52nd International Congress of Meat Science and Technology

52nd International Congress of Meat Science and Technology

13th-18th August 2006, Dublin, Ireland

Organising Committee
Mr. Declan Troy, Chairperson, Teagasc, Dublin
Prof. Joseph Buckley, University College Cork
Dr. Joseph Kerry, University College Cork
Mr. Patrick Moore, Bord Bia, Dublin
Dr. Tony Smith, Department of Agriculture and Food
Prof. Brian McKenna, University College Dublin
Mr. Cormac Healy, Meat Industry Ireland, Dublin
Dr. Briege Byrne, Teagasc, Dublin
Dr. Rachel Pearce, Teagasc, Dublin
Mr. Eric Donald, Teagasc, Co. Carlow
Ms. Lorraine Ryan, Teagasc, Co. Carlow
Dr. Lance O'Brien, Teagasc, Co. Carlow
Mr. Con Breen, Teagasc, Dublin

Scientific Committee
Prof. Joseph Buckley, Chairperson, University College Dublin
Mr. Declan Troy, Teagasc, Dublin
Dr. Joseph Kerry, University College Cork
Mr. Patrick Moore, Bord Bia, Dublin
Dr. Tony Smith, Department of Agriculture and Food
Prof. Brian McKenna, University College Dublin
Mr. Cormac Healy, Meat Industry Ireland, Dublin
Dr. Vivion Tarrant, Former Director of Operations, Teagasc, Dublin

The publication of these proceedings is kindly sponsored by Teagasc, the Irish Agriculture and Food Development Authority.

Teagasc
Agriculture and Food Development Authority

52nd
International Congress of
Meat Science and **Technology**

HARNESSING AND EXPLOITING
GLOBAL OPPORTUNITIES

edited by:

Declan Troy

Rachel Pearce

Briege Byrne

Joseph Kerry

Wageningen Academic
Publishers

ISBN-10: 90-8686-010-9
ISBN-13: 978-90-8686-010-4

First published, 2006

Wageningen Academic Publishers
The Netherlands, 2006

FOREWORD

On behalf of the scientific and organising committees, I am delighted to present the proceedings of the 52nd International Congress of Meat Science and Technology (ICoMST) which was held in Dublin, Ireland from 13th-18th August 2006. The proceedings contain over 300 offered papers presented at the congress in addition to 4 papers from invited speakers. The remaining 20 invited speaker papers are published in a special issue of Meat Science vol. 74 no.1, (2006).

This major event has been the main forum where the scientific, technical and economic developments in meat science have been discussed on an annual basis since 1954. The main theme of the Dublin ICoMST 2006 was "harnessing and exploiting global opportunities" now emerging through developments in key science based technologies and innovations. The congress focused on discussing the scientific basis of the opportunities and on ways to overcome major technological challenges in their implementation. Scientists from a broad range of sectors participated from around the world including research, academia, government agencies and industry.

The themes included the science of meat quality, scientific advances underpinning meat safety, new developments in meat packaging, understanding future consumer needs and wants, the latest technology responses from the global meat industry and finally the conference ended with a session on the latest scientific knowledge on the role of meat in our diet.

A new approach this year was to also address specific hot topics important to the industry and meat scientists. To this end, we achieved much interaction between the research community and the industry, a dialogue we hope will continue beyond this meeting.

Appreciation is expressed to all those who have contributed to the publication of these proceedings.

Finally, I would like to extend my utmost appreciation to all the delegates and sponsors for contributing to the success of the Congress.

Declan J Troy
Congress Chairman

TABLE OF CONTENTS

GLOBAL PERSPECTIVES AND CONSUMER NEEDS IN A CHANGING WORLD

L. Catlett

Regents Professor and Dean, College of Agriculture and Home Economics, New Mexico State University, USA.. Email: lcatlett@nmsu.edu

Keywords: consumer, global food market, healthy meat, nutragenomics, micronutrient isolation

Introduction

The post World War II years were dominated by the Green Revolution as scientists and policy makers were intent on "growing two blades of grass where one grew before". They succeeded, and continue to succeed, as agricultural productivity has gained faster than population growth for the last half of century. According to the United Nations Food and Agriculture Organization (UNFAO), the number of undernourished people in developing nations dropped from 920 million in 1980 to 799 million in 2000 while the world added 1.6 billion people, or a 43% gain in agricultural productivity/undernourishment reduction. Since 1990, in the developing nations of Asia and Africa, total food calories (excluding beer!) have increased on average by 5% to the point that every man, woman and child has on average 2900 calories/day. Undernourishment is not a food production issue, but rather a distribution issue. By the year 2050, when the world's population is forecasted to peak at between 7.5-8.0 billion people and then stabilise and begin to slowly decrease, the additional 1.2 to 1.7 billion new people can be added and provided for without increasing manufacturing production capacity, nor for that matter, agricultural production capacity since less than 65% of the world's arable land is currently in use.

The point is this—the world knows how to produce (both things and food).

Consumers in developed (developing) nations spend between 10-15% (25-30%) of their income on food (including meals taken away from home) which is approximately one half of what they spent in 1960. Furthermore, the world is awash in excess cash. In 2005, U.S. households held approximately 5 trillion euros in cash ($6.45 trillion), Canadian households 1.55 trillion euros ($2 trillion), Japanese households 3.1 trillion euros ($4 trillion), Australian households 770 billion euros ($1 trillion) and European households 6.2 trillion euros ($8 trillion).
China's middle class (331 million) is larger that the entire populations of both Canada and the U.S. At the end of 2005, China had 393 million mobile phone accounts becoming the world's largest user. In 2002, China had only 190 million mobile phone accounts. Mexico's middle class grew to 39 million during the last 15 years.
In the U.S., 63 percent of all people are overweight and 31% are obese. Thirty three percent of all normal weight Canadians in 1995 were overweight by 2003. According to the International Obesity Task Force, 300 million people are obese and 750 million are overweight worldwide. In Africa it is now four times more likely to find a child that is overweight or obese as it is to find one that is malnourished.

The point is this—the world is richer and knows how to consume (both things and food). The post World War II production world has been replaced with the modern consumer world where the food chain emphasis is top down (customer driven) not bottom up (producer driven). Kevin Roberts, CEO of Saatchi & Saatchi (part of Publicis) said it best—"For the first time the consumer is boss, which is fascinatingly frightening, scary and terrifying, because everything we used to do, everything we used to know, will no longer work."

Food is Free

It is important to now view food as essentially free. If all that can be offered by producers, wholesalers and retailers is a basic calorie rich diet, then they are operating with a production world mentality which has not existed for at least 30 years. A consumer mentality world, however, would concentrate on value added components of food (the big four are convenience, variety, health and safety—CVHS).
Coffee that is brewed at home costs approximately 6 cents per cup and is, for all intents and purposes—"free". A latte, cappuccino or Starbucks is not. Consumers pay for the value added of ***convenience/variety***. The developing nations of Asia have increased the portion of their diet that is meat from 10% in 1990 to 15% in 2003 and added another 800 million in population increase during the same time period (UNFAO). Africa followed a similar pattern from 9.4% to 14.3%. A latte in a plaza in Rome is about convenience versus a home brewed cup, while the increase in meat-rich diets for emerging nations is about diet variety.
Wine became the most popular alcoholic beverage in the U.S. in 2005 in part because of the value added ***health/safety*** component. Ditto the assault by policy makers on snack foods and soft drinks sold in schools in North America and Europe as childhood obesity soared from 13% in 1960 to 31% in 2000. One hundred years ago, the gastrointestinal infection mortality per 100,000 people was approximately 120 in most Western nations whereas today it is essentially zero. Thus prior to and immediately after World War II food safety issues were paramount as science began to understand disease vectors. However, by the late 1950s most food safety issues were conquered and safety issues faded

away. Only in the last few years with the advent of Bovine Spongiform Encephalopathy (BSE), or "mad cow" disease has the issue of safety of the food system been on the public's radar.

While food may be free—convenience, variety, health and safety are not and consumers worldwide are increasingly willing to pay. In economic terms, all of these factors have an income elasticity that is normal to luxury, *i.e.* for every 1% increase in incomes, there is a positive increase (or more than 1% increase if a luxury item) in the demand for the products/services that they represent. While certain cuts of meat and processed meat such as hamburger may in general have a negative income elasticity, certainly hamburger meat that can be certified to the source (have traceability) or be certified disease-free would have a positive income elasticity. In the U.S. the demand for the highest restaurant quality beef cuts (prime grade) by the general customer in non-restaurant venues is at least twice the available supply each week (approximately 3% of slaughter).

To emphasise the concept of "Food is Free", a closer look at another industry is important. The telecommunications industry during the last century reaped trillions of euros from customers as they built complex infrastructures to carry voice worldwide *via* telephones. As the wireless industry emerged in the late 1980s they copied their land line counterparts and focused on voice by concentrating on selling a certain number of minutes of talk time. However, by the late 1990s voice minutes were expanding to the point that they were packaged with certain other features such as music downloads, Internet access, pictures, location via global positioning, calendars, ringers, text messaging, instant messaging and games. Today, voice is free on most wireless systems worldwide. If all a wireless company can offer is voice, they are obsolete. Thirty percent of all voice communications are now "free" via an Internet protocol called VoIP (Voice over Internet Protocol). What most wireless companies do today is offer free talk time essentially *via* numerous minutes, but charge for the other wireless services that each individual customer wants.

In May, 2006, IGD listed it's "Ten to Watch" outstanding food retailers in the world as: 1) AS Watson (Hong Kong), 2) Central Retail Corporation (Thailand), 3) Couche-Tard (Canada), 4) eBay (U.S.), 5) Pantaloon (India), 6) Perekriostok (Russia), 7) Tchibo (Germany), 8) Wegmans (U.S.), 9) Whole Foods (U.S.) and 10) Wu Mart (China). All of these retailers concentrate on market segmentation and differentiation rather than mass marketing. It is interesting to note that the list includes the world auction market eBay which is not a food retailer *per se* but certainly has revolutionised the way consumers shop. Pantaloon in India has in-store kitchens so that customers can dine in, or have the chef prepare their food for take away. Many of the other retailers in the "Ten to Watch" list have in-store kitchens as well. Whole Foods concentrates on the organic market as well as the high-end customer and thus has the nickname "Whole Paycheck". Whole Foods' customers may have fun with the nickname, but they are loyal and the company is setting a blistering pace in store opening and earnings.

Market of One

Retailers worldwide are looking for ways to differentiate the marketplace by any means possible, whether it is *via* income, age, race or the latest rage "a market of one". The idea behind a "market of one" is that the generation of young people that were born in the mid to late 1980s is the most consumer driven "have it your way" group in history.

In the U.S., 30% of all college graduates have attended more than 3 universities before they graduate with a bachelor's degree. The typical student "picks and chooses" what courses and campuses they like best and/or how they want the class either on campus or *via* distance over the Web. The estimate from several sources puts new "blog" sites at approximately 30,000 per day as individuals express their own views and opinions on any and all issues. On demand publishing now allows anyone with only a nominal investment to fully publish and print a book, record an ISBN number and place it on Web ordering sites such as Amazon, Barnes and Noble and Chapters.

For the food industry, the concept of "a market of one" leads to three natural organisational changes. Two involve bricks and mortar—1) big box super stores that cater to a wide variety of choices that often involve more than food such as Wal Mart Supercenters, but include giants like Tesco which is the 9th largest clothing retailer in the U.K. Likewise, ASDA's concept of "fashion democracy" involves providing an extensive array of choices at reasonable prices. Safeway has moved to a lifestyle store concept that promotes consumer choice; and 2) special narrowly focused stores that either cater to neighbourhoods, cultural or ethnic channels or other characteristics. However, these narrowly focused stores have to be clustered so that consumers can easily select meats, breads, fruits and vegetables from individual stores or have the choices under one roof á la Tesco's neighbourhood stores or Trader Joe's limited array but complete stores to satisfy consumers' desire for convenience. The third area is, of course, the Internet. Estimates range all over the map, but there is no doubt that growth of all consumer items including food has been in double digits for more than a decade (doubling every 5 years on average). The convenience of ordering special food items *via* the Internet allows consumers to be more selective and certainly high-end in terms of quality as the meat industry has learned in the last 5 years. High-end special meats and meat products lend themselves to Internet sales as customers pick holiday and special event products. Food items that are considered commodities have limited success via Internet sales as transportation costs push them past local prices.

Food Is HealthCare

A luxury to one generation becomes the necessity of the next. Only during the post World War II era have individuals viewed longevity as part of their lives. Life expectancies waxed and waned around 50 years of age in developed nations prior to WWII. However, as ages began to creep up (79 in the U.S. and 80 in Canada and Europe), people began to assume that they could live longer and in so doing, began to understand that potentially how and what they eat might impact longevity. Broad studies emerged post WWII implicating almost every food as being bad for human health based on spurious science such as cranberries, red meat, mushrooms, apples, raw eggs, poultry, pork, coffee and a whole host of others. Fortunately, just as many studies emerged to counter each. Fads occur almost daily as blueberries are shown to be anti-carcinogenic or green tea imparts cardiovascular health.

Healthcare officials routinely use the model that 85% of healthcare issues faced by humans are behaviour related while only 15% involves non choice variables. In other words, human choices concerning lifestyles are the most important attributes about health--thus the focus that food is healthcare. Yet in the new world of food where CVHS and a Market of One dominate, broad studies that suggest a particular food may be either bad or good for humans holds little value. Two new fields of study point to the future of food—1) nutragenomics (NG) and 2) micro nutrient isolation (MNI).

NG is the study of how food reacts biochemically with each individual's own DNA. Lard may be good for some, but bad for others. Some individuals may be able to eat large quantities of red meat daily with no bad side effects for others may have to limit their intake because of their inability to biochemically to store, process or mitigate certain fats and/or proteins. Needless to say this involves not only the sciences of nutrition, chemistry, and biology but also statistics, mathematics and computer algorithms.

MNI takes NG to a more detailed level. Instead of concentrating on a certain food such as beef, MNI looks for specific elements and/or molecules that impart either a benefit or non-benefit to an individual's health. Certain amino acids, for example, have been shown recently to be very important to potential problems with obesity and diabetes.

Taken together, NG and MNI merge to a vernacular expression—***prescription food.*** Just as doctors prescribe a certain medicine for an individual, the future of food involves consumers knowing not only ***what*** food to eat, but ***when*** during the day and ***how*** much to consume to maintain the best possible healthcare.

The Big Two for the Meat Industry

The world's meat industry faces two great opportunities for unbelievable growth. The first entails the most tonnage and the second the largest investment in human capital.

Big Market One

As the world becomes richer, the demand for meat will increase in all emerging economies. Yet it will not be the same as the past. Consumers are savvy. If consumers do not buy *via* the Internet, they at least search for information on products *via* the Web. As the world becomes more mobile *via* cell phones, the information will be more readily available. Therefore, merely providing calories and or protein *via* commodity meat will not pass muster in a world where food is free. Even in emerging societies, meat producers, wholesalers and retails must increasingly be aware of convenience, variety, health and safety.

Big Market Two

As the world ages, especially in mature economies, healthcare is a major issue and food is an integral part. NG and MNI dominate. The future is not so distant whereby every individual will have a daily prescription broken down by the quarter hour on what micronutrients/vitamins/minerals are necessary for optimum health and likewise what quantities and types of foods are necessary to achieve the balance. Meat isn't meat in this world, but ***"smart meat"***. Human capital and human capital development is essential.

What is critical is that the meat industry worldwide will grow quantity-wise and just as importantly, quality-wise. Just as Starbucks is not about coffee but a lifestyle, the same is true for meat. Quantity and quality are moving targets that have to adjust to changes in lifestyles world wide. To be sure, we consume more coffee thanks to Starbucks, but only because we can get it "skinny", "with a twist", "to go", "no foam" all while hooked to the Internet *via* WiFi or WiMax in a comfortable environment listening to music or watching movies with our iPods streaming data while we Instant Message the person next to us in a chair and text message 23 others on our chat or party line. Such is the future of meat, be it in China, Ireland, U.K, U.S. or Cameroon.

Sources

Eurostat (www.europa.eu.int)
International Food Information Council Foundation (www.//ific.org)
Statistics Canada (www.statcan.ca)
The Economist (www.economist.com)
U.S. Bureau of Census (www.census.gov)
United Nations Food and Agriculture Organization (www.un.org)

INNOVATING MEAT PRODUCTS TO MEET CONSUMER WANTS AND NEEDS: KEEPING MEAT ON THE MENU

A.C. Haselgrove

Dawn Group, Cross Hands Business Park, Cross Hands, Carmarthenshire, SA14 6RB, Wales.
Email: alison.haselgrove@uk.dawnmeats.com

Keywords: meat, consumer, product development

Introduction

From the very earliest of times, and certainly well before meat science and technology were defined as disciplines - meat has been adapted for human consumption needs. In fact, by merely coining the term "meat", we are disassociating the basic raw material, from its animal origin –for example, cattle, lambs or pigs - and also distinguishing it from more medical or scientific terms such as "flesh" or "tissue". When we speak of "meat" - we immediately suggest a substance destined for consumption. Over human history, meat has always been adapted to facilitate consumption – with an ever-increasing body of knowledge and practice - building to the point at which we find ourselves today - such that meat science and technology has become a sophisticated discipline. The meat industry is now able to draw on a vast body of knowledge to adapt meat to best suit the needs of today's (and tomorrow's) consumers. However, when we consider current science and technology in the context of contemporary consumer needs, have we really travelled so far, for, whilst the way in which consumer requirements manifest themselves is, self-evidently, constantly evolving, have the core needs of consumers really changed so much over the centuries?

Those in the field of marketing will be familiar with the 4 P's – product, price, promotion and positioning – the elements that marketers believe are core to any marketing strategy. In the world of meat technology, we also have our own version of the 4 P's.

The 4Ps

Basic consumer needs, which, whilst patently manifesting themselves in ever changing ways, do stem back to some fundamentals in the interaction of meat with its consumers.

Palatability

This is an obvious need, but core to what always has and continues to drive consumers to come back to a food product is the critical question - does it taste good? A recent TNS report (TNS, 2005), which investigated home consumption in the UK, revealed that "enjoyment" was the primary driver of food selection in the home – especially for weekend eating. This was backed up in late 2005 by a 2 year study (Foodnavigator.com, 2006), by international ingredients company Tate and Lyle, who in both qualitative and quantitative research in Germany, France and Britain found that 75% of consumers felt that the taste of their choice of food was more important than, or equally important to, it's nutritional merits.

How did ancient man work out that cooking meat made it a whole lot tastier than raw hunks of flesh, and how closely was the discovery of fire and fire-making techniques followed by the idea of cooking meat? Long before Louis Maillard investigated and explained the complex chemical processes that were occurring when meat roasted, people were experimenting with roasting meat - just because it tasted great. It's also no surprise to dedicated carnivores that even in early civilisations meat became the aspirational centrepiece of the lavish meals of the wealthy classes. Homer writes of ancient Greek contemporaries feasting on sheep meat and several Roman authors - such as Petronius - satirised pretentious and upwardly mobile "plebs" for serving meat at every course – a wild extravagance considered to be in poor taste (Grant, 1999). Images of lavish Medieval and Tudor banquets with attendees gorging on meat and tossing the bones to the dogs beneath the tables are also familiar and there is documentary evidence of these menus. (Paston-Williams, year) Meat, therefore, enjoyed for its great flavour and social status, has been with us for centuries and taste continues to be paramount. From these beginnings, much research today in meat science and technology focuses on making meat as delicious as possible to the human palette. We will now move on to our second 'P'.

Portability

So you've nipped out of the cave with a few hunter friends and brought down your sabre tooth tiger or woolly mammoth – now what? You have to get that carcass turned into edible portions pretty pronto – not just before some creature bigger than you decides to feast on your kill – but so that it can be moved and stored in meal friendly portions. This isn't very different from the modern challenges of portioning, packing and transportation for today's sophisticated retail and food service environments.

So you and your hunting party have worked out how to portion and move your felled beast and you have your meat ready for to be prepared for eating. But even with a few friends over – it would be challenging to eat a whole woolly mammoth at once, and who knows when you might have another successful hunting expedition? This brings us to our third 'P'.

Preservation

Being able to preserve meat has always pre-occupied people involved in its handling. In earlier societies where seasonal availability dictated what was available in the daily diet, it was desirable to be able to store meat for later consumption, for example in the long snowy months of northern winters or in the intense summer heats of hot countries. Whilst the understanding of those nutritional requirements that make for a healthy balanced diet is a relatively modern discipline, just through empirical observation many societies have long valued meat for its contribution to growth and strength. It has also been valued for flavour and variety in the largely vegetable and cereal diets of the poorer members of society. Meat has long been required to be moved to where it is needed to be eaten and then stored there without it spoiling to the point at which it is a hazard to health to eat it – or is simply rank and rotten and thus extremely unappetising.

In the home over previous centuries, meat was smoked, dried, iced and pickled to retain its edibility. All these disciplines still find a place in the way in which we handle and present meat for eating today. After the industrialisation of Western society, the urban workforce could no longer live off the land - yet they still needed to be fed and meat was an aspirational part of the diet. Also, the desire for meat to be available under more challenging conditions such as long sea voyages and military campaigns has given us a legacy of meat innovation – from the Mongol horsemen of the 13th century storing and tenderising meat under their saddles – to the methods for the jarring and canning of meat being devised in 1795 by Nicolas Appert (www.answers.com, 2005), out to win Napoleon's 12k franc prize for effective ways to feed French troops with nutritious meat which had not spoiled in transit.

Coming to the present day, the prizes are greater still – with modern consumer markets for meat now worth enormous sums a year –Datamonitor (Datamonitor, 2005a) predict a 2006 market value for meat products at Euro 20.5 billion in Europe alone - delivering the "Meat P's" has become the focus of the vibrant industry in which we work. This brings us to our 4th 'P' of meat technology.

Profitability

In the commercial meat sector, achieving adequate levels of profitability pays wages – so this sector is keen to use the best of contemporary meat technology to improve the potential for profit in modern meat processing. However, consumers are a complex and increasingly individualistic bunch, so choosing those elements of meat technology that have most appeal to consumers is crucial to commercial success. For this, keeping up to date with what is driving consumer needs and behaviour is essential.

The Contemporary Consumer

This section discusses some of the trends which are currently influencing the European markets. Many of these will no doubt have relevance to other developed markets around the world. We will now consider how those basic elements of palatability, portability, preservation and profitability manifest themselves to those in the meat industry today. Palatability leads us to an exploration of contemporary consumer tastes and concerns, and if we extend the definition to include not only the organoleptic properties of meat but also the "intellectual palatability" of concepts in the minds of modern consumers, we can uncover a wealth of significant issues such as farming regimes, animal welfare and ethical production, which are highly relevant to the work of those of us here today. Consumer tastes vary by the cultural and societal context in which a person lives, and despite the effects of globalisation, many consumers still have much-loved dishes and culinary traditions which they hold dear to their hearts – even viewing them as defining elements of their cultural, ethnic or national identity. On high days and holidays, as the saying goes, many will be drawn back to these dishes, but with the busy lifestyles that many consumers in developed countries now lead, food is being eaten in increasing different ways. If meat is to stay on the menu for contemporary consumers, it needs to adapt and present itself as a relevant and desirable constituent of modern diets.

Much consumer research has been conducted to give insight into consumer wants and needs – how they are currently behaving and how this behaviour might evolve. In 2002 in Denmark, the team at the MAPP Centre published a review paper (Brunsø *et al.*, 2002), collating a large body of research into consumers' food choices and perceptions of quality. They constructed a Total Food Quality Model, and within this explored consumer attitudes to beef and pork in several European countries. Their findings demonstrated that quality perception in meat is influenced by both "objective" factors such as muscle conformation, fat distribution and colour and by "subjective" judgements such as the credibility of health statements or claims for superior farming and animal rearing techniques. In fact, their work further showed that the meat industry potentially has even greater hurdles to overcome than in other food sectors. This is because the measures that consumers use to judge a "good" piece of meat – that it will be tasty, tender and succulent – such as a "nice" colour or low levels of fat – are often spurious and actually leave them less likely to choose meat which won't deliver against their expectations. Even marbling, which some consumers felt was a better indication of tenderness, has been proven by tests at the Animal Research Centre, in Nebraska to be a less important indicator than the largely invisible factor of collagen content (Shackleford, 2005). The author has been involved in numerous consumer focus groups discussing purchasing behaviour and meat selection and would consequently agree that we seem to face higher levels of challenge than for other food stuffs.

The underlying messages are encouraging - consumers have strong views on the meat they buy because the area is important to them – meat is seen by those who include it in their daily diet as both a key source of good nutrition and an important focus for a meal occasion – good meat makes or breaks the enjoyment of the meal. The good news continues for those of us with an interest in the beef sector. The 2003 study (Datamonitor, 2005b) by the Food and Brand Lab based at the University of Illinois had American men listing steak as their number one "comfort food" or choice when indulging themselves. Yet, a 1997 consumer audit (Huston, 2005) by the National Cattlemen's Beef Association revealed that 25% of American consumer eating experiences with beef were poor.

For every positive message we have - we always seem to have this potential for consumer backlash and the author's experience of consumer listening groups is that on the whole, customers are very suspicious of those of us providing their meat for them. They have an underlying assumption that the industry is out to trick them, hiding poor quality, fatty and gristly meat under labels that are too large, and obscuring it in brightly coloured opaque trays. So, transparency of packaging – literally and metaphorically - is critical, as is building a reputation through ongoing adherence to high quality standards, which allows consumers to have full confidence in the credence of our quality claims.

Working with a natural raw material is always a challenge, our meat will vary from carcass to carcass to a lesser or greater degree and many researchers are engaged in the study of those factors, which ones trigger variation and what can be done to ameliorate those effects – from breeding, to feeding and slaughtering techniques. In the past we have also turned to processing methods to help us even out variability and achieve consistency – yet this route is also increasingly being challenged by contemporary consumer attitudes.

Food, being an important part of everyday life, is sensitive to the so-called mega-trends in society. In particular, 3 major well documented trends seem to be re-occurring in research into consumers' food choices in both Europe and the USA - convenience, health and wellbeing and indulgence or personal satisfaction – all of which are now manifesting themselves into the way in which people select and purchase foods. As a consequence, the meat products we offer for retail and food service sale need to demonstrate, through their composition, design and marketing positioning that these trends have been fully understood and addressed. In addition, it is becoming increasingly evident from research into consumer attitudes and actual behaviour that consumers are becoming progressively more polymathic in their psychological interaction with food. No longer can we safely just group consumers into broad generic behavioural groups based on social class, economic circumstances, educational status or postcode and assume that this will be as relevant for their choice of foods as it is for their taste in books or selection of shampoo. Within each contemporary consumer can exist an intricate set of interrelated, sometimes contradictory and ever-evolving mental filters that build to a complex individual psychographic profile. Each individual's experiences and how they recall and interpret this to build aspirations and perceived needs, will lead to lifestyle choices and behavioural patterns and their relationship to meat products will be no less influenced by these complex and converging trends.

Convenience is becoming an over-worked term, but at its core is a desire by consumers to seek out products which effectively and efficiently allow them to juggle the multiple tasks that they have to accomplish in the day, and to fulfil their desire to increase the time they have for the priorities in their lives – be that quality free, leisure and family time. Unless we in the meat supply chain cater to consumers' desire to eat meat in more convenient formats then our part in the meal selection will diminish in both relevance and volume. Convenience can be delivered in a variety of ways – prior preparation of meat items removes the need for consumers to have an in-depth knowledge of meat cuts – both reflecting and acknowledging the skills gap that exists in the general public. Meat that is partly or fully prepared and ready to pop in the oven in its tray even removes the need for the consumer to handle meat – a task that many find distasteful and off-putting.

In western societies we are now increasing eating in less formal situations. The trend towards informal and fragmented eating is well documented. We are experiencing the demise of formal "mealtimes", and the rise of phenomena such as S.A.D. – "Stuck At Desk" eating – nick-named "Al desko" eating, along with the trend towards eating more frequent smaller quantities of foods, eating whilst on the move – so called "Dash Board Dining" – a term for in-car eating. People are also more likely to skip meals altogether these days – maybe topping up on energy boosting snacks as they get hunger pangs later in the day. For example, a survey (Datamonitor, 2005c) commissioned by Smoothie King in 2004 in the USA found that lunch is skipped by 58% of US consumers in favour of on-the-go alternatives. Of those who do take lunch 82% regularly spend 30 mins or less and 43% as little as 15 mins on their lunch slot. All across Europe and the USA the average length of journey to work is rising, eroding free time at either end of the day. These busy consumers now want convenient food at their finger tips. At a recent focus group of working mothers that the author attended, 7 out of 10 stated that at 3pm they had no idea what they would be serving their families for the evening meal. They now "repertoire shop" - that is they load their shopping trolleys with a selection of items that they know the family will eat – and then mix and match these ad hoc as they pull a meal together in the home. Younger consumers and those older "empty nesters" with no responsibilities for feeding a host of "chicks" can afford to be and are even less planned, relying upon their fridges, freezers (where most have selected fresh meat and frozen it themselves) along with takeaway menus and local restaurants to keep them fed and watered.

So the industry has to rise to the challenge. The erosion of formal "at the table" and "meat and two veg" eating occasions could equate to the erosion of our own market for meat. From a home-cooked meal in which meat may be 50% or more of the meal's weight, its more convenient alternative - say a pizza or a pasta ready meal - may have less than 10% meat. It all has potential to diminish our "share of stomach" and our sales volumes unless we can make our meat products highly relevant to the consumer's new eating patterns.

Both product formats and their presentations have to accommodate this new consumer behaviour. Beyond the meat itself, the packaging solutions must meet the needs of a widening number of eating venues and situations – on the move (on the tube, in the car, at your desk), easy open (for an increasingly economically active grey market), non-spilling/ re-sealable, ovenable/ microwavable and even self-heating. Of course the industry hasn't stood still. When it comes to the portioning of meat – we've sliced it, diced it, wafer-thinned it, reformed it into burger patties and made bite-sized snacks. We have extended its shelf life through preservatives, anti-oxidants and modified atmosphere packs. We have further prepared it in oven-ready portions, we have cooked it and made it ready to microwave in minutes. Yet, you only have to cross-reference a market, such as the dairy sector, to see the accelerating trend to make food able to be consumed away from the traditional " table at home", solving the practical difficulties of mess and sticky hands and the social inhibitions of wafting smells etc.

A critical determinant in many peoples' choices of food is now also the contribution that food can make to their overall wellbeing. Whilst the number of people actually following a prescriptive weight-loss diet is falling - a Healthfocus International Survey in 2005 (Datamonitor, 2005b) found that only 8% of American consumers fell into this group - the number of those actively seeking out foods and drinks to "ensure good future health" has risen above 30%. Similarly, with 75% of US shoppers seeking to "eat light" rather than diet and over 50% of UK consumers feeling they follow a well-balanced diet, meat - as an acknowledged nutrient-rich food - has to find its successful role in this trend. On the plus side for those in the sector, meat is not grouped with the type of sugary, highly processed foods that the European Commission Green Paper (European Commission, 2005) on "promoting healthy diets", which was published in late 2005, dismissed as "energy-dense and micronutrient-poor". Most consumers do believe meat can be part of an inherently healthy diet.

Yet we have challenges here too. Where previously the natural variability of meat has been evened out through the addition of chemicals – through the use of brines, tumbles and injections – this is now increasingly coming into conflict with the desire for natural unadulterated foods. Our challenge will be to ensure the consistent eating quality of meat in an environment that is demanding reduced or even completely additive-free products. Maintaining taste and preservation without salt or at significantly reduced salt levels has been a hand's on pre-occupation of the author's team of Meat Technologists along with many others in the industry over the past 18 months. Salt's link with raised blood pressure and the life-threatening diseases attributed to hypertension, continues to be very much at the forefront of government and media attention and therefore salt reduction remains on all our agendas. Studies, such as those by Dr Thun and colleagues (Chao *et al.* 2005) at the American Cancer Society, into the increased incidence of colon cancer in people eating high levels of red meat may inhibit consumers from including red meat in their diets, yet this may be countered by the growing trend of "less of the best". This is the tendency – especially noted in older consumers – to meet their "self-indulgence" needs by still treating themselves to what they crave – just in smaller quantities. So the demand for high quality and premium positioned meat products – targeted at the "a little of the best" consumer may prove a profitable sector for us to explore – as long as we can deliver viable product concepts with claims that the consumers find to be credible.

May we even go further and add "positive" nutrition to our meat – either via pre- or post-slaughter methods, boosting its health credentials through the addition of desirable nutraceuticals? The Dawn Group is supporting a Masters Degree project investigating the consumer acceptance of this topic during 2006. The modern consumer is looking for even more. In meeting their needs for a convenient product, a healthy product or an indulgent product we have succeeded in raising their expectations; now delivering one of these attributes in isolation is likely to be no longer enough. Whilst we in the meat industry have worked hard to rise to the "mega-trend" challenges, it seems we must now go further still. Building on the Danish "Food Related Lifestyles" work by the MAPP Centre, Teagasc (Teagasc, 2003) undertook further research in 2003 in the UK and Ireland and has attempted to define some new demographic groups with specific attitudinal and behavioural characteristics in relation to food. This research has substantiated the interaction and overlap between these key trends. As consumers experience expands, this drives higher expectations of us as food manufacturers and food industry experts to deliver the ultimate combination of attributes. Further consumer studies such as the 2005 Mintel reports on "Eating and Work" (Mintel, 2005a) and "Cooked and Delicatessen Meats" (Mintel, 2005b) highlight the need for new meat products that span more than one mega-trend. For a product to succeed, it must " convey a bundle of intrinsic and extrinsic attributes that better meet the rational and emotional needs of the different consumer groups within the overall shopping base" (Mintel, 2005b). Attitudinal research from several Datamonitor reports (Datamonitor 2005b,d, 2004) supports this further, challenging us to resolve the dilemma that consumers believe the most tasty food will be likely to be the most unhealthy and that healthy foods will be lacking in taste (Datamonitor, 2005b).

Conclusions

The challenge for the meat industry is to create meat products that taste fantastic, deliver sound health and nutritional benefits, and are presented in a format deemed convenient in a wide selection of eating occasions. Our brands must only make those claims that consumers find credible, engaging and relevant to their lives today. Our reputations must go before us, building trust and securing consumer confidence. So, it may all be a long way from that original Woolly Mammoth - but with the skill, expertise and enthusiasm we have within the industry, we will be more than capable of rising to the challenge of keeping meat securely on the menu in the years ahead.

References

Brunsø, K., Ahle Fjord, T. and Grunert, K.G. (2002). Consumers' Food Choice and Quality Perception. Aarhus School of Business.

Chao, A., Thun, M.J., Connell, C.J., McCullough, M.L., Jacobs, E.J., Flanders, W.D., Rodriguez, C., Sinha, R. and Calle, E.E., (2005). Meat consumption and risk of colorectal cancer. Journal of American Medical Association, 293, 2, 172-182.

Datamonitor (2004). Global Consumer Trends. DMCM0683, July 2004.

Datamonitor (2005a). Datamonitor Consumer Markets Knowledge Centre (Statistics Database).

Datamonitor (2005b). Profiting from Consumers' Desires for Healthy Indulgences, DMCM2362, December 2005.

Datamonitor (2005c). New trends in snacking and drinking on the go. DMCM2385, December 2005.

Datamonitor (2005d). Overweight consumers and the future of foods and drinks. DMCM2384, December 2005.

European Commission (2005). Promoting healthy diets and physical activity: a European dimension for the prevention of overweight, obesity and chronic diseases. COM (2005) 637 final, Brussels.

Grant, M. (1999). In: Roman Cooking: Ancient recipes for modern kitchens. Serif Publishing. ISBN 1897959397.

Huston, J. (2005). Private communication from Dr. John Huston, National Cattlemen's Beef Association.

Mintel Market Intelligence (2005). Eating at work. December 2005. Mintel, London.

Mintel Market Intelligence (2005). Cooked and Delicatessen Meats. December 2005. Mintel, London.

Paston–Williams, S. (1993). The Art of Dining: A History of Cooking and Eating. National Trust Books, 0707801737.

Shackleford, S. (2005). Private communication from Dr Steve Shackleford – Research Food Technologist, US Meat Animal Research Centre.

Teagasc (2003). Convenience food lifestyles report. Teagasc.

TNS (2005). TNS Worldpanel Usage (In Home).

www.answers.com (2005). Biography of Nicolas François Appert. December 2005.www.foodnavigator.com, January 2006.

AUTOMATION –THE MEAT FACTORY OF THE FUTURE

N.T. Madsen, J.U. Nielsen and J.K. Mønsted

Danish Meat Research Institute, Maglegaardsvej 2, 4000 Roskilde,Denmark. Email:ntm@danishmeat.dk

Keywords: automation, pigs, beef, slaughter

Introduction

For many years, slaughtering has been industrialised with respect to organisation, work specialisation and use of production lines. Interestingly enough it was from seeing the conveyor systems used by the slaughter industry that Henry Ford got the inspiration to implement car assembly lines in the beginning of the last century. One could claim that the meat industry is now taking inspiration from the robot assisted and computer controlled manufacturing systems applied in the car industry.

So what is driving this development after decades of limited change in the technological development of slaughter processes? The industry benefits significantly from automation: one of the most important benefits after increased labour productivity is the improved working environment. Slaughtering, cutting and boning operations are very labour intensive and require hard and repetitive work. For many young people a job in a slaughterhouse is therefore not their first choice. This situation could be changed by automation, which will not only remove arduous work but will in addition introduce more interesting jobs in terms of planning, supervision and control of the new technology. Automation will also improve the hygiene and thereby the food safety by less manual product handling and cross contamination between carcasses because the tools in the machines can be cleaned more effectively between each carcass. Product quality and yield may also be improved, and the production recipes may be changed instantly in accordance with information on market demands.

The poultry and fish industries have implemented automated processes with success and, in New Zealand in particular, the lamb slaughter industry has been progressive in developing and using new technology. Only in recent years have many processes for pig slaughter and deboning been conducted by newly developed robots. The status of this development will be reviewed here with the Danish program for automated slaughter of pigs as a case. Regarding beef, the development in technology for the slaughter process has been quite limited partly due to the biological variation in animals and the cost/benefit of applying complex technology. Some of the potential options for R&D will be suggested here.

The option to replace operators with equipment may create benefits for the industry, particularly when the efforts put into the implementation of the machines are duly considered.

Automation development methods and procedures

Development of automated solutions for slaughter, cutting and boning is a complex task because of the biological variation, the complexity of the material to be processed and the variety of professional disciplines involved. In order to control project economy and progress the Danish Meat Research Institute (DMRI) has applied a staged development model comprising a number of distinct phases:

- Phase 0. Initiation
- Phase 1. Generation and evaluation of ideas
- Phase 2. Method development
- Phase 3. Experimental machine
- Phase 4. Prototype machine
- Phase 5. Production version machine
- Phase 6. Project evaluation

Projects for automation are carefully selected on the basis of their cost/benefit. During the idea generation phase, different teams are competing to identify the best solution. All projects are conducted individually by separate project teams with representatives from the industry and participating suppliers and manufacturers attached. The transition from one phase to the following phase includes both economic and technical criteria for decision of whether to proceed or not. One of the decision tools is a cost/benefit analysis combined with an investment analysis in order to justify whether the solution is still economically feasible for the industry or whether the project should be terminated.

The development work is supported by use of the most contemporary and advanced tools: CAD-tools for mechanical and electronics design, robots for initial testing of tool operations and simulation in combination with CT-scanned models of the products to be processed. The latter can be illustrated by the machine for round cutting of the hind legs.

The process involves cutting of the tail bone, trimming of fat at the tail head, groin meat and fat. A new CT-scanner is used for the development process. A large number of hind legs are scanned with high resolution (1x1x2mm) to form a catalogue, representing the variation in the Danish pig population. In the catalogue the variation of the joint is described in principal components. The analysis reveals the shape variation with respect to the anatomical fix points used for gripping the product. This shape variation can then be used to design the tool in an optimal sense, based on the catalogue.

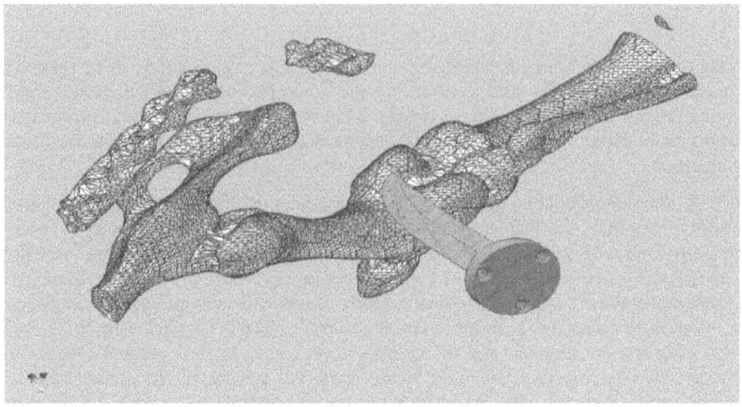

Figure 1: CT- scan of hind leg.

An automated pork slaughter line

The DMRI started development of new technology for optimal welfare during lairage in the late 80's and stunning of pigs in groups in the late 90's (Zink, 1995). During slaughterhouse lairage, the pigs can be moved by automatically driven gates. This means that today the pigs may be transported unstressed in homogeneous groups all the way from the farm until they are CO_2 stunned prior to sticking. The pigs are stunned in groups of 5-7. The groups are lead into a pater noster gondola which is lowered into the CO_2 where the pigs loose consciousness prior to sticking. The result of this, besides optimal animal welfare, is a mortality for slaughter pigs during transport and lairage which has remained below 0.2 $^{0}/_{00}$ for more than ten years. Furthermore, the drip loss in the meat is lower.

Most of the first part of the unclean slaughter line has been automated for decades with respect to scalding, dehairing, singeing and scraping. Only in recent years has it become possible to conduct more processes for slaughter and deboning using machines.

Modern pork slaughter lines in the EU run at 200-400/hour. In Denmark a line speed of 360/hour is usual whereas line speed in the USA is more than 1000/hour. The Danish pork industry launched an automation development strategy in 1998. The total investment in development exceeds €40 million and is organised in more than 30 individual projects which are in different stages of development. The aim is to automate a major part of the processes on the slaughter line and several cutting and boning operations. This was driven by expectations of better competitiveness, the high labour cost and an expectation of lower labour supply. In 2007 it will be possible to equip a slaughter line with 10 machines and a manning approximately 50% compared to the manual operation. The machines involved in such a slaughter line design are illustrated in Figure 2 with the present state of development indicated. In Figures 3, 4 and 5 photos of selected machines are given followed by the operations they perform. The hygienic advantage of the machine is that all tools are cleaned thoroughly between carcasses.

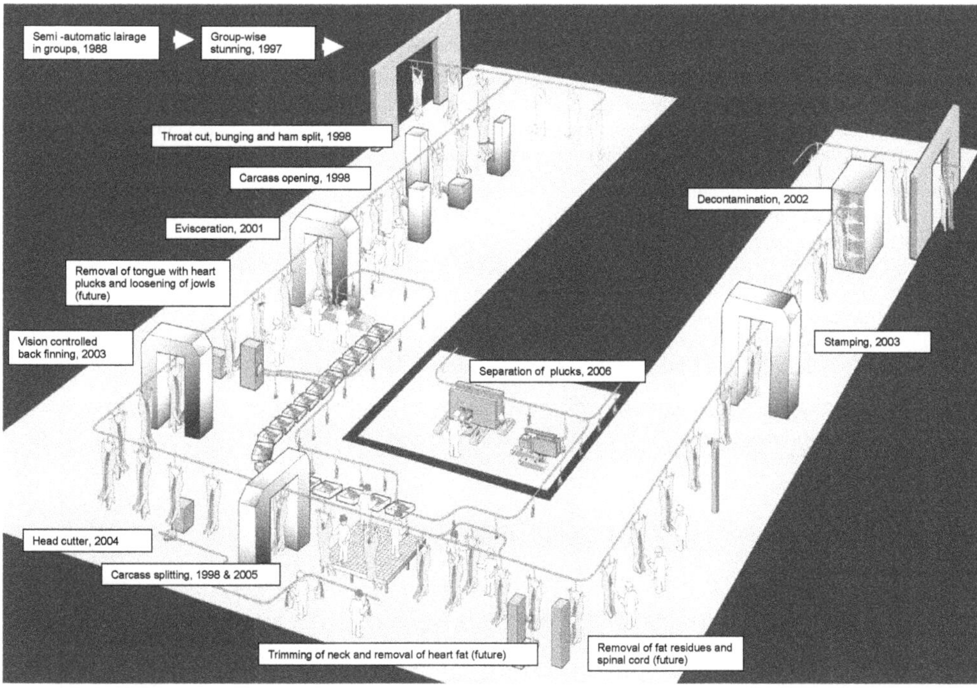

Figure 2: A slaughter line of 360 pigs/hour can be equipped with 10 automated machines and staffed with approximately 50% compared to a manual operation. The task of the machine is given in the textbox with the year of availability. Examples of machines are shown below.

Figure 3: Evisceration.

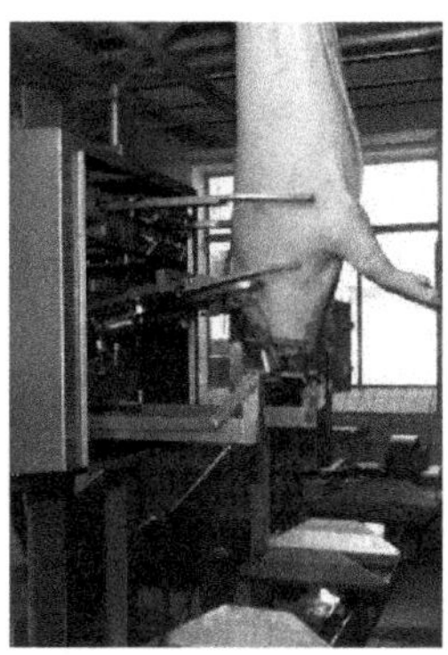

Figure 4: Head Cutting.

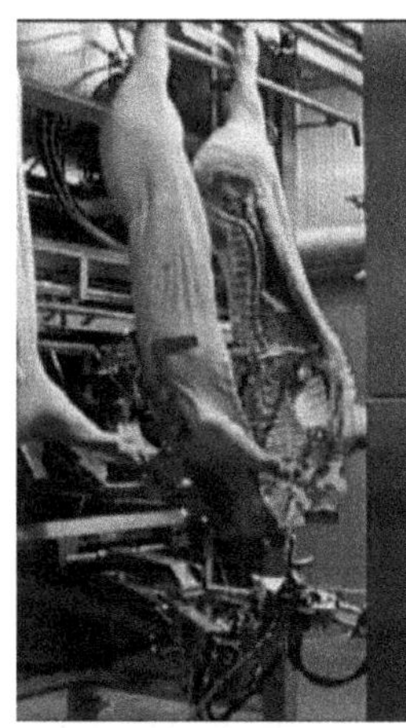

Figure 5: Trimming of neck and removal of heart fat (experimental machine).

Evisceration: The evisceration machine cuts the diaphragm, tears off the flare fat and loosens the intestinal tract such that the entire set of intestines and organs are removed from the carcass prior to separation, Anon. (2001). Because the separation takes place outside the carcass it can be performed more safely and the risk of faecal contamination is reduced.

Head cutting machine: A new machine for automatic cutting off of entire heads was recently introduced at some Danish slaughterhouses. The machine operates in two modes. It can either cut off the head entirely and leave it on a separate conveyor system or alternatively the machine can cut the neck only and leave the head hanging at the carcass by the jowls. Cutting off the heads before carcass splitting reduces the risk of contamination from the mouth and throat regions. The machine also provides better yield by leaving more neck meat on the carcass.

Trimming of neck, removal of heart fat and cutting off the foreleg: This machine, still in the experimental phase, will automate some of the critical cleaning of the carcasses after splitting. The machine automatically trims the neck region for glands and blood residues and from the breast part it will remove the heart fat. Finally the machine will cut off the forelegs. This cutting will be done more precisely than in a manual operation.

When the planned development program has been completed the only remaining manual tasks on a clean slaughter line will be removal of the sexual organs, separation of viscera, trichina sampling, freeing ends of tenderloins, final trimming and veterinarian inspection.

Automation of pig carcass cutting and deboning

As shown in Figure 6 below, the first process in the cutting department is usually to cut the tenderloin from the half carcass. A robot based machine for automation of this process is currently being developed.

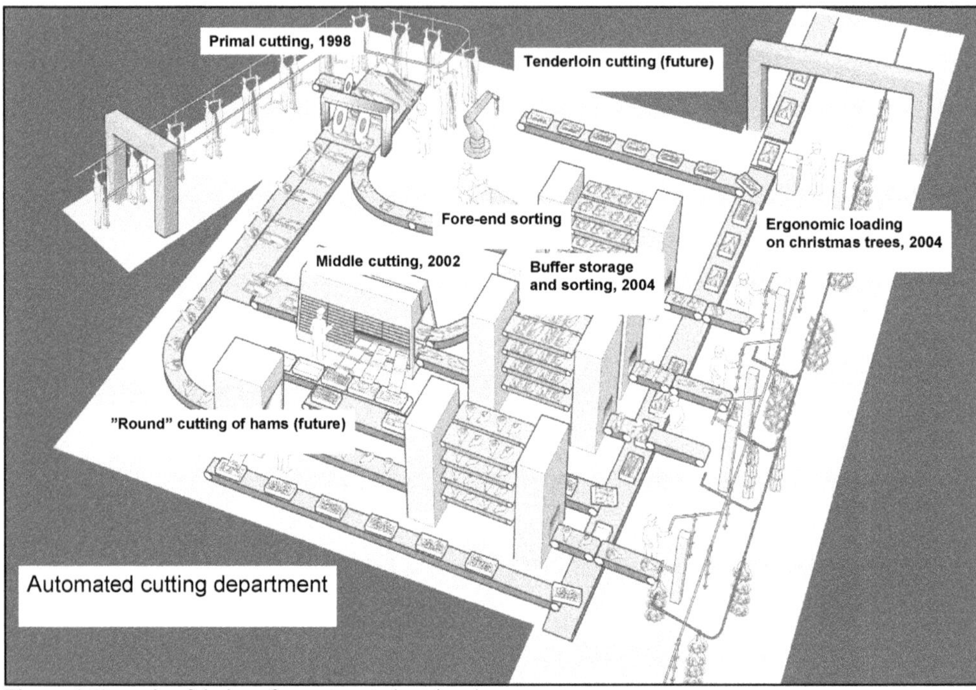

Figure 6: Example of design of an automated cutting department.

In 1998 a machine for *automatic carcass cutting* was introduced. The machine separates the half carcass into the three primal cuts: fore-end, middle and hind leg by sawing after precise measurements and positioning.

After carcass cutting the middles and the hind legs are conveyed to cutting machines while the fore-ends are moved directly to a buffer storage and sorting device.

The middles are processed in a machine finished in 2002 which cuts off the rib tops and separates the loin and the belly. The economic benefit of using the machine comprises savings on labour costs as well as added value by enhanced yield, because sawing of the rib tops can be more accurate and the division of the loin and the belly can be optimised according to the product specifications. From this machine the loins and bellies are moved to the buffer storage.

The hams will be processed by a new machine still in the experimental phase. The machine will automatically remove the tail bone and perform round cutting of the ham.

In the buffer storage the different cuts will be sorted according to customer specifications such as weight class, lean meat content, thickness of the fat layer *etc*. The cuts now arranged in quality classes will be loaded onto Christmas trees by means of a semi-automatic device, which will ensure that heavy lifts are avoided.

Automatic boning of fore-ends

Traditional boning of fore-ends requires a large number of operators and is a strenuous operation. In the future two types of machines – one for the removal of surface bones and one for the removal of interior bones - will be available for automatic boning of fore-ends and result in approx. 40% less operators, Figure 7.

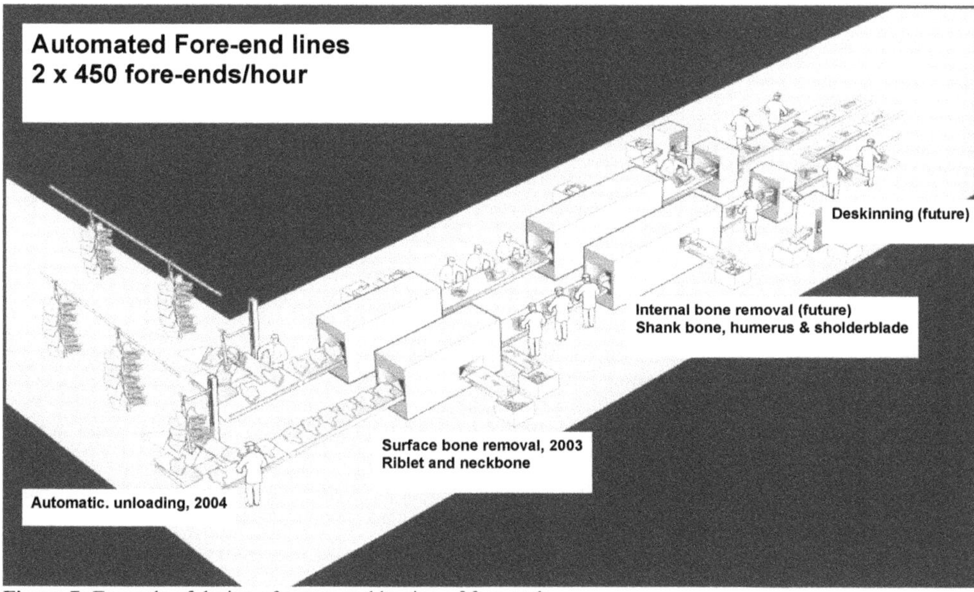

Figure 7: Example of design of automated boning of fore-ends.

The machine for removal of the surface bones - neck bone and riblet - was developed by DMRI and Townsend Engineering (Hansen, 2004). A number of these machines have been installed at Danish Slaughterhouses since 2004. The machine works at a rate of 450 fore-ends per hour and can be operated by one person (Figure 8). A further 2-3 operators carry out final trimming of the boned fore-ends. Very precise and powerful tools make it possible to cut closer and more uniformly to the bones. The machine thus improves the yield gain compared to manual boning.

Figure 8: Surface bone removal fore-ends.

Figure 9: Experimental model for internal bone removal in fore-ends.

The other machine for removal of the interior bones - shank bones, humerus and shoulder blade – is currently being developed by DMRI also in cooperation with Townsend Engineering. The machine is designed for flow boning, *i.e.* the fore-end is moved through the machine without stops at a rate of 450 fore-ends per hour.

The machine has been divided into three sections: entry, loader and processing. The fore-ends arrive at the machine with the shank end first and the rind downwards (Figure 9). After boning the fore-ends will leave the machine in one piece and continue for trimming, defattening and skinning. The machine replaces operators, and the yield is almost as good as that which can be achieved in manual boning.

Automation of beef slaughter

Slaughter procedures for cattle have been rather stable for decades and only limited technological development has taken place. Most of the development has been in the area of manually operated tools which have been improved to ease the physical work for operators or tools developed for improving the hygienic quality of slaughter.

In the USA, high line speed in excess of 300/hour is achieved by dividing processes across more operators and by ensuring that the animals slaughtered are relatively homogeneous in size, as slaughter lines are specialised for steers and heifers. Several pieces of equipments are duplicated and processes are divided across several machines, *e.g.* hide pulling. These plants run in several shifts.

The majority of cattle in the world are, however, still slaughtered at lower line speeds. In the EU, most of the larger plants run in single shifts at line speeds from 30-75/hour and the plants are seldom specialised but must slaughter all types of cattle as they are delivered (Figure 10). This implies that new technology must be flexible and match a large biological variation in dimensions.

Presently the situation is that the EU-beef industry is dominated by few new slaughter plants and many older slaughter lines on which a variable degree of upgrading and renovation are going on. A declining production, over capacity and ongoing structural changes imply that new investments in R&D have been limited.

Slaughter must mainly take place regionally due to welfare and zoonotic issues as well as geographical constraints. This implies that a high labour productivity is necessary, and to obtain this it is necessary to invest in new technology.

The benefits for the beef industry with more automation options are similar to that of the poultry, fish and pork industries with goals of increased efficiency, better food safety and hygiene as well as a better working environment. For retail packaging, the needs are identical.

Given that individual companies and industry organisations are simply too small to undertake a larger R&D task, it seems necessary to join R&D investments between interested parties internationally. This will reduce the risk and the individual financial contribution and ensure a better match with EU market requirements and better market access. Joint R&D must involve a critical selection of potential projects that have a reasonable payback time and/or result in other direct advantages (hygiene, ergonomics, food safety).

The processes most relevant for automation of beef slaughter depend mainly on labour costs and legislative requirement. In a joint Scandinavian study, the following areas suited for automation have been identified:

- Automatic splitting and removal of the spinal cord (SRM) in one process (spinal cord removal is likely to be a permanent requirement and involves several operators, Figure 11)
- Automatic bung cutting, neck and breast opening, and removal of head and tail. (more processes may be carried out with different tools in the same robot, Figure 12)
- Automatic separation (cutting) in hind and forequarter (*e.g.* ongoing R&D between e+v and Banss)
- Cleaning of dirty hides prior to slaughter (a manual equipment from SFK/MLA may be automated for cleaning of belly, Figure 13)
- Automatic hide pulling (this process is critical with respect to hygienic slaughter and carcass presentation)
- The area of boning still needs a thorough analysis, but preliminary evaluations show an interesting cost/benefit as well as documenting that the physical strain for boning operators doing piece work is extremely high.

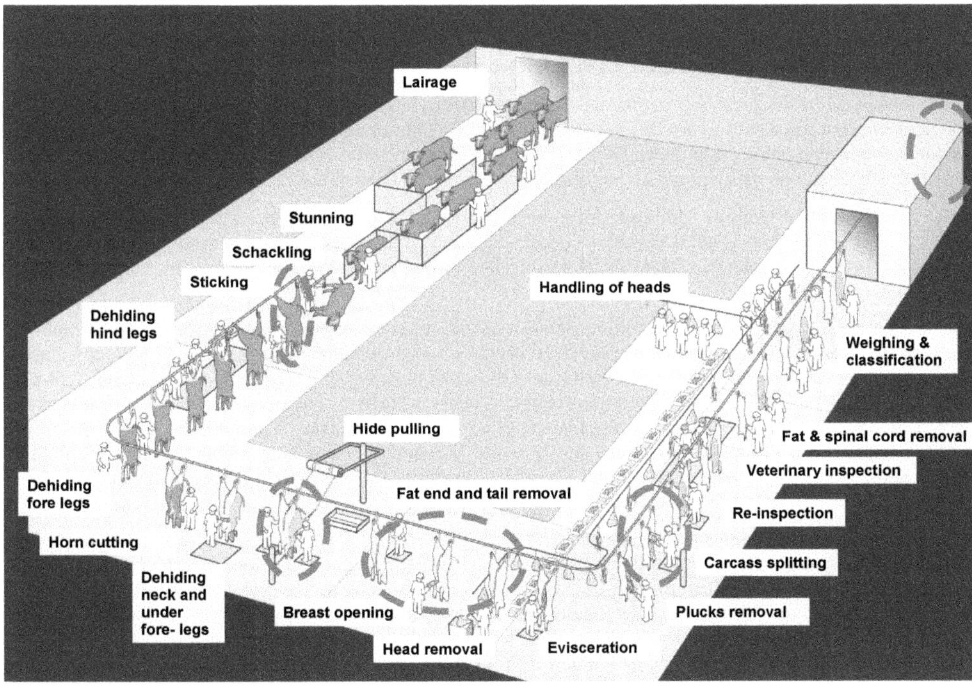

Figure 10: Traditional beef slaughter line with potential automation areas marked.

Some of the machines suggested are illustrated in Figures 11, 12, 13.

Figure 11: Splitting and spinal cord removal.

Figure 12: Automatic bung cutting, neck and breast opening, and removal of head and tail.

Figure 13: Automatic dedagger for cleaning the belly before dehiding.

Preliminary cost/benefit analysis of some of the machines above shows a pay back time of 3-6 years when doing one or two shifts per day. Considering that it is a complex and risky R&D process it is necessary that a sufficient market for machines is ensured and that the R&D cost can be carried by several interested parties. In addition the R&D project should aim to qualify for support by the technology platforms under the 7th EU framework.

Benefits and barriers and automation

It is important that automation is driven by an improvement in competitive strength. This means that it is a prerequisite for a successful commercialisation of the developed machines that they can be paid back within a short time frame, often as low as 3 years for meat industry investments if the production economy is the only parameter considered. The R&D investment to get a machine ready for the market is high, and market penetration and recovery of the R&D investment may take time.

Therefore substantial development will require a long term strategic decision, and international cooperation between the meat industry and the machine suppliers. Because of the risk and time profile of the R&D investments involved it will be of major importance to obtain public funding for this development.

The economic advantage of machines versus manual operations depends strongly on labour wages. For example, the economic benefit of automation looks very different with wages of €10/hour compared to €20/hour which implies that the introduction of new technology will depend strongly on the efficiency and labour market conditions. For example, access to low cost labour within the enlarged EC will influence the exploitation of automation technology.

An automated slaughter plant requires additional focus on monitoring the quality of the different processes to avoid quality defects or down time on the line. If deviations occur, these may affect further operations on the line on the next machines. This is solved by either increased visual monitoring by the remaining operators or by automatic vision systems.

A well run high tech slaughter line also requires intensified attention to organisational issues such as training, cooperation between professional groups and employee involvement, so production management will have to adjust to new values and policies. The management must emphasise a thorough planning of the commissioning phase. This plan must address both the technical issues and the organisational aspects in order to obtain optimal results. Unless skills, motivation and preparations for the technology are taken care of, the commissioning phase will be long or may even end up in failure. This is a well known experience *e.g.* from the introduction of milking robots for cows in the dairy industry, where some farmers returned to manual milking, while others were successful with identical machines.

Conclusions

Machines for automation of a major part of slaughter and boning operations for pigs are now available and provide competitive advantages and a better working environment, animal welfare, hygiene and food safety as well as yields. Their successful introduction requires careful planning of the implementation as well as attention to new organizational values and policies to reach the full benefits. The rate of exploitation internationally will depend on the labour market situation and international competition.

For beef slaughter, some potential areas for automation have been identified. As the beef industry structure is more fragmented and the development will be complex and resource demanding, a strategic R&D effort is likely to require cooperation between several partners to share the development investment and ensure industry relevance.

Acknowledgements

The machines presented in this paper have been developed by or in close cooperation with the following companies: ABB, ATTEC, Butina, e+v, Food Vision, Lodam, JARVIS, KJ INDUSTRIES, KUKA, SFK Meat systems, SFK-technology, Townsend Engineering. For some of the processes mentioned here there are other concepts and suppliers of technology. For example, STORK and JARVIS have several alternative equipment solutions.

The R&D program for pig slaughter was partly supported by the Danish Innovations law. The study for beef automation ideas was partly funded by Swedish Meats, Gilde Norsk Kjött and The Danish Livestock and Meat Board.

References

Anon. (2001). Slaughter-line for pigs is completed. An automated evisceration process is integrated into the slaughter-line at Danish Crown. *Fleischwirtschaft*,81,5, 111-115.

Folkmann, P. and Christensen, F. (2003). Innovation and advanced technology behind a new machine for automatic cutting of pork middles. *Fleischwirtschaft*, 83, 3, pp. 89-91.

Hansen, F. (2004). Robot ready for 47 million fore-ends. *Fleischwirtschaft*,84, 1, pp. 31-32.

Madsen, K.B. and Nielsen, J.U. (2004). Encyclopaedia of Meat Sciences I, 43-49. Amsterdam: Elsevier.

Madsen, K.B. and Nielsen, J.U. (2002). Automated meat processing. In: Kerry J and Ledward D (eds) *Meat Processing – Improving Quality*, pp. 283-296. Abington: Woodhead Publishing.

Nielsen, J.U. (1999). Automatic evisceration of pigs. In: *Meat Automation – Proceedings on Special User's Meeting on New Automation Technologies for Meat Production*, pp. 78-81. Brussels, Belgium.

Zink, J. (1995). Application of automation and robotics to pig slaughtering. In: *Proceedings of the 41st Congress of Meat Science and Technology*, pp. 420-424. San Antonio, Texas, USA.

NUTRITIONAL CONTRIBUTION OF RED MEAT IN THE DIET – THE STATE OF ART

M. Strong

Human Nutrition and Meat Science Manager, Meat and Livestock Commission, PO Box 44, Winterhill House, Snowdon Drive, Milton Keynes, MK6 1 AX, UK. E-mail address: maureen_strong@mlc.org.uk

Keywords: red meat, processed meat, nutrient composition, chronic disease, health

Introduction

The International Meat Secretariat (IMS) Human Nutrition and Health (HNH) Committee was formally established in 2004. Its membership includes a number of meat industry nutritionists from around the world. At the time it was agreed that certain synergies could be gained by working together, as individual industry organisations had limited expertise in this field, typically employing only one nutritionist in each country. Many countries have no formal organisation supporting their meat industry. The Committee's terms of reference are to advise the IMS on matters relating to human health and nutrition, to promote nutrition within the IMS and to act as a specialist group in contact with international organisations dealing with health and nutrition.

The HNH Committee recognised that if the meat industry was to effectively defend itself against ongoing negative criticism, it would need to amass a sound scientific evidence base. Thus, the Committee has established a regular dialogue, sharing the results of research and in some cases funding research where gaps exist. All this activity is based on the guiding principle that advice given is based on existing scientific evidence and appropriate dietary recommendation. The promotion of red meat globally is done within this context and in the context of a healthy balanced diet and lifestyle. However, it was identified at the time that there were gaps in the evidence base, and that the main health benefits associated with eating red meat relating to its nutritional composition, were being frequently overlooked in the literature.

In December 2004, the British Nutrition Foundation (BNF) was commissioned by the Meat and Livestock Commission (MLC) on behalf of the IMS HNH Committee to prepare a scientific review of red meat in the diet.

BNF is a scientific and educational charity, which promotes the well being of society through the impartial interpretation, and effective dissemination of evidence based nutritional knowledge and advice. It works in partnership with academic and research institutes, the food industry, educators and government. Its members include many of the UK's leading food industry companies.

The BNF was given a broad brief. The review should examine red meat consumption patterns, dietary and lifestyle factors associated with meat consumption and the effects of red meat intake on health and chronic disease outcomes. Both negative and positive aspects needed to be considered in order to present a balanced view which would withstand peer review. Individual organisations were able to contribute country specific data.

This resulted in a comprehensive review report which has now been published (BNF 2005). The purpose of this paper is to give a summary of the key findings. It will focus on the health aspects of red meat in relation to links with chronic disease (particularly colorectal cancer (CRC) and cardiovascular disease (CVD)). Issues such as the definition of red and processed meat, consumption, composition, contribution to the diet and dietary and lifestyle aspects of red meat consumers are discussed. The intention is to present a condensed version of the review to a wider meat science and meat industry audience who may have otherwise been unaware that this work has been done.

Defining Red and Processed Meat

Much of the evidence investigating associations between meat intake and health/disease outcomes is based on epidemiological studies. Not all studies define what is meant by meat and, where definitions are offered, they are not always the same, which can make comparisons between studies difficult. Some studies include poultry under the definition of meat, while others exclude it; some look at total meat consumption (including red meat and processed meat), while others analyse red meat and processed meat separately. In general, red meat refers to beef, pork and lamb in main dishes and processed meat refers to meat products, such as sausages, burgers and smoked, cured and tinned meats. Offal is also a form of meat, but there is little epidemiological evidence specific to this category of meat.

The following definitions for red and processed meat are based on definitions currently used in epidemiological studies looking at the health effects of meat consumption, such as the European Prospective Investigation into Cancer and Nutrition (EPIC) study.

Red meat includes beef, veal, pork and lamb (fresh, minced and frozen).
Processed meat includes meat that has been preserved by methods other than freezing, such as salting, smoking, marinating, air drying or heating *e.g.* ham, bacon, sausages, hamburgers, salami, corned beef and tinned meat.

Red Meat Consumption

Data from national dietary surveys indicate that intakes of meat vary widely between countries, and between subgroups of the population within countries. However, some similarities have been found; for example, women tend to consume less meat than men. As different methodologies are adopted to collect information on food intake, it is difficult to make direct comparisons between data from different countries. Table 1 shows average daily intakes of red and processed meat in a range of European countries, as determined by the EPIC calibration study.

Table 1: Mean daily intake (g/day) of total meat, red meat, processed meat and red + processed meat in selected countries participating in the EPIC calibration study.

	Total meat*		**Red meat**		**Processed meat**		**Red meat + processed meat**	
	Men	**Women**	**Men**	**Women**	**Men**	**Women**	**Men**	**Women**
Greece	78.8	47.1	45.3	25.5	10.0	5.8	55.3	31.3
Spain	170.4	99.2	74.0	37.8	52.8	29.6	126.8	67.4
Italy	140.1	86.1	57.8	40.8	33.5	19.6	91.3	60.4
Germany	154.6	84.3	52.2	28.6	83.2	40.9	135.4	69.5
Netherlands	155.6	92.7	63.8	41.0	72.4	37.9	136.2	78.9
UK	108.1	72.3	40.0	24.6	38.4	22.3	78.4	46.9
Denmark	141.1	88.3	69.6	44.1	51.9	25.3	121.5	69.4

*Total meat includes pork, beef, veal, lamb/mutton, poultry, game, rabbit, horse, goat and offal.
Source: Linseisen *et al.* (2002)

When considering the health effects of red meat consumption, it is useful to consider how meat is consumed as part of the diet (*e.g.* portion size) and the type of meat consumed. From the EPIC data it is evident that different European countries favour certain meats.

Data from the recent North/South Food Consumption Survey for Ireland found that, compared to all meats, portions of beef and veal were largest (on average 60g), and consumed by 55% of the sample (Irish Universities Nutrition Alliance, 2001). According to the survey, the most commonly consumed meats in Ireland were bacon, ham and processed meats (consumed by 71% of the sample). Average meat intake increased from 87g to 134g per day when intakes of all processed and carcass meats were included, although the average intake of processed meat was lower than for red meat alone.

In the UK, bacon and ham were also the most commonly consumed meats (77% men and 64% of women consuming these products) (Henderson *et al.* 2003a). Overall, carcass meat consumption in the UK has been recorded at just 32g per day (Department of Food and Rural Affairs, 2005). However, this does not take into account ready meals and convenience meat products that have become very popular in the UK. The popularity of these foods, along with other meat products such as meat pies, makes it difficult to accurately measure the average amount of red meat consumed. For example, data from the North/South Ireland Food Consumption Survey found that there was a 43% overestimation of meat intake, without disaggregation of composite foods (such as lasagne, sausage rolls) (Cosgrove, 2005a).

Nutrient Composition

Another challenging aspect of investigating patterns of red meat intake is that the composition of different types of meat varies widely. For example, lean red meat is low in total fat, saturated fatty acids (SFAs) and salt, whereas untrimmed meat is higher in both total fat and SFAs; processed meat is generally higher in salt and fat and often contains other components, such as nitrites, which are added during processing. Moreover, the composition of different processed meats and the types of processed meat consumed vary widely between countries. It is therefore important to distinguish between different types of meat when looking at patterns of intake, as well as considering the health effects of red meat. This is not always straightforward, however, as definitions vary and not all studies distinguish between different kinds of meat consumed.

Data on the nutrient composition of red meat are available in food composition tables and databases. There are over 150 food composition tables and electronic databases worldwide and undoubtedly meat and meat products will be listed in all of these, although slightly different nutrient values are likely to be found in different versions. Table 2 shows the nutrient composition of 100g of lean raw beef, lamb and pork according to the food composition tables of four different countries, and illustrates how these figures can vary.

Meat is principally composed of protein (which provides 17kJ/4kcal of energy per gram), but also contains fat in varying amounts (providing 37kJ/9kcal of energy per gram). The more fat that meat contains, the higher the energy content will be, as shown in Table 3.

Table 2: Comparison of selected nutrients in beef, lamb and pork (per 100g) according to food composition databases from four countries.

	Denmark	UK	Australia	USA
Beef, lean, raw				
Energy (kJ)	470	571	520	531
Protein (g)	22.3	22.5	23.0	22.3
Fat (g)	2.5	5.1	3.6	3.5
Niacin (mg)	10.1	5.0	3.0	6.5
Vitamin B_{12} (µg)	1.4	2.0	1.1	0.9
Iron (mg)	2.4	1.8	2.0	1.6
Zinc (mg)	4.7	4.1	4.2	4.0
Selenium (µg)	6.5	7.0	10	30.8
Lamb (leg), lean, raw				
Energy (kJ)	545	651	501	561
Protein (g)	20.2	20.2	20.4	20.5
Fat (g)	5.5	8.3	4.2	5.1
Niacin (mg)	7.5	5.4	5.6	6.3
Vitamin B12 (µg)	1.2	2.0	0.9	2.8
Iron (mg)	2.2	1.4	2.3	1.8
Zinc (mg)	3.3	3.3	3.4	3.8
Selenium (µg)	1.4	2.0	10.0	23.4
Pork, lean, raw				
Energy (kJ)	445	519	N/A	502
Protein (g)	21.6	21.8	N/A	20.9
Fat (g)	2.1	4.0	N/A	3.4
Niacin (mg)	7.3	6.9	N/A	4.4
Thiamin (mg)	0.8	1.0	N/A	1.0
Vitamin B_{12} (µg)	0.7	1.0	N/A	0.8
Iron (mg)	0.7	0.7	N/A	1.2
Zinc (mg)	3.6	2.1	N/A	2.0
Selenium (µg)	6.9	13.0	N/A	28.9

N/A: data was not available.

Sources: Danish Institute for Food and Veterinary Research (2005b); Chan *et al.* (1995); Red Meat and Health Expert Advisory Committee (2001); United States Department of Agriculture (2005).

Table 3: Energy, fat and protein content of lean and untrimmed cuts of red meat (per 100g; UK figures).

Meat (barbequed or grilled)	Energy kJ (kcal)	Fat (g)	Protein (g)
Rump steak – lean	741 (176)	5.7	31.2
Rump steak - lean and fat	849 (203)	9.4	29.5
Leg joint of lamb – lean	879 (210)	9.6	30.8
Leg joint of lamb – lean and fat	986 (236)	13.0	29.7
Loin chops of pork – lean	780 (186)	6.8	31.1
Loin chops of pork – lean and fat	1066 (255)	15.8	28.3

Source: Chan *et al.* (1995).

Advances in food processing technologies and breeding programmes, as well as modification of animal feeds and modern butchery techniques have all led to a reduction in the fat content of carcass meat over the past 15 years. For example, the fat content of carcass meat in the UK has been reduced by over 30% for pork, 15% for beef and 10% for lamb (Higgs, 2000; Lee *et al.* 1995). New Zealand has seen similar changes, and in particular has worked with butchers to encourage the trimming of fat from red meat before sale, through the introduction of a 'Quality Mark'. These efforts have resulted in a 48% reduction of fat in beef, and a 43% reduction in lamb (Laugesen, 2005).

Overall, lean red meat contains similar proportions of monounsaturated fatty acids (MUFAs) to saturated fatty acids (SFAs), although as illustrated in Table 4, the exact proportions vary depending on the type of meat (Chan *et al.* 1995).

Table 4: Typical fatty acid composition (g/100g) of different types of red meat (lean only, cooked) (UK figures).

	Beef	Lamb	Pork	Bacon (grilled)
Total SFA	3.26	5.36	2.31	7.91
Total MUFA	3.41	4.06	2.56	8.85
Total PUFA	0.38	0.59	1.15	2.71
n-6 PUFA	0.36	0.48	1.02	2.41
n-3 PUFA	0.09	0.23	0.12	0.31

Source: MAFF (1998).

Although red meat is seen as a contributor to SFAs intake, lean red meat actually contains a higher proportion of unsaturated fatty acids (see Table 4). Meat, primarily lean meat, also contains the long-chain *n*-3 fatty acids, eicosapentaenoic acid (EPA) and docosapentaenoic acid (DPA). Although levels of these are low, there are few other food sources (the richest source being oily fish), and therefore lean red meat can make a contribution to intake of these fatty acids. Indeed, an Australian study has shown that beef, lamb and pork provide 32% of these fatty acids to the diet (Howe *et al.*, 2003). Furthermore, much work is currently underway to identify methods through which to alter the fatty acid profile of foods, such as meat, in order to reflect a positive fatty acid profile in terms of heart health (Givens and Shingfield 2004).

Meat from ruminant animals is a source of another *naturally occurring* fatty acid - conjugated linoleic acid (CLA). CLA is a collective term used to describe a mixture of positional and geometric isomers of linoleic acid. CLA is found naturally in small amounts in products from ruminant animals *e.g.* lamb, beef, cheese and milk. Feeding practices influence the CLA content of meat. For example, concentrations of CLA in Irish and Australian beef can be 2 to 3 times higher than those in beef from the USA, reflecting the greater consumption of pasture throughout the year (Moloney, 2006). Based largely on animal studies in rodents, there is interest in the potential health benefits of CLA in humans in connection with cancer, blood lipids and an influence on the lean:fat tissue ratio (in favour of lean tissue) (Calder, 2002); however, no clear conclusions have yet been made with respect to human health.

Processed meats and meat products that contain lower amounts of lean meat are likely to have a lower micronutrient content per 100g, but may provide other nutrients not usually found in meat (*e.g.* carbohydrate and fibre). Also, the addition of some ingredients (*e.g.* soy, fibre) can be used to add functional properties to meat products that could offer potential benefits for health (Fernandez-Gines *et al.*, 2005).

Overall meat products and processed meats are more likely to have a higher content of sodium than lean meat. Sodium is added to meat products to enhance and modify the flavour, the physical properties and sensory attributes of the food, and to contribute to the preservation of the product (Matthews and Strong 2005). Owing to the adverse health effects associated with a high intake of sodium, work is underway within the UK to reduce the amount of salt in processed meat products and, in particular, products that contain the highest levels of sodium (Matthews and Strong 2005).

Contribution to Nutrient Intake

Despite the variations in composition with country, species, cut, fatty acid profile and in some cases the reformulation of processed meat products, meat and meat products can make an important contribution to nutrient intakes in the diet. They provide a number of essential nutrients, including protein, long-chain *n*-3 fatty acids, iron, zinc, selenium, vitamin D and vitamin B_{12}.

Red meat is a well recognised source of bioavailable iron, contributing around 20% of iron intake, on average, in developed countries (Henderson *et al.*, 2003b; Russell *et al.*, 1999). In the light of current low levels of iron intake in many developed countries, particularly among women, meat has the potential to make an important contribution to intakes. A number of studies have confirmed the positive effect of including meat in the diet, on intakes of iron. For example, a study by Gibson and Ashwell (2003) found an increased risk of low iron intake (below the UK LRNI) in those who ate the least meat and processed meat ($\leq$90g per day), compared with high consumers ($\geq$140g per day).

In some countries, for example the UK, there are also concerns over low intakes of selenium (BNF2001;SACN 2005b). Meat is one of the main providers of selenium in the diet, particularly in parts of the world where selenium levels in the cereals consumed are low.

Red meat is also classified as a *source* of zinc, in EU food labelling terms (BNF 2002). Red meat consumption provides 30-40% of the recommended zinc intake in some countries. As with iron, the zinc in meat is in a highly bioavailable form, making meat a more 'efficient' provider of this mineral. Concern exists, *e.g.* in the UK, about the zinc intakes of some population groups (Henderson *et al.*, 2003b).

Red meat also contains a range of B vitamins, although the levels vary between types of meat, in particular it is classified as a *rich source* of vitamin B_{12}, in EU food labelling terms. As this vitamin is only found naturally in foods of animal origin, sub-groups of the population who do not consume meat or animal products may have inadequate intakes (Herbert, 1988;1994).

Red meat is also a good source of vitamin D, which may be more easily utilised than the vitamin D present in other foodstuffs (Groff *et al.*, 1995). Furthermore, it has been suggested that components of meat protein may enhance the utilisation of vitamin D in humans, particularly where exposure to sunshine is limited (Dunnigan and Henderson 1997).

Diet and Lifestyle Factors Associated with Red Meat Consumption
There are relatively few large-scale studies that have looked at the dietary and lifestyle habits of meat-eaters compared to other dietary groups, such as vegetarians. However, this topic has recently been examined in two large UK cohorts: the EPIC-Oxford cohort (Davey *et al.*, 2003) and the UK Women's Cohort Study (Cade *et al.* 2004). Some distinct differences in the diet and lifestyle characteristics of meat-eaters, compared to other dietary groups have been highlighted. In particular, meat-eaters seem to be generally older, have a higher BMI (which may be linked) and are more likely to smoke compared to non meat-eaters. As might be expected, nutrient intakes were found to differ between meat-eaters and other dietary groups. Meat-eaters were found to have higher protein, total fat and SFA intakes and lower carbohydrate and fibre intakes than non meat-eaters, and their consumption of fruit and vegetables was also lower. Marked differences in micronutrient intakes were also found.
However, it is noteworthy that these cohort studies do not differentiate between consumers of unprocessed and processed meat. A recent cross-sectional study of Irish adults indicates that it is important to distinguish between meat groups as there is a large variation in dietary quality between consumers of red meat, white meat and processed meat. For example, increasing processed meat intake has been found to be associated with a lower intake of wholemeal bread, fruit and vegetable and fish intake and poorer overall dietary quality (Cosgrove *et al.* 2005b).
The variation in diet and lifestyle characteristics between meat-eaters and non meat-eaters may have important implications for associations between meat consumption and health and disease. However, vegetarians represent only a small minority of the population *e.g.* in the UK it is estimated that around 3-7% of the population are vegetarian (Phillips 2005). Furthermore, the mortality rates of vegetarians are similar to those of comparable non-vegetarians, suggesting that much of this benefit may be attributed to non-dietary factors, such as lower prevalence of smoking or to other aspects of the diet other than the avoidance of meat and fish *e.g.* higher intake of fruit and vegetables (Appleby *et al.*, 2002).

Health Aspects of Red Meat
Various research studies have been conducted to try to determine whether there is a link between red and processed meat intake and a number of conditions, including colorectal cancer (CRC), cardiovascular disease (CVD), obesity and type 2 diabetes.

Colorectal cancer (CRC)
Most of the published literature on meat in relation to cancer development has focused on CRC. There have been some studies investigating possible associations between meat and other types of cancer, including gastric, breast, prostate and kidney cancers and cancer of the pancreas, however the evidence in relation to these other types of cancer has been found to be weak or inconsistent (Department of Health, UK, 1998; Key *et al.*, 2004).
The scientific interest in the association between red and processed meat intake and CRC over recent years has generated a number of reviews, including two large reports, one from the World Cancer Research Fund (WCRF) and another from the UK Committee on Medical Aspects of Food and Nutrition Policy (COMA). Both panels agreed that the results of epidemiological studies were not consistent. The conclusion of the WCRF (1997) report was that "The evidence shows that red meat *probably* increases risk and processed meat *possibly* increases risk of colorectal cancer". A new analysis from the WCRF is expected in 2007.
The COMA report (Department of Health, UK, 1998) concluded that "there is moderately consistent evidence from cohort studies of a positive association between the consumption of red or processed meat and risk of colorectal cancer." This report recommended that the current, average level of red and processed meat intake in the UK should not increase. Those with high levels of intake (>140g per day) were recommended to reduce their consumption.

CRC is the third most common cancer in the world. In 2002, CRC was estimated to account for over a million new cancer cases worldwide (9.4% of all cancer cases) (Ferlay *et al.* 2004). Genetic changes associated with colorectal cancer are well established and inherited mutations in key genes are considered to be responsible for about 20% of cases. The remaining 80% of cases are sporadic (*i.e.* arise spontaneously) and appear to be influenced by environmental and lifestyle factors, such as diet and physical activity level. This is also supported by the findings of epidemiological studies (see below).
Incidence rates for CRC are approximately 10-fold higher in developed compared to developing countries (Ferlay *et al.*, 2004). Changes in incidence rates have also been noted in populations over time. CRC was very rare in Japan in the 1960s, however there has been an almost five-fold increase in CRC in Japanese men over the past 30 years, and incidence in Japanese men aged 55-60 is now twice that of men in the UK. This cannot all be attributable to genetic effects as there cannot have been such a rapid change in the Japanese gene pool over such a short period of time. Environmental factors such as diet, therefore, must play a part (Bingham and Riboli, 2004).
It has been suggested that the contribution of diet to CRC incidence could be as much as 80% (Willett, 1995). If this is the case, it means that the majority of CRC cases may be preventable. Few specific diet-related factors have been shown

unequivocally to contribute to the pathogenesis of CRC, however, there is a general consensus that some aspects of the western diet increase the risk of CRC, owing to the large variation in incidence between developed and developing countries (World Health Organisation, 2003).

Table 5 shows diet-related risk factors for which there is evidence of an association with CRC risk. Also, it has been shown that risk factors tend to cluster, so individuals who are obese will often be physically inactive, smoke and consume low levels of fruit and vegetables and high levels of meat (Potter, 1999).

Table 5: Diet-related risk factors for CRC (from Key *et al.*, 2004).

Diet-related risk factor	**Level of evidence**
Overweight/obesity	Increases risk (best established diet-related risk factor)
Physical activity	Consistent association with reduced risk
Fruit and vegetables	*Probably* decreases the risk
Red and processed meat	*Probably* increases risk
Alcohol	*Probably* contributes to a small increase in risk

There is some evidence that several other diet-related factors may be associated with a reduced risk of CRC, including dietary fibre, folate, calcium and vitamin D. Evidence in relation to folate, calcium and vitamin D intake has not been firmly established, however, recent findings from EPIC study have shown dietary fibre to be inversely related to CRC incidence (Bingham *et al.* 2003).

In the UK, the incidence of CRC has increased substantially over the past 35 years, yet red meat intake has declined by around 25% over the same period. A similar pattern has been seen in other European countries, such as Norway, where risk of CRC has increased by 50% over the same period (Hill, 2002).

Table 6 shows the mean daily intake of different types of meat for 10 European countries from the EPIC study (females only shown) together with CRC incidence for each country. Meat consumption in the UK was found to be less than in many Mediterranean countries, such as Spain and Italy, and yet CRC incidence is higher in the UK than several Southern European countries.

The two meta-analyses carried out by Sandhu *et al.* (2001) and Norat *et al.* (2002) provide evidence from the combined results of cohort (and case-control) studies of an association between high red and processed meat intake and CRC risk. This is supported by the results of some more recent large prospective studies, such as the EPIC cohort. EPIC results also indicate that processed meat intake may be a stronger risk factor for CRC than red meat intake (Norat *et al.* 2005).

The underlying mechanism for this association is still uncertain, although recent human intervention studies have suggested that the endogenous *N*-nitrosation arising from ingestion of haem iron (not inorganic iron or protein) could be the most likely mechanism (Cross *et al.* 2003). This would also help explain why poultry, which contains much lower amounts of haem, is not associated with increased CRC risk. It has been suggested that processed meat intake is more strongly associated with CRC risk than unprocessed red meat, however, there is currently no straightforward mechanism that could explain this. Nitrites or nitrates added to meat during processing could increase exogenous exposure to nitrosamines and other NOCs, but not all processed meats contain added nitrites (Norat *et al.* 2005).

Table 6: Variation in meat consumption and CRC incidence across 10 EPIC countries.

	Mean daily intake (g/day)					
	All meat	Red meat	Poultry	Sausages	Other processed	CRC Incidence (ASR)
Denmark	88.3	44.1	16.8	15.6	9.8	19.2
Norway	88.6	28.5	10.6	22.9	23.6	16.8
Germany	84.3	28.6	13.0	29.4	11.5	15.7
The Netherlands	92.7	41.0	12.4	16.4	21.5	14.4
UK	72.3	24.6	24.0	9.3	13.0	12.4
France	106.0	44.4	21.8	12.2	17.8	11.8
Spain	99.2	37.8	24.4	13.1	16.5	11.3
Sweden	92.3	32.3	9.3	21.3	22.0	11.1
Italy	86.1	40.8	20.2	9.52	10.1	10.9
Greece	47.1	25.5	11.8	3.1	2.8	8.0

Data shown taken from EPIC calibration study (females only). Data adjusted for age, day of the week, season (Linseisen *et al.* 2002). CRC incidence reported as age standardised (world) rate (ASR) per 100,000 (Ferlay *et al.* 2004)

It is important to put these findings in context, however. Red and processed meat intake is only one of a number of potential risk factors for CRC. Indeed, the most established diet-related risk factors for CRC are overweight and obesity and low physical activity level. Therefore reducing the risk of CRC involves modification of a number of dietary and lifestyle factors including maintaining a healthy body weight, increasing physical activity level and consuming more fruit and vegetables and dietary fibre (Slattery, 2000).

Moreover, whether the consumption of meat is an *independent* risk factor for CRC is still not certain. Colorectal carcinogenesis is a multi-step process involving many different factors; a number of biological pathways may be involved and an accumulation of alterations to DNA occur. It is therefore unlikely that factors determining CRC risk act in isolation. Furthermore, because not all studies have actually examined the independent effect of meat intake, it is not possible to exclude the possibility that the association may be confounded by other dietary, genetic or associated factors (Sandhu *et al.* 2001). Further studies with long-term follow-up, repeated measures of diet, more in depth consideration of dietary patterns, more detailed measures of cooking methods, and genetic markers of susceptibility may therefore be required (Willett, 2005).

Nevertheless, a picture seems to be emerging that patterns of food intake are important, *i.e.* any increased risk associated with a high intake of red and processed meat appears to be lessened if dietary fibre intakes are increased (Norat *et al.* 2005). Thus lean red meat is unlikely to significantly increase the risk of CRC when consumed in moderation and as part of a healthy balanced diet that includes plenty of fibre from fruit, vegetables, pulses and whole grain cereals.

Thus, currently available evidence suggests that the UK guidelines set by COMA with respect to reducing CRC risk are still appropriate. These suggest that an individual's consumption of red and processed meat should not rise and that higher consumers (> 140g per day or 12-14 portions per week) should consider a reduction in intake (Department of Health, UK, 1998). It is worth noting that recommendations in different countries vary somewhat; for example, the Australian Cancer Council of New South Wales (NSW) recommends that people consume moderate amounts of red meat (65-100g of cooked red meat, 3-4 times a week) and limit consumption of processed meats (The Cancer Council NSW, 2003). The American Cancer Society recommends limiting consumption of red meats, especially those high in fat and processed meat (American Cancer Society, 2001).

Average daily intakes of red and processed meat in most countries are still below the level thought to increase the risk of CRC. For example, average intakes of red meat in Europe (in men) range from 40g per day in the UK to 74g per day in Spain, while average intakes of processed meat (in men) range from 10g per day in Greece to 83.2g per day in Germany (see Table 1) and therefore it is only the small proportion of high consumers of meat and meat products that may need to consider a reduction in consumption.

Cardiovascular Disease (CVD)

CVD, which includes coronary heart disease (CHD) and stroke, is the leading cause of death worldwide, accounting for 18 million deaths each year. It has a multifactorial aetiology. Although there are a number of unmodifiable risk factors, including a genetic predisposition, many of the major causes such as hypertension, obesity, diabetes, high blood cholesterol and high blood triglyceride concentrations are modifiable by diet, along with other lifestyle factors (*e.g.* physical activity and smoking cessation) (Stanner 2005).

Meat has often been assumed to be a contributor to increased risk of heart disease because of its relatively high contribution to fat intakes and its perceived high content of SFAs, but as previously discussed advances in food processing technologies and breeding programmes, as well as modification of animal feeds and modern butchery techniques have all led to a reduction in the fat content of carcass meat over the past 15 years.

Prospective studies have demonstrated a positive association between meat and CVD risk, but any causal link remains to be established. For example, Fraser (1999) reported a significant association between beef consumption and fatal CHD in a male cohort of Seventh-Day Adventists, with those eating beef up to 3 times a week having a 1.9-fold increased risk, and those eating beef 3 or more times a week having a 2.3-fold increased risk, compared to vegetarians. However, the study found no association in women and no significant increase in risk of non-fatal myocardial infarction (MI). Subdivision of subjects in a pooled analysis of vegetarian studies into meat-eaters (those eating meat more than once a week), occasional meat-eaters (less than once a week), fish-eaters and vegetarians, demonstrated the death rate ratio to be significantly below 1 for occasional meat-eaters, fish-eaters and vegetarians, using the meat-eaters as the reference population (Key *et al.* 1998; 1999a).

In the Nurses' Health Study, consumption of red meat was associated with an increased risk of CHD, after adjustment for age (relative risk for one additional serving per day was 1.43, 95% CI:1.35-1.65) and this was attributed to its contribution to intake of SFAs (Hu *et al.* 1999a). However, the association was substantially attenuated and no longer significant in multivariate analyses.

In the Iowa Women's Health Study, higher consumption of red meat was significantly associated with CHD mortality (risk ratio 1.44, 95% CI: 1.06 - 1.94) (Kelemen *et al.* 2005). Amongst a cohort of men in the US, intake of haem iron, particularly from red meat, was significantly related to increased risk of fatal CHD or non-fatal myocardial infarction (MI) after adjustment for dietary cholesterol and fats (Ascherio *et al.* 1994). As meat- eaters vary from non or infrequent meat eaters in a number of ways (*e.g.* they are often more likely to smoke and have a higher BMI), it is very difficult for studies to isolate the effects of meat *per se,* even if they correct for a number of recognised potential confounders. Moreover, none of these investigations have attempted to distinguish between the effects of different types of meat, particularly lean versus untrimmed meat or processed versus unprocessed meat.

Dietary intervention studies have, however, suggested that whilst untrimmed meat is cholesterol-raising, this is not true of diets containing fat-trimmed lean meat. Most studies have demonstrated lean red meat to have similar effects on total, LDL- and HDL-cholesterol or triglyceride levels as white meat or soybean products (Li *et al.* 2005). In fact, a recent review of dietary intervention and cross-sectional studies concluded that diets low in SFAs and containing lean red meat are associated with a reduction in LDL-cholesterol levels in both healthy subjects and those with hypercholesterolaemia. Studies also suggest that lean meat does not have an adverse effect on blood lipids in patients with existing CVD (Watts *et al.* 1988). Lean red meat was also found to have no effect on thrombotic risk factors (Li *et al.* 2005).

Meat from ruminant animals contains some *trans* fatty acids as well as conjugated linoleic acid (CLA). *Trans* fatty acids are recognised to have a more potent effect on blood cholesterol than SFAs, by raising levels of LDL-cholesterol and lipoprotein (a) and decreasing HDL-cholesterol levels, although this may not occur with the natural *trans* fats in meat and milk (BNF, 1995b). Prospective studies, such as the Nurses' Health Study, have shown a high intake of *trans* fatty acids to increase risk of CHD (Oh *et al.* 2005, Hu *et al.* 1997) but current European diets are generally sufficiently low in these fatty acids not to warrant concern (Hulshof *et al.* 1999). In Britain, current intake is low, at 1.2% of energy (Henderson *et al.* 2003a) (Dietary Reference Value is 2% energy), whilst reductions in intake are less apparent in North America.

There have been reports of some interesting isomer-specific effects of CLA on the blood lipid profile in human subjects, however, findings have been inconsistent and this might be attributed to the variability of the dose level and/or the mix of CLA isomers used, particularly as results from animal studies show that specific isomers of CLA may be responsible for specific biological effects (Stanner, 2005). Unfavourable effects of a high dose of one of the CLA isomers (*t*10*c*12 CLA, 3.4g/day) has also been reported in relation to increased insulin resistance, oxidative stress and inflammatory biomarkers, and whilst such an intake is not consumed by diet alone, at present the effects of CLA on human health remain unclear (Tricon *et al.* 2005).

In the GISSI-Prevenzione Investigators 1999 trial, high doses of long-chain PUFAs from fish or supplements were given to patients who had already had a heart attack, and demonstrated a reduction in cardiac events (*e.g.* death, nonfatal MI and non-fatal stroke) (Burr *et al.* 1989). Prospective studies (*e.g.* the Nurses' Health Study) have also shown an inverse association between fish consumption and *n*-3 fatty acids and CHD deaths (Hu *et al.* 2002). Despite being present at low levels, particularly when compared with oil-rich fish, meat, primarily lean meat, contains medium (α-linolenic acid) and long-chain *n*-3 PUFAs (EPA, DPA and DHA) and is a significant source of these cardio-protective fatty acids for the average consumer.

The relative risk of both CHD and stroke increases as blood pressure rises. In a recent review of the effect of sodium on blood pressure, the Scientific Advisory Committee on Nutrition (SACN) in the UK concluded that although studies that have prospectively collected 24-hour urine (a good marker of sodium intake) suggest that a high salt (sodium chloride) intake has adverse effects on CVD mortality, there are insufficient reliable data on morbidity and premature mortality outcomes to reach clear conclusions (Scientific Advisory Committee on Nutrition, 2003). However, SACN concluded that reducing the average salt intake in Britain would confer significant public health benefits by contributing to a reduction in CVD burden. Similar recommendations have been made elsewhere (American Heart Association, 2000). Carcass meat contains very little sodium naturally, but salt is added to meat products for a variety of technical reasons.

Thus whilst red meat contains SFAs, a high intake of which can have adverse effects on CVD risk factors such as blood cholesterol levels, it also contains other fatty acids (*n*-3 PUFAs, MUFAs) and nutrients (*e.g.* B vitamins and selenium) that offer potential cardioprotective benefits. Processed meat products make a substantial contribution to total salt intake which is thought to influence blood pressure, a risk factor for CVD. To address this the meat industry in the UK has been working to reduce the salt content of meat products for some time (Matthews and Strong 2005).

Obesity

Meat-eaters have been shown to have a higher BMI than comparable vegetarians (Key *et al.*, 1999b); however, evidence of an association between meat consumption and obesity does not automatically indicate that there is a causal link. Furthermore there are problems with interpreting studies comparing meat eaters with vegetarians, as vegetarians tend to be more health-orientated generally *e.g.* consuming more fruit and vegetables and having higher levels of physical activity. Obesity is a complex disorder with a diverse range of causal factors and to identify one dietary factor as causal would be a gross oversimplification of a complex process.

Nevertheless, there is modest evidence that a high dietary fat to carbohydrate ratio in the diet is more likely to promote obesity development (British Nutrition Foundation 1999b). Meat makes a relatively high contribution to dietary fat intakes; a recent UK dietary survey indicated that meat and meat products contribute around 23% of total fat intake (Henderson *et al.* 2003a). However, with the progressive decline in the fat content of carcass meat, lean red meat can now contain as little as 2% fat. As dietary fat comes from a variety of sources, reducing the fat content of the diet *as a whole* (target recommendation 35% of food energy) is a key feature of health policy activities.

There is some evidence that it can be advantageous to include lean meat in weight loss diets, as high protein intakes have been found to lead to increased satiety (Stubbs, 1995). Halton and Hu (2004) recently conducted a systematic review of studies investigating the effects of high protein diets on thermogenesis, satiety, body weight and fat loss. Overall, they concluded that there is some evidence that diets higher in protein lead to increased weight loss and fat loss compared to diets lower in protein in the short-term (6-month period), but further longer-term studies are needed. Possible mechanisms to explain this effect include increased satiety from protein, decreased subsequent energy intake and the displacement of carbohydrate in higher protein diets. It seems likely that several of these mechanisms work together and are inter-related (Hu, 2005).

Type 2 Diabetes

The prevalence of type 2 diabetes is increasing rapidly worldwide and this is thought to be linked to the increasing prevalence of obesity. Currently an estimated 120 million people worldwide are affected by type 2 diabetes and the incidence has been predicted to double to 215 million by 2010. There are a number of risk factors for type 2 diabetes, the most important being genetic predisposition, obesity and physical activity level (British Nutrition Foundation 2004). Evidence from a small number of recent cohort studies in the US suggest that frequent consumption of processed meat (*e.g.* 5 times or more per week) may be associated with an increased risk of type 2 diabetes. However, to date there have been no intervention studies (the means of demonstrating cause and effect) and purported mechanisms are merely speculative. The evidence that exists points to any effect being linked with processed rather than lean red meat, and processed meat products can be high in fat. Hence the findings are in accord with current recommendations to reduce fat intake and to include lean meat in the diet, in moderation.

People with type 2 diabetes, like the general population, are recommended to follow a healthy balanced diet, low in fat (especially SFAs) and rich in fruit, vegetables and whole grain cereals, to maintain a healthy body weight and to stay physically active (British Nutrition Foundation 2004). There is no evidence to suggest that lean red meat cannot be recommended, in moderation, as part of a healthy balanced diet for people with type 2 diabetes. In fact, single-meal intervention studies suggest that an energy-restricted, high-protein, low-fat diet (that includes red meat) may actually help improve overall glucose control in type 2 diabetes (*e.g.* Gannon *et al.*,2003).

Conclusion

The main health benefits associated with eating red meat relate to its nutritional composition. Red meat contains high biological value protein and important micronutrients, all of which are essential for good health throughout life. The majority of the population in most developed countries consume meat and meat products, and therefore meat makes a significant contribution to nutrient intake for most individuals. In addition, meat can be a versatile food that adds variety to eating occasions and is enjoyed by many. Some people choose not to eat meat, for a variety of reasons, but as there is no evidence that a moderate intake of lean red meat has any negative effects on health, there is currently no real scientific justification for excluding it from the diet. Thus, as recommended in healthy eating advice around the world, lean red meat, consumed in moderation, can be promoted as part of a healthy balanced diet.

Acknowledgements

British Nutrition Foundation, London, UK

In particular C S Williamson, R K Foster, S A Stanner and J Buttriss who were the authors of the original review paper.

The International Meat Secretariat Human Nutrition and Health Committee:

Meat and Livestock Australia

The Beef Information Centre Canada and Canada Pork

Danske Slagterier

Centre d'information des Viande – France

The Irish Food Board

The Dutch Meat Board

The New Zealand Beef and Lamb Marketing Bureau

The Animal Nutrition and Animal Products Institute – South Africa

The National Cattlemen's Beef Association and National Pork Board – USA

Livestock and Meat Commission – Northern Ireland

The Meat and Livestock Commission

References

American Cancer Society (2001). *American Cancer Society Recommendations for Nutrition and PhysicalActivity for Cancer Prevention.* Accessed via www.cancer.org.

American Heart Association (2000). American Heart Association Dietary Guidelines. *Circulation.* 2000; 102, 2284.

Appleby, P.N., Key, T.J., Thorogood, M., Burr, M.L. and Mann, J. (2002). Mortality in British vegetarians. *Public Health Nutrition* 5,1, 29-36.

Ascherio, A, Willett, W.C., Rimm, E.B., Giovannucci, E.L., Stampfer, M.J. (1994). Dietary iron intake and risk of coronary disease among men. *Circulation.* 89, 969-74.

Bingham, S.A., Day, N.E., Luben, R., Ferrari, P., Slimani, N., Norat, T., Clavel-Chapelon, F., Kesse, E., Nieters, A., Boeing, H., Tjonneland, A., Overvad, K., Martinez, C., Dorronsoro, M., Gonzalez, C.A., Key, T.J., Trichopoulou, A., Naska, A., Vineis, P., Tumino, R., Krogh, V., Bueno-de-Mesquita, H.B., Peeters, P.H., Berglund, G., Hallmans,

G., Lund, E., Skeie, G., Kaaks, R. and Riboli, E. (2003). Dietary fibre in food and protection against colorectal cancer in the European Prospective Investigation into Cancer and Nutrition (EPIC): an observational study. *Lancet*, 361, 1496-501.
Bingham, S.A. and Riboli, E. (2004). Diet and cancer – the European Prospective Investigation into Cancer and Nutrition. *Nature*. 4, 206-215.
British Nutrition Foundation (BNF) (1995b). *Task Force Report; Trans fatty Acids*. British Nutrition Foundation, London.
British Nutrition Foundation (BNF) (1999b). *Obesity: The Report of the British Nutrition Foundation Task Force*. Oxford, Blackwell Science.
British Nutrition Foundation (2001). Selenium and Health. London, British Nutrition Foundation.
British Nutrition Foundation (2002). Nutrition Labelling and Health Claims. British Nutrition Foundation, London.
British Nutrition Foundation (2004). Diabetes. Accessed via www.nutrition.org.uk
British Nutrition Foundation (2005). Review Paper: Red Meat in the Diet. British Nutrition Foundation. London.
Burr, M.L., Fehily, A.M., Gilbert, J.F., Rogers, S., Holliday, R.M., Sweetnam, P.M., Elwood, P.C. and Deadman, N.M. (1989). Effects of changes in fat, fish and fibre intake on death and myocardial reinfarction: diet and reinfarction trial (DART). *Lancet ii*, 757-61.
Cade, J.E., Burley, V.J. and Greenwood, D.C. (2004). The UK Women's Cohort Study: comparison of vegetarians, fish-eaters and meat-eaters. *Public Health Nutrition*. 7, 871-8.
Calder, P.C. (2002). Conjugated linoleic acid in humans – reasons to be cheerful? *Current opinion in clinical nutrition and metabolic care* 5,2, 123-126.
Chan, W., Brown, J. and Church, S. (1995). *Meat, Poultry and Game*. Supplement to McCance and Widdowson's *The Composition of Foods*. MAFF, London.
Cosgrove, M., Flynn, A., Kiely, M. (2005a). Impact of disaggregation of composite foods on estimates of intakes of meat and meat products in Irish adults. *Public Health Nutrition*. 8,3, 327-37.
Cosgrove, M., Flynn, A., Kiely, M. (2005b). Consumption of red meat, white meat and processed meat in Irish adults in relation to dietary quality. *British Journal of Nutrition*, 93,6, 933-42.
Cross, A.J., Pollock, J.R. and Bingham, S.A. (2003). Haem, not protein or inorganic iron, is responsible for endogenous intestinal *N*-nitrosation arising from red meat. *Cancer Research* 63,10, 2358-60.
Danish Institute for Food and Veterinary Research (2005b). Food Composition Data. Accessed via www.Foodcomp.dk.
Davey, G.K., Spencer, E.A., Appleby, P.N. (2003). EPIC-Oxford: lifestyle characteristics and nutrient intakes in a cohort of 33 883 meat-eaters and 31 546 non meat-eaters in the UK. *Public Health Nutrition* 6,3, 259-69.
Department for Food and Rural Affairs (2005). *Family Food 2002/03*. The Stationery Office, London. Accessed via http://statistics.defra.gov.uk/esg/publications/efs/2003/default.asp.
Department of Health (1998). Report on health and social subjects 48. *Nutritional Aspects of the Development of Cancer: Report of the Working Group on Diet and Cancer of the Committee on Medical Aspects of Food and Nutrition Policy*. London: The Stationary Office.
Dunnigan, M.G., Henderson, J.B. An epidemiological model of privational rickets and osteomalacia. Proc Nutr Soc 1997;56:939–56.
Ferlay, J., Bray, F., Pisani, P. *et al.* (2004). GLOBOCAN 2002: Cancer Incidence, Mortality and Prevalence Worldwide IARC CancerBase No. 5. version 2.0. IARC *Press*, Lyon, 2004. Accessed via http://www-dep.iarc.fr/globocan/database.htm.
Fernandez-Gines, J., Fernandez-Lopez, J. and Sayas-Barbera, E. (2005). Meat Products as Functional Foods: A Review. *Journal of Food Science* 70, 2,R37-R43.
Fowler (2004). European Meat consumption. *Meat Demand Trends, Meat and Livestock Commission.*
Fraser, G.E. (1999). Associations between diet and cancer, ischemic heart disease, and all-cause mortality in non-Hispanic white California Seventh-day Adventists. *American Journal of Clinical Nutrition* 70,532S-8S.
Gannon, M.C., Nuttall, F.Q., Saeed, A., Jordan, K. and Hoover, H. (2003). An increase in dietary protein improves the blood glucose response in persons with type 2 diabetes. *American Journal of Clinical Nutrition* 78, 4, 734-41.
Gibson, S. and Ashwell, M. (2003). The association between red and processed meat consumption and iron intakes and status among British adults. *Public Health Nutrition* 6, 4, 341-350.
GISSI-Prevenzione Investigators (1999). Dietary supplementation with n-3 poly-unsaturated fatty acids and vitamin E after myocardial infarction: results from the GISSI-Prevenzione trial. *Lancet* 354, 447-455.
Givens, D. and Shingfield, K. (2004). Foods derived from animals: the impact of animal nutrition on their nutritive value and ability to sustain long-term health. *Nutrition Bulletin* 29, 4, 325-332.
Groff, J., Gropper, S. and Junt, S. (1995). *Advanced Nutrition and Human Metabolism*. 2nd edition. West Publishing Company, Minneapolis/St Paul.
Halton, T.L. and Hu, F.B. (2004). The effects of high protein diets on thermogenesis, satiety and weight loss: a critical review. *Journal of the American College of Nutrition*. 23, 373-85.
Henderson, L., Gregory, J., Irving, K *et al.* (2003a). The National Diet and Nutrition Survey: adults aged 19 to 64 years. Volume 2: Energy, protein, carbohydrate, fat and alcohol intake. London, HMSO.
Henderson, L., Irving, K., Gregory, J. *et al.* (2003b). The National Diet and Nutrition Survey: adults aged 19 to 64 years. Volume 3: Vitamin and mineral intake and urinary analytes. London, HMSO.

Herbert, V. (1988). Vitamin B_{12}: plant sources, requirements and assay. *American Journal of Clinical nutrition* 48 (3 Suppl): 852-8.
Herbert, V. (1994). Staging vitamin B_{12} (cobalamin) status in vegetarians. *American Journal of Clinical nutrition* 59 (5 Suppl): 1213S-1222S
Higgs, J. (2000) The changing nature of red meat: 20 years of improving nutritional quality. *Trends in Food Science and Technology*, 11, 85-95.
Hill, M. (2002). Meat, cancer and dietary advice to the public. *European Journal of Clinical Nutrition* 56 Suppl 1, S36-41.
Howe, P.R.C., Meyer, B.J., Record, S., and Baghurst, K. (2003). Relative Contributions of Meats and Fish to Very Long Chain Omega-3 (VLC ω3) Intakes. *Asian Pacific Journal of Clinical Nutrition, 27*,12 (Suppl) S27.
Hulshof, K.F., van Erp-Baart, M.A., Anttolainen, M. *et al.* (1999). Intake of fatty acids in western Europe with emphasis on trans fatty acids: the TRANSFAIR Study. *European Journal of Clinical Nutrition* 53: 143-57.
Hu, F.B., Stampfer, M.J., Manson, J.E., Rimm, E., Colditz, G.A., Rosner, B.A., Hennekens, C.H., Willett, W.C. (1997). Dietary fat intake and the risk of coronary heart disease in women. *New England Journal of Medicine*, 337:1491-9.
Hu, F.B., Stampfer, M.J., Manson, J.E., Ascherio, A., Colditz, G.A., Speizer, F.E., Hennekens, C.H. and Willett, W.C. (1999a). Dietary saturated fats and their food sources in relation to the risk of coronary heart disease in women. *American Journal of Clinical Nutrition*. 70, 6, 1001-8.
Hu, F.B., Bronner, L., Willett, W.C., Stampfer, M.J., Rexrode, K.M., Albert, C.M., Hunter, D., and Manson, J.E. (2002). Fish and omega-3 fatty acid intake and risk of coronary heart disease in women. *The Journal of the American Medical Association* 10, 14, 1815-21.
Hu, F.B. (2005). Protein, body weight and cardiovascular health. *American Journal of Clinical Nutrition* 82: (suppl) 242S-7S.
Irish Universities Nutrition Alliance (IUNA) (2001). The North/South Ireland Dietary Survey, Accessed via http://www.iuna.net/documents/Food%20Survey%202001.pdf.
Kelemen, L.E., Kushi, L.H., Jacobs, D.R. Jr, Cherhan, J.R. (2005). Associations of dietary protein with disease and mortality in a prospective study of postmenopausal women. American Journal of Epidemiology, 161, 239-49.
Key, T.J., Fraser, G.E., Thorogood, M., Appleby, P.N., Beral, V., Reeves, G., Burr, M.L., Chang-Claude, J., Frentzel-Beyme, R., Kuzma, J.W., Mann, J., McPherson, K. (1998). Mortality in vegetarians and non-vegetarians: a collaborative analysis of 8300 deaths among 76,000 men and women in five prospective studies. *Public Health Nutrition*, 1, 33-41.
Key, T.J., Fraser, G.E., Thorogood, M., Appleby, P.N., Beral, V., Reeves, G., Burr, M.L., Chang-Claude, J., Frentzel-Beyme, R., Kuzma, J.W., Mann, J., McPherson, K. (1999a). Mortality in vegetarians and non-vegetarians: detailed findings from a collaborative analysis of 5 prospective studies. *American Journal of Clinical Nutrition* 70 (3 Suppl): 516S-524S.
Key, T.J., Davey, G.K. and Appleby, P.N. (1999b). Health benefits of a vegetarian diet. *Proceedings of the Nutrition Society* 58, 2, 271-5.
Key, T.J., Schatzkin, A., Willett, W.C., Allen, N.E., Spencer, E.A., Travis, R.C. (2004). Diet, nutrition and the prevention of cancer. *Public Health Nutrition* 7, 1A, 187-200.
Laugesen, M. (2005). Decreased Red Meat Fat Consumption in New Zealand 1995-2002. *New Zealand Medical Journal,* 118(1226), 1751-61.
Lee, S., Buss, D., Holcombe, D. *et al.* (1995). Nutrient content of retail cuts of beef, pork and lamb - preliminary results. *Journal of Human Nutrition and Dietetics*, 8, 75-80.
Li, D., Siriamornpun, S., Wahlqvist, M.L., Mann, N.J., Sinclair, A,J. (2005). Lean meat and heart health. *Asian Pacific Journal of Clinical Nutrition* 14, 113-9.
Linseisen, J., Kesse, E., Slimani, N., Bueno-De-Mesquita, H.B., Ocke, M.C., Skeie, G., Kumle, M., Dorronsoro Iraeta, M., Morote Gomez, P., Janzon, L., Stattin, P., Welch, A.A., Spencer, E.A., Overvad. K., Tjonneland, A., Clavel-Chapelon, F., Miller, A.B., Klipstein-Grobusch, K., Lagiou, P., Kalapothaki, V., Masala, G., Giurdanella, M.C., Norat, T., Riboli, E. (2002). Meat consumption in the European Prospective Investigation into Cancer and Nutrition (EPIC) cohorts: results from 24-hour dietary recalls. *Public Health Nutrition* 5, 1243-1258.
Matthews, K. and Strong, M. (2005) Salt – its role in meat products and the industry's action to reduce it. *Nutrition Bulletin* 30, 1, 55-61.
Meat and Wool New Zealand (2004). Meat Consumption and Expenditure. *Meat and Wool New Zealand - Economic Service*. September 2004.
Moloney, A., Reducing Fats in raw meat. In: Improving the fat content of foods (CM Williams and J L Buttriss eds) Woodhead Publishing Ltd (2006).
Ministry of Agriculture Fisheries and Food (MAFF) (1998). *Fatty Acids. Supplement to McCance and Widdowson's The Composition of Foods*. HMSO, London.
Norat, T., Lukanova, A., Ferrari, P. and Riboli, E. (2002). Meat consumption and colorectal cancer risk: a dose-response meta-analysis of epidemiological studies. *International Journal of Cancer* 98, 2, 241-56.
Norat, T., Bingham, S., Ferrari, P., Slimani, N., Jenab, M., Mazuir, M., Overvad, K., Olsen, A., Tjonneland, A., Clavel, F., Boutron-Ruault, M.C., Kesse, E., Boeing, H., Bergmann, M.M., Nieters, A., Linseisen, J., Trichopoulou, A., Trichopoulos, D., Tountas, Y., Berrino, F., Palli, D., Panico, S., Tumino, R., Vineis, P., Bueno-de-Mesquita, H.B.,

Peeters, P.H., Engeset, D., Lund, E., Skeie, G., Ardanaz, E., Gonzalez, C., Navarro, C., Quiros, J.R., Sanchez, M.J., Berglund, G., Mattisson, I., Hallmans, G., Palmqvist, R., Day, N.E., Khaw, K.T., Key, T.J., San Joaquin, M., Hemon, B., Saracci, R., Kaaks, R., Riboli, E. (2005). Meat, fish, and colorectal cancer risk: the European Prospective Investigation into cancer and nutrition. *Journal of the National Cancer Institute* 97, 12, 906-16.

Oh, K., Hu, F.B., Manson, J.E., Stampfer, M.J., Willett, W.C. (2005). Dietary fat intake and risk of coronary heart disease in women: 20 years of follow-up of the nurses' health study. *American Journal of Epidemiology* 161: 672-9.

Phillips, F. (2005). Vegetarian Nutrition – *Nutrition Bulletin* 30, 2, 132-167.

Potter, J.D. (1999). Colorectal cancer: molecules and populations. *Journal of the National Cancer* Institute, 91, 11, 916-32.

Red Meat and Health Expert Advisory Committee (2001) *The role of red meat in healthy Australian diets.* Report prepared by Red Meat and Health Expert Advisory Committee.

Russell, D. *et al.* (1999). *National Nutrition Survey. New Zealand Food, New Zealand People. Key Results of the 1997 National Nutrition Survey.* Ministry of Health, Wellington.

Sandhu, M.S., White, I.R. and McPherson, K. (2001). Systematic review of the prospective cohort studies on meat consumption and colorectal cancer risk: a meta-analytical approach. *Cancer Epidemiology, Biomarkers and Prevention* 10, 5, 439-46.

Scientific Advisory Committee on Nutrition (SACN) (2003). *Salt and Health.* The Stationary Office, London.

Scientific Advisory Committee on Nutrition (SACN) (2005b). Paper for discussion: Selenium. http://www.sacn.gov.uk/meetings/committee/2005_06_15.html# (accessed 1/8/05).

Slattery, M.L. (2000). Diet, lifestyle and colon cancer. *Seminars in gastrointestinal disease* 11, 3, 142-6.

Stanner, S. (2005). *British Nutrition Task Force Report on Cardiovascular disease: Diet and Emerging Risk Factors.* Blackwells, Oxford.

Stubbs, R.J. (1995). Macronutrient effects on appetite. *International Journal of Obesity* 19, Suppl.5: S11-S5.

The Cancer Council NSW (2003). *Meat and Cancer.* Accessed via www.cancercouncil.com.au.

Tricon, S., Burdge, G.C., Williams, C.M., Calder, P.C., Yaqoob, P. (2005). The effects of conjugated linoleic acid on human health-related outcomes. Proceedings of the Nutrition Society 64, 171-82.

United States Department of Agriculture (USDA) (2005). http://www.nal.usda.gov/fnic/foodcomp/ (accessed 1/8/05).

Watts, G.F., Ahmed, W., Quiney, J., Houlston, R., Jackson, P., Iles, C., Lewis B. (1988). Effective lipid lowering diets including lean meat. *British Medical Journal,* 296 (6617): 235-7.

Willett, W.C. (1995). Diet, nutrition and avoidable cancer. *Environmental Health Perspectives* 103 (Suppl 8): 165-170.

Willett, W.C. (2005). Diet and cancer: an evolving picture. *Journal of the American Medical* Association 293, 2, 233-4.

World Cancer Research Fund (1997). *Food, Nutrition and the Prevention of Cancer: A Global Perspective.* Washington DC: American Institute for Cancer Research.

World Health Organisation (2003). *Diet, nutrition and the prevention of chronic diseases.* Geneva, WHO.
Accessed via (http://www.who.int/dietphysicalactivity/publications/trs916/en/gsfao_global.pdf).

SESSION 1: MEAT QUALITY

- GENOMICS AND BIOTECHNOLOGY

IDENTIFICATION OF AMERICAN BISON TISSUE BASED ON THE MITOCHONDRIAL *CYTOCHROME B* GENE

F. Schwägele, K. Spiegel and K.O. Honikel*

Federal Research Centre for Nutrition and Food, Institute for Chemistry and Physics, E.-C.-Baumann-Str. 20, 95326 Kulmbach, Germany. E-mail: fredi.schwaegele@bfel.de

Keywords: animal species, PCR, mitochondrial DNA

Introduction

Bison is reared in North America (USA and Canada) at a significant extent and exported into the European Union. Bison meat is classified as a "gourmet product" because of its uncommon and exceptional taste. However, with respect to its meat texture it is very similar to beef and for this reason it should be unequivocally labeled as bison. The exact taxonomy of the American bison (bison bison) is as follows: kingdom – animal; phylum – chordata; subphylum – vertebrates (having a spine); class – mammalians (having mammary glands); subclass – eutheria (having placenta); super order (ungulates); order - artiodactyls (having an even number of fingers); suborder – ruminants (having the rumen); family – bovidae, subfamily – bovinae; genus – bison; species – bison bison (Wilson and Reeder, 1993). Although buffalo (bubalus bubalis), cattle (bos taurus) and bison belong to the same subfamily (bovinae) they differ from each other by genus. Bison has to be legally treated in accordance with the regulation of the European Parliament (EC) 1760/2003 dealing with the labeling of beef and beef products. On this account the Federal Ministry for Nutrition, Agriculture, and Consumer Protection as well as the German Association for Meat Business argued for obligatory labeling of bison and bison products. As a consequence, there exists a need for molecular biological methods to differentiate between bison, buffalo and cattle. Based on mitochondrial DNA a polymerase chain reaction (PCR) system was developed to provide a possibility to distinguish between tissues of the mentioned animals.

Materials and Methods

DNA used for PCR was isolated from 50 mg meat in each case applying a modified CTAB protocol (Binke *et al.*, 2003). The DNA content was determined according to Warburg and Christian (1942). PAGE was performed as described by Altmann *et al.*, (2004) using the following molecular weight standards: λ-DNA/Eco RI + Hind III and pBR322/HaeIII. DNA amplification was carried out in a *Perkin Elmer Gene Amp System 9600*.

Primer design focusing on mitochondrial DNA was implemented by means of *HUSAR Data Base* (German Cancer Research Center, Heidelberg). The following primer pair was used for the amplification of a bison (bison bison) specific DNA fragment of a length of 138 bp:

CY-Bis/forward: 5′ - AAATCCACTCAATACACCTCCC - 3′ (22 bp);
CY-Bis/reverse: 5′ - CTAATCCTTGCCCTCATTCC - 3` (20 bp).

In addition a buffalo specific primer system was applied resulting in a buffalo (bubalus bubalis) specific DNA fragment consisting of 242 bp:

CY-Buf/forward: 5′ - TAGGCATCTGCCTAATTCTG - 3′ (20 bp);
CY- Buf/reverse: 5′ - ACTCCGATGTTTCATGTTTCT - 3′ (21 bp).

Results and Discussion

In order to identify the animal species and to investigate the specificity of the PCR system based on the *cytochrome b* gene for bison, various animal species were tested by PCR for cross similarity. No PCR product was amplified for goat, pig, horse, chicken, buffalo and cattle (Fleckvieh and Schwarzbunt) (Figure 1A: lane 2- 9). However, a distinct band with a length of 138 bp can be observed in the case of bison (Figure 1A: lane 10).

Further studies were performed to exclude a cross similarity between buffalo and bison, as well as the other above mentioned animals using a buffalo specific primer system, which is also based on the mitochondrial *cytochrome b* gene (Rajapaksha *et al.*, 2003). Figure 1B clearly shows, that there is only a distinct band in lane 7 reflecting an amplification product of 242 bp, that is specific for buffalo, but not for bison and all the other animal tissues tested. Only in the case of pig, a weak band was observed indicating a cross similarity between pig and buffalo. In Figure 2 the primer sequences in the mitochondrial *cytochrome b* gene region between bison, buffalo and cattle is depicted visualising the number of existing nucleobase mismatches.

Conclusions

The results demonstrate clearly, that the primer system designed for bison based on the mitochondrial *cytochrom b* gene provides the possibility for unequivocal differentiation between bison and the other animal species tested. As there was no animal material available originating from yak (bos grunniens) and the East European bison (bison bonasus), the so called wisent, further investigations have to be performed to prove that no cross similarity exists in the case of the chosen primer system between bison on the one hand and yak and wisent on the other hand.

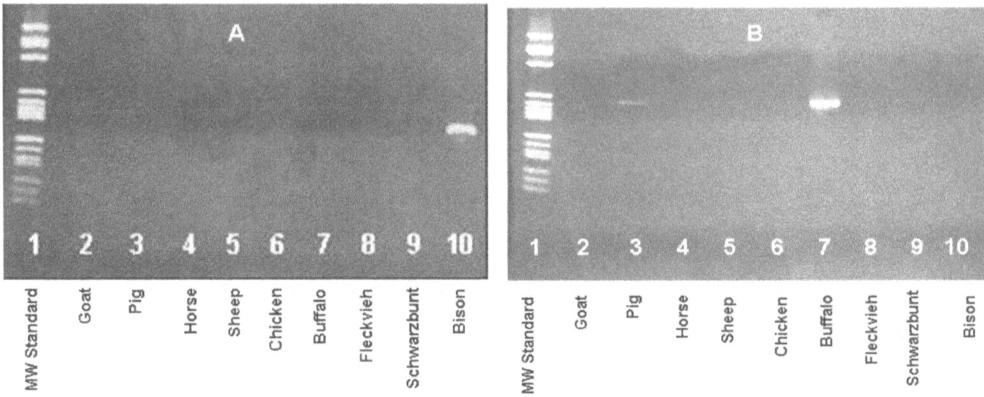

Figure 1: PAGE of various amplification products visualized by Ethidiumbromide intercalation: Gel A on the left shows only one distinct DNA band (lane 10) with 138 bp reflecting the specificity of the CY-Bis primer system for bison (bison bison) solely. There are no amplicons for tissue in the case of the other animals. On Gel B depicted on the right there is a clear DNA band (lane 7) with 242 bp for buffalo. This demonstrates the specificity of the CY-Buf primer system for buffalo (bubalus bubalis). Merely a weak band has been observed in lane 3 where DNA of pig tissue was isolated and amplified.

Forward primer:

Bison bison	**AAATCCACTC AATACACCTC CC**	
Bubalus bubalis	**AAACCCACTC AACACACCTC CC**	**2 Mismatches**
Bos taurus	**CAATCCACTC AACACACCCC CT**	**4 Mismatches**

Reverse primer:

Bison bison	**CTAATCCTTG CCCTCATTCC**	
Bubalus bubalis	**CTAATCCTCA TTCTCATGCC**	**5 Mismatches**
Bos taurus	**CTAATTCTTG CTCTAATCCC**	**4 Mismatches**

Figure 2: Forward and reverse primer sequences in comparison of Bison (bison bison), Buffalo (bubalus bubalis) and cattle (bos Taurus). Concerning taxonomic aspects the three animals belong to the subfamily bovinae. However, they are different with respect to genus. The specificity of the primer system for bison is reasonable because of the number of existing mismatches (underlined) in the case of buffalo and cattle.

References

Altmann, K., Binke, R., Schwägele, F. Qualitative determination of goat in meat products by means of the nuclear single-copy gene *beta-casein*. Innovations in Food Technology 22: 64 – 65 (2004).

Binke, R., Eichner, R., Zäh, M., Schwägele, F. Entwicklung eines leistungsfähigen Extraktionssystems zur Isolierung von Nucleinsäuren aus Fleisch und Fleischerzeugnissen für die PCR. Arch. Lebensmittelhyg. 3: 52 – 55 (2003).

Rajapaksha, W.R.A.K.J.S., Thilakaratne, I.D.S.I.P., Chandrasiri, A.D.N., Niroshan, T.D. Development of a PCR assay for identification of buffalo meat. Asian-Aust. Journal of Animal Science, 16: 1046 – 1048 (2003).

Warburg, O., Christian, W. Isolation and crystallisation of enolase. Biochem. 310: 384 – 421 (1942).

Wilson, D.E., Reeder, D.M. (eds.). Mammal species of the world (2nd edition). Washington: Smithonian Institution Press (1993).

POLYMERASE CHAIN REACTION IDENTIFICATION OF HORSE AND DONKEY MEAT IN HEAT-TREATED MEAT

H. Yetim[1], Z. Kesmen[1], F. Sahin[2] and H. W. Ockerman*[3]

[1.] *Erciyes Univ., Coll. of Eng., Food Sci. Dept., 38039, Kayseri, Turkey,* [2.] *Yeditepe Univ., Coll. of Eng. Bioeng. Dept. 34755, Istanbul, Turkey,* [3.] *The Ohio State Univ., Dept of Anim. Sci. 43210, Columbus, OH, USA.*
Email: ockerman.2@osu.edu

Keywords: species identification, PCR, horse, donkey, heat- processed meat products

Introduction

The differentiation of meat species in meat products has been considered important to preserve the consumers from the presence of unknown, less desirable meat types for economic, religious and health reasons (Koh *et al.*, 1998). However, species identification of cooked meat products to detect adulteration or fraudulent substitutions is a problem for the food control authorities (Meyer *et al.*, 1994). For this purpose, numerous analytical methods have been developed for meat species identification based on protein analysis; electrophoretic, chromatographic and immunological techniques (Patterson and Jones, 1990; Zerifi *et al.* 1992). Although most of these methods identified the species origin of raw meats, they were less sensitive when the meat had been thermally processed, because the temperatures during the heat treatment destroyed the species-specific proteins or the epitopes (Calvo *et al.*, 2001). Advances in DNA technology have led to rapid development of alternative approaches for species identification studies. Advantages of DNA analysis compared to protein analysis include stability at higher temperature, presence in all types of tissue and greater variation in genetic code (Wolf *et al.*, 1999). DNA methods have commonly been based on PCR analysis of species-specific mitochondrial DNA (mt-DNA) sequences. The superiority of mitochondrial based DNA analysis derives from the fact that each cell contains several copies of mt-DNA and have improved the possibility of amplifying template molecules of adequate size among the DNA fragments brought about by heat denaturation (Montiel-Sosa *et al.*, 2000). Also there is a relatively high mutation rate compared to nuclear genes to allow the differentiation between the closely-related species (Brodman *et al.*, 2001). PCR based methods exhibit high specifity, sensitivity, rapidity and suitability for complex identification. This study evaluated a method for species identification of horse and donkey meat in heat-processed meat by species-specific PCR of mt-DNA.

Materials and Methods

Preparation of Meat Samples: Bovine, ovine, swine, horse and donkey muscle tissues were used. A mixture of each specie was prepared from raw minced beef or sheep meats by adding donkey and horse meat at 5 different levels (0.0, 0.1, 0.5, 1.0 and 5.0%). Meat mixtures were subjected to heating for 30 min at 100°C and 120°C.

DNA extraction: DNA was extracted following the method of DeSalle *et al.* (1993) and Kesmen (2005).

Oligonucleotide Primers: Two sets of oligonucleotide primers were designed to amplify specific regions for the horse and donkey mt-DNA sequences. The first set amplifies a specific fragment of 153 bp which is part of the horse mitochondrial genome (ATPase8- ATPase6) using 5'-CTATCCGACACACCCAGAAGTAAAG-3' as forward primer and 5'-GATGCTGGGAAATATGATGATCAGA-3' as reverse primer. The second set detects a donkey species-specific amplicon of 145 bp from the donkey mitochondrial genome (NADH dehydrogenase subunit 2) using 5'-CATCCTACTAACTATAGCCGTGCTA-3' as forward pirmer and 5'-CAGTGTTGGGTT GTACACTAAGATG-3' as reverse primer.

PCR Amplification Specific Fragment: The PCR amplification reaction was performed in a total volume of 50 µl (Kesmen, 2005). The products of PCR amplification were analyzed by electrophoresis through a 2% agarose gel.

PCR Specificity and Sensitivity Test: 100ng/µl of bovine, ovine, horse and donkey genomic DNA and a negative control without DNA were used in PCR respectively, to evaluate the specificity of primers used in the study. Also serial 1:10 dilutions of raw horse and donkey DNA were made to 0.01 ng/µl. Each dilution was added separately to the reaction mixtures for the sensitivity test of the PCR assay.

Result and Discussion

The PCR results demonstrated that the specific primers for horse and donkey designed in this study hybridize on the species-specific DNA sequences of each specie and yield 153 and 145 bp fragments respectively in 0.01 ng DNA levels without any cross-reaction between the other species studied (Figure 1 and 2). Although detection of horse and donkey meat was successful in all mixtures heated at 100°C for 30min, regardless whether the product was prepared from beef or sheep (Figure 3). However, the same samples heated at 120°C for 30min could not be detected at the substitution level of 0.1% (Figure 4).

Conclusion

The use of a PCR method is shown to be a reliable, highly specific and sensitive method for the detection of horse and donkey meat in heat processed meat products. PCR assay was able to amplify DNA from the cooked and autoclaved meat since the heat treatment could not seriously destroy the DNA. Nevertheless, in autoclaved meat, DNA samples degrade to small fragments, averaging 300 bp (Meyer *et al.*, 1994). Therefore the PCR system developed in this research, targeting an even shorter DNA fragments (153 and 145 bp), could be used succesfully in the analysis of highly degraded DNA. These results confirm the appropriateness of mt-DNA for meat species detection containing high number of copies per cell. Also the mt-DNA accepts the mutations fast enough to allow differentiation between closely related species. Particularly mt-DNA sequence encoding for subunit 8 and subunit 6 of ATP synthase complex was chosen as a target sequence to design horse specific primer. Such region was preferred to other mt-DNA coding sequences because both ATPase-8 and ATPase-6 exhibit a relatively high degree of variation among the vertebrates (Tartaglia *et al.,* 1998). Although phylogenetically close animal species are studied, identifications appear to be highly species-specific. It is a simple and reliable method for meat species identification that can be adapted to many differentiation problems.

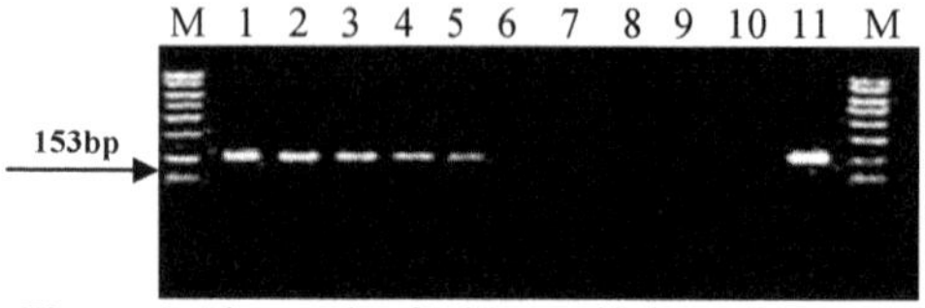

Figure 1: The result of PCR specificity and sensitivity test for the specific horse primers.M: Marker; 1: 100ng horse; 2: 10ng horse; 3: 1 ng horse; 4: 0.1 ng horse; 5: 0.01 ng horse; 6: donkey; 7: pig; 8:beef; 9: lamb; 10:Negative control; 11: Positive control .

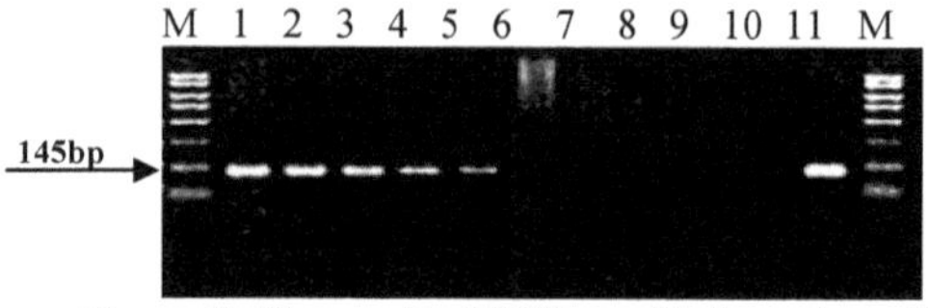

Figure 2: The result of PCR specificity and sensitivity test for the specific donkey primers. M: Marker; 1: 100ng donkey; 2: 10ng donkey; 3: 1 ng donkey; 4: 0.1 ng donkey; 5: 0.01 ng donkey; 6:horse; 7: pig; 8:beef; 9: lamb; 10:Negative control; 11: Positive control.

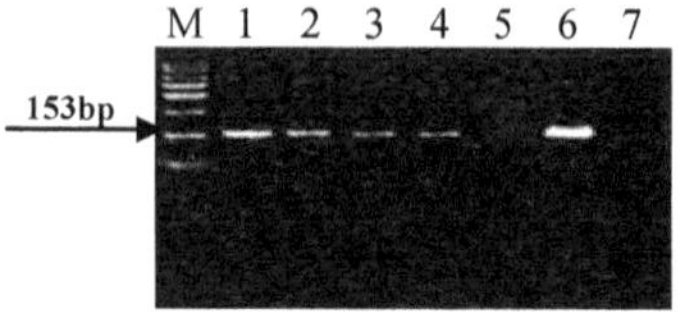

Figure 3: Electrophoretic gel of horse-beef mixtures heated at 100 °C for 30 min. M: Marker; 1: horse %5 + beef 95%; 2: horse %1 + beef 99%; 3: horse %0.5 + beef 99.5%; 4: horse %0.1 + beef 99.9%; 5:beef 100%; 6: Positive control; 7: Negative control.

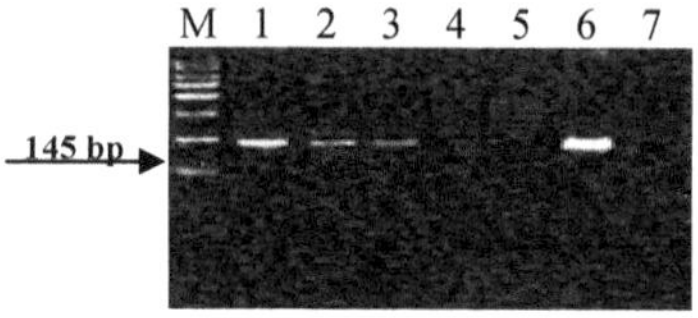

Figure 4: Electrophoretic gel of donkey-beef mixtures heated at 120°C for 30 min. M: Marker; 1: donkey %5 + beef 95%; 2: donkey %1 + beef 99%; 3: donkey %0.5 + beef 99.5%; 4: donkey %0.1 + beef 99.9%; 5: beef 100%; 6: Positive control; 7: Negative control.

References

Brodmann, P.D., Nicholas, G., Schaltenbrand, P. and Ilg, E.C. (2001). Identifying unknown game species: with nucleotid sequencing of the mitochondrial sitokrom b gene and a subsequent basic local alignment search tool search. Z. Lebensm. Unters Forsch. 212, 491-496.

Calvo, J.H., Zaragoza, P. and Osta, R. (2001). Technical note: A quick and more sensitive method to identify pork in processed food by PCR amplification of a new specific DNA fragment. J. Anim. Sci., 79, 2108-2112.

DeSalle, R., Williams, A. and George, M. (1993). Isolation and characterization of animal mitochondrial DNA. Meth. Enzymol. 224, 174-204.

Guoli, Z., Mingguang, Z., Zhijiang, Z., Hongsheng, O. and Qiang, L. (1999). Establishment and application of a polymerase chain reaction fort he identification of beef. Meat Sci., 51, 233-236.

Kesmen, Z. (2005). A research on the meat species identification in fresh and processed meat products by using pcr techniques. PhD Dissertation, Ataturk Univ. Erzurum, 85 p.

Koh, M.C., Lim, C.H., Chua, S.B., Chew, S.T. and Phang, S.T.W. (1998). Random amplified polymorhic DNA (RAPD) fingerprints for identification of red meat animal species. Meat Sci., 48, (3/4) 275-285.

Meyer, R., Candrian, U. and Lüthy, J. (1994). Detection of pork in heated meat products by Polymerase Chain Reaction (PCR). J. Assoc. Off. Anal. Chem. Int., 77, 617-622.

Montiel-Sosa, J.F., Ruiz-Pesini, E., Montoya, J, *et al.* (2000). Direct and highly species-specific determination of pork meat and fat in meat products by PCR amplification of mitochondrial DNA. J.Agric., Food Chem., 48 2829-2832.

Patterson,R. L. and Jones S.J. (1990). Review of current technique for verification of the species origin of meat. Analyst, 115, 501-506.

Tartaglia M., Ernestina, S., Pestalozza, S., Morelli, L., Antonucci, G. and Battaglia, P.A. (1998). Detection of bovine mitochondrial DNA in ruminant feeds: A molecular approach to test fort he presece of bovine derived materials. J. Food Prot., 61 (5), 513- 518.

Wolf, C., Rentsch, J. and Hübner, P. (1999). PCR-RFLP analysis of mitochondrial DNA: A reliable method for identification. J. Agric. Food Chem., 47, 1350-1355.

Zerifi, A., Labie, C. and Benard, G. (1992). SDS-PAGE Technique fort he species identification of cooked meat. Fleiswirtschaft 1. 54-59.

MUSCLE PROTEOME OF "BLONDE D'AQUITAINE" YOUNG BULLS AND MEAT EATING QUALITIES

M. Morzel[1]*, C. Terlouw[1], J. Groelly[1], C. Chambon[1], A. Listrat[2], C. Jurie[2], H. Dubroeucq[2] and B. Picard[2]

INRA, Centre de Theix, 63122 St Genès Champanelle, France. 1 Qualité des Produits Animaux 2Unité des Recherches sur les Herbivores. Email : morzel@clermont.inra.fr

Keywords: beef, Blonde d'Aquitaine, proteome, sensory quality, predictors

Introduction

Blonde d'Aquitaine is a bovine French breed renowned for its palatable meat. Sensory properties of meat depend largely on muscle composition in terms of proteins. For example, some proteins and fragments detected 72h *post mortem* correlate well with pork tenderness (Lametsch *et al.,* 2003). The present study was designed to determine whether muscle proteins present early after slaughter predict later eating qualities of aged and cooked meat of Blonde d'Aquitaine, using proteome techniques.

Materials and Methods

Ten 15-month old Blonde d'Aquitaine bulls were slaughtered and *Longissimus thoracis* samples were taken 1-h post-mortem. The *Longissimus thoracis* muscle was then separated from the carcass, aged for 14 days and grilled. Meat sensory properties (initial, overall and residual tenderness, flavour intensity, juiciness) were rated by a trained panel. Muscle samples taken at slaughter were characterised by measuring citrate synthase (CS, oxidative metabolism) activity, total hydroxyproline (OH-Prol, i.e. collagen) content and composition in myosin heavy chain (MyHC) types (I, IIa, IIx). Two-dimensional electrophoresis (2DE) was performed (5-8 pH gradient, 11% polyacrylamide gel, 800µg total protein per gel). Proteins of interest were identified by MALDI-ToF mass spectrometry.

Pearson and Spearman rank correlations were calculated between muscle traits or spot intensities and sensory ratings. Proteins significantly correlated in both correlations to sensory ratings were subjected to a multiple forward stepwise regression analysis. Principal Component Analysis (PCA) was performed for each sensory property using the 9 proteins identified as the most strongly correlated with the sensory trait.

Results and Discussion

OH-Prol was negatively correlated to CS (Table1), an oxidative metabolism marker. Muscle fibre composition, in terms of MyHC types, were not strongly associated with measured meat quality traits. In accordance with previous reports (Renand *et al.*, 2001), the higher the total collagen content, the less tender the meat initially and overall and the tougher ($P<0.10$) the bolus when swallowed (residual tenderness). This is consistent with the positive link between initial and overall tenderness, themselves negatively linked to residual tenderness.

Table 1: Correlation coefficients between muscle traits and sensory ratings. * $p<0.05$; ** $p<0.01$; *** $p<0.005$.

		CS	OH-Prol	MyHC			tenderness			flavour
				I	IIa	IIx	initial	overall	residual	
OH-Prol		-0.63*								
	I	0.45	-0.38							
MHC	IIa	0.57	-0.15	0.20						
	IIx	-0.37	0.43	-0.47	-0.49					
	initial	0.35	-0.87***	0.22	-0.04	0.51				
tenderness	overall	0.30	-0.87***	0.09	0.00	-0.42	0.97***			
	residual	-0.24	0.59	0.31	0.26	0.26	-0.67*	-0.68*		
flavour		-0.41	-0.16	0.05	-0.72*	0.12	0.31	0.27	-0.46	
juiciness		0.06	-0.59	0.31	-0.50	-0.30	0.59	0.53	-0.57	0.70*

Out of 250 separated by 2DE, more than 40 were linked to texture and juiciness, whereas only 3 spots could predict flavour (Table 2). The finding that only few proteins, within the limits of observed molecular weight and isoelectric point, predict flavour, is coherent with the report that complex mechanisms involving lipid content, *post mortem* oxidation and production of small peptides determine flavour of cooked beef (Spanier *et al.*, 2004).

Regarding texture attributes, only positive correlations were found between spot intensities and initial/overall tenderness, while mostly negative correlations were found between spot intensities and residual tenderness, in accordance with the negative significant correlation between initial/overall and residual tenderness (Table 1). Several spots were good predictors of initial tenderness: spot 5702 alone (identified as succinate dehydrogenase: SDH; Figure 1) explains 67% of variability. Using two spots (0802 and 1301, non identified), 74% of variability can be explained.

Table 2: Number of spots significantly ($p<0.05$) correlated (Pearson and Spearman) with meat sensory ratings.

	initial tenderness	overall tenderness	residual tenderness	flavour	juiciness
Positive correlation	31	27	8	2	18
Negative correlation	0	0	39	1	2

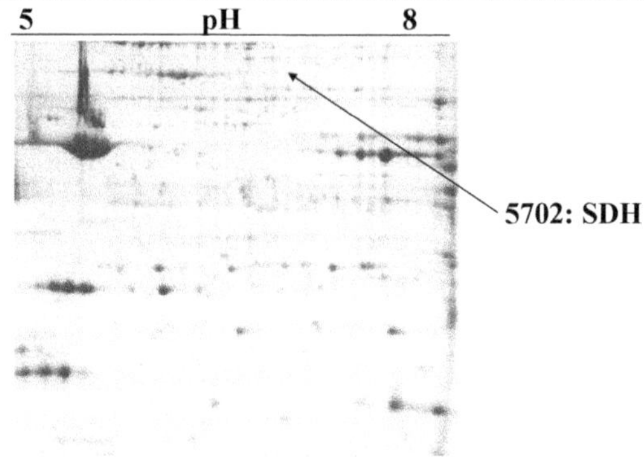

Figure 1: 2DE separation of bovine muscle total proteins and position of spot 5702.

PCA analysis (Figure 2) further illustrates the clear association of overall tenderness with several spots (e.g. 2307 and 5702) on the first axis of PCA (90.2% of variation explained). This axis identifies correctly animals according to tenderness (e.g. animals 6459 and 5863, were rated most and least tender, respectively). The second axis (8.0% of variation explained) was not correlated with any of the phenotypic measurements.

Figure 2: PCA analysis of 10 animals (dots) and the 9 spots (lines) most strongly correlated with initial tenderness.

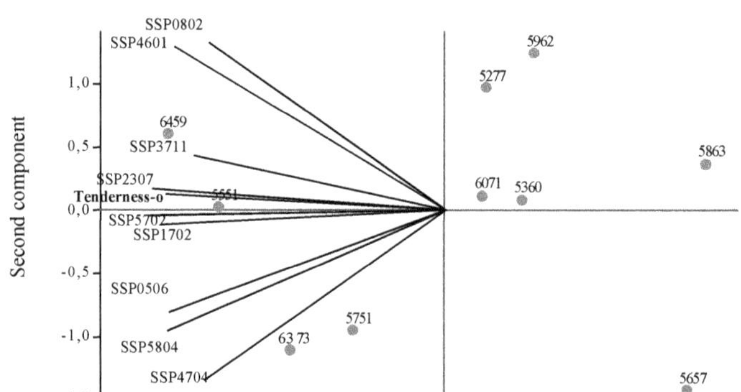

Conclusions

Presence of many early *post mortem* proteins, including SDH, predict later initial and overall tenderness. Results suggest that in Blonde d'Aquitaine young bulls, characterised by a mostly glycolytic muscle metabolism, animals with the "least glycolytic / most oxidative" muscles provide more tender meat.

References

Lamestch, R., Karlsson, A., Rosenvold, K., Andersen, H.J., Roepstorff, P. and Bendixen, E. (2003). Postmortem proteome changes of porcine muscle related to tenderness. *Journal of Agricultural and Food Chemistry, 51*, 6992-6997.

Renand, G., Picard, B., Touraille, C., Berge, P. and Lepetit, J. (2001). Relationships between muscle characteristics and meat quality traits of young Charolais bulls. *Meat Science, 59*, 49-60.

Spanier, A.M., Flores, M., Toldra, F., Aristoy, M-C., Bett, K.L., Bystricky P. and Bland, J.M. (2004). Meat flavor: contribution of proteins and peptides to the flavor of beef. In: "*Quality of fresh and processed foods*", ed. Shahidi *et al.*, Kluwer academic, p 33-49.

BOVINE MUSCLE 20S PROTEASOME: QUANTIFICATION IN TISSUE CRUDE EXTRACTS USING ELISA, RADIAL IMMUNODIFFUSION TECHNIQUES AND PRACTICAL APPLICATIONS

L. Aubry[1], C. Herrera-Mendez[1], G. Coulis[1], M.A. Sentandreu[2], D. Levieux[3], A. Ouali[1]* and D. Dutaud[1]

[1.] *Unité de biochimie, SRV, INRA-Theix, 63122 St Genès Champanelle, France.* [2.] *Departament of Food Science, Instituto de Agronomica y Tecnologia de Alimentos (C.S.I.C), Apt. 73, 46100 Burjassot, Valencia, Spain.* [3.] *Unité d'immunochimie, SRV, INRA-Theix, 63122 St Genès Champanelle, France. Email: aouali@clermont.inra.fr*

Keywords: proteasome, ELISA, RID, muscle, bovine

Introduction

In the food industry, it was shown that the 20S proteasome contribute significantly to the improvement in meat texture of slow-twitch oxidative muscles and as well to a high pH in meat (Ouali, 1999). Hence, the 20S proteasome might be used as a marker, not only for pathologies but also for meat quality. This work develops two accurate immunochemical quantitative tests, the ELISA method and the less known radial immunodiffusion (RID).

Materials and Methods

The 20S proteasome was purified from bovine *Diaphragma* muscle by Dutaud (1998). SDS-PAGE (Laemmli, 1970) was conducted using acrylamide concentration (12%). Immunoblot was carried out as described by Sentandreu *et al.*, (2003). The monoclonal antibody (mcp 20) directed against alpha subunits of human 20S proteasome was prepared as previously described (Hendil *et al.*, 1995). Preparation of the antiserums were carried out (Dutaud *et al.*, 2002). The polyclonal antibody obtained against proteasome was used at a dilution of 1/1000. *ELISA procedure:* For proteasome quantification, the procedure used was based on a sandwich ELISA technique. The capture antibody was the monoclonal antibody whereas the secondary antibody was Peroxidase-labelled IgG and HRP activity was revelated. *Proteasome quantification by RID:* Radial immunodiffusion (RID) (Mancini *et al.*,1965), is a precipitation reaction carried out by application of antigen solution to a gel into which a monospecific antibody solution has been added. It is an adaptation of gel-precipitation reactions that permits the quantitation of an antigen. From the known standard concentrations, a standard curve can be drawn by plotting the square root of the antigen concentration versus the ring diameter. From this linear calibration curve the concentration of the unknown antigen samples can be determined

Results and Discussion

When analysed by SDS-PAGE, the purified proteasome exhibited a series of bands, characteristic to the 20S proteasome complex (see Figure 1A, lane 2). Pure proteasome preparations were used for the production of polyclonal antibodies and as a standard for the quantification of this complex by ELISA and RID.

The specificity of the rabbit antiserum was tested by western blot against a crude muscle (Figure 1A, lane 1) and against the purified 20S proteasome (Figure 1A, lane 2). As shown in Figure 1B, the antibody recognised several major bands corresponding to the proteasome subunits (Figure 1B, lane 1). When tested against the whole muscle extract, the number of labelled bands were lower, a similar pattern was observed (Figure 1B, lane 2) and no other bands were detected suggesting a high specificity of the antibody.

Assesment of the optimun dilution for mcp 20 was perfomed. mcp 20 therefore used at a dilution of 1/4500 and a typical standard curve was established (Figure 2). As shown in Figure 3, a linear relationship was observed for all muscle extract dilutions tested (r^2= 0.998). For three dilutions (Figure 3) the plots of the measured concentration versus the added amount of purified proteasome were all linear with slope values close to 1 (from 0.99 to 1.05; see Fig 3). The initial proteasome concentration in the muscle extract tested was estimated to be 53.5 ± 3.1 µg/ml (6 replicates).

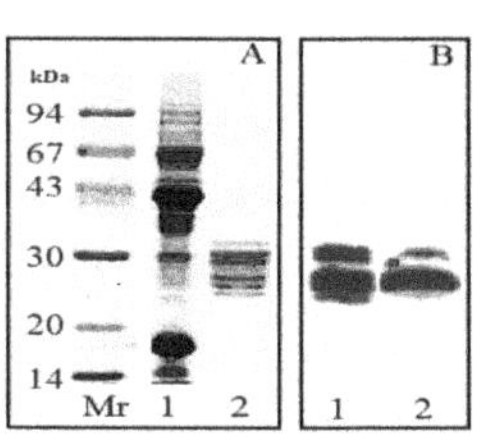

Figure 1: : Immunoblot analysis of the specificity of the polyclonal rabbit anti-20S proteasome (RABPs2) and the monoclonal (mcp 20 and mcp 21) antibodies. (A) Mr markers. Crude muscle extract from *Diaphragma pedialis* muscle stained with coomassie (lane 1); Purified proteasome stained with silver (lane 2). (B) Proteasome revealed by the polyclonal antibody on the purified proteasome (lane 1) and the muscle crude extract (lane 2).

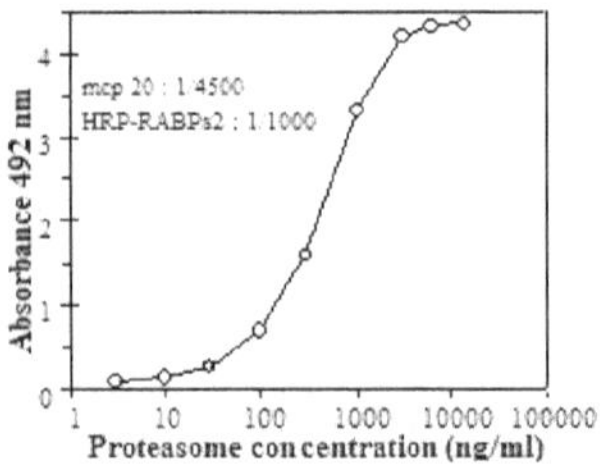

Figure 2: Standard curve obtained with optimum dilution of mcp 20 (capture antibody) and RABPs2.

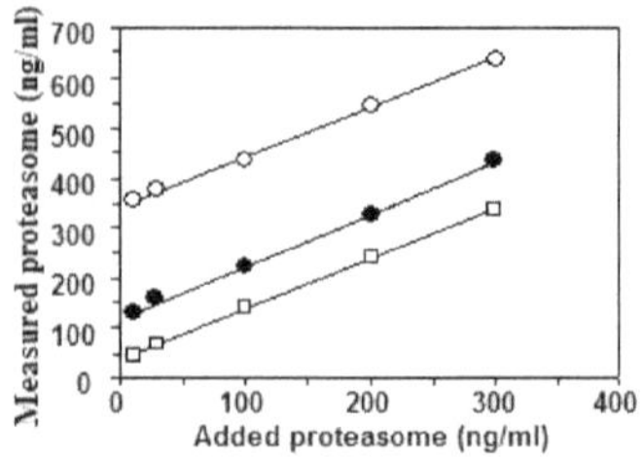

Figure 3: Dose-response curves of 20S proteasome measured by the sandwich-ELISA test. Proteasome concentration was measured on serial dilutions of a crude muscle extract (1/1500, open square; 1/500, closed circles; 1/150, open circles) supplemented with various amounts of purified proteasome. The initial proteasome concentration was determined by linear regression analysis of the plots.

The complex concentration in crude extracts was determined using radial immunodiffusion (RID). The results were compared to those obtained using the ELISA. The agarose gel with the wells containing either purified proteasome of known concentrations or crude extracts from *Longissimus* (L) and *Supraspinatus* (SS) muscles of unknown concentration is presented in Figure 4a. From the diameter of the precipitin rings corresponding to the purified proteasome, a standard curve was plotted (Figure 4b). From this curve the concentration of proteasome in the crude muscle extracts were estimated to be 265 and 200 µg/g for *Longissimus* and *Supraspinatus* muscles, respectively. These values are not significantly different from those obtained using the ELISA technique which yielded values of 275 and 209 µg/g, respectively. Both methods provided therefore very close values.

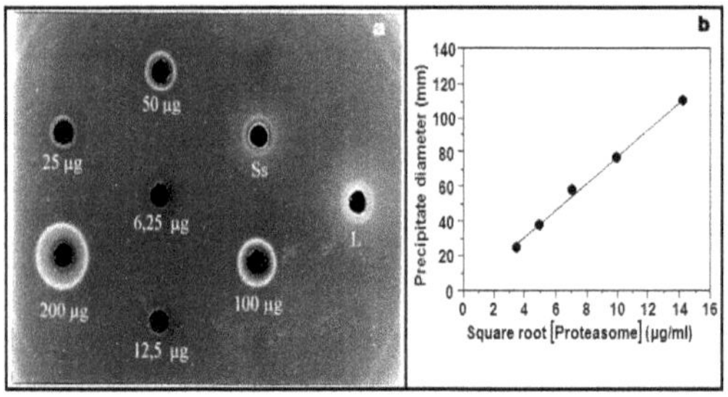

Figure 4: Quantitative determination of proteasome concentrations by radial immunodiffusion (RID). (a) The RABPs2 is incorporated into liquefied agar and allowed to gel. The purified proteasome is introduced at different concentration as indicated in the figure in the circular wells. Diluted crude extracts from *Longissimus* (L) and *Supraspinatus* (SS) muscles were introduced in two different wells as indicated. Note the radial diffusion of the antigen which is precipitated by the polyclonal antibody. (b) Plot of the ring diameter versus the square root of the proteasome concentration. The equation and r^2 value of the plot were: $y= 7.8x + 0.01$, $r^2= 0.99$.

Conclusion

In this study, we further provide strong evidence for a possible accurate measurement of proteasome content in bovine tissues using another immunochemical technique referred as radial immunodiffusion (RID). RID is a more easy-to-use method than ELISA and requires only low cost material. We showed that RID provided very similar proteasome concentration values compared to ELISA for all bovine tissues tested. The plot of concentration values obtained with ELISA versus values obtained with RID gave linear relationship with a correlation coefficient of 0.99 and a slope close to 1, suggesting that each of these methods can be used independently for proteasome quantification.

References

Dutaud, D. (1998). Quantification et caractérisation de protéasome 20S de muscle de bovin en relation avec l'attendrissage de la viande bovine. PhD Thesis, Blaise Pascal University, Clermont-Ferrand, France.

Dutaud, D., Aubry, L., Henry, L., Levieux, D., Hendil, KB., Kuehn, L., Bureau, J.P. and Ouali A. (2002). Development and evaluation of a sandwich ELISA for quantification of the 20S proteasome in human plasma. *Journal of Immunological Methods, 260*, 183-193.

Hendil, K.B., Kristensen, P. and Uerkvitz, W. (1995). Human proteasomes analysed with monoclonal antibodies. *Biochemical Journal, 305*, 245-252.

Laemmli, U.K. (1970). Cleavage of structural proteins during the assembly of the head of bacteriophage T4. *Nature, 227*, 680-685.

Mancini, G., Carbonara, A.O. and Heremans J. F. (1965). Immunochemical quantitation of antigens by single radial immunodiffusion. *Immunochemistry, 2*, 235-254.

Ouali A. (1999). Strructure and biochemistry of muscle as related to meat texture. Proc. XIV European Symposium on the Quality of Poultry Meat (Bologna, Italy). *1*, 91-121.

Sentandreu, M.A., Aubry, L. and Ouali, A. (2003). Purification of bovine cathepsine B: proteomic characterization of the different forms and production of specific antibodies. *Biochem. Cell. Biol., 81*, 317-326.

Q204X MYOSTATIN MUTATION EFFECTS ON CARCASS AND MEAT QUALITY TRAITS IN HETEROZYGOUS CHAROLAIS YOUNG BULLS

H. Levéziel[1]*, J.F. Hocquette[2], J. Lepetit[3], C. Denoyelle[4], V. Dodelin[5], N Payet[1], A. Malafosse[6] and G. Renand[7]

[1] INRA/Université de Limoges, Unité de Génétique Moléculaire Animale, Faculté des Sciences et Techniques, 87060 Limoges, France; [2] INRA Unité de Recherches sur les Herbivores, 63122 Saint Genès Champanelle, France; [3] INRA Unité Qualité des Produits Animaux, 63122 Saint Genès Champanelle, France; [4] Institut de l'Elevage, Laboratoire d'Analyses et Technologies des Viandes, Villers Bocage, 14310, France; [5] Institut de l'Elevage, Service Sélection, 75975 Paris Cedex 12, France; [6] UNCEIA, 75975 Paris Cedex 12, France; [7] INRA Station de Génétique Quantitative et Appliquée, 78352 Jouy en Josas Cedex, France. Email: hubert.leveziel@jouy.inra.fr

Keywords: beef, myostatin mutation, carcass, meat quality

Introduction

Double muscling in cattle is known to be controlled by mutations in the myostatin gene (Grobet *et al.*, 1998). Homozygous cattle with two copies of the mutated allele have a dramatically enhanced muscle development (Ménisier, 1982). Except in few breeds (Belgian Blue, Piedmontese) the double-muscled animals are generally discarded for replacement breeding due to the adverse effects on maternal traits. Nevertheless, the selective superiority of heterozygous animals may lead to the maintenance of a significant frequency of the myostatin mutations in some breeds. The availability of molecular tests allows the detection of the different mutations (Fahrenkrug *et al.*, 1999; Miranda *et al.*, 2002) that are segregating in these populations and consequently the estimation of the allele frequencies and/or the genotype effects on meat production traits. In a sample of 128 purebred cows Dunner (2001) estimated the frequency of the Q204x mutation (specific of the Charolais) to 16% in the French Charolais breed. In experimental crossbred populations Short *et al.* (2002) and Casas *et al.* (2004) estimated the effects of the C313Y and nt821 mutations (specific of the Piedmontese and Belgian Blue respectively) on live and carcass traits. The objective of the present study was to evaluate the frequency of the myostatin Q204x allele in the French Charolais breed and to evaluate its effect on live, carcass and meat quality traits when comparing heterozygous carriers with normal homozygous cattle.

Materials and Methods

In France, artificial insemination sires of the specialized beef breed are progeny tested for estimating breeding values of live and carcass traits. Four hundred and eight male calves, randomly procreated in pure bred Charolais herds by 22 Charolais sires, entered two feed lots in 2003, were fattened with corn silage and eventually were slaughtered in two slaughter houses when they reached 728 kg live weight on average. In addition to the carcass traits usually recorded (dressing percentage, fat and muscling scores) meat quality attributes and related muscle characteristics were also recorded. Twenty four hours post slaughter, a 6th to 9th rib section was excised. Muscle lightness was measured on the *Longissimus thoracis* surface. The rib eye area and the ratio of fat to muscle + fat weights in the dissected 6th rib were measured. Steaks sliced in the *Longissimus thoracis* muscle were placed for a 14 day ageing period and then frozen. Meat texture was evaluated by the measurement of Shear Force with a Warner-Bratzler device. For meat quality measurements, thawed steaks were cooked on a grill with a core temperature of 55°C, which is the usual cooking temperature for beef meat in France. Sensory attributes (Tenderness, Juiciness and Flavour) were measured by 12 trained panellists. Three muscle characteristics related to meat quality were also measured on muscle samples 24 hours post mortem: collagen (assessed via hydroxyproline), lipid (assessed via ether extract) contents and muscle fibre section mean area.

Blood samples were obtained from all the bull calves and sires, and most (n=326) of the dams. DNA was extracted and genotyped for the Q204x mutation (TaqMan 5' nuclease assay, 7900 HT Applied Biosystems). Two genotypes were found: homozygous (+/+) or heterozygous (+/mh). According to the sire and dam genotypes, the sire inherited allele (al_s = + or al_s = mh) was determined unless the three animals were heterozygous (al_s = ?).

The base model for analysing the performance included the feed-lot/slaughter house (F) fixed effects and the sire (s) random effects. For estimating the allele substitution effect, the sire inherited allele effect, pooled within sire ($al_s(s)$), was added to the base model. For estimating the average additive genetic effect (a), the calves' genotype (+/+ or +/mh) effect was added to the base model and (a) was computed as the difference between the heterozygous minus the homozygous calf effects.

Results and Discussion

Among the 22 sires, 7 (32%) were heterozygous for the Q204x mutation while 52 among the 326 genotyped cows were heterozygous (16%). The cow sample being more highly representative of the whole breed variability, the Q204x allele frequency in the French Charolais breed can be estimated to 9%.

Among the 138 calves from the 7 heterozygous sires, 68 inherited the a_s=+ and 40 the a_s=mh allele. For 30 calves, it was not possible to determine which allele was transmitted by the sire however. The substitution effects on the carcass muscling score averaged +1.05 standard deviation ($P<0.001$), ranking from 0.2 to 1.8 across sires.
Among the 408 calves, 86 were heterozygous, inheriting the mh allele either from the sire (n=40) or the dam (n=22). The origin of the mh allele could not be determined in 24 calves. The difference in carcass muscling score between calves with (+/mh) genotype and calves with (+/+) was estimated to **a**=+1.11 standard deviation ($P<0.001$), i.e. exactly the same estimate as the within sire substitution effect as expected for a causal mutation. The additive genetic effect (a) of the mutated allele was estimated and results are reported in the Table 1.

Table 1.

Trait		mean	± SD°	a =+/mh - +/+	a / SD	Prob[diff]
Dressing Percentage	%	57.7	1.9	1.9	0.98	<0.001
Muscling Score	/16	10.0	1.7	1.9	1.07	<0.001
Rib Eye Area	cm²	51.8	8.3	5.5	0.66	<0.001
Fat / (Fat+Muscle)	%	20.6	3.9	-3.0	-0.77	<0.001
Lipid content	%	1.62	0.96	-0.45	-0.46	0.001
Collagen content	%	0.38	0.07	-0.03	-0.40	0.005
Fibre Diameter	μ²	2824	678	-274	-0.40	0.003
L*	/100	35.8	4.7	2.1	0.43	0.001
Shear Force	kg	3.91	0.79	0.04	0.05	0.69
Tenderness	/100	62.2	8.6	2.2	0.26	0.06
Juiciness	/100	60.1	6.3	0.3	0.05	0.74
Flavour	/100	55.1	5.4	-0.5	-0.08	0.54

For the carcass composition and conformation, the effect of the mutated allele is important and of similar magnitude as the Piedmontese C313Y or Belgian Blue nt821 effects estimated by Short *et al.* (2002) and Casas *et al.* (2004). While no effect could be detected on Flavour, Juiciness and Shear Force, the presence of the Q204x has large effects on muscle characteristics: less lipids, less collagen, thinner muscle fibres. The mutated allele also gives lighter meat which tends to be tender.

Conclusions

A first estimation was obtained on the actual effect of the Q204x mutation inheritance on the carcass and meat quality of heterozygous Charolais in a population of commercial beef bulls. These effects are large enough to anticipate that genotyping of this mutation could allow breeders and/or producers to provide the market with beef cattle that fit part of the retailer or consumer requirements. Knowing the actual genotype of beef cattle may generate, therefore, an added value in comparison to the present situation where heterozygous cattle are marketed in the same market segment as normal homozygous cattle.

References

Casas E., Bennett G.L., Smith T.P.L., Cundiff L.V. (2004). Association of myostatin on early calf mortality, growth and carcass composition traits in crossbred cattle. Journal of Animal Science, 82, 2913-2918.

Dunner S. (2001). Detection and frequencies of the mutations in the myostatin gene causing muscular hypertrophy in European bovine breeds for marker assisted selection purposes. European demonstration project n° BIO4-CT98-0421 final report. 28pp.

Fahrenkrug S.C., Casas E., Keele J.W., Smith T.P.L. (1999). Direct genotyping of the double-muscling locus (mh) in Piedmontese and Belgian Blue cattle by fluorescent PCR. Journal of Animal Science, 77, 2028-2030.

Grobet L., Poncelet D., Royo L.J., Brouwers B., Pirottin D., Michaux C., Ménissier F., Zanotti M., Dunner S., Georges M. (1998). Molecular definition of an allelic series of mutations disrupting the myostatin function and causing double-muscling in cattle. Mammalian Genome, 9, 210-213.

Ménissier F. (1982). General survey of the effect of double muscling on cattle performance. In: King J.W.B., Ménissier F (Eds.) Muscle hypertrophy of genetic origin and its use to improve beef production. Martinus Nijhoff Publishers, London, pp. 23-53.

Miranda M.E., Amigues Y., Boscher M.Y., Ménissier F., Cortés O., Dunner S. (2002). Simultaneous genotyping to detect myostatin gene polymorphism in beef cattle breeds. Journal of Animal Breeding and Genetics, 119, 361-366.

Short R.E., MacNeil M.D., Grosz M.D., Gerrard D.E., Grings E.E. (2002). Pleiotropic effects in Hereford, Limousin and Piedmontese F2 crossbred calves of genes controlling muscularity including the Piedmontee myostatin allele. Journal of Animal Science, 80, 1-11.

Acknowledgements

This research was supported by the AGENAE consortium.

MYOSTATIN AND ADIPOGENIC TRANSCRIPTION FACTOR EXPRESSIONS DURING SKELETAL MUSCULAR GROWTH OF JAPANESE BLACK CATTLE

M. Shibata*[1], K. Matsumoto[1], K. Aikawa[1], T. Muramoto[1], S. Muroya[2], M. Ohe[2], I. Nakajima[2], K. Chikuni[2], S. Fujimura[3] and M. Kadowaki[3]

[1.] *National Agricultural Research Center for Western Region (WeNARC), Ohda, Shimane, 694-0013, Japan,* [2.] *National Institute of Livestock and Grassland Science, Tsukuba, Ibaraki, 305-0901, Japan,* [3.] *Faculty of Agriculture, Niigata University, Niigata, 950-2181, Japan, Email: shibatam@affrc.go.jp*

Keywords: gene expression, Japanese Black Cattle, *M. Semitendinosus*, myostatin, PPARγ2

Introduction

Myostatin is classified as a member of the transforming growth factor-β superfamily and identified as a specific negative regulator of skeletal muscle growth in myostatin gene null mice, which have reduced rates of body fat accumulation (McPherron *et al.*, 1997). Similarly, natural myostatin mutations in cattle have been shown to be associated with a double-muscling phenotype (Grobet *et al.*, 1997). Although we (Shibata *et al.*, 2003) and others have reported myostatin gene expression in fetal bovine and calf skeletal muscle, detailed long-term developmental changes in myostatin gene expression in cattle have not been characterized. The present study investigates the association of myostatin in the formation of skeletal muscle in Japanese Black Cattle (JBC). We report developmental changes in the myostatin gene and adipogenic transcription factor expression from the fetal to the fattening period.

Materials and Methods

Skeletal muscles were obtained by slaughter or by biopsy from JBC which had been bred at the WeNARC. Donor cattle at 8 months gestation were slaughtered and fetuses obtained. *Musculus (M.) Semitendinosus* tissue was obtained by biopsy, under local anesthesia, from JBC aged 4, 10, 16, 22, and 28 months. All collected samples were rapidly frozen in liquid nitrogen and stored at –80°C until RNA extraction. Total RNA was extracted from muscle tissues using the TRIZOL reagent according to the manufacturer's protocol. First strand cDNA was synthesized from 3 μg of total RNA using SuperScript II RNase H^- reverse transcriptase with oligo dT primer. After reverse transcription, expressions of target genes were determined by quantitative real time PCR with a TaqMan probe using an ABI 7700 detection system, or a semi-quantitative reverse transcription (RT)-PCR. Relative target genes mRNA levels were normalized with glyceraldehyde-3-phosphate dehydrogenase (GAPDH).

Results and Discussion

We report long-term developmental changes in myostatin gene expression in *M. Semitendinosus* of JBC up to the fattening period. The results showed two peaks: the highest in the fetal stage and the second at 16 months of age (Figure 1). *M. Semitendinosus*, *M. Semimembranosus* and *M. Biceps femoris*, all present in the hindlimbs, increase in weight in double-muscled cattle (Boccard and Dumont, 1974); further, the muscular hypertrophy that is characteristic of these cattle was more marked in hindlimbs than forelimbs (Dumont, 1982). Higher expression of the gene was always found in *M. Semitendinosus* compared to *M. Longissimus lumborum* of JBC (Shibata *et al.*, 2003). These results suggest that hindlimb development at middle of fattening period in bred cattle may need the potential control of muscle growth by myostatin.

To investigate the relationship between myostatin and fat accumulation in skeletal muscle, expressions of the adipogenic transcription factors, CCAAT/enhancer binding protein (C/EBPα) and peroxisome proliferator -activated receptor (PPARγ2), were measured at different stages of skeletal muscle development by semiquantitative RT-PCR. PPARγ2 expression at 16 months was higher than at any other stage, but C/EBPα level did not change between the different stages (Figure 2). The fat content of double-muscled cattle is lower than that of normal cattle (Shahin *et al.*, 1985) and, interestingly, study with myostatin null mice recently revealed that a decrease in body fat accumulation results from a reduction of adipogenesis and, accordingly, decreased leptin secretion (Lin *et al.*, 2002). This suggests that myostatin is involved in the regulation of adiposity. It is interesting that higher PPARγ2 gene expression was observed at 16 months, compared with other stages in JBC. A previous report indicated that fat deposition in dressed JBC carcasses was slow until 15 months, and then dramatically increased at 15 to 16 months. The increase in fat accumulation in skeletal muscle has been shown to occur at 15–20 months in JBC (Mitsuhashi *et al.*, 1987). These findings suggest that myostatin may be related to marbling formation at the middle of the fattening period in JBC.

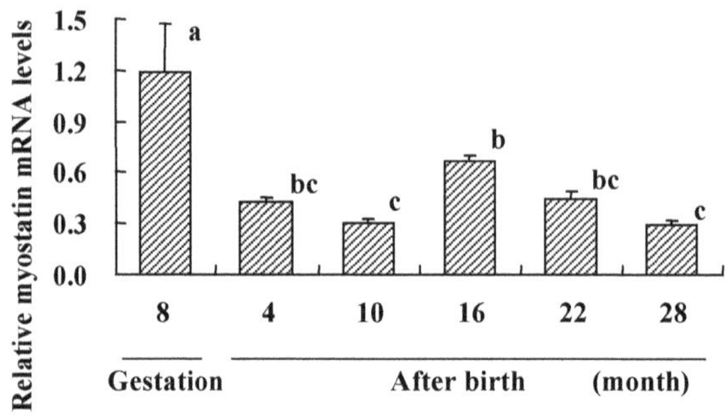

Figure 1: Developmental changes in myostatin gene expression in *M. Semitendinosus* of Japanese Black Cattle by real time PCR. Relative myostatin mRNA levels were normalized with GAPDH mRNA. Bars represent the means ± SEM of 3 to 8 animals. Values with different letters (a - c) are significantly different ($P<0.05$).

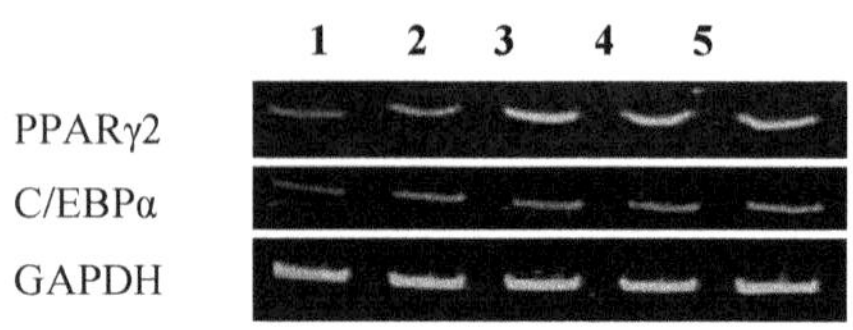

Figure 2: Polyacrylamide gel electrophoresis of RT-PCR products amplified from *M. Semitendinosus* with gene specific primer pairs. Lane 1: 4 months, Lane 2: 10 months, Lane 3: 16 months, Lane 4: 22 months, Lane 5: 28 months after birth.

Conclusion

The present study revealed high expressions of both myostatin and PPARγ2 genes in *M. Semitendinosus* at 16 months. This increase in genes may be related to marbling formation in the fattening period.

References

Boccard R., and Dumont, B. L. (1974). Effects of hereditary muscular hypertrophy on the musculature of cattle. Annales de Genetique et de Selection Animale. 6, 2, 177-186.

Dumont B. L. (1982). Carcass composition and muscle structure in hypertrophied animals. Pages 111-134 in Muscle Hypertrophy of Genetic Origin and its Use to Improve Beef Production. King J. W. B. and F. Menissier, eds. Martinus Nijhoff Publishers, Hague.

Grobet L., Martin, L. J. R., Pirottin, D., Brouwers, B., Riquet, J., Schoeberlein, A., Dunner, S., Menissier, F., Massabanda, J., Hanset, R. and Georges. M. (1997). A deletion in the bovine myostatin gene causes the double-muscled phenotype in cattle. Nature Genetics 17: 71-74.

Lin J., Arnold, H. B., Della-Fera, M. A., Azain, M. J., Hartzell, D. L. and Baile, C. A. (2002). Myostatin knockout in mice increases myogenesis and decreases adipogenesis. Biochemical Biophysical Research Communication 291: 701-706.

McPherron, A.C., Lawler, A. M. and Lee. S. J. (1997) Regulation of skeletal muscle mass in mice by a new TGF-β superfamily member. Nature 387: 83-90.

Mitsuhashi T., Mitsumoto, M. and Y.Yamashita. (1987). Age-associated changes in tenderness and ether extract content of beef muscle from Japanese Black Steers. The Bulletin of the Chugoku National Agricultural Experiment Station 1: 51-57 (in Japanese).

Shahin K. A. and Berg. R. T. (1985). Fat growth and partitioning among the depots in double muscled and normal cattle. Canadian Journal of Animal Science 65: 307-318.

Shibata M., K. Ohshima, T. Kojima, T. Muramoto, K. Matsumoto, M. Komatsu, K. Aikawa, Fujimura, S. and M. Kadowaki. (2003). Nucleotide sequence of myostatin gene and its developmental expression in skeletal muscles of Japanese Black Cattle. Animal Science Journal 74: 383-390.

CARCASS TRAITS AND MEAT QUALITY OF DIFFERENT PIG GENOTYPE

G. Kralik*, G. Kusec, V. Margeta and D. Hanzek

Josip Juraj Strossmayer University of Osijek, Faculty of Agriculture Department of Specialized Zootechnics, Trg svetog Trojstva 3, 31000 Osijek, Croatia, Email: gkralik@pfos.hr

Keywords: pigs, carcass traits, meat quality

Introduction

For the production of pork in the Republic of Croatia, different combination of crossbred (meat type) pigs are used, as well as imported hybrid pigs. The main objective is to evaluate pig genotypes characterized by both quantity and quality of produced meat. Intensive selection for fast growth and high lean meat proportion resulted in increased sensitivity of pigs to production and environmental conditions. This resulted in higher production of meat with poor technological quality traits. Poor meat quality decreases any positive economic effect which is one of the biggest problems in the slaughter industry in Croatia today. Because of the high heritability of pork quality traits, it is possible to influence on their improvement with selection. Edwards *et al.*, (2003) demonstrated that proper characterization of different breeds and lines for carcass and meat quality traits is essential to the choice of parents for pork production. The aim of our research was to compare the leanness and meat quality traits of the Croatian genotypes (CCB) with imported TOPIGS hybrids and to evaluate which genotype can be considered as more desirable regarding the carcass and meat quality traits.

Materials and Methods

Research was performed on pig carcasses with Croatian genotypes /(LWxSL)GL/ - CCB (n=20) and TOPIGS hybrids (n=20). At the slaughter line, the measurements of warm carcass weight, carcass length ("a" from *os pubis* to the 1st rib; "b" from *os pubis* to *atlas*), ham length and circumference were taken from which ham index was calculated. Initial pH values (pH_{45}) were measured 45 minutes after exsanguination. After 24 hours of cooling, cold carcass weight, muscle surface and (belonging) backfat surface of the *M. longissimus dorsi- MLD* (cm^2), ultimate pH (pH_{24}) values, water holding capacity (WHC, cm^2), drip loss (%) and colour of the *M. longissimus dorsi* were taken. Shares of muscle tissue, fatty tissue and bones were determined by total dissection of the carcasses by the method of Weniger *et al.* (1963). Backfat and MLD surface were measured by the geometric procedure (Comberg, 1978) and expressed as the fat/muscle area ratio; water holding capacity (w.h.c.) was determined using the compression method (Grau and Hamm, 1952) and by measuring drip loss according to Kauffmann *et al.* (1992). The lightness of meat was measured using a "Minolta CR-300" device on a *M. longissimus dorsi* cut and expressed as CIE L* value. Statistical analysis was performed using STATISTICA (6.0) for Windows program.

Results and Discussion

Table 1 presents carcass quality traits of investigated pig groups. Significant differences were identified for carcass length, ham index and fat / MLD surface ratio. CCB pigs had significantly ($P<0.05$) longer carcass length "a", and more significantly ($P<0.001$) longer carcass length "b". Ham index was significantly higher ($P<0.01$) for the CCB genotype compared to TOPIGS. TOPIGS hybrid pigs had more desirable fat / MLD surface ratio; the difference between examined groups was significant ($P<0.01$). It is evident that carcasses from examined groups had similar shares of lean meat, fatty tissue and bones of the main and less valuable carcass parts ($P>0.05$).

Meat quality trait measurements on investigated pigs are presented in Table 2. Results of pH values in MLD samples, according to border initial pH_{45} value 6.0, recommended by Hoffmann (1994), as well as the final pH value (pH_{24} 5.5) recommended by Forest (1998) showed that both pig groups are characterized with "normal" meat quality. Statistically significant differences in pH_{24} and meat colour ($P<0.05$) between Croatian crossbreeds and TOPIGS group suggest better meat quality traits of the former. The mean values of WHC >9 cm^2 (Blendl *et al.*, 1991), drip loss >5% (Kauffman *et al.*, 1992) and Minolta L* >53 (Hofmann, 1994) indicate PSE condition of meat originating from TOPIGS hybrid pigs.

Table 1: Carcass quality traits.

Indicator	CCB	TOPIGS	Significance of differences
Weight of warm carcasses (kg)	86.40±8.42	85.55±8.40	n.s.
Weight of refrigerated carcasses (kg)	84.60±6.11	83.84±7.78	n.s.
Lean meat percentage (%)	55.27±3.34	54.12±2.43	n.s.
Fatty tissue percentage (%)	24.61±2.61	24.74±2.48	n.s.
Bone percentage (%)	10.17±0.98	9.53±0.34	n.s.
Carcass length – a (cm)	87.55±2.39	85.65±2.37	*
Carcass length - b (cm)	105.85±3.17	102.40±2.28	***
Ham index	0.46±0.01	0.44±0.02	**
MLD surface (cm^2)	41.21±3.50	45.87±7.29	n.s.
Surface of MLD belonging fat (cm^2)	17.55±1.38	15.14±2.77	n.s.
Fat/MLD surface ratio	0.43±0.03	0.33±0.05	**
Ham (%)	30.26±1.00	30.30±0.38	n.s.
Loin (%)	17.18±1.03	16.39±1.33	n.s.
Belly-rib part (%)	16.64±1.30	16.91±0.91	n.s.
Shoulder (%)	13.50±0.23	13.93±0.68	n.s.
Neck (%)	10.73±1.00	10.80±0.62	n.s.
Less valuable parts (%)	11.69±0.37	11.67±0.36	n.s.

Table 2: Meat quality.

Indicator	CCB	TOPIGS	Significance of differences
pH_{45}	6.26±0.21	6.25±0.19	n.s.
pH_{24}	5.55±0.10	5.50±0.05	*
W.H.C. (cm^2)	8.76±1.70	9.94±2.03	n.s.
Drip loss (%)	3.73±1.24	5.92±2.66	n.s.
Colour (Minolta L*)	51.59±3.33	56.19±2.92	*

Conclusions

On the basis of the present study, the following conclusions can be derived:

- Croatian genotypes (CCB) had significantly longer carcasses and more favorable ham index than TOPIGS hybrid pigs. On the other hand, TOPIGS hybrids had better fat/MLD ratio than CCB pigs. Other carcass traits did not differ significantly.
- Although significant differences between investigated groups were found only for pH24 and Minolta L* values, Croatian genotypes of pigs may be considered as more desirable regarding the meat quality traits, because of high incidence of PSE meat in TOPIGS hybrid pigs.

References

Blendl, H., E. Kallweit, J. Scheper (1991) Qualitatanbieten: Schweine-fleisch. AID, 1103, Bonn.

Comberg, G. (1978) Schweinezucht. Verlag Eugen Ulmer, Stuttgart.

Edwards, D.B., R.O., Bates, W.N. Osburn (2003) Evaluation of Duroc-vs. Pietrain-sired pigs for carcass and meat quality measures. Journal of Animal Science 81:1895-1899.

Grau, R. and R. Hamm (1952) Eine einfache Methode zur Bestimmung der Wasserbindung im Fleisch. Die Fleischwirtschaft, 4:295-297.

Hofmann, K. (1994) What is quality? Definition, measurement and evaluation of meat quality. Meat Focus International, Vol. 3, Part 2, February 1994.

Kauffman R.G., R.G. Cassens, A. Sherer and D.L. Meeker (1992) Variations in pork quality. National Pork Producers Council Publication, Des Moines, IA, pp. 1-8.

Weniger, H.J., D. Steinhauf and G. Pahl (1963) Musculare Topography of carcasses. BLV Verlagsgesellschaft, München.

EFFECT OF IGF-II AND HALOTHANE GENOTYPE ON CURED COOKED LOIN PRODUCTION

K. Van den Maagdenberg[1], E. Claeys[1], A. Stinckens[2], N. Buys[2] and S. De Smet[1]*

[1.] *Laboratory for Animal Nutrition and Animal Product Quality, Department of Animal Production, Ghent University, Proefhoevestraat 10, 9090 Melle, Belgium,* [2.] *Laboratory of Livestock Physiology, Immunology and Genetics, KULeuven, Kasteelpark 30, 3001 Heverlee, Belgium. Email: Karijn.VandenMaagdenberg@UGent.be*

Keywords: halothane gene, PSE, IGF-II allele, stress genotype, ham

Introduction

During the last decades, pig breeding programs have strongly focused on selection for fast growth of lean meat. However, this may be accompanied by negative effects on certain meat quality characteristics. It is well known that the halothane gene increases the carcass lean content but equally increases the risk of development of PSE (Pale, Soft and Exudative) meat. Cooking and curing of PSE hams results in lower technological yields. This may be associated with aberrant colour development and texture defects (Fernandez *et al.*, 2002; Leach *et al.*, 1996), particularly in polyphosphate-free cooked ham production (Martoccia *et al.*, 1994). The gene in its heterozygous form (Nn) is believed to have mainly intermediate effects on carcass and meat quality.

On the other hand, Van Laere *et al.*, (2003) recently identified a paternally expressed mutation located in the regulatory sequence of the IGF-II gene (insulin-like growth factor-II). The mutation is known as the IGF-II intron3 G3072A mutation and has a similarly large effect on carcass lean content and apparently no negative effects on meat quality (Nezer *et al.*, 1999). However, the effects on curing and cooking of meat have not yet been assessed. Therefore, the aim of this study was to investigate the effect of the IGF-II allele (Apat vs. Gpat) in combination with the stress genotype (Nn vs. NN) on the properties of cured and cooked loin. For practical reasons, loin samples were used to simulate this type of processing, although cured cooked ham production from hind-leg muscles is more common

Materials and Methods

<u>Animals</u>: The animals (n = 53) were the progeny of one sire and 13 sows and consisted of castrates and females (n = 26 and 27 respectively). The sire (Rattlerow-Seghers) was heterozygous for IGF-II and homozygous negative for halothane. The sows were heterozygous for the halothane gene. Because of the paternal imprinting of the IGF-II gene, offspring were only considered with known paternal allele of interest. The number of Nn animals was 13 Apat and 17 Gpat, whereas there were 10 Apat and 13 Gpat NN animals. Apat animals inherited the IGF-II mutation which is related to the higer muscle mass versus the Gpat animals carrying the paternal wild type allele. Pigs were fed ad libitum on a conventional diet. After overnight fasting, they were slaughtered following CO_2 stunning in a private abattoir. The average live weight at slaughter was 109 ± 7 kg.

<u>Cured cooked loin preparation</u>: At 24 h post mortem, the left loin was taken from the carcass and backfat was removed. The dorsal end of the *Longissimus* between the 3rd and 4th vertebra (between 0.9 and 1.5 kg) was used for curing and cooking the day after. A protocol for the preparation of high-quality cooked ham was followed. With a multi-needle injection system, 15% brine (20.2% salt, 5.6% dextrose, 1.1% ascorbate and 1.1% flavourings) was injected, followed by 30 minutes of tumbling (15 rpm, 4°C, 85°, not-vacuum) in a Rhüle PR15. Afterwards, the meat was individually vacuum-packed in a shrink-bag and cooked until a nucleus temperature of 67°C. After 2 h of cooling, the samples were stored at 2°C for 48 h. The curing yield was calculated as the percentage difference between the weight after curing and tumbling and the initial weight. Similarly, cooking losses were calculated from the weight difference before and after cooking. The overall yield is defined as the final weight relative to the initial weight.

<u>Analyses</u>: Colour and colour stability was measured with a Hunterlab Miniscan apparatus (D65 light source, 10° standard observer, 45°/0° geometry, 1 in. light surface, white standard) after 2 h and 5 days of display. Samples were wrapped in oxygen-permeable foil and stored at 4°C under fluorescent light. L*, a*, b* values were measured in quadruplicate. Reflection values at 630 and 650 nm were measured in duplicate to estimate the concentration of nitrosylhemochrome pigment (% NH = R650/R630). A texture profile analysis was done with a Lloyd TA 500 Texture Analyser directly after cutting. Samples of 3 cm thickness were compressed two times at the same place to 70% of its initial thickness. Several texture traits (hardness, gumminess, chewiness and stiffness) were derived from the resultant time versus force (load) curve.

Data were analysed with a univariate general linear model with gender, IGF-II and halothane genotype as fixed factors. Preliminary analysis revealed no significance of the interaction terms, and no effect of carcass weight as covariate. Statistical analyses were done using SPSS 12.0 for windows.

Results and Discussion

No effect of IGF-II genotype and gender on curing yield, cooking losses and overall yield was found (Table 1). However, overall yield was significantly higher in Nn animals compared to NN animals. This is in accordance with literature data (Fernandez *et al.*, 2002; Leach *et al.*, 1996). It should be stressed that the effects of differences in fresh meat quality on processing yields is highly dependent on the type of processing, e.g. on the use of additives in the brine. Texture parameters of the cooked samples were not influenced by the IGF-II genotype. However, significant effects were found for the halothane genotype and gender. Gumminess and chewiness were significantly higher in samples from halothane carriers. Samples from gilts had significantly higher values for all texture traits. Texture traits were negatively related to overall yield and positively related to carcass lean meat content ($|r| \approx 0.4$). Carcass lean meat content and overall yield were negatively related ($r = -0.26$). In this study, the differences in lean meat content between the two IGF-II genotypes and gender were much larger (> 5%) than between the two halothane genotypes (< 2%). Hence, differences for yield and texture traits that were observed may be partly, but not entirely, associated with differences in fat content.

The IGF-II mutation and gender did not influence colour and colour stability of the cooked ham as well as the nitrosylhemochrome concentration. Nn pigs had slightly but significantly higher b* values compared to NN pigs after 2 hours of exposure to light.

Table 1: Least square means ± SE for technological yield, texture and colour traits.

	IGF-II genotype			Halothane genotype			Gender		
	A	G	P	Nn	NN	P	Barrow	Gilt	P
				Technological yield traits					
Curing yield (%)	14.1 ± 0.4	14.4 ± 0.3	ns	14.1 ± 0.3	14.4 ± 0.4	ns	14.1 ± 0.3	14.4 ± 0.3	ns
Cooking losses (%)	12.6 ± 0.6	12.0 ± 0.5	ns	13.5 ± 0.5	11.0 ± 0.6	**	12.0 ± 0.6	12.5 ± 0.5	ns
Overall yield (%)	99.7 ± 0.8	100.7 ± 0.7	ns	98.7 ± 0.7	101.7 ± 0.8	**	100.4 ± 0.8	100.0 ± 0.8	ns
				Texture traits					
Hardness (N)	49.1 ± 1.4	48.0 ± 1.3	ns	49.9 ± 1.3	47.2 ± 1.4	ns	45.0 ± 1.4	52.1 ± 1.3	**
Gummines (N)	13.7 ± 0.5	13.5 ± 0.4	ns	14.3 ± 0.4	12.9 ± 0.5	*	12.6 ± 0.4	14.6 ± 0.4	**
Chewiness (N.mm)	154.2 ± 5.3	151.5 ± 4.6	ns	161.0 ± 4.6	144.7 ± 5.3	*	140.3 ± 5.0	165.4 ± 4.8	**
Stiffness (N.mm)	5.3 ± 0.2	5.3 ± 0.2	ns	5.4 ± 0.2	5.3 ± 0.2	ns	4.8 ± 0.2	5.9 ± 0.2	**
				Colour traits					
L* 2h	76.14 ± 0.32	76.02 ± 0.28	ns	75.92 ± 0.28	76.24 ± 0.32	ns	76.31 ± 0.31	75.85 ± 0.30	ns
a* 2h	5.84 ± 0.11	5.99 ± 0.10	ns	5.86 ± 0.10	5.97 ± 0.10	ns	5.96 ± 0.11	5.87 ± 0.10	ns
b* 2h	12.39 ± 0.06	12.43 ± 0.05	ns	12.56 ± 0.05	12.23 ± 0.05	**	12.34 ± 0.06	12.48 ± 0.05	ns
% NH 2h	1.06 ± 0.01	1.05 ± 0.01	ns	1.06 ± 0.01	1.05 ± 0.01	ns	1.05 ± 0.01	1.06 ± 0.01	ns
Δ a*	3.91 ± 0.10	3.86 ± 0.09	ns	3.87 ± 0.09	3.90 ± 0.10	ns	3.81 ± 0.09	3.96 ± 0.09	ns
Δ Chroma	0.67 ± 0.08	0.62 ± 0.07	ns	0.66 ± 0.07	0.62 ± 0.08	ns	0.53 ± 0.08	0.75 ± 0.08	ns
Δ % NH	0.02 ± 0.01	0.03 ± 0.01	ns	0.03 ± 0.01	0.02 ± 0.01	ns	0.02 ± 0.01	0.04 ± 0.01	ns

*, ** = significantly different at $P < 0.05$ and 0.01 respectively, ns = not significant

Δ a*, Δ Chroma, Δ Hue and Δ % NH is calculated as the difference between 2 h and day 5 as measures for colour stability

Conclusions

The IGF-II genotype has no influence on the technological yield, texture and colour traits of cured cooked loin. Also no interaction effect with the halothane genotype is to be expected. However, the halothane genotype affected technological yield and texture, and gender also had an effect on the texture of cured cooked loin.

References

Fernandez, X., Gilbert, S., Vendeuvre, J. L. (2002). Effects of the halothane genotype and pre-slaughter treatment on pig meat quality. Part 2. Physico-chemical traits of cured-cooked ham and sensory traits of cured-cooked and dry-cured hams. Meat Science, 62: 439-446.

Leach, L. M., Sutton, D. S., McKeith, F. K., Wilson, E. R. (1996). The growth performance, carcass characteristics, and meat quality of halothane carrier and negative pigs. Journal of Animal Science, 74: 934-943.

Martoccia, L., De Smet, S., Van Damme, B., Demeyer D. (1994). Effect of fresh pork quality on the production of polyphosphate-free cooked ham. In: Proc. 40th International Congress of Meat Science and Technology, S-VIB.01.

Nezer, C., Moreau, L., Brouwers, B., Coppieters, W., Detilleux, J., Hanset, R., Karim, L., Kvasz, A., Leroy, P., Georges, M. (1999). An imprinted QTL with major effect on muscle mass and fat deposition maps to the IGF2 locus in pigs. Nature Genetics, 21: 155-156.

Van Laere, A., Nguyen, M., Braunschweig, M., Nezer, C., Collette, C., Moreau, L., Archibald, A. L., Haley, C., Buys, N., Tally, M., Andersson, G., Georges, M., Andersson, L. (2003). A regulatory mutation in IGF2 causes a major QTL effect on muscle growth in the pig. Nature, 425: 832-836.

Acknowledgements

This research is financially supported by the Institute for the Promotion of Innovation through Science and Technology, Brussels. The authors thank S. Lescouhier and T. Luyten for their excellent technical assistance.

EARLY AND RELIABLE DETECTION OF BOAR TAINT AND ITS GENETIC PREDISPOSITION

D.V. Byrne[3]*, G. Nijs[1], S. Millet[1], M.J. Van Oeckel[1], N. Warnants[1], B. Sonck[2], J. Vestergaard[3], J.-E. Haugen[4], K. Verheyden[5], H. De Brabander[5], F.A.M. Tuyttens[1] and D.L. De Brabander[1]

[1]Institute for Agricultural and Fisheries Research (ILVO-Ghent), Animal Science Unit, Scheldeweg 68, 9090 Melle (Belgium) [2]Institute for Agricultural and Fisheries Research (ILVO), Unit Technology and Food, Burgemeester Van Gansberghelaan 115, 9820 Merelbeke (Belgium), [3]Royal Veterinary and Agricultural University (KVL), Department of Food Science-Sensory Science, Rolighedsvej 30, 1958 Frederiksberg C (Denmark), [4]Norwegian Institute for Food Research (MATFORSK), Osloveien 1, 1430 Ås (Norway), [5]Ghent University, Research Group Veterinary Public Health and Zoonoses, Salisburylaan 133, 9820 Merelbeke (Belgium) E-mail: g.nijs@clo.fgov.be

Keywords: boar taint, detection, diet, breed, management

Introduction

Pork products originating from intact male pigs may have a distinct and unpleasant odour, flavour or taste, commonly known as boar taint. It is mainly perceived when the fat fraction in the meat is heated. Two principal compounds are presumed to cause this phenomenon, although the extent of their contribution to boar taint is still disputed: androstenone (male sex hormone) and skatole (metabolite of amino acid tryptophan).

As boar taint is an unwanted characteristic in the pork industry, it has to be avoided. However, welfare issues rise, as surgical castration is the most common and widespread method to eliminate this phenomenon. Governments have become more aware of these animal welfare concerns and are responding by adjusting legislation. This leads to a demand for the development of alternative solutions to surgical castration.

Several alternatives (e.g. immunocastration, chemical castration) are currently being investigated but are not ready yet for implementation in practice. As legislation prohibiting castration is pending in Belgium, focus is drawn on the development of reduction and detection methods to prevent boar tainted meat or meat products from reaching the consumer, rendering the production of intact males possible.

The purpose of this research project is to find a reliable method to reduce the occurrence of boar taint and to timely detect the presence of this abnormal odour, rendering surgical castration unnecessary and thereby enhancing animal welfare and making the production of entire male pigs possible.

Materials and Methods

The research will focus on three main strategies:

1. Reducing boar taint by altering management strategies. In a series of experiments the effects of various feed ingredients, of genotype in conjunction with slaughter weight, and of hygiene status on the presence of boar taint in the meat and fat will be investigated. Meat and fat samples will be taken, vacuum packed and stored at -18°C until analysis. Presence of boar taint will be assessed by a consumer panel, and linked to the results of both an expert panel and laboratory analyses of specific boar taint components using liquid chromatography-mass spectrometry (LC-MS).
2. Finding a predictor that will permit a swift detection of boar taint in live animals. This would make timely identification of pigs prone to develop boar taint possible, allowing us to take specific measures (immunocastration, early slaughtering) to prevent these animals from developing this taint. Hence, we will investigate whether the intensity of boar taint can be predicted by measuring the development of physical parameters, by observing the behaviour as well as by other measurements such as skin lesions and hygiene status.

For this purpose, a total of 102 animals will be observed for 24 h every fortnight using a camera line-up and a video-recording system. These video images will be analysed using The Observer (Noldus Information Technology) during which aggressive, sexual and social behaviour will be scored. Every fortnight, testis size of each boar will be measured using callipers and skin lesions and hygienic status will be scored.

3. Post-mortem detection of boar taint to prevent boar-tainted meat from reaching the consumer. Several detection systems will be implemented to determine whether or not boar taint is present in meat or fat samples. A threshold level for the acceptability of boar taint will be established for the Belgian consumer, by presenting the meat samples to a consumer panel. The reasons for acceptance or rejection of the meat by the consumer panel will be elucidated further by presenting the meat to an expert panel that is trained to detect and characterise boar taint compounds. The presence of skatole, indole and androstenone levels will be quantified at the Faculty of Veterinary Medicine of the University of Ghent (BE) using a LC-MS technique. The assessments by the consumer panel will also function to fine-tune electronic noses and to train sniffer pigs to detect boar taint in meat or fat samples. A test-setup has been build and clicker training is used to teach the sniffer pigs how to use the experimental set-up to indicate tainted meat. Different types of electronic noses will be investigated and fine-tuned for the post-mortem detection of boar taint in meat and fat samples or on the slaughter line by the Royal Veterinary and Agricultural University (KVL), Copenhagen (DK) in cooperation with Matforsk, Ås (NO). Finally, the soldering iron method will be implemented in the slaughter line and evaluated for provisioning of an indicative value of the intensity of boar taint.

Results and Discussion

This project started in June 2005 and will take 4 years to complete. The sniffer pigs are currently being trained. Observations of behaviour and related measurements in fattening pigs are being carried out in order to detect differences between boars that develop boar taint and boars that do not. The experiment with the different dietary ingredients has just finished and samples have been taken to perform analyses as described above. The expert panel has been trained and consumer and expert panel evaluation of meat and fat samples of the 111 pigs from this experiment will start soon.

Acknowledgments

Research funded by the Belgian Federal Government of Public Health, Food chain safety and Environment, project nr. R-04/007-Boartaint. Thanks to Myriam Audenaert, Koen Desomviele, Kristof Dierkens and Roland Limpens for technical assistance.

GENETIC DIVERSITY AND BREED COMPARISON OF CARCASS TRAITS IN TASMANIAN CORRIEDALE AND EAST FRIESIAN SHEEP BY RAPD MARKERS

A.E.O. Malau-Aduli*[12], C.W. Bignell[1], F. Tavassoli-Salardini[A], A.J. Smolenski[2], A. Palmer[3], J. Bignell[4], S. Burbury[5], R. Batchelor[6], B.S. Malau-Aduli[1], S.A. Adediran[1], P.A. Lane[1] and R.J. Clark[7]

[1.] Animal Production and Genetics, School of Agricultural Science, University of Tasmania, Private Bag 54 Hobart, TAS 7001, Australia, [2.] Molecular Genetics Laboratory, TIAR, University of Tasmania, Private Bag 54 Hobart, TAS 7001, Australia, [3.] Tasman Group Services Pty Ltd, Longford, TAS 7301, Australia, [4.] Thorpe Farm, Bothwell, TAS 7030, Australia, [5.] Kuranda Lambs, Woodbury, Tasmania, Australia, [6.] Roberts Pty Ltd Hobart, TAS 7001, Australia, [7.] Tasmanian Institute of Agricultural Research, University of Tasmania, Private Bag 98 Hobart, Tasmania 7001, Australia. E-mail: Aduli.MalauAduli@utas.edu.au

Keywords: RAPD markers, Corriedale, East Friesian, meat quality, genetic diversity

Introduction

The Australian lamb meat industry exploits crossbreeding in which some 40% of the annual slaughter of about 18 million lambs is first cross progeny from terminal sire meat rams mated to Merino ewes (Fogarty *et al.*, 2005a). However, research has shown that even though the Merino breed is leaner and has reasonably comparable eye muscle area with other meat breeds (Fogarty *et al.*, 2003), carcasses from Merinos have poorer acceptability and high meat pH due to their greater susceptibility to pre-slaughter stress (Fogarty *et al.*, 2000). As a result of the ever-increasing consumer demands for meat products that meet their specifications, the farmer needs to constantly explore diverse breed combinations that best meet such demands. In Tasmania, the Corriedale and East Friesian breeds constitute a significant proportion of the dual-purpose prime lamb breeds, hence an improvement in the genetic merit of their carcass and meat quality traits has the potential to substantially increase the productivity and value of the industry. The objective of this paper was to provide some preliminary information on carcass quality measurements in Corriedale and East Friesian prime lambs produced in Tasmania and to explore the utilisation of RAPD markers for studying the genetic variation between breeds.

Materials and Methods

Two hundred East Friesian and Corriedale prime lambs were commercially slaughtered in a Tasmanian abattoir and their DNA extracted and screened with nine RAPD markers whose primer sequences are shown on Table 1.

Table 1: RAPD primers, sequences and guanine + cytosine (GC) percentage.

Primer	5' to 3' sequence	GC%
A1	GCACTGAGTA	50
A2	ACGTCGAGCA	60
A5	GAATCGGCCA	60
P1	ACAACGCCTC	60
P2	GGGAACGTGT	60
P3	CTGGGCAACT	60
P4	CCGTGACTCA	60
T3	GCTGCTCGAT	64
T4	ACCGCGGTCT	64

Details of the laboratory procedures of genomic DNA extraction, polymerase chain reaction assays, gel electrophoresis, RAPD band scoring and analyses have been published elsewhere (Malau-Aduli *et al.*, 2006). The carcass data were subjected to a one-way analysis of covariance in which breed was fitted as a fixed effect and slaughter day as a covariate utilising the SAS statistical software. Pairwise t-test between means was computed.

Results and Discussion

At the same slaughter weight and body condition score, Corriedales had significantly ($P<0.01$) higher fat score, thicker subcutaneous fat (29.6 vs 23.3 mm) and wider eye muscle (49.8 vs 36.2 mm) than East Friesians (Figure 1). On the other hand, East Friesians had significantly ($P<0.01$) heavier hot carcass weight (23.8 vs 22.3 kg), larger eye muscle area (42.9 vs 39.5 cm^2) and longer eye muscle (67.1 vs 65.6 mm) than Corriedales, while shortloin length (SLL) in the two breeds did not significantly differ ($P>0.09$). These results are consistent with those of Fogarty *et al.*, (2005b) that reported significant sire breed differences in hot carcass weight, carcass fat and eye muscle area between Border Leicester, East Friesian, Corriedale, Finnsheep, Coopworth, White Suffolk and Booroola Leicester. RAPD marker assays are based on polymerase chain reaction amplification of random segments of the DNA with an identical pair of primers 8-10 bp in length consisting of arbitrary nucleotide sequence (Rabouam *et al.*, 1999; Liu and Cordes 2004). Genetic variation and divergence within and between breeds of interest are assessed by the presence or absence of each product which is dictated by the DNA sequence at each locus. The power to detect polymorphisms is very high given that 5-20 bands (Figure 2) can be produced using a given primer pair and multiple sets of random primers can be used to scan the entire genome for differential RAPD bands. RAPD has several advantages over other molecular markers because it can be used with uncharacterised genomes without prior knowledge of nucleotide sequence information and can be applied to problems in which only small quantities of DNA are available. It is also efficient and inexpensive.

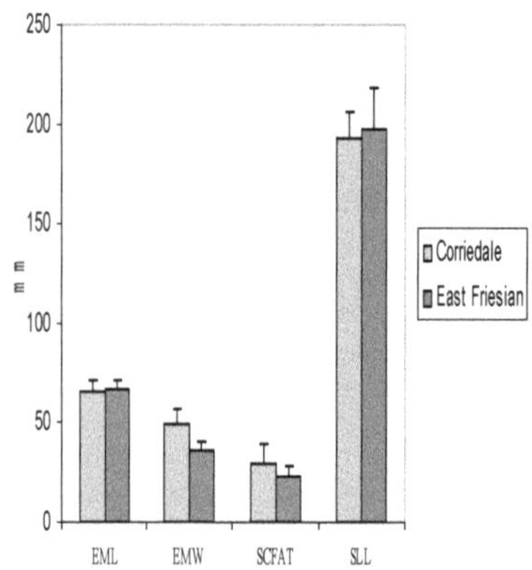

Figure 1: Variations in eye muscle length (EML), eye muscle width (EMW), shortloin length (SLL) and subcutaneous fat thickness (SCFAT) ± s.e. (mm) of East Friesian and Corriedale carcasses.

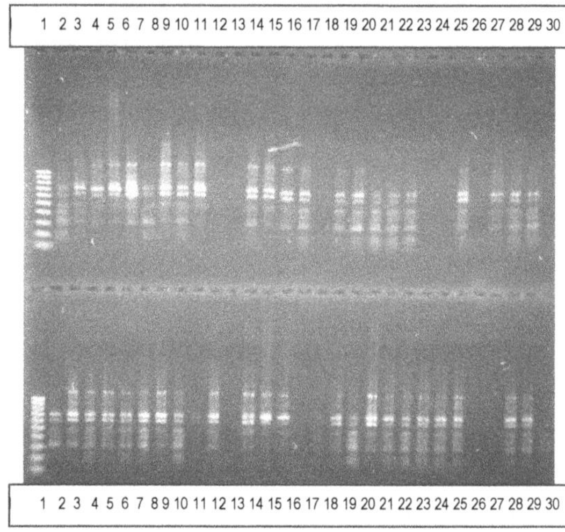

Figure 2: Differential RAPD bands utilising marker A2 in Corriedale (Upper Row) and East Friesian (Lower Row) sheep. Lane 1 is the 100bp DNA size ladder.

References

Fogarty, N. M., Hopkins, D. L., van de Ven, R. (2000). Lamb production from diverse genotypes 2. Carcass characteristics. *Animal Science*, 70: 147-156.

Fogarty, N. M., Safari, E., Taylor, P. J., Murray, W. (2003). Genetic parameters for meat quality and carcass traits and their correlation with wool traits in Australian Merino sheep. *Australian Journal of Agricultural Research*, 54: 715-722.

Fogarty, N. M., Ingham, V. M., Gilmour, A. R., Cummins, L. J., Gaunt, G. M., Stafford, J., Hocking-Edwards, J. E., Banks, R. G. (2005a). Genetic evaluation of crossbred lamb production 1. Breed and fixed effects for birth and weaning weight of first-cross lambs, gestation length and reproduction of base ewes. *Australian Journal of Agricultural Research,* 56: 443-453.

Fogarty, N. M., Ingham, V. M., Gilmour, A. R., Cummins, L. J., Gaunt, G. M., Stafford, J., Hocking-Edwards, J. E., Banks, R. G. (2005b). Genetic evaluation of crossbred lamb production 2. Breed and fixed effects for post-weaning growth, carcass and wool of first-cross lambs. *Australian Journal of Agricultural Research*, 56: 455-463.

Liu, Z. J., Cordes, J. F. (2004). DNA marker technologies and their applications in aquaculture genetics. *Aquaculture*, 238: 1-37.

Malau-Aduli, A. E. O., Bignell, C. W., Hegarty, R. S., Oddy, H., Johns, W., Tavassoli-Salardini, F., Smolenski, A. J., Malau-Aduli, B. S., Wells, B. B., Lane, P. A., Clark, R. J. (2006). Random amplified polymorphic DNA marker prospects for detecting genetic variation in growth and meat quality of crossbred lambs sired by rams with high EBVs for muscle or growth fed high or low levels of nutrition. *Australian Journal of Experimental Agriculture*, Volume 46: Accepted for publication

Rabouam, C., Comes, A. M., Bretagnolle, V., Humbert, J. F., Periquet, G., Bigot, Y. (1999). Features of DNA fragments obtained by random amplified polymorphic DNA (RAPD) assays. *Molecular Ecology*, 8: 493-503.

REALTIME PCR ANALYSIS OF MRNA TRANSCRIPTS IN POST-MORTEM BOVINE MUSCLE

P.C. Stapleton* and A.M. Mullen

Meat Technology Department, Ashtown Food Research Centre, (formerly The National Food Centre), Teagasc, Ashtown, Dublin 15, Ireland. Email:patricia.stapleton@teagasc.ie

Keywords: RNA stability, bovine, real-time PCR

Introduction

The application of approaches such as transcriptome and proteome analysis has opened up new possibilities for discovering genes, proteins or molecular polymorphisms that control muscle growth and meat quality traits (Hocquette *et al.*, 2001). Transcriptomics enables analysis of the complete set of RNA transcripts produced by the genome at a given time and provides a dynamic link between the genome, the proteome and the cellular phenotype. However to carry out such analysis, high quality undegraded RNA is necessary. The stability of mRNA has emerged as a key step in the regulation of eukaryotic gene expression with half-lives of mRNAs varying from a few minutes to more than 24 hours (Tourrière *et al.*, 2002). In human tissue, extensive RNA stability was observed up to 38-48 h post-mortem (Schramm *et al.*, 1999). Since access to muscle tissue is usually post-mortem, there can be an appreciable time lag in commercial abattoirs between time of slaughter and muscle excision online. Post-mortem stability in fibrous tissue, such as muscle, has not been extensively studied, and is of particular interest in respect of analyzing gene expression in meat-producing animals in relation to meat quality. The aim of this study was to investigate RNA integrity from bovine striploin muscle, collected 0.5 – 48 h post-mortem in an abattoir environment. Subsequently, GAPDH and Calpain II mRNA transcript levels were quantitatively determined using realtime PCR.

Materials and Methods

Heifers were slaughtered under controlled conditions at an abattoir facility at Ashtown Food Research Centre, Dublin. Samples were excised from the *longissimus dorsi* (LD) muscle at intervals of 0.5 – 48 h post-mortem. Samples were stored in RNAlater or flash frozen in liquid nitrogen and stored at –20°C or –80°C as appropriate. Total RNA was isolated using (1) TRIreagent or (2) the commercial Qiagen RNeasy mini-kit and all samples were subsequently DNAsed. The Agilent Bioanalyzer (Agilent Technologies, Germany) was used to calculate the integrity and concentration of total RNA. Samples with RNA Integrity Numbers (RIN) of >7 are considered to be of sufficient quality for downstream gene expression analysis.

Randomly primed reverse transcription was carried out for first strand cDNA synthesis. PCR was performed using primers designed specifically for glyceraldehyde-3 phosphate dehydrogenase (GAPDH 216 bp) (U85042) and calpain II (287 bp) (J05065). RT-PCR products were analysed by agarose gel electrophoresis. Realtime RT-PCR analysis was carried out for mRNA transcript detection using the Faststart SYBR Green 1 master mix kit (Roche Diagnostics, UK). Standard dilutions were used to generate crossing point values which were used to quantify the transcript levels of GAPDH and calpain II of bovine muscle samples collected from 0.5 – 48 h post-mortem.

Results and Discussion

The Agilent Bioanalyzer was used to assess total RNA integrity and concentration and is regarded as a quality control standard for all routine gene expression experiments. Meat samples stored in RNAlater at -20°C, -80°C or flash frozen, from which RNA was isolated using TRIreagent or the commercially available kit all gave high quality total RNA up to 2 days post-morte*m* (Table 1). Good RNA Integrity Numbers (RIN >7) were obtained for all of these samples. The RIN is a software tool designed to estimate the integrity of total RNA samples. The expert software automatically assigns an integrity number to a eukaryote total RNA sample, with RIN values >7 being considered high quality RNA. Therefore, these extracts are of sufficient quality for gene expression analysis using total RNA.

RT-PCR analysis showed that transcript intensity appeared to remain constant for GAPDH (housekeeping gene) and calpain II (involved in post-mortem meat tenderisation). From initial quantitative realtime RT-PCR analysis, which accurately determines transcript levels, it appears that there is no difference in transcript quantity for GAPDH and calpain II up to 48 h post-mortem, suggesting that transcripts are relatively stable up to 48 h post-mortem (Figure 1). Further samples are being tested over this post-mortem period and the numbers of candidate genes examined will be expanded.

Table 1: RNA Integrity Numbers (RIN) obtained from bovine muscle cellular RNA samples using the Agilent 2100 Bioanlayser. RIN >7 are indicative of high quality RNA. *pm* = time post-mortem, 1 = muscle samples stored in RNAlater at -20°C; 2 = muscle samples stored in RNAlater at 4°C for 24 hours followed by subsequent RNAlater removal and tissue storage at -80°C; 3 = muscle samples flash frozen in liquid nitrogen and stored at -80°C.

Extraction Method	TRIReagent			Qiagen RNeasy kit		
Storage	1	2	3	1	2	3
Sample	RNA Integrity Number (RIN)					
0.5h *pm*	8.5	8.5	8.5	7.4	8.4	7.6
1h *pm*	8.6	8.5	7.9	7.9	7.9	7.5
1.5h *pm*	8.7	8.5	7.9	7.6	7.5	7.5
2h *pm*	8.7	8.6	7.9	8.0	8.0	7.3
3h *pm*	8.3	8.3	8.2	7.7	7.6	7.7
4h *pm*	9.2	8.6	7.7	7.7	7.8	7.5
5h *pm*	8.7	8.2	7.5	7.9	7.9	7.4
24h *pm*	8.1	8.5	7.8	8.1	7.9	7.6
48h *pm*	8.2	8.6	7.6	7.5	7.4	7.5

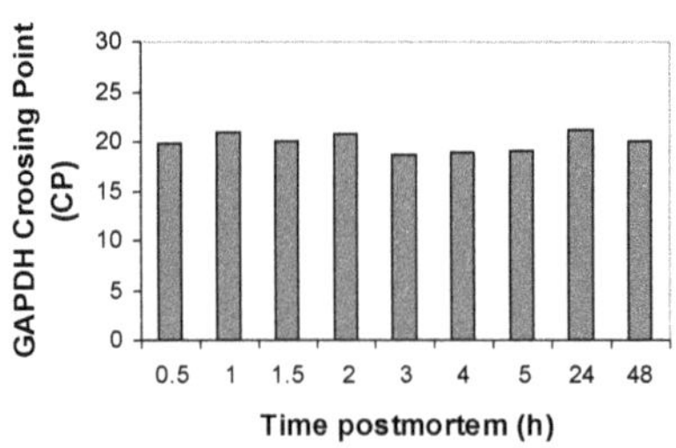

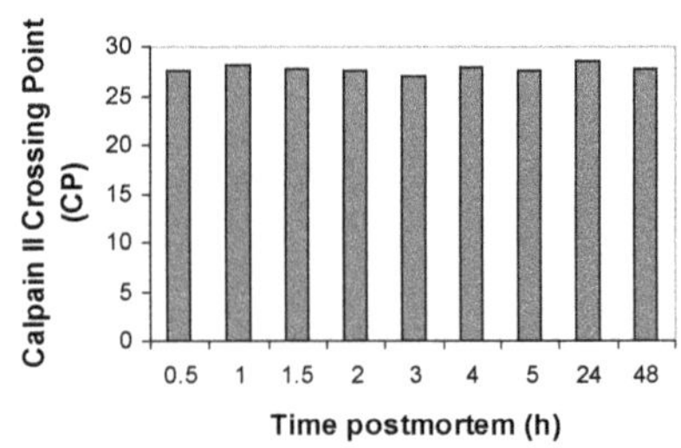

Figure 1: Realtime RT-PCR analysis of GAPDH and calpain II mRNA transcripts. Standard dilutions (of known copy number) generated crossing point values which were used to determine GAPDH and calpain II copy numbers in unknown bovine muscle samples collected from 0.5 – 48 h post-mortem.

Conclusions

The Agilent Bioanalyzer is a fast and reliable method of assessing RNA integrity in addition to making a quantitative measurement. Either flash freezing or storing muscle samples in RNAlater enabled the extraction of high quality total RNA from 0.5 – 48 h post-mortem, indicating high bovine muscle RNA stability post-mortem. High quality total RNA, suitable for gene expression analysis, was isolated from beef up to 2 days post-mortem demonstrating that a time delay between slaughter and sample collection should not adversely affect cellular RNA quality.

References

Hocquette, J.F., Cassar-Malek, I., Listrat, A., Jurie, C., Jailler, R. and Picard, B. (2001). Some recent advances in muscle biology and its regulation by nutrition: consequences for bovine meat quality. *Recent Advances in Animal Nutrition in Australia,* 13:135-143.

Schramm, M., Falkai, P., Tepest, R., Schneider-Axmann, T., Przkora, R., Waha, A., Pietsch, T., Bonte, W. and Bayer, T. A. (1999). Stability of RNA transcripts in post-mortem psychiatric brains. *Journal of Neural Transmisssion,* 106 : 329-335.

Tourrière, H., Chebli, K. and Tazi, J. (2002). mRNA degradation machines in eukaryotic cells. *Biochimie,* 84:821-837.

THE INFLUENCE OF PRKAG3 GENE ON MEAT QUALITY OF STRESS RESISTANT FATTENERS

M. Koćwin-Podsiadła*, E. Krzęcio, A. Zybert, H. Sieczkowska and K. Antosik

University of Podlasie, Chair of Pig Breeding and Meat Science, 08-110 Siedlce, 14 Prusa Str., Poland. Email: kocwin@ap.siedlce.pl

Keywords: PRKAG3, pigs, meat quality

Introduction

The presence of the RN gene effects pork meat quality. The meat from RN^- has 40 to 70% higher glycogen content, a low ultimate pH measured at 24 hours post slaughter (<5.5), paler meat, reduced technological yield and lower (about 1%) protein content. Initially, the phenotype of RN gene was estimated on the basis of Napole yield, then *in vivo* or *post mortem* measurements of muscle glycolytic potential (GP) (Fernandez *et al.*, 1990, Le Roy *et al.*, 1994, Lundstrom *et al.*, 1996). Milan *et al.*, (2000) found dominant RN^- mutation in PRKAG3 gene in codon 200 (Q200R) and shown that phenotypic effect of $Arg^{200}\rightarrow Gln$ reduces glycogen content in meat after slaughter. In addition, a new allele (V199I) in PRKAG3 gene affecting glycogen content, ultimate pH and colour, has been found (Milan *et al.*, 2000, Ciobanu *et al.*, 2001). The presence of Q200R mutation was found in pure or crossbred Hampshire pigs, while V199I polymorphism in breeds other than Hampshire (Milan *et al.*, 2000, Ciobanu *et al.*, 2001). The objective of this experiment was the assessment of the impact of the PRKAG3 gene on the meat quality of stress resistant fatteners.

Materials and Methods

The investigations covered 384 stress resistant fatteners [Landrace-79, Landrace x Duroc – 125, Landrace x Yorkshire – 67, (Landrace x Yorkshire)x Duroc – 83, (Landrace x Yorkshire)x(Duroc x Pietrain)-30]. The animals were kept under the same environmental conditions and fed a full bath feed. The animals were slaughtered 2-4 hours after transportation using electrical stunning method and recumbent bleeding out (Inarco system). The following meat quality characteristics were determined: pH of meat measured directly in *longissimus lumborum* (LL) muscle (45 minutes, 120 minutes and 24 hours after slaughter) using pH-Master apparatus produced by Draminski., electrical conductivity (EC) evaluated in 35 minutes, 120 minutes and 24 hours *post mortem* using LF-Star apparatus (Matthaus -Germany), R_1 indicator expressed as IMP/ATP ratio at 45 minutes *post mortem* according to Honikel and Fischer (1977) meat lightness (L*) measured Minolta CR-310 Chroma Meter in CIE L*a*b* system, water holding capacity (WHC) according to Grau and Hamm (1952) with Pohja and Niniivaara (1957) modification, technological yield in cured and cooked meat (24 hours after slaughter) according to Naveau *et al.*, (1985) with a modification of temperature in geometric centre of the probe (72°C), drip loss determined in 48, 96 and 144 hours *post mortem* according to Prange *et al.*, (1977) and shear force (in 48 and 144 hours after slaughter) using the Instron 1140 apparatus with Warner-Bratzler device. The analysis of protein, fat, water and dry matter content in LL muscle, was performed. The RYR1 genotypes were established according to Fujii *et al.*, (1991). At 45 minutes *post mortem*, samples from LL muscle were collected in tubes with 0.5M PCA for subsequent determination of the glycolytic potential (GP) according to formula proposed by Monin and Sellier (1985). The phenotype of RN gene was identified on the basis of glycolytic potential (GP) and its bimodal distribution: rn^+rn^+ (GP≤130µmol/g) $RN^-/?$ (GP>130µmol/g). The polymorphism of PRKAG3 gene was identified according to Milan *et al.*, (2000). The data was analysed using one-way analysis of variance in non-orthogonal scheme. The significance of differences between means was calculated using Duncan's test.

Results and Discussion

The analysis of variance showed the influence of PRKAG3 gene only on the following (from 15 measured in different time) meat quality parameters: glycolytic potential, lactate content, ultimate pH measured in 24 hours after slaughter and protein content (Table 1).

Meat from AA (RN^-/RN^-) animals compared with meat tissue from GG (rn^+/rn^+) fatteners had the lowest pH_{24} (lower about 0.1 pH units) and lactate content (Table 1). The pH_{24} range obtained in this experiment was lower then in the study of Le Roy *et al.*, (2000), where pH_{24}LD between RN^-/RN^- and rn^+/rn^+ ranged 0.2 units (5.5 vs. 5.7 respectively). No statistically significant differences between AA and GG animals (RN^-/RN^- and rn^+/rn^+ respectively) for GP level, glycogen content and protein content (typical for RN^- phenotype) were identified in the present study (Table 1). The data presented in Figure 1 showed a high share (55%) of animals with high glycolytic potential (GP>130µmol/g with phenotype $RN^-/?$) and high glycogen content in group of fatteners identified by PCR-RFLP DNA test as GG (rn^+/rn^+) animals. In the investigations of Fontanesi *et al.*, (2003) it was shown, that high GP in some animals, was not explained by the presence of the 200Q allele, this may suggest other genetic factors could influence this parameter in different pig populations. Also, Meadus *et al.*, (2002) reported, that PRKAG3 mutation was not found in 27% of retailed pork chops samples that had high GP values.

Table 1: Means and their standard deviations for meat quality traits from different PRKAG3 genotypes.

Trait	PRKAG3				F-emp. p-value
	AA (RN^-/RN^-) n=6	AG (RN^-/rn^+) n=27	GG (rn^+/rn^+) n=351	Average n=384	
Glycolytic potential (µmol/g)	145,63b ±21,87	120,65a ±22,04	134,55ab ±26,95	133,75 ±26,79	4,034 0,02
Glycogen	56,83 ±9,61	43,32 ±11,74	47,77 ±13,59	47,61 ±13,49	2,814 0,061
Lactate	31,98a ±5,28	33,94ab ±11,12	39,28b ±11,00	38,79 ±11,05	4,165 0,02
pH_{24}	5,50A ±0,11	5,54AB ±0,09	5,61B ±0,12	5,60 ±0,12	6,667 0,002
Protein (%)	22,32a ±0,52	22,68b ±0,55	22,38ab ±0,60	22,41 ±0,60	3,377 0,04

Means marked different letters a, b differ statistically at $p \leq 0.05$

Fig.1 Glycolytic potential and glycogen content in group of animals identify as GG(rn+/rn+) by PRKAG3 PCR-RFLP DNA test

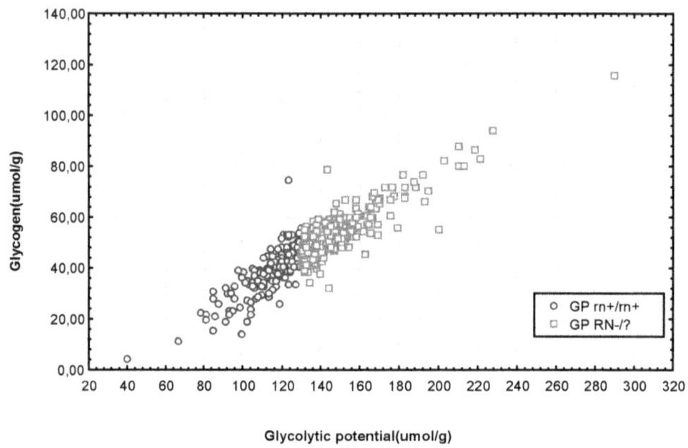

References

Ciobanu D., Bastiaansen J., Malek M., Helm J., Woollard J., Plastow G., Rothschild M. (2001). *Genetics*, 159, 1151-1162.

Fernandez X., Naveau J., Talmant A., Monin G. (1990). *J. Rech. Porcine en France*, 22, 97-100.

Fontanesi L., Davoli R., Nanni Costa L., Sotti E., Russo V. (2003). *Cytogenetic and Genome Research*, 102, 145-151.

Fujii J., Otsu K., Zorzato F., de Leon S., Khanna S., Weiler V. K., O'Brien P. J., MacLennan D., H. (1991). Science, 253, 448-451.

Grau R., Hamm R. (1952). *Fleischwirtschaft*, 4, 295 – 297.

Honikel K. O., Fisher H. (1977) *Journal of Food Science*, 42, 1633 – 1636.

Le Roy P., Elsen M.J., Cartiez J.C., Talmant A., Juin H., Sellier P., Monin G. (2000). *Genetics, Selection, Evolution*, 32, 165-186.

Le Roy P., Przybylski W., Burlot T., Bazin C., Lagant H., Monin G. (1994). *J. Rech. Porcine en France*, 26,311-314.

Lundstrom K., Andresson A., Hansson I. (1996) *Meat Science*, 42, 145-153.

Lundstrom K., Andresson A., Maerz S., Hansson I. (1994). Proc. 40th ICoMST, The Hague.

Meadus W.J., MacInnis R., Dugan M.E.R., Aalhaus J.L. (2002). *Canadian Journal of Animal Science*, 82, 449-451.

MilanD., Jeon J.T., Looft C., Amarger V., Robic A., Thelander M., Rogel-Gaillard C., Paul S, Iannuccelli N., Rask L., Ronne H., Lundstrom K., Reinsch R., Gellin J., Kalm E., Le Roy P., Chardon P., Andersson L. (2000). *Science*, 288, 1248-1251.

Monin G., Sellier P. (1985). *Meat Science*,13, 49-63.

Naveau J., Pommeret P., Lechaux P. (1985) *Techni. Porc*, 8, 7 – 13.

Prange H., Jugrrt L., Schrner E. (1977) *Arch. Exper. Vet. Med. Leipzig*, 31, 2, 235 – 248.

Pohja N.S., Ninivaara F.P. (1957) *Fleischwirtschaft*, 9, 193-195

Przybylski W., Koćwin-Podsiadła M., Kaczorek S., Krzęcio E. (1998). *Polish Journal of Food and Nutrition Science*, 7/48, 1, 83-88.

THE INFLUENCE OF PRKAG3 GENE ON CARCASS COMPOSITION TRAITS OF STRESS RESISTANT FATTENERS

M. Koćwin-Podsiadła*, E. Krzęcio, A. Zybert, H. Sieczkowska and K. Antosik

University of Podlasie, Chair of Pig Breeding and Meat Science, 08-110 Siedlce, 14 Prusa Str., Poland. Email: kocwin@ap.siedlce.pl

Keywords: PRKAG3, pigs, carcass composition traits

Introduction

The detrimental effect of the Rendement Napole (RN) gene on pork meat quality attributes has been demonstrated by a number of researches (Le Roy *et al.*, 1990, 2000, Koćwin-Podsiadła 1995). The RN^- allele is also known to positively influence the lean meat content, backfat thickness and the weight of lean meat joints (Enfalt *et al.*, 1997, Le Roy *et al.*, 2000, Hamilton *et al.*, 2003). Previously, the phenotype of RN gene was estimated on the basis of Napole yield, then *in vivo* or *post mortem* measurements of muscle glycolytic potential (GP) (Fernandez *et al.*, 1990, Le Roy *et al.*, 1994, Lundstrom *et al.*, 1996). Milan *et al.*, (2000) identified the dominant RN^- mutation in the PRKAG3 gene. The objective of this experiment was assessment of the impact of PRKAG3 gene on carcass composition of stress resistant fatteners.

Materials and Methods

The investigations covered 384 stress resistant fatteners [Landrace-79, Landrace x Duroc – 125, Landrace x Yorkshire – 67, (Landrace x Yorkshire)x Duroc – 83, (Landrace x Yorkshire)x(Duroc x Pietrain)-30]. The animals were kept under the same environmental conditions and fed a full bath feed. The animals were slaughtered at approximately 110kg body weight 2-4 hours after transportation using electrical stunning method and recumbent bleeding out. Measurements of pork meatiness and its components were made using Danish apparatus ULTRA-FOM 100 produced by SFK Technology. The estimation of carcass composition was performed in accordance with methodology using in Polish Pig Testing Stations (Różycki 1996). All data concerning carcass composition traits were standardised on 85 kg of hot carcass weight. The RYR1 genotypes were established according to Fujii *et al.*, (1991). At 45 minutes *post mortem*, samples from LL muscle were collected in tubes with 0.5M PCA for subsequent determination of glycolytic potential (GP) according to formula proposed by Monin and Sellier (1985). The polymorphism of PRKAG3 gene was identified according to Milan *et al.*, (2000). The data were analysed using one-way analysis of variance in non-orthogonal scheme. The significance of differences between means was calculated using Duncan's test.

Results and Discussion

The analysis of variance demonstrated the influence of PRKAG3 gene only on the following carcass composition traits: backfat thickness measured over *Gluteus medius* muscle in 3 points (I-st, II-nd and III-rd cross), average bacfat thickness, *Longissimus dorsi* muscle weight and belly weight (Table 1).

No statistically confirmed differences between AA and GG animals (RN^-/RN^- and rn^+/rn^+) for analysed carcass composition traits was found in the present study. AG (RN^-/rn^+) animals compared with homozygous AA and GG fatteners (RN^-/RN^- and rn^+/rn^+ respectively), had a thicker backfat measured over *Gluteus medius* muscle and average backfat. Differences were confirmed statistically between AA and AG genotypes (Table 1). Moeller *et al.*, (2003) showed that PRKAG3 gene had no effect on backfat thickness with the exception of last lumbar backfat depth. Other studies have reported, that the RN^- phenotype is associated with leaner carcasses and a decrease of backfat thickness (Le Roy *et al.*, 1996, 2000; Enfalt *et al.*, 1997; Hamilton *et al.*, 2003).

It was confirmed statistically that heterozygous AG (RN^-/rn^+) fatteners compared with homozygous AA and GG fatteners (RN^-/RN^- and rn^+/rn^+ respectively), had the significantly lowest *Longissimus dorsi* muscle weight and highest belly weight (Table 1). In contrast, Le Roy *et al.*, (2000) reported significant increase in the weight of lean joints (ham and loin) and a decrease in belly weight in RN^- allele (RN^-/rn^+) carriers identified by GP distribution. Olsson *et al.*, (2003) also identified the RN^- phenotype on the basis of GP distribution, and a higher percentage share of *Longissimus dorsi* muscle in carriers of RN^- allele (RN^-/rn^+).

The diverging results relating to the PRKAG3 gene on carcass composition traits were probably due to animals with high glycolytic potential who have been identified by PCR-RFLP DNA test as GG (rn^+/rn^+) animals.

Table 1: Means and their standard deviations for carcass composition traits from different PRKAG3 genotypes.

Trait	PRKAG3				F-emp. p-value
	AA RN⁻/RN⁻ n=6	AG RN⁻/rn⁺ n=27	GG rn⁺/rn⁺ n=351	Average n=384	
Lean meat content (%)	56,05 ±2,49	55,53 ±4,33	55,99 ±4,05	55,96 ±4,05	0,192 0,826
Backfat thicknes over *gluteus medius* muscle (cm): I cross	1,61A ±0,24	2,18B ±0,40	1,91A ±0,37	1,93 ±0,38	9,498 0,000
II cross	1,25a ±0,19	1,55b ±0,36	1,41ab ±0,33	1,42 ±0,33	3,433 0,033
III cross	2,01A ±0,31	2,50B ±0,49	2,27AB ±0,43	2,28 ±0,44	5,172 0,006
Average backfat thickness (cm)	1,82A ±0,19	2,18B ±0,32	2,06AB ±0,28	2,03 ±0,29	6,414 0,002
Longissimus dorsi muscle weight (kg)	3,13B ±0,25	2,73A ±0,30	2,94AB ±0,33	2,92 ±0,33	6,562 0,002
Belly weight (kg)	6,28A ±0,57	7,07B ±0,75	6,56A ±0,58	6,60 ±0,61	10,693 0,000

Means marked different letters A, B differ statistically at p≤0.01; a, b differ statistically at p≤0.05

References

Enfalt A.C., Lundstrom K., Karlsson A., Hansson I. (1997). Estimated frequency of the allele RN⁻ in Swedish Hampshire pigs and comparison of glycolytic potential, carcass composition and technological meat quality among Swedish Hampshire, Landrace and Yorkshire pigs. *Journal of Animal Science*, 75, 2924-2935.

Fernandez X., Naveau J., Talmant A., Monin G. (1990). Distribution du potentiel glycolytique dans une population porcine et relation avec le Rendment Napole. *J. Rech. Porcine en France*, 22, 97-100.

Fujii J., Otsu K., Zorzato F., de Leon S., Khanna S., Weiler V. K., O'Brien P. J., MacLennan D., H. (1991). Identification of a mutation in porcine ryanodine receptor associated with malignant hyperthermia. *Science*, 253, 448-451.

Hamilton D.N., Miller K.D., Ellis M., McKeith F.K., Wilson E.R. (2003). Relationship between longissimus glycolytic potential and swine growth performance, carcass traits and pork quality. *Journal of Animal Science*, 81, 2206-2212.

Koćwin-Podsiadła M., Przybylski W., Kurył J., Talmant A. Monin G. (1995). Muscle glycogen level and meat quality in pigs of different halothane genotypes. *Meat Science*, 40, 121-125.

Le Roy P., Elsen J.M., Caritez J.C., Talmant A., Juin H., Sellier P., Monin G. (2000). Comparison between the three porcine RN genotypes for growth, carcass composition and meat quality traits. *Genetics, Selection, Evolution*, 32, 165-186.

Le Roy P., Monin G., Elsen J.M., Caritez J.C., Talmant A., Lebret B., Lefaucheur L., Mourot J., Juin H., Sellier P. (1996). Effect of the RN genotype on growth and carcass traits in pigs. Proc. 47th Mtg. Eur. Assoc. Anim. Prod., vol.2, Lillehamer, Norway, 311.

Le Roy P., Naveau J., Elsen J.M., Sellier P. (1990). Evidence for a new major gene influencing meat quality in pigs. *Genet. Res.*, 55, 33-40.

Le Roy P., Przybylski W., Burlot T., Bazin C., Lagant H., Monin G. (1994). Etude des relations entre le potentiel glycolytique du muscle et caracteres de production dans les lignes Laconie et Penshire. *J. Rech. Porcine en France*, 26,311-314.

Lundstrom K., Andresson A., Hansson I. (1996). Effect of RN gene on technological and sensory meat quality in crossbred pigs with Hampshire as terminal sire. *Meat Science*, 42, 145-153.

Milan D., Jeon J.T., Looft C., Amarger V., Robic A., Thelander M., Rogel-Gaillard C., Paul S, Iannuccelli N., Rask L., Ronne H., Lundstrom K., Reinsch R., Gellin J., Kalm E., Le Roy P., Chardon P., Andersson L. (2000). A mutation in PRKAG3 associated with excess glycogen content in pig skeletal muscle. *Science*, 288, 1248-1251.

Monin G., Sellier P. (1985). Pork of low technological quality with a normal rate of muscle pH fall in the immediate post-mortem period: The case of the Hampshire breed. *Meat Science*, 13, 49-63.

Ollson V., Andersson K., Hansson I., Lundstrm K. (2003). Differences in meat quality between organically and conventionally produced pigs. *Meat Science*, 64, 287-297.

Różycki M. (1996). Zasady postępowania przy ocenie świń w Stacjach Kontroli Użytkowości Rzeźnej trzody Chlewnej. Stan Hodowli i Wyniki Oceny Świń. Instytut Zootechniki, Kraków, 505-541 (in polish).

THE EFFECT OF GENOTYPE, HANGING METHOD AND AGING PERIOD ON MEAT QUALITY

F.O. Lively[1,3], T.W.J. Keady[2], *B.W. Moss[1,3], D.C. Patterson[1,3], E.L.C Tolland[1], D.J. Devlin[1], I.J. Tollerton[1], G. Kirkpatrick[1] and A. Gordon[1]

[1]*Agricultural and Food Biosciences Institute, Newforge Lane, Belfast BT9 5PX, UK.*
[2]*Teagasc, Livestock Research Centre, Athenry, Co. Galway, Ireland.*
[3]*Queen's University of Belfast, School of Agriculture & Food Science, Newforge Lane, Belfast BT9 5PX. Email: bruce.moss@afbini.gov.uk*

Keywords: meat quality, genotype, hanging method, ageing, tenderness

Introduction

Currently around 47% of the beef production in Northern Ireland originates from the dairy herds. Sinclair *et al.* (2001) showed the *m. longissimus lumborum* from Aberdeen Angus breed was more tender and juicier than the Holstein breed and there was no difference between the Charolais and Holstein breeds. Sinclair *et al.* (2001) used the tenderstretch hanging method and comparisons were made at both 7 and 14 days aging. Lively *et al.* (2005) studied only Charolais and Holstein breeds and found an interaction between breed and hanging method for Warner-Bratzler shear force (WBSF) after 7 days aging in the longissimus dorsi. The WBSF values were higher in Charolais (than Holsteins) when hung by Achilles tendon (4.3 and 2.8kg/cm² respectively), however the difference was less when carcases were hung tenderstretch (Charolais 3.0 and Holstein 2.3kg/cm²). Given that increased aging leads to a reduction in WBSF values, the present experiment was designed to further investigate the interaction between aging period, genotype and hanging method.

Materials and Methods

The experiment involved a total of 41 steers, consisting of two breeds, Holstein (HOL) and Charolais (CH). All animals were offered grass silage and libitum supplemented with 4.5 kg concentrate for 98 days prior to slaughter. All cattle were slaughtered at the same time, resulting in a range of carcass weights. At slaughter, one side of each carcase was randomly allocated to one of two hanging techniques, Achilles Tendon (AT) or Tenderstretch (TS, carcase suspended through aiteh bone at 45 min post mortem until 48 h post mortem). At 3 days post slaughter the sirloin was removed, and the *longissmus dorsi* (LD) dissected out free of external fat. A slice 1 inch thick (2.5 cm) was removed from the anterior portion of the dissected LD, vacuum packed and blast frozen for further analysis. Sequential slices of similar thickness were vacuum packed and stored at 2°C until blast frozen at 7, 14 and 21 days post slaughter. Prior to measurement of cooking loss and WBSF, samples were allowed to thaw out in cold room at 4°C for 24 hours. Meat samples were removed from the vacuum bags, weighed and placed in new vacuum pack bags, for cooking in a water bath at 75°C for a period of 50 minutes. Samples were allowed to cool overnight, reweighed and ten cores (1.3cm) diameter removed parallel to fibre direction from each steak for measurement of WBSF. Shear force was measured using a Instron Universal testing machine, using a Warner-Bratzler shear force attachment and cross head speed 100mm/min.

Results and Discussion

Cooking loss and Warner-Bratzler shear force values (WBSF) were significantly affected by period of aging (see Table 1). Cooking loss was higher at 14 and 21 days aging than at 3 and 7 days aging (Table 1). WBSF values generally decreased as aging period increased (Table 1). Both cooking loss and WBSF values were significantly higher in Charolais than Holstein breed. WBSF values were significantly lower in carcases hung by the tenderstretch method. The aging by genotype interaction shows that for the Holstein breed there was no significant change in cooking loss with aging (33.3, 31.6, 32.5, 32.1% for 3, 7, 14 and 21d respectively) whereas for the Charolais breed, cooking loss at 3 days aging (32.4%) was significantly lower than at other times of aging (34.5, 35.7, 35.1% for 7, 14, 21 days respectively). When hung by the Achilles tendon, WBSF values decreased with aging for 3 to 21 days (4.65, 4.48. 4.33. 4.04 respectively), however when hung TS WBSF values did not change with aging from 3 to 21 days (3.57, 3.69, 3.62, 3.50 kg/cm²). WBSF values for Charolais were significantly higher than for Holstein when hung AT (5.11 and 3.64 kg/cm² respectively), but not when hung TS (3.83, 3.36 kg/cm²respectively). The method of tenderstretch hanging had a much greater effect for the Charolais breed than Holstein breed. The current results show the increase in tenderness, *i.e.* lower shear force, due to the tenderstretch method of hanging, which is well documented in the literature (Sørheim *et al.*, 2001). Of particular interest is the hanging by genotype interaction. Sinclair *et al.*, (2001) showed that when hung

Table 1: Effect of genotype, hanging method and aging on cooking loss and shear force values of *Longissimus dorsi*

Treatment		Cooking loss (%)	WBSF (kg /cm²)
Period Aged	3	32.3a	4.11a
	7	33ac	4.09a
	14	34.1b	3.98a
	21	33.6bc	3.77b
	Significance	*	**
	sed	0.52	0.074
Genotype	CH	34.4	4.47
	HOL	32.1	3.50
	Significance	**	***
	sed	0.50	0.181
Hanging technique	AT	33. 7	4.38
	TS	32.8	3.60
	Significance	NS	***
	sed	0.50	0.181
Interactions	Aging X Genotype	**	ns
	Aging X Hanging	ns	**
	Hanging X Genotype	**	ns

Within a column means with a common superscript are not significantly different. ns not statistically significant; * $p<0.05$; ** $p<0.01$ *** $p<0.001$.

tenderstretch there was no difference in eating quality between Charolais and Holstein breed. The current results also show the genotype WBSF values of the two genotypes to be closer when tenderstretch hung. Although the average carcase weight at slaughter of the Holsteins in the current experiment was lower than the Charolais (CH 360, HOL 281kg), covariate adjustment for carcass weight did not change the pattern of results reported in Table 1. The cooking loss and WBSF values for 7 days aging in the current studies on meat frozen after aging are higher than those reported by Lively *et al.*, (2005) when undertaken on fresh *L. dorsi.* It should be noted that the pattern of results for 7 days aging (Lively *et al.*, 2005) was similar to the current studies, with biggest genotype differences for WBSF being shown when hung AT rather than TS. Lively *et al.* (2006a) considered that the greater marbling of the Holstein breed may account for the greater tenderness in the LD muscle. The contribution of other factors such as pH/temperature decline, or proteolytic activity need further investigation.

Conclusion

The current results show the importance of post slaughter procedures e.g. hanging method, aging when undertaking breed comparisons.

References

Sinclair, K.D., Lobley, G.E., Horgan, G.W., Kyle, D.J., Porter, A.D. and Matthews, K.R. (2001). Factors influencing beef eating quality 1. Effects of nutritional regime and genotype on orgonoleptic properties and instrumental texture. Animal Science. **72**, 269-277.

Lively, F.O., Keady T.W.J., Moss, B.W., Patterson, D.C. and Kilpatrick, D.J. (2005). The effect of genotype and pelvic hanging technique on meat quality. Proc. BSAS. 8th Annual Langford Food Industry Conference. 18-19 May 2005. p.59.

Lively, F.O. Moss, B.W., Keady, T.W.J., and Gordon, A. (2006). Proceedings of the Agricultural Research Forum, p 84.

Sørheim, O. Idland, J., Halvorsen, E. C., Frøystein, T. Lea, P. and Hildrum K. I. (2001). Influence of beef carcass stretching and chilling rate on tenderness of *m. longissimus dorsi. Meat Science, 57, I79-85.*

ANALYSIS OF BOVINE MUSCLE DNA POLYMORPHISMS IN RELATION TO MEAT QUALITY, IN PARTICULAR INTRAMUSCULAR FAT

L. Pannier*[1,2], P. Stapleton[1], T. Sweeney[2] and A.M. Mullen[1]

[1] *Department of Meat Technology, Ashtown Food Research Centre Teagasc, Ashtown, Dublin 15, Ireland.*
[2]*Department of Animal Husbandry and Production, University College Dublin, Belfield, Dublin 4, Ireland.*
Email: liselotte.pannier@teagasc.ie

Keywords: bovine candidate genes, intramuscular fat, bovine diversity, DNA marker

Introduction

Intramuscular fat (IMF) deposition, also subjectively assessed as marbling, can be defined as the percentage of lipid content of muscle. It is a continuously varying trait in cattle exhibiting heritabilities of up to 0.65 (Marshall, 1999) and it is hypothesed to contribute to the juiciness and flavour of cooked meat (Schmutz, 2003). The identification of genes that contribute to continuous variation of meat quality, including IMF levels, is a challenge for molecular genetics and advances in molecular genetics have led to the identification of loci and DNA variations (polymorphisms) that affect traits of interest in livestock (Anderson, 2001). Some of these DNA polymorphisms may have the potential to serve as markers of meat quality. These genetic markers could help to improve the quality of beef, by marker-assisted selection (MAS) (Dekkers, 2004) by selecting these animals with a high genetic potential for a certain trait and by so producing better tasting, more valuable and more consistent beef. Three candidate genes were chosen for this study and were hypothesed to be one of the many genetic factors influencing IMF. Known single nucleotide polymorphisms (SNPs) in these genes have been determined. The leptin gene, also called the obese gene, has been assigned to bovine chromosome 4. Its product, a 16 kDa hormone called leptin, which functions in the regulation of adiposity, may be involved in controlling IMF levels. The thyroglobulin (TG) gene is located on chromosome 14 and codes for a protein TG, which is a precursor molecule of the thyroid hormones, which affect lipid metabolism. Finally the diacylglycerol-O-acyltransferase1 (DGAT1) gene, also located on chromosome 14, codes for DGAT1, a microsomal enzyme that catalyses triglyceride synthesis. Two known SNPs in the leptin gene in exon 2 (Buchanan *et al.*, 2002) and exon 3 (Haegeman *et al.*, 2000), one SNP in the TG gene in the 5' untranslated region (Barendse, W.J., 1999) and two SNPs in the DGAT1 gene in exon 8 (Winter *et al.*, 2002) have been determined. The objectives of this study were to investigate the frequency of the SNPs genotypes in the Irish cattle population and to determine if an association exists between these SNPs and IMF values. Only a few studies have been performed to show a significant effect of the SNPs on IMF values in cattle.

Materials and Methods

Blood samples were collected from pedigree bulls from a number of artificial insemination (AI) stations in Ireland. This sample set was used to investigate the frequency of the genotypes of the four polymorphisms in different breeds. To date, breeds tested include Aberdeen Angus, Charolais, Friesian, Limousin and Simmental. Secondly, bovine *M. longissimus dorsi* (LD) (249) were collected from slaughter-weight commercial cattle. Heifers were slaughtered under controlled conditions from a number of commercial abattoirs in Ireland. Meat quality measurements were analysed and the fat content for every sample was determined using the CEM analysis system, which is a hot solvent extraction (Bostian *et al.*, 1985). Genomic DNA was isolated from blood and muscle samples using the QIAamp® DNA mini kit. Diagnoses of the SNPs of leptin, TG and DGAT1 gene were detected by Polymerase Chain Reaction-Restriction Fragment Length Polymorphism analysis (PCR-RFLP). PCR was carried out, using primers specific for the chosen candidate gene and the amplified fragments were a 545bp fragment for TG gene (Barendse, W.J., 1999), 411bp for DGAT (Winter *et al.*, 2002), and a 94bp and a 331bp fragment for two SNPs of the leptin gene (Haegeman *et al.*, 2000; Buchanan *et al.*, 2002). RFLP analysis was carried out by using specific restriction enzymes. Digests were separated and visualised on 3% agarose gels using gel electrophoresis and genotypes were assigned to each sample depending on its restriction pattern. For the DGAT1 gene, the two SNPs are positioned adjacent to each other, therefore, samples were cloned into a pCR®4-TOPO® vector and sequenced to distinguish between the different alleles. The genotype frequencies were performed using the FREQ procedure of SAS allowing differences between breeds and genotype to be discerned. Associations between the genotypes of the SNPs and IMF values were assessed using the GLM (one-way ANOVA) procedure of SAS.

Results and Discussion

Three genotypes CC, CT and TT, were detected for the leptin and TG genes and AA, KA and KK for the DGAT1 gene as previously reported (Haegeman *et al.*, 2000; Buchanan *et al.*, 2002; Barendse, W.J., 1999; Winter *et al.*, 2002). For the frequency study, some genotypes were present at a low frequency for certain breeds and not every genotype occurred in every breed (Table 1). Breed differences and geological aspects could partly influence the frequency values. For the two loci in the leptin gene, both the CC and CT genotypes are equally distributed while the TT genotype and also the KK genotype for DGAT1 occurs at a very low frequency or is even not present. The CC genotype of the TG gene is more frequent then the CT or TT, and the TT genotype does not occur in 76 Friesians and 40 Limousins. Genotyping of a larger number of animals from each breed would establish genotype frequencies in Irish breeds, allowing breeders to increase the frequency if favourable genotypes have been associated with meat quality traits.

Table 1: Frequency (%) study of the SNPs genotypes in the Irish cattle population. 56 Friesians were used for the DGAT1 study.

Gene	TG gene			leptin exon 3			leptin exon 2			DGAT1		
Breed (n)	**CC**	**CT**	**TT**	**CC**	**CT**	**TT**	**CC**	**CT**	**TT**	**AA**	**KA**	**KK**
Friesian (76)	73.7	26.3	-	47.4	51.3	1.32	31.6	61.8	6.58	76.8	17.9	5.36 (56)
Limousin (40)	77.5	22.5	-	70.0	25.0	5.50	60.0	30.0	10.0	80.0	20.0	-
Charolais (34)	61.8	29.4	8.80	47.1	50.0	2.94	44.1	52.9	2.94	67.6	29.4	2.94
A. Angus (16)	6.30	56.3	37.5	87.5	12.5	-	12.5	75.0	12.5	62.5	37.5	-
Simmental (15)	46.7	46.7	6.70	46.7	53.3	-	60.0	40.0	-	93.3	6.67	-

Results to date show a non-significant (P>0.05) association between genotypes and IMF values in Irish beef for the chosen candidate genes (data not shown). Thaller *et al.* 2003 identified an associated with higher levels of IMF for T/T homozygous German Holstein cattle in *M. longissimus dorsi* for the SNP of the TG gene and in *M. semitendinusos dorsi* for the DGAT1 gene. In our study genotype TT for TG gene, is only present for 4 animals. This number of TT homozygous animals may have been too small for association detection. Buchanan *et al.*, (2002) identified the C allele of the SNP in exon 2 of the leptin gene, as the allele to be associated with fatter carcasses. While no association was found in this study, it might be that other SNPs in the candidate genes are correlated to IMF and it is also important to note that there are many other genes that are thought to contribute towards the quality trait of marbling (IMF) or to effect fat metabolism. Marbling is quite a complex trait that is affected by many biological pathways of different genes. It is important to also consider other unmarked genes, the interaction of genes and the production environment, which may all, have an effect on fat metabolism.

Conclusion

The frequency study doesn't allow us to make a final conclusion due to the small number of animals for some breeds. Although association were observed in other breeds in other studies, no association was detected so far in the Irish cattle population for the chosen candidate genes.

References

Anderson, L. (2001). Genetic dissection of phenotypic diversity in farm animals. Nature Reviews Genetics 2,130-138.

Barendse, M.R. (1999). Assessing lipid metabolism. Patent, International Publication Number: WO 99/23248. World International property organisation.

Bostian, M. L., Fish, D. L., Webb, N. B. and Arey, J.J. (1985). Automated methods for determination of fat and moisture in meat and poultry products: collaborative study. Journal of the Association of official analytical Chemists, 68, 876-880.

Buchanan, Fiona C., Fitzsimmons, Carolyn J., Van Kessel, Andrew G., Thue, Tracey D., Winkelman-Sim, Dianne C. and Schumtz, Sheila M. (2002). Associations of a missense mutation in the bovine leptin gene with carcass fat content and leptin mRNA levels. Genetics Selection Evolution, 34, 105-116.

Dekkers, J.C.M., (2004). Commercial application of marker- and gene- assisted selection in livestock: Strategies and lessons. Journal of Animal Science, 82, 313-328.

Haegeman, A., Van zeveren, A. and Peelman, L.J. (2000). International Society for Animal Genetics, Animal Genetics 31, 68-79.

Marschall, D.M. (1999). Genetics of meat quality. In: The Genetics of Cattle (Ed. by R. Fries, A. Ruvinsky) CABI Publishing, Wallingford.605-636.

Schmutz, S. (2003). DNA studies of meat quality. http://skyway.usask.ca/~schmutz/leptin.html.

Thaller, G., Kühn, C., Winter, A., Ewald, G., Bellman, O., Wegner, J., Zühlke, H. and Fries, R. (2003). DGAT1, a new positional and functional candidate gene for intramuscular fat deposition in cattle. Animal Genetics 34:354-357.

Winter, A., Krämer, W., Werner, F.A.O., Kollers, S., Kata S., Durstewitz, G., Buitkamp, J., Womack, J.E., Thaller, G., and Fries, R. (2002) Association of a lysine-232/alanine polymorphism in a bovine gene encoding acyl CoA:diacylglycerol acyltransferase (DGAT) with variation at a quantitative trait locus for milk fat content. Proceedings of the National Academy of Science of the United States of America, 99,9300-5.

IDENTIFICATION OF HANWOO (KOREAN NATIVE CATTLE) BEEF BY REAL-TIME PCR IN SOUTH KOREA

J.O. Nam, B.R. Nam, J.M. Park, J.H. Choi, C.J. Kim, and J.M. Kim*

Department of Food Science and Biotechnology of Animal Resources, Konkuk University, Seoul 143-701, South Korea. Email: jinmkim@konkuk.ac.kr

Keywords: MCIR gene, real-time PCR, genotype, Hanwoo beef, Korean cattle

Introduction

Korean consumers prefer beef from Korean native cattle (Hanwoo) to imported dairy cattle because they believe that juiciness and flavour of Hanwoo is better. Therefore, Hanwoo beef has been regarded as the most expensive and high quality beef in South Korea (Kim *et al.*, 2002). Real time-polymerase chain reaction (RT-PCR) is currently considered as the most sensitive method to detect low abundant DNAs in samples. Compared to conventional PCR, real-time PCR has a high reliability because of excluding false-positive results and can allow a simultaneous faster detection and quantification of target DNAs (Madani *et al.*, 2005). This study was carried out to identify the Hanwoo (Korean native cattle) beef by genotyping after DNA extraction of 295 commercial beef samples. Since Hanwoo, Holstein and imported cattle meat have different patterns in the MC1R gene associated with the coat colours of cattle (C-type, C/T-type or T-type), we could identify the genotype using real-time PCR (Kim *et al.*, 2002).

Materials and Methods

Beef samples were purchased from butcher shops in South Korea, designated 1 to 295 (Seoul; 1 to 100, *Jung-Bu* area; 101 to 170, *Young-Nam* area; 171-210, *Ho-Nam* area; 211-245, *Chung-Cheong* area; 246 to 295). Samples (10mg of tissue) were added to a 1.5ml tube containing 400 µl Lysis A buffer solution and 40µl Lysis B buffer solution with 5µl Proteinase K solution added to the lysate. The solution was then incubated at 65°C for 30 minutes, mixed with 300 µl Chloroform, and vortexed for 1 minute. After vortexing it was centrifuged at 12,000 rpm for 10 minutes and 300µl supernatant was poured into a clean 1.5ml micro tube containing 300µl DNA binding buffer and 300µl Isopropanol. Then it was centrifuged at 10,000 rpm for 1 minute and passed through the DNA binding column. 650µl 75% Ethanol was added into a DNA binding column, centrifuged at 10,000rpm for 1 minute and the Ethanol was removed. It was then centrifuged at 10,000 rpm for 3 minutes to dry. 100µl distilled water or TE buffer was added to the column tube and the sample stored at room temperature over 5 minutes, and centrifuged at 10,000 rpm for 3 minutes. The DNA extraction stored at -20°C.

The MC1R gene was amplified by PCR with a GeneAmp PCR System 9600(Perkin-Elmer, Cetus, USA). The PCR reaction contained 10µl of final solution consisting of: 10~100ng of template DNA, 2µl of primer mixture, 1µl of probe mixture, 5µl of 2X master mixture. The thermocycle program was as follows: 50°C for 3 minutes (initial denaturation), 94°C for 5 minutes; 40 cycles of 94°C for 30 seconds (denaturation), 60°C for 30 seconds (annealing) and a final extension step at 72°C for 30 seconds (extension). Real-time PCR reactions performed with the Real-Time PCR 7700 (ABI Prism™ 7700 Sequence Detector, Applied Biosystem, Singapole), PTC-200 (Peltier thermal cycler, MJ RESEARCH, USA) for SNP analysis (Figure 1).

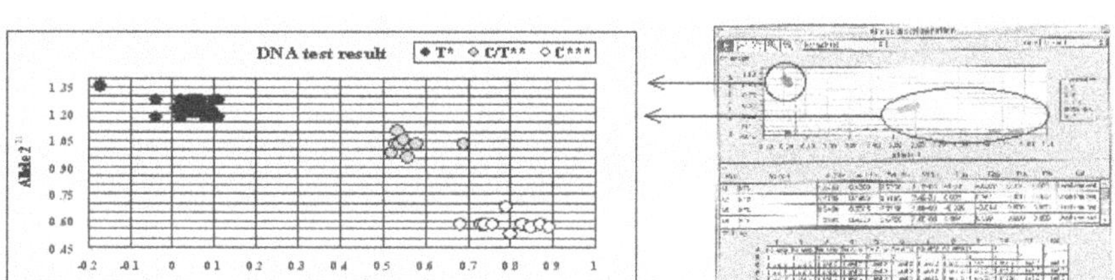

Figure 1: Real-time PCR analysis of MC1R gene in Hanwoo (Korean cattle), Holstein or imported cattle meat.
* T : MC1R gene of Hanwoo (Korean cattle) meat.
** C/T : MC1R gene possessed of Hanwoo (Korean cattle), Holstein and imported cattle meat.
*** C : MC1R of Holstein or imported cattle meat.
1) Control samples of Holstein and imported cattle meat.
2) Control samples of Hanwoo (Korean cattle) meat.

Results and Discussion

MC1R genotyping may be used to reveal Hanwoo (Korean cattle) meats that are purebred.

The result of real-time PCR assay for the proportions of Hanwoo beef were 84%, 85.7%, 95%, 91.4%, and 90% in Seoul, *Jung-Bu*, *Young-Nam*, *Ho-Nam* and *Chung-Cheong* area, respectively. Thus, beef samples in 295 butcher shops which are asserted to sell Hanwoo beef only, showed that 259 of 295 samples were Hanwoo beef gene type (T-type) and 36 of 295 samples were Holstein or imported dairy cattle gene type (C-type or C/T-type). In conclusion, the proportion of Hanwoo beef was 87.8% and the proportion of Holstein or imported dairy cattle meat was 12.2% (C-type; 9.8%, C/T-type; 2.4%).

Generally, most consumers could not differentiate imported meat from Hanwoo beef. Therefore, it will be necessary to make it mandatory for identification of Hanwoo beef and imported dairy cattle meat using MC1R genotyping based on Real-time PCR in butcher shops.

Conclusions

This study was carried out to identify the Hanwoo (Korean native cattle) beef by genotyping after DNA extraction of commercial beef of 295 samples. The proportion of Hanwoo beef was 87.8% and the proportion of Holstein or imported cattle meat was 12.2%.

References

Kim, C.J. and Lee, E.S. (2002) Effects of quality grade on the chemical, physical and sensory characteristics of Hanwoo (Korean native cattle) beef. Meat Science, 63(3): 397-405

Kim, J.M., Nam, Y.S., Choi, J.H., Lee, M.A., Jeong, J.U. and Kim, C.J. (2005) Identification of Hanwoo (Korean Native Cattle) Beef in Restaurants using Real-time PCR. Korean Food Science Animal Resources, 25(2): 203-209

Madani, M., Subbotin, S. A. and Moens, M. (2005) Quantitative detection of the potato cyst nematode, *Globodera pallida*, and the beer xyst nematode, *Heterodera schachtii*, using real-time PCR with SYBR green I dye. Molecular and Cellular Probes, 19: 81-86

Park, S.D., Kim, T.J. and Lee, J.I. (2005) Identification of Hanwoo and Holstein meat using MGB probe based real-time PCR associated with single nucleotide polymorphism (SNP) in Melanocortin 1 receptor (MC1R) gene. Korean Vet Resources, 45(1): 25-28

RAPD MARKER VARIATION IN MEAT QUALITY TRAITS OF POLL DORSET SECOND-CROSS LAMBS SELECTED FOR MUSCLE OR GROWTH

A.E.O. Malau-Aduli*[1], C.W. Bignell[1], R.S. Hegarty[2], H. Oddy[3], W. Johns[2], F. Tavassoli-Salardini[1], A.J. Smolenski[1], B.S. Malau-Aduli[1], B.B. Wells[1], P.A. Lane[1] and R.J. Clark[1]

[1.] *Animal Production and Genetics, School of Agricultural Science, University of Tasmania, TAS 7001, Australia,* [2.] *Beef Industry Centre of Excellence, NSW Department of Primary Industries, Armidale NSW 2351, Australia,* [3.] *Meat and Livestock Australia, 165 Walker St, North Sydney, NSW 2060, Australia. Email: Aduli.MalauAduli@utas.edu.au*

Keywords: RAPD markers, Poll Dorset crossbred lambs, meat quality, growth, muscle

Introduction

Crossbreeding has long been recognised as a powerful tool for exploiting genetic differences between two or more breeds to harness hybrid vigour in the offspring as a means of improving traits of economic importance. The growth and carcass merit of crossbred lambs is significant to the efficiency of production in the Australian meat sheep industry because the national flock is predominantly based on Merino. The development of DNA-based genetic markers has had a revolutionary impact on animal genetics in that genetic variation can be observed and exploited in the entire genome at the molecular level. Applications of DNA markers include investigations of genetic variability and inbreeding, species and strain identification, parentage assignment, the construction of high-resolution genetic linkage maps and identification of major genes for use in marker-assisted selection (Liu and Cordes 2004). Molecular markers can be classified into type I and type II markers. The former are markers associated with genes of known function, while the latter are associated with anonymous genomic segments. Random amplified polymorphic DNA (RAPD) markers are type II markers because their bands are amplified from anonymous genomic regions via the polymerase chain reaction. This research focuses on the prospects of RAPD markers distinguishing between Poll Dorset crossbred progeny carrying the genes for muscle or growth on the basis of their rams selected for high estimated breeding values (EBV) for the two traits.

Materials and Methods

One hundred and ninety six Poll Dorset x (Border Leicester x Merino) lambs from the Kirby Research Station in Armidale, New South Wales were used in a 3 x 2 experimental design for this study. They were progeny from 3 sire types selected on the basis of their EBVs for growth, muscle or control and accorded two levels of nutrition; low or high. Details of the pasture compositions of the nutrition levels, EBVs, sire selection, lambing, post lambing management, pasture assessment, live animal assessment, slaughter and abattoir assessment have been described (Hegarty *et al.* 2006). DNA was extracted from the muscle tissue and screened with nine RAPD markers whose primer sequences had been published (Malau-Aduli *et al.* 2006). Standard laboratory procedures of genomic DNA extraction, polymerase chain reaction assays, gel electrophoresis, RAPD band scoring and analyses were utilised (Sambrook and Russell, 2001).

Results and Discussion

Fat depth at the GR and C sites was significantly ($P<0.05$) influenced by level of nutrition and nutrition level x siretype interaction (Figure 1) in that fat depths were greater in crossbred lambs fed at high levels of nutrition than those fed low nutrition levels. Lambs selected for growth and fed high level of nutrition produced carcasses with the least KNIFE GR fat depth of 11.4mm, a significant reduction from 18.1mm in the control group fed low level nutrition (Figure 1). KNIFE GR fat depth was not affected by low levels of nutrition in crossbred lambs whose sires had been selected for growth or muscle. In Figure 2, all muscle measurements were similar except for slight differences in eye muscle depth (EMD) where high level of nutrition in groups selected for muscle and growth led to higher values than the control groups. All organisms are subject to mutations as a result of normal cellular operations or interactions with the environment, leading to genetic variation (polymorphism). In conjunction with selection and genetic drift, there arises genetic variation within and among individuals and species. For this variation to be useful to geneticists, it must be heritable and discernable whether as a recognisable phenotypic variation or as a genetic mutation distinguishable through molecular techniques. PCR amplification and gel electrophoresis resolution of products using RAPD primer A_2 in Figure 3 shows polymorphism of up to 5 bands was evident. It also demonstrates that at the molecular level, there are banding differences that can be picked up between crossbred progeny sired by rams with high genetic merit for growth or muscle. It would be interesting in further association studies to investigate if these differences can be significantly linked to QTL for inclusion in marker-assisted selection.

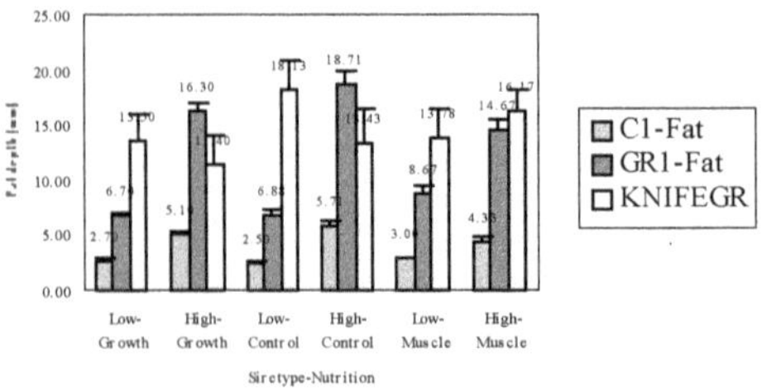

Figure 1: Fat depth variation at the C1 (CFAT), GR1 (GR1-Fat) sites and KnifeGR in mm.

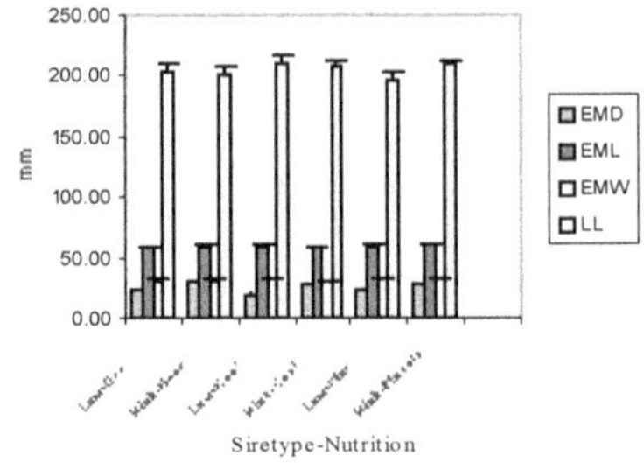

Figure 2: Siretype-nutrition variation in eye muscle depth (EMD), length (EML), width (EMW) and loin length (LL) in mm.

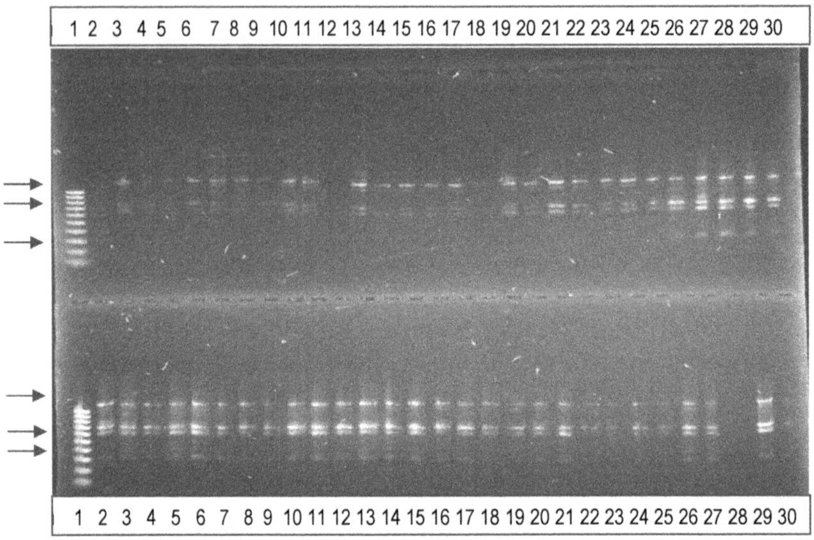

Figure 3: PCR amplification and gel electrophoresis resolution of products using RAPD primer A_2 in Poll Dorset x (Border Leicester x Merino) crossbred sheep. Upper Row Lane 12 and Lower Row Lane 28 controls without any band. 600-700bp and 1.1kb fragments produced bright fragments in all samples (except in the control lanes) indicating they were all crossbreds. At 300bp fragment in the upper row, a band is present in Lanes 26-29 (crossbreds sired by rams selected for high muscle EBVs) as distinct from Lanes 1-25 that did not have any band (crossbreds sired by either unselected rams or those selected for high growth EBVs).

References

Hegarty, R.S., Shands, C., Marchant, R., Hopkins, D.L., Ball, A. and Harden, S. (2006). Effects of available nutrition and sire breeding values for growth and muscling on the development of crossbred lambs 1. Growth and carcass characteristics. *Australian Journal of Agricultural Research* in press.

Liu, Z J. and Cordes, J.F. (2004). DNA marker technologies and their applications in aquaculture genetics. Aquaculture, 238: 1-37.

Malau-Aduli, A.E.O., Bignell, C.W., Tavassoli-Salardini, F., Smolenski, A.J., Palmer, A., Bignell, J., Burbury, S., Batchelor, R., Malau-Aduli, B.S., Adediran, S.A., Lane, P.A. and Clark, R.J. (2006). Genetic diversity and breed comparison of carcass traits in Tasmanian Corriedale and East Friesian sheep by RAPD markers. Proceedings of the 52nd International Congress on Meat Science and Technology, University College Dublin, Ireland.

Rabouam, C., Comes, A.M., Bretagnolle, V., Humbert, J.F., Periquet, G. and Bigot, Y. (1999). Features of DNA fragments obtained by random amplified polymorphic DNA (RAPD) assays. Molecular Ecology, 8: 493-503.

Sambrook, J.and Russell, D.W. (2001). Molecular Cloning: A laboratory manual, Cold Spring Harbor Laboratory Press, New York.

SESSION 1: MEAT QUALITY

- ANIMAL PRODUCTION AND PRODUCTION SYSTEMS

EFFECT OF WEANING ON FAT QUALITY IN SPANISH MERINO LIGHT LAMBS

M. Juárez[1], A. Horcada[1], M.J. Alcalde[1], A.M. Mullen[2], N. Brunton[2], M. Valera[1] and A. Molina[3]

[1]*Departamento de Ciencias Agroforestales. E.U.I.T.A. Universidad de Sevilla. Spain* [2]*Ashtown Food Research Centre. Teagasc. Dublin. Ireland* [3]*Departamento de Genética. Universidad de Córdoba. Spain*

Keywords: fat depot, CLA, Spanish Merino, light lambs

Introduction

Generally, the Mediterranean consumer looks for lamb meat with a reduced amount of fat (Castro *et al.*, 2005). For this reason, the Spanish Merino breed is oriented toward light lamb production, based on extensive production systems.
Fat depot amount and composition influence lamb carcass and meat quality (Pérez *et al.*, 2002). Intramuscular (IM), intermuscular (IN) and Subcutaneous (SC) fat depots have an important effect on meat quality due to their influence on firmness or softness (Wood *et al.*, 2004).
Conjugated linoleic fatty acid (CLA) is a collective term indicating a group of octadecenoic acids that are geometric and positional isomers of linoleic acid (C18:2) (Pariza *et al.*, 2001; Schmid *et al.*, 2006). Several health effects (anticarcinogenic, antiatherosclerosis, hypocholesterolemic, modulation of inmune system and reduction of body fat) have been reported for CLA in animals (Pariza and Hargraves, 1985; Lee *et al.*, 1994; Park *et al.*, 1999; Miller *et al.*, 2001).
Most authors agree about there are several factors that influence fat characteristics and CLA levels (breed, slaughter weight, fatness degree,...) and of these feeding is one of the most important factors (Vatansever *et al.*, 2000; Arsenos *et al.* 2006).
The main objective of this study was to characterise fat quality of Spanish Merino light lambs bred following a traditional Southern Spain production system (no weaning and concentrate availability from 45 days after birth) and following a standard system (weaning at 45 days and fattening with concentrate).

Materials and Methods

Sixteen pure-bred Spanish Merino lambs were selected. Eight lambs were assigned to **Treatment 1** (weaning at 45 days and fattening with concentrate) and eight to **Treatment 2** (not weaning and concentrate availability from 45 days after birth). Lambs were slaughtered at 21.90±0.24kg live weight. Samples of SC and IN fat were collected in the slaughterhouse within the first hour *post mortem*, vacuum packed and frozen. *Longissimus dorsi* samples were collected 24h post slaughter from the left side of the carcass and aged for 72 hours at 4°C prior to vacuum packing and freezing them. Total fatty acids were extracted, methylated and analysed by an adaptation of the method described by Elmore *et al.* (1999). Separation and quantification of the FAMEs (fatty acid methyl esters) was carried out using a gas chromatograph (GC, Varian Star 3400CX, Varian Associates, Inc., California, USA) equipped with a flame ionisation detector (FID) and fitted with a BPX-70 capillary column (120m, 0.25mm i.d., 0.2µm film thickness, SGE, Australia). Tricosanoic acid methyl ester (C23:0 ME) was used as an internal standard. Fatty acid profiles were expressed as a percentage with respect to overall fatty acids identified and grouped as follows: Saturated fatty acids (SFA), Monounsaturated fatty acids (MUFA), Polyunsaturated fatty acids (PUFA), w3 and w6 fatty acids and CLA, PUFA/SFA and w3/w6 ratios and desaturase activity Δ^9C16 and Δ^9C18 indexes were also calculted (Malau-Aduli *et al.*, 1998). Statistical analyses were performed using SPSS 12.0 for Windows.

Results and Discussion

The effect of weaning on fatty acid profiles of Spanish Merino light lambs SC, IN and IM fat deposits is presented in Table 1. The fatty acid profile of IM fat, as reported by Castro *et al.* (2005), has been shown to have higher amounts of PUFA than SC and IN fat depots ($p<0.001$). Furthermore, MUFA and SFA levels were higher for IM fat ($p<0.001$). CLA levels were not affected by the depot ($p>0.05$), showing similar levels in the three kinds of fat studied.
No difference between treatments ($p>0.05$) was observed for SFA or MUFA levels. However, PUFA/SFA, w3/w6 ratios and PUFA levels were higher for IN and SC fat from Treatment 2 lambs. Furthermore, MUFA and SFA levels were lower for IM fat lambs for all fat types. Therefore, longer suckling periods may improve healthy fatty acid profiles. This may be due to the presence of CLA in ewes milk and their grass-based production system.Δ^9C18 enzyme activity was higher in weaned lambs (Treatment 1), which would result in a greater capacity to desaturate C18:0 fatty acid, due to a higher ruminal activity. Δ^9C16 enzyme activity was higher in weaned lambs (Treatment 1) in IN fat only.

Table 1: The effect of weaning on the fatty acid groups in Spanish Merino light lambs Subcutaneous, Intra and Intermuscular fat depots.

Depot	*Intramuscular fat*			*Intermuscular fat*			*Subcutaneous fat*		
Treatment	1	. 2	SD[A]	1	. 2	SD[A]	. 1	2	SD[A]
SFA	43.29±1.26	43.68±0.78	ns	47.47±1.17	47.36±1.03	ns	46.7±1.288	45.8±1.157	ns
MUFA	43.62±1.20	41.75±0.57	ns	46.66±1.10	45.11±1.21	ns	47.1±1.210	46.6±1.141	ns
PUFA	13.0±1.25	14.5±0.91	ns	5.85±0.21	7.52±0.34	***	6.13±0.18	7.46±0.35	**
P/S[B]	0.30±0.03	0.33±0.02	ns	0.12±0.00	0.15±0.00	**	0.13±0.00	0.16±0.00	*
w3/w6	0.29±0.02	0.33±0.01	ns	0.07±0.00	0.10±0.01	*	0.09±0.00	0.12±0.00	*
CLA	0.89±0.07	1.74±0.19	***	0.87±0.06	1.97±0.25	***	1.02±0.08	2.04±0.18	***
Δ^9C16[C]	0.08±0.00	0.08±0.00	ns	0.08±0.00	0.07±0.00	*	0.10±0.00	0.10±0.00	ns
Δ^9C18[D]	0.75±0.00	0.73±0.00	*	0.74±0.00	0.69±0.00	***	0.79±0.01	0.77±0.01	*

[A]SD: Significant differences. ns = p>0.05; * = p<0.05; **= p<0.01; *** = p<0.001

[B]P/S: PUFA/SFA ratio

[C]$\Delta^9C16 = 100*[(C16:1)/(C16:0+C16:1)]$; [D]$\Delta^9C18 = 100*[(C18:1)/(C18:0+C18:1)]$

Conclusions

IM fat had higher levels of PUFA, however, no fat deposition effect was observed for CLA levels between IM, SC and IN fat depots.

Fat from those lambs that stayed with their mothers until slaughter had a healthier fatty acid composition in comparison to fat from those lambs weaned 45 days after birth.

References

Arsenos, G., Kufidis, D., Zygoyiannis, D., Katsaounis and Stamataris C. (2006). Fatty acid composition of lambs of indigenous dairy Greek breeds of sheep as affected by post-weaning nutritional management and weight at slaughter. Meat Science, 73, 55-65.

Castro, T., Manso, T., Mantecón, A.R., Guirao, J. and Jimeno, V. (2005). Fatty acid composition and carcass characteristics of growing lambs fed diets containing palm oil supplements. Meat Science, 69, 757-764.

Elmore, J.S., Mottram, D.S., Enser, M. and Wood, J.D. (1999). Effect of the Polyunsaturated Fatty Acid Composition of Beef Muscle on the Profile of Aroma Volatiles. Journal of Agricultural and Food Chemistry, 47, 1619-1625

Lee, K. N., Kritchevsky, D. and Pariza, M. W. (1994). Conjugated linoleic acid and atherosclerosis in rabbits. Atherosclerosis, 108, 19-25.

Malau-Aduli, A.E.O., Siebert, B.D., Bottema, D.K. and Pitchford, W.S. (1998). Breed composition of muscle phospholipids in Jersey and Limousin cattle. Journal of Animal Science, 76, 766-773.

Miller, A., Stanton, C. and Devery, R. (2001). Modultion of arachidonic acid distribution by conjugated linoleic acid isomers and linoleic acid in MCF-7 and SW480 cancer cells. Lipids, 36, 1161-1168.

Park, Y., Albright, K. L., Storkson, J.M., Liu, W, Cook, M. E. and Pariza, M. W. 1999. Changes in body composition in mice during feeding and withdrawal of conjugated linoleic acid. Lipids, 34, 243-248.

Pariza, M. W. and Hargraves, W. A. (1985). A beef derived mutagenesis modulator inhibits initiation on mouse epidermal tumors by 7,12-dimethylbenz(a)anthracene. Carcinogenesis, 6, 591-594.

Pariza, M. W., Park, Y. and Cook, M. E. (2001). The biologically active isomers of conjugated linoleic acid. Progress in Lipid Research, 40, 283-298.

Pérez, P., Maino, M., Tomic, G., Mardones, E. and Pokniak J. (2002). Carcass characteristics and meat quality of Suffolk Down suckling lambs. Small Ruminant Research, 44, 233-240

Schmid, A., Collomb, M., Sieber, R. and Bee, G. 2006. conjugated linoleic acid in meat and meat products: A review. Meat Science, 73, 29-41.

Vatansever, L., Kart, E., Enser, M., Nute, G.R., Scollan, N.D., Wood, J.D. and Richardson, R.I. (2000). Shelf life and eating quality of beef from cattle of different breeds given diets differing in n-3 polyunsaturated faty acid composition. Animal Science, 71, 471-482.

Wood, J.D., Richardson, R.I., Nute, G.R., Fisher, A.V., Campo, M.M., Kasapidou, E., Sheard, P.R. and Enser, M. (2004). Effect of fatty acid on meat quality: a review. Meat Science, 63 (1), 21-32.

THE INFLUENCE OF SOURCE AND LEVEL OF SELENIUM IN CHICKEN FEED ON BREAST MEAT QUALITY

V. Tomović*[1], N. Džinić[1], Lj. Petrović[1], L. Perić[2], and D. Žikić[2]

[1]*Faculty of Technology;* [2] *Faculty of Agriculture, University of Novi Sad, 21000 Novi Sad, Serbia and Montenegro. E-mail: tomovic@uns.ns.ac.yu*

Keywords: chicken, inorganic selenium, organic selenium, breast meat quality

Introduction

Over the past few years, consumer demands regarding aspects of meat quality have substantially increased. Therefore, a challenge to the meat industry is to enhance the image of meat purchased at the supermarket. There are many meat quality characteristics that attract consumer attention (appearance, colour, texture, tenderness, juiciness, aroma) (Janssens, 1998). Consumers prefer fresh meat with mimimum loss of water during handling and cooking. Therfore the water-holding capacity of the meat (Mahan and Kim, 1999) and colour are considered among the most important meat quality characteristics. It is well known that quality and especially chemical composition of produced meat are highly affected by feed composition. In order to evaluate chicken meat quality, as well as its quantity, different additives (vitamin E and C, α-tocopherol, Se, ...) have been introduced to standard chicken feed, and regarding the published data, positive effects have been achieved (Sheehy *et al.*, 1997; Ahn *et al.*, 1998; Edens, 1996; 1997). The aim of this work was to investigate the influence of different sources and levels of selenium in chicken feed on technological quality characteristics (pH, drip loss and colour) of breast meat.

Materials and Methods

Cobb 500 hybrids, both sex, were used for the investigations. The trial was designed as 4 x 8 test (4 groups and 8 replicates). This includes 32 pens with 78 broilers in each pen. Stocking density was 15 birds/m^2. Air temperature was adjusted in accordance to the technological demands. Feed and water supply were *ad libitum.* Feed composition was the same for all four groups, based on corn, soya and sunflower, except added premixes. Experimental treatments were the following: (A) Feed containing 0.3ppm Se from sodium selenite (NaSe); (B) Feed containing 0.2ppm Se from NaSe and 0.1ppm Se from Sel-Plex; (C) Feed containing 0.1ppm Se from NaSe and 0.2ppm Se from Sel-Plex; (D) Feed containing 0.3ppm Se from Sel-Plex.

After 42 days of fattening, two groups of broilers containing 24 birds each, of similarly same final mass were slaughtered, processed and chilled by standard technological procedure. Breast meat was taken from each bird for the determination of pH value, colour and drip loss. pH_u was determined 24 h *post mortem* (pm) on the cranial part of breast using the portable pH-meter ULTRA X. Drip loss (DL) was measured on breast meat. Breast meat was separated from the bone and both parts were separately hung the hook and covered with plastic bag (Honikel, 1998). All the samples were put into a refrigerator at 4°C. After 24 hours, the first measurement was carried out (on one half of the breast meat), and after 48 hours the second measurement was done on the other part of the meat of the same bird. Colour of breast meat was determined on fresh cut 24h and 48h *post mortem,* after chilling, using the Minolta Chroma Meter CR-400, and the colour characteristics were expressed in CIELa*b* system (Robertson, 1977).

Results and Discussion

The results of technological breast muscle characteristics determination are presented in Table 1. The highest average pH_u value – 5.89, was determined in breasts of birds of experimental group A, and the lowest average value – 5.81, was in breast muscles of experimental group C. The differences between the average pH_u values of muscles of experimental groups are not statistically significant ($P > 0.05$), and the mean values respond to values of normal quality of meat. Statistically significantly lower ($P < 0.01$) mean value of DL was determined in breast meat of experimental group D (0.60%) compared to DL of breast meat of experimental groups A and B (0.91%), measured 24 h pm. The lowest DL value was measured 48h (0.84%) in breast meat of experimental group D, and it was significantly smaller ($P < 0.05$ or $P < 0.01$) compared with DL of breast meat of experimental groups A and B (1.13%, 1.25%, respectively). The results of DL determination show that average values of breast meat DL, measured 24 h pm, are statistically significantly lower ($P < 0.05$ or $P < 0.01$) or have a tendency of statistical significance, than the DL values measured 48 h pm. The obtained results are in accordance with the results of Edens (2001), Naylor *et al.* (2000) and Mahan *et al.* (1999), who found that organic Se (0.3ppm), introduced to chicken feed, affects decrease of DL, and that meat quality can be preserved during storage replacing the inorganic Se with organic Se in feed. The lightness (L*) of breast cut was determined instrumentally (Table 2) 24h pm. Breast muscles of experimental group D were the darkest, but not statistically significantly ($P > 0.05$).

Table 1: Influence of source and level of Se on drip loss and pH value of breast meat.

Traits	Groups	*24 h pm* *Mean ± SD*	*48 h pm* *Mean ± SD*	*24-48 h pm* Paired t-test
pH_u	A	5.89±0.17		
	B	5.86±0.19		
	C	5.81±0.15		
	D	5.87±0.18		
DL (%)	A	0.91^{A}±0.32	1.13^{aAB}±0.23	0.049
	B	0.91^{A}±0.27	1.25^{aA}±027	0.003
	C	0.79^{AB}±0.27	1.05^{abAB}±0.28	0.070
	D	0.60^{B}±0.21	0.84^{bB}±0.32	0.047

The average lightness values (L*) are in the range from 50.95 (D) to 51.43 (B) and respond to values of normal meat quality (48 < L* < 53). The lightness (L*) of breast muscle of all investigated groups determined 48 h pm is statistically significantly (P < 0.05 or P < 0.01) higher compared with lightness of breast muscles determined 24 h pm; however, the difference between average lightness values (L*) measured after 48 h pm was not statistically significant (P > 0.05).

Table 2: Influence of source and Se level on breast meat colour 24 and 48 h *post mortem.*

Traits	Groups	*24 h pm* *Mean ± SD*	*48 h pm* *Mean ± SD*	*24-48 h pm* Paired t-test
CIE*L** (lightness)	A	51.23±3.46	53.67±2.88	<0.001
	B	51.43±3.23	53.65±2.85	0.036
	C	51.31±3.83	53.83±3.01	<0.001
	D	50.95±5.39	53.00±4.80	0.007
CIE*a** (redness)	A	3.29±1.51	3.88^{ab}±1.45	<0.001
	B	3.13±1.13	3.03^{b}±1.04	0.712
	C	3.36±0.92	3.34^{ab}±1.02	0.898
	D	3.72±0.91	4.20^{a}±0.75	0.008
CIE*b** (yellowness)	A	6.20±1.68	7.27±1.90	<0.001
	B	7.29±1.62	7.89±1.92	0.231
	C	6.70±1.12	7.51±1.03	0.004
	D	6.53±1.78	7.90±2.21	0.001

ab – P<0.05; AB – P<0.01

The portion of red colour (a*) in breast meat of experimental group D, determined 48 h pm is significantly higher (4.20) compared to breast meat of experimental group B (3.03). The portion of yellow colour (b*) determined 48 h pm (7.90) is significantly higher (P < 0.01) in breast meat of experimental group D compared to breast meat of the same group, but determined 24 h pm (6.53).

Conclusions

The addition of organic Se to chicken feed (containing 0.3 ppm Se from Sel-Plex) significantly affected the decrease in drip loss of breast meat. The colour (lightness – L*) is the darkest, and the portion of red colour (a*) the highest in the same group (group D) of investigated breast muscles, but the stated differences are not statistically significant.

References

Ahn, C.N., Chae, H.S., Kim, D.W., Yoo,Y.M., Kim, Y.K.and Rhee, Y.C. (1998). Effects of full fat flax seed, α-tocopherol, ascorbic acid and selenium on the storage of broiler meats. Journal of Liverstock Science 40: 96-102.

Edens, F.W. (1996). Organic selenium: from feathers to muscle integrity to drip loss. Five years onward:no more selenit. In: Biotechnology in the Feed industry.Proceedings of Alltech' s 12th Annual Symposium. Nottingham University Press, Nottingham, UK pp.165-185.

Edens, F.W. (1997). Potential for organic selenium to replace selenite in poultry diets. Zootecnica International 20: 28-31.

Edens, F.W. (2001). Involvement of Sel-Plex in physiological stability and performance of broiler ckickens. In Biotechnology in the Feed industry.Procceedings of Alltech's 17th Annual Symposium. Nottingham University Press, Nottingham, UK pp.349-376.

Honikel, K.O. (1998). Reference methods for the assessment of phisical characteristics of meat. Meat Science, 49: 447-457.

Janssens, G. (1998). Vitamin E improves chicken meat quality. Meat Processing International Edition. Septermber/October 42:44-46.

Mahan, D.C. and Kim.Y.Y. (1999). The role of vitamins and minerals in the production of high quality pork. Review.Asian-Australian Journal of Animal Sciences 12: 287-294.

Naylor, A.J., Chost, M and Jacques, K.A. (2000). Effects of selenium source and level on performance and meat quality in male broilers. Poultry Science 79 (Suppl) 117.

Sheehy, P.J.A., Morrissey, P.A., Buckley, D.J.and Wen, J. (1997). Effects of vitamins in the feed on meat quality in farm animals: Vitamin E. In: Recent advances in animal nutrition. Nottingham University Press, Nottingham, UK pp.3-27.

Robertson, A.R. 1977. The CIE 1976 Color-Difference formulae, Color Research Applied, 2, 7.

EFFECT OF ACCOMMODATION SYSTEM ON COLOUR AND QUALITY OF *LONGISSIMUS DORSI* AND *EXTENSOR CARPI RADIALIS* MUSCLES OF STEERS

J. Rogalski[1]*, T. Moreno[2], P.G. Dunne[1], F.J. Monahan[3], P. French[4] and A.P. Moloney[1,5]

[1]*Meat Technology Department, Teagasc, Ashtown Food Research Centre, Ashtown, Dublin 15, Ireland,* [2] *Centro de Investigaciones Agrarias de Mabegondo, Apartado 10, 15080 A Coruña, Spain,* [3] *Department of Food Science, University College Dublin, Dublin 4, Ireland,* [4] *Teagasc, Moorepark Research Centre, Fermoy, Co. Cork, Ireland,* [5] *Teagasc, Grange Research Centre, Dunsany, Co. Meath, Ireland.*
E-mail: jaroslaw.rogalski@teagasc.ie

Keywords: beef, steers, accommodation, housing, muscle, colour

Introduction

The factors that affect beef quality have been widely studied and can be broadly categorised as either intrinsic, e.g. breed, gender, age, muscle, or extrinsic. Extrinsic factors relate both to environmental ('on-farm') conditions under which animals are produced, e.g. nutrition, accommodation, climate and welfare, and also post-mortem technological factors that affect muscle tissue properties and thus meat quality, e.g. electrical stimulation, chilling rate and packaging. Currently, other than nutritional effects, there is a paucity of information with regard to effects of extrinsic 'on-farm' factors which may cause variation in meat quality of cattle. Foremost among these factors is the accommodation system under which cattle are housed.

In Ireland, approximately 38% of cattle are slaughtered annually between January and May and the majority of these cattle are finished indoors, usually in 'conventional' slatted floor accommodation (SLA). Provision of such accommodation can represent a substantial capital and overhead cost and thus, cheaper alternatives such as out-wintering pads (OWP) have been examined. From an animal welfare perspective, straw bedding (STR) is often perceived to be preferable to a slatted floor. When operated optimally, these accommodation options differ in environment, space allowance and floor type, all of which could influence beef quality from a consumer perspective. Since there is a deficit of information concerning this issue, the objective of this study was to document the effect of type of cattle accommodation on the quality of two muscles.

Materials and Methods

Forty-five Charolais crossbred steers were blocked on bodyweight and within block accommodated at random either outside on OWP at 18m^2/head or inside on SLA at 2.5m^2/500kg bodyweight or on STR at 4m^2/500kg bodyweight . Animals on each treatment were penned in groups of five and offered a total-mixed ration (500g of concentrates dry matter (DM) and 500g of grass silage DM/kg total DM) *ad libitum*. Post-slaughter after a 132-day feeding period, carcass weight and fat score were recorded. Two days post-mortem, pH, colour and drip loss of *M. longissimus dorsi* (LD) and *M. extensor carpi radialis* (ECR) were measured. Following vacuum storage at <4°C for a further 12 days, colour was again measured and samples stored frozen for subsequent assessment of tenderness (as Warner-Bratzler shear force, WBSF), organoleptic characteristics and lipid content. Data were analysed using models appropriate to a randomised complete block or a randomised block split plot design, as required. Composition and pigment concentrations were measured in samples collected at 2days post-mortem and stored frozen.

Results and Discussion

Carcass weight averaged 351, 362 and 372 (s.e.d. 6.63) kg ($P<0.05$) for SLA, STR and OWP, respectively. Type of accommodation had no effect on the redness, saturation or hue of either LD or ECR muscles at 2 or 14 days post mortem (Table 1). Type of accommodation did not affect drip loss, shear force or lipid content of either muscle examined (Table 1) or the sensory attributes of tenderness, texture, flavour, juiciness, chewiness or acceptability of LD muscle (data not shown). As previously observed (Dunne *et al.*, 2005), the ECR muscle was less red and had lower colour intensity (saturation) but also had lower drip loss, less lipid and required greater shear force than LD muscle (Table 1). The pH of ECR muscle was higher ($P<0.05$) for animals accommodated on OWP compared to that for animals accommodated on SLA or STR, while LD muscle from OWP was darker at 2 days post-mortem than LD from animals accommodated on STR. The effect of muscle on haem pigments approached significance ($P=0.065$) but there was no effect of accommodation system.

In general, there were no important effects of accommodation system on any meat quality characteristic; muscle was a more important source of variation. Previously, Jensen and Oksama (1996) reported that accommodation system influenced colour (lightness, hue and saturation), pigmentation and shear force of LD from Danish Friesian and Danish Jersey young bulls. However, in their study loose-housed and tie-stall bulls were compared. Thus, effects may have been mediated by differences in the level of physical activity of both groups. Dunne *et al.*, (2005) reported that exercise increased redness of ECR but not LD. In the present study, by contrast, LD of OWP steers was darker at 2days post-mortem which may relate to metabolic differences mediated by physical activity. The OWP steers had more space and thus more opportunity at least, if not more inclination, to exercise although definitive conclusions are difficult to draw in this regard. Any marginal differences in LD colour were absent at day 14. This was as expected since during ageing mitochondrial decay leads to a reduction in competition for available oxygen by mitochondrial respiratory enzymes. Coupled with structural changes, this leads to increased penetration by oxygen and a deeper oxymyoglobin layer (Hood, 1980), thus attenuating any treatment differences present at day 2. Differences in meat quality indices due to muscle rather than treatment variability, as observed in the present study are not surprising since muscles, particularly those chosen here, are known to exhibit contrasting histochemical, biochemical and metabolic characteristics due to their location, and thus function within the animal (Klont *et al.*, 1998).

Tenderness is a critical component of eating quality, particularly for beef. While there remains some conjecture regarding the exact molecular basis for post-mortem tenderisation of beef, it is generally accepted that genetic factors and post-slaughter technologies have a greater effect on tenderness than 'on-farm' environmental factors, such as accommodation system (Tatum *et al.*, 1999). Tenderness differences between LD and ECR most likely relate to compositional differences, particularly in collagen content, reflecting differences in anatomical location and muscle function but also histochemical and biochemical differences (Klont *et al.*, 1998).

Table 1: Colour (2 and 14 days), pH and meat quality indicators of *M. longissimus dorsi* (LD) and *M.extensor carpi radialis* (ECR) of steers accommodated in slatted-floor housing (SLA), in straw-bedded housing (STR) or on out-wintering pads (OWP).

	LD			ECR				Significance	
Variable	SLA	STR	OWP	SLA	STR	OWP	SED	Muscle	System
Colour (2d)									
Lightness	36.2^{ab}	37.5^{b}	35.6^{a}	40.3^{c}	39.3^{c}	39.2^{c}	0.73	***	NS [1]
Redness	13.8	14.1	13.9	13.0	13.0	13.0	0.47	***	NS
Hue	29.4	30.2	29.0	32.2	31.4	31.3	0.72	***	NS
Saturation	15.8	16.3	15.9	15.4	15.3	15.2	0.53	*	NS
pH	5.56^{a}	5.56^{a}	5.56^{a}	5.81^{b}	5.81^{b}	5.90^{c}	0.030	***	* [2]
Colour (14d)									
Lightness	37.2	36.7	36.5	37.3	36.7	36.4	0.62	NS	NS
Redness	16.4	16.8	16.5	15.3	15.0	14.6	0.53	***	NS
Hue	30.0	29.6	29.8	30.8	30.3	30.5	0.61	***	NS
Saturation	19.0	19.3	19.0	17.8	17.4	17.0	0.55	*	NS
Drip loss (g/kg)	15.7	14.3	18.4	4.9	5.1	5.2	1.76	***	NS
WBSF (N)	26.8	27.3	25.9	58.0	60.1	64.7	2.88	***	NS
Lipid (g/kg)	31.7	30.5	28.6	9.4	7.9	8.7	3.07	***	NS
Pigments (mg/g)	7.02	7.24	7.53	6.56	6.91	6.57	0.570	NS [3]	NS

[a,b,c] Within rows, means assigned different superscripts differ significantly (P<0.05).
[1] Muscle × system (P=0.04), [2] Muscle × system (P=0.07), [3] Muscle P=0.065

Conclusions

Under the conditions of the study, compared to conventional indoor housing, accommodating cattle on OWP had a minor transient effect on beef colour and no impact on beef lipid content or eating quality, simulated by WBSF. No evidence emerged of an effect of type of indoor housing on major meat quality characteristics.

Acknowledgements

J. Rogalski was in receipt of a European Union Marie Curie Fellowship.

References

Dunne, P.G., O'Mara, F.P., Monahan, F.J., French, P. and Moloney, A.P. (2005). Colour of muscle from 18-month-old steers given long term daily exercise. Meat Science, 71: 219-229.

Hood, D.E. (1980). Factors affecting the rate of metmyoglobin accumulation in pre-packaged beef. Meat Science, 4: 247-265.

Jensen, L.R. and Oksama, M. (1996). Influence of different housing systems on carcass and meat quality in young bulls. Proceedings of the 42nd ICoMST, p.436-437.

Klont, R.E., Brocks, L. and Eikelenboom, G. (1998). Muscle fibre type and meat quality. Meat Science, 49:S219-S229.

Tatum, J.D., Belk, K.E., George, M.H. and Smith, G.C. (1999). Identification of quality management practices to reduce the incidence of retail beef tenderness problems: development and evaluation of a prototype quality system to produce tender beef. Journal of Animal Science, 77:2112-2118.

EFFECTS OF FINISHING DIET AND FASTING TIME ON THE GLYCOLYTIC POTENTIAL AND QUALITY OF LONGISSIMUS MUSCLE IN CROSSBRED PIGS

K. Partanen*[1], M. Honkavaara[2], H. Siljander-Rasi[1], N. Kivistö[2] and M. Ruusunen[3]

[1.] *MTT Agrifood Research Finland, Animal Production Research, Tervamäentie 179, FI-05840 Hyvinkää, Finland,* [2.] *Finnish Meat Research Institute, P.O. Box 56, FI-13101 Hämeenlinna, Finland,* [3.] *Department of Food Technology, P.O. Box 66, FI-00014 University of Helsinki. Email:kirsi.partanen@mtt.fi*

Keywords: pork, feeding, fasting, pH, colour, drip loss, glycogen, lactic acid

Introduction

Colour and water-holding capacity are important attributes of pork quality. Both quality traits are affected by biochemical processes during the post-slaughter conversion of muscle to meat, where pH is an important factor. The extent of post-mortem pH fall and the ultimate pH of pork are mainly determined by the muscle glycogen content at slaughter. High muscle glycogen levels can result in pork with low ultimate pH, pale colour and decreased water-holding capacity. Muscle glycogen stores are generally expressed as glycolytic potential (GP), which is a measure of the compounds present in the muscle that can be converted to lactic acid and thus contribute to pH decline (Monin and Sellier, 1985). There are some differences in the GP of muscles and pork quality between pig breeds, but differences found between purebred pigs can differ when cross-breeding is applied (Monin and Sellier, 1985; Enfält *et al.*, 1997). The GP of muscles and pork quality can be influenced by the length of fasting before slaughter (Leheska *et al.*, 2003) and by feeding pigs diets of low starch and sugar content (Rosenvold *et al.*, 2001). This study focused on the effects of finishing diet's carbohydrate composition and fasting time on the GP and quality of longissimus muscle in crossbred pigs.

Materials and Methods

One hundred and sixty 25kg pigs were assigned in pairs and sexes separately to a performance experiment to study how the carbohydrate composition of finishing diet and fasting time influence the GP, ultimate pH, colour and drip loss of longissimus muscle in crossbred pigs. Half of the pigs were crosses of Finnish Landrace and Yorkshire (LY) and the other half were crosses of Duroc, Finnish Landrace and Yorkshire (DLY, 25% Duroc). The pigs were fed a barley-soybean meal diet up to the 80kg body weight. Thereafter half of the pigs received a barley-soybean meal control diet and another half received a high-fibre diet which contained barley, barley fibre, faba beans and rapeseed cake. Control and high-fibre diets contained 529 and 405g starch, 35 and 42g sugars, and 183 and 257g neutral detergent fibre per kg dry matter, respectively. The pigs were slaughtered in a commercial slaughterhouse in 106kg body weight. Fasting times averaged 25 and 41h and they were obtained by giving pigs their last meal either 7am on the day of transportation or 3pm the previous day. The pigs were slaughtered the morning after the transportation. Sixteen pigs had been slaughtered during the transportation day and they were removed from the data.

The ultimate pH and colour measurements (L*, Minolta DP301 device) and samples for GP determination were taken from the longissimus muscle 24h after slaughter. Drip loss was determined from a 100g sample of meat before and after the storage in sealed plastic bag at 4°C for three days. Glycogen and lactic acid contents were determined from one randomly selected pig per pen. Glycogen content was analysed as glucose: 10 µl of homogenate (sample in 0.1 M phosphate buffer, pH 7.0) were hydrolysed in 200µl of 0.1M HCl at 100°C for 2h, after which pH was adjusted to 6.5-7.5 and glucose was determined via NADP+ reduction with linked assay involving hexokinase and glucose-6-phosphate dehydrogenase (Glucose (HK) 16-50, Sigma Diagnostics). Lactate concentration was determined from the homogenate via NAD+ reduction with a linked assay involving lactate dehydrogenase and glutamate pyruvate transaminase (Boehringer-Mannheim no. 139 084). The GP was calculated as 2 × (glycogen + glucose + glucose-6-phosphate) + lactic acid (Monin and Sellier, 1985).

The data were analysed separately for LY- and DLY-crosses using the MIXED procedure of SAS and the model which had the fixed effects of sex, fasting time and finishing diet and their interactions and for meat quality measurements the random effect of pen within sex × fasting time × finishing diet interaction.

Results and Discussion

In LY-crosses, several significant or moderate ($P < 0.20$) interactions were found, and therefore the results are presented for each sex and treatment combination in Table 1 and main effects are given in the text when appropriate. There was a sex × finishing diet interaction in the GP ($P = 0.07$) and glycogen ($P = 0.15$) and lactate contents ($P < 0.05$) of longissimus muscle. In gilts, control and high-fibre diets resulted in fairly similar GP (151 vs. 156µmol/g; $P > 0.05$) and glycogen (29 vs. 28µmol/g; $P > 0.05$) and lactate contents (93 vs 99µmol/g; $P > 0.05$). However, the barrows that were fed high-fibre diet had lower GP (134 vs. 162µmol/g; $P < 0.05$) and glycogen (21 vs. 32; $P < 0.05$) and lactate contents (90 vs. 97µmol/g; $P < 0.05$) than those fed the control diet. Longer fasting decreased the GP (153 vs. 148µmol/g; $P < 0.05$) and glycogen content (32 vs. 24µmol/g; $P < 0.05$). However, there was a moderate sex × fasting time interaction in the lactate content ($P = 0.15$). Increasing fasting time decreased muscle lactate content in barrows (97 vs. 90µmol/g;

P < 0.05), but not in gilts (96 vs. 96μmol/g; P > 0.05). There were moderate sex × fasting time (P = 0.06) and finishing diet × fasting time interactions (P = 0.17) in the ultimate pH of longissimus muscle. Longer fasting increased the ultimate pH in barrows (5.56 vs. 5.64; P < 0.05), but not in gilts (5.55 vs. 5.54; P > 0.05). Finishing diet did not affect the ultimate pH in 25h fasting (5.57 vs. 5.57; P > 0.05), but in 41h fasting high-fibre diet resulted in higher ultimate pH than control diet (5.61 vs. 5.54; P < 0.05). The lightness of meat was not affected by the investigated factors (P > 0.05). A moderate sex × fasting time interaction was found in the drip loss (P = 0.11). Longer fasting decreased the drip loss in barrows (4.5 vs. 3.1%; P < 0.05), but not in gilts (3.7 vs. 3.7%; P > 0.05).

Table 1: Effects of sex, finishing diet and fasting on the glycogen metabolism and meat quality in LY-crosses.

	Gilts				Barrows				
	Control		High-fibre		Control		High-fibre		
	25 h	41 h	25 h	41 h	25 h	41 h	25 h	41 h	SEM
GP, μmol/g$^{F,\ S×D}$	159	143	158	153	179	145	143	124	12
Glycogen, μmol/g$^{F,\ S×D}$	33	25	30	27	30	25	24	19	5
Lactate, μmol/g$^{S×D,\ S×F}$	92	92	94	99	99	95	95	85	3
Ultimate pH$^{S,\ S×F,\ D×F}$	5.56	5.51	5.55	5.57	5.58	5.63	5.53	5.66	0.04
L*	51.3	50.1	50.8	51.3	50.8	50.1	51.6	49.6	1.4
Drip loss, %$^{S×F}$	3.9	3.6	3.5	3.9	4.6	2.7	4.4	3.6	0.6

^{S} The effect of sex is significant (P < 0.05). ^{F} The effect of fasting time is significant (p < 0.05). $^{S×F}$ There is a sex × fasting time interaction (P < 0.20). $^{D×F}$ There is a finishing diet × fasting time interaction (P < 0.20).

In DLY-crosses, interactions were scarce and therefore the main effects of sex, finishing diet and fasting time are presented in Table 2. Longer fasting decreased the glycogen content (P < 0.05) and tended to decrease the GP of longissimus muscle (P = 0.06). However, lactate content was not affected by fasting time (P > 0.05). Sex and finishing diet did not affect the GP and glycogen and lactate contents (P > 0.05). The DLY-crosses that were fed high-fibre diet had higher ultimate pH and darker meat colour (P < 0.05) than those fed the control diet. Longer fasting tended to increase the ultimate pH (P = 0.09), but did not affect meat colour (P > 0.05). There was a moderate sex × finishing diet interaction (P = 0.13) in the drip loss. The control diet resulted in similar drip losses in both gilts and barrows (3.8 vs. 3.7%; P > 0.05), whereas barrows that were fed high-fibre diet had lower drip loss than the respective gilts (3.1 vs. 4.4%; P < 0.05).

Table 2: Effects of sex, finishing diet and fasting on the glycogen metabolism and meat quality in DLY-crosses.

	Sex			Finishing diet			Fasting time		
	Gilts	Barrows	SEM	Control	High-fibre	SEM	25 h	41 h	SEM
GP, μmol/g	150	148	5	152	146	5	157	142	5
Glycogen, μmol/g	29	29	2	30	28	2	32^{b}	26^{a}	2
Lactate, μmol/g	93	90	2	92	91	2	93	90	2
Ultimate pH	5.55	5.57	0.02	5.53^{a}	5.60^{b}	0.02	5.54	5.58	0.02
L*	50.6	51.2	0.6	51.7^{b}	50.0^{a}	0.06	50.9	50.8	0.06
Drip loss, %	3.7	3.8	0.3	3.7	3.8	0.3	4.1	3.4	0.3

Conclusions

A high-fibre finishing diet and 41h fasting can lower the glycolytic potential of longissimus muscle and have positive effects on pork quality. However, these effects seem to depend on the crossbreed and sex of pigs.

References

Enfält, A.-C., Lundström, K., Hansson, I., Lundeheim, N. and Nyström, P.E. (1997). Effects of outdoor rearing and sire breed (Duroc or Yorkshire) on carcass composition and sensory and technological meat quality. Meat Science, 45: 1-15.

Leheska, J.M., Wolf, D.M. and Maddock, R.J. (2003). Effects of fasting and transportation on pork quality development and extent of post-mortem metabolism. Journal of Animal Science, 81: 3194-3202.

Monin, G. and Sellier, P. (1985). Pork of low technological quality with a normal rate of muscle pH fall in the intermediate post-mortem period: The case of the Hampshire breed. Meat Science, 13: 49-63.

Rosenvold, K., Lærke, H.N., Jensen, S.K., Karlson, A.H., Lundström, K. and Andersen, H.J. (2001a). Strategic finishing feeding as a tool on the control of pork quality. Meat Science, 59: 397-406.

ADDITION OF LIQUID FLAVOUR VITAMIN D_3 IN DRINKING WATER BEFORE PIG TRANSPORT: QUALITY CARCASS MENSURATIONS

E.T.F. Silveira*[1], A.P. Tonietti[1], N.C. Azambuja[1], C.B. Ludtke[2], R.O. Roça[2], and O. Dalla Costa[3]

[1] *Institute of Food Technology, Meat Technology Center, Av. Brasil, 2880, 17070-178, Campinas, SP, Brasil,* [2] *Universidade Estadual Paulista, UNESP, FCA, Botucatu, SP, Brasil,* [3] *EMBRAPA, Centro Nacional de Pesquisa de Suínos e Aves, Concórdia, SC, Brasil. Email: tadeu.s@uol.com.br*

Keywords: pigs, vitamin D_3, carcass quality, meat quality

Introduction

Pre-slaughter operations are recognised worldwide as a key animal welfare issue. The pig meat production chain is investing in new technologies in order to minimise economic losses due to stress, injury and mortality. In this way, levels of blood splashed, skin damage, bone fractures and meat quality traits have been considered to better understand their interactions with handling before slaughter, pig welfare and meat quality. Levels of skin damage are also associated with plasmatic levels of cortisol and creatine phosphokinase, two important blood indices of stress, and the DFD condition. Scores of skin damage seem to be more associated with the conditions of psychological stress and physical efforts in relation to a short period of stress (Warris *et al.*, 1998).

This study was conducted to evaluate the efficiency of liquid flavour vitamin D_3 in the drink water of pigs for seven days before slaughtering and its effects on carcass (bruise, broken bones and skin damage) and the meat (pH, colour, water holding capacity and tenderness) quality.

Materials and Methods

Animals. A total of 522 pigs from the same farm and genetic origin were allocated systematically according to presentation order into three experimental groups (T01, control, n=174, T02, 1.5 ppm, n=174 and T03, 3.0 ppm, n=174). Liquid flavour vitamin D_3 was administered into the drinking water for 7 days before slaughter.

Slaughter. The animals were transported at night for 1 hour and slaughtered in a commercial abattoir. Water was available to the animals during the fasting period in the farm (12h) and lairage time (6h).

Measurements. The skin damage levels were obtained visually in the ham, shoulder and body of the carcasses in the slaughter line using standard pictures with a scale from 0 to 3 (0 = absence, 1 = light, 2 = moderate and 3 = severe) as described by Barton Gade (1993). Rate of pH decline in *Semimembranosus* muscle measured at 1 and 24h after bleeding; colour evaluation in *M.longíssimus dorsi* at 24h postmortem using Minolta colourimeter (CR300); water-holding capacity (WHC) in *Semimembranosus* evaluated at 24h postmortem according to the pressing method described by Hoffmann *et al.*, (1982), drip loss in *M.longíssimus dorsi* using the methodology reported by Honikel, (1998) and Warner Bratzler shear force (WBS) of 1.27x1.27x2.54 cm core samples sheared perpendicular to the fibre direction in three steaks of *M.longíssimus dorsi* were determined with a TAXT2i Texture Analyzer (full scale load 5kg/crosshead speed 200mm/min) attached to a Warner Bratzler accessory. The statistical design was based on randomised block design blocked on pen location. Mean of skin damage, pH, colour, WHC, drip loss, tenderness were analysed using general linear models with terms including the fixed effect of treatment group and the random effect of block, assessed at 5% significance level.

Results and Discussion

Table 1 contains the results of the skin damage for the treatments studied. Larger levels of skin damage in the areas of the shoulder ($p < 0.0001$), body ($p < 0.0332$) and ham ($p < 0.0011$) were obtained for the group control (T0) when compared to the liquid flavour vitamin D_3 (T1=1.5 ppm; T2=3.0 ppm) groups. This represents a good profit to the pig abattoir because these carcasses are normally transported to the export market. Pereira *et al.* (2005) developed similar experimental work with chicken and concluded that the carcasses from the control group were downgraded significantly (51.23%, $p = 0.03$) compared to the group treated with liquid flavour vitamin D_3 (T1=2.0 ppm). Warris *et al.* (1998) reported that elevated levels of skin damage are associated with animal welfare, levels of CPK, lactate, cortisol and incidence of the DFD condition.

Table 1: Skin damage values.

Variable	T0 n=150	T1 n=150	T2 n=150	*P* Values
Shoulder	1.399	0.792	0.745	<0.0001
Body	1.145	0.825	0.815	0.0332
Ham	0.889	0.642	0.602	0.0011

T0. Control; T1. Liquid flavour vitamin D3 (1.5ppm); T2. Liquid flavour vitamin D3 (3.0ppm)

Considering the pH values (1 and 24h *postmortem*), drip loss, WHC and tenderness, there was no significant difference in relation to the treatments studied. The ultimate pH found in this study is in the range (5.49 to 5.78) classified as normal meat. The colour (L* values, 43.55 to 45.10) and drip loss (4.98 to 5.48%) obtained in this experiment can be classified from reddish-pink, soft, exudative (RSE, L* values 52 to 58 and drip loss > 5%) to reddish-pink, firm, non-exudative (RFN, L* values 45 to 52 and drip loss < 5%), as proposed by Silveira (1997). Water holding capacity values (0.33 to 0.37) obtained in this work falls in a range considered RFN (WHC values, 0.36 ± 0.04) meat according to Silveira (1997).

Table 2: Quality characteristics evaluated in *Longissimus dorsi* (LD); *Semimembranosus* (SM) pig muscles according to the treatments studied.

Meat Quality Traits	T0 N=40	T1 n=40	T2 n=40	*P* Values
pH_1 (SM)	6.03	6.10	5.90	0.289
pH_{24} (SM)	5.64	5.65	5.69	0.270
L* (LD)	44.80	43.55	45.10	0.032
a* (LD)	11.44	11.72	11.24	0.255
b* (LD)	-2.14	-2.00	-1.09	0.268
Drip Loss (LD, %)	4.98	5.11	5.48	0.229
Water Holding Capacity (SM)	0.30	0.31	0.30	0.371
Tenderness (LD) kgf	5.86	5.87	5.64	0.326

T0. Control; T1. Liquid flavour vitamin D_3 (1.5 ppm); T2. Liquid flavour vitamin D_3 (3.0ppm)
L*. Luminosity; a*. Redness; b*. Yellowness.

Conclusions

The efficiency of liquid flavour vitamin D_3 has been proven in this study since there was a reduction in carcass damage and no negative influence on meat quality characteristics. It is important to point out that it was observed that the pigs were less agitated during the loading/unloading operations, which contributed positively to this fact. When the production of pig meat in Brazil (3.140.000 tones of meat) is considered, the use of liquid flavour vitamin D_3 will provide more pig carcasses which could be exported. Moreover, this represents economical advantages for the pork industry since both markets, fresh and processed products, are reached simultaneously.

References

Barton-Gade, P. A. (1993) Methods of assessing meat quality, Danish Meat Research Institute, Roskilde.

Honikel, K.O. (1998). Reference methods for the assessment of physical characteristics of meat. Meat Science, 49, 447-457.

Pereira, L.L.N.V; Silveira, E.T.F.; Beraquet, N.J.; Petenate, A.; Andrade, J.C., Buzelli, M.L. (2005).Addition of liquid flavor vitamin D_3 in drinking water before chicken transport and its effects on the stress, carcass and meat quality. In: ICOMST, 51. Proceedings of the 51st International Congress of Meat Science and Technology, Baltimore, USA,

Sakata, R.; Deguchi, T.; Nagata, Y. (1989). Assessment of rapid method for determining water holding capacity of meat. In: ICOMST, 39. Proceedings..., Copenhagen, Denmark,. l 3: 181-183.

Silveira, E.T.F. (1997). Slaughtering techniques and its effects on pork quality. PhD thesis. UNICAMP.

Warris, P.D., Brown, S.N., Barton-Gade, P., Santos, C., Costa, L.N., Lambooij, E. and Geers, R. (1998). An analysis of data relating of pig carcass quality and indices of stress collect in the European Union. Meat Science, 49, 137-144.

EFFECT OF VITAMIN E SUPPLEMENTATION ON EVOLUTION OF MEAT FATTY ACID COMPOSITION DURING STORAGE IN HIGH OXYGEN PACKS

I. Álvarez*[1], V. Cañeque[1], J. De La Fuente[2], S. Álvarez[1], C. Pérez[3], S. Lauzurica[2], M.T. Silva[1], and M.T. Díaz[4]

[1.] *Departamento de Tecnología de los Alimentos, INIA, 28040 Madrid, Spain.* [2.] *Departamento de Producción Animal, Facultad de Veterinaria, UCM, 28040 Madrid, Spain.* [3.] *Departamento de Fisiología Animal, Facultad de Veterinaria, UCM, 28040 Madrid, Spain,* [4.] *Estación Tecnológica de la Carne, Instituto Tecnológico Agrario de Castilla y León, 37770 Guijuelo, Salamanca, Spain. Email: acero@inia.es*

Keywords: lamb meat, fatty acids, vitamin E, modified atmosphere packaging.

Introduction

Muscle foods, by virtue of the fact that they contain unsaturated lipid and prooxidant components, are prone to lipid oxidation, whereby molecular oxygen reacts with unsaturated lipids to form lipid peroxides. Because of this, high oxygen packs (MAP) can promote lipid oxidation. Vitamin E (VE) content of meat, a widely studied antioxidant, has been shown to be a critical determinant of the susceptibility of meat stored in MAP to lipid oxidation. It is now well established that lipid oxidation produces the lost of essential fatty acids reducing the nutritive value of meat. Numerous studies have been devoted to oxidative changes in lamb meat during its storage under refrigeration by an increase in the amount of Thiobarbituric Acid Reactive Substances (TBARS) (Lauzurica *et al.*, 2005). On the contrary, few works have studied fatty acid (FA) changes in lamb meat during refrigerated storage. The aim of this research was to study the evolution of total FA composition of intramuscular lipids of VE supplemented lamb meat during refrigerated storage in MAP.

Materials and Methods

Thirty-six weaned male Manchego breed lambs were randomly assigned to one of the four dietary treatments (9 lambs per treatment) consisting of one non-supplemented (E0) and three VE-supplemented diets: 250 (E250), 500 (E500) and 1000 (E1000) mg/kg feed. The fattening period lasted an average of 37 ± 1.5 days, from an initial live weight of 13.2 ± 0.5 to a slaughter weight of 26.2 ± 0.3kg. The lambs were slaughtered in a commercial abattoir. After a 24h chilling period (4°C) the right m. *longissimus dorsi* (LD) was dissected and cut in 4 slices which were randomly assigned to 0, 14, 21 and 28 days of MAP storage (70% O_2; 30% CO_2). Muscle slices were kept in darkness at 2 ± 1°C during storage. The concentration of VE (α-tocopherol) in LD was determined at initial time according to Cayuela *et al.* (2003). Intramuscular fat was extracted from LD at each time of storage (Hanson and Olley, 1963). FA methyl esters were formed according to Morrison and Smith (1964) and then gas chromatographic analysis of them was performed. Data were statistically analysed using the MIXED procedure of the Statistical Analysis System package (SAS, 1996).

Results and Discussion

Table 1 shows α-tocopherol concentration, intramuscular fat proportion and FA profile of E0, E250, E500 and E1000 in LD. α-Tocopherol deposited in muscle increased as the concentration of VE in feed rise (Lauzurica *et al.*, 2005). Initial intramuscular fat proportion and FA profile were not significantly different among experimental groups. Significant interaction was observed between level of VE supplementation and storage time for saturated FA (SFA), monounsaturated FA (MUFA), polyunsaturated FA (PUFA), PUFAn6 and PUFA/SFA ratio. Non-supplemented lambs had the highest change in FA profile during storage, increasing SFA (10%) and MUFA (7%) and decreasing PUFA (40%), principally PUFAn6, after 28 days of storage respect to initial values. In the other groups, there was also a change in FA profile, but it was moderate. E250 group displayed an increase of MUFA (4%) and a reduction of PUFA (13%) after 28 days of storage respect 0 day. E500 and E1000 only showed a change in MUFA proportion, which after 14 and 21 days of storage increased, and a week later decreased to a similar level to day 0. During lipid oxidation, oxygen attacks the double bond in fatty acids, increasing this sensitivity when increasing unsaturation. Therefore, PUFA were more prone to oxidation compared with MUFA and SFA during storage in MAP. Besides, higher α-tocopherol concentration in meat, which acts as an effective antioxidant by acting as a free radical quencher, implied lower changes in fatty acid composition during storage. From a nutritional point of view, due to a decrease in PUFA, the PUFA/SFA ratio decreased significantly in E0 from 0.43, which came closest to the nutritional recommendation of 0.45 (Department of Health, 1994), to 0.23, whereas VE-supplemented groups did not modify their ratio during refrigerated storage.

Conclusions

Non-supplemented VE lamb meat had the highest change in FA profile during storage, increasing SFA and MUFA and decreasing PUFA after 28 days of MAP storage with respect to initial values. FA profile in VE-supplemented lamb meat was more stable from a level of supplementation of 250 mg of VE / kg in feed. VE supplementation reduced changes in FA profile and the nutritional lipid value of meat was better conserved.

Table 1: α-Tocopherol concentration in LD, intramuscular fat percentage and FA profile of E0, E250, E500 and E1000 during lamb storage in MAP.

	Time	Level of supplementation (VE)													Sig.		
	(T)	E0	SE		E250	SE		E500	SE		E1000	SE		S_{ve}	VE	T	VE*T
αTocopherol (μg/g) [1]	0	0.95	0.09	d	2.17	0.20	c	2.67	0.18	b	3.57	0.15	a	-	***	-	-
IM Fat (%)	0	1.82	0.11		1.89	0.13		1.78	0.11		1.66	0.11		-	ns	-	-
SFA (%)	0	41.67	0.71	y	42.17	0.86		41.43	0.68		40.93	0.68		ns	**	**	**
	14	42.56	0.75	ay	43.30	0.86	a	39.91	0.74	b	39.87	0.70	b	**			
	21	43.53	0.78	axy	43.33	0.81	ab	39.63	0.77	bc	40.66	0.68	c	***			
	28	45.99	0.83	ax	42.44	0.86	ab	41.60	0.70	b	42.03	0.70	b	***			
S_t		***			ns			*			ns						
MUFA (%)	0	40.61	0.84	y	41.35	0.97	y	40.39	0.79	y	42.09	0.79	y	ns	ns	***	**
	14	43.77	0.85	x	43.74	0.97	x	41.87	0.82	x	43.85	0.81	x	ns			
	21	44.62	0.87	x	42.58	0.95	x	42.30	0.83	x	43.51	0.79	x	ns			
	28	43.40	0.89	x	43.15	0.97	x	41.48	0.81	xy	42.49	0.81	xy	ns			
S_t		***			**			**			**						
PUFA (%)	0	17.72	1.05	x	16.50	1.23	x	18.17	0.99		16.98	0.99		ns	**	***	***
	14	13.67	1.08	by	13.02	1.23	by	18.25	1.05	a	16.25	1.02	ab	**			
	21	11.89	1.12	ayz	14.09	1.19	abxy	18.06	1.09	b	15.83	0.99	ab	**			
	28	10.56	1.17	az	14.34	1.23	abxy	16.93	1.02	b	15.46	1.02	b	**			
S_t		***			*			ns			ns						
PUFAn6 (%)	0	14.34	0.89	x	14.14	1.05	x	16.01	0.85		14.32	0.85		ns	**	***	**
	14	11.68	0.92	by	10.94	1.05	by	15.97	0.89	a	13.73	0.87	ab	**			
	21	10.27	0.95	byz	11.93	1.01	bxy	15.89	0.93	a	13.42	0.85	ab	***			
	28	8.94	0.99	bz	12.19	1.05	abxy	14.75	0.87	a	13.16	0.87	a	***			
St		***			**			ns			ns						
PUFAn3 (%)	0	1.41	0.18		1.88	0.21		1.78	0.17		1.48	0.17		ns	ns	ns	ns
	14	1.67	0.19		1.59	0.21		1.99	0.18		1.97	0.17		ns			
	21	1.30	0.19		1.71	0.20		1.89	0.19		1.87	0.17		ns			
	28	1.04	0.21		1.71	0.21		1.83	0.17		1.72	0.17		*			
St		ns			ns			ns			ns						
PUFA/SFA	0	0.43	0.03	x	0.40	0.04		0.44	0.03		0.42	0.03		ns	**	***	**
	14	0.32	0.03	by	0.30	0.04	b	0.46	0.03	a	0.41	0.03	ab	**			
	21	0.27	0.03	byz	0.33	0.03	b	0.46	0.03	a	0.39	0.03	ab	***			
	28	0.23	0.03	bz	0.34	0.04	ab	0.41	0.03	a	0.37	0.03	a	**			
St		***			ns			ns			ns						

SFA: C10:0, C12:0, C14:0, C15:0, C16:0, C17:0, C18:0, C20:0; MUFA: C17:1, C16:1, C18:1; PUFA: C18:2n6, C18:2 *Cis-9 Trans-11*, C18:3n3, C20,3n6, C204n6, C20:5n3,C225n3, C226n3. SE: Standard error; Sig.: [1]Significance of the Model $y_{ijk} = \mu + VE_i + T_j + VE_i \times T_j + \varepsilon_{ij(k)}$; S_{ve}: Significance of dietary supplementation of VE within each storage period; S_t: Significance of storage period within each dietary supplementation VE level; ns: no significant; * $P < 0.05$; ** $P < 0.01$; *** $P < 0.001$; a.b.c Means in the same row with different superscripts differ significantly ($P < 0.05$).; x.y.z Means in the same column with different superscripts differ significantly ($P < 0.05$). [1]Also published in Lauzurica *et al.* (2005)

References

Cayuela, J.M., Garrido, M.D., Bañón, S. J. and Ros, J.S. (2003). Simultaneous HPLC Analysis of α-tocopherol and Cholesterol in fresh pig meat. Journal of Agricultural Food and Chemistry, 51, 1120-1124.

Department of Health. (1994). Report on health and social subjects. No. 46. Nutritional aspects of cardiovascular disease. London: HMSO.

Hanson, S.W.F. and Olley, J. (1963). Application of the Bligh and Dyer method of lipid extraction to tissue homogenates. Biochemistry Journal, 89: 101-102.

Lauzurica, S., De la Fuente, J., Díaz, M.T., Álvarez, I., Pérez, C. and Cañeque, V. (2005). Effect of dietary supplementation of vitamin E on characteristics of lamb meat packed under modified atmosphere. Meat Science, 70, 639-646.

Morrison, W.R. and Smith, L.M. (1964). Preparation of fatty acid methyl esters and dimethyl acetals from lipids with boron fluoride-methanol. Journal of Lipids Research, 5: 600-608.

SAS/STAT, (1996). SAS User´s Guide, Version 8. SAS Institute, Cary, NC.

EFFECT OF NUTRITION ON CARCASS AND PORK QUALITY IN TWO GENOTYPES

G. Mas*[1], C.E. Realini[2], M. Llavall[1], M.A. Oliver[2], M. Gispert[2], N. Panella[2], J. Tibau[2], J. Soler[2], E. Fabrega, R. Roca[3] and D. Coll[1]

[1.] UPB España S.A., Ctra Viver, Km 6,3, Viver i Serrateix, Barcelona 08673, Spain, [2.] IRTA, Institute for Food and Agricultural Research and Technology, Granja Camps i Armet, Monells 17121 Spain, [3.] Grupo Omega de Nutrición Animal, SL. Email: carolina.realini@irta.es

Keywords: pork, quality, carcass, genotype, diet

Introduction

Characterisation of breeds for carcass and meat quality traits is essential to the choice of terminal boar breeds within pork production, as different pork markets or pork chains require different product specifications (Edwards *et al.*, 2003). In addition, modern society is concerned about the importance of meat in the diet, and the consequences of consuming pork fat in human health. High levels of fat consumption and particularly of saturated fatty acids have been associated with increased serum LDL-cholesterol, and increased risk of coronary heart disease in humans (Department of Health, 1994). Meats containing greater contents of the highly unsaturated lipids may be more prone to lipid oxidation than those more saturated (Yang *et al.*, 2002). Vitamin C has been used in different ways to enhance lean colour stability of meat during display. The aim of this study was to evaluate the effect of feeding a diet rich in oleic acid and supplementation with antioxidants (Biocitro®) on carcass and pork quality of two genotypes.

Materials and Methods

One hundred and sixty eight pigs from 2 sexes (barrows and gilts) and 2 genotypes were fed one of three finishing diets: 1) grain and soy diet (CONTROL); 2) grain and soy plus 3.83 % high oleic acid supplement (HO); and 3) grain and soy plus 3.83% HO plus 200 ppm Biocitro® (HO-ANTX). Animals (60 kg live weight) were fed CONTROL (n=53) and HO (n=115) diets until they reached 90 kg live weight. During the finishing period animals were fed the 3 diets (CONTROL n=53, HO n=55, and HO-ANTX n=60) until slaughter. Diets had similar nutrient content and diet composition which is shown in Table 1. Genotypes were crosses of Landrace*Large White with Pietrain (negative for the halothane gene, NN) or Large White and were slaughtered at 115 (fresh pork market) and 130 kg (processed ham market), respectively. The high oleic acid supplement is a natural byproduct of the olive industry composed of a mixture of Ca-salts rich in oleic acid. Biocitro® is a natural byproduct of the citric juice industry and a rich source of vitamin C and flavonoids. Carcass weight and fat and muscle depths measured using the Fat-O-Meat'er probe was recorded within 1 h post-mortem. At 24 h post-mortem ultimate pH and muscle colour (Minolta CR400) were recorded. The left side of each carcass was cut and the loin dissected into lean, subcutaneous and intermuscular fat and bone following the procedure of Walstra and Merkus (1995). Intramuscular fat of the *longissimus thoracis* (LT) was determined by Near Infrared Transmittance (NIT, Infratec 1265). There were no interactions (P>0.05) between sex and diet.

Table 1: Diet composition.

Composition, %	CONTROL	HO	HO-ANTX
Corn	25.0	5.0	5.0
Barley	26.9	47.2	47.2
Wheat	20.0	20.0	20.0
Soy	22.3	20.3	20.3
Fat	2.90	1.39	1.39
High oleic acid supplement	-	3.83	3.83
Biocitro®	-	-	0.02
Lysine	0.24	0.27	0.27

Results and Discussion

Dietary treatments HO and HO-ANTX had no effect (P>0.05) on carcass and pork quality of Pietrain and Large White crosses (Tables 2 and 3, respectively). Gilts showed lighter carcasses and were leaner than barrows showing greater lean percent and lower fat depth in both Pietrain and Large White crosses. Pietrain gilts had greater percent of carcass as ham and lower percent as loin and bacon than barrows. There were no differences in percent of ham and loin between males and gilts for Large White, but gilts had lower bacon percent than males. Lean percent from loin dissections was greater and subcutaneous and intermuscular fat percent lower for gilts than barrows for both Pietrain and Large White crosses. There were no differences in bone percent of the loin between barrows and gilts for Pietrain crosses while Large White gilts had greater bone percent than barrows. There were no differences in marbling percent, Minolta L* or ultimate pH between sexes for Pietrain crosses. L* and ultimate pH did not differ between barrows and gilts for Large White crosses, while marbling percent of LT was lower for Large White gilts compared with barrows.

Table 2: Carcass and pork quality characteristics for Pietrain crosses.

	Sex		Diet			
Carcass Characteristics	Barrows	Gilts	CONTROL	HO	HO-ANTX	RMSE†
Carcass weight, kg	89.2^{a}	83.7^{b}	86.4	86.2	86.7	3.937
Fat depth^{b}, mm	16.7^{a}	13.1^{b}	14.6	14.9	15.1	2.439
Lean %, FOM	55.9^{a}	59.6^{b}	58.0	57.8	57.4	2.442
Ham, % of carcass	24.2^{a}	25.3^{b}	24.8	24.8	24.6	0.740
Loin, % of carcass	17.7^{a}	17.1^{b}	17.5	17.1	17.5	0.838
Bacon, % of carcass	9.3^{a}	8.7^{b}	9.0	9.0	8.9	0.590
Loin Dissection, %						
Lean	55.0^{a}	62.9^{b}	58.7	59.5	58.7	3.794
Subcutaneous fat	28.6^{a}	22.0^{b}	25.6	24.3	25.9	3.608
Intermuscular fat	5.40^{a}	3.94^{b}	4.46	4.68	4.88	0.942
Bone	11.0	11.4	11.1	11.4	11.0	1.064
Pork Quality						
Marbling LT^{c}, %	1.24	0.95	1.07	1.06	1.16	0.323
L*, lightness Minolta	50.3	48.8	49.7	49.3	49.8	2.956
pH ultimate LT	5.53	5.52	5.52	5.54	5.51	0.110

Table 3: Carcass and pork quality characteristics for Large White crosses.

	Sex		Diet			
Carcass Characteristics	Barrows	Gilts	CONTROL	HO	HO-ANTX	RMSE†
Carcass weight, kg	95.1^{a}	92.9^{b}	92.4	94.7	95.0	4.107
Fat depth^{b}, mm	21.5^{a}	17.7^{b}	20.2	19.0	19.6	3.909
LEAN % FOM**	51.3^{a}	54.9^{b}	52.3	53.8	53.2	3.722
Ham, % of carcass	23.5	23.7	23.8	23.5	23.5	0.827
Loin, % of carcass	17.8	17.6	17.9	17.5	17.6	0.910
Bacon, % of carcass	9.7^{a}	9.3^{b}	9.5	9.4	9.6	0.626
Loin Dissection, %						
Lean	49.4^{a}	54.7^{b}	51.4	52.3	52.4	4.748
Subcutaneous fat	34.9^{a}	29.3^{b}	33.0	32.0	31.2	4.823
Intermuscular fat	5.2^{a}	4.6^{b}	5.1	4.78	4.90	0.954
Bone	10.4^{a}	11.3^{b}	10.5	10.9	11.3	1.132
Pork Quality						
Marbling LT***, %	1.44^{a}	1.13^{b}	1.30	1.27	1.28	0.110
L*, lightness Minolta	47.2	48.4	47.3	48.0	48.2	2.632
pH ultimate LT	5.50	5.47	5.51	5.48	5.47	0.315

†RMSE: root mean-square error, **Fat-O-Meat'er probe between 3-4 last ribs at 60 mm off the midline, ***NIT, Means within the same row with different letters (a,b) differ $P<0.05$

Conclusions

Dietary supplementation with 3.83% of high oleic acid supplement and 200 ppm of Biocitro® did not alter carcass and pork quality characteristics in Pietrain and Large White crosses. Feeding diets rich in oleic acid to finishing pigs can achieve greater content of monounsaturated fatty acids (data presented in a separate paper at ICoMST 2006) without modifying carcass traits and offering the consumer a healthier product.

References

Department of Health. (1994). Report on health and social subjects. No. 46. Nutritional aspects of cardiovascular disease. London:HMSO.

Edwards, D.B. and Bates, R.O. (2003).Evaluation of Duroc vs. Pietrain–sired pigs for carcass and meat quality measures. Journal of Animal Science, 81:1895-1899

Yang, A., Lanari, M.C., Brewster, M.J. and Tume, R.K. (2002). Lipid stability and meat colour of beef from pasture and grain-fed cattle with or without vitamin E supplement. Meat Science, 60:41-50.

Walstra, P. and Merkus, G. S. M. (1996). Procedure for assessment of the lean meat percentage as a consequence of the new EU reference dissection method in pig carcass classification. Report ID-DLO 96.014, March, p. 22.

INFLUENCE OF GRAZING SYSTEMS ON VEAL FATTY ACID PROFILE

T. Moreno, E. Bispo*, D. Franco, N. Pérez, and L. Montserrat

Centro de Investigaciones Agrarias de Mabegondo, Apartado 10, 15080 A Coruña, Spain. E-mail address: teresamoreno@ciam.es

Keywords: calves, pasture, CLA, ω-3, *longissimus thoracis*

Introduction

Beef is often criticised for having higher proportions of saturated fatty acids (SFA) than plant food. However, beef usually contains an array of fat molecules that have beneficial effects on human health (Moloney *et al.*, 1999). Grass-based diets are the main sources of C18:3 ω-3 for animal feeding. Consumption of C18:3 ω-3 has been recommended to improve nutritionally the ω-6: ω-3 ratio in humans. Moreover, in recent years, it has been demonstrated that the conjugated linoleic acid (CLA) has important biological effects, including anticarcinogenic activity (Visonneau *et al.*, 1997) and the ability to reduce atherosclerosis (Ip *et al.*, 1994). CLA is found naturally in a variety of ruminant fats, as a result of incomplete biohydrogenation of linoleic and linolenic acid by the rumen bacterium. Diets rich in these fatty acids increase the concentration of the CLA in the fat deposit of the animal. Therefore, grazing-based diets of cattle increase the concentration of CLA and the ω-3 polyunsaturated fatty acids (PUFA) when compared to concentrate-based diets (French *et al.*, 2000). The objective of this study was to investigate the effect of the different beef production systems from the North West of Spain (Galician region) on the fatty acid profile of the intramuscular lipid.

Materials and Methods

Thirty male Rubia Gallega-breed calves were assigned at random to three different finishing treatments (10 calves/treatment): (1) a concentrate and hay *ad libitum* diet was provided to calves housed indoors for 165 days, following a previous grazing period of 57 days (C); (2) Calves were fed on pasture for 197 days complemented by concentrate during the last month of grazing, after which they were finished indoors with concentrate and hay *ad libitum* for an additional 50 days (PC); and (3) Calves were grazing only on a rotational system followed by a pasture-finishing program for a total of 250 days (P). In all treatments, calves kept suckling their mothers up to the slaughter time. During the indoor period, PC and C animals suckled twice a day. Calves under the P and PC treatment were slaughtered four and three weeks later respectively than those under the C treatment, to allow the animals to reach the same slaughter weight (about 330 kg). Carcasses were chilled for 48h at 4°C. Samples of the *Longissimus thoracis* muscle were collected, split into steaks, vacuum packed and stored at –20°C, which were later used for the assessment of nutritional composition (fatty acid content). Intramuscular fat was extracted and methylated. Fatty acids (FA) were analysed by gas chromatography with a 100 m CP-Sil 88 column and the concentration of individual fatty acids in the muscle was determined. Data were statistically analysed using the General Linear Model (GLM) of the SAS package.

Results and Discussion

Fatty acid (FA) data are summarised in Table 1. Concentration of C18:3 ω-3 showed significant differences (P<0.001) among treatments, reaching the highest value in animals which were only grazing for 250 days (P).
However, no differences were found for this ω-3 FA between animals which were finishing on concentrate for 165 and 50 days, C and PC respectively. This reveals that the relative increase in the C18:3 ω-3 FA was due to the effect of finishing animals on a pasture program compared to finishing on concentrate. The same data pattern was found for total ω-3 PUFA content in intramuscular fat.

Table 1: Average concentration (mg/100 g muscle) of main FA in *Longissimus thoracis* muscle.

FA	Grazing Days (Treatments) 250 (P)	197 (PC)	57 (C)	F-test ‡
C18:1 trans 11 (TVA)	14.08 ± 0.85	12.53 ± 1.34	16.35 ± 1.63	+
C18:2 ω-6c (linoleic)	81.62 ± 7.12	80.32 ± 6.57	110.65 ± 15.73	+
C18:3 ω-3 (linolenic)	60.48 ± 4.67	28.33 ± 3.38	26.99 ± 1.93	***
CLA c9t11	8.05 ± 0.36	5.50 ± 0.67	5.30 ± 0.39	**
PUFA	266.57 ± 19.68	185.36 ± 16.69	223.48 ± 24.12	*
MUFA	747.39 ± 41.35	840.97 ± 101.77	1035.14 ± 74.58	+
SFA	1109.04 ± 56.69	1181.61 ± 132.58	1341.19 ± 64.74	NS
ω-6 PUFA	107.93 ± 9.43	96.92 ± 8.24	136.32 ± 18.73	NS
ω-3 PUFA	150.59 ± 10.51	82.94 ± 9.15	81.85 ± 6.19	***
PUFA:SFA Ratio	0.23 ± 0.03	0.17 ± 0.02	0.17 ± 0.02	NS
ω-6: ω-3 Ratio	0.73 ± 0.02	1.24 ± 0.09	1.62 ± 0.12	***

†Values after ± refer to the standard deviation of the mean.
‡ +=<0.1, *=<0.05, **=<0.01, ***=<0.001, NS= Not Significant.

Animals under the P treatment showed significantly (P<0.001) higher concentration in total ω-3 PUFA than animals under C and PC treatments, whereas no differences were found between C and PC. It seems that content of ω-3 FA in veal fat is highly influenced by the type of diet on which animals were fed during the finishing period. Significant differences (P<0.01) were also found between calves under the P treatment and those under the C and PC treatment for the CLA content, whereas no differences were found between C and PC. This may be explained by the same reasons as mentioned above.

Data of associations of grazing levels and pertinent nutritionally parameters were fitted to linear regression straights. These linear regressions represent graphically the data discussed above.

Figure 1
Association of grazing levels and CLA

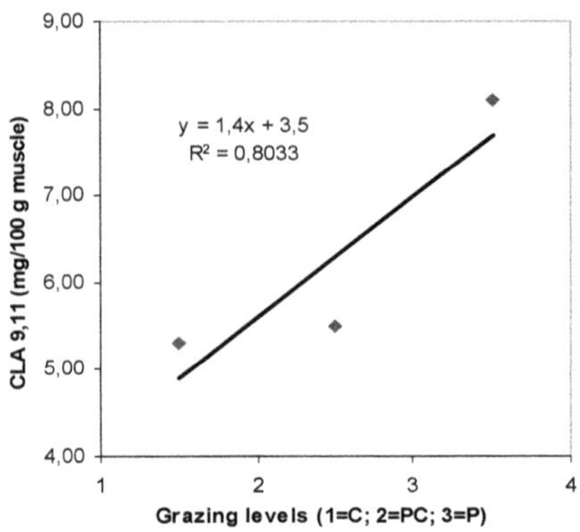

Figure 2
Association of grazing levels and ω-3 PUFA

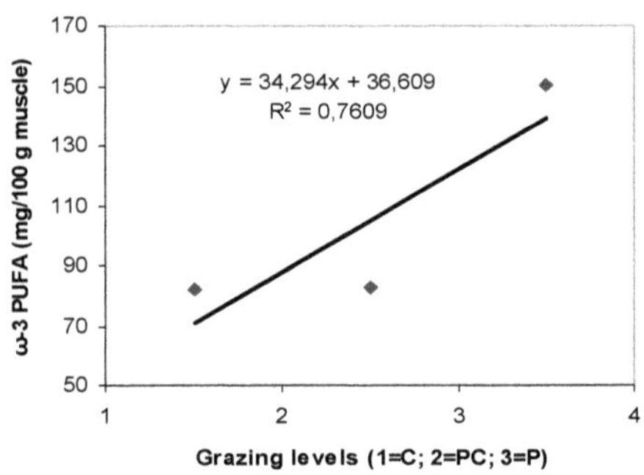

Figure 3
Association of grazing levels and ω-6:ω-3 PUFA

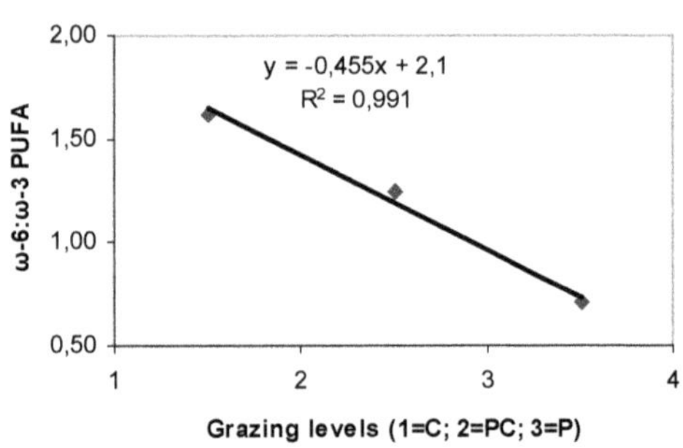

Conclusions

Diets based on pasture finishing improved the content of some healthy fatty acids in veal, such as C18:3 ω-3, ω-3 PUFA, PUFA, CLA and ω-6: ω-3 ratio, compared to concentrate finishing for young calves.

References

French, P., Stanton, C, Lawless, F., O´Riordan, E.G., Monahan, F.J., Caffrey, P.J. and Moloney, A.P. (2000). Fatty acid composition, including conjugated linoleic acid, of intramuscular fat from steers offered grazed grass, grass silage, or concentrate based diets. *Journal of Animal Science* 78: 2849-2855.

Ip, C., Scimeca, J.A. and Thompson, H.J. (1994). Conjugated linoleic acid cancer. Supp.1994; 74(3): 1050-1054.

Moloney, A., French, P. and Stanton, C. (1999). Grass-fed beef: A natural health food,. *The Irish Scientist.*

Visonneau, S., Cesano, A., Tepper, S.A., Scimeca, J.A., Santali, D., and Kritchevsky, D. (1997). Conjugated linoleic acid suppresses the growth of human breast adenocarcinoma cells in SCID mice. Anticancer Research 17, 969-974.

STRAW INTAKE AFFECTS CARCASS QUALITY AND PORK QUALITY

M. Trezona*[1], B.P. Mullan[1], J.R. Pluske[2], D.W. Pethick[2] and D.N. D'Souza[3]

[1.] Department of Agriculture and Food, Locked Bag No. 4 Bentley DC, 6983 Western Australia, [2.] Division of Veterinary and Biomedical Sciences, Murdoch University, Murdoch 6150 Western Australia, [3.] Alltech Biotechnology, 68 Nissan Drive Dandenong South, 3175 Victoria Australia. Email: mtrezona@agric.wa.gov.au

Keywords: pig housing, straw, growth performance, pork quality, carcass quality

Introduction

In Australia, the majority of pigs are raised in either conventional (indoor) or low cost, deep-litter housing systems. Different types of pig housing systems have different effects on pig carcass quality and objective pork quality (Lambooij, 2004). It has been reported that pigs raised in deep-litter systems and conventional systems that have been enriched with straw have higher growth rates, live weights, carcass weights and backfat thickness compared to pigs of the same age raised in barren conventional housing systems (Beattie *et al.*, 2000; Gentry *et al.*, 2002). Comparisons of objective pork quality between pigs raised in conventional systems and deep-litter or straw-enriched systems have indicated there can be differences in colour, drip loss and shear force (Beattie *et al.* 2000, Maw *et al.* 2001, Lambooij *et al.*, 2004).

van Barneveld *et al.*, (2003) demonstrated that when growing pigs are housed (bedded) on rice hulls, up to 10% of the dietary intake is bedding. These authors suggested that pigs grown on other types of bedding, such as cereal straw, could have similar levels of intake. The consumption of bedding material is likely to alter the digestible amino acid: digestible energy ratio of the diet, and alter the pigs amino acid requirements, changes which may not be accounted for in the formulated diet. This in turn could cause variation in growth performance and carcass quality (van Barneveld *et al.*, 2003).

The aim of this experiment was to determine whether the differences in carcass and pork quality found between pigs housed conventionally and on deep-litter could be replicated by including straw in the diet of growing pigs.

Materials and Methods

This experiment was a 2x2 factorial design. There were 2 dietary treatments: control diets (CD) which were commercial cereal-based grower finisher rations; and the control diets + 10% wheat straw (SD). The diets were pelleted to ensure the incorporation of the added straw. The calculated analyses for the grower and finisher diets were: CD grower: digestible energy (DE)=13.5 MJ/kg, available lysine (LYS)=9.3 g/MJ DE, 68 g/kg crude fibre (CFIB), 22.6 g/kg acid detergent fibre (ADF) and 183.0 g/kg neutral detergent fibre (NDF); SD grower: DE=13.3 MJ/kg, LYS=9.3 g/MJ DE, 103.0 g/kg CFIB, 53.1 g/kg ADF, 183.0 g/kg NDF; CD finisher: DE=13.0 MJ/kg, LYS=6.9 g/MJ, 68.9 g/kg CFIB, 200.0 g/kg NDF; SD finisher: DE=12.8 MJ/kg, LYS=6.9 g/MJ DE, 105.0 g/kg CFIB, 32.5 g/kg ADF, 200.0 g/kg NDF. There were 2 flooring treatments in the otherwise identical conventional grower-finisher pens. The first flooring treatment was partially slatted concrete flooring (CF), common to conventional pig production systems, and the second flooring treatment was straw bedding (SF) to mimic deep-litter production systems. The pens were bedded thickly with straw to completely cover the concrete floor, and were cleaned out twice a week after which approximately 14kg of fresh straw/pen was added. Pigs had *ad libitum* access to feed and fresh cool water.

Ninety-six Large White x Landrace female pigs were stratified by weight at 8 weeks of age, into groups of 6. Each group was then randomly allocated across pens and treatments. There were 4 pens, and therefore 24 pigs, per treatment. Individual live weight and feed supplied per pen were recorded weekly. Pigs were slaughtered at a commercial abattoir at 24 weeks of age (~120 kg LW). Twenty-four hours post slaughter a sample of the *longissimus dorsi* from the carcasses of 3 pigs per pen were collected. A 20mm steak was cut from the sample and pHu, colour and drip loss were measured on the freshly exposed surface. Drip loss was measured using the filter paper method and surface lightness (L*), redness (a*) and yellowness (b*) were measured using a Minolta Chromameter CR-400. A 100g block of *l. dorsi* was used to determine cook loss and measure shear force (Warner Bratzler shear blade fitted to an Instron Universal Testing Machine). A further 20g block of *l. dorsi* was used to determine the percentage of intramuscular fat (%IMF) via ether extraction. Data were analysed by using Genstat 8th Edition 2005 (Lawes Agricultural Trust) to conduct two-way analyses of variance.

Results and Discussion

Significant main effects on live weight were found for diet and floor type. Pigs fed the 10% straw diet and/or housed on straw had the highest live weights compared to pigs that had no access to straw ($P<0.05$). There was a similar trend for carcass weight where pigs with access to straw either in the diet or as bedding had heavier carcass weights ($P<0.100$) than pigs without straw. For both measures there was no interaction between diet and floor ($P>0.05$). There was no effect of treatment dressing percentage or P2 backfat depth ($P>0.05$) (Table 1).

Table 1: The effect of diet and floor type on carcass quality and objective pork quality of the *longissimus dorsi* in 24 week old female pigs.

	Treatment				SED	P-value		
	SD -CF	SD-SF	CD-CF	CD-SF	Diet*Floor	Diet	Floor	Diet*Floor
Live weight (kg)	115.1^{bc}	119.0^{c}	110.4^{a}	114.1^{ab}	2.26	0.005	0.023	0.963
Carcass weight (kg)	78.53^{ab}	81.53^{b}	76.41^{a}	78.53^{ab}	2.05	0.085	0.084	0.762
Dressing %	68.17	68.52	69.17	68.78	0.827	0.288	0.970	0.529
P2 (mm)^	13.7	14.8	14.0	14.6	0.996	0.906	0.243	0.724
Shear Force (kg)	6.12	5.23	6.03	5.87	0.607	0.521	0.230	0.396
Colour L*	53.24	53.04	51.53	53.15	1.515	0.457	0.508	0.399
a*	6.07	6.51	5.79	5.61	0.506	0.107	0.719	0.393
b*	4.21	4.66	3.99	4.15	0.401	0.182	0.257	0.660
pH	5.72	5.67	5.73	5.71	0.050	0.445	0.321	0.745
Drip loss (%)	2.03^{a}	2.59^{b}	2.08^{a}	1.88^{a}	0.191	0.017	0.200	0.007
Cook Loss (%)	33.96^{b}	33.3^{ab}	33.78^{ab}	32.73^{a}	0.585	0.369	0.045	0.634
%IMF	1.14	1.62	1.54	1.43	0.262	0.577	0.330	0.131

^{a,b,c} different superscripts within rows indicate significant difference, LSD 5%.

There was a main effect of diet (P=0.017) and an interaction between diet and floor (P=0.007) on percent drip loss. Pigs with access to straw either via the straw diet or straw bedding had a significantly higher percent drip loss compared to pigs that had no access to straw. Johnston *et al.* (2005) found drip loss was higher in pork from pigs housed in deep-litter systems compared to pigs in conventional systems, however other researchers have found percent drip loss and percent cook loss to be higher in pigs housed in a barren conventional environment (Gentry *et al.* 2002, Lambooij *et al.*, 2004). The results for percent cook loss concur with the findings of Gentry *et al.*(2002) and Lambooij *et al.*, (2004) where pork from pigs housed on concrete had a higher percent cook loss compared to pork from pigs housed on straw (P=0.045). The difference in %IMF of the *l. dorsi* between treatments was not significant (P>0.05). However, the difference in average %IMF between pigs fed the straw diet and housed on concrete (1.14%) compared to pigs fed the straw diet and housed on straw (1.62%) may have positively affected eating quality. Fernandez *et al.*, (1999) reported that taste panellists comparing pork from different categories of %IMF identified the texture of pork with 1.5-2.5% IMF as being more favourable than pork with <1.5% IMF.

Conclusions

The results from this experiment indicate that the ingestion of straw affects growth performance, carcass quality and pork quality and may contribute to the differences that are found between pigs housed conventionally (indoors) and on deep-litter. Similarities in the results between pigs that were fed the straw diet and pigs fed the control diet and housed on straw confirm that pigs consumed bedding in sufficient quantities to affect growth, carcass and meat quality parameters.

References

Beattie, V.E., O'Connell, N.E., Moss, B.W. (2000). Influence of environmental enrichment on the behaviour, performance and meat quality of domestic pigs. Livestock Production Science, 65:71-79.

Fernandez, X., Monin, G., Talmant, A., Mourot, J., Lebret, B. (1999). Influence on intramuscular fat content on the quality of pig meat. 2. Consumer acceptability of *m. longissimus lumborum*. Meat Science, 53:67-72.

Gentry, J.G., McGlone, J.J., Blanton, Jr. J.R and Miller, M. F. (2002). Alternative housing systems for pigs: Influences on growth, composition, and pork quality. Journal of Animal Science, 80:1781-1790.

Johnston, L., Morrison, R., Rudstron, M. (2005). Welfare, Meat quality and performance of pigs housed in large-group, deep-litter systems. 7th International Livestock Environment Symposium, Beijing, China. American Society of Agricultural Engineers.

Lambooij, E., Hulsegge, B., Klont, R.E., Winkelman-Goedhart, H.A., Reimert, H.G.M. and Kranen, R.W. (2004). Effects of housing conditions of slaughter pigs on some post mortem muscle metabolites and pork quality characteristics. Meat Science, 66:855-862.

Maw, S.J., Fowler, V.R., Hamilton, M., Petchey, A.M. (2001). Effect of husbandry and housing of pigs on the organoleptic properties of bacon. Livestock Production Science, 68: 119-130.

van Barneveld, R.J., Dove, H., Cadogan, D.J., Ru, Y.J., Edwards, A.C., Choct, M. (2003). Diet composition of growing pigs housed in a deep-litter (rice hulls) system. In: Manipulating Pig Production IX. p.122, ed. J.E. Paterson (Australasian Pig Science Association: Werribee).

VITAMIN D_3 SUPPLEMENTATION FEEDING IMPROVES TENDERNESS OF BEEF BUT NOT OF PORK

R. Lahucky*[1], I. Bahelka[1], U. Kűchenmeister[2], K. Vasickova[1], J. Mojto[1] and K. Ender[2]

[1.] *Slovak Agriculture Research Centre, 94992 Nitra, Slovak Republik,* [2]. *Research Institute for the Biology of Farm Animals, 18196 Dummerstorf, Germany. Email: lahucky@scpv.sk*

Keywords: beef, pigs, vitamin D_3, vitamin E, tenderness, longissimus dorsi

Introduction

Improving and stabilizing meat quality parameters is an area attracting increasing attention from all segments of the meat chain. Results from a number of studies showed that feeding of supra/nutritional levels of some nutrients may improve the antioxidative capacity and meat quality parameters. These nutrients include magnesium, vitamin E and others (Buckley *et al.*, 1995; Swigert *et al.*, 2004). A major factor in determining consumer satisfaction with meat is tenderness. The association of calcium with meat tenderness is well defined and increasing muscle calcium increases the activity of calpains, which are intracellular proteases responsible for postmortem meat tenderness (Koohmaraie, 1992). Montgomery *et al.* (2000) have reported that feeding very high concentrations of vitamin D_3 to cattle results in increased levels of plasma and muscle calcium. But vitamin D_3 supplementation did not affect quality characteristics (Swigert *et al.*, 2004) or tenderness of pork but changes in L* and a* values were positively influenced (Wiegand *et al.*, 2002). Results of the further study (Wilborn *et al.*, 2004) indicated that L* values can be reduced, and subjective colour and firmness scores increased. The objective of this research was to characterize the effects of dietary supplementation with vitamin D_3 individually, and in combination with vitamin E on meat quality characteristics of beef and pork.

Materials and Methods

Thirty bulls (*Spotted breed*) were fed standard feed (control group, group C, n=10) and experimental groups (each n=10) were supplemented with vitamin D_3 (7.5 x 10^6/animal/day, for 7 days, D group) or with vitamin D_3 (7.5 x 10^6/animal/day, for 7 days) and vitamin E (2.500 α – tocopherylacetate/animal/day, 6 weeks, DE group) before slaughter. In total 36 pigs *(Slovak White Meaty)* were genotyped *(RYR1*, DNA based test*)*. The control group (group C, n=12) received a standard diet. The experimental groups (each n=12) received a supplemental level of vitamin D_3 (500,000 IU/kg of feed intake for 5 days, D group) and a supplemental level of vitamin D_3 (500,000 IU/kg of feed intake for 5 days) and vitamin E (ROVIMIX® E-50 SD, 500 mg α-tocopherol/kg of feed for 30 days, DE group) before slaughter.

Animals (bulls at average weight of 550 kg, pigs 110 kg) were slaughtered according to standard commercial procedures in an Institute abattoir. Blood was collected and calcium determined (atomic absorption). The pH value (45 min) of the carcass, colour (24 h and 5 days, by Miniscan) and drip loss (24 h) were measured. The *Longissimus dorsi* (LD) was removed for further analyses. Vitamin E (α-tocopherol) level was measured by HPLC. Lipid oxidation and antioxidant capacity (muscle homogenate stimulated by Fe^{2+}/ascorbate) were assessed by the 2-thiobarbituric acid method (TBARS, expressed as malondialdehyde – MDA). Shear force was determined in cooked samples (internal temperature 75°C) with Warner-Bratzler (W-B) manufacturer apparatus. Statistical analyses were calculated as mean values and standard deviations and differences were evaluated by t-test.

Results and Discussion

Vitamin D_3 supplementation exhibited plasma calcium concentration in bulls (2.60 vs 2.78 $mmol.l^{-1}$) and was greater in pigs (2.71 vs 3.90 $mmol.l^{-1}$) as was shown also by Wiegand *et al.* (2002). Feeding supplemental vitamin E increased ($P<0.05$) levelsof α-tocopherol in beef (2.81 vs. 4.18 $mg.kg^{-1}$) and in pork (2.01 vs. 3.21 $mg.kg^{-1}$). The results are in agreement with those published earlier (Buckley *et al.*, 1995). Meat quality characteristics (pH, drip loss) were not significantly ($P>0.05$) influenced. Colour data (a^* value) were higher ($P<0.05$) for LD muscle (5 d) from vitamin D_3 and vitamin E treated pigs (Table 1). Similar results were reported also by Wiegand *et al.* (2002).

Table 1: Meat quality values ($\overline{x} \pm s$) of pigs.

Trait	Control		Vit. D_3		Vit. D_3 + E	
pH_1	6.24	0,33	6.33	0,16	6.32	0,28
Drip loss, %	3.42	1,25	3.20	1,06	3,04	1,00
Colour L^*	48.85	3,12	46.72	2,32	46.59	4,15
24 h a^*	2.19	2,09	2.44	1,21	2.51	0,44
Colour L^*	49.20	2,05	49.16	2,15	48.30	4,47
5d a^*	2.73^a	0,76	3.74^b	1,15	3.63^b	0,81
W-B, kg	4.86	1,04	4.77	0,83	5.00	0,72

$^{a, b}$ ($P<0.05$)

Results from measures of tenderness (W-B, kg) indicated that feeding of vitamin D_3 to bulls improves shear force value ($P<0.05$) for 14 d (5.98 vs.4.22 and 4.11 kg) but not in pigs for 5 d (Table 1). The results obtained in this study are similar to those reported by other workers (Montgomery *et al.,* 2000, Wiegand *et al.,* 2002). Antoxidant capacity measured as MDA (nM/mg protein) after muscle homogenate incubation (120 min) with Fe^{2+}/ascorbate was improved ($P<0.05$) by vitamin E and partially by vitamin D_3 in beef (0.09 and 0.27 vs. 0.57) and in pork (0.11 and 0.28 vs. 0.43) when compared to control animals.

Conclusions

Vitamin D_3 supplementation increases blood calcium and improves tenderness of beef but not in pork. Colour (a^* value) can be improved in pork. Supplementation vitamin E increases α-tocopherol in beef and pork. Antioxidant capacity was improved in beef and pork by vitamin E and partially by vitamin D_3 supplementation.

References

Buckley, D. J., Morrissey, P. A., Gray, J. I. (1995). Influence of dietary vitamin E on the oxidative stability and quality of pig meat. Journal of Animal Science, 73: 3122-3131.

Koohmaraie, M. (1992). The role of Ca^{2+} - dependent proteases (calpains) in postmortem proteolysis and meat tenderness. Biochemie (Paris), 74: 239-245.

Montgomery, J. L, Parish, F. C. Jr., Beitz D, C., Horst, R.L., Huff-Lonergan, E.J., Trenkle, A.H. (2000). The use of vitamin D_3 to improve beef tenderness. Journal of Animal Science, 78: 2615-2621.

Swigert, K.S., McKeith, F.K., Carr, T.C., Brewer, M.S., Culbertson, M. (2004). Effects of dietary vitamin D_3, vitamin E, and magnesium supplementation on pork quality. Meat Science, 67: 81-86.

Wiegan, B. R, - Sparks J. C., Beitz D. C., Parish, F.C.,Jr., Horst, R.L., Trenkle, A.H., Ewan, R.C. (2002). Short-term feeding of vitamin D_3 improves color but does not change tenderness of pork-loin chops. Journal Animal Science, 80: 2116-2112.

Wilborn, B.S., Kerth, C.R., Owsley, W.F., Jones, W.R., Frobish, L.T. (2004). Improving pork quality by feeding supranutritional concentrations of vitamin D_3. Journal of Animal Science, 82: 218-224.

BETA-ADRENERGIC FEEDING DIFFERENTIALLY ALTERS FIBRE TYPE-SPECIFIC GENE EXPRESSION IN PORCINE MUSCLE

A. Gunawan, B.T. Richert, A.P. Schinckel, A.L. Grant and D.E. Gerrard*

Department of Animal Sciences, Purdue University, West Lafayette, Indiana 47907 USA
Email:dgerrard@purdue.edu

Keywords: fibre type, myosin heavy chain isoform, qRT-PCR

Introduction

The primary goal of animal agriculture is to produce the greatest quantity of high quality protein possible for the least amount of input. Much progress has been made over the past few decades, mainly as a result of improved nutrition and genetics, and from the use of growth promotants in meat-animals. However, as incremental increases in muscle protein deposition become more difficult to achieve through traditional means, we must delve further into the biological mechanisms controlling muscle growth. One aspect of muscle biology that has been largely overlooked by most in the area of enhancing muscle growth is by understanding how muscle-specific contractile genes differentially respond to hypertrophy signals. Although understandable, many scientists have erroneously assumed that when muscle experiences hypertrophy all contractile proteins are up-regulated. However, muscle consists of a heterogeneous population of muscle cells (fibres), which collectively dictates the overall biochemical and contractile nature of muscle. Depending on the physiological status of the animal, (i.e., growing, maintaining, senescing, etc.) muscle has the ability to modify its functional characteristics to accommodate these 'states'. Molecular mechanisms controlling this phenomenon are unknown, but are particularly intriguing to those interested in animal agriculture because the abundance of each type of fibre in muscle are associated with changes in animal performance and ultimate meat quality development. One commercially used swine growth promotant, Paylean®, dramatically increases the amount of type IIB myosin heavy chain (MyHC) in skeletal muscle at the expense of slower-specific MyHC isoforms (Depreux *et al.*, 2002). The active ingredient in Paylean®, ractopamine, is a beta-adrenergic agonist that repartitions nutrients in growing pigs toward muscle and increases protein accretion (Beermann, *et al.*, 1987). Ultimately, the feeding of such a growth promotant results in animals that grow faster and more efficient, and yield carcasses with more muscle and less fat. The exact mechanisms underlying this improved efficiency are not well understood but may be a result of differentially changing muscle fibre type-specific gene expression. Therefore, the objective of this experiment was to determine the relative expression of muscle fibre type-specific genes after feeding ractopamine.

Materials and Methods

Ractopamine (RAC, Paylean®, Elanco Animal Health, IN) was administered at 20 mg/kg daily in feed for either 0, 1, 2 or 4 wk to 44 pigs (90 kg) obtained from Purdue University Animal Sciences Research and Education Center. Control pigs were slaughtered after a 4 wk period of growth, whereas pigs that were fed RAC for 1 or 2 wk followed a 3 or 2 wk period of growth, respectively. This was an attempt to achieve the predetermined duration of feeding, and target an average slaughter weight of 115 kg. Pigs were processed according to normal industry procedures and LD muscle samples were taken immediately post-exsanguination. Samples were frozen in liquid nitrogen and stored at -80°C. Total RNA was extracted and quantified using the Ribogreen quantification kit (Molecular Probes, Eugene, OR). RNA was reverse transcribed and subjected to real time polymerase chain reaction (qPCR). Primer sequences for myosin heavy chain isoforms I, IIA, IIX, and IIB, GS, CS, and β-actin have been described previously (de Costa, *et al.*, 2002). Additional primer sequences for PPARα, and β-ARs 1 and 2 were used as well. The first cycle in the log linear region of amplification where a significant increase in fluorescence was detected above background was designated the threshold cycle (Ct). Therefore, a low Ct indicates high gene expression. Data were analyzed using the GLM procedure of SAS with Ct as the independent variable. Data were reported as least square means and an analysis of variance was used to determine differences among those means.

Results

Real time PCR was used to determine the effect of RAC on gene expression in porcine skeletal muscle. The control to test for experimental variation, β-actin, did not differ over the course of the experiment. Expression of the type I MyHC gene was not altered by RAC during the entire length of the study (data not shown). Type IIA MyHC gene expression was decreased ($P < 0.0001$) 7-fold by 1 wk of RAC administration, remained 4-fold lower ($P < 0.001$) at 2 wk, but was not different from controls by 4 wk (Figure 1). RAC did not influence MyHC type IIX gene expression by 1 wk, but decreased ($P < 0.001$) expression by 2 and 4 wk (Figure 1). MyHC type IIB expression was increased ($P < 0.0001$) 2-fold by 1 wk, and remained elevated throughout the remainder of the study (Figure 1). Citrate synthase (CS) gene expression was not different from controls at 1 wk, but decreased ($P < 0.01$) by 2 and 4 wk (data not shown). Glycogen synthase (GS) mRNA abundance was greater ($P < 0.05$) by 1 wk of RAC administration, returned to the level of control at wk 2, and decreased ($P < 0.01$) by 4 wk (data not shown). Lactate dehydrogenase (LDH) gene expression was not affected by RAC. Expression of the PPARα gene decreased ($P < 0.05$) by 1 wk of RAC administration, but was not different from controls by 2 and 4 wk. RAC had no effect on β_1-AR gene expression (data not shown), however, β_2-AR gene expression decreased ($P < 0.05$) by 2 and 4 wk (data not shown).

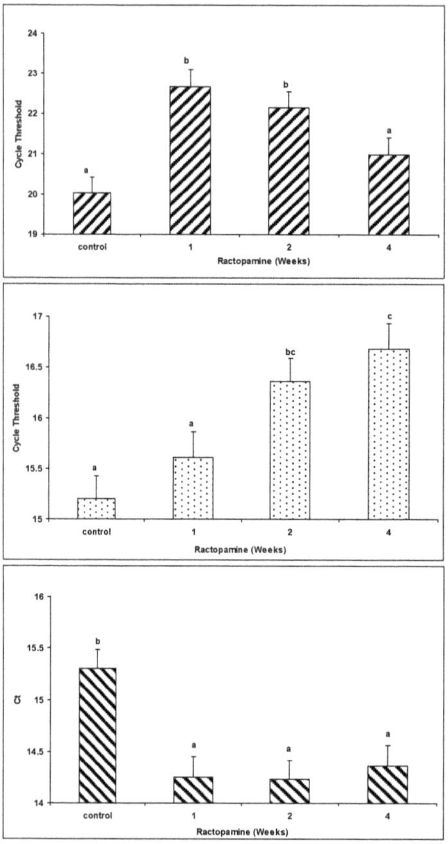

Discussion

Although MyHC genes are homologous, expression of each isoform was affected differently by RAC in our study. Furthermore, changes in gene expression reported in our study mirrors RAC-induced changes in fibre type frequency as well as MyHC protein abundance (Aalhus *et al.*, 1992; Depreux *et al.*, 2002). Therefore, RAC-induced muscle hypertrophy is likely a result of transcriptional regulation of MyHC genes. Effects of RAC administration over a 4 wk period are shown in Figure 1 and, although the effect on type IIA MyHC gene expression returns to the level of controls at wk 4, we found no attenuation of MyHC type IIX and IIB gene expression, suggesting our study concluded before RAC-induced signal transduction was attenuated. Together, results presented herein provide evidence indicating RAC differentially affects MyHC gene expression, resulting in an acute transition between MyHC isoforms (1 wk), and suggest a chronic adaptation of energy metabolism (4 wk) is likely after prolonged stimulation by a β-agonist.

Conclusions

Data presented herein suggest RAC differentially induces expression of type II MyHC genes. However, RAC was administered to pigs during the finishing phase where pigs normally exhibit muscle hypertrophy. Control pigs in the current study experienced muscle growth, therefore, it cannot be concluded that RAC causes muscle hypertrophy. The possibility remains that muscle fibre type-specific gene expression was altered because muscle was growing more rapidly. Further research is needed to determine if RAC alters gene expression in porcine skeletal muscle in the absence of hypertrophy to more clearly define a cause-and-effect relationship between altered gene expression and muscle hypertrophy.

Figure 1: Differential expression of type IIA (top), IIX (middle) and IIB (bottom) myosin heavy chain isoform in pig longissimus muscle. Pigs were fed 20 ppm ractopamine for 0 (control), 1, 2 and 4 wks prior to slaughter. Muscle samples were taken and subjected to real time PCR analysis to determine relative MyHC gene expression. Note: Cycle threshold is inversely related to mRNA abundance, as lower amounts of mRNA require more cycles to cross the detection threshold of the detector.

References

Beermann, D.H., Butler W.R., Hogue D.E., Fishell V.K., Dalrymple R.H., Ricks C.A., C.G. Scanes (1987) Cimaterol-induced muscle hypertrophy and altered endocrine status in lambs. Journal of Animal Science. 65:1514-1524.

Depreux F.F.S., A.L. Grant, D.E. Gerrard (2002) Influence of halothane genotype and body-weight on myosin heavy chain composition in pig muscle as related to meat quality. Livestock Production Science 73:265-273.

Depreux F.F.S., A.L. Grant, D.B. Anderson, D.E. Gerrard (2002) Paylean® alters myosin heavy chain isoform content in pig muscle. Journal of Animal Science 80:1888-1894.

Aalhus, J.L., A.L. Schaefer, A.C. Murray, S. D. M. Jones. (1992). The effect of ractopamine on myofiber distribution and morphology and their relation to meat quality in swine. Meat Science 31: 397-409.

THE EFFECT OF FINISHING DIET ON BEEF QUALITY TRAITS

G. Holló*[1], J. Seregi[1], K. Ender[2], I. Repa[1] I. Holló[1] and K. Nuernberg[2]

[1] *Institute of Diagnostic Imaging and Radiation Oncology, University of Kaposvár, Guba S. 40.Hungary* [2] *Research Institute for Biology of Farm Animals, Dummerstorf, Wilhelm-Stahl-Allee 2, Germany. Email: hollo.gabriella@sic.hu*

Keywords: beef quality, fatty acid composition

Introduction

Beef quality in terms of chemical composition and sensory aspects is affected by many factors, including breed and diet. The ratio of *n*-6 and *n*-3 fatty acids of beef from grain-based-diets is shown higher than the beef from forage fed animals (Enser *et al.*, 1998, cit. Elmore *et al.*, 2004). The reason for this is the high level of linoleic acid (C 18:2 *n*-6) in the grain but the forages are rich in linolenic acid (C 18:3 *n*-3) the base of the long chain *n*-3 fatty acids. The widespread fattening system used in Hungary (the indoor production system), is based on maize silage, hay and high concentrate diets. Beef produced from this feeding regime has undesirable fatty acid composition from a human nutritional point of view (Holló *et al.*, 2001). Therefore different methods should be aimed to modify the fatty acid profile. The objective of this study was to compare the beef quality traits and fatty acid profile of intramuscular fat from Hungarian Simmental (HS) and Holstein (HF) bulls fed two concentrate types in the final finishing period.

Materials and Methods

The experimental design and the nutrient content of different feedstuffs is shown in Table 1 and Table 2. Concentrates for groups III and IV were supplemented with 20% linseed meal containing 44.77 % linolenic acid and fed in the last 67 days of growing-finishing period. The animals were slaughtered at the target live weight of approximately 550 kg. After a 24hrs chilling, *longissimus* (LD) muscle samples were taken at the 12th ribs from the right half carcasses. The right half carcasses were dissected for carcass composition. The fatty acid composition of samples was determined according to method of Nuernberg *et al.*, (2002). SAS package was used for statistical analysis.

Table 1: Experimental design.

Group	Breed	n	Feeding
I.	HF	6	
II.	HS	5	Maize silage, hay, 2-4 kg concentrate
III.	HF	6	Maize silage, hay 2-4 kg concentrate with
IV.	HS	5	linseed

Table 2: Chemical composition of feed.

	Maize silage	Hay	Concentrate	Concentrate with linseed
Dry matter	29.73	91.6	89.36	90.19
Crude protein	6.27	7.1	15.27	14.80
Crude fat	1.87	1.7	3.53	11.98
Crude ash	4.86	5.1	6.93	6.13

Results and Discussion

The fattening and slaughter records are shown in *Table 3*. The HS bulls and the linseed supplemented groups had significant higher ($P<0.05$) daily weight gain. No significant differences between the two diets were found for the slaughter and dressing data. The dressing percentage and lean meat as well as fat content of HS were significantly higher than that of HF bulls. The meat quality traits of the intramuscular fat content of LD for HS and HF bulls were 1.57 and 1.45, respectively. The intramuscular fat of HS contained significantly higher C 20:0, C 18:1 *trans*-9 and C 22:2 *n*-6 compared to HF bulls (Table 4). In the linseed supplemented groups the CLA content in LD was 1.5 times higher, however the meat of HS contained about 0.1% higher CLA percentage. The proportion of linolenic acid increased by linseed supplemented concentrate. The long chain *n*-3 PUFA (C 20:5 C 22:5) are significantly increased by linseed feeding. The amount of *n*-6 fatty acids were not affected. The ratio of *n*-6/*n*-3 fatty acids can be altered from 14:1 and 13:1 with linseed supplementation 5.2:1.

Table 3: Animal performance and carcass traits.

Traits	Without linseed suppl.		With linseed suppl.	
	HF (I)	HS (II)	HF (III)	HS (IV)
Initial age, d	170.83±13.04	191.80±27.52	178.17±10.78	168.40±14.28
Initial weight, kg	205.17±24.77	208.60±40.25	205.25±26.35	207.10±22.53
Final age, d	443.83 ±13.04	437.80±27.52	451.17±10.78	414.40±14.28
Final live weight, kg	528.50± 29.32	556.80±49.69	538.67±56.10	584.20±25.12
Daily gain, g/d	1184±92	1415±183	1221±167	1533±123
Dressing percentage, %	57.13±0.92	58.20±0.60	56.66±1.09	59.64±1.54
Lean meat in right half carcass, kg	92.67±7.20	105.90±6.09	94.19±10.57	114.20±5.85
Bone in right half carcass, kg	30.81±3.95	26.68±2.90	29.25±2.21	26.44±1.37
Fat in right half carcass, kg	8.40±3.12	11.70±2.52	9.58±2.82	11.80±3.98
Tendon in right half carcass, kg	5.15±0.81	4.53±1.32	5.44±0.74	6.42±2.19

Table 4: The fatty acid composition of LD.

Fatty acids, %	Without linseed suppl.				With linseed suppl.				*Sign*
	HF (I)		HS (II)		HF (III)		HS (IV)		
	mean	SE	mean	SE	mean	SE	mean	SE	
Intram. fat	1.48	0.15	1.57	0.17	1.42	0.16	1.57	0.17	
C12:0	0.08	0.01	0.06	0.02	0.09	0.01	0.07	0.02	
C14:0	1.90	0.19	2.36	0.22	2.14	0.20	2.12	0.22	
C16:0	21.66	0.90	23.53	1.06	23.12	0.97	21.85	1.06	
C16:1	2.88	0.31	3.79	0.36	3.22	0.33	3.16	0.36	
C18:0	15.61	0.79	15.07	0.93	15.29	0.85	14.76	0.93	
C18:1*trans*-9	4.20	0.29	5.39	0.3	5.24	0.31	5.79	0.34	B; D
C18:1*cis*-9	28.47	1.95	27.12	2.31	24.98	2.11	27.43	2.31	
C18:2*trans*	0.20	0.02	0.24	0.02	0.39	0.02	0.37	0.02	D
C18:2*n*-6	11.43	1.06	9.33	1.26	10.11	1.15	10.23	1.26	
C18:3*n*-6	0.06	0.01	0.07	0.01	0.04	0.01	0.03	0.01	D
C18:3*n*-3	0.59	0.12	0.54	0.14	1.60	0.13	1.79	0.14	D
C20:0	0.03	0.01	0.12	0.02	0.05	0.01	0.11	0.02	B
C20:3*n*-6	0.71	0.07	0.63	0.08	0.71	0.07	0.49	0.08	
C20:4*n*-6	3.79	0.42	2.92	0.50	3.67	0.46	2.83	0.50	
C20:5*n*-3	0.14	0.02	0.10	0.03	0.26	0.03	0.22	0.03	D
C22:2*n-6*	0.02	0.01	0.06	0.01	0.04	0.01	0.07	0.01	B
C22:5*n*-3	0.33	0.04	0.28	0.05	0.47	0.04	0.35	0.05	D
C22:6*n*-3	0.07	0.02	0.06	0.02	0.09	0.02	0.08	0.02	
c-9,*tr*-11CLA	0.38	0.04	0.52	0.04	0.57	0.04	0.66	0.04	B; D
SFA	40.99	1.41	42.73	1.67	42.35	1.52	40.47	1.67	
UFA	59.01	1.41	57.27	1.67	57.65	1.52	59.53	1.67	
PUFA	18.10	1.68	15.02	1.99	18.55	1.82	17.34	1.99	

B-breed, D-diet

Conclusions

With the linseed supplementation in concentrate at the finishing phase of the maize silage fattening, the *n*-6 and *n*-3 fatty acid ratio can be altered more beneficial whereas the slaughter and the carcass quality traits are not significantly changed.

References

Elmore J.S., Warren, H.E. Mottram, D.S. Scollan, N.D. Enser, M. Richardson, R.I. Wood. J.D. (2004) A comparison of the aroma volatiles and fatty acid compositions of grilled beef muscle from Aberdeen Angus and Holstein-Friesian steers fed diets based on silage or concentrates. Meat Science, 68 27–33.

Holló G., Csapó J., Szűcs E., Tőzsér J., Repa I., Holló I. (2001) Influence of breed, slaughter weight and gender on chemical composition of beef. Part 2. Fatty acid composition of fat in rib samples. Asian-Australasian Journal of Animal Sciences, 14. (11) 1719-1723.

Nuernberg, K., Nuernberg, G., Ender, K. Lorenz, S., Winkler, K., Rickert, R., Steinhart, H. (2002) n-3 fatty acids and conjugated linoleic aid content of longissimus muscle in beef cattle. Eur. J. Lipid Sci. Techn. 104, 463-471.

THE USE OF SODIUM CHLORIDE AND BETAINE TO IMPROVE HYDRATION STATUS OF LAMBS AT SLAUGHTER

K.L. Pearce[1,2]*, D.G. Masters[1], R.H. Jacob[3], D.L. Hopkins[4] and D.W. Pethick[2]

[1.] CSIRO Livestock Industries, Private Bag 5, Wembley, WA 6914, Australia, [2.] Division of Veterinary and Biomedical Science, Murdoch University, Murdoch, WA 6150, Australia. [3.] Department of Agriculture Western Australia, Baron-Hay Court, South Perth, WA 6151, Australia. [4.] NSW Department of Primary Industries, Centre for Sheep Meat Development, PO Box 129, Cowra NSW 2794, Australia., Email: k.pearce@murdoch.edu.au

Keywords: lamb, sodium chloride, betaine, dehydration

Introduction

In Australia lambs are often routinely denied access to water for about 15 h prior to transport from farms to abattoirs. Although water is made available in lairage it is common for sheep not to drink due to the high stress environment (Jacob *et al.*, 2006b). Dehydration can be common at slaughter and corresponds to reductions in the fluid content of meat causing liveweight and carcass weight loss (Cole, 1995; Jacob *et al.*, 2006a). Importantly, Jacob *et al.*, (2006a) showed that dehydration can be averted if sheep drink in lairage. The priority of a dehydrated animal is to actively maintain blood volume which impacts on blood pressure and activates homeostatic mechanisms. Blood volume is maintained by drawing fluid from the muscle interstitial and intracellular fluid stores. The strategic feeding of high levels of sodium chloride and/or betaine in a finishing system prior to low or minimal water intake in lairage may prevent dehydration by increasing blood volume during transport and lairage (Kirkendall *et al.*, 1976). An elevated plasma volume would reduce the need to withdraw fluid from the muscle fluid spaces to maintain homeostatic blood volume and pressure. Less depletion of the muscle fluid stores may result in higher levels of fluid present at slaughter and therefore less muscle fluid loss. The high salt intake on farm will also increase water intake and result in high levels of fluid in the gastrointestinal tract which may act as a water reserve during times of water deprivation. Where sheep do have access to water in lairage, the high salt diet may also stimulate water intake above normal levels whilst in lairage and prevent dehydration. The consumption of betaine in either scenario may also be beneficial. In animals, betaine may have an osmoregulatory function similar to that in plants (Kettunen *et al.*, 2001) and assist in the retention of water in muscle tissue cells. This study evaluated the addition of sodium chloride and or betaine to a diet fed to lambs prior to a period of water deprivation and prior to commercial slaughter where water was made available in lairage.

Materials and Methods

Two separate experiments were conducted. Both experiments were a 2 x 2 factorial design with two levels of sodium chloride added (0 and 5%) and two levels of betaine (0 and 0.6%) in the diets (+salt+betaine, +salt-betaine, -salt+betaine, -salt-betaine). In both experiments, the sheep were allowed *ad lib* access to low salt water during the feeding period. For experiment 1, 96 Poll Dorset-Merino cross wether lambs (mean liveweight of 45.9 ± 0.17kg) were used. The sheep were split into 4 groups of 24 so that the experiment had 4 time replicates (n = 4 x 24 = 96). The sheep were fed 1.5kg/head d. of pellets for 7 days (6 per treatment) after which the 24 sheep were randomly allocated to one of 2 group pens. All sheep remained in these rooms for 48 hours without food or water, after which they were commercially slaughtered. Water intake was measured over the last 3 days of the 7 day feeding period. The lambs were weighed immediately prior to the 48 h period of water deprivation and prior to slaughter. For experiment 2, 204 Merino wether lambs (mean liveweight of 43.1 ± 0.35 kg) were used. The lambs were allocated to 1 of 12 group pens in groups of 17 with each pen being randomly allocated to 1 of the 4 experimental diets. The feed was allocated at 1.5kg/head.d and fed for 7 days. At 10am on the final day (day 8), all sheep were held in one pen with no feed or water prior to transport to the abattoir. Upon arrival at the abattoir the sheep were held in covered lairage yards with *ad lib* access to water available in 1 trough per pen. The lambs were slaughtered at 10am on day 10 after a total period of 48 h off feed and with a 24 h period immediately prior to slaughter with free access to water. The lambs were weighed immediately prior to transportation to the abattoir. For each animal in both experiments the hot carcass weight and GR fat depth were recorded and a 2 g sample taken of the *M. semimembranosis* (SM) and *M. semitendinosis* (ST) muscles within 1 h of slaughter for muscle dry matter content (muscle dry matter = muscle dry weight / wet weight * 100).

Results and Discussion

Under two different commercial scenarios there was no benefit of feeding sodium chloride and or betaine for 7 days prior to farm curfew, lairage and slaughter on liveweight or carcass loss or muscle dry matter. However in both experiments the consumption of the high salt diet did result in a higher (but not significant) fluid content in plasma and gut at the start of the 48 hour water deprivation/transport and lairage period [see (Pearce *et al.*, 2006)]. When no water was made available prior to slaughter (experiment 1) the high salt group excreted all the excess fluid in the plasma to achieve normal blood volume and pressure. In addition, the high salt group did not begin conserving fluid until later than the low salt group. A necessary trigger for fluid conservation is a decrease in blood volume below a homeostatic threshold. As a result of these processes, both groups conserved similar fluid levels in the muscle over the 48 h period.

Had the stimulus to conserve fluid been triggered earlier then it is possible that the fluid content of the muscles at slaughter may have been higher and consequently treatment differences in carcass weight and dressing percentage may have occurred. When water was made available from 24-48 h whilst in lairage (experiment 2), the high salt group drank proportionately more fluid than the low salt group. However at 24 h the high salt group had excreted all the excess fluid so that the high and low salt groups were at a similar hydration status and any water intake in lairage had a similar outcome in both treatment groups and is likely to have maintained blood volume. Betaine did not show any benefit for preventing liveweight and carcass loss. This may be because the levels of betaine in the diets were not high enough or because the osmotic role of betaine is specifically for non-ruminants.

Table 1: The effect of treatment on liveweight, carcass and muscle parameters (Mean ± SEM).

Parameter	Treatment				Significance		
	-salt -betaine	+salt -betaine	-salt +betaine	+salt +betaine	Salt	Betaine	Interaction
Experiment 1							
Water intake (Litres/lamb.day)	5.4±0.33	8.3±0.69	5.7±0.36	9.1±0.92	<0.001	0.39	0.71
Liveweight prior to 48 h water deprivation period (kg)	46.1±0.27	45.5±0.38	45.9±0.38	45.8±0.38	0.27	0.88	0.56
Liveweight loss after 48 hour water deprivation period (kg)	4.6±0.42	4.9±0.18	4.5±0.38	4.9±0.22	0.31	0.14	0.52
Hot carcass weight (kg)	20.9±0.18	20.5±0.20	20.9±0.22	21.1±0.16	0.88	0.61	0.86
GR fat depth (mm)	12.1±0.41	12.2±0.60	12.7±0.67	13.2±0.49	0.57	0.04	0.02
Dressing percentage (%)	46.0±0.4	45.5±0.5	45.6±0.6	45.2±0.5	0.66	0.86	0.91
SM muscle dry matter (%)	27.8±0.2	27.9±0.2	27.9±0.2	28.1±0.2	0.42	0.47	0.83
ST muscle dry matter (%)	27.6±0.31	28.3±0.31	27.8±0.31	27.6±0.31	0.56	0.56	0.21
Experiment 2							
Liveweight prior to leaving farm (kg)	43.5±0.77	43.9±0.77	43.8±0.77	44.2±0.77	0.48	0.94	0.55
Hot carcass weight (kg)	16.7±0.33	16.7±0.33	16.7±0.33	16.7±0.33	0.95	0.97	0.88
Dressing percentage (%)	38.4±0.65	38±0.65	37.8±0.65	37.8±0.65	0.56	0.63	0.22
GR Depth (mm)	5.9±0.38	7.2±0.38	6.2±0.38	4.8±0.38	0.95	0.05	0.13
SM muscle dry matter (%)	25.0±0.3	24.7±0.3	24.8±0.3	25.2±0.3	0.78	0.13	0.53
ST muscle dry matter (%)	24.3±0.3	24.2±0.3	24.6±0.3	24.3±0.3	0.35	0.74	0.41

Conclusions

There was no benefit of feeding sodium chloride and or betaine for 7 days prior to farm curfew, lairage and slaughter on liveweight or carcass loss with and without access to water in lairage. In both experiments, the high salt groups excreted all the excess fluid and were at a similar physiological and hydration status to the low salt groups at slaughter.

References

Cole, N.A. (1995). Influence of three-day feed and water deprivation period on gut fill, tissue weights, and tissue composition in mature wethers. *Journal of Animal Science* 73, 2548-2557.

Jacob, R.H., Pethick, D.W., Clark, P., D'Souza, D., Hopkins, D.L. and White J (2006a). Quantifying the hydration status of lambs in relation to carcass characteristics. *Australian Journal of Experimental Agriculture* 46, 429-437.

Jacob, R.H., Pethick, D.W., Ponnampalam, E. Speijers J, Hopkins DL (2006b). The hydration status of lambs after lairage at 2 Australian abattoirs. *Australian Journal of Experimental Agriculture* 46, (in press).

Kettunen H, Peuranen S, Tiihonen K (2001). Betaine aids in the osmoregulation of duodenal epithelium of broiler chicks, and affects the movement of water across the small intestinal epithelium in vitro. *Comparative Biochemistry and Physiology Part A* 129, 595-603.

Kirkendall, W.M., Connor, W.E., Abboud, F., Rastogi, S.P., Anderson, T.A., Fry, M. (1976). The effects of dietary sodium on blood pressure, body fluids, electrolytes, renal function and serum lipids in normosensitive man. *Journal of Clinical Medicine* 87, 418-434.

Pearce, K.L., Jacob, R.H., Masters, D.G., Hopkins, D.L., Pethick, D.W. (2006). The use of sodium chloride and betaine to improve the hydration status of lambs at slaughter. *Livestock Production Science* (Submitted).

EFFECT OF THE FINISHING DIETS ON BEEF NUTRITIONAL COMPOSITION OF SIX COMMERCIAL MUSCLES

T. Moreno, D. Franco, E. Bispo, N. Pérez and L. Monserrat*

Centro de Investigaciones Agrarias de Mabegondo, Xunta de Galicia, Apartado 10, 15080 A Coruña, Spain. Email: lorenzo.monserrat.bermejo@xunta.es

Keywords: PUFA/SFA, n-6/n-3, CLA, commercial muscles, correlation

Introduction

Nowadays beef consumers and human nutritionists are very much concerned about total fat consumption and its nutritional composition rates, such as PUFA/SFA and n-6/n-3 ratios, due to their incidence on obesity and cardiovascular problems. Conjugated linoleic acid (CLA) is also a subject of interest because of its anticarcinogenic action (Parodi, 1994). It has been verified that grazing diets on animals have beneficial effects on PUFA/SFA and n-6/n-3 ratios, increase in CLA content and decrease in SFA content of beef (Enser *et al.*, 1998; French *et al.*, 2000). Beef production systems in the region of Galicia, North West of Spain, have been based on herds of the late maturating breed Rubia Gallega (RG), where calves are slaughtered very small (7-9 months of age), and animals are usually reared and/or finished with concentrate feed, which increase fat deposition. The Protected Geographical Indication (PGI) "Ternera Gallega" commercialises mainly the "Ternero" type, slaughtered before 10 months of age, and the "Añojo" type, slaughtered between 10-18 months, which represents only 3% of total commercialised calves. However, it should be interesting to encourage beef production based on grazing systems with or without a concentrate finishing for economical and technical reasons. In addition, analyses of the intramuscular fatty acid composition are usually studied on the *Longissimus dorsi* muscle, although it is known that there are differences between muscles (Enser *et al.*, 1998). Since there is a paucity of information in this regard, the objective of this study was to document the effect of grazing finishing diets on the fatty acid composition of main commercial muscles in the "Añojo" type.

Materials and Methods

Twenty male calves of RG breed were randomly assigned to either grass (GF) or concentrate finishing (CNF). Calves were weaned at 8-9 months and grazed on a rotational system until 11 months of age. At this time, one group of ten calves continued grazing for 65 days in a meadow sown with *Lolium perenne* and white and red clover, whereas the other group of ten calves, were fed 65 days on a concentrate *ad libitum*, composed of a mixture of barley (81%), soybean meal (14%) and mineral/vitamin mix (5%) and hay. Post-slaughter, weight conformation (EUROP) and fat score (1 to 5) of the carcasses were recorded. Two-days post–mortem samples of the *longissimus thoracis* (LT), *Semimembranosus* (SM), *Semitendinosus* (ST), *Gluteus medius* (GM), *Cuadriceps femoris* (CF) and *Gluteus Biceps* (BF) were collected and used for fatty acid analysis by gas chromatography. Data were statiscally analysed using a correlation procedure and an analysis of variance of the SAS package (SAS, 1999).

Results and Discussion

Weight, conformation and fat score of the carcasses averaged respectively 251 kg, R-O, and 2.5 for GF and 300 kg, U-R, and 2.8 for CNF. The correlation coefficients between LT muscles and the other muscles (Table 1) were high for n-3 content, moderate for n-6 content and low or very low for the other parameters studied.

Table 1: Correlation coefficients (signification level) between *L. thoracis* (LT) and *gluteus biceps* (GB), *Cuadriceps femoris* (CF), *gluteus medius* (GM), *semitendinosus* (ST) *semimembranosus* (SM) muscles.

		Muscle				
Fatty acid	Muscle	GB	CF	GM	ST	SM
SFA	LT	0.46 (*)	0.41 (+)	0.09 (n.s)	0.03 (n.s)	0.21 (n.s)
MUFA	LT	0.37 (+)	0.19 (n.s)	-0.26 (n.s)	0.06 (n.s)	0.01 (n.s)
PUFA	LT	0.05 (n.s)	0.54 (**)	0.28 (n.s)	0.19 (n.s)	0.59 (**)
n-3	LT	0.74 (***)	0.79 (***)	0.78 (***)	0.71 (***)	0.83 (***)
n-6	LT	0.32 (n.s)	0.67 (**)	0.49 (*)	0.62 (**)	0.77 (***)
CLA	LT	0.65 (***)	0.52 (*)	0.45 (*)	0.23 (n.s)	0.27 (n.s)

Muscles of GF had more n-3 content and less n-6/n-3 ratio than CNF calves, as previously reported by Enser *et al.*, (1998) and French *et al.*, (2000), but the differences were smaller in this study than in the other studies. Also as these authors reported, GF calves had a higher value of CLA and PUFA/SFA and lower value of n-6 than CNF calves (Table 2), although the differences were not significant for all muscles. Discrepancy about smaller differences between GF and CNF treatments in this study and the other studies could be due to that intramuscular fat is deposited very late in the RG breed (Brea *et al.*, 1997).

There were significant differences among muscles of the GF calves in all parameter studies (Table 2). The LT muscle had lower n-3, n-6 content and PUFA/SFA ratio and higher n-6/n-3 ratio than the other muscles. The SM and ST muscles had the lowest fat and CLA content, likely because of the difference in phospholipid concentration, which is the greatest in red oxidative muscle fibres (Turkki and Campbell, 1967). In CNF calves the LT muscle had also lower n-3 content and PUFA/SFA ratio than the other muscles, and its fat content is the highest. In the other parameters there were not differences between muscles.

Table 2: Some and total fatty acid (FA) content and pertinent FA ratios of *longissimus thoracis* (LT), *semimembranosus* (SM), *semitendinosus* (ST), *gluteus medius* (GM), *cuadriceps femoris* (CF) and *gluteus biceps* (GB) muscles from calves fed on grazing (GF) and concentrate (CNF) finishing diets.

FA (mg/100g muscle)	Finishing treatment	muscle						sed†
		LT	SM	ST	GM	CF	GB	
n-3	GF	19b	25a	26a	27a	27a	27a	4.64
	CNF	11b	15a	15a	14ab	15a	15a	3.23
	Sig. Level‡	***	***	***	***	***	***	
n-6	GF	17c	21ab	20bc	25a	23a	23a	4.35
	CNF	24	27	28	28	30	27	7.59
	Sig. Level	**	*	**	n.s	+	+	
CLA	GF	0.89ab	0.62b	0.75b	1.09a	0.92ab	0.78ab	0.34
	CNF	0.69	0.59	0.63	0.64	0.75	0.58	0.28
	Sig. Level	*	n.s	n.s	**	n.s	+	
TOTAL	GF	460a	368b	370bb	512a	458a	452a	95.6
	CNF	498a	389b	432ab	433ab	444ab	412ab	105.6
	Sig. Level	n.s	n.s	*	*	n.s	n.s	
FA ratios								
PUFA/SFA	GF	0.15c	0.26a	0.25a b	0.19b c	0.24ab	0.24ab	0.07
	CNF	0.13b	0.22a	0.20a	0.19a	0.22a	0.22a	0.06
	Sig. Level	n.s	+	+	n.s	n.s	n.s	
n-6/n-3	GF	0.90a	0.86ab	0.77b	0.90a	0.86ab	0.87aa	0.10
	CNF	2.24	1.87	1.87	2.03	2.07	1.87	0.49
	Sig. Level	***	***	***	***	***	***	

†Means in the same row bearing different letters differ at $P<0.05$ (Duncan test); sed refers to standard error deviation of mean.

‡ +,*,**,*** refer to significant difference between treatments in some column at 10,5,1 and 0.1% level; n.s. refers to no significant difference.

Conclusions

Composition of the LT muscle relative to the other commercial muscles studied from the carcass could be a good reference for the n-3 PUFA content but only moderately acceptable for SFA, MUFA and PUFA content. In RG breed the low intramuscular fat deposition affects the differences between the fatty acid compositions of calves fed on grass and concentrate finishing diets.

References

Brea, T., García, J., Monserrat, L., Carballo, J. A.and Sánchez, L. (1997). Análisis alométrico de la composición química de los machos de raza Rubia Gallega a edades comerciales. ITEA, vol extra, nº 18: 754-756.

Enser, M., Hallet, K., Hewitt, B., Fursey, G. A. J., Wood, J. D.and Harrington, G. (1998). Fatty acid content and composition of U.K. beef and lamb muscle in relation to production system and implications for human nutrition. Meat Sci. 49: 329-341.

French, P., Stanton, C., Lawless, F., O`Riordan, E. G., Monahan, F. J., Caffrey, P. J. and Moloney, A. P. (2000). Fatty acid composition, including conjugated linoleic acid, of intramuscular fat from steers offered grazed grass, grass silage, or concentrate-based diets. J. Anim. Sci. 78: 2849-2855.

Parodi, P. W. (1994). Conjugated linoleic acid: An anticarcinogenic fatty acid present in milk fat. Aust. J. Dairy Technol. 49: 93-97.

Turkki, P. R.and Campbell, A.M. (1967). Relation of phospholipids to other lipid components in two beef muscles. Journal of Food Science. 32: 151-154.

THERMAL MEAL TREATMENT IMPROVES SOME MEAT QUALITY TRAITS IN CHICKEN

A. Dalle Zotte

Department of Animal Sciences, University of Padova, Agripolis, Viale dell'Università, 16, Legnaro (PD), Italy; Email: antonella.dallezotte@unipd.it

Keywords: chicken, feed, thermal treatment, gender, meat quality

Introduction

Treatment of feed components may modify both physical and chemical characteristics. The pelleting conditions greatly influence enzyme inactivation and pelleting at temperatures over 85°C negatively affect bird performances (Inborr and Bedford, 1994). Pelleting has been shown to increase digestibility and weight gain (Petterson *et al.*, 1991) more than the extrusion process (Vukić Vranješ *et al.*, 1994), but high temperature short time extrusion and expansion processes of whole feeds has been proved to enhance AME of feeds for broilers (Plavnik and Sklan, 1995). Recently, a process of thermal meal treatment has been developed by Bühler with the aim to produce contamination-free premixes. This research was designed to determine whether this thermal meal treatment could enhance the meat quality traits in the broiler.

Materials and Methods

Three hundred and twenty hybrid chicks (Ross 508) were, at 1 day of age, divided into 2 groups. One group were fed a commercial diet (Control), while the other group received the control diet submitted to a further thermal process (Bühler technology, DTMT4000) which hygenize the meal by the addition of steam (up to the mass temperature of 85°C for 180 sec, followed by drying for 90 sec) obtaining the Sanix diet of Natcor s.n.c.. In order to cover the energy and nutrient requirements of birds along their growth, 3 diets were formulated. In each phase, Control and Sanix diets were prepared (Table 1). Chicks were reared indoor in pens up to 3 weeks of age and randomly allocated to cages by 4 weeks (38 kg liveweight/m^2). At 56 days of age 133 chickens, equally representing the 2 diets, were slaughtered and the breast and thigh meats dissected and used for meat quality analysis. Diets were analysed for chemical composition (dry matter, protein, fat, crude fibre, ash), for starch, lysine, macrominerals, xanthophylls and carotenoids, gross energy and peroxydes contents (Table 1). Meat measurements included the pH (pHu) and the L*a*b* colour values (CIElab, 1976) of *Pectoralis major* (PM) and *Iliotibialis lateralis* (ITL) muscles at 24 hours *post mortem*. Water losses and the Warner-Bratzler shear force of PM and thigh were measured after 2 weeks of frozen storage at -20°C. The L*a*b* colour was also determined 24 hours *post mortem* on the skin of the thigh. ANOVA was performed using the proc GLM of the SAS (1990) program, by including the diet D (Control, Sanix), the gender G and the DxG interaction as fixed effects.

Results and Discussion

Thermal heat treatment increased the content of crude fat but reduced the content of starch in all the diets, resulting in a moderate increase of the gross energy content (Table 1). Also the xanthophyll and carotenoid content was slightly increased by the heat treatment.

Table 1: Composition of the experimental diets (g/100g, as-fed basis).

	Starter phase (1-21d)		Grower phase (22-49d)		Finisher phase (50-56d)	
	Control	Sanix	Control	Sanix	Control	Sanix
Dry matter (DM), g/100g	88.35	88.22	88.00	88.38	88.07	88.39
Crude protein	20.85	22.02	20.06	19.65	20.64	20.50
Crude fat	6.27	6.50	7.17	8.67	7.36	8.25
Crude fibre	2.45	3.14	2.76	2.70	1.96	1.97
Crude ash	5.57	5.29	5.47	5.68	6.02	5.87
Starch	36.2	34.0	35.0	33.6	35.3	33.9
Lysine	1.07	1.29	0.97	1.37	0.89	1.01
Xanthophylls and carotenoids	0.66	0.67	0.41	0.55	0.53	0.68
Ca	1.17	1.19	1.43	0.91	0.91	1.52
K	0.80	0.85	0.83	0.80	0.78	0.80
P	0.58	0.57	0.48	0.59	0.50	0.58
Na	0.13	0.11	0.26	0.18	0.18	0.14
Peroxydes, mg active O_2/kg fat	41.68	43.28	26.40	32.00	47.41	36.42
Gross Energy, MJ/kg	16.86	17.03	17.05	17.30	17.08	17.29

The results of the effects of the two experimental diets on the chickens' weight and meat quality are summarized in Table 2. No difference in slaughter weight between diets was observed. Steam heating treatment of the Sanix diet did not affect the water loss or the Warner-Bratzler shear force of the breast meat. On the contrary, the birds fed with the Sanix diet exhibited less cooking losses of the thigh than those fed the Control diet. Colour parameters were strongly affected by the diets. L* was significantly reduced by the Sanix diet both in breast and thigh meats. The greatest effect of the Sanix diet concerned the b* value (yellowness), significantly higher in breast meat and in thigh skin, compared to the Control diet. Sex effect was significant for slaughter weight, for meat water losses and for some colour values of meat and skin.

Table 2: Quality traits of chicken meat measured 24h *post-mortem* (water losses and Warner-Bratzler shear force measured after 2 weeks frozen storage at -20°C).

	Diet (D)		Gender (G)		Significance[1]		
	Control	Sanix	F	M	D	G	DSR
animals for SW, pHu, L*a*b* colour, n	68	65	52	81			
animals for WHC and W-B shear force, n	32	32	32	32			
Slaughter Weight, g	3648	3574	3345	3877	ns	***	221
Breast:							
Cooking losses, %	25.3	25.4	24.8	25.8	ns	*	1.6
Warner-Bratzler shear force, kg/cm^2	4.50	5.12	5.06	5.05	ns	ns	0.30
pHu m. *Pectoralis major* (PM)	5.76	5.78	5.76	5.78	ns	ns	0.12
Lightness (L*) m. PM	56.0	55.1	55.3	55.8	*	ns	2.2
redness (a*) m. PM	-0.45	-0.24	-0.28	-0.41	ns	ns	1.07
yellowness (b*) m. PM	10.80	12.30	12.70	10.50	**	***	2.68
Thigh:							
Thawing losses, %	1.36	1.29	1.50	1.15	ns	**	0.46
Cooking losses, %	20.6	18.6	20.6	18.7	**	*	2.9
total losses, %	21.9	19.9	22.0	19.8	**	**	2.9
Warner-Bratzler shear force, kg/cm^2	3.72	3.87	3.78	3.81	ns	ns	0.37
pHu m. *Iliotibialis lateralis* (ITL)	6.20	6.25	6.22	6.23	ns	ns	0.16
L* ITL	54.7	53.6	53.4	54.9	**	***	2.3
a* ITL	0.55	0.84	0.66	0.72	ns	ns	1.23
b* ITL	11.70	12.40	12.40	11.70	ns	ns	2.86
L* skin	71.9	71.5	72.2	71.2	ns	**	2.0
a* skin	-1.20	-1.01	-0.81	-1.39	ns	*	1.54
b* skin	27.6	30.3	30.9	27.0	***	***	4.1

(1) *: P<0.05; **: P<0.01; ***: P<0.001; No interaction DxG was found.

Research supported by NATCOR s.n.c., Via Fondo Muri, 43 - S. Tomio di Malo (VI), Italy.

Conclusions

Steam heating treatment of the Sanix diet modify the content of some nutrients. From the meat quality point of view, the highest content of xanthophylls and carotenoids in the Sanix diet increased meat and skin yellowness.

References

Inborr, J.and Bedford, M.R. (1994). Stability of feed enzymes to steam pelleting during feed processing. Animal Feed Science and Technology, 46: 179-196.

Petterson, D., Graham, H. and Åman, P. (1991). The nutritive value for broiler chickens of pelleting and enzyme supplementation of a diet containing barley, wheat and rye. Animal Feed Science and Technology, 33: 1-14.

Plavnik, I.and Sklan, D. (1995). Nutritional effects of expansion and short time extrusion on feeds for broilers. Animal Feed Science and Technology, 55: 247-251.

Vukić Vranješ, M., Pfirter, H.P.and Wenk, C. (1994). Influence of processing treatment and type of cereal on the effect of dietary enzymes in broiler diet. Animal Feed Science and Technology, 46: 261-270.

EFFECTS OF COMPENSATORY GROWTH FEEDING STRATEGY ON FATTY ACID COMPOSITION OF PORCINE MUSCLE

S.A. Hogan* and J.P. Kerry

Department of Food & Nutritional Sciences, University College Cork, Cork, Ireland. Email: s.hogan@ucc.ie

Keywords: compensatory growth, pigs, fatty acids, grass

Introduction

Compensatory growth may be defined as a physiological process whereby growth is accelerated after a period of restricted development, usually due to reduced energy intake, such that the animal's weight equals or approaches that of animals whose growth was never reduced (Hornick *et al.*, 2000). Increased growth rates and carcass leanness have been demonstrated in pigs using such production strategies.

Foods containing high n-6:n-3 PUFA ratios have been implicated in the onset of certain cancers and heart disease. It is feasible that the ratio of n-6:n-3 PUFA of pork might be improved by feeding grass-based diets high in α-linolenic acid (as in ruminants). Grass is also low in metabolisable energy (restricted energy feed source) compared to conventional pig concentrate diets. Feeding grass-based diets to pigs could, therefore, produce leaner pork whilst improving fatty acid profiles in accordance with contemporary health guidelines. The purpose of this study was to determine the effects of compensatory growth diets on the fatty acid composition of muscle from Landrace and Duroc pigs.

Materials and Methods

Landrace (L) or Duroc (D) × (Large White × L) female pigs (n=72) were assigned to one of six diets: (1) Control – concentrate weaner and finisher diets to slaughter (C), (2) grassmeal diet (GM – included at 100 and 200 g/kg) of weaner and finisher diets respectively) to slaughter (GM105), (3) GM to 50 kg followed by finisher diet (GM50), (4) GM to 80 kg followed by finisher diet (GM80), (5) as (4) except α-tocopheryl acetate (200 mg/kg feed) and 5% rapeseed oil added to finisher diet (α-TOC), and (6) as (5) except green tea catechins (200 mg/kg feed) as antioxidant (GTC). Grass was included in pelleted form. Pigs were slaughtered at 105 kg and stored (24 h at 3°C) prior to removal of *M. longissimus dorsi* for subsequent analysis.

Muscle lipid extract was fractionated (neutral (NL) and polar lipid (PL) fractions) by the solid phase extraction (SPE) procedure of Lauridsen *et al.*, (1999). Lipid fractions were transesterified using 0.05 M sodium methylate solution. Fatty acid methyl esters (FAMEs) were quantified using a Varian CP-3800 gas chromatograph equipped with FID and Varian CP-Sil 88 column (50 m × 0.25 mm i.d.). Helium was used as carrier gas at 30 psi. Oven temperature was programmed at 150°C for 35 min, increased at 4°C/min to 240°C and held for 2 min. Injector and detector temperatures were 270°C and 300°C respectively. FAMEs were identified by comparison with pure standards and expressed as a percentage of total fatty acids. Statistical analysis was carried out by GLM procedure (ANOVA) of SPSS (Chicago, IL, USA).

Results and Discussion

Dietary treatments significantly affected SFA and MUFA contents of NL and MUFA and PUFA contents of PL (Table 1). Inclusion of rapeseed oil during re-alimentation (α-TOH and GTC diets) increased MUFA levels of both NL and PL fractions and improved (lowered) the n-6:n-3 balance of porcine muscle. The higher oleic acid (C18:1 n-9) levels observed in both lipid fractions resulting from dietary rapeseed oil inclusion is supported by the findings of Rhee *et al.* (1988). Animals fed rapeseed oil, however, also had significantly lower PL PUFA content than other dietary treatments (with the exception of GM50). Levels of arachidonic acid (C20:4 n-6) were highest in pigs fed GM105 diets, an effect supported by Bee *et al.*, (2002), who reported an increase in this fatty acid in pigs fed low energy diets. It would appear that the increased levels of α-linolenic acid (C18:3 n-3) in PL fractions of α-TOH and GTC fed pigs were due to addition of rapeseed oil rather than the effects of grassmeal consumption.

With the exception of the palmitoleic acid (C16:1) content of NL, all MUFA contents of NL and PL fractions from Duroc pigs were higher than those from Landrace animals (Table 1). This effect is probably related to the higher IMF content of the former breed (Suzuki *et al.*, 2003). Arachidonic acid (C20:4 n-6) contents were significantly higher in lipids from Landrace pigs. Pig genotype did not affect either SFA content or n-6:n-3 ratios of NL or PL fractions.

Table 1: Fatty acid compositions (%) of neutral and polar lipids of *M. Longissimus dorsi* muscle from pigs fed experimental diets.

	Neutral lipid									
	Dietary treatment							*Breed*		
	C	GM105	GM50	GM80	α-TOH	GTC	*sig*	L	D	*sig*
C14:0	1.36	1.50	1.38	1.26	1.40	1.43	ns	1.45	1.33	**
C16:0	22.16	21.39	21.80	22.31	21.55	21.75	ns	21.56	22.10	ns
C16:1	3.65^{a}	3.49^{ab}	3.38^{ab}	3.25^{b}	3.46^{ab}	3.38^{ab}	*	3.46	3.41	ns
C18:0	12.39^{a}	11.63^{ab}	11.17^{bc}	11.41^{bc}	10.61^{c}	10.76^{c}	***	11.23	11.43	ns
C18:1 (n-9)	42.93^{a}	43.06^{a}	42.34^{a}	42.95^{a}	44.67^{b}	44.94^{b}	***	43.03	43.93	*
C18:2 (n-6)	7.32	7.30	7.68	7.08	7.38	7.37	ns	7.36	7.36	ns
C18:3 (n-3)	0.75	0.76	0.73	0.67	0.73	0.74	ns	0.73	0.74	ns
C20:4 (n-6)	0.35^{a}	0.46^{c}	0.34^{a}	0.42^{bc}	0.37^{ab}	0.38^{ab}	***	0.42	0.35	***
C20:5 (n-3)	0.10	0.10	0.10	0.10	0.10	0.09	ns	0.10	0.10	ns
C22:6 (n-3)	0.24	0.26	0.29	0.28	0.32	0.26	ns	0.25	0.30	**
Σ SFA[1]	36.65^{a}	35.21^{ab}	35.01^{ab}	35.63^{ab}	34.23^{b}	34.64^{b}	**	34.95	35.51	ns
Σ MUFA[1]	47.36^{a}	47.25^{a}	46.67^{a}	46.96^{a}	49.01^{b}	49.25^{b}	***	47.30	48.19	**
Σ PUFA[1]	9.45	9.60	9.78	9.23	9.60	9.54	ns	9.55	9.52	ns
n-6/n-3[1]	7.42	7.28	7.51	7.49	7.07	7.35	ns	7.57	7.14	ns
	Polar lipid									
	Dietary treatment							*Breed*		
	C	GM105	GM50	GM80	α-TOH	GTC	*sig*	L	D	*sig*
C14:0	0.42	0.41	0.42	0.40	0.43	0.44	ns	0.45	0.40	**
C16:0	19.49	19.67	19.90	19.43	19.49	19.71	ns	19.49	19.74	ns
C16:1	0.76^{a}	0.65^{ab}	0.66^{ab}	0.50^{b}	0.58^{ab}	0.60^{ab}	*	0.58	0.67	*
C18:0	8.68	9.99	8.75	9.20	9.27	9.86	*	9.19	9.40	ns
C18:1 (n-9)	12.08^{a}	11.94^{a}	11.30^{a}	11.83^{a}	19.92^{b}	20.55^{b}	***	14.12	15.09	**
C18:2 (n-6)	28.93^{ab}	27.97^{b}	28.50^{ab}	29.49^{a}	25.44^{c}	25.21^{c}	***	28.10	27.08	***
C18:3 (n-3)	0.74^{ab}	0.75^{ab}	0.75^{ab}	0.70^{a}	0.84^{c}	0.84^{c}	*	0.77	0.77	ns
C20:4 (n-6)	11.82^{a}	13.29^{c}	11.16^{a}	12.10^{b}	11.77^{ab}	11.89^{ab}	***	12.35	11.66	***
C20:5 (n-3)	0.92	0.95	1.04	0.95	1.00	0.90	ns	0.99	0.93	ns
C22:6 (n-3)	1.26	0.97	1.07	1.11	1.16	1.25	ns	1.16	1.11	ns
Σ SFA[1]	31.25	32.99	31.96	32.07	31.98	32.88	ns	31.99	32.38	ns
Σ MUFA[1]	13.24^{a}	12.92^{a}	12.43^{a}	12.85^{a}	20.93^{b}	21.59^{b}	***	15.11	16.21	**
Σ PUFA[1]	45.97^{a}	46.07^{a}	44.47^{b}	46.44^{a}	42.40^{c}	42.46^{bc}	***	45.51	43.76	***
n-6/n-3[1]	14.73^{ab}	16.51^{a}	15.56^{ab}	16.10^{ab}	13.05^{c}	13.14^{c}	**	14.95	14.75	ns

[1] Sum of fatty acids: Not all data shown. Means within the same row with different superscripts significantly different. * = $P < 0.05$, ** = $P < 0.01$, *** = $P < 0.001$, ns = non significant, $P > 0.05$.

Conclusions

Lipid composition of porcine muscle was most affected by inclusion of rapeseed oil to re-alimentation diets. Such treatments enhanced the nutritional quality of pork through augmentation of MUFA content and improvement of the n-6:n-3 balance of PL fractions. Inclusion of grassmeal to pig diets or pig genotype did not affect n-6:n-3 ratios. Overall, diet and breed affected the PL fraction of muscle lipids to an extent greater than NL.

References

Bee, G., Gerbert, S., Messikommer, R. (2002). Effect of low energy supply and fat source on the fatty acid pattern of adipose and lean tissue and lipogenesis in the pig. Journal of Animal Science, 80: 1564–1574.

Hornick, J.L., Van Eenaeme, C., Gerard, O., Dufrasne, I., Istasse, L. (2000). Mechanisms of reduced and compensatory growth. Domestic Animal Endocrinology, 19: 121-132.

Lauridsen, C., Nielsen, J.H. Henckel, P., Sørensen, M.T. (1999). Antioxidative and oxidative status in muscles of pigs fed rapeseed oil, vitamin E, and copper. Journal of Animal Science, 77: 105-115.

Rhee, K.S., Ziprin, Y.A., Ordonez, G., Bohac, C.E. (1988). Fatty acid profiles of the total lipids and lipid oxidation in pork muscles as affected by canola oil in the animal diet and muscle location. Meat Science, 23: 201-210.

Suzuki, K., Shibata, T., Kadowaki, H., Abe, H., Toyoshima, T. (2003). Meat quality comparison of Berkshire, Duroc and crossbred pigs sired by Berkshire and Duroc. Meat Science, 64: 35-42.

EFFECT OF FRESH CHICORY (*CICHORIUM INTYIBUM L.*) INCLUSION ON RABBIT MEAT QUALITY

L. Basso*[1], M.E. Cossu[1], G. Graziotti[2], C. Ríos[2], C. Gauna[3], C. Castellini[4] and A. Dal Bosco[4]

[1] Producción Animal, Facultad Agronomía, Universidad de Buenos Aires, Av. San Martín 4453, (1417), Argentina. [2] Facultad Ciencias Veterinarias, Univ. de Buenos Aires, Av. Chorroarín 280, (1417), Argentina. [3] Fac. Ciencias Veterinarias de La Pampa, La Pampa, Argentina. [4] Dip. Biología Vegetale e Biotecnologie Agroambientali e Zootecniche. Univ. di Perugia, Borgo 20 Giugno, 74. Italia. Email: lbasso@agro.uba.ar

Keywords: rabbit, chicory, meat quality

Introduction

The energy level of the rabbit fattening diet is a direct way to modify the growth performance and meat characteristics. In order to satisfy most rabbit meat consumers, diet fatty acid composition and manipulation has been well investigated (Gondret *et al.*, 1998). Rabbit meat possesses a relatively high content of polyunsaturated fatty acids (Ouhayoun *et al.*, 1987) and low FA-n6/FA-n3 ratio (Bernardini *et al.*, 1999) but forage addition increases FA-n3 (Facchin, 2001). Chicory plant is a concentrated source of fructo-oligosaccharides and inulin. Fresh chicory addition might influence growth performance and consequently meat quality traits of rabbits. The purpose of the study was to assess the effect of feeding fresh chicory (Cichorium intyibum L.) on the chemical, nutritional properties and muscle fibre type of rabbit meat.

Materials and Methods

A trial was conducted at an experimental station belonging to the Faculty of Agronomy, Buenos Aires University. From weaning to 52days of age, all NZW young rabbits were fed a standard feed (DE: 2463kcal/kg, CP:15,5%), from 53d to slaughter a finishing diet was used (DE:2571 kcal/kg, CP:16.01%). Litters of 26 NZW does (13 per group) were used. After weaning a total of 156 rabbits were allocated in collective cages until 53 days of age, then redistributed in bi-cellular cages until 80days of age. Animals were fed either a commercial (Control) or commercial diet plus *Cichorium intyibum L.* (chicory) (100g chicory dry matter/d/rabbit; 9.34%DM, CP: 28.8%DM, DE:2558 kcal/kgDM). Twenty six (13/group) 53d-of age rabbits were slaughtered to determine fatty acid (FA) composition in early age; the rest were slaughtered at 80 days. Samples of hindleg were obtained to determine composition in FA extracted according to Folch *et al.* (1957) and analysed by gas chromatography (Shimadzu GC-14B; capillary column: Ulbon HR-SS-10; 0,32 I.D.x 50 mL) and using Helium as carrier gas. Chemical analysis followed AOAC (1985). Ten minutes *post mortem* a sample of *Longissimus dorsi* from the left side of the carcass was obtained and histochemical treatment (freeezing in isopentane cooled by liquid nitrogen). Six serial cross-sections (10μm thick) from each muscle sample were obtained with a cryostat at 20°C. One was stained with azorubine (reference staining), four were processed according to the myofibrillar ATPase (Guth and Samaha, 1970) and one was stained for succino-dehydrogenase activity (Nachlas *et al.*, 1957). Fibres were classified as I, IIA y IIB (Brooke and Kaiser, 1970). Mean cross-sectional area was determined with a computerised image analysis system (Buche, 1990). For the variables, analysis was carried out using GLM procedure of SAS and mean values were compared using Tukey test (5%).

Results and Discussion

The fibre type distribution (Table 1) was slightly but significantly influenced by diet. Chicory rabbits presented a high level of I type fibres. Only IIA fibre cross-sectional area varied between diets ($P<0.01$), with less area for chicory fed . However, when all fibre type were analysed, the Chicory group presented the thinner fibres of all types (N.S.). It is known, that dietary energy level represents a tool to modify carcass composition, to a lesser degree, this factor influences muscle biology (fibre cross-sectional area); the major energy level determined the increase in anaerobic glycolytic energy metabolism. Chicory rabbits were significantly leaner at slaughter than control, presented less intramuscular fat and more protein content ($P<0.01$; Table 2). At 53 days of age, hindleg fatty acid composition differed slightly between treatments: only for C18:3, and n-6/n-3 ratio with a better value for the chicory group (more linolenic acid content and less n-6/n-3 ratio). At slaughter, fatty acid composition differed significantly. The level of C18:3 increased with age but, although difference between treatments was not significant, chicory group showed higher levels than control group. The high level of chicory saturated fatty acid was due to differences in C16:0 and C18:0 concentration. The lower content of polyunsatured fatty acids was due to the C18:2 level (32.8 and 26.8 %Agtot for control and chicory group respectively) and explains the lower n-6/n-3 ratio for the chicory group, closest to the recommendation of the U.S. Department of Health.

Conclusions

The results of this study show a good response to the inclusion of chicory on carcass quality and meat traits. Carcasses were leaner and meat presented less intramuscular fat and more protein content. The chicory group showed high linolenic acid content and a better n-6/n-3 ratio.

Table 1: Effect of diet on histochemical traits of *Longissimus dorsi* muscle.

	Control	Chicory	Probability	Rsd[1]
Fibre type distribution (% of the total number of fibres)				
I	14,2[b]	17,1[a]	P=0,0220	6,2
IIA	11,8	12,0	P=0,1267	4,5
IIB	74,0	70,9	P=0,1528	21,7
Fibre cross-sectional area (μm^2)				
I	2179	1868	P=0,0875	453
IIA	1868[A]	1435[B]	P=0,0100	403
IIB	4397	4126	P=0,3776	781

(1)Rsd= residual standard deviation a, b= P<0,05; A, B=<0,01

Table 2: Effect of diet on hindleg chemical and fatty acid composition.

	Control	Chicory	Probability	Rsd[1]
Final weight (g)	2232	2268	P=0,3760	105
Total fat (% RC[2])	3,67[A]	2,70[B]	P=0,007	0,88
Humidity (%)	72,5[b]	73,6[a]	P=0,0471	1,5
Protein (%)	19,5[B]	20,9[A]	P=0,0018	1,1
Ether extract (%)	3,20[A]	2,03[B]	P=0,0012	0,85
Fatty acid composition (%Agtot) of 53d-of age rabbits				
C18:3	2,50[b]	3,03[a]	P=0,0109	0,35
SFA[3]	36,6	37,6	P=0,4621	2,58
MUFA[4]	23,2	24,5	P=0,3636	2,64
PUFA[5]	40,2	37,9	P=0,2888	4,03
n-6/n-3	13,8[A]	10,2[B]	P=0,0022	1,82
Fatty acid composition (%Agtot) of 80d-of age rabbits				
C18:3	2,81	3,24	P=0,1817	0,61
SFA[3]	33,7[B]	39,4[A]	P<0,0001	1,79
MUFA[4]	25,4	25,9	P=0,6775	2,10
PUFA[5]	40,9[A]	34,7[B]	P<0,0001	1,68
n-6/n-3	12,5[A]	8,78[B]	P=0,0010	1,77

(1)Rsd= residual standard deviation; (2) RC= chilled carcass – (liver, kidneys, organs of chest and neck); (3) saturated fatty acids; (4) monounsaturated fatty acids; (5) polyunsaturated fatty acids;

References

A.O.A.C. (1985). *Official Methods of Analysis*. Ed.XXX AOAC Publ., Washington, DC. USA.

Bernardini, M., Castellini, C.and Dal Bosco, A. (1997). Livello di ω-3 nella carne di coniglio in relazione al contenuto della dieta. Atti XII Congresso Nazionale ASPA, Pisa 23-26 Giugno, 381-382.

Buche, P. (1990). RACINE: Un système d'analyse multi-images de coupes sériées. Application à la caractérisation de fibres musculaires. Ph.D.thesis, Université RennesI. France.

Brooke, M. and Kaiser, K. (1970). Muscle fiber types:how many and what kind?. Arch. Neurol., 23: 369

Folch, J., Lees, M. and Stanley, G.H.S. (1957). A simple method for the isolation and purification of lipids from animal tissues. J. of Biological Chemistry, 226: 497-509.

Facchin, E. (2001). Resultados preliminares de crianza rural en Brasil. Seminario Internacional para el desarrollo sostenible de cunicultura en América Latina. 6-10 agosto . Santa Rosa, La Pampa. Argentina

Gondret, F., Mourot, J., Lebas, F. and Bonneau, M. (1998). Effects of dietary fatty acids on lipogenesis and lipid traits in muscle, adipose tissue and liver of growing rabbits. Anim. Sci., 66: 483-489

Guth, L. and Samaha, F.J. (1970). Procedure for the histochemical demonstration of actomyosin ATPase. Research Note. Experimental Neurology, 28: 365-367.

Nachlas, M.M., Tsou, K., De Souza, E., Chang, C. and Seligman, A.M. (1957). Cytochemical demonstration of succinic dehydrogenase by the use of a new p-nitrophenyl substitued ditetrazole. J. of Histochemistry and Cytochemistry, 41: 1151-1162.

Ouhayoun, J., Kopp, J., Bonnet, M., Demarne, Y. and Delmas, D. (1987). Influence de la composition des graisses alimentaires sur les propriétés des lipides périrénaux et la qualité de la viande du lapin. Sci. Aliments, 7, 521-534.

EFFECT OF THE REARING SYSTEM ON MEAT QUALITY OF CROSS-BRED F_2 (WILD BOAR X PIG)

S. Dimatteo*, M. Ragni, S. Tarricone, L. Melodia, A. Zarrilli and G. Marsico

Department of Animal Production, University of Bari, 70100 Bari, Italy. Email: s.dima@agr.uniba.it

Keywords: pig hybrid, housing, quality, welfare

Introduction

Several researches have shown that rearing system affects behaviour and productive performances of pigs. Morrison *et al.*, (2003) have shown that pigs housed in large groups on deep-litter have longer feeding events with less disruption from other pigs compared to pigs maintained in conventional housing systems. In fact the absence of welfare, as a consequence of a non optimal rearing system, leads to stressed conditions. If these conditions are long term they may cause negative reactions such as alterations of social and feeding behaviour, growth reduction, reproductive ineffectiveness, but also pathological conditions such as gastric and intestinal damage, hypertension and immune suppression. This aspect is particularly important when we consider wild animals or rustic genotypes. For this kind of animals the space allowance is more consistent. The rearing of wild animals for human feeding supplies high quality and nutrition productions. It is however necessary to consider their welfare which is related to a greater demand of space, to a higher environmental quality and to the possibility of behavioural expression. The hybrid animals can also allow the improvement of some quantitative and qualitative traits of the animal productions. The rearing of the hybrid wild boar x pig, as any other wild or rustic genotype, can supply niche products with high added value (Vicenti *et al.*, 1991; Pugliese *et al.*, 2002). The aim of the current experiment is to examine the effects of different rearing systems (indoor *vs* outdoor) on the meat quality traits of pig hybrids, testing the positive effect of space allowance on animals and, potentially, on welfare and meat quality.

Materials and Methods

Fourteen second generation hybrids (wild boar x (wild boar x pig)), weaned at 50 days of age, have been subdivided into 2 groups, the first reared in indoor pigsty and the second one in an outdoor pen. Both groups have been fed on a complete feed (M.E. 3,008.00 Kcal/Kg). The chemical composition of diet, measured in accordance with A.S.P.A. indications (1980), is: Moisture 10.80; Protein 12.45; Fat 5.36; Ash 8.13; Crude fibre 5.02; N-free extract 22.61; NDF 23.01; ADF 7.56; AIA 0.76. The pH value has been measured at slaughtering (pH_1) and after 24 hours of storage at +4°C (pH_2). Meat colour has been estimated by the Hunter Lab system using a colorimeter (illuminant D 65), which measures the values of Lightness (L), Redness (a) and Yellowness (b). Tenderness has been measured using a Warner Bratzler shear device applied to an Instron 5544 and expressed as the cutting force (kg/cm2). The chemical composition of *Longissimus dorsi* (LD) and *Quadriceps femori* (QF) muscles has been measured in accordance with A.S.P.A. indications (1980), and the fatty acid composition has been determined, as extracted/methylated using a gas chromatography system (Chromopack CP9000) with a capillary 60 m silica-glass column and cyanopropyle film at 100%, by measuring the contents of saturated fatty acids (SFA), unsaturated fatty acids (UFA), mono unsaturated fatty acids (MUFA) and polyunsaturated fatty acids (PUFA). The atherogenicity and thrombogenicity indexes have been calculated accordingly to Ulbricht and Southgate (1991). The PCL/PCE (plasma cholesterol lowering/plasma cholesterol elevating) ratio was also determined (Reiser and Shorland, 1990). The data have been analysed for variance and the significance between the means evaluated using Student's "t" test (SAS, 1996).

Results and Discussion

No significant differences of the colour parameters have been measured on the raw meat, except for the redness index "a" of the QF muscle, significantly ($P < 0.05$) higher for the meat of the animals reared in the outdoor pen (Table 1). The pH value has been significantly higher ($P < 0.01$) for the QF muscle, after refrigeration, of the indoor reared subjects (6.28 vs 6.02). The shear force has not been affected by the rearing system. The physical measurements have shown some differences, although not significant, of the pH and colour. This last parameter is mainly stronger and less bright in the outdoor reared animals, probably as a consequence of the higher concentration of myoglobin caused by the higher rate of movement in the outdoor pen (Mayoral et al., 1999). The chemical composition of meat has been significantly affected by the rearing system, the ash content was higher (1.50% vs 1.80%; $P<0.05$) for the cooked QF muscle of the animals reared outdoor (Table 2). Space allowance has significantly reduced the undetermined percentage of the cooked LD muscle of the indoor reared pigs (0.94% vs 1.71%). Meat produced from the outdoor rearing system has shown specific traits, in agreement with other studies on improved genotypes (Sather et al., 1997; Enfalt et al., 1997). Fatty acid composition of the muscles has not been affected by the rearing system, except the for MUFA content of cooked LD muscle, significantly higher (47.41% vs 43.95%; $P<0.05$) for indoor animals (Table 3).

Conclusions

The substantial equivalence of the data from the two kinds of rearing systems suggests the possibility that the stress produced by the smaller space allowance could bring to the production of lean meat as well as the facility of movement of the animals reared outdoor. Data confirm that meat of outdoor reared animals has some dietary traits, such as the fatty acidic composition, more useful for human feeding (Pugliese *et al.*, 2002).

Table 1: Colour, pH and tenderness parameters.

		Indoor	Outdoor
Colour			
LD:	L	45.07	49.08
	a	8.34	9.88
	b	11.95	14.23
QF:	L	43.99	43.84
	a	7.64 b	9.39 a
	b	8.57	9.24
pH			
pH_1 : LD		7.03	6.80
	QF	6.95	6.75
pH_2 : LD		6.31	6.01
	QF	6.28 A	6.02 B
WBS (Kg/cm^2)			
LD raw		2.90	2.85
QF raw		1.54	1.96
LD cooked		3.23	3.36
QF cooked		3.44	3.31

A, B: $P<0.01$; a, b: $P<0.05$

Table 2: Chemical composition of cooked meat (%).

	QF		LD	
	Indoor	Outdoor	Indoor	Outdoor
Moisture	67.59	66.34	66.18	65.76
Protein	27.49	28.13	28.79	28.46
Fat	2.27	2.30	2.46	2.27
Ash	1.50 b	1.80 a	1.62	1.80
Undetermined	1.16	1.44	0.94 b	1.71 a

Table 3: Fatty acids composition of cooked meat (%).

	QF		LD	
	Indoor	Outdoor	Indoor	Outdoor
MUFA	45.17	44.82	47.41 a	43.95 b
PUFA	8.92	10.22	8.04	9.57
ω6	8.42	9.57	7.67	8.98
ω3	0.50	0.64	0.37	0.60
ω6/ω3	17.96	15.57	26.97	16.83
Atherog. index	0.64	0.62	0.60	0.67
Thrombog. Index	1.49	1.45	1.45	1.53
PCL/PCE	1.07	1.12	1.10	1.05

References

A.S.P.A., (1980). Zoot. Nutr. Anim. 6:19-34

Enfalt, A.C., Lundstrom, K., Hansson, I., Lundeheim, N. and Nystrom, P.E. (1997) – Effect of outdoorrearing and sire breed (Duroc or Yorkshire) on carcass composition and sensory and technological meat quality. Meat Science, 45, 1-15.

Mayoral, A. I., Dorado, M., Guillen, M. T., Robina, A., Vivo, J. M., Vazquez, C. and Ruiz, J. (1999) – Development of meat carcass quality characteristics in Iberian pigs reared outdoors. Meat Science, 52, 315-324.

Morrison, R.S., Hemsworth, P.H., Cronin, G.M. and Campbell, R.G. (2003). The effect of restricting pen space and feeder availability on the behaviour and growth performance of entire male growing pigs in a deep-litter, large group housing system. Applied Animal Behaviour Science 83 (2003) 163–176

Pugliese, C., Badii, M., Bozzi, R., Acciaioli, A., Campodoni, G. and Franci, O. (2002) – Fatty acid composition of raw and cured ham fat of Cinta Senese and Large White x Cinta Senese pigs as affected by rearing system. 48th International Congress Meat Science and Technology, Roma.

Reiser and Shorland, (1990). Meat fats and fatty acids. In A.M. Pearson T.R. Dutson (Ed): Meat and Health. Advances in Meat Research, 6:21-62. Elsevier Science Publishers Ltd. London

Sather, A.P., Jones, S.D.M., Schaefer, A.L., Colyn, J. and Robertson, W.M. (1997) – Feedlot performance, carcass composition and meat qualità of free-range reared pigs. Can. J. Anim. Sci., 77, 225-232.

SAS (1996), SAS users guide. Ed. SAS Inst. Inc. Cary, NC, USA.

Ulbricht, T.L.V. and Southgate, D.A.T. (1991). Coronary heart disease: seven dietary factors. Lancet, 338, 985-992.

Vicenti, A., Marsico, G., Centoducati, P., Moramarco, V., Braghieri, A.and Ciruzzi, B. (1991) - Confronto fra cinghiali, suini ed F1 (cinghiale x suino), castrati e macellati a 200 giorni di età. IV. Composizione chimica della carne ed acidità dei giorni di deposito. Atti I Conv. Europeo "Allevamenti di Selvaggina", 219.

DEHYDRATION DOES NOT CHANGE THE EATING QUALITY OF LAMB MEAT

R.H. Jacob[1]*, K.L. Pearce[2] and D.W. Pethick[2]

[1.] *Department of Agriculture Western Australia, Baron-Hay Court, South Perth, WA 6151, Australia.* [2.] *Division of Veterinary and Biomedical Science, Murdoch University, Murdoch, WA 6150, Australia. Email: rjacob@agric.wa.gov.au*

Keywords: lamb, dehydration, consumer panel, urine specific gravity

Introduction

A recent study suggested that the prevalence of subclinical dehydration in lambs at the time of slaughter in Australian abattoirs could be high (Jacob *et al.*, 2006). This was attributed to lambs failing to drink water in lairage rather than to extreme climatic conditions. Whilst dehydration can reduce carcass weight, meat yield and be an indicator of compromised animal welfare during the pre-slaughter period, little is known about the potential for dehydration to reduce the eating quality of lamb meat. Eating quality dominates the decision by consumers to repurchase meat so is an important part of the strategy to build demand for Australian lamb meat products worldwide (Russell *et al.*, 2005).
A study by Warner *et al.* (2002) suggested that water deprivation may actually improve tenderness but did not examine other sensory factors. Jacob *et al.* (2006) found urine specific gravity (USG) to be a simple indicator of water deprivation suitable for use in a commercial scenario. The upper physiological limit for USG in sheep is 1.045 (Kaneko *et al.*, 1997) above which tissue hydration can be reduced (Jacob *et al.*, 2006). In this study we used USG to identify lambs of different hydration status for the purpose of conducting consumer taste panel evaluations of the eating quality of lamb meat.

Materials and Methods

First cross merino lambs were selected from 9 different vendor consignments, and slaughtered under commercial conditions at an abattoir in Western Australia. The lambs were slaughtered by electrical head stunning followed by exsanguination. Urine was collected from the bladder of all castrated male lambs in each consignment soon after slaughter and USG measured immediately using a Reichert veterinary refractometer (Model 10436, Cambridge Instruments Inc., Buffalo, USA). From each consignment 6 lambs were then selected consisting of 3 lambs with a low USG (USG<1.030) and 3 lambs with a high USG (USG>1.045). This procedure was done over 2 days with 4 consignments being sampled on October 18 2005 and 5 consignments on October 25 2005._The lamb carcasses were electrically stimulated using a post dressing medium voltage stimulation unit with settings of 2.5ms pulse duration, 1A current, and 15Hz frequency.
One backstrap *(m. longisimuss thoracis et lumborum)* and 2 silversides (*m. biceps femoris)* were dissected from each carcass 48 hours post slaughter then aged for 6 days at a temperature of 2°C. Ultimate pH (pHu) was measured in both muscles 48 hours after slaughter. Samples of loin BKS045 (Anonymous 1998) and silverside OUT005 (Anonymous 1998) were then prepared and frozen in readiness for consumer panel evaluation.
Consumer panel sensory evaluation was conducted at a centralised location in Sydney, using untrained consumer panels. Consumers were asked to score steaks for tenderness, liking of flavour, liking of smell, juiciness and overall liking from 0-100 and also to rate the meat as either unsatisfactory, good every day or better than every day. The panels were conducted according to the method outlined in Thompson *et al.* (2005) whereby 1.5cm thick chops were grilled to a medium degree of doneness (71°C internal temperature) following thawing. They were then halved and presented to consumers giving a total of 10 consumer responses for each experimental muscle making a total of 1080 steaks tested this way. Panel members were allocated across treatments such that each consumer tasted 3 samples from high USG lambs and 3 samples from low USG lambs. Samples were allocated to a testing layout using a Latin square design and the 10 portions of each cut were served in a different order to avoid bias due to order of tasting.
Eating quality data were analysed using analysis of variance (ANOVA) for the effects of USG class (low and high) for each cut of meat (loin BKS045 and silverside OUT005). Data was blocked for the fixed effects of consignment within each sample date. Carcass data was also analysed using ANOVA with tissue depth (GR) being adjusted using carcass weight as a covariate. These analyses were performed using Genstat 7.0. (VSN International Ltd, www.vsn-intl.com). Eating quality data was "clipped" prior to analyses to reduce consumer variation. This involved removing the highest and lowest values in each group of 10 scores.

Results and Discussion

A significant difference (P<0.01) occurred between the GR tissue depth in the low and high USG categories (Table 1). This may have been due to a selection bias caused by smaller numbers of low USG lambs being available than high USG lambs rather than an effect of USG category on tissue depth. In all consignments the majority of lambs had a high USG.
There was no effect (P>0.05) of USG category on the eating quality of loin or silverside for overall liking, liking of flavour, juiciness, liking of smell, and tenderness (Table 2). However, the percentage of consumers rating the loin

unsatisfactory was higher for low USG compared to high USG lambs. Although significant, the GR tissue depth difference between groups was small and may not have accounted for this difference in failure rate.

Table 1: Carcass and urine specific gravity data.

Carcass measurement	Low USG	High USG	5% LSD	Significance
GR tissue depth(mm)	13.9	15.6	1.31	**
Carcass weight adjusted for shrinkage (kg)	22.8	23.4	0.83	NS
pHu *m. longisimuss thoracis et lumborum*	5.59	5.62	0.04	NS
pHu *m. biceps femoris*	5.63	5.64	0.03	NS
USG	1.009	1.055	0.0013	**

* P<0.05, ** P < 0.001, NS not significant P>0.05, 5% LSD least significant difference at the 5% level

Table 2: Consumer panel data.

Attribute	Low USG	High USG	5% LSD	Significance
		Loin (BKS045)		
Overall liking	66.7	69.6	4.94	NS
Liking of flavour	65.6	68.5	4.65	NS
Juiciness	62.2	63.9	6.15	NS
Liking of smell	67.7	68.9	3.71	NS
Tenderness	72.5	72.7	5.62	NS
% Unsatisfactory	9.9	1.9	7.08	*
		Silverside (OUT005)		
Overall liking	64.4	66.3	4.83	NS
Liking of flavour	64.0	65.8	4.82	NS
Juiciness	66.1	65.7	4.32	NS
Liking of smell	67.4	71.0	3.61	NS
Tenderness	62.1	64.0	5.58	NS
% Unsatisfactory	6.2	6.8	6.08	NS

* P<0.05, ** P < 0.001, NS not significant P>0.05, 5% LSD least significant difference at the 5% level

Conclusions

Dehydration of lambs within the range found under commercial Australian conditions does not change the eating quality of lamb meat.

References

Anonymous (1998). Handbook of Australian Meat. (Ausmeat: Brisbane).

Jacob, R., Pethick, D., Clark, P., D'Souza, D., Hopkins, D. and White, J. (2006). Quantifying the hydration status of lambs in relation to carcass characteristics. *Australian Journal of Experimental Agriculture*, 46, 1-9.

Jacob, R., Pethick, D. W., Ponnampalam, E., Speijers, J. and Hopkins, D. (2006). The hydration status of lambs after lairage at 2 Australian abattoirs. *Proceedings of the Australian Society of Animal Production*. Accepted for publication.

Kaneko, J., Harvey J. and Bruss, M. (1997). Clinical biochemistry of domestic animals.' (Academic Press: San Diego, California).

Russell, B., McAlister, G., Ross, I. and Pethick, D. (2005). Lamb and sheep meat eating quality- industry and scientific issues and the need for integrated research. *Australian Journal of Experimental Agriculture,* 45, 465-467.

Thompson, J., Gee, A., Hopkins, D., Pethick, D., Baud, S. and O'Halloran, W. (2005). Development of a sensory protocol for testing palatability of sheep meats. *Australian Journal of Experimental Agriculture,* 45, 469-476.

Warner, R., Cotrell, J., D'Souza, D., Kerr, M., Bond, J. (2002). The effect of time off water pre-slaughter on lamb meat quality. In: Proceedings of the 48th International Congress of Meat Science and Technology. Rome. (Ed.s R Chizzolini) pp. 368-69.

INFLUENCE OF GENDER, GENETIC BACKGROUND AND REARING METHOD ON THE RELATIONSHIP BETWEEN PROTEIN EXPRESSION AND MEAT QUALITY

A. Kwasiborski*[1], T. Sayd[1], C. Chambon[1], V. Santé-Lhoutellier[1], D. Rocha[2] and C. Terlouw[1]

[1] *INRA, Unité Qualité des Produits Animaux, Centre de Theix, 63122 St Genès Champanelle, France*
[2] *Genus plc, Genus Cambridge Research Laboratory, Department of Pathology, University of Cambridge CB2 1QP Cambridge, United kingdom. Email: anthony@clermont.inra.fr*

Keywords: pigs, proteome, meat quality, *longissimus dorsi*

Introduction

With a mean consumption of 43.5kg per year and per capita in 2001, pork is the most consumed meat in Europe (Devine, 2003), but its technological and sensory qualities are still insufficiently controlled. Rearing method, breed and gender influence raw material and end-product qualities. Meat quality depends on post-mortem metabolism, involving many proteins: enzymes, transport, stress or chaperone proteins. This study evaluated effects of rearing method, genetic background and gender on protein expression and its relationship with meat quality in pigs.

Materials and Methods

Twelve castrated male and 12 female pigs, sired by Duroc or Large White PIC purebred, were reared either outdoors or indoors in a 2 x 2 x 2 factorial design. Dams were Large White x Landrace. Immediately after slaughter, *Longissimus lumborum* (LL) samples were taken to determine various meat quality parameters including ultimate pH, colour (L*, a*, b*), and drip loss at 3 days and cooking loss. LL proteomic maps were established with 2-dimensional electrophoresis, using an immobilised pH 5-8 gradient for the first dimension and a SDS-PAGE gel for the second dimension. After staining with Coomassie blue, each gel was scanned and intensity of spots was evaluated using PDQuest software. The identifications were made by MALDI-ToF mass spectrometry.

Results and Discussion

ANOVA found that for 122 of a total of 248 spots intensity varied significantly: 18 were influenced by rearing method, 10 by breed, 90 by gender and 32 by various interactions between these factors (Table 1). Eighty spots were identified and could be classified in different categories including glycolysis enzymes, stress and chaperone proteins.

Table 1: Number of spots of which intensity varied due to rearing method, sire breed or gender(left) and due to their interactions (right).

	Factors		
	rearing	sire	gender
Total	18	10	90
Spots affected by one factor	12	6	74

	Interactions			
	rearing x sire	rearing x gender	sire x gender	rearing x sire x gender
Total	8	10	10	4
Spots affected by interactions of factors	4	4	7	2

ANOVA (Table 2) found that ultimate pH was influenced by gender and by the interaction between rearing method and breed. Drip loss was influenced by breed. The other meat qualities were not influenced by the treatment factors.

Table 2: Means and significant treatment effects on meat quality parameters.

	Mean by factors						influencing factors	
	rearing		sire		gender			
Variable	Outdoor	Indoor	Duroc	Large White	Male	Female		
Ultimate pH	5,56	5,53	5,56	5,53	5,63	5,47	Gender p=0.035	Rearing x sire p= 0.050
L*	53,01	51,74	52,07	52,68	51,48	53,27		
a*	8,10	8,50	8,35	8,24	7,96	8,64		
b*	4,84	4,02	4,37	4,49	4,10	4,76		
Drip day 3	4,52	4,17	3,38	5,21	3,59	5,15	sire p=0.020	
Cooking loss	18,48	19,85	18,11	20,23	18,31	20,03		

Multiple forward stepwise regression analysis was used to identify proteins most strongly correlated with meat quality parameters. When means of explanatory or explained variables varied according to treatment group, or when their regression coefficients differed (Figure 1), multiple regressions were carried out separately for each group. Explanatory variables were retained when $p<0.05$. In castrated males, 72% of ultimate pH variability was explained by 2 unidentified proteins. In females, a fragment of the creatine kinase and one other unidentified protein explained 76% of ultimate pH variability (Table 3). Significant multiple regression models were also found for the other meat quality parameters (Table 3).

As these parameters are also influenced by ultimate pH (Renerre, 1990; Offer and Cousins, 1992), the multiple regression models were tested again adding ultimate pH as a possible explanatory variable. Ultimate pH was retained in 3 regressions in addition to the proteins (Table 3). For example, in outdoor pigs, intensity of a myoglobin chemical form explains 57% of L* variability, addition of ultimate pH improves the model with 14% (Table 3). The relationship between L* and myoglobin is well known and related to the proportions of oxymyoglobin MbO_2 and metmyoglobin MetMb, which depend on the activity of oxygen-consuming enzymes and on MetMb reducing activity (Lindhal et al, 2001). L* is low when myofilament spacing is low due to high ultimate pH (Bendall and Swatland, 1988). For indoor pigs, L* is uncorrelated with this myoglobin chemical form, suggesting that the latter is not correlated with relative amounts of MbO_2 and MetMb.

Table 3: Proteins explaining the meat quality variations. Arrows indicate model improvement after inclusion of ultimate pH.

variables	experimental group	retained proteins	Correlation coefficient R	% variability explained
ultimate pH	Male	5021= unidentified	0.73	49
		4305= unidentified	0.73	72
		7220=Creatine kinase	-0.39	81
		8026=unidentified	0.46	90
	Female	7220=Creatine kinase	-0.81	62
		3101= unnidentified	0.74	78
L*	Indoor	None		0 → 56
	Outdoor	6004= Myoglobin	-0.78	57 → 71
a*	Male	7107= Creatine kinase	0.72	46
		5021=unidentified	-0.67	81
	Female	38= Ca2+ binding protein	0.69	42
		3303= Guanosine diphosphate dissociation inhibitor	-0.01	62
		5021= unidentified	-0.49	79
		7107= Creatine kinase	-0.63	87
		4408= Leucine aminopeptidase	-0.58	93
b*	Male	6211= unidentified	0.83	67 → 86
	Female	6117= unidentified	-0.75	51
Drip day 3	Large White sire	6106= Creatine kinase	0.83	62
		1425= unidentified	0.66	94
		8026= unidentified	0.48	99
	Duroc sire	None		0
Cooking loss	Male	2004= Dj1 protein	-0.74	51
		2615= Heat shock protein 70	-0.43	73
	Female	5613= Transferrin	-0.68	41
		7019= Glyceraldehyde-3-phosphate dehydrogenase	0.18	60

Figure 1: Scatter plot and regression lines for cooking loss and transferrin expression.

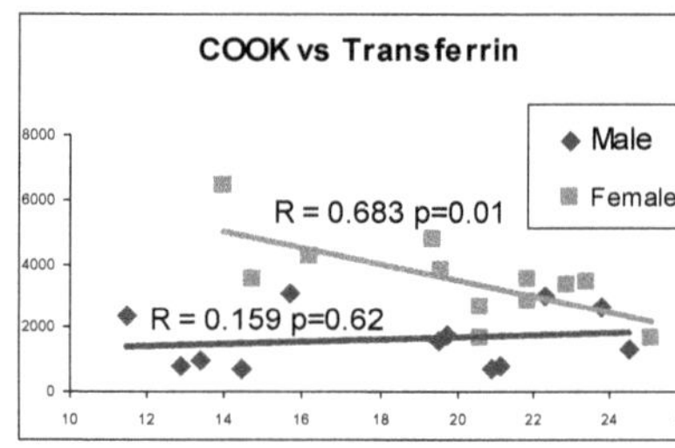

Figure 2: Localisation of the spots implicated in the meat quality variability

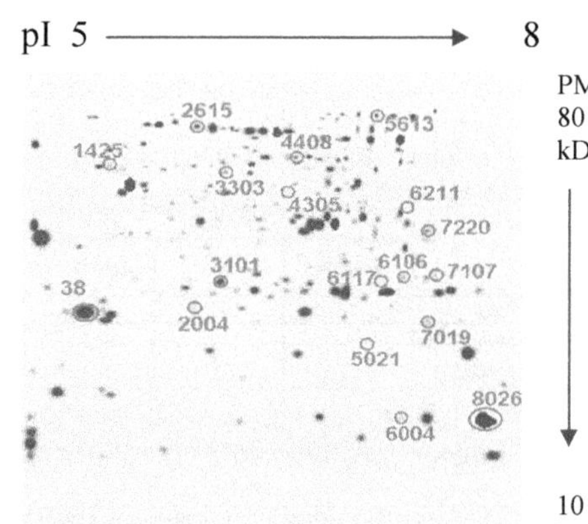

Conclusion

Protein expression may explain up to 99% of the variability in certain meat quality parameters. Relationships between protein expression and meat quality depend on gender, breed and rearing method. This study is a first step towards a better understanding of the role of certain proteins in meat quality.

References

Bendall JR. and Swatland HJ (1988). A review of the relationships of pH with physical aspects of pork quality. *Meat Science*, 24: 85-126.

Devine R. (2003). La consommation des produits carnés. INRA production Animal, 16 (5) : 325-327.

Ledward DA. (1992). Colour of raw and cooked meat. In Johnston DE, Knight MK & Ledward DA. The chemistry of muscle-based foods (pp 128-144). Cambridge: Royal Society of chemistry.

Lindhal G., Ludstrom K. and Tornberg E. (2001). Contribution of pigment content, myoglobin forms and internal reflectance to the colour of pork loin and ham from pure breed pigs. *Meat Science*, 59: 141-151.

THE IMPROVEMENT OF LAMB MEAT SHELF LIFE THROUGH FEEDING WITH SUB-PRODUCTS OF *ROSMARINUS OFFICINALIS, L.*

G. Nieto*, B. Linares, P. Díaz, S. Bañón and M.D. Garrido

Department of Food Technology and Science and Human Nutrition. Veterinary Faculty, University of Murcia, Campus Espinardo, 30071 Murcia, Spain. Email: gnieto@um.es

Keywords: lamb, colour, lipid oxidation, rosemary

Introduction

The purpose of this study is to examine the possibility of producing lamb meat rich in components from the sub-products of rosemary which allow improved health measures, quality and shelf life. Lamb meat is usually commercialised without undergoing any transformation. Refrigerated lamb steaks packed in a protective atmosphere to prolong shelf life predominate the market. This delays microbiological and colour spoilage but increases lipid oxidation deterioration. Recent studies show the antioxidant, anti-microbial, anti-viral and anti-mutagenic effect of some natural antioxidants (rosemary, thyme, sage, etc) incorporated by means of animal feed (Tang *et al.*, 2001; O'Grady *et al.*, 2006). After the extraction of essential oils, the transformation of aromatic medicinal plants (rosemary, thyme, ..) generates a distilled excedant. This subproduct constitutes a potential source of natural antioxidants, most of which are directed at the food industry as protective agents against lipid oxidation (Hagerman *et al.*, 1998).

Materials and Methods

Thirty lambs of the Segureña breed were studied. Lambs were fed with milk and their normal foodstuff with added rosemary distilled leaf (w:w) until they reached 25kg live weight. Three groups were formed; R1 (10% rosemary), R2 (20% rosemary) and C (control). At 24 h *post mortem*, the *Longissimus dorsi* muscle was filleted, packed in a modified atmosphere (70%O_2:30%CO^2), and stored for a maximum of 21 days at 4°C in natural lighting conditions. Meat spoilage was determined by measuring colour (Minolta Chromameter Reflectance II), the TBARs index (mg MDA/kg meat), microbiological activity (total viable, total psycrophiles, mould and yeast; log u.f.c/g), and sensory analysis at 0, 7, 14 and 21 days of storage.

Result and Discussion

TBARs values (Table 1) show delayed lipid oxidation in meat from lambs receiving rosemary dietary supplement as compared to the C group. This antioxidant effect was similar in R1 and R2 as has also been described by McCarthy *et al* (2001). L, a, b values showed that rosemary (R1 and R2) increased colour stability. From day 14 of storage higher a* values ($p<0.05$), implies less pigment oxidation and a better appearance, as has also been described by O'Grady et al (2006). This was contrasted with the sensory analysis (Table 2). Independently of the dosage used, a positive effect of rosemary was found on lean and fat colour and rancid odour. Neither putrid nor acid odour was detected in the samples. The antimicrobial effect of rosemary (Table 3) was moderate. A difference from C group with regard to R1 and R2, was observed only in the total mesophiles .A similar effect has been seen by Govaris *et al*, (2005) in studies on turkey breast.

Table 1: Average values and SD of TBArs (TBA), L*, a* and b* coordinates in lamb at 0, 7, 14 and 21 days under retail conditions.

	Level	Day 0	Day 7	Day 14	Day 21
TBA	C	0.02 ± 0.01	4.32 ± 2.24 [a]	8.43 ± 1.86 [a]	10.29 ± 0.75 [a]
	R1	0.01 ± 0.01	1.18 ± 1.31 [b]	3.58 ± 3.53 [b]	4.85 ± 3.51 [b]
	R2	0.02 ± 0.01	0.88 ± 0.51 [b]	3.82 ± 2.90 [b]	5.23 ± 3.65 [b]
L*	C	41.50 ± 2.89	44.82 ± 2.58	49.84 ± 3.94	53.60 ± 2.93
	R1	40.71 ± 1.92	44.28 ± 2.72	45.77 ± 5.62	48.02 ± 7.51
	R2	42.21 ± 1.42	44.76 ± 1.97	48.34 ± 3.69	48.98 ± 4.81
a*	C	17.35 ± 1.50	16.383 ± 1.52	8.49 ± 4.01 [b]	4.16 ± 2.22 [b]
	R1	16.84 ± 1.11	17.68 ± 1.45	13.93 ± 4.43 [a]	10.71 ± 5.9 [a]
	R2	17.68 ± 0.90	16.61 ± 1.56	12.14 ± 4.45 [a]	9.10 ± 4.73 [a]
b*	C	6.72 ± 1.14	12.39 ± 2.29	13.99 ± 2.26 [a]	15.34 ± 2.07 [a]
	R1	6.64 ± 1.06	11.94 ± 0.78	11.64 ± 1.72 [b]	13.50 ± 3.17 [b]
	R2	7.24 ± 0.47	11.70 ± 0.68	12.55 ± 1.10 [ab]	13.99 ± 2.53 [ab]

C: Control; R1: 10% Rosemary; R2: 20% Rosemary. Means with different superscripts are significantly different ($p<0.05$).

Table 2: Average values and SD of sensory attributes in lamb at 0, 7, 14 and 21 days under retail conditions.

	Level	Day 0	Day 7	Day 14	Day 21
RO	C	1.00 ± 0.00	1.65 ± 1.41	4.05 ± 1.06 [a]	4.70 ± 0.78 [a]
	R1	1.00 ± 0.00	1.37 ± 0.74	1.87 ± 1.15 [b]	1.93 ± 1.32 [b]
	R2	1.00 ± 0.00	1.14 ± 0.37	2.07 ± 1.53 [b]	1.64 ± 0.80 [b]
CM	C	5.85 ± 0.33	4.35 ± 1.27	2.95 ± 1.18	1.05 ± 1.11 [b]
	R1	4.68 ± 0.45	4.75 ± 0.65	3.56 ± 1.01	3.37 ± 1.06 [a]
	R2	6.00 ± 0.00	5.35 ± 0.62	3.71 ± 1.15	3.21 ± 1.46 [a]
FC	C	5.85 ± 0.33	4.55 ± 0.36 [b]	3.90 ± 0.81 [b]	2.35 ± 0.94 [b]
	R1	5.75 ± 0.46	4.93 ± 0.32 [ab]	3.87 ± 0.58 [a]	3.31 ± 0.70 [ab]
	R2	6.00 ± 0.00	5.21 ± 0.56 [a]	3.64 ± 0.62 [ab]	3.21 ± 0.56 [a]

C: Control; 10: 10% Rosemary; 20: 20% Rosemary.
Scoring scale: (1: minimum; 6: maximum). RO: rancid odour; CM: meaty colour; CF: fat colour

Table 3: Average total viable count (TVC), total psycrophile count (TPC) and mould and yeast (MLC) (log ufc/g) in lamb at 0, 7, 14 and 21 days under retail conditions.

	Level	Day 0	Day 7	Day 14	Day 21
	C	4.12 ± 0.51 [a]	4.23 ± 0.55 [a]	4.45 ± 0.59 [a]	5.24 ± 0.61 [a]
TVC	R1	2.42 ± 0.85 [b]	2.09 ± 0.99 [b]	2.76 ± 0.91 [b]	3.74 ± 0.76[b]
	R2	2.28 ± 0.49 [b]	2.20 ± 1.08 [b]	3.38 ± 0.41 [b]	3.39 ± 0.46 [b]
	C	3.06 ± 0.85	3.71 ± 0.72	4.57 ± 0.91	6.48 ± 0.55 [a]
TPC	R1	2.57 ± 0.48	2.84 ± 0.59	3.75 ± 0.76	5.23 ± 0.82 [ab]
	R2	2.10 ± 0.90	3.15 ± 0.90	3.62 ± 0.22	4.55 ± 0.39 [b]
MLC	C	1.60 ± 1.10	2.02 ± 0.72	2.41 ± 1.42	3.94 ± 1.17 [a]
	R1	0.88 ± 0.79	1.12 ± 0.75	1.46 ± 0.91	2.77 ± 0.92 [b]
	R2	1.70 ± 0.56	1.08 ± 0.39	1.11 ± 0.50	1.72 ± 0.79 [b]

C: Control; 10: 10% Rosemary; 20: 20% Rosemary.
Means with different superscripts are significantly different ($p<0.05$).

Conclusions

The feeding of lambs with 10% or 20% rosemary distilled leaf improves the quality and shelf life of raw lamb meat packed in a modified atmosphere (70%O_2:30%CO_2) due to the fact that this delays the oxidation of fat and colour and microbiological spoilage, improving the appearance of the meat. As such this dietary supplementation of lambs, is advantageous.

References

Govaris, A., Florou-Paneri, P., Botsoglou, E., Giannenas, I., Amvrosiadis, I., Botsoglou, N. (2005). The inhibitory potential of feed supplementation with rosemary and/or α-tocopheryl acetate on microbial growth and lipid oxidation of turkey breast during refrigerated storage. Food Science and Technology.

Hagerman, A.E.; Riedl, K.M.; Jones, G.A.; Sovik, K.N.; Richard, N.T; Hartzfeld, P.W.; Riechel, T.L.(1998). High molecular weight plants polyphenolics (tannins) as biological antioxidants. Journal of Agricultural and Food Chemistry. 46: 1887-1892

O'Grady M.N., Maher M., Troy D.J., Moloney A.P., Kerry J. P. (2005). An assessment of dietary supplementation with tea catechins and rosemary extract on the quality of fresh beef. Meat Science, 73: 132-143.

McCarthy T. L.; Kerry J. P.; Kerry J. F.; Lynch P. B.; Buckley D. J. (2001b). Evaluation of the antioxidant potential of natural food/plant extracts as compared with synthetic antioxidants and vitamin E in raw and cooked pork patties. Meat Science, 57: 45-52.

Tang, S., Kerry, J.P., Sheehan, D., Buckley, D.J., Morrissey, P.A. (2001). Antioxidative effect of dietary tea catechins on lipid oxidation of long-term frozen chicken meat. Meat Science, 57: 331-336.

INFLUENCE OF MAIZE GRAIN SUPPLEMENTATION ON CHEMICAL COMPOSITION OF MEAT CALVES

E. Dal Molin, G. Destefanis, A. Brugiapaglia, P. Barge* and M.T. Barge

Dipartimento di Scienze Zootecniche, Università di Torino, Facoltà di Agraria, via Leonardo da Vinci 44, 10095 Grugliasco, Torino, Italia. Email: paolo.barge@unito.it

Keywords: calves, breed type, maize grain supplementation, meat quality

Introduction

Veal meat production is economically important since calves represent about a quarter of the cattle slaughtered in Italy. Veal meat is appreciated for its pale colour and high tenderness, which is the result of a liquid diet, based on milk replacer with a restricted iron content and slaughter at an early age. In order to promote rumination, health and welfare of these animals, the 97/2/CE Directive (EU Council, 1997) set a minimum daily amount of fibrous feed to be provided to calves. Fibrous feed supplementation, however, could affect meat quality. The aim of the present study is to determine how breed type and feeding plan, characterized by a different maize grain supplementation, influences chemical composition of veal meat.

Materials and Methods

During the entire fattening period, thirty-two male calves, 16 Friesian (F) and 16 Crossbreed (C), were fed on commercial milk replacer (350 kg milk powder /calf), according to the standard feeding program, and maize silage (50 kg/calf). In addition, calves received an individual amount of 65 kg (T1 group; 8F+8C) or 100 kg (T2 group; 8F+8C) of maize grain.

Crossbreeds were from different breeds and were considered of average quality by the market.

The average age of the animals was 19 days at the start of the trial and the trial lasted 178 days.

After slaughtering, two samples of *longissimus thoracis et lumborum* were taken, between the 8th T.V. and 1st L. V., from the right and the left side of each carcass.

The meat samples were analysed for chemical composition water, protein and ether extract content according to AOAC procedures (AOAC, 1970), as well as haem iron content (μg/g muscle) according to Hudzik (1990).

Data were analysed by ANOVA, considering the breed type (2 levels) and the feeding plan (2 levels) as factors as well as their interaction.

Results and Discussion

The results of the chemical analysis are reported in Table 1.

Table 1: Chemical composition of *longissimus thoracis et lumborum* of calves.

	Water (%)			Protein (%)			Ether extract (%)			Fe (μg/g)		
	F	C	mean	F	C	mean	F	C	mean	F	C	mean
T1	75.55	76.24	75.89	21.40	21.19	21.29	1.56^A	1.09^B	1.32	4.35	3.93	4.14
T2	76.17	75.86	76.01	21.04	21.52	21.28	1.22^B	1.14^B	1.18	3.63	4.16	3.89
mean	75.86	76.05		21.22	21.35		1.39	1.12		3.98	4.04	

Means with different letters differ significantly (P<0.01)

For all the variables taken into account, the interaction breed type x feeding plan was statistically significant (P<0.05 for the ether extract; P<0.01 for the other variables).

The meat of Friesian calves, fed on T1 diet, had lower water content and higher protein, ether extract and haem iron content, while Crossbred calves showed the contrary.

Since ANOVA showed a significant effect of the breed type on the ether extract as well as the interaction breed type x feeding plan being significant, ANOVA was performed considering each factor separately (within each level of the other factor) in order to draw a conclusion on its main effects.

Consequently, the meat of Friesian calves, fed on T1 diet had a significant higher ether extract content (P<0.01) in comparison with both Crossbred or Friesian calves, fed onT2 diet.

In particular, it is important to point out that the increase of the ether extract of the meat of animals of the FT1 group compared to FT2 was equal to 28%.

Moreover, for calves fed the T2 diet, the ether extract content of Crossbred veal meat was not statistically different from that of Friesian calves.

Due to different experimental condition, it is not easy to compare the results of this study with others. However, in general, some authors (Cozzi *et al.*, 2001, Cozzi *et al.*, 2002 ; Xiccato *et al.*, 2001) reported that the use of fibrous feeds (dried beet pulp, wheat straw and maize grain), in partial replacement or in addition to the milk diet, produced no differences or increased the ether extract content of the meat.

Conclusions

In this study, we must stress the importance of the feeding plan choice in relation to the breed type. Indeed a higher amount of maize grain in the diet did not modify the veal meat characteristics of Crossbreds, while it improved those of Friesians. In the latter breed type, this supplementation seems to have influenced the whole utilization of the diet to the extent that intramuscular fat deposition was limited. From a qualitative point of view, it must be borne in mind that the higher fat deposition of Friesians on the T1 diet could negatively influence both the visual meat appraisal, considering the importance at retail sale of packaged meat, and the nutritional characteristics since consumers prefer meat with a low fat content.

Acknowledgments

This research was supported by Auchan, Gruppo Rinascente.

References

AOAC (1970). Official methods of analysis (11th Ed.). Washington, DC, USA: Association of Official Analytical Chemist.

Council Directive 97/2/EC, (1997). Official Journal of the European Communities of the 28 January 1997, No. L 25/24.

Cozzi, G., Gottardo, F., Fregolent, G., Ravarotto, L., Segato, S., Dalle Zotte, A. (2001). The provision of solid feeds for veal calf welfare: effects on growth performance and meat quality. Proceedings of the ASPA XIV Congress, Firenze, 12-15 June.

Cozzi, G., Gottardo, F., Mattiello S., Canali, E., Scanziani, E., Verga, M., Andrighetto, I. (2002). The provision of solid feeds to veal calves: I. Growth performance, forestomach development, and carcass and meat quality. Journal of Animal Science, 80, 357-366.

Hudzik, E. (1990). Comparaison de la détermination du fer héminique dans les viandes crues par la méthode de Hornsey avec l'absorption d'extraits aqueux à 410 nm. Sciences des aliments, 10, 209-213.

Xiccato, G., Trocino, A., Queaque, P.I., Sartori, A., Carazzolo, A. (2001). Rearing veal calves with respect to animal welfare: effects of group housing and solid feed supplementation on growth performances and meat quality. Livestock Production Science, 75, 269-280.

EFFECT OF FEED RESTRICTION ON THE FAT QUALITY OF MEAT FROM IBERIAN PIGS DESTINED TO DRY-CURED INDUSTRY

J. Peinado*[1], T. Zaballos[2], J. Viguera[1], J. Sánchez[1], J. Ruiz[3] and C. López-Bote[4]

[1.] Imasde Agropecuaria, S.L., 28224 Pozuelo de Alarcón, Madrid, Spain, [2.] CRAPE, S.Coop.Ltda., 37300 Peñaranda de Bracamonte, Salamanca, Spain, [3.] Dpto. Tecnología de los Alimentos, Universidad de Extremadura, 10071 Cáceres, Spain, [4.] Dpto. Producción Animal, Universidad Complutense de Madrid, 28040 Madrid, Spain. Email: jpeinado@e-imasde.com

Keywords: Iberian pig, fatty acid profile, feed restriction, subcutaneous fat, intramuscular fat

Introduction

Sensory characteristics of Iberian dry-cured products are highly appreciated. The fat content and fatty acid (FA) profile of raw meat take a key role in the flavour compound formation during the ripening process and the oily appearance of Iberian dry cured-products (Ruiz *et al.*, 2000). Further, an insufficient fat content enhances dehydration and the products would ripen too quickly, impairing the appropriate development of quality (López-Bote, 1998). In order to increase fat deposition, Iberian pigs are traditionally slaughtered between 150 and 160 kg BW. The feeding program used in most farms consists of a restricted feeding from 30 to 100 kg for development of the bones and the muscle tissue, and subsequently, *ad libitum* access to a high caloric diet rich in oleic acid until to slaughter. However, the restricted feed intake (FI) level may affect the FA deposition because the endogenous lipid synthesis is reduced (Nürnberg *et al.*, 1998), and consequently the quality of final product could be modified. The aim of this study was to evaluate the influence of the feed intake level on fatty acid composition and melting point of fat from meat of Iberian pigs destined to dry-cured product industry.

Table 1: Composition of the experimental diets.

Item	34 to 97 kg BW	97 kg BW to slaughter
Main ingredients, %		
Barley	56	58
Wheat	15	30
Soybean meal, 44% CP	14	7
High oleic sunflower oil	1.7	2.5
Others	13.3	2.5
Calculated nutrient composition[a], %		
NE (kcal/kg)	2.332	2.394
Lysine	0.82	0.61
Starch	44.1	47.2
C 16:0	0.36	0.40
C 18:0	0.08	0.11
C 18:1	1.59	2.13
C 18:2	0.94	0.89
C 18:3	0.07	0.08
Analysed composition[b], %		
Crude protein	14.8	12.8
Crude fibre	4.2	4.5
Ether extract	3.3	4.0

[a]According to FEDNA (2003); [b]According to AOAC (2000).

Materials and Methods

A total of 96 castrated male crossbreed Iberian x Duroc pigs 125d of age with an average weight of 34.8 kg were used. Pigs were separated into two experimental groups according to the FI level during the growing phase from 34 to 97 kg BW: 1.6 (LOW) *vs* 2.1 (HIGH) kg feed/d. Pigs were placed in two barns of 90 m^2 in a semi-intensive controlled-environment and had *ad libitum* access to water. When each experimental group reached 97 kg BW, pigs were fed *ad libitum* a finishing diet rich in oleic acid up to slaughter (Table 1). At the preplanned slaughter age of 350 d (with an average weight of 156.4 and 174.9 kg BW for the LOW and HIGH groups) all pigs were electrically stunned, exsanguinated and eviscerated. Twenty carcasses from each treatment were randomly selected and subcutaneous fat (SF) and *longissimus dorsi* muscle (LM) samples were taken at the level of the last rib. All samples were vacuum-packed and kept at –20°C for subsequent analyses. Lipids from inner and outer SF layers were extracted independently by method proposed by Bligh and Dyer (1959), while neutral and polar intramuscular lipids were obtained according to the method developed by Marmer and Maxwell (1981). Fat extracts were methylated and analysed by gas chromatography as described by Rey and López-Bote (2001). Additionally, the melting point of both SF layers was determined using AOAC 920.157 procedure (AOAC, 2000).

The experimental unit for data was a SF or LM sample from 20 carcasses sampled at random per treatment. Data were analysed as a completely randomized design using the GLM procedure of SAS (SAS Inst., Inc., Cary, NC, USA).

Results and Discussion

The SF from pigs with a LOW FI level showed higher C18:2, C18:3, and C20:5 content in the inner layer but lower C18:1 content in the outer layer than the SF from pigs with a HIGH FI level (Table 2). As expected, the outer layer had higher content of unsaturated FA and lower melting point than the inner layer in all experimental groups. However, FI level did not modify the melting point of inner and outer layer of SF, and similar values for water migration during the ripening and for the brightness or oiliness of the final product could be expected among primal cuts from all experimental pigs. Furthermore, no FI levels effect was observed for the main FA composition either from the neutral or the polar muscle lipids, which are the major source of the volatile aroma and flavour compounds produced during the curing process.

Table 2: Effect of feed intake level (LOW *vs* HIGH) on the fatty acid profile and the melting point of the subcutaneous (SF) and intramuscular (IF) fat.

Item	LOW	HIGH	SEM (n=20)	*P*<	LOW	HIGH	SEM (n=20)	*P*<
SF fatty acid content, %	Inner layer				Outer layer			
C 16:0	24.81	24.64	0.19	0.55	23.75	23.43	0.20	0.28
C 16:1 *(n-7)*	2.03	1.99	0.05	0.63	1.98	2.07	0.05	0.28
C 18:0	15.04	14.41	0.32	0.17	13.76	12.37	0.30	0.01
C 18:1 *(n-9)*	46.62	47.53	0.37	0.09	49.04	50.75	0.33	0.01
C 18:2 *(n-6)*	6.53	6.20	0.08	0.01	5.97	5.78	0.11	0.27
C 18:3 *(n-3)*	0.343	0.324	0.005	0.01	0.302	0.296	0.006	0.47
C 20:5 *(n-3)*	0.087	0.078	0.003	0.04	0.103	0.103	0.004	0.99
C 22:6 *(n-3)*	0.035	0.019	0.006	0.07	0.063	0.065	0.008	0.89
SF melting point, ºC	33.2	33.7	0.7	0.67	29.9	29.7	0.2	0.46
IF fatty acid content, %	Neutral lipids				Polar lipids			
C 16:0	26.72	27.18	0.24	0.19	25.00	25.52	0.39	0.35
C 16:1 *(n-7)*	4.10	4.44	0.15	0.11	3.31	3.79	0.17	0.04
C 18:0	13.34	12.72	0.37	0.24	11.62	11.35	0.36	0.60
C 18:1 *(n-9)*	49.36	49.73	0.50	0.60	42.18	44.54	1.26	0.19
C 18:2 *(n-6)*	2.35	2.14	0.11	0.17	10.47	7.88	1.34	0.17
C 18:3 *(n-3)*	0.110	0.099	0.006	0.21	0.233	0.184	0.018	0.05
C 20:5 *(n-3)*	0.060	0.057	0.004	0.51	0.469	0.281	0.063	0.04
C 22:6 *(n-3)*	0.054	0.063	0.008	0.47	0.132	0.090	0.018	0.11

Conclusions

The fatty acid content of intramuscular fat and the melting point of subcutaneous fat were not affected by feeding restriction before fattening at the studied levels. Therefore, no effect of feeding restriction level from 34 to 97 kg BW on the profile of volatile compounds and on the curing losses of dry-cured products is expected.

Acknowledgements

Financial support was provided by the Ministerio de Industria, Comercio y Turismo (CDTI Project 040387; PROFIT Project FIT060000200441), and by the Agencia de Desarrollo Económico de la Junta de Castilla y León (ADE Project 0404SA0005).

References

AOAC (2000). Official Methods of Analysis (17th Ed.). Association of Official Analytical Chemists. Arlington, VA, USA.

Bligh, E.G. and Dyer, W.J. (1959). A rapid method of total lipid extraction and purification. *Canadian Journal of Biochemistry and Physiology, 37*: 911-917.

FEDNA (2003). Normas FEDNA para la formulación de piensos compuestos. De Blas, C., Mateos, G. G. and García, P., ed. Fundación Española para el Desarrollo de la Nutrición Animal, Madrid, Spain.

López-Bote, C.J. (1998). Sustained utilization of the Iberian pig breed. *Meat Science, 49 (1)*: S17-S27.

Marmer , W.N. and Maxwell, R.J. (1981). Dry column method for the quantitative extraction and simultaneous class separation of lipids from muscle tissue. *Lipids, 16*: 365-371.

Nürnberg, K., Wegner, J. & Ender, K. (1998). Factors influencing fat composition in muscle and adipose tissue of farm animals. *Livestock Production Science, 56*:145-156.

Rey, A. and López-Bote, C.J. (2001). Effect of dietary copper and vitamin E supplementation, and extensive feeding with acorns and grass on *longissimus* muscle composition and susceptibility to oxidation in Iberian pigs. *Journal of Animal Physiology and Animal Nutrition, 85*: 281-292.

Ruiz, J., Ventanas, J., Cava, R., Andrés, A.I. and García, C. (2000). Texture and appearance of dry cured ham as affected by fat content and fatty acid composition. *Food Research International, 33*: 91-95.

EFFECTS OF GRASS SILAGE, GRAZED GRASS AND A CONCENTRATE DIET ON THE FATTY ACID COMPOSITION OF BEEF SUBCUTANEOUS FAT

J.D Wood*[1], H.E. Warren[3], G.G Stonehouse[1], K.G. Hallett[1], F.M. Whittington[1], N.D. Scollan[2] and R.I. Richardson[1]

[1] *Division of Farm Animal Science, School of Veterinary Science, University of Bristol, Langford, Bristol, BS40 5DU, UK.* [2] *Institute of Grassland and Environmental Research, Aberystwyth, Wales, SY23 3EB, UK.* [3] *Nottingham Trent University, Southwell ,Notts ,NG25 0QF, UK. Email: Jeff.Wood@bristol.ac.uk*

Keywords: beef, fatty acids, conjugated linoleic acid, grass, grass silage

Introduction

Fatty acid composition is of major importance in the healthiness of meat, especially in the ruminant species cattle and sheep, which have relatively saturated fat. However, ruminants deposit significant amounts of 'beneficial' fatty acids in meat, particularly the n-3 (omega-3) polyunsaturated fatty acids (PUFA) and conjugated linoleic acid (CLA, 18:2-9c 11t).

Diet is a major factor in ruminant meat fatty acid composition, with grass-based diets increasing levels of α-linolenic acid (18:3n-3) and the longer-chain n-3 PUFA. Grass, however, is a variable feed stuff, according to the time of year and whether it is fed fresh or conserved as silage or hay. In this work, we contrast fresh grazed grass, grass conserved as silage and a standard concentrate in terms of their effects on the fatty acid composition of subcutaneous fat in beef steers.

Materials and Methods

Twenty-four steers from each of two breeds, Holstein-Friesian (HF) and Aberdeen Angus x Holstein-Friesian (AA) were used in this study, part of a bigger study into the effects of age, breed and diet on fatty acid composition and meat quality in beef (Warren *et al.*, 2006). From 6 months of age, 8 steers from each breed were fed either a concentrated diet (based on barley and full fat soyabean meal) or a grass silage diet and reared to 19 months so that the growth rate of these two groups was similar. A third group of 8 steers from each breed was fed grass silage from 6 to 12 months then grazed fresh grass from 12 to 19 months during the Spring and Summer. Fresh and conserved grass was perennial ryegrass. At 19 months the cattle were slaughtered and samples of subcutaneous fat were removed from the loin region. Lipids were extracted into chloroform and hydrolysed with 2M KOH in methanol containing an internal standard, C21:0. Fatty acids were extracted into petroleum ether and methylated with diazomethane. They were then analysed by gas liquid chromatography (GLC).

Breed affected the results for weight and carcass composition but fatty acid composition was overwhelmingly influenced by diet so only the results for diets (breeds pooled) are presented.

Results and Discussion

Results for live weight and carcass characteristics are in Table 1. Steers fed occentrate and grass silage were similar in weight as intended, although AA were slightly heavier than HF. The groups fed fresh grass were lighter, with no difference between breeds. The animals fed grass silage were fatter than those fed concentrate, with the fresh grass group being the least fat. AA were fatter than HF.

Table 1: Live weight and carcass fat cover.

	Concentrate		Grass silage		Fresh grass		SED		Sig
	AA	HF	AA	HF	AA	HF			
Final live weight (kg)	519	501	526	492	467	473	14.9	ns	***
Fat score[1]	71	30	113	74	52	32	7.5	*	*

[1] Estimated subcutaneous fat % x 10. Score (20-145) given by carcass assessor.

The results for the fatty acid composition of subcutaneous fat are in Table 2. The proportion of linoleic acid (18:2n-6), which is the major fatty acid in grain and soya, was 2.5 times higher in the concentrate group. The proportion of 18:3n-3 was about 3 times higher in the fresh grass and grass silage groups than in concentrates and was significantly higher in thefresh grass than in the grass silage group. These results for 18:2n-6 and 18:3n-3 reflect differences in the composition of the diets. The proportion of trans vaccenic acid (18:1trans) was about 2.8 times higher in concentrate and fresh grass than grass silage. CLA followed the same pattern, being about two times higher in Concentrate and the fresh grass than Grass silage. The relationship between 18:1trans and CLA is shown in Figure 1. The linear relationship between the proportions of these two fatty acids reflects the production of CLA from 18:1trans, most of which occurs in adipose tissue in cattle (Tanaka, 2005). The implication is that rumen fermentation of the concentrate and the fresh grass diets resulted in similar amounts of 18:1trans production in the rumen which was absorbed from the blood stream and synthesised into CLA in adipose tissue. The pattern of ruminal fermentation of grass silage was apparently different from that of fresh grass, as has been shown previously (Schmid, *et al.*, 2006). This led to a much smaller production of 18:1 trans and also stearic acid (18:0), the main product of complete biohydrogenation of 18C PUFA.

Table 2: Fatty acid composition of subcutaneous fat (%).

	Concentrate	Fresh grass	Grass silage	SED	Sig.
C12:0	0.12^{c}	0.08^{a}	0.09^{b}	0.007	***
C14:0	3.86^{a}	3.46^{a}	4.51^{b}	0.214	***
C16:0	26.20^{b}	24.76^{a}	28.89^{c}	0.649	***
C16:1	5.26^{a}	5.79^{b}	7.15^{c}	0.372	***
C18:0	14.04^{b}	15.20^{b}	10.09^{a}	1.069	***
C18:1tr	4.07^{b}	4.32^{b}	1.48^{a}	0.272	***
C18:1cis9	31.17	32.06	32.83	0.909	ns
C18:1cis11	1.08^{b}	0.86^{a}	1.03^{b}	0.062	**
C18:2	2.39^{b}	0.97^{a}	0.90^{a}	0.067	***
CLA,9c11t	0.95^{b}	0.93^{b}	0.43^{a}	0.052	***
C18:3	0.22^{a}	0.68^{c}	0.59^{b}	0.024	***

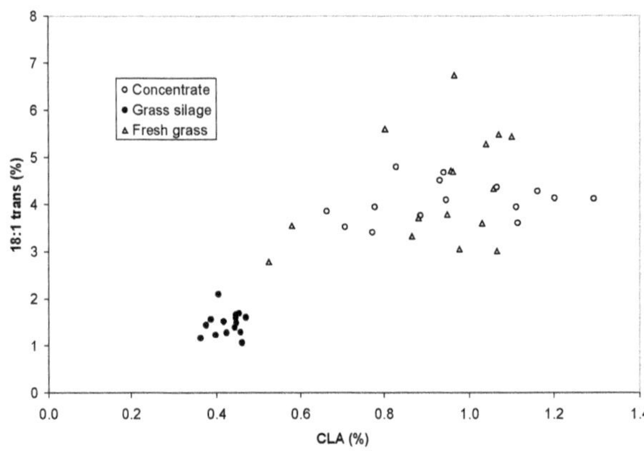

Figure 1: Relationship between 18:1 trans (%) and CLA (%). Linear regression : y=4.27x, $r^{2=}$0.67.

Conclusions

Feeding fresh grass (grazing), grass silage and concentrate to beef cattle resulted in different fatty acid profiles in subcutaneous fat. Compared with grass silage, fresh grass produced higher proportions of 18:3 n-3, 18:1 trans and CLA and also a higher proportion of 18:0. The concentrate diet produced more 18:2 n-6 and similar proportions of 18:1 trans and CLA as fresh grass.

References

Schmid, A., Collomb, M., Sieber, R. and Bee, G. (2006). Conjugated linoleic acid in meat and meat products: a review. Meat Science, 73: 29-41.

Tanaka, K. (2005). Occurrence of conjugated linoleic acid in ruminant products and its physiological functions. Animal Science Journal, 76: 291-303.

Warren, H.E., Scollan, N.D., Enser, M., Hallett, K.G., Hughes, S.I., Richardson, R.I. and Wood, J.D. (2006). Effects of age, breed and a concentrate or grass silage diet on beef quality. 1. Animal performance, carcass quality and muscle fatty acid composition. Journal of Animal Science, (Submitted).

Acknowledgements

This was a Sustainable Livestock Production LINK project, funded by Defra, MLC, Tesco Stores Ltd., JSR Farms and southern Counties Fresh Foods Ltd.

EFFECTS OF DAIDZEIN FEEDING OF PREGNANT SOWS ON CARCASS AND MEAT QUALITY AND SKELETAL MUSCLE CELLULARITY OF THE PROGENY

C. Rehfeldt*[1], I. Adamovic[2], M. Mau[1] and G. Kuhn[1]

[1.] *Research Institute for the Biology of Farm Animals, Research Unit Muscle Biology and Growth, Wilhelm-Stahl-Allee 2, D-18196 Dummerstorf, Germany,* [2.] *University of Belgrade, Faculty of Agriculture, Nemanjina 6, 11000 Belgrade, Serbia. Email: rehfeldt@fbn-dummerstorf.de*

Keywords: isoflavone, pigs, sow feeding, carcass quality, meat quality, muscle fibre

Introduction

Biologically active compounds in the maternal circulatory system may influence not only the maternal organism but also the foetus. In this context dietary phytoestrogens are of interest, as they are able to cross the placenta. Phytoestrogens are plant-derived compounds with structural and functional similarities to natural estrogens (Ren *et al.*, 2001a). Soy and soy products are the most significant dietary sources of isoflavonic phytoestrogens, mainly of genistein and daidzein (Barnes *et al.*, 1994). In pigs, the level of circulating isoflavones and thus their bioavailability depends on the isoflavone content of the soy-product in the diet (Kuhn *et al.*, 2004). Isoflavones have been shown to exhibit multi-biological properties including estrogenic and anti-estrogenic effects, inhibitory effects on protein tyrosine kinases (e.g. growth factor receptors), modulation of gene transcription, and antioxidant actions (Ren *et al.*, 2001a). From these various features both promoting and inhibiting properties of isoflavones on cellular growth and development are to be expected. Results on the influence of isoflavones on growth and differentiation of porcine skeletal muscle are not yet available. To better understand the significance of isoflavones contained in feed on pig performance, the aim of this study was to investigate the effects of maternal dietary daidzein supplementation during late gestation on the progeny at birth and at market weight with focus on body composition, carcass and meat quality, and associated skeletal muscle cellularity.

Materials and Methods

Animals and feeding. Sixteen multiparous sows (German Landrace) were bred to the same German Landrace boar and housed individually. All sows were fed twice daily with a special soybean-free triticale-wheat based diet (2.6 up to 5 kg/daily) containing 13.0 MJ metabolizable energy per kg dry matter, 16% crude protein, and 0.94% lysine. Sows were divided into experimental (n=8) and control (n=8) groups with balanced live weights and parity. From day 85 of gestation until parturition the diet of the experimental group was supplemented with 1 mg of pure daidzein (Nanjing Agricultural University, China) per kg body weight and day. This approximates a concentration of daidzein (c50 mg/kg feed) as it has been measured in a common triticale-soybean meal diet (Kuhn *et al.*, 2004a). After birth two male piglets per litter were collected for detailed analysis. The remaining piglets were reared until market weight and slaughtered at 180.4±5.3 days of age (n=66 control; n=55 daidzein).

Piglet body composition, carcass and meat quality were analyzed as described previously (Kuhn *et al.*, 2004b). *Muscle histology and biochemistry.* Samples from *semitendinosus* (ST) muscle were collected from newborn piglets and slaughter pigs. Neonatal muscle was further analyzed according to Rehfeldt *et al.* (2001). In adult pigs, muscle fibre characteristics were measured in a sample of 29 controls and 25 experimental (daidzein) pigs only. Samples from the ST midbelly were taken, mounted on cork-chucks and snap-frozen in liquid nitrogen. Serial sections of 10 µm were stained for cytoplasm/nuclei by haematoxylin/eosin or exposed to a combined reaction for NADH-TR and ATPase at pH 4.2, which enables to classify muscle fibre types (STO, FTO, FTG) (Rehfeldt *et al.*, 1993). Microscopic evaluation was carried out by image analysis (AMBA, IBSB, Berlin, Germany). Homogenates of neonatal ST muscle were used to quantify DNA, protein, and creatine kinase (CK) activity as described previously (Rehfeldt *et al.*, 2001); RNA was quantified with SYBR®Green II (Oksbjerg *et al.*, 2000). *Statistical analyses.* Data were subjected to ANOVA by the GLM or mixed classification models of SAS (SAS Inst. Inc., Cary NC). Significance of LSmeans differences was tested by the Student's t-test ($P<0.05$).

Results and Discussion

Effects on newborn piglets. Maternal performance, such as litter size, percentages of live born piglets and runts, average birth weight per litter and litter weight, was not different between control and daidzein-supplemented sows (data not shown). In contrast, Ren *et al.* (2001b) found increased birth weights in males in response to maternal daidzein, but at a very low doses supplemented to a normal soy-based diet. In our study, maternal daidzein elevated the concentration of fat in the whole piglet body suggesting that overall development was positively influenced (Kuhn *et al.*, 2002) without being apparent as higher birth weight. Average tissue proportions of the piglets were not changed in response to daidzein feeding (Table 1). Likewise, neither muscle weight nor histological and biochemical properties were significantly affected, even though numerical decreases in fibre number, protein and CK activity were seen. However, the response to maternal daidzein depended on litter size as seen by significant interactions. The percentage of muscle tissue tended to decrease by maternal daidzein feeding in piglets belonging to large litters ($P=0.09$), whereas the percentage of skin increased ($P=0.03$). Accordingly, muscular CK activity was lower ($P=0.06$) suggesting that myogenesis was slightly impaired by daidzein in piglets that highly compete for maternal nutrients. Recent results of an *in vitro* study revealed that daidzein and genistein inhibit the growth of porcine muscle satellite cells in a dose-dependent manner (Mau *et al.*, 2006). Litter size ($\geq$15 vs. <15), but not sex, was an influential factor for birth weight and maturity as seen by lower body fat percentage, ST muscle weight, myofibre numbers, and CK activity in piglets from large litters.

Effects on postnatal growth, carcass and meat quality. Postnatal growth, carcass and meat quality at market weight were not changed by maternal daidzein feeding (not shown). Likewise, average meat quality measured on the *longissimus* muscle of the progeny, was not influenced. However, the pigs originating from large litters exhibited higher muscle pH_{45} (P=0.02) indicating a slight improvement of meat quality by maternal daidzein, but the mechanism remains unclear. Carcass weight and composition were sex-dependent. Male castrates showed greater live and carcass weights, back fat thickness, but lower meat percentage and loin area (see also Correa *et al.*, 2006), and different meat quality and muscle structure than females.

Table 1: Body composition and *semitendinosus* (ST) muscle characteristics of male neonatal piglets (LSmeans ± SE) born to sows supplemented with dietary daidzein of 1 mg/kg body weight/d from d 85 of gestation to parturition as compared to untreated controls.

Item	Control	Daidzein	*P* : T[a]	L[b]	T x L
Birth weight, g	1.266 ± 50	1.272 ± 46	0.93	0.01	0.53
Meat, %	43.4 ± 0.60	42.9 ± 0.66	0.64	0.68	0.07[e]
Subcutaneous fat, %	8.6 ± 0.35	9.0 ± 0.39	0.50	0.32	0.79
Bones, %	36.6 ± 0.75	36.4 ± 0.83	0.84	0.56	0.20
Skin, %	10.6 ± 0.16	11.0 ± 0.18	0.14	0.26	0.10[f]
Protein, %	14.75 ± 0.24	15.76 ± 0.27	0.99	0.63	0.96
Fat, %	1.19 ± 0.04	1.30 ± 0.04	0.04	0.01	0.85
ST Muscle weight (g)	2.77 ± 0.20	2.63 ± 0.22	0.65	0.05	0.28
ST Protein (mg/g)	91.2 ± 4.3	86.7 ± 4.7	0.50	0.14	0.79
ST CK (IU/mg protein)[c]	3.76 ± 0.28	3.58 ± 0.22	0.56	0.06[d]	0.76
ST Total fibre number	458.3 ± 29.6	409.4 ± 32.4	0.30	0.03	0.51

[a]T – treatment; [b]L – litter size group; [c]CK – creatine kinase activity; [d]lower in large litters (*P*=0.06); [e]Decreases in large litters (*P* = 0.09); [f]Increases in large litters (*P* = 0.03)

Conclusions

The isoflavone compound daidzein when supplemented to a soy-free sow diet during late gestation in an amount typical for soy-based diets marginally affects growth, carcass and meat quality, and skeletal muscle cellularity of the progeny. Slight detrimental effects on foetal muscle development in piglets from large litters can be compensated for during postnatal growth. Higher doses and/or other kinds of isoflavones might have greater effects on muscle growth and meat quality.

References

Barnes, S., Kirk, M. and Coward, L. (1994) Isoflavones and their conjugates in soy foods - extraction conditions and analysis by HPLC mass-spectrometry. Journal of Agricultural and Food Chemistry 42: 2466-2474

Correa, J. A., Faucitano, L., Laforest, J. P., Rivest, J., Marcoux, M. and Gariépy, C. (2006) Effects of slaughter weight on carcass composition and meat quality in pigs of two different growth rates. Meat Science 72: 91-99

Kuhn, G., Hennig, U., Kalbe, C., Rehfeldt, C., *et al.* (2004a) Growth performance, carcass characteristics and bioavailability of isoflavones in pigs fed soybean based diets. Archives of Animal Nutrition 58: 265-276.

Kuhn, G., Kanitz, E., Tuchscherer, M., Nürnberg, G., Ender, K. and Rehfeldt, C. (2004b) Growth and carcass quality of offspring in response to porcine somatotropin (pST) treatment of pregnant sows. Livestock Production Science 85: 103-112.

Mau, M., Viergutz, T., Rehfeldt, C. (2006). Influence of estrogens and isoflavones on porcine muscle cell growth. Archives of Animal Breeding 49: Special issue, 81-85.

Oksbjerg, N., Petersen, J. S., Sørensen, I. L., Henckel, P., Vestergaard, M., Ertbjerg, P., Møller, A. J., Bejerholm, C. and Stoier, S. (2000) Long-term changes in performance and meat quality of Danish Landrace pigs: a study on a current compared with an unimproved genotype. Animal Science 71: 81-92.

Rehfeldt, C. and Ende,r K. (1993) Skeletal muscle cellularity and histochemistry in response to porcine somatotropin in finishing pigs. Meat Science 34: 107-118.

Rehfeldt, C., Kuhn, G., Vanselow, J., Fürbass, R., Fiedler, I *et al.* (2001) Maternal treatment with somatotropin during early gestation affects basic events of myogenesis in pigs. Cell & Tissue Research 306: 429-440.

Ren, M.Q., Kuhn, G., Wegner, J. and Chen, J. (2001a) Isoflavones, substances with multi-biological and clinical properties. European Journal of Nutrition 40: 135-146.

Ren, M.Q., Kuhn, G., Wegner, J., Nürnberg, G., Chen, J. and Ender, K. (2001b) Feeding daidzein to late pregnant sows influences the estrogen receptor beta and type 1 insulin-like growth factor receptor mRNA expression in newborn piglets. Journal of Endocrinology 170: 129-135.

COMPENSORY GROWTH RESPONSE IN PIGS ON PERFORMANCE, COMPOSITION OF WEIGHT GAIN AT CARCASS AND MUSCLE LEVEL AND MEAT QUALITY TRAITS

A. Heyer and B. Lebret*

Institut National de la Recherche Agronomique (INRA), UMR SENAH, 35590 Saint-Gilles, France. Email: Benedicte.Lebret@rennes.inra.fr

Keywords: pigs, compensatory growth, tissue deposition rates, intramuscular fat, meat quality

Introduction

Compensatory growth in growing/finishing pigs is a well-known physiological phenomenon of accelerated final growth rate induced by a restriction - re-alimentation feeding. This feeding strategy influences the rates of lean and adipose tissue deposition at the carcass level (Campbell *et al.*, 1983), however composition of weight gain at the muscle level was hardly described in pigs. High level of intramuscular fat (**IMF**) is often reported to have beneficial effects on pork eating quality. Previous results showed that increasing age at slaughter of pigs increases the potential for IMF deposition through elevated intramuscular adipocyte number, whereas actual IMF deposition rises with energy intake (Gondret and Lebret, 2002). Thus, a feeding strategy that leads to both higher final age and daily feed intake might positively influence IMF, and therefore pork quality. Besides, it was shown that increased final growth rate might improve pork tenderness through higher muscle protein turn-over (Therkildsen *et al.*, 2002). This study aimed to increase age at slaughter of pigs by a strong feed restriction until 70 kg live weight (**LW**) followed by *ad libitum* feeding up to 110 kg LW in order to 1) increase age at slaughter of pigs and potential for IMF deposition, 2) evaluate composition of weight gain during restriction and re-alimentation and 3) increase final daily feed intake and IMF, and thereby improve pork eating quality.

Materials and Methods

At the average LW of 30 kg, 112 Duroc x (Large White x Landrace) pigs (females and castrated males) were equally allocated to either group **AL** (*ad libitum* feeding during growing (30-70 kg LW) and finishing (70-110 kg LW) with conventional growing and finishing diets), or **CG** (compensatory growth: feed restriction at 65% of AL feed intake during growing (growing diet), and *ad libitum* thereafter (finishing diet)). Pigs were kept in individual pens and slaughtered at 70 kg (n=30) or 110 kg (n=82). Fourteen pigs issued from the same litters were slaughtered at 30 kg LW for determination of carcass composition at start of the experiment. Carcass weight, average backfat thickness ($3^{rd}/4^{th}$ lumbar vertebra and $3^{rd}/4^{th}$ last rib levels), muscle depth ($3^{rd}/4^{th}$ last rib level) and lean meat content (calculated for 110 kg LW only, from linear measurements) were measured on the day of slaughter. The day after, the ham was partially dissected into muscles, subcutaneous fat, and bones. Samples of *m. longissimus* (**LD**) were taken, trimmed of external fat and freeze dried before IMF (=petroleum ether extract) and protein (=6.25*nitrogen, Dumas method) determinations. Deposition rates of lean and fat tissues and of LD components during both growing and finishing periods were calculated from tissue weights and LD chemical composition. Data were submitted to an analysis of variance (proc GLM, SAS), including the fixed effects of feeding regimen and sex and the random effect of sow within sire.

Results and Discussion

Feed restriction at 65% of the *ad libitum* level during growing phase resulted in a 35% decrease in average daily gain and a 22 d increase in age at 70 kg LW for CG compared with AL pigs (Table 1). During re-alimentation average daily feed intake (**ADFI**) of CG pigs was strongly increased (+255%) compared to their growing period, and corresponded to 108% of the ADFI of the AL pigs during finishing. This demonstrated a compensatory feed intake for the CG pigs, which had additionally a better finishing feed conversion ratio. This led to a 13% higher growth rate of these pigs during finishing. Hence, CG pigs expressed a compensatory growth, and were 19 d older at 110 LW than AL pigs. Thus, the objective of the study to produce older pigs with same market weight and higher final growth rate through a specific feeding strategy was reached.

Feed restriction reduced carcass fatness, as already reported (Gondret and Lebret, 2002). After re-alimentation, carcass fatness and lean meat content did not differ between CG and AL pigs, indicating that compensatory growth modifies the composition of weight gain with higher lipid than protein deposition at the carcass level. It was already demonstrated that a restriction-re-alimentation feeding strategy might improve pig performance and influence carcass traits. However, the effects depend on the onset, intensity and duration of the restriction, and onset and duration of the re-alimentation (Campbell *et al.*, 1983; Therkildsen *et al.*, 2002).

IMF was slightly lower for CG than AL pigs at 70 kg LW, and was still lower for CG pigs at 110 kg LW (Table 2). LD protein content was not affected by feeding regimen. Feed restriction reduced muscle (-30%) and lipid (-50%) deposition rates in both ham and LD. During re-alimentation, muscle deposition rates in ham and LD were similar in CG and AL pigs, but LD protein deposition tended to be reduced in CG pigs. Subcutaneous fat deposition was increased, but IMF deposition did not differ between CG and AL pigs during re-alimentation, explaining the similar carcass fatness and lower IMF of the CG pigs at 110 kg LW. Thus, CG pigs did not catch up on the previous retardation of IMF deposition during a 33 d re-alimentation period, which favoured external fat deposition. Hence, an increased IMF content through CG feeding strategy could not be obtained. Meat quality traits (pH1, pHu, drip loss) were not affected by the feeding strategy (data not shown).

Conclusions

A 65% feed restriction until 70 kg LW followed by *ad libitum* feeding during finishing resulted in an expected higher age of pigs at commercial market weight compared to controls. Re-fed pigs expressed a compensatory growth during finishing, and similar carcass traits at 110 kg LW due to higher deposition rate of adipose than muscle tissues at the carcass level during re-alimentation. However, at muscle level, composition of weight gain was not modified. Consequently, the initial hypothesis of increased IMF deposition in CG pigs during re-alimentation could not be verified. Technological pork quality was not influenced by the feeding strategy. It can be hypothesized that an elevated IMF content in re-fed pigs, that would improve pork quality, might be achieved through forwarding the pattern of restriction – re-alimentation periods.

Table 1: Growth performance and carcass composition of pigs slaughtered at either 70 or 110 kg LW and allocated to different feeding regimes.

	LW70				LW110			
	AL	**CG**	**RSD**[1]	***P*-value**	**AL**	**CG**	**RSD**	***P*-valu**
No. of animals	56	55			41	40		
Live weight, kg	72.0	71.3	2.10	0.074	110.9	112.8	2.21	**0.001**
Age at LW 70/110, d	118	140	5.48	**0.001**	154	173	6.70	**0.001**
Daily feed intake, kg/d	2.32	1.53	0.13	**0.001**	3.62	3.90	0.30	**0.001**
Daily weight gain, g/d	980	634	48.0	**0.001**	1113	1257	101.3	**0.001**
Feed conv. ratio, kg/kg[2]	2.37	2.41	0.13	0.131	3.27	3.10	0.19	**0.001**
No. of animals	15	15			41	40		
Hot carcass, kg	55.3	56.4	1.85	0.142	90.1	90.2	2.10	0.797
Dressing, %	78.0	78.3	1.11	0.157	81.2	80.0	1.01	**0.001**
Backfat thickness, mm	14.4	11.4	1.65	**0.039**	21.0	20.6	3.20	0.603
Muscle depth$_{\text{last rib}}$, mm	47.0	48.6	5.59	0.460	61.8	60.1	5.33	0.183
Lean meat content, %					58.9	58.8	2.08	0.867

Table 2: Chemical composition of *m. longissimus* and muscle, protein and fat deposition rates in ham and *m. longissimus* of pigs slaughtered at either 70 or 110 kg LW and allocated to different feeding regimes

	LW70				LW110			
	AL	**CG**	**RSD**	***P*-value**	**AL**	**CG**	**RSD**	***P*-value**
No. of animals	15	15			41	40		
IMF of LD, %	1.50	1.24	0.35	0.074	2.53	2.20	0.66	**0.032**
Protein of LD, %	22.4	22.1	0.45	0.115	22.8	22.6	0.53	0.094
No. of animals	13	13			13	13		
Deposition rates in ham								
Muscle, g/d	66.3	48.7	5.74	**0.001**	65.3	60.8	11.7	0.414
Subcutan. fat, g/d	11.6	6.5	2.53	**0.001**	20.9	26.6	4.64	**0.033**
Deposition rates in m. longissimus								
Muscle, g/d	24.8	17.3	2.63	**0.001**	23.9	19.9	5.43	0.100
Protein, g/d	5.8	4.0	0.61	**0.001**	5.8	4.8	1.33	0.054
IMF, g/d	0.46	0.24	0.13	**0.001**	1.09	0.92	0.48	0.547

[1]Residual standard deviation. [2]Interaction between feeding regimen and sex was found for the 70-110 kg LW period ($P<0.05$).

References

Campbell, R., Taverner, M. and Curic, D. (1983). Effects of feeding level from 20 to 45 kg on the performance and carcass composition of pigs grown to 90 kg live weight. Livestock Production Science, 10:265-272.

Gondret, F. and Lebret, B. (2002). Feeding intensity and dietary protein level affect adipocyte cellularity and lipogenic capacity of muscle homogenates in growing pigs, without modification of the expression of sterol regulatory element binding protein. Journal of Animal Science, 80: 3184-3193.

Therkildsen, M., Riis, B., Karlsson, A., Kristensen, L., Ertbjerg, P., Purslow, P.P., Aaslyng, M.D. and Oksbjerg, N. (2002). Compensatory growth response in pigs, muscle protein turn-over and meat texture: effects of restriction/realimentation period. Animal Science, 75: 367-377.

MUSCLE TRAITS, PRESLAUGHTER STRESS AND MEAT QUALITY INDICATORS AS INFLUENCED BY PIG REARING SYSTEM

B. Lebret*[1], A. Foury[2], P. Mormède[2], E.M.C. Terlouw[3], A. Vautier[4] and P. Chevillon[4]

[1.] Institut National de la Recherche Agronomique (INRA), UMR SENAH, 35590 Saint-Gilles, France; [2.] INRA UMR Neurogénétique et Stress, 33077 Bordeaux, France; [3.] INRA QuaPA, Theix, 63122 St-Genès-Champanelle, France; [4.] Institut Technique du Porc, BP35104, 35651 Le Rheu, France. Email: Benedicte.Lebret@rennes.inra.fr

Keywords: pigs, rearing system, muscle metabolism, stress hormones, meat quality

Introduction

Consumers often associate alternative pig rearing systems like outdoor housing or straw bedding with better animal welfare and meat quality, and lower environmental impact (Ngapo *et al.*, 2003). However, effects of alternative rearing systems on pork quality often differ between studies, possibly due to differences in housing conditions, pig genotype, and behavioural and physiological responses of pigs to preslaughter handling. An additional advantage of alternative systems may be the larger group sizes, due to increased space allowance, that reduces the need for mixing pigs before slaughter, thus avoiding fights between animals and therefore reduces variability in pork quality. This study aimed to evaluate the effects of different alternative rearing systems and preslaughter handling procedures on muscle metabolism, preslaughter stress and meat quality indicators, compared to the conventional system.

Materials and Methods

On 3 different rearing sites, 3 alternative pig rearing systems, free range (**FR**, 150 m²/pig, 1 group of 40 pigs), outdoor courtyard with a shed (**OUT**, 1.3 m²/pig, 4 groups of 10 pigs), and indoor housing on straw bedding (**ST**, 1.3 m²/pig, 2 groups of 25 pigs) were each compared with the conventional or control system (1 per site) (**CONT**, 0.65 m²/pig, fully slatted floor, 4 pens of 10 pigs). FR and ST pigs and their controls were (Large White X Pietrain) X (Large White X Landrace) crossbreeds, and OUT pigs and their controls were synthetic line X (Large White X Landrace) crossbreeds. Pigs were reared during the winter and fed *ad libitum* with conventional diets. Pigs from the same rearing site were slaughtered on the same day in a commercial slaughterhouse 80-km away. With the exception of FR pigs, animals within comparisons and systems were mixed to form a single group during lairage. Blood and urine samples were taken at evisceration to determine levels of plasma glucose, lactate, creatine kinase (**CK**) and cortisol, as well as of urine cortisol, adrenaline and noradrenaline (Foury *et al.*, 2005). Thirty min after slaughter, **pH1** of *Semimembranosus* muscle (**SM**) was measured and an SM sample was taken to determine lactate dehydrogenase (**LDH**) and citrate synthase (**CS**) activities and glycolytic potential (**GP**) (Lebret *et al.*, 2006). The next day, the number of skin lesions (>3 cm) on the right carcass side and SM ultimate pH (**pHu**) were recorded. Within comparisons, the effect of rearing system was tested using analysis of variance (proc GLM, SAS). Halothane genotype effect (NN or Nn) was added for FR-CONT and ST-CONT comparisons. Urine and plasma hormones and CK levels were analysed after logarithmic transformation to fit a normal distribution. Correlations between biological parameters were calculated using the "Pooled Pearson" method to correct for treatment effects.

Results and Discussion

The rearing system influenced muscle metabolism and quality traits (Table 1). Compared to their controls, FR and ST pigs had a higher CS activity and FR pigs had a lower LDH activity, whereas the OUT system had no effect on muscle metabolism. The increased SM oxidative capacity with alternative rearing is in agreement with the findings of Gentry *et al.*, (2004). However, the present study shows that effects depend on rearing system, and possibly genotype. Muscle pH1 and lactate levels (not shown) were not influenced by rearing conditions, in accordance with Lebret *et al.*, (2006). The OUT system did not affect GP or pHu of SM, but FR and ST pigs had higher GP and lower pHu, suggesting higher resting muscle glycogen stores and/or lower glycogen consumption during preslaughter for these pigs. A similar but lower effect of alternative (straw bedding + outdoor access) or free range rearing was previously found on SM GP and pHu (Terlouw *et al.*, 2004; Lebret *et al.*, 2006). FR pigs had fewer skin lesions, lower plasma CK (indicator of cell suffering consecutively to high muscle activity) and lower urine adrenaline and noradrenaline, and higher plasma glucose, than their controls. However plasma lactate (not shown), and cortisol and urine cortisol levels were unaffected. OUT pigs had fewer skin lesions and lower CK activity than controls, but similar levels of other plasma and urine components. The ST system did not have a significant effect on skin lesions, plasma and urine metabolites, or hormone levels.

Strong correlations were found between biological parameters (Table 2). Plasma CK activity was positively correlated with urine catecholamines and the number of skin lesions. Skin lesions were correlated with urine adrenaline levels. GP was negatively correlated with plasma CK and urine cortisol and catecholamines, whereas strong correlations were found between pHu and plasma CK, urine catecholamines, and skin lesions ($r=0.38$, $P<0.001$). These correlations, the lesser skin damage and lower levels of plasma CK and urine catecholamines indicate that the higher GP and lower pHu of the FR pigs are at least in part explained by their lower physical exercise and/or stress during the preslaughter period (no mixing) (Foury *et al.*, 2005). In accordance, we observed fewer aggressive events (> 5 s) during the first hour in lairage for the pigs reared in alternative systems than for controls, independently to mixing in lairage (13 vs 25 aggressive events for FR and CONT pigs; 32 vs 70 for OUT and CONT pigs; 71 vs 180 for ST and CONT pigs, respectively). In agreement with present results, Terlouw *et al.* (2004) reported less aggressive behaviour of outdoor pigs during mixing and higher muscle glycogen content at slaughter compared with conventionally reared pigs.

Conclusions

The largest differences between alternative and conventional pig rearing systems were found in the free-range system. However, outdoor courtyard or straw bedding showed similar, albeit less strong effects. Alternative rearing systems increased muscle oxidative capacity and glycolytic potential at slaughter, resulting in lower meat pHu in 2 of the 3 systems investigated. This is partly explained by reduced physical activity including less aggressive behaviour of pigs during the preslaughter period.

Table 1: Influence of rearing systems on muscle traits, skin lesions, and plasma and urine components.

	FR *vs* conventional			OUT *vs* conventional			ST *vs* conventional		
	FR	CONT	Sign.	OUT	CONT	Sign.	ST	CONT	Sign.
No. of pigs	39	39		30	36		50	39	
Semimembranosus traits									
LDH, µmol substrate/min/g	2294	2478	***	2456	2476	ns	2408	2389	ns
CS, µmol substrate/min/g	12.4	10.0	***	8.9	9.0	ns	11.4	8.5	***
Glycolytic pot., µmol/g	151	131	***	150	148	ns	138	126	*
pH1	6.28	6.30	ns	6.30	6.22	ns	6.29	6.35	ns
pHu	5.58	5.85	***	5.63	5.69	ns	5.69	5.90	***
No skin lesions	5.8	22.6	***	9.4	19.0	*	12.1	9.4	ns
Plasma components									
Glucose, µmol/ml	11.1	8.8	***	7.1	6.1	ns	9.4	8.2	ns
Creatine kinase, UI/ml	9.1	28.3	***	2.8	7.1	***	26.7	25.0	ns
Cortisol, ng/ml	33.6	31.2	ns	42.5	35.0	ns	55.8	49.9	ns
Urine hormones									
Cortisol, ng/mg creatinine	29.7	32.2	ns	8.0	11.1	ns	36.0	25.7	ns
Adrenaline, ng/mg creatinine	8.4	17.8	***	3.8	5.1	ns	16.8	17.3	ns
Noradrenaline, ng/mg creat.	17.1	26.7	**	13.1	17.4	ns	22.0	33.7	ns

Table 2: Correlation coefficients between physiological and meat quality traits, and skin lesions.

	Plasma CK	Nb skin lesions	GP SM	pHu SM
Plasma CK		0.37 ***	-0.28 ***	0.32 ***
Urine cortisol	0.16 ns	0.17 ns	-0.25 *	0.22 *
Urine adrenaline	0.42 ***	0.37 ***	-0.31 **	0.55 ***
Urine noradrenaline	0.31 ***	0.18 ns	-0.32 **	0.49 ***

* : P<0.05; ** : P<0.01; *** : P<0.001.

References

Foury, A., Devillers, N., Sanchez, M.P., Griffon, H., Le Roy, P.and Mormède, P. (2005). Stress hormones, carcass composition and meat quality in Large White X Duroc pigs. Meat Science, 69: 703-707.

Gentry, J.G., McGlone, J.J., Miller, M.F. and Blanton Jr., J.R. (2004). Environmental effects on pig performance, meat quality, and muscle characteristics. Journal of Animal Science, 82: 209-217.

Lebret, B., Meunier-Salaün, M.C., Foury A., Mormède P., Dransfield E. and Dourmad J.Y. (2006). Influence of rearing conditions on performance, behavioral and physiological responses of pigs to preslaughter handling, carcass traits and meat quality. Journal of Animal Science, *in press*.

Ngapo, T.M., Dransfield, E., Martin, J.F., Magnusson, M., Bredahl, L. and Nute, R.G. (2003). Consumer perceptions: pork and pig production. Insights from France, England, Sweden and Denmark. Meat Science, 66: 125-134.

Terlouw, E.M.C., Astruc, T. and Monin, G. (2004). Effect of genetic background, rearing and slaughter conditions on behaviour, physiology and meat quality of pigs. Proc. EU Workshop "Sustainable Pork Production: Welfare, Quality, Nutrition and Consumer Attitudes", Copenhagen, Denmark, 17-18/06.13 Pp.

EFFECT OF AD-LIBITUM FEEDING OF FRESH GUINEA GRASS, WILD SUNFLOWER, IPILIPIL AND ERYTHRINA LEAF MEAL ON CARCASS QUALITY OF RABBITS

R.W. Seneviratne*, P.D.G.E. Perera and H.W. Cyril

Department of Animal Science, Faculty of Agriculture, University of Peradeniya, Peradeniya 20400, Sri Lanka
Email: ruwani_ws@yahoo.co.uk

Introduction
Protein malnutrition in humans is a major problem in third world countries. Therefore it is important to exploit the availability of non traditional meat sources as proteins for humans. Rabbit meat could play an important role in this context (Lang, 1981). In most animal production systems feed is one of the most costly and important factors which controls the cost of production (Ibrahim, 1988). Introduction of forages in rabbit diets could cut down the cost of feeding as well as amount of concentrate fed to rabbits. The main objective of this study was to find out the effect of *ad libitum* feeding of Guinea grass, Wild sunflower, Ipilipil and Erythrina Leaf Meal (ELM) on growth and carcass quality of rabbits.

Materials and Methods
This study was carried out as two experiments at the Livestock field Station at the University of Peradeniya, Sri Lanka. Experiment 1 was conducted using fresh Guinea grass, Wild sunflower and Ipilipil. Each treatment consisted of 25 grams of broiler finisher mash. *Ad-libitum* amount of fresh Guinea (*Panicum maximum*), Wild sunflower (*Tithonia diversifolia*) and Ipilipil (*Leucena leucocephala*) were added to each treatment separately. A total of 24 New Zealand White rabbits were used. For each treatment eight rabbits were randomly assigned and fed individually.

Experiment 2 was carried out using *Erythrina indica*. *Erythrina* leaf was stripped and initially wilted in the shade for two days for the production of leaf meal. The wilted material was oven dried and ground through a mill (Model Christy Lab Mill 47455). Control diet consisted of corn 25%, rice bran 25%, coconut poonac 40%, di-calcium phosphate 2.5%, salt 0.25% and premix 0.25%. Treatment diet 1(T1) was formulated by replacing 50% of coconut poonac with ELM. Treatment diet 2 (T2) was formulated by replacing (100%) coconut poonac with ELM. A total of 18 New Zealand white rabbits were used. For each treatment six rabbits were randomly assigned and fed individually.

During the trial period of both experiments feed intake and weekly weight gain were recorded. At the week 14 all rabbits were fasted for 24 hours and slaughtered. Offal weights and carcass quality characteristics were recorded. Data was analysed according to the completely randomised design using analysis of covariance technique. Dunnet test was used for mean separation.

Results and Discussion
Experiment 1

Table1. Effect of *ad libitum* Feeding of Fresh Guinea Grass (GG), Wild Sunflower (WS) and Ipilipil (IP) on Growth and Carcass Quality of Rabbits.

Parameter	Feed Type		
	GG	WS	IP
Average daily weight gain (g/day)	29.0	26.3	25.85
Average daily feed intake (g/day)	73.4	79.8	75.1
Dressing %	50.4	46.92	46.02
Abdominal fat%[a]	0.62	0.16	0.16
Eye muscle area (cm^2)	7.2	6.9	7.4

a. Live weight basis%

Average daily weight gain was highest in rabbits fed with Guinea Grass and it was lowest for Ipilipil. Average daily feed intake was highest for Wild Sunflower and lowest for Guinea Grass. Dressing % and abdominal fat % was highest for Guinea Grass. Lowest dressing % and abdominal fat % was recorded for Ipilipil. Eye muscle area was highest for Ipilipil and lowest for Guinea Grass. The differences were significant for average daily weight gain, dressing % and abdominal fat % but not for average feed intake and eye muscle area.

Experiment 2

Table 2: Effect of Feeding of Erythrina Leaf Meal (ELM) on Growth and Carcass Quality of Rabbits.

Parameter	Diet		
	Control	20%ELM(T1)	40% ELM(T2)
Average daily weight gain (g/day)	13.6	10.10	6.58
Average daily feed intake (g/day)	64.9	69.98	75.78
Dressing %	49.76	45.52	47.26
Abdominal fat%[a]	1.06	0.00	0.06
Eye muscle area (cm^2)	4.14	3.86	4.61

a. Live weight basis%

Average daily weight gain was highest in control diet and lowest in T2. Average daily feed intake was highest in T2 and lowest in control diet. Dressing % was highest in control diet and lowest in T1. Abdominal fat % was highest in control and no abdominal fat was found in T1. Eye muscle area was highest in T2 and lowest in T1. The differences were significant for dressing % and abdominal fat % but not for average daily weight gain, average daily feed intake and eye muscle area.

Table 3: Effect of Feeding of ELM on Meat : Bone Ratio.

Part of the Carcass	Diet		
	Control	20%ELM (T1)	40%ELM (T2)
Fore half	3.54	2.04	2.95
Hind Half	2.43	2.00	3.50
Whole Carcass	2.82	2.55	3.90

Meat to bone ratio (Table 3) was significantly lowest in the fore half for T1 but differences were not significant for the hind half and whole carcass.

Conclusions

Use of fresh guinea grass in rabbit feed rations is economical in terms of reducing cost of production by reducing feed intake and increasing profit margin by increasing dressing percentage and daily weight gain. By incorporating ELM into the rabbit rations abdominal fat percentage can be reduced.

References

Ibrahim, M.N.M. 1981. *Feeding tables for ruminant in Sri Lanka.* Kandy: Kandy Offset Printers Ltd.

Lang, J. 1981. The nutrition of commercial rabbits, physiology, digestibility and nutrient requirements. *Nutrient Absttr.* Rev. 51: 197 - 225

LOIN MARBLING IS ASSOCIATED WITH PIG GROWTH RATE AND OVERALL CARCASS LEANNESS

S. Jungst, B. Fields, A. Williams, A. Sosnicki* and C. Booher

PIC USA, P.O. Box 348, Franklin KY, 42135-0348 USA. Email: Andrzej.Sosnicki@pic.com

Keywords: pork, meat quality, marbling, growth, lean

Introduction

There is increasing interest in the degree of marbling in fresh pork loins in the pork industry. This is being led by the ever increasing amount of fresh pork exported to Japan and used by the food service industry. Although contradictory studies have been published on the impact of increased marbling level on consumer acceptability, most agree pork should have 2-4% intramuscular fat (IMF). Fernandez *et al.*, (1999b) and Brewer *et al.*, (2001) established that more highly marbled chops were juicer, more tender, and more flavoursome, but they appeared paler and were less likely to be purchased by consumers. Both Casteels *et al.*, (1994) and Rincker *et al.*, (2005) agreed IMF levels had little effect on the sensory properties of fresh pork. A desirable level of 2.5% to 3.0% was called for by DeVol *et al.*, (1988) while Fernandez *et al.*, (1999a) determined that flavour and juiciness were enhanced when IMF levels were over 2.5%. Others have suggested a threshold of 2% IMF to ensure desirable eating quality (Bejerholm and Barton-Gade, 1986). A survey of the US pork industry claimed the average marbling score of fresh pork loins in the retail case was 2.37 (AMSA, 2003), suggesting there is still a need to increase the amount of marbling in fresh pork loins.

While debating the importance of increased marbling levels is simple, putting this into practice is not. The amount of marbling in pork loins is known to be directly correlated with the amount of backfat (Huff-Lonergan, 2002). Furthermore, studies that have looked at the impact of nutrition on IMF and marbling levels demonstrated IMF can be increased by decreasing the protein:digestible energy ratio or feeding increased levels of fat during finishing, but at the detriment of growth rate (Cisneros *et al.*, 1996; Eggert *et al.*, 1998).

Without a premium payment system or a pre-defined bonus structure for producers delivering pigs with increased levels of marbling, the economic impact seems to be negative. However, if there were a trade off with certain production factors such as fatter carcasses but with faster growth, the additional costs of producing pigs with increased marbling may be nullified. Objectives of this study were to look beyond the well documented eating quality parameters associated with marbling and to investigate the production factors that affect the overall performance of the pig, including the marbling levels.

Materials and Methods

PIC commercial sires of predominantly Duroc (19 boars), RN^- Hampshire (23 boars), and Synthetic (21 boars) genetic backgrounds were used in this study. A total of 1,903 pigs were farrowed by Camborough 22 or 23 sows at a PIC commercial research farm and were harvested at a commercial processor in the U.S. Market weights ranged from 88 kg to 170 kg with a mean of 127.7 kg. Feed intake was measured on individual pigs using FIRE (Feed Intake Recording Equipment) feeders. The pigs were fed corn-soy based diets formulated to meet PIC's nutrient recommendations for grow – finish pigs. Paylean™ was not fed. Carcasses were measured for backfat and loin depth using a Fat–O–Meater. Subjective marbling scores were assigned using the National Pork Board score of 1 (devoid) to 10 (abundant) (NPB, 1999) following visual observation of the sirloin end of each boneless loin. Data were analyzed using SAS PROC MIXED (SAS Inst., Inc., Cary, NC).

Results and Discussion

Growth rate and carcass leanness are major determinants of the amount of marbling in boneless loins. Least square means are presented in Table 1 for the growth, carcass, and meat quality traits by subjective loin marbling score. All genotypes performed in a similar manner as there was no interaction between genotype and marbling score. Days to 125 kg showed a linear decrease of -.97 days for each 1 unit increase in subjective marbling score. For average daily gain, there was a curvilinear relationship indicating that the pigs grew faster but the rate of the increase changed at the higher marbling scores. Feed conversion was better at the higher levels of marbling and improved by .017 for each 1 unit increase in marbling score. Carcasses that displayed more backfat thickness, less loin depth, and a lower percentage of carcass lean had greater amounts of marbling. Furthermore, there was a linear relationship between subjective marbling score and loin pH, with loin pH increasing by .022 for each 1 point increase in marbling score. There was a curvilinear relationship between subjective marbling score and subjective loin colour score. The subjective loin colour score increased, but at a faster rate at the higher levels of marbling.

Table 1: Production, carcass and loin meat quality means by subjective marbling score.

	Subjective Loin Marbling Score[a]						
Trait	1	2	3	4	5	Linear	Quadratic
Production traits	n = 646	n = 538	n = 234	n = 173	n = 98		
Average daily gain, g /day	889	903	907	925	907	.012**	- .007*
Age at 125 kg, days	173.0	172.1	171.0	168.7	169.8	- .970***	.207
Feed conversion, kg feed / kg gain	2.51	2.51	2.50	2.42	2.47	- .017*	.001
Carcass traits (adj. for carcass wt.)	n = 681	n = 574	n = 225	n = 187	n = 101		
Hot carcass weight, kg	95.8	97.4	96.1	95.0	93.3	- 1.639***	- 1.031**
Backfat thickness[b], mm	18.5	18.5	19.1	19.8	20.6	.021***	.004†
Loin depth[b], mm	59.4	58.4	58.7	57.9	57.9	- .016*	.003
Lean[b], %	54.7	54.5	54.3	53.7	53.3	- .357***	- .049
Loin meat quality traits	n = 694	n = 618	n = 258	n = 192	n = 103		
$pH_{24\text{-hours}}$	5.64	5.67	5.67	5.69	5.73	.022***	.003
Subjective colour score, (1 to 6)	2.94	3.07	3.10	3.26	3.55	.141***	.032***
Minolta L*	48.41	48.58	49.02	48.92	48.61	.076	- .106

[a] Subjective marbling scores based on the National Pork Board scale of 1 (devoid) to 10 (abundant).
[b] Carcass traits were adjusted to a 96.1 kg hot carcass weight.
†P < .10, *P < .05, **P < .01, *** P < .001

Conclusions

Loin marbling is significantly affected by growth rate and overall leanness. Increased levels of marbling will result in a lower lean percentage, but this will be counteracted by faster growth rates and better feed conversion. Muscle colour and pH will also be improved in these carcasses. Based on the known relationships between pH and muscle colour it is expected that water holding capacity and tenderness may be improved as well.

References

American Meat Science Association (2003). Benchmarking Value in the Pork Supply Chain: Quantitative Strategies and Opportunities to Improve Quality – Final Report. AMSA, Savoy, IL, USA.

Bejerholm, C. and Barton-Gade, P.A. (1986). Effect of intramuscular fat level on eating quality of pig meat. In: Proceedings of 32nd Meeting of European Meat Research Workers (pp. 389-391). Gent, Belgium.

Brewer, M.S., Zhu, L.G. and McKeith, F.K. (2001). Marbling effects on quality characteristics of pork loin chops: consumer purchase intent, visual and sensory characteristics. Meat Science, 59: 153-163.

Casteels, M., Van Oeckel, M.J., Boschaerts, L., Spincemaille, G. and Boucque, Ch.V. (1994). The relationship between carcass, meat and eating quality of three pig genotypes. Meat Science, 36: 253-269

Cisneros, F., Ellis, M., Baker, D.H., Easter, R.A. and McKeith, F.K. (1996). The influence of time of feeding of amino-acid deficient diets on the intramuscular fat content of pork. Animal Science, 63: 517-522.

DeVol, D.L., McKeith, F.K., Bechtel, P.J., Novakofski, J., Shanks, R.D. and Carr, T.R. (1988). Variation in composition and palatability traits and relationships between muscle characteristics and palatability in a random sample of pork carcasses. Journal of Animal Science, 66: 385-395.

Eggert, J.M., Farrand, E.J., Mills, S.E., Shinckel, A.P., Forrest, J.C., Grant, A.L. and Watkins, B.A. (1998). Effects of feeding poultry fat with supplemental beef tallow on pork quality and composition. http://ansc.purdue.edu/swine/swineday/sday1998/psd06-98.htm.

Fernandez, X., Monin, G., Talmant, A., Mourot, J. and Lebret, B. (1999a). Influence of intramuscular fat content on the quality of pig meat – 1. Composition of the lipid fraction and sensory characteristics of *m. longissimus lumborum*. Meat Science, 53: 59-65.

Fernandez, X., Monin, G., Talmant, A., Mourot, J. and Lebret, B. (1999b). Influence of intramuscular fat content on the quality of pig meat – 2. Consumer acceptability of *m. longissimus lumborum*. Meat Science, 53: 67-72.

Huff-Lonergan, E., Baas, T.J., Malek, M., Dekkers, J.C.M., Prusa, K. and Rothschild, M.F. (2002). Correlations among selected pork quality traits. Journal of Animal Science, 80: 2-10.

National Pork Board, (1999). Pork Quality Standards. Des Moines, IA, USA.

Rincker, P., Killefer, J., Matzat, P., McKeith, F., (2005). The effect of ractopamine and intramuscular fat content on sensory attributes in ork from pigs with similar genetics. In: Proceedings of 51st International Congress of Meat Science and Technology (pp. 77-78). 7-12 August 2005. Baltimore, Maryland, USA.

SUPPLEMENTARY FEEDING TO IMPROVE MILK AND MEAT PRODUCTION OF CATTLE UNDER SMALLHOLDER FARM CONDITIONS IN BANGLADESH

M.A.S. Khan*, M.A.R. Chowdhury, M.N. Islam and M.S.R. Siddiki

Department of Dairy Science, Bangladesh Agricultural University, Mymensingh-2202, Bangladesh
Email: s-khan@royalten.net.bd

Keywords: supplementation, meat, milk, body weight

Introduction

Cattle on smallholder farms produce milk, manure and meat that are used to improve the food security of the farm household. The smallholder farmers of Bangladesh cannot feed their cattle adequately. Most fibrous feeds available to animals are deficient in some nutrients and are generally poor in digestibility. Straw is the most important crop residue contributing more than 90% of the basal dry matter available to the cattle in Bangladesh (Saadullah *et al.,* 1994). Recently considerable efforts of many animal nutritionists have been directed to improve the nutritive value of rice straw and its utilisation (Hock *et al.*, 1988). Their results suggest that the nutritional limitation can be overcome by physical and chemical treatment or by supplementation with specific nutrients to provide an optimum ruminal condition for rumen microorganisms. It has been observed from on-station trials that the supplementation of urea molasses block (UMB) with straw-based diet could increase feed intake, daily milk yield, lactation period and increase live weight gain in cows and calves in Bangladesh (Saadullah, 1991). The most common system for keeping cattle in Bangladesh on which most resource-poor people depend, is the mixed crop livestock farming system. However, the feeding system commonly used in smallholder farms in Bangladesh is different to that of organised large farms in urban areas including farms of peri-urban areas. It was observed in an earlier on station experiment that 500g/d/cow UMB is required for maximum benefit. But it is important to determine the amount of UMB for small farmer conditions at village level. The main objective of the experiment was to determine the optimum level of UMB required for maximum benefit regarding meat and milk production under smallholder farm conditions in Bangladesh.

Materials and Methods

The experiment was conducted at some selected villages in the Mymensingh district. One hundred and eight lactating crossbred (Holstein x local zebu) cows were randomly divided into four groups each consisting of 27 cows whose average live weight was 302 (SE 53.93) kg and all the cows were of 2nd lactation. Four treatments were 0.00, 0.35, 0.50 and 0.65 kg/d/cow UMB fed with basal diets designated as T_0, T_1, T_2 and T_3 respectively. The basal diet consisted of rice straw and mixed grass (at a ratio of 4:1) available to smallholder farmers and concentrates mixture made from locally available feed ingredients (36% wheat bran, 36% rice polish, 27% sesame cake and 1% common salt). The UMB consisted of 39% molasses, 20% wheat bran, 20% rice polish, 10% urea, 6% lime powder (CaO) and 5% common salt (NaCl). The detailed composition of diets are given in Table 1.

Table 1: Composition of diets for different treatment groups (Fresh basis).

Ingredients (Fresh basis)	Treatments (kg/d)			
	T_0	T_1	T_2	T_3
Roughages				
a. Straw	*ad libitum*	*ad libitum*	*ad libitum*	*ad libitum*
b. Mixed grass	3.00	3.00	3.00	3.00
Concentrate	2.75	2.75	2.75	2.75
UMB	-	0.35	0.50	0.65

The feed and milk samples were collected randomly from study area and were analysed for proximate components according to the method of AOAC (2003). Milk yield of each individual cow was taken daily and body weight of the cow was recorded monthly after calving to confirm pregnancy of non-pregnant cow-180 days of lactation. Body weight of each calf was recorded monthly up to 6th months of age. Data for milk yield, body weight changes of cows and calves was statistically analysed by one factor completely randomized block design using the MSTAT statistical package program. Significant differences among the treatment means were compared by DMRT.

Results and Discussion

As shown in Table 2, there were increasing values for DM and ME ($P>0.05$) intake. On the other hand, CP intake increased significantly ($p<0.001$) as the level of UMB increased with basal diets. Analysis indicated that there was a positive relationship of DM, CP and ME intake with different levels of UMB intake.

Table 2: Nutrient intake of unsupplemented and UMB supplemented cows.

Ingredients (DM basis)	Diets (kg/d)				SEM	Significance
	T_0	T_1	T_2	T_3		
Total DM (kg/d)	8.63	8.98	9.69	9.77	0.38	NS
CP (g/d)	587^b	671^b	793^a	825^a	41.68	***
ME (MJ/d)	62.67	65.74	72.11	97.74	14.37	NS

Values with different superscripts in the same row differ significantly
NS= Non-Significance (P>0.05), ***=P<0.001

Table 3: Composition of milk at unsupplemented and at different levels of UMB supplemented cows.

Parameters	Treatments				SEM	Significance
	T_0	T_1	T_2	T_3		
Milk fat (g/kg)	41.2	44.5	47.1	47.1	0.528	NS
Protein (g/kg)	34.3	34.4	35.7	36.0	0.131	NS
Lactose (g/kg)	41.4	40.2	40.7	40.6	0.139	NS
Total Solids (g/kg)	122.6	126.1	130.6	131.8	0.625	NS
Ash (g/kg)	6.7	7.0	7.1	7.1	0.098	NS

Table 4: Mean values of milk yield and body weight gain of cows and calves.

Parameters	Treatments				SEM	Significance
	T_0	T_1	T_2	T_3		
180 days av. yield (kg/d)	5.42^b	5.49^b	6.81^a	6.83^a	0.47	*
Lactation average (kg/d)	3.68 (306 days)	5.08 (303 days)	5.95 (336 days)	5.99 (335 days)	0.81	NS
Body weight changes of cows (g/d)	9.37^c	65.89^{ab}	88.04^a	88.44^a	22.35	*
Calf weight gain (g/d)	159^b	167^b	215^a	228^a	14.12	***

Values with different superscripts in the same row differ significantly
NS= Non-Significance (P>0.05), *=P<0.05, ***=P<0.001

The average milk yield of 180 days in 4 groups differed significantly (P<0.05). The highest milk yield was observed in T_3 but the difference between T_2 and T_3 was negligible and the difference was not statistically significant (P>0.05). The lactation average was also increased linearly by the increasing level of UMB intake by cows. The body weight gain of cows up to the 6th month of lactation was higher in supplemented cows when compared to unsupplemented cows. Body weight of calves up to the 6th month of age was significantly (P<0.001) increased in the groups of calves of those dams receiving UMB with higher milk composition (Table 3). Cattle fattening before Eid-ul-Azha (Muslim festival) for beef production has become an important business for smallholder farmers. Of the calves born, very few are reared as bulls for natural breeding. Most of the male calves are therefore available for rearing as beef animals.

Conclusions

The addition of 0.50 kg/d/cow UMB supplement to the basal diets improved milk yield, milk composition and meat of cows and calves. Cattle fattening can also be used as a path out of poverty.

References

AOAC, (2003). Official Method of Analysis of AOAC International. 17th Edition, (edt. Horwitz, W.) Revision 2, Volume 1 & 2, AOAC International, Suite 500, 481 North Frederick Avenue Gaithersburg, Maryland 20877-2417, USA.

Hock, R. D. V., Muttetuwegama, G.S. and Sciere, J.B. (1988). Overcome the nutritional limitations of rice straw for ruminants. Urea ammonia treatment and supplementation with rice bran and glicirida for lactating Surti buffaloes., Asian Australasian Journal of Animal. Science., 1 (4):201-208.

Saadullah, M. (1991). The importance of urea molasses blocks and by-pass protein in animal production: the situation in Bangladesh. Isotope and related techniques in animal production, IAEA-SM-318/28 pp.145-156.

Saadullah, M., Sarkar, D.R.D., Haque, M.A., Samad, M. A. and Islam, S. (1994). Study on the effect of replacement of wheat bran by urea molasses block lick in straw base diet on the performance of draught cows. Bangladesh Journal of Animal Science., 23(1-2):163-168.

INTRAMUSCULAR FATTY ACID COMPOSITION OF EARLY AND LATE LAMBS IN SICILY

A. Lanza[1], M. Bella[1], V. Vasta*[2] and A. Priolo[1]

[1.] *DACPA Sez. Scienze delle Produzioni Animali, University of Catania, via Valdisavoia 5, 95123 Catania, Italy,* [2.] *Dottorato di Ricerca in Produzioni Foraggere Mediterranee, University of Palermo, viale delle Scienze 13, 90128, Palermo, Italy. Email: vvasta@unict.it*

Keywords: lamb, fatty acid composition, feeding system, Sicily

Introduction

In recent years meat quality has been seen as a main target in animal production. In particular, consumers demand for healthier food has encouraged research focussing on the manipulation of meat fatty acid (FA) composition in order to improve benefits to human health. It has been demonstrated that lowering the *n*-6 fatty acids content and increasing the *n*-3 fatty acids contents in the human diet reduces the incidence of cardiovascular diseases (Enser *et al.*, 1998). Several studies focussed on the meat content of the *cis*-9, *trans*-11 isomer of linoleic acid (conjugated linoleic acid or CLA), as it has been shown that this fatty acid exerts favourable effects against carcinogenesis in mammals (McGuire and McGuire, 1999). It is known that animal species, breed, sex, age and feeding system are all factors influencing meat fatty acid composition. Conversely to other European countries, in Sicily two types of lamb are produced from the same breed: early lambs, fed exclusively maternal milk, and late lambs, allowed to graze and also receiving concentrate supplements. The intramuscular fatty acid profile of these two types of meat were compared. This research was developed in the framework of a three-year collaboration between "Ente di Sviluppo Agricolo Regione Siciliana" (ESA) and the University of Catania.

Materials and Methods

Forty-two male Comisana lambs were used in this trial. Twenty-two lambs were fed exclusively maternal milk and were slaughtered at 40 days of age (early lambs). The remaining 20 lambs were allowed to graze on natural pasture and received also a daily ration of concentrate and hay until slaughter at 8 months of age (late lambs). After slaughtering, carcasses were stored at 4°C and 24hrs post-mortem the muscle *longissimus dorsi* (LD) was excised from the carcasses. Meat intramuscular fatty acids (FA) were extracted according to Folch *et al.* (1957), as modified by French *et al.* (2000) from 5g of ground meat removed at the level of the 13th thoracic rib. Fatty acid methyl esters were separated by gas chromatography. Meat FA were identified according to retention time with reference to fatty acid standards. Meat FA are expressed as g/100 g of total fatty acid methyl esters. Data were analysed by ANOVA general linear model (GLM) including "lamb type" (early or late lamb) in the model.

Results and Discussion

Medium chain saturated fatty acids (C12:0, C14:0 and C15:0) were present at higher amount ($P < 0.0005$) in the intramuscular fat of early lambs (Table 1). A similar pattern was found for palmitoleic acid (C16:1 9-*cis*; $P < 0.05$). The percentages of these fatty acids in early lamb meat were higher compared to data reported by Valvo *et al.* (2005) on meat fatty acid composition of lambs fed exclusively maternal milk and with their dams fed green grass or concentrate. Oleic acid did not differ between the two groups of animals and was the predominant FA in lambs intramuscular fat, confirming results reported by Sañudo *et al.* (2000). *Trans*-vaccenic acid (C18:1 11-*trans*) and linolenic acid (C18:3 *n*-3) were present at higher amount ($P < 0.005$) in the meat of late lambs compared to early lambs. Late lambs had free access to pasture and according to Santos-Silva *et al.* (2002) a green herbage diet enhances the inclusion of these two fatty acids into lamb fat. The content of CLA in meat was not affected by lamb type. Very long chain FA, total *n*-6 and *n*-3 FA were not different among the two groups of animals. Only C22:6 *n*-3 (DHA) was found at significantly higher percentage in early lamb meat.

Conclusions

This study has shown that some differences occur in meat fatty acid composition between early and late lambs in Sicily. In particular, early lamb meat contains higher percentages of medium chain saturated fatty acids and lower *trans*-vaccenic acid and linolenic acid. These differences were probably caused by the different rearing conditions and age at slaughter of the two groups of animals. Conjugated linoleic acid, total *n*-6 and *n*-3 FA contents did not vary among the two types of lamb. The *n*-6 / *n*-3 ratio was similar in the meat of the two groups of lambs and its low value was favourable for consumers' health.

Table 1: Effect of lamb typology on meat intramuscular fatty acid composition (g/100g FA methyl esters).

	Early lamb	Late lamb	SEM	*P* - value
No of lambs	*n* = 22	*n* = 20		
C12:0	0.82	0.23	0.062	<0.0005
C14:0	5.42	2.20	0.309	<0.0005
C14:1	0.26	0.16	0.014	<0.0005
C15:0	0.48	0.38	0.015	<0.0005
C15:1	0.11	0.33	0.025	<0.0005
C16:0	20.93	19.60	0.373	0.075
C16:1 9-*cis*	1.42	1.06	0.071	0.010
C17:0	0.74	0.97	0.025	<0.0005
C17:1	0.31	0.66	0.062	0.004
C18:0	9.38	13.61	0.391	<0.0005
C18:1 11-*trans*	1.08	2.29	0.139	<0.0005
C18:1 9-*cis*	27.67	26.46	0.478	0.211
C18:1 11-*cis*	1.70	2.11	0.080	0.008
C18:2 12-*trans*	0.56	0.65	0.037	0.242
C18:2 *n*-6 (linoleic)	12.87	13.17	0.394	0.712
C18:3 *n*-6 (γ-linolenic)	0.14	0.15	0.007	0.934
C18:3 *n*-3 (α-linolenic)	1.47	2.46	0.178	0.004
C18:2 9-*cis*,11-*trans* (CLA)	0.65	0.52	0.049	0.953
C18:2 10-*trans*,12-*cis*	0.03	0.04	0.005	0.383
C20:2 *n*-6	0.10	0.13	0.023	0.516
C20:3 *n*-3	0.12	0.13	0.011	0.770
C20:4 *n*-6	7.93	7.63	0.230	0.901
C20:5 *n*-3	1.67	1.67	0.118	0.989
C22:5 *n*-3	2.10	2.16	0.090	0.750
C22:6 *n*-3	1.43	0.74	0.071	<0.0005
n-3	6.79	7.16	0.365	0.625
n-6	21.09	21.21	0.660	0.928
n-6 / *n*-3	3.20	3.77	0.239	0.236

References

Enser, M., Hallett, K.G., Hewett, B., Fursey, G.A.J., Wood, J.D. and Harrington, G. (1998). Fatty acid content and composition of UK beef and lamb muscle in relation to production system and implications for human health. Meat Science, 49, 329-341.

Folch, J., Lees, M. and Stanley, S.G.H. (1957). A simple method for the isolation and purification of total lipides from animal tissues. Journal of Biological Chemistry, 226, 497-509.

French, P., Stanton, C., Lawless, F., O'Riodan, E.G., Monahan, F.J., Caffrey, P.J. and Moloney, A.P. (2000). Fatty acid composition, including conjugated linoleic acid, of intramuscular fat from steers offered grass, grass silage, or concentrate-based diets. Journal of Animal Science, 78, 2849-2855.

McGuire, M.A. and McGuire, M.K. (1999). Conjugated linoleic acid (CLA): a ruminant fatty acid with beneficial effects on human health. Proc Am Soc Anim Sci, http://.asas.org/jas/symposia/proceedings/0938.pdf.

Santos-Silva, J., Bessa, R.J.B. and Santos-Silva, F. (2002). Effect of genotype, feeling system and slaughter weight on the quality of light lambs. II. Fatty acid composition. Livestock Production Science, 77, 187-194.

Sañudo, C., Enser, M.E., Campo, M.M., Nute, G.R., María, G., Sierra, I. and Wood, J.D. (2000). Fatty acid composition and sensory characteristics of lamb carcasses from Britain and Spain. Meat Science, 54, 339-346.

Valvo, M.A., Lanza, M., Bella, M., Fasone, V., Scerra, M., Biondi, L. and Priolo, A. (2005). Ewe feeding system (grass v. concentrate) strongly affects intramuscular fatty acids of lambs fed exclusively maternal milk. Animal Science, 81, 431-436.

PRINCIPAL COMPONENT ANALYSIS TO CHARACTERISE THE CHEMICAL COMPOSITION OF BEEF ACCORDING TO AGE AND GENDER

S. Uzcátegui-Bracho[1], L. Arenas de Moreno*[2], N. Jerez-Timaure[2], N. Huerta-Leidenz[2], M. Giuffrida-Mendoza[3] and J. Ortega[4]

[1.] Facultad de Ciencias Veterinarias. Departamento de Biología Animal, [2.] Facultad de Agronomía. Instituto de Investigaciones Agronómicas, [3.] Facultad de Medicina. Departamento de Ciencias Fisiológicas. Universidad del Zulia. Aptdo. 15252. Maracaibo, Venezuela. Email: sojanuzcategui@cantv.net

Keywords: *longissimus dorsi*, chemical composition, beef cattle, principal components analysis

Introduction

Meat chemical composition has been well characterised in other countries, where research has demonstrated that intrinsic factors (gender, species) and extrinsic factors (plane of nutrition, growth regulation, castration, *etc.*) are largely responsible for the variation found in beef nutrient composition (Byers *et al.*,1988). However, little information exists about the chemical composition in Venezuelan ruminant meat produced under tropical conditions. The purpose of the present investigation was to assess the relative ability of principal components analysis (PCA) to correctly characterise Venezuelan beef, considering different ages and gender according to their nutrient content (proximate, mineral composition and fatty acids profile). This statistical approach will allow us to convert a large number of related variables into a smaller set of factors, composed of a weighed set of the original variables, to describe specific patterns of behaviour.

Materials and Methods

A randomly selected group of 145 cattle raised under tropical conditions in Venezuela (mostly grass-fed) were slaughtered according to typical industry procedures at a commercial packing house. Animals were classified by age (2.5, 3.0, 3.5 and 4.0 yr estimated by dentition) and gender (61 bulls, 64 steers and 20 heifers).

After 48h *post-mortem,* two steaks (2.5cm thick) were obtained from the *longissimus dorsi* (LD) muscle of each carcass, vacuum-packaged and stored at -30°C. Duplicates of ground samples were analysed for proximate composition and mineral analyses, as described by Huerta-Leidenz *et al.* (2003). Total lipids content was determined according to Folch *et al.* (1957). Fatty acids (FA) were determined by gas chromatography, following the methodology used by Uzcátegui-Bracho *et al.* (1999). Experimental data were subjected to multivariate analysis by using the PROC PRINCOMP procedure of the Statistical Analysis System (SAS, 2000). The number of factors retained from each PCA was determined by variance value explained by each factor, and by factor interpretability. Labelling of the factors was primarily descriptive and based on our interpretation of the pattern structures. The highest-weight variables of the principal component 1 (PC1) were: total lipids, pentadecilic (C15:0), palmitic (C16:0), stearic (C18:0), oleic (C18:1 *cis*), elaidic (C18:1 *trans*), linoleic (C18:2) acids, total saturated (SFA), unsaturated (UFA), monounsaturated (MUFA), polyunsaturated fatty acids (PUFA), and *cis* and *trans* fatty acids. The variables for PC2 were: UFA/SFA, MUFA/SFA and PUFA/SFA ratio, while in PC3 the variables were: Ca, Mg and P.

Results and Discussion

Multivariate analysis showed that PC1 explains 51.0% of the total variability, due to the presence of relatively high concentrations of the selected variables in beef samples. PC2 and PC3 explain 13.1% and 9.8%, respectively. Together, PC1, PC2 and PC3 explain 73.9% of the total variability. It was not possible to group the variables by age with PCA, this could be due to the narrow range of ages considered in the study.

PCA allowed the separation of original variables into four groups. Animals of group 1 represented most of the bull samples showing the strongest relationship with UFA/SFA, MUFA/SFA and PUFA/SFA, with low contents of SFA (C16:0, C18:0), MUFA (C15:1, C18:1 *cis*, C18:1 *trans*) and PUFA (C18:2). In relation to unsaturated level, our results showed that most bulls presented more UFA than SFA. Equally, other studies (Mitchell *et al.*1991; Yang *et al.* 1999) have reported similar results in beef.

Group 2 included bulls and a small proportion of steers and heifers. Animals from this group showed low levels of the relationships UFA/SFA, MUFA/SFA, PUFA/SFA. The amount of SFA (C16:0, C18:0) (C18:1 *cis*, C18:1) PUFA (C18:2) (*trans* C15:1) and MUFA, were also low. These results can be attributed to the low total content of lipids in the samples. Group 3 was represented by most steers, heifer samples, showing low levels of the relationships UFA/SFA, MUFA/SFA, PUFA/SFA, but a high content of C15:0, C16:0, C18:0, *cis* C18:1, C18:1, *trans* C18:2, total UFA SFA, MUFA PUFA. Group 4 represented the majority of steers showing high concentrations of total lipids, C15:0, C16:0, C18:0, C18:1 *cis*, C18:1 *trans*, C18:2, total SFA, UFA. This group also showed the highest contents of MUFA PUFA, but UFA/SFA, MUFA/SFA PUFA/SFA. Similar results in fresh meat were reported by Valero-Leal (2000) who demonstrated that steers presented 0.24g more lipids than bulls. Huerta-Leidenz (1993) attributed the fact that steers tend to accumulate, more intramuscular fat than bulls to hormonal effects.

Conclusion

Beef samples used in this study represent the national beef herd; animals were selected in the slaughterhouse that gathers animals from several farms in the country. PCA allows us to better characterise meat samples by gender than by age, maybe due to the small variability by age presented in the animals.

Variables related to fat and minerals were grouped in different PCs. Meat from steers and heifers showed the highest concentration of total lipids, SFA, UFA, MUFA PUFA and low levels of the relationships UFA/ SFA, MUFA/ SFA PUFA/SFA.

References

Byers, F.M., Cross, H.R. and G.T. Schelling. (1988). Integrated Nutrition, Genetics and Growth Management Programs for Lean Beef Production. In: Designing Foods. Animal product Options in the Marketplace. National Research Council. National Academy Press. Washington DC.

Folch, J.; Lees, M. & G. H. Sloane Stanley. (1957). A simple method for the isolation and purification of total lipides from animal tissues. Journal of Biological Chemistry, 226, 497-509.

Huerta-Leidenz, N.; Arenas de Moreno, L.; Morón-Fuenmayor, O.; Uzcátegui-Bracho, S. (2003). Composición mineral del músculo *longissimus* crudo derivado de canales bovinas producidas y clasificadas en Venezuela. Archivos Latinoamericanos de Nutricion, 53, 1, 96-101.

Huerta-Leidenz, N.O. and G. Ríos. (1993). La castración del bovino a diferentes estadios de su crecimiento. II. Las características de la canal. Una revisión. Rev. Fac. Agron. (LUZ) 10:163-187.

Mitchell, G.E., Reed, A.W. and Rogers, S.A. (1991). A research note: Influence of feeding regimen on the sensory qualities and fatty acid contents of beef steaks. Journal of Food Science, 56, 4, 1102.

Statistical Analysis System (S.A.S). (2000). SAS Institute, Inc; Cary, N.C. U.S.A.

Uzcátegui-Bracho, S.; Huerta-Leidenz, N.; Arenas de Moreno, L.; Colina, G. and N. Jerez-Timaure. (1999). Contenido de humedad, lípidos totales y ácidos grasos del músculo *longissimus* crudo de bovinos en Venezuela. Archivos Latinoamericanos de Nutricion , 49, 2, 171-180.

Yang, A., Larsen, T.W., Powell, V.H. and Tume, R.K. (1999). A comparison of fat composition of Japanese and long-term grain-fed Australian steers. Meat Science, 51, 1-9.

Valero-Leal, K. (2000). In: Proceedings of the 46th International Congress of Meat Science and Technology. Análisis comparativo de la composición proximal y del contenido de ácidos grasos de la carne de Búfalo de agua (*Bubalus Bubalis*) y la carne de vacunos acebuados de 24 meses de edad. Buenos Aires, Argentina.Vol. (1) 74-75.

DIETARY MANIPULATION OF PORCINE INTRAMUSCULAR FAT LEVELS AND SUBSEQUENT PROTEOMIC ANALYSIS

D. Corcoran*[1], B. Mee[1], P. Sullivan[1], E. O'Doherty[1], K. O'Reilly[1], A. Reynolds[2], J. O'Doherty[2] and A.M. Mullen[1]

[1]*Teagasc, Ashtown Food Research Centre, Ashtown, Dublin 15, Ireland.* [2] *University College Dublin, Lyons Research Farm, Newcastle, Co. Dublin. *Email: deirdre.corcoran@teagasc.ie*

Keywords: pork, intramuscular fat, *longissimus dorsi*, protein profile, bioanalyzer

Introduction

The quality and nutritive value of meat are critical traits for today's consumers with fat content becoming increasingly important to the consumer. Although pork meat with low fat content is desirable for reducing human caloric intake, an intramuscular fat level (IMF) below 2.5% is related to lower sensory quality traits (Fernandez *et al.* 1999). Studies suggest that dietary conditions can modulate IMF content at slaughter. Restriction of energy intake results in a decreased IMF content (Candek-Potokar *et al.*, 1998; Gondret and Lebret 2002; Wood *et al.*, 1996) and a low-protein diet results in an increased IMF content when compared to adequate diets (Adeola and Young 1989; Karlsson *et al.*, 1993). In the current study we were interested in exploiting this fact to produce pigs divergent for IMF content and to investigate if changes induced at the cellular level could be monitored through examination of protein profiles early post slaughter. Parallel studies will consider other factors such as differential gene expression. Therefore, the objective of this study was to administer dietary treatments to alter the IMF status in pork *M. longissimus dorsi* and to examine if differences are reflected at a proteomic level.

Materials and Methods

Forty pigs were offered one of 5 diets ad-libitum containing 13.8 MJ digestible energy and varying lysine levels: **Treatment 1**; 12.5 g/kg lysine from 40 kg to 65 kg and 10.5 g/kg lysine from 65 kg to slaughter, **Treatment 2**:10.5 g/kg lysine from 40 kg to slaughter, **Treatment 3**: 8.5 g/kg lysine from 40 kg to 65 kg and 10.5 g/kg lysine from 65 kg to slaughter, **Treatment 4**: 8.5 g/kg lysine from 40 kg to slaughter, **Treatment 5**:12.5 g/kg lysine from 40 kg to slaughter. Animals were then slaughtered according to standard commercial procedures in an Irish abattoir and samples of *M. longissimus dorsi* were taken for protein analysis at 3 hours post mortem. Compositional analysis was performed (Bostian *et al.*, 1985) to determine IMF content. Soluble protein extracts were prepared using 5.0% TCA following a modification of the method of Stoeva *et al.*, (2000). Extracts, from treatment groups which were significantly different for IMF, were analysed on an Agilent 2100 Bioanalyzer using a Protein 200 Plus Assay (Agilent Technologies, 76337 Waldbronn, Germany). These extracts were also run on a one dimensional sodium dodecyl sulphate-polyacrylamide gel (1-DE SDS-PAGE gel; 12% acrylamide).

Results and Discussion

Animals produced on diets with varying lysine levels showed altered IMF levels at slaughter. Those on treatment 5 showed a significantly lower level of intramuscular fat than those on treatment 3 and 4 (0.39% Vs 1.49%, $P<0.05$; Table 1). The greatest divergence in IMF level 5 was observed between treatment groups 3 and 5, therefore these groups were chosen for further protein analysis as these samples were most likely to have markedly different protein profiles. Protein extracts from Treatment 3 and 5 were investigated using a 'lab on a chip' technology (Bioanalyzer, Agilent Technologies). This novel approach enables a reduction in labour and time for protein analysis and possesses the added advantage of having a minute sample requirement. However this novel approach did not conclusively identify protein differences between IMF extremes as the protein profiles of both treatment 3 and 5 were similar. The only variation of note occurred in the lower molecular weight region of the samples (6-21 kDa), however the existence of inter group variation within this region precluded meaningful analysis (Figure 1). Comparable results were obtained by analyzing the protein samples from each treatment electrophoretically using a 1-DE gel (Figure 2). As both 1-DE electrophoresis and the Bioanalyzer act on the same principle of separating proteins according to their molecular weight, 2-DE electrophoresis may prove a more comprehensive method for identifying subtle deviations in protein profiles.

Table 1: Intramuscular fat levels (IMF) in pigs *post mortem*.

Treatment	Mean IMF (%)	Std Dev	Sample size(n)
1	0.64^{ab}	0.341	7
2	1.22^{ab}	0.744	6
3	1.54^{ac}	0.906	7
4	1.49^{a}	0.727	7
5	0.39^{b}	0.219	8

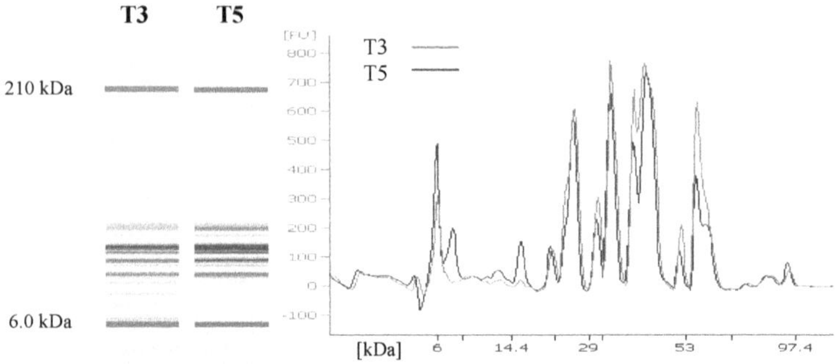

Figure 1: Bioanalyzer gel and electropherogram images of TCA extracts from samples with extremes of IMF levels.

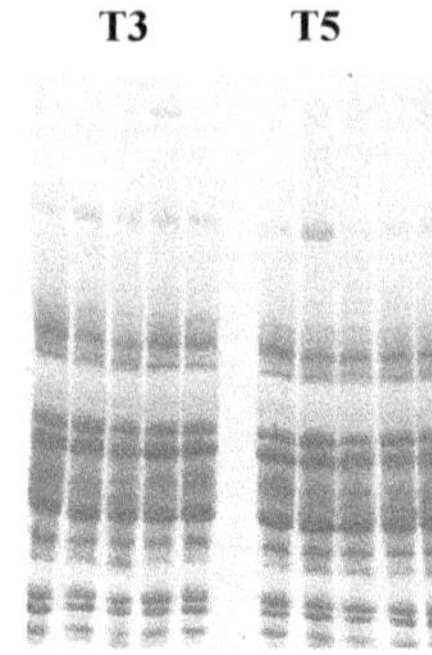

Figure 2: SDS-PAGE analysis of proteins from treatments 3 & 5.

Conclusion

Dietary treatments with varying lysine levels resulted in variation in IMF content in pork *longissimus dorsi* muscle. Protein profiles were investigated using both Bioanalyzer protein chip and 1-DE SDS-PAGE techniques. We have shown that protein deviations which may occur between pork samples of different IMF levels are not detected in proteins isolated by TCA extraction. It is possible that differences in protein profiles are being masked by co migration of the proteins of interest. The use of 2-DE electrophoresis would overcome this problem. Also more pronounced proteomic differences earlier in the live animal may be evident in myofibrillar and/or connective tissue. While results proved inconclusive in this instance, the Bioanalyzer did prove a useful tool for analysis of muscle protein species.

References

Adeola, O. and Young, L.G. (1989), Dietary protein-induced changes in porcine muscle respiration, protein synthesis and adipose tissue metabolism. Journal of Animal Science, 67, 664-673.

Bostian, M.L., Fish, D.L. Webb, N.B. and Arey, J.J. (1985). Automated methods for determination of fat and moisture in meat and poultry products: collaborative study. Journal - Association of Official Analytical Chemists, 68, 876-80.

Candek-Potokar, M., Lefaucheur, L., Zlender, B. and Bonneau, M. (1998). Effects of age and/or weight at slaughter on longissimus dorsi muscle: Biochemical traits and sensory quality in pigs. Meat Science, 48, 287-300.

Fernandez, X., Monin, G., Talmant, A., Mourot, J. and Lebret, B. (1999). Influence of intramuscular fat content on the quality of pig meat. Meat Science, 53, 59-72.

Gondret, F. and Lebret, B. (2002). Feeding intensity and dietary protein level affect adipocyte cellularity and lipogenic capacity of muscle homogenates in growing pigs, without modification of the expression of sterol regulatory element binding protein. Journal of Animal Science, 80, 3184-3193.

Karlsson, A., Enfalt, A.C., Essen-Gustavsson, B., Lundstrom, K., Rydhmer, L. and Stern, S. (1993). Muscle histochemical and biochemical properties in relation to meat quality during selection for increased lean tissue growth rate in pigs. Journal of Animal Science, 71, 930-938.

Stoeva, S., Byrne, C., Mullen, A.M., Troy, D.J. and Voelter, W. (2000). Isolation and identification of proteolytic fragments from TCA soluble extracts of bovine *M.longissimus dors.* Food Chemistry, 69, 365-370.

Wood, J.D., Brown, S.N., Nute, G.R., Whittington, F.M., Perry, A.M., Johnson, S.P. and Enser, M. (1996). Effects of breed, feed level and conditioning time on the tenderness of pork. Meat Science, 44, 105-112.

GRAZING BOTANICALLY DIVERSE PASTURES INFLUENCES TISSUE LONG-CHAIN FATTY ACID PROFILE IN LAMBS

M. Lourenço, G. Van Ranst, K. Raes, V. Fievez and S. De Smet*

LANUPRO, Ghent University, Proefhoevestraat 10, 9090 Melle, Belgium. Email: Stefaan.DeSmet@UGent.be

Keywords: botanical diversity, fatty acids, grazing, meat

Introduction

It is well known that feeding factors, particularly the feed fatty acid (FA) profile, affect the FA composition of animal products. The botanical composition of pastures may also influence rumen and intermediary FA metabolism, although less is known in this respect for meat producing ruminants, compared to dairy cattle. However, Ådnøy *et al.*, (2005) reported an increased intramuscular (IM) polyunsaturated FA (PUFA) content in lambs grazing diverse mountain pastures, compared to monocultures of lowland pastures. The objectives of this trial were to study the effect of grazing on 3 botanically different pastures on the FA metabolism and pattern in intramuscular and subcutaneous fat of lambs.

Materials and Methods

Twenty-one male lambs of similar genetic background were assigned to 3 different pastures of 7 lambs each: Botanical diverse pasture (BD), Leguminosa rich pasture (L) and Intensive English ryegrass pasture (IR). Animals started at an average live weight of 22.3 ± 3.1 kg and an age of 86 ± 7 days (not significantly different between groups). Animals grazed the pastures *ad libitum* for 12 weeks, and did not receive supplementary feed. The stocking density was low and on average less than 850 kg live weight ha^{-1} for the BD pasture and less than 1200 kg live weight ha^{-1} for the IR and L pasture. The botanical composition of the 3 pastures (Table 1) was determined on 9 occasions during the trial, according to De Vries (1933).

Table 1: Overview of the botanical composition of the 3 pastures grazed by the animals (n=9).

Plant Family	BD	L	IR
Poaceae	*Agrostis stolonifera* (38%)	*Lolium perenne* (19%)	*Lolium perenne* (69%)
	Bromus hordeaceus (18%)	*Phleum pratense* (14%)	*Bromus hordeaceus* (17%)
	Phleum pratense (9%)		*Lolium multiflorum* (5%)
Asteraceae	*Carduus nutans* (12%)		
	Taraxacum officinale (4%)		
Ranunculaceae	*Ranunclus acris* (5%)		
Fabaceae		*Trifolium repens* (41%)	
		Medicago sativa (20%)	

During the trial, weekly representative samples of the 3 pastures were taken for FA analysis and DM determination. After collection, the fresh samples were stored immediately in liquid N_2 for 4h, after which the FA extraction was started immediately. After 12 weeks, lambs were slaughtered at an average live weight of 24.8 ± 4.2 kg; 36.4 ± 3.5 kg and 35.6 ± 3.8 kg for the BD, L and IR group, respectively (significantly lower for the BD group compared to the L and IR group). *Longissimus thoracis* (left side of carcass between rib 7 and 8) and subcutaneous (SC) fat samples were taken 24h post-mortem from chilled carcasses (4°C), vacuum packed and stored at -20°C until FA analysis. FA of fresh pasture samples were extracted in triplicate with chloroform/methanol (2/1, v/v), as described by Lourenço and Fievez (2005). Meat and adipose tissue samples were extracted in duplicate as described by Raes *et al.* (2001). Fatty acids methylesters (FAME) were prepared according to Raes *et al.* (2001) and analyzed by GC (Raes *et al.*, 2004). For grass FA the temperature program used was: 150°C for 2 min, followed by an increase at a rate of 1°C/min until 200°C. A one-way ANOVA was used to evaluate the effect of the different pastures on grass, meat and fat FA, with Duncan as post-hoc test (SPSS for windows, version 11.0).

Results and Discussion

Table 2 presents the FA composition of the pastures. It is clear that pasture L had the highest total amount of FA, although the proportions of the sum of C18:2 n-6 and C18:3 n-3 were similar across groups with, however, pasture IR being richer in C18:3 n-3 and pastures BD and L having a higher content of C18:2 n-6.

There were significant effects of the type of pasture on the FA profile of SC and IM fat. The CLA c9t11 content in the SC fat significantly differed between the three pasture groups (BD > IR > L), but was not significantly different in the IM fat.

Table 2: Total fatty acid content (mg/g tissue) and proportions of FA (% of FAME) of pasture samples (n=12), and subcutaneous fat and meat of animals (n=7) according to pasture group.

Sample	Fatty acids	BD	L	IR	SEM	Sign.
Pasture	Total (mg/g DM)	18.5[c]	29.8[a]	25.5[b]	1.23	<0.001
	C18:2 n-6	18.2[a]	17.3[a]	13.9[b]	0.778	0.001
	C18:3 n-3	51.7[b]	52.1[b]	57.2[a]	1.67	0.050
Fat	Total (mg/g fat)	609	669	660	18.8	0.098
	CLA c9t11	1.32[a]	0.676[c]	1.01[b]	0.091	0.001
	C18:2 n-6	1.19[b]	2.35[a]	0.862[c]	0.078	<0.001
	C18:3 n-3	1.30[b]	3.53[a]	1.50[b]	0.166	<0.001
Meat	Total (mg/g meat)	16.0	24.4	19.6	2.58	0.094
	C18:2 n-6	7.06[a]	5.28[b]	3.37[c]	0.564	0.001
	C18:3 n-3	2.64[b]	3.99[a]	2.59[b]	0.235	0.001
	CLA c9t11	0.897	0.738	0.903	0.097	0.415
	C20:4 n-6	4.16[a]	1.17[b]	1.33[b]	0.549	0.002
	C20:5 n-3	2.76[a]	1.09[b]	1.33[b]	0.315	0.003
	C22:5 n-3	2.69[a]	1.08[b]	1.26[b]	0.296	0.002
	C22:6 n-3	0.427	0.293	0.340	0.037	0.059
	Indices of desaturation and elongation					
	C20:4 n-6/C18:2 n-6	0.543[a]	0.393[b]	0.219[c]	0.043	<0.001
	C20:5 n-3/C18:3 n-3	1.02[a]	0.518[b]	0.273[b]	0.103	<0.001
	C22:5 n-3/C18:3 n-3	0.993[a]	0.487[b]	0.271[b]	0.094	<0.001
	C22:6 n-3/C18:3 n-3	0.161[a]	0.076[b]	0.131[a]	0.012	<0.001

The proportions of C18:2 n-6 and C18:3 n-3 in SC fat of animals grazing pasture L were significantly higher than for the other groups. A similar effect was found in IM fat for C18:3 n-3, but not for C18:2 n-6, which significantly differed between the three pasture groups in the order BD > L > IR. Although the differences between tissues cannot be readily explained, the higher C18:3 n-3 content in meat and fat from L pasture lambs might be related to differences in rumen outflow rate. This has been shown to be higher for animals fed clover rich diets (Lee *et al.*, 2003).

Significantly higher proportions of C20:4 n-6, C20:5 n-3 and C22:5 n-3 in IM fat of pasture BD animals were also observed. Yet, no significant differences were found for C22:6 n-3. A confounding effect of the lower IM fat content in this group and thus a higher phospholipid/triglyceride ratio cannot be ruled out. However, after correction for differences in IM fat content, the proportions of long-chain FA remained higher for pasture BD lambs (data not shown). This is also reflected in the higher indices for elongation and desaturation activity, suggesting some stimulation of elongation and desaturation of long-chain FA in muscles of pasture BD animals.

Conclusions

From this study, it is clear that the botanical composition of pastures may affect the CLA c9t11 and the PUFA profile of the subcutaneous and intramuscular fat of exclusively grazing lambs. It seems that other factors than the FA supply interfere in the rumen and/or intermediary FA metabolism.

References

Ådnøy, T., Haug, A., Sørheim, O., Thomassen, M.S., Varszegi, Z., Eik, L.O. (2005). Grazing on mountain pastures – does it affect meat quality in lambs? *Livestock Production Science* 94, 25-31.

De Vries, D. (1933). The ranking method, an estimation method for botanical grassland research with determination of ranking order. *Verslagen van Landbouwkundige Onderzoekingen*, 'S Gravenhage, The Netherlands, n° 39A (in Dutch).

Lee, M.R.F., Harris, L.J., Dewhurst, R.J., Merry, R.J. and Scollan, N.D. (2003). The effect of clover silages on long chain fatty acid rumen transformations and digestion in beef steers. *Animal Science* 76, 491-501.

Lourenço, M. and Fievez, V. (2005). Content, pattern and esterification of fatty acids in fresh grass in relation to extraction solvents and sample storage conditions. In *Proceedings of the XX International Grassland Congress*, p. 183, Dublin, Ireland, June 2005.

Raes, K., De Smet, S. and Demeyer, D. (2001). Effect of double-muscling in Belgian-Blue young bulls on the intramuscular fatty acid composition with emphasis on conjugated linoleic acid and polyunsaturated fatty acids. *Animal Science* 73, 253-260.

Raes, K., Haak. L., Balcaen, A., Claeys, E, Demeyer, D. and De Smet, S (2004). Effect of linseed feeding at similar linoleic acid levels on the fatty acid composition of double-muscled Belgian-Blue young bulls. *Meat Science* 66, 307-315.

SUPPLEMENTATION OF HEIFERS WITH RUMINALLY-PROTECTED POLYUNSATURATED FATTY ACIDS: EFFECTS ON COLOUR STABILITY OF RETAIL-PACKAGED MINCED BEEF

P.G. Dunne*[1], J. Rogalski[1], S. Childs[2], F.J. Monahan[3], D.A. Kenny[3] and A.P. Moloney[1,4]

[1]*Teagasc, Ashtown Food Research Centre, Ashtown, Dublin 15, Ireland,* [2]*Teagasc, Athenry Research Centre, Athenry, Co. Galway, Ireland,* [3] *Department of Food Science, University College Dublin, Dublin 4, Ireland and* [4]*Teagasc, Grange Research Centre, Grange, Dunsany, Co. Meath, Ireland. E-mail: peter.dunne@teagasc.ie*

Keywords: ruminally-protected, heifers, polyunsaturated fatty acids, beef, colour

Introduction

Meat and milk of ruminant origin contain a substantial proportion of saturated lipids which result from lipolysis and biohydrogenation of dietary unsaturated lipids by rumen microorganisms (MacRae *et al.*, 2005). If a more unsaturated lipid composition is desired rumen hydrogenation must be prevented. Strategies to achieve this usually involve chemical modification of dietary lipids or feeding a seed source of dietary lipids where the oilseed itself provides some degree of protection from rumen microbial activity. Supplementation of cattle with ruminally-protected polyunsaturated fatty acids (RP-PUFA) and a consequent increase in PUFA in muscle lipids, would be expected to decrease the colour stability of the meat. Dietary PUFA are preferentially deposited in membrane phospholipids in ruminants (DeSmet *et al.*, 2004) and the fatty acid composition of phospholipids rather than triglycerides has been shown to be largely responsible for the susceptibility of meat to lipid oxidation.

Lipid oxidation and oxymyoglobin oxidation, manifested as meat discolouration, are linked (O'Grady *et al.*, 2001). Additionally, mincing of beef destroys cellular compartmentalisation and thus brings pro-oxidant components into close proximity with labile PUFA. There is also evidence that packaging of meat in a high oxygen atmosphere typical of beef retail packs provides an additional challenge to the oxidative stability and thus colour stability of meat, particularly if mincing has occurred (Monahan, 2000). The objective of the present study was to determine the effect of supplementation of heifers with a source of long chain ω-3 RP-PUFA on the colour stability of minced beef. We hypothesised that such supplementation would decrease the colour stability of beef under simulated retail display conditions.

Materials and Methods

Continental crossbred heifers were individually offered a daily bolus ration of 1kg (freshweight, approximately 850g dry matter (DM)) that contained the ω-3 RP-PUFA supplement of eicosapentaenoic acid (EPA, C20:5ω-3) and docosahexaenoic acid (DHA, C22:6ω-3) (Nutreco Holding N.V., 3800 AG Amersfoort, The Netherlands) at 0 (control), 69, 138 or 275g PUFA per kg (PU00, PU69, PU138 and PU275, respectively). This was followed by 1.5kg (1.278kgDM) of a high crude protein (26.2g/100g) balancer ration. Each afternoon 3.5kg (3.028kgDM) of another balancer ration was offered to all heifers with 1.5kg (1.24kgDM) of straw. Diets were formulated to be isoenergetic, isolipidic and isonitrogenous. Vitamin E, as *dl*-α-tocopheryl acetate was added to the RP-PUFA supplement at a rate of 5,000mg/kg and to the vitamin/mineral supplement included in the balancer rations at a rate of 10,000 I.U./kg (1mg *dl*-α-tocopheryl acetate = 1 I.U. (international units)). After 8 weeks of supplementation, heifers were humanely harvested, samples of neck muscle were recovered and were stored at -80°C (6 months) prior to mincing.

Eighteen hours before mincing samples were placed in a dark chill room at <2°C to thaw. Sections of neck muscle were minced by passing through a mincer with 3mm holes. When minced, samples were divided into six approximately equal portions, formed into patties of least 2.54cm thickness and dispensed into styrofoam trays with absorbent pads. Trays were sealed under oxygen impermeable barrier film (oxygen transmission rate: $8cm^3O_2/m^2/24h$ at 23°C and 75% relative humidity) following evacuation and flushing with $80\%O_2$:$20\%CO_2$ in a modified atmosphere packaging machine. Trays were randomly positioned in an open-fronted retail display cabinet under permanent fluorescent lighting (2800 lm) and permanently shielded by an insulating blind, to maintain a uniform temperature distribution. Cabinet temperature was monitored using three needle thermocouples.

Approximately 3hours after packaging, the 'L' (lightness), 'a' (redness) and 'b'(yellowness) values of the minced beef were measured (day 0) using a benchtop Hunter lab UltraScan XE spectrocolorimeter. Saturation and hue angle were calculated as $\sqrt{(a^2+b^2)}$ and $[\tan^{-1}(b/a)][180/\pi]$, respectively. Colour was measured again on days 1, 3, 5, 7 and 10. Data were analysed using ANOVA appropriate to a split-plot design. Treatment (4 levels) was in the main plot and display time (6 levels) in sub-plot.

Results and Discussion

There was no effect of treatment or treatment × display time interaction for any of the measured colour variables except for 'b' value (treatment × display time, $P = 0.038$). Thus while the PU00 'b' value tended to be lowest up to day 5, it was equal to PU275 on day 7 but higher ($P<0.05$) on day 10. Between days 7 and 10, the PU00 'b' value tended to decrease (Δ= - 0.46 b units, P>0.05) whereas the PU275 'b' value decreased (P<0.05) by 1.04 b units. This may prove useful in providing an index of colour stability for display periods in excess of 7 days but future trials are required to reinforce this suggestion. Redness and saturation are normally used as indices of colour stability (Figure 1), although Insausti *et al.* (1999) and Lindahl *et al.* (2001) claimed that changes in the 'b' value could be valuable in interpreting changes in pigment proportions, which are responsible for the phenomenon seen as unstable surface colour of meat, as red oxymyoglobin is converted to brown metmyoglobin. Critical considerations in the present study may be the muscle concentration of vitamin E as well as PUFA and also the low mean cabinet temperature (2.6°C).

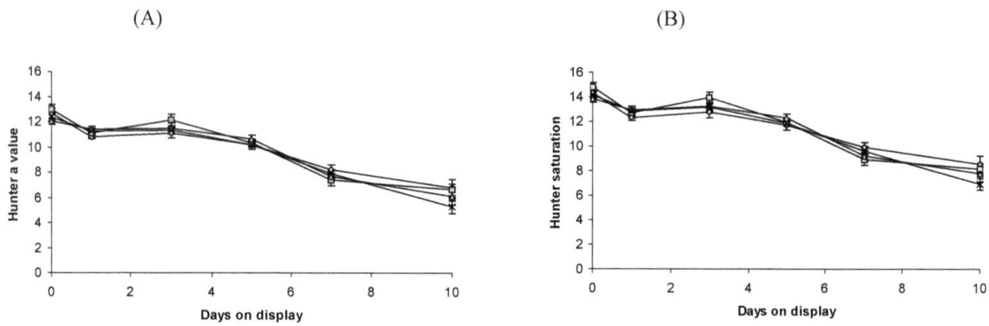

Figure 1: (A) Redness (Hunter 'a' value) and (B) saturation of minced beef from heifers offered ruminally-protected PUFA at 0 (PU00 (control)◇), 69 (PU69, □), 138 (PU138, △) or 275 (PU275, ×) g/kg and packaged in 80%O_2:20%CO_2 under permanent fluorescent lightning (2800lm).

Conclusions

Supplementing heifers with RP-PUFA did not have a deleterious effect on colour stability of minced beef stored for 1 week in high O_2 packs relative to control heifers, when indicated by redness or saturation. Thus, it is concluded that there were no negative effects of RP-PUFA supplementation based on usual measures but that the b value might have potential as a more sensitive index of colour stability.

Acknowledgements

J. Rogalski was in receipt of a European Union Marie Curie Fellowship. S. Childs was funded under the Walsh Fellowship Scheme of Teagasc, the Irish Agriculture & Food Development Authority. The authors thank Ms. Paula Reid, Ashtown Food Research Centre for statistical analysis.

References

DeSmet, S., Raes, K. and Demeyer, D. (2004). Meat fatty acid composition as affected by fatness and genetic factors: a review. Animal Research, 53: 81-98.

Insausti, K. Beriain, M.J., Purroy, A., Alberti, P., Lizaso, L. and Hernandez, B.(1999). Colour stability of beef from different Spanish native cattle breeds stored under vacuum and modified atmosphere. Meat Science, 53:241-249.

Lindahl, G., Lundstrom, K. and Tornberg, E. (2001). Contribution of pigment content, myoglobin forms and internal reflectance to the colour of pork loin and ham from pure sire breeds. Meat Science, 59: 141-151.

MacRae, J., O'Reilly, L. and Morgan, P. (2005). Desirable characteristics of animal products from a human health perspective. Livestock Production Science, 94: 95-103.

Monahan, F.J. (2000). Oxidation of lipids in muscle foods: fundamental and applied concerns, in: *Antioxidants in muscle foods – Nutritional strategies to improve quality*. Ed by Eric A. Decker, Cameron Faustman & Clemente J. Lopez-Bote. John Wiley & Sons, Inc., New York, pp 3-24.

O'Grady, M.N., Monahan, F.J. and Brunton, N.P. (2001). Oxymyoglobin oxidation and lipid oxidation in bovine muscle-mechanistic studies. Journal of Food Science, 66: 386-392.

EFFECT OF DIETARY FATTY ACID PROFILE ON PORK CARCASS FAT CONTENT AND DISTRIBUTION

C.E. Realini[1]*, P. Duran[2], R. Lizardo[2], M. Gispert[1], M.A. Oliver[1] and E. Esteve-Garcia[2]

[1.] *Institut de Recerca i Tecnologia Agroalimentàries (IRTA), Meat Technology Center, Granja Camps i Armet, Monells 17121 Spain.* [2.] *IRTA, Department of Animal Nutrition, Apartat 415, Reus 43280 Spain Email: carolina.realini@irta.es*

Keywords: diet, pork, carcass, lipid, fatty acids

Introduction

The pork industry has made significant efforts to decrease total amount of fat deposited in pigs through genetics, nutrition and management. There is also a trend in human nutrition towards a higher intake of polyunsaturated fatty acids (PUFAs). It is well known that the amount and fatty acid composition of the diet in monogastric species such as chickens and pigs influences carcass fat quality. Previous research has shown that diets rich in PUFAs reduce abdominal fat in broilers compared with saturated or monounsaturated fats (Crespo and Esteve-Garcia, 2002). In addition, Shimomura *et al.*, (1990) reported less body fat accumulation in rats fed a safflower oil diet than in rats fed a beef tallow diet. The aim of this study was to evaluate the effect of dietary fatty acid profile on pork carcass fat content and distribution.

Materials and Methods

Sixty crossbred (Duroc x Landrace) gilts (61 kg average live weight) were fed a basal diet (barley and soybean meal) supplemented with 10% of the following fats and oils: 1) tallow (T); 2) high oleic sunflower oil (HOSF); 3) sunflower oil (SFO); 4) linseed oil (LO); 5) fat blend: 55% tallow, 35% sunflower oil, 10% linseed oil (FB); or 6) fish oil blend: 40% fish oil, 60% linseed oil (FO). Diets had similar nutrient content differing in the fatty acid (FA) profile (Table 1). Feed lipids were extracted following the procedure of Folch *et al.*, (1957), converted to fatty acid methyl esters using BF_3 and methanolic KOH (Morrison and Smith, 1964) and analyzed using GC (Agilent 6890N, Wilmington, DE, USA). Gilts were slaughtered (97 kg average live weight) using CO_2 stunning and carcass and flare fat weights were recorded within 1 h postmortem. Fat and muscle depths were measured using the Fat-O-Meat'er probe and carcass lean percentage estimated according to Gispert and Diestre (1994). The left side of each carcass was cut and the loin and belly dissected into lean, fat and bone following the procedure of Walstra and Merkus (1995). The right side of each carcass was reduced into small pieces, frozen, ground (Grinder Cato-PA160, Spain), homogenized and subsamples (500 g) collected. Fat content of the ground carcass was determined following the ether extraction method 960.39 (AOAC, 1990). Intramuscular fat of *longissimus thoracis* (LT) and *semimembranosus* (SM) muscles was determined by Near Infrared Transmittance (NIT-Infratec 1265).

Table 1: Fat content and fatty acid composition of the experimental diets (g/kg feed).

	T	HOSF	SFO	LO	FB	FO
Fat content	124.5	134.2	130.2	113.9	120.9	115.6
Fatty acids						
Palmitic	28.7	7.87	11.2	8.21	20.0	13.0
Stearic	22.9	4.6	5.6	3.6	12.8	4.1
Oleic	34.8	86.8	32.5	20.7	29.5	19.0
Linoleic	14.2	21.5	73.3	26.2	26.0	21.5
Linolenic	1.9	1.3	1.3	47.1	18.9	31.2
Long chain PUFAs	2.5	3.2	9.9	13.5	8.0	19.7
Sum PUFAs	16.4	22.8	74.5	73.4	45.3	71.0

T=tallow; HOSF= high oleic sunflower oil; SFO= sunflower oil; LO=linseed oil; FB= fat blend; FO=fish oil blend

Results and Discussion

Dietary fat sources differed in the fatty acid profile showing higher values of PUFAs for SFO, LO and FO followed by FB, with lower values for HOSF and T (Table 1). Although fat content of ground carcass and intramuscular fat of LT and SM muscles were numerically lower for LO compared with other diets, values did not differ ($P>0.05$) among dietary treatments (Table 2). Diets SFO and LO showed lowest values of fat content of ground carcass and SM intramuscular fat, while HOSF showed highest values for carcass and SM fat percentage. There were no differences ($P>0.05$) among treatments in carcass characteristics with the exception of estimated carcass lean percentage which was higher ($P<0.05$) for T than HOSF and SFO (Table 3). There were no differences ($P>0.05$) among treatments in ham and shoulder percentage. Dietary SFO fed animals showed higher ($P<0.05$) percentage of loin compared with FB, and higher ($P<0.05$) percentage of belly than T. Results from primal dissections showed no differences ($P>0.05$) in lean, fat and bone percentages of the loin among treatments (Table 4). Lean percentage of the belly was higher ($P<0.05$) and fat percentage was lower ($P<0.05$) for T and LO compared with HOSF.

Table 2: Fat content of ground carcass and intramuscular fat of *longissimus thoracis*, and *semimembranosus* muscles.

Fat content, %	T	HOSF	SFO	LO	FB	FO	RMSE
Carcass	23.7	24.4	21.5	21.2	23.1	23.7	2.803
Longissimus thoracis	1.53	1.65	1.51	1.37	1.46	1.64	0.418
Semimembranosus	2.85	3.29	2.60	2.56	2.73	2.64	1.100

RMSE: root mean-square error

Table 3: Carcass characteristics and percentage of major primal cuts.

Carcass characteristics	T	HOSF	SFO	LO	FB	FO	RMSE
Carcass weight, kg	77.9	78.7	76.3	74.8	79.1	73.5	6.798
Flare fat %	1.50	1.85	1.83	1.61	1.63	1.69	0.300
Last rib fat depth, mm	13.5	15.8	15.8	14.3	14.1	14.8	2.071
Estimated lean %	56.2^{b}	53.4^{a}	53.6^{a}	55.3^{ab}	55.4^{ab}	54.7^{ab}	1.947
Major primal cuts, % of carcass							
Ham	24.5	24.4	24.2	24.7	24.6	24.3	0.815
Loin	18.3^{ab}	17.7^{ab}	18.4^{b}	17.9^{ab}	17.3^{a}	18.0^{ab}	0.891
Shoulder	13.9	13.7	13.6	13.6	13.8	13.9	0.536
Belly	9.07^{a}	9.83^{ab}	9.87^{b}	9.48^{ab}	9.82^{ab}	9.47^{ab}	0.659

Means within the same row with different letters differ $P<0.05$

RMSE: root mean-square error

Table 4: Dissection of loin and belly into lean, fat and bone.

Loin dissection, %	T	HOSF	SFO	LO	FB	FO	RMSE
Lean	54.7	51.4	51.0	54.2	53.8	52.0	3.197
Fat*	34.3	37.8	38.4	34.9	34.8	36.6	3.567
Bone	11.1	10.9	10.6	11.0	11.4	11.5	0.998
Belly dissection, %							
Lean	49.0^{b}	44.9^{a}	47.2^{ab}	49.3^{b}	48.3^{ab}	47.9^{ab}	2.798
Fat*	42.9^{a}	47.3^{b}	44.7^{ab}	42.6^{a}	43.9^{ab}	43.9^{ab}	3.172
Bone	8.15	7.85	8.12	8.06	7.75	8.17	1.094

*Skin-on

RMSE: root mean-square error

Conclusions

Dietary supplementation with fat sources differing in the fatty acid profile showed minor changes in pork carcass fat content and distribution. Preliminary results from this study suggest that diets rich in polyunsaturated fatty acids may affect lipid content differently in swine compared to other monogastric species such as chickens.

References

Crespo, N. and Esteve-Garcia, E. (2002). Nutrient and fatty acid deposition in broilers fed different dietary fatty acid profiles. *Poultry Science*, 81:1533-1542.

Folch, J., Lees, M. and Stanley, G.H.S. (1957). A simple method for the isolation and purification of total lipids from animal tissues. *Journal of Biological Chemistry*, 226:497-509.

Gispert, M. and Diestre, A. (1994). Classement des carcasses de porc en Espagne: un pas vers l'harmonization communautaire. *Techni-Porc*,17(2):29–32.

Morrison, W.R. and Smith, L.M. (1964). Preparation of fatty acid methyl esters + dimethylacetals from lipids with boron fluoride-methanol. *Journal of Lipid Research*, 5:600-608.

Shimomura, Y., Tamura, T. and Suzuki, M. (1990). Less body-fat accumulation in rats fed a safflower oil diet than in rats fed a beef tallow diet. *Journal of Nutrition*, 120:1291-1296.

Walstra, P. and Merkus, G.S.M. (1996). Procedure for assessment of the lean meat percentage as a consequence of the new EU reference dissection method in pig carcass classification. Report ID-DLO 96.014, Zeist, The Netherlands, 22pp.

EFFECT OF DIETARY FAT QUALITY ON THE LIPID QUALITY AND STABILITY OF A CURED AND FERMENTED PORK SAUSAGE

A. Hugo*[1], S.J. Madisa[1], E. Roodt[1], G. Osthoff[1] and P.J. Jooste[2]

[1.] Department of Microbial, Biochemical and Food Biotechnology, University of the Free State, PO Box 339, Bloemfontein, 9300, South Africa, [2.] Department of Biotechnology and Food Technology, Tshwane University of Technology, Private Bag X680, Pretoria, 0001, South Africa. E-mail: HugoA.sci@mail.uovs.ac.za.

Keywords: oxidized oil, pigs, salami, stability

Introduction

An estimated 50,000 tonnes of oils and fats (mostly sunflower oil) are used annually in fried foods in South Africa (Kock *et al.*, 1997). As a result of regulations relating to quality of edible oils and fats for human consumption (Government Gazette, 1996), approximately 60 % of these oils and fats should accumulate as non-toxic used oils that need to be discarded in an appropriate manner. These non-toxic oils should contain less than 25 % polar compounds and less than 16 % polymerized compounds. They should also have the potential to be used as high energy food sources for animal production (Kock *et al.*, 1997). The aim of this study consequently was to evaluate the effect of the inclusion of such oils in pig diets on the lipid quality and stability of a high value consumer product such as salami.

Materials and Methods

Twenty four Large White gilts weighing ± 32 kg, were randomly assigned to each of four dietary treatments. These consisted of a control diet (CL), a diet supplemented with 3 % fresh sunflower oil (FSO), a diet supplemented with 3 % used sunflower oil (USO) and a diet supplemented with 3 % used sunflower oil plus 200 mg/kg of α-tocopheryl acetate (USOE). Pigs were slaughtered after a 85 day feeding period at an average live weight of ± 100 kg. The left loin including the last three ribs was removed and deboned. Total lipid was extracted from the backfat and muscle (Folch *et al.*, 1957). Fatty acid methyl esters were prepared using methanol-BF3 (Slover and Lanza, 1979) and identified and quantified using a Varian GX 3400 gas chromatograph. A colorimetric method (AOAC nr. 971.30, 1990) was used to determine the α – tocopherol content in backfat and muscle. The lean as well as backfat from the pigs of each dietary treatment was pooled and used to manufacture salami according to good manufacturing practices using a commercial starter culture (Floracarn SL Hansen). Salamis were sampled directly after stuffing, after fermentation and at the end of ripening for pH and stability measurements. Extracted lipids were subjected to testing for thiobarbituric acid reactive substances (TBARS) (Raharjo *et al.*, 1992), peroxide value (PV) (AOAC nr. 965.33, 1990), free fatty acids (FFA) (Pearson, 1968) and total carbonyl compounds (TCC) (Berry and McKerrigan, 1958). Analysis of variance (ANOVA) and the Tukey-Kramer multiple comparison test was used to determine whether measurements from different treatments differed significantly (NCSS, 2004).

Results and Discussion

The backfat and muscle of animals fed sunflower oil contained significantly more linoleic acid (C18:2) and less palmitic acid (C16:0) compared to those receiving the control diet. Dietary supplementation with α-tocopheryl acetate resulted in signifcant increases in the α-tocopherol content of both backfat and muscle (Table 1).

Table 1: Chemical properties and fattty acid composition of gilts in the four dietary groups.

Dietary groups	**CL** (n = 12)	**FSO** (n = 12)	**USO** (n = 12)	**USOE** (n = 12)	**Sign.**
Properties of backfat					
C16:0 (%)	21.24 ± 1.70^b	19.55 ± 2.02^a	19.61 ± 1.18^a	18.89 ± 1.89^a	*
C18:0 (%)	11.14 ± 0.90	10.17 ± 1.29	10.94 ± 1.66	10.32 ± 1.60	NS
C18:1 (%)	36.47 ± 1.00^b	32.35 ± 1.47^a	31.82 ± 1.80^a	32.03 ± 0.99^a	***
C18:2 (%)	18.65 ± 2.69^a	27.42 ± 4.15^b	27.39 ± 3.56^b	28.26 ± 3.07^b	***
α-tocopherol (µg/g)	8.36 ± 2.41^a	8.29 ± 2.48^a	6.91 ± 2.06^a	24.16 ± 7.61^b	***
Properties of muscle:					
C16:0 (%)	22.42 ± 0.79^b	21.11 ± 0.94^a	21.64 ± 0.81^a	20.97 ± 0.99^a	*
C18:0 (%)	11.39 ± 0.65	11.61 ± 0.68	11.57 ± 0.59	11.70 ± 0.78	NS
C18:1 (%)	37.14 ± 2.18^b	35.86 ± 2.23^a	34.69 ± 2.59^a	33.48 ± 2.06^a	*
C18:2 (%)	12.99 ± 2.07^a	15.64 ± 2.36^b	16.96 ± 1.58^b	17.17 ± 2.55^b	*
α-tocopherol (µg/g)	1.92 ± 0.66^a	1.97 ± 0.58^a	1.48 ± 0.89^a	4.77 ± 1.23^b	***

Sign. = Significance level; NS = Not significant; * = $P < 0.05$; ** = $P < 0.01$; *** = $P < 0.001$

Significantly higher pH values in salami were observed in the USO group and lower pH values in salami of the USOE group compared to the CL and FSO groups at the end of fermentation and drying (Table 2). This can possibly be ascribed to the negative effect of oxidation products (transferred from the used oil to the meat), on starter growth, or to the more conducive environment created for starter growth by α-tocopherol. This pH difference was, however, not large enough to induce significant differences in weight loss from salamis in the different treatment groups. Although salami usually contains nitrite (an antioxidant), oxidative differences were observed in salami made from meat of the different treatments. Salami made from the meat of pigs receiving the sunflower oil supplemented diets had significantly higher FFA and TCC contents than salami made from meat of the CL diet throughout the manufacturing process. No significant differences were, however, observed between the FSO, USO and USOE groups in terms of FFA and TCC at any point of manufacture. The same significant trend was observed with PV's except that salami manufactured from the USO treatment had a significantly higher initial PV than salami manufactured from the other sunflower oil treatments. It would seem that FFA production (lipolysis), peroxide formation (oxidation products of FFA) and carbonyl formation (oxidation products of peroxides) are related to lipid unsaturation and that oil quality and α-tocopherol level had little effect on these values. Contrary to FFA content, PV and TCC content TBARS values were significantly influenced by dietary oil quality and α-tocopherol level towards the end of the manufacturing process of salami. At the end of the drying period, the TBARS values of the salami from the CL and USOE groups were well below the critical value of 1.0 considered to be the cut-off point in terms of rancid tastes, while salami from the FSO and USO groups exceeded this value.

Table 2: Lipid stability in salami after manufacture (Day 0), fermentation (Day 2) and drying (Day 21).

Parameter:	**Treatments:** Day	**CL**(n=12)	**FSO**(n=12)	**USO**(n=12)	**USOE**(n=12)	**Sign.**
Weight loss (%)	2	1.54	1.89	1.71	1.83	-
	21	21.56	22.04	21.67	21.83	-
pH	0	5.77 ± 0.01	5.77 ± 0.01	5.78 ± 0.01	5.76 ± 0.02	NS
	2	5.18 ± 0.02^b	5.19 ± 0.01^b	5.24 ± 0.02^c	5.07 ± 0.07^a	***
	21	4.75 ± 0.02^b	4.76 ± 0.03^b	4.78 ± 0.01^b	4.68 ± 0.09^a	***
Free Fatty Acids (%)	0	1.91 ± 0.42^a	2.19 ± 0.33^b	2.29 ± 0.17^b	2.31 ± 0.37^b	*
(g oleic/100g total fatty acids)	2	2.70 ± 0.34^a	3.07 ± 0.18^b	2.95 ± 0.48^b	3.33 ± 0.40^b	**
	21	4.41 ± 0.34^a	5.04 ± 0.26^b	4.86 ± 0.38^b	5.12 ± 0.47^b	***
Total carbonyl compounds	0	11.90 ± 1.99^a	14.74 ± 1.97^b	14.53 ± 3.39^b	14.79 ± 3.27^b	*
(mmol/kg lipid)	2	13.06 ± 17.01^a	17.01 ± 2.63^b	17.43 ± 1.73^b	15.99 ± 2.26^b	***
	21	13.42 ± 1.17^a	15.31 ± 2.74^b	15.82 ± 1.88^b	16.42 ± 2.73^b	*
Peroxide value	0	5.72 ± 2.20^a	11.18 ± 2.79^b	15.01 ± 3.17^c	12.01 ± 3.84^b	***
(milliequiv. peroxide/kg fat)	2	6.11 ± 1.08^a	17.14 ± 5.67^b	18.80 ± 7.26^b	18.61 ± 6.09^b	***
	21	8.28 ± 2.43^a	13.10 ± 5.79^b	14.45 ± 6.70^b	14.53 ± 5.51^b	*
TBARS value	0	0.05 ± 0.03	0.05 ± 0.02	0.05 ± 0.02	0.04 ± 0.02	NS
(mg molonaldehyde/kg meat)	2	0.09 ± 0.05	0.08 ± 0.03	0.10 ± 0.03	0.09 ± 0.01	NS
	21	0.48 ± 0.35^a	1.22 ± 0.30^b	1.14 ± 0.24^b	0.40 ± 0.38^a	***

Sign. = Significance level; NS = Not significant; * = $p < 0.05$; ** = $p < 0.01$; *** = $p < 0.001$

Conclusions

This research indicated that high levels of polyunsaturated fatty acids and oxidized oils in the diets of pigs had a negative effect on lipid stability of high value products like salami. It was, however, demonstrated that dietary supplementation with α-tocopheryl acetate may be efficient in retarding oxidative rancidity in salami manufactured from fatty tissue originating from animals fed oxidized oils or feed rich in polyunsaturated fats.

References

AOAC. 1990. *Official Methods of Analysis,* 15th ed. Association of Official Analytical Chemists, Virginia.

Berry, B.N. and Mc Kerrigan, A.A. (1958). Carbonyl compounds as a criterion of flavour deterioration in edible fats. Journal of the Science of Food and Agriculture 9: 693-696.

Folch, J., Lees, M. and Sloane-Stanley, G.H. (1957). A simple method for the isolation and purification of total lipids from animal tissue. Journal of Biological Chemistry 226: 497-509.

Government Regulation No. R. 1316. (1996). Regulations relating to edible fats and oils. Government Gazette No. 17365, 16 8th 1996:3-4.

Kock, L., Botha, A. and Jeffery, J. (1997). Fryer oil initiative for SA: restaurant waste oils now available in S.A. for animal feeds. AFMA MATRIX June: 23-25.

NCSS. (2004). Statistical System for Windows. Number Cruncher Statistical Systems, Kaysville, Utah, USA.

Pearson, D. (1968). Application of chemical methods for the assessment of beef quality III – Methods related to fat spoilage. Journal of the Science of Food and AgricultureAgric. 19: 553-556.

Raharjo, S., Sofos, J.N.S. and Schmidt, G.R. (1992). Improved speed specificity, and limit of determination of an aqueous acid extraction thiobarbituric acid-C_{18} method for measuring lipid peroxidation in beef. Journal of Agricultural and Food Chemistry 40: 2182-2185.

Slover, H.T. and Lanza, E. (1979). Quantitative analysis of food fatty acids by capillary gas chromatography. Journal of the American Oil Chemistry Society 56: 933-943.

EFFECTS OF DIETARY GROUND FLAXSEED ON COLOUR AND LIPID STABILITY IN BEEF *LONGISSIMUS DORSI*

J.P. Grobbel*, M.C. Hunt, M. Seyfert, and J.S. Drouillard

Department of Animal Sciences and Industry, Kansas State University, Manhattan, KS 66506 USA
Email: jgrobbel@ksu.edu

Keywords: beef, colour and lipid stability, flaxseed

Introduction

Flaxseed (*Linum usitatissimum*) is high in omega-3 fatty acids, especially α-linolenic acid (C18:3n3). Many potential health benefits may be possible through the consumption of more omega-3 fatty acids (Alexander, 1998; Harbige, 1998). Flax-fed beef had increased polyunsaturation (Scollan *et al.*, 2001), decreased lipid stability (Wood *et al.*, 1999), and decreased colour stability (Hutchins *et al.*, 1967). Flax-fed beef without vitamin E in the diet had a ten-fold increase in C18:3n3 but decreased colour stability when compared to the control (LaBrune *et al.*, 2002). Drouillard *et al.* (2002) found that including vitamin E in the diet for flax-fed beef still resulted in increased C18:3n3 but the meat had similar colour stability to that of non-flax fed beef. The objectives of this study were to determine the effects of dietary flax on fatty-acid composition and lipid and myoglobin oxidation in beef *longissimus*.

Materials and Methods

Beef *longissimus dorsi* steaks (2.54 cm thick) from grain-fed Holstein steers fed 0% ground flaxseed (control, n=15) or 10% ground flaxseed dry matter basis with vitamin E (n=15) for 180 d were packaged in polyvinyl chloride film over wrap 11 d post-mortem. Steaks were displayed under continuous fluorescent lighting at 1°C for 5 d with instrumental CIE L*a*b* measurements taken daily. Trained visual panellists (n=6) evaluated muscle colour each day of display. Initial beef colour was evaluated using an eight-point scale, where 1=bleached red, 4=cherry red, and 8=very dark red. Muscle colour stability was measured using a five-point scale where 1=very bright cherry red, 2=bright cherry red, 3=slightly dark red to tannish red, 3.5=borderline acceptable, 4=moderately grayish tan to brown, and 5=tan, to brown to the nearest half-point. Metmyoglobin reducing activity (MRA) was determined on d 0 and 5 of display by reduction of nitric oxide metmyoglobin reduction using a modified procedure of Watts *et al.* (1966), where MRA = (observed decrease in metmyoglobin concentration / initial metmyoglobin concentration) × 100. 2-Thiobarbituric acid reactive substances (TBARS) were measured on d 0 and 5 of display as an indication of lipid oxidation by using a slightly modified method to that described by Witte *et al.* (1970). Fatty acid composition was determined from *longissimus dorsi* samples using gas chromatography (Sukhija and Palmquist, 1988). The PROC MIXED procedure of SAS® (SAS Institute, Inc., Cary, NC) was used to analyze the data.

Results and Discussion

Carcasses from both feeding regimens graded USDA Select, and those fed 0% flax had a yield grade of 2.9 and pH of 5.7 compared with 3.0 and 5.8, respectively, for the 10% flax group. Initial beef colour (cherry red) was the same ($P > 0.05$) for both 0 and 10% flax treatments. Visual colour during display remained the same ($P > 0.05$) for both treatments, but scores decreased throughout display (Table 1). There were no diet × day interactions for any additional colour traits including L*, a*, or b*, a*/b*, hue angle, and saturation index (data not shown). Metmyoglobin-reducing activity decreased ($P < 0.05$) throughout display as expected with d 0 having 90.7% metmyoglobin reduced and d 5 having 65.5% reduced but there were no differences ($P > 0.05$) between treatments.

Table 1: Means for visual colour scores[a] from 0 to 5 d of display of steaks from steers fed 0 and 10% flax.

	Time (days)						
	0	1	2	3	4	5	SE
Visual colour scores	2.5[b]	2.9[c]	3.0[c]	3.1[c]	3.4[d]	3.6[e]	0.12

[a]1=very bright cherry red, 2=bright cherry red, 3=slightly dark red to tannish red, 3.5=borderline acceptable, 4=moderately grayish tan to brown, and 5=tan to brown
[bcde]Means with different superscript letters are different ($P < 0.05$)

Fatty acid composition was affected by flax in the diet; steaks from cattle fed 10% flax had more ($P < 0.05$) omega-3 fatty acids (C18:3 *n*-3, C20:5 *n*-3, and C22:5 *n*-3; Figure 1) and fewer ($P < 0.05$) total *n*-6 to total *n*-3 fatty acids (Figure 2) than those steaks from cattle fed 0% flax. Regardless of differences in fatty acid composition, TBARS values were very low (less than 0.12 for both treatments throughout display), and although statistically different, of no practical importance.

Figure 1: Total n-3 fatty acids from cattle fed 0 or 10% flax

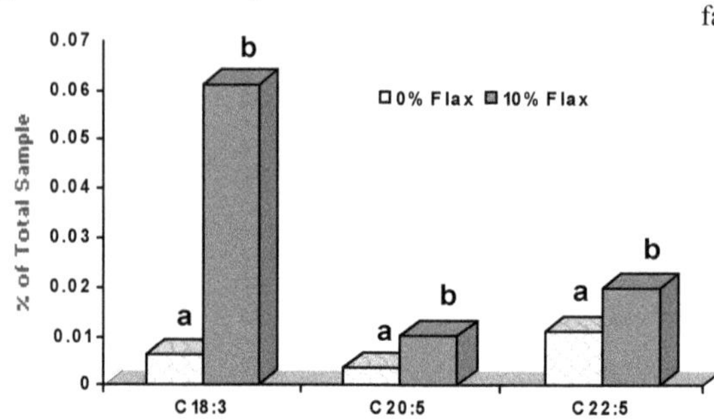

Figure 2: Comparison of total n-6 to n-3 fatty acids from cattle fed 0 or 10% flax

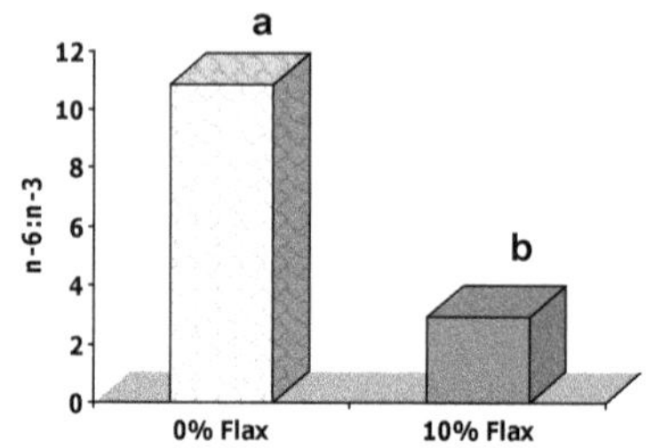

[ab]Means with different superscript letters differ $P < 0.05$

LaBrune *et al.* (2002) found that steaks from cattle fed flax without vitamin E had lower colour stability than steaks from control cattle. When vitamin E was included in the diet of cattle fed flax, colour stability was equal to that of steaks from control animals (Drouillard *et al.*, 2002). The current study confirms that vitamin E should be included in the diet of cattle fed flax and will not compromise the change in fatty acid composition occurring from including flax in the diet. In addition, it appears that the MRA of muscle does not decrease with the increase in fatty acid unsaturation as colour stability between the two groups was similar.

Conclusions

Including 10% flax and vitamin E in beef-cattle diets will increase the percent of unsaturated fatty acid and notably increase the omega-3 fatty acids. Including flax in the diet of cattle would have positive effects for the health profile of fatty acids without compromising colour or lipid stability.

References

Alexander, J.W. (1998). Immunonutrition: The role of omega-3 fatty acids. Nutrition. 14: 627-633.

Drouillard, J.S., Good, E.J., Gordon, C.M., Kessen, T.J. and Sulpizio, M.J. (2002). Flaxseed and flaxseed products for cattle: Effects of health, growth performance, carcass quality and sensory attributes. In: Proceedings of 59th Flax Institute of the United States. Flax Institute. Fargo, North Dakota.

Harbige, L.S. (1998). Dietary n-6 and n-3 fatty acids in immunity and autoimmune disease. In: Proceedings of the Nutrition Society. 57: 555-562.

Hultquist, D.E. (1978). Methemoglobin reduction system of erythrocytes. In: Fleisher S., L. Packer, editors. Methods in enzymology. Biomembranes LII. New York: Academic Press. p. 465-466.

Hutchins, B.K., Liu, T.P. and Watts, B.M. (1967). Effect of additives and refrigeration on reducing activity, metmyoglobin, and malonaldehyde of raw ground beef. Journal of Food Science. 32: 214-217.

LaBrune, H.J. (2000). Effects of grain processing and lipid addition on blood glucose, performance, carcass characteristics, and meat quality in finishing cattle. MS Thesis, Kansas State University, Manhattan, KS, USA.

Scollan, N.D., Choi, N.J., Kurt, E., Fisher, A.V., Enser, M. and Wood, J.D. (2001). Manipulating the fatty acid composition of muscle and adipose tissue in beef cattle. British Journal of Nutrition. 85: 115-124.

Sukhija, P.S. and Palmquist, D.L. (1988). Rapid method for determination of total fatty acid content and composition of feedstuffs and feces. Journal of Agricultural and Food Chemistry. 36: 1202-1206.

Watts, B.M., Kendrick, J., Zipser, M.W., Hutchins and B. Saleh, B. (1966). Enzymatic reducing pathways in meat. Journal of Food Science. 31: 855-861.

Witte, V.C., Krause, G.F. and Bailey, M.E. (1970). A new extraction method for determining 2-thiobarbituric acid values of pork and beef during storage. Journal of Food Science. 35: 582-585.

Wood, J.D., Enser, M., Fisher, A.V., Nute, G.R., Richardson, R.I. and Sheard, P.R.. (1999). Manipulating meat quality and composition. In: Proceedings of the Nutrition Society. 58: 363-370.

Acknowledgements

This is Contribution No. 06-305-A from the Kansas Agricultural Experiment Station, Manhattan, Kansas.

LIPOGENIC ACTIVITY AND FATTY ACID COMPOSITION OF 3 DIFFERENT IBERIAN X DUROC PIG GENOTYPES

R. Cava* and M.R. Ramírez

Tecnología de los Alimentos. Facultad de Veterinaria. Avda Universidad s.n. 10071 Cáceres. Spain. E-mail: rcava@unex.es

Keywords: lipogenic enzymes, glucose-6-phosphate dehydrogenase, malic enzyme, Duroc, Iberian

Introduction

Iberian is a rustic breed with a high adipogenic potential. Endogenous synthesis of fatty acids mainly takes place *in situ*, in the adipose tissue, from acetyl CoA and malonyl CoA to produce palmitic acid (C16:0), from which can be synthesized stearic acid (C18:0) by elongation. Malic enzyme (ME) and glucose-6-phosphate dehydrogenase (G6PDH) are the main enzymes involved in supplying NADPH for these reactions (Wise and Ball, 1964).On the other hand, one of the alternatives more often applied to improve productive parameters of Iberian pigs is the cross with Duroc at 50%. A specific law for Iberian products passed in 2001 in Spain (B.O.E., 15th October 2001). It regulates the genotype for the elaboration of dry-cured meat products labelled as "Iberian" being allowed the use of pure Iberian pigs as well as Iberian x Duroc crosses, but in the crosses obliges producers always to employ Iberian females. Therefore, it is of prime importance to assess the consequences of the use of different Duroc paternal lines in Iberian x Duroc crosses as well as the differences between Iberian x Duroc reciprocal crosses on the adipogenic character and fatty acid composition of subcutaneous and intramuscular fat (IMF).

Materials and Methods

Animals. Three groups of 10 pigs each were studied (5 males and 5 females) from the genotypes: GEN1: ♂ Iberian x ♀ Duroc1, GEN2: ♂ Duroc1 x ♀ Iberian; GEN3: ♂ Duroc2 x ♀ Iberian. GEN1 and GEN2 are reciprocal crosses, while the difference between GEN2 and GEN3 is the Duroc sire line. Duroc1 corresponded to pigs selected for the production of dry-cured meat products, while Duroc2 corresponded to animals selected for meat production. Pigs were slaughtered after 316 days at 150-165Kg. Backfat thickness (BFT) and ham fat thickness (HFT) were measured in the 5th rib and in the *Biceps femoris* muscle in the carcass and ham, respectively. *Longissimus dorsi* (LD) muscles were dissected and stored at -80°C until their analyses.

Methods. Intramuscular lipids were extracted according to Bligh and Dyer (1959), fractionated by SPE (Monin *et al.*, 2003) Fatty acid methyl esters (FAMEs) from IMF and subcutaneous fat (SCF) were prepared by acidic esterification in presence of sulphuric acid. Fatty acids (g/100g) were analysed using a Hewlett-Packard model HP-5890A gas chromatograph. Lipogenic enzyme activity was measured spectrophotometrically and expressed on the basis of nmol NADP reduced to NADPH per min per mg protein, using the extinction coefficient of NADPH at 340 nm of 6.22×10^6 M^{-1} cm^{-1} following the method of Bautista *et al.*, (1988) for the determination of G6PDH activity and the method described by Spina *et al.* (1970) for the determination of ME activity. The protein content was determined by Bradford method (1976).

Results and Discussion

GEN2 had significantly higher BFT and HFT than GEN3, while the level of fatness of GEN1 was intermediate (Table 1). The fatty acid composition of subcutaneous fat showed the highest levels of PUFA in the GEN3.

Table 1: Fat depths and fatty acid composition of subcutaneous fat.

		Genotype			Sex			Probability		
		GEN1	GEN2	GEN3	♂	♀	sem	gen	sex	Interaction
Fat depths(cm)	BFT	6.3ab	6.4a	5.3b	5.9	6.1	0.18	*	ns	Ns
	HFT	3.1ab	3.4a	2.7b	3.2	2.9	0.11	**	ns	Ns
Fatty acids (FA)	ΣSFA	38.05	39.00	37.68	38.33	38.18	0.22	ns	ns	Ns
(g/100g FA)	ΣMUFA	50.90	49.65	50.30	50.24	50.29	0.25	ns	ns	Ns
	ΣPUFA	11.05b	11.34ab	12.01a	11.44	11.53	0.14	*	ns	Ns

a,b: Different letters in the same row indicate significant statistical differences (Tukey Test, $p<0.05$). ns: non significant. SFA: saturated fatty acids, MUFA: monounsaturated fatty acids, PUFA: polyunsaturated fatty acids. * $p<0.05$, ** $p<0.01$, ***$p<0.001$.

The IMF content of LD was significantly affected by the genotype (Table 2); GEN2 had significantly higher IMF content than GEN1 and GEN3. An adequate level of IMF is a determinant factor for the elaboration of dry cured meat products (Gandemer, 2002), so GEN2 would be more suitable for the production of meat products. In the IMF of the LD muscle, PUFA level was significantly higher in GEN1 and GEN3 than in GEN2. The differences between reciprocal crosses (GEN1 vs. GEN2) are caused by the lowest IMF content of GEN1, which should have increased the relative proportion of unsaturated fatty acids in this genotype, since neutral lipids –NL- did not showed differences in the fatty acid composition between reciprocal crosses. However, GEN3 showed the highest percentages of unsaturated fatty acids, which could cause undesirable technological and sensory consequences, since PUFA are extremely sensitive to oxidation, leading to meat texture, flavour and colour alterations (Morrissey *et al.*, 1998).

Table 2: IMF content and fatty acid composition of IMF and NL of *Longissimus dorsi* muscle.

	Genotype			Sex			Probability		
	GEN1	GEN2	GEN3	♂	♀	sem	gen	sex	Interaction
IMF (g/100g)	3.84b	5.87a	3.32b	4.39	4.33	0.28	***	ns	Ns
FA imf (g/100g FA)									
ΣSFA	38.27b	40.60a	36.97b	38.33	38.90	0.40	***	ns	**
ΣMUFA	53.64ab	53.03b	54.41a	53.82	53.57	0.23	*	ns	Ns
ΣPUFA	8.09a	6.37b	8.63a	7.85	7.52	0.27	***	ns	Ns
FA NL (g/100g FA)									
ΣSFA	40.16a	41.54a	38.10b	39.40	40.41	0.38	***	*	*
ΣMUFA	54.58b	53.57b	55.98a	54.99	54.45	0.28	***	ns	*
ΣPUFA	5.25ab	4.89b	5.93a	5.60	5.13	0.18	*	ns	Ns

a,b: Different letters in the same row indicate significant statistical differences (Tukey Test, $p<0.05$). IMF: intramuscular fat. NL: neutral lipids. ns: non significant. SFA: saturated fatty acids, MUFA: monounsaturated fatty acids, PUFA: polyunsaturated fatty acids. * $p<0.05$, ** $p<0.01$, *** $p<0.001$.

The lipogenic enzyme activity of subcutaneous fat and LD muscle showed similar results between reciprocal crosses (GEN1 vs. GEN2) whereas it was strongly influenced by the Duroc sire genotype, as GEN3 showed lower enzyme activity of G6PDH and ME than GEN1 and GEN2 in both locations (Figure 1 and 2). The higher lipogenic activity of GEN1 and GEN2 is in accordance to the higher SFA levels in these genotypes, since ME and G6PDH the main enzymes involved in supplying NADPH for the endogenous synthesis of fatty acids.

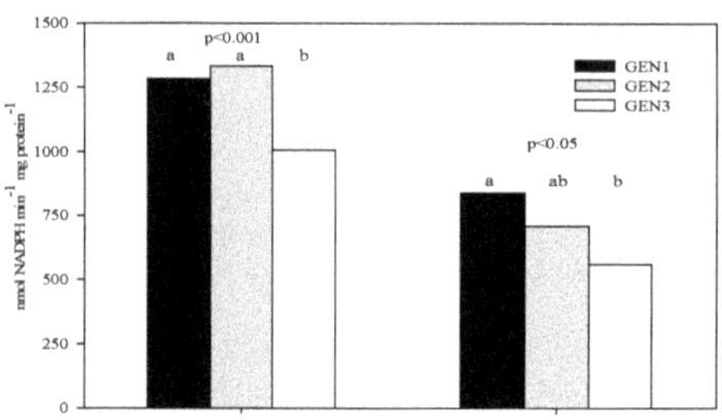

Figure 1. G6PDH and ME activities (nmol of NADPH/min/mg protein) in SCF

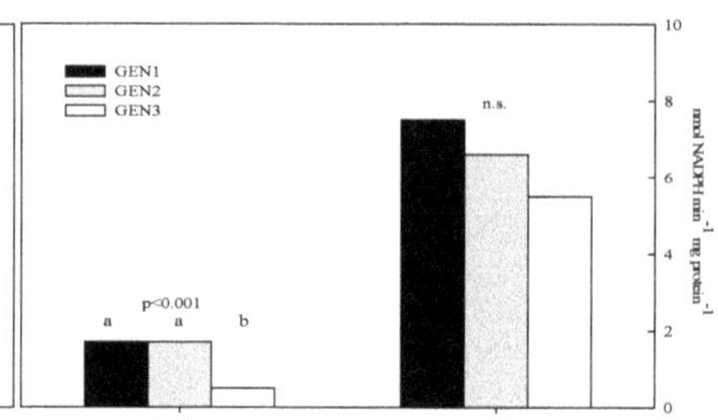

Figure 2. G6PDH and ME activities (nmol of NADPH/min/mg protein) in LD

Conclusions

Iberian x Duroc reciprocal crosses (GEN1 vs. GEN2) showed a similar adipogenic character. Nonetheless, the differences found between the genotypes from the 2 paternal lines of Duroc (GEN2 vs. GEN3) indicate that crossing a selected Duroc genotype (Duroc2) with Iberian pigs reduces the adipogenic level and as consequence IMF level and it also modifies fatty acid profile, increasing unsaturated fatty acids, which could have negative effects on the production of fresh meat and dry-cured meat products.

References

B.O.E., 15 de Octubre del 2001. Real Decreto 1083/2001, de 5 de Octubre. Norma de Calidad para el jamón ibérico, paleta ibérica y caña de lomo ibérico elaborados en España.

Bautista, J.M., Garrido-Pertierra, A., and Soler., G. (1988). Glucose-6-phosphate dehidrogenase from Dicentrarchus labrax liver: kinetic mechanism and kinetics of NADPH inhibition. *Biochim et Biophys Acta, 967,* 3, 354-363.

Bligh, E. G., and Dyer, WJ. (1959). A rapid method of total lipid extraction and purification. *Can. J of Biochem. Physiol., 37,* 911–917.

Bradford, M. (1976). A rapid and sensitive method for the quantification of microgram quantities of protein utilizing the principle of protein-dye binding. *Anal. Biochem., 72*: 248-254.

Gandemer, G. (2002). Lipids in muscles and adipose tissues, changes during processing and sensory properties of meat products. *Meat Sci., 62*, 309-321.

Monin, G., Hortós, M., Díaz, I., Rock, E., and García-Regueiro, J. A. (2003). Lipolysis and lipid oxidation during chilled storage of meat from Large White and Pietrain pigs. *Meat Sci., , 64,* 7–12.

Morrissey, P.A., Sheehy, P.J.A., Galvin, K., Kerry, J.P, and Buckley, D.J. (1998). Lipid stability in meat and meat products. *Meat Sci., 49,* Supplement 1, S73-S86.

Spina, J., Bright, H.J., and Rosenbloom, J. (1970). Purification and properties of L-Malic Enzyme from Escherichia coli. *Biochemistry, 9,* 3794-3801.

Wise, E., and Ball, E.G. (1964). Malic enzyme and lipogenesis. *Proceedings of the National Academic of the Science of the United State of America,* 52, 1255-1263.

FAT CONTENT AND FATTY ACID PROFILE OF STEERS REARED IN ITALY: CHANGES ALONG WITH TIME

P. Salvatore, S. Nicoli, M. Lucarini*, L. D'Evoli, P. Gabrielli, L Marletta and G. Lombardi-Boccia

National Institute for Food and Nutrition Research – Via Ardeatina 546, 00147 Rome (Italy)
Email: lucarini@inran.it

Keywords: fatty acids, total fat, Italian steers

Introduction

Bovine meat is one of the most consumed in Italy (Turrini and Lombardi-Boccia, 2002). Data on the composition of the Italian Total Diet reported that bovine meat accounted for 39.4% of the total meat and meat products daily consumption (Lombardi-Boccia *et al.*, 2004). In recent years, much attention has been paid on the nature of fat in meat because it is considered a risk factor in the developing of cardiovascular diseases. Consumer concerns led the meat industry toward efforts aimed at reducing carcass fatness. The present study was undertaken to update and broaden data on the fat content and composition of meat from steers reared in Italy. The objective of this study was therefore to evaluate the total fat content, fatty acids profile and cholesterol content of five meat cuts (fillet, sirloin-steak, eye-round, top-side and knuckle) of steers reared in Italy. Furthermore present data were compared with those coming from previous studies carried out in Italy and changes in fatty acids composition along with time are discussed.

Materials and Methods

Five crossbred steers (Charolais and Limousine) were reared on 5 farms and raised on commercial pellets (UNIFEED). The steers were slaughtered at 21 months of age, and five retail cuts (fillet, sirloin steak, eye-round, top-side and knuckle) from each steer were trimmed away of external fat, packaged in vacuum packs and delivered to the laboratory. In order to obtain representative meat samples for analyses, each meat cut was sliced in pieces of about 20g and subdivided in several equal aliquots: some aliquots were immediately taken for analysis, the others were frozen at -30°C under vacuum and stored for subsequent analyses, each being carried out in triplicate.

Lipid: Intramuscular fat was extracted by means of a modification of the method of Folch, Lees and Stanley (1957) using chloroform/methanol (2/1, v/v). *Fatty acid composition:* Fatty acids were esterified using boron trifluoride in methanol as esterification reagents. The recovered fatty acids methyl ester (FAME) were analysed using a HP 5890 II series gas chromatograph equipped with a flame ionisation detector (FID) and operated with a split ratio of 100:1. Separations were accomplished for saturated fatty acids on a Supelcowax 10 TM (60mx 0.25mm i.d., 0.25μ film thickness) column; detector temperature was 260°C, injector temperature was 240°C and the following temperature program was used: from initial temperature of 120°C to 180°C at 8°/min and to 240°C at 3°/min. Separation of trans fatty acids was carried out using a SP 2560 column (100mx25mm i.d., 0.2μ film thickness), detector and injector temperature was 200°C, the separations were accomplished with an isotherm at 170°C. Identification of the components was based on comparison of retention time with standard and using Standard Reference Material: NIST 1546 (meat homogenate) and BCR 163 (beef-pork fat blend). *Cholesterol:* this was quantified by enzymatic assay (Boehringer Mannheim/R-Biopharm).

Results and Discussion

The five meat cuts showed a low total fat content, from about 1.1% detected in eye-round and top-side, which displayed about half the amount found in the other cuts, to a maximum of 2.9% detected in sirloin-steak. When the total intramuscular fat content detected in this study was compared with that reported in the Italian Food Composition Table (Carnovale and Marletta, 2000), which describes the composition of Italian meat of about ten years ago, a marked decrease in total fat content along with time was observable: the reduction in intramuscular fat content was up to 56% for fillet, the minimum reduction was found for knuckle (21%), with the mean value of total fat reduction (comprehensive of the five meat cuts analysed in this study) being about 44%. The fatty acid profile of total extractable lipids from the five cuts analysed is given in Table 1. Among the SFA, palmitic and stearic acids were the most represented in all the cuts analysed, sirloin steak showing the highest amounts of both, in contrast, the lowest amounts of palmitic acid were found in top-side and stearic acid in eye-round. Among the MUFA, oleic acid was present in the highest amount and sirloin steak and knuckles were the cuts with the highest concentration of it (84 and 80%, respectively). The most abundant PUFA were linoleic and arachidonic acids. Top-side was the cut with the highest concentration of linoleic acid (19.2%) and the fillet had the highest content of arachidonic acid (about 4.2%).

Table 1: Fatty acid profile (% total fatty acids) of the five raw cuts analysed (f.w.)

Fatty ac.	fillet		sirloin-steak		eye-round		top-side		knuckle	
	mean	s.d.	mean	s.d.	mean	s.d.	mean	s.d.	mean	s.d.
14:0	1.92	0.27	2.71	0.21	0.97	0.24	1.80	0.42	2.17	0.26
15:0	2.13	1.05	1.80	0.15	0.99	0.34	2.21	0.05	1.91	0.22
16:0	22.41	1.31	25.51	0.58	22.44	0.35	22.13	0.83	21.59	0.49
17:0	2.37	0.40	2.16	0.06	1.94	0.07	1.60	0.18	1.72	0.10
18:0	18.16	1.15	18.42	0.63	16.91	0.41	17.16	0.55	17.32	0.03
21:0	0.86	0.02	0.72	0.04	3.84	0.3	1.34	0.01	1.09	0.04
14:1	0.82	0.19	0.65	0.08	0.99	0.07	1.42	0.12	0.88	0.25
16:1 n7	2.24	0.07	2.44	0.08	2.18	0.54	1.97	0.36	2.68	0.54
17:1 n8	1.79	0.68	0.97	0.18	1.18	0.64	2.39	0.51	1.56	0.19
18:1 n9	24.99	1.70	28.82	0.32	23.89	1.94	21.55	2.21	28.53	1.94
18:1 n7	2.24	0.17	1.43	0.47	2.23	0.59	2.28	0.28	2.06	0.82
18:2 n6	14.53	1.11	10.61	0.84	16.34	1.23	19.21	1.44	13.09	1.04
18:3 n3	0.86	0.16	0.65	0.43	0.81	0.04	0.77	0.01	0.92	0.97
20:4 n6	4.16	0.45	2.51	0.11	3.99	0.48	3.76	0.24	3.87	0.03
20:5 n3	0.51	1.66	0.65	0.05	1.26	1.48	1.24	0.10	0.52	0.05
SFA	47.8		51.3		47.0		46.2		45.8	
MUFA	32.0		34.3		30.5		30.0		35.7	
PUFA	20.1		14.4		22.4		25.0		18.4	

Sirloin steak, the cut with the highest proportion of intramuscular fat, had the highest percentage of SFA (about 51.3%). MUFA were detected in the highest amount in knuckles (35.7%). The two cuts lowest in fat were found to be the cuts richest in PUFA: 25% and 22%, respectively. Compared with data reported in previous studies (Lintas *et al.*, 1977; Carnovale and Marletta, 2000) which describe the composition of Italian meat from 30 years ago (Figure1) a marked decrease in total fat content as well as changes in fatty acid content along with time was observed.

Fig.1 Changes in fatty acid profile in Italian beef (% total fatty acids).

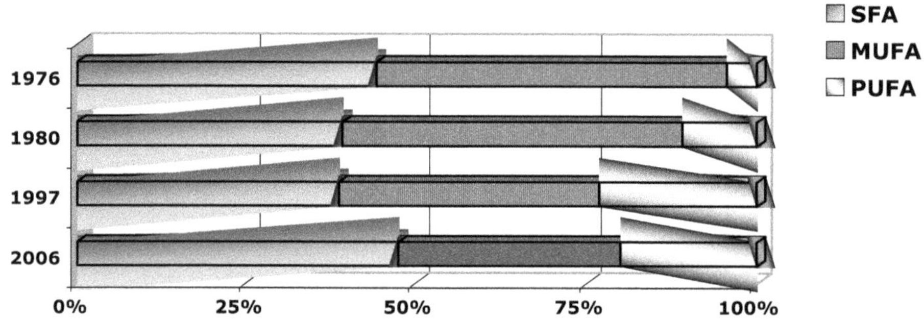

References

Carnovale, E. and Marletta, L. (2000). Tabelle di Composizione degli Alimenti. EDRA ed: Milano

Lintas, C., Balduzzi, MA., Bernardini, MP. and Massi, O. (1977). Indagini sulla composizione delle carni alimentary. Notal Acidi grassi di bovini adulti

Lombardi-Boccia, G. Lanzi, S., Lucarini. M. and Di Lullo, G. (2004). Meat and Meat product Consumption in Italy: Contribution to trace Elements and selected Vitamin B supply. International Journal Vitamin and Nutrition Research, 74: 247-251

Turrini A. and Lombardi-Boccia G. (2002). The formulation of the market basket for evaluating the Italian total diet. Nutrition Research 22: 1151-1162.

FATTY ACID COMPOSITION OF SUCKLING LAMBS AS AFFECTED BY INCREASING CALCIUM SOAP ON EWE'S DIET

M.A. Lurueña-Martínez*[1], C. Palacios-Riocerezo[2], I. Revilla[1] and A.M. Vivar-Quintana[1]

[1.] Área de Tecnología de los Alimentos, E.P.S. de Zamora, Universidad de Salamanca, Avda. Requejo 33, 49022 Zamora, Spain. [2.] Área de Producción Animal. Facultad de Ciencias Agrarias y Ambientales, Universidad de Salamanca, Avda. Filiberto Villalobos 119-129, 37007 Salamanca, Spain. Email: luru74@hotmail.com

Keywords: suckling lamb, calcium soap, fatty acids, adipose tissue

Introduction

In Mediterranean areas, meat from suckling lambs is traditionally consumed, especially in specific periods (Christmas, Easter); commanding high market prices (Sañudo *et al.*, 2000). In Spain, suckling lambs are usually slaughtered immediately after weaning (10-12 kg live weight) and generally have been exclusively fed with milk. The Spanish market demands light carcasses (under 13 kg) with a characteristic degree of fatness that decides, to great extent, their commercial value. The optimum level of fatness is that which maintains a balance between the minimum amount needed to satisfy consumer tastes and that which guarantees carcass conservation, together with the flavour and aroma of the meat (Castro *et al.*, 2005). It is generally accepted that the fatty acid composition of the adipose depots of suckling animals depends on the composition of the milk they consume (Velasco, 1999; Zygoyiannis *et al.*, 1992) just as the fatty acid composition of this milk depends on the composition of ewe's diet (Goulas *et al.*, 2003). Palm oil, marketed as such and in the form of calcium soaps made from palm oil fatty acids, is one of the principal vegetable fats used in animal feeds (Castro *et al.*, 2005).

Our aim was to investigate the different effects of two ewe's diets (with and without calcium soaps) on the fatty acids of suckling lambs.

Materials and Methods

Sixty 25-30-day-old suckling lambs were studied. Two groups of thirty lambs from three different breeds (assaf, castellana and churra) were slaughtered at 11 kg live weight. One of these groups was fed with a basic diet composed of 18% beetroot pulp, 26% alfalfa, 22% barley, 12% corn, 12% soy, 10% cotton. The other group was fed with an added calcium soap diet composed of 18% beetroot pulp, 26% alfalfa, 18% barley, 12% corn, 12% soy, 10% cotton, 4% calcium soap. Lipids from adipose tissue were extracted using a standard chloroform/methanol procedure (Folch *et al.*, 1957). Fatty acid composition of lipids was methylated (Murrieta *et al.*, 2003) and analysed by gas chromatography (GC 6890 N, Agilent Technologies, USA) using a capillary column of 100 m × 0.25 mm × 0.20 μm (Supelco, Inc., Bellefonte, PA, USA). One microlitre was injected into the chromatograph, equipped with a split/splitless injector and a FID detector. The oven temperature program was 150 °C increasing at 1°C/min up to 165°C then increasing at 0.20°C/min up to 167°C and then increasing 1.50°C up to 225°C where it was maintained for 15 min. Injector and detector temperatures were 250°C. The carrier gas was helium at 1 ml/min and split (20:1). Fatty acids were expressed as a proportion by total weight. Data were compared by analysis of variance according to the General Linear Model Procedure of Statgraphics Plus package (Statgraphics Plus for Windows 2.1, Statistical Graphics Corp., 1996).

Results and Discussion

Table 1 displays data on the fatty acid content of adipose tissue (g/100g). In concordance with data found in the bibliography (Castro *et al.*, 2005), the most abundant fatty acids in all depots studied were oleic acid (C18:1 n-9), palmitic acid (C16:0), stearic acid (C18:0), butyric acid (C4:0) and myristic acid (C14:0).

As table 1 shows, breed and ewe's diet had an important influence over fatty acids of suckling lambs. Breed was more influential than diet, but the aim of this study was the latter.

With respect to lambs from ewes with calcium soap diet, the average content of C16:0, C16:0 ramified and C18:0 (saturated fatty acids) was significantly higher than those raised with basic diet ewe's milk. This increasing was caused by palm oil fatty acids that compose calcium soap.

On the other hand, calcium soap diet lambs presented lower monounsaturated and polyunsaturated acid values than basic diet lambs, like oleic acid (C18:1 n-9), linoleic acid (C18:2 n-6) and α-linolenic acid (C18:3 n-3). Other fatty acids that were significantly decreased in calcium soap diet lambs were caprylic acid (C8:0), myristoleic acid (C14:1 n-5), palmitoleic acid (C16:1 n-7), heptadecenic acid (C17:1), elaidic acid (9 *trans* C18:1 n-9) and tricosanoic acid (C23:0). This decrease was caused by the increase of saturated fatty acids in the diet (calcium soap) and the decrease of unsaturated fatty acids (22% to 18% of barley).

Table 1: Effect of diet and breed on fatty acid composition expressed as g/100 g.

	Assaf		Castellana		Churra			
	Basic	Soap	Basic	Soap	Basic	Soap	BREED	FEED
C4:0	9.04	6.10	15.89	20.24	14.00	8.07	**	ns
C6:0	0.07	0.05	0.08	0.05	0.10	0.07	ns	ns
C8:0	0.11	0.09	0.15	0.09	0.10	0.06	ns	**
C10:0	1.74	1.81	1.70	1.86	1.45	1.44	ns	ns
C11:0	0.11	0.13	0.11	0.07	0.09	0.06	**	ns
C12:0	2.34	1.58	2.50	2.89	2.04	2.08	*	ns
C14:0 ramified	0.09	0.09	0.05	0.07	0.07	0.09	*	ns
C14:0	12.93	11.51	11.03	10.94	11.83	12.46	ns	ns
C15:0 ramified	0.55	0.41	0.39	0.38	0.45	0.49	**	ns
C14:1 n-5	0.37	0.38	0.46	0.27	0.45	0.31	ns	**
C15:0	0.85	0.84	0.69	0.65	0.68	0.76	**	ns
C16:0 ramified	0.19	0.20	0.14	0.16	0.16	0.22	*	*
C16:0	23.82	28.21	20.58	22.96	22.29	25.31	**	**
C16:1 n-7	2.21	2.00	2.34	1.23	2.48	1.51	*	**
C17:0	0.79	0.93	0.83	0.63	0.79	0.90	**	ns
C17:1	0.36	0.39	0.40	0.25	0.45	0.37	**	**
C18:0	9.20	12.63	9.67	10.38	8.60	12.32	ns	**
9 *trans* C18:1 n-9	1.82	0.85	1.06	0.72	0.85	1.18	**	**
C18:1 n-9	24.67	24.51	25.64	20.56	26.27	24.53	ns	**
C18:1	1.04	0.75	0.66	0.52	0.79	0.68	**	**
C18:2 n-6	3.45	2.18	1.95	1.63	2.12	2.07	**	**
C18:3 n-3	0.49	0.65	0.57	0.29	0.71	0.46	**	**
C18:2 CLA	0.60	0.35	0.58	0.44	0.45	1.05	*	ns
C23:0	0.26	0.12	0.16	0.01	0.17	0.09	**	**
C22:5 n-3	0.14	0.11	0.10	0.01	0.16	0.08	**	**

* *P*<0,1 ***P*<0,05 ns: not significant

Conclusions

The results obtained in this study indicate that fatty acid composition of adipose depots of suckling lambs were affected by the ewe's diet composition. Supplementation with calcium soap made from palm oil fatty acids increased some saturated fatty acids, like C16:0, and decreased some unsaturated fatty acids, like C18:1 n-9.

Acknowledgements

The authors are grateful for financial support from the Fundación Caja Rural. M. A. Lurueña-Martínez was supported by doctorate grant from Consejería de Educación de la Junta de Castilla y León and Fondo Social Europeo.

References

Castro, T., Manso, T., Mantecón, A. R., Guirao, J., Jimeno, V. (2005). Fatty acid composition and carcass characteristics of growing lambs fed diets containing palm oil supplements. Meat Science, 69: 757-764.

Folch, J., Lees, M., Sloane Stanley, G. H. A. (1957). Simple method for the isolation and purification of total lipids from animal tissues. Journal of Biological Chemistry, 226: 497-509.

Goulas, C., Zervas, G., Papadopoulos, G. (2003). Effect of dietary animal fat and methionine on dairy ewe's milk yield and milk composition. Animal Feed Science and Technology, 105: 43-54.

Murrieta, C. M., Hess, B. W., Rule, D. C. (2003). Comparison of acidic and alkaline catalysts for preparation of fatty acid methyl ester from ovine muscle with emphasis on conjugated linoleic acid. Meat Science, 65: 523-529.

Sañudo, C. Alfonso, M., Sánchez, A., Delfa, R. Teixeira, A. (2000). Carcass and meat quality in light lambs from different fat classes in the EU carcass classification system. Meat Science, 56: 89-94.

Velasco S. (1999) Caracterización de los depósitos adiposos de corderos lechales en función de diversos parámetros productivos. Tesis Doctoral. Facultad de Veterinaria de Madrid.

Zygoyiannis, D., Kufidis, D., Katsaounis, N., Phillips, P. (1992). Fatty acid composition of carcass fat of indigenous (*Capra prisca*) suckled Greek kids and milk of their does. Small Ruminant Research, 8: 83-95.

INTERRELATIONSHIPS IN FATTY ACID COMPOSITION BETWEEN SUCKLING LAMBS AND THE FEEDING EWE MILK

M.A. Lurueña-Martínez*[1], C. Palacios-Riocerezo[2], I. Revilla[1] and A.M. Vivar-Quintana[1]

[1] *Área de Tecnología de los Alimentos, E.P.S. de Zamora, Universidad de Salamanca, Avda. Requejo 33, 49022 Zamora, Spain.* [2] *Área de Producción Animal. Facultad de Ciencias Agrarias y Ambientales, Universidad de Salamanca, Avda. Filiberto Villalobos 119-129, 37007 Salamanca, Spain. Email: luru74@hotmail.com*

Keywords: suckling lamb, fatty acid composition, ewe milk, multivariate analysis

Introduction

In Spain, suckling lambs are usually slaughtered immediately after weaning (10-12 kg live weight) and generally have been exclusively fed with milk. It is generally accepted that the fatty acid composition of the adipose depots of suckling animals depends on the composition of the milk they consume (Velasco, 1999). In fact, the impact of a milk-feeding regime on lamb meat quality could be dramatic because milk-fed lambs are to be considered "functional monogastrics" and their rumen underdevelopment protects dietary polyunsaturated fatty acids from dehydrogenation (Napolitano *et al.*, 2002).

On the other hand, it is well known that yield and composition of ewe milk undergoes marked changes throughout the year, depending on the feed availability and metabolic and endocrine changes related to the climate and the advancement of lactation (Hassan, 1995). Breed is another important factor that is closely related to milk quality. At present, there are not many multivariate studies in the literature about this subject. The aim of the present study, was to examine the correlation between fatty acids from ewe milk and suckling lambs.

Materials and Methods

Sixty 25-30 day-old suckling lambs were studied. Two groups of thirty lambs from three different breeds (assaf, castellana and churra) were slaughtered at 11 kg live weight. One of these groups was fed with a basic diet composed of 18% beetroot pulp, 26% alfalfa, 22% barley, 12% corn, 12% soy, 10% cotton. The other group was fed with an added calcium soap diet composed of 18% beetroot pulp, 26% alfalfa, 18% barley, 12% corn, 12% soy, 10% cotton, 4% calcium soap.

Lipids from adipose tissue were extracted using a standard chloroform/methanol procedure (Folch *et al.*, 1957). Lipids from ewe milk were extracted using the International Standard Method described in ISO 14156:2001. Fatty acid composition of lipids were methylated (Murrieta *et al.*, 2003) and analysed by gas chromatography (GC 6890 N, Agilent Technologies, USA) using a capillary column of 100 m × 0.25 mm × 0.20 µm (Supelco, Inc., Bellefonte, PA, USA). One microlitre was injected into the chromatograph, equipped with a split/splitless injector and a FID detector. The oven temperature program was 150°C increasing at 1 °C/min up to 165 °C then increasing at 0,20 °C/min up to 167 °C and then increasing 1,50 °C up to 225°C where it was maintained for 15 min. Injector and detector temperatures were 250°C. The carrier gas was helium at 1 ml/min and split (20:1). Fatty acids were expressed as a proportion by total weight.

Data were compared by a multivariate method according to the Multiple-Variable Analysis Procedure of Statgraphics Plus package (Statgraphics Plus for Windows 2.1, Statistical Graphics Corp., 1996).

Finally, only the twelve most important fatty acids were considered: butyric acid (C4:0), caproic acid (C6:0), caprylic acid (C8:0), capric acid (C10:0), lauric acid (C12:0), myristic acid (C14:0), palmitic acid (C16:0), palmitoleic acid (C16:1), stearic acid (C18:0), oleic acid (C18:1 n-9c), linoleic acid (C18:2 n-6) and α-linolenic acid (C18:3 n-3).

Results and Discussion

Significant correlations between ewe milk and suckling lambs fatty acids were found after multivariate analysis. Data are given in Table 1 and show that the most influential milk fatty acids were caproic, caprylic, palmitoleic (saturated fatty acids) and oleic. In suckling lambs, the most influenced fatty acids were palmitoleic, linoleic and α-linolenic (unsaturated fatty acids). Other fatty acids from lamb highly correlated with fatty acids from milk were butyric, lauric and myristic.

Butyric, caproic, caprylic and capric from ewe milk (short-chain saturated fatty acids) were in negative correlation with nine fatty acids from lamb. An increase in these fatty acids from milk caused a decrease in correlated fatty acids from lamb. On the other hand, palmitoleic acid, stearic acid, oleic acid and linoleic acid from milk (long-chain fatty acids) were in positively correlated with palmitoleic acid, linoleic acid and α-linolenic acid from lamb.

Capric acid from lamb had not been significantly influenced by any milk fatty acid. Only oleic acid from lamb was significantly correlated with one fatty acid from milk (butyric acid).

Table 1: Correlations between ewe milk and suckling lamb fatty acids.

		Suckling lamb fatty acids											
		A	B	C	D	E	F	G	H	I	J	K	L
Ewe milk fatty acids	A	0.37**	ns	ns	ns	-0.25	-0.32*	ns	-0.41**	ns	-0.24	-0.21	-0.42**
	B	0.32*	-0.19	ns	ns	-0.28	-0.29*	ns	-0.40**	0.20	-0.22	-0.28*	-0.38**
	C	0.29*	-0.18	ns	-0.15	-0.27	-0.26	ns	-0.33*	0.18	-0.18	-0.32*	-0.31*
	D	0.22	ns	-0.16	-0.17	-0.26	-0.23	ns	-0.25	0.20	ns	-0.36**	-0.23
	E	ns	ns	ns	ns	ns	ns	ns	ns	ns	ns	-0.37**	ns
	F	ns	0.28*	ns	ns	ns	ns	ns	ns	ns	ns	ns	ns
	G	-0.32*	ns	ns	ns	0.26	0.29*	0.20	ns	ns	ns	0.18	ns
	H	ns	0.33*	0.46**	ns	ns	ns	-0.35*	0.56**	-0.30*	0.22	ns	0.49**
	I	ns	ns	ns	ns	ns	0.17	ns	0.22	-0.29*	ns	0.37**	0.22
	J	ns	ns	0.29*	ns	ns	ns	-0.16	0.42**	-0.29*	0.18	0.40**	0.41**
	K	-0.23	ns	0.16	ns	ns	0.20	ns	0.41**	-0.17	0.20	0.44**	0.33*
	L	ns	0.20	ns	ns	ns	ns	ns	ns	ns	ns	-0.21	0.18

* $P<0.05$ ** $P<0.01$ ns: not significant (P>0.3)

A: butyric acid, B: caproic acid, C: caprylic acid, D: capric acid, E: lauric acid, F: myristic acid, G: palmitic acid, H: palmitoleic acid, I: stearic acid, J: oleic acid, K: linoleic acid, L: α-linolenic acid. Negative values indicate an inverse correlation between fatty acids.

On the other hand, α-linolenic acid from milk did not significantly affect any lamb fatty acid meanwhile only lauric acid and myristic acid were each in correlation with only one lamb fatty acid.

Linoleic acid from ewe milk was positively correlated with linoleic acid and α-linolenic acid from lamb. These results mean, according to Napolitano *et al.*, (2002), that lamb rumen underdevelopment protects dietary polyunsaturated fatty acids from dehydrogenation.

Conclusions

An interrelationship was found between ewe milk and suckling lamb fatty acids. The most influential milk fatty acids were caproic, caprilic, palmitoleic and oleic. The most influenced suckling lamb fatty acids were palmitoleic, linoleic and α-linolenic.

Acknowledgements

The authors are grateful for financial support from the Fundación Caja Rural. M. A. Lurueña-Martínez was supported by doctorate grant from Consejería de Educación de la Junta de Castilla y León and Fondo Social Europeo.

References

Folch, J., Lees, M., Sloane Stanley, G. H. A. (1957). Simple method for the isolation and purification of total lipids from animal tissues. Journal of Biological Chemistry, 226: 497-509.

Hassan, S. H. (1995). Effects of crossing and environmental factors on production and some constituents of milk in Osimi and Saidi sheep and their crosses with Chios. Small Ruminant Research, 18: 165-172.

International Standard. (2001). Milk and milk products. Extraction methods for lipids and liposoluble compounds. ISO 14156:2001 (E), IDF 172:2001 (E).

Murrieta, C. M., Hess, B. W., Rule, D. C. (2003). Comparison of acidic and alkaline catalysts for preparation of fatty acid methyl ester from ovine muscle with emphasis on conjugated linoleic acid. Meat Science, 65: 523-529.

Napolitano, F., Cifuni, G. F., Pacelli, C., Riviezzi, A. M., Girolami, A. (2002). Effect of artificial rearing on lamb welfare and meat quality. Meat Science, 60: 307-315.

Sañudo, C. Alfonso, M., Sánchez, A., Delfa, R. Teixeira, A. (2000). Carcass and meat quality in light lambs from different fat classes in the EU carcass classification system. Meat Science, 56: 89-94.

Velasco S. (1999). Caracterización de los depósitos adiposos de corderos lechales en función de diversos parámetros productivos. Doctoral Thesis. Facultad de Veterinaria de Madrid.

THE EFFECT OF TRANSPORT ON PH EVOLUTION OF DIFFERENT MUSCLES IN SUCKLING LAMBS

J. De La Fuente[2], C. Pérez[3], C. Vieira*[1], M. Sánchez[1], E. González de Chávarri[2], M.D. García[1], I. Álvarez[4] and M.T. Díaz[1]

[1]*Estación Tecnológica de la Carne, ITACyL, 37770 Guijuelo, Salamanca. *Email: vieallce@itacy.es* [2]*Departamento de Producción Animal, Facultad de Veterinaria, UCM, 28040 Madrid.* [3]*Departamento de Fisiología (Fisiología Animal), Facultad de Veterinaria, UCM, 28040 Madrid.* [4]*Departamento de Tecnología de los Alimentos, INIA, 28040 Madrid.*

Keywords: transport time, stocking density, suckling lamb, pH

Introduction

Animal transport is an important and necessary phase in all systems of meat production. There are great number of factors involved in animal transport such as stocking density, duration of transport, mixing of animals etc., which can affect animal welfare and also carcass and meat quality. The impact of transport on animals can be minimized ensuring that transport is carried out in the best of possible conditions (Warriss, 1998). pH is one of the main parameters used to measured meat quality and it can be affected by stress due to transport (Devine *et al.,* 1993). In fact, parameters as important in meat quality as colour, texture, and sensory properties are strictly related to pH. There are very few studies related to lamb transport and mainly related to transport of suckling lambs. Therefore, the aim of this study was to evaluate the effect of transport time and stocking density on pH evolution in several muscles of suckling lambs transported to slaughterhouse.

Materials and Methods

Forty-eight male suckling lambs (mean live weight 13.08 ± 0.17) were used in a 2 x 2 factorial design (two transport time: 30 minutes and 5 hours x two stocking densities: $0.12m^2$/lamb and $0.20m^2$/lamb). Lambs remained with their dams until they were loaded up into an experimental lorry to be transported to the slaughterhouse. Lambs were randomly assigned to four groups of transport; two groups were transported for 30 minutes with the two densities and two groups for 5 hours with the two densities. All the journeys were done on two consecutives weeks in April and occurred early in the morning with an environmental temperature of 10-12°C. After transport, the lambs were unloaded and held in lairage less than 1 hour. Animals were, after electrical stunning, slaughtered and the carcasses were chilled at 4°C. pH were measured immediately after slaughter (initial), and at 45 min, 3 h and 24 h postmortem using a penetrating electrode adapted to a portable pH meter in the *Semitendinosus* (ST), *Longisimus* (LD) and *Psoas major* (PM) muscles. All analyses were performing using Statgraphics Plus. 5.0, the pH values and drops were analysed statistically by analysis of variance.

Results and Discussion

The table 1 shows pH values and drops of the muscles *Semitendinosus*, *Longisimus* and *Psoas major.* In all muscles studied, the pH values found were in a normal range for lamb, and only one case of DFD meat was detected, pH at 24 h of 6.13, in ST muscle of a lamb transported for 30 min at low density ($0.20m^2$/lamb). According to Brown *et al.,* (1990) the incidence of DFD is low for young ruminants.

The pH at 3h and pH at 24 h in ST muscle were significantly affected by transport time and stocking density respectively ($P<0.05$). The transport time did not affect significantly to pH values and pH drop in LD muscle, however stocking density affected significantly pH values initially ($P<0.05$) and at 24 hours ($P<0.01$). On the other hand, transport time and stocking density affected pH values and pH drop in PM muscle. The different response of the muscles to the transport could be due to the different fibre types of the studied muscles. Muscles involved in movements (walking, standing up) such as ST and LD have a high proportion of alpha-white fibres with high glycogen storage (Gregory, 1998). Therefore the transport time did not affect them. However, PM has a high proportion of beta-red fibres with low glycogen storage (Gregory, 1998). In this muscle, lambs transported 30 min had a high initial pH, maintaining this high pH at 24 h. According to Sanz *et al.,* (1996), short journeys can be more stressful than longer ones, since in short journey animals had less time to adapt to the new situation and may have arrived with lower glycogen reserves and therefore high pHs.

Ultimate pH (pH 24 h) of ST, LD and PM were affected by stocking density showing animals transported at low stocking density ($0.20m^2$/lamb) higher values than those transported at $0.12m^2$/lamb. Low stocking density increases the capacity of lambs for movement, and factors such as lorry movements might have caused them to move to retain their balance, producing a reduction on muscle glycogen and consequently high ultimate pH. This result agrees with Ibañez *et al.,* (2002) who found that suckling lambs transported at low stocking density (4 lambs/m^2) had higher lactate dehydrogenase (LDH) activity than lambs transported at high stocking density (8 lambs/m^2) because of the greater amount of exercise of the former lambs.

Table 1: Mean and ± standard error of pH values and drops of *Semitendinosus*, *Longisimus* and *Psoas major* initially, at 45 min, 3 h and 24 h postmortem according to transport time (T) and stocking density (De).

	Transport time (T)		Stocking density (De)		Significance		
	30 min	5 hours	0.12m²/lamb	0.20m²/lamb	T	De	T*De
Semitendinosus							
pH value							
Initial	6.50 ± 0.04	6.40 ± 0.04	6.46 ± 0.04	6.44 ± 0.04	NS	NS	NS
45 min.	6.28 ± 0.03	6.24 ± 0.04	6.27 ± 0.04	6.25 ± 0.04	NS	NS	NS
3 h.	6.03 ± 0.03	6.12 ± 0.04	6.09 ± 0.04	6.06 ± 0.03	*	NS	NS
24 h.	5.60 ± 0.03	5.62 ± 0.04	5.56 ± 0.03	5.66 ± 0.04	NS	*	NS
pH drop							
0-45 min.	0.22 ± 0.04	0.16 ± 0.04	0.19 ± 0.04	0.19 ± 0.04	NS	NS	NS
45-3 h.	0.26 ± 0.04	0.12 ± 0.04	0.19 ± 0.04	0.19 ± 0.04	*	NS	NS
3-24 h.	0.42 ± 0.04	0.74 ± 0.05	0.52 ± 0.04	0.64 ± 0.05	NS	NS	NS
Longisimus							
pH value							
Initial	6.70 ± 0.04	6.61 ± 0.04	6.71 ± 0.04	6.60 ± 0.04	NS	*	NS
45 min.	6.56 ± 0.04	6.48 ± 0.03	6.52 ± 0.03	6.52 ± 0.04	NS	NS	NS
3 h.	6.27 ± 0.04	6.25 ± 0.04	6.28 ± 0.05	6.24 ± 0.04	NS	NS	NS
24 h.	5.56 ± 0.02	5.50 ± 0.05	5.46 ± 0.03	5.59 ± 0.04	NS	**	NS
pH drop							
0-45 min.	0.14 ± 0.04	0.13 ± 0.06	0.19 ± 0.05	0.08 ± 0.05	NS	NS	NS
45-3 h.	0.29 ± 0.05	0.23 ± 0.04	0.25 ± 0.05	0.28 ± 0.04	NS	NS	NS
3-24 h.	0.70 ± 0.04	0.99 ± 0.07	0.81 ± 0.06	0.88 ± 0.05	NS	NS	NS
Psoas major							
pH value							
Initial	6.25 ± 0.05	6.08 ± 0.03	6.20 ± 0.05	6.16 ± 0.04	**	NS	NS
45 min.	5.96 ± 0.04	6.03 ± 0.04	6.06 ± 0.04	5.93 ± 0.04	NS	*	NS
3 h.	5.92 ± 0.04	5.84 ± 0.05	5.85 ± 0.04	5.91 ± 0.06	NS	NS	NS
24 h.	5.63 ± 0.02	5.52 ± 0.05	5.52 ± 0.03	5.63 ± 0.03	*	*	NS
pH drop							
0-45 min.	0.30 ± 0.05	0.05 ± 0.04	0.14 ± 0.05	0.21 ± 0.06	**	NS	NS
45-3 h.	0.03 ± 0.05	0.19 ± 0.04	0.20 ± 0.04	0.02 ± 0.05	*	**	NS
3-24 h.	0.29 ± 0.04	0.55 ± 0.06	0.33 ± 0.05	0.51 ± 0.06	NS	NS	NS

*P<0.05; **P<0.01; NS: not significance

Conclusions

PM muscle was more affected by transport time and stocking density than the other studied muscles, showing lambs transported short time the highest pH, initially and at 24 h. However the three muscles studied were affected by stocking density, displaying highest ultimate pH lambs transported at low stocking density.

References

Brown, S.N., Bevis, E.A. and Warriss, P.D. (1990) An estimate of the incidence of dark cutting beef in the United Kingdom. Meat Science, 27, 249.

Devine, C.E., Graafhuis, A.E., Muir, P.D. and Chrystall, B.B. (1993) The effect of growth rate and ultimate pH on meat quality of lambs. Meat Science. 35, 63-77.

Gregory, N.G. (1998) Animal welfare and meat science. 1st Ed. CABI Publishing, Wallingford, pp 298.

Ibañez, M., De La Fuent, J., Thos, J. and Gonzales De Chavarri, E. (2002) Behavioural and physiological responses of suckling lambs to transport and lairage. Animal Welfare. 11, 223-230.

Sanz, M.C., Verde, M.T., Saez, T. and Sañudo, C. (1996) Effect of breed on muscle glycogen and dark cutting incidence in stressed young bulls. Meat Science. 43, 37-42.

Warriss, P.D. (1998) Choosing appropriate space allowance for slaughter pigs transported by road. A review. The Veterinary Record. 142, 449-454.

THE EFFECT OF TRANSPORT VEHICLE, LAIRAGE, AND HARVEST DAY ON PORK QUALITY

D.J. Newman*, C.C. Carr, M.H. Ryan, J. Meisinger and E.P. Berg

University of Missouri, Division of Animal Science, Columbia, MO, 65211, USA. Email: djn837@mizzou.edu

Keywords: pork, quality, stress, lairage, transport

Introduction

Preslaughter handling of pigs, including stockman interaction (loading, unloading, driving animals), transport, and lairage, has been designed to meet the pork industry's requirements rather than the pig's needs. Various preslaughter handling such as transport (Warris and Bevis, 1986), lairage (Brown *et al.*, 1999), and preslaughter stress (Hambrecht *et al.*, 2004) can trigger a stress response and may have negative consequences for pork quality aspects such as colour, water-holding capacity, and other sensory characteristics. Warris (1998) and Gispert *et al.*, (2000) found that pigs transported for long periods of time are more likely to form dark-firm-dry pork because of exhaustion and fatigue. Bradshaw *et al.*, (1996) found that short (≤1h) compared with long transports were more detrimental because the pigs did not recover from the stress of loading and could not adapt to the conditions of transport. The effect of lairage duration depends on factors such as environmental temperature and mixing. Aaslyng and Gade (2001) found that a lairage time of less than one hour does not allow for sufficient resting and may negatively impact both animal welfare and meat quality. However, extended lairage times of six hours and more may increase carcass damage and stress caused by fighting and prolonged periods without feed (Warris, 1998). The negative factors associated with transport, lairage, and preslaughter stress increase the chances of economic losses to producers and packers.

Materials and Methods

Mixed market hogs (Pig Improvement Company, Franklin KY) from the same commercial farm were harvested via electrical stunning in four groups on two days. Animals were harvested during a week in February at a processing plant in the northern Midwest. Pigs (n = 639) were randomly assigned to one of eight treatments in a 2 x 2 x 2 factorial arrangement, with two transport vehicle animal locations (top deck or bottom deck), two transport times (short (<3 h) or long (6 h), and two lairage durations (long (6 h) or short (<3h)). Four groups of pigs were processed over one week during the winter at a commercial plant. At 24h postmortem, loins were collected for ultimate meat quality. Measurements were taken at the $10^{th}/11^{th}$ rib interface of the longissimus muscle (LM). L* (lightness), a* (redness), and b*(yellowness) colour values were assessed in the LM on a freshly cut surface after a 10 minute blooming period with a Minolta Portable Chroma Meter (Model CR 210) calibrated to a black and white tile standard with a D_{65} illuminant. Both colour saturation (chroma; $[a^{*2}+b^{*2}]^{.05}$) and hue angle (true red; arctangent (b*/a*) x [360° /2x3.14)]) were calculated. pH was measured at the 10th rib location of the LM using an MPI pH meter (Meat Probes). Drip loss samples were collected from a 2.54cm thick loin chops using a 2.54cm diameter coring device. Samples were weighed, placed in drip loss tube (meat juice containers, C. Christensen Laboratory, Denmark) and stored at 2°C. After 24h of storage samples were re-weighed for calculation of drip loss percentage. Data were analyzed by the General Linear Model (GLM) procedure of SAS (v. 9.1, SAS Inst., Inc., Cary, NC).

Results and Discussion

Results are printed in Table 1. Pigs from the coldest collection day (Thursday) had lower ($P < 0.01$) ultimate pH values and tended ($P=0.06$) to be lighter in colour than pigs from the Tuesday harvest. However, pigs harvested on Thursday had an unusually lower ($P < 0.01$) drip loss percentages than pigs harvested on Tuesday. Additionally, drip loss was significantly ($P < 0.01$) increased for pigs on the bottom deck versus bottom deck. Short transport and short lairage increased ($P < 0.01$) paleness (i.e. high L* values) and decreased a* (redness) values in the longissimus. Colour saturation was decreased ($P < 0.01$) in the short lairage treatment and also on the cold day collection group. Hue Angle was decreased ($P < 0.01$) in the short transport group. These data suggest that longer transport or lairage times can allow pigs to adjust to the increased stress of loading and mixing animals.

Table 1: Least squares means (standard errors) for fresh pork loin quality measurements associated with time in transport, trailer deck level, length of lairage and harvest day.

	Time in Transport[a]			Deck Level[b]			Time in Lairage[c]			Harvest Day[d]		
Variable	**Long**	**Short**	**P-val**	**Bottom**	**Top**	**P-val**	**Short**	**Long**	**P-val**	**Tues**	**Thurs**	**P-val**
pH	5.64 (0.009)	5.63 (0.009)	0.57	5.63 (0.009)	5.64 (0.009)	0.32	5.62 (0.008)	5.65 (0.009)	< 0.01	5.72 (0.010)	5.55 (0.008)	< 0.01
Drip, %	2.97 (0.166)	3.18 (0.168)	0.37	3.39 (0.159)	2.76 (0.174)	< 0.01	3.07 (0.159)	3.08 (0.174)	0.98	3.34 (0.181)	2.81 (0.152)	< 0.05
L*	53.08 (0.159)	53.65 (0.159)	0.01	53.51 (0.153)	53.22 (0.164)	0.20	53.69 (0.151)	53.04 (0.165)	< 0.01	52.95 (0.173)	53.79 (0.145)	0.06
a*	16.53 (0.055)	16.45 (0.056)	0.30	16.54 (0.053)	16.44 (0.057)	0.21	16.38 (0.053)	16.59 (0.058)	< 0.01	16.61 (0.060)	16.37 (0.050)	< 0.01
Saturation	17.89 (0.056)	17.88 (0.056)	0.92	17.94 (0.054)	17.83 (0.058)	0.18	17.77 (0.054)	18.00 (0.059)	< 0.01	18.01 (0.061)	17.75 (0.051)	< 0.01
Hue Angle	22.34 (0.155)	22.96 (0.156)	< 0.01	22.64 (0.149)	22.66 (0.160)	0.94	22.69 (0.148)	22.62 (0.161)	0.75	22.66 (0.169)	22.65 (0.141)	0.96

[a]Long = 6 h transport duration; Short = 3 h transport duration
[b]Trailer description
[c]Short lairage = <3h; Long lairage = 6h
[d]Tuesday February 14 and Thursday February 16.

Conclusions

The preslaughter time period is crucial in preserving ultimate meat quality. Environment, transport, and lairage all play a significant role in contributing to factors that decrease both ultimate pork quality and good economic returns. In the present study improved meat quality can be obtained from animals that are allowed adequate recovery time from either short transport or short lairage. Furthermore, because of the decrease in meat quality seen from the coldest collection day, it may be necessary to modify the transport vehicle or lairage environment so that animals are not exposed to harsh environmental factors that ultimately result in poor meat quality.

References

Aaslyng, M.D., and Gade, P.B. (2001). Low stress pre-slaughter handling: effect of lairage time on the meat quality of pork. *Meat Science*, 57 (1): 87-92.

Bradshaw, R. H., R.F. Parrott, J.A. Goode, D.M. Lloyd, R.G. Rodway, and D. M. Broom. (1996). Stress and travel sickness in pigs: effects of road transport on plasma concentrations of cortisol, beta-endorphin and lysine vasopressin. *Animal Science*, 63: 507-516.

Brown, S.N., Knowles, T.G., Edwards, J.E., and Warriss, P.D. (1999). Behavioural and physiological response of pigs to being transported for up to 24 hours followed by six hours recovery in lairage. *Veterinary Record*, 145, 421-426

Gispert, M., Faucitano, L., Oliver, M.D., Coll, C., Siggers, K., Harvery, K., and Diestre, A. (2000). A survey of pre-slaughter conditions, halothane gene frequency, and carcass and meat quality in 5 Spanish pig commercial abattoirs. *Meat Science*, 55, 97-101.

Hambrecht, E., Eissen, J.J., Nooijen, R.I.J., Ducro, B.J., Smits, C.H.M., den Hartog, L.A., and Verstegen M.W.A. (2004a). Preslaughter stress and muscle energy largely determine pork quality at two commercial processing plants. *Journal of Animal Science*, 82:1401-1409.

Warriss, P.D., and Bevis, E.A. (1986). Transport and lairage times in British slaughter pigs. *British veterinary Journal,* 142 (2): 124-130.

Warriss, P.D. (1998). The welfare of slaughter pigs during transport. *Animal Welfare*, 7: 365-381.

THE STRESS RESPONSIVENESS OF THREE DIFFERENT BEEF BREED TYPES AND THE EFFECT ON ULTIMATE PH AND MEAT COLOUR

H.A O'Neill*[1], E.C. Webb[1], L. Frylinck[2] and P. Strydom[2]

[1.] *Department of Animal and Wildlife Sciences, University of Pretoria, Pretoria, 0002,* [2.] *Agricultural Research Council, South Africa. Email: adri.oneill@up.ac.za*

Keywords: meat tenderness, urinary catecholamine's, beef breeds, ultimate pH, meat colour

Introduction

Determining the factors that result in a low ultimate pH (pHu) and establishing the limits of residual glycogen levels obtained under various conditions (Immonen and Puolanne, 2000), are central to stress and meat quality research. A contributing factor towards meat tenderness is animal breed type but this has not been quantified in terms of breed stress responsiveness and the effect on meat tenderness. Meat tenderness was identified as one of the major contributing factors to the perception of meat quality (Koohmaraie, 1994). The tenderness of meat varies considerably and therefore prevents meat producers from marketing their produce on the basis of consistent quality (Maher *et al.*, 2004). The present study focuses on the urinary catecholamine output of three different beef breeds (Brahman-x, Simmental-x and Nguni-x), the pHu and the colour of the meat. The measurement of catecholamines in animals is usually highly invasive, but their measurement in excreted urine is not (Hay and Mormede, 1998). In this study the methodology was extended to the post-mortem situation. Previous studies have reported that the measurement of urinary catecholamine's indicated that despite the massive release of catecholamines associated with slaughter, urinary concentrations are unaffected by this owing to the delay between elevation of concentrations in the plasma and subsequent elevation in the urine. The concentration of nor-epinephrine is the result of neuronal washout from tissues with sympathetic nerves and thus an important indicator of sympathetic nervous system activity (Young *et al.*, 1984) and therefore an indicator of stress responsiveness. The two hormonally active catecholamines are the amine-containing derivatives of catechol, 1,2-dihydroxybenzene, nor-epinephrine and its methyl derivative epinephrine. These compounds are synthesised in the adrenal medulla from tyrosine and stored in granules to await their exocytotic release under the control of the sympathetic nervous system (Voet and Voet, 1995).

Materials and Methods

Animals (n=180) were slaughtered at the specified A-age (no permanent incisors) and fatness-class 2-3 (5-7mm). Three different beef breeds Brahman-x (Br-x), Simmental-x (Sm-x) and Nguni-x (Ng-x) were treated in exactly the same manner during the growth and feeding period, and fed the same feed. Sixty Br-x, 60 Sm-x crosses and 60 Ng-x crosses were divided into 2 different feed withdrawal groups each. Thirty animals from each breed group were transported 40 km to the abattoir after which feed were withdrawn for 24 hours pre-slaughter or 3 hours pre-slaughter. Thirty animals from each breed group were transported 40 km to the abattoir after which feed was withdrawn for 3 hours pre-slaughter. pHu was measured 20 hours post-mortem with a digital handheld meat pH meter (Sentron, Model 1001) fitted with a polypropylene spear type gel electrode. pH measurements were done on the *M. longissimus thoracis* of the carcasses. At 20 hours post-mortem the colour of the meat was measured with a Minolta colorimeter as described by Dunne *et al.* (2005). Catecholamine's were measured from a urine sample that was collected from the bladder with a syringe and needle approximately 12 minutes post-mortem, immediately after evisceration. The urine was preserved as described by Lowe *et al.* (2000) and analysed by HPLC-method as described by Gouarne *et al.* (2004). Concentrations of catecholamine's are volume related and for this reason only creatinine related concentrations were considered in the statistical analysis. The two different feed withdrawal groups were pooled together and a general linear model analysis were done on the pHu, colour 20 hours post-mortem and the urinary catecholamine output.

Results and Discussion

Table 1: Urinary catecholamine outputs from Brahman-x (Br-x), Simmental-x (Sm-x) and Nguni-x (Ng-x) breeds.

Catecholamine (ng/μmol creatinine)	Br-x	Sm-x	Ng-x
Nor-epinephrine	2.5633±1.04228^{a}	2.47700±.88997^{a}	4.7228±2.10624^{b}
Epinephrine	1.1642±0.66371^{a}	1.7074±1.17029^{a}	3.3679±2.2891^{b}

$^{a\,b}$ $p<0.05$

There were statistically significant differences for the nor-epinephrine and epinephrine concentrations ($p<0.05$) with Ng-x having the highest output and the Br-x and Sm-x having lower outputs. Catecholamine's are often implied as the cause of the depletion of glycogen in the pre-slaughter period. This is based on their action of rapid mobilisation and depletion of glycogen. If livestock encounter stressors causing the release of catecholamine's in the immediate pre-slaughter period, there is a possibility for depletion of muscle glycogen (Lacourt and Tarrant, 1985), because of the fact that epinephrine activates muscle adenylate cyclase and thereby stimulates glycogen breakdown (Voet and Voet, 1995).

In turn, lowered glycogen prevents an acceptable decrease in pH and attainment of pHu for optimal conversion of muscle to meat (Purchas *et al.*, 1999; Warriss, 1990). The decrease in carcass pH was marginally slower for Ng-x compared to Br-x and Sm-x. The pHu values also tended to be higher in the Ng-x compared to Br-x and Sm-x. Colour measurements for Ng-x indicated significantly darker scores for Ng-x compared to Br-x and Sm-x. The results suggest that the Ng-x has a lower glycolytic potential and this agrees with the results obtained for indigenous goats in South Africa (Webb *et al.*, 2005).

Conclusion

Certain cattle breeds indigenous to Africa, such as the Nguni, have shown an ability to adapt to harsh extensive conditions. (Strydom *et al.*, 2000). Ng-x cattle had a higher urinary catecholamine output as well as significantly higher pHu values. This correlates with the fact that catecholamine's will have an effect on the glycogen profile of the animal, its glycolytic potential and the colour of the meat. One can therefore conclude that the Nguni breed is more stress responsive with a subsequent lower glycolytic potential. This does not mean that this breed will result in tougher meat as other parameters with regard to meat tenderness may be involved. This aspect is currently under investigation.

References

Dunne, P.G., O'Mara, F.P., Monahan, F.J., French, P. and Moloney, A.P. (2005). Colour of muscle from 18-month-old steers given long-term daily exercise. Meat Science, 71: 219-229.

Gouarne, C., Foury, A. and Duclos, M. (2004). Critical study of common conditions of storage of glucocorticoids and catecholamines in 24-h urine collected during resting and exercising conditions. Clinica Chimica Acta, 348: 207–214.

Hay, M. and Mormede, P. (1998). Urinary excretion of catecholamine's, cortisol and their metabolites in mesishan and large white sows: Validation as a non-invasive and integrative assessment of adrenocortical and sympathoadrenal axis activity. Veterinary Research, 29: 119-128.

Immonen, K. and Puolanne, E. (2000). Variation of residual glycogen-glucose concentration at ultimate pH values below 5.75. Meat Science, 55: 279-283.

Koohmaraie, M. (1994). Muscle proteinases and meat ageing. Meat Science, 36: 93-104.

Lacourt, A. and Tarrant, P.V. (1985). Glycogen depletion patterns in myofibres of cattle during stress. Meat Science, 15: 85-100.

Maher, S.C., Mullen, A.M., Moloney, A.P., Buckley, D.J. and Kerry, J.P. (2004). Quantifying the extent of variation in the eating quality traits of the *M. longissimus dorsi* and *M. semimembranosus* of conventionally processed Irish beef. Meat Science, 66: 351-360.

Purchas, P.W., Yan, X. and Hartley, D.G. (1999). The influence of a period of ageing on the relationship between ultimate pH and shear values of beef *M. longissimus thoracis*. Meat Science, 51: 135-144.

Strydom, P.E., Naude, R.T., Smith, M.F, Scholtz, M.M. and van Wyk, J.B. 2000. Characterisation of indigenous African cattle breeds in relation to meat quality traits. Meat Science. 55: 79-88.

Voet, D. and Voet J.G. (1995). Biochemistry, Publisher New York: Wiley C.

Warriss, P.D. (1990). The handling of cattle pre-slaughter and its effect on carcass and meat quality. Applied animal behaviour science, 21: 171-186.

Webb, E.C., Casey, N.H. and Simela, L. 2005. Goat meat quality. Small Ruminant Research, 60: 153-166.

Young, J.B., Rosa, R.M. and Landsberg, L. (1984). Dissociation of sympathetic nervous system and adrenal medullary responses. American Journal of Physiology, 247: E35-E40.

EVALUATION OF SLAUGHTER CONDITIONS ON THE QUALITY OF BEEF CARCASSES PRODUCED IN SLAUGHTERHOUSES OF NORTHERN MÉXICO

E. Díaz[1]*, J. Anaya[2], H. González[1], A. Sánchez-Escalante[1] and G. Torrescano[1]

[1] *Department of Meat and Seafood Products. Meat Science Laboratory. Centro de Investigación en Alimentación y Desarrollo, A.C. PO Box 1735. Hermosillo, Sonora, 83000, México.*
[2] *UNISON-DAG, Blvd. Luis Encinas y Rosales, Col. Centro, Hermosillo, Sonora, México.*
Email: gtorrescano@cascabel.ciad.mx

Keywords: slaughterhouse, pH, colour, sarcomere length

Introduction

The state of Sonora is located in the northern part of Mexico and the land is arid and cattle are fattened on concentrates. Slaughterhouses are subject to health inspections through the Agriculture Secretary (SAGARPA). Slaughterhouses that meet federally approved standards for inspection (or Federal Inspection Type Plants; TIF) have high sanitary standards and advanced technological processing levels. The beef quality is extremely variable, depending on *antemortem* factors such as age, sex, breed, fasting and transportation. Besides *antemortem* conditions, chilling conditions may also have a marked influence on tenderness. Fast chilling of carcasses may induce a rapid temperature decline in superficial muscles, leading sometimes to cold shortening which affects the meat tenderness.

The objective of this investigation was to evaluate the *antemortem* influence (sex, race, fasting and transportation) and chill conditions on some important quality indicators such as temperature, pH, colour and tenderness of beef carcasses produced at Sonora's TIF slaughterhouses.

Materials and Methods

Two TIF slaughterhouses were evaluated in the summer; plant A is located in the north of the state, and plant B is situated in the south. The experimental material consisted of 135 beef carcasses in each processing plant. Cooler room conditions (temperature, relative humidity and air velocity) were determined using a Dickson temperature and humidity chart recorder and a thermo anemometer. Temperature and pH values were measured in three muscles of the carcasses: *Semimembranosus (SM)*, *Longissimus thoracis* (*LT*) and *Pectoralis profundus (PP)*. The pH and temperature values were measured at the end of the slaughter line 45 min *post mortem* (pH_0 and T_0) and after cooling (pH_{24} and T_{24}); also colour values were determined after 24h. The pH and temperature were measured with a portable pH meter equipped with a combined (glass/reference) electrode (Hanna model HI 98140) and a digital thermometer (Oakton Acorn Series 400). Colour was measured at 24h over the 12th rib eye area using a colorimeter Minolta BC-10 that directly transmits the measured values to the CIE system: lightness L* and coordinates a* and b*. Sarcomere length was evaluated at 45 min and 24h *post mortem*, according to Torrescano *et al.* (2003). Statistical analysis of measured data was performed using SPSS (version 10.0.6), and an analysis of variance test was carried out at a significance level $P < 0.05$.

Results and Discussion

The processing plants (A and B) were working at 50% and 60% of their capacity and the time of fasting was 2h for plant A and 16h for plant B. The distance from the feedlot places to the slaughterhouse was less than 20km. The beef breed slaughtered in A and B plants were *Bos Taurus* (50% and 70%), 50% *Bos indicus* (plant A) and 30% *B.taurus* x *B. indicus* cross (plant B). Cooling room conditions were similar in both plants; the relative humidity and wind speed average values were 78 % and 0.35 m/s, respectively. The time–temperature (Figure 1) decline curves for the chilling room reveal a rapid rate of cooling.

Figure 1. Chilling rooms temperature

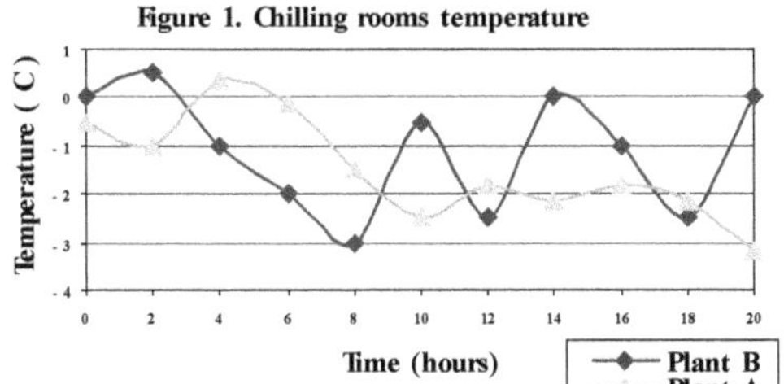

Results of physical carcass traits are given in Table 1. Significant differences ($p<0.05$) were found for temperature and pH on both plants, LT (3.83) presented the lowest T_{24} and SM (10.04) the highest in the B plant, which was probably due to low air velocity registered below the 0.5 m/s according to USMEF (2005). The pH values indicate proper acidification of the three muscles, the final pH_{24} value was 5.42 – 5.76 and indicate a proper process of glycolysis as confirmed by numerous publications (MSA, 2004).

Table 1: Indicator of beef carcass at 45 min. and 24 h.

	45 MINUTES		24 HOURS	
Temperature °C	A Plant	B Plant	A Plant	B Plant
P. profundus	38.98^{bx} ± 1.82	39.59^{ay} ± 1.22	4.69^{ay} ± 1.72	6.91^{by} ± 2.43
L. thoracis	39.23^{ax} ± 1.91	39.24^{ay} ± 1.09	4.78^{ay} ± 1.52	3.83^{bz} ± 1.74
Semimembranosus	39.28^{bx} ± 1.61	40.71^{ax} ± .85	9.89^{ax} ± 3.87	10.04^{ax} ± 3.0
pH				
P. profundus	6.48^{ax} ± .27	6.67^{bx} ± .17	5.74^{cx} ± .14	5.76^{cx} ± .2
L. thoracis	6.43^{ax} ± .27	6.61^{by} ± .18	5.50^{dy} ± .12	5.58^{cy} ± .15
Semimembranosus	6.50^{ax} ± .19	6.57^{by} ± .16	5.42^{dz} ± .11	5.54^{cy} ± .17
Colour				
L*			36.39^{a} ± 2.16	35.0^{a} ± 3.29
a*			21.59^{a} ± 2.07	21.09^{a} ± 3.35
b*			12.59^{a} ± 1.15	11.84^{b} ± 1.64
Sarcomere length, µm				
P. profundus	2.29^{ax} ± 0.3	2.24^{ay} ± .18	1.96^{by} ± 0.3	1.95^{by} ± 0.2
L. thoracis	2.34^{ax} ± 0.3	2.3^{ax} ± 0.15	2.13^{bx} ± 0.2	1.99^{cx} ± 0.2
Semimembranosus	2.20^{ay} ± 0.2	2.22^{ay} ± 0.14	1.93^{ay} ± 0.1	1.96^{ay} ± 0.2

Means within a column $^{x\text{-}y}$ and row $^{a\text{-}d}$ with a common letter are not significantly different (P<0.05).

The colour values L* and a* were an average of 36 and 21 in both plants, respectively. This confirms an absence of DFD in the meat, as DFD may be present in the meat when L*<33 (Mullen *et al.*, 2000); the results indicate a darker red in agreement with Wulf and Wise (1999), Levrino (2004). The sarcomere length of muscles from carcasses suspended vertically did not showed differences (p>0.05) in both plants; according to Hostetler *et al.* (1972) all muscles were above 2µm that is the minimum value of a relaxed muscle. The shortenings were considered as normal (< 15%) in all abattoirs evaluated (Smulders *et al.*, 1990).

Conclusions

The pH_{24} and the colour value of L* > 33 confirm an absence of DFD in the meat. The rapid rate of cooling confirms the absence of cold shortening in the meat.

References

Hostetler, R.L., Link, B.A., Landmann, W.A. and Fitzhugh, H.A. (1972). Effect of carcass suspension on sarcomere length and shear force of some major bovine muscles. Journal of Food Science, 37,132-135

Levrino, G.M. (2004). Transporte de ganado bovino, bienestar animal y calidad de la carne *Producción animal,* 19, Nº. 202 pags. 35-46.

Meat Standards Australia (2004). www.mla.com.au

Mullen, A.M., Murray, B. and Troy, D. (2000). Predicting the eating quality of meat. Report No. 29. The National Food C., Dublin. Teagasc.

Smulders, F.J.M., Marsh, B.B. Swartz, D.R., Russell, R.L. and Hoenecke, M.E. (1990). Beef Tenderness and Sarcomere Length. Meat Science, 28,349-363.

Torrescano, G., Sánchez, A., Giménez, B., Roncalés, P., Beltrán, J.A. (2003). Shear values of raw samples of 14 bovine muscles and their relation to muscle collagen characteristics. Meat Science, 64, 85-91.

USMEF. (2005). Cold Change Management. http://www.usmef.org/TradeLibrary/index_animalhealth.asp

Wulf, D.M & Wise, J.W. (1999). Measuring muscle color on beef carcasses using the L*a*b* Color Space. Journal of Animal Science, 77, 2418-2427.

EFFECT OF TYPE OF STUNNING ON LIPID OXIDATION AND COLOUR OF LIGHT LAMB MEAT

M.B. Linares[1,2], R. Bórnez[1,2] and H. Vergara*[1,2]

[1] Departamento de Ciencia y Tecnología Agroforestal, Universidad de Castilla-La Mancha, 02071, Albacete, Spain, [2] Sección Calidad Alimentaria, Instituto de Desarrollo Regional, Universidad de Castilla-La Mancha, 02071, Albacete, Spain. Email: herminia.vergara@uclm.es

Keywords: lamb, stunning, lipid oxidation, colour, modified atmosphere

Introduction

It is well known that pre-slaughter stress affects meat quality in small ruminants (Kannan *et al.*, 2003). In addition, some authors (Buckley *et al.*, 1995; Juncher *et al.*, 2003) have pointed out that some parameters such as colour and lipid oxidation could be more affected by stress. In the EU, stunning previous to slaughter is a legal procedure to ensure that animals remain in an adequate state of unconsciousness up to time of slaughter (Directive 93/119/EU). But the electrical stunning system, used for sheep, is found to have many disadvantages, such as hemorrhages, bone fractures or intense muscular activity (Vergara and Gallego, 2000), which can make stunning unadvisable. In some species (broilers or pigs) electrical stunning has been replaced by a gas method due to the numerous advantages both for meat quality and animal welfare (Gregory, 2005). In sheep, the effects of gas systems (mainly CO_2) are practically unknown, especially as regards lipid oxidation and colour coordinates, which are the aim of this study.

Materials and Methods

Twenty male lambs of the Manchega Spanish breed from the flock of the Experimental farm of Castilla-La Mancha University were slaughtered at 25 kg of weight. Animals were distributed in two groups (n= 10 each) according to type of stunning. In the first group, the animals were stunned using the gas method (GSL, 90%CO_2 for 90 sec). The other group was stunned electrically (ESL, 110 V 50 Hz for 5 sec). Immediately after stunning, lambs were slaughtered using standard commercial procedures. Carcasses were chilled at 4°C in a conventional chiller. The *Longissimus dorsi* (LD) muscle was removed (at 24 h post-mortem) from each carcass and cut into three samples that were placed in clear rigid trays and packed in a conventional atmosphere (30%CO_2+70%O_2). Samples were chilled at 2°C and were analysed (twenty samples per sampling time, ten from each group) at 7, 14 and 21 post-slaughter. Sample rancidity was determined in duplicate from 2 g of the LD muscle as described by Botsoglou *et al.*, (1994) by determining 2-thiobarbituric acid-reactive substances (TBARS). Colour measurements (L^*, a^*, b^*) were taken immediately after the packs were opened and the measurements were taken on the surface of the LD using a Minolta CR400 colorimeter. The data were analysed using an analysis of variance to determine the effects of type of stunning on colour coordinates and TBARS values.

Results and Discussion

Table 1 shows TBARS values for the two different types of stunning (GSL *vs* ESL). Rancidity level increased ($p < 0.001$) in both groups throughout time of storage, in agreement with Berruga *et al.*, (2005). The GSL group showed lower oxidation values than the ESL at all times ($p < 0.001$). The reason could be that CO_2 used in the gas stunning system may be a determinant in the reduction of O_2 availability in muscle cells, which then reduces the oxidative processes (Bekhit and Faustman, 2005).

Table 1: Effect of type of stunning of lamb on TBARS values (mg malondialdehyde/kg^{-1} of meat) in meat packed in modified atmosphere (30%CO_2+70%O_2).

Time post-packaging (days)	GSL	ESL	ANOVA
7	0.82±0.3^a	4.96±0.42^a	***
14	1.18±0.33^a	4.2±0.22^a	***
21	3.48±0.39^b	7.57±0.45^b	***
ANOVA	***	***	

GSL: gas stunned lambs; ESL: electrically stunned lambs. *** Indicates significant differences ($p < 0.001$)
^{a,b} values in the same column with different superscripts are significantly different with time ($p < 0.05$) according to a Tukey test.

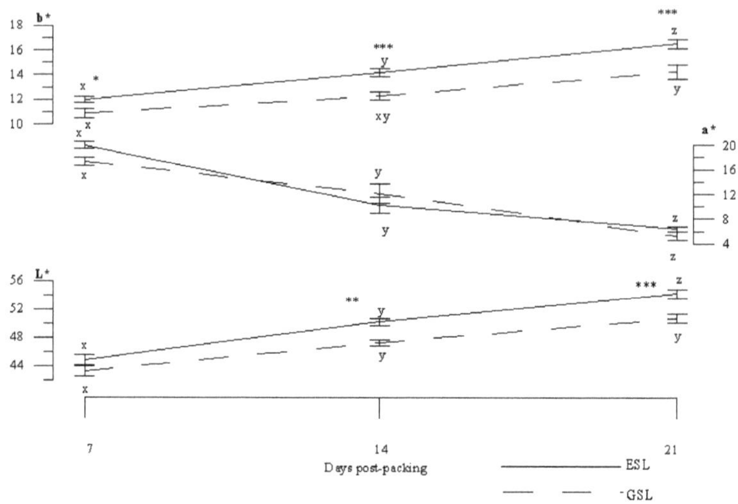

Figure 1: Colour coordinates (L*, a*, b*) in gas-stunned lambs (GSL) and electrically stunned lambs (ESL). x,y,z: indicate significant differences due to the time of storage (7,14 and 21 days) ($p < 0.05$). *, **,***: indicate significant levels of 0.05, 0.01 and 0.001 respectively for the same time post-packing among the two groups of lambs from the different types of stunning.

L* and b* values increased (Figure 1) with time in both groups whereas the a* values showed an opposite trend, in agreement with the results shown by Vergara and Gallego (2001). On the other hand, the oxidative process of meat (Table 1) was related to the instability of colour with time, according to Bekhit and Faustman (2005). In this sense, L* and b* colour coordinates were higher in the ESL than in the GSL samples, although there were no significant differences between groups with regard to the a* parameter at any time of storage. As some authors mentioned (Guidera *et al.,* 1997), lipid and pigment oxidation in lambs are strongly related.

Conclusions

The type of stunning affects lipid oxidation as well as colour coordinates values. The stunning of lambs using gas can increase the shelf-life of the meat packed in a conventional modified atmosphere (rich in oxygen) since the TBARS values were less than in samples of lambs stunned electrically and this was related to a higher stability in colour.

References

Bekhit, A. E. D. and Faustman, C. (2005). Metmyoglobin reducing activity. Meat Science, 71(3), 407-439.

Berruga, M. I., Vergara, H., Gallego, L. (2005). Influence of packaging conditions on microbial and lipid oxidation in lamb meat. Small Ruminant Research, 57: 257-264.

Bostoglou, N. A., Fletouris, D. J., Papageorgiu, G. E., Vassilopoulus, V. N., Mantis, A. J., Trakatellis, A. G. (1994). Rapid, sensitive and specific thiobarbituric acid method for measuring lipid peroxidation in animal tissue, food and feedstuff samples. Journal of Agricultural and Food Chemistry, 42: 1931-1937.

Buckley, D. J., Morrissey, P. A., Gray, J. I. (1995). Influence of dietary vitamin E on pig meat. Journal of Animal Science, 73: 3122-3130.

EU Council Directive 93/119. (1993). On the protection of animals at time of slaughter or killing. Official Journal. European Commission. No. L340/21.

Gregory, N. G.(2005). Recent concerns about stunning and slaughter- a review. Meat Science, 70:481-491.

Guidera, J., Kerry, J. P., Buckley, D. J., Lynch, P. B., Morrissey, P. A. (1997). The effect of dietary vitamin E supplementation on the quality of fresh and frozen lamb meat. Meat Science, 45(1): 33-43.

Juncher, D., Ronn, B., Hansen, T. B., Henckel, P., Karlsson, A., Skibsted, L. H., Bertelsen, G. (2003). Effect of pre-slaughter physiological conditions on the oxidative stability of colour and lipid during chill storage of sliced, retail packed roast ham. Meat Science, 63:151-159.

Kannan, G., Kouakou, B., Terrill, T. H., Gelaye, S. (2003). Endocrine, blood metabolite and meat quality changes in goats as influenced by short-term, pre-slaughter stress. Journal of Animal Science, 81:1499-1507.

Vergara, H., Gallego, L. (2000). Effect of electrical stunning on meat quality of lamb. Meat Science, 56:345-349.

Vergara, H., Gallego, L. (2001). Effects of gas composition in modified atmosphere packaging on the meat quality of Spanish Manchega lamb. Journal of the Science of Food and Agriculture, 81:1353-1357.

INFLUENCE OF BREED, GENDER AND SLAUGHTER WEIGHT ON HISTOCHEMICAL TRAITS AND LONGISSIMUS QUALITY OF LAMB

C. Gariépy[1], L. Dragomir[2], C. Avezard[1], M. Thériault[3] and F. Castonguay[3]

[1]*Agriculture and Agri-Food Canada, 3600, Casavant Blvd West, Saint-Hyacinthe, Quebec, Canada J2S 8E3.* [2]*Animal Science Dept., Laval University, Québec, Canada G1K 7P4.* [3]*Agriculture and Agri-Food Canada, Lennoxville, Quebec, Canada J1M 1Z3. Email: gariepyc@agr.gc.ca*

Keywords: lamb, meat quality, muscle fibre

Introduction

Any increase in slaughter weight of lamb that could translate in improved carcass yield without affecting its fat cover represents a definitive economic advantage to producers, processors and consumers alike. Such gain however must not be realized at the expense of the quality of the meat as observed in other species. In lamb, a large genetic variability prevails and, relative to other species, there is a paucity of information with respect to muscle structure and meat quality. As muscle size increase, assessment of its chemical and histochemical composition may help in understanding and documenting the effects of production factors on meat quality which is the purpose of this study.

Materials and Methods

A total of 108 lambs (27 males and 27 females of each of the Suffolk (Su) and Dorset (DP) breeds) were allocated to one of three targeted slaughter weight (SW) intervals: 41-44kg, 46-49kg, 51-54kg in a 2x2x3 factorial design. These breeds were selected for their opposite pattern in fat deposition. Lambs were selected at weaning (approx. 55 d old) and were then raised under controlled conditions with an 18% protein concentrate fed *ad libitum* up to 35 kg BW. They were then fed a 15 % protein concentrate up to slaughter. High-quality hay and water were also provided *ad libitum.* There were 8 lambs of the same gender per pen, with 4 of each Su and DP. As they reached their SW, lambs were electrically stunned and slaughtered in a commercial abattoir after a 12 h feed withdrawal. Within 1h post mortem, a 5 g *Longissimus* (L) sample was taken and frozen in liquid nitrogen-cooled isopentane for further histochemical analyses. Carcasses were chilled for 24 h at 1°C and were graded according to Canadian regulations. Backfat (BF) thickness and loin eye area (LEA) were then measured between the 12th-13th ribs. The right L muscle (rack) was removed and frozen for ulterior quality analyses. After thawing at 4°C, a 3.5 cm thick chop was used for ultimate pH and colour (L*a*b*) measurements of the bloomed surface. This chop was then freeze dried for the determination of water, fat (3-AOAC 991.36) and protein (3-AOAC 992.15) content. Another 2.5 cm chop was used for drip loss measurement after a 48 h period at 4°C. Shear forces were measured across 6-9 (5 x 1 x 1 cm) cores prepared along the fibre axis of the remaining portion of the L muscle cooked to an internal 68°C with thermocouples. Cooking losses were also measured. Cryostat prepared transverse serial sections were stained for myosin ATPase following an alkaline pre incubation (pH 10.4) (1) for the determination of fibre twitch. Their oxidative capacity was assessed with SDH stain (2). Five serial bundles were analyzed per muscle with an image analysis system. Fibre cross sectional area was also measured. Fibres were classified as slow twitch oxidative (SO), fast twitch oxidative glycolytic (FOG) or fast twitch glycolytic (FG).

Results and Discussion

BF was thicker in DP than in Su. BF thickness was also superior in females than in males and an increase with slaughter weight was measured. LEA however increased with SW only (Table 1).

Only fibre distribution (% and relative area) are reported in table 2 as fibre size did not change. FOG fibres however were overall smaller (1796 μm^2) compared to SO (2352 μm^2) and FG (2273 μm^2). Contradictory results are found with respect to lamb fibre size (6, 8). Except for the RA of fibres from males Su, we report a proportion of FOG larger than that of FG compared to others (4, 5, 6, 7). These results could suggest uncompleted transformation of some FOG fibres into FG as growth occurs. However, no significant change in LEA was observed between lambs of the last two SW groups. In addition, comparison between breeds did show an increase in proportion of FG at the expanse of FOG population. In DP, FOG decreased and FG increased with SW. Much smaller variations were observed in Su. These breed x SW interactions for % FG and FOG were not significant in terms of relative area. With respect to gender, a similar pattern was observed. Su males which had the fastest growth rate (results not shown) had smaller % and RA of FOG and larger proportion and RA of FG than females while no gender effect were observed in FOG and FG distribution in DP. The smaller proportion of SO fibres was higher in male than in female.

Consequent with backfat thickness, females had greater intramuscular fat (IMF) and lower moisture content than males while DP had greater IMF and lower moisture than Su. Treatments had no effect on protein content and on both drip and cooking losses (results not shown). Observed colour differences were not of practical magnitude. Meat from male had greater shear force however there was no effect of breed. Unexpectedly, shear force decreased with SW which might be associated with possible cold shortening attenuation in larger and more insulated carcasses.

Table 1: Effects of breed, gender and slaughter weight on backfat thickness and loin eye area.

Parameters	Female						Male							Effect[2]			
	Dorset			Suffolk			Dorset			Suffolk							
	41-44[1] (n=10)	46-49 (n=9)	51-54 (n=9)	41-44 (n=9)	46-49 (n=9)	51-54 (n=9)	41-44 (n=8)	46-49 (n=9)	51-54 (n=8)	41-44 (n=7)	46-49 (n=9)	51-54 (n=8)	SEM	G	B	SW	X
Backfat (mm)	7.0	8.6	10.7	6.2	6.5	7.9	5.3	5.9	6.5	3.8	4.0	4.2	0.8	***	***	***	NS
LEA (mm^2)	1542.7	1612.8	1702.4	1584.5	1720.2	1639.9	1452.4	1665.4	1763.7	1534.1	1653.0	1693.7	68.1	NS	NS	***	NS

[1] Slaughter weight (kg); [2] G: Gender; B: Breed; SW: Slaughter weight; X: Interactions; [3] Backfat thickness and loin eye area (LEA) measured between the 12th -13th ribs; * $P < 0.05$; ** $P < 0.01$; *** $P < 0.001$

Table 2: Effects of breed, gender and slaughter weight on Longissimus histochemical traits.

Parameters[3]	Female						Male							Effect[2]			
	Dorset			Suffolk			Dorset			Suffolk							
	41-44[1] (n=8)	46-49 (n=6)	51-54 (n=7)	41-44 (n=3)	46-49 (n=5)	51-54 (n=8)	41-44 (n=4)	46-49 (n=8)	51-54 (n=8)	41-44 (n=1)	46-49 (n=4)	51-54 (n=6)	SEM	G	B	SW	X
SO (%)[3]	6.9	6.5	5.7	6.6	7.8	7.8	7.9	7.7	7.6	9.3	8.7	8.0	2.1*	0.1	NS	NS	
FOG (%)[3]	65.5	61.6	58.7	62.4	62.1	62.8	64.4	63.3	58.4	51.2	51.9	54.7	4.4***	***	NS	BxSW	BxG
FG (%)[3]	27.6	31.9	35.5	31.0	30.1	29.4	27.8	29.0	34.0	39.5	39.4	37.3	4.1**	**	NS	BxSW	BxG
RA SO (%)[4]	7.7	8.3	6.5	6.8	8.7	9.7	8.2	8.9	9.2	10.8	10.1	9.3	2.60.06	NS	NS	NS	
RA FOG (%)	60.2	54.4	56.7	56.6	54.5	56.0	60.5	59.3	52.6	45.9	43.9	47.8	5.5**	***	NS	BxG	
RA FG (%)	32.1	37.4	36.7	36.6	36.8	34.2	31.2	31.8	38.2	43.3	46.0	42.9	5.5*	**	NS	BxG	

[1] Slaughter weight (kg); [2] G: Gender; B: Breed; SW: Slaughter weight; X: Interactions; [3] Muscle fibre type: SO (slow oxidative), FOG (fast oxidative glycolytic) FG (fast glycolytic); [4] RA: relative area; * $P < 0.05$; ** $P < 0.01$; *** $P < 0.001$

Table 3: Effects of breed, gender and slaughter weight on meat quality parameters.

Parameters[3]	Female						Male							Effect[2]			
	Dorset			Suffolk			Dorset			Suffolk							
	41-44[1] (n=10)	46-49 (n=9)	51-54 (n=8)	41-44 (n=9)	46-49 (n=8)	51-54 (n=9)	41-44 (n=7)	46-49 (n=8)	51-54 (n=8)	41-44 (n=5)	46-49 (n=6)	51-54 (n=8)	SEM	G	B	SW	X
pH	5.51	5.48	5.48	5.52	5.61	5.57	5.59	5.58	5.52	5.48	5.56	5.540.06		NS	NS	NS	BxG
Shear force (kg)	2.7	3.9	3.2	4.1	3.2	3.0	4.2	3.6	3.4	4.6	4.0	3.60.4		*	NS	*	NS
*a**	18.9	19.4	20.1	19.1	18.7	18.4	18.0	20.1	18.6	18.7	16.7	17.00.8		0.07	**	NS	BxSW
*b**	10.5	9.5	9.7	9.9	9.0	8.9	8.2	11.1	9.7	9.6	8.0	7.80.8		NS	NS	NS	BxGxSW
*L**	40.2	37.4	39.4	40.8	39.1	38.4	38.2	42.4	38.9	41.9	40.3	39.41.3		NS	NS	NS	NS

[1] Slaughter weight (kg); [2] G: Gender; B: Breed; SW: Slaughter weight; X: Interactions; * $P < 0.05$; ** $P < 0.01$; *** $P < 0.001$.

Conclusions

Large contribution of small FOG fibres in the L could suggest further growth potential and perhaps increasing risk for cold shortening. Overall, increasing slaughter weight had no negative effect on meat quality. However, high lean yield genotype such as Su should be preferred in order to maintain carcass yield. Histochemical profile from other muscles should complete such study.

References

Guth L. and Samaha J. (1970). Exper. Neur. 28: 365-367.

Nachlas M.M., Kwan-Chung Tsou K., De Souza E., Chang C., and Seligman A.M., (1957). J. Histochem. Cytochem. 5: 420-436.

AOAC, (1995). Official method of analysis, Washington D.C. Association of Official Analytical Chemist.

Solomon M.B., Moody W.G., Kemp J.D., and Ely D.G. (1981). J. Animal Science 52: 1019-1025.

Pinkas A., Marinova P., Tomov I., and Monin, G. (1982). Meat Science 6: 245-255.

Shackelford S.D., Wheeler T.L. and Koomaraie M. (1995). J. Animal Science 73: 2986-2993.

Carpenter C.E., Rice O.D. Cockett N.E. and Snowder G.D. (1996). J. Animal Science 74: 388-393.

Peinado B., Latorre R., Vaquez-Auton J.M., Poto A., Ramirez G., Lopez-Albors., Moreno F. and Gil F. (2004). Anat. Histol. Embryol. 33, 236-243.

CARCASS CHARACTERISTICS AND MEAT QUALITY OF MERINO EWES CROSSED WITH SIX SIRE BREEDS

J.J.E. Cloete*[1,2], L.C. Hoffman[1] and S.W.P. Cloete[1,3]

[1]Department of Animal Sciences, University of Stellenbosch, Private Bag X1, Matieland 7602, South Africa; [2]Cape Institute of Agricultural Training: Elsenburg, Private Bag X1, Elsenburg 7607 South Africa; [3]Animal Production Division, Elsenburg Agricultural Centre, Private Bag X1, Elsenburg 7607, South Africa; Email: jasperc@elsenburg.com

Keywords: cooking loss, colour, drip loss, *longissimus dorsi*, sheep

Introduction

Fluctuations in the ratio between wool and meat prices and low wool prices have resulted in the emphasis on the two products changing markedly. Commercial producers often seek other ways to exploit short-term benefits resulting from an increase in the price of meat than within flock selection, without compromising the wool-producing capacity of their ewe flocks (Cloete *et al.*, 2004b). Crossbreeding of Merino ewes with mutton type rams in a terminal crossbreeding system may be considered to realize this goal (Kleeman *et al.*, 1983).
The production of commercial crossbred slaughter lambs results in lambs with a higher birth weight, better growth performance up to the age of weaning and these lambs should reach mature slaughtering weight in an intensive fattening system faster (Doloksaribu *et al.*, 2000, Snowder and Duckett, 2003).
In the South African scenario little research has been done using Merino ewes as dam line in a crossbreeding program. The purpose of this study was to evaluate the meat quality of Merino and Merino crossbred lambs. Ram breeds that were used were Dormer, Ile de France (IdF), SA Mutton Merino (SAMM), Merino Landsheep (ML), Suffolk and Dorper rams, while purebred Merinos were used as control.

Materials and Methods

The experiment was carried out on the Tygerhoek experimental farm in the Southern Cape of South Africa. The ewes were randomly allocated to single sire-mating groups (equal number of ewes were assigned to each breed). Approximately 20 ewes were mated with two rams of each sire breed (i.e. 40 in total for each ram genotype). One ram of each breed was retained for breeding, to provide sire links across the two years.
Data included 274 slaughter and carcass records of lambs. It was intended to slaughter all lambs at a live weight of 45 kg but this objective was not reached and the lambs were slaughter at a live weight of 42 kg. Lambs were slaughtered at a commercial abattoir using standard South African slaughter techniques for this species (Cloete *et al.*, 2004a).
The following carcass measurements were taken (according to the procedures described by Cloete *et al.*, 2004a): Cold carcass weight, fat depth at the 13th rib 25 mm from the midline and between the 3rd and 4th lumbar vertebrae 25 mm from the midline. The pH of the *M. longissimus dorsi* was measured at two different times, namely 45 minutes and 48 hours after slaughter (Cloete *et al.*, 2005). The *M. longissimus dorsi lumborum* was dissected from the 1st to the 6th lumbar vertebrae and used for the following analyses as described by Cloete *et al.* (2005): cooking loss, drip loss, colour and meat tenderness.
The random effect of sire within breed group was generally not significant for the respective slaughter traits, and these effects are not presented. For slaughter traits and carcass characteristics, all the traits were statistically analysed with and without slaughter weight as a covariant, in the cases where slaughter weight had no significant effect the data were presented without correction for difference in slaughter weight.

Results and Discussion

Purebred Merino lambs were slaughtered at 340±7 days of age, which was on average between 40 - 50 days later than the crossbred lambs (Table 1). The crossbred lambs were also between 9 - 11 % heavier at slaughter than purebred Merinos. The fat cover of the purebred Merino and ML cross lambs were lower than that of the other crossbred lambs. Dorper and Dormer crossbred lambs had the thickest ($P < 0.05$) fat cover at the 13th rib and between the 3rd and 4th lumbar vertebrae (Table 1). These two breeds are regarded as early maturing breeds, which tend to put on more fat at the same live weight than late maturing breeds.
Initial pH measurements (45 min after slaughter) did not differ between the different breed combinations. However, final pH values (measured 48 hours after slaughter) of Merino, IdF cross and ML cross lambs were higher than those of Dormer and Dorper sired lambs. Cooking loss of the muscle from Merino lambs was higher than that of the crossbred lambs, but no differences were found for drip loss between the different breed combinations.

Table1: Least square means (±se) depicting carcass traits and fat depths of Merino and Merino crossbred lambs (with slaughter mass as co-variant).

Trait	Sire breed					
	Merino	Dormer	IdF	ML	Suffolk	Dorper
Slaughter age (days)	340±7^{a}	289±5^{cd}	278±7^{d}	282±7^{cd}	300±7^{b}	296±6^{bc}
Slaughter weight (kg)	38.7±0.8^{b}	43.0±0.6^{a}	43.2±0.8^{a}	43.5±0.8^{a}	42.7±0.8^{a}	43.2±0.6^{a}
Backfat depth (13th rib) (mm)	1.73±0.20^{c}	2.76±0.14^{a}	2.15±0.19^{b}	1.78±0.18^{c}	2.14±0.18^{b}	2.67±0.14^{a}
Backfat depth (3rd and 4th lumbar) (mm)	2.13±0.21^{c}	3.29±0.15^{a}	3.01±2.05^{a}	1.88±0.20^{c}	2.64±0.19^{b}	3.40±0.16^{a}

^{abc} Means in the same row with different superscripts differ ($P < 0.05$)

No differences were found in the L* colour reading, but the a* value of the muscle from purebred Merino lambs were higher than that of all the other breed combinations. This higher a* value could be linked to the higher ultimate pH of the muscle from the purebred Merino lambs (Cloete *et al.*, 2005b). The *M. longissimus dorsi* did not differ between the different crossbred combinations for shearing value.

Table2: Least square means (±se) depicting meat quality traits of Merino and Merino crossbred

Trait	Sire breed					
	Merino	Dormer	IdF	ML	Suffolk	Dorper
pH45	6.47±0.07	6.49±0.05	6.47±0.07	6.41±0.07	6.59±0.07	6.47±0.05
pH48	5.61±0.02^{a}	5.54±0.01^{b}	5.58±0.02^{a}	5.59±0.02^{a}	5.56±0.02^{ab}	5.56±0.01^{b}
Cooking loss %	30.5±0.7^{a}	27.2±0.7^{b}	27.9±0.7^{b}	29.9±0.7^{ab}	27.9±0.7^{b}	28.2±0.7^{b}
Drip loss %	0.95±0.08	0.98±0.07	1.00±0.07	1.08±0.07	1.02±0.07	1.13±0.08
Colour L*	35.2±0.4	34.9±0.3	35.8±0.4	36.0±0.3	35.4±0.3	34.8±0.4
a*	15.1±0.5^{a}	12.8±0.4^{c}	12.9±0.4^{c}	13.7±0.4^{b}	13.8±0.4^{b}	14.2±0.5^{b}
b*	8.33±0.33^{ab}	8.31±0.3^{ab}	8.29±0.31^{ab}	8.49±0.30^{ab}	7.94±0.29^{b}	9.33±0.33^{a}
Shearing value (N)	80.7±4.9	69.3±4.5	77.2±4.6	83.8±4.5	85.2±4.4	74.4±5.0

Conclusions

Although the number of rams used per breed in this investigation was low and these results should thus be interpreted with due caution, the variation between sires within breed group was generally not significant. The sampling of sires within crossbred group thus does not appear to have influenced the outcomes of the study markedly and any differences could thus be attributed to breed and not to the sampling of sires between breeds. This study shows that crossbred lambs reached slaughter weight at an earlier age than purebred Merino lambs. No conclusive advantages as pertaining to the physical meat quality characteristics evaluated in favour of any of the terminal sire breeds on Merino ewes were noted.

References

Cloete, J.J.E., Cloete, S.W.P. and Hoffman, L.C. 2005. Behaviour of Merinos divergently selected for multiple rearing ability in response to external stimuli. Small Rum Res. 60, 227-236.

Cloete, J.J.E., Hoffman, L.C., Cloete, S.W.P. and Fourie, J.E. 2004a. A comparison between the body composition, carcass characteristics and retail cuts of South African Mutton Merino and Dormer sheep. S. Afr. Journal of Animal Science 34, 44-51.

Cloete, S.W.P., Cloete, J.J.E., Herselman M.J. and Hoffman, L.C. 2004b. Relative performance and efficiency of five Merino and Merino type dam lines in a terminal crossbreeding system with Dormer or Suffolk sires. S. Afr. Journal of Animal Science 34 135-143.

Dolosaribu, M., Gatenby, R.M., Subandriyo and Bradford, G.E. 2000. Comparison of Sumatra sheep crossbreds. III. Reproductive performance of F_2 ewes and weights of lambs. Small Rum. Res. 38, 115-121.

Kleeman, D.O., South, M.H.L., Dolling, C.H.S. and Ponzoni, R.W. 1983. Survival, growth and wool production of South Australian strong-wool Merino and first-cross Merino lambs from birth to 16 months of age. Aust. J. Exp. Agric. Anim. Husb. 23, 271-279.

Notter, D. R. 1987. The crossbred sire: Theory. Journal of Animal Science 65, 99-109.

Snowder G.D. and Duckett S.K. 2003. Evaluation of the South African Dorper as a terminal sire breed for growth, carcass, and palatability characteristics. Journal of Animal Science 81, 368-375.

EFFECT OF ANATOMICAL LOCATION ON THE CHEMICAL COMPOSITION OF *LONGISSIMUS THORACIS ET LUMBORUM* MUSCLE IN CATTLE

M. Oliván[1], V. Sierra*[1], A. Coto-Montes[2], N. Aldai[1], M.J. Martínez[1], M. Mocha[1] and K. Osoro[1]

[1.] *S E R I D A, Apdo. 13, 33300 Villaviciosa, Asturias, Spain.* [2.] *Departamento de Morfología y Biología Celular, Universidad de Oviedo, 33006 Oviedo, Asturias, Spain. Email: veroniss@serida.org*

Keywords: *Longissimus* muscle, cattle, moisture, fat, protein

Introduction

Most studies concerning meat quality in ruminants are based on the analysis of the *Longissimus thoracis et lumborum* (LTL) muscle mainly due to its big size, easy access and great importance in the carcass. Furthermore, a high homogeneity of physico-chemical characteristics along the muscle is supposed (Jeremiah and Murray, 1984, Belew *et al.*, 2003), although some authors have shown that there is significant variability along this muscle in quality traits such as tenderness (Gariepy *et al.*, 1990).

However, there are relatively few studies which have investigated the possible variation in the chemical composition of meat within this muscle. The objective of this work was to study the differences of chemical composition (moisture, intramuscular fat and protein contents) at two different anatomical locations (cranial and medial sections) of bovine LTL muscle.

Materials and Methods

Thirty three LTL muscles were taken from yearling bulls of two local breeds from northern Spain, Asturiana de los Valles (AV, n=23) and Asturiana de la Montaña (AM, n=8), and their crosses (AV x AM, n=2). Animals of the AV breed were homozygous (*mh/mh*), heterozygous (*mh/+*) or normal (+/+) for muscular hypertrophy gene. Calves were weaned after 8 months suckling, fattened by feeding concentrate meal and barley straw *ad libitum* and slaughtered at a live weight around 500 kg and with 14 to 18 months age in a commercial abattoir following approved EU procedures. Twenty four hours post-slaughter the left half carcass was quartered between the 5th and 6th thoracic vertebrae with a circular saw, and the LTL muscle was extracted and transported to the laboratory. Two loin steaks of 2.5 cm thick were sampled at two different locations in the muscle, at the 6th rib level (cranial) and at the 9th rib level (medial), aged at 4°C for 7 days and frozen at –20°C for subsequent determinations. Muscle chemical composition (percentage of moisture, intramuscular fat and protein) was determined in ground meat by near infrared spectroscopy transmittance (NIT) using calibrations developed by Oliván *et al.* (2002). Beef samples were scanned by duplicate in a Meat Analyzer 1265 of Infratec (FOSS), which operates from 850 to 1050 nm at 2nm intervals. Samples were placed into a glass cup of 130 mm diameter and the average spectra of 15 scan locations was recorded as log $1/T$ (T= transmittance). Analysis of variance was carried out to test the effect of sampling location on the chemical composition of meat. Lineal regressions between variables were assessed by Pearson correlation coefficient (r). All statistical analyses were performed using SPSS (SPSS Inc., Ireland) version 11.5 (2002).

Results and Discussion

Table 1 shows the chemical composition of the loin at two different sampling locations: cranial and medial. In spite of the short distance between these two ribs (6th and 9th), location affected significantly the meat composition. Samples taken in the cranial section had lower moisture (73.4 *vs* 74.8, P<0.001), higher intramuscular fat (2.9 *vs* 1.9, P<0.001) and higher protein content (22.7 *vs* 22.5, P<0.05) than samples obtained from the medial section.

Table 1: Chemical composition of meat at the cranial and medial sections of *Longissimus thoracis et lumborum* muscle

	CRANIAL SECTION (6th rib)			MEDIAL SECTION (9th rib)			EFFECT
	Mean	range	sd	mean	range	sd	LOCATION
Moisture (%)	73.4	72.0-74.9	0.72	74.8	73.2-75.7	0.61	***
Fat (%)	2.9	0.9-4.9	1.18	1.9	0.5-3.5	0.85	***
Protein (%)	22.7	21.8-23.9	0.51	22.5	21.5-23.2	0.43	*

These results agree with those reported by O'Neill *et al.*, (2004), who found that moisture and intramuscular fat contents did vary along the location in the LTL muscle (P<0.05), with a lower percentage of moisture and higher of fat in the cranial than in the medial sections, although these authors considered this variation not important due to the small difference between the average values (74.73 *vs* 74.84 for moisture; 1.85 *vs* 1.59 for fat).

Our results showed higher differences, probably due to the wider range of intramuscular fat content in the animals studied (0.5 to 4.9%). Furthermore, we found that the relationships between the chemical composition (moisture, fat and protein) at the cranial and medial sections were lineal (Figure 1). It can be seen that for moisture (Figure 1a) differences between both sections (cranial and medial) decreased and that regression points moved towards the line of equality as the amount of moisture in muscle increased, while for fat (Figure 1b) and protein (Figure 1c), opposite effect was found, with differences between locations increasing as the amount of fat or protein in muscle was higher. These data show that the magnitude of variation in the chemical composition between the medial and cranial sections of the LTL muscle depends on the composition of meat, being higher in animals or breeds with higher capacity for intramuscular fat deposition.

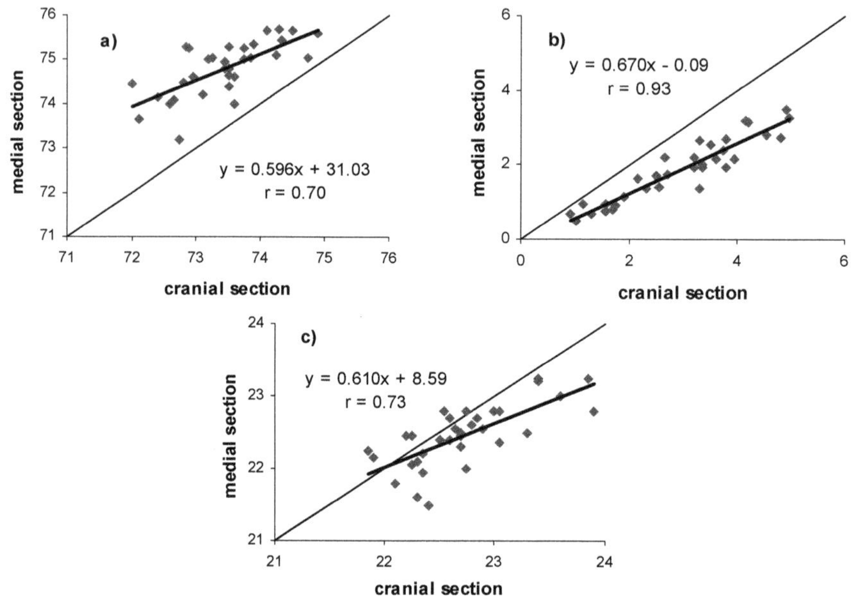

Figure 1: Relationship between the moisture **(a)**, intramuscular fat **(b)** and protein **(c)** content (%) of LTL measured at the cranial and medial sections. Solid line is the line of equality.

Conclusions

There was a significant variation in the chemical composition of meat along the LTL muscle, which should be taken into account by researchers when obtaining meat samples for assays. This means that meat samples for chemical analysis should always be extracted at the same region within the muscle in order to avoid confusion when comparing results from different studies.

References

Belew, J.B., Brooks, J.C., McKenna, D.R. and Savell, J.W. (2003). Warner-Bratzler shear evaluations of 40 bovine muscles. Meat Science, 64, 507-512.

Gariepy, C., Jones, S.D.M. and Robertson, W.M. (1990). Variation in meat quality at three sites along the length of the beef longissimus muscle. Canadian Journal of Animal Science, 51, 35-42.

Jeremiah, L.E. and Murray, A.C. (1984). The influence of anatomical location within the *Longissimus dorsi* muscle on everall tenderness and on the concentration and solubility of intramuscular collagen. Canadian Journal of Animal Science, 64, 1045-1047.

Oliván, M, de la Roza, B., Mocha, M. and Martínez, M.J. (2002). Prediction of physico-chemical and texture characteristics of beef by near infrared transmittance spectroscopy. Proc. 10th International Conference of Near Infrared Spectroscopy, Korea, pp. 197-202.

O'Neill, D.J., Troy D.J. and Mullen, A.M. (2004). Determination of potential inherent variability when measuring beef quality. Meat Science 66, 765-770.

SPSS 11.5. (2002). Guía breve. SPSS Inc. Ireland.

ENDOCRINE PROFILE SEPARATION OF UNITED STATES DEPARTMENT OF AGRICULTURE BEEF CARCASS GRADES

E.P. Berg*, D.H. Keisler, M.M Brandt, D.L. Meyer and T.B. Schmidt

University of Missouri, Division of Animal Science, Columbia, MO, 65211, USA; E-mail: BergEP@missouri.edu

Keywords: beef, carcass, fat, GH, IGF-I, leptin

Introduction

Geary *et al.*, (2003) reported correlations between serum leptin and 12th rib back fat, USDA yield grade, marbling score, kidney-pelvic-heart fat (KPH), and USDA quality grade in beef cattle. McFadin *et al.*, (2003) found positive correlations between serum leptin and 12th rib back-fat, USDA yield grade, and marbling score in beef cattle. The results of these trials suggest that serum leptin concentrations could be used as a means to predict beef carcass merit prior to harvest. Likewise, bovine GH and IGF-I have been documented as endocrine regulators for lean muscle growth in cattle. Trenkle and Topel (1978) reported correlations between GH, percent carcass fat, and percent carcass muscle. Anderson *et al.*, (1988) reported negative correlations with percentage of carcass fat, GH, and IGF-I in beef bulls; however, the true role of these hormones relative to carcass merit, merits further study. The objectives of this project were to 1) evaluate endocrine hormone (leptin, IGF-I, GH) concentrations from a random allotment of United States commercial beef cattle relative to beef carcass composition and quality and 2) evaluate gender differences of leptin, IGF-I, and GH concentrations across carcass compositional endpoints.

Materials and Methods

Cattle were randomly selected from a commercial abattoir immediately after exsanguination on March 30, 2004, May 17, 2004, August 17, 2004, and January 3, 2005. The total number of young, A-maturity steers and heifers used in the final analysis was 995 and 757, respectively. No pre-harvest information was gathered for any of the cattle prior to initiation on the project. The random selection of cattle was assumed to represent the general population of market cattle sold through the commercial abattoir (based on McKenna *et al.*, 2002).

Blood samples were collected at exsanguination in a 15 ml sterile polypropylene conical tube, allowed to clot for 30 min at room temperature, and then placed on ice and transported 3.5 h to the University of Missouri. Samples were centrifuged at 2,500 X g for 45 min. After centrifugation, serum was pippetted from tubes, placed in 48 well plates (5 mL/well; ABgene, Inc., Rochester, NY), and stored at -20° C until analyses. Leptin concentrations were determined by a double-antibody RIA as described by Delavaud *et al.*, (2000). Leptin assay inter- and intra-assay coefficients of variation were 2 and 3% (n = 5 and 35, respectively). Insulin Growth Factor-I and GH concentrations were assayed as described by Lalman *et al.*, (2000). The IGF-I assay inter-and intra-assay coefficients of variation were 3 and 3% (n = 9 and 27, respectively). Bovine GH assay inter- and intra-assay coefficients of variation were 2 and 2% (n = 7 and 38, respectively).

Hot carcass weight (HCWT) was recorded prior to entry into the chill for approximately 24 h at 2°C. After chilling, carcasses were ribbed at the 12th/13th rib interface and allowed to bloom. Marbling score and skeletal maturity were recorded for determination of USDA quality grade and ribeye area, subcutaneous fat thickness, and the percentage kidney, pelvic, and heart fat were combined with HCWT to determine USDA yield grade. The GLM procedure of SAS (SAS Inst. Inc., Cary, NC) was utilized to test the fixed effects of kill day and gender. Least squares means of leptin, IGF-I, and GH were evaluated for least significant differences across USDA quality and yield grade categories.

Results and Discussion

Sample population. Yield grade within the present trial was distributed as follows: Yield Grade 1 (15.3 %), Yield Grade 2 (40.1 %), Yield Grade 3 (35.6 %), Yield Grade 4 (8.0 %) and Yield Grade 5 (0.9 %). The average USDA quality grade for the present study was low choice (not shown in tabular form) with a marbling score of small10.

Serum leptin. The average serum leptin concentration was less ($P = 0.008$) for steers (10.9 ng/mL) than heifers (11.9 ng/mL; Figure 1). Heifers had lighter weight carcasses (331.9 vs. 352.2 kg, $P < 0.0001$), greater 12th rib fat depth (1.3 vs. 1.1 cm, $P < 0.001$), greater KPH (2.5 vs. 2.4 %, $P < 0.001$), and more marbling (Small40 vs. Small10, $P < 0.001$) than steers. Statistical evaluation of the gender heifer leptin concentration across yield grade classification. Due to the low number of steers (n = 3) and heifers (n = 13) representing yield grade 5 carcasses, yield grade 4 and 5 classifications were combined. As carcass cutability declined (increase in numeric yield grade) in steer carcasses, leptin concentration continued to rise; each yield grade category differed ($P < 0.05$) from the previous. A different pattern emerged for heifers. Yield grade 1 carcasses differed from yield grades 2 and 3, however, yield grades 3, 4, and 5 did not differ ($P > 0.05$). The observance of this leptin plateau indicates the possibility that in young heifers as the proportion of whole body fat increases, leptin concentration will elevate in a like manner until it reaches a break point at or near the yield grade 3 classification.

Bovine growth hormone. Steers had greater ($P < 0.001$) serum GH concentrations than heifers (Figure 3). With the steer carcasses having numerically lower yield grades and heavier hot carcass weights it would be an obvious factor attributing to gender based differences. Evaluation of the gender × yield grade interaction term in the GH model approached a trend ($P = 0.11$). Figure 4 displays steer and heifer GH concentration across yield grade classification (pooling yield grades 4 and 5). There appears to be a larger numeric decline in GH concentration as young steers of the same carcass maturity decline in carcass cutability (higher yield grade classification and more total carcass fat). Young heifers of similar age possess lower concentrations of GH at each yield grade category and appear to have less of a decline in GH concentration with increased adiposity.

<u>Insulin-like growth factor-I.</u> Mean IGF-I concentration in steers (16.70 ng/mL) were greater ($P < 0.001$) than heifers (12.0 ng/mL; Figure 3). Yield grade ($P < 0.001$) and gender ($P < 0.001$) were associated with IGF-I concentrations at the point of exsanguination, but the interaction term (gender × yield grade) was not ($P = 0.24$). Although the interaction term of gender × yield grade was not significant ($P = 0.24$), an interesting numeric distribution is presented for steers versus heifers (Figure 5).

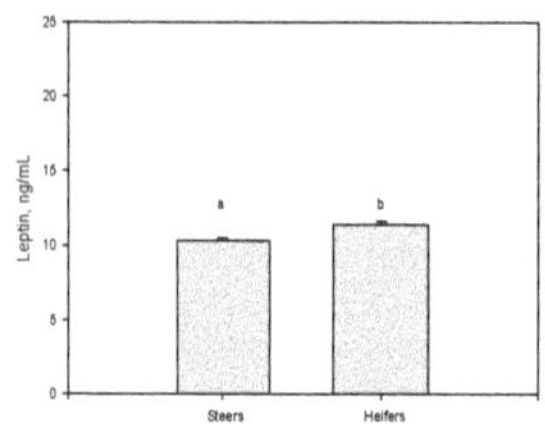

Figure 1: Main effect of gender on serum leptin concentration. Bars differ at $P<0.05$).

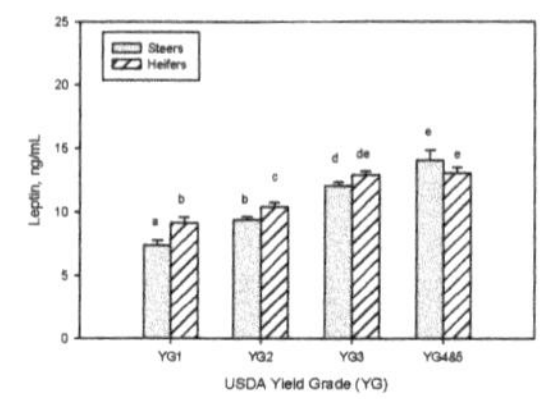

Figure 2: Interactive effect of gender and yield grade ($P= 0.08$) on serum leptin concentration. Bars differ at $P<0.05$).

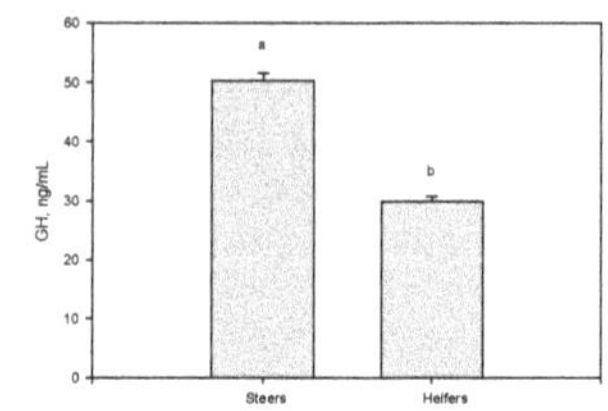

Figure 3: Main effect of gender on serum GH concentration. Bars differ at $P < 0.05$).

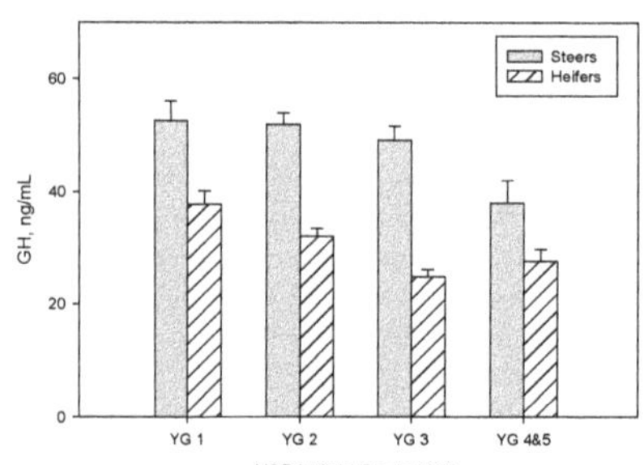

Figure 4: Interactive effect of gender and yield grade (YG; $P = 0.11$) on serum GH levels.

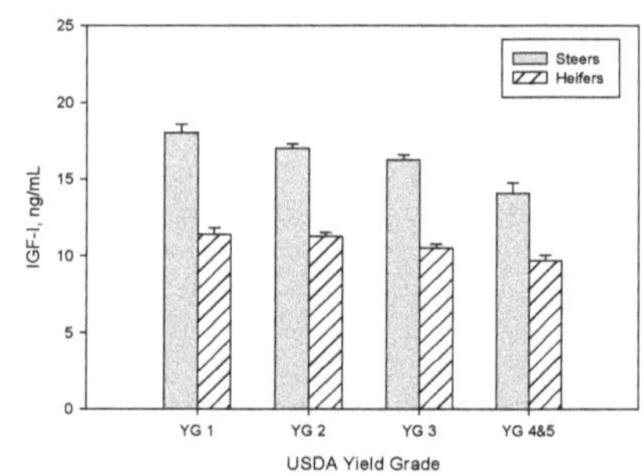

Figure 5: Interactive effect of gender and yield grade (YG; $P = 0.24$) on serum IGF-I levels.

Conclusions

As discussed above, previous research clearly demonstrates the decline of anabolic hormones with age. In the present study, it can be assumed that all animals used in the investigation were of similar chronological age based on carcass maturity scores. Heifers of the same age appear to possess more stable GH (Figure 4) and IGF-I (Figure 5) levels across carcass cutability (yield grade) classification, while their castrated counterparts have higher serum concentrations of these anabolic hormones yet appear to decline to a greater degree with declining carcass cutability.

References

Anderson, P. T., W. G. Bergen, R. A. Merkel, *et al.* (1988). The relationship between composition of gain and circulating hormones in growing beef bulls fed three dietary crude protein levels. J. Anim. Sci. 66:3059-3067.

Delavaud, C., F. Bocquier, Y. Chilliard, *et al.* (2000). Plasma leptin determination in ruminants: effect of nutritional status and body fatness on plasma leptin concentration assessed by a specific RIA in sheep. J. Endocrinol. 165:519-526.

Geary, T. W., E. L. McFadin, M. D. MacNeil, E. E. Grings, R. E. Short, R. N. Funston, and D. H. Keisler. (2003). Leptin as a predictor of carcass composition in beef cattle. J. Anim. Sci. 81:1-8.

Lalman, D. L., J. E. Williams, B. W. Hess, M. G. Thomas and D. H. Keisler. (2000). Effect of dietary energy on milk production and metabolic hormones in thin primiparous beef heifers. J. Anim. Sci. 78:530-538.

McFadin, E. L., D. H. Keisler, T. B. Schmidt, C. L. Lorenzen, and E. P. Berg. (2003). Correlations between serum concentrations of leptin and beef carcass composition and quality. J. Muscle Foods, 14:81-87.

McKenna, D. R., D. L. Roeber, P. K. Bates, *et al.* (2002). National Beef Quality Audit-2000: Survey of targeted cattle and carcass characteristics related to quality, quantity, and value of fed steers and heifers. J. Anim. Science, 80:1212-1222. Trenkle, A. and D. G. Topel. (1978). Relationships of some endocrine measurements to growth and carcass composition of cattle. J. Anim. Sci. 46:1604.

MUSCLE PROFILE: BOVINE MUSCLE PROPERTIES OF HANWOO

Y.K. Kim*, P.N. Sung, B.Y. Park, S.H. Cho, H.S. Choi, Y.M. Yoo, J.N. Ahn and B.H. Baek

National Livestock Research institute, RDA. 564 Omokchun-dong, Kwonsun-gu, Suwon, Korea.
Email: yke7e7@rda.go.kr

Keywords: Hanwoo beef, muscle, meat quality, sensory property

Introduction

Different bovine muscles have different functions to contribute their various quality traits. Ramsbottom *et al.* (1945) compared 25 major muscles from heifer carcasses and found significant differences in tenderness and Warner-Bratzler shear force (WBS) between muscles. Many researchers reported differences in tenderness (as measured by shear force and/or sensory property) among muscles (Prost *et al.,* 1975). Muscles vary greatly in proteolysis, rigor shortening, or cold shortening, potentially attributing to their quality variation. Thus, this variation results in the relative contribution of various factors in the meat quality traits of individual muscle. An intramuscular fat content has been thought to affect palatability (Tatum, 1981); however that theory has been difficult to prove. Jennings *et al.* (1978) and Tatum *et al.* (1980) have reported positive relationships between marbling (or percent fat) and tenderness. It appears that there are consumers willing to pay a premium for guaranteed tender beef. However, inconsistency in meat tenderness has been identified as one of the major problems in beef industry (Boleman *et al.*, 1998; Rhee *et al.*, 2003). Therefore, the objective of this study was to determine the extent of variation and relationship among meat quality properties and palatability traits in major beef muscles of Hanwoo beef.

Materials and Methods

A total of 6 Korean Hanwoo steers (approximately 550-650kg live weight) were obtained at the National Livestock Research Institute (NLRI), and slaughtered. Chemical compositions were determined by Food scan^{TM}Lab (FOSS). WB-shear force was measured on cooked steaks (25mm thick) according to the method described by Wheeler *et al.* 2000). Water-holding capacity (WHC) was determined by the centrifugation method (Kristensen and Purslow, 2001). On the following day of slaughter, a total of 60 muscles were dissected, and frozen until analysis, WHC was expressed as a percentage of weight loss of sample tissue during the centrifugation. Five trained sensory panellists evaluated sensory characteristics of tenderness, juiciness and flavour intensity on a six-point scale. The panellists were asked to score from one point for extremely dislike to six point for extremely like. Thin slice samples (*ca.* 4mm in thickness, and 50 x 70mm in size) were cooked by placing on the tin plate (*ca.* 245-255^{oC}) with turning at the first pooling of liquid on the surface of the strip, or at the start of shrinkage. The cooked strip was immediately served to each panellist for evaluation.

Results and Discussion

The weight and intramuscular fat contents (%) of individual muscle are in Table 1. *Longissimus dorsi* muscle weighed highest (6.21±1.2kg) among the muscles followed by *Biceps femoris* muscle (5.43±1.30kg). Intramuscular fat contents were high in the *Seratus ventralis* muscle (25.81±8.88%), Intercostal muscle, *Transverse thoracis* muscle and *Semispinalis* muscle while relatively low fat was accounted in Sartorius muscle(2.13±0.93%). As shown in Table 1 and Figure 1, the proportion of intramuscular fat contents in the muscle originated from the same retail cut. This difference could be attributed to animal to animal variation and location effects of the portion (Rhee *et al.*, 2003). Intramuscular fat contents were relatively high in the muscles around chuck and loin when compared to the other muscles. The WBS, WHC, intramuscular fat, protein and sensory properties of twenty one major muscles among sixty muscles were investigated the physico-chemical properties (Table 2). The Warner-Bratzler shear force (WBS) values were significantly low for *Seratus ventralis* muscle and Intercostal muscle, and followed by *Infraspinatus* muscle and *Psoas major* muscle, while *Semimembranosus* muscle and *longissimus dorsi* muscle had significantly high WBS values ($p<0.05$). Across all muscles, tenderness rating of sensory evaluation was highly correlated with WBS scores. Intercostal muscle, *Seratus ventralis* muscle, External abdominal oblique muscle and *Psoas major* muscle had the most favourable scores in flavour whereas the muscles around the hindquarter had the lowest flavour scores when compared to the other muscles. There was no significant difference in WHC among different muscles ($p>0.05$) however, juiciness scores of sensory evaluation were significantly different depending on the muscles ($p<0.05$).

Conclusion

The result indicated that meat quality traits such as intramuscular fat contents, tenderness and juiciness are highly variable within and among different beef muscles. This variation might be based on the complex interaction of various biochemical traits that changes from muscle to muscle depending on the animals. This result will provide the information to facilitate the development of cut-specific strategies by targeting specific muscle characteristics.

References

Boleman, S.L. *et al.* (1998). J. Anim. Sci. 76:96-103.

Jennings, T.G. *et al.* (1978). J. Ani. Sci. 46; 658.

Kristensen, L., Purslow. P.P. (2001). Meat Sci. 58:17.

Prost, E. *et al.* (1975). J. Ani. Sci. 41:534.

Ramsbottom, J.M. *et al.* (1945) Food Res.10; 497.

Rhee, M.S. *et al.* (2003). J. Anim. Sci. 82:534

Tatum, J.D. *et al.* (1980). J. Ani. Sci. 50; 833

Wheeler, T.L. *et al.* (2000). J. Anim. Sci. 78, 2856

Table 1: Weight and intramuscular fat contents(%) of major beef muscle of Hanwoo beef

Muscle	Weight (kg)	Intramus-cular Fat	Muscle	Weight (kg)	Intramus-cular Fat	Muscle	Weight (kg)	Intramus-cular Fat
Cutaneous trunci m.	1.62±0.30	6.30±3.13	Triceps brachii m.	3.52±1.08	7.43±3.53	Seratus ventralis m.	1.83±0.95	25.81±8.88
Trapezius m.	1.51±0.39	10.82±5.88	Rhomboideus m.	1.05±0.78	9.45±8.58	intercostal m.	3.20±0.51	24.96±10.72
Longus capitis m.	1.15±0.47	6.72±2.77	Teres major m.	0.40±0.11	4.06±1.24	Deep gluteal m.	0.35±0.10	8.13±4.24
Brachiocephalicus m.	1.51±0.47	6.47±2.60	Biceps brachii m.	0.44±0.19	8.83±4.69	Diaphragm	0.67±0.12	19.95±7.12
Sternocephalic m.	0.58±0.13	6.03±3.64	Brachialis m.	0.37±0.06	5.99±4.96	Supraspinatus m.	1.40±0.28	6.52±3.05
Splenius m.	1.48±1.07	10.83±9.33	Subscapularis m.	1.25±0.34	5.52±2.20	Infraspinatus m.	2.06±0.40	16.28±9.24
Pectineus m.	0.34±0.06	7.75±3.87	Deltoid muscle	0.33±0.21	4.93±2.44	Semispinalis m.	1.77±0.32	21.95±1.58
Longus colli m.	0.93±0.30	6.26±1.77	Teres minor m.	0.32±0.15	7.39±2.23	Spinalis muscle	2.38±0.88	17.54±9.69
Seratus cervicis m.	3.23±1.04	13.41±8.25	Longissimus m.	6.21±1.25	10.77±5.13	deep pectoral m.	3.45±0.57	8.42±3.41
Semispinals m.	2.46±1.01	17.54±9.38	Psoas major m.	1.90±0.35	9.54±3.15	Rectus femoris m.	1.69±0.31	7.10±0.33
Omohyoid m.	0.74±0.34	4.24±0.73	Psoas minor m.	0.48±0.25	11.20±5.09	Vastus lateralis m.	1.96±0.26	4.45±1.64
Omotransverse m.	0.67±0.20	6.76±3.57	Iliocostalis m.	0.49±0.12	13.28±5.13	Gracialis m.	1.08±0.30	5.12±2.63
Rectus abdominis m.	1.69±0.25	18.38±8.52	Multifidus m.	1.20±0.30	14.53±6.43	Middle gluteal m.	3.13±0.61	5.85±2.39
Superficial pectoral m.	1.47±0.14	14.24±7.28	Biceps femoris m.	5.43±1.30	8.51±4.25	Sartorius m.	0.32±0.09	2.13±0.93
Latissimus dorsi m.	2.20±0.39	7.48±4.32	Iliacus m.	0.42±0.18	7.75±2.80	Adductor m.	1.28±0.56	4.67±2.27
Semimembranosus m.	5.34±1.37	4.39±1.02	Semitendinosus m.	1.99±0.39	5.67±2.59	Vastus medialis m.	0.54±0.05	6.85±2.10

Table 2. Meat quality properties of major 21 muscles of Hanwoo beef.

Muscle	WBS kg/0.5inch^2	Water Holding Capacity (%)	Crude Protein(%)	Sensory panel score(Point/6Point)		
				Juiciness	Tenderness	Flavour
Deep pectoral m.	4.23±0.86c	54.30±4.17	19.25±0.91	3.25±0.27c	3.62±0.75bc	3.90±0.34bc
Superficial pectoral m	3.11±0.35b	49.56±3.48	17.86±1.86	3.97±0.82abc	4.08±1.08abc	4.15±0.42ab
Seratus ventralis m.	2.07±0.39a	43.79±2.85	15.90±2.21	4.83±0.62a	5.37±0.61a	4.53±0.25ab
Semispinals m.	2.63±0.48ab	48.75±5.23	17.06±2.26	4.60±0.58a	5.02±0.58a	4.70±0.23a
Supraspinatus m	3.47±0.22b	48.78±2.66	19.26±1.00	3.32±0.34bc	3.55±0.47bc	4.00±0.30bc
Infraspinatus m.	2.66±0.29ab	48.80±4.09	16.86±2.43	4.25±0.93ab	5.08±0.49a	4.35±0.57ab
Triceps brachii m	3.16±0.11b	50.87±1.56	19.99±0.82	3.93±0.64abc	4.18±0.76bc	4.25±0.43ab
intercostal m	2.08±0.48a	49.31±5.20	17.73±2.60	4.85±0.52a	4.68±0.79ab	4.65±0.32ab
Rectus abdominis m.	3.34±0.33b	51.80±3.61	16.75±2.37	4.05±0.64ab	4.55±0.66ab	4.03±0.30b
Longissimus m.	4.45±2.18c	54.19±5.69	20.05±0.94	3.80±0.77ab	4.43±0.83ab	4.38±0.46ab
Latissimus dorsi m.	3.76±0.4bc	53.28±3.07	20.20±0.81	3.55±0.47bc	3.45±0.72bc	4.05±0.35b
Psoas major m	2.85±0.16ab	56.88±3.35	19.91±0.75	4.04±0.45ab	5.00±0.28a	4.50±0.36ab
Biceps femoris m.	3.80±0.96abc	53.55±2.43	19.48±1.09	3.53±0.61bc	3.75±0.69bc	3.62±0.53bc
Tensor faciae latae m	3.13±0.34ab	52.15±3.72	19.83±0.97	3.70±0.32bc	4.08±0.44b	4.12±0.25bc
Rectus femoris m.	3.23±0.46ab	52.41±5.46	19.68±0.88	3.34±0.42bc	3.88±0.60bc	3.96±0.44bc
Semimembranosus m.	4.51±0.87c	54.08±4.29	21.22±0.85	3.15±0.51c	3.37±0.58bc	3.65±0.48bc
Middle gluteal m.	3.64±0.49bc	50.35±3.44	20.63±0.89	3.07±0.27c	3.90±0.41bc	3.90±0.31bc
Semitendinosus m.	3.46±0.22b	54.04±2.75	20.75±0.49	3.05±0.45c	3.48±0.37c	3.97±0.34bc

[a,b,c] Means in the same vertical column with different superscripts are significantly different ($P<0.05$). WBS; Warner-Bratzler shear force.

FACTORS DETERMINING THE CHEMICAL COMPOSITION OF PORK MEAT

S. Tyszkiewicz*, J. Strzelecki, A. Borys and M. Wawrzyniewicz

Meat and Fat Research Institute, Jubilerska 4, 04-190 Warszawa, Poland Email: styszkiewicz@ipmt.waw.pl

Keywords: pork meat, lean meat percentage, fat, water, protein

Introduction

The basic components of meat are: fat (F), water (H) and proteins (P), which aggregate to 100%, if we disregard mineral substances of minor importance for this balance. The fat content may be treated as the main independent factor of chemical composition variety, since fat appears in the volume (quantity) from several percent in lean meat up to over 90% in fat tissue. The water to proteins ratio in meat is more or less stable and it amounts to *ca* 4 grams of water to 1 gram of proteins. This ratio is called the Feder Number (FN). In previous work (Tyszkiewicz *et al.*, 2005) we confirmed that FN is different for various carcass elements and for lean meat it depends on its leanness /lean meat percentage (LMP), and not on the post mortem carcass weight (PCW). In the present work we would like to present the results of tests which took into account the role of proteins of fat tissue and connective tissue in the forming of the chemical composition of pork meat.

Materials and Methods

The subject of the test comprised 57 pork carcasses of animals of both sexes (27 gilts and 30 barrows) and of the different body weight (PCW from 61kg to 99kg) and different lean meat percentage (LMP from 42% to 59%). The carcasses have been parted into elements and dissected using the Walstra and Markus method (Walstra and Markus, 1995). From three carcass elements, loin, ham and shoulder, three tissue fractions have been separated namely: meat, intra-muscle fat and external fat with skin, their weights have been measured and samples for chemical trials taken. The fat content (F) has been determined with the Soxhlet method, that of water (H) – with the dryer method, then of protein nitrogen (N) - with the Kjeldahl method and that of collagen (C) - with the hydroxyproline method. Assuming that for collagen the recalculation ratio of nitrogen into proteins is 5.36, the collagen nitrogen (NC) quantity has been calculated and then deducted from protein nitrogen N. The difference (NP), on multiplying by 6.25, resulted in the muscle protein (PM) content. All the analytical methods were in compliance with ISO standards. The derivative parameters were also calculated: total protein (PT)=PM+C and Feder Number FN=H/PT. Multiplying the chemical component content by the weight of particular tissue fractions of particular carcass elements, the chemical composition of all the tested meat was calculated (balanced).

Results and Discussion

The influence of the carcass weight PCW and its leanness LPM on the content of particular chemical components and on the calculated Feder Number has been checked. Since no substantial influence of sex of the animals on the chemical composition of meat has been found, the total material has been treated as uniform. In Table 1 the values of coefficients of the linear correlation have been given, marking their significance with the following numbers: at 1 $p \leq 0.05$, at 2 $p \leq 0.01$ and at 3 $p \leq 0.001$. The mark 0 means significance at $p \leq 0.05$ in multiple correlation.

Table 1.

Factors	Chemical Components					
	Fat F	Water H	Muscle Protein PM	Collagen C	Total protein PT	Feder number FN
Carcass Leanness LPM	-0.875^3	0.836^3	0.911^3	-0.058	0.883^3	-0.329^1
Carcass Weight PCW	0.408^2	-0.450^2	-0.169	-0.124	-0.200	-0.231^0
Both agents LPM and PCW	0.922^3	0.902^3	0.912^3	0.145	0.885^3	0.432^3

From the regression equations describing the dependence of the particular chemical component or of the FN on carcass leanness LPM, the values corresponding to limit LPM values in EUROP classification have been calculated and presented in Table 2.

The dependence of water content on the fat free mass of fat tissues (HFT/100-FFT) 100% on the ratio of the collagen content to total protein of fat tissues (CFT/CFT+ PFT)100% has been analysed and, on that basis, the degree of hydration of collagen and other proteins of fat tissue has been determined, obtaining highly significant correlation coefficients – 0.677 for fat tissue on loin, - 0.715 for fat tissue of ham and – 0.660 for fat tissue of shoulder. See Figures.

Table 2.

Factors	Carcass Leanness (Lean Meat Percentage – LMP)				
	40%	45%	50%	55%	60%
Fat F [%]	37.7	32.6	28.0	23.4	18.8
Water H [%]	48.0	51.4	54.9	58.3	61.7
Muscle Protein PM [%]	11.1	12.4	13.7	15.0	16.4
Total Protein PT [%]	13.2	14.5	15.7	17.0	18.3
Feder Number FN [-]	3.64	3.56	3.49	3.43	3.38

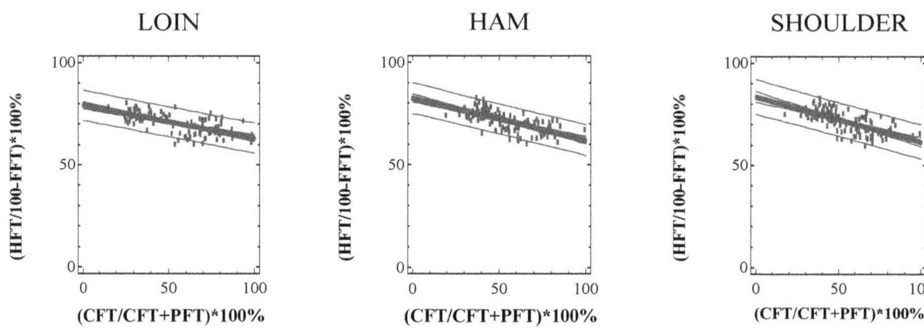

The calculated water to protein ratios in fat tissues is presented in Table 3.

Table 3.

Factors	Degree of protein hydratation g H_20/g		
	Loin A	Ham B	Shoulder C
Muscle Protein (1)	3.36 ^{BC}	3.66 ^{AC}	3.80 ^{AB}
Collagen of Fat Tissue	1.71	1.64	1.60
Other Then Collagen Protein of Fat Tissue	3.80 ^{BC}	4.73 ^{AC}	5.07 ^{AB}

ABC Significant difference at $p \leq 0.01$

Conclusions

The fat and water content depend mostly on carcass leanness and, to the smaller extent – on carcass weight. The content of muscle proteins and total proteins correlates and not with carcass weight. The content of collagen depends neither on carcass leanness nor on carcass weight. The protein content in meat determined with the Feder Number depend on carcass leanness and carcass weight. Both the factors result in a decrease of the Feder Number, *i.e.* on increasing the relative protein content in meat. Neither chemical composition nor protein content in meat depends on the sex of the pigs. The protein content in meat depends on the differentiated degree of hydration of proteins contained in muscle tissue, intra-muscular fat tissue and outer fat tissue. The degree of hydration of collagen is more or less stable and it amounts to about 1.6 grams of water to 1 gram of proteins. The degree of hydration of other proteins of fat tissue is variable and different for different carcass elements. Performed investigations allowed for estimation of the mechanism of observable depletion of degree proteins' hydration, accompanied by an increase in lean meat in the pig carcass. It was stated that the content of low-hydrated collagen in fat-tissue is constant, independent of fat-tissue content. Therefore, together with diminishing of fat-tissue content in the carcass, the content of highly-hydrated other proteins must become lower.

References

Tyszkiewicz, S., Wawrzyniewicz, M., Strzelecki, J., Borys, A. and Foltyn M. (2005). Effect of carcass leanness in pig on protein content and chemical composition of prime cuts. 51st ICoMST, Baltimore USA.

Walstra, P. and Marcus, G.S.M. (1995). Procedure for assessment of the lean meat percentage as a consequence of the new EU-reference dissection method in pig carcass classification. DLD-Research Institute for Animal Science and Health (ID-DLD), February.

COMPARISON OF EVALUATION RESULTS OF PORCINE CARCASS MEATINESS OBTAINED WITH THE ASSISTANCE OF DIFFERENT METHODS OF FULL AND SIMPLIFIED DISSECTION

D. Lisiak[1]*, A. Borys[1], A. Łyczyński[2] and K. Borzuta[1]

Meat and Fat Research Institute in Warsaw,, 04-190 Warszawa, ul. Jubilerska 4, Agricultural University in Poznań, Chair of Raw Materials of Animal Origin, 60-637 Poznań, ul. Wołyńska 33 Email: andrzejl@jay.au.poznan.pl, dsi.poznan@ipmt.waw.pl

Keywords: porcine carcass, dissection, methods

Introduction

Full dissection, which can be defined as the mechanical separation of the basic tissue components of all half-carcass cuts - with the exception of its peripheral parts - is still considered as the most precise method of determination of the tissue composition of a porcine carcass. In Poland, detailed methodology of full dissection was first published in 1957 (Kielanowski *et al.*, 1957). The method came to be known as the SKURTCh method which is an acronym formed from the name – **S**tacja **K**ontroli **U**żytkowości **R**zeźnej **T**rzody **Ch**lewnej (Control Station of Swine Slaughter Performance) and is still used in experiments even today. In Germany, another method of detailed dissection was elaborated known as the DLG method. After some modifications introduced in 1980s, the above method was adopted as the procedure to be applied in all EU member states (Borzuta 1998). Due to the considerable labour investment as well as the high costs of the full dissection, several methods of simplified dissection were elaborated in which the content of meat or fat is assessed with the assistance of regression equations. These methods are also known as reference methods. Among the best known methods of this type worth mentioning here are the SKURTCh method (Różycki, 1996) and the Walstra and Merkus (1996) method which was adopted as the reference method in all EU countries (Commission Regulation EC No 3127/1994).

Materials and Methods

The experiment was carried out on 30 pork carcasses (15 gilts and 15 hogs) randomly selected on the slaughter line in different meat processing plants. The after slaughter processing of carcasses was carried out in accordance with the method adopted in the meat industry but the kidney fat was left with the carcasses. After chilling, each half carcass was weighed separately to within 100g accuracy. Then the left half-carcass was dissected following the DLG methodology (Borzuta, 1998), whereas the right half carcass – according to the SKURTCh method (Kielanowski *et al.*, 1957). In addition, the obtained results were also used to calculate the meat content in the carcass according to the equations designed for the simplified methods, *i.e.* the method of Walstra and Merkus (Walstra and Merkus 1996) as well as the SKURTCh method (Różycki, 1996). In the case of the first method, data from the dissection of the left half-carcass were used, while in the case of the second one – those obtained from the right half-carcass. The obtained research results were elaborated statistically, assessing the significance of differences with the aid of the correlated pair method (Snedecor, 1956).

Results and Discussion

It is evident from Table 1 that the content of meat obtained from the dissection by the SKURTCh method is smaller than when the DLG method was applied (47.32 and 48.60%, respectively). This fact can be explained by different lines of cutting the carcass into cuts, different number of cuts subjected to dissection (*e.g.* in the case of the Polish method, the head is not dissected), different treatment of the carcass before dissection (absence of kidney fat in the carcass in the DLG method) as well as different tissue components obtained from the dissections (in the SKURTCh method, membranes and tendons are not separated). Furthermore, the above-mentioned factors also contributed to significant differences in the content of skin and bones (in the SKURTCh method, the whole head is included in bones) between the compared methods of full dissection. On the other hand, no significant differences were found in the content of dissection fat. The performed investigations also showed that meatiness assessed using the two simplified methods did not differ significantly from the results obtained using the DLG full dissection and was significantly higher in comparison with the full SKURTCh dissection method.

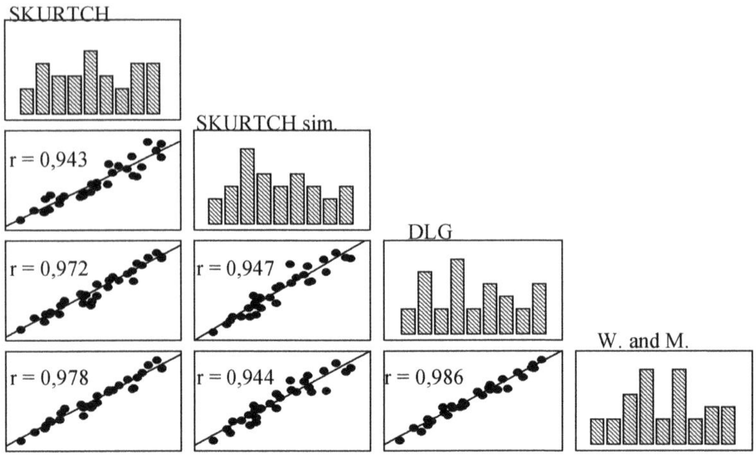

Figure 1: Correlations between the meat content in the carcass obtained using different dissection methods.

Table 1: Mean dissection results of carcasses obtained using different methods (%).

Tissue elements (%)	Full dissection DLG		Full dissection SKURTCH		Simplified dissection SKURTCH		Simplified dissection of Walstra and Merkus	
	$\bar{x}$	sd	$\bar{x}$	sd	$\bar{x}$	SD	$\bar{x}$	sd
Meat	48.60^{B}	5.86	47.32^{ACD}	5.97	48.23^{B}	6.21	48.60^{B}	6.02
Fat	28.43	6.65	28.33	6.98	-	-	-	-
Bones	13.90^{B}	1.09	19.08^{A}	1.22	-	-	-	-
Skin	6.26^{B}	0.92	4.80^{A}	0.89	-	-	-	-
membranes and tendons	2.21	0.45	-	-	-	-	-	-

A – B – the difference is statistically significant at P≤0.01

The results of meatiness assessment determined with the assistance of the examined methods were characterised by a very high correlation coefficient (Figure 1). However, the highest correlation was observed between the DLG method and the simplified method of Walstra and Merkus with r = 0.986. In combination with the fact that the average meatiness determined using these two methods was identical, it points to a very precise match of the method currently in force in the EU to the full carcass dissection with the aid of the DLG method.

Conclusions

The content of meat determined on the basis of full dissection carried out according to the Polish SKURTCh method differed, on average, by approximately 1.28% when compared with the DLG method of dissection. The main cause of these differences was the inclusion, in the case of the SKURTCh method, of the kidney fat to the carcass weight and, at the same time, the exclusion of the head which was treated as part of bones. The performed investigations showed that, out of the examined methods of simplified dissection, the simplified method of Walstra and Merkus revealed higher compliance with the content of meat in the carcass determined using the full dissection method of DLG.

References

Borzuta K. (1998). Studies on usefulness of different methods of meatiness evaluation for the classification of porcine carcasses in the Europ system. Annals of the Meat and Fat Research Institute, vol. XXXV/2, 1-84

Kielanowski J., Osińska Z. and Chomyszyn W. (1957). Sprawozdanie z działalności Stacji Kontroli Użytkowości Rzeźnej Trzody Chlewnej za lata 1951 – 1954 i za rok 1955, PWRiL Warszawa

Różycki M. (1996). Raport on pig breeding in Poland in 1995, XVI, 69-82

Snedecor G.W. (1956). Statistical methods. Er.5. The Iowa State Press.

Walstra P. and Merkus G.S.M. (1996). Procedure for assessment of the lean meat percentage as a consequence of the new UE reference dissection method in pig carcass classification - Raport ID-DLO 96.014, 1-22 - Inst. for Animal Sci. and Health, Lelystad.

FAT QUALITY OF PIGS FROM DIFFERENT PRODUCTION SYSTEMS

L. Basso*[1], M.E. Cossu[1], S. Moisá[1], J. Brunori[2], D. Campagna[3], G. Alleva[1] and R. Franco[2]

[1.] *Animal Production Department, Faculty of Agronomy at the University of Buenos Aires, addressed at Av. San Martín 4453, zip code (1417), Buenos Aires, Argentina.* [2.] *Pork Group, at E.E.A. INTA located in Marcos Juárez, Córdoba, Argentina.* [3.] *Faculty of Agrarian Sciences at the National University of Rosario, Rosario, Argentina. Email: lbasso@agro.uba.ar*

Keywords: pigs, production systems, fat quality

Introduction

Alternatives exist to introduce factors of differentiation in meat production such as the production system and feeding. There are very few research works on the effects of outdoor systems on the quality of pork, in some cases with contradictory results due to the fact that there exists a great number of factors that influence them. (Gentry *et al.*, 2002). Thus, the exercise that pigs do when pasturing has a direct and indirect effect on said quality, by the reduction of the growing speed, less consumption of supplementary food and the change in the proportions of the different types of muscular fibres (Stoll, 1992).

On the other hand, the quality of adipose tissue as regards its nutritional value, organoleptic and conserving properties, is related to the composition in fatty acids (Lizardo *et al.*, 2002). Furthermore, its composition is highly influenced by different factors, such as the genotype, sex, age, live weight and fattening grade of pigs (Girard *et al.*, 1988; Lebret and Mourot, 1998), the place of deposition (Marchello *et al.*, 1983; Miller *et al.*, 1990), the thermic environment (Katsumata *et al.*, 1995) and especially by nutrition. The energy and the content of lipids in the diet, its composition in fatty acids, the daily consumption of food and the division of energy between protein and lipids in the animal body may modify the composition of fatty acids of the adipose tissue. (Wiseman *et al.*, 1998, Lebret and Mourot, 1998).

The purpose of the experimental work is to assess the effect of the supplement with high quality pastures and the exercise done by the animals during growth and finishing in outdoor systems on the intramuscular fat, the quality of pork and its nutritious properties.

Materials and Methods

A summer-autumn trial was conducted at Experimental Station INTA Marcos Juárez and the Meat Quality Laboratory of the Agronomy Faculty at the University of Buenos Aires. Fifty four pigs INTA-MGC were used (27 barrows and 27 gilts), with an initial average weight of 26.4 ± 0.7 kg, aleatory distributed in three treatments: T1: pigs in outdoor conditions on a 1.4 ha lot, in a pasture with alfalfa (*Medicago sativa*) and white clover (*Trifolium repens*); T2: pigs in outdoor conditions on a 1.4 ha lot with no implanted pasture and with a coverage of cebadilla (*Bromus unioloides*); T3: pigs in indoor conditions located in stall with concrete floor. The pigs were fed ad-libitum with standard feed from 25 to 60 kg of live weight (DE: 3.30 Mcal/kg, CP: 18%, Lysine: 1.05%) and from 60 to slaughter (111.6 ± 7.09 kg) with finishing diet (DE: 3.28 Mcal/kg, CP: 17.5%, Lysine 1.0%) on a maize and soya basis. After slaughter samples of *Longissimus dorsi* muscle (last rib level) were obtained to determine composition in fatty acids from intramuscular fat, extracted according to Folch *et al.* (1957) and analysed by gas chromatography (Shimadzu GC-14B; capillary column: Ulbon HR-SS-10; 0.32 I.D.x 50 mL) and using helium as carrier gas. For the variables analysed, GLM procedure of SAS (SAS, 1998) was used and mean values were compared using Tukey test with a 5% significance.

Results and Discussion

The growing-finishing pig in outdoor systems and pasture produced better pork quality than traditional indoor systems, along with more favourable health characteristics. Fatty acid composition presented significantly more C18:1 and C18:3 ($p<0.05$; Table 1). The level of C18:3 increased by high quality pasture consume, with relation to the treatment without pasture and on concrete, probably due to the fact that pastures are characterized by a high level content of polyunsaturated fatty acids and mainly of C18:3 (more than 50% of the total fatty acids), as well as n-6/n-3 ratio of 0.20 (Morand-Fehr and Tran, 2001). The lower level content of polyunsaturated fatty acids was due to the C18:2 level (9.83% and 7.26% for indoor-concrete and outdoor-pasture group respectively) ($p<0.05$). This explains the low n-6/n-3 ratio for pasture group (12.77), which is nearest to the recommendation of the U.S. Department of Health (4,0).

Conclusions

The results of this study show a good response of pasture inclusion and outdoor conditions on intramuscular fat quality. Pasture group showed high oleic and linolenic acid content and a better n-6/n-3 ratio.

Table 1: Least-squares means and residual standard deviation (RSD) of fatty acids composition of intramuscular fat in different treatment.

Fatty acids	Outdoor		Indoor	RSD
	Pasture	Non pasture		
C16:0	24.58	24.60	25.15	0.76
C16:1	3.59	3.25	3.16	0.56
C18:0	12.07	12.32	12.34	1.04
C18:1	43.27 ***a***	42.81 ***a***	40.73 ***b***	1.90
C18:2	7.26 ***b***	8.56 ***ab***	9.83 ***a***	1.51
C18:3	0.58 ***a***	0.34 ***b***	0.25 ***b***	0.17
SFA	40.15	39.50	41.00	1.46
MUFA	49.80 ***a***	48.66 ***ab***	46.69 ***b***	2.07
PUFA	10.05 ***b***	11.84 ***a***	12.30 ***a***	1.51
n-6/n-3	12.77 ***b***	24.21 ***a***	30.30 ***a***	2.45

SFA: saturated fatty acids; MUFA: monounsaturated fatty acids; PUFA: Polyunsaturated fatty acids. Different letters indicate significant differences ($p<0.05$).

References

Folch, J., Lees, M. and Stanley, G.H.S. (1957). A simple method for the isolation and purification of lipids from animal tissues. J. of Biological Chemistry, 226: 497-509.

Gentry, J.G., McGlone, J.J., Blanton J.R. and Miller, M.F. (2002). Alternative housing systems for pigs: Influences on growth, composition and pork quality. J. Animal Science 80: 1781-1790.

Girard, J.P., Bout, J. and Salort, D. (1988). Lipides et qualités des tissus adipeux et musculaires de porc, facteurs de variation. I. Lipides et qualités des tissus adipeux, facteurs de variation. II. Lipides et qualités des tissus musculaires, facteurs de variation. Journéss Rech. Porcine France 20, 255-278.

Katsumata, M., Hirose,I. and Kaji, Y. (1995). Influence of high ambient temperature and dietary fat suppementation on fatty acid composition of depot fats in finishing pigs. Anim. Sci. Technol. (Jpn) 66, 225 - 232.

Lebret, B. and Mourot, J. (1998). Caractéristiques et qualité des tissus adipeux chez le porc. Facteurs de variation non génétiques. INRA. Prod. Anim. 11, 131-143.

Lizardo, R., van Milgen, J., Mourot, J., Noblet, J. and Bonneau, M. (2002). A nutritional model of fatty acid composition in the growing-finishing pig. Livestock Production Science 75, 167-182.

Marchello, M.J., Cook, N.K., Slanger, W.D., Johnson, V.K., Fischer, A.G. and Disusson, W.E. (1983). Fatty acid composition of lean and fat tissue of swine fed varius dietary levels of sunflower seed. Journal of Food Science 48, 1331-1334.

Miller, I.J., Shackelford, S.D., Hayden, K.D. and Reagan, J.O. (1990). Determination of the alteration in fatty acid profiles, sensory characteristics and carcass traits of swine fed elevated levels of monounsaturated fats in the diet. J. Animal Science 68, 1624-1631

Morand-Fehr, P., Tran, G. (2001). La fraction lipidique des aliments et les corps utilisés en alimentation animale. INRA Prod. Anim. 14 (5): 285-302.

SAS. (1998). SAS Users Guide: Statistics. SAS Institute Inc., Cary, NC (USA).

Stoll, P. (1992). Comparaison de différents modes d´engraissement du porc. I. Comportement au pâturage, performances d´engraissement et d´abattage. Revue Suisse Agric. 24(6): 381-385.

Wiseman, J. and Agunbiade, J.A. (1998). The influence of changes in dietary fat and oils on fatty acid profiles of carcass fat in finishing pigs. Livestock Production Science 54, 215-225.

TEXTURAL AND SENSORY PROPERTIES OF MEAT FROM PODOLIAN CATTLE

R. Marino *[1], M. Albenzio[1], M. Caroprese[1], S. Pelosi[2], A. Santillo[1], A. Sevi[1] and A. Braghieri[3]

[1]*Dipartimento PRIME, Facoltà di Agraria– Via Napoli, 25 –71100 Foggia, Italy* [2]*CNR-ISMAR, Lesina, Italy*
[3]*Dipartimento di Scienze delle Produzioni Animali, Università degli Studi della Basilicata. Via dell'Ateneo Lucano, 10 - 85100 Potenza, Italy* * *E-mail: r.marino @unifg.it*

Keywords: podolian, rearing system, ageing time, texture profile, sensory characteristics

Introduction

Production factors including breed, age, sex, feeding, rearing system and technological factors, such as refrigeration or ageing time, can affect a number of meat characteristics. Research has shown evidence that concentrate-fed animals produce more tender meat and with better flavour than forage-fed animals. Inadequate tenderness is the most frequent motivation of consumer dissatisfaction, so any improvement in this parameter would increase the value of the final product. Prolonged ageing time may improve meat tenderness (Campo *et al.*, 1999; Marino *et al.*, 2006). During ageing, muscles undergo a series of physical and biochemical changes which are responsible of their conversion to meat. The present research aims to evaluate the effects of rearing system (indoor vs outdoor), protein level (high and low protein supplementation) and ageing time on the textural and sensory properties of meat from Podolian young bulls.

Materials and Methods

Twenty-four Podolian young bulls were divided into 3 groups (8 subjects each): indoor (IND), pasture with high protein level supplementation (PHP-16% D.M.) and pasture with low protein level supplementation (PLP-12% D.M.). Diet was composed of oat hay as forage plus durum wheat flour shorts and field beans as concentrate, the amount in each group was calculated according to INRA (1988) to satisfy the growth requirement. Longissumus dorsi lumborum (LD) was removed from the right carcass side and aged in vacuum-packaging at 4°C until 11 and 18 days post-mortem. Rheological properties were tested on ten parallel pipes for each raw sample (1 cm^2 in cross-section) using two different instrumental measurements (Instron 4301 universal testing machine). The shear force (WBSF) was assessed using a Warner-Bratzler device (100 mm/min crosshead speed using 100 kg load cell). Texture profile was analysed using a modified compression device that avoids transversal elongation of the samples. Each sample underwent two cycles of 80% compression, force by time data were used to calculate the following parameters: hardness, cohesiveness, springiness, chewiness. The sensory properties were evaluated on steaks grilled at an internal temperature of 75°C by a panel of 10 assessors, previously trained to develop a communal sensory vocabulary and to evaluate the intensity of attributes (odour, savoury, acidic, bitter, sweet, flavour, tenderness, juiciness, chewiness, fatness). Attributes were rated on the basis of 100 mm unstructured lines with anchor points at each end (0: absent and 100: very strong). Data were subjected to an analysis of variance, using the GLM procedure of the SAS statistical software (1999).

Results and Discussion

Rearing system and protein level did not account for WBSF and texture profile (Table 1) in agreement with Keane and Allen *et al.* (1998). The specific effects of dietary constituents on meat quality are not easy to evaluate. The feeding regime can have an influence on animal growth and it is difficult to establish whether the meat characteristics are due to dietary components for their intrinsic properties or to diet effects on animal growth rate and body composition (Muir *et al.*, 1998). Nevertheless, significant effects of protein level was found in relation to the sensory properties (Figure 1), where meat from grazing animals supplementated with high protein level showed lower flavour intensity ($P<0.01$), tenderness ($P<0.001$) and juiciness ($P<0.05$) then grazing animals supplementated with low protein level and outdoor animals. Berge *et al.* (1993) observed lower meat tenderness with increased amounts of muscle production in steers fed a high level of protein rich soybean-rapeseed meal compared with a low level of soybean-rapeseed meal.

Ageing time significantly improved, WBSF and texture profile of Podolian meat (Table 1); after 18 days lower WBSF, hardness, chewiness ($P<0.001$), cohesiveness ($P<0.01$) and springness ($P<0.05$) was found when compared to those after 11 days of ageing.

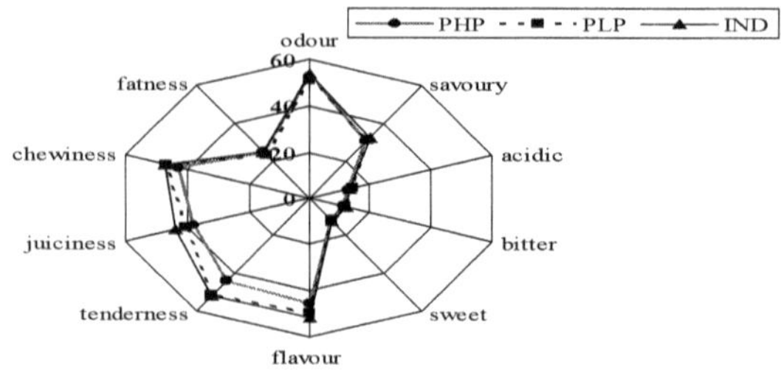

Figure 1: Sensory properties of Podolian meat as affected by rearing system.

Table 1: Rheological properties of Podolian meat as affected by rearing system and ageing (means± SE).

			Rearing system			*mean*	Effects. P		
		Ageing	**PHP**	**PLP**	**IND**		*Rearing system*	*Ageing*	*Ag.x Rear.system*
WBSF	kg	11	6.39±0.46	6.72±0.44	7.64±0.50	6.92 ±0.27	NS	***	NS
		18	5.9±0.46	5.69±0.44	6.09±0.50	5.89 ±0.27			
Hardness	kg	11	9.45±0.53	8.97±0.58	9.74±0.60	9.39 ±0.44	NS	***	NS
		18	7.79±0.53	7.94±0.58	8.18±0.60	7.97 ±0.44			
Cohesiveness		11	0.16±0.01	0.16±0.008	0.17±0.009	0.17 ±0.01	NS	**	NS
		18	0.12±0.01	0.12±0.008	0.13±0.009	0.13 ±0.01			
Springiness	cm	11	0.71±0.025	0.66±0.027	0.69±0.028	0.69 ±0.02	NS	*	NS
		18	0.64±0.025	0.61±0.029	0.61±0.029	0.62 ±0.02			
Chewiness	kg*cm	11	1.04±0.075	0.96±0.081	1.09±0.084	1.03 ±0.08	NS	***	NS
		18	0.70±0.075	0.69±0.089	0.73±0.087	0.71 ±0.08			

NS= not significant. *= P<0.05; **= P<0.01;***=P<0.001 Means followed by different letters differ significantly at P < 0.05.

During ageing meat is subjected to gradual softening processes. Indeed, muscle undergo a series of physical and biochemical changes which are responsible for their conversion to meat; in particular, these modifications may be due to the action of endogenous protelytic enzymes (Koohmaraie, 1996). Extending the ageing time from 11 to 18 days improved flavour intensity, tenderness, juiciness and chewiness (P<0.001) of meat (data not shown). Some authors (Campo *et al.*, 1999; Monson *et al.*, 2005) found that flavour intensity increased after a longer ageing time, probably due to post-mortem processes, such as proteolysis and lipolysis, which result in the development of flavour precursors. Indeed, over 1000 volatile compounds responsible for meat flavour have been identified (Mottram, 1998).

Conclusions

Our results suggest that extending ageing time from 11 to 18 days results in increased meat tenderness and improved chewing texture and flavour intensity. High level of protein supplementation did not change meat texture profile and impaired some sensory properties. This result reveals that, in formulating finishing diets, good meat eating quality could be achieved optimising feed availability (pasture) and reducing the protein level of diet supplementation.

References

Berge, P., Culioli, J., Renerre, M., Touraille, C., Micol, D. and Geay,Y. (1993). Effect of feed protein on carcass composition and meat quality in steers. Meat Science, 35, 79–92.

Campo, M.M., Sañudo, C., Panea, B., Alberti, P. and Santolaria P., (1999). Breed type and ageing time effects on sensory characteristics of beef strip loin steaks. Meat Science, 51, 383-390.

I.N.R.A. (1988). Alimentation des Bovins Ovins and Caprins, ed. R. Jarrige. Institut National de la Recherche Agronomique, Paris.

Keane, M.G. and Allen, P., (1998). Effects of production system intensity on performance, carcass composition and meat quality of beef cattle. Livestock Production Science, 56, 203-214.

Koohmaraie, M., 1996. Biochemical factors regulating the toughening and tenderization processes of meat. Meat Science 43, (suppl.1), S193-201.

Marino R., Albenzio M., Braghieri A., Muscio A., Sevi A. 2006.Organic farming:effects of forage to concentrate ratio and ageing time on meat quality of Podolian young bulls. Livestock Production Science (in press)

Monson F., Sañudo C., Sierra I. (2005). Influence of breed and ageing time on the sensory meat quality and consumer acceptability in intensively reared beef. Meat Science, 71:471–479.

Mottram, D.S., (1998). Flavour formation in meat and meat products: a review. Food Chem. 62, 415-424.

SAS. 1999. SAS/STAT User's Guide (Version 8.1). Statistical Analysis System Inst, Cary, NC.

CONTEXTUAL VOLUME GRADING VS. SPECTRAL CALIBRATION

L.B. Christensen[1], A. Lyckegaard[2], C. Borggaard[1], R. Romvari[3], E.V. Olsen[1], W. Branscheid[4] and M. Judas[4]

[1]Danish Meat Research Institute, DK-4000 Roskilde. [2]Informatics and Mathematical Modelling, Technical University of Denmark, DK-2800 Lyngby. [3]University of Kaposvar, Faculty of Animal Science, H-7400 Kaposvar. [4]Bundesforschungsanstalt für Ernährung und Lebensmittel, Standort Kulmbach, D-95326 Kulmbach. Email: lbc@danishmeat.dk

Keywords: CT scans, contextual volume grading, Hounsfield spectrum, lean meat weight, virtual dissection

Introduction

The EU reference for determination of the content of lean meat in a pig carcass is based on a manual dissection of the carcass tissues into fat, bone and lean meat. The result of the grading is expressed as the lean meat weight (LMW), i.e. the weight of the total volume of lean meat in the carcass, as a percentage of carcass weight. In the EUPIGCLASS project, computer tomography scanning (CT) was investigated as a potential objective grading reference technique. Dobrowolski *et al.*, (2004), analysed the Hounsfield spectra of CT scanned carcasses to predict the dissected lean meat weight . One advantage of the spectral approach is, despite the simplicity, it achieved a high correlation (r<0.996) to the manual dissection reference. One drawback is, however, the influence from tissues like bone marrow and cartilage that have a comparable attenuation of x-rays to fat and lean meat. As the method includes no spatial information of individual tissue contributions, the presence of these tissues will limit the correlation.

Very recently, a new approach has emerged (Lyckegaard *et al.*, 2006) including the spatial information of individual voxels and their neighbours. The present paper compares this new approach to the conventional spectral analysis of CT generated data with respect to the determination of lean meat weight.

Materials and Methods

57 pig half carcasses were CT scanned and dissected according to the EU-recommendation (data set A). A contextual volume grading model was made to estimate the total weight W_A with a linear model of the form: $W_A = \beta_{fat}V_{fat} + \beta_{meat}V_{meat} + \beta_{bone}V_{bone}$, where V_i denotes the volume [cm^3] of the i-th tissue class. From a physical point of view, β_i is assumed to correspond to a mass density [g/cm^3] of each tissue class. The three volumes were found by binning all voxels into one of three classes using the Owen-Hjort-Mohn algorithm. From the Hounsfield value of each voxel, the algorithm optimises the probability of binning the voxels into one of the three classes: bone, fat or lean meat. The probability is conditional depending on the surrounding voxels. Further details on contextual volume grading may be found in (Lyckegaard *et al.*, 2006).

As contrast to the contextual volume grading, an alternative statistical model was made from the same data to estimate the LMW. The Hounsfield spectrum between -500 and 1500 HU was scaled to unit variance and then used as input for the PLS analysis to model the dissected lean meat content. Both models were validated (leave-one-out) and finally tested on a second unknown data set consisting of 25 CT scanned (in another CT scanner) and dissected carcasses (data set B).

Results and Discussion

Data set A was generated with a voxel size of 0.0095 cm^3 so the volume was found by simple scaling of the number of voxels in the three tissue classes. The contextual volume grading procedure lead to the following estimation model: $W_A = 0.6542 \cdot V_{fat} + 0.7519 \cdot V_{meat} + 1.1215 \cdot V_{bone}$. The performed validation of the model resulted in a high correlation of r=0.9909 with a residual error RMSEP=584 g. The values for predicting LMW with the same model was found to be r=0.9863 with a RMSEP=1408 g. The surprisingly small coefficients for density of fat and lean meat were investigated on a second data set B, but first the alternative statistical analysis was performed on set A.

The statistical PLS model estimated the LMW with a similar high correlation of r= 0.994 and a residual error SEP = 339 g in the leave-one-out validation. The PLS analysis used the spectral range between -500 and 1500 HU. The model for predicting LMW used 3 factors. No scaling was applied to the Hounsfield spectra prior to the PLS modelling. The two analysis strategies are thus producing similar prediction quality.

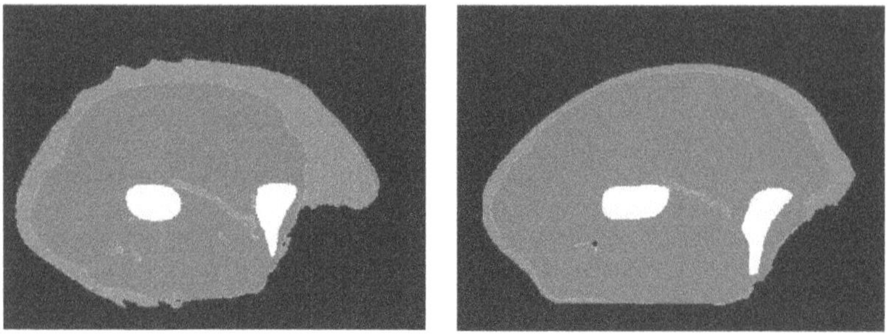

Figure 1: The result of the contextual volume grading procedure on a CT image from data set A (left) and data set B (right). By comparing the images, small deviations in the classification of the skin are apparent, a difference that will affect the estimated density coefficients. One other difference is the occurrence of classified intramuscular fat in the left image. The presence may be caused by artefacts from the image reconstruction algorithms, used in the specific CT scanning.

The two models were then tested on the second data set B for prediction of the LMW. The data set was generated with a slightly smaller pixel size of 0.0081cm^3 on a different CT scanner, so before applying the above models, both have to be scaled accordingly to achieve the tissue volumes expressed in SI units. Apart from this adaptation, no alteration whatsoever was made to any of the models.

The statistical PLS model attained a correlation between the predicted LMW and the reference value of r=0.994 with a SEP=1837g. The contextual volume grading model managed a correlation r=0.9967 but with a surprisingly high RMSEP=6991 g.

To investigate this matter, in comparison, the contextual volume grading was repeated on the second data set of 25 animals. The analysis lead to a second model $W_B = 0.9836 \cdot V_{fat} + 1.1084 \cdot V_{meat} + 1.4628 \cdot V_{bone}$ with a RMSEC=121 g. These coefficients are somewhat closer to fulfilling our expectations of the density of fat and lean meat tissue. The calculated coefficient for the bone tissue represents a mean value with a larger variation due to the inclusion of bone marrow in the bone tissue bin and to the natural density difference between cortical and trabecular bone structures.

Due to the very high correlation between the contextual volume grading result when performed on the two data sets, we believe there must be a systematic rather than random difference between the CT scanning of the two samples A and B. The difference is yet to be investigated.

Conclusions

We have demonstrated two fundamentally different analysis of CT scanning for grading of pigs. Both the contextual volume grading and the statistical PLS approach demonstrate very high correlation to a dissected reference LMW with quite small residual error contributions.

Despite the large, yet unresolved, difference (offset) between the density constants representing the two data sets, we find the contextual volume grading a very interesting alternative to the conventional spectral analysis of the Hounsfield spectrum. Further improvements of the technique seem possible in order to avoid misclassification of skin as meat instead of fat. Figure 1 left supports the assumption.

Acknowledgements

Data set A was provided by the EC project G6RD-CT-1999-00127 EUPIGCLASS of the Measurement and Testing activity (Competitive and Sustainable Growth programme).

References

Dobrowolski, A., Branscheid, W., Romvari, R., Horn, P., Allen, P. (2004): X-ray computed tomography as possible reference for the pig carcass evaluation. Fleischwirtschaft 84, (3) pp. 109-112.

Lyckegaard, A., Larsen, R., Christensen, L. B., Vester-Christensen, M. and Olsen, E. V. (2006): Contextual analysis of CT scanned pig carcasses. ICoMST52, Dublin.

CONTEXTUAL ANALYSIS OF CT SCANNED PIG CARCASSES

A. Lyckegaard[1], R. Larsen[1], L.B. Christensen*[2], M. Vester-Christensen[1] and E.V. Olsen[2]

[1] Informatics and Mathematical Modelling, Technical University of Denmark, DK-2800 Kgs. Lyngby [2] Danish Meat Research Institute, DK-4000 Roskilde. Email: lbc@danishmeat.dk

Keywords: contextual analysis, tissue classification, virtual dissection, computed tomography, weight prediction

Introduction

Knowledge of the weight of tissue types in pig carcasses is generally only available after manual dissection. The use of computed tomography (CT) has been demonstrated to be a promising approach to gain knowledge on the lean meat weight (Romvari, 2005), but less effort has been put into gaining knowledge about the weight of other tissue types from CT. Knowing the weight of individual tissue types will directly give access to other measures such as the weight of the carcass and the Lean Meat Percentage (LMP).

Until now, most analyses of CT scans have been based on the Hounsfield spectra that do not consider the spatial context in the CT scan. By applying contextual methods from the field of image analysis we hope to make a virtual dissection of pig carcasses.

Materials and Methods

57 CT scanned and manually dissected left half carcasses were used. Each CT scan consisted of approximately 140 slices(z-direction) of the size 512x512 pixels (xy-direction). The resolution in the xyz-directions were 1 mm, 1 mm, 10 mm respectively. The carcass weight was also available.

To classify each slice of the CT scans, we used the Owen-Hjort-Mohn algorithm contextual Bayesian classifier (Larsen, 2000). We classified each voxel to belong to one of the classes, c_{fat}, c_{meat} and c_{bone}, on the basis of its own and the neighbours' voxel values. For each voxel we denoted its value by X and the value of its neighbours to the north, south, east and west by X_N, X_S, X_E, X_W which lead to the feature vector $\mathbf{D}=(X, X_N, X_S, X_E, X_W)^T$. Given this feature vector we want to make a classification, that is we want to find the $v \in \{\text{fat, meat, bone}\}$ that maximizes $P(C=c_v|\mathbf{D}=\mathbf{d})$. Using Bayes Theorem and the law of total probability we have:

$$P(C=c_v \mid \mathbf{D}=\mathbf{d}) = \frac{P(\mathbf{D}=\mathbf{d} \mid C=c_v)P(C=c_v)}{P(\mathbf{D}=\mathbf{d})} = \frac{P(C=c_v)\sum_{a,b,c,d} P(\mathbf{D}=\mathbf{d} \mid \mathbf{C}=(c_v,c_a,c_b,c_c,c_d))g(c_a,c_b,c_c,c_d \mid c_v)}{h(\mathbf{d})} \quad (1)$$

where the prior $P(C=c_v)$ can be estimated from the Hounsfield spectra, $h(\mathbf{d})$ is the unconditional density for $\mathbf{d}$ and the index a,b,c,d is one of the 3^4 different class configurations of the neighbours. g(.) is especially interesting in the sense that we only allow 2 different classes within the neighbourhood and only in the spatial pattern shown in Figure 1. When postprocessing the result from the Bayesian classifier using mathematical morphology it is possible to correct some of the misclassification of marrow as meat or fat. The final result of this virtual dissection is a class label on each of the voxels in the CT scan, i.e. when knowing the voxel volume, the volume of each tissue type can be estimated.

Figure 1: Only these spatial patterns are allowed for the neighbourhood.

For the carcass weight we assume it can be modelled as a weighted sum of the tissue volumes:

$$W = \beta_{fat}V_{fat} + \beta_{meat}V_{meat} + \beta_{bone}V_{bone} \quad (2)$$

The β's in this model can be interpreted as tissue densities, so estimating the β's from known examples makes it possible to predict carcass weight from a CT scan.

Results and Discussion

Using the method described above, the 57 CT scanned left half carcasses were virtually dissected. An example of the virtual dissection of two different slices is shown in Figure 2. The method is demonstrated to be robust to noise and artefacts but this also means that finer structures disappear in the virtually dissected image. The postprocessing step works well and the marrow inside the bone is not found to be either fat or meat.

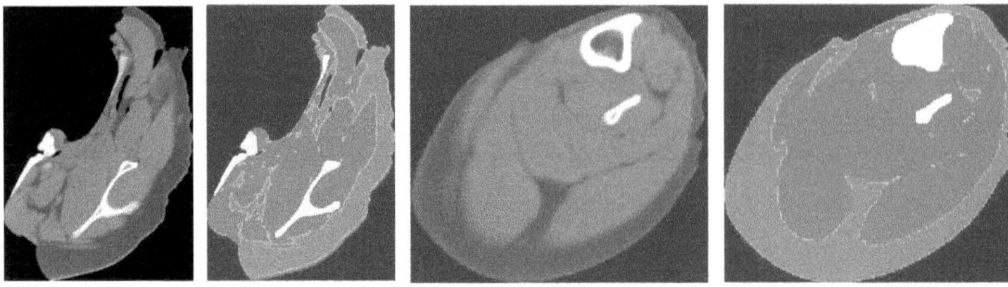

Figure 2: Result of the virtual dissection of two different slices. From left to right: CT slice, virtually dissected slice, CT slice, virtually dissected slice.

Estimating β's in equation (2) from the 57 pig half carcasses we obtained the correlation between measured and predicted half carcass weight shown in Figure 3 (left) with R=0.9918 and RMSEC=0.5537 kg. The slope of the regression line is 0.9878, and it has an offset of 0.4638. Performing leave-one-out cross validation (Figure 3, right) gave R=0.9909 and a residual sum of squares of RMSEP= 0.5840 kg. The regression line has slope 0.9856 and offset 0.5468.

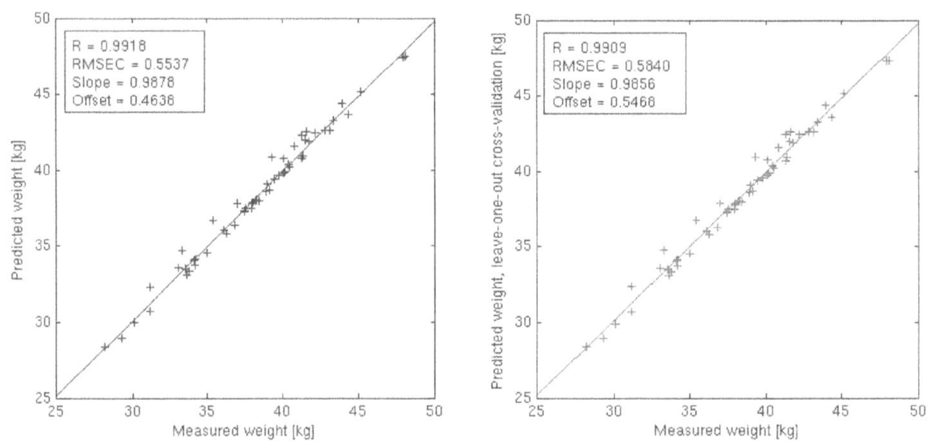

Figure 3: Left shows the correlation between measured and predicted half carcass weight. Right shows the correlation between measured and predicted weight half carcass when performing cross validation.

Conclusions

A contextual analysis method, the Owen-Hjort-Mohn algorithm, combined with a postprocessing step using mathematical morphology was developed for performing a virtual dissection of pig carcasses from CT scans. The virtual dissection was performed on 57 CT scanned left half carcasses and a model of the carcass weight based on virtual dissections was suggested and evaluated against known weight.

Acknowledgements

R. Romvari is gratefully acknowledged for providing data from the EC project G6RD-CT-1999-00127 EUPIGCLASS of the Measurement and Testing activity (Competitive and Sustainable Growth programme).

References

Larsen, R. (2000). 3D Contextual Bayesian Classifiers, IEEE Transactions on Image Processing, vol. 10(3), pp. 518-524.

Romvari, R., Dobrowolski, A., Repa, I., Allen, P., Olsen, E., Szabo, A. and Horn, P. (2006). Development of a computed tomography calibration method for the determination of lean meat content in pig carcasses, Acta Veterinaria Hungarica 54(1), pp. 1-10.

CARCASS AND MEAT QUALITY OF YOUNG BULLS IN SLOVAK REPUBLIC

J. Mojto[1], J.Kožuch[2], K. Zaujec[1] and R. Lahucky*[1]

[1.] *Slovak Agriculture Research Centre, 949 92 Nitra, Slovak Republic,* [2.] *Ministry of Agriculture of the SlovakRrepublic, 812 66 Bratislava, SlovakRrepublic, Email: mojto@scpv.sk*

Keywords: beef, quality audit, young bulls

Introduction

Improvement of meat quality in slaughter cattle necessitates knowledge of its real quality immediately after killing at the slaughterhouse. Results can effectively help authorities of the country to take important decisions, and they also bring significant information to meat processors and farmers. In some countries it is usual to do a quality audit of beef (Brooks *et al.*, 2000; McKenna *et al.*, 2002). Numbers of scientific workers from various research institutions in the country take part in it. This study presents the results of the first such audits that took place in the Slovak Republic in 2003 and 2005.

Materials and Methods

The quality audit was performed on young bulls to the age of two years. The following number of carcasses was evaluated: 13,805 carcasses in 2003 and 14,111 carcasses in 2005. Conformation and fat cover of carcasses were evaluated by the classification system SEUROP (Regulation EEC No 1208/81, Regulation EEC No 1026/91).

Data were obtained from the most important slaughterhouses in the country. In addition, samples were taken of the muscle *M. longissimus dorsi* (LD) from 1,360 carcasses to assess quality parameters. The following was assessed; water content, proteins and intramuscular fat by the Infratec 1265 apparatus, pH value, colour of meat (% remission) spectrophotometrically, content of free water, LD area and marbling in meat by 10 points scale (1 = very intensive marbling, 10 = without traces of i.m. fat) 48 hours post mortem. On Day 7 we detected shear force in cooked meat by Warner-Bratzler method. Variation-statistical values were calculated with each of qualitative parameters (Statgraphic).

Results and Discussion

Analysis of results (Table 1) showed that there are important differences between the compared years. While in 2003, most carcasses were in conformation classes U (46.14 %) and R (44.52 %), in 2005 they were in classes R (52.50 %) and O (31.07 %) according to the degree of meatiness. It represents a shift by 1 class towards worse meatiness on average. We found significant differences also when evaluating the degree of fat cover in 2003 and 2005. The number of carcasses increased in classes 2 and 3, and decreased in class 1. Knowledge and experience from a number of European countries show that processors of slaughter cattle are more interested in carcasses in classes of fat cover 2 and 3. Mentioned changes are due more to a stricter system of classification and to better control of classifiers than to real changes in carcass quality. Therefore it is necessary to take the results in 2005 as accurate.

Table 1: Percentual representation of slaughter carcasses.

Confor-	Fat class											
mation	1		2		3		4		5		Total	
	2003	2005	2003	2005	2003	2005	2003	2005	2003	2005	2003	2005
E	0.09	0.42	1.32	0.36	0.04	0.25	0.91	-	-	-	2.36	1.03
U	35.26	1.16	8.82	7.51	1.95	1.16	0.12	0.13	0.01	0.01	46.14	9.97
R	3.95	10.88	32.15	30.03	6.77	10.93	1.56	0.60	0.09	0.06	44.52	52.50
O	3.35	10.04	2.25	10.40	0.29	9.52	0.01	0.96	0.16	0.15	6.06	31.07
P	0.47	2.98	0.71	1.55	0.28	0.57	-	0.29	-	0.03	0.90	5.42
Total	43.11	25.49	44.71	49.85	9.33	22.43	2.59	1.98	0.25	0.25	100.00	100.00

A nutritional and physical-technological characteristic of LD is in table 2. There is low content of i.m. fat in meat of young bulls (2.22 %) which corresponds also with results of marbling in LD. We found the highest variability with shear force of meat (v = 53.47 %) and with content of i.m. fat (v = 51.79 %). We found extremely dark meat (DCB) with high end value of pH in 21.45 % carcasses, it is markedly less than in our first large study in 1982 (Mojto *et al.*, 1982), when we found DCB in 56 % carcasses of young bulls. Decrease in DCB occurrence is a result of significant changes in manipulation of animals before slaughter (shorter transport times, social stabilisation of animals, etc.). LD area was on average 94.86 cm^2, within the span from 60 to 135 cm^2. In the USA they found span from 50 to 150 cm^2 (McKenna *et al.*, 2002) during the audit.

Table 2: Mean values and variability of qualitative parameters.

Parameters	$\bar{x}$	s	v %	min.	max.
Total water, %	74.80	1.62	2.17	60.09	78.87
Total protein, %	21.90	1.02	4.66	17.87	26.06
Intramuscular fat, %	2.22	1.15	51.79	0.49	9.91
Free water, %	32.35	4.48	13.85	20.17	40.88
pH value	5.90	0.48	8.14	5.20	7.10
Meat color, %	8.56	2.25	26.36	3.80	21.00
Cooking loss, %	40.84	6.25	15.31	17.17	59.35
W-B shear force, kg	6.43	3.44	53.47	0.60	18.30
Marbling score	7.55	1.27	16.89	3.00	9.00
Longissimus muscle area, cm^2	94.86	11.28	11.89	66.20	135.01

Conclusion

The first clear and representative results of carcass and meat quality in the category of young bulls were obtained in the Slovak Republic. They create the basis for observation of trends in future years that will be evaluated from periodically repeated audits of beef quality. Also, further slaughter categories will be incorporated gradually.

References

Brooks,J.C., Belew,J.B., Griffin,D.B., Gvartney,B.L., Hale,D.S., Henning,W.R., Johnson,D.D., Morgan,J.B., Parrish,F.C., Reagan,J.O., Savell,J.W. (2000) National Beef Tenderness Survey-1998, Journal of Animal Science, 78:1852-1860.

McKenna,D.R., Roeber,D.L., Batest,P.K., Schmidts,T.B. Hale,D.S., Griffin,D.B., Savell,J.W., Brooks,J.C.,Morgan,J.B., Montgomery,T.H., Belk,K.E., Smith,G.C. (2002) National Beef Quality Audit – 2000: Survey of targeted cattle and carcass characteristics related to quality, quantity, and value of fed steers and heifers, Journal of Animal Science, 80:1212-1222.

Mojto,J., Jedlička,J., Foltys,V., Páleník,Š. (1982) Význam a možnosti určovania kvality hovädzieho mäsa podľa hodnoty pH. In: Zborník Produkcia a kvalita mäsa hovädzieho dobytka a problémy speňažovania, ČSVTS Dom techniky Košice, 68-72.

IMPROVAC® IMMUNIZED BOARS COMPARED TO SURGICAL CASTRATES: CONTROL OF BOAR TAINT AND GROWTH PERFORMANCE

E.T.F. Silveira*[1], E. Poleze[2], O. Umehara[2], A.P. Tonietti[1], M.L.T. Buzelli[1], M.M.H Haguiwara[1], L. Miyagusku[1] and D. Hennessy[3]

[1.] Institute of Food Technology, Meat Technology Center, Av. Brasil, 2880, 17070-178, Campinas, SP, Brasil, [2.] Laboratorios Pfizer Ltda. Divisão de Saúde Animal, Condomínio Edifício Birmann 11, Rua Alexandre Dumas, 1711, 8º. Andar 04717-000, São Paulo, Brasil, [3.] Pfizer Australia, 231/1 Queens Road Melbourne, Vic Australia 3004. Email: tadeu.s@uol.com.br

Keywords: pigs, Improvac® vaccine, boar taint, growth performance, meat yields

Introduction

International competition is driving the pig meat production chain to invest more in food safety, animal welfare, traceability and leaner animals. Raising intact male pigs for meat production eliminates the need for surgical castration and enables better production efficiency when compared to castrated pigs. However, consumer acceptance of fresh pork from intact male pigs is hindered by the taint compounds. Androstenone (a steroid produced by testes) and skatole (produced by bacteria in the gut) have been recognised to be the major causes of taint (Babol *et al.*, 2002). This study was conducted to evaluate the efficiency of Improvac®, a vaccine directed against gonadotrophin – releasing factor (GnRF), to control boar taint and to allow improved growth performance by rearing entire male pigs rather than surgically castrated male pigs.

Materials and Methods

Animals. A total of 48 male pigs, 18-22 days old, initial weight 4.2 – 8.5 kg, from the same farm and genetic origin were allocated systematically according to presentation order into two experimental groups (T01-24 surgically castrated and T02-24 Improvac® vaccinated). Improvac® was administered subcutaneously as two 2mL doses; the first dose at about 15 weeks and the second dose at about 19 weeks of age. Two critically ill pigs from group T01 were withdrawn from the study.

Production performance. At the end of the experimental period the following parameters were evaluated: mean weight (MW), average daily weight gain (ADG), feed efficiency (FE), and meat yields (MY) in half carcass.

Slaughter. The animals were kept in the evaluation station for 20 weeks then slaughtered in a commercial abattoir at 23 weeks. The chilled carcasses were transported to the Meat Technology Centre and prepared for boning purposes according to the methodology described by Walstra and Merkus (1996).

Evaluations. Serum testosterone analysis was performed according to Nunes (1982). Skatole and androstenone fat assays evaluations followed the methodology modified by Oliveira *et al.*, (1999).

Statistical analysis. The statistical design was based on randomised block design blocked on pen location. Mean of weights, feed efficiency, average daily weight gain, carcass evaluations, and androstenone levels were analysed using general linear models with terms including the fixed effect of treatment group and the random effect of block, assessed at 5% significance level.

Results and Discussion

Table 1 contains the results of the boar taint compounds for both treatment groups. Only two animals (both in group T02) had a testosterone value above the limit of detection (0.1 pg/mL). No animals in either group had a value for skatole above the accepted sensory threshold limit of 0.2 μg/g (Desmoulin *et al.*, 1982). There were no significant differences between Improvac® vaccinated and surgically castrated pigs for androstenone levels, which were under the accepted sensory threshold limit of 1.0 μg/g (Bonneau *et al.*, 1993), confirming the efficiency of the Improvac® vaccine. Improvac® vaccinated pigs had a significantly higher mean weight, higher average daily weight gain and better feed efficiency, and higher meat yields than surgically castrated pigs (Table 2). This agrees with the results of Mccauley *et al.*, (2003) who reported that Improvac vaccinated pigs may be a means of maintaining high weight gain in the period while maximizing growth performance in finishing boars. Carcass weight was higher in T02 but not significant. Moreover, there was an increase in the amount of meat in some commercial cuts such as leg (P=0.052), tenderloin (P<0.05), shoulder, belly, and ventral part of the belly (P<0.05), representing economical advantages for the pork industry since both markets, fresh and processed products, are reached simultaneously.

Conclusions

Improvac® was highly effective in controlling boar taint compounds and in suppressing testosterone to levels comparable to those normally found in surgically castrated pigs. Improvac® boars had a significantly better mean weight, average daily weight gain, feed efficiency, and total meat yield compared to surgically castrated pigs.

Table 1: Serum testosterone, skatole and androstenone fat values.

Evaluations	Surgically Castrated (n=22)*	Improvac Vaccinated (n=24)	*P* Values
Testosterone (pg/ml):			
< 0.1 n (%)	22 (100%)	22 (92%)	N/A
0.4 – 2.2 n (%)	0 (0%)	2 (8%)	
Skatole (μg/g):			
< 0.2 n (%)	22 (100%)	24 (100%)	N/A
Androstenone (μg/g):			
Geometric mean	0.32	0.28	0.37

* Two critically ill pigs from the surgically castrated group were withdrawn from the study.

Table 2: Growth performance values.

Production Parameters	Surgically Castrated	Improvac Vaccinated	*P* Values
Mean Weight (kg)	124.41	137.76	0.0025
Average Daily Gain (kg/day)	0.851	0.944	0.0032
Feed Efficiency	2.467	2.237	<0.0001
Carcass Weight (kg)	103.82	108.93	0.09
Meat Yields	26.640	29.057	0.0052

References

Babol, J., Squires, E.J. and Gullet, E.A. (2002) Factors affecting the level of boar taint in entire male pigs as assessed by consumers' sensory panel. Meat Science, 61: 33-40.

Bonneau, M. (1993) Effects of different compounds on boar taint. 44TH Ann. Mtg. Europ. Assoc. Anim. Prod., Aarhus.

Desmoulin, B., Bonneau, M., Frouin, A. and Bidadrd, J. P. (1982) Consumer testing of pork and processed meat from boars: the influence of fat androstenone level. Livestock Production Science, 9, 707- 715.

Mccauley, I.; Watt, M., Suster, D., Kerton, D. J., Oliver, W. T., Harrell, R.J. and Dunshea, F. R. A (2003) GnRF vaccine (Improvac®) and porcine somatotropin (Reporcin®) have synergistic effects upon growth performance in both boars and gilts. Australian Journal of Agricultural Research, 54: 11-20.

Oliveira, J.J.V., Silveira, E.T.F. and Viana, A. G. (1999) Determinação de escatol e androstenona em toucinho e glândula salivar submaxilar de suínos. Revista Brasileira de Análise de Alimentos, 5: 12-20.

Walstra, P. and Merkus, G.S.M. (1996) Procedure for assessment of the lean meat percentage as a consequence of the new EU reference dissection method in pig carcass classification. Report ID-DLO 96.014. Lelystad: DLO- Institute for Animal Science and Health, p. 22.

A COMPARISON OF MEAT CHARACTERISTICS BETWEEN DUCK AND CHICKEN BREAST

S.-T. Joo*, M.D. Shawkat Ali, G.-H. Kang, Y.-H. Hwang and G.-B. Park

Division of Applied Life Science, Gyeongsang National University, Jinju, Gyeongnam 660-701, Korea.
Email: stjoo@gnu.ac.kr

Keywords: duck, chicken, muscle fibre, breast meat

Introduction

Duck is a waterfowl unlike chicken and has a different physiology compared with other poultry. More recently, duck cuts such as breast and leg have become more available, which offer more options for diet conscious consumers. Normally the slaughter procedure for duck is the same as for chicken in spite of duck having more red muscle fibres compared to chicken (Smith *et al.*, 1993) and is considered a red meat. Therefore, there is a need for different slaughter, processing and preservation methods for better quality duck meat. The objective of this study was to investigate the pH decline pattern postmortem, composition of muscle fibre type, meat characteristics and fatty acid composition of duck and chicken breast.

Materials and Methods

Twenty-four broilers (Ross) and 24 ducklings (Cherry berry) of 45 days were stunned and killed by conventional neck cut. Breast meat (Pectoralis major) was removed from each carcass at the following times post-mortem: 15 min (3 birds from each species), 30 min (3 birds from each species), and 1 h (complete processing of the rest of the birds). The breast meat was then placed in a plastic bag and kept in a cold storage room at 4°C. The pH was measured by a pH-meter (MP230, Mettler, Switzerland). The surface colour (CIE L* and a*) of meat samples was measured using a Minolta Chromameter (Minolta CR 301, Tokyo, Japan). Shear force was measured by using the Instron Universal Testing Machine (Model 3343) from each cooked breast meat sample. For fatty acid analysis, lipids were extracted with chloroform and methanol and fatty acid methyl esters were analysed on a gas chromatography (Agilent, 6890, USA). For muscle fibre type, the serial frozen sections (8 μm thick) were stained by histochemical reactions for observing the distribution of myofibre types. Myosin adenosine triphosphatase (ATPase) activities (Padykula and Herman, 1955) were detected after acid (pH 4.3) and alkaline (pH 10.5) pre-incubation (Brooke and Kaiser, 1969), as was reduced nicotinamide adenine dinucleotide dehydrogenase (NADH-DH) activity (Okamoto *et al.*, 1976). On the 2 series of photographs (×250) of each tissue taken at the same location on the different specimens, the myofibres divided into types I, IIA and IIB according to the nomenclature of Brooke and Kaiser (1969).

Results and Discussion

Significant differences were found in muscle pH at 1 and 3h postmortem between two species. The pH of chicken meat was significantly lower than duck meat at 1 and 3 h postmortem, but the ultimate pH at 24h was not different. Protein content was significantly higher in chicken breast whereas fat content was higher in duck breast. Muscle fibres of duck breast contained 23.4% type I, 48.5 % type IIA and 28.1 type IIB, whereas chicken breast contained 100% type IIB muscle fibre only. Smith *et al.* (1993) also found 16% and 100 % white fibres in *Pectoralis* muscle of duck and chicken, respectively. These results suggested that the rapid pH decline and lower fat content of chicken breast would be due to its higher content of white muscle fibre compared to duck breast. Because of the red muscle fibres in duck breast, as expected, significant higher redness (a*) and lower lightness (L*) values were observed in duck breast. During cold storage for 7 days, the a* value of duck remained the same but the L* value was increased. Also duck breast showed significantly higher TBARS values and cooking loss (%) compared to chicken breast during storage period. Shear force decreased when storage period was increased in both chicken and duck breast meat, and it decreased rapidly in duck breast compared to chicken breast. Alvarado and Sams (2000) also found higher cooking loss and shear force value in duck breast compare to chicken breast at different postmortem time. The fatty acids (%) C14:0, C16:0, C16:1, C18:2 and C18:3 were significantly higher while C18:0 was significantly lower in duck breast compared to chicken. Significant changes in fatty acids composition were found in both chicken and duck breast meat at 1 and 7 days of cold storage, and the change was severe in duck breast compared to chicken breast. Results suggested that the higher TBARS values in duck breast might be due to higher content of unsaturated fatty acids and severe changes in fatty acid composition during cold storage. Data also suggested that duck should be slaughtered and processed in a different way from chicken for a desirable meat quality.

Conclusions

Duck had higher red muscle fibres compared to chicken, resulting in different pH decline pattern at postmortem, proximate composition and meat characteristics. Higher TBARS values and cooking loss % were observed in duck breast compared to chicken breast during 7 days of cold storage. Data suggested that duck would be slaughtered and processed with different way from chicken for a desirable meat quality.

Table 1: The pH, proximate composition (%) and muscle fibre type (%) in chicken and duck breast meat.

Meats	pH at postmortem times					Proximate composition				Muscle fibre type		
	15 m	30 m	1 h	3 h	24 h	Moisture	Protein	Fat	Ash	I	IIA	IIB
Chicken	6.62^{A}	6.42^{B}	6.19^{CX}	6.06^{DX}	5.87^{E}	75.47	22.04^{X}	1.05^{Y}	1.07^{X}	0	0	100
Duck	6.55^{A}	6.42^{B}	6.32^{CY}	6.17^{DY}	5.94^{E}	76.41	20.06^{Y}	1.84^{X}	0.92^{Y}	23.4	48.5	28.1

$^{A-E}$Means with different superscripts within a raw in pH at different postmortem time differ significantly ($p<0.05$); $^{X-Y}$Means with different superscripts within same column differ significantly ($p<0.05$).

Table 2: The meat quality characteristics of chicken and duck breast meat during cold storage at 4°C.

Day	L*		a*		[1]TBARS		Cooking loss (%)		[2]Shear force	
	Chicken	Duck	Chicken	Duck	Chicken	Duck	Chicken	Duck	Chicken	Duck
1	57.1^{AX}	39.7^{BY}	1.7^{ABY}	18.2^{X}	0.14^{CY}	0.24^{CX}	29.2^{AY}	34.5^{X}	3.5^{A}	3.8^{A}
3	54.1^{BX}	41.8^{ABY}	2.5^{AY}	19.1^{X}	0.21^{BY}	0.32^{BX}	27.2^{ABY}	35.5^{X}	3.3^{A}	3.4^{B}
5	53.5^{BX}	41.7^{ABY}	2.0^{ABY}	18.8^{X}	0.22^{BY}	0.35^{BX}	24.8^{BCY}	35.6^{X}	3.4^{A}	3.4^{B}
7	57.0^{AX}	43.2^{AY}	1.3^{BY}	19.0^{X}	0.29^{AY}	0.40^{AX}	22.2^{CY}	35.6^{X}	2.7^{B}	3.1^{C}

$^{A-C}$Means with different superscripts within same column differ significantly ($p<0.05$); $^{X-Y}$Means with different superscripts within a row in the same parameters differ significantly ($p<0.05$); [1]mg malonaldehyde/ kg sample; [2]kg/cm^2.

Table 3: Fatty acid composition of chicken and duck breast meat.

Fatty acid	Chicken breast		Duck breast	
	1 day	7 days	1 day	7 days
C14:0	0.36^{BY}	0.78^{A}	0.91^{AX}	0.38^{B}
C16:0	17.22^{BY}	22.53^{A}	21.83^{X}	22.01
C 16:1	2.17^{Y}	3.27	4.16^{AX}	2.14^{B}
C 18:0	18.17^{AX}	10.54^{B}	10.46^{AY}	14.48^{B}
C 18:1	34.32	36.67	35.66	31.48
C 18:2	14.15^{BY}	16.74^{A}	19.34^{AX}	15.14^{B}
C 18:3	0.53^{Y}	0.60	0.84^{AX}	0.52^{B}
C 20:4	11.23	7.06	5.49^{B}	12.12^{A}
C22:5	0.89	0.94	0.68	0.83
C22:6	0.96	0.88	0.61	0.91
SFA	35.75	33.85	33.21^{B}	36.86^{A}
USFA	64.25	66.15	66.79^{A}	63.14^{B}
MUSFA	36.49	39.93	39.82^{A}	33.62^{B}
PUSFA	27.75	26.22	26.97	29.51
MUSFA/SFA	1.03	1.18	1.20^{A}	0.91^{B}
PUSFA/SFA	0.77	0.77	0.81	0.80

$^{A-B}$Means with different superscripts in a row within chicken breast or duck breast differ significantly ($p<0.05$); $^{X-Y}$Means with different superscripts in a row within 1 day chicken and duck breast differ significantly ($p<0.05$).

References

Alvarado, C. Z. and Sams, A. R. 2000. The influence of post-mortem electrical stimulation on rigor mortis development, calpastatin activity, and tenderness in broiler and duck pectoralis. Poultry Science, 79: 1364-1368.

Brooke, M. H. and Kaiser, K. K. 1969. Some comments on the histochemical characterization of muscle adenosine triphosphate. Journal of Histochemistry and Cytochemistry, 17: 431-432.

Padykula, H. A. and Herman, E. 1955. Factors affecting the activity of adenosine triphosphatase and other triphosphatase as measured by histochemical techniques. Journal of Histochemistry and Cytochemistry, 3: 161-169.

Smith, D. P., Fletcher, D. L., Buhr, R. J. and R. S. Beyer. 1993. Pekin ducklings and broiler chicken pectoralis muscle structure and composition. Poultry Science, 72: 202-208.

FEMUR STRENGTH OF PIGS

M. Ruusunen[1], K. Partanen[2] and E. Puolanne*[1]

[1.] *Department of Food Technology, University of Helsinki, P.O. Box 66 (Agnes Sjöbergin katu 2), FI-00014 Helsinki, Finland,* [2.] *MTT Agrifood Research Finland, Animal Production Research, Pigs, Tervamäentie 179, FIN-05840 Hyvinkää, Finland. Email: marita.ruusunen@helsinki.fi.*

Keywords: pigs, femur strength

Introduction

Bones provide the main structural support for the body. Bony tissue is classified as either compact bone on the bone surface or spongy bone in the deeper layers. The main structural elements of bones are the epiphysis at both ends, with a joint surface, and the diaphysis in the middle. In the core of the diaphysis is the medulla containing soft material. Bones contain both inorganic mineral and organic components. The inorganic components, mostly calcium and phosphorous, constitute 65-70 percent of bone weight (Ross *et al.*, 1995). The inorganic salts form a solid porous material able to resist pressure. The other constituents, connective tissue, bone cells, nerves, arteries and veins etc. are embedded in bone matrix. The organic part, mostly collagen, increases the tensile properties, especially resistance to bending. There is, however, a need for better understanding of bone strength of pigs because their bones break easily during transport to the abattoir and during the slaughter process, resulting in loss of carcass weight. The joint defects, e.g. osteochondrosis, are also rather common. Bone and joint defects have also been linked to high growth rate of pigs. Bone weakness and other bone problems also constitute significant animal welfare issues.

This study investigated the effect of femur dimensions, mineral content and mineral density on femur strength in pigs slaughtered at the same age.

Materials and Methods

The study consisted of 18 pigs from nine litters (9 gilts and 9 barrows): 7 Landrace, 7 Yorkshire and 4 crossbreed. The pigs were housed individually in pens. A barley-soybean based diet was used. The pigs were slaughtered at the age of 165 ± 2 d at the experimental abattoir of University of Helsinki, Department of Food Technology. One day after slaughter the bones were dissected from the carcasses and weighed. The femur was weighed separately.

Four rings (thickness 7 mm) were cut from the diaphysis of femur. The bone area and wall thickness of the ring were measured with an image analysis system using Zeiss stereo microscope (Stemi DV4, Carl Zeiss, Germany), AxioCam camera (Carl Zeiss AxioCamera MRc, Germany) and KS300 image analysis computer program (KS300, Carl Zeiss Vision GmbH, Germany). Bone mineral content (BMC) and bone mineral density (BMD) were measured from the femur rings by scanning all femur rings at the same time with dual-energy X-ray absorptiometry (DXA) and using a Lunar PIXImus 2 densitometer (software version 1.42.006.010; Lunar Corporation, Madison, WI). After measuring the femur ring dimensions, the rings were broken with Instron compressometer (Instron TM-M, Instron Ltd., UK). The maximum force required to break the ring of bone was recorded.

The statistical analysis was performed with SAS Program version 8.02 (SAS, 1999). Linear regression analysis (REG procedure) was used to explain the variation in femur ring strength.

Results and Discussion

The weight of pigs at slaughter was on average 108 kg and the carcass weight was on average 83.9 kg. The bone content in the carcass was on average 14.9% (Table 1).

Table 1: The live and carcass weights of the pigs and the measured parameters of femur.

N	18
Live weight, kg	108.0 ± 7.0
Carcass weight, kg	83.9 ± 5.6
% bone in the carcass	14.9 ± 1.1
Femur weight, g	325 ± 44
Femur ring area, cm^2	3.2 ± 0.40
Femur wall thickness, mm	4.7 ± 0.5
Femur ring strength, kg	38.0 ± 10.2
Bone mineral content per femur ring, g (BMC)	2.45 ± 0.31
Bone mineral density, g/cm^2 (BMD)	0.76 ± 0.06

The femur ring area was on average 3.2 cm^2, but the variation was large ranging from 2.5 to 4.1 cm^2, although the pigs were slaughtered at the same age. Femur wall thickness was on average 4.7 mm ranging from 3.9 to 5.4 mm. Bone mineral content (BMC) and density (BMD) measurements were made from the femur rings. The correlation between

BMC values and femur weight was weakly positive (r=0.458, P=0.0558). The reason for this was the larger femur ring area in heavier femurs (r=0.808, P<0.0001).

Femur ring strength was on average 38 kg varying from 27.6 to 71.4 kg. Keller (1994) reported high positive correlations between bone mineral mass and bone strength. In this study, however, several measured traits explained femur ring strength. Live weight, carcass weight, femur ring area, femur wall thickness and BMC accounted for 84% of the variation in femur ring strength while femur ring area, femur wall thickness and BMC accounted for 73% of the variation in femur ring strength (Table 2). The results showed that the bone mass in pigs must be large enough with high mineral content to guarantee strong bones. Thick femur wall strengthens the bone.

Table 2: Coefficients of determination (R^2) for simple linear regressions of measured traits on femur strength.

	R^2
Live weight, carcass weight, femur weight, femur ring area, femur wall thickness, BMD, BMC	0.85
Live weight (P<0.013), carcass weight (0.021), femur ring area (P<0.0001), femur wall thickness (P<0.0004), BMC (P<0.0001)	0.84
Femur ring area (P<0.0003), femur wall thickness (P<0.0059), BMC (P<0.0001)	0.73

Conclusions

For strong bones in pigs it is required that the bone mass is large with a high mineral content.

Acknowledgements

The authors wish to thank the Finnish Ministry of Agriculture and Forestry and Finnish meat industry for funding this study.

References

Keller, T.S. (1994). Predicting the compressive mechanical behaviour of bone. Journal.of Biomechanics 27: 1159-1168.

Ross, M.H., Romrell, L.J. and Kaye, G.I. (1995). Histology: a text and atlas. Williams & Wilkins.

SAS. (1999). SAS, Version 8.2, SAS Institute Inc., Cary, NC.

MEAT QUALITY IN LOMBARDIA REGION ITALY

L. Bertocchi*[1], R. Bravo[1], E. Monti[1], R. Tadeo [2], S. Benicchio [1], and C. Sgoifo Rossi [3]

[1.] *Istituto zooprofilattico Sperimentale della Lombardia ed Emilia Romagna B.Ubertini, via Bianchi 9, 25124 Brescia Italy.* [2.] *Centro Miglioramento del Latte e della Carne Bovina di Brescia via Bianchi 9, 25124 Brescia Italy,* [3.] *Dipartimento di Scienze e Tecnologie Veterinarie per la Sicurezza Alimentare, Facoltà di Medicina Veterinaria, via Celoria, Milano Italy E mail: lbertocchi@bs.izs.it*

Keywords: meat quality, beef, organoleptic analysis

Introduction

Lombardia is the most developed region of Italy for meat technology. This area is the best indicator of economic productivity from agriculture and the highest profitability from food. Lombardia produces 14% of the national meat production. After the negative affect of BSE on meat consumption up to 2001, livestock breeders relaunched the market starting from the development and certification of production regulations . The application of Reg CE 1760 made it possible to give the consumer additional information about meat. This information concerns the place and name of the producer, animal place of origin and the stock farming characteristics with particular reference to meat quality.

This study presents data on quality and organoleptic analysis on 605 meat samples obtained from an equal number of male cattle belonging to three different French breeds bred in Lombardia for meat production.

Materials and Methods

Beef cattle were imported from France at the age of 10-16 months, weighing 200-400 kg. They were fattened intensively for 6\12 months and then butchered at the age of 18-24 months, weighing 600-700 kg. Feeding was controlled to prevent the animals from growing fat and producing fat meat, which consumers don't like. To monitor the results of this stock-farming system and to inform consumers about meat quality characteristics, the production and labelling regulations provided the basis for the collection of meat samples to be sent for laboratory analysis. Samples 2.5 cm. thick, taken from *Longissimus dorsi* muscle between the 5th and 7th thoracic vertebra, were collected in the slaughter-house three days after butchery. Laboratory analysis to define chemical, quality and organoleptic characteristics of the meat followed ASPA methodologies.

The following parameters were considered:

- pH: 24 hours post-mortem, using a HANNA pHmeter HI 98150 (Hanna Instruments Italia Srl., PD)
- colour, using a Minolta, Chroma Meter CR-300 colorimeter (Instruments Lab Control s.n.c., RE)
- tenderness, as resistance to cutting, expressed as the kgf required to cut 5 cylinders of cooked meat with a diameter of 12.7 mm and thickness of 25 mm, with longituadinal axis parallel to the muscular fibres, using an Instron 5544 and Warner Bratzler device.
- weight loss after cooking, using gravimetric method on meat cylinders with the same size as the samples used for tenderness measurement, with double weighing before and after bain-marie cooking at 75°C.
- proteins, total fat and ash with NIR.

Results and Discussion

Both 24h pH post mortem and meat colour appears to have been influenced by sensitivity to stress and a nervous disposition in males, compared to females. This was also observed in animals with hypertrophic muscles, compared to animals with a normal muscle growth (Table l).

Differences in lipid content and colour between cattle breeds were observed in relation to nutrition, organoleptic properties and shelf life, in meat for retail sale. Traditionally, Blonde D'aquitaine have less fat which is characteristic of this genetic type (Table 1). In Charollais beef there is less resistance to cutting, though not significantly, probably because of a higher lipid content. There is no significant difference between the other parameters. This result is related not only to correct feeding during breeding, but also to proper management of pre-slaughter loading and unloading of animals at the slaughter house and with stressful events. Cattle with muscle hypertrophy are more susceptibility to stress and consequently to producing dark, firm and dry beef (DFD).

Table 1: Effect of genetic breed on quality and organoleptic meat characteristics.

Values expressed in %	Limousine	Charolaise	B. D'aquitaine	P
Weight carcass, kg	376.86 [a]	413.84 [b]	461.17 [c]	0.02
Yield in weight after slaughter %	61.78 [a]	59.12 [b]	64.23 [c]	0.03
pH at 45 minuts	6.73	6.71	6.89	NS
pH at 24 hours	5.77 [a]	5.65 [a]	5.86 [b]	0.04
Weight loss after cooking , %	30.30	29.30	31.78	NS
Weight loss of raw meat after 3 days	5.11	5.14	5.98	NS
Shelf-life	5.80	5.70	6.00	NS
Colorimetric characteristics,				
L	42.18	43.21	41.01	NS
a*	21.63 [a]	22.04 [a]	19.13 [b]	0.04
b*	13.31	13.78	12.04	NS
Cutting resistence, kgf	3.38	3.19	3.56	NS
Protein %	22.85	21.55	23.08	NS
Fat %	2.87 [a,]	3.59 [a,]	1.91 [b]	0,05

[a,b,c]($P<0.05$). Different letters on the same line mean significant difference

Conclusions

Fat content is an important meat quality factor and the results of this study showed very low meat fat levels, which is not what consumers want. Post-mortem meat pH after 24 hours and colour were influenced by susceptibility to stress and aggression in male cattle as well as animals with muscular hypertrophy. These factors must be considered in relation to proper animal management.

The results of this study have shown the excellent quality and nutritional characteristics of meat obtained from animals that were produced using the Reg Ce 1760/2000 meat labelling system. In Brescia in 2001, 10 livestock breeders, 5 slaughter houses and 14 butchers used these regulations, and a total of 190 animals were slaughtered, labelled and certified. In 2005, 38 breeders, 9 slaughter houses and 27 butchers were involved and a total of 2318 animals were slaughtered and certified.

SENSORY CHARACTERISTICS OF SUCKLING LAMBS. INFLUENCE OF BREED AND SLAUGHTER WEIGHT

I. Revilla*, G. Rodríguez-López, and A.M. Vivar-Quintana

Area de Tecnología de Alimentos. Universidad de Salamanca, E.P.S de Zamora, Avda Requejo 33, 49022 Zamora, Spain. Email: irevilla@usal.es

Keywords: colour, taste, aroma, texture, sensory panel

Introduction

Suckling lamb is a typical product of the Mediterranean areas associated to the production of ewe's milk for cheese-making. Many lamb products are also associated with Protected Geographical Indications, these quality brands include some specific breeds but also foreign breeds, introduced due to its higher milk production, have an increasing relevance. When the influence of breed on physico-chemical and sensory quality is considered, the effect is unclear, making difficult differentiation between foreign and local breeds. Indeed, although the range of carcass weights in light lambs is short (9 to 12 kg of live weight for suckling lambs of *Lechazo de Castilla y León* PGI), there are high differences in the tissues composition and meat quality between them (Díaz *et al.*, 2003) but for the sensory characteristics the effect is unclear. In some studies, they appear to affect some organoleptic scores (Jeremiah *et al.*, 1998, Martínez-Cerezo *et al.*, 2005), but in others they have no effect (Solomon *et al.*, 1980). These differences may be explained in some cases by differences in slaughter weights when breed effect was compared and because, at the same slaughter weight, breeds have different levels of maturity. Taking this into account the aim of this work was to determine if there are differences between Spanish and foreign breeds at different slaughter weights and to decide the optimum slaughter weight for the studied breeds.

Materials and Methods

Nineteen suckling lambs (animals that were raised exclusively on maternal milk from birth to slaughter at one month age), from three breeds: 30 Churra lambs, 30 Castellana lambs and 30 Assaf lambs were slaughtered according to standard commercial procedures in Magnus abattoir (Arcenillas, Zamora, Spain). Within breed, three slaughter live weights were considered: 9.5 kg, 10.5 and 11.5 kg. At 72 h post-mortem, carcasses were cut into two halves and the *Longissimus dorsi* from left half of carcasses was taken to laboratory

Sensory analysis was developed by a fifteen member trained panel using the quantitative descriptive analysis (Murray and Delahunty, 2001) on 2 cm-thick slices of *Longissimus lumborum* wrapped in aluminium foil and cooked on a pre-heated double hot plat grill at 200ºC until the internal temperature reached 70ºC. Then, they were cut into portions, each one uniform and free from obvious connective tissue and fat streaks. The sub-samples were immediately wrapped in aluminium foil, coded and kept at 60ºC. Meat was served following a randomised design for order and carry over effects. Within the descriptors generated by the panel, the following were defined as being the most discriminating and related to the consumer perception: brown, pink, aroma intensity, taste intensity, liver odour, hardness, juiciness, fat sensation, springiness and fibrous sensation. Reference scales (Severiano *et al.*, 2005) to assess the intensity of texture parameters were used. The anchored reference standards, terminology definitions and evaluation techniques were agreed upon by the assessors during training. For quantification of the intensity of each attribute, 9-point scales were employed, in which "1" referred to the minimum intensity and "9" the maximum for each parameter.

Statistical analysis was performed using the Statgraphic Plus program for Windows (Manugistics Inc. 1995). Outliers were detected by means of Box and Wisker plot, the significance of effects and their respective interactions were obtained using General Linear Model procedures. The LSD Fisher-test was employed to test for statistically significant differences between samples, differences were considered significant at the $P<0.05$ level.

Results and Discussion

The first step in sensory evaluation of suckling lamb meat was to define the parameters most significant in order to describe the samples. Initially, more than twenty parameters were generated by the panel but after several reductions, taking into consideration the frequency and intensity assigned to each parameter for the panel, the following parameters and definitions were selected. The colour parameters chosen were brown: characteristic brown colour on the cooked meat surface and pink: characteristic pink-red colour in the cooked meat centre. Among the texture parameters the panellist selected hardness: the effort to bite the sample, juiciness: initial juiciness perceived during the firsts 2-3 chews, fat sensation: fat perception in the mouth springiness: recover of the initial form after suppression of initial deformation force, fibrous sensation: sensation due to the low extensibility of meat fibre. Indeed some flavour and odour parameters were used as aroma intensity: sensations perceived by the olfactory organ, taste intensity: sensations perceived during tasting and liver odour: odour associated with the liver.

Regarding the effect of breed and slaughter weight on sensory characteristics of suckling lambs, results show (Table 1) that both factors affected the organoleptic properties studied. The outer colour was affected by the slaughter weight showing the lower values the medium weight animals (10.5 kg) for the three breeds and there were no statistical differences for inner colour.

All the texture parameters were affected by breed. The Assaf breed was considered harder and with higher springiness, lower fat sensation and slightly lower juiciness, while Churra breed had higher juiciness and Castellana produced lower fibrous sensation. The slaughter weight had influence on almost all the texture parameters. In Spanish breeds, animals of 10.5 kg of live weight had lower hardness and fibrous sensation, but for Assaf breed this slaughter weight presented the higher values for these two parameters. Juiciness tended to increase with slaughter weight for the three breeds but the fat sensation was higher at 10.5 kg live weight. Finally, springiness was the only texture parameter not affected by slaughter weight.

Table 1: Means of sensory parameters for the three breeds (B) (Castellana, Churra and Assaf) and the three live slaughter weights (W).

	Castellana			Churra			Assaf			Significance	
	9.5 kg	10,5 kg	11,5 kg	9.5 kg	10,5 kg	11,5 kg	9.5 kg	10,5 kg	11,5 kg	B	W
Brown	4.48^{b}	3.19^{a}	4.97^{b}	4.51^{b}	3.69^{a}	4.53^{b}	4.77^{b}	3.65^{a}	4.51^{b}	n.s.	***
Pink	3.89^{bc}	3.60^{ab}	3.26^{a}	$3.97^{b,c}$	3.78^{abc}	3.97^{bc}	3.48^{ab}	4.20^{c}	$3.62^{a,b}$	n.s.	n.s.
Hard.	4.14^{bcd}	3.68^{a}	3.74^{ab}	4.44^{de}	3.48^{a}	3.81^{abc}	4.29^{bcd}	4.82^{de}	4.29^{bcd}	***	*
Juiciness	4.18^{ab}	4.24^{ab}	4.45^{ab}	4.30^{ab}	4.47^{bc}	4.86^{c}	4.30^{ab}	3.97^{a}	4.46^{bc}	*	**
Fibrous	4.91^{cd}	3.82^{a}	4.61^{bc}	5.07^{cd}	4.29^{b}	5.06^{cd}	4.73^{c}	5.28^{d}	4.67^{c}	***	***
Fat sens	3.80^{bcd}	4.19^{d}	3.61^{bc}	3.84^{bcd}	3.97^{cd}	3.88^{cd}	3.38^{ab}	3.76^{bcd}	3.34^{ab}	***	**
Springiness	4.37^{b}	3.75^{a}	4.21^{ab}	4.37^{b}	4.46^{b}	4.59^{bc}	4.12^{ab}	4.43^{b}	4.44^{b}	***	n.s.
Taste int.	5.58^{d}	4.59^{a}	5.44^{d}	5.22^{bcd}	5.22^{bcd}	4.64^{ab}	5.27^{bcd}	4.76^{abc}	5.16^{abcd}	n.s.	**
Odour int	5.74^{cd}	4.56^{a}	5.82^{d}	5.23^{bc}	5.21^{bc}	5.31^{bcd}	5.35^{bcd}	4.93^{ab}	5.16^{b}	n.s.	***
Liver odour	3.81^{bcd}	3.81^{bcd}	3.17^{abc}	3.45^{abcd}	3.42^{abcd}	3.81^{cd}	3.24^{abcd}	2.98^{ab}	2.84^{a}	***	n.s.

$^{a-b}$ Different letter means differences statistically significant at α=0.05 within the row.
n.s.= no significant effects;* p< 0.1, ** p<0.05;*** p<0.01

Regarding olfactive and gustative parameters liver odour was influence only by breed, showing Assaf breed the lower value that is a positive characteristic of this breed. Finally, taste and odour intensity was affected by slaughter weight but not by breed showing, in general, the lower values for these two parameters in the suckling lambs with 10.5 kg live weight.

Conclusions

Spanish consumers prefer light colour, taste and odour for meat lamb differing from other countries, where lambs are slaughtered at heavier weights. On the other hand, tenderness, juiciness and a moderate fat sensation that contributes to sustained juiciness are desirable characteristics for meat. Taking into account that the Churra and Castellana breeds presented better sensory characteristics than the Assaf breed, the optimum slaughter weight for suckling lambs of three breeds was 10.5 kg.

Acknowledgments

Authors are grateful for financial support of Asovino Cooperative.

References

Díaz, M.T., Velasco, S., Pérez, C., Lauzurica, S., Huidobro, F. and Cañeque, V. (2003). Physico-chemical characteristics of carcass and meat Manchego-breed suckling lambs slaughtered at different weights. Meat Science, 65: 1085-1093.

Jeremiah, L.E., Tong, A.K.W. and Gibson, L.L. (1998). The influence of lamb chronological age, slaughter weight and gender. Flavor and texture profiles. Food Research International, 31: 227-241.

Martínez-Cerezo, S., Sañudo, C., Medel, I. and Olleta, J.L. (2005). Breed, slaughter weight and ageing time effects on sensory characteristics of lamb. Meat Science, 69: 571-578.

Murray, J.M. and Delahunty, C. M. (2001) Descriptive sensory analysis: past, present and future. Food Research International, 34: 461-471.

Severiano-Pérez, P., Vivar-Quintana, A.M. and Revilla, I. (2006). Determination and evaluation of the parameters affecting the choice of veal meat of the "Ternera de Aliste" quality appellation. Meat Science, in press.

Solomon, M.B., Kemp, J.D., Moody, W.G., Ely, D.G.and Fox, J.D. (1980) Effect of breed and slaughter weight on physical, chemical and organoleptic properties of lamb carcasses. Journal of Animal Science, 51: 1102-1107.

CHARACTERISTICS OF BEEF STEAKS FROM "CARNE DE LA SIERRA DE GUADARRAMA" QUALITY LABEL

M.T. Silva[1], S. Velasco*[1], O. Mozun[2], P.F. Rouco[3], J.F. Calahorra[3] and L. Ruiz[3]

[1.] *IMIDRA. Departamento de Agroalimentación, Comunidad de Madrid, 28800 Alcalá de Henares, Madrid, Spain,* [2.] *INVAC, Madrid, Spain.* [3.] *Departamento de Producción Animal, Facultad de Veterinaria, Universidad Complutense de Madrid, 28080 Madrid, Spain. Email: susana.velasco@madrid.org*

Keywords: beef, gender, texture, *longissimus dorsi,* ageing

Introduction

The term meat quality includes a sum of parameters that in short are: pH, colour, texture, water holding capacity, quantity and composition of fat, quantity and types of collagen and sensory attributes. Tenderness and colour are particularly important. While the first determines the consumer's satisfaction, the second causes a consumer's response and decision to buy or not to buy the product. Researchers have found a relationship between ultimate muscle pH (Watanabe *et al.*, 1995), muscle colour (Wulf *et al.*, 1996) and meat tenderness. Because of its importance in the carcass and its high commercial value, M. *longissimus dorsi* (thoracic part) (LD) is considered as an indicator of tenderness and is used in several studies like this. The objective of this study was to assess the effects of sex and ageing time on the tenderisation process using both instrumental and sensory tests.

Materials and Methods

Nineteen male and eleven female beef animals with the label de PGI "Carne de la Sierra de Guadarrama" (a Protected Geographical Indication from Madrid) were used in this study. These animals were entire males and females of 13-15 months old, with near 241 kg HCW for females and 399 kg for males and carcass classification values from U to R, and levels of fatness from 1 to 4. They were slaughtered in a commercial abattoir and processed according to the rules. Carcasses were weighed and classified by an official of the PGI and chilled under commercial conditions during three ageing intervals: 1 (from 3 to 6 days), 2 (from 7 to 9 days) and 3 (since 10 to 14 days). After the ageing period, M. *longissimus dorsi* (LD) was removed and sliced into 2 cm thick steaks that were vacuum packaged and randomly assigned to subsequent analyses. Meat quality parameters measured were: pH, colour, TPA (texture profile analysis), water-holding capacity, total collagen content and its solubility, and intramuscular fat quantity. To measure texture, we used the TPA (texture profile analysis) test on raw samples and steaks cut in pieces of 1 cm^2 with muscle fibres parallel to the longitudinal axis in order to apply the force transversal to the muscle fibres. Meat samples were analysed in a texturometer with a cylindrical 10 mm-diameter probe of ebonite and a deformation force of 75%. Sensory analysis was carried as described by Onega *et al.,* (2000) on samples for hardness, springiness and juiciness, complete with the parameters fat sensation (in mouth), flavour intensity, number of chewings until swallowing and overall. Scales used to assess intensity of every parameter were interval scales 10 cm long. Meat steaks were grilled to 80ºC (internal temperature) and cut into 1 cm^3 samples. 12-15 trained assessors tasted each muscle/animal/ageing combination. Each assessor tasted 6 portions of meat per session (2 replicates for each muscle). Statistical analysis of the data was made using the analysis of covariance of the GLM procedure of Statgraphics Plus (1994), with hot carcass weight used as covariate. The Duncan test was used to analyse differences between means.

Results and Discussion

The effects of gender and the three ageing intervals in LD muscle on pH, water-holding capacity, intramuscular fat, collagen composition, and colour parameters are shown in Table 1.

The percentage of intramuscular fat was higher in females than males and shows a significant effect (as expected) and lightly higher values for females than those found by Ruiz de Huidobro *et al.,* (2003) for the same PGI meat label. Total collagen content, insoluble collagen and solubility were highest in males although only with significant effects in solubility and insoluble collagen ($P<0.05$). Ageing had no significant effect on any collagen parameters. pH had quantitative interaction between studied factors. In water-holding capacity it was observed that early ageing (3 to 6 days) showed a higher liquid expulsion than half or long ageing. Ruiz de Huidobro *et al.*, (2003) found that WHC increased significantly during ageing one to six days in heifers, while in bulls there was no significant variation. Expelled water (%) in heifers' meat was reduced significantly during the first and third days but then remained unchanged until the sixth day. The factors studied had no influence on product colour.

Table 1: Arithmetic means and mean squares of error for pH, WHC, intramuscular fat, collagen and colour parameters of LD at different ageing periods and in both sexes.

	Gender (G)			Ageing (A)					
	Male	Female	Sig	1	2	3	Sig	G x A	MSE
pH	5.44	5.47	NS	5.44[a]	5.51[b]	5.41[a]	*	*	0.005
WHC (%)	14.33	12.71	NS	15.48[a]	12.66[b]	12.41[b]	*	NS	8.62
Intramuscular fat (%)	2.57	4.38	*	3.70	3.79	2.94	NS	NS	1.64
TC (mg/g muscle)	6.48	6.06	NS	6.09	6.34	6.38	NS	NS	2.02
IC (mg/g muscle)	5.42	4.45	*	4.59	5.19	5.03	NS	NS	1.71
Solubility (%)	22.67	10.00	*	15.71	16.36	16.94	NS	NS	70.29
L*	37.52	38.18	NS	37.93	36.76	38.86	NS	NS	5.83
a*	17.11	17.73	NS	15.72	17.31	19.22	NS	NS	22.67
b*	8.58	10.08	NS	8.41	9.11	10.48	NS	NS	4.52
C*	19.19	20.56	NS	17.94	19.62	22.06	NS	NS	24.81
H*	26.71	30.61	NS	28.69	27.76	29.53	NS	NS	14.98

Means in the same row not followed by a common letter differ significantly ($p \leq 0.05$). Sig: Significance; NS (non significance). MSE: mean square of the error. Ageing 1 (3 to 6 days), 2 (7 to 9 days) and 3 (since 10 to 14 days).

Effects of gender and the three ageing intervals on muscle on textural and sensorial parameters in LD are shown in Table 2. Aged meat during a half period interval (7 to 9 days) displayed the lowest hardness and springiness (by TPA) than the other intervals in both sex. Chewiness by compression measurement presented a qualitative interaction between gender and ageing interval. From early ageing (3 to 6 days) to medium ageing (7 to 9 days) chewiness decreased in both sex, but at long ageing (10 to 14 days) the female meat samples had higher values of chewiness than the male.
According to sensorial analysis, female's meat was less hard and springy than male's, due to its higher intramuscular fat. Also, females showed higher flavour intensity. By consequence, this meat was best when evaluated by assessors. No significant differences were found in juiciness, fat sensation and number of chewings. Sensorial parameters, including overall, were similar in all three ageing intervals.

Table 2: Arithmetic means and mean squares of the error of textural and sensorial parameters of LD at different ageing period and in both sex.

	Gender (G)			Ageing (A)					
	Male	Female	Sig	1	2	3	Sig	G x A	MSE
TPA Hardness (N)	175.77	207.45	NS	204.93[a]	158.14[b]	211.75[a]	*	NS	1981.13
TPA Springiness (g)	75.43	76.97	NS	76.27[ab]	74.71[a]	77.63[b]	*	NS	6.26
TPA Chewiness	3714.4	5085.7	*	4795.8[a]	3522.8[b]	4881.6[a]	*	*	896404
Hardness	4.84	3.45	*	4.11	4.04	4.31	NS	NS	0.80
Springiness	4.38	3.70	*	4.00	3.88	4.22	NS	NS	0.63
Juiciness	2.48	2.86	NS	2.74	2.60	2.67	NS	NS	0.59
Fat sensation	2.49	2.59	NS	2.65	2.58	2.39	NS	NS	0.26
Flavour intensity	4.12	4.96	*	4.64	4.32	4.65	NS	NS	0.46
Chewings (number)	22.27	20.43	NS	23.16	20.24	20.64	NS	NS	20.89
Overall	4.50	5.26	*	4.65	4.90	5.08	NS	NS	0.41

Means in the same row not followed by a common letter differ significantly ($p \leq 0.05$). Sig: Significance; NS (non significance). MSE: mean square of the error. Ageing 1 (3 to 6 days), 2 (7 to 9 days) and 3 (since 10 to 14 days).

Conclusions

There was a tendency for "better overall" with increase in ageing period but the effect was not significant. The samples of female meat were less hard and springy and had a higher flavour intensity than males, probably due to the higher intramuscular fatness in the former. However all meats studied were considered poor for hardness.

References

Ruiz de Huidobro F., Miguel E., Blázquez B. and Onega E. (2003). A comparison between two methods (Warner-Bratzler and texture profile analysis) for testing either raw meat or cooked meat. Meat Science, 69: 527-536.

Watanabe A., Daly C.C and Devine C.E (1995). The effects of the ultimate pH of meat on tenderness changes during ageing. Meat Science, 42: 67-68.

Wulf D.M., Morgan J.B., Tatum J.D. and Smith G.C. (1996). Effects of animal age, marbling score, capastatin activity, subprimal cut, calcium injection, and degree of doneness on the palatability of steaks from Limousin steers. Journal of Animal Science, 74, 569-576.

COMPARISON OF 3 PIG BREEDS IN TERMS OF MEAT QUALITY AND EVALUATION OF CUT OFF CRITERIA FOR ESTABLISHING A QUALITY PROGRAM

D. Mörlein*, G. Link and M. Wicke

Research Centre for Animal Production and Technology, University of Göttingen, Driverstrasse 22, D-49377 Vechta, Germany. Email: daniel.moerlein@agr.uni-goettingen.de

Keywords: pork, drip loss, PSE, IMF, electrical conductivity, *longissimus dorsi*

Introduction

Until today the meat industry is facing the variability of meat quality regardless of numerous concerns for its improvement, e.g. breeding programs, feeding regimes, slaughter handling etc. Despite those efforts, post slaughter there is still variation in meat quality. Currently, polarisation between discount and premium segments has been observed. Thus, in order to satisfy the premium consumer's need for constant & guaranteed quality it is necessary to select carcasses with quality beyond the standard. This research focuses on evaluating the status quo of carcass traits, meat composition and sensory properties of pigs currently found at the market in Germany. Simulating a meat quality program, we also aimed to evaluate property thresholds that may guarantee better eating quality bearing in mind the percentage of pigs meeting this program's conditions.

Materials and Methods

This research included in total 606 pigs originating from 3 commercially available breeds (both castrated male & female). The pigs were slaughtered in a commercial abattoir within regular conditions (electric stunning) at 6 dates distributed within one year. Thus, we realised a nearly fully balanced design (6 dates * 3 breeds * 2 sexes). The carcasses for the research were randomly selected except for design conditions. Slaughter weight and carcass traits were provided by the company using SFK Fat-O-Meater. Electrical conductivity (EC) at *longissimus* muscle was measured at $2^{nd}/3^{rd}$ last rib 24 hours post mortem with Matthäus LF-star. Drip loss of samples taken 24 hours post mortem was measured according to the bag-method (Honikel, 1987) after 48 hours storage time at 4°C. From an adjacent cutlet intramuscular fat was extracted (petrol ether) with prior HCl pre-treatment. Another sample was cooked on a contact grill to 73°C core temperature to further measure cooking loss and shear force. Sensory properties of 328 samples were determined by a 6 member instructed panel. Statistical analyses were performed with SAS PROC MIXED including fixed effects (breed, sex, date of slaughter) and all interactions.

Results and Discussion

In Table 1 least squares means of selected parameters are presented with relation to breed and sex, respectively.

Table 1: LS means of selected traits dependent on breed and sex, respectively (n = 606).

	breed			sex	
trait	Breed G	Breed H	Breed S	castrates	females
Slaughter weight [kg]	95.18^{b}	93.51^{a}	95.61^{b}	94.67	94.87
Meat percentage [%]	58.20^{b}	57.32^{a}	57.41^{a}	56.87^{a}	58.43^{b}
Backfat thickness [mm]	15.18	14.99	14.81	15.83^{b}	14.15^{a}
Muscle thickness [mm]	65.56^{b}	62.58^{a}	63.42^{a}	63.55	64.16
LD muscle area [cm²]	54.80^{c}	51.15^{a}	52.34^{b}	51.35^{a}	54.17^{b}
Intramuscular fat [%]	1.03^{a}	1.20^{b}	1.39^{c}	1.32^{a}	1.10^{b}
E. conductivity [mS/cm]	6.30^{b}	5.86^{a}	6.45^{b}	6.08	6.33
Drip loss [%]	6.38^{b}	6.70^{b}	5.92^{a}	6.43	6.24
Cooking loss [%]	29.79^{ab}	30.15^{b}	29.25^{a}	29.91	29.55
Shear force [N]	47.84^{b}	44.35^{a}	45.52^{a}	45.98	45.83

^{a,b,c} different superscripts indicate significant differences between breeds and sexes, respectively ($\alpha < 0.05$)

The IMF content of most of the samples did not reach the level of 1.5 – 2.0 percent as is usually recommended with respect to optimal eating quality. Clearly, breed differences exist with breed S outperforming the other breeds under investigation. This was most probably due to the 25% Duroc in that breed. Besides, that breed shows significantly less drip loss. Interestingly, genotype S showed the highest mean EC values despite the lowest drip loss. Contrarily, breed G performs best in terms of meat percentage and muscle area. At the same time IMF was lowest compared to the other breeds.

In Figure 1 relative frequency distribution of Electric conductivity (EC) values (left) and mean drip loss (48 hrs.) dependent on EC values (right) are depicted for all samples regardless the breed.

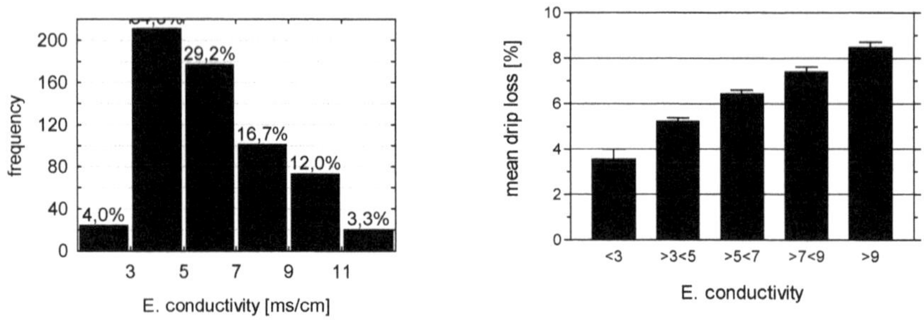

Figure1: Relative frequencies of Electric conductivity (EC) values (left) and mean drip loss dependent on conductivity class (right; LS means ± standard error of mean); n = 606.

Applying EC values above 7 and 9 ms/cm as indicators for pale, soft, exudative (= PSE) conditions of the meat, we must state that 32.0 and 15.3 percent, respectively of the samples under investigation were PSE like. From a recent monitoring in 6 commercial slaughter facilities in German that covered more than 20,000 carcasses (Altmann *et al.*, 2005), the authors applied the more strict EC values greater than 7 ms/cm. Thus, they similarly concluded up to 27% of the carcasses show PSE - in accordance with Warriss *et al.* (1998) for carcasses in the EU. The relationship between drip loss and EC can also be seen from Figure 1 (right). Above breeds, the mean drip loss significantly increased between EC classes.

Table 2 shows the results, *i.e.* yield of carcasses in % and corresponding traits, after application of 2 cut off criteria (meat percentage and conductivity) in order to select carcasses with favoured traits, especially less drip and more IMF. Among all breeds, a slight increase in IMF and remarkable decrease of drip was noted as well as an enormous decrease of the yield: more than 50% were cut off. This is partly due to breed S because of the higher EC values. Only 37 % of breed S carcasses meet the criteria. However, with respect to the favoured traits this breed performs best showing least drip and mean IMF around 1.50%. Nevertheless a big variation both of drip loss and instrumental tenderness remained (data not shown).

Table 2: Application of EC and lean meat percentage cut off values to discriminate carcasses.

	original data,	Quality program criteria: lean meat > 54 < 59 %, EC < 7 ms/cm			
trait	no cut off	All carcasses	Breed G	Breed H	Breed S
program carcasses [%]	100	43.6	41.1[+]	51.0[+]	37.1[+]
Slaughter weight [kg]	94.91	94.12	95.2^{b}	92.7^{a}	94.9^{b}
Meat percentage [%]	57.6	56.82	57.0^{b}	56.8^{ab}	56.5^{a}
E. conductivity [mS/cm]	6.24	4.82	5.04^{b}	4.87^{b}	4.48^{a}
IMF [%]	1.21	1.24	1.05^{a}	1.22^{b}	1.50^{c}
Drip loss [%]	6.37	5.49	5.47	5.68	5.19
Shear force [N]	45.92	44.75	46.4	44.2	43.8
Liking score*	52.15	53.03	51.3^{a}	51.6^{a}	56.3^{b}

[+] relative proportion within breed.

* sensory liking score: 0 = worst … 100 = best.

[a,b,c] different superscripts indicate significant differences between breeds ($\alpha < 0.05$).

Conclusion

The strategy of choice for implementation of a meat quality program shall include both the selection of a certain breed and discrimination of "premium carcasses" at slaughter. The data presented here suggest the application of EC measurement since it is more reliable and less dangerous (no glass electrodes that can break) compared with pH measurements. The threshold carcasses with lower drip loss and higher IMF content were selected.

References

Altmann, M. *et al.* (2005). Incidence of PSE condition in market hogs - Monitoring results for various abattoirs in Germany. Fleischwirtschaft, 85: 101-104.

Honikel, K. O. (1987). The water binding of meat. Fleischwirtschaft, 67: 1098-1102.

Warriss, P. D. *et al.* (1998). An analysis of data relating to pig carcass quality and indices of stress collected in the European Union. Meat Science, 49: 137-144.

COMPARISON OF INTRAMUSCULAR FAT, DRIP LOSS, CIE A* VALUE, FATTY ACIDS AND AROMA PATTERN IN PORK FROM THREE DIFFERENT BREEDS

S.K. Lee*, S.M. Kang and C.G. Kang

Department of Animal Products and Food Science, Kangwon National University, Chunchon, Kangwon Province, 200-701, Republic of Korea. Email: skilee@kangwon.ac.kr

Keywords: pork, breed, CIE a*, fatty acids, aroma pattern

Introduction

Korean native black pigs originated from wild small species among black pigs from Manturia and settled in the Korean Peninsula almost 2,000 years ago as domesticated animals. Kang *et al.* (2005) suggested Korean native black pork had greater water-holding capacity, colour stability and sensory quality during refrigerated storage. There are also wild pigs (*Sus scrofa creanus*) that are native pigs in Korea. Because of the importance of original gene sources, many Korean researchers are actively studying native livestock as well as wild animals in several fields (Choi *et al.*, 2005; Chung *et al.*, 2001; Kim *et al.*, 2001; Lee *et al.*, 2005). Therefore, this study was carried out to compare the intramuscular fat, CIE a* value, fatty acids and aroma pattern in *M. longissimus* from pigs of 3 different breeds such as wild pigs, Korean native black pigs and modern genotype pigs.

Materials and Methods

M. longissimus from 5 wild pigs (*Sus scrofa creanus*, WP) of 111kg, 5 Korean native black barrows (KNP) of 64kg and 5 modern genotype barrows (Landrace×Yorkshire, MGP) of 114kg were utilised for quality measurements. The wild pigs consisted of 1 boar of 100%, 1 barrow and 2 gilts of 50% (100% wild boar×100% modern genotype gilt) and 1gilt of 25% (50% wild boar×50% modern genotype gilt). After slaughter, carcasses were chilled at 2°C for 24 hr and then deboned. The 3cm thick samples were packaged in low density polyethylene zipper bags and stored at 2±0.3°C refrigerator (CRF1021D, Samsung, Korea) for 12 days. Intramuscular fat (IMF, %) and drip loss (%) were performed as a described by AOAC (1995) and Honikel (1998), respectively. The CIE a* value (redness) was measured using a colour difference meter (CR-400, Konica Minolta Sensing Inc., Japan). The discrimination of meaty aroma pattern by principal component analysis (PCA) was analyzed using an electronic nose (FOX 3000, Alpha MOS, France). The fatty acid methyl esters (AOAC, 1995) were analyzed using the Agilent 6890N GC equipped with a HP-Innowax column (30 m length×0.32 mm id×0.25 μm film thickness). Data was analysed using the General Linear Model procedure of the SAS (1999) program.

Results and Discussion

IMF was significantly higher in KNP than in WP and MGP ($P<0.05$) (Figure 1). Drip loss was significantly lower in MGP than in the others during storage ($P<0.05$) (Figure 2). The CIE a* value (redness) was significantly higher in KNP than in the others during storage ($P<0.001$) (Figure 3). The PCA by electronic nose of *M. longissimus* from WP, KNP and MGP is presented in Figure 4. Aroma pattern of all breeds was perfectly discriminated by electronic nose. Therefore, we could discriminate between pork of 3 breeds by using PCA method from electronic nose. The fatty acid composition of *M. longissimus* from WP, KNP and MGP is presented in Table 1. SFA was significantly lower in WP than in KNP ($P<0.05$) but UFA and PUFA n3 were significantly higher in WP than in KNP ($P<0.05$). MUFA and MUFA/SFA ratio were significantly higher in WP and MGP than in KNP ($P<0.05$, $P<0.01$, respectively) and UFA/SFA ratio was significantly highest in WP ($P<0.05$). However, there were no significant differences in PUFA, PUFA/SFA ratio and PUFA n6 between the breeds. PUFA n6/n3 ratio was significantly lowest in MGP and significantly lower in WP than in KNP ($P<0.0001$).

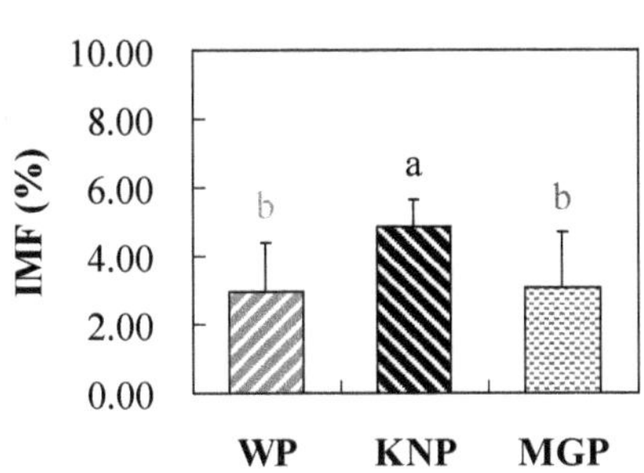

Figure 1: Comparison of intramuscular fat of *M. longissimus* from wild pigs, Korean native black pigs and modern genotype pigs.

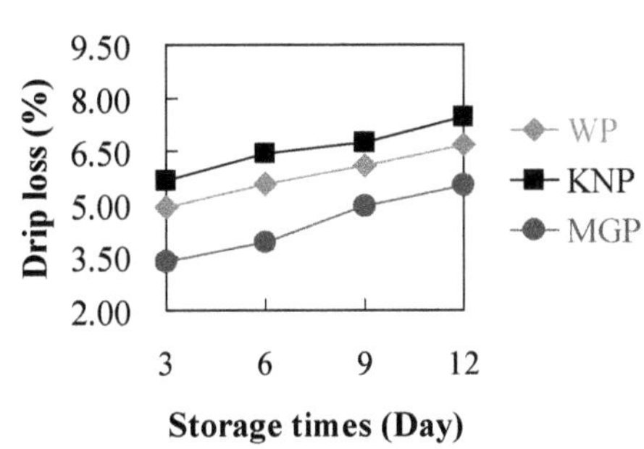

Figure 2: Comparison of drip loss of *M. longissimus* from wild pigs, Korean native black pigs and modern genotype pigs during refrigerated storage.

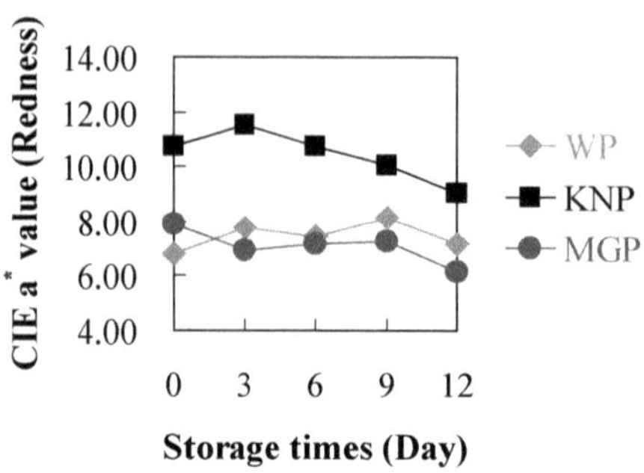

Figure 3: Comparison of CIE a* value (redness) of *M. longissimus* from wild pigs, Korean native black pigs and modern genotype pigs during refrigerated storage.

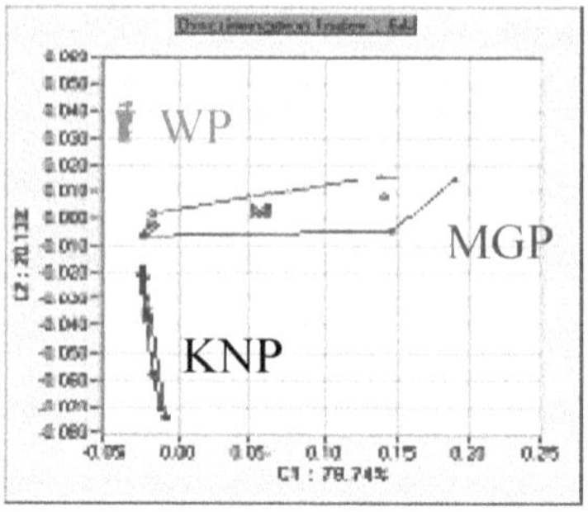

Figure 4: Comparison of PCA by electronic nose of *M. longissimus* from wild pigs, Korean native black pigs and modern genotype pigs.

Table 1: Comparison of fatty acid composition (%) of *M. longissimus* from wild pigs, Korean native black pigs and modern genotype pigs.

Items	Breed WP	KNP	MGP	SEM[A]	P^B
Saturated fatty acid (SFA)	38.44[b]	42.05[a]	40.58[ab]	0.491	*
Unsaturated fatty acid (UFA)	61.56[a]	57.95[b]	59.42[ab]	0.494	*
Total	100.00	100.00	100.00		
Monounsaturated fatty acid (MUFA)	48.43[a]	46.69[b]	48.52[a]	0.326	*
Polyunsaturated fatty acid (PUFA)	13.13	11.26	10.90	0.529	ns
UFA/SFA	1.60[a]	1.38[b]	1.46[b]	0.029	*
MUFA/SFA	1.26[a]	1.11[b]	1.20[a]	0.017	**
PUFA/SFA	0.34	0.27	0.27	0.016	ns
PUFA n6	12.33	10.65	10.14	0.499	ns
PUFA n3	0.80[a]	0.60[b]	0.76[ab]	0.036	*
PUFA n6/n3	15.41[b]	17.75[a]	13.34[c]	0.559	****

[a,b,c] Means in the same row with different superscripts are significantly different.

[A] Standard error of the mean.

[B] ns, not significant; *, $P<0.05$; **, $P<0.01$; ***, $P<0.001$; ****, $P<0.0001$.

Conclusions

KNP had the highest IMF and the reddest colour value during refrigerated storage. MGP had the best water-holding capacity during refrigerated storage and wild pork had the highest UFA/SFA ratio. The pork from 3 different breeds could be discriminated by different aroma patterns using an electronic nose.

References

AOAC. (1995). Official methods of analysis. 16th ed, Association of Official Analytical Chemists, Washington, DC, USA.

Choi, Y. S., Park, B. Y., Lee, J. M. and Lee, S. K. (2005). Comparison of carcass and meat quality characteristics between Korean native black pigs and commercial crossbred pigs. Korean J. Food Sci. Ani. Resour. 25: 322-327.

Chung, E. R., Kim, W. T., Kim, Y. S., Lee, J. K. and Han, S. K. (2001). Genetic diversity and breed identification of Korean native pig using AFLP markers. J. Anim. Sci. Technol. (Kor.) 43: 777-788.

Honikel, K. O. (1998). References methods for assessment of physical characteristics of meat. Meat Sci. 49: 447-457.

Kang, S. M., Chae, B. J., Kim, Y. S., Kang, C. G., and Lee, S. K. (2005). Comparison of the meat quality of Korean native black pork and modern genotype pork fed by different dietary lysine levels during refrigerated after thawing. Annals of Animal Resources 16: 49-59.

Lee, S. K., Ju, M. K., Kim, Y. S., Kang, S. M. and Choi, Y. S. (2005). Quality comparison between Korean native black ground pork and modern genotype pork during refrigerated storage. Korean J. Food Sci. Ani. Resour. 25: 71-77.

SAS. (1999) SAS User's Guide. SAS Institute, Inc., NC, USA.

EFFECT ON MEAT QUALITY AND STORAGE CHARACTERISTICS DEPENDING ON PSE STATUS OF BROILERS BREAST MEAT

H.S. Chae*, N.K. Singh, Y.M. Yoo, C.N. Ahn, S.G. Jong, J.S. Ham and D.H. Kim

Nationa Liivestock Research Institute Rural Development Administration, 77 Chooksan-gil, Kwonsun-gu, Suwon,441706, Korea. Email: hs6226@rda.go.kr

Keywords: broilers breast meat, PSE, storage characteristics, meat quality

Introduction

During the recent past, increases in the consumption of parted chicken have been noticed in the Korean consumption trade. Due to the increase in company franchises, a simultaneous increase has also been noticed in the consumption of parted chickens. However, there is growing concern over the PSE status of parted chicken among consumers because of the appearance of chicken parts. Certain advanced countries have been practicing regulations for parted chickens. United States follows USDA, 1998 (3 step system) for the quality grading of poultry parts whereas Japan and UK (MAFF, 1999) follow a 2 step system for the same. Whole chicken has basically 3 large parts i.e. Breast, legs and wings. Consumers who preferred drum sticks previously have recently started to like breast parts because of its low fat percentage. The present study focuses on the PSE status of broiler's breast to investigate the effects of different degrees of PSE on its quality and storage characteristics.

Materials and Methods

Samples were collected from the large scale slaughter house (70,000 birds/day). Thirty five-day raised broiler chickens were considered for the study. Prepared chicken averaging 1251-1350g was selected for the study. Breast was separated and skin was removed. Thereafter, analysis was done on the basis of pH, meat color, storage characteristics (until 3 days). Thiobarbituric acid reactive substances (TBARS), volatile base nitrogen (VBN), shear force, cooking loss and water holding capacity were also measured.

Results and Discussion

Breast PSE examination revealed that PSE meat either slight or serious had a low pH range below 6.0 in 80-100% of the samples compared to normal PSE (Table 1). Similar findings have also been reported for PSE meat (Petracci *et al.*, 2004). In 70 to 100% of samples either slight or serious PSE was observed to have lightness more than 67 compared to normal. Only 20% of the slight PSE was below 64 whereas none of the sample in serious PSE was either in the range of 64 to 67 or less than 64. (Table 2). Almost 100% of the samples showed the redness range of 3 to 6 (data not shown). Fifty percent of the slight PSE was less than 3, whereas 20% was in the range of 3-6, another 20% was between 6-9 and only 10% was above 9. The average redness for slight PSE was 4.23±4.01. 73.3% was found to be below 3 and 3.3% was in the range of 3-6. Ten per cent was between 6-9 and 13.3% of samples were observed to be above 9. The average redness calculated for normal PSE was 3.54±3.53. The average yellowness was 12.68±4.09 (serious), 11.93±3.70 (slight) and 8.39±3.01 (normal). The average yellowness followed an increasing pattern with the increase in the degree of PSE. TBARS values showed increasing trend with the degree of PSE during storage. The values were found to be significantly different ($P<0.05$) for slight and serious PSE's compared to normal (Table 3). VBN values for slight and serious PSEs were significantly different on day 1 compared to normal (data not shown).

Table 1: Relationship between PSE and pH of meat breast; (Unit: %).

Items	Average pH	pH range		
		< 6.0	6.0 ~ 6.2	>6.2
Normal	6.10±0.13	23.3	53.3	23.3
PSE(Slight)	5.95±0.06	80.0	20.0	-
PSE(Serious)	5.87±0.12	100	-	-

* Survey numbers: 43 heads.

The cooking loss % during storage showed an increasing trend with increase in the degree of PSE. Slight and serious PSEs were significantly different ($p<0.05$) compared to the normal. A decreasing trend was also observed in shear force during storage (Table 4). However, there was no significant difference observed. No significant variation was noticed in terms of water holding capacity.

Table 2: Relationship between PSE and lightness (L*) of breast meat; (Unit: %).

Items	Avar. L*	CIE(L*) range			
		< 64	64 ~ 67	>67 ~ 70	>70
Normal	65.04±4.08	36.7	30.0	23.3	10.0
PSE(Slight)	67.93±4.43	20.0	10.0	50.0	20.0
PSE(Serious)	69.29±2.25	-	-	33.3	66.6

Table 3: Relationship between PSE and TBARS of breast meat during storage; (Unit: mg MA/kg).

Items	1day	3day
Normal	0.032 ± 0.006^{b}	0.052 ± 0.007^{b}
PSE(Slight)	0.023 ± 0.002^{c}	0.065 ± 0.006^{ab}
PSE(Serious)	0.041 ± 0.004^{a}	0.075 ± 0.011^{a}

[ab]Means±SE in the same row with different superscripts differ significantly. ($p<0.05$)

Table 4: Physical characteristics of breast meat during storage.

Items	Cooking loss (%)		Shear force(kg/0.5inch^2)		Water holding capacity (%)	
	1day	3day	1day	3day	1day	3day
Normal	15.54 ± 2.06^{c}	18.60 ± 1.12^{b}	1.73 ± 0.19^{a}	1.66 ± 0.44^{a}	61.17 ± 1.94^{b}	61.44 ± 2.14^{b}
PSE(Slight)	18.19 ± 1.02^{b}	20.31 ± 2.12^{b}	1.68 ± 0.44^{a}	1.35 ± 0.46^{a}	65.18 ± 0.72^{a}	65.14 ± 1.01^{a}
PSE(Serious)	23.49 ± 1.08^{a}	24.04 ± 0.95^{a}	1.38 ± 0.18^{a}	1.21 ± 0.38^{a}	64.42 ± 1.43^{a}	60.26 ± 0.84^{b}

[abc]Means±SE in the same column with different superscripts differ significantly. ($p<0.05$).

Conclusion

Increase in the degree of PSE of broiler breast meat during storage causes a fall in pH and shear force and an increase in lightness, cooking loss, TBARS and VBN. Breast meat PSE status confirms the actual pale soft and exudative nature of meat.

References

Petracci, M.; Bianchi,M.; Cavani,C. (2004). Color Variation and characteristics of broiler breast meat during processing in Italy. Poultry Science, 83(12): 2086-92

UK MAFF (1999). Enforcement guide to EC Poultrymeat marketing standards regulations.

USDA (1998). United States Classes, Standards and Grades for poultry.

INTRAMUSCULAR COLLAGEN CHARACTERISTICS OF DOUBLE-MUSCLED AND NORMAL JAPANESE SHORTHORN STEERS

N. Shiba*[1, 2], A. Watanabe[1], T. Suzuki[3], E. Tsuneishi[4], M. Matsuzaki[4], M. Higuchi[1] and H. Iwamoto[5]

[1.] National Agricultural Research Center for Tohoku Region, Morioka 020-0198, Japan, [2.] Graduate School of Bioresource and Bioenvironmental Science, Kyushu University, Fukuoka 812-8581, Japan, [3.] Animal Industry Research Institute, Iwate Agricultural Research Center, Takizawa 020-0173 Japan, [4.] National Agricultural Research Center for Kyushu Okinawa Region, Koshi 861-1192, Japan, [5.] Faculty of Agriculture, Graduate School, Kyushu University, Fukuoka 812-8581, Japan. E-mail: nshiba@affrc.go.jp

Keywords: Japanese Shorthorn, double-muscled, collagen solubility, cross-link

Introduction

Cattle displaying a heavy or extreme muscular appearance express what is known as 'double muscling' or muscular hypertrophy. This is a genetically controlled characteristic, and the manifestation of double muscling varies with the breed depending on the genetic background, environment, sex and stage of maturity. Ngapo *et al.*, (2002) indicated that double-muscled cattle have lower collagen content in muscle and higher pyridinoline cross-links in collagen than normal cattle. Mature collagen has more pyridinoline cross-links and should be less heat-labile with greater mechanical strength (Bailey, 1984). It is generally accepted that collagen is a major component in connective tissues of muscle, and the texture of meat is controlled by its quantity and quality. The Japanese Shorthorn is a breed of Japanese beef cattle and is bred mainly in the northern area of the Tohoku region of Japan. The meat has less marbling, but the number of Japanese consumers who prefer less marbling or lean meat has been increasing recently. The efficiency of lean meat production would be improved, when the merit of double-muscling could be used in the production of Japanese Shorthorn beef; but the quality of double-muscled Japanese Shorthorn beef has not been well studied. The aim of the present study was to compare the characteristics of intramuscular collagen in double-muscled Japanese Shorthorn beef cattle with normal cattle in order to evaluate the meat quality of double-muscled beef.

Materials and Methods

Three normal (22 months of age, average body weight 695.2kg) and 3 double-muscled (22 months of age, average body weight 686.8kg) castrated male Japanese Shorthorn steers were used. Steers were slaughtered and samples of *m. longissimus dorsi* (LD muscle) and *m. biceps femoris* (BF muscle) were excised. Some muscle pieces were used to prepare intramuscular connective tissue (IMCT) and the remaining samples were frozen at -20°C until analysis. IMCT samples were prepared by the method of Nishimura *et al.* (1998). The shear force value of IMCT preparations embedded in acrylamide gels was measured using a rheometer (RT-3010D-CW, RHEOTECH, Tokyo, Japan) with a shearing wire and a crosshead speed of 30cm/minute. For chemical analysis, meat samples were thawed at 4°C and trimmed of external fat and epimysial connective tissue. The muscle samples were minced and prepared to determine moisture, fat and protein contents (%), total and soluble collagen contents (mg/g of muscle) and pyridinoline contents (moles/mole of collagen). Soluble collagen was extracted from samples by heating for 70 min at 77°C in 0.25-strength Ringer's solution, using a water bath (Hill, 1966). The supernatant samples and minced meat samples used to measure the total amount of collagen in each muscle were hydrolyzed in 6 N HCl for 12 hours at 121°C. The hydroxyproline concentration of each hydrolyzate was determined by a colorimetric method (Bergman and Loxley, 1963). The collagen content was calculated using conversion factors of 7.25 (total collagen) and 7.52 (soluble collagen), respectively (Cross *et al.*, 1973). Collagen solubility was expressed as the percentage soluble collagen of the total collagen content. Pyridinoline was purified by ion-exchanged cellulose according to the method of Skinner (1982), and was determined by the liquid chromatographic procedure of Arakawa *et al.*, (1992). Pyridinoline concentration was determined by monitoring fluorescence intensity using excitation and emission wavelengths of 295 and 395 nm, respectively and compared with the intensity of an external standard purchased from Wako (Osaka, Japan).

Results and Discussion

In LD and BF muscle, moisture and protein contents were higher while fat content was lower in double-muscled cattle than in normal cattle and double-muscled cattle had a lower content of total, soluble and insoluble collagen than normal cattle. But collagen solubility and pyridinoline in collagen did not show clear differences (Table 1). These results indicate that the muscles of double-muscled cattle had lower intramuscular collagen content from the dilution of collagen with other proteins and moisture, but the solubility and pyridinoline cross-links of collagen were not different. The shear force value of IMCT preparations from double–muscled cattle was lower than that from normal cattle (Figure 1). This may be attributed to different collagen proportions.

Table 1: Chemical composition and collagen characteristics of muscles of normal and double-muscled cattle.

	M. longissimus dorsi						*M. biceps femoris*		
Cattle	Normal			Double-muscled			Normal	Double-muscled	
	1	2	3	1	2	3	3	1	2
Moisture (%)	65.1	66.7	67.2	72.0	71.1	72.3	69.1	73.0	72.8
Fat (%)	14.2	11.3	10.9	4.6	4.3	4.3	8.5	2.1	3.7
Protein (%)	19.6	20.9	20.6	21.6	22.4	22.1	20.7	22.6	22.1
Total collagen (mg/g)	3.26	3.10	2.77	2.46	1.86	3.17	5.21	3.32	3.39
Soluble collagen (mg/g)	0.39	0.54	0.43	0.27	0.22	0.51	0.45	0.41	0.38
Insoluble collagen (mg/g)	2.86	2.56	2.34	2.19	1.64	2.66	4.76	2.91	3.01
Collagen solubility (%)	12.1	17.5	15.7	11.0	12.0	16.0	8.6	12.4	11.3
Pyridinoline (moles/mole of collagen)	0.16	0.14	0.16	0.11	0.14	0.17	0.16	0.16	0.15

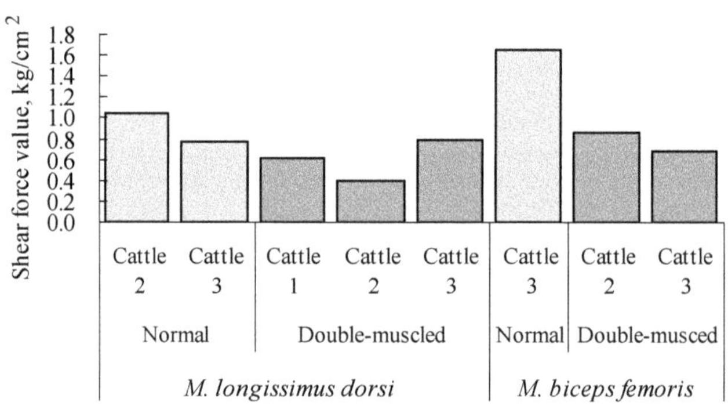

Figure 1: The shear force value of IMCT preparations embedded in acrylamide gels of muscles of normal and double-muscled cattle.

Conclusions

Muscle from double-muscled Japanese Shorthorn beef cattle had a high protein and low collagen content and low shear force value of IMCT, but collagen solubility and pyridinoline cross-links did not differ from normal cattle. From these results, it seems that double-muscling of Japanese Shorthorn is efficient to produce tender beef.

References

Arakawa, N., Kim, M. and Otsuka, M. (1992) An improved high-performance liquid chromatographic assay for the determination of pyridinoline in connective tissues. Journal of Nutritional Science Vitaminology, 38: 375-380.

Bailey, A.J. (1984). Recent Advances in the Chemistry of Meat. Special Publications No. 47, 22-40. The Royal Society of Chemistry, London.

Bergman, I. and Loxley, R. (1963). Two improved and simplified methods for the spectrophotometric determination of hydroxyproline. Analytical Chemistry, 35: 1961-1965.

Hill, F. (1966). The solubility of intramuscular collagen in meat animals of various ages. Journal of Food Science, 31: 161-166.

Ngapo, T.N., Berge, P., Culioli, J. and De Smet, S. (2002) Perimysial collagen crosslinking in Belgian Blue double-muscled cattle. Food Chemistry, 77: 15-26.

Nishimura, T., Liu, A., Hattori, A. and Takahashi, K. (1998). Changes in mechanical strength of intramuscular connective tissue during postmortem aging beef. Journal of Animal science, 76: 528-532.

Skinner, S.J.M. (1982) Rapid method for the purification of the elastin cross-links, desmosine and isodesmosine. Journal of Chromatography, 299: 200-204.

ESTABLISHMENT OF SUBCUTANEOUS PREADIPOCYTE CLONAL LINE FROM CHINESE MEISHAN PIG

I. Nakajima*, M. Oe, S. Muroya, N. Kanematsu, and K. Chikuni

National Institute of Livestock and Grassland Science, Norindanchi P.O.Box 5, Tsukuba, Ibaraki 305-0901, Japan. Email: ikuyon@affrc.go.jp

Keywords: pigs, backfat, cell culture, preadipocyte, differentiation

Introduction

Chinese pigs such as the Meishan breed provide a particularly interesting genetic model to study. The Meishan can be characterised by its wrinkled face and skin (Figure 1a,1b), and is highly known as one of the most prolific breeds of pig where females begin puberty at only 60 days of age, and have a large litter size of 15-16 piglets. Meishan pigs also have the characteristics of slow growth and higher fat deposits than conventional breeds of pig (Legault, 1985; White *et al.*, 1995). Also, their backfat can become quite thick (Figure 1c).

Since subcutaneous adipose tissue in pigs represents a major source of both cost inefficiency and consumer concerns, numerous investigations have been carried out to regulate the carcass fat for the meat quality traits through nutritional controls. On the other hand, approach from the comparative study among greatly different genotypes such as generating crosses between Meishan and Western pigs, have been set up to identify the quantitative trait loci associated with backfat thickness (Rohrer, 2000; Sato *et al.*, 2003). However, there are some difficulties to monitor adipose development *in vivo* studies, due to the complicated interactions of all kinds of neural and hormonal signals involved in adipocyte metabolism. Therefore, investigators have established *in vitro* cell culture systems as useful tools for studying adipose development.

We have already established a preadipocyte clonal line from Western crossbred subcutaneous tissue (PSPA) (Nakajima *et al.*, 2003) for the study of pig fat. The present study was designed to establish another pig preadipocyte clonal line from the Meishan pigs (MSPA).

Figure 1: (a) Adult Meishan pig. (b) Meishan piglets. (c) Meishan meat. Note the thickness of the subcutaneous adipose tissue.

Materials and Methods

Isolation and cloning of Meishan pig preadipocytes.

Meishan pig preadipocytes were obtained according to Nakajima *et al.* (2003). Dorsal subcutaneous tissue was dissected from fetuses (85 days gestation) that were produced at the National Institute of Livestock and Grassland Science. Briefly, the tissue was minced and digested in Dulbecco's modified Eagle's medium (DMEM, 1 g/l glucose) containing 1 mg/ml collagenase. The digestion proceeded for 1 hr in a sterile plastic tube at 37°C for 30 min with agitation. After filtration through a 75-mm mesh and centrifugation for 7 min at 300 g, the supernatant was discarded. The pellet fraction, composed primarily of stromal-vascular cells, was resuspended and seeded into a tissue culture flask in growth medium, DMEM supplemented with 10% fetal bovine serum (FBS), 1,000U/ml penicillin, and 1 mg/ml streptomycin. The cell monolayer was tripsinized, and cells were then cloned by a limiting dilution. Two to four weeks later, the different clones were grown separately, and MSPA clone was selected upon the basis of its high growth rate and frequency of adipose conversion.

Differentiation of preadipocyte to adipocytes.

In order to produce mature adipocytes, MSPA cells were plated at 2.1×10^4 cells/cm^2 and grown for 3 days to obtain confluency. After reaching confluence (0 day), adipose conversion was induced in high glucose (4.5 g/l) DMEM containing 10% FBS in addition to various combinations of 5 μg/ml insulin, 0.25 μM dexamethasone, 33 μM biotin, 17 μM pantothenate, and 5 mM octanoate. The medium was changed every other day and the cells were allowed to differentiate for 10 more days. Control cultures were grown after confluency in non-adipogenic growth medium as the preadipose state.

Triglyceride assay.

Triglyceride (TG) in the cell lysate was extracted with chloroform-methanol and quantified enzymatically using a Triglyceride G Test Wako Kit.

Results and Discussion

Exponentially growing MSPA cells derived from Meishan subcutaneous tissue resembled fibroblasts in their typical spindle shape (Figure 2). The cells were passed every 4 days in non-adipogenic growth medium at a constant density of inoculation of 1×10^4 cells/cm^2. As shown in Figure 3, the ratio N/N_0 (N, cell number after 4 days over N_0, the initial number of inoculated cells) of 15-20th passage was nearly 1 which means the cells were quiescent. This phase was overcome after 20 passages, and then constantly, the cell numbers were doubly increased after 4 days culture ($N/N_0>2$). After achieving confluence and in the presence of insulin, dexamethasone, biotin, pantothenate, and octanoate, MSPA cells exhibited large amounts of TG at 10 days (data not shown). While confluent cells were cultured in growth medium, they did not accumulate any intracellular lipid droplets. Furthermore, adipose conversion was not induced upon exposure of MSPA cells to a standard hormonal cocktail of 1-methyl-3-isobutylxanthine, dexamethasone, insulin and fetal bovine serum, which is the medium to differentiate one of the most popular preadipocyte cell lines, the mouse 3T3-L1 cells (Student *et al.*, 1980). This reaction of MSPA cells was consistent with the result of PSPA cells, which strongly supported the idea that there is species specificity in response to adipocyte inducers between pig and mouse.

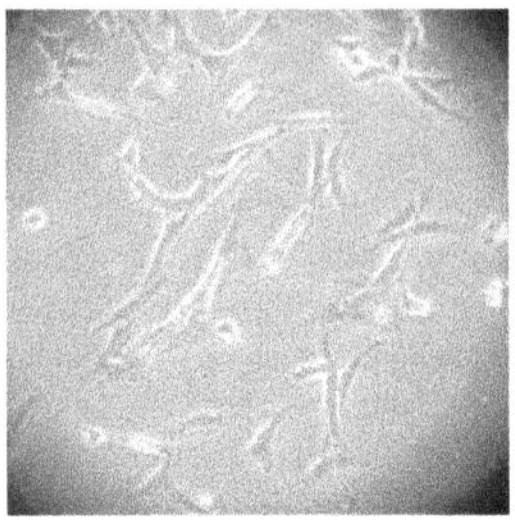

Figure 2: Exponentially growing MSPA cells with a fibroblastic appearance after 28 passages. (Original magnification ×100).

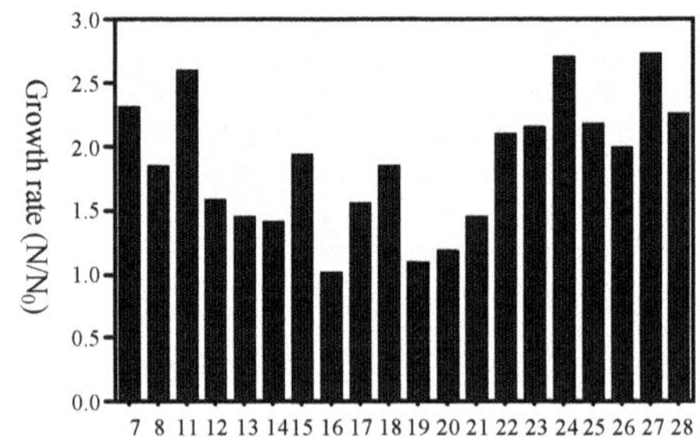

Figure 3: Growth rate of MSPA cells in 4days passage. N, cell number after 4 days; N_0, the initial number of inoculation cells.

Conclusions

We have established a clonal subcutaneous preadipocyte cell line derived from the Meishan pig. So far, MSPA cells have undergone at least 28 passages with no detectable loss of phenotypic properties or the ability to proliferate and develop into mature adipocytes. The study of comparing MSPA cells with PSPA cells may give us an answer as to why Chinese breeds deposit so much fat, and this will give us a clue to develop methods for manipulating the backfat content of pigs in the near future.

References

Legault, C. (1985) Selection of breeds, strains and individual pigs for prolificacy. Journals of Reproduction and Fertility, 33: 151-166.

Nakajima, I., Muroya, S.and Chikuni, K. (2003) Growth arrest by octanoate is required for porcine preadipocyte differentiation. Biochemical and Biophysical Research Communications, 309: 702-708.

Rohrer, G.A. (2000) Identification of quantitative trait loci affecting birth characters and accumulation of backfat and weight in a Meishan-White composite resource population. Journal of Animal Science, 78: 2547-2553.

Sato, S., Oyamada, Y., Atsuji, K., Nade, T., Sato, S., Kobayashi, E., Mitsuhashi T., Nirasawa, K., Komatsuda, A., Saito, Y., Terai, S., Hayashi, T. and Sugimoto, Y. (2003) Quantitative trait loci analysis for growth and carcass traits in a Meishan × Duroc F2 resource population. Journal of Animal Science, 81: 2938-2949.

Student, A.K., Hsu, R.Y. and Lane, M.D. (1980). Induction of fatty acid synthetase in differentiating 3T3-L1 preadipocytes. Journal of Biological Chemistry, 255: 4745-4750.

White, B.R., Lan, Y.H., McKeith, F.K., Novakofski, J., Wheeler, M.B. and McLaren, D.G. (1995) Growth and body composition of Meishan and Yorkshire barrows and gilts. Journal of Animal Science, 73: 738-749.

EVALUATION OF CARCASS COMPOSITION IN LIVE PIGS WITH ON-FARM ULTRASOUND MEASUREMENT

M. Honkavaara*[1], K. Partanen[2] and H. Siljander-Rasi[2]

[1] Finnish Meat Research Institute, Box 56, FI-13101 Hämeenlinna, Finland, [2] MTT Agrifood Research Finland, Tervamäentie 179, FI-05840 Hyvinkää, Finland. Email: markku.honkavaara@ltk.fi

Keywords: pigs, ultrasound measurement, subcutaneous fat, cutting yield, lean meat percentage

Introduction

Evaluation of the lean meat percentage of live animals before slaughter is very important from an economical and technical point of view. It enables farmers to produce different crossbreeds for desired purposes. Evaluation of the lean meat percentage of pigs before slaughter with ultrasound measurements could facilitate co-operation between pork production and meat products manufacture. As a result, the meat industry would get optimal raw material for its process. As is well known, the lean meat percentage is based on the yields of pork cuttings.

In this research the ultrasonic on-farm performance test was used to evaluate the lean meat percentage in pig carcasses.

Materials and Methods

The on-farm performance test includes measurement of fat thickness with ultrasound (Scanoprobe II, Scanco Inc. USA). In this study, the thickness of subcutaneous fat was measured in 75 live pigs with ultrasound at four points a couple of days before slaughter. The points were the shoulder (just behind the scapula, 6 cm from the midline of the back), the loin (in the midline of the back, between the 12th and 13th ribs), the S1 point (between the 15th rib and 1st lumbar vertebra, 8 cm from the midline of the back) and the ham (10 cm from the pigtail in the midline of the back). The test animals included 36 two-breed crosses (Landrace x Yorkshire) and 39 three-breed crosses (Duroc x Landrace x Yorkshire). Furthermore, there were 38 castrates and 37 gilts. On the slaughter line, the lean meat percentage of carcasses was measured with the Hennessy GP4. A day after slaughter the carcasses were commercially cut and the cuts were weighed (Honkavaara 2002, Swatland 2000). Cutting yields were calculated as a percentage of the carcass without head and feet. Regression analysis was used to evaluate the variation in the lean meat percentage of pig carcasses against the ultrasound values.

Results and Discussion

The average live weights of tested pigs and their shoulder, loin, S1 point and ham fat thicknesses are shown in Table 1. Ultrasound measurement indicated shoulder fat to be thickest, followed by ham fat and S1 point fat, while loin fat was thinnest.

Table 1: The average fat thickness values measured with ultrasound in 75 live pigs.

Variable	Mean	Std dev	Min	Max
Live weight, kg	104.7	4.6	93.0	114.5
Shoulder fat, mm	16.9	0.9	14.0	20.0
Loin fat, mm	10.8	0.9	8.0	13.0
S1 point fat, mm	11.0	1.1	8.0	13.0
Ham fat, mm	11.8	1.1	9.0	15.0

In this study, the average slaughter weight of pig carcasses was 78.1 $\pm$ 4.0 kg. Their lean meat percentage was 58.8 $\pm$ 1.9 %. Table 2 shows the average cutting yields of tested animals. The heaviest primal cut was the ham, followed by the forequarter, belly and back and tenderloin.

Forequarter cuts accounted for 28.3 % of the variation in the thickness of shoulder fat while foreshank meat explained 13.3 %, boneless shoulder 5.2 %, subcutaneous fat 5.0 % and fatty trimmings 2.9 %. Cross breeding had some effect on the variation in shoulder fat. In two-breed crosses, forequarter cuts explained 43.2 % and in three-breed crosses 28.5 % of the variation in shoulder fat thickness. Sex also had an effect on the variation in shoulder fat. In castrates, forequarter cuts accounted for 55.4 % of the variation while in gilts they accounted for only 18.6 % of the variation.

It was found that retail back cuts could explain only 11.0 to 11.4 % of the variation in loin fat and S1 point fat thickness. Fatty trimmings and subcutaneous fat explained most of the variation in back fat thickness. Cross breeding had no significant effect on the variation in loin fat and S1 point fat thickness only 9.1 to 17.8 % while sex affected the variation in back fat thickness. Back cuts of gilts explained 39.5 % and those of castrates only 11.5 % of the variation in loin fat thickness. In addition, back cuts of gilts explained 33.2 % of the variation in S1 point fat thickness whereas those of castrates could only explain 25.7 %

Furthermore, retail ham cuts explained only 19.0 % of the variation in ham fat thickness: hindshank meat explained the most at 6.1 %, followed by fatty trimmings at 4.0 % and abdominal cut at 3.3 %. Cross breeding had an effect on the variation in ham fat thickness. In two-breed crosses ham cuts explained 31.7 % and with three-breed crosses 20.8 % of the variation in ham fat thickness. Sex also had some effect on the variation in ham fat. Ham cuts of castrates explained 29.0 % while those of gilts explained 24.1 % of the variation in ham fat thickness.

In this study, the measured fat thickness values explained only 7.8 % of the variation in the lean meat percentage of pig carcasses. S1 point fat thickness explained 3.4 %, loin fat thickness 2.5 %, shoulder fat thickness 0.9 % and ham fat thickness 0.9 % of the variation in the lean meat percentage. Fat thickness values of two-breed crosses explained only 4.6 %, and those of three-breed crosses 15.8 % of the variation in the meat percentage of carcasses. Fat thickness values of castrates explained 20.2 %, whereas those of gilts explained only 10.8 % of the variation in the lean meat percentage of carcasses.

Table 2: Average yield of retail cuts of 75 pig carcasses (mean ± standard deviation without head and feet).

Retail cuts	Yield, %	
Forequarter:	30.1 ± 1.1	
Anterior back		5.7 ± 0.4
Boneless shoulder		10.2 ± 0.8
Fatty trimmings		3.7 ± 0.7
Foreshank meat		1.1 ± 0.2
Skin		1.5 ± 0.3
Subcutaneous fat		2.7 ± 0.5
Bones		5.2 ± 0.5
Back:	16.9 ± 0.8	
The *longissimus dorsi* muscle		8.8 ± 0.6
Fatty trimmings		1.6 ± 0.3
Skin		0.8 ± 0.1
Subcutaneous fat		2.3 ± 0.6
Bones		3.4 ± 0.3
Tenderloin:	1.8 ± 0.1	
Trimmed tenderloin		1.4 ± 0.1
Trimmings		0.4 ± 0.1

Retail cuts	Yield, %	
Belly:	17.1 ± 0.9	
Boneless belly		10.9 ± 0.9
Fatty trimmings		1.4 ± 0.4
Skin		1.4 ± 0.2
Bones		3.4 ± 0.3
Ham:	31.9 ± 0.9	
Inside joint		5.5 ± 0.4
Outside joint		6.8 ± 0.5
Corner joint		3.5 ± 0.3
Roasting joint		2.8 ± 0.3
Hindshank meat		2.0 ± 0.2
Fatty trimmings		2.9 ± 0.6
Skin		1.6 ± 0.2
Subcutaneous fat		3.1 ± 0.8
Abdominal cut		2.1 ± 0.4
Bones		3.8 ± 0.3

Conclusions

The results show that the yields of retail cuts of pig carcasses were nearly insignificant in explaining the variation of fat thickness in the shoulder, back and ham when measured with ultrasound on the farm. Furthermore, the results show that it is not possible to evaluate the lean meat percentage of pig carcasses by ultrasound fat thickness measurement on the farm a couple of days before slaughter. This suggests that ultrasonic estimation of muscle dimensions, like the loin muscle area in live pigs, should be included in the procedure in addition to fat thickness measurement.

Acknowledgements

The authors wish to thank Leena Hintsanen, Finnish Breeding Association, for ultrasound measurements of pigs. This study was financed by the Ministry of Agriculture and Forestry, Finland.

References

Honkavaara, M. (2002). Pork quality and cutting yields of breeding pigs in Finland. In: Proceedings of the 48th International Congress of Meat Science and Technology, Rome, Italy, pp. 678-679.

Swatland, H.J. (2000). Meat Cuts and Muscle Foods, Nottingham University Press, Nottingham, United Kingdom, pp. 68-69.

CARCASS AND MEAT QUALITY OF DUROC-BOAR OFFSPRING

F.J. Tan[1], T.H. Huang[1], Y.Y. Tseng[1] and H.W. Ockerman*[2]

[1.] *Department of Animal Science, National University of Science and Technology, Pingtung, 912, Taiwan.* [2.] *Department of Animal Sciences, The Ohio State University, Columbus, 43210, OH, USA. Email: ockerman.2@osu.edu*

Keywords: Duroc, carcass characteristic, quality, pork

Introduction

The swine industry has taken great efforts to produce high quality pork. Various parameters, including genetic, nutrition, management, cooking etc. influence eating quality of pork. Several studies have investigated breed effects on carcass composition and pork quality in pigs (Lo *et al.*, 1992; Lonergan *et al.*, 2001; D'Souza and Mullan, 2001). This study examined the effect of breed on the carcass characteristics and meat quality of offspring pigs from Duroc boars mated with sows of either Landrace (L), Yorkshire (Y) or hybrid (LY).

Materials and Methods

Two hundred and ninety-one Duroc-boared offspring pigs which came from Duroc (D) boars mated with sows of Landrace (L), Yorkshire (Y) or hybrid (LY), were slaughtered at a live weight of approximately 120 kg, according to standards of commercial procedures, in a Taiwanese abattoir. Carcass characteristics including live weight, carcass weight, carcass length, backfat thickness, loin eye area, dressing percentage, lean percentage, fat percentage, bone percentage, and meat qualities of loin and ham muscles including colour score, marbling score, cooking loss, drip loss, shear force, colour values, pH values, and chemical compositions of loin and shoulder muscles were evaluated (Huang, 2005). Data were analysed using SAS software.

Results and Discussion

Table 1 shows that the live weights and carcass weights of offspring pigs from Duroc boars mated with LYD sow were significantly ($P<0.05$) lower than those pigs from Duroc boar mated with LD or YD sow. The offspring pigs from D × LD had significantly longer carcass length than those from D × LYD hybrids. D × YD hybrids had significantly thicker backfat thickness and smaller loin eye areas. D × YD hybrids had significantly lower lean percentage and higher fat percentage. When considering meat quality, the loin muscles from D × YD hybrids had significantly higher NPPC colour scores by sensory evaluation (Table 2). This darker colour could also be illustrated by the significantly higher a values of those samples. D × LYD hybrids had significantly higher shear values which indicated a tougher tissue. No significant difference of NPPC marbling scores, cooking loss, purge, L and b values, and pH of loin muscles and pH values of the ham muscles were observed among the various breeds of pigs. The proximate compositions of loin muscles of the three breeds were not significantly different (Table 3). D × YD hybrids had significantly higher content of tasty amino acids, sweet amino acid, and MSG-line amino acids in loin muscles when compared with pigs of the other two hybrids. D × LYD hybrids had significantly higher total saturated fatty acid percentage and lower total polyunsaturated fatty acid percentage than the others.

Table 1: Carcass characteristics of offspring pigs from Duroc boar mated with various sow breeds.

Item	Breed		
	LD (n = 64)	LYD (n = 216)	YD (n = 11)
Live weight, kg	123.05±13.15^{a}	117.88±11.32^{b}	121.65±11.80^{a}
Carcass weight, kg	105.65±11.25^{a}	101.96±9.91^{b}	105.28±10.38^{a}
Carcass length, cm	87.66±3.07^{a}	85.85±3.33^{b}	86.61±2.52^{ab}
Backfat thickness, cm	1.95±0.45^{b}	2.10±0.44^{b}	2.38±0.39^{a}
Loin eye area, cm^2	60.92±7.59^{a}	59.97±8.26^{ab}	56.98±11.23^{b}
Dressing percentage, %	85.88±1.42^{b}	86.50±1.14^{a}	86.53±0.90^{a}
Lean percentage, %	57.16±2.96^{a}	56.98±2.75^{a}	54.48±3.04^{b}
Fat percentage, %	11.28±3.07^{ab}	10.18±2.69^{b}	12.24±2.70^{a}
Bone percentage, %	15.07±1.25	15.05±1.24	15.32±1.65

Mean±SE; ^{a,b} Values in the same row with different superscripts are significantly different ($P < 0.05$)

Table 2: Meat quality of loin and ham muscles of offspring pigs from Duroc boar mated with various sow breeds.

Item	Breed		
	LD (n = 64)	LYD (n = 216)	YD (n = 11)
Loin muscle			
Colour score	2.57±0.84[b]	2.92±0.79[ab]	3.14±0.55[a]
Marbling score	1.60±0.65	1.81±0.71	1.82±0.64
Cooking loss, %	26.46±2.69	28.08±3.48	28.08±1.92
Dripping loss, %	4.05±1.99	4.68±2.60	5.24±2.60
Shear force, kg	4.65±01.90[b]	6.08±1.77[a]	4.01±0.85[b]
L-value	40.78±5.15	41.92±3.47	40.15±2.24
a-value	8.09±0.98[b]	7.95±0.99[b]	9.04±0.95[a]
b-value	6.86±0.86	6.95±0.89	6.80±0.54
pH_1	5.89±0.32	5.78±0.27	5.89±0.24
pH_{24}	5.67±0.25	5.61±0.15	5.65±0.10
Ham muscle			
pH_1	6.15±0.28	6.07±0.27	6.08±0.20
pH_{24}	5.86±0.31	5.78±0.26	5.78±0.25

Mean±SE; [a,b] Values in the same row with different superscripts are significantly different (P0.05)

Table 3: Chemical composition of loin and shoulder muscles of offspring pigs from Duroc boar mated with various sow breeds.

Item	Breed		
	LD (n = 64)	LYD (n = 216)	YD (n = 11)
Loin muscle			
Crude protein, %	24.06±1.45	24.35±1.25	24.23±1.37
Crude fat, %	2.97±1.66	2.84±1.36	3.51±1.70
Moisture, %	72.06±1.69	72.06±1.93	71.07±1.32
Ash, %	1.07±0.25	1.05±0.25	1.01±0.23
Tasty amino acid, mg/100g	8993±1927[b]	9373±2547[b]	10512±2243[a]
Sweet amino acid, mg/100g	5109±1521[b]	5876±2027[ab]	6187±2052[a]
MSG-like amino acid, mg/100g	7194±1255[b]	7148±1991[b]	8146±1243[a]
Glycogen, mg/g	1.74±0.76	1.59±0.66	1.88±0.67
Shoulder muscle			
Total saturated fatty acid, %	37.20±2.75[b]	39.17±2.86[a]	36.72±3.43[b]
Total monounsaturated fatty acid, %	37.91±2.55	39.40±2.45	39.19±3.49
Total polyunsaturated fatty acid, %	23.41±4.16[a]	19.96±3.77[b]	22.86±6.55[a]

Mean±SE; [a,b] Values in the same row with different superscripts are significantly different ($P < 0.05$)

Tasty amino acid = Asp + Gly + Glu + Ser + Thr; Sweet amino acid = Ala + Gly + Pro + Ser + Thr; MSG-like amino acid = Asp + Glu + Met + Ser (Huang, 2005).

Conclusions

Carcass characteristics and meat quality traits from the offspring pigs which came from a Duroc boar mated with sows of either Landrace, Yorkshire or hybrid were varied. Among the breeds, the results showed that D × LYD hybrids had significantly lighter live and carcass weights; D × YD hybrids had significantly thicker backfat thickness and smaller loin eye areas, thus leading to a significantly lower lean percentage and higher fat percentage. The loin muscles from D × YD hybrids had significantly higher NPPC colour scores. D × LYD hybrids had significantly higher shear values. D × LYD hybrids had significantly higher total saturated fatty acid percentage.

References

D'Souza, D.N. and Mullan, B.P. (2001). The effect of genotype, sex and management strategy on the eating quality of pork. Meat Science, 60: 95-101.

Huang, T.H. (2005). Expression on carcass characteristics and meat quality of Duroc-sired offspring pig. National Pingtung University of Science and Technology, Master's thesis. Pingtung, Taiwan.

Lo, L.L., McLaren, D.G., McKeith, F.K., Fernando, R.L. and Novakofski, J. (1992). Genetic analyses of growth, real-time ultrasound, carcass, and pork quality traits in Duroc and Landrace pigs: 1. Breed effects. Journal of Animal Science, 70: 2373-2386.

Lonergan, S.M., Huff-Lonergan, E., Rowe, L.J., Kuhlers, D.L. and Jungst, S.B. (2001). Selection for lean growth efficiency in Duroc pigs influences pork quality. Journal of Animal Science, 79: 2075-2085.

EFFECT OF FAT SCORE ON THE QUALITY OF VARIOUS MEAT PRODUCTS

R. Hadorn*, P. Eberhard, D. Guggisberg, P. Piccinali, H. Schlichtherle-Cerny, D. Scherrer and G. Bee

Swiss Federal Research Station for Animal Production and Dairy Products, Agroscope Liebefeld-Posieux (ALP), Berne, Switzerland. Email: ruedi.hadorn@alp.admin.ch

Keywords: fat score, quality, processing, sensory, meat products

Introduction

During the last decades, breeding strategies aimed to increase lean tissue and to decrease adipose tissue deposition in pig carcasses. Because of the close relationship between the intake and the concentration of polyunsaturated fatty acids (PUFA) in porcine adipose tissue, both factors, lower fat deposition and dietary PUFA-intake, are often followed by increased PUFA concentrations in the body fat. Increased PUFA tissue levels are associated with higher incidence towards oxidation (→ rancid) and impaired texture (→ soft, greasy, oily fat). Both quality traits are of great importance for the meat processing industry (Prabucki, 1991). In larger Swiss slaughter houses, fat quality is determined by fat score (FS), which is similar to the iodine number and related to the degree of unsaturation of the fat (Scheeder *et al.*, 1999). The carcass fat quality requested by the abattoirs is a FS of < 62. Pig producers not meeting those requirements experience marked price reductions.

The aim of this work was to investigate possible effects of FS on different quality traits (nutrient composition, fatty-acid profile, sensory characteristics, texture and fat oxidation parameters) of four different meat products (salami, raw-cured bacon, pork hamburger, Vienna sausage). Correlations were tested on a level of significance of $P \leq 0.05$.

Materials and Methods

The meat and fat tissue of 47 Large White gilts were used. The pigs were from the trial of Bee (2005), in which the effects of four different feeding regimes on the FS during the last weeks prior to slaughter were tested (one animal died during the trial). The pigs were slaughtered at the ALP-slaughter house in two series within 6 weeks.

According to the procedure of Proviande (2003), a fat sample from each pig was collected from the back fat close to the hips and the FS was determined the same day at UFAG laboratories (Sursee, Switzerland) using the method described earlier (Scheeder *et al.*, 1999). The following day, the carcasses were dissected according to Rebsamen *et al.*, (1985) and then prepared for the production of four different meat products. At the Education Centre of the Swiss Meat Industry (ABZ, Spiez, Switzerland), salami (SAL), raw-cured bacon (RCB), pork hamburger (PHB) and Vienna sausage (VIS) were produced following traditional recipes. With the exception of RCB (n = 47), raw material was pooled by FS-classes for each series (SAL: n = 20) or over both series (PHB and VIS, n = 14 each), respectively (Table 1). Raw-material quality was also described by the ABZ-specialists.

The investigated parameters include main nutrients, fatty acid profile (Bee *et al.*, 2004), fat oxidation by GC-MS (aroma components as markers) and texture (sliceability, fat firmness) by a Warner-Bratzler-apparatus (WB). The four meat products were also assessed for their sensory profile (14 to 17 product-dependent odour, flavour and texture criteria on a 10-point intensity scale) by ten trained panelists. Half of the PHB were heated twice in order to detect potential flavour defaults. Additionally, the meat to fat ratio was assessed in the RCB at the 8 and 11 rib by planimetry and the a_w-value for SAL.

Results and Discussion

A good FS-variation resulted from the trial of Bee (2005), whereas the number of pigs within the individual FS-classes was not uneven (Table 1).

Table 1: Distribution of animals in the individual fat-score classes.

Fat-score class	series		Fat-score class	series		Fat-score class	series	
	1	2		1	2		1	2
< 58.0	-	1	62.1 – 62.5	2	1	64.6 – 65.0	-	3
58.1 – 59.0	1	-	62.6 – 63.0	3	3	65.1 – 65.5	1	1
59.1 – 60.0	5	2	63.1 – 63.5	1	-	65.6 – 66.0	3	-
60.1 – 61.0	5	1	63.6 – 64.0	-	-	66.1 – 67.0	-	1
61.1 – 62.0	2	7	64.1 – 64.5	1	1	> 67.0	-	2

Raw material for SAL- and RCB-production were characterized by the ABZ-specialists as wet and greasy when FS-values were above 65, whereas no FS-related differences were observed for VIS- and PHB-raw materials. Except for RCB (r^2 = 0.11 to 0.17), no significant correlations were found between FS and dry matter, protein, total fat and ash content, respectively. FS was correlated with the saturated-fatty acid content of the RCB (r = -0.55), SAL (r = -0.68) and VIS (r = -0.70). Only a few sensory characteristics of the RCB like spicy (r = 0.40), smoky (r = 0.33) and fibrous (r = 0.41) were linked to the FS whereas no significant sensory correlations were observed for the other meat products. Regardless of the FS, reheating of cooled PHB was followed by negative effects on the attributes rancid, juicy, tender and animal-like. Only the 1-octen-3-ol signal (peak height) in RCB (r = 0.51) and in SAL (r = 0.46), but not in the other meat products was correlated with the FS. Reheating of PHB was followed by significantly increased signals for pentanal, 1-pentanol and 1-octen-3-ol. Texture of SAL (WB-total work: r = -0.66, Figure 1; max. WB-force: r = -0.60) and RCB-fat firmness (needle penetration force, 10-15mm: r = -0.31) were negatively correlated with FS. The FS-related differences in texture were similar to the ones, which were seen between the two series. The a_w-values of SAL were positively correlated with the FS (Figure 2). Interestingly, a significant positive correlation between FS and meat:fat-ratio occurred in the RCB (Figure 3), which was mainly due to changes in the fat area (r = -0.46) and may be related to the occurrence of fatty breasts.

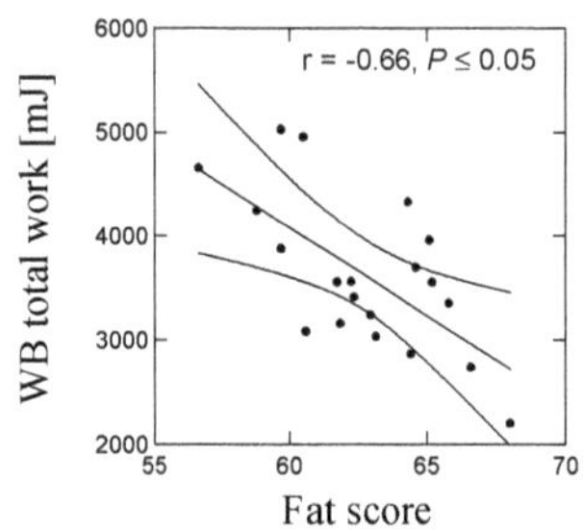

Figure 1: Correlation between the FS and WB total work in the SAL.

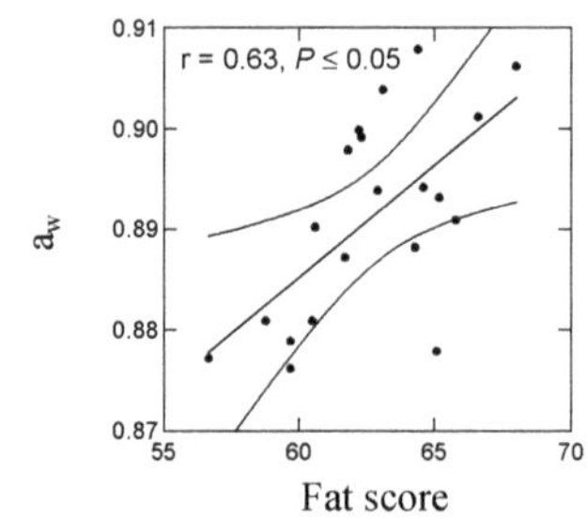

Figure 2: Correlation between FS and a_w-value in the SAL.

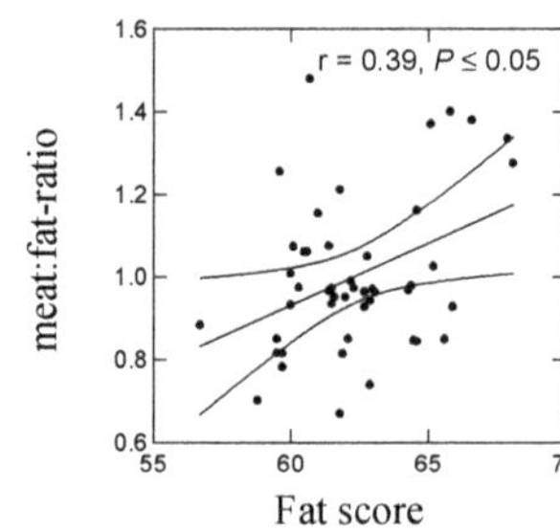

Figure 3: Correlation between FS and meat:fat-ratio in the RCB.

Conclusions

Several FS-related effects were observed for the cured and air-dried products like SAL and RCB, but not for the VIS and PHB. The FS was related to raw material characteristics, texture and shelf life (a_w-value) for SAL, and to raw material characteristics, meat:fat-ratio, fat oxidation and fat firmness for RCB.

Due to the relevance of the FS to the profitability of pig production the threshold of 62 is always questioned by pig producers and feeding manufacturers. However, it is not possible to redefine the actual FS-limit based on the present data. Thus, redefining the FS-limit should be a matter of market-related considerations between stakeholders.

References

Bee G. (2005). Unterschiedliche Fütterungsstrategien in der Endmast und deren Einfluss auf die Fettzahl. Tagungsbericht vom 13. Mai 2005, Schriftenreihe des Institutes für Nutztierwissenschaften, ETH Zürich, Heft 26: 155-158.

Bee, G., Guex, G. and Herzog, W. (2004). Free-range rearing of pigs during the winter: Adaptations in muscle fiber characteristics and effects on adipose tissue composition and meat quality traits. Journal of Animal Science, 82: 1206-1218.

Prabucki A.L. (1991). Qualitätsanforderungen an Schweinefleisch. Schweinefleischqualität - Qualitätsschweinefleisch. Tagungsbericht vom 22. Mai 1991, Schriftenreihe aus dem Institut für Nutztierwissenschaften, ETH Zürich, Heft 5: 5-10.

Proviande (2003). Richtlinien über die Erfassung der Fettqualität bei Schweinen vom 5. Dezember 2003. Proviande Klassifizierungsdienst, Bern.

Rebsamen, A., Schwörer, D. und Lorenz, D. (1995). Die Schlachtkörperzerlegung beim Schwein in der MLP Sempach. Der Kleinviehzüchter 43: 223-259.

Scheeder, M.R.L., Bossi, H. und Wenk, C. (1999). Kritische Betrachtungen zur Fettzahl-Bestimmung. Agrarforschung 6: 1-8.

SESSION 1: MEAT QUALITY

- MUSCLE BIOLOGY AND BIOCHEMISTRY

EFFECTS OF *POST-MORTEM* TEMPERATURE ON ISOMETRIC TENSION, SHORTENING AND PH IN OSTRICH MUSCLE

L.C. Hoffman[1]*, S.St.C. Botha,[1,2] and T.J. Britz[1]

[1] Department of Animal Sciences, [2] Department of Food Science, University of Stellenbosch, Private Bag X1, Matieland 7602, South Africa. E-mail: lch@sun.ac.za

Keywords: hot deboning, isometric tension, shortening, ostrich, rigor shortening

Introduction

Cold-shortening occurs when muscle is exposed to low temperatures (<10-15°C) early *post-mortem* (p-m), when ATP and pH (above 6.20) levels are still high (Nuss and Wolfe, 1980-81). On the other hand, rigor tension occurs much later and at any temperature between 0 and 37°C, reaching maximum values when ATP levels have been depleted and pH is at a minimum value (Nuss and Wolfe, 1980-81).

The aim of this study was to investigate the development of isometric tension and shortening in ostrich *M. gastrocnemius, pars interna* during *rigor mortis* at 7 and 37°C respectively in an attempt to determine the time course of rigor, pH decline, and degree and extent of shortening.

Materials and Methods

Ten rested ostriches were slaughtered as described by Wotton and Sparrey (2002). The right leg *M. gastrocnemius, pars interna* were removed within 20 min to 1 h p-m. Two muscle strips were cut (with dimensions of 10x10x30 mm, weighing between 1.5 and 3 g) parallel to the fibre axis, from each muscle sample. The muscle strips were put into two separate rigometers (Rigotech®, ©ReoLogica, 1999, Sweden) at constant temperatures of 7 and 37°C, respectively. Isometric tension, shortening and pH were recorded during *rigor mortis* every 15 min for 23 h p-m. The change in pH and temperature were also measured continuously every 10 min for 23 h p-m in the intact left leg *M. gastrocnemius, pars interna* from the same ostrich carcasses.

The data were subjected to factorial analysis of variance (ANOVA) using SAS version 8.2 statistical software (SAS, 1999). Shapiro-Wilk tests were performed for testing non-normality (Shapiro & Wilk, 1965).

Results and Discussion

Maximum shortening was significantly ($P<0.0001$) higher at 37°C (33.39±3.57%) than at 7°C (10.69± 2.63%), while the pH values at the time of maximum shortening did not differ significantly ($P=0.7539$) for 7°C (6.11±0.25) and 37°C (6.09±0.37) respectively. This indicated that the muscle strips at 7 and 37°C from the same ostrich carcass were at the same initial energy levels early p-m leading therefore to similar pH values at the point of maximum tension development. Maximum shortening was reached sooner ($P=0.0006$) in the muscle strips at 37°C (5.59±3.65 h) than the muscle strips at 7°C (23.00±0.45 h).

As illustrated in Figure 1, maximum tension and shortening were reached sooner in muscle strips at 37 than at 7°C. Maximum shortening was reached very rapidly in the muscle strip at 37°C (5.59±1.53 h), while the shortening in the muscle strip at 7°C increased at a slow rate, tending towards an asymptotic value, although not yet reaching a constant value.

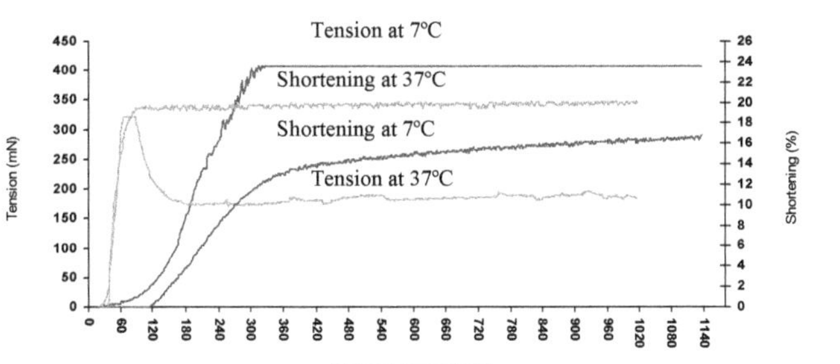

Figure 1: Development of muscle isometric tension and shortening from the time (20 min p-m) the muscle strips from an individual ostrich were placed in the rigometers, maintained at 7 and 37°C, respectively.

The intact muscles reached a minimum pH value (5.85±0.22) within 2 h after slaughter, while muscle strips at 7°C reached a minimum pH of 5.90±0.21 at 6.41±4.51 h p-m ($P<0.05$). This indicated that p-m glycolysis and anaerobic production of ATP were faster in muscles with higher muscle temperatures during the rigor process, which in turn could explain the higher percentage of maximum shortening obtained at 37 (33.39±3.57%) than at 7°C (10.69±2.63%).

As illustrated by the pH and temperature data in Figure 2, the change in pH for the muscle strip at 37°C showed a similar initial fall in pH similar to the intact muscle refrigerated at <4°C, followed by an increase in pH after a minimum value had been reached. In general, the pH for intact *M. gastrocnemius, pars interna* decreased from 1 h p-m to a mean minimum value of 5.85±0.22 at 2 h p-m, indicating a rapid fall in pH during the first 2-3 h p-m, after which the pH then increased to a mean value of 6.14±0.19 at 24 h p-m. Muscle strips at 7°C reached a minimum pH at 6.42±4.51 h p-m. This clearly indicates that the change in pH is highly temperature dependent.

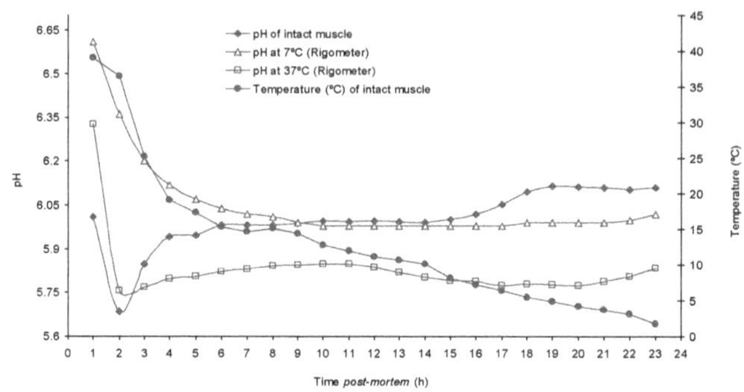

Figure 2: Hourly temperature (°C) and pH values indicating the decline in temperature (•) and pH change (♦) during the first 23 h p-m for intact *M. gastrocnemius, pars interna* refrigerated at <4°C, as well as for muscle strips from the *M. gastrocnemius, pars interna* maintained at 7 (△) and 37°C (□) in the rigometers from the same ostrich carcass.

Conclusions

It was concluded that the rate of rigor development, the course of *rigor mortis*, and the rate of change in pH in ostrich *M. gastrocnemius, pars interna*. The degree of maximum tension and shortening were also temperature dependant, where maximum tension was higher at 7 than at 37°C, while, in contrast, maximum shortening was significantly higher at 37 than at 7°C. Muscle strips at 37°C reached a maximum tension value within 4.08±3.89 h p-m, while the minimum pH was also reached within 4.83±3.82 h p-m in muscle strips at 37°C, indicating that full *rigor mortis* in ostrich muscles occurred at the point of minimum pH. The intact muscles reached a minimum pH at approximately 2 h p-m, while muscle temperature was still relatively high (32.08±4.29°C). It is therefore suggested that ostrich muscles not be hot-deboned within the first 2 h p-m.

Acknowledgments

This project was supported by Swartland Ostrich Abattoir, Malmesbury, South Africa, who contributed the ostrich carcasses and who made it possible to conduct research at the abattoir premises. This study was also made possible by the two year prestige scholarship from the National Research Foundation. The Norwegian Food Research Institute (MATFORSK, Ås, N-1430 Norway) is thanked for the use of the RigoTech equipment.

References

Nuss, J.I. and Wolfe, F.H. (1980-81). Effect of *post-mortem* storage temperatures on isometric tension, pH,

Reologica Instruments, AB. (1999). User's manual for RigoTech Meat Analyzer, Version 3.0. Sweden.

SAS Institute, Inc. (1999). SAS/STAT User's Guide, Version 8, 1st printing, Volume 2. SAS Institute Inc, SAS Campus Drive, Cary, North Carolina 27513.

Shapiro, S.S. and Wilk, M.B. (1965). An analysis of variance test for normality (complete samples). *Biometrika*, 52, 591-611.

Wotton, S. and Sparrey, J. (2002). Stunning and slaughter of ostriches. *Meat Science*, 60, 389-394.

STUDIES ON ATP RELATED COMPOUNDS, FREE AMINO ACIDS AND FREE SUGARS OF PORK FROM DIFFERENT BREEDS OF PIG IN TAIWAN

M.T. Chen* and Y.W. Huang

Department of Bioindustry Technology, Da-Yeh University, 112 Shan Jeau Rd., Da Tsuen, Changhua, Taiwan 515.
Email:michen@mail.dyu.edu.tw

Keywords: pork, ATP, related compounds, free amino acid, free sugar

Introduction

The aim of this study was to understand the water-soluble flavour compounds of pork, of five different breeds of pigs (Duroc, Yorkshire, Landrace, Black and hybrid (LYD)) which are popular breeds in Taiwan.

Materials and Methods

Duroc (9), Yorkshire (8), Landrace (8), Black (11) and Hybrid (18) were fed for 6.5 to 7.5 months of age with the same composition of feeds, and slaughtered at a local slaughterhouse. The loin was taken and cooled at 2°C for analyses. The ATP and related compounds, free amino acids, free sugars, pH and moisture content were determined.

The ATP and related compounds were analysed by a procedure modified from the methods of Boyle *et al.* (1991) and Ryder (1985). The samples were treated by the method described by Ryder (1985). Free amino acids were extracted using the method modified from Chu *et al.*, (1996) and Konosu *et al.*, (1974), and analysed by HPLC chromatography. Free sugars were extracted by the same method as free amino acids, and pre-column derivatised with PMP, then analyzed by HPLC. The pH value was determined with a pH-meter (HANNA, Italy) and moisture content was determined by the method of AOAC (1984). The data was analysed by SAS software (1994).

Results and Discussion

The pH values and moisture content of pork from different breeds of pigs at the age of 6.5 to 7.5 months were found to range between pH 5.4 and 6.22, and 71.3 to 73%, respectively. Tables 1 and 2 indicate the concentration of ATP and its related compounds of pork from different breeds of pigs. The result found was that ATP depleted more rapidly from Black pig than others, and completely depleted after 24 hr postmortem, while Inosine and Hypoxanthin increased remarkably. It was found that inosine content in pork of hybrid was lower than the others 1 hr postmortem. After 24 hr postmortem change in inosine level from different breeds compared as follows: Duroc and Black>Landrace>Hybrid and Yorkshire in descending order. IMP content which is associated with meat flavour, was highest in pork from Duroc among the five breeds whether 1 hr or 24 hr postmortem. So Duroc is often used as the base for crossbreeding of the meat herd such as LYD. Table 3 shows the free amino acids and total free amino acids in pork from different breeds. It was found there were no significant differences in free amino acids and total amino acids contents among the different breeds, except the methionine was higher in hybrid pork. Table 4 shows free sugar contents in loin and ham from different breeds. Glucose of loin was the highest for Duroc, then Black>Landrace and Hybrid>Yorkshire in descending order, and ribose content was the highest in Hybrid pork and lowest in Black among the different breeds. However, free sugar content from ham was different from the result of loin. It was found that glucose contents in Duroc and Black were the highest and Yorkshire and Landrace were the lowest among the different breeds, while ribose content was found higher in Hybrid and Duroc than other breeds. The concentration of glucose in loin was higher when compared to the ham, but no significant difference in ribose was found between different parts.

Table 1: Comparison of ATP related compounds (mg/g) in pork loin** from different breeds* 1hr post-mortem.

Items\Breed	Black	Hybrid	Duroc	Yorkshire	Landrace
IMP	0.005 ± 0.105^{c}	0.936 ± 0.079^{b}	1.344 ± 0.121^{a}	0.215 ± 0.134^{c}	0.851 ± 0.105^{b}
ATP	0.043 ± 0.112^{c}	0.589 ± 0.085^{ab}	0.478 ± 0.130^{b}	0.676 ± 0.143^{ab}	0.833 ± 0.112^{a}
ADP	0.538 ± 0.100^{a}	0.474 ± 0.076^{a}	0.193 ± 0.116^{b}	0.481 ± 0.128^{ab}	0.194 ± 0.100^{b}
AMP	1.092 ± 0.105^{a}	0.175 ± 0.080^{c}	0.339 ± 0.122^{bc}	0.572 ± 0.135^{b}	0.523 ± 0.105^{b}
Hx	0.005 ± 0.0015^{bc}	0.0013 ± 0.001^{d}	0.012 ± 0.002^{a}	0.008 ± 0.002^{ab}	0.003 ± 0.001^{cd}
Ino	0.088 ± 0.012^{a}	0.031 ± 0.009^{b}	0.105 ± 0.014^{a}	0.097 ± 0.016^{a}	0.070 ± 0.012^{a}

* : Mean±S.E.
** : pigs were fed for 6.5~7.5 months.
a,b,c,d : means within the same row without the same superscript letters are significantly different ($P < 0.05$).

Table 2: Comparison of ATP related compounds (mg/g) * in pork loin * * from different breeds 24hr post- mortem.

Items\Breed	Black	Hybrid	Duroc	Yorkshire	Landrace
IMP	0.005 ± 0.114^{d}	0.624 ± 0.108^{c}	1.615 ± 0.116^{a}	1.115 ± 0.116^{b}	1.193 ± 0.205^{ab}
ATP	0.007 ± 0.048^{c}	0.132 ± 0.045^{bc}	0.274 ± 0.049^{a}	0.401 ± 0.049^{a}	0.252 ± 0.086^{ab}
ADP	1.590 ± 0.121^{a}	0.647 ± 0.114^{b}	0.343 ± 0.123^{b}	0.560 ± 0.123^{b}	0.748 ± 0.218^{b}
AMP	0.097 ± 0.009^{ab}	0.077 ± 0.009^{b}	0.112 ± 0.010^{a}	0.080 ± 0.010^{b}	0.115 ± 0.017^{a}
Hx	0.014 ± 0.007^{b}	0.005 ± 0.007^{b}	0.014 ± 0.007^{b}	0.006 ± 0.007^{b}	0.063 ± 0.013^{a}
Ino	0.093 ± 0.008^{ab}	0.047 ± 0.008^{c}	0.105 ± 0.008^{a}	0.050 ± 0.008^{c}	0.077 ± 0.015^{bc}

* : Mean±S.E.
* * : pigs were fed for 6.5~7.5 months.
a,b,c, d : means within the same row without the same superscript letters are significantly different ($P < 0.05$).
ND : Not detected. ; TR : Trace.

Table 3: Comparison of free amino acid (mg/100g) * in pork loin * * from different breeds *.

Items\Breed	Black	Hybrid	Duroc	Yorkshire	Landrace
Asp	4.65±1.48	5.87±1.21	4.13±1.70	5.36±1.32	4.30±1.48
Thr+Ser	28.87±4.82	20.60±3.77	19.11±5.33	21.10±5.65	24.96±5.65
Glu	21.05±4.65	14.31±3.63	10.43±5.14	13.99±4.45	12.83±5.45
Gly	12.29±1.65	9.72±1.29	8.94±1.82	9.73±1.93	8.49±1.93
Ala	23.28±3.54	19.54±2.76	19.30±3.91	18.11±4.15	16.46±4.15
Cys	ND	TR	TR	TR	TR
Val	9.67±1.62	6.91±1.27	6.28±1.79	7.65±1.90	6.73±1.90
Met	3.91 ± 1.12^{ab}	6.91 ± 1.12^{a}	2.68 ± 1.29^{b}	1.98 ± 1.42^{b}	3.25 ± 1.29^{b}
Ile	7.82±1.55	5.04±1.21	4.07±1.71	4.78±1.81	4.43±1.81
Leu	13.41±2.65	8.62±2.07	7.14±2.93	7.51±3.11	7.76±3.11
Tyr	8.65±1.63	6.08±1.27	4.43±1.91	5.66±1.91	5.29±2.04
Phe	10.20±2.08	8.64±2.08	4.16±2.36	5.65±2.55	7.88±2.80
His	7.60±1.36	5.55±1.07	6.69±1.51	5.10±1.60	5.09±1.60
Lys	12.33±2.82	8.40±2.20	6.69±3.11	10.00±3.30	9.95±3.30
Arg	10.07±2.10	6.29±1.64	4.37±2.32	6.86±2.46	7.11±2.46
Total	167.93±31.14	120.59±23.34	103.61±34.43	119.66±36.51	117.94±36.51

* : Mean±S.E.
* * : pigs were fed for 6.5~7.5 months.
a,b : means within the same row without the same superscript letters are significantly different ($P < 0.05$).

Table 4: Comparison of free glucose and ribose (mg/100g) * in pork loin * * and ham from different breeds.

Part	Items\Breed	Black	Hybrid	Duroc	Yorkshire	Landrace
Loin	Glucose	114.36 ± 7.45^{b}	93.91 ± 6.00^{b}	145.74 ± 8.48^{a}	36.29 ± 8.48^{d}	104.11 ± 8.67^{bc}
	Ribose	3.80 ± 0.44^{c}	6.99 ± 0.49^{a}	6.00 ± 0.77^{ab}	4.11 ± 0.69^{bc}	4.12 ± 0.65^{bc}
Ham	Glucose	80.41 ± 6.92^{ab}	64.78 ± 5.37^{b}	92.35 ± 7.83^{a}	30.71 ± 7.83^{c}	38.40 ± 7.75^{c}
	Ribose	3.42 ± 0.42^{b}	6.95 ± 0.43^{a}	6.93 ± 0.92^{a}	4.18 ± 0.59^{b}	3.64 ± 0.55^{b}

* : Mean±S.E.
* * : pigs were fed for 6.5~7.5 months.
a,b,c : means within the same row without the same superscript letters are significantly different ($P < 0.05$).

Conclusions

Changes in ATP and its related compounds, free amino acids and free sugars, and pH value of pork are different among different breeds postmortem. They are affected by not only breed, but also sex, age and location of carcass and nutrition. Therefore, the factors affecting palatability of meat such as flavour and taste are very complicated.

References

A.O.A.C. (1984). Official Methods of Analysis. 14th ed. Washington. D.C., USA.

Boyle, J.L., Lindsay, R.C. and Stuiber, D.A. (1991). Adenine nucleotide degradation in modified atmosphere chill-stored fresh fish. Journal of Food Science, 56: 1267-1270.

Konosu, S., Watanabe, K. and Shimizu, T. (1974). Distribution of nitrogenous constitutes in the muscle extracts of eight species of fish. Bulletin of the Japanese Society of Scientific Fisheries, 40: 909- 915.

Ryder, J. M. (1985). Determination of adenosine triphosphate and its breakdown products in fish muscle br high-performance liquid chromatography. Journal of Agricultural and Food Chemistry, 33: 678-680.

SAS. 1994. “SAS User’s Guide: Statistics.” Cary, NC. SAS Institute, Inc.

THE CALPAIN SYSTEM IN CHICKEN SKELETAL MUSCLE AND CHANGES IN PROTEOLYTIC ACTIVITY DURING AGEING

H.L. Lee[1§], V. Santé-Lhoutellier[2§], S. Vigouroux[1], Y. Briand*[1] and M. Briand[1]

[1]*Laboratoire de Génie Chimique et Biochimique, Unite Biochimie - CUST - Université Blaise Pascal, 63174 Aubière, France and* [2] *INRA QuaPA, Biochimie et Protéines du Muscle, 63122 Saint Genès Champanelle, France. Email: yves.briand@univ-bpclermont.fr* §HSL and VS contributed equally to this work

Keywords: chicken, muscle, *post mortem*, proteolysis, calpains

Introduction

Post mortem tenderisation of meat depends largely on the action of proteases, which weaken myofibrillar structures. Among these proteases, calpains probably constitute the proteolytic system that plays a key role in meat ageing, at least in the first few hours. The role of calpains in *post mortem* proteolysis and tenderisation is well documented, particularly in cattle (Koohmaraie, 1996; Geesink *et al.*, 2000), where μ-calpain is the active form able *in vitro* to reproduce observations made *in vivo*, unlike m-calpain, which requires high calcium concentrations. However, as shown by Boehm *et al.*, (1998) in cattle and Veiseth *et al.*, (2001) in sheep, μ-calpain activity drops very quickly after death and probably cannot account for all the phenomena observed. In cattle, the proteasome may also participate in tenderisation (Robert *et al.*, 1999) as it damages myofibrils *in vitro* and remains active for eight days after slaughter (Lamare *et al.*, 2000). Data on meat ageing are incomplete and contradictory in chickens (Shreurs, 2000), in which 80% of maximum tenderness is reached within 10 hours after slaughter, compared with 10 days in cattle (Dransfield, 1994). The nature of calpains in chicken is unclear, and one, two or even three ubiquitous calpains of differing calcium sensitivity have been reported (Ishiura *et al.*, 1978; Wolfe *et al.*, 1989; Sorimachi *et al.*, 1995). Our aim was to identify the special features of calpains in chicken, in comparison with calpains in cattle, and to study *post mortem* changes in the calpain system.

Materials and Methods

Samples of about 2 grams of *Pectoralis superficialis* muscle were collected at different times (5 min, 1 h, 3 h, 6 h, 12 h, 24 h, 48 h, 72 h) after death, from nine 7-week-old female chickens whose carcasses were kept at 4°C in the slaughterhouse of the Theix INRA research centre. Samples were frozen in liquid nitrogen, powdered and stored at -80°C until use. Calpain activity was measured by zymography. Samples were analysed by non-denaturing electrophoresis in gels copolymerised in the presence of casein, EDTA and EGTA, according to a slightly modified version of the method of Raser *et al.*, (1995). Gels were incubated in the presence of calcium and a reducing agent that activates calpains. In the region of a band of activated calpain, the casein was digested into small fragments which diffused out of the gel. The gels were then stained with Coomassie blue, and a clear band was noted in the presence of calpain. The bands were digitised using a scanner (Umax) and Photoshop software. The signals obtained were quantified using QuantityOne software (Bio-Rad). Under our experimental conditions, the signal was proportional to the amount of protein loaded on the gel.

Results and Discussion

Specificity of chicken calpains: Calpain activities were measured in raw muscle extracts from cattle and chicken. Zymograms were incubated in the presence of different calcium concentrations. Figure 1 shows that four types of calpains appeared successively. In the absence of calcium, no activity was visible, thus demonstrating the method's specificity. In the presence of 10 μM calcium, a calpain of low mobility was seen in chicken (1), and bovine μ-calpain appeared at 30 μM calcium (2). The second form in chicken was seen in the presence of low calcium concentration (100 μM calcium) (3) whereas in bovine, the m-calpain was activated at 1 mM and 5 mM calcium (4). In addition, the proportions of μ- and m-calpains differed greatly, since in cattle they were present in roughly equal quantities, whereas in chicken μ-calpain was found in small amounts and accounted for under 15% of total calpain activity.

Post mortem changes in calpain activity in chicken muscle: Calpain activity was measured by zymography using chicken muscle extracts obtained at different times *post mortem* (see Materials and Methods). μ-calpain activity was very low at the outset and rapidly diminished until it was negligible 12 hours *post mortem* (Figure 2). μ/m-calpain activity was stable and from 24 hours onwards there was a slight decrease, which was less than 10% after 72 hours. A third form of calcium-dependent protease appeared after 6 hours and progressively increased up to 72 hours. Preliminary work suggests that this was a phosphorylated form of μ/m-calpain.

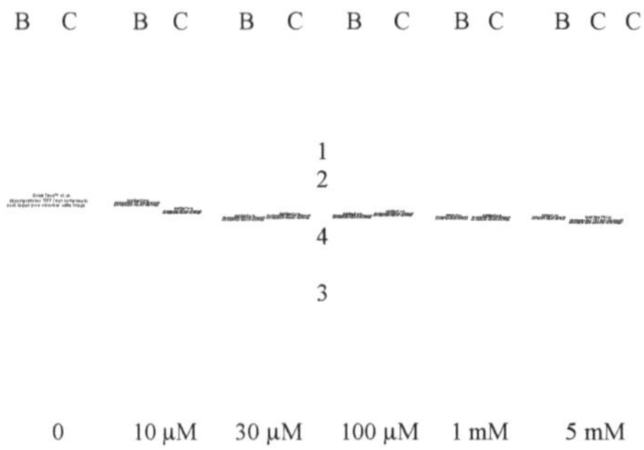

Figure 1: Activities of calpains from bovine *semi Tendinosus* muscle and/or chicken muscle *pectoralis superficialis* measured by zymography, as a function of calcium concentration B = bovine; C = chicken.

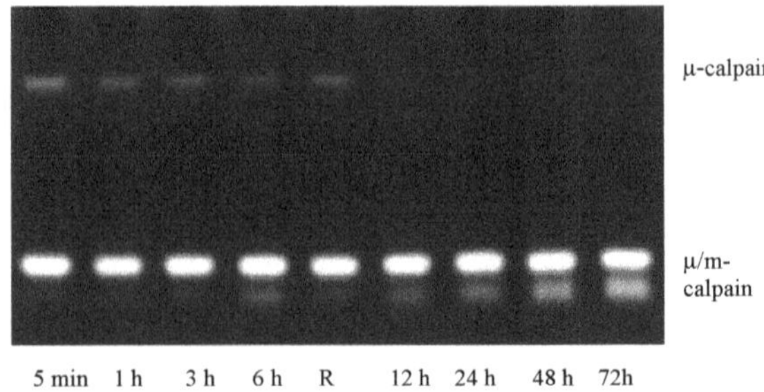

Figure 2: *Post-mortem* changes in calpain activities in chicken *pectoralis superficialis* muscle. R= reference.

Chicken and mammalian calpain isoforms differ in electrophoretic mobility and the calcium sensitivity of the former is much greater than that of the latter, justifying the name µ/m-calpain for the form with the greatest electrophoretic mobility. In contrast to the two mammalian forms, which are present in roughly equivalent amounts, in chicken the µ/m-calpain form predominates and µ-calpain accounts for less than 15% of the total activity. The calcium sensitivity of these chicken calpains means that they are active soon after death, and µ-calpain activity decreases sharply in the first few hours (and fit the pH decline in muscle) due to autolysis. It is possible that, as in mammals, chicken µ-calpain is in the active form *post mortem* and causes early destabilisation of myofibrillar structures, thereby permitting the action of other proteolytic systems like µ/m-calpain or the proteasome. Tenderisation of chicken muscle does indeed take place in the first few hours *post mortem* (Dransfield, 1994). Moreover, we have shown in the laboratory that 50% of proteasome activity in chicken muscle remains 48 hours after slaughter.

References

Boehm, M.L., Kendall T.L., Thompson V.F. and Goll D.E. (1998). Changes in the calpains and calpastatin during post-mortem storage of bovine muscle. *Journal of Animal Science,* 76, 2415-34.

Dransfield, E. (1994). Optimisation of tenderisation, ageing, and tenderness. *Meat Science,* 36, 105–121.

Geesink G.H., Ilian M.A., Morton J.D. and Bickerstaffel R. (2000). Involvement of calpains in postmortem tenderization. In *Proceedings of the New Zealand Society of Animal Production*, 60, 99-102.

Ishiura S., Murofushi H., Suzuki K. and Imahori K. (1978). Studies of a calcium-activated neutral protease from chicken skeletal muscle. I. Purification and characterization. *J. Biochem. (Tokyo),* 84, 225-30.

Koohmaraie M. (1996) Biochemical factors regulating the toughening and tenderization process of meat. *Meat Science*, 43, 193-201.

Lamare M.C., Taylor R.G., Farout L., Briand Y. and Briand M. (2002). Changes in proteasome activity during postmortem aging of bovine muscle. *Meat Science.,* 61, 199-204.

Raser K.J., Posner A. and Wang K.K. (1995). Casein zymography: a method to study m-calpain, m-calpain, and their inhibitory agents. *Archives Biochemistry and Biophysics,* 319, 211-6.

Robert N., Briand M., Taylor R.G., Briand Y. (1999). The effect of proteasome on myofibrillar strucutures in bovine skeletal muscle. *Meat Science,* 51, 149-53.

Schreurs F.J.G. (2000). Post-mortem changes in chicken muscle. *World's Poultry Science Journal,* 56, 319-46.

Sorimachi H., Tsukahara T., Okada-Ban M., Sugita H., Ishiura S. and Suzuki K. (1995). Identification of a third ubiquitous calpain species: chicken muscle expresses four distinct calpains. *Biochimica Biophysica Acta.,*1261, 381-93

Suzuki A. and Sorimachi H. (1998). A novel aspect of calpain activation. *FEBS Letters,* 433(1-2), 1-4.

Veiseth E., Shackelford S.D., Wheeler T.L. and Koohmaraie M. (2001). Effect of postmortem storage on mu-calpain and m-calpain in ovine skeletal muscle. *Journal of Animal Science,* 79(6), 1502-8.

Wolfe F.H., Sathe S.K., Goll D.E., Kleese W.C., Edmuns T. and Duperret S.M. (1989). Chicken skeletal muscle has three Ca2+-dependent proteinases. *Biochim. Biophys. Acta.,* 998, 236-50.

THE POST MORTEM ACTIVITY OF GLYCOGEN DEBRANCHING ENZYME

M. Ylä-Ajos* and E. Puolanne

Department of Food Technology, P.O. Box 44, University of Helsinki, 00014 University of Helsinki, Finland. Email: maria.yla-ajos@helsinki.fi

Keywords: glycogen debranching enzyme, RN genotype, pH decline, pigs, *longissimus dorsi*

Introduction

Shortly after slaughter the enzymes in meat operate in conditions quite similar to prevailing in a living muscle. To maintain the ATP level constant, energy is produced by glycogen degradation. Glycogen phosphorylase together with glycogen debranching enzyme (GDE) catalyse the breakdown of it to glucose-1-phosphate and free glucose (Brown and Illingworth-Brown, 1966). Glucose-1-phosphate is further broken down to lactate resulting in an acidification of muscles. When measuring the enzyme activities in meat, the samples are usually taken as soon as possible after slaughter. The *post mortem* pH decrease, however, continues for 24 or 48 hours in pig muscles. Thus, the glycolytic enzymes must maintain their activity a fairly long period of time. Both the rate and the extent of pH decrease differ between animals and between muscles. In Hampshire breed pigs a dominant RN^- allele in PRKAG3 gene is common (Milan *et al.*, 2000). The allele is associated with an increased glycogen content, faster pH decrease and lower ultimate pH in glycolytic muscles from RN^- carriers compared with the wild type animals (Fernandez *et al.*, 1992; Enfält *et al.*, 1997; Lindahl *et al.*, 2004). Furthermore, the ultimate pH is attained not until 48 h after slaughter in RN^- carries while in the wild type animals the pH decrease ceases 24 h *post mortem* (Lindahl *et al.*, 2004). This research focuses on the change in the *post mortem* GDE activity during the period of 0.5 h to 48 h after slaughter in *longissimus dorsi* muscle from Hampshire crossbred animals with and without presence of the RN^- allele.

Materials and Methods

Fourteen Hampshire x (Swedish Landrace x Yorkshire) crossbred pigs were stunned in CO_2 and slaughtered at a commercial slaughterhouse in Sweden. A DNA test (Milan *et al.*, 2000) was applied to identify RN^- allele carriers (RN^- carrier, n=7) and wild type animals (n=7). Samples from *M. longissimus dorsi* were obtained 0.5, 3, 5, 24 and 48 hours after bleeding, frozen in liquid nitrogen and stored at -80°C until analyzed. The pH was measured (Knick portable pH-meter equipped with a combination gel electrode, SE104, Knick Berlin, Germany) from *M. longissimus dorsi* at the last thoracic vertebra at 0.5, 3, 5, 24 and 48 hours after bleeding.

The GDE activity was measured by a method following the change in the iodine complex spectrum of phosphorylase limit dextrin described more detailed (Kylä-Puhju *et al.*, 2005). The method enables the use of crude meat extract in the GDE activity analyses. Limit dextrin is the natural substrate for GDE and it was available in our laboratory. The temperature of the reaction mixture was 39 °C and pH 6.3±0.05. The reproducibility of the GDE activity assay was determined by using an internal standard. A large muscle sample was ground in liquid nitrogen, diced, stored at -80°C and analyzed with the samples. The coefficient of variation (CV) for the internal standard was ±18.3%. The differences in the GDE activity and in pH between the genotypes were tested by using the MIXED procedure of SAS v. 8.02. The model included the fixed effects of genotype, sex, time and interaction between genotype and time and a random animal effect to account for repeated measurements performed within the same animal.

Results and Discussion

The pH 0.5 h *post mortem* was similar in the RN^- carriers (6.53±0.06) and in wild type animals (6.65±0.06). However, as reported earlier (Enfält *et al.*, 1997; Lindahl *et al.*, 2004), the rate of pH decrease was faster and the ultimate pH was lower in the RN^- carriers (5.21±0.03) than in wild type animals (5.36±0.03). The GDE activity was similar in both RN genotypes up to 5 h after slaughter (Figure 1). Estrade *et al.* (1994) found that 5 min after slaughter, both the GDE activity and the glycogen phosphorylase activity are uniform between the RN genotypes. Thus, it seems that immediately after slaughter, the glycogenolytic enzymes are not more active in RN^- carriers than in wild type animals despite the higher glycogen content in RN^- carriers. Thereby the GDE activity does not explain the faster pH decrease early *post mortem* in RN^- carriers than in wild type animals.

The GDE activity decreased slowly as a function of time *post mortem*, and the decrease was statistically significant until after 5 hours *post mortem*. At 24 h *post mortem* the GDE activity was significantly reduced in both genotypes, but the decrease was more pronounced in wild type animals. The GDE activity 24 h *post mortem* was higher in RN^- carriers compared with wild type animals, and this difference was maintained up to 48 h *post mortem*. In RN^- carries, despite the higher rate of pH decrease, the ultimate pH is attained not until 48 h after slaughter (Lindahl *et al.*, 2004). Thus, the extended pH decrease in RN^- carriers might be a result of longer maintenance of the high GDE activity in these animals. Furthermore, in some situations (like when very fast chilling is concerned), the GDE may catalyze a rate-limiting step in *post mortem* glycogenolysis and thus glycolysis (Taylor *et al.*, 1975; Kylä-Puhju *et al.*, 2005).

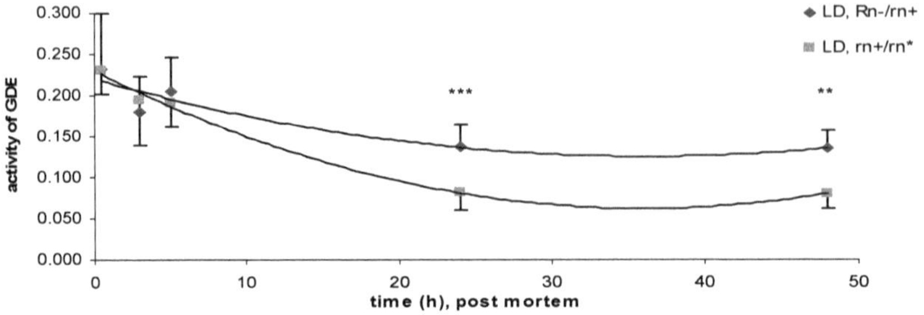

Figure 1: Activity of GDE (Δabs/Δmin) with standard deviations in RN⁻ carriers (♦) and in wild type pigs (■, rn^+). Asterisks indicate significant differences between the two genotypes (*** P<0.001, ** P<0.01).

The RN^- mutation is located in the gene coding for adenosine monophosphate-activated protein kinase (AMPK), which is a key metabolic enzyme (Milan *et al.*, 2000). Activated AMPK inhibits the ATP-consuming pathways and stimulates ATP-generating pathways (Hardie and Carling, 1997) and is important for maintaining the activity of glycogen phosphorylase in *post mortem* muscle (Shen and Du, 2005). Thus the mutation in the AMPK gene might explain why RN^- carriers exhibit a prolonged high GDE activity, as found in the present study. The GDE activity was measured in laboratory, in fixed conditions. In carcass, however, the temperature decreases concurrent with pH decrease. The pH decrease has a minor effect on the GDE activity, but the effect of temperature decrease is significant (Kylä-Puhju *et al.*, 2005). Thus, carcass chilling may further rise the difference in GDE activity between RN genotypes after 5 hours *post mortem*.

Conclusions

The decrease in the GDE activity with increasing *post mortem* time is slower and less pronounced, the pH decrease is faster and the ultimate pH lower in RN^- carriers than in wild type animals. The long period of high GDE activity in RN^- carriers may enable the extended pH decrease.

Acknowledgements

This study is a part of a larger project studying the effect of the RN-gene on meat colour stability and glycogen metabolism. The authors wish to thank Dr. G. Lindahl, Dr. J. Young, Dr. P. Theil, research director H. Andersen and Prof. N. Oksbjerg from DIAS, Research Centre Foulum, Denmark for their contribution to this work.

References

Brown, D.H. and Illingworth-Brown, B.I. (1966). Enzymes of glycogen debranching: Amylo-1,6-glucosidase (I) and oligo-1,4→1,4-glucantransferase (II). New York: Academic Press.

Enfält, A.C., Lundström, K., Karlsson, A. and Hansson, I. (1997). Estimated frequency of the RN allele in swedish hampshire pigs and comparison of glycolytic potential, carcass composition, and technological meat quality among swedish hampshire, landrace, and yorkshire pigs. J. Animal Sci., 75: 2924-2935.

Estrade, M., Ayoub, S., Talmant, A. and Monin, G. (1994). Enzyme activities of glycogen metabolism and mitochondrial characteristics in muscles of RN- carrier pigs. Comp. Biochem. Phys., 108: 295-301.

Fernandez, X. *et al.*, (1992). Bimodal distribution of the muscle glycolytic potential in french and swedish populations of hampshire crossbred pigs. J. Sci. Food Agric., 59: 307-311.

Hardie, D.G. and Carling, D. (1997). The AMP-activated protein kinase--fuel gauge of the mammalian cell? Eur. J. Biochem./ FEBS, 246: 259-273.

Kylä-Puhju, M., Ruusunen, M. and Puolanne, E. (2005). Activity of porcine muscle glycogen debranching enzyme in relation to pH and temperature. Meat Science, 69: 143-149.

Lindahl, G. *et al.*, (2004). A second mutant allele (V199I) at the PRKAG3 (RN) locus. I. effect on technological meat quality of pork loin. Meat Sci., 66: 609-619.

Milan, D. *et al.* (2000). A mutation in PRKAG3 associated with excess glycogen content in pig skeletal muscle. Science, 288: 1248-1251.

Shen, Q.W. and Du, M. (2005). Role of AMP -activated protein kinase in the glycolysis of postmortem muscle. J. Sci. Food Agric., 85: 2401-2406.

Taylor, C., Cox, A.J., Kernohan, J.C. and Cohen, P. (1975). Debranching enzyme from rabbit skeletal muscle. Purification, properties and physiological role. Eur. J. Biochem., 51: 105-115.

COMPOSITION OF TROPOMYOSIN ISOFORMS AND MYOSIN HEAVY CHAIN ISOFORMS IN BOVINE SKELETAL MUSCLES

M. Oe*, I. Nakajima, S. Muroya and K. Chikuni

National Institute of Livestock and Grassland Science, Tukuba, Ibaraki 305-0901, Japan. Email: mooe@affrc.go.jp

Keywords: tropomyosin, myosin heavy chain, isoform, skeletal muscle

Introduction

Tropomyosin (TPM) and Myosin heavy chain (MyHC) are myofibril proteins that work together on muscle contraction. In skeletal muscle, TPM has three isoforms: TPM1, TPM2 and TPM3 (Perry, 2001). TPM1 is expressed in fast skeletal muscle and TPM3 is expressed in slow skeletal muscles, forming a dimer with TPM2. MyHC expressed three isoforms in bovine skeletal muscles (Chikuni *et al.*, 2004). MyHC-slow is a slow-type isoform and MyHC-2a and -2x are fast-type isoforms.

Although there are one TPM and two MyHC fast-type isoforms, the relationships between the fast type TPM and MyHC isoforms in muscle fibres are not clear. The aim of this study was to analyse the presence of MyHC and TPM isoforms in ten different types of skeletal muscles using real time polymerase chain reaction (real time-PCR).

Materials and Methods

Skeletal muscle samples for real time-PCR were obtained from the tongue, *masseter*, *pectralis profundus*, diaphragm, *psoas majoir*, *longissimus thoracis*, *semispinalis*, *semitendinosus*, *semimembranosus* and *biceps femoris* of three Holstein cows. The muscle pieces were excised within 1 h after slaughter and immediately frozen in liquid nitrogen. The frozen samples were crushed into fine powder, and total RNA was extracted from the frozen muscle powder by an ISOGEN total RNA extraction kit (NipponGene, Tokyo, Japan). Single-stranded cDNA was synthesized by RNase H minus M-MLV reverse transcriptase (TOYOBO, Osaka, Japan). The cDNAs from the ten muscles were used as a template of multiplex PCR (Table 1). PCR products were electrophoresed on 4% NuSieve GTG agarose gel and then stained with ethidium bromide. The image of the PCR products was scanned by FluorChem™8900 with AlphaEase FC software (Alpha Innotech, San Leandro, CA, USA).

Results and Discussion

The TPM2 isoform comprised about 50% of the total TPM in all muscles examined, while the amounts of TPM1 and TPM3 isoforms differed among the muscles. Comparing the TPM isoform composition with the MyHC isoform composition, there was high positive correlation between MyHC-slow and TPM3 ($r = 0.92$) (Table 2). The sum of MyHC-2a and -2x (fast-type MyHC isoforms) had high positive correlation with TPM1 ($r = 0.85$), but the correlations between MyHC-2a or -2x alone and particular TPM isoforms were not as high.

In the tongue and *semispinalis* muscles, the MyHC-2a isoform was the only fast-type MyHC isoforms expressed, while both TPM1 and TPM3 isoforms were expressed in these muscles. Therefore, the IIA type fibre, which expresses MyHC-2a, would be regulated delicately by changing the TPM isoform composition.

Conclusions

TPM isoforms affect muscle contraction together with the MyHCs during post mortem rigor development.

References

Chikuni, K., Muroya, S. and Nakajima, I. (2004). Myosin heavy chain isoforms expressed in bovine skeletal muscles. Meat Science, 67(1); 87-94.

Perry, S. V. (2001). Vertebrate tropomyosin: distribution, properties and function. J Muscle Res Cell Motil, 22(1); 5-49.

Table 1: Nucleotide sequence of the PCR primers used for RT-PCR.

	Forward	Reverse
TPM 1	5'ATGGAGGCCATCAAGAAGAAGATGCAGATG 3'	5'GGAGTATTTGTCCAGTTCATCTTCGGTGG 3'
TPM 2	↑	5'TCCCTTTTAGCTTTTTCTGGAGGGCCTG 3'
TPM 3	↑	5'CTTACTTCTTTCTTCTGCCTGCTTCTGC 3'
MyHC-slow	5'TGCTGCTCTCAGGCCCCTGCCACCTT 3'	5'GTCATCATGGCCATGTCCTCGATCTTGTC 3'
MyHC-2a	5'CACTTGCTAACAAGGACCTCTGAGTTCA 3'	↑
MyHC-2x	5'CTTTCCTCATAAAGCTTCAAGTTCTGACC 3'	↑

Table 2: Correlation coefficients (*r*) between content of TPM isoforms and content of MyHC isoforms in skeletal muscle (n = 30).

	MyHC-slow	MyHC-2a	MyHC-2x	MyHC-2a + -2x
TPM 1	- 0.85	0.48	0.69	0.85
TPM 3	0.92	- 0.65	- 0.62	- 0.92

ANTITHROMBIN III FROM BOVINE SKELETAL MUSCLE: PURIFICATION, CHARACTERISATION AND CELLULAR LOCATION

C. Herrera-Mendez, L. Aubry, G. Coulis and A. Ouali*

Unité de biochimie, SRV, INRA-Theix, 63122 St Genès Champanelle France. Email : aouali@clermon.inra.fr

Keywords: antithrombin, bovine skeletal muscle, purification

Introduction

In blood, thrombin activity is mainly driven by antithrombin III (AT-III), a specific inhibitor. Extravascular thrombin functions must be regulated and AT-III is a potential candidate. The enzyme expressed by muscle cells is located in the vicinity of the external leaflet of the plasma membrane at the neuromuscular junction level. This feature is in good agreement with a thrombin regulatory function at the neuromuscular junction (Liu *et al.*, 1994). AT-III was found inside all fibres of mouse skeletal muscles suggesting that, as thrombin, this protein is also expressed in muscle cells (Businaro *et al.*, 1995). In cattle, the presence of a signal peptide is undoubted (Mejdoub *et al.*, 1991). Indeed, because thrombin is located extracellularly, the control of its activity by muscle AT-III means that serpin must be exported. Despite the fact that AT-III is most likely expressed in skeletal muscle, the protein has never been purified and characterized. The present work is therefore intended to purify and characterize AT-III from bovine skeletal muscle.

Materials and Methods

Purification of AT-III was carried out from a crude extract prepared as in Bige *et al.*, (1985) from bovine *diaphragma* muscle within 1h post exanguination. The procedure comprises of four to five chromatography steps. *Inhibitory activity* against trypsin was measured as described in Tassy *et al.*, 2005. *Titration* of bovine pancreatic trypsin, bovine plasmin and thrombin were carried out as in Chase and Shaw (1967). Chymotrypsine was titrated as described in Kezdy and Kaiser (1969). *Stoichiometry* of the enzyme–inibitor interaction was determined: predefined concentrations of enzyme were incubated with increasing amounts of muscle AT-III and the residual activity was measured. *Second order association rate constant* (K_{ass}) of the muscle AT-III for the peptidases tested: thrombin, trypsin, chymotrypsin and plasmin were determined as described in Schechter and Plotnick (2004). *SDS/PAGE* was perfomed as described previously (Laemmli, 1970). Molecular masses were estimated by using the Pharmacia low M_r calibration kit. *Immunoblots* were carried out as described previously (Sentandreu *et al.*, 2003). *pH stability* of the purified inhibitor was tested in the pH range 2 to 12. *Heat stability* of the AT-III was determined at different temperatures ranging from 40°C to 100°C. *Sequencing N-terminal* of muscle AT-III was perfomed using an Applied Biosystems 477A pulsed liquid sequencer. *MALDI-TOF mass spectrometry analysis*: peptide map of the trypsin digest was perfomed as previously described (Sentandreu *et al.*, 2003). *Immunolocalization* of AT-III was perfomated on traverse sections of bovine onglet muscle using the rabbit anti-human AT-III polyclonal antibody.

Results and Discussion

Identity of the purified serpin was assessed using different technical approaches. SDS-PAGE analysis of purified AT-III showed only one band with an estimated molecular mass of 58 kDa (Figure 1, lane b). The rabbit anti-human AT-III polyclonal antibody specificity was analysed using muscle crude extract and the purified protein. As shown in Figure 2a, after SDS-PAGE, the antibody labelled only one band in the crude extract (lane 1) running similarly to the purified AT-III (lane 2). In non denaturing conditions, this antibody recognized also one band in the crude extract (Figure 2b lane 1) running as the purified AT-III (Figure 2b, lane 2). These findings suggest a high specificity of the antibody which will be used for the AT-III localisation in muscle tissue.

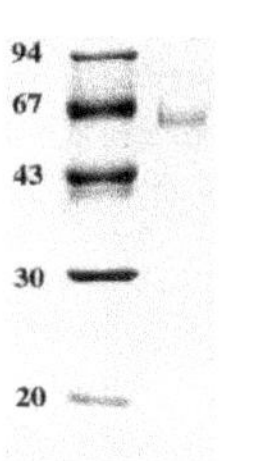

Figure 1: The inhibitor was run in denaturating conditions on a 12.5% polyacrylamide slab gel and revelated by silver staining. (a) Mr markers. (b) Purified AT-III.

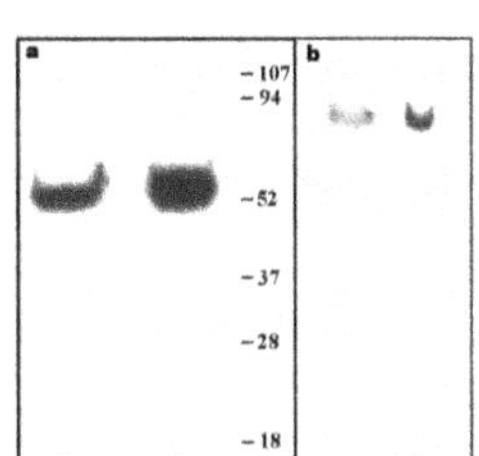

Figure 2: Analysis by immunoblotting using the rabbit anti-human AT-III polyclonal antibody. (2a, line 1): crude extract; (2a line 2): purified AT-III. In non denaturanting conditions (2b line 1): crude extract; (2b line 2): purified AT-III.

Primary sequence analysis using the Edman procedure provided the following sequence: $H^{1}RSPVEDVCTAKPR^{14}$. This sequence showed 100% similarity with the cDNA deduced sequence of the mature bovine AT-III (SwisProt accession number: P41361), a serpin inhibiting specifically thrombin (Mejdoub *et al.*, 1991). The strong labelling of the purified muscle AT-III by a polyclonal antibody raised against human antithrombin III supported our findings. Finally, mass spectrometry trypsin peptide map identified this protein as bovine AT-III, the peptide finger print covering about 60 to

70% of the bovine protein sequence (not shown). Several sets of evidence therefore stressed that the purified serpin is undoubtedly bovine AT-III.

Treatment of AT-III for 15 min at different temperatures ranging from 35 to 100°C confirmed the low thermal stability of normal bovine AT-III (Figure 3a). A decrease of about 2 to 5% in the activity was already observed after 15 min at 40°C. At 45 °C an activity loss of 15% is generally achieved. Analysis of the data using a sigmoid curve fitting programme gave a temperature of transition Tm of 52.7 °C. Bovine AT-III is relatively stable only at pH 10. Above this value the activity falls sharply to about 30% of its maximal activity at pH 12.

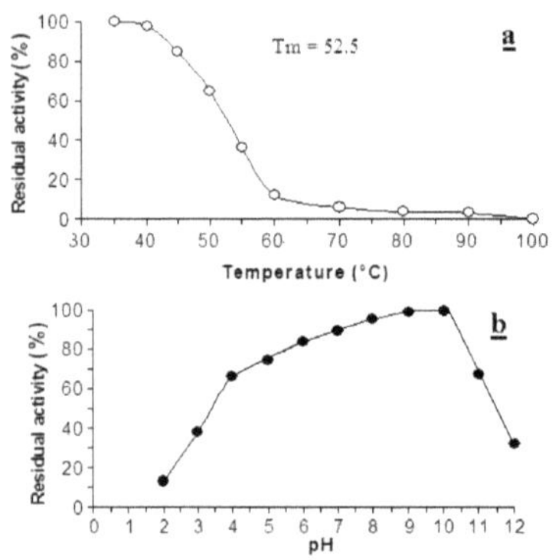

Figure 3: (a) Residual activity (percentage of the control) of AT-III: (a) Heated 15 min at different temperatures ranging 40° to 100°C. (b) Pre-incubated for 1h at different pH ranging from 2 to 12.

Table 1: Association rate constants for inhibition of serine proteinases.

	Cofactor	k_{ass} ($M^{-1}s^{-1}$)
Trypsin	-	5 x10^5
Chymotrypsin	-	6,8 x10^4
Plasmin	-	1,7 x10^4
Thrombin	- heparine -	1,8 x10^5
	+ heparine (5 µM)	2,3 x10^7

Stoichiometry of inhibition (SI): These results stressed forward that AT-III inhibition of both trypsin and thrombin occurs through a 1/1 enzyme/inhibitor ratio (SI=1). Inhibitory pattern of bovine muscle AT-III and association rate constant: the results depicted in Table 1 indicate that, though to various extent, all peptidases are sensitive to the inhibitory action of AT-III. In the presence of heparin, the k_{ass} for thrombin increased about 100 fold whereas other peptidases did not show any sensitivity to this cofactor (not shown). Several other serine peptidases were also tested including Human leukocyte elastase, cathepsin G Human kidney urokinase, porcine pancreatic kallikrein and tissue-plasminogen activator and none of them seemed to be inhibited by bovine muscle AT-III. The muslce AT-III is highly concentrated between the plasma membrane and the myofibrils, indicating that muscle AT-III is exclusively intracellular with a preferential peripheral localization.

Conclusion

Antithrombin III has so far never been purified from muscle tissue and this is the first time that a purification procedure was proposed. The present work further stressed that AT-III must definitely be included as a new partner in the studies on the role of thrombin in muscle development. Using intact skeletal muscle and primary muscle cells, examination, at the gene and protein levels, of the AT-III functions in muscle cell differentiation and muscle development will be therefore the next step in our view of a better understanding of the role of thrombin/antithrombin in these complex processes.

References

Bige, L., Ouali, A. and Valin, C. (1985). Purification and characterization of a low molecular weight cysteine proteinase inhibitor from bovine muscle. Biochim Biophys. Acta 843, 269–275.

Businaro R., Nori S.L., Toesca A., *et al.*. (1995). Immunohistochemical detection of three serum protease inhibitors in mouse skeletal muscle by confocal laser scanning microscopy. Ital J Anat Embryol. 1995;100 Suppl 1:123-30.

Chase, T. and Shaw, E. (1967) p-nitrophenyl-p_-guanidinobenzoate HCl: a new active site titrant for trypsin. Biochem. Biophys. Res. Comm. 29, 508–514

Kezdy, F. J. and Kaiser, E. T. (1969). Principles of active site titration of proteolytic enzymes. Methods Enzymol. 29, 3–27

Laemmli, U.K. (1970). Cleavage of structural proteins during the assembly of the head of bacteriophage T4. *Nature, 227*, 680-685.

Liu Y., Fields R. D., Festoff B. W. and Nelson P. G. (1994). Proteolytic action of thrombin is required for electrical activity-dependent synapse reduction Proc. Natl. Acad. Sci. U. S. A. 91, 10300–10304.

Mejdoub H., Le Ret M., Boulanger Y., Maman M., Choay J. and Reinbolt J. (1991). The complete amino acid sequence of bovine antithrombin (AT3). J Protein Chem. 10, 205-212.

Schechter N.M. and Plotnick M.I. (2004). Measurement of the kinetic parameters mediating protease-serpin inhibition. Methods. 32, 159-168.

Sentandreu, M. A., Aubry, L. and Ouali, A. (2003). Purification of bovine cathepsine B : proteomic characterization of the different forms and production of specific antibodies. *Biochem. Cell. Biol.*, *81*, 317-326.

Tassy, C., Herrera-Mendez, C.H., Sentandreu, M.A., *et al.* (2005). Muscle endopin 1, a muscle intracellular serpin which strongly inhibits elastase: purification, characterization, cellular localization and tissue distribution. Biochem. J. 388, 273-280.

TIGHTLY BOUND CALCIUM WITHIN I BAND: POSSIBLE FUNCTION IN TITIN AGGREGATION AND CALPAIN 1 BINDING TO TITIN

G. Coulis*, L. Aubry, C. Herrera-Mendez and A. Ouali

QuaPA, Muscle biochemistry group, INRA-THEIX, 63122 Saint Genès Champanelle,France.
Email: gcoulis@clermont.inra.fr

Keywords: titin, calpain 1, calcium, N-line regions, titin polymerization

Introduction

Meat tenderisation is based upon an enzymatic process implying several proteolytic systems (Sentandreu *et al.*, 2002). Amongst them, calpains have been and are still extensively investigated by meat scientists. Besides the implication of ubiquitous calpains in the meat tenderising process, some recent investigations provide contradictory findings about a possible contribution of muscle specific calpain 3 (Parr *et al.*, 1999; Ilian *et al.*, 2000; Geesink *et al.*, 2005).
Within myofibrils, calpain 1 and calcium are colocalised at N1- and N2-line levels (Raynaud *et al.*, 2005; Vignon *et al.*, 1989). Calpain 3 has been also localised at the N1 and N2 levels of the sarcomere (unpublished data). Hence, the question arising is: what relationship between these different partners, i.e. calcium, calpains and titin may exist? The purpose of the present study was to provide partial answers to that question by investigating the identification of the calcium binding site at the N1 line region of titin and the consequences of calcium binding on titin itself and titin/calpain 1 interactions.

Materials and Methods

Recombinant fragment preparation:
Recombinant titin fragment spanning Z9 to I1 domains of titin and located in the N1-line region of the protein was expressed in *Escherichia coli* using the pET vectors.
Calcium titration of the recombinant Z9-I1 titin fragment:
Calcium titration of the recombinant fragment was carried out as described previously by Johnson and Tikhunova (2002). Increasing concentrations of calcium were added, from a 2M stock solution to the recombinant fragment and changes in the intrinsic tryptophan fluorescence recorded using a Perking Elmer LS50 spectrofluorometer (λexc 305 nm λ Em 358nm). The fluorescence data were fit to the one-site model of the non-linear Hill equation: y= ymax /1+10(n*(Kd-pCa))
Polymerization of the titin fragment in the presence of calcium:
After overnight treatment of the Z9-I1 recombinant fragment with Chelex-100 resin, the mixture was run on a Superose 12 HR 10/30 column. The monomeric form eluted at a Mr of about 53000 Da was collected and used for the calcium-induced polymerization study. The fragment was incubated at room temperature without calcium (control) or with 0.1 and 1μM calcium for 30 min. An aliquot of each mixture was then loaded on a Superose 12 10/30 column. Proteins were eluted at a flow rate of 0.3 ml/min and fractions of 0.3 ml were collected.
Titin/calpain 1 interaction
Binding assays were carried out with both bovine or porcine calpain 1 and Z9-I1 titin fragment by using solid phase ELISA whether in presence 4mM calcium or 4mM EGTA.

Results and Discussion

Calcium titration of the recombinant Z9-I1 titin fragment:

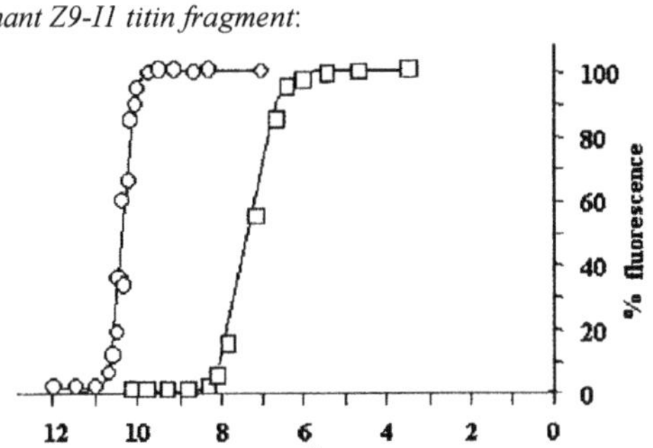

Figure 1: Calcium titration of Quin-2 (open squares) and of the Z9-I1 titin fragment (open circles).

As shown in Figure 1, increasing concentrations of calcium causes a sharp increase in the Trp fluorescence and half-maximal binding was found to occur at pCa 10,31±0.03 giving a Kd value of $4,92\times10^{-11}$ M.

The free calcium concentration was calibrated by titrating Quin-2 in similar conditions (λexc 330 nm λem 495nm). The Kd value of 61nM obtained for Quin-2 was identical to the value of 60nM reported by the supplier suggesting that the free calcium concentrations are accurate up to the nM range.

Calcium induced polymerization of the Z9-I1 titin fragment

Calcium induced the spontaneous polymerization of the recombinant fragment at the expense of the monomer which disappears almost totally at a higher calcium concentration. These findings suggest a calcium-dependant association of titin strands to form the end-filaments package at least in the N1-line region.

Calpain/titin interaction

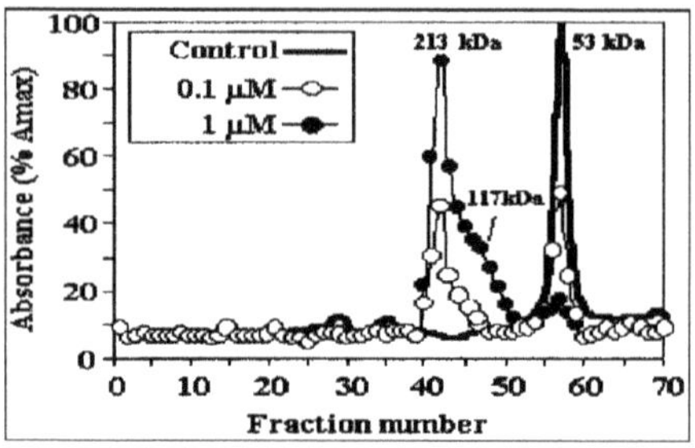

Figure 2: Elution profile of the recombinant fragment from a Superose 12 HR column after incubation for 30 min at room temperature, without calcium (inset and thick line) or with either 0.1 µM free calcium (open circles) or 1 µM free calcium (close circles). Fractions of 0.3 ml were collected.

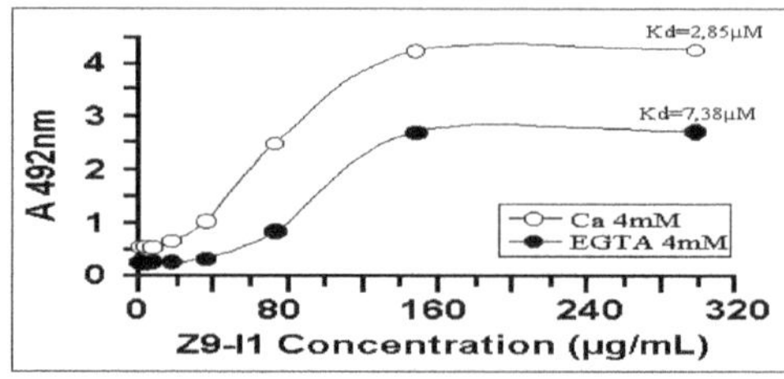

Figure 3: ELISA binding assays carried out with porcine calpain 1and Z9-I1 titin fragment whether in presence of 4mM calcium or 4mM EGTA.

Calcium increases the amount of titin fragment bound to coated calpain however with a comparable affinity in the presence (Kd=2,85µM) or absence(Kd=7,38µM) of calcium. Theses constants affinities are weaker than those observed with a 150kDa titin fragment corresponding to Z8-I5 domains (Raynaud *et al.*, 2005). Thus, calpain 1 probably interacts strongly in a calcium-dependent manner with a region including I2-I5 domains.

Conclusions

This study provides possible answers to the physiological significance of tightly bound calcium in skeletal muscle fibres. A calcium-dependent aggregation of titin molecules would allow a better coordination of the stretching and shortening of each titin strand and would ensure a better force transmission between adjacent sarcomeres as well. The calcium induced-aggregation could have beneficial effect upon titin-calpain interactions increasing the amount of calpain bound to titin.

References

Geesink, G. H., Taylor, R. G and Koohmaraie, M. (2005). Calpain 3/p94 is not involved in postmortem proteolysis. *J. Anim. Sci.* 83 (7):1646-52.

Ilian, M.A., Morton, J.D., Kent, M.P., Le Couteur, C.E., Hickford, J, Cowley R and Bickerstaffe, R (2000). Intermuscular variation in tenderness: association with the ubiquitous and muscle-specific calpains. *J. Anim. Sci.* 79, 122-132.

Johnson, J.D. and Tikhunova, S.B. (2002) In: Calcium-Binding protein protocols, Vol. 2: Methods and techniques pp.89-102, Humana Press Inc., Totowa, NJ.

Parr, T., Sensky, P.L, Scothern, G.P., Bardsley, R.G., Buttery, P.J., Wood, J.D. and Warkup, C. (1999). Relationship between skeletal muscle-specific calpain and tenderness of conditioned porcine longissimus muscle. *J. Anim. Sci.* 77, 661-668.

Raynaud, F., Fernandez, E., Coulis, G., Aubry, L., Vignon X., Bleimling, N., Gautel, M., Benyamin, Y. and Ouali, A. (2005). Calpain 1-titin interactions concentrate calpain 1 in the Z-band edges and in the N2-line region within the skeletal myofibril. *FEBS J.* 272(10):2578-90.

Sentandreu, M.A., Coulis, G. and Ouali, A. (2002). Role of muscle endopeptidases and their inhibitors in meat tenderness. *Trends in food science and technology* 13, 400-421

Vignon, X., Beaulaton, J. and Ouali, A. (1989). Ultrastructural localization of calcium in post-mortem bovine muscle: a cytochemical and X-ray microanalytical study. *J. Histochem* 21(7):403-11

COMPARISON OF HISTOCHEMICAL CHARACTERISTICS AND MEAT QUALITY TRAITS IN DIFFERENT PIG BREEDS

Y.C. Ryu*, Y.M. Choi, S.H. Lee, G.W. Go, H.G. Shin and B.C. Kim

Division of Food Science, Korea University, 136713 Seoul, Korea. Email: ycryu@korea.ac.kr

Keywords: histochemical characteristics, *longissimus dorsi*, meat quality, pig breeds

Introduction

Mechanisms controlling pork quality development are often associated with altered postmortem muscle metabolism. Specifically, changes in the rate of glycolysis can create unfavorable muscle pH. A high rate of pH decline and a low ultimate pH result in muscle protein denaturation and diminish quality parameters (Ryu *et al.*, 2005). One of the main factors determining muscle biochemical pathways is fibre type composition: skeletal muscle is composed of different types of fibres, which are the results of co-ordinated expression of distinct sets of structural proteins and metabolic enzymes (Chang *et al.*, 2003). Because muscle fibres contain different myosin heavy chains, which are responsible for their different ATPase activity (Picard *et al.*, 1999), it is possible that fibre type composition may be associated with postmortem changes in the conversion of muscle to meat, and subsequently meat quality (Ryu and Kim, 2006). Therefore, the variation in fibre type characteristics can explain part of the variation in some meat quality traits.

Some recent studies found correlations between muscle fibre characteristics and meat quality traits in beef (Ozawa *et al.*, 2000) and in pigs (Ryu and Kim, 2005). For the practical application of this knowledge to improve and control meat quality, more information of the effects of the fibre type characteristics on meat quality is necessary. Therefore, this research focuses on comparing muscle histochemical characteristics between Berkshire, Landrace, Yorkshire, and crossbred pigs.

Materials and Methods

This study involved 594 pigs from four different breeds as follows: Berkshire, n=105; Landrace, n=102; Yorkshire, n=221 and crossbred pigs (Landrace×Yorkshire×Duroc, LYD), n=166. Pigs were slaughtered according to standard commercial procedures in a Korean abattoir. Within 45 min postmortem, muscle samples for histochemical analysis were taken from the *longissimus* muscle at the 8th *thoracic vertebrae.* Samples were cut into 0.5 × 0.5 × 1.0 cm pieces, promptly frozen in isopentane cooled by liquid nitrogen, and stored at -80°C until subsequent analyses. Serial transverse muscle sections (10μm) were obtained from each sample with a cryostat at -20°C and mounted on glass slides. The myosin adenosine triphosphatase activities were detected after acid (pH 4.7) preincubation (Brooke and Kaiser, 1970). About 600 fibres per sample were examined by an image analysis system and the muscle fibres were divided into type I, IIa, and IIb according to the nomenclature of Brooke and Kaiser (1970).

Muscle pH was measured directly on carcass at the 7th/8th *thoracic vertebrae* using a spear type electrode at 1 h (pH_1) and 24 h postmortem (pH_u). Following 24 h of chilling, *M. longissimus dorsi* was taken to evaluate the meat quality traits. Drip loss was determined by suspending muscle samples standardised for surface area in an inflated plastic bag for 48 h at 2°C. The colour of the meat was measured at the 8th/9th *thoracic vertebrae* at 24h postmortem with a chromameter after exposing the surface to the air for 30 min at 2°C. The average of triplicate measurements was recorded and the results were expressed as C.I.E. L*, a*, and b*. The data were analysed statistically using the SAS program. The differences between the breeds were studied using analysis of variance and Duncan's test was used to locate the differences.

Results and Discussion

Muscle fibre number and cross-sectional area were not significantly different between pig breeds, which could be explained by the differences in weight and age of the pigs (Table 1). A clear difference in histochemical property was observed in the results of fibre type composition. The area percentage of type I fibres was higher ($P< 0.001$), and that of type IIb fibres was lower ($P < 0.05$) in Berkshire pigs than those of other breeds. The muscle pH_1 and pH_u were significantly higher in Berkshire pigs (Table 2). Water-holding capacity measured as drip loss and colour parameters were significantly different between breeds ($P < 0.001$). Ryu and Kim (2005) reported that the percentage of type IIb fibres was negatively related to the pH at 45 min postmortem and positively related to drip loss and lightness. Solomon *et al.*, (1998) also suggested that high proportion of type IIb fibres may be more prone to PSE pork because of its anaerobic nature, higher glycogen content, and lower ultimate pH. In this study, Berkshire pigs, which showed the highest muscle pH, the lowest drip loss, and L* values, contained a significantly higher percentage of type I fibres and a lower percentage of type IIb fibres than those of other breeds.

Table 1: Cross-sectional area, the number of fibres per mm^2, and fibre type composition (means ± s.d.) of *longissimus dorsi* muscle in different breeds

	Breed				Levels of significance
	Berkshire	Landrace	Yorkshire	LYD	
Cross-sectional area (μm^2)	4262 ± 743.8	4063 ± 732.1	4170 ± 691.5	3993 ± 734.1	NS
Fibre number/mm^2	241 ± 40.6	253 ± 45.8	246 ± 39.6	258 ± 44.4	NS
Area percentage (%)					
Type I	7.51 ± 3.12^a	6.21 ± 2.15^{bc}	6.80 ± 2.57^{ab}	5.43 ± 2.53^d	***
Type IIa	5.80 ± 3.26^a	5.60 ± 2.85^a	6.42 ± 2.81^{ab}	7.13 ± 2.48^b	**
Type IIb	85.03 ± 11.80^a	88.19 ± 3.87^b	86.78 ± 3.71^b	87.45 ± 3.30^b	*
Number percentage (%)					
Type I	10.69 ± 4.39^a	9.14 ± 3.32^b	8.92 ± 3.64^b	8.13 ± 3.79^b	***
Type IIa	9.02 ± 4.36^a	9.40 ± 3.72^{ab}	10.61 ± 4.36^b	11.44 ± 4.24^c	***
Type IIb	80.29 ± 6.49	81.53 ± 4.98	80.47 ± 5.21	80.43 ± 4.89	NS

Levels of significance: NS = not significant; * $P < 0.05$; ** $P < 0.01$; *** $P < 0.001$.
Means within a row with no common superscript differ significantly ($P < 0.05$).

Table 2: Meat quality traits (means ± s.d.) of the porcine *longissimus dorsi* muscle in different breeds

	Breed				Levels of significance
	Berkshire	Landrace	Yorkshire	LYD	
Muscle pH_1	6.19 ± 0.24^a	5.86 ± 0.33^c	6.03 ± 0.36^b	6.02 ± 0.27^b	***
Muscle pH_u	5.66 ± 0.22^a	5.50 ± 0.14^d	5.56 ± 0.11^c	5.61 ± 0.12^b	***
Drip loss (%)	2.97 ± 1.55^c	5.99 ± 2.81^a	4.66 ± 2.69^b	4.24 ± 2.14^b	***
Lightness (L^*)	45.29 ± 2.47^a	49.72 ± 3.63^d	47.66 ± 2.36^c	46.60 ± 3.43^b	***
Redness (a^*)	7.87 ± 1.69^a	6.60 ± 1.42^b	6.85 ± 1.33^b	6.83 ± 1.23^b	***
Yellowness(b^*)	3.10 ± 1.16^a	3.77 ± 1.3^b	3.36 ± 0.91^a	3.69 ± 1.13^b	***

Levels of significance: *** $P < 0.001$.
Means within a row with no common superscript differ significantly ($P < 0.05$).

Conclusions

Comparing the fibre type composition observed from the different breeds, the results imply that the *longissimus dorsi* muscle of Berkshire pigs is more oxidative than that of other breeds. A high pH value in Berkshire pigs is due to a high percentage of type I fibres and a low percentage of type IIb fibres. Based on these results, it is concluded that muscle fibre composition is useful parameter to explain the variation of the meat quality.

References

Brooke, M. H., and K. K. Kaiser. (1970). Three myosin adenosine triphosphatase system: the nature of their pH liability and sulphydryl dependence. *Journal of Histochemystry and Cytochemystry*, 18: 670–672.

Chang, K. C., N. Da Costa, R. Blackley, O. Southwood, G. Evans, G. Plastow, J. D. Wood, and R. I. Richardson. (2003). Relationships of myosin heavy chain fibre types to meat quality traits in traditional and modern pigs. *Meat Science*, 64: 93–103.

Ozawa, S., T. Mitsuhashi, M. Mitsumoto, S. Matsumoto, N. Itoh, K. Itagaki, Y. Kohno, and T. Dohgo. (2000). The characteristics of muscle fiber types of longissimus thoracis muscle and their influences on the quantity and quality of meat from Japanese Black steers. *Meat Science*, 54: 65–70.

Picard, B., C. Barboiron, M. P. Duris, H. Gagniere, C. Jurie, and Y. Geay. (1999). Electrophoretic separation of bovine muscle myosin heavy chain isoforms. *Meat Science*, 53: 1–7.

Ryu, Y. C., and B. C. Kim. (2005). The relationship between muscle fiber characteristics, postmortem metabolic rate, and meat quality of pig longissimus dorsi muscle. *Meat Science*, 71: 351–357.

Ryu, Y. C., and B. C. Kim. (2006). Comparison of histochemical characteristics in various pork groups categorized by postmortem metabolic rate and pork quality. *Journal of Animal Science*, 84: 894–901.

Ryu, Y. C., Y. M. Choi, and B. C. Kim. 2005. Variations in metabolite contents and protein denaturation of the longissimus dorsi muscle in various porcine quality classifications and metabolic rates. *Meat Science*, 71: 522–529.

Solomon, M. B., R. L. J. M. van Laack, and J. S. Eastridge. (1998). Biophysical basis of pale, soft, exudative (PSE) pork and poultry muscle: A. review. *Journal of Muscle Foods*, 9: 1–12.

PRERIGOR SKELETAL ALTERATION TO IMPROVE BEEF MUSCLE TENDERNESS

P. Tribot Laspière and Y.M. Chatelin

Institut de l'Elevage, Service Qualité des Viandes, Route d'Epinay, 14310 Villers-Bocage, France.
Email : paul.tribot-laspiere@inst-elevage.asso.fr

Keywords : beef, meat, tendercut, tenderness, sensory analysis

Introduction
For the consumer, tenderness is one of the most important properties of beef meat. The main factors controlling meat tenderness occur after slaughtering (Dransfield, 1994). If the ageing effect on meat tenderness is widely known and usually used in the meat industry, others factors like Pelvic Suspension (PS), despite its efficiency on tenderness in different hind limb muscles, is not as widely used as ageing is. Perhaps because of its constraints (equipment and extra time requirement, heavy carcass fall risk), many countries do not use it at all. The effects of prerigor cutting bone techniques on tenderness have been investigated in the past. This treatment, the Tender Cut (TC) technology, could be an interesting alternative to stretch muscles without using Pelvic Suspension. But, the results on tenderness were various according the authors (Wang *et al.*, 1993; Ludwig *et al.*, 1997; Shanks *et al*, 2002). The first objective of this investigation was to evaluate the effect of TC at 2 ageing stages on meat tenderness of eight muscles, and secondly, to evaluate the relative importance of TC and ageing on tenderness.

Materials and Methods
Fifteen cows were slaughtered (3.5 to 9 years old, weight carcasses from 231 to 363kg). The mean carcass score was P+3 according to the EUROP grading system. Electrical stimulation was not used. Carcass sides of these animals were randomly allocated to one of the two pre rigor treatments: classical treatment, without any tender stretch alteration (NTC) – or tender stretch by the Tender Cut process (TC). One side from each carcass was skeletal cut approximately 60 minutes after bleeding while the other side served as control (NTC) under industry conditions. For implementing TC technology, 3 skeletal cuts were given (Claus *et al.*, 1991): at the 12th vertebrae, through the *ischium* on the pelvic bone and at the junction between the 4th/5th sacral vertebrae. The carcasses were chilled (at 2°C) until cutting 2 days after slaughtering. pH was measured in the centre of the *Longissimus dorsi* at the 13th rib.

The 8 major muscles of the hindquarter were removed: *Longissimus dorsi* (LD), *Tensor fascia latae* (TF), V*astus lateralis* (VL), *Semi tendinosus* (ST), 2 parts of the *Semi membranosus* (*Ischium* side SMI and the *Gastrocnemius* side SMG) and 2 parts of the *Biceps femoris* (*Ischium* side BFI and the *Gastrocnemius* side BFG). All 8 muscles removed from each carcass sides were cut in 2 parts before vacuum-packaging. One side was directly frozen while the other side was aged for 7 additional days at 2°C before freezing (-24°C). After thawing, each muscle was cubed and cooked in an oven at 310°C for 7 minutes, until the internal end-point temperature of approximately 57°C was reached. A selected and trained panel of 12 judges performed the sensory analysis. The samples (4 in each plate) were served to the judges who compared and scored them for tenderness on a scale from 0 (tough) to 100 (tender). From the same carcass and the same muscle each plate contained the 4 samples corresponding to all treatments: non-tender cut with 2 days ageing (NTC2), non-tender cut with 9 days ageing (NTC9), tender cut with 2 days ageing (TC2) and tender cut with 9 days ageing (TC9).
Statistical evaluation was performed using the procedure MIXED in SAS.

Results and Discussion
The sensory evaluation results showed as found by Claus, JR *et al.* (1997), that tenderness was improved by TC as compared with the NTC treatment on LD, and to a lesser degree on BFI and BFG whatever the ageing duration (Figure 1). The tenderness of TF, VL, ST, SMI and SMG was not affected.
According to previous works, as compared with NTC treatment, TC resulted in stretching or not stretching muscle fibres and respectively more tender or unchanged meat, depending upon the muscle anatomical position. The particular positive effect performed on LD became from the especially strong stretching muscle, which was due to the weight of the fore quarter.

As Ludwig *et al.* (1997) showed, the TC process, combined with ageing, resulted in an additional effect on tenderness. TC plus 9 days ageing gave a higher gain in tenderness than this obtained with one of the two treatments (TC or ageing) applied alone (Figure 1).

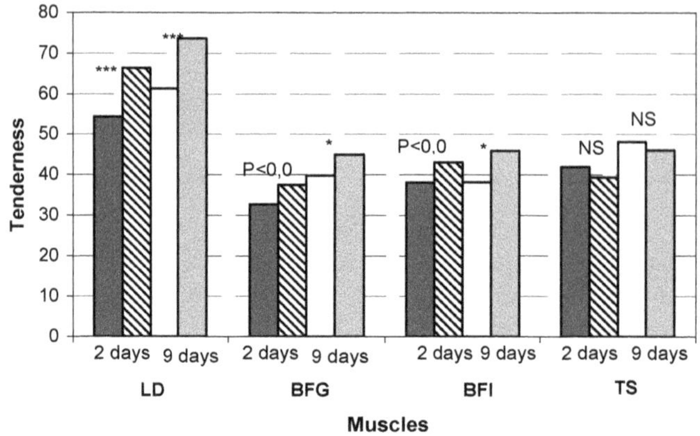

* P<0,05 ** P<0,01 *** P<0,001 NS : non significant

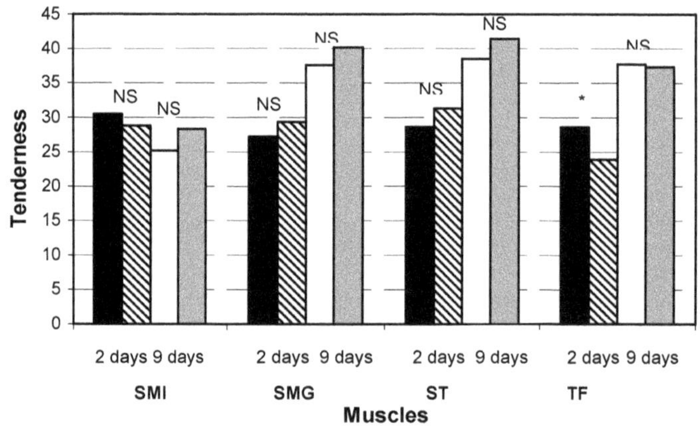

Figure 1: Comparative effects of TC and NTC on meat tenderness from different muscles after 2 and 9 days ageing.

Conclusion

TC treatment significantly improved tenderness in LD similar to 1.7 week ageing. This process could easily be used in the meat industry to improve tenderness in this muscle or to reduce the ageing period whilst preserving tenderness.

References

Claus, J. R. and Marriott, N. G. (1991). Selected skeletal alteration to improve muscle tenderness through a process now refereed to as the Tendercut^{TM} process. CIT Docket 226-VPI: VTIP Disclosure No. 91-001. Virginia Center for Innovative Technology. Herndon, VA.

Claus, J. R., Wang, H. and Marriott, N.G. (1997). Prerigor carcass muscle strectching. Effects on tenderness of grain-fed beef under commercial conditions. Journal of Food Science, Vol 62, No 6, 1231-1234.

Dransfield, E. (1994). Optimization of tenderisation, ageing and tenderness. Meat Science, 36, 105-121.

Ludwig, C.J., Claus, J.R., Marriott, N.G., Johnson, J. and Wang, H. (1997). Skeletal Alteration to Improve Beef Longissimus Muscle Tenderness. Journal of Animal Science, 75, 2404-2410.

Schanks, B. C., Wulf D. M., Reuter B. J. and Maddock R. J. Increasing tenderness of beef round and sirloin muscles through prerigor skeletal separations. Journal of Animal Science.

Wang, H., Claus J. R. and Marriott N.G. (1994). Selected skeletal alterations to improve tenderness of beef round muscles. Journal of Muscle Foods 5, 137-147

INFLUENCE OF MYOSIN ISOFORMS ON HISTOCHEMICAL PROPERTIES AND MEAT QUALITY IN PIGS

Y.M. Choi*, Y.C. Ryu, S.H. Lee, G.W. Go, H.G. Shin and B.C. Kim

Division of Food Science, College of Life and Environmental Sciences,Korea University, 5-1 Anam-dong, Sungbuk-gu, Seoul 136-713, South Korea. Email: bckim@korea.ac.kr

Keywords: myosin heavy chain, myosin light chain, histochemical properties, pork quality

Introduction

Myosins are hexamers, which are composed of two myosin heavy- (MHC) and four light chains (MLC) isoforms. The variety of possible myosin isoforms is reflective of the structural and functional diversity inherent in the skeletal muscles, such as contractility and energetic properties. Myosin content is significantly lower in slow-twitch than in fast-twitch fibres (Reggiani *et al.*, 2000). In addition, MHC isoform contents affect the post-mortem metabolic rate and meat quality (Choi *et al.*, 2006). Therefore, we examined the influence of MHC and MLC isoform content on histochemical properties and meat quality in the porcine *longissimus dorsi* muscle.

Materials and Methods

A total of 69 crossbred (Yorkshire × Landrace × Duroc) pigs (45 gilts and 24 castrated male pigs) were evaluated. Within 45 min postmortem, muscle samples were obtained from the porcine *longissimus dorsi* muscle for myosin and histochemical analyses. At 24h postmortem, muscle samples were taken in order to evaluate meat quality. Both MHC and MLC contents were determined using SDS-PAGE (Choi *et al.*, 2006; Joo *et al.*, 1999, respectively). The muscle sections were stained for myofibrillar ATPase reactivity after preincubation at pH 4.7 (Brooke and Kaiser, 1970). Meat colour was evaluated, using a Minolta chromameter (CR-300, Minolta Camera Co., Japan). Drip loss was also assessed in accordance with the procedure of Honikel (1987), and filter paper fluid uptake was measured as described by Kauffman *et al.* (1986). Protein solubility was assessed in accordance with the method developed by Joo *et al.* (1999). The Pearson correlation coefficients were evaluated using the partial correlation coefficients. The data were classified into two groups (low and high) on the basis of the MHC fast/slow isoform ratios. Least squares analyses were conducted using the GLM procedure of the SAS statistical package (2001).

Results and Discussion

The MHC 1 isoform content was positively correlated with the density of type I fibres (r = .71) (Table 1). Moreover, the MHC fast/slow ratio evidenced a correlation with fibre type I and IIB density (r = -.64 and .35, respectively). The MLC 1s isoform content was positively correlated with the mean of fibre area and density of type I fibres (r = .26 and .32, respectively), whereas the MLC fast/slow isoform ratios evidenced the opposite tendency (r = -.25 and -.34, respectively). Muscles with higher MHC fast/slow ratios were found to have lower ultimate pH values and also showed an excessive degree of protein denaturation, hence exhibiting a paler surface and higher drip loss than was observed in muscles with lower MHC fast/slow ratios (P < 0.05) (Table 2).

Table 1: Correlations between contents of the myosin isoforms and histochemical properties in the porcine *longissimus dorsi* muscle.

	Myosin heavy chain isoform			Myosin light chain isoform			
	1	2	Fast/slow ratio	1s	1f	3f	Fast/slow ratio
Muscle fibre area (μm^2)							
Mean	-.07	0.07	-.06	.26*	.03	-.08	-.25*
Type I	-.13	0.12	.17	.32**	-.35*	.04	-.34**
Type IIA	.21	-0.25*	-.33**	.20	-.03	-.05	-.18
Type IIB	-.01	0.05	-.13	.20	.08	-.07	.20
Muscle fibre density per mm^2							
Sum	.08	-0.07	.03	-.26*	.02	.12	.25*
Type I	.71***	-0.70***	-.64***	-.07	.22	-16	.05
Type IIA	-.01	0.01	-.11	-.07	.14	-.24*	.10
Type IIB	-.12	0.12	.35**	-.25*	-.08	.01	.28*

Levels of significance: * $P < 0.05$, ** $P < 0.01$, *** $P < 0.001$.

Table 2: Meat quality measurements of the porcine *longissimus dorsi* muscle in groups categorized by myosin heavy chain fast/slow ratio

	Myosin heavy chain fast/slow ratio		Levels of Significance
	Low (N = 46)	High (N = 23)	
Meat quality traits			
Drip loss (%)	5.91[a] (0.39)[1]	3.98[b] (0.27)	***
Filter-paper fluid uptake (mg)	62.32[a] (4.91)	48.85[b] (3.44)	*
Lightness (L^*)	47.80[a] (0.64)	46.28[b] (0.45)	*
Muscle $pH_{24\,h}$	5.55[b] (0.01)	5.66[a] (0.01)	*
Protein solubility (mg/g)			
Total protein solubility	179.16[b] (4.47)	194.05[a] (3.00)	**
Sarcoplasmic protein solubility	69.20 (2.10)	73.13 (1.41)	NS
Myofibrillar protein solubility	109.96[b] (3.17)	120.92[a] (2.13)	**

[1] Standard error of least-square means.
Levels of significance: NS = not significance, * $P < 0.05$, ** $P < 0.01$, *** $P < 0.001$.
[a, b] Least-square means with different superscripts in the same row significantly differ ($P < 0.05$).

Conclusions

In this study, the results indicate that both the MHC and MLC isoforms may potentially serve as determinants of certain muscle fibre characteristics in the porcine *longissimus dorsi* muscle. Moreover, MHC isoform contents appear to influence both extent of protein denaturation and overall quality in pork.

References

Brooke, M. H. and Kaiser. K. K. (1970). Muscle fiber types: how many and what kind? Archives of Neurology, 23: 369-379.

Choi, Y. M., Ryu, Y. C. and Kim, B. C. (2006). Effect of myosin heavy chain isoforms on muscle fiber characteristics and meat quality in porcine *longissimus* muscle. Journal of Muscle Foods, in press.

Joo, S. T., Kauffman, R. G. Kim, B. C. and Park, G. B. (1999). The relationship of sarcoplasmic and myofibrillar protein solubility to colour and water-holding capacity in porcine *longissimus* muscle. Meat Science, 52: 291-297.

Honikel, K. O. (1987). Evaluation and control of meat quality in pigs. pp. 129-142, Martinus Nijhoff, Dordrecht, The Netherlands.

Kauffman, R. G., Eikelenmoom, G., Van Der Wal, P. G. Merkus, B. G. and Zaar. M. (1986). The use of filter paper to estimate drip loss of porcine musculature. Meat Science, 18: 191-200.

Reggiani, C., Bottinelli, R. and Stience, G. J. (2000). Sarcomeric myosin isoforms: Fine tuning of molecular motor. News in Physiological Science, 15: 26-33.

SAS Institute. (2001). SAS user's guide, version 8.2. Cary, NC: SAS Institute.

THE RELATIONSHIP BETWEEN COLLAGEN CHARACTERISTICS IN MUSCLE AND MARBLING OF BEEF

A. Listrat[1]*, A.L. Pissavy[1], C. Jurie[1], C. Lethias[2], D.W. Pethick[3], and J.F. Hocquette[1]

[1] *INRA, Herbivore Research Unit, Theix, 63122 Saint-Genès Champanelle, France,* [2] *CNRS, IBCP, Passage du Vercors, 69367 Lyon, Cedex 07, France,* [3] *Murdoch University, Division of Veterinary and Biomedical Sciences, Perth, Australia.*
Corresponding author: Anne Listrat, E-mail: listrat@clermont.inra.fr

Keywords: beef, collagen isoform, marbling, muscle metabolism

Introduction

Marbling is a factor determining the quality grade of beef meat in North America, Asia and Australia. It refers to the appearance of white flecks or streaks of adipose tissue within connective tissue of muscles between the bundles of muscle fibres. It is thus closely linked to intramuscular fat (IMF) content. In continental Europe, IMF content in beef is lower, but it plays a major role in determining flavour. Indeed, flavour score markedly increases with increasing IMF content (Goutefongea and Valin, 1978).

The number and the diameter of intramuscular adipocytes are major factors determining IMF content and hence marbling (Cianzo *et al.,* 1985). In addition, metabolic activity of muscles was shown to differ between bovine breeds (Angus *vs* Limousin) with different ability to develop marbling (Hocquette *et al.,* 2003). The hypothesis of this work was that the characteristics of the connective tissue in which intramuscular adipocytes are included may differ between muscles of Angus and Limousin breeds as well as between muscle types. To test this hypothesis, we have determined muscle contents of type XII and XIV collagens which were previously thought to play a role in adipocyte differentiation (Ruehl *et al.,* 2005; Tahara *et al.,* 2004).

Materials and Methods

Two groups of 10 steers from two different breeds were used as previously described (Hocquette *et al.,* 2003): Angus which produces a marbled meat and Limousin which produces a weakly marbled meat. All animals had a long (6 months) finishing period with a similar cereal-rich diet [rolled wheat (47-50%), triticale (17-18%), hay (14-18%), lupins (9%), Metabolizable energy: 12 KJ/kg DM, Crude Protein: 15%], which allowed them to express their genetic potential for IMF deposition.

Animals were slaughtered at 23 months of age and samples of two muscle types were taken: *Rectus abdominis* [RA] (oxidative), and *Semitendinosus* [ST] (glycolytic). In all samples, activities of oxidative enzymes (isocitrate dehydrogenase [ICDH], cytochrome-*c* oxydase [COX]), glycolytic enzyme (phosphofructokinase [PFK]) as well as protein contents were measured as described by Hocquette *et al.* (2003). The muscle contents of two isoforms of fatty acid-binding proteins specific of muscle fibres (H-FABP) or intramuscular adipocytes (A-FABP) were determined by ELISA (Hocquette *et al.,* 2003). Total fat and triacylglycerol (TAG) contents was measured as described (Hocquette *et al.,* 2003). All results were expressed per g tissue wet weight (for TAG and FABP contents) or per mg of protein in muscles (for enzyme activities).

Muscle collagen content was determined from the hydroxyproline concentration (Listrat *et al.,* 2004). The relative contents of type XII and XIV collagens within muscle were measured following western-blotting analysis. Each sample was measured in triplicate and results were expressed in arbitrary units.

Differences between genotypes and muscles were analyzed by variance analysis using SAS. Fixed effects included genotype, animal nested within genotype, muscle type and interaction between muscle and genotype. The effect of genotype was tested against animals within genotypes. A correlation analysis between the different parameters was also realized followed by a regression test to determine if the slope of the regression line was significantly different of zero.

Results and Discussion

In a previous paper (Hocquette *et al.,* 2003), we showed that the Angus muscles had a higher IMF content, as well as a more oxidative and a less glycolytic muscle metabolism that the Limousin muscles. In this paper on the same animals, we showed that the breed effect was highly significant for both total collagen and type XIV collagen contents ($P < 0.001$). For the latter, there was no muscle effect. These two parameters were indeed higher (x 1.6 and 1.7, respectively) for the Angus breed that for the Limousin breed. For type XII collagen, there was no breed effect but a significant muscle effect ($P < 0.01$), type XII collagen content being higher in RA than in ST (x 2.5). The interaction between muscle and breed effects was never significant.

High and positive correlations between protein contents of the two fatty acid binding protein (FABP) isoforms and the relative amounts of type XII and XIV collagens were observed (Table 1) when all 40 samples were included in the analysis (two muscles from two groups of ten steers). The correlations were higher for collagen XII than for collagen XIV.

	Collagen XIV		Collagen XII	
TAG	R	**+0.34***		NS
A-FABP	R	**+0.39***	R	**+0.48****
H-FABP	R	**+0.42****	R	**+0.54*****
COX	R	+0.28 t	R	**+0.42****
ICDH	R	**+0.38***	R	**+0.42****
PFK	R	**-0.32***		NS
LDH	R	**-0.43****		NS

Table 1: Coefficients of correlation between collagen type XII and XIV relative contents and metabolic characteristics of muscles (NS: not significant; t: $P < 0.10$; *: $P < 0.05$; **: $P < 0.001$).

Some positive correlations were also observed between the contents of the two studied collagen types and the activities of oxidative enzymes.

We can thus conclude that type XII collagen is a relevant indicator of muscle type, especially associated with intramuscular fat differentiation (from its correlation with A-FABP content) and oxidative metabolism (from its correlation with H-FABP content, COX and ICDH activities), but not with IMF content. By contrast, type XIV collagen is breed specific and hence associated with muscle parameters (including IMF content) which discriminate breeds.

Recent *in vitro* studies with preadipocytes from a bovine intramuscular cell line (Tahara *et al.,* 2004) and with 3T3-L1 preadipocytes (Ruehl *et al.,* 2005) suggested that collagen types may affect adipogenesis. Our *in vivo* results support this idea since type XII and XIV collagen contents were both correlated with A-FABP content, a marker of intramuscular adipocyte differentiation. But, although both type XII and XIV collagens were associated with the metabolic activity of muscles, only type XIV collagen was associated with IMF content. This is consistent with the fact, that these two collagen types could have distinct functions.

The meat of Angus is considered to be tender and tasty due to its greater ability to lay down intra-muscular marbling fat during the finishing phase. The development of adipose tissue has been proposed to disorganise the structure of the intramuscular connective tissue (Nishimura *et al.,* 1999). In addition, both type XII and XIV collagens could participate in extracellular matrix deformability, by making it more rigid or more flexible according to the constraints on the tissue (Nishiyama *et al.,* 1994). The higher proportion of type XIV collagen in the Angus muscles could favour the development of IMF or could be due to the disorganisation of the connective tissue by IMF

Conclusions

From our *in vivo* results, we can hypothesize that type XII and XIV collagen contents may be associated with metabolism of intramuscular lipids. We thus confirm previous *in vitro* studies (Tahara *et al.,* 2004; Ruehl *et al.,* 2005). Both type XII and XIV collagen may contribute to IMF accumulation, but probably by different biological mechanisms since type XII collagen content mainly differs between muscle types and type XIV collagen content mainly differs between breeds.

References

Cianzo, D.S., Topel, D.G., Whitehurst, G.B., Beitz, D.C. and Self, H.L. (1985). Adipose tissue growth and cellularity: changes in bovine adipocyte and number. Journal of Animal Science, 60: 970-976.

Goutefongea, R. and Valin, C. (1978). Quality of beef. 2. Comparison of the organoleptic properties of beef from cows and from young bulls. Annales de Technologie Agricole, 27: 609-627.

Hocquette, J.F., Jurie, C., Ueda, Y., Boulesteix, P., Bauchart, D. and Pethick, D.W. (2003). The relationship between muscle metabolic pathways and marbling of beef. In: « Progress in Research on Energy and Protein Metabolism», Souffrant W.B., Metges C.C. (Editors), 2003. EAAP Publication N° 109, Wageningen Pers. (The Netherlands), pp. 513-516.

Nishimura, T., Hattori, A. and Takahashi, K. (1999). Structural changes in intramuscular connective tissue during the fattening of Japanese Black cattle: effect of marbling on beef tenderization. Journal of Animal Science, 77: 93–104.

Tahara, K., Aso, H., Yamasaki, T., Rose, M.T., Takasuga, A., Sugimoto, Y., Yamagushi, T. and Takano, S. (2004). Cloning and expression of type XII collagen isoforms during bovine adipogenesis. Differentiation, 72: 113-122.

Ruehl, M., Erben, U., Schuppan, D., Wagner, C., Zeller, A., Freise, C., Al-Hasani, H., Loesekan, M., Notter, M., Wittig, B.M., Zeitz, M., Dieterich and W., Somasundaram, R. (2005). The elongated first fibronectin type III domain of collagen XIV is an inducer of quiescence and differentiation in fibroblasts and preadipocytes. Journal of Biological Chemistry, 280: 38537-38543.

Listrat, A. and Hocquette, J.F (2004). Analytical limits of total and insoluble collagen content measurements and of type I and III collagens analysis by electrophoresis in bovine muscles. Meat Science, 68: 127-136.

Nishiyama, T., McDonough, A.M., Bruns, R.R. and Burgeson, R.E. (1994). Type XII and XIV collagens mediate interactions between banded collagen fibers in vitro and may modulate extracellular matrix deformability. Journal of Biological Chemistry, 269: 28193-28199.

MYOGLOBIN LAYERS IN MEAT: A MULTISPECTRAL IMAGING APPROACH

B. Hernández*, C. Sáenz, C. Alberdi, S. Alfonso, and J.M. Diñeiro

Departamento de Física. Universidad Pública de Navarra. Campus Arrosadía. 31006 Pamplona, Spain. Email: bego@unavarra.es

Keywords: multispectral imaging, myoglobin, beef

Introduction

Meat colour depends on several productive factors like breed, sex, animal age or pH (Seideman *et al.*, 1984) but once they are fixed, the main factor responsible of colour and colour changes is the evolution of the surface concentrations of haeminic pigments, particularly of myoglobin. Prior to being exposed to air, myoglobin is found in the reduced form (Mb). When meat is exposed to air, oxygen diffusion through the surface converts Mb to the oxygenated form or oxymyoglobin (MbO_2) and, with time, metmyoglobin (MMb) will appear after the oxidation of MbO_2. If enough time of exposure to air is allowed, a layered structure appears where MMb is found immediately beneath the surface, followed by a deeper layer of MbO_2 and an almost oxygen free region dominated by Mb. However, even for long times of exposure to air, layer thickness is of the order of a few millimetres (Renerre, 2004) and therefore beyond the capabilities of standard spectrophotometric instrumentation.

Presently a single method is available to determine position and width of myoglobin layers. The so called 'squash plate method' consists in squashing a meat sample between two transparent glass plates. Glass prevents blooming of the observed surface and oxygenation takes place along the direction parallel to the plates. In this way the development of pigment layers can be observed through the glass and their position and thickness measured (O'Keeffe and Hood, 1982; Powell and Macdougall, 1993; Gutzke *et al.*, 1997). Boundaries between pigments layers are determined by visual inspection, using a magnifier lens and their positions and widths measured with a precision ruler or a micrometer.

However, the overall process depends on the visual capabilities of the observer and several factors will limit the accuracy and usefulness of the results. Pigment layers are identified by their colour since each of the three pigment forms has a distinct colour. However boundaries between pigment layers are not sharp and colour transition between them is gradual, making difficult the determination of the end of one layer and the beginning of the next one. Pigment layers need to be completely developed to be discernible and the method does not work if a mixture of pigments coexists in the same region. In addition, colour differences between pigments may be subtle and measurements biased by the colour discrimination capabilities of the observer. Furthermore colour differences will depend on the illumination used during the measurement. This data is not provided in existing works but meat colour is known to be greatly affected by illumination.

In this work we use standard spectrophotometric methods combined with a multispectral imaging acquisition system to determine the layered structure of myoglobin derivatives during meat oxygenation.

Materials and Methods

The multispectral imaging acquisition system consists of an objective lens, a CRI VariSpec^{TM} Liquid Crystal Tunable Filter (LCTF), a relay lens and a 1392 x 1040 pixels, 12 bit, monochrome digital camera Evolution VF. The LTCF clear aperture is 20 mm and works in the wavelength range 400-720nm with 10nm of bandwidth. All the system is mounted in a Hama Repro XZ adjustable column with 4x100 W tungsten lamps for illumination. System optics was selected to work in macro mode to increase the spatial resolution. Reflectance calibration using a lambertian white diffuser allowed raw images to be converted in reflectance images and spatial calibration was performed to convert pixel values to distances.

Two 20 x 20cm square highly transparent glass plates formed the 'squash plate'. Plates were kept fixed and parallel using four 2cm long spacers. Both plates were screwed to the spacers making a solid and stable 'sample holder' where samples were squeezed and imaged through.

Meat samples were obtained from beef *longissimus dorsi* muscle. Samples of 2.2cm thick were cut from the inner part of the muscle to ensure that initially only Mb was present. Samples were immediately introduced between the squash plates. The border of the squash plate was covered with oxygen permeable film to allow oxygenation and prevent excessive dry of the sample.

Relative concentrations of myoglobin forms were obtained using the standard procedure (A.M.S.A., 1991) from reflectance images using (K/S) ratios at specific isobestic points (474, 525, 572 and 610 nm). In this way, images corresponding to pigment concentrations can be constructed.

A final false colour image is constructed assigning each concentration image to each of the RGB colour channels. In this way pigments can be easily identified.

Results and Discussion

As an example of the resulting pigment concentration images, pigment layers after 1 hour and 8 hours of oxygenation time at room temperature (20ºC) are shown in Figure 1. These images are B/W versions of the original colour coded images. In these images Mb appears darker (Mb covers most of the imaged area in this sample) and MbO_2 appears as a whitish band, broad and close to the image surface in the left image and as a somewhat narrower band, shifted towards the inner part of the sample, after 8 hours. MMb appears dark (slightly lighter than Mb) and immediately beneath the surface after 8 hours. MMb can not be distinguished in the B/W image corresponding to 1 hour of oxygenation, because concentration values of MMb are small, but not zero, at this time.

Both images have 1040 (width) x 1392 (height) pixels. The imaged area after spatial calibration is 1.28 (width) x 1.72 cm (height). The scanned area is therefore 2.20 cm^2 and the spatial resolution is better than 0.0125 mm/pixel (80 pixel/mm).

Figure 1: Images of a sample of beef *longissimus dorsi* muscle showing the development of pigment layers after 1 hour (left) and 8 hours (right) of exposure to air. The meat sample covers the upper 2/3 of each picture. The slightly curved line that delimits the end of the meat sample defines the surface in contact with air.

Conclusions

Multispectral techniques can be used to study myoglobin layer development during meat oxygenation. It is an instrumental method, not biased by the observer or the illuminating source. It provides high spatial resolution and should permit in the future accurate analysis of pigment concentration distributions in meat. It is a non destructive method and pigment concentrations can be studied for long periods of time for the same sample.

References

S.C. Seideman, H.R. Cross, G.C.Smith and P.R. Durland. "Factors associated with fresh meat color: a review". Journal of Food Quality, 6, 211-237 (1984).

M. Renerre (2004). Myoglobin oxidation and meat colour. VII Congreso Nacional de Color. Pamplona. Spain.

O'Keeffe, M. and Hood, D. E. (1982). Biochemical factors influencing metmyoglobin formation on beef from muscles of differing colour stability, Meat Science 7, 3, 209-228.

Powell, V. H., Macdougall, D.B. (1993). In: Proceedings of the 39th Congress Meat Science and Technology. Calgary (Canada), 141.

Gutzke, D.A., Trout, G.R. and D'Arcy, B.R. (1997). "The relationship between the colour stability and the formation of the metmyoglobin layer at the surface of beef, pork and venison muscles". In Proceedings 43rd International Congress of Meat Science and Technology (pp. 652-653) 27 July-1 August 1997. Auckland, New Zealand.

A.M.S.A. (1991). Guidelines for meat color evaluation. National Live Stock and Meat Board. 444 North Michigan Avenue. Chicago Illinois.

CHANGES IN ENZYMES ASSOCIATED WITH ENERGY METABOLISM IN PRE AND *POST-MORTEM* LONGISSIMUS THORACIS BOVINE MUSCLE ANALYSED BY PROTEOMICS

X. Jia[1,2], K.I. Hildrum[1], F. Westad[1], E. Kummen[3], L. Aass[4] and K. Hollung*[1]

[1] *Matforsk AS, Norwegian Food Research Institute, Osloveien 1, N-1430 Ås* [2] *Department of Chemistry, Biotechnology and Food Science, Norwegian University of Life Sciences, P.O. Box 5003, N-1432 Ås,* [3] *GENO Øyer, 2636 Øyer and Department of Animal and Aquacultural Sciences,* [4] *Norwegian University of Life Sciences, P.O.Box 5003, N-1432 Ås, Norway. Email: Kristin.hollung@matforsk.no*

Keywords: cattle, proteomics, energy metabolism, muscle metabolism, *longissimus thoracis*

Introduction
Many factors can affect final meat quality (Maltin *et al.*, 2003), of which the rate of early post mortem glycolysis seems to be an important determinant of eating quality (O'Halloran *et al.*, 1997). However, the biochemical mechanisms behind post mortem glycolysis in post mortem muscles are poorly understood. Soluble muscle proteins include many sarcoplasmic and mitochondrial enzymes, which are involved in biochemical metabolic processes in living animals. Hence, to turn muscles into high quality meat, an improved understanding of which metabolic proteins are involved and how the biochemical and energy metabolism is regulated during the early post mortem period, is a prerequisite. Proteomics is an important cornerstone in post-genome sciences and has also been applied in meat science in recent years (Bendixen, 2005; Bouley *et al.*, 2004; Jia *et al.*, 2006; Lametsch *et al.*, 2003). The objective of this study was to investigate the changes in metabolic proteins in bovine muscles between pre and post mortem muscles.

Materials and Methods
The experiment included 9 NRF (Norwegian Red) dual-purpose young bulls slaughtered (approximately 13 months of age/450 kg live weight) at a performance test station (GENO-Breeding and AI Association) in 2004. The procedure for excision of biopsies had been approved by the Norwegian Animal Research Authority (Immonen *et al.*, 2000). Muscle samples were homogenized in TES buffer (10 mM Tris (pH 7.6), 1 mM EDTA and 0.25 M sucrose). 2-DE and silver staining was performed according to the method described previously (Jia *et al.*, 2006). The protein spots were matched across all 36 gels (2 time points × 9 animals × 2 technical replicates) using Image Master 2D Platinum v. 5.0 (GE Healthcare) and the spot report was imported into Unscrambler version 9.2 (CAMO A/S, Norway) and Matlab version 7.0.4. Significant spots were identified by cross-model validation. Proteins were identified by peptide mass fingerprinting (PMF) and MS/MS of at least 3 protein fragments.

Results and Discussion
Figure 1A shows a representative 2-DE pattern of the proteins extracted from muscle samples in the pH range between 4 and 7. Most of the soluble proteins were located within the range pH 5 to 7. Image analysis allowed matching of 833 spots across all 36 gels. The PCA score plot of the complete data set of spot intensities from

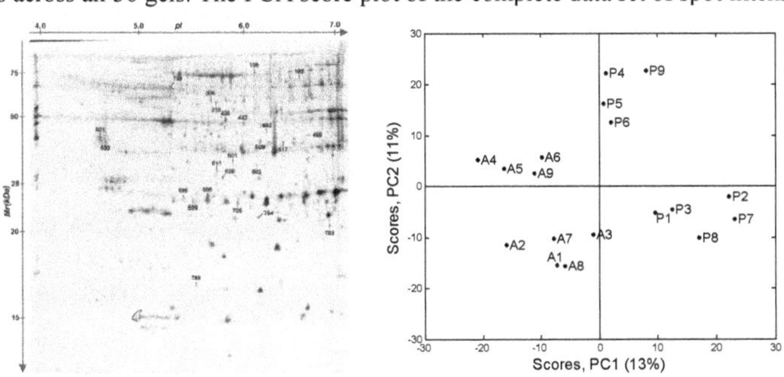

(A) (B)

Figure 1: A. Silver-stained 2-DE master gel of bovine *longissimus thoracis*. Protein spots that are changed in the post mortem samples are numbered. **B.** Principal component analysis (PCA) score plot of protein spots. A1-A9: biopsies taken from nine animals. P1-P9: Samples taken after slaughter from the same nine animals.

comparative image analyses reveals that two clusters related to pre and post slaughter samples are formed by the first two components (Figure 1B). In the first set of technical replicates 49 protein spots were significantly changed, and in the second technical replicate 56 proteins were significantly changed. The 24 protein spots numbered in Figure 1A were

significantly changed in both data sets and underwent further protein identification. Among the selected 24 protein spots, 18 spots increased and 6 decreased in intensity in the post slaughter samples, respectively.
In this study, several glycolytic enzymes involved in energy metabolism, in addition to 3-hydroxyisobutyrate dehydrogenase and two enzymes in the TCA cycle, increased in intensity after slaughter (Figure 2). The finding that several glycolytic enzymes increased in intensity after slaughter may suggest increased rate of glycolysis, in order to support and maintain the ATP production.
The goal of this study was to investigate the changes in metabolic proteins from live animals to samples taken immediately after slaughter. Several processes such as transportation, lairage, stunning, exsanguination, dehiding and electrical stimulation could contribute to the changes in the protein profile. In this study we did not address these questions, but rather investigated the consequence of all these processing parameters. Further studies are needed to separate the individual effects of the processing steps on the protein profile.

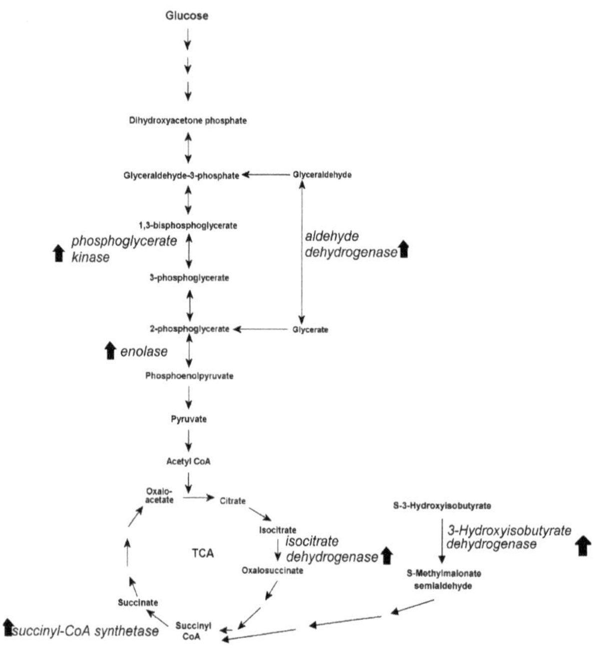

Figure 2: Biochemical pathways in energy metabolism that are influenced in bovine *longissimus thoracis* post mortem. The identified metabolic enzymes by MALDI-TOF MS/MS are shown. The arrows show increased enzyme abundance.

Conclusions

We observed a clear shift in energy metabolism in the muscle post mortem with an increase in enzymes involved in both the glycolytic pathway as well as in the TCA cycle. These findings suggest that an increased aerobic energy metabolism occurs in the first hour after slaughter. The increased aerobic energy metabolism will probably affect the rate of glycolysis in muscles after slaughter and eventually lead to variation in meat quality. However, additional studies including more animals are required to reveal how changes in protein expression in pre and post mortem muscles are related to meat quality traits in cattle.

Acknowlegements

The work was supported by a grant from The Fund for the Research Levy on Agricultural Products.

References

Bendixen, E. (2005). The use of proteomics in meat science. *Meat Science,* 71:138-149.

Bouley, J. Chambon, C. and Picard, B. (2004) Mapping of bovine skeletal muscle proteins using two-dimensional gel electrophoresis and mass spectrometry. *Proteomics,* 4:1811-1824.

Immonen, K. Schaefer, D.M., Puolanne, E., Kauffman, R.G. and Nordheim E.V. (2000) The relative effect of dietary energy density on repleted and resting muscle glycogen concentrations. *Meat Science,* 54:155-162.

Jia, X., Hollung, K., Therkildsen, M., Hildrum, K.I. and Bendixen, E. (2006) Proteome analysis of early post-mortem changes in two bovine muscle types: M. longissimus dorsi and M. semitendinosis. *Proteomics* 6:936-944.

Lametsch, R., Karlsson, A., Rosenvold, K., Andersen, H.J., Roepstorff, P. and Bendixen, E. (2003) Postmortem proteome changes of porcine muscle related to tenderness. *J Agr Food Chem,* 51:6992-6997.

Maltin, C., Balcerzak, D., Tilley, R. and Delday, M. (2003) Determinants of meat quality: tenderness. *Proc Nutr Soc,* 62:337-347.

O'Halloran, G.R., Troy, D.J. and Buckley, D.J. (1997) The relationship between early post-mortem pH and the tenderisation of beef muscles. *Meat Science,* 45:239-251.

DETECTION OF MULTIPLE PROTEASE ACTIVITIES IN BOVINE SKELETAL MUSCLES BY TWO-DIMENSIONAL ZYMOGRAPHY

M. Higuchi*[1], N. Shiba[1], A. Watanabe[1] and Y. Ueda[2]

[1.] *Grazing and Meat Production Research Team, National Agricultural Research Center for Tohoku Region, Morioka 020-0198, Japan,* [2.] *Intensive Grazing Research Team, National Agricultural Research Center for Hokkaido Region, Sapporo 062-8555,Japan. Email: mikito@affrc.go.jp*

Keywords: cattle, skeletal muscle, proteases, zymography

Introduction

It is widely recognized that proteases in skeletal muscles play principal roles in the degradation of myofibrillar proteins in *post mortem* muscles. Two groups of proteases are known to be involved in this process, namely calpains and cathepsins (Ouali *et al.*, 1987), although several proteases other than these two groups may be involved in the meat ageing process. It is important to examine the quantity and activity of these proteases for the investigation of their effects on *post mortem* muscles and numerous studies have been performed with beef samples. However, most studies deal with only one or a small number of proteases and require large beef samples (e.g. from 10 to over 50 g). It is necessary to detect multiple proteases simultaneously in beef samples to comprehensively understand the cooperation of these proteases in the degradation of myofibrillar proteins in *post mortem* muscles. In addition, when the sample size is limited, it is desirable to apply a method to accurately detect protease activities in these muscle samples (Kent *et al.*, 2005). Zymographic analysis is widely used for the detection of multiple protease activities in relatively small volumes of animal tissues and body fluids. One-dimensional casein zymography has been applied for the detection of neutral μ- or m-calpain activities in meat samples (Veiseth *et al.*, 2001). However, as far as we know, there are no reports using this technique for the detection of acidic protease activities in beef samples.

It is well known that two-dimensional electrophoresis (2-DE) is more effective in separating large number of proteins than one-dimensional electrophoresis, and the two-dimensional (2-D) zymography method, a combination of casein zymography and 2-DE was reported to detect multiple proteolytic enzymes (Kaino *et al.*, 1998). The purpose of this study is to clarify whether 2-D zymography gives better results for detecting multiple protease activities in beef samples, eliminating the need for large muscle samples.

Materials and Methods

One Japanese Shorthorn steer was slaughtered at 24 months of age and 640 kg bodyweight in our experimental slaughterhouse. All of the slaughter procedures were performed following The Guidelines for Experimental Animal Handling, National Agricultural Research Center for Tohoku Region.

Longissimus lumborum (LL), *Biceps femoris* (BF), *Psoas major* (PM) and *Semitendinosus* (ST) muscles were collected within two hours of slaughter and snap frozen in liquid nitrogen. The samples were then stored at -80°C until further analysis.

One gram of frozen sample was homogenized with 3 ml ice-cooled homogenization buffer (50 mM Tris-HCl, 10mM EDTA-2Na and 1/100 concentration of protease inhibitor cocktail (SIGMA), pH 7.5). The homogenate was centrifuged at 4°C and the aliquot was dispensed into two microtubes. The microtubes were centrifuged at full speed for 10 min at 4°C to purify the aliquot which was vacuum condensed and desalted by passing through a Sephadex® G25 gel column, followed by quantification of the protein content. An immobilized pH gel strip (BioRad, 11 cm length, for pI 5-8) was rehydrated with 185 μl of rehydration solution containing 2mg of desalted sarcoplasmic protein. First dimension isoelectric focusing was performed, applying 6,000 V voltages to the rehydrated strip for 5 hours. Rehydration and isoelectric focusing were performed at 10°C.

A polyacrylamide slab gel 12.5% (w/v) containing 0.25% (w/v) casein without SDS was prepared for two- dimensional electrophoresis, with buffer containing 30 mM Tris base, 200 mM glycine and 10 mM EDTA. The focused gel strip was put on the top of the slab gel with 1% (w/v) agarose as an adhesive. The gel was run at 120 mA current for 16 h, maintaining the buffer temperature below 10°C throughout electrophoresis.

Fifty mM sodium-acetate buffer was prepared for gel incubation to restore the protease activities. The pH was adjusted to 4.0, 4.5, 5.0, 5.5, 6.0 and 6.5. Calcium chloride was added at a concentration of 10 mM or 100 mM to investigate whether the proteases were Ca-dependently activated. Upon completion of electrophoresis, the gel was washed twice by gentle shaking in incubation buffer for 15 min at room temperature to wash off EDTA and protease inhibitors. The gel was then transferred into fresh incubation buffer and incubated for 48 h at 37°C with gentle shaking, changing the buffer after 24 hrs of incubation. The gel was fixed and stained with Coomassie Brilliant Blue R-250 (Waco Chemicals) dissolved in fixation buffer (10% acetic acid, 30% methanol and 70% ion-exchanged water) and destained with fixation buffer.

Results and Discussion

Figure 1 shows the typical two-dimensional zymographic pattern using bovine LL and ST muscles. More than ten protease activity spots could be detected under the conditions mentioned in the Legend. Western blot analysis suggested that two of the spots corresponded to m-calpain and cathepsin-B. PM exhibited stronger activities for most of the spots than BF, LL and ST muscles. However, several spots exhibited higher intensity in muscles other than PM, suggesting that the composition of the proteases differed by muscle.

At pH 6.5 and 6.0, no activity spots were detected. In contrast, most spots showed strong caseinolytic activities at pH 5.5 and 5.0. At pH 4.5 and 4.0, proteolytic activities were generally weaker than at pH 5.0 and 5.5, however, several spots retained relatively strong activities. These results indicate that the pH of the incubation buffer should be adjusted corectly when particular protease activities are to be investigated with this method. The concentration of calcium ions apparently affected caseinolytic activity. Although proteolytic activities were seen at 10 mM of Ca^{2+}, most activity spots were brighter at Ca^{2+} concentrations of 100 mM than at 10 mM. It is unlikely that all activated spots are calpains and one possible explanation for the phenomenon is that the substrate casein in the gel was affected by the increase in Ca^{2+} concentration and became prone to degradation by the proteases in muscles.

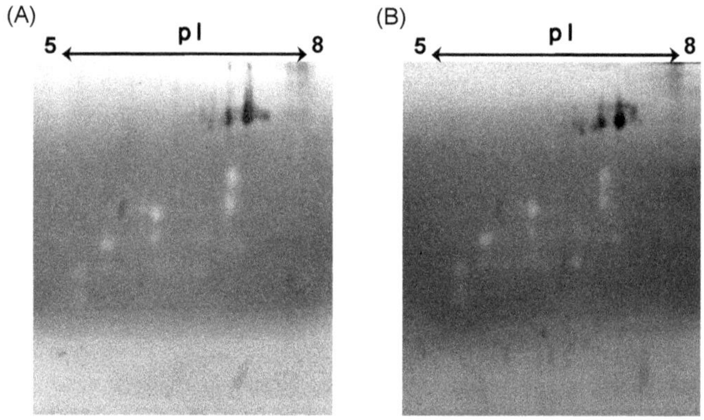

Figure 1: Two-dimensional zymography of the water-soluble proeins of bovine LL (A) and ST (B). Caseinolytic activity spots are seen as CBB-unstained bright spots. Protein content: 2.0 mg, incubation buffer: 50 mM sodium acetate, 100 mM calcium chloride, pH=5.5, incubation: 48 hrs at 37 °C.

Conclusions

Although the two-dimensional casein zymography used in this study needs to be significantly improved (e.g., identification of the proteases, longer incubation time and difficulty in quantifying protease activities), it seems to be a good method of detecting multiple protease activities with a small amount of skeletal muscle, enabling us to investigate proteases in muscle not only *post mortem* but also in living animals, facilitating knowledge on the differences in muscular protease activities influenced by factors such as cattle breed, sex, age, feed and exercise.

References

Kaino, S., Furui, T., Hatano, S., Kaino, M., Okita, K. and Nakamura, K. (1998). Two-dimensional zymography for analysis of proteolytic enzymes in human pure pancreatic juice. *Electrophoresis*, 19: 782-787.

Kent, M.P., Veiseth, E., Therkildsen, M. and Koohmaraie, M. (2005). An assessment of extraction and assay techniques for quantification of calpain and calpastatin from small tissue samples. *Journal of Animal Science*, 83:2182-2188.

Ouali. A., Garrel, N., Obled, A., Deval, C. and Valin, C. (1987). Comparative action of cathepsins D, B, H, L and of a new lysosomal cysteine proteinase on rabbit myofibrils. *Meat Science*, 19:83-100.

Veiseth, E., Shackelford, S.D., Wheeler, T.L and Koohmaraie, M. (2001). Effect of post-mortem storage on mu-calpain and m-calpain in ovine skeletal muscle. *Journal of Animal Science*, 79:1502-1508.

DRIP LOSS INFLUENCES MUSCLE STRUCTURAL ANALYSIS OF THE PORCINE *LONGISSIMUS DORSI* POST MORTEM

G. Link, D. Mörlein* and M. Wicke

Research Centre for Animal Production and Technology, Georg-August-University of Goettingen, Driverstrasse 22, D– 49377 Vechta, Germany. E-mail: daniel.moerlein@agr.uni-goettingen.de

Keywords: pork, drip loss, shrinkage, *longissimus dorsi,* muscle fibre

Introduction

It is generally accepted, that proteolytic processes influence meat quality during time of ageing. These processes are also responsible for drip formation. In beef it was observed by Offer and Cousins (1992) that fibre bundles separate 4 to 6h *post mortem* from perimysium and within rigor mortis muscle fibres separate from epimysium while shrinking. That is caused by the degradation of the costameres, i.e. the cell membrane penetrating link between cytoskeleton and extracellular connective tissue (Taylor *et al.*, 1995). The emerging space between sarcolemma and cell body was identified as drip channels (Bertram *et al.*, 2004). M-calpain seems to be associated with β1-integrin and plays a significant role in their *post mortem* degradation (Lawson, 2004).
Considering this, the timing of sampling for histological analysis will effect morphometric analysis. The objective of this study was to quantify the shrinking of porcine *longissimus dorsi* muscle samples and how that affects the total muscle fibre count and cell density.

Materials and Methods

54 carcasses of commercially slaughtered pigs were investigated. They genetically derived from 3 genotypes. The sample includes both castrates and female pigs. Electrical conductivity (EC) was measured 24 h *post mortem* at $2^{nd}/3^{rd}$ last rib. Drip loss of 2.4 cm thick slices excised 30 h post mortem was measured after storage for 48 h at 4°C in polyethylene bags as described by Honikel (1987). *Longissimus* muscle cross sectional area was recorded. Muscle samples for histological investigations were taken 30 h *post mortem* and stored in liquid nitrogen. 12 μm thick cross sections were stained according to Horak (1983) to simultaneously identify fast twitch glycolytic (FTG), fast twitch oxidative (FTO) and slow twitch oxidative (STO) muscle fibres. Morphometric analysis was performed with the LUCIA software (Nikon). To yield the total amount of fibres per muscle as usual, single fibre diameters were recorded within a region of interest (ROI) followed by extrapolation to the *longissimus* cross sectional area. Shrinkage was determined measuring the area of primary muscle fibre bundles and subsequent subtraction of the summed area of the single measured fibres within that bundle. At 10 subsamples the actual fibre density was determined by counting all fibres within a ROI (9.87 mm²) also followed by extrapolation to the cutlet area to estimate the total fibre number of the muscle. Statistics were computed with SAS.

Results and Discussion

To illustrate shrinking, images of samples with varying drip loss are given in Figure 1. All samples show gaps between fibre bundles. The space between single fibres increases with rising drip loss. A high cellular damage must be assumed that is capable to cause shrinking up to 25 % and higher.

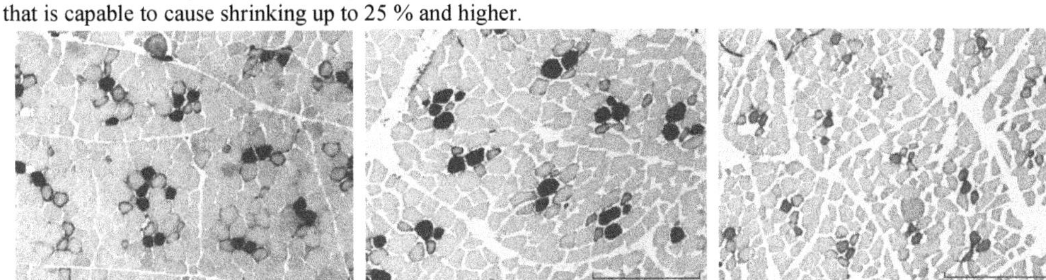

Shrinking:	5.2 %	15.0 %	25.6 %
Drip loss:	3.0 %	8.8 %	11.0 %

Figure 1: Porcine *longissimus* muscle cross sections stained for fibre typing (STO = dark; FTG = bright, FTO = bright with dark border) showing shrinkage related to the drip loss of the muscle sample (bar length = 500 μm).

Mean carcass traits and corresponding drip loss and shrinking can be seen in Table 1 demonstrating a great variation of the samples, e.g. drip loss or electric conductivity (EC). Shrinking extends from 1.1 % to 25.6 %.

Table 1: Basis statistics of carcass data, drip and shrinking (mean and standard deviation; n=54).

	Carcass weight, kg	Lean meat, %	*L. d.* area, cm²	EC, mS/cm	Drip loss, %	Shrinking, %
Mean	94.96	57.66	52.0	6.66	6.73	11.44
St.dev.	6.35	2.49	6.66	2.57	2.54	5.19

Correlations suggest FTG fibres to be more susceptible to shrinking compared with STO and FTO fibres (Table 2). With rising content of FTG fibres in muscle, exudativity would more likely occur. Muscle with higher frequencies of FTO and STO fibres would be more resistant in terms of generating purge. There is a tight relationship between drip loss and shrinkage of the cells (r = 0.66). Electrical conductivity is closely related both to the rate of shrinking and to drip loss and thus may be a good predictor.

Table 2: Correlations between histological parameters and carcass traits (n=54).

	Conductivity, mS/cm	Drip loss, %	Carcass weight, kg	Lean meat, %	Shrinkage, %
% FTG	0.17	0.41*	-0.03	-0.17	0.34*
% FTO	-0.09	-0.30*	-0.11	0.31*	-0.21
% STO	-0.17	-0.27*	0.22	-0.16	-0.29*
µm FTG	0.17	-0.30*	0.32*	0.29*	-0.31*
µm FGO	0.03	-0.10	0.05	0.16	-0.13
µm STG	0.29*	0.01	0.24	0.04	0.09
E. Conductivity		0.48*	0.31*	0.03	0.57*
Drip loss			-0.08	0.03	0.66*

* $p<0.05$

In Table 3 calculations of the total amount of muscle fibres and the cell density has been as usual and with consideration of the shrinking effect, respectively. If the counted mean cell density of 187 cells per mm² is declared as the gold standard, an overestimation occurs by simple calculating without considering the shrinking effect. The total amount of muscle fibres would be overestimated with 111.2 %.

Table 3: Estimated total amount of muscle fibres and cell density with and without considering shrinkage (mean and standard deviation; n=10).

	Shrinkage considered		Shrinkage not considered	
	counted cells / mm²	amount of muscle fibres	estimated cells / mm²	Total amount of muscle fibres
Mean	187	962.915	207	1,071,061
St.dev.	33.3	141389	38.3	182.932
Relation	100%	100%	110.9 %	111.2 %

Conclusions

Muscle fibre types susceptibility to shrinking is as follows: FTG>FTO>STO.
A high relationship between drip loss and shrinkage was observed.
Overestimation of the total amount of muscle fibres occurs if shrinkage was not considered.
A sampling within 45 min *post mortem* is recommended or biopsy has to be taken to most accurately measure fibre diameters and to calculate the most correct cell density.
Fibre bundles may be used to quantify shrinking and subsequently correct morphometric analyses, if autopsy sampling is inescapable.

References

Bertram, H. C., Whittaker, A.K., Andersen, H. J. and Karlsson, A. H. (2004). Visualization of drip channels in meat using NMR microimaging. Meat Science, 68, 667-670.

Honikel, K. O. 1987. The water binding of meat. Fleischwirtschaft, 67, 1098-1102.

Horák, V. (1983). A successive histochemical staining for succinate dehydrogenase and "reversed"-ATPase in a single section for the skeletal muscle fibre typing. Histochemistry, 78, 545-553.

Lawson, M. A. (2004). The role of integrin degradation in post mortem drip loss in pork. Meat Science, 68, 559-566.

Offer, G. and Cousins, T. (1992). The Mechanism of Drip Production: Formation of Two Compartments of Extracellular Space in Muscle Post Mortem. Journal of the Science of Food and Agriculture, 5, 107-116.

Taylor, G. R., Geesink, G. H., Thompson, V. F., Koohmaraie, M. and Goll, D. E. (1995). Is Z-Disk degradation responsible for post-mortem tenderisation? Journal of Animal Science, 73, 1351-1367.

DIFFERENTIAL REGULATION OF CALPASTATIN PROMOTERS BY AGENTS THAT AFFECT PORK TENDERNESS

P.L. Sensky*, K.K. Jewell, T. Parr, R.G. Bardsley and P.J. Buttery

Division of Nutritional Sciences, School of Biosciences, University of Nottingham, Sutton Bonington Campus, Loughborough, Leicestershire, LE12 5RD. Email: paul.sensky@nottingham.ac.uk

Keywords: pork, calpastatin, cells, regulation, growth

Introduction

Variation in meat tenderness is a major factor affecting consumer choice in the UK. Despite efforts to control nutrition and breeding, and the stress associated with handling and slaughter processes, the tenderness of pork is not always of consistent quality. An important biochemical factor during conditioning is undoubtedly the calpain-specific inhibitor calpastatin. Variable quantities of calpastatin at slaughter are known to suppress calpain-mediated postmortem proteolysis to different degrees and contribute to the variation in ultimate tenderness. In support of this, high levels of calpastatin are induced by β-agonist feeding over several weeks in several livestock species and are clearly associated with increased toughness (Koohmaraie, 1996). The biochemical mechanism underlying the β-agonist-induced increase in calpastatin expression remains unproven, although cAMP responsive elements (CRE) have been identified in bovine and porcine calpastatin gene promoters that suggest β-adrenergic signalling by cAMP-mediated protein kinases and phosphorylation of CRE binding protein (CREB) (Cong *et al.*, 1998; Parr *et al.*, 2001). Adrenergic pathways are also activated in the short term by physiological stress, which is well known to affect tenderness and it has been shown that relatively short term adrenaline administration can affect calpastatin expression (Sensky *et al.*, 1996). There are at least three different promoter regions in the porcine calpastatin gene (1xa, 1xb and 1u, Parr *et al.*, 2004) that, via differential promoter usage and alternative splicing, give rise to three isoforms of calpastatin (Types I, II and III) that vary in their N terminal sequences. These promoter regions contain a number of transcription factor motifs that could affect the expression of calpastatin under different environmental conditions. Specifically, there are elements in all the promoter regions that suggest susceptibility to stimuli mediated by β-agonists, Ca^{2+} or calcineurin (Figure 1). In this study, the responsiveness of different porcine calpastatin promoter constructs to stimuli that simulate these pathways is investigated *in vitro* using a non-differentiating skeletal muscle cell line.

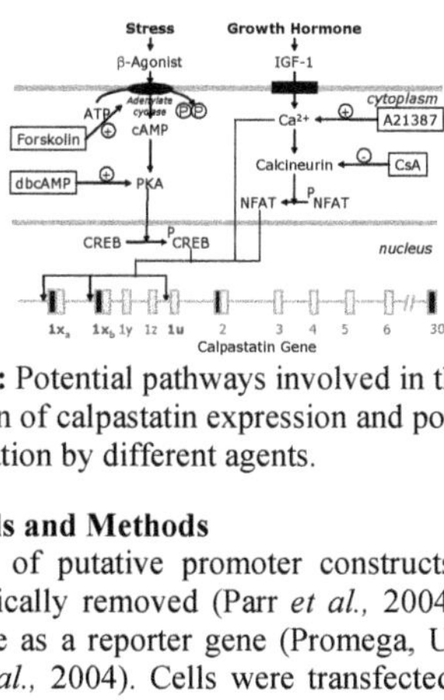

Figure 1: Potential pathways involved in the regulation of calpastatin expression and potential manipulation by different agents.

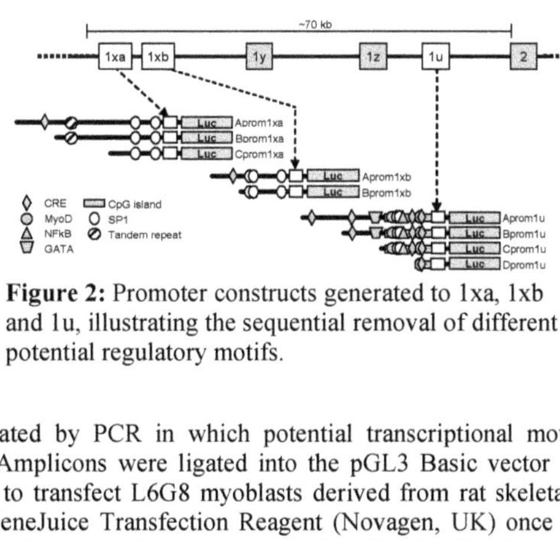

Figure 2: Promoter constructs generated to 1xa, 1xb and 1u, illustrating the sequential removal of different potential regulatory motifs.

Materials and Methods

A series of putative promoter constructs were generated by PCR in which potential transcriptional motifs were systematically removed (Parr *et al.*, 2004, Figure 2). Amplicons were ligated into the pGL3 Basic vector linked to luciferase as a reporter gene (Promega, UK) and used to transfect L6G8 myoblasts derived from rat skeletal muscle (Parr *et al.*, 2004). Cells were transfected using the GeneJuice Transfection Reagent (Novagen, UK) once they had reached 50 - 70% confluence and, immediately after transfection, cells were treated with either 2 mM dibutyryl cAMP (dbcAMP), 10 μM forskolin (adenylate cyclase activator), 1 μM A23187 (Ca^{2+} ionophore) or 10 μM cyclosporin A (CsA, calcineurin inhibitor). Firefly and *Renilla* Luciferase activities were determined using the Dual-Glo Luciferase Assay System (Promega, UK) with *Renilla* luciferase activity used to correct for the transfection efficiency. Data are expressed as the change in transfection efficiency relative to the untreated constructs and significance of treatments was determined by Student's unpaired t-test.

Results and Discussion

All promoter constructs except for Dprom1u exhibited full functionality as previously reported (Parr *et al.*, 2004). Activity of the 1xa promoter series was reduced by dbcAMP and forskolin to differing degrees depending on the degree of truncation of the construct (Table 1). A23187 had the opposite effect, increasing the activity of the 1xa promoter constructs, whilst CsA had no effect. The only treatment to affect expression of the 1xb promoter activity was A23187 which increased the activity of the Bprom1xb construct. Treatment of the 1u constructs with the different reagents generally produced the opposite effect to that seen with the 1xa promoters, with dbcAMP and forskolin having positive effects on promoter activity and A23187 suppressing activity.

Table 1: Change in percentage expression of different porcine calpastatin promoter constructs following treatment with different agents. * $P < 0.05$, ** $P < 0.01$, *** $P < 0.001$.

	2 mM dbcAMP	10 μM forskolin	1 μM A23187	10 μM CsA
Aprom1xa	-20.2%**	-12.9%*	+8.1%	-8.7%
Bprom1xa	-12.8%	-5.4%	+11.5%*	+5.0%
Cprom1xa	-5.2%	-1.8%	+8.7%**	-1.1%
Aprom1xb	-2.7%	-5.5%	+3.7%	-4.3%
Bprom1xb	-0.6%	-4.9%	+17.7%**	+7.6%
Aprom1u	+22.8%***	+0.1%	-8.4%	+5.2%
Bprom1u	+18.2%	+19.9%*	-31.2%**	-8.2%
Cprom1u	+15.5%	+11.6%*	-42.1%**	+1.8%
Dprom1u	-6.9%	-3.1%	+2.9%	-9.0%

The results support whole animal studies indicating that all three promoters are functional and that different mRNA transcripts and protein isoforms of calpastatin are present in different muscle types and respond to β-adrenergic stimulation (Parr *et al.*, 2004). In porcine longissimus the abundance of Type I and III mRNA transcripts is comparable, while Type II is barely detectable. However, at the protein level the Type III isoform mainly derived from the 1u promoter completely dominates in porcine longissimus. The effects seen on the 1u promoter series support the known link between β-adrenergic stimulatory pathways, including β-agonist feeding and physiological stress, and calpastatin in a way that could account for the increased toughness in treated animals. The opposite effect on the 1xa promoter series is unexpected, given that CRE motifs are also found in this promoter, and this warrants further attention. Growth hormone (pST) treatment in pigs has been reported to reduce calpastatin mRNA expression in longissimus (Ji *et al.*, 1998) and is believed to act by increasing IGF-1 thereby increasing Ca^{2+} concentration in muscle. A23187, which raises intracellular Ca^{2+}, had strong effects on transcriptional efficiencies with a differential effect on 1xa and 1u promoters. Ca^{2+}-dependent effects on gene expression can act by both calcineurin-dependent and independent signaling pathways, but the lack of an effect of CsA on any promoter construct suggests that calcineurin-mediated signaling can be discounted.

Conclusions

The close relationship between elevated calpastatin expression and increased toughness in livestock species indicate that calpastatin plays a critical role in determining the ultimate tenderness of meat. Reduction of muscle calpastatin pre-slaughter by selective breeding or other interventions would provide a means of reducing the incidence of toughness. This study demonstrates that the expression of calpastatin in a model system can be manipulated by factors associated with physiological stress or changes in intracellular calcium in whole animals that might account for variable rates of tenderisation. The most striking and novel result is the large differential effect of Ca^{2+} which may suggest strategies for controlling the activities of the 1xa and 1u promoters of the calpastatin gene pre-slaughter that could have a major impact on meat quality.

References

Cong, M., Thompson, V.F., Goll, D.E.and Antin, P.B. (1998). The bovine calpastatin gene promoter and a new N-terminal region of the protein are targets for cAMP-dependent protein kinase activity. Journal of Biological Chemistry, 273: 660-666.

Ji, S.Q., Frank, G.R., Cornelius, S.G., Willis, G.M. and Spurlock, M.E. (1998) Porcine somatotropin improves growth in finishing pigs without altering calpain 3 (p94) or alpha-actin mRNA abundance and has a differential effect on calpastatin transcription products. Journal of Animal Science, 76: 1389-1395.

Koohmaraie, M. (1996) Biochemical factors regulating the toughening and tenderization processes of meat. Meat Science, 43: S193-S201.

Parr, T., Jewell, K.K., Sensky, P.L., Brameld, J.M., Bardsley, R.G. and Buttery, P.J. (2004) Expression of calpastatin isoforms in muscle and functionality of multiple calpastatin promoters. Archives of Biochemistry & Biophysics, 427: 8-15.

Parr, T., Sensky, P.L., Bardsley, R.G.and Buttery, P.J. (2001) Calpastatin expression in porcine cardiac and skeletal muscle and partial gene structure. Archives of Biochemistry and Biophysics, 395: 1-13.

Sensky, P.L., Parr, T., Bardsley, R.G. and Buttery, P.J. (1996) The relationship between plasma epinephrine concentration and the activity of the calpain enzyme-system in porcine longissimus muscle. Journal of Animal Science, 74: 380-387.

THERMALLY-INDUCED CONFORMATIONAL CHANGES OF NATURAL ACTOMYOSIN EXTRACTED FROM NORMAL AND PSE PORK *LONGISSIMUS*

H. Wang, M. Pato, Z. Pietrasik and P.J. Shand*

Department of Applied Microbiology and Food Science, University of Saskatchewan, Saskatoon, SK, S7N 5A8, Canada. Email: zeb.pietrasik@usask.ca

Keywords: gelation, hydrophobicity, dynamic rheology, pork proteins

Introduction

Thermal unfolding and gelation properties of meat proteins, including myosin, salt soluble protein and myofibrillar protein with different species, pH and ionic strength have been intensively studied. Myosin plays a very important role in the gelation process in meat and meat products. Actin is important in reinforcing the gel structure of myosin (Yasui *et al.*, 1982). Myosin and actin are present as an actomyosin complex in full and post rigor meat. Natural actomyosin (NAM) can be extracted under mild conditions and is thought to represent the native state of proteins in post mortem pork meat. We have previously (Rathgeber *et al.*, 2002) reported that pale, soft, and exudative (PSE) pork produced meat gels with lower strain values compared to those from normal pork. The cause of poor gelation may be related to denaturation and degradation of myofibrillar proteins, but the biochemical and physicochemical properties of proteins from normal and PSE pork during heating and their gelation mechanisms are not fully understood.

The objective of the present study was to investigate the gelation behavior of NAM extracted from normal and PSE pork. Protein conformational changes during heating were studied by examining alterations of tertiary and/or quaternary structures (surface hydrophobicity and surface sulfhydryl and disulfide contents), while protein unfolding was monitored by determining alterations of secondary structures (α–helical content).

Materials and Methods

Pork loins from seven PSE and five normal pigs were selected based on pH at 2 and 24 h post mortem and on amount of drip loss (Normal: $pH_2 > 5.8$, $pH_u < 5.8$, drip loss < 5%; PSE: $pH_2 < 5.8$, $pH_u < 5.8$, drip loss > 5%). Meat was frozen at 72 h postmortem and later used for extraction of pork NAM as described by Wang *et al.* (2005). Heat treatment of NAM involved dilution of extracted NAM to 5.0 mg protein/mL in 6 M KCl, 20 mM KH_2PO_4/K_2HPO_4 buffer, pH 7.0 and then heating from 25 to 90°C at a constant rate of 1.3°C/min. An unheated solution kept at 4°C served as a control.

The α–helical content was estimated as described by Wang *et al.* (2005). Protein surface hydrophobicity was determined with the fluorescent probe 1-anilinonaphthalene-8-sulfonic acid (ANS) according to Li-Chan *et al.*, (1985). Total and surface sulfhydryl (S-SH) (also called reactive sulfhydryl) contents of heat-treated NAM were determined in the presence and absence of 8 M urea, respectively, by colorimetric assay using Ellman's reagent 5, 5'-dithio-bis (2-nitrobenzoic acid). The sulfhydryl plus disulfide (SH+SS) content of thermally-treated NAM was measured by using 2-nitro-5-thiosulfobenzoate according to Damodaran (1985) with slight modifications as described by Wang *et al.*, (2003). Disulfide content was calculated as the difference between (SS+SH) and total sulfhydryl contents.

Dynamic rheological properties of NAM (20 mg/mL in 0.6 M KCl, 20 mM KH_2PO_4/K_2HPO_4 buffer, pH 7.0) from PSE and normal pork were measured by using an AR-1000 advanced rheometer (TA Instruments, New Castle, DE, USA), which was equipped with a parallel plate measuring system (40 mm diameter). The sample was heated at a rate of 1 °C/min from 20 to 80°C, using a programmable temperature-controlled water bath.

Results and Discussion

The α–helical content of NAM from both normal and PSE pork decreased to about 3% upon heating above 70°C, indicating that approximately 90% of the α–helical structure of the actomyosin molecules were unfolded at 70°C and higher temperatures (Figure 1A). The decline of α–helical content of NAM from PSE meat occurred at a lower temperature than that of NAM from normal meat (~54 °C versus 57°C).

The surface hydrophobicity (S_0-ANS) of NAM from PSE meat was higher than that from normal meat below 40°C, indicating greater exposure of hydrophobic groups on the surface (Figure 1B). This result was in agreement with our previous report for pork stored at –20°C for up to 6 months (Wang *et al.*, 2005). The S_0-ANS of NAM from both meat groups rapidly increased at temperatures above 40°C and reached maximum values at 80°C. Interestingly, S_0-ANS of NAM from normal pork was higher than that from PSE meat at temperatures above 55°C, showing greater changes with heating in hydrophobicity of NAM from meat of normal quality. Surface hydrophobicities of NAM from normal and PSE pork at 70°C were 3.6 and 2.4 times greater than that at 40°C.

Possibly, the lower change in development of hydrophobicity of PSE meat would contribute to less structure development during heating.

Concentrations of free sulfhydryl groups declined (disulfide content increased, Figure 1C) during heat treatment of pork proteins, but quantity varied with meat quality. About 44% of available sulfhydryl groups formed disulfide bonds or other permanent chemical bonds at 80°C in NAM from normal, while this number was only 26% for thermally-treated NAM from PSE pork.

Dynamic thermal rheology of NAM from normal and PSE pork during heating from 25 to 85°C gave very similar rheological patterns. However, there were a few subtle differences. Namely, changes of G′ of NAM from PSE pork always occurred at a lower temperature during thermal treatment compared with NAM from normal pork (Figure 1D). The temperature at the lowest storage modulus (G′) during thermal rheology was in accordance with that at maximum α–helical content decline of NAM. In addition, the NAM from normal meat had a slightly higher initial G′ and G″, and δ at peak than corresponding values for NAM from PSE pork. Although the NAM from normal and PSE pork seemed to have different gelation kinetics during heating from 50-60°C, their final elastic response was similar.

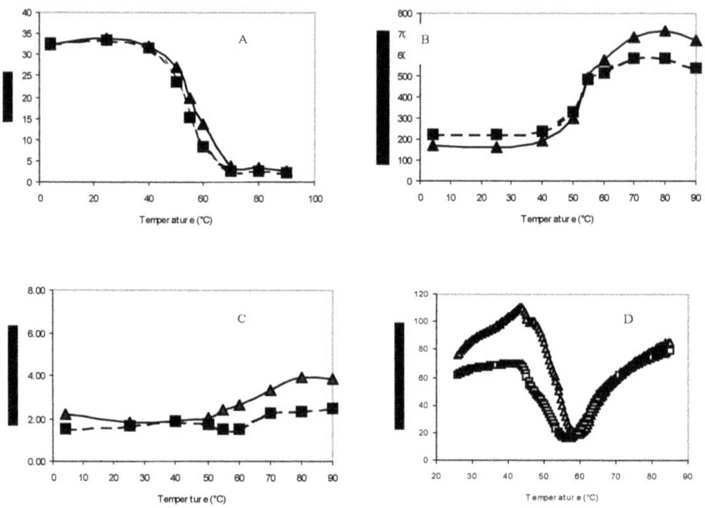

Figure 1: Alpha-helical content (A), surface hydrophobicity (B), disulfide content (moles per 10^5 g of protein) (C) and storage modulus (D) of NAM from normal (▲) and PSE (■) pork during heating. Heating rate was 1°C/min for storage modulus (dynamic rheological profile); 1.3°C/min for all other parameters.

Conclusions

Overall, NAM from normal pork underwent aggregation with a higher extent of hydrophobic interaction and disulfide bonds, as well as higher temperatures at maximum velocity for conformational change and unfolding than that from PSE pork. As a consequence, NAM from normal pork had superior rheological properties.

References

Damodaran, S. (1985). Estimation of disulfide bonds using 2-nitro-5-thiosulfobenzoic acid: Limitations. *Analytical Biochemistry, 145,* 200-204.

Li-Chan, E., Nakai, S., and Wood, D. F. (1985). Relationship between functional (fat, heating, emulsifying) and physicochemical properties of muscle proteins. Effects of heating, freezing, pH and species. *Journal of Food Science, 50,* 1034-1038.

Rathgeber, B. M., Boles, J. A., Pato M, and Shand, P. J. (2002). Myosin degradation and impaired gelation observed in PSE pork. In *48th International Congress of Meat Science and Technology* (pp. 590-591). Volume II.

Wang, H., Liceaga-Gesualdo, A., and Li-Chan, E. C. Y. (2003). Biochemical and physicochemical characteristics of muscle and natural actomyosin isolated from young Atlantic salmon (*Salmo Salar*) fillets stored at 0 and 4°C. *Journal of Food Science, 68,* 784-789.

Wang, H., Pato, M. D., and Shand, P. J. (2005). Biochemical properties of natural actomyosin extracted from normal and pale, soft and exudative pork loin after frozen storage. *Journal of Food Science, 70,* C313-320.

Yasui, T., Ishioroshi, M., and Samejima, K. (1982). Effect of actomyosin on heat induced gelation of myosin. *Agricultural and Biological Chemistry, 46,* 1049-1059.

TWO-DIMENSIONAL GEL ANALYSIS OF TROPONIN T ISOFORMS AND FRAGMENTS DURING POSTMORTEM AGING

S. Muroya*[1], M. Oe[1], I. Nakajima[2] and K. Chikuni[1]

[1.] *Meat Protein Research Team and [2]Animal Product Research Team, National Institute of Livestock and Grassland Science, Tsukuba, Ibaraki 305-0901, Japan E-mail: muros@affrc.go.jp*

Keywords: ageing, beef, isoform, troponin T, two-dimensional gel electrophoresis (2DE)

Introduction

Troponin T (TnT) is degraded mainly into the 30-kDa peptide, during postmortem tenderisation of beef. A peptide generated by TnT degradation was found to suppress sourness of lactic acid and is useful as a conditioning indicator in pork. Despite the significance of elucidating postmortem TnT degradation, the details of TnT degradation are still not understood to date. In bovine muscles, at least eight fast-type TnT (fTnT) (fTnT1/16 and 17, fTnT2/16 and 17, fTnT3/16 and 17, and fTnT4/16 and 17) and two slow-type TnT (sTnT) (sTnT1 and sTnT2) isoforms have been found to be expressed so far. These fTnT and sTnT isoforms are generated by the alternative splicing of the mRNA originated from fTnT and sTnT genes, respectively. The molecular weight (MW) and isoelectric point (pI) of the isoforms are estimated to range widely from 32.1 to 29.8 kDa, and from 5.61 to 9.23, respectively (Muroya *et al.*, 2003). To clarify degradation of multiple TnT isoforms, the analysis is required to be highly resolutive, by which the isoform proteins can separate from one another. Taking advantage of high resolution, 2DE analysis provides the benefits of being able to analyze complex mixtures of myofibrils and sarcoplasm. In the present study, in order to comprehend the changes in TnT isoforms accompaning beef aging, whole muscle proteins including all TnT-related polypeptides were applied to 2DE analysis. To identify TnT spots, western blotting, using isoform-specific anti-TnT antibodies on 2DE gels were performed.

Materials and Methods

The longissimus thoracis (LT) for 0 d postmortem samples were excised from a Holstein and a Japanese Black steer within 1 hr after slaughter. After overnight hanging of the carcasses at 2°C, the muscle blocks were excised from the carcass. At 1 and 14 d postmortem, the muscle samples were prepared from the central part of the block, and bagged during aging at 2°C. The sample for whole muscle proteins was prepared by homogenizing the muscle sample in lysis buffer containing 8 M urea, 50 mM dithiothreitol (DTT), 4% CHAPS, and 0.2% Bio-Lyte 3/10 Ampholyte (Bio-Rad, CA, USA.), to prepare 2DE samples. For myofibril preparation, muscle samples were homogenized and extracted with homogenizing buffer, 20 mM potassium-phosphate buffer (pH 6.8) containing 100 mM KCl, 1 mM $MgCl_2$, 1 mM EDTA, 1mM 2-mercaptoethanol, and 0.05% sodium azide. Then, the samples were solubilized in lysis buffer. After determining protein concentration, 100-200 μg of protein was loaded onto a Bio-Rad ReadyStrip, 11 cm, pH 3-10 non-linear and immobilized pH gradient (IPG) strips. Isoelectric focusing (IEF) was performed in a Protean IEF cell (Bio-Rad). Strips were rehydrated overnight. For the subsequent IEF, voltage was increased gradually up to 8,000 V for 2.5 h, and then was maintained at a total voltage of 60,000 or 80,000 Vh. The second dimensional SDS-PAGE was performed in a Criterion cell (Bio-Rad) on 12.5% polyacrylamide gel at 100 V for 1 h. Overall proteins in preparative gel were stained in SYPRO Ruby (Bio-Rad). Experimental pI was determined using a 2DE protein standard (Bio-Rad). After 2DE, the proteins on the gel for detecting phosphorylation were fixed and stained with Pro-Q® Diamond phosphoprotein gel stain (Invitrogen, Eugene, OR, USA) according to manufacture's protocol. After destaining, the fluorescence of the gel was scanned at excitation 540 nm and emission 595 nm with Ettan DIGE Primo (GE Healthcare, Fairfield, CT, USA). As for western blot analysis, the proteins on the gel were transferred to a polyvinylidenefluoride membrane. The electroblotted membrane was incubated with commercial anti-fTnT or -sTnT polyclonal goat antibodies (Santa Cruz Biotechnology, Inc., CA, USA). The positive spots were detected with avidin-DH and biotin-conjugated HRP, and DAB substrate kit (VECTORSTAIN® ABC kit, Vector Laboratories, CA, USA.) as performed previously (Muroya *et al.*, 2006).

Results and Discussion

The 2DE western blot image showed that at least nine and three anti-fTnT antibody-positive spots were distributed on the 0 and 14 d gel, respectively. The MW of the fTnT spots were 36.5, 35.4, 34.8, and 32.8 kDa on the 0 d gel, and were 32.1, 28.3, and 26 kDa on the 14 d gel. The spots of four different MW isoforms on the 0 d gel were of fTnT1, fTnT2, fTnT3, and fTnT4 isoforms, respectively (Muroya *et al.*, 2004, 2006). In the overview of 2DE map of intact fTnT isoforms, the highest MW isoform was acidic, and in contrast, the lowest MW isoforms tended to be basic, with intermediate MW isoforms having broad pI range. As for the N-terminal variants, fTnT spots consisted of fTnT2 and fTnT3 isoforms in most part, plus fTnT1 and fTnT4 isoforms in minor part, in the case of bovine LT muscle. On the other hand, 32.1, 28.3, and 26 kDa polypeptides on the 14 d gel were degradation products generated during postmortem aging, since they appeared with aging, showing lower MW than that of intact TnT isoforms. The fTnT isoform spots on the 0 d gel were distributed within a range of pI 5.7-10, while on the 14 d gel, spots of fTnT fragments were exclusively located at pI of more than 9.6. These results clearly demonstrated that, in spite of the broad pI range, all of the fTnT isoforms were degraded into basic polypeptides during postmortem aging of beef. Similarly, two and one anti-sTnT antibody-positive spots were detected on the 0 and 14 d gel, respectively. The sTnT-related spots on the 0 d gel were of pI 5.6 and 5.7, while the degraded sTnT-spot was at pI of more than 9.6 on the 14 d gel. The MW of the sTnT peptides were 34.5 kDa on the 0 d gel, and 31.0 kDa on the 14 d gel. This result indicates that sTnT isoforms were also degraded exclusively into basic polypeptides. Moreover, we detected spots of phosphorylated myofibril protein by gel staining specific to phosphorylated protein. Since no difference in the TnT spot pattern between beef whole muscle and myofibril samples was observed by western blot analysis, the myofibril fraction was used to improve the content of TnT proteins. Consequently, three of the fTnT-related spots on the 0 d gel were shown to be phosphorylated, but none of the phosphorylated spots matching anti-TnT antibody-positive spots was found on the 14 d gel. It is likely that TnT proteins are phosphorylated in an isoform-specific manner in live bovine skeletal muscle, and that changes in phosphorylated spots were caused by a cut-off of the N-terminal region of TnT during beef aging.

Conclusion

Whole beef longissimus muscle proteins were developed on 2DE gel and multiple TnT-related spots were identified by western blotting. The results showed that more than ten fTnT isoform spots (pI 5.7-9.6<) and two sTnT spots (pI 5.6-5.7) present at slaughter were degraded exclusively into basic spots (pI 9.6<) at day 14 postmortem, indicating that TnT is cleaved primarily in the glutamic acid-rich amino-terminal region.

References

Muroya, S., Kitamura, S., Tanabe, S., Nishimura, T., Nakajima, I., and Chikuni, K. (2004). N-terminal amino acid sequences of troponin T fragments, including 30 kDa one, produced during postmortem aging of bovine longissimus muscle. Meat Science, 67, 19-24.

Muroya, S., Nakajima, I., and Chikuni, K. (2003). Amino acid sequences of multiple fast and slow troponin T isoforms expressed in adult bovine muscles. Journal of Animal Science, 81, 1185-1192.

Muroya, S., Nakajima, I., Oe, M., and Chikuni, K. (2006). Difference in postmortem degradation pattern among troponin T isoforms expressed in bovine longissimus, diaphragm, and masseter muscles. Meat Science, 72, 245-251.

CONVERSION OF METHMYOGLOBIN INTO RED MYOGLOBIN DERIVATIVE BY *PSEUDOMONAS FRAGI*

M. Motoyama*, M. Kobayashi, K. Sasaki and M. Mitsumoto

Animal Products Research Team, National Institute of Livestock and Grassland Science, 305-0901 Tsukuba, Ibaraki, Japan. Email: mmichiyo@affrc.go.jp

Keywords: *Pseudomonas fragi*, *Brochothrix thermosphacta*, lipid oxidation, colour, myoglobin

Introduction

Meat colour largely results from the chemical state of myoglobin. Undesirable discolouration of meat during preservation is due to the accumulation of a myoglobin derivative, metmyoglobin (Renerre, 1990). Generally, bacteria increase metmyoglobin ratio. An increase of metmyoglobin occurs concomitantly with an undesirable increase in lipid oxidation. A common meat bacterium, *Pseudomonas* sp., decreased a red myoglobin derivative, oxymyoglobin (Bala *et al.*, 1977), via increased oxygen consumption (Chan *et al.*, 1998). On the other hand, Faustman *et al.* (1990) reported that ground beef homogenates inoculated with *Pseudomonas* sp. showed a decrease in metmyoglobin and a colour reversion from brown to red. In this study, we isolated *Pseudomonas fragi* from beef, and investigated these bacterial effects on meat colour and lipid oxidation.

Materials and Methods

Bacterial Preparation

The bacteria used in this study were obtained from slimy beef preserved at 4°C for 2 weeks. Single colonies were purified on nutrient agar (1.5% w/v) plates after 2 days of incubation at 4°C under aerobic conditions. Isolates were identified by the following criteria: cell morphology, motility, growth temperature, and 16S rRNA sequences. Regarding *Pseudomonas* sp., further identification was performed by sequencing *rpoB* (Tayeb *et al.*, 2005). Pure culture were maintained at 4°C in nutrient broth and subcultured at a weekly interval.

Inoculation test on Sterile Meat

Semitendinosus and *Longissimus* muscles of Japanese Black beef steers were used. The outsides of the muscles were sterilized by dipping them into boiling water for 5 s. Each sample was formed into a circular disk 5 cm in diameter and 1 cm in thickness. The pure cultures were inoculated meat upper surface (20 cm^2) with 1.0×10^4 cfu/cm^2. Each sample was placed in an autoclaved 60 ml weighing boat, overwrapped with oxygen-permeable PVC film and stored for 10 days at 4°C under fluorescent lighting. The light intensity was 2000 lx. Lipid oxidation was evaluated in terms of 2-Thiobarbituric acid reactive substances (TBARS) values. TBARS were measured in samples displayed for 0 and 10 days by spectrophotometric analysis (Witte *et al.*, 1970, Mitsumoto *et al.*, 1993) using Shimadzu UV-2400PC spectrophotometer (SHIMADZU CORPORATION, Kyoto). The TBARS values were expressed as mg of malonaldehyde (MDA) equivalents per kg of muscle. Meat colour was measured by reflectance spectrometry using the Shimadzu UV-2400PC, and the redness was expressed as an a^* value. The values were analyzed by the general linear model procedure of SAS version 9.1.3 (SAS Institute Inc., Cary, NC).

Inoculation Test with 0.05% Myoglobin Containing Nutrient Broth

0.05% horse skeletal myoglobin (Sigma Chemical Co., St Louis, Mo) containing nutrient broth was prepared. After filter-sterilization with a 0.45 µm pore-size filter (Dismic-25CS, Toyo Roshi Kaisha, Tokyo), the pure cultures were inoculated with 1.0×10^7 cfu/ml. After incubation for 92 h at 25°C, the cells were removed by centrifugation (10,000 rpm, 5 min) and absorption spectra of the supernatant were measured from 650 to 350 nm at room temperature (23-25°C) using the Shimadzu UV-2400PC spectrophotometer.

Results and Discussion

Seven different type colonies were identified, five were identified as *Pseudomonas* sp. (including *Pseudomonas fragi*; *P. fragi*) and two were identified *Brochothrix thermosphacta* (*B. thermosphacta*). Both bacteria are commonly observed in chilled meat (Dainty and Mackey, 1992). *B. thermosphacta* did not change meat colour and lipid oxidation (Table 1). *P. fragi* significantly improved redness ($P < 0.05$), and showed a tendency to lower the increase in TBARS. This observation agreed with Bala *et al.* (1977). Nutrient broth with the presence of *B. thermosphacta* as well as no-inoculated nutrient broth showed the metmyoglobin spectrum pattern, with characteristic peaks at 503 nm (Sakata *et al.*, 1996) (Figure 1). The α and β absorption bands were 580 and 543 nm, respectively. The myoglobin derivative(s) converted by *P. fragi* had a spectrum

Table 1: Effect of growth of *B. thermosphacta* and *P. fragi* on meat redness and TBARS

Treatment	Redness (a^*)		TBARS	
	Day0	Day10	Day0	Day10
Control	21.8	8.1	0.1	2.1
B. thermosphacta	19.3	6.1	0.1	2.0
P. fragi	18.7	12.4 †	0.1	0.8

† Significantly different from the control ($P < 0.05$). n=6.

that was very similar to that of oxymyoglobin (Sakata *et al.,* 1996). Therefore, oxymyoglobin was presumably responsible for high a^* value of the *P. fragi* inoculated beef at day10.

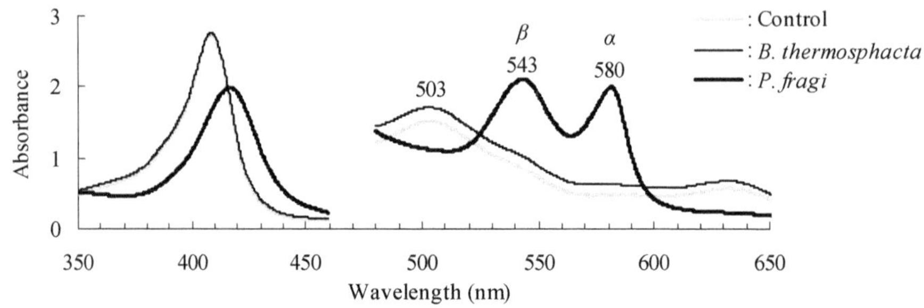

Figure 1: Spectra of bacteria in inoculated/non-inoculated 0.05%Mb-containing nutrient broth after 92h culture. The control contained sterile nutrient broth in place of cultures. The α and β absorption bands observed in *P. fragi*-inoculated-broth were 580 and 543 nm, respectively.

Conclusions

We isolated *P. fragi* from chilled beef. *P. fragi* improved meat colour during preservation. The improvement meat colour was probably due to the formation of oxymyoglobin. Today, a colouring agent for meat products, nitrite, is currently being questioned from the standpoint of human health. Nitrite may be a precursor of carcinogenic N-nitroso compounds (nitrosamines). Therefore, this bacterial metmyoglobin conversion system may be a key to a new technology for preserving and/or improving the appearance of meat and meat products.

References

Bala, K., Marshall, R.T., Stringer, W.C., Nuimann, H.D. (1977). Effect of *Pseudomonas fragi* on the Color of Beef. Journal of Food Science, 42(5): 1176-1179.

Chan, W.K.M., Joo, S., Faustman, C., Sun, Q., Vieth, R. (1998). Effect of *Pseudomonas fluorecens* on Beef Discoloration and Oxidation In Vitro. Journal of Food Protection, 61(10): 1341-1346.

Dainty, R.H., Mackey, B.M. (1992). The relationship between the phenotypic properties of bacteria from chill-stored meat and spoilage processes. Journal of Applied Bacteriology Symposium Supplement, 73: 103S-114S.

Faustman, C., Johnson, J.L., Cassens, R.G., Doyle, M.P. (1990). Color reversion in beef. Influence of psychrotrophic bacteria. Fleischwirtsch. 70: 676-679.

Mitsumoto, M., Arnold, R.N., Schaefer, D.M., Cassens, R.G. (1993). Dietary versus postmortem supplementation of vitamin E on pigment and lipid stability in ground beef. Journal of Animal Science, 71: 1812-1816.

Renerre, M. (1990). Review: Factors involved in the discoloration of beef meat. International Journal of Food Science and Technology, 25: 613-630.

Sakata, R., Honikel, K.O., Morita, H., Nagata, Y. (1996). Physikalisch-chemische Merkmale von Nitrosomyoblobin: Wasserextrahierbarkeit und Stabilität von Myoglobinderivaten bei Verwendung eines Oxidationmittels. *Fleischwirtschaft*, 76: 1046-1048.

Tayeb, L.A., Ageron, E., Grimont, F., Grimont, P.A.D. (2005). Molecular phylogeny of the genus *Pseudomonas* based on rpoB sequences and application for the identification of isolates. Research in Microbiology, 156: 763-773.

Witte, V.C., Krause, G.F., Bailey, M.E. (1970). A new extraction method for determining 2-Thiobarbituric acid value of pork and beef during storage. Journal of Food Science, 35: 582-585.

SOL-GEL IMMOBILIZED LUCIFERASE-BASED ATP BIOSENSOR FOR MEAT QUALITY DETERMINATION IN POSTMORTEM PIG MUSCLE

A.D. Weaver[1], S.S. Jedlicka[2,4], J.L. Rickus[2,3,4] B.C. Bowker[3] and D.E. Gerrard[1]*

[1.] Departments of Animal Sciences and [2.] Agricultural and Biological Engineering, Weldon School of Biomedical Engineering, Physiological Sensing Facility at Bindley Bioscience Center, West Lafayette, Indiana USA, [3.] USDA-ARS, Food Technology and Safety Laboratory, Beltsville, MD USA. Email: dgerrard@purdue.edu

Keywords: pork, metabolism, ATP, luciferase

Introduction

Pale, soft, exudative (PSE) and associated meat quality defects represent a serious problem facing the pork industry. Not only do these undesirable quality attributes negatively impact further processing potential of a valuable commodity, they also create a level of indecisiveness on the part of consumers wishing to purchase a highly satisfying and nutritious food item. In addition, production of pork with reduced quality robs the industry of potential revenues. R.G. Kauffman (Univ. of Wisconsin) has suggested in his "*Report Assessing the Definable and Potential Losses Due to Quality Problems in the US Pork Industry*" that lowered pork quality costs the industry \$100 million annually. Granted, the industry has responded to an increased incidence of adverse quality development and has instigated new processing procedures to alleviate some of the negative attributes of this problem by using various injection or "pumping" strategies; however, the bottom line is that pork possessing these abhorrent traits represents an inefficiency in the food production chain, as muscle protein functionality, and therefore product quality, is less than optimal in some pork carcasses. Unfortunately, the industry has perpetuated this problem by aggressively selecting for improved meat-producing genotypes. As a result, the incidence of pork quality problems has and will continue to increase. Classically, PSE pork results from a rapid rate of muscle metabolism (energy) that is stimulated by ATP depletion during the early postmortem period (Kastenschmidt *et al.*, 1968). This accelerated metabolism results in a low muscle pH while the carcass temperature is still high and leads to severe muscle protein denaturation and thus, diminished pork quality attributes. The other major pork quality aberration, known as 'acid meat', which possesses quality characteristics similar to PSE pork (a pale color, soft texture and an exudative cut lean surface) results from an exaggerated or prolonged energy metabolism postmortem, which results in a greater than normal pH decline in muscle tissues at 24 hr after slaughter. The enzyme luciferase is commonly used to measure ATP in biological samples. This enzyme catalyzes the ATP-dependant oxidation of luciferin and produces light as a consequence:

$$\text{ATP} + \text{luciferin} + O_2 \xrightarrow[\text{Mg}^{2+}]{\text{luciferase}} \text{oxyluciferin} + \text{AMP} + \text{PP}_i + CO_2 + \text{light}(\lambda_{max}=560\text{nm})$$

In an excess of co-substrates, the light produced correlates with the ATP concentration present. Despite the widespread use of luciferase as a lab-based bioassay for ATP, there are few practical ATP biosensors available, due to the challenge of luciferase immobilization and the instability of the luciferase enzyme in solution. Recently, a biologically friendly method of immobilizing luciferase in a silica-based sol-gel was discovered by Cruz-Aguado, *et al.* (2004). The objective of this study was to explore the use of a silica immobilized luciferase-based optical biosensor in monitoring muscle metabolism postmortem and correlating these results to pork quality.

Materials and Methods

To test the hypothesis that muscle ATP could be used for predicting fresh pork quality, ten crossbred pigs averaging 120 kg were delivered to the Purdue Meat Science Research Laboratory and slaughtered according to normal processing practices, except some carcasses were randomly subjected to varying amounts of electrical stimulation at 10 min to simulate adverse pork quality development (Bowker *et al.*, 1999). At various times postmortem (15, 30, 60, 90 and 120 min), muscle exudates were sampled using cotton swabs. These swabs were immediately subjected to a firefly luciferase assay and luminescence was recorded. After a 24 hr chill, muscle exudates were analyzed using the same protocol and traditional pork quality characteristics were measured. An optical luciferase based biosensor was developed using the immobilization technique of Cruz-Aguado, *et al.* (2004). This method, using a biocompatible sol-gel precursor and a covalently linked stabilizing sugar moiety allows for luciferase activity retention very close to the activity of the enzyme in solution form. Diglycerylsilane (DGS) was synthesized by heating neat tetramethyl orthosilicate (TMOS) and anhydrous glycerol in a 1:2 molar ratio at 110 °C for 15 hours, followed by complete removal of the evolved methanol by distillation at 140 °C (Brook, *et al.*, 2004). DGS was hydrolyzed and mixed at an 80% molar ratio with N-(3-triethoxysilylpropyl) gluconamide (Gelest, Inc.). Luciferase (Sigma, 1:30 volumetric ratio to the sol), D-luciferin (Sigma, 2 mM final concentration), and magnesium chloride (10 mM) were added to the hydrolyzed sol in phosphate buffer, pH 8.0. Bovine serum albumin (0.1 mg/mL), EDTA (0.5 mM) and DTT (0.5 mM) were added as stabilizing agents. The resulting sol mixture was pipetted into capillary tubes containing 1000 µm optical fibres (PolyMicro Technologies). The fibre/silica/enzyme probe was allowed to sit at 4 °C overnight prior to use. To test the efficacy of a luciferase probe, the probes were connected to a jacketed fibre via a bare fibre adaptor. Jacketed fibres were connected to a hand-held USB 2000 spectrometer containing a linear CCD detector (Ocean Optics, Inc.). The spectrometer was operated from a laptop PC using standard operating software (Ocean Optics, Inc.). Silica-based luciferase probes were tested as 'real-time' biosensors using five crossbred pigs. Pigs were slaughtered under normal processing conditions except one side of each carcass was electrically stimulated (400V, 20 pulses) at approximately 30 min postmortem. Immediately prior to and after stimulation and hourly up to 4 hr postmortem, luciferase containing probes were placed in the longissimus muscle at random locations. Luminescence was recorded using a 5 second integration time for several minutes post-insertion to ensure probe stability, and the luminescence output averaged across the given time period. A new probe was used for each time point and each side of the carcass. Probe reusability will be the focus of future work.

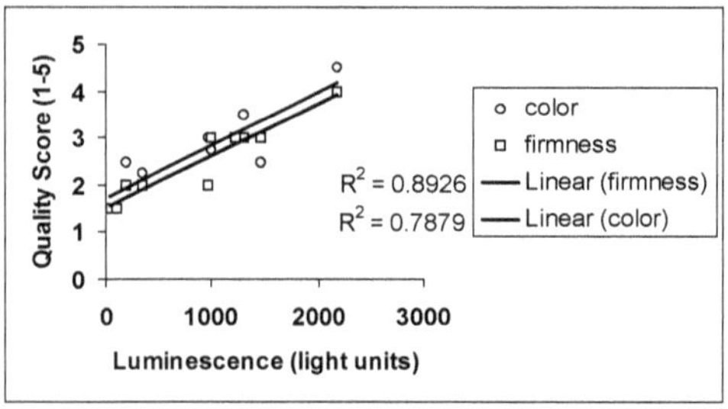

Figure 1: Relationship between luciferase-based luminescence and fresh pork colour and firmness scores. (1=pale or soft; 5=dark or firm).

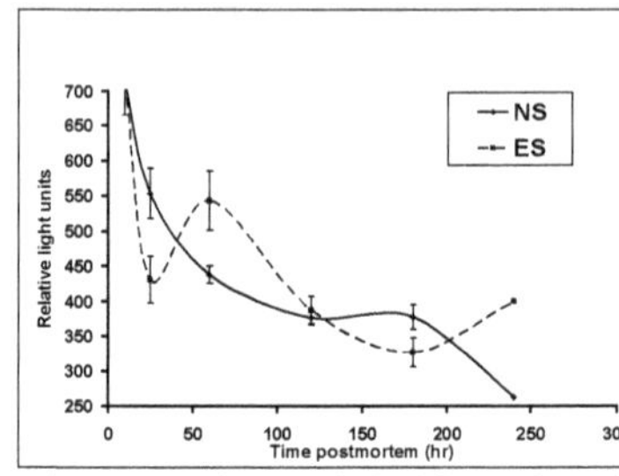

Figure 2: Changes in luciferase-based luminescence (relative light units) early postmortem in electrically stimulated (ES) and control (NS).

Results and Discussion

Figure 1 shows that luminescence in the postmortem exudates correlated with subjective fresh pork colour and firmness scores (0.79 and 0.89, respectively). These data suggest that higher levels of ATP postmortem are useful predictors of pork quality attributes. Curiously, muscle pH around this time postmortem normally correlates well with these traits. Changes in muscle luminescence postmortem parallel pH values using this model. Electric stimulation rapidly increases the consumption of ATP by the muscle (Figure 2) and as a result, anaerobic glycolysis is increased and lactate accumulation is exaggerated compared to non-stimulated carcasses. Although events leading to adverse pork quality development in this manner may not exactly simulate classic PSE development, it is a good model for rapid postmortem muscle metabolism. Additionally, the ability of the luminescence based optical biosensor for ATP to detect the changes in the postmortem muscle tissue indicates that the sensor is a potential option for rapid meat quality testing. Although more work to optimize the probes must still be done, it is a demonstration that a traditional lab-based assay can be translated into an inexpensive, hand-held device for industrial use.

Conclusions

These data suggest that the relative amount of ATP in muscle cells early postmortem is related to pork quality. Based on the results shown, the development of a hand-held real-time biosensor using a luminescence based platform is a viable possibility for the rapid detection of ATP. Additional experiments will be necessary to further support these claims.

References

Bowker, B.C., E.J. Wynveen, A.L Grant, D.E. Gerrard, (1999) Effects of electrical stimulation on early postmortem pH and temperature declines in pigs from different genetic lines and halothane genotypes. Meat Science 53:125-133.

Brook, M.A., Y. Chen, K. Guo, Z. Zhang, J.D. Brennan. (2004) Sugar-modified silanes: precursors for silica monoliths. J. Mater. Chem. 14:1469-1479.

Cruz-Aguado, J.A., Y. Chen, Z. Zhang, N.H. Elowe, M.A. Brook, J.D. Brennan. (2004) Ultrasensitive ATP detection using firefly luciferase entrapped in sugar-modified sol-gel derived silica. J. Am. Chem. Soc. 126: 6878-6879.

Kastenschmidt, L.L., W.G. Hoekstra, E.J. Briskey. (1968) Glycolytic intermediates and co-factors in "fast-" and "slow-glycolyzing" muscles of the pig. J. Food Sci. 33:151.

THE CALPAIN SYSTEM IN 15 EUROPEAN CATTLE BREEDS AND RELATIONSHIP TO SHEAR FORCE

P. Ertbjerg*[1], M. Christensen[1], S. Failla[2], S. Gigli[2], J.L. Olleta[3], C. Sañudo[3], B. Panea[4], P. Albertí[4], J.F. Hocquette[5], R.I. Richardson[6], G.R. Nute[6] and J.L. Williams[7]

[1]Department of Food Science, Royal Veterinary and Agricultural University, 1958 Frederiksberg C, Denmark. [2]Istituto Sperimentale di Zootecnia, 00016 Monterotondo, Rome,Italy. [3]Departamento de Producción Animal y Ciencia de los Alimentos, Universidad de Zaragoza, 50.013 Zaragoza, Spain. [4]Centro de Investigación y Tecnología Agroalimentaria, 50016 Zaragoza, Spain. [5]Unité de Recherches sur les Herbivores, Croissance et Métabolisme du Muscle, Theix, 63122 Saint-Genès-Champanelle, France. [6]Division of Farm Animal Science, University of Bristol, Langford BS40 5DU, United Kingdom. [7]Roslin Institute, EH25 9PS Roslin Midlothian, United Kingdom. Email: peer@kvl.dk

Keywords: beef, μ-calpain, m-calpain, calpastatin, shear force

Introduction

This study, which is part of the European Project GEMQUAL (Genetics of Meat Quality), examined young bulls from 15 different breeds (both dairy and beef) representing the genetic diversity of European cattle. Post-mortem (p.m.) storage of beef at refrigerated temperatures is known to increase meat tenderness. The calpain system is believed to be involved in tenderisation of beef. Moderate correlations between calpastatin activity and shear force have been reported in *Bos indicus* crosses (r = 0.43, Whipple *et al.*, 1990), whereas lower correlations were reported for *Bos taurus* breeds (r = 0.27, Shackelford *et al.*, 1994). Here, we report some characteristics of the calpain system in European cattle breeds and correlations to Warner Bratzler shear force in *longissimus thoracis* muscle.

Materials and Methods

The experiment included 437 young bulls from 15 European cattle breeds. Muscle samples (ca. 20 grams) were taken within 45 minutes post-mortem (p.m.) from the right side of *M. longissimus thoracis*, chopped into very small pieces, immediately frozen in liquid nitrogen and stored at -80°C until use. *M. longissimus thoracis* (between 8th and 13th rib) from the left side of the carcass was stored at 3°C ± 1°C until 10 days p.m., a sample was cut, vacuum packaged and frozen for Warner-Bratzler (WB) shear force. Meat slices for determination of WB shear force were heated in a water bath at 80°C until a 75°C internal temperature was reached. Peak force was determined on 10 cores (1x1x2 cm) per slice cut perpendicularly to the fibre axis. Calpain and calpastatin activities were determined in one gram muscle representing a mixture of the very finely chopped muscle of 20 grams. Briefly, an extract was prepared by homogenisation in 10 volumes of extraction buffer. After centrifugation at 10,000 x *g* for 30 min, one aliquot for calpastatin determination was heated to 95 °C for 5 min, centrifuged, and the supernatant was assayed for calpastatin activity as described in Ertbjerg *et al.* (1999). Then, one aliquot of the supernatant was taken for calpain activity determination by casein zymography. Briefly, casein minigels with 26 wells were run at 80 V for three hours at 4 °C, removed, and incubated with calcium for 1 hour at room temperature. After staining and destaining, the density of each band was measured and quantified relative to reference standards within each gel and relative to activity of the standards in a casein assay (Ertbjerg *et al.*, 1999).

Results and Discussion

From Table 1, it appears that the two dairy breeds Danish Red Cattle and Holstein and the specialised beef breed Limousin contained significantly higher μ-calpain activity than the other breeds, whereas the beef breeds Piemontese and Marchigiana contained the lowest μ-calpain activity. The m-calpain activity was the highest in Danish Red Cattle and the lowest in the beef breed Pirenaica. The calpastatin activity was the highest in Danish Red Cattle and the lowest in the double-muscled Piemontese breed. Although the Asturiana de los Valles breed is also a double-muscled breed,, the calpastatin activity in this breed appeared to be intermediate as compared to the other breeds.

Partial correlations (i.e. corrected pearson correlations for the breed differences) between μ-calpain or calpastatin activities and WB shear force were significant ($P < 0.01$), but explained only 4 and 2% of its variation, respectively (data not shown). The poor relationship between the calpain system and WB shear force were breed-dependent. There was indeed a high negative correlation between μ-calpain and WB shear force in Casina (r = -0.70, $P < 0.001$) (Figure 1) and Jersey (r = -0.39, $P < 0.05$) but no significant correlation for other breeds. Piemontese showed a positive correlation between m-calpain and WB shear force (r = 0.41, $P < 0.05$). A significant correlation between calpastatin activity and WB shear force was found in Highland (r = 0.52, $P < 0.05$) and Pirenaica (r = 0.60, $P < 0.001$) (Figure 1), but not in any of the other breeds studied.

Table 1: Differences in µ-calpain, m-calpain and calpastatin activity between 15 European cattle breeds. Values are expressed as means ± standard deviations. Jersey (n=30), South Devon (n=27), Aberdeen Angus (n=30), Highland (n=29), Holstein (n=29), Danish Red Cattle (n=29), Simmental (n=20), Asturiana de los Valles (n=30), Casina (n=31), Avileña (n=30), Pirenaica (n=31), Piemontese (n=30), Marchigiana (n=29), Limousin (n=31) and Charolais (n=31).

	µ-Calpain (U/g meat)	**m-Calpain** (U/g meat)	**Calpastatin** (U/g meat)
Danish Red Cattle	9.62 ± 1.04^{a}	5.74 ± 0.96^{a}	34.74 ± 4.88^{a}
Holstein	9.30 ± 0.84^{a}	5.60 ± 0.84^{ab}	32.87 ± 4.08^{ab}
Limousin	9.07 ± 0.59^{ab}	5.23 ± 0.93^{bcde}	31.47 ± 4.66^{bcd}
Jersey	8.50 ± 1.57^{b}	4.67 ± 1.00^{fg}	32.16 ± 3.74^{bcd}
Charolais	8.43 ± 0.77^{b}	5.38 ± 0.56^{abcde}	33.08 ± 5.38^{ab}
Simmental	8.43 ± 0.62^{b}	5.52 ± 0.94^{abc}	32.98 ± 3.17^{ab}
Asturiana de los Valles	7.47 ± 1.35^{c}	5.43 ± 0.63^{abcd}	30.24 ± 4.13^{def}
Casina	7.28 ± 1.31^{cd}	5.48 ± 0.74^{abc}	31.26 ± 3.52^{bcd}
Highland	7.23 ± 1.23^{cd}	5.51 ± 0.66^{abc}	32.53 ± 2.76^{bc}
Avileña	7.04 ± 2.03^{cd}	4.99 ± 0.59^{efg}	32.92 ± 3.57^{ab}
Pirenaica	7.03 ± 1.66^{cd}	4.59 ± 0.97^{g}	30.67 ± 5.12^{cde}
South Devon	7.03 ± 1.44^{cd}	5.21 ± 0.74^{bcde}	28.90 ± 4.08^{efg}
Aberdeen Angus	6.93 ± 0.91^{cd}	5.04 ± 0.49^{def}	30.51 ± 3.32^{cdef}
Piemontese	6.68 ± 1.76^{d}	5.43 ± 0.88^{abcd}	27.19 ± 3.37^{g}
Marchigiana	6.68 ± 1.25^{d}	5.16 ± 1.13^{cde}	31.15 ± 3.31^{bcd}

[a-g] Within columns, values with the same letter are not significantly different

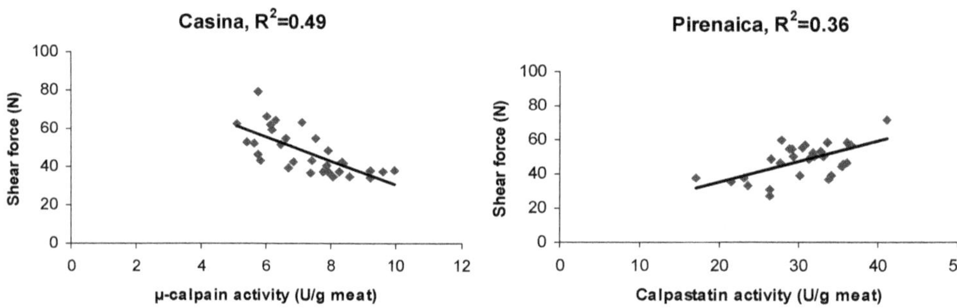

Figure 1: Correlation between µ-calpain activity and shear force in the Casina breed (left) and between calpastatin activity and shear force in the Pirenaica breed (right).

Conclusions

In conclusion, the dairy breeds showed relatively higher values for all components of the calpain system than beef breeds, whereas the double-muscled Piemontese breed had the lowest calpastatin activity. Overall, the 45 min post-mortem µ-calpain and calpastatin activity showed poor correlation to day 10 shear force. However, this relationship was strongly breed-dependent. Thus, 49% of the variation in shear force could be explained by µ-calpain in the Casina breed. Also, 36% of the variation in shear force could be explained by the calpastatin activity in the Pirenaica breed and 27% in the Highland breed. However, no strong significant relationship was revealed in other breeds.

References

Ertbjerg, P., Henckel, P., Karlsson, A., Larsen, L. M., Møller, A. J. (1999). Combined effect of epinephrine and exercise on calpain/calpastatin and cathepsin B and L activity in porcine *longissimus* muscle. *Journal of Animal Science*, 77: 2428-2436.

Shackelford, S.D., Koohmaraie, M., Cundiff, L.V., Gregory, K.E., Rohrer, G.A., Savell, J.W. (1994). Heritabilities and phenotypic and genetic correlations for bovine postrigor calpastatin activity, intramuscular fat content, Warner-Bartzler shear force, retail product yield, and growth rate. *Journal of Animal Science*, 72: 857-863.

Whipple, G., Koohmaraie, M., Dikeman, M.E., Crouse, J.D., Hunt, M.C., Klemm, R.D. (1990). Evaluation of attributes that affect Longissimus muscle tenderness in Bos taurus and Bos indicus cattle. *Journal of Animal Science*, 68: 2716-2728.

CALPAIN-1 ACTIVITY IS RELATED TO DRIP LOSS IN COMMERCIAL PORK PRODUCTION

P.P. Purslow*[1], M.C. Cha[1], S. Wang[1], I.B. Mandell[2], T. Widowski[2], C.F.M. de Lange[2], A. Robinson[2], J. Brown[2] and E.J. Squires[2]

[1.] *Department of Food Science, University of Guelph, Guelph N1G 2W1 Ontario, Canada,* [2.] *Department of Animal and Poultry Science, University of Guelph, Guelph N1G 2W1 Ontario, Canada.*
Email: ppurslow@uoguelph.

Keywords: meat quality, pork, drip loss, proteolysis, tenderisation, *longissimus dorsi, semimembranosus*

Introduction

Unexplained variability in meat quality (tenderness, appearance and water-holding capacity) is a consistent concern for the meat industry in both domestic and export markets. Specific to research on pork, there is a great deal of existing knowledge on the effect of individual factors such as animal stress (Hemsworth *et al.,* 2002), nutrition (Warriss *et al.,* 1998), genotype (Lonergan *et al.,* 2001) and a number of post-mortem factors (Schäfer *et el.,* 2002). However, individual factors do not completely explain variability in quality due to substantial and complex interactions between a number of factors, such as genotype, management factors, nutrition and stress (D'Souza *et al.,* 1998a, Warris *et al.,* 1998). Figure 1 represents the experiments we have undertaken to assess the interactions of genotype, behaviour, handling, nutrition and post-mortem conditions in determining phenotypic measures of meat quality (colour, drip loss, tenderness) in commercial practice in Ontario and assessing underlying mechanisms by biochemistry and gene expression studies (both in muscle and brain tissue). This paper describes a subset of the data, specifically relating calpain activity in ham and loin muscles immediately post mortem to drip loss and shear force.

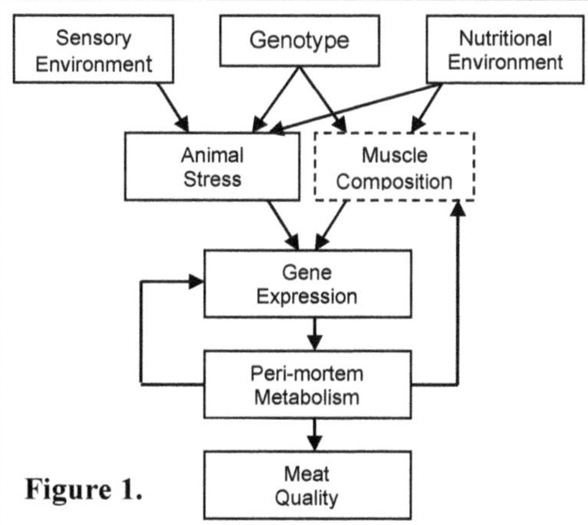

Figure 1.

Materials and Methods

12 barrows and 12 gilts from 24 commercial producers and 2 university farm-based herds (n=624) were subjected to on-farm behavioural tests and behaviour studies on shipping and reception at a common commercial slaughter plant. Producers were selected to represent a range of management and nutritional practices and inherently chosen to give a reasonable snapshot of variation seen in, commercial practice in Ontario. At the slaughter plant, blood samples were collected immediately at sticking for measurement of cortisol, glucose, lactate and CPK. The pH and temperature in *longissimus* and *semimembranosus* muscles were monitored at 1, 2, and 24 hours post mortem (pH_{1h}, pH_{2h}, pH_{24h}). Loin and ham samples were shipped to the University of Guelph Meat Laboratory for measures of backfat thickness, loin eye area, total lipid content, colour (La*b*), drip loss, cooking loss, processing yield and cooked meat Warner-Bratzler (WB)shear force measurements.

Subsamples of *longissimus* and *semimembranosus* muscles taken immediately post-mortem were snap-frozen in liquid nitrogen and subsequently stored at -80°C. Following analysis of the meat quality results, a sub-set of 65 loin and ham samples representatively spanning the range of drip loss and shear force values obtained were selected for measurement of net calpain-1 (micro-calpain) and calpain-2 (milli-calpain) activity by zymography, based upon the methods of Melody *et al.,* 2004. Four repeats of each activity determination were conducted per muscle sample. Internal controls of calpain-1 and -2 were used in each zymogram and the activity in each sample related to mg of pure enzyme standard. Statistical analysis of simple correlations and partial correlations (with one controlled variable) were performed using SPSS.

Results and Discussion

Variations seen in phenotypic measures on meat quality in commercial practice were consistent with the results of a previous US survey (Kauffman *et al.,* 1992). Colour (L –value) ranged from 35-62 and drip loss from 2.61% to 12.54% across both loin and ham samples. WB shear force values ranged from 1.07-5.43 Kg in ham samples and from 2.06-7.66 Kg in loin samples. Given the (intentionally) highly multivariate nature of the study, full multivariate analysis and careful reconciliation of effects through gene expression differences are underway. However, it is instructive that significant effects can still be seen in simple correlations of two variables.

Figure 2 shows the correlation between the % drip loss and calpain-1 (micro-calpain) activity of all 65 ham and loin samples in the sub-set studied to date. There is a clear negative relationship; higher calpain-1 activity is correlated to lower drip loss. This negative correlation is highly significant ($0.01<p<0.02$). This result is consistent with the implications from previous studies (Kristensen and Purslow, 2001; Melody *et al.*, 2004) that degradation of cytoskeletal proteins by calpain reduces drip loss. Connections between myofibrils and the cell surface seem to be required in order to translate reduction in myofibrillar volume caused by the post-mortem fall in pH into reduction in cellular, and hence meat volume. Increased cytoskeletal degradation from higher levels of net activity of calpain-1 apparently reduces the ability to force water from the muscle cells out into extracellular spaces and drip channels.

Fig.2. Correlation between calpain 1 activity and % driploss of ham and loin

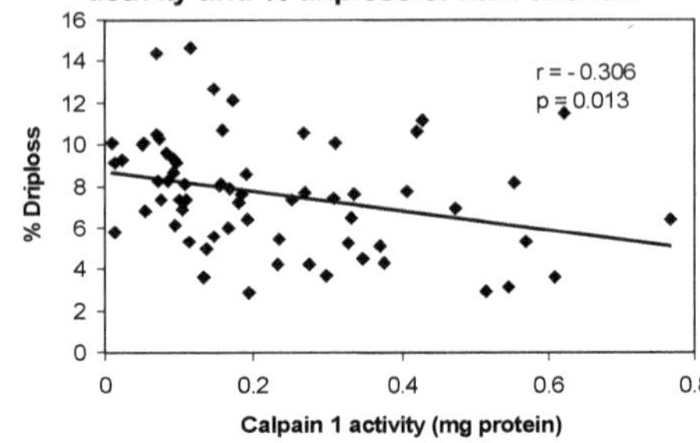

Figure 3 demonstrates that the negative correlation between drip loss and calpain-1 activity is significant (P=0.029) in semimembranosus samples (n=32). The correlation in loin samples only, although again showing a negative trend, is non-significant (P= 0.259).

Fig. 3. Correlation between calpain 1 activity and % driploss of ham only

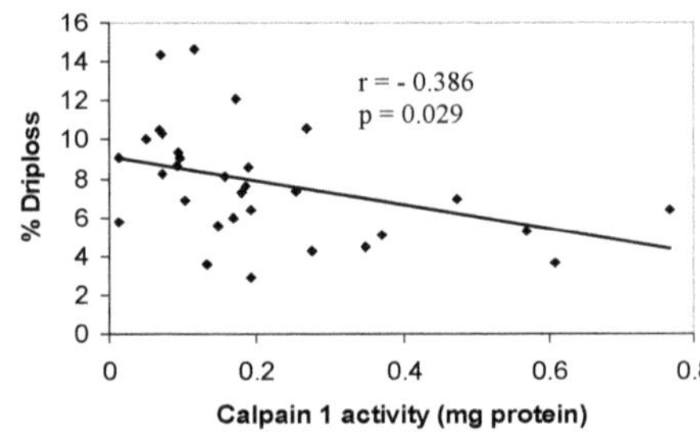

Although calpain-2 (milli-calpain) was highly correlated with calpain-1 activity, no significant correlations with shear force or drip loss were seen in either ham or loin samples Simple correlations between WB shear force and calpain-1 activity were poor (P=0.756) and similarly, shear force was poorly correlated with drip loss (P=0.570).

WB shear force shows a significant (P=0.016) negative correlation with pH_{24h}. We therefore carried out partial correlation analysis, holding pH as the controlled variable, and examining the correlation between calpain-1 activity and WB shear force. No significant correlation was found (P=0.666).

Drip loss in both loin and ham samples showed a significant negative correlation (P=0.036) with pH_{1h} at and a highly significant (P=0.005) negative correlation with pH_{24h}. (I.e. lower pH corresponded to higher drip loss). Because calpain activity is pH-dependent, pH is a co-factor in any simple correlation between drip loss and calpain activity. Partial correlation analysis of calpain -1 activity against drip loss with pH_{1h} or pH_{24h} as the variable held controlled shows significant negative correlations, with P=0.035 and P=0.013, respectively. Thus, calpain-1 activity is significantly related to drip loss even when pH variations are taken out of the picture.

Conclusions

The considerable interactions between genotype, management practices in commercial production, stress physiology, nutrition and post-mortem metabolism mean that there are highly multivariate influences on meat quality in commercial practice. Nevertheless, a clear and significant negative correlation exists between calpain-1 activity and drip loss, suggesting that this enzymic pathway is a common mechanism on which many of the influences may act.

References

D'Souza DN *et al.*, (1998). *Aust J Agric Res*, 49: 1021-1025.

Kauffman RG *et al.*, (1992). Variations in pork quality. National Pork Producers Council, Des Moines, IA, USA

Hemsworth PH *et al.*, (2002). *Aust. J. Agric. Res*, 53:493-501.

Kristensen L and Purslow PP (2001). *Meat Science*, 58, 17-23.

Lonergan SM *et al.*, (2001). *Journal of Animal Science*, 79:2075-2085.

Melody JL *et al.*, (2004). *Journal of Animal Science*, 82:1195-1205.

Schäfer A *et al.*, (2002). *Meat Science*, 64:355-366.

Warriss PD *et al.*, (1998). *Meat Science*, 49:137-144.

MMP-9 ACTIVITY IS INCREASED IN THE FIBROBLASTS BUT NOT MYOBLASTS IN RESPONSE TO GLYCOPROTEIN COATING

M.C. Cha and P.P. Purslow*

Department of Food Science, University of Guelph, Guelph, Ontario, Canada N1G 2W1
Email: ppurslow@uoguelph.

Keywords: muscle turnover, connective tissue, MMP-9, MMP-3, fibroblasts, myoblasts

Introduction

Tenderness is an important meat quality and a common cause of unacceptability of meat purchases is toughness (Jeremiah, 1982). Both myofibrillar and intramuscular connective tissue (IMCT) components of meat contribute to cooked meat toughness. Whilst is generally accepted that meat tenderisation post-mortem is primarily a result of proteolysis of myofibrils by endogenous calpain enzymes (Koohmaraie, 1991), both myofibrillar and IMCT components are turned over during muscle growth. A large family of at least 26 structurally-related zinc metalloproteinases (MMP's) is primarily responsible for connective tissue catabolism (Nagase *et al.*, 2006). Skeletal muscles contain many enzymes of the MMP family and their inhibitors (Balcerzak *et al.*, 2001). There is good evidence that MMP expression in skeletal muscle is influenced by nutritional status and mechanical stimulation (e.g. exercise). IMCT is structurally and functionally complex; it is divided in to perimysium and endomysium, both with different collagen and proteoglycan components, and different structural organization. It is not clear which cell types within muscle are responsible for the maintenance and turnover of which IMCT components. As part of a basic cell culture study of MMP expression by a variety of cell types found in muscle, the current research was conducted to examine the activity of MMPs secreted by the cultured fibroblasts and myoblasts and how activity is modulated by the IMCT components that the cell types are interacting with. We studied activity of two important MMP enzymes; MMP-9 (also called gelatinase B) is most active in hydrolysing gelatin, native type IV collagen and elastin (Murphy and Crabbe, 1995) and MMP-3 (stromelysin-1) is capable of degrading proteoglycans, gelatin, type IV collagen, laminin, fibronectin, and type IX collagen (Okada *et al.*, 1989).

Materials and Methods

Fibroblasts (Rat 2, R2) and myoblasts (C2C12) were cultured in Dulbecco's Modified Eagle's Medium (DMEM) with 10% fetal bovine serum. Substrates diluted in PBS were coated to the tissue culture plates by incubating overnight at 4°C. The plates were air-dried before use. Equal number of cells was loaded into the coated and the non-coated plates for each substrate coating experiment. The cells were then growth at 37°C in 5% CO_2. After 24 hours, the culture medium from each plate was collected and freeze-dried to 3:1 concentrated and the cells in the plate were counted. The concentrated culture medium was used to test the MMP activity.

The MMP-9 and MMP-3 activities in the culture medium were determined by gelatin or casein zymography, respectively. Gelatin zymography was carried out with minor modification of the method of Kleiner and Stetler-Stevenson (1994). Casein zymography was performed according to .t(he method described by Fernandez-Resa *et al.*, (1995). Data were analysed by one-way ANOVA using an SPSS general linear model program. Means were separated by Fisher's protected least significant difference (LSD) test.

Results and Discussion

The rate of C2C12 myoblast differentiation into myotubes is known to be affected by the type of extracellular matrix component they are in contact with (Lawson and Purslow, 2001) To look at effects of interactions on the activity of MMPs expressed from the cells, we chose collagen I fibronectin, laminin as test IMCT components and used bovine serum albumin (BSA) as the negative control.

Figure 1 shows a gelatin zymogram gel showing MMP-9 activity in fibroblasts and myoblasts grown on laminin. (MMP-9 standard; R2 fibroblasts on no coating; L+R2 fibroblasts on laminin coating, C2 myoblasts on no-coating, and L+C2 myoblasts on laminin coating in lane 1 to lane 5, respectively. Generally, fibroblast show more MMP activity than myoblasts. Laminin coating did not modify MMP-9 activity in myoblasts (no difference in band densities between the Lane 4 and Lane 5), but, laminin coating increased MMP-9 activity in fibroblasts (increased band density in the Lane 3 as compared to that in Lane 2).

Figure 2 shows that, compared to control (non-coated cultures) interactions with glycoprotein (fibronectin (FN), laminin) substrates increased mean MMP-9 activity (n=4 cultures) in the fibroblasts as compared with that of in the myoblasts ($p<0.05$). However, the proteins collagen I and BSA have no such effect. Myoblasts are likely to interact with laminin in the basement membrane far more than Collagen I in the intracellular matrix

In contrast with MMP-9, MMP-3 activity in either the fibroblasts or the myoblasts as examined by casein zymography was not influenced by coating any of the substrates tested (Data not shown).

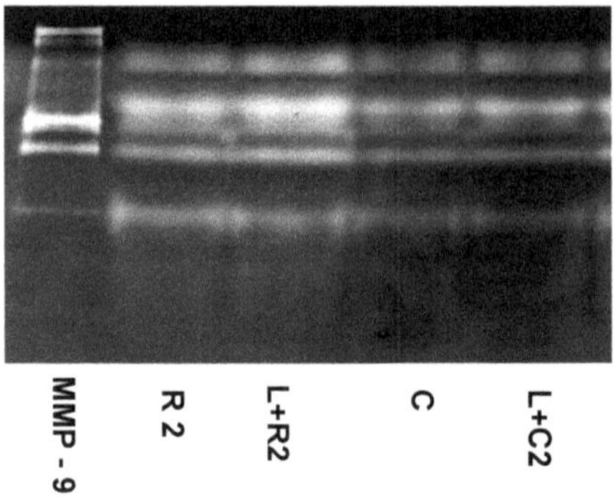

Figure 1.

Figure 2.

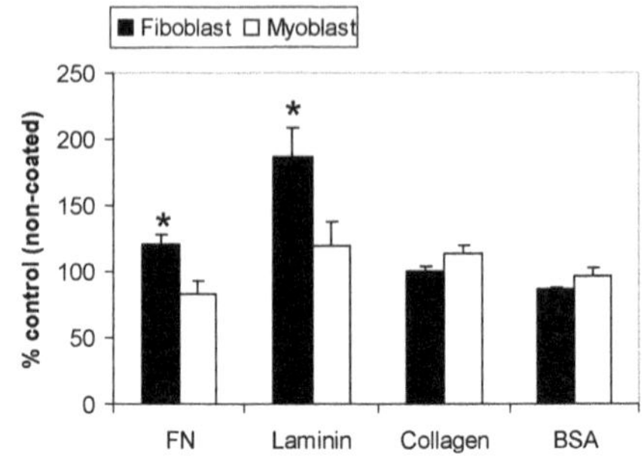

Conclusions

This research demonstrates that MMP activity expression from fibroblasts and myoblasts in culture is different. Interactions between cells and IMCT components also interactively influenced MMP-9 activity. Laminin coating increased MMP-9 activity in fibroblasts but not in myoblasts. MMP-9 activity increased in fibroblasts in response to glycoprotein coating as compared with that in myoblasts. These interactions may be important in controlling the expression of MMPs by fibroblasts in the endomysium and perimysium during muscle growth and MCT turnover.

References

Balcerzak, D. *et al.*, (2001) *Journal of Animal Science*, 79, 99-107.

Fernandez-Resa, P. *et al.*, (1995) *Anal. Biochem*, 224, 434-435.

Jeremiah, L.E. (1982) *Journal of Consumer Studies and Home Economics*, 6,137-154.

Koohmaraie, M *et al.*, (1991). *Journal of Animal Science*, 69, 617-624.

Kleiner, D.E. and Stetler-Stevenson, W.G. (1994) *Anal. Biochem*, 218,325-9.

Lawson, M.A. and Purslow, P.P. (2001) *Br. J. Poultry Science*, 42, 315-320.

Murphy, G. and Crabbe, T. (1995) *Methods Enzymol*, 248:470-84.

Nagase, H. *et al.*, (2006) *Cardiovasc. Res.* 69, 562-573.

Okada, Y, *et al.*, (1989) *FEBS Lett*, 244,473-6.

DIFFERENTIATION OF ACID MEAT FROM OTHER QUALITY GROUPS (RFN AND PSE) THROUGH THE ANALYSIS OF PROTEINS FROM ITS DRIP

E. Pospiech*[1,2], B. Mikołajczak[1], E. Iwańska[1], B. Grześ[1], A. Łyczyński[3] and W. Kapelański[4]

[1]Institute of Meat Technology, Agricultural University of Poznań, 31 Wojska Polskiego Str., 60-624 Poznań, Poland, pospiech@au.poznan.pl; [2]Meat and Fat Research Institute, 239 Głogowska Str., 60-111 Poznań, Poland, [3]Department of Raw Materials of Animal Origin, Agricultural University, 33 Wołyńska Str., 60-637 Poznań, Poland, [4]Department of Pig Breeding, University of Technology and Agriculture, 28 Mazowiecka Str., 85-084 Bydgoszcz, Poland

Keywords: protein, 2D electrophoresis, mass spectrometry, meat quality, acid meat

Introduction

Processes taking place in the muscle tissue directly after slaughter cause protein changes resulting in specific sensory and functional properties of the muscle tissue. Changes in proteins in the case of muscle defects are of particular interest. The observed reduced water holding capacity in the case of PSE meat is associated primarily with protein denaturation, especially myosin (Stabursvik *et al.*, 1984) as well as with a slower degradation of cytoskeletal proteins, including, in particular, titin (Pospiech *et al.*, 2004). In the case of acid meat, which is also characterised by low water holding capacity, the degradation of titin is faster than in PSE meat (Pospiech *et al.*, 2004), but is distinguished by usually greater acidification. There is no information whether this process is associated with other proteins or if they can differentiate muscle in regard to their quality. In order to find the answer to this question, a comprehensive analysis of proteins of meat drip was performed using 2-dimensional electrophoresis (2DE) and mass spectrometry (MS).

Materials and Methods

The experimental material comprised *longissimus thoracis* and *lumborum* muscles derived from 12 pigs of known origin and pre-slaughter weight. Experiments were carried out on muscles of varying quality, which was assessed on the basis of the following measurements: pH value 45 m and 24 h after slaughter, determination of the R value which corresponds to the ratio of IMP to ATP nucleotides (Honikel and Fischer 1977) as well as the electrical conductivity (EC) measured 90 minutes and 24 h after slaughter. Employing criteria recommended by Borzuta and Pospiech (1999), the following three quality muscle groups were distinguished: RFN, PSE and ASE. Samples from the carcasses of the slaughtered pigs were collected four times: 45 m, 3, 48 and 144 h after slaughter. The samples were then frozen to allow subsequent assessment of protein changes in the drip from meat which was obtained by centrifugation of slightly defrosted meat samples, employing a method typically used to determine water holding capacity (Honikel 1998). Simultaneously, 24 hours after slaughter, muscles were cut from carcasses and divided into portions to estimate water holding capacity and meat tenderness 48 and 144 h after slaughter. Samples were vacuum packed and stored in refrigerated conditions at 2-4°C. Tenderness and water holding capacity were assessed instrumentally employing methods suggested by Honikel (1999) and Grześ *et al.* (2005) respectively. Electrophoretic analysis of muscle proteins was performed using the method of 2DE (Mikołajczak *et al.*, 2005) and the separated proteins were evaluated with the assistance of a scanning densitometer of the Image Master® VDS type. Some proteins whose presence varied in relation to the muscle quality were subjected to further analysis with the aid of a mass spectrophotometer using the MALDI-TOF method (Lametsch *et al.*, 2001) and the Bruker Autoflex mass spectrometer. Selected spots were cut from the polyacrylamide gel, digested using trypsin and their MS MALDI TOF spectra were registered. Proteins were identified on the basis of comparison of the obtained peptide maps with the maps developed for protein registered in databases. The final analysis took into consideration only those proteins for which the protein score was greater than 76 and was significant at $p<0.05$. All the results obtained were subjected to statistical single and two-factorial analysis of variance using the STATISTICA program.

Results and Discussion

Data from the Tables 1 and 2 showed that the measurement results of the pH, PE and R values differentiated the examined muscles with regard to their quality. The slowest *post mortem* changes occurred in RFN meat. The best tenderness characterised the acid meat, but the largest drip loss was observed in case of this and PSE meat (Table 3). Depending on the type of sample, 2D gels of separated proteins were revealed from about 70 to 80 spots. It was found that 14 of them were not found in each sample. However, protein "D" drew special attention (Table 4) as it was characteristic only for acid meat. It was characterised by the molecular weight ranging from 25÷30kDa and pI 9.5. It was found only in those samples which showed acid meat properties and derived from pigs which were crosses of [(Polish Landrace x Polish Large White) x (Duroc x Pietrain)]. It was not found in the meat of the Danish Landrace crossed with the Danish Yorkshire. Investigations carried out with the aid of the mass spectrometer revealed that the domains of this protein corresponded to the basic cytosolic protein (Schoentgen *et al.*, 1992) (protein score of 86). It takes part in processes associated with the transmission of neurohormonal signals, although its precise role has not been elucidated completely. The finding that the basic cytosolic protein belongs to the group of protein kinases is of particular interest. The AMP-stimulated protein kinase responsible for the development of acid meat also belongs to this group. The observed lack of significant differences in the amount of the D protein at different times after slaughter

(Table 4) may point to the fact that its occurrence is associated with the influence of a factor of pre-slaughter nature. Protein D, being characteristic for acid muscles, sets them apart from other quality groups. However, this differentiation is by no means perfect since only 67% of all muscles showing ASE meat properties contained this protein.

Table 1: Changes of pH value of various quality muscles during their storage in chilled room.

Meat quality	No. of samples	Term of measurement 45'	3 h	24 h	48 h	144 h
RFN	3	6.66^{c}±0.20*	6.35^{b}±0.31	5.70^{b}±0.21	5.47±0.01	5.57^{bc}±0.04
PSE	3	5.75^{a}±0.06	5.39^{a} ±0.10	5.44^{a}±0.05	5.45±0.03	5.49^{ab}±0.04
ASE	6	6.17^{b}±0.22	5.76^{a} ±0.32	5.39^{a}±0.03	5.48±0.06	5.48^{a}±0.04

* standard deviation; ^{a,b,c} - means marked with various letters differ significantly in regard to the muscle quality

Table 2: Electrical conductivity values (mS/cm) and the ratio of IMP/ATP of investigated muscles.

Meat quality	Electrical conductivity 1.5h	24h	IMP/ATP 45'
RFN	2.72^{a} ±0.64	1.56^{a}±0.07	0.98^{a}±0.01
PSE	9.07^{c}±3.36	10.10^{b}±3.13	1.11^{b}±0.04
ASE	5.18^{b}±1.51	9.47^{b}±1.57	1.02^{a}±0.03

Table 4: Protein „D" content (pixels) in the centrifugal drip of acid meat.

Term of measurement	Protein D content
45 min.	52.67±60.12
3h	42.17±56.58
48h	57.00±83.91
144h	39.67±54.16

Table 3: Drip loss (%) and tenderness (N/cm^2) of meat of investigated pigs.

Meat quality	Drip loss 48h	144h	Meat tenderness 48h	144h
RFN	0.86^{a}±0.22	1.50^{a}±0.47	62.88^{b}±32.87	45.77^{a} ±5.65
PSE	3.00^{b}±0.51	5.09^{b}±1.30	55.73^{ab}±16.61	44.73^{a}±13.16
ASE	3.94^{b}±1.46	6.77^{b}±2.06	41.94^{a}±9.17	32.18^{a}±6.61

Conclusions

The results obtained show that the muscle protein composition variation may be associated with the character of changes that take place after slaughter. The proteins "D" characterised by the molecular weight ranging from 25÷30kDa and pI 9.5 sets acid meat apart from other quality groups. The occurrence of this protein may be associated with pre-slaughter factors. Differentiation of acid meat is not absolute and only 67% of all muscles showing ASE meat properties contained this protein. The results obtained indicate that it is necessary to continue investigations in this area in order to find a more convincing explanation of the role of this protein in the development of acid meat.

References

Borzuta K. and Pospiech E. (1999). Analysis of benefits associated with increased meatiness of pigs and losses caused by deterioration of meat quality. [Analiza korzyści związanych ze wzrostem mięsności tuczników oraz strat spowodowanych pogorszeniem jakości mięsa]. Gospodarka Mięsna 9, 36–40.

Enfält A-C, von Seth G., Lundström K., Josell Å., Lindahl G., Hedebro – Velander I., Braunschweig M. and Anderson L. (2002). A third allele (v1991) at the RN locus with effect on carcass composition and meat quality. 48th ICoMST, Italy, Rome, 2002, Proc. vol. 2, 618–619.

Grześ B., Pospiech E., Koćwin-Podsiadła M., Krzęcio E., Łyczyński A., Zybert A., Mikołajczak B. and Iwańska E. (2005). Analysis of the causes of meat quality variations of pigs with the identical genotype. Pol. J. Food Nutr. Sciences 14/55, SI 1, 59–64.

Honikel K.O. (1998). Reference methods for the assessment of physical characteristics of meat. Meat Sci. 49, 4, 447-457.

Honikel K.O. and Fischer H., (1977). A rapid method for the detection of PSE and DFD porcine muscles. J. Food Science 42, 1633-1636.

Lametsch R., Roepstorff P. and Bendixen E. (2001). Identification of post mortem protein degradation of pork with two-dimensional gel electrophoresis and mass spectrometry. 47th ICoMST Poland, Kraków, Proc. Vol. 1, 4-P2, 246–247.

Mikołajczak B., Pospiech E., Greaser M.L., Grześ B., Iwańska E. and Łyczyński A. (2005). Two dimensional electrophoresis of proteins from centrifugal drip of various quality meat. Annals of Animal Sci. Suppl. 2, 123–126.

Pospiech E., Szalata M., van Laack R.L.J.M., Sosnicki A.A. and Greaser M.L. (2004). Tenderness and protein changes of pork in gelation to pig genotype (RN$^-$) and rate of postmortem glycolysis. Ann. Anim. Sci. Suppl. (2004), 2, 257-261.

Schoentgen F, Seddiqi N, Bucquoy S, Jolles P, Lemesle-Varloot L, Provost K. and Mornon JP. (1992). Main structural and functional features of the basic cytosolic bovine 21 kDa protein delineated through hydrophobic cluster analysis and molecular modelling. Protein Eng. Jun. 5 (4): 295-303.

Stabursvik E., Fretheim K. and Frøystein T. (1984). Myosin denaturation in pale, soft, and exudative (PSE) porcine muscle tissue as studied by differential scanning calorimetry. J. Sci. Food Agric. 35, 240–252.

Acknowledgements: This study was partly financed by the Ministry of Science and Information Technologies in Poland, grant no. 2 P06T 079 27 and by the European Commission, project ΣChain, Contract no. FP6 – 518451.

SINGLE NUCLEOTIDE POLYMORPHISMS IN THE CALPAIN AND CALPASTATIN GENES AND BEEF TENDERNESS

R.M. Hamill*[1], S.A. Costello[1], E. O'Doherty[1], C.W. Ernst[3], T. Sweeney[2] and A.M. Mullen[1]

[1]*Ashtown Food Research Centre, Teagasc, Ashtown, Dublin 15, Ireland. Email ruth.hamill@teagasc.ie*
[2]*Department of Animal Husbandry and Production, Faculty of Veterinary Medicine, National University of Ireland Dublin, Belfield, Dublin 4, Ireland* [3]*Department of Animal Science, Michigan State University, East Lansing, MI 48824, USA*

Keywords: tenderisation, SNP, beef, *longissimus dorsi*

Introduction

Currently, the biochemical pathway with the most evidential support for involvement in post-mortem tenderisation is that of the calpain family of proteases. Calpain I and II act to break down the myofibrillar proteins via calcium-dependent proteolytic degradation post-mortem and thus increase meat tenderness. Calpastatin (CAST) is the inhibitor of calpain I and II activity, regulating post-mortem proteolysis. Increased CAST activity is associated with less tender meat (Koohmaraie *et al.*, 1995). If single nucleotide polymorphisms in these genes are found to display an association with beef tenderness, they could provide the basis of a diagnostic test for tenderness or improve breeding programmes through marker assisted selection. SNP variants exist in exons 9 and 14 of the calpain I gene that alter the protein sequence of calpain I (Page *et al.*, 2002). These have been found to be associated with a measure of beef tenderness, i.e. Warner-Bratzler shear force (WBSF) in resource populations (Page *et al.*, 2002, Casas *et al.*, 2006). A restriction fragment length polymorphism (RFLP) exists in Calpain II (Zhang *et al.*, 1996) that may also be associated with beef tenderness. These markers have been tested for association with WBSF in bovine *M. longissimus dorsi* in the Irish herd and a SNP in exon 9 was significantly associated with this parameter (Costello *et al.*, 2004). Following on from this work, our aim was to determine the frequencies of SNP variants at candidate loci in the calpain family for beef tenderness in different breeds in the Irish herd. A recently-identified SNP allele in the CAST gene has been shown to have a significant association with reduced WBSF values and a reduction in the proportion of unacceptably tough steaks (Schenkel *et al.*, 2006). The aim of this research was also to test this marker for association with WBSF, in Irish bovine *M. longissimus dorsi*.

Materials and Methods

DNA was isolated from 25mg of *M. longissimus dorsi* muscle (on which tenderness was also characterised) and thawed blood samples using the QIAamp® DNA minikit. Muscle samples were used in both the association analysis and the frequency study while the blood samples which were procured from Irish AI stations were used solely to determine genotype frequencies. Tenderness scores were measured by WBSF on steaks from the posterior end of the LD using a modified method of Shackelford *et al.* (1991). Polymerase Chain reaction (PCR) was carried out using primers specific for calpain I exon 9 and 14 and the partial calpastatin gene (Schenkel *et al.*, 2006). Restriction digestions were carried out using the restriction enzymes *Btg*I (calpain I, exon 9), *Dpn*II (calpain I, exon 14) and *Rsa*I (CAST). Digests were subjected to electrophoresis on 2% agarose gels, photographed under UV transillumination and genotypes scored. Association analysis was performed between the observed CAST genotypes and WBSF at day 14 post mortem using the GLM procedure of GenStat.

Results and Discussion

Genotype frequencies for all breeds tested are presented in Table 1. The GA genotype at the Calpain I, exon 9 locus, which is significantly associated with reduced WBSF (Costello *et al.*, 2004), was the minor genotype at that locus in the Irish herd.

Table 1: Breed specific frequencies (%) for Calpain I and Calpain II genotypes- genotypes were assigned following published designations.

Breed (n)	Calpain I, exon 9		Calpain I, exon 14		Calpain II		
	GG	GA	VV	VI	AA	AB	BB
Friesian (64)	95.3	4.7	80.8	19.2	16.7	64.8	18.5
Belgian Blue (11)	100.0	-	100	-	27.3	54.5	18.2
Limousin (11)	81.8	18.2	81.8	18.2	45.5	54.5	-
Charolais (9)	77.8	22.2	100	-	22.2	44.4	33.3
Aberdeen Angus (7)	100	-	83.3	16.7	-	85.7	14.3
Others (17)	82.3	17.7	93.3	6.7	31.3	50.0	0.063

The GA genotype at the Calpain I, exon 9 genotype varied in frequency among breeds and was absent from the Belgian Blue and Aberdeen Angus samples tested. This indicates the potential of marker-assisted selection, based on the A allele as a functional marker, to significantly increase beef tenderness in the Irish herd by selectively breeding animals with favourable genotypes in breeding programmes. There was variation among genotype frequencies for the VI genotype at the exon 14 locus which was absent from the Belgian Blue and the Charolais breeds, although this allele was not significantly associated with tenderness, while there was less variation in genotype frequency among breeds for the Calpain II genotypes.

A significant association was observed between WBSF and genotypes at the calpastatin locus in the Irish herd (p= 0.015). The CC genotype had a LSM shear force of 41.8 and the GG genotype had a LSM shear force of 51.6: the GC genotype had an intermediate LSM shear force of 45.5. This association concurs with results of recent studies (Schenkel *et al.* 2006). The overall frequency of the CC genotype was 0.41 (Table 2), again indicating the potential for incorporation into marker assisted breeding programmes. However, samples with potentially confounding factors such as extreme values of sarcomere length and intramuscular fat (IMF) should be excluded in a controlled sample set for future analysis. Care must also be taken when selecting against the alternative alleles at these loci, as they may contribute positively to other quantitative traits not examined here.

Table 2: Observed genotypic frequencies (%) in Irish bovine *M. longissimus dorsi.*

Gene	Calpain I, exon 9 (n=284)			Calpain I, exon 14 (n=284)		Calpain II (n=284)			Calpastatin (n=196)		
	GG	GA	AA	VV	VI	AA	AB	BB	CC	CG	GG
Frequency	76.5	23.2	0.4	75.6	24.4	15.2	53.0	31.9	0.41	0.47	0.13

Conclusion

The calpastatin CC genotype was observed to have an effect on tenderness in the Irish herd. Both this genotype and the A allele at the calpain 1, exon 9 locus may have potential for incorporation into marker-assisted breeding programmes to improve the consistency of beef quality in the Irish herd. In particular, the calpain 1 allele is currently at low frequency in the Irish herd, thereby offering considerable potential as a functional marker to enhance beef tenderness in the Irish population.

References

Casas, E., White, S.N., Wheeler, T.L., Shackelford, S.D., Koohmaraie, M., Riley, D.G., Chase, C.C., Johnson, D.D. and Smith, T.P.L. (2006). Effects of calpastatin and mu-calpain markers in beef cattle on tenderness traits. Journal of Animal Science 84,520-525.

Costello, S., O'Doherty, E., Mullen, A.M., Kim, K.-S., Ernst, C. and Sweeney, T. (2004). Analysis of the relationship between DNA polymorphisms in candidate genes and beef tenderness. In: Proceedings of the 50th International Congress of Meat Science and Technology. Helsinki, Finland.

Koohmaraie, M., Killifer, J., Bishop, M.D., Shackleford, S., Wheeler, T.L. and Arbona, J. (1995). Calpastatin-based methods for predicting meat tenderness, In: A. Ouali, *et al.*, eds. Expression of Muscle Proteinases and Regulation of Protein Degradation as Related to Meat Quality, 395-412. Audet Tijdschriften, Nijmegen, The Netherlands.

Page, B.T., Casas, E., Heaton, M.P., Cullen, N.G., Hyndman, D.L., Morris, C.A., Crawford, A.M., Wheeler, T.L., Koohmaraie, M., Keele, J.W., and Smith, T.P.L. (2002). Evaluation of single-nucleotide polymorphisms in CAPN1 for association with meat tenderness in cattle. Journal of Animal Science, 80, 3077-3085.

Schenkel, F., Miller, S. Jiang, Z. Mandell, I. Ye, X. Li, H. and Wilton, J.W. (2006). Association of a single nucleotide polymorphism in the calpastatin gene with carcass and meat quality traits of beef cattle. Journal of Animal Science 84, 291-299.

Shackleford, S.D., Wheeler, T.L. and Koohmaraie, M. (1991). Identification of threshold levels for Warner Bratzler shear force in beef top loin steaks. Journal of Muscle Foods 2, 289-296.

Zhang, H.M., S.K. DeNise, and R.L. Ax (1996). A novel DNA polymorphism of the bovine calpain gene detected by PCR-RFLP analysis. Journal of Animal Science 74,1441.

MANIPULATION OF THE RATE OF PROTEOLYSIS IN BOVINE M. *LONGISSIMUS DORSI*

P. Sullivan[1,2]*, A. White[1], M. Cairns[2] and A.M. Mullen[1]

[1]*Ashtown Food Research Centre (formerly known as The National Food Centre), Teagasc, Ashtown, Dublin 15, Ireland.* [2] *The National Diagnostic Centre, National University of Ireland, Galway, Ireland. Email: paul.sullivan@teagasc.ie*

Keywords: beef, tenderisation, *longissimus dorsi*, SDS-PAGE

Introduction

Tenderness is seen as one of the most important attributes in eating quality of beef (Koohmarie, 1998). Understanding the biochemical basis of meat quality, particularly tenderness, is of great importance to the meat industry. Production of tender meat has led to a number of procedures being put in place to increase the quality of meat, through breeding programmes, slaughtering methods, post-mortem intervention techniques and storage conditions. However variation is still found in beef tenderness (Maher *et al.*, 2004). Several studies have shown the rate of *post-mortem* glycolysis to be an important contributor to this variation (White *et al.*, 2006b; O'Halloran *et al*, 1997; Takahashi *et al.*, 1984), which is also affected by the rate of cooling, which affects rigor onset (White *et al.*, 2006a; Locker and Hagyard, 1963). Interactions between pH and temperature during the onset of rigor directly influence meat tenderness, water-holding capacity and meat colour through their effects on proteolysis, protein denaturation, and myofibrillar shrinkage (White *et al.*, 2006a; Bertram *et al.*, 2004). Close examination of the protein profiles in meat over the ageing period reveals novel indicators of tenderness (Sanchez *et al.*, 2005) which may lead to a greater understanding of the processes contributing to quality.

Meat samples collected through a non-research abattoir can result in samples with variation in sarcomere length are difficult to categorise for tenderness. Manipulation of post-mortem handling of the muscle can allow for divergence in proteolysis, which in turn may result in protein profiles with novel quality indicators. Therefore, the objective of this study was to manipulate the rate of proteolysis, through the application of chilling regimes, to enable closer analysis of products of proteolysis which may be related to quality.

Materials and Methods

A model system was established to produce beef which was divergent for proteolysis through alteration of the rate of glycolysis (White *et al.*, 2004) Hot-boned *M. longissimus dorsi* (LD) muscles (n=5) were cut in half and wrapped by Pi-Vac (PiPatente, Germany) to prevent shortening. Each half was immersed in a water bath set at either 5°C or 20°C for 8 hours *post-mortem* and then placed in a chill set at 2°C for 14 days.

pH and temperature were recorded up to 48hrs *post mortem*. Both sarcomere length (day 2 *post-mortem*) and Warner Bratzler shear force (day 2, 7, 14 and 21 day *post mortem*) were measured according to the methods of Cross *et al.* (1980) and Shackleford *et al.* (1991). Both myofibrillar and TCA extractions were performed on samples taken at 1hr, 3hr, 8hr, 1d, 7d and 14d and these were run on 12% 1-dimensional sodium dodecyl sulphate polyacylamide gel electrophoresis (SDS-PAGE) (Laemmeli *et al.*, 1970).

Results and Discussion

The difference in holding temperature was shown to have a significant effect on the pH of the LD from 3-8hrs *post mortem* ($P < 0.001$) and also at 24hrs and 48hrs ($p<0.05$). LD muscles held at 5°C were found to have a slower rate of pH decline than those muscles held at 20°C (Figure 1).The temperature of storage was found to have no significant effect ($P > 0.05$) on the sarcomere length of the LD and was also shown to have no significant effect ($P >0.05$) on the Warner Bratzler shear force values of the muscle. This may have been due to the muscles being wrapped by Pi-Vac. The mean sarcomere length was 1.81 ± 0.17 for the 5°C treatment and 1.85 ± 0.16 for the 20°C treatment. Hence the impact of pH/temperature alone was not enough to cause a significant difference in the tenderness of these samples.

Initial results from qualitative analysis of the SDS-PAGE gels for both TCA (Figure 2) and myofibrillar extractions show little or no difference between the proteolytic profiles of the two treatments. However, further semi-quantitative analysis is required to verify this. Also use of 2D SDS-PAGE may identify differences which are not evident using 1D SDS-PAGE.

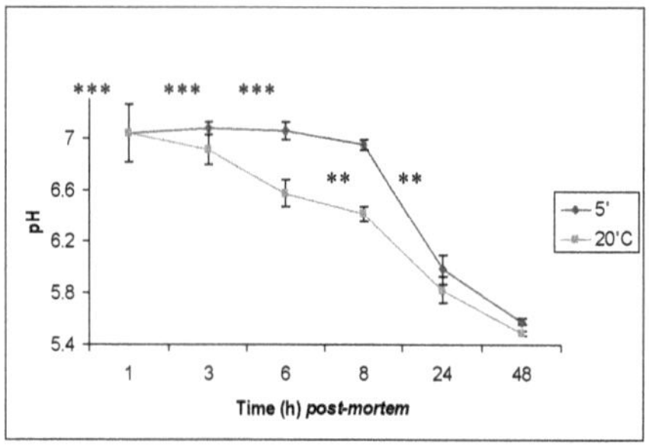

Figure 1: Mean pH value of hot-boned *M. longissimus dorsi* muscle held at 5° and 20 ℃ (bars represent standard error of the mean).

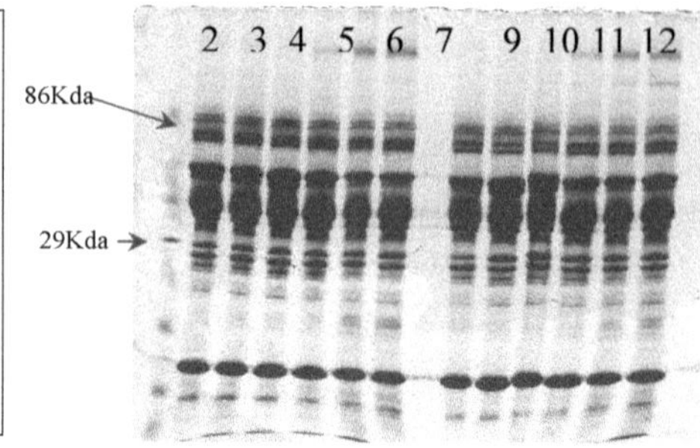

Figure 2: TCA extractions loaded onto 12% 1D SDS-PAGE. Lanes 2-7: 5°C samples taken at 1h, 3h, 8h, 1d, 7d and 14d, respectively. Lanes 9-14: 20°C samples taken at 1h, 3h, 8h, 1d, 7d and 14d, respectively.

Conclusions

Pre-rigor temperature regimes altered the glycolytic pathway of the muscles. Use of the Pi-Vac machine restricted sarcomeres from shortening. This may have contributed to the lack of impact of temperature regimes on Warner Bratzler shear force values. Further work, such as 2-D gel electrophoresis may provide further insight in to the proteolytic profile.

References

Bertram, H.C., Schafer, A., Rosenvold, K. and Andersen, H.J., (2004). Physical changes of significance of post-mortem water distribution in porcine M. *longissimus*, *Meat Science* 66, 915–924.

Cross, H.R., West, R.L., and Dutson, T.R. (1980). Comparison of methods for measuring sarcomere length in beef *semitendinosus* muscle, *Meat Science,* 5, pp. 261–266.

Koohmaraie, M. (1998). The role of the endogenous proteases in meat tenderness. In *Proceedings reciprocal meat conference* (Vol. 41, pp.89-100)

Laemmeli, U.K. (1970). Cleavage of structural proteins during assembly of the head of the bacteriophage T4. *Nature*, 227: 680-685.

Locker, R.H. and Hagyard, C.J. (1963). Acold shortening effect in beef muscles. *Journal of Science Food and Agriculture*, 14, 787-793.

Maher, S.C., Mullen, A.M., Moloney, A.P., Buckley, D.J., Kerry, J.P. (2004). Quantifying the extent of variation in eating quality traits of the M. *longissimus dorsi* and M. *semimembranosus* of conventionally processed Irish beef. *Meat Science*, 66, 351-360.

O'Halloran, G.R., Troy, D.J., and Buckley, D.J. (1997). The relationship between early post-mortem pH and the tenderisation in beef muscles. *Meat Science*, 45, 239-251.

Sanchez V., O'Reilly, K., White, A., Troy, D. and Mullen, AM. (2005). Proteolytic fragments in bovine exudate as potential tenderisation markers. In *Proceedings 51st international congress of meat science technology.*

Shackleford, S.D., Koohmarie, M., Whipple G., Wheeler, T.L., Miller, M.F. and Crouse, J.D. (1991). Predictors of beef tenderness: development and verification. *Journal of Food Science*, 56, 1130-1135.

Takahashi, G., Lochner, J.V., and Marsh, B.B. (1984). The effects of low frequency electrical stimulation on beef tenderness. *Meat Science*, 11, 207-225.

White, A. P., O'Sullivan, A., O'Neill, E. E., Troy, D. J. (2004). Manipulation of pre-rigor glycolytic behaviours to produce consistent beef tenderness. In *Proceedings of the 50th International Congress of Meat Science and Technology* (pp. 62–63). Helsinki, Finland.

White, A., O'Sullivan, A., Troy. D.J., and O'Neill, E.E. (2006a). Manipulation of the pre-rigor glycoytic behaviour of bovine M. *longissimus dorsi* in order to identify causes in inconsistencies in tenderness. *Meat Science*, 73, 151-156.

White, A., O'Sullivan, A., Troy. D.J., and O'Neill, E.E. (2006b). Manipulation of the pre-rigor phase to investigate the significance of proteolysis and sarcomere length in determining the tenderness of bovine M. *longissimus dorsi*. *Meat Science*, 73, 204-208.

FIBRE TYPE OF KOREAN NATIVE PORK AND ITS EFFECTS ON POST-MORTEM PROTEOLYSIS

I. Hwang[1], B.Y. Park[2] and C.S. Lee[3]

[1] Department of Animal Resources and Biotechnology, Chonbuk National University, JeonJu, Korea [2] National Livestock Research Institute, RDA. Suwon, Korea, [3] Konkuk University, Chungju, Korea Email: inho.hwang@chonbuk.ac.kr

Keywords: pork, fibre type, proteolysis, *longissimus dorsi*

Introduction

The reddish feature of Korean native pork is a favorable characteristic to Korean consumers, and likely related to fibre composition. Fibre type is a significant component affecting meat quality due to its relation to postmortem glycolytic rate, proteolytic rate and water-holding capacity (Eggert *et al.,* 2002). pH/temperature window during rigor development has a significant effect on meat quality though its effect on proteolysis, shortening and protein denaturation (Hwang *et al.,* 2004). As Korean native black pig (KNBP) has two carcass traits of red meat colour and small carcass size of which effect on pH/temperature window is great, the fibre type of KNBP has great effect on postmortem proteolysis and meat quality. The current study was conducted to characterize fibre type of Korean native black pig (KNBP) and its relation to glycolysis, proteolysis, and objective meat quality with reference to Landrace.

Materials and Methods

Twenty market-weighted male pigs (10 landrace, 118 kg, and 10 KNBP, 72 kg) were sampled from the NLRI breeding program. The pigs were assigned to a 2 x 3 factorial which was composed of two chilling regimes (-3 and 5°C) and three ageing times (1, 7, and 14 d at 1°C). Pigs were conventionally slaughtered, and placed in a 1°C chiller until the following day. pH, temperature, WB-shear force, and meat colour were determined similar to those described by Hwang *et al.* (2004). Relative composition of MyHC-I isoform in myofibril was determined by applying an indirect enzyme-linked immunosorbent assay (ELISA) following Picard *et al.,* (1994). Longissimus muscle tissue was biopsied during bleeding, immediately frozen in liquid nitrogen, and stored at –70°C until analysis. Primary and secondary antibodies were human MyHC-I monoclonal antibody (F36.5B9, 2C8, isotype mouse IgG2a, Biocytex biotechnology) and rabbit anti-mouse IgG (conjugated with alkaline phosphatase, Bethyl, Lab. Inc). *p*-nitrophenyl phosphate solution (Sigma, SL, USA) was used for colour development and absorbance was measured at 405 nm using a plate reader (MicroScreener LB 9260, EG and E BERTHOLD, Germany). Relative percentage of MyHC-I was calculated against a standard curve of *m. masseter* tissue. Postmortem proteolytic rate was quantified by a tricine-SDS-PAGE, and peptides were identified by a LC/MS/MS procedure described by Hwang *et al.,* (2005). For relative quantification, a horse myosin peptide cocktail (2.5 – 16.9 kDa, Amersham Biosciences) was run in triplicate, and changes in relative percentage of proteolytic products during ageing were calculated against 16.9 kDa peptide.

Results and Discussion

Table 1 and Figure 1 describe fibre composition, objective meat quality and rate of proteolysis during chiller ageing. The results demonstrated that KNBP longissimus muscle had a higher level in hunter a* value (red dimension), and that was related to a higher proportion of slow myosin heavy chain (MyHC-I). Given the result of an early study (Depreux, *et al.,* 2002) which reported that MyHC-I was negatively related to carcass weight, we could not exclude that the lighter carcass weight of KNBP, with similar age, was a possible factor for the higher proportion of the slow fibre type. However, our previous study (Hwang *et al.,* 2004) found that old and heavy KNBP (ca. 100 kg and 13 month old) also showed a similar colour characteristic. The results collectively implied that genetic components were involved in the distinct colour feature of KNBP. It has been well documented that carcass temperature of 10-15°C during rigor development could minimize muscle shortening and maximize proteolysis (Hwang *et al.,* 2003). Based on longissimus temperature at pH 6.2, pH/temperature decline of KNBP during rigor development was greatly favorable for resulting in tender meat (Table 1). This could be a consequence of slower glycolytic rate due to the higher proportion of slow myosin heavy chain (Bowker *et al.,* (2004), and faster chilling rate for the small carcasses. As previous studies demonstrated, this was reflected by a fast appearance of proteolytic peptides (Figure 1) but the presence of which was not significant on WB-shear force (Table 1). While bearing this fact in mind the proteolytic rate is significantly faster in white type muscle, thus one might expect a faster proteolytic rate for Landrace due to the higher frequency of white type fibres (Bowker *et al.,* 2004). However, KNBP showed a significantly faster degradation rate for some proteins (ca. creatine kinase, GAPDH, myosin light chain, titin and troponin I).

Table 1: Differences in objective meat qualities and postmortem proteolysis between landrace and Korean native black pig(KNBP).

	Breed		Av.se	F value		df[d]
	Landrace	KNBP		Breed	Ageing	
pH at 3 h	6.3	6.5	0.06	6.12^{*}		1(1)/18
Temperature at 3 h	26.0	23.1	0.87	5.77^{*}		1(1)/18
Temperature at pH 6.2	21.5	11.4	2.52	7.93^{*}		
pH at 24 h	2.3	3.0	0.87	0.3		
Temperature at 24	5.5	5.6	0.05	2.58		
MyHC-I (%)	11.3	14.2	0.71	7.91^{*}		1(1)/18
WB-shear force (kg)	5.8	5.4	0.22	2.77	52.0^{***}	1(2)/56
Hunter a*	6.8	10.2	0.32	57.27^{***}	7.12^{**}	1(2)/56
Hunter L*	45.9	42.1	0.64	17.44^{***}	11.47^{***}	1(2)/56
Band A (%)[f]	157.7	133.4	9.17	3.5		1(1)/18
Band B (%)[f]	11.2	12.6	0.71	1.85	30.55^{***}	1(2)/56
Band C (%)[f]	8.2	10.2	0.51	7.37^{**}	26.64^{***}	1(2)/56
Band D (%)[f]	8.8	10.5	0.47	6.96^{*}	17.86^{***}	1(2)/56
Band E (%)[f]	9.4	11.4	0.74	3.87	13.88^{***}	1(2)/56
Band F (%)[f]	7.6	8.9	0.86	1.24	19.54^{***}	1(2)/56
Peak A	7.1	6.8	0.35	0.39	203.5^{***}	1(2)/17

*$P<0.05$, **$P<0.01$, ***$P<0.001$, [d] Numerator/denominator degree of freedom for breed (ageing), [f] Relative percentage of 16.9 kDa horse myosin peptide.

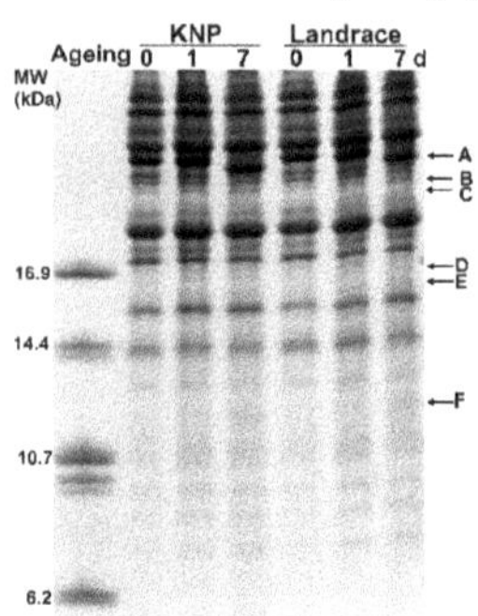

Figure 1: Changes in small molecular proteins during chiller ageing for Landrace and Korean native black pig(KNP), and 2DE-chromathgraphic profile (cation exchange, SCX, and reverse phase, RP) during ageing. Band A: Adenylate Kinase, Troponin I fast skeletal muscle, Glyceraldehyde-3-phosphate dehydrogenase; B: Creatine kinase, glyceraldehyde-3-phosphate dehydrogenase, Triosephosphate isomerase; C: Creatine kinase, GAPDH, Myosin light; D: Titin, Superoxide dismutase 1, GAPDH, Troponin I; E: Myosin light chain 1, Glyceraldehyde-3-phosphate dehydrogenase, Cytochrome c oxidase subunit IV isoform 1, Creatine kinase; F: Myosin light chain 2, myozenin 1, Creatine kinase, Glyceraldehyde-3-phosphate dehydrogenase , KIAA0613 protein.

Conclusions

The more a reddish colour of KNBP was related to higher frequency of slower fibre. KNBP showed more favorable pH/temperature profile during rigor development and was coincided with a faster proteolytic rate.

Acknowledgements

The BioGreen21 Research Promotion funded by RDA is thanked for financial support.

References

Bar, A and D. Pette. (1998). *FEBS letters*. 235, 153-155.
Bowker, B.C. *et al.* (2004). *Meat Science*. 68, 587-594.
Depreux, F.F.S. *et al.* (2002). *Meat Science*. 73, 265-273.
Eggert, J.M. *et al.* (2002). *Meat Science*. 61, 117-126.
Hwang, I. H. *et al.* (2004). *AJAS*. 17, 1599-1607.
Hwang, I.H. *et al.* (2003). *Meat Science*. 65, 677-691.
Hwang, I.H. *et al.* (2004). *Meat Science*. 68, 497-505.
Picard, B. *et al.* (1994). *Meat Science*. 36, 333-343.

DESMIN, 30KDA POST-MORTEM PATTERNS AND THE CALPAIN SYSTEM AS INDICATORS OF TENDERNESS IN BEEF ANIMALS

G.L. Marais[1], L. Frylinck[1], P.E. Strydom*[1], E.C. Webb[2], and J.D. Snyman[1]

[1.] *Agricultural Research Council of South Africa, Private Bag X2, Irene, South Africa, 0062.* [2.] *Department of Animal and Wildlife Sciences, University of Pretoria, 0002. E-mail: lorinda@arc.agric.za*

Keywords: beef tenderness, calpain system, SDS-PAGE, desmin, 30 kDa protein

Introduction

One of the most important meat quality attributes is tenderness (Ouali, 1991). Techniques such as sodium dodecyl sulphate-polyacrylamide gel electrophoresis (SDS-PAGE) and Western-blotting could be used for assessing meat tenderness. The degradation of proteins such as titin, nebulin, and desmin have been shown to play a role in the development in meat tenderness (Huff-Lonergan *et al.* 1996) and act as substrates for the calpain system known to be involved with the *post mortem* ageing of meat. These techniques are gradually becoming more sensitive, rapid and reliable as technologies develop in the form of capillary electrophoresis combined with mass spectroscopy. The purposes of this paper is to illustrate the correlation between the degradation patterns of desmin and the development of the 30kDa protein on the individual level and correlate them with different aspects of the calpain system and to determine how the data corresponds to the measured shear force on the *M. longissimus*.

Materials and Methods

The four animals (Brahman-144; Nguni-39; Simmental-130; Nguni-49) chosen as examples for this article form part of a larger project as described by Strydom and Frylinck (2005). These carcasses were not electrically stimulated (NS) as electrical stimulation (ES) influences the processes of meat tenderness. This paper focussed on the expression of the inherent tenderness characteristics without external *post mortem* influences. To study the degradation of desmin and appearance of the ~30 kDa degradation products as a result of tenderisation the *M. longissimus* proteins from samples taken at 1, 7, and 14 days *post mortem*, were separated using 30%T 0.5%C SDS-PAGE separation gels with 12%T, 12.5%C stacking gels (Fritz *et al.*, 1989). The separated muscle protein bands by means of SDS-PAGE electrophoresis were analysed by densitometry with the help a video image analyser equipped with the ImageMaster 1D Software (Amersham Pharmacia Biotech). The transfer and Western blotting procedures were carried out in accordance with the procedure of Huff-Lonergan (1996). The electrophoresis and Western immuno-blot patterns were analysed densitometrically with the ImageMaster 1D Software (Amersham Pharmacia Biotech). Calpains and calpastatin were extracted from frozen samples (Dransfield 1996) and separated by means of a two-step gradient ion-exchange chromatography-method (Geesink and Koohmaraie 1999). Calpain assays were done using azo-casein as substrate (Dransfield 1996) and one unit of calpain activity was defined as an increase in absorbance at 366 nm of 1.0 per hour at 25 °C. One unit of calpastatin activity was defined as the amount that inhibited one unit of m-calpain activity. Myofibril fragment lengths were measured using a Video Image Analyzer and tenderness was measured by Warner Bratzler shear force (SF) measurements (Strydom and Frylinck, 2005).

Results and Discussion

The intensities of the diminishing desmin protein (Figure 1) and the developing intensities of the 30 kDa protein bands as revealed by Western immuno-blotting and SDS-PAGE techniques, at 1-, 7-, and 14d *post mortem* in Brahman-144, Nguni-39, Simmental-130, and Nguni-49 are shown in Table 1. High correlations existed between these proteins and the calpain system analysed by means of the myofibril fragmentation at 1 day (r=0.68 to 0.92 for desmin; r=-0.71 to -0.81 for 30 kDa), 7 d *post mortem* (r=0.70 to 0.91 for desmin; r=-0.50 to -0.80 for 30 kDa), and rate of ageing (MFL at 1 d p.m – MFL at 14 d p.m) (r=0.68 to 0.92 for desmin; r=-0.70 to -0.85 for 30 kDa). Calpastatin and μ-calpain activity levels measured at 1 h post slaughter as well as the calpastatin/μ-calpain ratio (Table 2) correlated with rate of desmin degradation (r=0.80, –0.55, and 0.76 respectively) and 30 kDa formation (r=-0.83, 0.65, and -0.65 respectively). In contrast to these results, SF only correlated with diminishing desmin and formation of 30 kDa in respect with the tempo of tenderisation (SF at 1 d p.m. – SF at 14 d p.m.) (r=0.86 and 0.84 respectively). The actual shear force measured at 1 d and 14 d did not correlate well with the intensities of these proteins. The Br-144 and Sm-130 animals (Table 2) were the tougher of the four animal, but the MFL at 14 d p.m. were similar for al the animals. Also, the calpastatin level of Br-144 was unexpectedly the most favourable of the four animals. This show that one should judge all animals on an individual level, and that the variation between animals within breeds must be kept in mind.

Figure 1: Immuno-blots showing the degradation of desmin over time at 1, 7, and 14 days *post mortem* in *M. longissimus* of the Brahman (Bh-144), Nguni (Ng-39), Simmental (Sm-130) and Nguni (Ng-49).

Table 1: Densitometric scans of SDS-PAGE and Western blots of nebulin, desmin and the 30 kDa degradation protein (µg/µl) of examples of four animals representing three genotypes; Brahman-X, Nguni and Simmental-X.

Protein	Br-144	Ng-39	Sm-130	Ng-49
• 30 kDa (SDS-PAGE)				
- 1 d p.m.	0.028	0.017	0.011	0.000
- 7 d p.m.	0.062	0.019	0.018	0.039
- 14 d p.m.	0.153	0.052	0.059	0.138
- (1-14) d p.m.	0.125	0.035	0.048	0.138
• Desmin (Western blot)				
- 1 d p.m.	8.339	9.896	10.185	8.512
- 7 d p.m.	7.417	9.147	8.743	7.936
- 14 d p.m.	5.571	8.051	8.685	7.878
- (1-14) d p.m.	2.768	1.845	1.500	0.634

Table 2: Tenderness (SF) and calpain system measurements on examples of M. longissimus of four individual animals representing three genotypes; Brahman-X, Nguni, and Simmental-X.

Characteristic	Br-144	Ng-39	Sm-130	Ng-49
SF – 1 d p.m. (kg)	10.48	6.61	9.54	6.25
SF – 14 d p.m. (kg)	6.17	4.08	7.30	4.33
MFL – 7 d p.m. (µm)	25.9	34.0	41.3	31.8
MFL – 14 d p.m. (µm)	23.1	23.2	24.5	21.8
MFL – (1 – 14) d p.m. (µm)	9.48	26.8	37.5	16.8
Calpastatin (U/g meat)	2.13	2.12	3.13	3.66
µ-Calpain (U/g meat)	1.26	1.21	1.16	2.11
Calpastatin/µ-calpain	2.00	2.24	2.76	1.74

SF = Warner Bratzler shear force; MFL = myofibril fragment length; d p.m. – days *post mortem*

Conclusions

The tenderisation of meat is a multi-factorial event, therefore it will not be enough to only measure the contribution of the tenderisation mechanism involving the calpain system on the individual animal level in order to predict its potential meat tenderness. New technologies must include the potential contribution of the other mechanisms, such as the level of contraction of actin and myosin, as well as the contribution of the connective tissue characteristics.

References

Fritz, J.D., Swartz, D.R., and Greaser, M.L. (1989). Factors affecting polyacrylamide gel electrophoresis and electroblotting of high-molecular-weight myofibrillar proteins. *Analytical Biochemistry*, 180, 205-210.

Dransfield, E. (1996). Calpains from thaw rigor muscle. *Meat Science*, 43, 311-320.

Geesink, G.H., and Koohmaraie, M., (1999). A rapid method for quantification of calpain and calpastatin activities in muscle. *Journal of Animal Science,* 77, 3225-3229.

Huff-Lonergan, E., Mitsuhashi, T., Beekman, D.D., Parish, F.C., Olson, D.G., and Robson, R.M. (1996). Proteolysis of specific muscle structural proteins by µ-calpain at low pH and temperature is similar to degradation in post-mortem Bovine muscle. *Journal of Animal Science,* 74, 993-1008.

Ouali, A. (1991). Sensory quality of meat as affected by muscle biochemistry and modern technologies. In *Animal Biotechnology and the Quality of Meat Production*. Eds. Fiems, L.O., Cottyn, B.G., & Demeyer, D.I. Elsevier Science Publishers B.V., Amsterdam. pp 85-105.

Strydom, P.E. and Frylinck, L. (2005). The effect of genotype, duration of feed withdrawal and electrical stimulation on meat quality. 51st International Congress of Meat Science and Technology. August 7-12, 2005 – Baltimore, Maryland USA.

SESSION 2: MEAT SAFETY

RISK ANALYSES BASED CONTROL OF *SALMONELLA* ON PORK CUTS ON THE ISLAND OF IRELAND

D.M. Prendergast*, S.J. Duggan and G. Duffy

Food Safety Department, Ashtown Food Research Centre, Teagasc, Ashtown, Dublin 15, Ireland.
Email: deirdre.prendergast@teagasc.ie

Keywords: *Salmonella*, pork, prevalence, abattoir, risk assessment

Introduction

Salmonella spp. are the second most common cause of bacterial food borne illness and pork is now recognised as one of the most important food borne sources of *Salmonella* (Berends *et al.*, 1998; Beloeil *et al.*, 2004). In the Republic of Ireland, every pig herd is tested on an on-going basis. Twenty four pigs from each herd are tested three times a year and herds are assigned a category (1-3) based on a calculated weighted average of the three most recent tests. A certificate is issued grading the herd as Category 1 (≤ 10% positive), category 2 (> 10% ; ≤ 50% positive) or category 3 (> 50% positive). At slaughter, pigs from category 3 herds are slaughtered separately from other pigs and in a manner that minimises the risk of contamination. While there is a considerable amount of information available on the occurrence of *Salmonella* in pork on the island of Ireland this has not been amalgamated and there are many gaps in knowledge. This study is employing a quantitative risk assessment approach to determine the contribution of processing to pork contamination during slaughter. This will involve a quantitative assessment of the contamination levels, types of *Salmonella* and risk factors contributing to its transmission via pork leading to an assessment of how well current controls are operating. To fill in key gaps in data needed for the exposure assessment, microbiological studies on *Salmonella* in pork will be conducted at pork plants and at retail level in the Republic of Ireland and Northern Ireland to investigate the prevalence, numbers and types of *Salmonella* on pork cuts. This will allow a direct comparison of *Salmonella* contamination on pork cuts in both jurisdictions to be fed into a quantitative risk assessment model. This study reports the findings of the prevalence of *Salmonella* spp. in three commercial pork abattoirs in the Republic of Ireland.

Materials and Methods

Samples were collected in the boning halls of three commercial pork abattoirs during two visits between October 2005 and March 2006. To ensure that samples were representative of the factory throughput, the day on which sampling was carried out was altered during each sampling visit. A total of sixty samples from each oyster cut (primal leg cut) were taken over the entire working day, thirty in the morning and thirty in the afternoon. Sampling took place 2 hrs after work commenced. A 25-50g sample was excised from the oyster cut using a sterile scalpel and tweezers and transferred to sterile stomacher bags. Samples were transferred to a coolbox with ice packs and transported to the laboratory under chilled conditions 0-4°C. The method employed for the isolation and enumeration of *Salmonella* spp. is described in Figure 1.

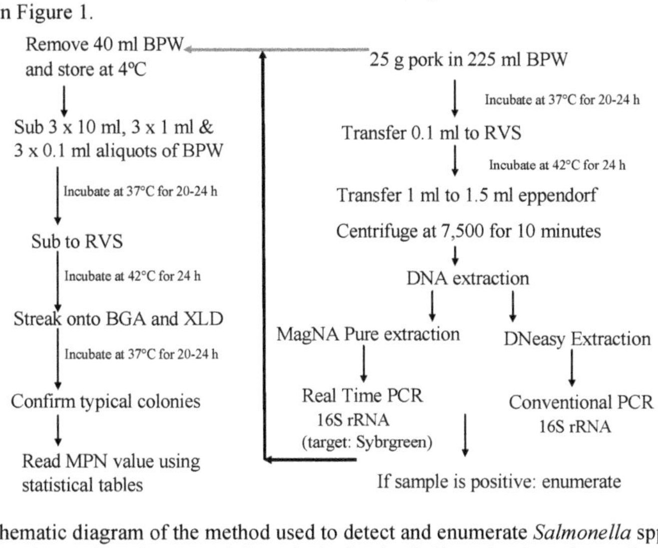

Figure 1: Schematic diagram of the method used to detect and enumerate *Salmonella* spp. in pork pieces

Information on the product that passed through the boning hall on the day of sampling i.e., slaughter date, number of animals slaughtered, number of suppliers and category of herd was obtained from each abattoir after each visit.

Results and Discussion

The number of samples positive for *Salmonella* spp. over two visits in each abattoir is shown in Table 1. In total, 360 samples were examined for the presence of *Salmonella* and 1.11% of these samples were found to be positive. *Salmonella* was not recovered in Abattoirs A or C. In abattoir A, the category of herd that passed through the boning hall on each sampling day was 1 and 2 only. In abattoir C, the category of herd that passed through the boning hall was 1, 2 and 3. However, further information revealed herds were classified as category 3 as the supplier did not have any certificate for these animals and therefore it is unknown if these animals were category 3. In abattoir B, 4 of the 120 samples were positive for *Salmonella* spp. These positive samples were all obtained from samples taken during the second visit. The category of herd that passed through the boning hall on the two sampling days was 1, 2 and 3. The percentage of category 3 animals on the first and second sampling days was 17.0 and 9.0% respectively.

Table 1: Incidence of *Salmonella* spp. on pork cuts (oyster) in the boning halls of Abattoirs A, B and C.

Abattoir	Time of sampling		No. positive		% positive		MPN g^{-1} ($\log_{10}$ cfu g^{-1})	
	Morning	Afternoon	Morning	Afternoon	Morning	Afternoon	Morning	Afternoon
A	60	60	0	0	0	0	-	-
B	60	60	1	3	1.70	5.00	<0.30 (-0.52)	<0.30 (-0.52)
C	60	60	0	0	0	0	-	-

Conclusions

In this present study, the incidence of *Salmonella* on pork cuts in the boning halls of the three abattoirs was 1.11%. It has been reported that the prevalence of *Salmonella* on raw pork in the Republic of Ireland has decreased from 9% in 2000 to 2% in 2003 (Anon, 2004; 2005) and the results of this present study are in keeping with this trend. Future work will increase the number of visits in each abattoir from two to five bringing the total number of samples analysed from 360 up to 900.

References

Anon (2004) Report on zoonoses in Ireland, 2000 and 2001. Published by FSAI http://www.fsai.ie/publications/index.asp

Anon (2005) Report on zoonoses in Ireland, 2002 and 2003. Published by FSAI http://www.fsai.ie/publications/index.asp

Berends, B. R., Van Knapen, F., Mossel, D. A. A., Burt, S. A. and Snijders, J. M. A. (1998) Impact on human health of Salmonella spp. on pork in The Netherlands and the anticipated effects of some currently proposed control strategies. *International Journal of Food Microbiology*, 44, 219-229.

Beloeil, P. A., Fravalo, P., Fablet, C., Jolly, J. P., Eveno, E., Hascoet, Y., Chauvin, C., Salvat, G. and Madec, F. (2004) Risk factors for *Salmonella* enterica subsp. enterica shedding by market-age pigs in French farrow-to-finish herds. *Preventive Veterinary Medicine,* 63, 103-120.

Catarame, T.M.G., O' Hanlon, K.A., McDowell, D.A., Blair, I.S. and Duffy, G. (2006) Comparison of a Real-Time Polymerase Chain Reaction assay with a culture methods for the detection of *Salmonella* in retail meat samples. *Journal of Food Safety*, 26, 1-15.

Acknowledgements

Safe*food*, The Food Safety Promotion Board and the Food Institutional Research Measure (FIRM) administered by the Irish Department of Agriculture and Food are acknowledged for funding this project. The co-operation and support of a stakeholder forum including regulatory authorities and the pork industry in the Republic and Northern Ireland is also gratefully acknowledged.

EVALUATION OF MICROBIOLOGICAL PERFORMANCE OBJECTIVES FOR THE BEEF INDUSTRY

P. Nally[1], E. Cummins[1], F. Butler[1]*, S. O'Brien[2], E. Carney[2], G. Duffy[2] and J.J. Sheridan[2]

[1.] Biosystems Engineering, UCD School of Agriculture, Food Science and Veterinary Medicine, University Collage Dublin, Earlsfort Terrace, Dublin 2, Ireland. [2.] Ashtown Food Research Centre, Teagasc, Ashtown, Dublín 15, Ireland. Email: Francis.butler@ucd.ie

Keywords: beef burgers, performance objectives, risk assessment

Introduction

Considerable advances in the development of quantitative risk assessment modelling of microbial pathogens in food have occurred in the past ten years. Simultaneously, risk-based management approaches to food safety have been introduced. These involve the application of food safety objectives (FSO), performance objectives (PO), and performance criteria (PC) in order to relate public health goals to the level of stringency required for food safety measures and systems.

FSO for microbiological criteria can be defined as the maximum prevalence and/or numbers of a microbial hazard in a food considered tolerable for consumer protection at the time of consumption. Setting the FSO at the time of consumption requires setting of Performance Objectives (PO) and Performance Criteria (PC) at critical points further back in the food chain. A PO is the maximum level (frequency and/or concentration) of a hazard in a food at a specified point in the food chain that should not be exceeded in order to achieve an FSO. A PC is the outcome of one or more steps in the food safety management system that must be met in order to achieve an FSO. From these parameters, food safety controls such as process criteria, product criteria and microbiological criteria may be derived. This paper presents an example of deriving risk-based criteria for *Escherichia coli* O157:H7 using risk assessment information. Food, especially undercooked beef burgers, has been implicated in many outbreaks of *E. coli* O157:H7 worldwide. For the present work, a previously developed risk assessment model for *E. coli* O157:H7 in Irish minced beef (Duffy, 2006) was used to generate a series of performance objectives and food safety objectives and demonstrate how these values influence the baseline risk per serving.

Materials and Methods

In this case study, potential performance objectives were selected based on the results of surveys of *E. coli* O157:H7 prevalence and numbers on beef trimmings and beef burgers in Ireland (Cagney 2004; Carney, 2006). A wide range of different combinations of prevalence and concentration values at the point of boxed trim or for freshly formed 100g beef burgers were selected as potential PO (Table 1).

Table 1: Potential performance objectives to be evaluated using the Duffy *et al.* (2006) risk model.

PO	Location	Initial prevalence (%)	Initial numbers (CFU/g)
1	Trimming	0.25	1
2	Trimming	0.5	1
3	Trimming	0.5	100
4	Trimming	2	1
5	Trimming	2	100
6	Trimming	3	100
7	Trimming	0.25	0.01
8	Trimming	0.1	1
9	Trimming	0.1	0.01
10	Trimming	0.05	1
11	Trimming	0.05	0.01
12	Trimming	0.25	0.001
13	Burger	0.25	1
14	Burger	0.5	1
15	Burger	1	100
16	Burger	2	100
17	Burger	3	1,000
18	Burger	3	10,000
19	Burger	0.1	1
20	Burger	0.01	1

These values were substituted into the Duffy *et al.* (2006) risk model for a fresh refrigerated beef burger and keeping all other post-production assumptions (for storage, handling, preparation, etc.) the same as in the original assessment, the model was re-simulated to generate outcomes for concentration (dose) and prevalence at time of consumption (after cooking), and the expected risk of illness.

Results and Discussion

Duffy *et al.* (2006) predict that the mean risk of *E. coli* O157:H7 illness from the consumption of a 100 g portion of a fresh, refrigerated beef burger in Ireland was approximately 1 in 1 million (exact value -5.94 log).

The results of re-simulating the Duffy *et al.* (2006) risk model with the relevant PO are presented in Table 2. These results show that the risk of illness is dependent on both the prevalence and numbers of *E. coli* O157:H7 on beef trimmings and in beef burgers. For example, in the case of a prevalence of *E. coli* O157:H7 on beef trimmings of 0.25% (potential PO 1 and 7) but with different counts (1 and 0.01 CFU/g), the resulting (mean) estimates for the probability of illness (in log units) per serving are -3.77 and -7.35 respectively. Similarly, for the same average *E. coli* O157:H7 count on trimmings, for example, 0.01 CFU/g, (potential PO 9 and 11) but with different prevalence values (0.1%, and 0.05%), the resulting estimate for the mean probability of illness (in log units) are -7.29 and -8.33 respectively.

Table 2: Results of re-simulating the Duffy *et al.* (2006) risk model with the potential PO.

Potential PO	Log probability of illness		
	5th percentile	Mean	95th percentile
1	-3.77	-6.38	-2.28
2	-3.68	-6.21	-1.96
3	-3.90	-6.18	-1.89
4	-3.71	-5.42	-1.49
5	-3.20	-5.59	-1.29
6	-3.07	-5.44	-1.10
7	-7.35	-7.80	-6.99
8	-4.33	-6.77	-2.61
9	-7.29	-7.42	-7.06
10	-5.71	-7.34	-4.29
11	-8.33	-9.12	-7.89
12	-7.69	-7.82	-7.58
13	-6.30	-8.54	-4.26
14	-6.13	-8.22	-4.01
15	-5.63	-7.99	-3.62
16	-5.34	-7.67	-3.32
17	-5.34	-7.56	-3.11
18	-5.84	-7.56	-3.10
19	-6.77	-8.93	-4.68
20	-7.21	-8.99	-5.06

Conclusions

If a mean probability of illness per serving of one in one million is considered to be a tolerable risk, then potential PO 13, 14, 19 and 20 would be suitable choices as a PO for freshly formed beef burgers, while potential PO 7, 9, 11 and 12 would be suitable choices as a PO for beef trimmings. It would appear that the PO could be "not more than 0.01 CFU/g for beef trimmings" or "not more than 1 CFU/g for beef burgers". The PO need not be more stringent; likewise, a less stringent PO would not provide a tolerable level of protection. Consideration also must be given to other sources of *E. coli* O157:H7 that result in disease plus whether the goal is to reduce over time or eliminate the number of cases from ground beef. The methodology illustrated in this paper demonstrates how risk assessment models can be used in the setting of PO for improved human health.

References

Duffy, G., Cummins, E., Nally, P., O'Brien, S. and Butler, F. (2006). A review of quantitative microbial risk assessment in the management of *Escherichia coli* O157:H7 on beef. Meat Science (accepted).

Cagney, C., Crowley, H., Duffy, G., Sheridan, J. J., O'Brien, S., Carney, E., Anderson, W., McDowell, D. A., Blair, I. S. and Bishop, R. H. (2004). Prevalence and numbers of *Escherichia coli* O157 in minced beef and beef burgers from butcher shops and supermarkets in the Republic of Ireland. Food Microbiology, 21, 2, 203–212.

Carney, E., O'Brien, S. B., Sheridan, J. J., McDowell, D. A., Blair, I. S. and Duffy, G. (2006). Prevalence and numbers of *Escherichia coli* O157 on beef trimmings, carcasses and head meat at a beef slaughter plant. Food Microbiology, 23, 1, 52–59.

SALMONELLA MIGRATION INTO MARINATED TURKEY BREAST AFTER CONTAMINATION DURING THREE STEPS OF THE MARINATION PROCESS

V. Tuntivanich, A. Orta-Ramirez and A.M. Booren*

Department of Food Science and Human Nutrition, Michigan State University, East Lansing, Michigan 48824, USA, Email: booren@msu.edu

Keywords: bacterial migration, *Salmonella*, turkey muscle, marination

Introduction

Consumer trends show an increasing demand for ready-to-eat (RTE) products, which include marinated whole-muscle products (Russell, 2002). Many meat and poultry products sold in retail markets are value-added by means of marination and mechanical tenderisation. It is often assumed that the interior of intact, undamaged whole-muscle is sterile (Elmossalami and Wassef, 1971). However, several studies have demonstrated that the inner portion of whole-muscle can be contaminated during processing. Warsow *et al.*, (2002) evaluated the potential for *Salmonella* migration into intact, whole-muscle turkey breasts during marination. *Salmonella* counts increased with the application of vacuum to the process and decreased with depth below the inoculated surface.

Migration is affected by several factors, including bacterial motility (Thomas *et al.*, 1987), proteolytic activity (Gill and Penney, 1982, Gupta *et al.*, 1983), water availability within the tissue, and fibre orientation (Maxcy, 1981). Tumbling is a process that increases uptake of marinade ingredients and promotes tenderisation (Xiong and Kupski, 1999), while vacuum is often applied to improve marinade penetration into the product. If a product surface and/or marinade were contaminated with a pathogenic microorganism, tumbling, vacuum and/or marinade composition could increase the potential for bacterial migration into the whole-muscle raw product. The objective was to study *Salmonella* migration into marinated turkey breast meat after contamination during three steps of the marination process.

Materials and Methods

This study focused on three different treatments: "pre-", "during", and "post-marination", depending on the time period at which the contamination occurred. For "pre-marination", 20 mL of *Salmonella* (8 strains)-inoculated marinade (10^8 CFU/mL) was added drop wise onto a sterile tray. An irradiated whole-muscle turkey breast was tossed over on the contaminated tray, then placed into a sterile bag before subjecting to 20 min vacuum tumbling marination. Turkey breast muscle for "post-marination" treatment was tumbled with sterile marinade for 20 min before turning over on the contaminated tray. For the "during marination" sample, whole-muscle turkey breast was tumbled with *Salmonella*-inoculated marinade.

After marination, three 1 cm-thick slices (cranial, middle, and caudal) were removed from the turkey breast longitudinal to the muscle fibre (Figure 1a). The middle slice was taken from the thickest part of the breast muscle. Cranial and caudal slices were sectioned at ~5 and 10 cm, respectively, measuring from both ends. Within these three slices, 1 cm^3 cubes were excised along the axis of the muscle fibre (Figure 1b) and enumerated for surviving *Salmonella*. All of the dissection was done using a self-cauterizing electrosurgical unit.

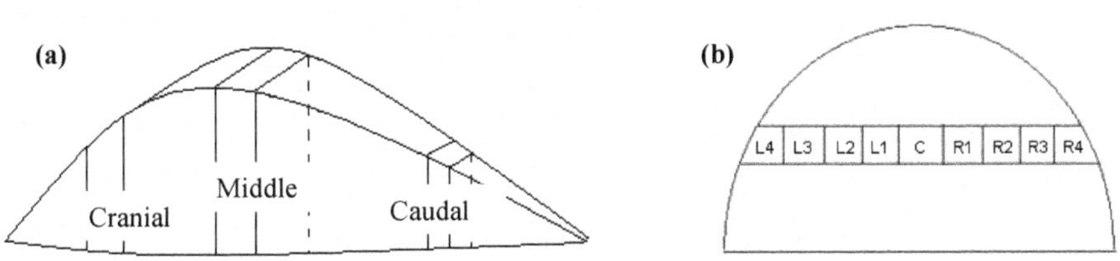

Figure 1: Sampling diagram (a) 3 slices excised from marinated whole-muscle turkey breast (b) 1-cm^3 segments dissected from cranial, middle, and caudal slices. Segments labeled R1, R3, L1, and L3 were dissected from cranial and caudal slices and segments labeled C, R2, R4, L2, and L4 were dissected from middle slice.

Results and Discussion

The results suggest that penetration of *Salmonella* into whole-muscle turkey breasts does not follow a random pattern. *Salmonella* counts decreased as the depth below the inoculated surface from four directions (right, left, top, and bottom) increased. The highest counts were observed in the pieces next to the inoculated surface. The center segment (C) from the middle slice for all treatments showed the lowest *Salmonella* migration among all segments. *Salmonella* plate counts at the center (C) from the middle slice for "during marination" treatment was 3.43 Log CFU/mL (Figure 2b), which was higher ($P<0.05$) than those of "pre-" and "post-marination" treatments (Figures 2a and 2c, respectively). This means

Salmonella penetrated more when turkey meat was tumbled with a contaminated marinade solution. There was a slight difference in *Salmonella* distribution for the cranial and caudal slices, as compared to the middle slice. This may be due to the actual structure of turkey breast muscle.

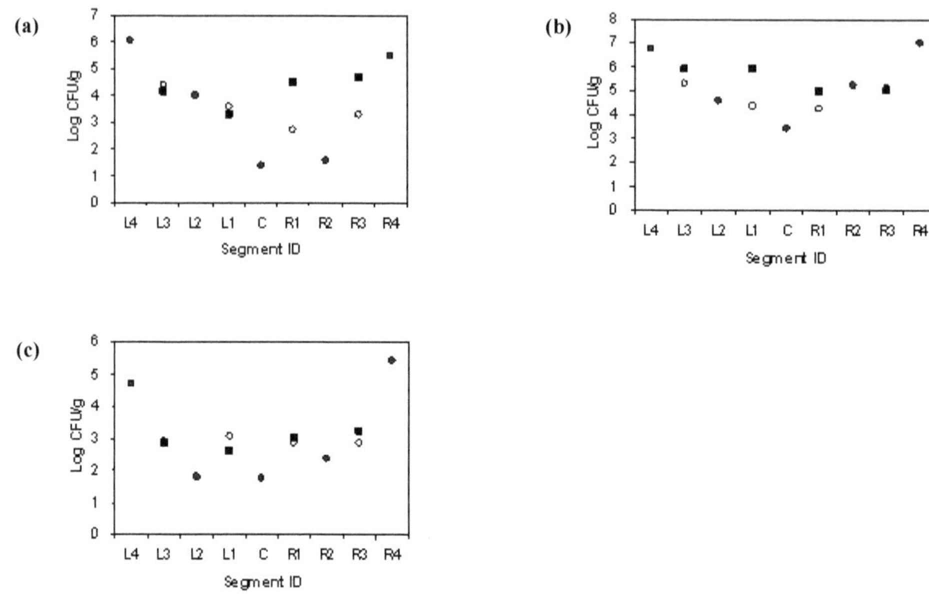

Figure 2: *Salmonella* migration pattern for (a) pre-marination (b) during marination and (c) post-marination. Open circles, solid circles, and solid squares represent sample cubes from cranial, middle, and caudal slices, respectively.

Conclusions

These results suggest that *Salmonella* can enter and survive in value-added or marinated turkey products. Our results provide important information for product and process safety, which is necessary to develop a quantitative model for pathogen migration into whole-muscle poultry products during the marination process. In addition, our research provides preliminary information and feasibility in developing 3-D profiles of bacterial penetration into whole-muscle products during the marination process.

References

Elmossalami, E. and Wassef, N. (1971). Penetration of some microorganisms in meat. Zentralblatt-fuer-Veterinaermedizin-Reihe B., 18:329-336.

Gill, C.O. and Penney, N. (1982). Bacterial penetration of muscle tissue. *Journal of Food Science*, 47:690-691.

Gupta, L., Kalra, M.S., Singh, A., and Singh, B. (1983). Penetration of bacteria into muscles of goat, pork and poultry meat. *Journal of Food Science and Technology*, 20:13-16.

Maxcy, R.B. (1981). Surface microenvironment and penetration of bacteria into meat. *Journal of Food Protection*, 44:550-552.

Russell, J. (2002). Annual Deli Report: Fresh Start. *The National Provisioner*. Oct. 2002. 216(7):14.

Thomas, C.J., O'Rourke, R.D., and McMeekin, T.A. (1987). Bacterial penetration of chicken breast muscle. *Food Microbiology*, 4:87-95.

Warsow, C.R., Marks, B.P., Ryser, E.T., Orta-Ramirez, A., and Booren, A.M. (2002). *Salmonella* penetration into whole muscle turkey during vacuum marination. Abstracts of the Institute of Food Technologists Annual Meeting, Abstract No. 88-4.

Xiong, Y.L. and Kupski, D.R. (1999). Monitoring phosphate marinade penetration in tumbled chicken filets using a thin-slicing, dye-tracking method. *Journal of Poultry Science*, 78:1048-1052.

QUANTIFICATION OF TOTAL VIABLE COUNTS (TVC'S) ON FRESH MEAT CARCASSES BY A NOVEL REAL-TIME POLYMERASE CHAIN REACTION

A. Dolan*[1], C. Burgess[1], T. Barry[2], S. Fanning[3] and G. Duffy[1]

[1.] *Food Safety Department, Ashtown Food Research Centre, Teagasc, Ashtown, Dublin15, Ireland,* [2.] *Department of Microbiology, National University of Ireland, Galway, Ireland,* [3]. *School of Agriculture, Food Science, and Veterinary Medicine, University College Dublin, Belfield, Dublin* [4.] *Email: anthony.dolan@teagasc.ie*

Keywords: total viable count, real-time PCR, beef, DNA, RNA

Introduction

The safety of food must be assured by a preventative approach based on the application of a food safety management system such as Hazard Analysis Critical Control Points (HACCP) and setting of microbiological criteria. Testing of samples against these criteria can be used to provide assurance that process controls, performance criteria or food safety objectives have been achieved (Stewart *et al.*, 2002). The total viable count can be performed by a cultural method (ISO 4833 which takes 3 days) or by an alternative rapid method if shown to be equivalent (ISO 16140:2003). The aim of this project is to develop and apply highly specific and user friendly nucleic acid based (DNA and RNA) rapid methods based on Real-Time PCR for the determination of total viable counts on fresh meat carcasses. Subsequently a procedure for demonstration of equivalence based on that outlined in EN/ISO:16140 will be followed. As it is essential that only viable organisms be detected, Reverse Transcriptase PCR which detects specific RNA sequences (only expressed by viable cells) will be employed. It will be applied in a Real-Time (Lightcycler) automated PCR format, which uses fluorescence to detect the presence and quantity of a particular gene in real time thus making it make more rapid and user friendly than conventional PCR (Yaron and Matthews, 2002).

Materials and Methods

In order to establish the microflora of a beef carcass, 90 slaughtered steers were sampled over a 5 week period in an Irish abattoir. Each carcass was sampled using the sponge swab technique of four sites (neck, flank, brisket and rump) before chilling as outlined in the EU 471 directive. The samples were stored at 4°C until further processing on the same day. Samples were plated onto Plate Count Agar (PCA) to establish the total viable count (TVC) and on a range of different agar bases in order to identify particular species that were typical or dominant. Following sampling, 40 bacterial colonies were selected at random for identification by 16S rDNA sequencing.

Using the information obtained from the sequencing, pure bacterial cultures typical of those occurring on beef carcasses were selected. DNA was extracted from pure cultures grown overnight at the appropriate temperature using the DNeasy® Tissue Kit (Qiagen, Hilden, Germany), according to the manufacturers instructions. For the PCR, a master mix of total volume 45µl was prepared, to which 5µl of DNA was added. The master mix contained novel primers for a specific gene target common to all bacteria. The PCR was carried out on a DNA Engine Dyad™. At present, the assay is being transferred to a real-time PCR platform on the Roche LightCycler®.

Results and Discussion

The predominant microbial flora at the pre-chilling stage of the slaughter process have been shown to be Lactic Acid Bacteria and *Enterobacteriaceae* spp. In addition, a broad range of Gram-positive and Gram-negative bacteria were detected, including *Acinetobacter* spp., Staphylococcae and *Pseudomonas* spp. In the PCR reaction, designed primers were tested against 40 bacterial isolates, and the expected region was amplified in all species (Figure 1). The selected gene is highly expressed in all bacteria and therefore will be very amenable to an RNA based reverse transcriptase PCR assay. Once this target has been fully characterised, it will be assessed for the ability to be used as a working assay for total viable counts using the above outlined technology.

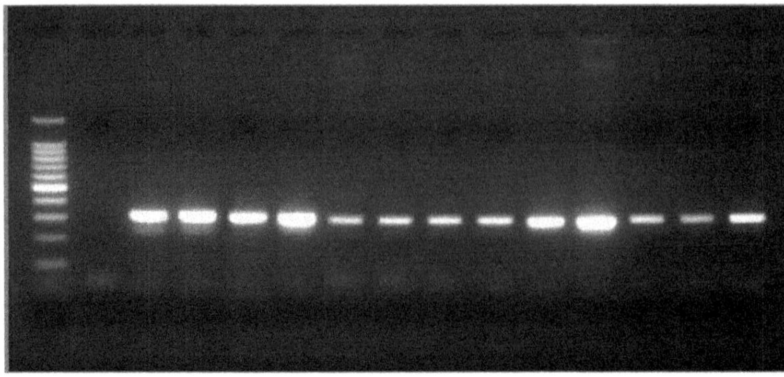

Figure 1: 2% agarose gel showing the amplicon obtained in the PCR reaction. Lane: 1. 100 base pair ladder (Promega Corporation, WI, USA); 2. Negative control; 3. *Acinetobacter calcoaceticus*; 4. *Aeromonas hydrophila*; 5. *Aeromonas salmonicida*; 6. *Enterobacter aerogenes*; 7. *Enterococcus faecium*; 8. *Enterococcus gallinarum*; 9. *Lactobacillus plantarum*; 10. *Leuconostoc mesenteroides*; 11. *Pseudomonas flourescens*; 12. *Pseudomonas putida*; 13. *Staphylococcus epidermis*; 14. *Staphylococcus haemolyticus*; 15. *Staphylococcus saprophyticus*.

Conclusions

Based on these results it may be concluded that this method seems to be adaptable to a real-time PCR assay to quantify total bacteria on fresh meat carcasses. An alternative rapid method, which gives results within a few hours, would be of immense benefit to the meat industry, allowing real time assessment of product quality and reduced holding time of the product.

References

Stewart, C.M., Bruce-Tompkin, R. and Cole. M.B. (2002). Food safety: new concepts for the new millenium. *Innovative Food Science and Emerging technologies,* 3: 105-112.

Yaron S. and Matthews K. R. (2002). A reverse transcriptase-polymerase chain reaction assay for detection of viable *Escherichia coli* O157:H7: investigation of specific target genes. *Journal of Applied Microbiology,* 92:633-40.

Acknowledgements

This project is funded by the Food Institutional Research Measure (FIRM) administered by The Department of Agriculture and Food, Ireland.

TOWARDS THE DEVELOPMENT OF A RAPID NUCLEIC ACID BASED TECHNIQUE FOR THE SENSITIVE AND SPECIFIC DETECTION OF *SALMONELLA* SPP. ON FRESH MEAT

E. McCabe[1], C. Burgess[1], S. Fanning[3], T. Barry[2] and G. Duffy*[1]

[1]Food Safety Department, Ashtown Food Research Centre, Teagasc, Ashtown, Dublin 15, Ireland; [2]Department of Microbiology, National University of Ireland, Galway, Ireland [3]School of Agriculture, Food Science, and Veterinary Medicine, University College Dublin, Belfield, Dublin 4. Email: Geraldine.duffy@teagasc.ie

Keywords: *Salmonella*, real-time PCR, meat

Introduction

The safety of food must be assured by a preventative approach based on the application of a food safety management system such as Hazard Analysis Critical Control point (HACCP). The setting of microbiological criteria and testing of food samples against the criteria, can be used to support food safety management systems and provide assurance that process controls and performance criteria or food safety objectives have been achieved (Stewart *et al.*, 2002). To date EU microbiological criteria have not addressed the presence of pathogens such as *Salmonella* on fresh meat carcasses, however this is now being addressed and is included in the EU Commission "Microbiological criteria for foodstuffs". Testing may be carried out by a cultural method (ISO 6579) which takes 3-4 days for positive clearance of samples, or by an equivalent rapid method in EN/ISO 16140:2003. The requirement to test for *Salmonella* means that industry has an immediate need for an alternative rapid method as the delays incurred in waiting for test results will otherwise involve a considerable sized product holding area and delayed product release. The aim of the project is to develop and apply highly specific and user friendly nucleic acid based (DNA and RNA) rapid methods based on real time PCR or fluorescent in situ hybridisation (FISH) approaches for the detection of Salmonella on fresh meat carcasses and chicken skin. The equivalence of the developed method with the traditional method will be carried out according to the procedures outlined in EN/ISO:16140

Materials and Methods

Initial studies included further validation of a 16S rRNA DNA target method (Catarme *et al.*, 2006., Trkov *et al.*, 2003) using a *Salmonella* enterica serovar bank and a range of non-*Salmonella* species by real time and conventional PCR. DNA was extracted from 1 ml of overnight pure cultures of *Salmonella* using the DNAeasy extraction kit (Qiagen), and real time PCR was carried out using 16S rRNA target specific primers and the SYBR Green detection format using the Lightcycler (Roche) platform.

Further studies investigated gene targets identified from the literature in addition to looking for novel gene targets. Identified targets hilA (Pathmanathan *et al.*, 2003), sip b/c (Ellingson *et al.*, 2004), and the ttr (Malorny *et al.*, 2004) were carried out in real time DNA polymerase chain reaction (PCR) protocol using a panel of *Salmonella* serovars and non-*Salmonella* species. When a suitable gene was identified RNA was extracted from twenty three overnight pure cultures of *Salmonella* spp. and twelve non-*Salmonella* strains using the SV total RNA isolation kit (Promega) and RNA was reverse transcribed into cDNA using the ImProm-II reverse transcription system (Promega). Real time PCR was carried out on both the DNA and the cDNA for the twenty three strains and the twelve non-Salmonella strains with the newly designed primer set. The newly developed real time protocol using the SYBR Green detection format was assessed using inocula as low as 10 CFU ml^{-1}.

Results and Discussion

A 16S rRNA DNA target was tested in conventional and real time PCR format using a *Salmonella* enterica serovar bank and a range of non-*Salmonella* strains. It was found that although this method was highly sensitive and specific for *Salmonella* spp. that there was some cross-reactivity with these primers and some of the non-*Salmonella* species. It was concluded that there was a need for a more specific DNA/RNA target for *Salmonella* spp. detection.

From the number of genes tested the target chosen was against the hilA gene. The nucleic acid from twelve typed strains of *Salmonella* were sequenced and aligned and new primers designed against the hilA gene. Using the newly designed primers RNA was successfully reverse transcribed into cDNA and real time PCR was carried out using both DNA and cDNA from all of the twenty three strains. The method was found to be both sensitive and specific for *Salmonella* spp. and there was no cross reactivity with the primers and any of the twelve non-*Salmonella* strains. Further development of the method would be to design a probe towards the PCR product and an internal amplification control (IAC), these design measures would ensure specificity, sensitivity and efficiency of the method.

Conclusions

The requirement for testing for *Salmonella* means that industry has an immediate need for an alternative rapid method as the delays incurred in waiting for test results will otherwise involve a considerable sized product holding area and delayed product release and thus have a significant economic impact. The potential DNA/RNA target for the hilA gene has been identified for the detection of *Salmonella* spp. and is now being investigated for use in either a real time reverse transcriptase polymerase chain reaction (RT-PCR) protocol or a fluorescent in situ hybridization (FISH) format for the detection of *Salmonella* spp. in meat samples based on specificity, limit of detection and user friendliness.

References

Catarame,T. Duffy ,G., O'Hanlon, K., Sheridan, J.J., Blair, I.S. and McDowell, D.A. (2006).

Comparison of a culture method with a real time PCR. Method for detection of *Salmonella* in retail meat samples. *Journal of Food Safety,* 26, 1-15

Ellingson JL, Anderson JL, Carlson SA, Sharma VK. (2004) Twelve hour real-time PCR technique for the sensitive and specific detection of *Salmonella* in raw and ready-to-eat meat products. *Molecular and Cellular Probes,* 18(1):51-7.

International Organisation for standardization (ISO) no. 16140:2003. Microbiology of food and animal feeding stuffs-Protocol for the validation of alternative methods.

International Organisation for standardization (ISO) no.6579:2003. Microbiology of food and animal feeding stuffs-Horizontal method for the detection of *Salmonella.*

Malorny B, Paccassoni E, Fach P, Bunge C, Martin A, Helmuth R. (2004). Diagnostic real-time PCR for detection of *Salmonella* in food. *Applied and Environmental Microbiology,* 70(12):7046-52.

Pathmanathan SG, Cardona-Castro N, Sanchez-Jimenez MM, Correa-Ochoa MM, Puthucheary SD, Thong KL. (2003). Simple and rapid detection of Salmonella strains by direct PCR amplification of the hilA gene. *Journal of Medical Microbiology,* 52:773-6.

Stewart, C.M., Bruce-Tompkin, R. and Cole. M.B (2002). Food safety: new concepts for the new millennium. *Innovative Food Science and emerging technologies,* 3: 105-112.

Trkov M, Avgustin G. An improved 16S rRNA based PCR method for the specific detection of *Salmonella enterica* (2003). *International Journal of Food Microbiology,* 80(1):67-75.

Acknowledgements

This project is funded by the Food Institutional Research Measure (FIRM) administered by The Irish Department of Agriculture and Food.

DEVELOPMENT OF METHODS FOR THE ISOLATION AND DETECTION OF *CAMPYLOBACTERACEAE* IN THE MEAT FOOD CHAIN

Ó.A. Lynch, C. Cagney and G. Duffy*

Ashtown Food Research Centre, Teagasc, Ashtown, Dublin 15, Ireland. Email: Geraldine.Duffy@teagasc.ie

Keywords: *Campylobacter*, meat, detection methods

Introduction

Campylobacter spp. are recognised as one of the leading causes of food-related illness and are the most common cause of gastric-infections in Ireland. In 2004 alone, there were 1711 reported cases of human *Campylobacteriosis* in the Republic (CIR 43.7/100,000), with males under the age of five carrying the highest burden of infection (HPSC, 2004). The isolation and identification methods used to examine clinical and food samples for *Campylobacter* are designed to primarily detect *Campylobacter jejuni* and *Campylobacter coli* and are biased towards these species. However, an isolation procedure incorporating the use of Hydrogen enriched atmosphere together with filtration onto blood agar and incubation at a lower temperature (37 °C) reportedly facilitates recovery of a much wider range of *Campylobacteraceae* for clinical specimens (Lastovica and Le Roux, 2000). Using these new methodologies, evidence is accumulating from studies carried out in South Africa and Europe that there are other *Campylobacter* strains of importance associated with gastric illness, including *Campylobacter concisus* and *Campylobacter upsaliensis*. These emerging species are being isolated at high rates and now appear to be as clinically significant as *C. jejuni* and *C. coli* (Enberg *et al.*, 2000; Abenhus *et al.*, 2002; Maher *et al.*, 2002).

The distribution and reservoirs of emerging *Campylobacteraceae* in the food chain have not yet been established. It is therefore essential to modify current detection methods for the isolation of a wider range of *Campylobacteraceae* from food and environmental samples. This will contribute to effective monitoring and surveillance and allow for a better understanding of the epidemiology and transmission routes of this pathogen. The following summarises the methods and techniques investigated, to establish new isolation methods capable of recovering a wide range of *Campylobacter* from meat.

Materials and Methods

Campylobacter strains (n=59) of human, clinical and animal origin were used in this study. All strains were supplied by the Danish Food and Veterinary Research Institute in Copenhagen, Denmark, and the Red Cross children's Hospital in Cape Town, South Africa and stored in 80% Glycerol at – 80 °C.

Enrichment: *Campylobacteraceae* spp. were inoculated into a range of different enrichment media {Hunts Enrichment Broth (Hunt *et al.*, 1992), Bolton broth (Oxoid, Basingstoke, U.K) and *Campylobacter* Enrichment Broth (CEB) (Lab M, Surrey, UK). Each broth was incubated in an atmospheric cabinet (Don Whitley) supplied with 2.5% O_2, 7% H_2, 10% CO_2, and 80.5% N_2 (personal communication, Keevil, W. University of Southampton) for 0-72 h at 37 °C. During the enrichment period, *Campylobacter* were plated out at various time intervals onto two agar types; Tryptose Blood Agar with 5% lysed horse blood (TBA) and Anaerobe Basal Agar with 5% lysed horse blood (ABA) (Oxoid). Agar plates were incubated under the same conditions for 48 h to allow for enumeration. Morphology and motility of each sample was assessed at every time interval using NanoOrange Staining (Molecular Probes, the Netherlands) (Alonso *et al.*, 2002).

Isolation: Meat pieces (25 g) were inoculated with control *Campylobacter* strains at levels of 1.0 $\log_{10}$ cfu ml $^{-1}$ and 2.0 $\log_{10}$ cfu ml $^{-1}$. Samples were placed in a filter stomacher bag with 225 ml CEB and 5% lysed horse blood and incubated for 24 or 48 h in the gas atmosphere described above at 37 °C. A range of options were investigated for recovery of *Campylobacter* from the enrichment broth, alone and in combination, including stomaching versus pulsification (Microgen Bioproducts, U.K); centrifugation of enrichment broth to remove food debris, at various times and speeds; membrane filtration (various filter types and pore sizes: Cellulose nitrate 0.45μm, cCellulose acetate 0.45 μm, Cellulose nitrate 0.6 μm, Mixed ester 0.6 μm and Polycarbonate 0.6 μm) and the volume of enrichment broth to be transferred to the membrane surface (0.2, 0.3, 0.4 ml).

Validation: Using the optimised enrichment and isolation method described above, beef, pork and chicken (25g) were inoculated individually with 59 *Campylobacteraceae* strains to levels of $\log_{10}$ 1.0 cfu ml $^{-1}$ and $\log_{10}$ 2.0 cfu ml $^{-1}$ and recovery was tested.

Identification: Cultures were confirmed as *Campylobacteraceae* by use of the new O.B.I.S L-ala biochemical identification kits (Smith *et al.*, 2006) (Oxoid, Basingstoke, U.K) in conjunction with the KOH test to determine Gram reaction (Halebian *et al.*, 1981). *Campylobacter* latex agglutination kits (Microgen Bioproducts, UK) were used to establish the thermophillic/non-thermophillic status of each strain. Catalase, oxidase and hippurate testing were also carried out.

Polymerase Chain Reaction: PCR protocol is currently under optimization which allow for speciation of a wide range of known and emerging *Campylobacters*. The method allows for differentiation of species based on nucleotide sequence differences in a common house keeping gene (Klena *et al.,* 2004). Once complete the procedure will allow for broad speciation using a minimum of multiplex reactions.

Results and Discussion

The method developed involved placing a meat samples (25g) in a filter stomacher bag with 225 ml CEB + 5% lysed horse blood and pulsifying it for 15 s. The sample was then placed in an atmospheric cabinet supplied with 2.5% O_2 , 7% H_2 , 10% CO_2 , and 80.5% N_2 for 24 h at 37 °C. After incubation, 10 ml of broth was centrifuged at 2,500 rpm for 10 s followed by a dilution series. 200µl of supernatant was transferred to a mixed ester membrane (pore size 0.6 µm) on the surface of an ABA plate (with 5% lysed horse blood). The filters were removed after 15 s and plates were incubated for up to 6 days in the above described atmosphere and examined at intervals throughout the incubation period. Suspect colonies were confirmed using a biochemical matrix developed by Lastovica *et al.,* (2000) with the addition of the OBIS L-ala biochemical test and Latex agglutination. Speciation of *Campylobacter* positive isolates was achieved using an optimised PCR protocol (Klena *et al .,* 2004).

Conclusion

The above described protocol has been applied to examine the *Campylobacter* populations in fresh meat in Ireland over a 12 month period and the initial results indicate that in addition to *C. jejuni* and *C. coli,* a wider range of *Campylobacter* species are recoverable from these meats.

References

Alonso J.L., Mascellaro S., Moreno Y., Ferrus M. and Hernandez J. (2002). Double-staining method for differentiation of morphological changes and membrane integrity of *Campylobacter coli* cells. *Applied and Environmental Microbiology,* 68(10): 5151-4.

Aabenhus R., Permin H., On SL. and Andersen L.P. (2002). Prevalence of *Campylobacter concisus* in diarrhoea of immunocompromised patients. *Scandanavian Journal of Infectious Disease,* 34(4): 248-52.

Engberg J., On S.L.W., Harrington C.S. and Gerner-Smidt P. (2000). Prevalence of *Campylobacter, Arcobacter, Helicobacter* and *Sutterella* spp. in human faecal samples as estimated by a re-evaluation of isolation methods for *Campylobacter. Journal Clinical Microbiology,* 38: 286-291.

Halebian, S, Harris, B, Finegold, S.M, Rolfe, R.D (1981). Rapid methods that aid in distinguishing Gram-negative anaerobic bacteria. *Journal of Clinical Microbiology,* 13(3): 444-448

Health Protection Surveillance Centre (2004). Annual Report 2004.

Hunt, J.M (1992). "*Campylobacter*". FDA Bacteriological Analytical Manual, 7th Edition. P 77-94

Lastovica A.J. and Le Roux E. (2000). Efficient isolation of *Campylobacter* from Stools. *Journal of Clinical Microbiology,* 2798-2799.

Klena, J.D, Parker, C.T, Knibb, K, Ibbitt, C, Devane, P.M.L, Horn, S.T, Miller, W.G, Knonkel, M.E. (2004). Differentiation *of Campylobacter coli, jejuni, lari* and *upsaliensis* by a multiplex PCR developed from nucleotide sequences of the lipid A gene IpxA. *Journal of Clinical Microbiology* 42(12): 5549-5557.

Maher M., Finnegan C., Collins E., Ward B., Carroll C. and Cormican M. (2002). Evaluation of culture methods and a DNA probe-based PCR assay for detection of *Campylobacter* species in clinical specimens of faeces. *Journal of Clinical Microbiology,* 41(7): 2980-2986.

Smith, C.M, Colbourne, N.R, Stephens, P.J, Druggan, P (2006). A simple and rapid biochemical screening test for the differentiation of *Campylobacter* species from other contaminating organisms. Presented at Conference on Emerging *Campylobacter* in the Food Chain. Croke Park, Dublin, Feb 8th 2006

Acknowledgements

This work was funded by the EC FP5 CAMPYCHECK project (QLK CT 2002 02201).

THE EFFECT OF MEAT SAFETY SYSTEMS AT HIGH THROUGHPUT ABATTOIRS ON THE MICROBIOLOGICAL QUALITY OF MEAT

P.A Tshabalala*[1], E. Buys[2] and H.L. de Kock[2]

[1.] *Department of Agriculture, Private Bag x343, Pretoria, South Africa.* [2.] *Department of Food Science, University of Pretoria, Pretoria, South Africa. Email: papisot@nda.agric.za*

Keywords: beef, food safety systems, *E. coli* serotypes

Introduction

In South Africa, there are two classes of high throughput red meat abattoirs, the export and the non-export abattoir. South Africa uses the Meat Safety Act (Act 40, 2000) and its regulations to administer the South African meat industry by providing a framework for developing meat hygiene and safety programmes, resulting in the assurance of wholesomeness and supply of quality meat to consumers. Standards and requirements at export abattoirs are more stringent and monitored by the Department of Agriculture (DoA).

Although all abattoirs have to meet structural and hygiene requirements, export abattoirs in addition have to meet the following:

1. A local high throughput abattoir wishing to export meat and meat products would have to apply and, if successful, would be registered as an export establishment.
2. The Department of Agriculture (National /Provincial) assigns a full-time veterinarian to be stationed at each export abattoir. The EU requires that such a veterinarian carries out ante-mortem inspection. The ante-mortem records will be incorporated into a traceability system.
3. Exporting abattoirs implement auditable HACCP systems.
4. Exporting abattoirs have a traceability system that covers the whole chain from farm-to-table.

The non-export abattoirs utilise an inspection system, which removes diseased and dirty animals from the food supply. Visual inspection by sensory methods such as smell, sight and sense of touch are relied upon to identify unsafe food. Such a system does not target and reduce microbial pathogens on raw meat. Since bacteria such as *E. coli* O157:H7 or *Salmonella* cannot be detected by visual inspection, they remain present in meat and poultry products delivered to distributors and consumers. This research focuses on the determination of total bacterial counts, and the incidence of *Pseudomonas* spp. *Enterobacteriaceae*, lactic acid bacteria (LAB) *E. coli* serotypes, *Salmonella*,spp and *Staphylococcus aureus* on fresh beef after chilling, produced from high throughput abattoirs with food safety systems and beef from high throughput abattoirs without food safety systems.

Materials and Methods

Two high throughput abattoirs participated in the project. Each abattoir was visited two times, where ten carcasses were randomly selected followed by the collection of 20 swab samples i.e. 40 samples per abattoir. Envirosponges (USDA) were used to swab a 25 cm^2 area on the right side of the carcass and a corresponding left fore quarter was swabbed for *Salmonella* analysis. The swab samples were kept cool (below 5°C) during transportation. 20 swab samples were processed at the University of Pretoria, whereas the remaining 20 were send to Onderstepoort Veterinary Institute for *Salmonella* analysis.

Laboratory evaluation: Each swab was homogenized in 225 ml 0.1 % Buffered Peptone Water (BPW) for 2 minutes. Serial dilutions were prepared with the use of a 0.1 % BPW (oxoid). The spread plate method was further applied (taking the dilution factor into consideration) to quantify the various microbial groups (Herbert, 1990) using the following agars: Standard Plate Count Agar (Oxoid-CM0463), (Martley *et al.*, 1970) followed by incubation at 25°C for 72 h; Pseudomonas Agar Base (Oxoid-CM0559) with added glycerol, followed by incubation at 25°C for *Pseudomonas* spp.; Bair-Parker Agar (Bio-Rad SA), followed by incubation at 35°C for 48 h for *Staphylococcus* sp.; MRS Agar (Biolab) followed by incubation for 48 h for Lactic Acid Bacteria and Violet Red Bile-Glucose (Oxoid) for *Enterobacteriaceae* followed by incubation at 37°C for 24 hours.

Results and Discussion

Table 1: Incidence of *Staphylococcus aureus* and *Salmonella* spp on fresh beef carcasses from high throughput abattoirs with (abattoir A) and without (abattoir B) food safety systems (n=20) after chilling (ND= not detected).

Abattoir	Visit	Number of samples positive (n=10)	*Staphylococcus aureus* (log cfu/g)	*Salmonella* spp.
A (HACCP)	1	6	2.09	ND
	2	7	1.11	ND
B	1	0	0	ND
	2	3	0.32	ND

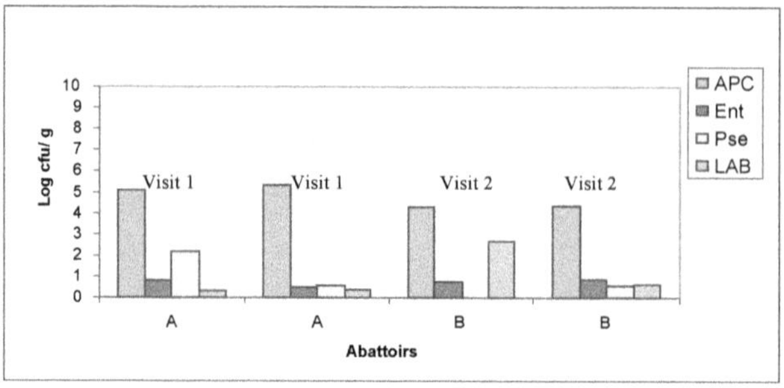

Figure 1: Incidence of target bacteria on fresh beef carcasses from high throughput abattoirs with (abattoir A) and without (abattoir B) food safety systems (n=20) (APC=Aerobic Counts, Ent = Enterobacteriaceae, Pse = *Pseudomonas* spp. and LAB = Lactic Acid Bacteria).

Table 2: Incidence of *E. coli* serotypes isolated during processing of fresh beef carcasses in high throughput abattoirs with or without food safety systems.

Abattoir	Processing step	*E. coli* O157:H7	*E. coli* serotypes
A (HACCP)	AE	+ (1/5)*	RPN
	AF	+ (1/5)*	RPN
	AC	-	RPN
B	AE	-	RPN
	AF	-	RPN
	AC	-	RPN

AE= After Evisceration; AF = After the final wash; AC = After Chilling; RPN = Rough non pathogenic.
*samples positive for *E. coli* O157:H7.

The Aerobic Plate Counts from abattoir A were high compared to abattoir B. A difference was observed in the Enterobacteriaceae counts from abattoir A compared to abattoir B (Figure 1), an indication that the Hygiene Management Systems used at abattoir B are effective in controlling carcass contamination during slaughtering. *Staphylococcus aureus* was present in 65 % of the samples collected from abattoir A compared to 30 % presence from abattoir B. Since abattoir A has an auditable food safety system, the high *Staphylocouss aureus* incidence indicates that the food safety system is not managed optimally. Twenty percent of samples at abattoir A tested positive for *E. coli* O157:H7 compared to none at abattoir B and higher APC and *Pseudomonas* spp. counts were recorded at abattoir A compared to B. Some researchers have suggested that *E. coli* O157:H7 has always been present in the abattoirs but is probably out competed for survival by other organisms and as such has not been able to grow sufficiently to cause foodborne outbreaks. It was further noted that in the US there has been more food borne outbreaks implicating *E. coli* O157:H7 since the inception of the Pathogen Reduction Programme. This programme involves the use of decontaminants which are not selective but may also eradicate the protective organisms. In South Africa only chlorinated water is used in the final wash. *Salmonella* was not detected in all the samples. In the FSIS's National beef microbiological baselines data collection programme (FSIS, 1994) the mean number of *Salmonella* was recovered from 1 % of 2,089 carcasses. In a study carried by the provincial department of agriculture in Northern Cape (not published), *E. coli* O157:H7 was isolated mostly from sheep and pork carcasses compared to beef carcasses.

Conclusion

The effectiveness of the food safety system at abattoir A for controlling microbial contamination of carcasses does not seem to be optimal.

References

Food Safety Inspection Services (1994). National beef microbiological baseline data collection programme:Steers and heifers. October 1992-September 1993. Document dated January 1994 compiled by Microbiological Division, FSIS, United States Department of Agriculture.

Herbert, R. A. (1990). Methods for enumerating micro-organisms and determining biomass in natural environments. In: Grigorova, R. & Norris, J. R. editors, Methods of microbiology, Academic Press, New York. Pp 1-39.

Martley, F. G., Jayashankar, S. & Lawrence,R. C. (1970). An improved agar medium for the detection of proteolytic organisms in total bacterial counts. *Journal of Applied Bacteriology,* 33, 129-133.

Modisane, B. M. (1999). An evaluation of the extent of which export status red meat abattoirs in the North West Province of South Africa comply with the international requirements for the hygienic production of good quality meat. Master of Business Administration Thesis. Buckinghamshire Chilterns University College. Pp 34.

DISSEMINATION OF *SALMONELLA* SPP. IN A LARGE POULTRY SLAUGHTER PLANT

L. Briceño,[1] C. Narváez-Bravo,*[2] A. Rodas-González,[2] T.E. Wittum,[3] and A.E. Hoet[3]

[1]*Facultad Experimental de Ciencias, Universidad del Zulia, Maracaibo, Venezuela.* [2]*Facultad de Ciencias Veterinarias, Universidad del Zulia, Maracaibo, Venezuela,* [3]*Department of Veterinary Preventive Medicine, The Ohio State University, Columbus, Ohio, 43210, USA. Email: claudianarvaez519@yahoo.es*

Keywords: *Salmonella*, poultry, processing, environment, antibiotics resistant

Introduction

Poultry and its derivated products are considered one of the most important food sources to act as a vehicle for foodborne *Salmonella* (Bean and Griffin, 1990). Therefore, the control of this microorganism in poultry has been the focus of numerous reports worldwide. Several studies have shown that live birds can introduce this pathogen into the slaughter plant and be an important source of contamination of the final product (Fuzihara *et al.,* 2000; Boscan *et al.,* 2005). Additionally, the number of *Salmonella* strains with resistant to antibiotics has increased (Threlfall, 1992) in human and farm animals, becoming an important public health concern. In Venezuela, there is little information on how *Salmonella* spread throughout the slaughter and packing process and their antimicrobial susceptibility. The main objective of this research was to study the spread of *Salmonella* spp. in a large poultry slaughter plant by sampling carcasses at different stages during the processing, as well as edible by-products and the environment. Additionally, tests of antimicrobial susceptibility of the strains isolated were conducted.

Materials and Methods

Location and samples: The present study was performed in a large poultry slaughter plant located in Zulia State, Venezuela. Eighty five (n=85) birds approximately 6 week-old were randomly selected at their arrival to the plant to be tagged and followed through of the slaughter process. A total of 332 carcass samples were colleted in four different phases: carcasses after defeathering (DEF) (n=85), evisceration (EVIS) (n=85), chilling (CHI)(n=81), and final packed product (FPP)(n=81). Intestines (duodenal and colon) and internal organs (livers and spleens) of each selected bird (n=85) were also collected. Additionally, thirty five (n=35) samples from edible by-products such giblet pack (neck, n=9; liver, n=9; gizzard, n=9) and legs, n=8; as well as 103 environmental samples such as water, ice, and diverse surfaces were also collected.
Sampling procedure: Intestine and organs: These were collected by the researchers immediately after hocking and opening the carcasses using sterile gloves and placing the sample into prelabed sterile bags. Carcasses and packed product: Immediately after DEF, EVIS and CHI, the tagged carcasses were removed from the production line to be sampled. Each individual carcass was introduced in a sterile plastic bag containing 225 ml of Peptone Water. Once the bag was properly sealed, it was vigorously shaken and manually massaged for 1 minute. Then, the carcass was placed again in the production line. A similar process was performed with the FPP. Edible by-products: Necks, livers and gizzards from other birds that were being processed at the same time with the tagged birds, were pooled and collected in individual sterile bags. Eight legs from the storage room (during the process) were chosen and pooled into sterile plastic bags containing 225 ml of Peptone Water. Environmental samples: Surface samples from conveyor belts in the evisceration and packing room, chiller's slide (2 mt^2), hooks (total surface) from the evisceration and packing room, and baskets used in the packing process (50 cm^2) were collected throughout the slaughter process. Sterile swabs were used for sampling hooks and baskets, and for larger surfaces, such as the conveyor belts and chiller, sterile gauze pads were used. Water and ice samples: Using sterile flasks, 100 ml of potable water (used during processing), water from scalding/washing area, water from the chiller, and ice (commercially and in-plant produced for the chiller) were individually collected.
Microbiological analysis: Bacteriological methods recommended by the Food and Drug Administration were used (FDA, 2003) for *Salmonella* detection in the peptone water from carcasses and legs rinses, blending from edible subproducts and the environment. For the detection of *Salmonella* spp in the intestine and internal organs (liver and spleen) the bacteriological methods described by Animal Disease Diagnostic Laboratory, Ohio Department of Agriculture (2000) were used. For antimicrobial susceptibility testing, agar diffusion methods of Kirby and Bauer were used (Bauer *et al.,* 1996). Statistical analysis: Chi-squared analysis was used to test differences among frequencies for each slaughter process, edible by-products, environmental samples, water, ice samples and antimicrobial susceptibilities (SAS, 1996).

Results and Discussion

Forty five percent (38/85) of the intestine and internal organs sampled were positive to *Salmonella* spp. Therefore, almost half of the selected birds were already positive to *Salmonella* at their arrival to the plant. When carcass samples from the different phases of the process were analysed, 50.6% (168/332) were positive and distributed as follows: DEF 46/85 (54.1%), EVIS 37/85 (43.5%), CHI 43/81 (53%), and in FPP 42/81 (51.9%). No significant differences were observed (P=0.3). Further analysis revealed that from the initial 46 carcasses with positive surface contamination after the scalding and defeathering, 61% (28/46) were still positive at packing. Indeed, 43.6% (17/39) of the initial 39 carcasses negative after scalding and defeathering were contaminated at packing. This could indicate that the carcasses that initiate the slaughter process with surface contamination will be 2.01 (OR) times more likely to be contaminated at the end of the process (packed product). However, because the sample size was small, it was not possible to find a significant difference between such values (P =

0.11). When analyzing the possibility that infected birds (positive intestinal content for *Salmonella* at arrival) will be more likely to end as a contaminated packaged product, it was observed that from the 38 positive birds at arrival, 55% (21/38) of them ended up positive at packing, compared to 44% (21/47) of those birds that were negative at arrival but ended up positive at packing. This result may also suggest that infected birds are more likely (OR 1.53) to end as a contaminated product than negative birds at arrival; however, the difference between groups was not significant (P =0.33). Thirty four percent (n=12) of the 35 samples from edible subproducts (giblet pack and legs), were positive for *Salmonella*, and were distributed as follows: neck 4/9 (44.4%), liver 2/9 (22.2%), gizzard 3/9 (33.3%), and legs 3/8 (37.5%). No significant differences were observed (P= 0.8) among this group of samples. These results point out the widespread presence of *Salmonella* in the different edible by-products. With regard to the environmental samples, from the 103 collected samples, 15 (14.6%) were positive and distributed as follows: conveyor belt in the evisceration room, 1/9 (11.1%); conveyor belt in the packing room, 1/8 (12.5%); Chiller's slide 1/10 (10%); hooks at the evisceration room, 1/9 (11.1%); hooks at the packing room, 2/8 (25%); baskets for packing (beginning), 2/9 (22.2%); baskets for packing (end), 4/9 (44.4%); scalding water , 2/8 (25%); chiller's water, 0/9 (0%); potable water, 0/9 (0%); ice (commercial source) for chiller, 0/6 (0%); ice (in-plant source) for chiller, 1/9 (11.1%). The results highlight the presence of *Salmonella* in the environment, which could be an important source of contamination. With reference to antimicrobial susceptibility tests, the following results were obtained: 7.4% resistance to chloramphenicol (11/148), 10.8% to ciprofloxacin (16/148), 54.1% to neomycin (80/148), 84.5% to nalidixic acid (125/148), 62.2% to nitrofurantoin (92/148), 73% to tetracycline (108/148) and 53.4% to trimethoprim (79/148).

Conclusions

The high initial prevalence of infected poultry, the occurence of environmental contamination, and the possibility of cross contamination of the final product with *Salmonella* positive giblet packs are major factors that could explain the high number of contaminated packaged product, which is a major public health concern. Although it was expected to observe an increase in carcasses positive for *Salmonella* as they went through the process, especially in the last step (packing), in this study such an increase was not observed and the total prevalence of contaminated carcasses was steady through the four phases analysed. A possible explanation was the use of chloride (at least 20 ppm) in the chiller water, which could decrease the dissemination of *Salmonella*. Nevertheless, active dissemination and cross contamination of *Salmonella* did occur throughout the slaughter process, because some animals or carcasses that were negative at the beginning of the process were positive at the end of it. In this study it was not possible to show the association between infected poultry (or poultry with surface contamination) at arrival and the increase in the likelihood that the final product from those animals will be positive for *Salmonella*. It was, however, possible to observe a trend indicating that this could be a major risk factor; and therefore, possibly an important point of intervention. Additionally, a high level of resistance was found against antimicrobials of common use in humans, which also represents an important public health concern.

References

Bauer, A., Kirby, V., Sherris. J. and Turck. M. (1996). Antibiotic susceptibility testing by a single disk method. Am, J. Clin. Pathol. 45, 493-496.

Bean, N.H. and Griffin, P.M. (1990). Food -borne diseases outbreaks in the United States, 1973-1987: Pathogens and trends, J. Food Prot. 53, 804.

Boscán, D. Arzalluz, A.M. Ugarte, C.I. Sánchez, D. Diaz, D. Wittum, T.E and Hoet, A.E. (2005). Isolation of Salmonellas with zoonotic importance in viscera from broiler chickens in Zulia stste, Venezuela. Rev. Cientifica Facultad de Ciencias Veterinarias. 15 (6), 576-582.

Food and Drugs Administration. Bacteriologycal Analytical Manual. (2001). 8th Edition. http://www.cfsanfda.gov/~ebam/bam-toc.htlm

Fuzihara, T.O. Fernandes, S.A. and Franco, B.D.G. (2000). Prevalence and dissemination of *Salmonella* serotypes along the slaughtering process in Brazilian samall poultry salughterhouses. J. Food. Prot. 63 (12), 1749-1753.

Ohio Department of Agriculture, Animal Diseases Diagnostic Laboratory (ODA-ADDL). Microbiology Methods Manual.(2000). Reynoldsburg, OH, USA. 30-35pp.

Statistic Analysis System (S.A.S.). (1996). User's Guide. Institute Inc. Cary, NC.

Threlfall, E. J. (1992). Antibiotics and the selection of food-borne pathogens. J. Appl. Bacteriol. Symp. Suppl. 73, 96S-102S.

PRESENCE OF *SALMONELLA* SPP. IN PIG CARCASSES DURING SLAUGHTER: DIRECT OR CROSS CONTAMINATION?

M. Vieira-Pinto[1,*], R. Tenreiro[2] and C. Martins[1]

[1]*Universidade de Trás-os-Montes e Alto Douro. Departamento de Ciências Veterinárias. CECAV. Vila Real, Portugal.*
[2]*Universidade de Lisboa, Faculdade de Ciências, Centro de Genética e Biologia Molecular and Instituto de Ciência Aplicada e Tecnologia, Lisboa, Portugal. E-mail: mmvpinto@utad.pt*

Keywords: *Salmonella,* PFGE, carcass, contamination, pork, slaughterhouse

Introduction

Salmonella spp. infection is one of the main causes of human gastroenteritis worldwide and pork is an important source of contamination. At the slaughterhouse swine can carry *Salmonella* spp. in several tissues, which can be a source of contamination of closely related pigs in the slaughter process, and may be responsible for carcass contamination. The tissues mostly affected by *Salmonella* are those from the digestive tract and the corresponding lymphatic tissue such as tonsils and mesenteric- and mandibular lymph nodes (Berends *et al.*, 1996). Molecular typing methods are useful for defining clonal relationships between strains and provide more epidemiological information about the nature of contamination found at the slaughterhouses. Macrorestriction profiling by pulsed-field gel electrophoresis analysis (RFLP-PFGE) is now recognized to be one of the most reliable and discriminating tools for epidemiological typing of *Salmonella* isolates belonging to various serovars (Giovannacci *et al.*, 2001). This research focuses on the use of numerical analysis of macrorestriction profiles resolved by PFGE to identify clonal relationships between strains of the same *Salmonella* serotype, isolated from the carcass and from pig samples (ileum, tonsils, mandibular and ileocolic lymph nodes) in order to define the source of carcass contamination (direct or cross contamination).

Materials and Methods

From June 2003 to September 2004, several visits were made to an abattoir in the North of Portugal. In each visit a randomly selected group of pigs were sampled during the slaughter procedure. A total of 101 pigs were sampled. For each pig, samples of the: ileum (25g), ileocolic lymph nodes (25g), mandibular lymph nodes (10g) and tonsils (10g) were collected. In the corresponding half carcass an internal surface swab was performed, with a cotton sterilised gauze (hydrated in 25 ml of Buffered Peptone Water with 0.1% Tween).

At the laboratory all samples, except the carcass swab, were submerged in boiling water for 10 seconds to decontaminate the surface. Afterwards culture method was performed according to ISO 6579:2002 typical and suspicious *Salmonella* isolates were serotyped according to the Kauffmann-White scheme in the Laboratório Nacional de Investigação Veterinária (Lisbon, Portugal), the Portuguese Reference Laboratory for *Salmonella.*

Two main groups of samples were considered: the pig samples (ileum, ileocolic lymph nodes, tonsils and mandibular lymph nodes), which reflects the pig status as an asymptomatic carrier of *Salmonella* and the carcass, where the presence of *Salmonella* can be associated to pig infection and/or to carcass contamination. To perform RFLP-PFGE genotyping, BHI broth was inoculated with several typical colonies of *Salmonella* and incubated overnight at 37°C. Cells immobilised in agarose plugs were obtained from a bacterial suspension with $5x10^9$ cells/ml, according to the method previously described by Smith *et al.*, (1988). DNA fragments were separated in 1% agarose in 0.5X TBE (50 mM Tris, pH 8; 50 mM boric acid; 0.02 mM EDTA). Pulsed-field gel electrophoresis of the DNA fragments was performed on the Gene Navigator system, with 2,5 L of 0.5X TBE buffer for 22 hours at 200 V and 14°C, with a linear increase in switching time from 10.3 to 64.0 seconds, as described before by Bender *et al.*, (2001). After PFGE, the gel was stained with ethidium bromide (0.2 µg/ml), photographed under UV transilumination and the image digitalisation was processed with KODAK 1D 2.0 software. Macrorestriction profiles (MRPs) were analysed visually and each band position was carefully determined. Profiles of different strains were compared over several gels to ensure the reliability of the band matching process. The percentage between positive matched samples was estimated, as well as the association between two different positive samples using the Fisher-exact test.

Results and Discussion

Internal surface contamination with *Salmonella* spp. was detected on 13 (12.9%) carcasses. Table 1 summarises the information related to the positive carcasses: date of sample collection, pig identification, *Salmonella* serotype and genotype identified in the positive carcasses and in the positive pig samples related to carcass contamination.

Table 1: Positive carcass information: date of sample collection, pig identification, serotype and genotype identification.

Date of collection	Pig	*Salmonella* serotype	*Salmonella* genotype	
			Carcass	Pig samples
11/06/2003	22	Rissen	R_1	
	24	Rissen		R_1
2/07/2003	40	Typhimurium	Ty_1	Ty_1
	41	Typhimurium	Ty_1	Ty_1
	42	Typhimurium		Ty_1
	43	Typhimurium	Ty_1	Ty_1
	44	Typhimurium	Ty_1	Ty_1
	45	Typhimurium	Ty_1	Ty_1
	46	Typhimurium		Ty_1
	47	Typhimurium	Ty_1	Ty_1
12/07/2004	75	Rissen	R_2	
	76	Rissen	R_1	
	77	Rissen	R_3	R_1
19/07/2004	81	Rissen		R_1
	83	Rissen	R_1	R_1
13/09/2004	94	Rissen	R_1	
	95	Rissen	R_1	R_1

Among the positive carcasses, 8 (69%) belonged to *Salmonella*-positive pigs where the same genotype was identified in, at least, one corresponding pig sample, suggesting that direct contamination or internal infection had occurred (highly significant association; Fi = 0.356; $p < 0.001$). *Salmonella* spp. isolated in the remaining 5 (31%) positive carcasses was related to cross-contamination processes. From these, 3 presented the same genotype as the one identified in pig samples from animals slaughtered in the same day. These results emphasise the importance of the slaughter of *Salmonella*-infected pigs as an important risk factor for carcass contamination and, consequently, to the introduction of this foodborne pathogen into the human food chain.
In addition, it was observed that R_1 genotype was identified in several *Salmonella* Rissen isolated at different dates along the study period from pigs that belonged to different farms and, in some cases, that were transported in different vehicles (data not shown). For these reasons, it was hypothesised to consider it a resident strain in the slaughterhouse lairage, the only possible common source of *Salmonella* contamination for the pigs. With respect to the other MRPs (R_2 and R_3) identified in two *Salmonella* Rissen isolates, it was considered that they could represent clonal descendants from the resident strain (R_1) since these isolates were identified one year after the identification of R_1 and a high similarity among them was observed (data not shown). The lairage was identified as an important source of pig infection, that could occur 2-3 hours prior to slaughter, therefore more attention should be paid to the hygienic management of the lairage area and encourage the implementation of corrective measures.

Conclusions

The study indicates the usefulness of RFLP-PFGE to define, with more accuracy the possible dynamic routes of carcass contamination with *Salmonella* spp. and its relationship to the slaughter of positive pigs, that, according to the results, may be related to the pigs infection in lairage prior to slaughter. This study indicates that corrective measures could be more efficiently implemented in order to reduce pig infection and consequently carcass contamination.

References

Bender, J. B., Hedberg, C. W., Boxrud, D. J. and Besser, J. M. (2001). Use of molecular subtyping in surveillance for *Salmonella enterica* serotype Typhimurium. *The New England Journal of Medicine*, 344(3): 189-195.

Berends, B. R., Urlings, H., Snijders, J. M. A. and Van Knapen, F. (1996). Identification and quantification of risk factors in animals management and transport regarding *Salmonella* in pigs. *International Journal of Food Microbiology*, 30: 37-53.

Giovannacci, I., Queguiner, S., Ragimbeau, C., Salvat, G., Vendeuvre, J. L., Carlier, V. and Ermel, G. (2001). Tracing of *Salmonella* spp. In two porks slaughter and cutting plants using serotyping and macrorestriction genotyping. *Journal of Applied Microbiology*, 90: 131-147.

Smith, C.L., Klco, S.R. and Cantor, C.R., (1988). Pulsed-field gel electrophoresis and the technology of large DNA molecules. In: Davies, K.E. (Ed.), Genome Analysis: a Practical Approach. IRL Press, Oxford, pp. 41-72.

STEAM VACUUM VERSUS KNIFE TRIMMING FOR BEEF SLAUGHTER

B. Steenberg*, B. Dalsgaard, J.P. Teilmann, and H. Christensen

Danish Meat Association, Danish Meat Research Institute, DK-4000 Roskilde, Denmark.
Email: bsb@danishmeat.dk

Keywords: beef, decontamination, steam vacuuming

Introduction

Cattle arriving for slaughter can be very dirty, depending on the season and the stabling conditions both on the farm and at the meat plant. Traditionally, cattle are not cleaned before dehiding. This may cause a risk of pathogenic micro organisms spreading to the carcass during slaughter, unless careful slaughter processes are applied. Another source of contamination of the carcass with pathogenic bacteria is the faecal contamination that can occur in connection with evisceration. In the USA, steam vacuum is frequently used at several points along the beef slaughter lines. However in EU member states the method needs national approval if it is to be used to remove contamination before the veterinary inspection. This study was designed to document whether steam vacuuming is more effective in removing dirt and bacteria from the surface of a carcass than knife trimming.

Materials and Methods

The Danish Meat Research Institute (DMRI) has developed a light (approx 300 g) steam vacuum handle (now marketed by SFK Meat Systems, Denmark). The admission of steam is controlled through the handle of the equipment in order to prevent the operator from burning himself.

The temperature at the head of the steam vacuum handle was measured four times a day over six testing days.

To evaluate the risk of cross-contamination via the steam vacuum handle, some swabs of the head of the handle were taken 9-10 times a day over six testing days. The swabbing area was the area of the head of the handle that comes into contact with the meat surface, and 1 cm on the inside and outside of the head, about 20 cm^2 in all. The swabbing was done just after closing of vacuum and steam. All swabbings were done on days on which the steam vacuum equipment was used continuously at normal slaughtering speed (about 45 head/hour), so the swabs could represent the microbiological load under realistic conditions.

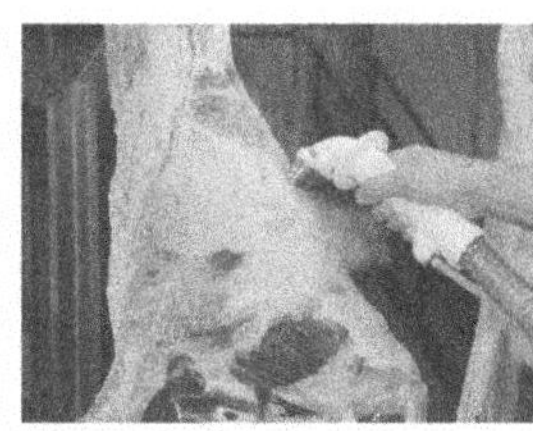

Figure 1: *Steam vacuum equipment in function.*

To evaluate the effect of the equipment on visual contamination, 19 carcasses with visual contamination were steam vacuumed and subsequently inspected by the local veterinary inspection authorities. Furthermore 35 carcasses were steam vacuumed on the outside round on one half of the carcass and trimmed with knife on the other half, and the carcasses were afterwards inspected by the local veterinary inspection authorities just before chilling.

The documentation of the microbiological effect of using the steam vacuum system took place over a period of six slaughter days under normal production conditions in a representative Nordic meat plant with a slaughter line speed of 40-50 head/hour. During the testing, there were two operators on a platform just after the splitting of the carcass. On one half the outside round was trimmed with a knife (operator 1), and on the other half the outside round was steam vacuumed (operator 2). Steam vacuuming was evaluatedon the outside round for 10 sec. and 30 sec. because these periods of time represent two scenarios: 1) 30 sec. is the treatment time it is possible to use at the current slaughtering speed if the operator only steam vacuums all of the outside round on both halves of the carcass, and 2) 10 sec. is a reasonable treatment period if steam vacuuming is only a part of the work the operator does.

Swabs were taken as listed in Table 1.

Table 1 Trial design.

Area	Treatment	Swabbing area	No. of samples/day	No. of repetitions	Total no. of samples
Outside round	Before trimming	100 cm^2 in the treated area	30	6	180
	With knife		30	3	90
	10 sec steam vac.		30	3	90
	30 sec steam vac.		30	3	90

The swabs 'before trimming' were taken just before the splitting of the carcass. The other swabs were taken just prior to chilling.

Before swabbing, the sterile gauze swabs were moistened with 0.85% NaCl buffered peptone water. Each swab was suspended in 25 ml of 0.85% NaCl buffered peptone water, stomached for 1 min. and then analysed for *E.coli* and

Aerobic Plate Count (APC). *E. coli* was obtained on Petri-film ™ EC, incubated at 37°C for 48 hours. APC was obtained in Plate Count Agar, incubated at 30°C for 3 days (NMKL no. 86, 3rd edition, 1999).
All counts were log transformed. The statistical analysis was performed with Proc. GLM (SAS Institute). Means were compared using Duncan's multiple range test (level of significance 0.01).

Results and Discussion

The temperature of the head of the steam vacuum handle was measured at 8 points on the outside of the head (Figure 2). The temperature was above 68.5°C in all the tests. At this temperature, pathogenic bacteria on the head of the equipment are killed instantly.
Fifty-eight swabs were taken from the head of the steam vacuum handle. There were no samples positive for *E. coli*, which indicates that the risk of cross-contamination with pathogenic bacteria such as *Salmonella* and *Campylobacter* is very small.

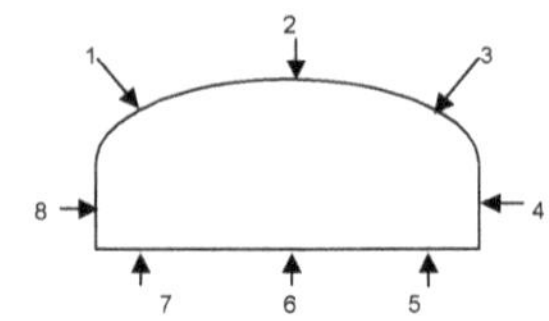

Figure 2:*Temperature measurement points at steam vacuum head.*

The evaluation of the effect of the equipment on visual contamination showed that steam vacuuming removed dirt and hair on fairly smooth surfaces, but colour traces from the contamination were not fully removed. If the contamination was located on cut surfaces, it was not possible to remove all of the contamination with steam vacuum without the meat getting a slightly 'cooked' appearance.
The microbiological analysis showed that there was a significant difference in the APC between knifetrimming and steam vacuuming (Figure 3). Figure 4 shows the percentage of samples that were positive for *E. coli*. Only a few samples were positive - and those samples had very low counts - making it impossible to estimate average counts. However, there was a tendency towards a reduction in *E. coli* after steam vacuuming of the outside round.

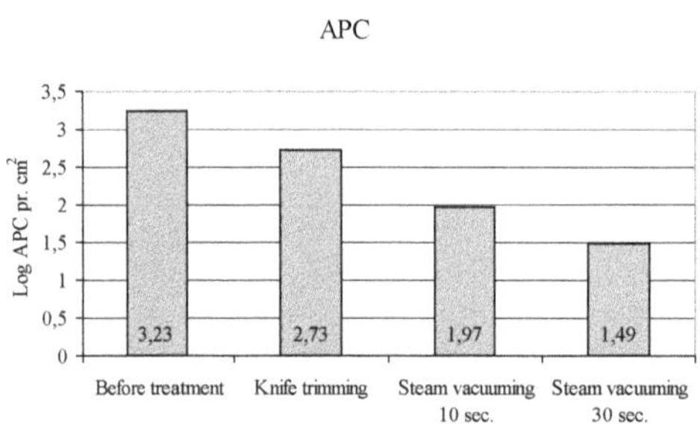

Figure 3: The average level of aerobic count (per cm²).

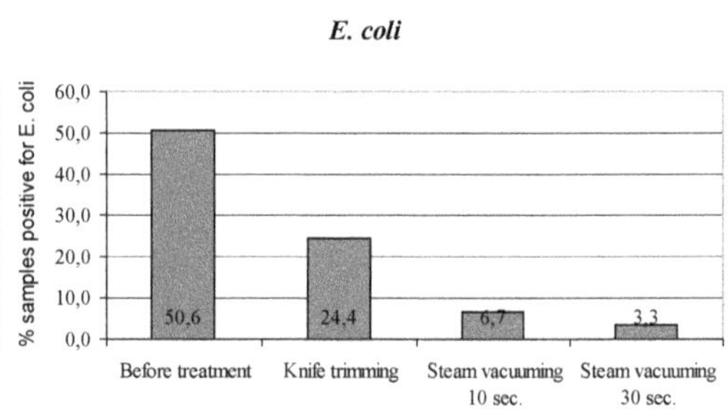

Figure 4: % samples positive for E. coli

There was also a statistically significant difference in the APC between steam vacuuming for 10 sec. and 30 sec. It was found that the longer the treatment time, the larger the reduction of the bacterial count.

Conclusions

The steam vacuum system developed by the Danish Meat Research Institute (DMRI) worked very well during the tests. High temperatures and even temperature distribution all over the head of the steam vacuum handle were recorded on all testing days. No swabs from the head of the handle were positive for *E. coli* so the risk of cross-contamination between carcasses with pathogenic bacteria is minimal. Steam vacuuming removed visual contamination effectively, and steam vacuuming gave a statistically significantly better microbiological result than knife trimming. The steam vacuum equipment has a large potential to increase product safety on the slaughter lines, but at the same time it should be pointed out that the equipment should only be used as a supplement to maintain a high level of slaughter hygiene. On the basis of this study, the Danish Veterinary and Food Administration allowed the method to be use to remove contamination on carcasses in Danish cattle abattoirs.

References

Dalsgaard, B., Teilmann, J. Christensen, H., (2003). Steam Vacuum System. Test and documentation of steam vacuuming at the slaughter line at a Danish cattle abattoir. Danish Meat Research Institute. Document no. 12912.
Danish Veterinary and Food Administration: Permission to use the Steam Vacuum system at cattle abattoirs. 17. November (2003). Journal no.2002-20-29-00202.
Steenberg B., Tørngren, M.A, Madsen, N.T, (2005). Efficiency of several decontamination hurdles on hygienic quality of beef carcasses. 51st International Congress of Meat Science and Technology. August 7-12, 2005 – Baltimore, Maryland USA

THE PRODUCTION AND MICROBIAL QUALITY OF SKIN-ON SHEEP CARCASSES

A.V. Fisher*[1], C.A. Wilkin[1], G. Purnell[2] and M. Howell[3]

[1] Division of Farm Animal Science, School of Veterinary Science, University of Bristol, Langford, Somerset BS40 5DU, UK, [2] Food Refrigeration and Process Engineering Research Centre, Faculty of Engineering, University of Bristol, Langford, Somerset, BS40 5DU, UK, [3] Meat Hygiene and Veterinary Division, Food Standards Agency, Aviation House, 125 Kingsway, London WC2B 6NH, UK, Email: alan.fisher@bristol.ac.uk

Keywords: sheep, skin-on carcass, singeing, microbiology

Introduction

In some regions of the world, such as parts of West Africa, it is common practice to place a carcass of a food animal directly on, or over, a fire to burn off the hair and singe the skin. In addition to removing unwanted hair, this burning imparts a smoked flavour to the meat and browns the surface, these being regarded as desirable changes. Demand for these skin-on, singed products by several ethnic groups resident in the United Kingdom is evident and may well occur in other developed countries. However, current EU legislation requires the skinning of ruminant carcasses and this conflict between demand and legality has resulted in a black market for these products, produced primarily from sheep. There are concerns that the illegally produced carcasses may pose a risk to human health for a variety of reasons and an approved procedure for the production of skin-on sheep carcasses that had acceptable microbiological status would be beneficial to the sheep industry and would be welcomed by specific consumer groups.

The fleece or skin is a primary vehicle for the introduction of contamination to the slaughterhouse (Koohmaraie *et al.*, 2005) but there is little information on the number of microorganisms remaining on the skin of a sheep carcass after singeing. This study aimed to devise a sequence of procedures (protocol) which resulted in the best carcass microbiological quality and which achieved a desired product, evaluated using assessors with prior experience of skin-on meat.

Materials and Methods

The animals used were purebred and crossbred Shetland ewes, from one farm. Coat colour ranged from all white through white/grey to black. The sheep were shorn within a week of slaughter so that the wool length was approximately 5 mm. They were conventionally stunned electrically and slaughtered. The hind feet were removed at the tarsus prior to further processing.

In order to achieve consistent singeing, purpose-built, automated equipment was constructed. This consisted of eight inwardly directed gas burners attached to a supporting microswitch-controlled octagonal ring that was chain driven and moved up and down around a suspended carcass. A hood was built above the rig to collect rising heat and fumes and was connected to a large displacement extractor fan using flexible ducting. Preliminary trials showed that a satisfactory carcass finish was achieved by three complete cycles of burner ring travel followed by removal of charred wool from the carcass surface with a pressure washer and water at a nominal 50^0C. This was followed by evisceration and, after completion of all dressing and inspection procedures, the carcass was returned to the singeing rig for a final single pass of the burners ('toasting'). These steps and their sequence were based on microbial skin counts and defined the protocol used to produce the skin-on carcasses. The microbiology of ten carcasses produced according to this protocol was compared with that of ten carcasses produced conventionally (skin removed) in the same abattoir on the same day. Skin samples from the rump, belly, flank, brisket, shoulder and neck, excised from the singed carcasses, were matched by samples of superficial tissues excised from the conventional carcasses and *Enterobacteriacea* and APC counts (CFU/cm^2) were made on both sets after culturing using standard EN/ISO methods.

Five Nigerian and two Ghanaian male postgraduate students, with previous handling and eating experiences of skin-on, smoked products, assessed the appearance of three skin-on carcasses from sheep with white, white/grey and brown fleeces, produced according to the standard protocol. They also assessed the aroma of boiled, skin-on meat. For each assessment, acceptability was scored using eight-point category scale ranging from extremely acceptable to extremely unacceptable and for carcass appearance there was an additional two-choice question: was it typical/atypical?

Results and Discussion

The conventional carcasses had APC counts consistently above two Logs, values being similar to those reported by Zweifel and Stephan (2003), but for skin-on carcasses all counts were less than one Log (Figure 1). Of the 60 counts (site x animal), 35 were below the detectable level on the skin-on carcasses compared with 5 on the conventionally dressed carcasses. Likewise, the conventional carcasses had higher *Enterobacteriaceae* numbers on all sites sampled, with a minimum site mean value of 0.44 Log (rump) compared with a maximum of 0.14 Log for the skin-on carcasses (neck site). The belly and brisket sites were the most heavily contaminated by *Enterobacteriaceae* in the conventionally dressed carcasses, a finding in agreement with Zweifel and Stephan (2003), but corresponding counts on the skin-on carcasses were not at detectable levels (Figure 2). In total there were 58 of the possible 60 counts of *Enterobacteriaceae* below detectable levels in the skin-on carcasses compared with 34 of 60 in the conventionally dressed carcasses.

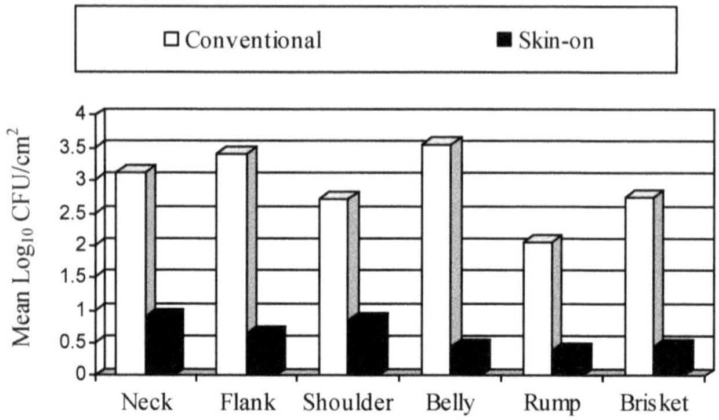

Figure 1: Comparison of APCC counts between conventionally dressed and skin-on carcasses produced according to the standard protocol on six sites on the carcass, immediately after preparation.

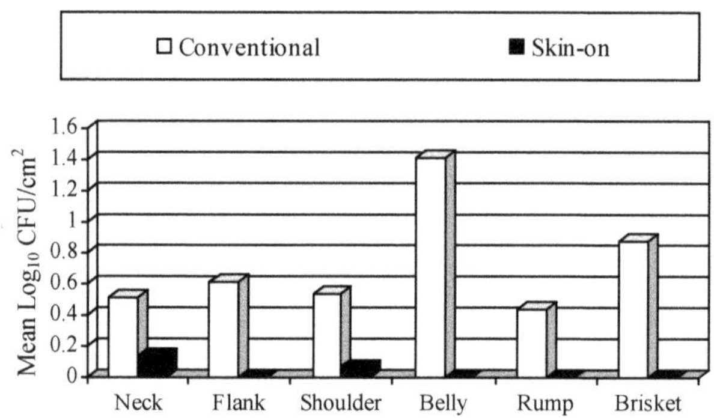

Figure 2: Comparison of *Enterobacteriaceae* counts between conventionally dressed and skin-on carcasses produced according to the standard protocol on six sites on the carcass, immediately after preparation.

The carcass from the white sheep was unanimously considered to be typical of smoked, skin-on product (hair-free with a good overall colour). Three assessors rated that carcass as being extremely acceptable and four as very acceptable. For the other two carcasses, six and four of the responses indicated a carcass in the acceptable half of the range and, overall, the production protocol adopted resulted in a desirable product. Similarly, the aroma of the meat from the carcass of the white sheep was judged to be extremely acceptable by three assessors and very acceptable by the other four.

Conclusions

Skin-on sheep carcasses can be produced, using singeing, washing and toasting interventions, to have a lower microbial loading than conventionally dressed carcasses and to meet consumer requirements. White fleeced sheep provide the visually most acceptable carcass.

References

Koohmaraie, M., Arthur, T.M., Bosilevac, J.M., Guerini, M., Shackelford, S.D. and Wheeler, T.L. (2005). Post-harvest interventions to reduce/eliminate pathogens in beef. Meat Science, 71: 79–91

Zweifel C. and Stephan, R. (2003). Microbiological monitoring of sheep carcass contamination in three Swiss abattoirs. Journal of Food Protection, 66: 946-952

REGULATORY HACCP – CARCASS AND ENVIRONMENTAL SURFACE BACTERIAL INDICATOR NUMBERS IN UK RED MEAT SLAUGHTERHOUSES

M. Howell*[1], S.M. Avery[2], C-A. Wilkin[2], D.J.I. Thomas[3], S. Buncic[2] and. M.L. Hutchison[2,3]

[1]Meat Hygiene and Veterinary Division, Food Standards Agency, Aviation House, 125 Kingsway, London, WC2B 6NH, England, [2]Department of Clinical Veterinary Science, Division of Farm Animal Science, University of Bristol, Langford, Bristol, BS40 5DU, England; [3]Hutchison Scientific Services Ltd, Units G and H, Mid Craigie Industrial Estate, Dundee, DD4 7RH, Scotland. Email: mary.howell@food.gsi.gov.uk

Keywords: carcass, slaughterhouse, environmental surfaces, microbiological testing, HACCP

Introduction

In keeping with an international trend of using a system to control hazards for meat processing, EU Commission Decision EC/471/2001 specified that all European meat slaughterhouses were to operate according to the seven HACCP principles specified by the Codex Alimentarius Commission (Anon, 1999). Verification that the slaughter and dressing processes were under control was to be undertaken in red meat plants by the weekly determination of microbial indicator numbers on carcases. The cleanliness of environmental surfaces was also assessed by periodic testing before the start of processing. Both sets of tests are undertaken in a highly standardised manner. Almost 4 years worth of standardised bacterial indicator test results have been collected directly from 89 UK slaughterhouses and this paper presents details of how numbers of these indicators have changed over that period.

Materials and Methods

Data collection and entry: A SQL server relational database was used to hold the microbiological test results that were collected. Laboratory test result certificates were collected from participating plants using a commercial courier. Data entry into the database was using a custom-coded data entry form constructed in Microsoft Access 2000 or an internet-based data entry system. These forms validated the results that were entered to ensure sensible entries were inputted (e.g. dates were verified to be in the correct format). Independent dual entry of randomly selected portions of the entered data was undertaken to help assure the quality of data entry.

Statistical analysis of test results: Subsets of data were imported directly into a statistical analyses program (StatsDirect, Cheshire, UK). For normally distributed data, t-tests, and ANOVA were used. A P value of <0.05 was used for significance.

Results and Discussion

Summary overviews of the collected data are shown in Table 1. To date, 46,852 carcass test results have been collected from 89 uniquely-licensed plants along with 86,692 environmental and food contact surface test results. Collectively these plants represent over 77% of the total national slaughtering capacity for the UK. Collections for study year 3 and 4 datasets (June 2004 to May 2006) are ongoing and thus comments on these data relate to our preliminary analyses only.

Table 1: The number of commercial slaughterhouse test results collected each year. The numbers of plants that supplied these results each year or part of the year are shown in parenthesis.

	Number test results collected (number of plants[a])			
Year	Cattle	Sheep	Pig	Surfaces
June 2002 to May 2003	6590 (42)	4597 (41)	5190 (35)	28995 (65)
June 2003 to May 2004	5714 (41)	4174 (43)	3908 (34)	30938 (56)
June 2004 to May 2005	4050 (36)	2906 (34)	3307 (23)	15468 (47)
June 2005 to April 2006	2313 (24)	1695 (21)	2408 (20)	11291 (35)
Total	18,667	13,372	14,813	86,692

[a]Multi-species plants have been counted once for each species that they process.

Percentile summaries of the test results were prepared for each year. The results are shown as Figure 1. Statistical analyses of the datasets used to generate Figure 1 showed that measured bacterial numbers recovered from cattle, sheep and pig carcasses and slaughterhouse environmental surfaces in UK plants have decreased significantly (P<0.05) over the last 4 years. In general, the largest observed decreases towards the higher end of the distribution of counts occurred in the first and the second years of the legislation being introduced. In the UK, many plants did not routinely test carcasses or surfaces prior to June 2002, and so initially, some staff may not have been adequately trained for the aseptic collection of samples. Consequently, it is possible that the initially-measured bacterial numbers were artificially-inflated by poor sampling technique. As staff improved their sampling technique, the microbiological counts fell to more accurately estimate the bacterial numbers on carcasses and surfaces.

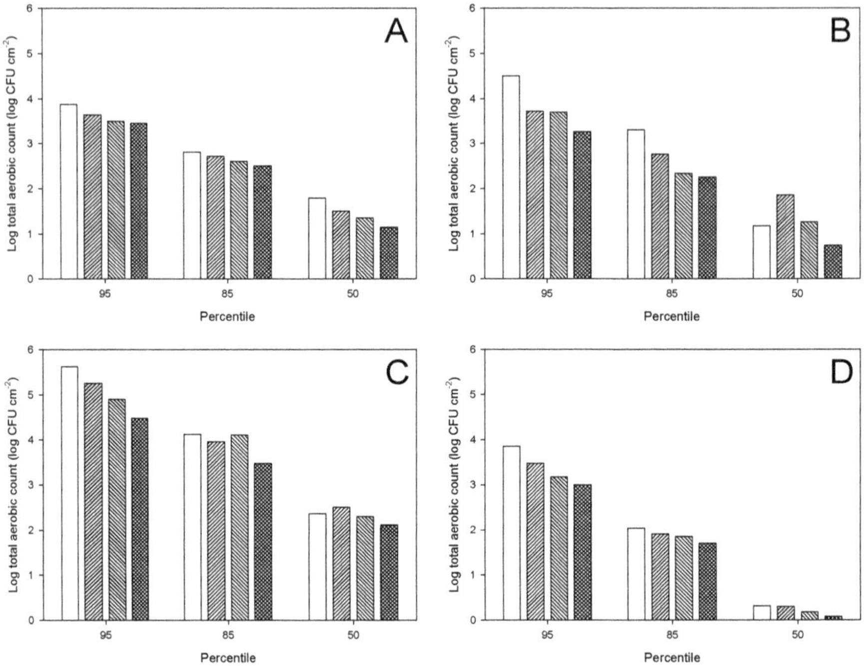

Figure 1: Percentile summaries of total aerobes on cattle (A), sheep (B), pig (C) carcasses and environmental surfaces (D) in UK slaughterhouses. Data were sorted into 12 month groups before performing calculations. Groups commenced on June 2002 (□), June 2003 (▨), June 2004 (▧), June 2005 (▩).

Balanced against the initially-rapid decreases in bacterial numbers is the influence of reduced testing frequencies. After a period of satisfactory microbiological test results plants can switch to a reduced testing frequency. Although incomplete, increased numbers of plants changing to a reduced testing schedule is one of the reasons for the diminishing size of the data sets summarised in Table 1. Since the relative contribution to the total dataset of those plants with satisfactory results decreased over the four year period that the test results were collected it is a likely reason for the less-pronounced decreasing trends in bacterial numbers between 2004 and 2006.

Conclusion

Commission Decision EC/471/2001 was a forerunner for the revision of hygiene legislation in Regulation EC 852/2004, and microbiological criteria in Regulation EC 2073/2005 covering all food of animal and non-animal origin. The observation that bacterial numbers in slaughterhouses declined after implementation of the required HACCP based procedures partly justifies the creation of these new regulations. Any reduction in the bacterial numbers associated with food is a positive thing in terms of potential reductions to foodborne illness. However, any positive affect of reduced bacterial numbers on meat as a consequence of monitoring should not be confused with the appropriateness of microbiological testing for process control purposes in red meat plants. Our previous conclusion (Hutchison *et al.*, 2005) that testing is of questionable value because of large variation in the test results remains unchanged.

Acknowledgements

The authors gratefully acknowledge the slaughterhouses who donated their microbiological test results and the support from the British Meat Processors Association. This study was jointly funded by the Reduction in Foodborne Disease Strategy and the Meat Hygiene Research programmes of the UK Food Standards Agency.

References

Anonymous. (1999) The evaluation of microbiological criteria for food products of an animal origin for human consumption. Opinion of the Scientific Committee on Veterinary Measures Relating to Public Health. Brussels: European Union.

Hutchison M. L., L. D. Walters, S. M. Avery, C. A. Reid, D. Wilson, M. Howell, A. M. Johnston and S. Buncic. (2005). A comparison of wet-dry swabbing and excision sampling methods for microbiological testing of bovine, porcine, and ovine carcasses at red meat slaughterhouses. *Journal of Food Protection,* 68:2155-2162.

PCR MOLECULAR TYPING TO TRACE *SALMONELLA* SPP. IN ALHEIRA PRODUCTION LINES

A. Esteves[1], T. Aymerich[2], M. Garriga[2], L. Patarata[1], M.C. Fontes[1] and C. Martins[1]*

[1] *Center of Studies in Animal and Veterinary Science (CECAV), University of Trás-os Montes e Alto Douro (UTAD), Apartado 1013, 5000-801, Vila Real, Portugal.* [2] *Institute for Food and Agriculture Research and Technology (IRTA), Meat Technology Center, Granja Camps i Armet. 17121, Monells, Spain.*
E-mail: cmartins@utad.pt

Keywords: *Alheira*, *Salmonella*, source, tracing, molecular PCR typing

Introduction

Alheira is a traditional Portuguese cooked meat product with specific characteristics, with respect to its raw material and its manufacturing technology. Previous studies showed that this product might contain some pathogens that could affect the safety of the product. In Esteves (2005), 96 samples of alheira from different production lines were analysed and *Salmonella* spp. was detected in 12.5% of them.

The study of the occurrence of a microorganism throughout a production line can be used to determine the main source and the spread of food product contamination (Samelis and Metaxopoulus, 1999). Several authors established that molecular typing methodologies of typing could be used to identify the major sources and routes of contamination of specific microorganisms in processing plants (Aguado *et al.*, 2001; Peccio *et al.*, 2003). The aim of the present study was to determine the source(s) and the type (s) of *Salmonella* contamination in alheira, by examining the occurrence of the microorganism throughout the production line of the product, and by PCR typing *Salmonella* isolates obtained from different locations at the production lines.

Materials and Methods

Experimental design: The production of 16 batch of Alheira was examined along the process. *Salmonella* spp. was isolated from various locations along the production line: samples of raw materials and seasoning (poultry P, pork R; bread B; casing T; casing desalting water WI; seasoning E); swabs of different contact surfaces (table I; reservoir Q; equipment S; handler's hands M); product in several stages of manufacturing (cooked meat Z; batter before the stuffing L); water from final washing (WII) and finished product (H), with a total of 14 different sampling points. For each sampling point, four independent samples were considered. 85 strains were isolated for further molecular typing.

Primer selection: A random Amplified Polymorphic DNA (RAPD) method using the primer M13R2 (5' GGA AAC AGC TAT GAT CAT GA- 3') and a Repetitive-Element PCR (rep-PCR) method using the primers ERIC1R (5'-ATG TAA GCT CCT GGG GAT TCA G-3'), ERIC 2 (5' AAG TAA GTA ACT GGG GTG AGT G- 3'), BOX1R (5' CTA CGG CAA GGC GAC GCT GAC G- 3') and BOXA2R (5' ACG TGG TTT GAA GAG ATT TTG G- 3') were evaluated for typing purposes. A preliminary study was carried out, testing all 5 primers in a sub-collection of 15 isolates of *Salmonella* spp, to choose the primer(s) which yielded the best results, in order to fulfil the objectives of the current work. The remaining *Salmonella* spp. collection was typed by the chosen PCR primers.

Salmonella spp. detection: The presence of *Salmonella* was investigated in 25 g, 25ml and 100 square cm of meat samples, water and contact surfaces respectively. Detection was automatically performed by an Enzyme Linked Fluorescent assay (MiniVIDAS-BioMérieux 99088), using the test sets and sample preparation as recommended by the manufacturer. Preenrichment was performed in a non selective medium (Buffered Peptone Water) followed by enrichment procedures in two different selective media Selenite Cystine and Rapapport-Vassiliadis (Oxoid), incubated at 35-37°C and 41-42°C for 6-8h, respectively. Presumptive *Salmonella* positive samples were streaked on plates of selective media, SM ID and Hektoen Enteric agar (BioMérieux). The confirmation/ identification of *Salmonella* was done by API 20E system (BioMérieux).Antigenic analysis by the agglutination test using polyvalent antisera (Bacto *Salmonella* O Poly A-I and Vi antisera, Difco) was also performed.

Molecular typing: DNA was extracted from pure cultures using Dneasy tissue kit (Quiagen), according to the manufacturer indications. DNA was quantified and the purity evaluated by spectophotometry using GeneQuant pro RNA/DNA Calculator (Amersham Pharmacia Biotech., 80-2109-99). PCR mixtures (25 µl) consisted of: 2.5 µl of Buffer (10x) (200mM Tris-Hcl, pH 8.4, 500m MKCl) (Invitrogen); 1.5 mM of $MgCl_2$ for the primer M13R2 and 3.5 mM for the others (Invitrogen); 200 µM dNTP (Roche Molecular Biochemical); 20 picomoles each primer (Roche); 1 U of Taq DNA polymerase (Invitrogen); 100 ng of genomic DNA. PCR reactions were performed in a Thermal Cycler (Perkin-Elmer Corp. 2700) as follows: initial denaturation at 95°C for 5 min; 40 cycles with denaturation at 95°C for 0.30 s, primer annealing for 1 min (temperatures for each primer: M13R2 38°C; ERIC1R 65°C; ERIC2- 68°C; BOXA2R 65°C; BOX1R- 68°C), and primer extension at 72°C for 1.30 min; followed by a final extension at 72°C for 5 min and cooling to 4°C. PCR products were resolved by agarose gel (1.5%) electrophoresis (80 mA, 90 minutes), in 0.5x Tris-acetate- EDTA buffer stained with 0.1 µg/ml of ethidium bromide (Sigma). The profiles obtained were observed under UV light, and the image was captured by a video camera.

Data analysis: Band profiles were analysed using Bio Profile software. Reproducibility was tested on 15% of the isolates. The similarity was estimated using Dice coefficient /UPGMA algorithm (NtsysPC 2.1). A cluster was defined with at least a group of three isolates and a limit a 95% similarity.

Results and Discussion

Salmonella spp. was detected in 4% of the samples along the production lines. Among the different incoming materials, casing and raw meat samples presented positive *Salmonella* samples. As occurrence data was not enough to identify sources to trace the pathogen in the processing plants, molecular typing methodologies such as RAPD and rep-PCR were considered for this purpose. Composite RAPD with M13R2 and ERIC 2 profiles were used to interpret the molecular typing results. Nine different clusters were obtained. In order to trace the *Salmonella* spp. contamination throughout the production line, three distinct areas were considered: area 1, reception; area 2, processing; area 3, finished product. Table 1 shows different profiles presented in each area, as well as the different origins of the typed isolates.

Table 1: Distribution of clusters in different areas of the alheira production lines.

Areas in processing plant	Isolates origin Sampling point	Clusters*
1 Reception	Casing, casing desalting water	*A, B, D, E, I*
	Pork, poultry	*A,B,C, D, E, F, G, H*
2 Processing	Table surface	*B, D*
	Hands	*B, D*
	Washing water after stuffing	*B*
3 Finished product	Alheira	*A, B, D, I*

* Clusters (A, B, C, D, E, F, G, H, I).

Clusters ***C, E, F, G*** e ***H*** included isolates exclusively from reception area - from raw meat and from casing. Clusters ***A, B, D*** and ***I*** consist of by isolates from two or more of the defined areas in production lines. These clusters included isolates from final product and from casing related samples. The occurrence of isolates from these clusters in alheira could be explained by the use of contaminated casings, as a direct vehicle for *Salmonella* spp.. The presence of isolates from cluster ***B*** and ***D***, in the processing area, respectively in table surfaces and in hands of workers suggests the possibility of cross contamination between intermediate manufacturing stages throughout contaminated surfaces. Considering the production lines and the production batches, it was possible to identify the sources and to trace the spread of the isolates from these clusters.

Conclusion

The presence of *Salmonella* spp. in alheira could be attributed to the use of contaminated casings, as a direct vehicle, or by cross contamination through contaminated table surfaces and hands of workers. The knowledge of *Salmonella* spp. contamination sources and spreading along the production line will allow us to define new strategies towards the control of *Salmonella* spp. in the alheira production lines.

References

Aguado, V., A.I.Vitasa and I. Garcia-Jalon. (2001). "Random amplified polymorphic DNA typing Apllied to the study of cross-contamination by *Listeria monocytogenes* in processed food products." *Journal of Food Protection.* 64:716-720.

Esteves, A. (2005). Perigos Microbiológicos em Alheiras. Principais vias de contaminação por *Staphylococcus aureus, Clostriudium perfringens* e *Salmonella* spp. Dissertação de Doutoramento UTAD. 257 pgs

Peccio, A., T. Autio, H. Korkeala, R. Rosmini and M. Trevisani. (2003). "*Listeria monocytogenes* occurrence and characterization in meat-producing plants." *Letters in Applied Microbiology.* 37:234-238.

Samelis, J. and J. Metaxopoulos. (1999). "Incidence and principal sources of *Listeria* spp. and *Listeria monocytogenes* contamination in processed meats and a meat processing plant." *Food Microbiology.* 16:465-477.

Acknowledgments

This work was financially supported by Agro Program (8.1) from Portuguese Agriculture Ministry.

SHIGA TOXIN-PRODUCING *ESCHERICHIA COLI* (STEC) IN DRY SAUSAGES

R. Pichner*[1], H. Hechelmann[1], H. Steinrück[2] and M. Gareis[1]

[1.] *Institute for Microbiology and Toxicology, Federal Research Centre for Nutrition and Food, E.-C.-Bauman-Str. 20, 95326 Kulmbach, Germany. *Email: Rohtraud.Pichner@bfel.de.* [2.] *Federal Institute for Risk Assessment, Diedersdorfer Weg 1, 12277 Berlin*

Keywords: STEC/EHEC, nitrite, fermented dry sausage

Introduction

Using good manufacturing technologies long fermented dry sausages do not pose a health hazard to the consumer (Kofoth, 1998). In September 2004 a recall was started for STEC-contaminated salametti products (EU-rapid alert, 2004). The recall affected plant produced mould-ripened salami products under ecological points of view (without nitrite). The aim of this study was to investigate the reasons for the recall action by analyzing products of the recall affected plant and products from retail and studying the survival of STEC in salami products with and without nitrite.

Materials and Methods

Product samples (35) from the recall-affected manufacturer and 100 salami products bought during October and December 2004 in Bavarian retail markets were analyzed for *E. coli* counts, STEC and other parameters as shown in Table 1.

Table 1: Material and methods for analysis of samples used in this study.

Materials	
Samples from recall affected plant	35 different kinds of dry sausages (mould-ripened, without nitrite)
Samples from retail markets	100 long fermented dry sausages (no products without nitrite available)
Samples from inoculation study	Sausage samples from produced batches (see table 2) at different stages of ripening (day 0, 2, 3, 6, 9, 14, 21, 28, 35, 42)
Methods	
Analysis according to German food legislation (LMBG, 1997)	pH-value
	a_W-value (only in control batch)
	Enterobacteriaceae counts (DHL-Agar [Merck, Germany], 24h, 37°C)
	E. coli counts (ECD [Merck, Germany], 24h 37°C)
	Qualitative STEC detection by PCR (Karch and Meyer, 1989)

To study the survival of *Enterobacteriaceae* and STEC in salami processed with and without nitrite, three batches of mould-ripened salami were produced using standard manufacturing technologies: batch 2 with 2.8% sodium chloride (NaCl) and batch 3 (with 2.8% nitrite curing salt) were inoculated with pool of *Enterobacteriaceae* and STEC. The nitrite curing salt was composed of 99.5–99.6% sodium chloride and 0.4-0.5% sodium nitrite. The *Enterobacteriaceae* inoculum was a pool of *Escherichia coli, Enterobacter liquefaciens* and *Proteus mirabilis*. For the STEC inoculum three strains were used, which have been isolated from products of the recall affected plant (serotypes: Ont:H42, O22:H8, O100:H-). The survival of the inocula was observed during the ripening of the sausages (see Table 2). Batch 1 was produced with 2.8% NaCl and taken as negative control.

Table 2: Materials, composition, inoculation doses and ripening programme for the inoculation study.

Composition:	Meat	Spices and additives				
	33% pork	0.2% Glucose		0.04% Nutmeg		0.03% Cardamom
	33% beef	0.3% Pepper		0.03% Ascorbate		0.03% Coriander
	30% pork lean fat	0.05% Paprica				
Starter cultures:	Biostart [Raps, Germany], Bactoferm™ (*Penicillium nalgiovense* [CHR HANSEN])					
Produced batches	Batch 1	Batch 2			Batch 3	
	+ 2.8% NaCl	+ 2.8% NaCl			+ 2.8% nitrite curing salt**	
	(control batch)	10 cfu/g STEC			10 cfu/g STEC	
		4 log_{10} cfu/g *Enterobacteriaceae*			4 log_{10} cfu/g *Enterobacteriaceae*	
Ripening programme						
Temperature (°C)	Day 0-1: 23	Day 2: 22	Day 3-5: 20	Day: 6-21: 18		Day22-42: 18
Rel. humidity (%)	Day 0: 95	Day 1-2: 93	Day 3-4: 90	Day 5-19: 88	Day 20-26: 85	Day 27-42: 80

**nitrite curing salt: composed of 99.5 – 99.6% NaCl and 0.4 – 0.5% $NaNO_2$

Results and Discussion

About 9% of the samples from the recall affected plant were found to be STEC positive. In 3% of the samples, *Enterobacteriaceae* up to 5 log_{10} cfu/g were counted. *E. coli* counts were up to 3 log $_{10}$ cfu/g. The microbiological analysis of salami products from retail markets showed counts below 10 cfu/g and no STEC in the samples. The inoculation study resulted in a survival of STEC over six weeks in sausages produced without nitrite.

Enterobacteriaceae counts decreased until the end of ripening to 100 cfu/g (see Figure 1).

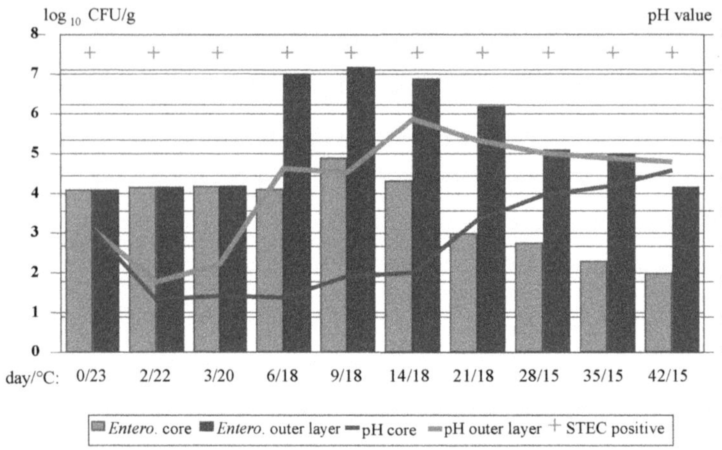

Fig. 1: Survival of STEC and *Enterobacteriaceae* in mould-ripened sausages, produced with 2.8% NaCl (composition and inoculation doses see tab. 2).

In sausages with 2.8% nitrite curing salt, STEC were detected until to the third week of ripening and *Enterobacteriaceae* counts fell under 10 cfu/g (see Figure 2).

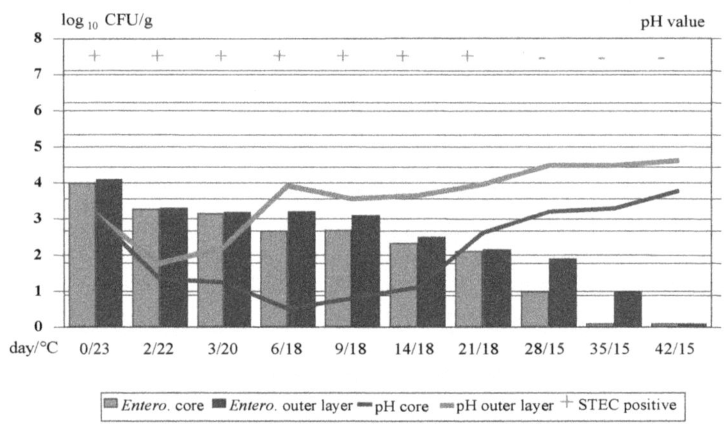

Fig. 2: Survival of STEC and *Enterobacteriaceae* in mould-ripened sausages, produced with 2.8% nitrite curing salt (composition and inoculation doses see tab. 2).

Conclusions

According to these results, the cause of the recall for STEC contaminated salametti products in 2004 in Germany was a deficient hygiene-management in the affected plant. Using good manufacturing technologies long fermented dry sausage does not bear a risk to consumer's health. The processing of salami products without nitrite should be done under strict compliance of hygiene regulations.

References

EU-rapid alert (2004): Alert notifications week 2004/41.-.

Karch, H., T. Meyer (1989): Single primer pair for amplifying segments of distinct shiga-like toxin genes by polymerase chain reaction. Journal of Clinical Micriobiology 27, 2751-2757.-.

Kofoth (1998): Überlebens- und Wachstumsfähigkeit von enterohämorrhagischen *Escherichia coli* (EHEC) in Rohwursterzeugnissen. Diss. med. vet. LMU München.-.

LMBG (1997): Amtliche Sammlung von Untersuchungsmethoden von Lebensmitteln nach § 35 des LMBG.

MICROBIAL SAFETY OF "SALSICCIA SARDA", A TYPICAL ITALIAN SEMI-DRY NATURALLY FERMENTED SAUSAGE

D. Meloni*, A. Mureddu , M.M. Colleo, R. Diana and R. Mazzette

Dipartimento di Biologia Animale, Sezione di Ispezione degli Alimenti di Origine Animale, Università degli Studi di Sassari, Via Vienna 2, 07100 Sassari, Italy. Email: melonidome@hotmail.com

Keywords: dry fermented sausage, meat processing plants, food microbiology

Introduction

Dry fermented sausages are ready-to-eat products, whose safety is essentially gained by a "hurdle technology" concept (Barbuti and Parolari, 2002). The main hurdles used during the processing of dry fermented products are nitrite and salt content, the decrease of redox potential and water activity (a_w), which inhibit many aerobic bacteria in the early stages of production (*Pseudomonas* spp. and other Gram-negative bacteria), and select the Lactic acid bacteria which cause the pH to fall. These hurdles are essential for the microbial safety and stability of quick-ripened fermented sausages, which are not fully dried. Although dry and semi-dry sausages have rarely been implicated in food poisoning, greater risk should be linked to the consumption of traditional products, because of an empirical application of hurdle technology (Leistner and Gould, 2002). "Salsiccia Sarda" is a typical Italian semi-dry sausage (a_w ranging from 0.90 to 0.95), naturally fermented, and included in the National list of the traditional products (D.M.18/07/2000). The physicochemical and microbiological profiles are unhomogeneous. The presence of spoilage bacteria and hygiene indicators, are strongly linked to the length of ripening, and at times to an incorrect acidification and drying (Mazzette *et al.*, 1998; Greco *et al.*, 2005). The aim of the present study was to evaluate the microbial safety of the various types of "Salsiccia Sarda".

Materials and Methods

Fifty-five samples from different types of ripened "Salsiccia Sarda" were collected from twenty-two processing plants, with high and low production capacity (Directive 92/5/EEC): thirty-six "*traditional*" and five "*smoked*" (horse-shoe shape, 30-35 mm∅, ripening of 10-25days), two "*cylindrical*" (30-35 mm∅, ripening of 10-15days), and twelve "*thin*" (horse-shoe shape, about 20 mm∅, ripening of less than 10 days). All the samples were analysed for the following parameters: a) *pH*, b) *aw* (Greco *et al.*, 2005), c) *microbiological parameters*, were determined using the A.P.H.A. methods (2001) or other specific protocols: Coliforms and *Escherichia coli*; *Pseudomonas* spp.; Mesophilic Anaerobic Sporeformers; Coagulase negative staphylococci (CNS); Lactic acid bacteria (LAB); Yeasts and Molds; *Clostridium perfringens*; *Salmonella* spp.; *Staphylococcus aureus*; *Listeria monocytogenes (L.m.),* (ISO 11290-1:1996 and ISO 11290-2:1998). Data were subjected to analysis of variance using the GLM procedures. The LSD test was used to evaluate differences between means (Statgraphics Plus, 5.1, Manugistics, MD, USA). Biochemical characterisation and presumptive identification were performed on all the *Listeria spp.* isolates, using the API Listeria identification system (Bio Mérieux). All the *Listeria monocytogenes* isolates were further characterised by a PCR-based method, aimed at *prfA* gene fragment detection (Jofré *et al.*, 2005) and by serotyping (Ueda *et al.*, 2002).

Results and Discussion

The pH, a_w (numeric values) and microbial ($\log_{10}$ CFU g^{-1}) results are presented in Table 1 and in the text as mean ± sd. a) pH: more than 90% of the sausages showed values ranging from 5.5 to 6, with no significant ($p>0.05$) differences between the sausage types. b) aw: the lower values were observed in the "*thin*" type sausage (0.87±0.01). More than 70% of the sausages showed values ranging from 0.90 to 0.92, while 20% ("*traditional*" type) showed values up to 0.94. c) Microbiological parameters: Coliforms were detected in 80% of samples with no significant differences in levels found between the sausage types. More than 80% of the positive samples, showed values ranging from 3 to 5 $\log_{10}$ CFU g^{-1}, and in the 15% of "*traditional*" sausages higher than 5 $\log_{10}$ CFU g^{-1}. E.coli was detected in 25.5% of samples tested. The highest contamination level ($p<0.05$) was found in samples with high pH values. Pseudomonas spp. was presence in all samples (4.58±0.87). In more than 30% of "*traditional*" and "*thin*" sausages, *Pseudomonas* spp. levels higher than 5 $\log_{10}$ CFU g^{-1} were detected. Mesophilic Anaerobic Sporeformers were detected in 10.9% of samples, and detected mainly in "*traditional*" (17.2%) and "*cylindrical*"(50%) sausage types. A significantly higher prevalence ($p<0.05$) of Mesophilic Anaerobic Sporeformers were found in sausages with a ripening period of less than 15 days (< 15days). LAB was the most prevalent microflora in all samples. Sausages with extended ripening period had the highest ($p<0.01$) levels of LAB. In 10% of "*traditional*" sausages, values higher than 8 $\log_{10}$ CFU g^{-1} LAB were found. CNS was found in more than 80% of samples, with some samples having values higher than 6 $\log_{10}$ CFU g^{-1}. Yeasts and Molds high levels of yeasts and moulds were found in all sample types. Cl. Perfringens was recovered from 14.5% of samples (all types), samples with high pH value were the most contaminated ($p<0.05$). S. aureus was detected in 14.5% of samples, and was strongly associated with lower ripening period ($p<0.05$) and high pH value ($p<0.01$). Listeria spp and L isteria monocytogenes, only twenty-one samples were analysed for *Listeria* spp. Of the 21 samples tested, *Listeria* spp. was detected in 76.1% of samples while *L. monocytogenes* was detected in 56.2% of samples. The contamination rate was <100cfu/g for all samples. Fifty-nine strains were identified as belonging to the following species: forty-two were *L. innocua*, fourteen were *L. monocytogenes* and three were *L. welshimeri*. All the *L. monocytogenes* isolates showed the same electrophoretical pattern (amplification product of 464*bp*). Serotyping revealed that fourteen *L.monocytogenes* strains belonged to the following serotypes: 1/2a (57.1%), 4b (28.5%), 3b (7.1%), 1/2c (7.1%). Thevenot *et al.*, (2005) reported a high heterogeneity in serotypes collected in sausages. This may allow non specific serotype to survive sausage processing and maturation. *Salmonella* spp. were not found in any of the samples.

Table 1: Physicochemical and microbiological parameters (mean±s.d.) of the different types of "Salsiccia Sarda" at the end of ripening.

Type Parameters	Traditional 36[1]	Smoked 5[1]	Cylindrical 2[1]	Thin 12[1]	Total 55[1]
pH	5.54± 0.37	5.46±0.29	5.50±0.41	5.69±0.35	5.56±0.35
a_w	0.90± 0.02	0.92±0.03	0.92±0.00	0.89±0.02	0.90±0.02
Coliforms	3.47± 1.28 (28)[2]	3.60±0.23 (4)[2]	3.11± 1.37	3.13± 1.26 (10)[2]	3.39 ± 1.18
E.coli	2.51± 1.23 (3)[2]	2.78 (1)[2]	2.60 (1)[2]	2.57± 1.32 (3)[2]	2.55±1.09
Pseudomonas spp	4.70±0.64	4.42 ± 0.17	4.48±0.47	4.74± 0.61	4.58±0.87
Anaerobes sporeformers	1.73±0.83	-	4.48 (1)[2]	-	2.19 ± 1.34
LAB	7.27± 0.73	7.03±0.71	7.23±1.19	7.48±1.05	7.29±0.80
CNS	6.83± 0.79	6.78±0.22	6.18±1.21	6.71±0.47	6.77±0.70
Yeasts and Molds	5.07± 0.49	5.39±0.14	5.39±0.10	5.16±0.34	5.13±0.43
Cl.perfringens	2.34±0.62(4)[2]	1.30 (1)[2]	5.0±2.46 (1)[2]	1.0 (1)[2]	2.14 ± 1.20
S.aureus	5.82±0,70 (4)[2]	4.0 (1)[2]	5.0 (1)[2]	4.17 ± 0.56 (2)[2]	5.07±0.98
L. monocytogenes	7/14[3]	-	1/1[3]	1/2[3]	9/14[3]

[1]= number of samples; [2]= mean±s.d referred to the number of positives in bracket; [3]=number of positive samples; - =below the sensibility limit of the method

Conclusions

The results indicate an incorrect management of the acidification (more than 90% of the samples showed values of pH higher than 5.5) and drying steps (the a_w value was higher than 0.90 in more than 70% of samples) during processing. These processing conditions may be ineffective in pathogen and spoilage bacterial control. Although the low contamination rate (<100cfu/g), the presence of *L. monocytogenes* in fermented sausages at the time of consumption still remain a major concern for manufacturers (Thevenot *et al.*, 2005). To increase the safety of these products, efforts should be encouraged to achieve the effectiveness of the following hurdles: a_w reduction (extended ripening) and pH reduction (Barbuti and Parolari, 2002). Effective hygiene control strategies should also be ensured to prevent cross contamination of products.

References

American Public Health Association (2001) Compendium of methods for the microbiological examination of foods (fourth edition) Edited by F. Pouch Downes, K.Ito, A.P.H.A., Washington DC, 2001.

Barbuti S. and Parolari, G. (2002) Validation of manufacturing process to control pathogenic bacteria in typical dry fermented products. *Meat Science*, 62, 323–329.

Greco M., Mazzette R., De Santis E.P.L., Corona A., Cosseddu A.M. (2005) Evolution and identification of lactic acid bacteria isolated during the ripening of Sardinian sausages. *Meat Science*, 69, 733-739.

Jofré A., Martin B., Garriga M., Hugas M., Pla M., Rodríguez-Lázaro D., Aymerich T. (2005) Simultaneous detection of *Listeria monocytogenes* and *Salmonella* by multiplex PCR in cooked ham. *Food Microbiology*, 22(1),109-11.

Leistner L. and Gould, G. (2002) Applications in industrialized countries, In: Hurdle technologies, combination treatments for food stability, safety and quality. Edited by Kluwer Academic Publisher, New York, 66-69.

Mazzette, R., De Santis, E.P.L., Greco, M., Bean, V. and Pisanu S. (1998) Critical Control Points (CCPs) in processing ripened sausages. In: Proceedings of the 52nd national congress of Italian Society of Veterinary Science (SISVet), 329-330.

Thèvenot, D., Delignette-Muller, Christieans, S., Vernozy-Rozand, C. (2005) Prevalence of *Listeria monocytogenes* in 13 dried sausage processing plants and their products. *International Journal of Food Microbiology*, 102, 85-94.

Ueda, F., Sugamata, M., Aota, M., Mochizuki, M., Yamada, F., Hondo, R. (2002) "Swift and definite serotyping for isolated *Listeria monocytogenes* strains" *Microbiologica*, 25, 165-171.

SCREENING OF BACTERIOCIN-PRODUCING BACTERIA ASSOCIATED IN TRADITIONAL THAI FERMENTED BEEF (MUM)

S. Piayura, P. Pinsirodom, Y. Surapantapisit and A. Swetwiwathana*

Department of Agro-industry, Faculty of Agricultural Technology, King Mongkut's Institute of Technology Ladkrabang (KMITL) Bangkok, 10520 Thailand. E-mail : adisorns@hotmail.com

Keywords: bacteriocin, lactic acid bacteria, fermented beef

Introduction

Lactic acid bacteria (LAB) are widely used as starter cultures for reliable and consistent acid production in various fermented foods. The inhibition of other microorganisms may also occur by the formation of various compounds, which are produced during fermentation. Among the variety of the inhibitory compounds synthesized by these LAB, bacteriocins have received much attention in the past decade. Mum is a kind of traditional thai fermented meat, which is normally made from minced beef, raw cow spleen and cow liver, cooked together with salt, garlic and cooked rice, mixed well and stuffed into cow's intestine. The product is left to ferment at room temperature for 3 days. The most important microorganisms during the spontaneous fermentation of this product belong to the LAB genera *Lactobacillus plantarum* and *Pediococcus cerevisiae* (Somathiti and Surapantapisit, 1987). Numerous reports about bacteriocin-producing LAB isolated from another Thai fermented meat products, such as Nham (Swetwiwathana, 2005), have applied these LAB as starter cultures to harm various pathogens during fermentation. The objective of the study was to isolate bacteriocin-producing LAB with potential to use as starter cultures to increase the microbiological safety of Mum. The brief characterisation of bacteriocins, such as inhibitory spectrum, identification using carbohydrate fermentation kit and microscopic identification, are also reported in this paper.

Materials and Methods

<u>Isolation of bacteriocin-producing bacteria from Mum:</u> In order to search for bacteriocinogenic bacteria, a total of 100 strains were randomly isolated from 10 samples of Mum sold in Khonkaen province by spread plate technique on MRS agar + 0.5 % calcium carbonate and incubated under micro-aerobic condition (candle jar) at 30°C for 48 h. Each colony with a clear zone was selected for the detection of antagonistic activity. All strains were precultured overnight in MRS broth (Oxoid) at 30°C before using to screen for bacteriocin production (De Man *et al.*, 1960).
<u>Bacteriocin screening medium:</u> A special bacteriocin screening medium (BSM), which was developed on the basis of MRS medium, was used as the bacteriocin screening medium for all isolates (Tichaczek *et al.*, 1992).
<u>Determination of antagonistic activity</u>: The direct method was performed as described by Fleming *et al.*, (1985). Evaluation of bacteriocin-producing strains was studied using the methods described by Tichaczek *et al.*, (1992) and Ennahar *et al.*, (1999) with 6 indicators (Table 1). Antimicrobial producers were examined after 24 hours for zones of inhibition. The most potent strains, which showed the best inhibitory spectrum to all 6 tested indicators, were selected for confirmation of their bacteriocins production in MRS broth with various indicators as described by Ennahar *et al.*, (1999) (Table 3) and compared to known bacteriocin-producer strains such as pediocin PA-1 producer strain of *P. pentosaceus* TISTR 536 (Swetwiwathana, 2005).
<u>Identification of suspect bacteriocin-producing strains:</u> Suspect bacteriocin-producing isolates were identified based on carbohydrate fermentation patterns using API 50 CHL kit test (bioMorieux Vitek, Inc., Hazelwood, Mo.). Cell morphology of each isolate was studied using a Gram stain. A catalase test for each strain was performed as recommended by Schillinger and Luecke, (1989).

Results and Discussion

Eight of 100 strains were found to produce antagonistic compounds against several indicators (Table 1). Only 2 strains M 13-5 and M 20-4 showed bactericidal board spectrum on all 6 tested indicators. Among these potent strains, M 13-5 was the only strain that exerted the best bactericidal board spectrum on *Lis. innocua.* Thus, these 2 aforementioned strains were selected for further characterisation. From the results of the catalase test, cell morphology and carbohydrate fermentation using API 50 CHL kit test of 2 selected isolates (Table 2), it can be concluded that there are at least 2 groups of suspected bacteriocin-producers of LAB isolated from Mum. One is M 13-5 which showed 99.8 % of identity to *P. pentosaceus* , and the other is M 20-4 which showed 99.9 % of identity to *Lb. plantarum* in API 50 CH database. These 2 groups of isolates, however, are currently being identified by 16S rDNA sequencing method.
In order to confirm the 'bacteriocin' definition of the product of these 2 isolates, inhibitory spectrum profiles of the antagonistic products were compared to the spectrum of a known pediocin PA-1 producer (Table 3). Most results were similar to the known pediocin PA-1 producer, therefore it can be assumed that M 13-5 is *P. pentosaceus* a group of bacteriocin-producers whose product is related to pediocin. The products from the other group of *Lb. plantarum* (M 20-4) exhibited a more narrowed spectrum than M 13-5 and might be an antilisterial bacteriocins. Both of these isolated LAB, are currently been investigated further by 16S rDNA sequences, for strain confirmation by bacteriocin

purification and identification. Further studies are being conducted to improve the quality and safety of these products in Mum production.

Table 1: Preliminary screening results of antagonistic substances produced by 8 of 100 strains isolated from Mum against 6 indicators using direct method.

Indicator strain	M 13-2	M 13-5	M 15-3	M 15-4	M 15-8	M 16-6	M 17-5	M 20-4
Lb.sakei (JCM1157^T)	+	+	+	+	+	-	+	+
Lb.plantarum ATCC14917^T	+	+	-	-	+	-	+	+
Lis. innocua ATCC33090^T	+	++	+	+	+	+	+	+
Staph. aureus ATCC12600	-	+	+	-	-	+	-	+
E. coli JM 109	+	+	+	+	+	+	+	+
Ent. faecalis JCM 5803 ^{T}	+	+	-	+	+	+	+	+

- = no inhibition (0-0.9 mm), + = inhibition zone 1-5 mm., ++ = inhibition zone 6-10 mm
ATCC, American Type Culture Collection, Rockville, Md; JCM, Japanese Culture of Microorganisms, Japan;
JM, commercial strain from Toyobo, Osaka, Japan

Table 3: Antimicrobial spectrum of BLIS from isolated LAB from Mum.

Indicator ^{a}	TISTR 536	M 13-5	M 20-4
P. pentosaceus JCM 5885	0	0	0
Lb. plantarum ATCC 14917^T	800	800	200
Lb. sakei subsp. *sakei* JCM 1157^T	3,200	3,200	800
Lc. lactis subsp. *lactis* ATCC 19435 ^{T}	0	0	0
Lc. lactis subsp. *lactis* JCM 7638	0	0	0
Lc. lactis subsp. *cremoris* TUA 1344L	800	800	0
Leuconostoc mesenteroides JCM 6124^T	400	400	0
Micrococcus luteus IFO 12708	800	800	0
Listeria innocua ATCC 33090 ^{T}	12,800	25,600	400
Enterococcus faecalis JCM 5803 ^{T}	1,600	1,600	100
Staph. aureus ATCC 12600	0	0	0
Bacillus circulans JCM 2504^T	0	0	0
B. coagulans JCM 2257^T	0	0	0
B subtilis JCM 1465^T	0	0	0
Escherichia coli JM 109	0	0	0
TISTR536	0	0	0
A 5	0	0	0
M 13	0	0	0

^{a} ATCC, American Type Culture Collection, Rockville, Md; JCM, Japanese Culture of Microorganisms, Japan; JM, commercial strain from Toyobo, Osaka, Japan; TUA, Tokyu University of Agriculture, Japan; IFO, Institute for Fermentation, Osaka, Japan; TISTR 536, the pediocin PA-1 producer strain (*P. pentosaceus*) [8)]

Table 2: Catalase test, morphology and carbo-hydrate fermentation (API 50 CHL) of the most potent strains.

Test	M 13-5	M 20-4
Catalase	-	-
Cell morphology	tetrad cocci	long rod
Gram stain	+	+
API 50CHL (percentage of identity)	*P. Pentosaceus* (99. 8 %)	*Lb. plantarum* (99. 9 %)

References

Schillinger, U., and Luecke, F-K. (1989) *Appl. Environ. Microbiol.* 55(8) : 1901-1906.

Tichaczek, P.S., Nissen-Meyer, J., Nes, I.F., Vogel, R.F., and Hammes., (1992) W.P. *System. Appl. Microbiol.* 15 : 460-468.

Ennahar, S., Zendo, T., Sonomoto, K., and Ishizaki, A. (1999) *Japanese J. of Lactic Acid Bacteria.* 10(1) : 29-37.

Somathiti, S., and Surapantapisit, Y. J. (1987) Fac. of Science., Kasetsart University, Thailand.

Swetwiwathana, A. (2005) Ph.D. Thesis of Department of Bioscience and Biotechnology, Kyushu University, Japan.

De Man, J.C., Rogasa, M., and Sharp, M.E. (1960) *J. Appl. Bact.* 23 : 130-135.

Fleming, H.P., Etchells, J.L., and Costilow, R.L. (1985) *Appl. Microbiol.* 30 : 1040-1042.

LISTERIA MONOCYTOGENES FROM MEAT AND MEAT-PROCESSING ENVIRONMENTS: BIOFILM FORMATION, MOTILITY AND HYDROPHOBICITY

A. Vergara*[1], V. D'Orio[1], D. Paludi[1], G. Di Bonaventura[2], R. Piccolomini[2], G. Colavita[3], and A. Ianieri[4]

[1.] *Dipartimento di Scienze degli Alimenti, Università degli Studi di Teramo;* [2.] *Centro Scienze dell'Invecchiamento, Fondazione Università "G. d'Annunzio", Chieti;* [3.] *Dipartimento DISTAAM, Università del Molise, Campobasso;* [4.] *Dipartimento di Sicurezza degli Alimenti, Università di Parma, ITALY. Email: dpaludi@unite.it*

Keywords: *Listeria monocytogenes*, meat, biofilm, motility, hydrophobicity

Introduction

When bacteria attach to surfaces and colonise them, their phenotypes change. They show, for instance, a decreased susceptibility to antimicrobials, increased exopolysaccharides production, and the community they form is called a biofilm (Costerton *et al.*, 1987). Biofilms create major problems in the food industry, since they, represent an important source of contamination of materials or foodstuffs coming into contact with them (Mattila-Sandholm and Wirtanen, 1992). *Listeria monocytogenes* is part of the microflora most commonly found in the meat industry (Roberts and Wiedmann, 2003). It has been shown that *L. monocytogenes* strains that persist in food industry environments form thicker biofilms than isolates found only sporadically (Lunden *et al.*, 2000), indicating that biofilm formation is important for the survival of *L. monocytogenes*. The quantity of microrganisms transferred to an inhert surface or food depends on the physicochemical properties of both surfaces and microrganisms and factors such as temperature (Fletcher, 1977), and hydrophobicity (Hood and Zottola, 1995). Further, it has been shown that flagella can positively modulate adhesion to both living and non-living surfaces (Vatanyoopaisarn *et al.*, 2000). The aim of this study was to evaluate the ability of 23 strains of *L. monocytogenes*, from meat and meat industry-environments, to form biofilms on three different surfaces commonly used in food-processing plants (polystyrene, glass, and stainless steel) with regard to different temperatures (4, 12, 22, 37°C), hydrophobicity and motility.

Materials and Methods

Bacterial strains. Twenty-three *L. monocytogenes* strains were tested: 19 from meat, and 4 from meat-processing plants by swabbing food and other contact surfaces and equipment. Biofilm production assay. Overnight cultures of *L. monocytogenes* were prepared, from o/night Trypticase Soy Agar-growth, in Trypticase Soy Broth (TSB) by incubating at selected temperature (4, 12, 22, or 37°C) under dynamic conditions. Cultures were then washed twice with phosphate-buffered saline (PBS; pH 7.3) and diluted with fresh TSB to reach a concentration of about 1 x 10^8 CFU/ml. Food-processing surfaces (glass, polystyrene, and stainless steel) were exposed to a standardized inoculum in 35-mm-diameter polystyrene tissue culture plates, and incubated at selected temperatures. After 24h-incubation nonadherent cells were removed by three washings in sterile PBS. *L. monocytogenes* biofilm biomass was determined spectrophotometrically (OD_{492}) by the crystal violet-based colorimetric method of Christensen *et al.* (1985). Considering the different growth surfaces of test materials, results were normalized by calculating the "biofilm production index" (BPI) = [$OD_{meanbiofilm}$/surface (mm^2)] x 1.000. Cell surface hydrophobicity assay. *L. monocytogenes* hydrophobicity was evaluated by Microbial Adherence to n-Hexadecane (MATH) test. Briefly, 4 ml of OD_{550} = 0.8 standardized inoculum in PBS were overlayed with 0.4 ml of n-hexadecane. After agitation for 1 min by vortexing, the phases were allowed to separate for 15 min at room temperature. The results were expressed as the proportion of the cells which were excluded from the aqueous phase, determined by the equation: $[(A_0 - A)/A_0]$ x 100, where A_0 and A are the initial and final optical densities of the aqueous phase, respectively. Strains were classified as: highly hydrophobic, for values >50%; moderately hydrophobic, for values ranging from 20 and 50%; and hydrophilic, for values <20%. Cell motility. Swimming, swarming, and twitching assays were performed at selected temperatures by inoculating a single colony from an agar growth. i) Swimming. Media used for assay was tryptone broth (10g/liter tryptone, 5 g/liter NaCl) that contained 0.3% agar. After incubation for 24 h, results were expressed as diameter (mm) of growth zone. ii) Swarming. Media used for assay consisted of 0.5% agar with 8 g/liter nutrient broth and 5 g/liter glucose. Swarm plates were typically allowed to dry at room temperature o/night before being used. Results were expressed as diameter (mm) of growth zone. iii) Twitching. Assay was performed with LB broth (10 g/liter tryptone, 5 g/liter yeast extract, and 10 g/liter NaCl) solidified with 1% agar. Twitch plates were briefly dried and strains were inoculated on the bottom of the Petri dish plate. After incubation for 24 h, the zone of motility at the agar/Petri dish interface was measured by staining with crystal violet. Statistical analysis. Assays were carried out in triplicate and repeated on two different occasions. Data were shown as mean ± standard deviation (SD). The significance of differences in biofilm formation was assessed by ANOVA, followed by Newman-Keuls Multiple Comparison Test. Relationship between biofilm formation and MATH% or motility was evaluated by calculating correlation Pearson r coefficient (r_P). Statistical significance was evaluated at a P value of < 0.05.

Results and Discussion

Biofilm, expressed as OD_{492}, was produced at significantly ($P<0.05$) higher levels on glass at 4°C (0.210±0.119), 12°C (0.265±0.124), and 22°C (0.392±0.167). At 37°C biofilm levels were comparable, regardless of substrate (0.571±0.262, 0.491±0.226, and 0.454±0.279 on glass, stainless steel, and plastic, respectively). According to previous studies (Blackman and Frank, 1996; Lunden *et al.*, 2000), our results showed that *L. monocytogenes* is able to adhere and form biofilm on commonly used surfaces in food-industry, regardless of the temperature tested. However, the ability to form biofilms was extremely variable in the population studied, making it difficult to select strong biofilm-producers strains with increased probability to persist in a food-processing environment. Bacterial adhesion to substrate is a consequence of complex physicochemical interactions, such as cell surface hydrophobicity (Hood and Zottola, 1995). Cell surface hydrophobicity level increased with temperature, although hydrophobicity at 37°C (mean MATH%, 32.3±10.1) was significantly ($P<0.001$) higher than those produced at 22°C (16.6±17.6), 12°C (12.4±6.0), and 4°C (8.7±10.6). It is plausible to hypothesize that temperature influences cell surface proteins and, consequently, the resulting hydrophobicity level (Chavant *et al.*, 2002). At 4°C, hydrophobicity levels negatively correlated with biofilm formation on both plastic (Pearson $r=-0.491$; $P<0.05$) and glass (Pearson $r=-0.499$; $P<0.05$). No strains were mobile by swarming or twitching, regardless of temperature. Conversely, 12 (52.2%) out of 23 strains tested exhibited flagellar motility (swimming) at 22°C. Motility by flagella, overcoming electrostatic repulsive forces, could facilitate the interaction between microrganism and substrate and so improve biofilm formation (Vatanyoopaisarn *et al.*, 2000). However, our results showed that swimming motility never correlated with biofilm produced, regardless of substrate considered.

Conclusions

The present work suggests that *L. monocytogenes* is able to organise itself as biofilm on surfaces commonly used in the food industry, regardless of temperature considered. This ability was exhibited also at 4°C, so raising a relevant hygienic/sanitary issue. Polystyrene is in use as a packaging material but is not much used in equipment in the food industry (Moretro *et al.*, 2003). However, our results suggest the use of polystyrene in food-processing plants since it might minimize biofilm formation, regardless of temperature considered.

References

Blackman, I.C., Frank, J. (1996). Growth of *Listeria monocytogenes* as biofilm on various food processing surfaces. J. Food Prot., 59: 827–831.

Chavant, P., Martinie, B., Meylheuc, T., Bellon-Fontaine, M., Hebraud, M. (2002). *Listeria monocytogenes* LO28: Surface physicochemical properties and ability to form biofilms at different temperatures and growth phases. Appl. Env. Microbiol., 68: 728–737.

Christensen, G.D., Simpson, W.A., Younger, J.J., Baddour, L.M., Barrett, F.F., Melton, D.M., Beachey, E.H. (1985). Adherence of coaugulase-negative staphylococci to plastic tissue culture plates: a quantitative model for the adherence of staphylococci to medical devices. J. Clin. Microbiol., 22: 996–1006.

Costerton, J.W., Cheng, K.J., Geesey, G.G., Ladd, T.I., Nickel, J.C., Dasgupta, M., Marrie, T.J. (1987). Bacterial biofilms in nature and disease. Annu. Rev. Microbiol., 41: 435–464.

Fletcher, M. (1977). The effects of culture concentration and age, time, and temperature on bacterial attachment to polystyrene. Can. J. Microbiol., 23: 1–6.

Hood, S.K., Zottola, E.A. (1995). Biofilms in food processing. Food Control, 6: 9–18.

Lunden, J.M., Miettinen, M.K., Autio, T.J., Korkeala, H.J. (2000). Persistent *Listeria monocytogenes* strains show enhanced adherence to food contact surface after short contact times. J. Food Prot., 63**:** 1204–1207.

Mattila-Sandholm, T., Wirtanen, G. (1992). Biofilm formation in the industry: a review. Food Res. Int., 8: 573–603.

Moretro, T., Hermansen, L., Holck, A.L., Sidhu, M.S., Rudi, K., Langsrud, S. (2003). Biofilm formation and the presence of the intercellular adhesion locus *ica* among staphylococci from food and food processing environments. Appl. Environ. Microbiol., 69:5648-5655.

Roberts, A.J., Wiedmann, M. (2003). Pathogen, host and environmental factors contributing to the pathogenesis of listeriosis. Cell. Mol. Life Sci., 60: 904–918.

Vatanyoopaisarn, S., Nazli, A., Dodd, C.E.R., Rees, C.E.D., Waites, W. M. (2000). Effect of flagella on initial attachment of *Listeria monocytogenes* to stainless steel. Appl. Env. Microbiol., 66: 860–863.

REDUCTION OF CAMPYLOBACTERS ON POULTRY USING STEAM OR HOT WATER HEAT TREATMENTS AND SUBSEQUENT CHILLING

G. Purnell*[1], J.E.L Corry[2] and C. James[1]

[1.] Food Refrigeration and Process Engineering Research Centre, Faculty of Engineering, University of Bristol, Langford, Somerset, BS40 5DU, UK, [2.] Division of Farm Animal Science, School of Veterinary Science, University of Bristol, Langford, Somerset, BS40 5DU, UK. Email: graham.purnell@bristol.ac.uk

Keywords: poultry, *Campylobacter*, appearance, steam, hot water

Introduction

Poultry is well-known to be colonised by *Campylobacter* spp. on the farm, and raw poultry carcasses are therefore often contaminated with campylobacters and are a major source of infection for humans (ACMSF, 2005). It has been predicted that a 2-log reduction in the numbers of campylobacters on poultry could lead to a 30-fold decrease in human campylobacteriosis (Rosenquist *et al.*, 2003). The studies presented here investigated methods of treating raw poultry carcasses to reduce or eliminate surface contamination with campylobacters without adversely affecting carcass appearance.

Materials and Methods

Equipment for controlled heat treatment (Figure 1) was constructed by FRPERC. The hot water immersion system provided controlled treatment temperatures (set point ±1°C) between room ambient and 100°C. The steam treatment cabinet provided a steam environment at atmospheric pressure (≈100°C). Carcasses were transferred to/from the heat treatment zones by pneumatic arms under electronic timer control. The rapid chilling system was an experimental unit (Figure 1) developed by Air Products plc (UK). This was an impingement air chilling system, designed to freeze the surface of the carcass for a limited period during the normal chilling process, and capable of operating at temperatures down to –40°C.

Figure 1: Experimental equipment: Hot water immersion, Steam and Rapid Chilling.

Laboratory studies used an inoculum consisting of equal volumes of *Campylobacter jejuni* (AR6 poultry isolate) grown in biphasic culture medium, and *Escherichia coli* (K12 Laboratory strain resistant to 200 ppm nalidixic acid) grown in BHI broth. Inoculum (3ml) was spread over the breast of each carcass, and at least 15min was allowed at room temperature before application of any decontamination treatments. Pieces of breast skin (10cm^2) were excised before and after treatment for enumeration of the bacteria. Industrial studies used naturally occurring bacteria. Carcasses were taken off the line immediately before chilling, heat treated, and returned to normal production or experimental rapid chilling within 5min. 10g neck skin samples were taken for enumeration of the bacteria. In all cases campylobacters were enumerated on mCCD agar (42°C for 48h in microaerobic atmosphere), and *E. coli* K12 (laboratory studies only) on MacConkey agar no. 3 with 100ppm nalidixic acid (37°C for 24h). During industrial studies, additional carcasses and portions derived from them were assessed by a panel on a 5 point scale for appearance acceptability after chilling.

Results and Discussion

The effect of a variety of treatment time/temperatures on carcass appearance was evaluated. Initial appearance trials established maximum severity heat treatments that could be tolerated without completely unacceptable damage. Steam treatment (≈100°C) for 5s and 10s, and hot water treatment at 80°C for 10s and 20s gave the best compromises between short processes (for industrial practicability) and treatment severity (expected bacterial kill) (Table 1). It was seen that surface appearance improved over time in chilled storage.

Table 1: Surface Appearance scores for carcasses and portions stored at 1°C for 5 or 11 days after treatment. A: acceptable; B: borderline; U: unacceptable.

	Control	Steam		Hot Water Immersion	
		(10s, 100°C)	(5s, 100°C)	(20s, 80°C)	(10s, 80°C)
Shelf day 5					
Whole bird	A	A	U	U	B
Thigh	A	U	B	U	A
Drumstick	A	A	A	A	A
Breast	A	U	B	A	A
Skinless breast	A	A	A	A	A
Wing	A	U	U	U	A
Shelf day 11					
Whole bird	A	B	B	A	A
Thigh	A	U	B	B	A
Drumstick	A	A	A	A	A
Breast	A	A	U	A	A
Skinless breast	A	A	A	A	A
Wing	A	U	A	B	A

The selected heat treatments were combined with three chilling regimes on inoculated carcasses (Table 2). Numbers of campylobacters were reduced during chilling even without any heat treatment, but crust freezing during chilling was more effective than other methods. Steam for 5 or 10 s was more effective than water (80°C for 10 or 20s). Crust freezing or 'chilling' at 15°C (probably with a drying effect) was most effective in combination with steam or hot water. Similar results were obtained for *E. coli K12*.

Table 2: Laboratory based reductions in inoculated C. jejuni ($\log_{10}$ CFU/cm^2).

	No heat treatment	Steam		Hot Water Immersion	
		(10s, 100°C)	(5s, 100°C)	(20s, 80°C)	(10s, 80°C)
Crust Freezing	1.5	3.2	2.3	2.8	2.4
0°C chilling	0.3	n/a	1.8	1.8	n/a
15°C chilling	0.4	n/a	3.2	2.6	n/a

Attempts to replicate the inoculated carcass laboratory results in industrial plant trials with naturally occurring bacteria were less successful. Firstly campylobacter positive flocks were difficult to identify, and even then positive numbers were low. Reductions from plant A (Table 3) were variable and substantially less than expected from laboratory results.

Table 3: Plant A; reductions in naturally occurring campylobacter ($\log_{10}$ CFU/cm^2).

	No heat treatment	Steam		Hot Water Immersion	
		(10s, 100°C)	(5s, 100°C)	(20s, 80°C)	(10s, 80°C)
Crust Freezing	0.8	0.7	ND	0.5	ND
Plant chilling	-0.3	0.2	ND	0.2	ND

Studies were moved to plant B where carcasses were taken from older flocks after previous thinning to increase likelihood of campylobacter positive flocks. Four experiments, each of 20 treated and 20 control carcasses gave mean campylobacter reductions of 1.15, 0.68, 1.37, and 0.97 $\log_{10}$ CFU/cm^2 respectively for the 20s, 80°C heat treatment, 0°C chill process. Whilst higher than in plant A reductions were still below the 1.8 $\log_{10}$ CFU/cm^2 seen in the laboratory studies.

Conclusions

Steam at atmospheric pressure or hot water can be used to reduce numbers of viable campylobacters on chicken carcasses. The effect on naturally-contaminated carcasses was less than expected from results using carcasses inoculated with high numbers of campylobacters. A controlled crust freezing regime gave better reductions than normal plant chillers. The reason for this could be that some of the campylobacters on naturally-contaminated carcasses are more protected against the heat treatment, being located in feather follicles or folds in the skin.

Acknowledgements

This work was funded by the UK Food Standards Agency. We are grateful for the assistance and goodwill of the two UK poultry processing companies who allowed access to their plants for these studies.

References

ACMSF 2005 Second report on Campylobacter, Advisory Committee on Microbiological Safety of Food, Food Standards Agency, London <http://www.food.gov.uk/science/ouradvisors/microbiogsafety/55817/>

Rosenquist, H., Nielsen, N. L., Sommer, H. M., Nørrung, B., and Christensen, B. B. (2003). Quantitative risk assessment of human campylobacteriosis associated with thermophilic *Campylobacter* species in chickens. *International Journal of Food Microbiology*, 83, 87-103.

SAFETY OF TRADITIONAL FERMENTED SAUSAGES: RESEARCH ON PROTECTIVE CULTURES AND BACTERIOCINS

J. Gasparik-Reichardt*[1], Sz. Tóth,[1] L. Cocolin[2], E.H. Drosinos[3], M. Hadziosmanovic[4], F. Caklovica[5], L. Turubatovic[6] and I. Petrohilou[7]

[1.] Hungarian Meat Research Institute, H-1097 Budapest, Gubacsi út 6/b, Hungary, [2.] University of Udine, Via Palladio 8 I-33100 Udine, Italy, [3.] The Agricultural University of Athens, 75 lera Odos str. GR-118 55 Athens, Greece, [4.] Zagreb University Veterinary Faculty, Heinzelova 55, 10000 Zagreb, Croatia, [5.] University of Sarajevo Veterinary Faculty, Zmaja od Bosne 90, 71100 Sarajevo, Bosnia and Herzegovina, [6.] Institute of Meat Hygiene and Technology, Kacanskog 13, 11000 Belgrade, Serbia and Montenegro, [7.] ETAT S.A., 80 Vouliagmenis Ave., 117 43 Athens, Greece Email: mikrobiologia@ohki.hu

Keywords: fermentation, sausages, protective starter cultures, bacteriocin, safety

Introduction

The main objective of the project was to isolate bacteriocin producing starter cultures from the endogenous lactic acid microflora of naturally fermented sausages and to select protective starter cultures which are active against pathogenic microbes in order to improve product safety and without any adverse effect on the sensorial characteristics of the sausage.

Materials and Methods

Traditional microbiological methods, API 50 CHL, API Staph kits and PCR-based methods were used for identification of isolated strains. AWDA method was applied for screening bacteriocin production and critical dilution method for determination of bacteriocin activity. Indicator strains for testing bacteriocin production were *Listeria monocytogenes* (NCTC 10527), *Staphylococcus aureus* (NCBF 1499) and *Escherichia coli* O157:H7 (NCTC 1207). For challenge tests *Listeria monocytogenes* was inoculated into sausage batter under laboratory conditions. The inoculated sausages were fermented and ripened at factories as usual.

Results and Discussion

For the purposes of the project, local factories were asked to produce a typical traditional sausage without starter cultures. Variations exist in the composition of the mix, the dimensions, as well as the fermentation-ripening process and time. The sausages were considered ready for consumption on the 28th day. Physicochemical and microbiological analyses of the raw materials, intermediate and final products (days 0, 2, 4, 7, 14 and 28) were carried out, in order to determine the hygienic parameters and the changes occurring as a result of the fermentation and ripening. The most important result, from the point of view of product safety, is that no *Listeria monocytogenes*, *Salmonella* or *Staphylococcus aureus* were detected in the final product in any of the countries. Enterobacteria, *Escherichia coli* and *Pseudomonas*, when present, were progressively eliminated regardless of their initial population. As expected, the main populations detected throughout the experiment were Lactic Acid Bacteria (LAB) and Micrococci. The physicochemical changes can be summarized as follows: decrease of the pH, decrease of a_w, decrease of nitrates and nitrites and increase of NaCl. Finally, sensorial analysis of the final products resulted in an overall acceptability of the product above 70% for all partners except for Croatia, possibly due to the extension of the regular ripening time for project harmonization purposes.

One hundred and fifty strains of LAB and 150 strains of catalase-positive cocci from the sausage fermentations were isolated in each country. These strains were characterized biochemically (API Systems) and genetically (sequencing of partial 16S rDNA). It was shown that the API system is insufficient for the identification of LAB, while it gives better results for staphylococci (Gasparik-Reichardt *et al.*, 2004). The main population of catalase-positive cocci belonged to the *Staphylococcus xylosus* and *S. saprophyticus* species and the main populations of LAB belonged to the *Lactobacillus sakei* and the *L. curvatus* species. The screening of all the isolated strains for the production of bacteriocins resulted in only 6 strains of *Lactobacillus sakei* (isolated by the Italian partners and designated as I151, I152, I153, I154, I155 and I156) and one strain of *Leuconoctoc mesenteroides* (isolated by the Greek partners and designated as E131), showing antibacterial activity towards *L. monocytogenes* only and not against *S. aureus* or *E. coli*. Based on the technological characteristics as well as the ability for bacteriocin production, three of the six *L. sakei* strains (I151, I154 and I155) were suggested as potential bio-protective cultures. *Lc. mesenteroides* E131, as an obligate heterofermentative strain produces carbon dioxide and slime, which are undesirable during the fermentation of sausages. Subsequently, it could be used only for the production of bacteriocin.

RAPD-PCR using three different primers was performed for the *L. sakei* strains and the results showed a coefficient of similarity higher than 86% suggesting that the strains could be considered as the same from a genetic point of view (Rantsiou *et al.*, 2005). The identification and the sequencing of the genes responsible for the bacteriocin production revealed that *L. sakei* I151 was producing sakacin P, since at both the DNA and protein levels, a 100% homology to the *sppA* gene and the SakP protein was observed, while *Lc. mesenteroides* E131 showed 100% similarity at both DNA and protein level to the *mesY* gene and MesY protein, respectively, thereby revealing the production of mesenterocin Y.

An ammonium precipitation protocol was developed for the production of semi-purified bacteriocins from *Lc. mesenteroides* E131 and *L. sakei* I154 (Drosinos *et al.*, 2005). Purified bacteriocin of *L. sakei* I154 was stable after 1 h at 40-100 °C, but after 15 min of heat treatment, it retained only 40% of the initial activity and was completely inactivated after 30 min. It retained activity for only two weeks at the temperature of -20 °C, while the activity was retained only 1 week at 30°C. Purified bacteriocin of *Lc. mesenteroides* E131 lost 40% of activity after incubation at high temperatures (90 and 100°C). Storage at temperatures ≤4°C did not affect the activity even after 4 weeks. At 30°C, however, activity was decreased after 2 weeks. The addition of Tween 80 had no effect on the activity of *Lc. mesenteroides* E131, while on the activity of *L. sakei* I154 it had a beneficial effect. At 30 °C, however, it retained activity for 3 weeks (20% of activity was retained).

The laboratory experiments to determine the optimal conditions (pH and temperature) for the highest antimicrobial activity against *L. monocytogenes* have shown that: the bacteriocins have greater effect at lower temperatures and pH values, the highest antilisterial activity was detected with bacteriocin of *Lc. mesenteroides* at pH 5.0 at all (12, 18 and 24 °C) temperatures. In spite of the inhibition or reduction of *L. monocytogenes,* the applied treatments did not kill the bacteria totally.

Using the protective cultures (*L. sakei* I151, I154 and I155) and semi purified bacteriocins, reduction of *L. monocytogenes* was measured in challenge tests during the production of fermented sausages (Caklovica *et al.*, 2005). The best results were obtained by *L. sakei* I151 and the bacteriocin of *Lc. mesenteroides* E131. The *in situ* validation of the selected culture and bacteriocins, suggests that *L. sakei* I151 and *Lc. mesenteroides* E131 bacteriocin are effective against *L. monocytogenes* and could be used as protective starter culture and bioprotective agent, respectively, in the production of naturally fermented sausages.

The evaluation of the effect of the addition of the selected protective culture or the bacteriocin, alone or in combination, on the technological and traditional quality characteristics of the products, was checked in a pilot production. The results of both microbiological and physicochemical analyses, showed a similar trend to that of the naturally fermented sausages with no additions. Generally, no essential differences in the organoleptic parameters among the samples in all fermentations were observed. All the samples showed acceptable flavour and taste without off-odours, but in some cases (Serbia, Croatia), the inoculated samples were graded better than the control. In order to determine the critical technological parameters, industrial scale production of sausages with the addition of *L. sakei* I151 as starter/protective culture and the bacteriocin of *Lc. mesenteroides* E131 as protective agent, was undertalen. The production of sliced and packaged products under modified atmosphere or vacuum, was studied in order to evaluate the influence of packaging on the microbial flora. It was revealed that both MAP and vacuum-packaging may be considered as effective packaging techniques with regard to the preservation of desirable traditional technological, sensorial and safety properties of the final product. Standard Operating Procedures (SOPs) were developed for the industrial scale production in each country.

Conclusions

The innovative features of the project are as follows: isolation of new protective cultures, purification of bacteriocins and their experimental implementation in fermented sausages, genetic characterization of genes and SOP's development for each country.

References

Caklovica, F., Kozacinski, L., Cvrtila Z., Veskovic-Moracanin, S., Gasparik-Reichardt, J., Zdolec, N.., Smajlovic, M., Alagic, D. (2005) Influence of selected LAB on L. monocytogenes during production of traditionally fermented sausages, Technologija mesa, 2005, 46:3-4, p: 185-193.

Drosinos, E.H., Mataragas, M., Nasis, P., Galiotou, M , Metaxopoulos, J. (2005). Growth and bacteriocin production kinetics of *Leuconostoc mesenteroides* E131. Journal of Applied Microbiology, 99:1314-1323.

Gasparik-Reichardt, J., Cocolin, L., Drosinos, E.H., Gaitis, F. (2004). Comparison of microflora of traditional fermented sausages identified with traditional and molecular methods. 50th ICoMST, 7-14 Helsinki, Finland, Abs. p:134.

Rantsiou, K., Urso, R., Cantoni, C., Cocolin, L. (2005). Sequencing of a sakacin P gene cluster of a *Lactobacillus sakei* isolated from Italian sausages and expression analysis in vitro and in situ. 8th Symposium on Lactic Acid Bacteria, Egmond aan Zee, NL, August 28 - September 1, Poster.

Acknowledgements

This research project is funded by the EU (Contact No: ICA4-CT-2002-10037).

THE GROWTH CHARACTERISATION OF *LISTERIA MONOCYTOGENES* IN DIFFERENT MEAT PRODUCTS UNDER VARIOUS CONDITIONS

J. Gasparik-Reichardt*[1], S. Pércsi[2], E. Horváth,[2] S. Tóth[1] and G. Szabó[1]

[1.] *Hungarian Meat Research Institute, H-1097 Budapest, 6/b Gubacsi út, Hungary,* [2.] *Campden & Chorleywood, 1107 Budapest, 21. Szállás utca, Hungary Email: mikrobiologia@ohki.hu*

Keywords: food safety, *Listeria monocytogenes*, predictive modeling, challenge test, meat products

Introduction

One of the main objectives of the project was to determine the effects of different storage temperatures and chemical composition of meat products most frequently consumed in Hungary on the growth of *Listeria monocytogenes* using predictive models. The effects of different storage temperatures and packaging technologies (vacuum and modified atmosphere) on the growth of *L. monocytogenes* in Bologna type sausage were also determined by challenge test. Furthermore, the results of challenge tests were also compared with the findings of predictive models.

Materials and Methods

Chemical (salt-, moisture- and nitrite content and pH) and microbiological parameters (TVC, *Listeria monocytogenes*) of 78 pre-packed, sliced and heat-treated meat products from supermarkets and retail shops were determined with standard chemical (AOAC) and microbiological methods (EN ISO 4833, EN ISO 11290-2). The chemical parameter data (Table 1) were used for the growth simulation of *Listeria monocytogenes* with predictive models (http://www.ifr.ac.uk/safety/growthpredictor/).

For challenge tests, *L. monocytogenes* 4ab No. 10 was grown in TGE broth for 24 hours at 37°C, then the cultures were placed in refrigerators at 3, 5, 8 and 10°C for adaptation to lower temperatures. Bologna type sausage was sliced with a Bizerba A 301-type machine and the samples (100-150 g) were aseptically placed in plastic bags. *L. monocytogenes* was inoculated on the surface of sliced Bologna type sausages with an initial *Listeria*-count of about 10^2 CFU/g. The samples were packaged under modified atmosphere and vacuum. The packages were stored at 3, 5, 8 and 10°C and samples in triplicate were taken periodically to determine total viable count and *Listeria*-counts.

Results and Discussion

The microbiological analyses of meat products purchased at retail level (Bologna type sausage, ham, Frankfurter) showed that *L. monocytogenes* was detected in 2 out of 78 samples (peeled Frankfurters).

Based on the chemical results (Table 1) the simulation of *L. monocytogenes* (Baranyi and Roberts, 1994) was carried out in Bologna type sausage, in Frankfurter and in ham. Figure 1 shows that with a predicted low inoculum (log 0.1 CFU/g) at 5°C, the time for a 2 fold increase of *L. monocytogenes* is between 3.5-9.5 days in the case of Bologna type sausage.

Table 1: Results of chemical analyses.

Product	Bologna type sausage			Ham			Frankfurter		
	Salt%	Nitrite mg/kg	pH	Salt%	Nitrite mg/kg	pH	Salt%	Nitrite mg/kg	pH
Mean	2.17	30	6.27	2.35	21	5.96	2.10	26	6.22
SD	0.26	12	0.2	0.48	24	0.31	0.33	15	0.32
Minimum	1.51	3	5.66	1.36	1	5.10	1.61	2	5.37
Maximum	2.44	54	6.54	2.94	2	6.33	2.91	65	6.69

The effects of different packaging technologies and different storage temperatures on the growth of microflora and *L. monocytogenes* were also determined with challenge tests. Differences were found at 3, 5, 8 and 10°C and between vacuum and modified atmosphere packaging. As expected, at higher temperature the growth was faster (both TVC and *Listeria*) and the modified atmosphere (30:70, CO_2:N_2) inhibited the growth of *L. monocytogenes*.

Figure 1 Simulation of growth of *Listeria monocytogenes* in Bologna type sausage at different temperature, composition and initial contamination.

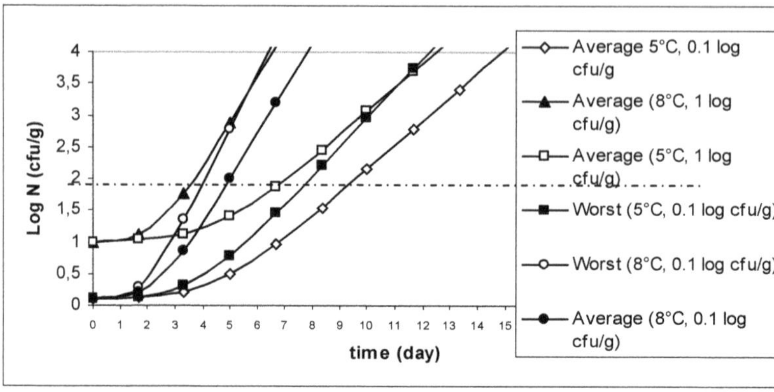

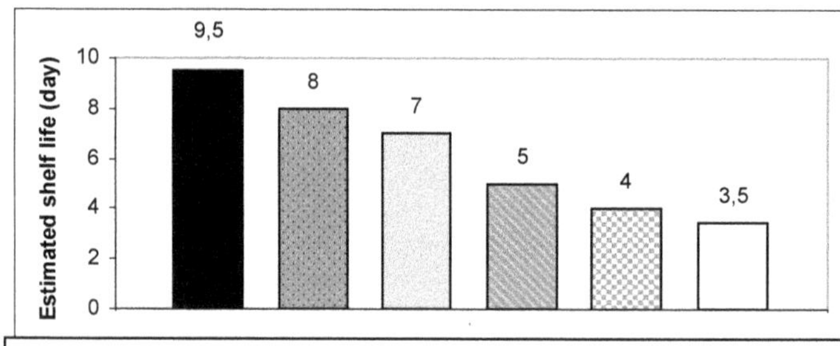

T[°C]	5°C	5°C	5°C	8°C	8°C	8°C
[log CFU/g]	0.1	0.1	1	0.1	0.1	1
composition*	average	worst	average	average	worst	average

*average composition: average compositional condition for *Listeria monocytogenes* to grow: average pH, average nitrite and salt content from Table 1.
worst composition: most favourable compositional condition for *Listeria monocytogenes* to grow: highest pH, lowest nitrite and salt content from Table 1.

Conclusions

We found that the suggested shelf life (when TVC reached 10^6 CFU/g) at 3°C was 41 days in the case of modified atmosphere packages and 28 days in the case of vacuum packages. The shelf life was shorter at 10°C, 6 days in case of vacuum packaging and 19 days in case of modified atmosphere packaging. The modified atmosphere packaging inhibited the growth of *L. monocytogenes* at 3°C for 16 days and at 10°C for 6 days compared with vacuum packaging. There were no differences for 2 log cycles increase of *L. monocytogenes* between the values determined with GrowthPredictor and vacuum-packaging (12-13 days at 3°C and 3 days at 10°C) but significant difference was found at modified atmosphere packaging (the time for 2 log cycle increase of *L. monocytogenes* was 37 days at 3°C and 12 days at 10°C).

Acknowledgements

This research project is co-funded by the National Office of Research and Technology (NKTH) of the Hungarian Republic.

References

Baranyi, J., Roberts, T. A. (1994): A dynamic approach to predicting bacterial growth in food. International Journal of Food Microbiology 23, p:277-294.
http://www.ifr.ac.uk/safety/growthpredictor/

PREVALENCE OF *ESCHERICHIA COLI* O157:H7 AND *SALMONELLA* SPP. IN FRESH MEAT AND MEAT PRODUCTS IN CATALUNYA, SPAIN

L. Cabedo*, L. Picart i Barrot, and A. Teixidó i Canelles

*Servei de Veterinaria, Direcció General de Salut Pública. Generalitat de Catalunya. Calàbria, 169, 7a. 08015 Barcelona. Spain. Email:*laura.cabedo@gencat.net*

Keywords: fresh meat, meat products, *E. coli* O157:H7, *Salmonella*, prevalence

Introduction

Escherichia coli O157:H7 and *Salmonella* spp. are pathogenic bacteria that colonise the intestines of farm animals. They can contaminate meat and meat products, due to cross contamination during slaughtering or meat processing in packing plants. *Salmonella* is the most reported zoonotic pathogen in Europe (EFSA, 2006), as well as in the U.S. (CDC, 2003).

One of the first steps in risk assessment is exposure assessment, which includes an assessment of the actual or anticipated human exposure to the pathogen. Frequency of contamination of foods and the initial contamination of the raw material, including considerations of regional differences, are factors that must be considered for exposure assessment (JEMRA, 1999). Therefore, in order to carry on the risk assessment for these pathogens, there is a need for the acquisition of data on the frequency of contamination of foods at the point of consumption.

The aim of this study is to survey the presence of *E. coli* O157:H7 and *Salmonella* spp. in fresh meat and meat products commonly consumed in Catalunya.

Materials and Methods

Collection of samples for the detection of *Salmonella* spp.: A total of 1025 samples consisting of 6 refrigerated fresh duck meat samples, 120 refrigerated fresh beef samples, 19 refrigerated fresh goat meat samples, 85 refrigerated fresh rabbit meat samples, 50 refrigerated fresh turkey meat samples, 30 refrigerated fresh quail meat samples, 187 refrigerated chicken samples, 121 refrigerated fresh pork samples, 121 refrigerated fresh lamb meat, 115 fresh pork sausage samples, 30 ground beef samples, 8 ground beef and pork samples, 74 refrigerated beef hamburger samples and 59 refrigerated chicken hamburger samples were collected.

Collection of samples for the detection of *E. coli* O157:H7: A total of 943 samples consisting of 6 refrigerated fresh duck samples, 119 refrigerated fresh beef samples, 18 refrigerated fresh goat meat samples, 54 refrigerated fresh rabbit meat samples, 29 refrigerated fresh turkey meat samples, 30 refrigerated fresh quail meat samples, 168 refrigerated chicken samples, 121 refrigerated fresh pork samples, 113 refrigerated fresh lamb samples, 114 fresh pork sausage samples, 30 ground beef samples, 8 ground beef and pork samples, 73 refrigerated beef hamburger samples and 60 refrigerated chicken hamburger samples were collected. All samples weighed 300g approximately, were representative of the most consumed fresh meat and meat products in the area, and were collected from retail stores or food industry in Catalunya, Spain, from 1998 to 2004. These samples were brought to the laboratory in a cooler, and immediately analyzed for the presence of *E. coli* O157:H7 or *Salmonella* spp. For the detection of *Salmonella*, 25 g of sample were placed in a sterile plastic bag with 225ml of buffered peptone water, and blended for 2 minutes in a stomacher lab blender. After incubation at 37°C for 18 ± 2h, 0.1ml of the blend was placed in Muller Kauffmann tetrathionate-novobiocin (MKTTn) broth and was incubated at 37 ± 1°C for 24 ± 3h. Ten ml of the blend was placed in Rappaport Vassiliadis soya (RVS) broth and was incubated at 41.5 ±1 °C for 24 ± 3h. Aliquots from MKTTn broth and RVS broth were surface plated on xylose lysine deoxycholate (XLD) agar and Rambach agar and were incubated at 37 ±1°C for 24 ± 3h. One to five typical colonies from each plate were inoculated in Kligler agar slants, and incubated at 37 ± 1°C for 24 ± 3h. Typical colonies were further confirmed with API 20E strips, or were grown on nutritive agar at 37°C for 24h for serological confirmation.

For the detection of *E. coli* O157:H7, a 25-g sample was placed in a sterile plastic bag with 225 ml of trypticase novobiocin broth, and blended for 2 minutes in a stomacher lab blender. After incubation at 37°C for 23 ± 1 h, it went through an immunomagnetic separation using Dynabeads coated with polyclonal antibodies against *E. coli* O157. The obtained cell suspension was plated on tellurite-cefixime-sorbitol McConkey agar and incubated at 37°C for 24h. Typical colonies were inoculated on Kliger agar slants and further incubated at 37°C for 24h. Positive presumptive isolates where tested for β-D-glucuronidase activity. Negative isolates were further confirmed by latex agglutination test and positive cultures to latex agglutination test were confirmed using API 20E strips, or were grown on nutritive agar at 37 °C for 24h for serological confirmation.

Results and Discussion

Results of analysed samples are shown in Table 1. *E. coli* O157:H7 was isolated from 4 samples out of 943 samples of fresh meat and meat products. This pathogen was present in 3.3 % of refrigerated quail samples, but only in 0.6% of refrigerated chicken samples and 0.8 % of refrigerated pork and lamb samples. *E. coli* O157:H7 was not isolated from ground beef, a mix of ground beef and pork, beef and chicken hamburgers or fresh pork sausage.

Salmonella was isolated from 91 samples out of 1025 samples. This pathogen was present 0.8% of refrigerated beef samples, 2% of refrigerated turkey meat samples, 2.5% of refrigerated pork samples, 2.7% of refrigerated beef hamburgers, 3.3% of ground beef samples, 6.1% of fresh pork sausage, 11.8% of refrigerated rabbit samples, 12.5% of mixed ground beef and pork samples, 16% of refrigerated chicken samples, 22% of refrigerated chicken hamburger and 73.3% of refrigerated quail samples (Table 1).

Table 1: Prevalence of *E. coli* O157:H7 and *Salmonella* in fresh meat and meat products in Catalunya, Spain.

Meat sample	*E. coli* O157:H7		*Salmonella*	
	Positive samples/analyzed samples	% positive samples	Positive samples/analyzed samples	% positive samples
Refrigerated duck	0/6	0	0/6	0
Refrigerated beef	0/119	0	1/120	0.8
Refrigerated goat	0/18	0	0/19	0
Refrigerated rabbit	0/54	0	10/85	11.8
Refrigerated turkey	0/29	0	1/50	2
Refrigerated quail	1/30	3.3	22/30	73.3
Refrigerated chicken	1/168	0.6	30/187	16
Refrigerated pork	1/121	0.8	3/121	2.5
Refrigerated lamb	1/113	0.8	0/121	0
Fresh pork sausage	0/114	0	7/115	6.1
Ground beef	0/30	0	1/30	3.3
Ground beef/pork	0/8	0	1/8	12.5
Refrigerated beef hamburger	0/73	0	2/74	2.7
Refrigerated chicken hamburger	0/60	0	13/59	22

Results show that prevalence of *E. coli* O157:H7 and *Salmonella* in refrigerated quail meat was the highest of the analysed meat samples. Quails are consumed as fresh meat, they are not used to elaborate meat products and undergo a thorough cooking process before consumption. Given the small size of quails, it is very likely that the heat treatment reaches all the carcass points, and therefore the health significance of the high prevalence of these pathogens is reduced. On the other hand, results also show a high prevalence of *Salmonella* in fresh rabbit, fresh chicken and chicken hamburgers, all of which are food products that are cooked before consumption, and therefore, the probability of the actual intake of *Salmonella* is reduced. However, the probability of cross contamination in the processing plant should be taken into account.

Conclusions

Microbial risk assessment involves exposure assessment (CAC, 2004), which is the degree of pathogen intake likely to occur. It also involves estimating the population at risk. Obtaining data on the prevalence of pathogens in different food products and different geographical areas should be a previous step in risk analysis. In Catalunya, *E. coli* O157:H7 was detected only in quail, chicken, pork and lamb, when analyzing fresh meat and raw meat products. However, the prevalence of *Salmonella* on fresh meat and raw meat products detected over 7 years, ranges from 0.8% to 73.3%. The pathogen was not isolated from fresh duck, goat and lamb.

References

Centers for Disease Control and Prevention. 2003.Foodborne diseases Active Surveillance Network (FoodNet) annual report 2003. Available at http://www.cdc.gov/foodnet/annual/2003/2003_report.pdf. Accessed 24 April 2006.

Codex Alimentarius Commission. 2004. Procedural Manual 14th Edition. Joint FAO/WHO Food Standards Programme. Food and Agricultural Organization & World Health Organization. Rome.

EFSA. 2006. The Community summary report on trends and sources of zoonoses, zoonotic agents and antimicrobial resistance in the European Union in 2004. The EFSA Journal 2005 – 310. European Food Safety Authority. Available at http://www.efsa.eu.int/science/monitoring_zoonoses/reports/1277/zoonoses2004-levels1-2-part11.pdf Accessed 24 April 2006.

JEMRA, 1999. Principles and guidelines for the conduct of microbiological risk assessment. CAC/GL-30. Available at www.codexalimentarius.net/download/standards/357/CXG_030e.pdf Accessed 24 April 2006.

SEROTYPES AND ANTIMICROBIAL RESISTANCE OF *SALMONELLA ENTERICA* ON RED MEAT CARCASSES AND IN THE LAIRAGES AFTER CLEANING, IN THE SOUTH-WEST OF ENGLAND

A. Small*[1,6], C. James[2], G. Purnell[2], S. James[2], R. Davies[3], M. Howell[4], M. Hutchison[5] and S. Buncic[1]

[1.] Division of Farm Animal Science, University of Bristol, Langford House, Bristol BS40 5DU, UK. [2.] Food Refrigeration and Process Engineering Research Centre (FRPERC), University of Bristol, Churchill Building, Langford House, Bristol BS40 5DU, UK. [3.] Veterinary Laboratories Agency, Weybridge, Addlestone, Surrey, UK. [4.] Food Standards Agency, Aviation House, 125 Kingsway, London WC2B 6NH, UK. [5.] Direct Laboratories, Wergs Road, Tettenhall, Wolverhampton, UK. [6.] Current address: Food Science Australia, PO Box 3312, Tingalpa DC, QLD 4173, Australia; Email: alison.small@csiro.au

Keywords: abattoir, lairage, carcass, *Salmonella*, pathogen

Introduction

Foodborne pathogens such as *Salmonella enterica* are carried in the gut of domestic livestock and are shed into the environment with the faeces (Fedorka-Cray *et al.*, 1998). This contamination may build up on animal coats and in the lairage prior to slaughter, and may be transferred through cross-contamination between animals and between animals and the environment, and ultimately be found on the resultant carcasses (Collis *et al*, 2004). In order to further understand this cross-contamination process, isolates of *Salmonella enterica* that had been taken from lairages and carcasses at five red meat abattoirs were compared with regard to serotype, Plasmid Profile and PFGE banding pattern.

Materials and Methods

One hundred and thirty-eight isolates of *Salmonella enterica* were serogrouped and tested for resistance to a panel of 16 antimicrobial agents (Amikacin {AK}; Ampicillin {AM}; Amoxycillin/Clavulanic Acid {AMC}; Apramycin {APR}; Chloramphenicol {C}; Ceftazidime {CAZ}; Ciprofloxacin {CIP};Cefotaxime {CTX}; Furazolidone {F}; Gentamicin {GN}; Neomycin {N}; Nalidixic Acid {NAL}; Streptomycin {S}; Compound Sulphonamides {SU}; Sulphamethoxazole/Trimethoprim {SXT} and Tetracycline {TET}). Selected isolates from each of the broad groupings thus determined were further subtyped and analysed by Pulsed-Field Gel Electrophoresis (PFGE) and Plasmid Profiling. The 137 isolates originated from survey samples taken from carcasses (n=102) and lairages following routine cleansing and prior to the onset of processing (n=35) of five red meat abattoirs (Small, *et al* 2005, 2006).

Results and Discussion

Seven serogroups of *Salmonella enterica* were found, of which serogroup C1 was the most predominant, comprising 46 isolates in total. All of these were sensitive to all 16 antimicrobials tested, and they originated from the lairages of two abattoirs, and from carcasses at one. Thirteen were fully typed and shown to be *Salmonella* Mbandaka. These all belonged to the same PFGE cluster (4), and fell within a group of Plasmid Profile clusters differing by only one band (4a, 4b,4c). The second most predominant serogroup was the ROUGH serogroup, comprising 32 isolates, all of which originated from carcasses at three abattoirs. Eight of these were sensitive to all 16 antimicrobials tested, and the two isolates undergoing further typing were identified as Orough:z10:e,n,z15, and belonged to the same PFGE cluster (4) and Plasmid Profile cluster (4b) as the *S.* Mbandaka isolates described above. The remaining 24 ROUGH isolates were Nalidixic Acid resistant, and a representative set of 6 were typed as Orough:i;l,w, carried no plasmids, and belonged to PFGE cluster 3.

There were 29 isolates in serogroup B: *S.* Kimuenza was found on one carcass at one abattoir, sensitive to all antimicrobials tested; Tetracycline resistant *S.* Typhimurium DT208 was associated with 11 pig carcass isolates from a second abattoir, and *S.* Derby was associated with carcasses and the lairage environment in another. These last belonged to PFGE cluster 2 and Plasmid Profile clusters 2a and 2b, and some isolates were resistant to Sulphamethoxazole/Trimethoprim and Compound Sulphonamides. *S.* Dublin (serogroup D) made up 16 isloates, taken from calf and sheep carcasses at a single abattoir. These were sensitive to the antimicrobials tested, and belonged to PFGE clusters 1A, 1B and 1C, and to Plasmid Profile clusters 1a and 1b, all of which differed by only one band. *S.* Anatum (serogroup E1) was identified in 3 lairage isolates, from two abattoirs (all fully sensitive), and from 6 carcass isolates from a third abattoir. The carcass isolates came from bobby calves, and were Chloramphenicol resistant. There was a single isolate of serogroup C3 (8,20:-:z6) from a lairage, resistant to Sulphamethoxazole/Trimethoprim and Compund Sulphonamides, and a single isolate of IIIb O61:-:1,5,7 which was fully sensitive, and originated from a lairage.

On only two occasions could lairage isolates be related to carcass isolates. In both instances, *S.* Mbandaka was found in the lairage before start and on beef carcasses later in the day. Of the 138 isolates, 90 (65.7%) were sensitive to all 16 antimicrobials in the panel, 24 (17.5%) were resistant to Nalidixic Acid, 11 (8%) to Tetracycline, 8 (5.8%) to Compound Sulphonamides and Sulphamethoxazole/Trimethoprim, and 4 (3%) to Chloramphenicol.

There seemed to be little similarity between abattoirs with regard to serotypes present, each abattoir yielding a particular population of *Salmonella enterica* isolates. For example, *S.* Derby and *S.* Dublin were found only at an abattoir processing sheep, calves and pigs. The former was the second most commonly isolated serotype from pigs, and the fourth most common in sheep in 2004 in Great Britain, while the latter is the most common serotype in cattle and the third most common in sheep (Kidd and Papadopoulou 2005). Similarly, *S.* Typhimurium DT208 was found on a single occasion on pig carcasses, and is probably an indication of the *Salmonella* carriage of that particular group of pigs (Botteldoorn *et al* 2004).

S. Mbandaka was isolated from both carcasses and lairage at a cattle abattoir, and was found in the lairage on more than one occasion, suggesting that the organism may be persisting in the lairage, or may be repeatedly introduced with cattle from a particular source. Its presence in the lairage of another cattle abattoir, and the genetic similarity of the isolates may suggest that both abattoirs receive cattle from the same supplier, but this was not possible to ascertain in the present study. The similarity between the *S.* Mbandaka isolates from the lairage and beef carcasses from one abattoir on the related processing day may suggest that the organism has been transferred from the lairage to the slaughterline through cross-contamination, as no cattle were present in the lairage at the time of sampling. Rostagno and others (2003) found similar evidence of environmental contamination matching carcass contamination in pig processing.

A number of isolates were resistant to Nalidixic Acid, an antimicrobial commonly used in marker organism studies. The finding of resistance in a large proportion of field isolates may have implications in such research.

Conclusions

Individual abattoirs yield distinct populations of *Salmonella enterica*, which may be harboured in the lairage environment, or may be repeatedly brought in by the animals processed. It is possible that a persistent strain may develop a stable population within the lairage and could lead to contamination of carcasses, through mechanical transfer onto the animals passing through the facility. Further work is required to explore this possibility. In England, it would appear that antimicrobial resistance of *Salmonella enterica* in the red meat species is quite low, but not zero, and it would be desirable to reduce this further.

References

Botteldoorn, N., Herman, L, Rijpens, N. and Heyndrickx, M. (2004). Phenotypic and molecular typing of *Salmonella* strains reveals different contamination sources in two commercial pig slaughterhouses. *Applied and Environmental Microbiology* 70: 5305-5324

Collis, V. J., C.-A. Reid, M. L. Hutchison, M. H. Davies, K. P. A. Wheeler, A. Small and S. Buncic (2004). Spread of marker bacteria from the hides of cattle in a simulated livestock market and at an abattoir. Journal of Food Protection, 67, 11, 2397-2402.

Kidd, S. and Papadopoulou (2005). Salmonella in Livestock Production in GB 2004, VLA, Addlestone, UK.

Rostagno, M. H., Hurd, H. S., McKean, J. D., Zeimer, C. J., Gailey, J. K. and Leite, R. C. (2003). Preslaughter holding environment in pork plants is highly contaminated with *Salmonella enterica. Applied and Environmental Microbiology* 69, 4489-4494.

Small, A. and Buncic, S. (2005). Prevalence of *Salmonella* spp. In red meat abattoir lairages in the South-west of England, prior to animals entering after routine cleaning operations. In: *Proceedings of the 51st International Congress of Meat Science and Technology,* Baltimore, USA, 7-12 August.

Small, A., James, C., Purnell, G. James, S., Davies, R., Howell, M., Hutchison M. and Buncic S. (2006). Presence of *Salmonella enterica* in the red meat abattoir lairage after routine cleansing and disinfection, and on carcasses. *Journal of Food Protection* submitted.

Acknowledgements

This project was funded by the Food Standards Agency, UK, and would not have been possible without the cooperation of the participating abattoirs.

MOISTURE-ENHANCED PORK – MICROBIAL SAFETY AND EATING QUALITY

A. Gunvig*, and M.A. Tørngren

Danish Meat Association, Danish Meat Research Institute, DK-4000 Roskilde, Denmark.
E-mail: agu@danishmeat.dk

Keywords: *Yersinia enterocolitica*, *Listeria monocytogenes*, texture, taste, odour

Introduction

There is an increasing demand for moisture-enhanced pork in Denmark, as well as in Europe and North America. The moisture-enhancement process guarantees a uniform, tender and juicy product. However, during the brine injection process, the sterile deep tissue can become contaminated with bacteria from the meat surface. Greer *et al.*, (2004) have shown that the number of *Listeria monocytogenes* increased by 2.34 log cfu/100 ml brine after 2.5 h production. This indicates that there is a risk of contamination with *L. monocytogenes* in the deep tissue. *L. monocytogenes* and *Yersinia enterocolitica* are both capable of growing in pork containing 0.8% sodium chloride and stored at 5°C. In order to inhibit these pathogens, it is necessary to add sodium lactate or sodium acetate to the brine.The addition of lactate and acetate salts can increase the tenderness and juiciness of enhanced pork and can significantly reduce the aerobic plate counts (APC) during display compared with untreated controls (Jensen *et al.* 2003).

The aim of the experiment was to evaluate the ability of brines containing sodium lactate and sodium acetate to inhibit the growth of *L. monocytogenes* and *Y. enterocolitica* during 21 days of storage at 5°C. Furthermore, the eating quality of the brine with inhibitory effect was evaluated.

Materials and Methods

Microbial safety experiment:

Nine pigs (sows, weight 74.0-78.0 kg and lean meat percentage 57.0-60.0%) were slaughtered according to standard commercial procedures in a Danish abattoir. pH was recorded at 24 h postmortem in the factory. *M. longissimus dorsi* (LD) and *M. biceps femoris* (BF) were removed, transported to the Danish Meat Research Institute (DMRI) and stored at 2°C. The meat was 2 days old when injected.

At DMRI, the LD and BF were moisture-enhanced and obtained weight gains of 12.1-22.5% (desired gain = 20%) with three different brines: (1) basic brine containing salt, starch and maltodextrin, (2) basic brine with the addition of sodium acetate (0.14%), sodium sesquicarbonate, sodium ascorbate and trisodium citrate, and (3) basic brine with the addition of sodium lactate (2.5%), sodium acetate (0.14%), sodium sesquicarbonate, sodium ascorbate and trisodium citrate. To obtain approximately 10^3 cfu/g in the product, the brines were inoculated with a 3-strain cocktail of *L. monocytogenes* or a 2-strain cocktail of *Y. enterocolitica* O3. The moisture-enhanced LD and BF were cut into approx. 15 cm-long pieces, vacuum-packed and stored for 21 days at 5°C.

After 0, 7, 14 and 21 days' storage, three meat samples for each combination of bacteria, muscle and brine were evaluated for odour and appearance and were then analysed for the presence of listeria or yersinia bacteria on the surface (cfu/cm^2) and in the deep tissue (cfu/g) on Oxford agar (Oxoid) or CIN agar (Oxoid). The APC and the number of *Pseudomonas* were analysed after 0 and 21 days.

The bacterial counts were log transformed and analysed by PROC GLM and grouped using the Duncan procedure (SAS version 9.13). The model included muscle type and brine composition and interactions as fixed effects. Levels of significance: $p > 0.05$= non-significant (ns), $0.05 > p > 0.01$ = *, $0.01 > p > 0.001$ = ** and $p < 0.001$ = ***.

Eating quality experiment:

Pigs were selected and processed in the same way as in the microbial experiment, but only brines 1 and 3 were used. After injection, the pork was frozen to -18°C during storage and thawed for 24 h before sensory analysis. The pork was cut into 2 cm-thick boneless chops and cooked on a frying pan to a core temperature of 75°C. An untreated pork chop was used as a control. The odour, tenderness and juiciness of the untreated and the brine injected chops were evaluated on a 15-point unstructured scale (0 = low intensity, 15 = high intensity) by an 8-member trained sensory panel.

Data were analysed using mixed models (SAS 8.2, 1999-2001). The model included the ageing period, muscle type, cooking temperature, brine composition and interactions as fixed effects, and assessors and pig number as random effects. Non-significant interactions were deleted from the model. Least squares means (Lsmeans) were calculated and separated using the probability of difference. The same significance levels as in the microbial experiment were used.

Results and Discussion

Microbial safety experiment:

The statistical analysis showed no significant difference between the muscles. On days 7 and 21, moisture enhancement with brine 3 resulted in a significantly lower number of *L. monocytogenes* in the deep tissue compared with brines 1 and 2. On day 14, moisture enhancement with brine 3 resulted in a significant lower number of *L. monocytogenes* in the deep tissue compared with brine 1 (control). Enhancement with brine 3 prevented growth of *L. monocytogenes* during 21 days of storage at 5°C.

The difference between brine 3 and brines 1 and 2 was more pronounced in the surface samples. On all days, the number of *L. monocytogenes* on the surface of brine 3 treated pork was significantly lower than surface samples from moisture-enhanced pork injected with brines 1 and 2. There was no significant difference between brines 1 and 2.

Identical results for moisture-enhancement with brine 3 were obtained in the experiment with *Y. enterocolitica*. Moisture enhancement with brine 2 resulted in an inhibition of the growth of 14 days on the surface and 21 days in the deep tissue.

On day 0, the number of *Pseudomonas* in the deep tissue was 1.5 log cfu/g, and after 21 days, the number of *Pseudomonas* in the moisture-enhanced pork injected with brines 1-3 was 6 log cfu/g, 3 log cfu/g and 3 log cfu/g, respectively. This indicates that the addition of sodium lactate and sodium acetate in brines 2 and 3 delayed the growth of *Pseudomonas*. The same tendency was seen in surface samples.

Eating quality experiment:

Uninjected LD chops were statiscally significantly more tender (8.2 points) than BF chops (6.2 points). Brine injection with the two brines resulted in significantly more tender meat than control samples: brine 3 (11.7 points) > brine 1 (10.5 points) > controls (7.7 points). Furthermore, the brine injection had no effect on the uniformity between muscle types, because the difference between LD and BF was maintained.

The juiciness of the meat was the same in uninjected LD and BF chops, and brine injection with brine 1 had only a small effect on BF chops, while LD chops were significantly juicier after brine-injection. However, brine injection with brine 3 resulted in juicier BF chops, while the juiciness of LD chops was the same as the juiciness obtained with brine 1.

The meat flavour was not affected by brine injection with brines 1 and 3, while salt taste and bitter taste were increased by brine injection, especially with brine 3. Finally, vinegar odour was not detected in the control samples and pork injected with brine 1, but was detected (3.6 points) in pork brine injected with brine 3.

Conclusions

The addition of 0.14% sodium acetate and 2.5% sodium lactate to a brine containing salt (0.8% in product), starch and maltodextrin inhibited the growth of *L. monocytogenes* and *Y. enterocolitica* during 21 days of storage at 5°C. Furthermore, the growth of *Y. enterocolitica* was inhibited when the pork was moisture-enhanced with a brine containing salt (0.8% in product), starch and maltodextrin and 0.14% sodium acetate during14 days of storage at 5°C.

The addition of sodium lactate (2.5%) and sodium acetate (0.14%) to brines delayed the growth of *Pseudomonas* in moisture-enhanced pork. Furthermore, this brine composition resulted in improved tenderness, juiciness and salt taste, as expected. Unfortunately, it also resulted in a more intense bitter taste and vinegar odour compared with an untreated control sample. Compared with brine 1, the addition of sodum acetate and sodium lactate resulted in improved tenderness and a higher intensity of salt taste and vinegar odour.

References

Jensen, J.M., Robbins, K.L., Ryan, K.J., Homco-Ryan, C., McKeith, F.K., Brewer, M.S. (2003). Effects of lactic and acetic acid salts on quality characteristics of enhanced pork during retail display. Meat Science 63: 501-508.

Greer, G.G., Nattress, F, Dilts, B., Baker, L. (2004). Bacterial contamination of recirculating brine used in the commercial production of moisture-enhanced pork. Journal of Food Protection, vol. 67, no. 1: 185-188.

PHYSICOCHEMICAL AND MICROBIOLOGICAL CHANGES IN SPANISH CHORIZO DURING RIPENING: EFFECT OF FAT AND LEAN ORIGIN AND PAPRIKA TYPE

C. Vieira,* B. Martínez, M.T. Díaz, M.J. Sanchez, M. Sanchez and M.D. Garcia

Estación Tecnológica de la Carne, Instituto Tecnológico Agrario de Castilla y León, Apdo. 58, 37770 Guijuelo, Spain. Email: vieallce@itacyl.es

Keywords: chorizo, paprika, ripening process

Introduction

Chorizo is one of the most typical Spanish meat products. Pork meat and pork backfat are the raw materials used in a normal recipe. However, occasionally, beef meat, especially that from lower price retail cuts from adult animals, could be used. In order to produce a high quality product, beef meat could be mixed with high quality fat tissues as that from Iberian pig. Iberian pig fat is characterized by a high percentage of unsaturated fatty acids and its beneficial effect on our health has been widely supported (Petrón *et al.*, 2004) but it has greater susceptibility to oxidation than saturated fat. Although synthetic antioxidants can retard the development of undesirable odours and flavours, questions regarding safety of these compounds together with consumer preference for "natural" products has provided an interest in natural antioxidants. The antioxidant effect of spices, garlic and paprika, has been widely demonstrated (Mateo *et al.*, 1997; Aguirrezábal *et al.*, 2000). Few studies have been done in fermented sausages without artificial additives (Mateo *et al.*, 1997; Aguirrezábal, *et al.*, 2000). Furthermore, no research has been carried out on using mixtures of beef meat and fat and lean from Iberian pigs to manufacture traditional Spanish chorizo. The aim of this study was to analyse the possibilities of using different mixtures of raw materials including shoulder beef meat and different fat tissues from Iberian pigs to make a high quality chorizo. The addition of sweet or spicy paprika was also studied.

Materials and Methods

Four different mixtures were produced as described in Table 1. Two batches were prepared of each mixture depending on the type of paprika used: one with 70% of sweet paprika and 30% of hot paprika (Batch A) and the other one with 70% of sweet paprika and 30% "ocal" paprika (Batch B), resulting in eight different batches. All mixtures had 20g/kg of salt, 30g/kg paprika, 0.35 g/kg oregano and 2g/kg fresh garlic.

Table 1: Raw materials and paprika type used in sausages.

	Beef meat	Cow fat	Iberian pig lean-fat	Iberian pig fat tissue	Paprika
Mixture 1A	60%	40%	-	-	Ocal/sweet
Mixture 2A	60%	-	40%	-	Ocal/sweet
Mixture 3A	50%	-	50%	-	Ocal/sweet
Mixture 4A	60%	-	-	40%	Ocal/sweet
Mixture 1B	60%	40%	-	-	Hot/sweet
Mixture 2B	60%	-	40%	-	Hot/sweet
Mixture 3B	50%	-	50%	-	Hot/sweet
Mixture 4B	60%	--	-	40%	Hot/sweet

After stuffing, the conditions during ripening were, 5 ± 2°C and 75% relative humidity for the first ten days and 7± 2°C and 65% relative humidity until the end of the ripening process (26 days). Three samples of each mixture were taken just before stuffing (day 0), and further samples were taken after 5, 8, 14, 20 and 26 days. The following analyses were performed on each sample: weight loss, pH, TBARS, Aw and bacteria counts (Aerobic psychotrophs, Enterobacteriace, Lactic acid bacteria).

Results and Discussion

Figures 1-3 show the changes in weight, water activity and pH of the different mixtures in relation to the origin of the lean and fat tissues used. No significant effect was found for type of paprika used for weight loss, pH, Aw, TBARs, and microbiological counts. Source of raw materials used significantly affected (P<0.001) weight loss along the ripening process (Figure 1), with the highest values for sausages made from lean and fat from cows, and the lowest for sausages formulated with Iberian pig internal fat, the rest being intermediate.

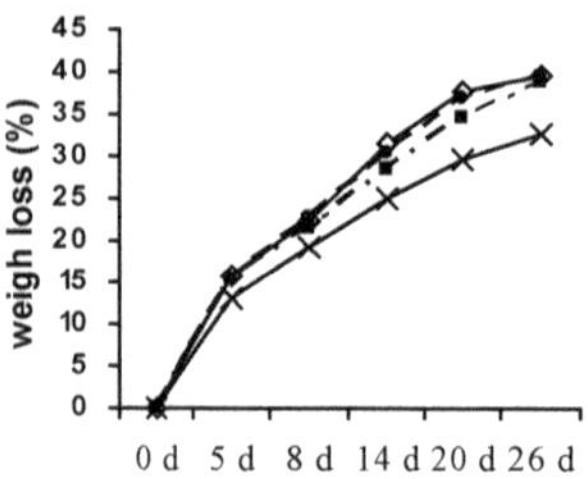

Figure 1: Weigh loss during ripening period

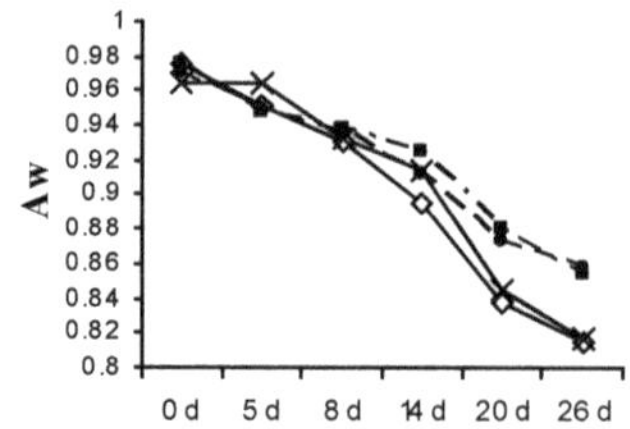

Figure 2: Changes in water activity during ripening period

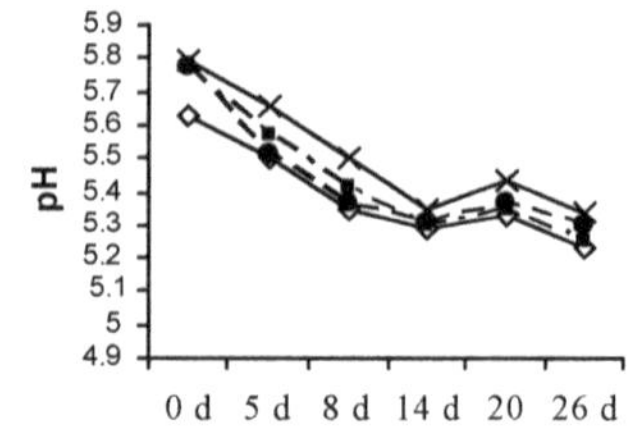

Figure 3: pH changes during ripening period

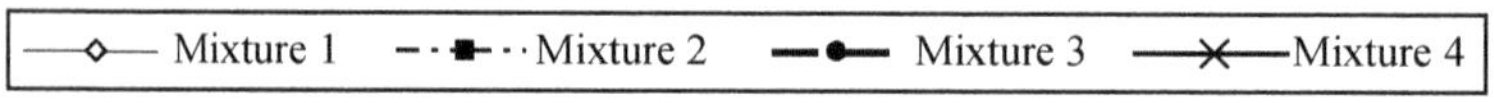

In relation to Aw (Figure 2), as expected, a gradual decrease throughout the ripening process was observed, but statistically significant differences ($p<0.01$) were also found between samplings up to day 8 of ripening. Mixture 1 and Mixture 4 showed a faster decrease ($P<0.05$) in Aw values than Mixtures 2 and 3, which did not differ ($P>0.05$). As is usual in fermented products, a gradual decrease in pH occurred during the first two weeks of ripening ($P<0.01$) (Figure 3). After this, an increase in the pH was observed, followed by a slight decrease. pH value of sausages including Iberian pig fat tissue (mixtures 4) was higher up to day 5 of ripening, without differences among the rest of mixtures manufactured.

The TBARs values decreased during the 5 first days of ripening, increased after until day 14, and then decreased gradually until the end of cured process. Mixture 1 showed the highest TBARs values ($P<0.01$) throughout the ripening process. In any case, values ranged from 1.86 to 4.5 mg/kg, which are in the line to that observed by Aguirrezábal *et al.*, (2000), and they can be considered acceptable. In this sense, several authors have observed that during the ripening of dry fermented sausage lipid oxidation did not occur, probably because the antioxidative effect of the species, curing agents and smoke (Domínguez and Zumalacarregui, 1991; Revilla and Quintana, 2006).

No significant differences were found for all microbiological groups tested ($P>0.05$) regarding raw materials used. The $\log_{10}$ value of total aerobic mesophilic bacteria was around 8 throughout the ripening period but *Enterobacteriaceae* counts decreased gradually throughout the cured process from a $\log_{10}$ value around 6 to a mean value around 4. Lactic acid bacteria counts increased gradually from a mean $\log_{10}$ value of 5 just after stuffing to a mean value of 8 at the end of the cured process, but statistically significant differences ($P>0.05$) were only found between days 0, 5 and 8.

Conclusion

From our results, physicochemical and microbiological changes during ripening process guaranteed the safety of the products elaborated. Since TBARs values can be considered acceptable, rancidity flavours are not expected. So, beef meat can be used as an alternative raw material to elaborate a quality product but it should be mixed with Iberian pork fat to decrease weigh loss.

Acknowledgements

The present work has been supported by Instituto Tecnológico Agrario de Castilla y León (ITACyL).

References

Arriguezabal, M.M., Mateo, J., Domínguez, M. C., and Zumalacarregui, J.M. (2000). The effect of paprika, garlic and salt on rancidity in dry sausages. Food Chemistry, 54, 77-81.

Domínguez, M. C., and Zumalacarregui, J.M. (1991). Lipolytic and oxidative changes in “chorizo” during ripening. Meat Science, 29, 99-107.

Mateo, J., Aguirrezábal, M., Domínguez, C.and Zumalacarregui, J.M. (1997). Volatile compounds in Spanish paprika. Journal of Food Composition and Analysis, 10, 225-232

Petrón, M. J., Muriel, E., Timón, M. L., Martín, L. and Antequera T. (2004). Fatty acid and triacylglycerol profiles from different types of Iberian dry-cured hams. Meat Science, 68 (1) 71-77.

Revilla, I. and Quintana, A.M.V. (2006). The effect of different paprika types on the ripening process and quality of dry sausages. International Journal of Food Science and Technology, 40, 411-417.

INFLUENCE OF PRE-SLAUGHTER HOLDING ON THE BACTERIAL QUALITY OF MARKET HOGS AND CARCASSES

F.J. Tan[1], C.C. Fang[1], C.Y. Cheng[1], Y.Y. Tseng[1] , Y.C. Lin*[2] and D.C. Liu[2]

[1.] *Department of Animal Science, National University of Science and Technology, Pingtung, 912, Taiwan.* [2.] *Department of Animal Science, National Chung Hsing University, Taichung, 402, Taiwan. Email: dcliu@dragon.nchu.edu.tw*

Keywords: pre-slaughter holding, bacterial quality, pigs, carcass

Introduction

During the slaughtering process, if animal's intestinal tracts contain heavy loads of ingesta, not only are they more difficult to handle during evisceration, but also more likely to be accidentally cut, thus contaminating the carcass. Therefore, for sanitary reasons, in modern slaughtering facilities, hogs are normally held off feed for 16 to 24 hours prior to slaughter. During preslaughter holding, water supplying and resting also reduce animal stress. Properly cleaning of the hogs at holding pens during this period of time can also reduce contamination. Several studies have investigated preslaughter holding treatment effects on the animal and carcass quality (Stalder *et al.*, 1998; Hurd *et al.*, 2001; Hurd *et al.*, 2002). The objective of this study was to investigate the influence of preslaughter holing on the bacterial qualities of the surfaces of pigs and carcasses.

Materials and Methods

Pigs were slaughtered at market weight (approximate 100±20 kg) in 4 abattoirs in Taiwan. In this study, "before preslaughter holding" referred to the time when hogs arrived at the abattoirs, which was about 6-7 pm; "after preslaughter holding" referred to the time when pigs were held for approximate 12 hour and right before being slaughtered; "after slaughtering" referred to the time, after slaughtering and right before the final carcass washing. Swab samples of 100 cm^2 were collected from surfaces on the pigs or carcasses at the location on the belly side 10 cm below central back lines. Each swab was stomached with 10 ml of 0.1% w/v peptone water. Serial dilutions were then made with 0.1% peptone as the diluents, and 1 ml portions of each homogenate were added and mixed with the selected pre-autoclaved melt agar on duplicate plates. The plates were incubated, and the plates evaluated for total plate counts, *Pseudomonas*, *Listeria*, *Enterococcus*, *Bacillus cereus*, *Staphylococcus aureus*, *Salmonella* and *Escherichia coli* which were enumerated and identified accordingly. Microbial counts in this study were expressed as $\log_{10}$ CFU per cm^2 of swabbed area. Data were analyzed using the SAS software.

Results and Discussion

In Taiwan, there are two major ways of dealing with hogs after auction. In this study, hogs in the "meat market slaughter plant" were first moved to holding pens at the meat market slaughter plant, washed, rested, held, and then slaughtered; hogs going to "abattoirs A, B, or C" were first mixed after auction, transported to the appropriate abattoir to be held, washed, rested, and then slaughtered. It was found that the microbial counts including total plate, *Pseudomonas*, *Enterococcus*, *Bacillus cereus*, *S. aureus*, *Salmonella* and *E. coli* on the surfaces of hog in abattoir A before preslaughter-holding were significantly higher than those at the other three abattoirs (Table 1). After approximately 12 hrs holding, the microbial counts on the surfaces of hogs in abattoir A were still higher than those of the hogs in other abattoirs (Table 2).

After slaughter by common commercial procedures, the microbial counts detected on the carcass surfaces decreased probably due to scalding and dehairing. In addition, most of the microbial counts on the surfaces of the carcasses slaughtered at abattoir A were still significantly higher than those in other abattoirs (Table 3). Mixing animals during holding could contribute more stress. Hurd *et al.* (2002) stated that rapid infection during transportation and during holding was a major cause of increased *Salmonella* prevalence in swine and suggested that holding pen should be chosen as an important control point for *Salmonella*. Also, in this study, a non-functionally water spraying and cleaning system in abattoir A, thus led to non-thoroughly and not-effectively washing and cleaning might be the major reason which led to higher microbial counts detected on the surfaces of hogs in abattoir A. It was also observed that the hog-driving workers in these abattoirs also helped in other tasks such as dehairing, eviscerating, and etc. due to lack of sufficient labors, thus led to higher likelihood of cross-contamination of carcasses.

Table 1: Microbial quality of pig surfaces before preslaughter-holding.

Location / Microorganisms	Meat market slaughtering plant (n=54)	Abattoir A (n=71)	Abattoir B (n=71)	Abattoir C (n=71)
	log CFU/cm^2			
Total plate count	$5.99^b \pm 0.83$	$6.48^a \pm 1.49$	$5.34^c \pm 0.70$	$5.54^c \pm 1.03$
Pseudomonas	$6.00^a \pm 1.17$	$5.78^{ab} \pm 1.08$	$4.65^c \pm 1.21$	$5.54^b \pm 0.96$
Listeria	$4.85^a \pm 0.76$	$4.19^b \pm 0.57$	$3.91^c \pm 0.66$	$4.15^b \pm 0.76$
Enterococcus	$4.08^b \pm 1.39$	$4.62^a \pm 0.70$	$4.14^b \pm 0.52$	$3.75^c \pm 1.00$
Bacillus cereus	$5.54^b \pm 1.53$	$6.94^a \pm 0.84$	$5.28^b \pm 1.84$	$5.56^b \pm 1.26$
Staphylococcus	$5.68^b \pm 1.28$	$6.39^a \pm 1.37$	$5.20^b \pm 1.23$	$5.16^b \pm 2.24$
E. coli	$1.50^b \pm 0.77$	$2.02^a \pm 0.87$	$1.57^b \pm 0.85$	$1.44^b \pm 0.93$

$^{a-c}$Means within a row that have different superscript are significantly different ($p<0.05$).

Table 2: Microbial quality of pig surfaces after preslaughter-holding.

Location / Microorganisms	Meat market slaughtering plant (n=60)	Abattoir A (n=72)	Abattoir B (n=72)	Abattoir C (n=72)
	log CFU/cm^2			
Total plate count	$5.67^b \pm 1.66$	$7.06^a \pm 1.27$	$5.71^b \pm 1.08$	$5.99^b \pm 1.06$
Pseudomonas	$5.05^c \pm 1.10$	$6.07^a \pm 0.05$	$5.54^b \pm 1.08$	$6.01^a \pm 0.77$
Listeria	$4.51^c \pm 0.65$	$5.07^a \pm 0.70$	$4.67^{bc} \pm 0.75$	$4.86^b \pm 0.69$
Enterococcus	$4.13^c \pm 1.48$	$4.78^a \pm 0.57$	$4.23^b \pm 0.46$	$4.40^b \pm 0.93$
Bacillus cereus	$5.61^b \pm 1.23$	$6.31^a \pm 1.54$	$5.69^b \pm 2.07$	$6.71^a \pm 0.79$
Staphylococcus	$5.52^c \pm 1.04$	$6.67^a \pm 1.02$	$5.56^c \pm 1.35$	$6.24^b \pm 0.70$
E. coli	$1.18^c \pm 0.90$	$2.28^a \pm 0.99$	$1.77^b \pm 0.71$	$1.96^b \pm 0.87$

$^{a-c}$Means within a row that have different superscript are significantly different ($p<0.05$).

Table 3: Microbial quality of carcass surfaces after slaughtering.

Location / Microorganisms	Meat market slaughtering plant (n=60)	Abattoir A (n=72)	Abattoir B (n=72)	Abattoir C (n=72)
	log CFU/cm^2			
Total plate count	$1.76^c \pm 1.81$	$3.84^a \pm 1.05$	$2.39^b \pm 0.74$	$2.49^b \pm 1.39$
Pseudomonas	$4.84^a \pm 1.41$	$3.49^b \pm 0.88$	$1.72^c \pm 0.83$	$2.03^c \pm 1.00$
Listeria	$1.67^b \pm 0.76$	$2.63^a \pm 0.99$	$1.51^b \pm 0.61$	$1.66^b \pm 1.11$
Enterococcus	$0.92^b \pm 0.51$	$2.80^a \pm 0.82$	$2.08^b \pm 1.57$	$0.72^c \pm 0.65$
Bacillus cereus	$3.35^b \pm 0.68$	$3.67^a \pm 1.35$	$2.55^b \pm 1.10$	$2.51^b \pm 0.87$
Staphylococcus	$1.61^a \pm 1.32$	$1.54^a \pm 0.91$	$1.11^b \pm 0.79$	$0.80^b \pm 0.93$
E. coli	$2.13^a \pm 1.45$	$1.15^b \pm 0.83$	$0.96^b \pm 0.76$	$1.02^b \pm 1.05$

$^{a-c}$Means within a row that have different superscript are significantly different ($p<0.05$).

Conclusions

In conclusion, preslaughter holding affected the microbial qualities of market hogs and carcasses. Moreover, proper handling during holding, and cleaning, transport, skinning and prevention of cross-contamination could lead to a decrease in microorganisms detected on carcasses.

References

Hurd, H.S. McKean, J.D., Wesley, I.V., Karriker, L.A. (2001). The effect of larirage on *Salmonella* isolation from market swine. Journal of Food Protection, 64: 939-944.

Hurd, H.S. McKean, J.D., Griffith, R.W., Wesley, I.V., Rostagno, M.H. (2002). *Salmonella enterica* infections in market swine with and without transport and holding. Applied and Environmental Microbiology, 68(5): 2376-2381.

Stalder, K.J., Maya, J., Christian, L.L., Moeller, S.J., Prusa, K.J. (1998). Effects of preslaughter management on the quality of carcasses from porcine stress syndrome heterozygous market hogs. Journal of Animal Science, 76:2435-2443.

EVALUATION OF *TOONA SINENSIS* ETHANOL EXTRACT AGAINST SELECTED MICROORGANISMS

F.J. Tan[1], R.W. Chen[1] and M.T. Chen*[2]

[1.] *Departament of Animal Science, National University of Science and TEchnology, Pingtung, 912, Taiwan,* [2.] *Department of Bioindustry Technology, Da-Yeh University, Changhua, 515, Taiwan. Email: tanfj@mail.npust.edu.tw*

Keywords: *Toona sinensis*, ethanol extract, microorganism

Introduction

Due to health concerns, consumers tend to accept "natural" antimicrobial agents rather than the artificial ones. Recently, antimicrobial properties of plant and spice extracts have been widely reported. *Toona sinensis* is a perennial tree that has become widely grown in Taiwan and China (Edmonds and Staniforth, 1998). Its tender buds have a special aroma and they are often consumed in Taiwan. Even though *Toona sinensis* was reported to have some antimicrobial activity (Shi, 2003), limited information regarding the antimicrobial effect of this plant was available. The objective of this study was to evaluate the antimicrobial effect of *Toona sinensis* ethanol extract against some selected common spoilage and pathogenic microorganisms in meat.

Materials and Methods

Fresh tender buds of *Toona sinensis* were purchased from a local farm in Pingtung, Taiwan, washed, dried in an oven at 45-50°C until totally dried (approximately 24 hrs), pulverized into powder, mixed with ethanol at a 1:20 (w/v) ratio, heated in a water bath of 50°C for 2 hrs, and filtered. The same procedure was conducted twice. The filtered residues of extract were then vacuum condensed at 50°C until dry and stored at 4°C. A 100μl of diluted bacterial suspension of *Staphylococcus aureus* CCRC 12657 (10^9 CFU/ml) or *Escherichia coli* CCRC 11509 (10^9 CFU/ml) was thoroughly mixed with approximately 15 ml TSA, which was pre-autoclaved and cooled to 50°C. When the agar was solidified, 30μl *Toona sinensis* ethanol extract of various concentrations from 0 to 5.0% was added into a well with a diameter of 7 mm on the solidified agar and incubated at 37°C for 24hr. Instead of TSA, a 100μl of diluted bacterial suspension of *Pseudomonas fragi* CCRC 10939 (10^7-10^9 CFU/ml) was mixed with pre-autoclaved NA (Nutrition agar), and incubated at 26°C for 24 hr. The inhibition zones (mm) were then determined.

Minimum inhibitory concentration (MIC)

The MIC of ethanol extract was determined, in triplicate, by an agar dilution method. Stock cultures of *S. aureus* CCRC 12657 and *E. coli* CCRC 11509 were grown in nutrient broth for 12-15 hr, and *P. fragi* CCRC 10939 for 24 hr, and then diluted to 10^5 CFU/ml with autoclaved water containing 0.1% peptone. A 100μl diluted bacteria culture and *Toona sinensis* ethanol extract of 0-4% were mixed thoroughly with approximately 15ml pre-autoclaved 50°C TSA (for *S. aureus* and *E. coli*) or NA (for *P. fragi*), solidified and incubated at 37°C for 24 hr for *S. aureus* and *E. coli* and at 26°C for 24 hr for *P. fragi*. Then the MIC was determined.

Results and Discussion

Concentration effect of Toona sinensis ethanol extract on the selected microorganisms

Table 1 shows that *Toona sinensis* ethanol extract had some antibacterial effects on the microorganisms tested in this study including S. *aureus* CCRC 12657, *E. coli* CCRC 11509, and *P. fragi* CCRC 10939. The higher concentration of the extract that was added, the larger the inhibition zone, which indicated a higher antimicrobial effect. When the concentration of extract was between 0.1 to 1.0%, the diameters of the inhibition zones of *S. aureus* CCRC 12657 were between 10.50 and 11.58 mm without significant ($P < 0.05$) difference in diameters. Adding extract up to 3 and 5%, increased the diameters of the inhibition zone significantly ($P < 0.05$) to 13.92 and 14.00 mm. Similarly, higher concentration of *Toona sinensis* ethanol extract had a larger antimicrobial effect on both *E. coli* CCRC 11509 and *P. fragi* CCRC 10939, which could be observed with larger inhibition zones. In addition, when higher than 0.5% of extract were added, S. *aureus* CCRC 12657 tended to have smaller diameters of inhibition zones when comparing with *E. coli* CCRC 11509 and *P. fragi* CCRC 10939 at the same levels of extract added.

Table 1: Antibacterial effects of *Toona sinensis* ethanol extract against the microorganism tested.

Extract conc. (%)	Inhibition zone diameter (mm)		
	S. aureus CCRC 12657	*E. coli* CCRC 11509	*P. fragi* CCRC 10939
0	–[1]	–	–
0.1	$10.50^{b}\pm1.32$[2]	$10.17^{f}\pm0.29$	$10.33^{c}\pm0.58$
0.3	$11.00^{b}\pm1.32$	$11.42^{e}\pm0.52$	$11.00^{bc}\pm0.00$
0.5	$11.58^{b}\pm2.24$	$12.67^{d}\pm0.58$	$13.50^{ab}\pm0.87$
0.7	$11.08^{b}\pm0.63$	$14.00^{c}\pm0.00$	$14.17^{a}\pm0.29$
0.9	$10.67^{b}\pm1.18$	$14.00^{c}\pm0.00$	$15.17^{a}\pm2.02$
1.0	$10.75^{b}\pm0.43$	$14.08^{c}\pm0.14$	$15.17^{a}\pm2.02$
3.0	$13.92^{a}\pm0.52$	$15.33^{b}\pm0.29$	$16.17^{a}\pm2.25$
5.0	$14.00^{a}\pm0.50$	$16.67^{a}\pm0.29$	$15.83^{a}\pm1.60$

[1]'–': Inhibition zones cannot be detected.
[2]a-f: Means within a column that have different superscripts are significantly different ($P < 0.05$).

The minimum inhibitory concentration of Toona sinensis ethanol extract against the microorganisms tested
The minimum inhibitory concentration (MIC) values obtained for *Toona sinensis* ethanol extract against the three selected microorganisms are shown in Table 2. The MIC value of *Toona sinensis* ethanol extract against *S. aureus* CCRC 12657 was the lowest which was 1.2%, followed by 3.2% of *E. coli* CCRC 11509 and the highest of 4.0 % of *P. fragi* CCRC 10939.

Table 2: Minimum inhibitory concentration (MIC) of *Toona sinensis* ethanol extract against the microorganism tested.

Microorganism	MIC (%)
S. aureus CCRC 12657	1.2
E. coli CCRC 11509	3.2
P. fragi CCRC 10939	4.0

Conclusions
In conclusion, *Toona sinensis* ethanol extract had antibacterial effect against *S. aureus* CCRC 12657, *E. coli* CCRC 11509, and *P. fragi* CCRC 10939.

References
Edmonds, J.M. and Staniforth, M. (1998). *Toona sinensis*. Curtis's Botanical Magazine, 15: 186-193.
Shi, Y.F. (2003). Herb knowledge guide. Suncolor Culture Co., Ltd., Taipei, Taiwan.

Acknowledgements
This work was supported by grant NSC 93-2313-B-020-028 from the National Science Council, Taiwan, ROC.

"SREMSKA" SAUSAGE WITH THE ADDITION OF PROTECTIVE CULTURES AND BACTERIOCINS

S. Saicic*[1], D. Karan and S. Veskovic-Moracanin

Institute of Meat Hygiene and Technology, 11000 Belgrade, Serbia and Montenegro.
Email: meatinst@beotel.yu

Keywords: Sremska sausages, protective starter cultures, bacteriocin, physico-chemical analysis

Introduction

"Sremska" sausage belongs to the group of very desirable dry fermented products of high quality (Tojagic, 1980). In order to ensure product safety and traditional sensory properties, it is useful to add to the product, components with some protective cultures, isolated from naturally fermented sausages as well as bacteriocins. With these additions, growth of some pathogen microorganisms and spoilage microorganisms is inhibited extending shelf-life (De Vuyst and Vandamme, 1994).

The main task of the research was to determine if the protective cultures and bacteriocins have any influence on physico-chemical changes and sensory characteristics of "sremska" sausage during production (smoking, fermentation, ripening and drying).

Materials and Methods

For production of "sremska" sausages pork and frozen pork-back fatty tissue, as well as, nitrite and NaCl mixture as additives and spices were used. The stuffing mass was divided into four parts: control sample (c); with protective cultures (*Lactobacillus sakei* I 154) (I); with protective cultures and bacteriocins (*L. sakei* I 154 and *Leuconostoc mesenteroides* E 131) (II) and with bacteriocins (*Ln. mesenteroides* E 131) (III). Sausage diameter was 32 mm. Smoking and ripening of sausages were carried out under controlled conditions, 21 days at 14-16°C with relative humidity 90-85% and 85-75%. Samples were taken from each batch at day 0, 3, 4, 7, 14 and 21during production process and subjected to physicochemical analysis (moisture and NaCl content, pH and a_w). Standard analytical methods were used. At the end of production, samples were tested by 10 panelists using quantitative descriptive test (JUS ISO 6658/2002) for evaluation of sensory characteristics of the product.

Results and Discussion

During production of "sremska" sausages, moisture content was gradually reduced, except between day 3 and 4, when a slight stagnation was noticed (Figure 1). The reason probably lies in temperature changes, from 16°C (drying and fermentation) to 14-15°C (ripening and drying), what probably decreased water evaporation. Moisture content of sausages was reduced slightly on average from 50% (day 0) to 20% (day 21). It is obvious that the samples with protective cultures and bacteriocins had lower water content compared with control samples. Faster decrease of pH was observed in experimental groups of the product, compared with control sample (Figure 3). Due to the influence of protective cultures and bacteriocins, faster and complete decomposition of added sugar can be expected, resulting in faster production of lactic acid and lower pH of sausages. a_w and NaCl values (Figure 2 and 4) were changed according to water content reduction of sausages. Protective culture and/or bacteriocins showed no undesirable influence on sensory characteristics of "sremska" sausages (Figure 5).

Conclusions

On the basis of the research results, addition of protective cultures and/or bacteriocins had no influence on physicochemical characteristics of "sremska" sausages. On the contrary, they had a positive effect on sensory characteristics of experimental sausages.

References

De Vuyst, L. and Vandamme, E. (1994) (Eds). Bacteriocins of lactic acid bacteria. Blackie, London, 151- 221.

Tojagic, S. Promena biohemijskih i organoleptickih svojstava sremske kobasice tokom proizvodnje. (1980) Ph.D. thesis, Tehnological faculty, Novi Sad.

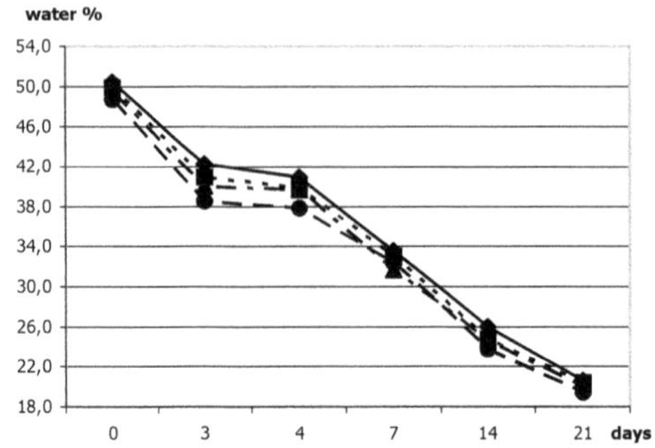

Figure 1: Changes of moisture content (%) during production of "sremska" sausages.

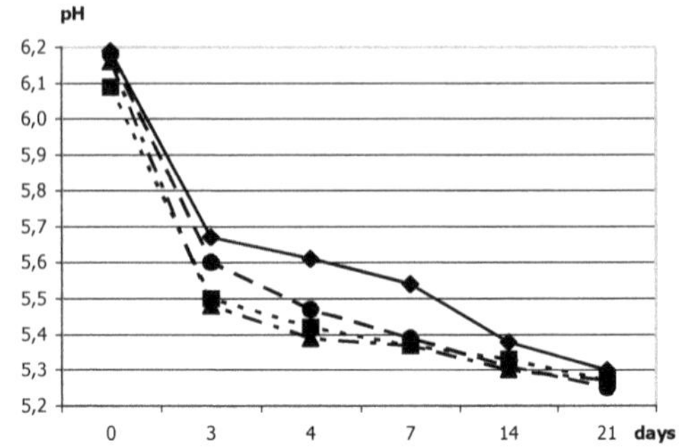

Figure 2: Changes of NaCl content (%) during production of "sremska" sausages.

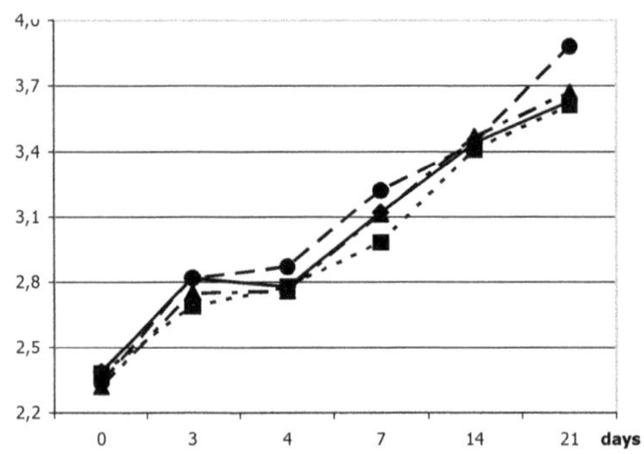

Figure 3: Changes of pH during production of "sremska" sausages.

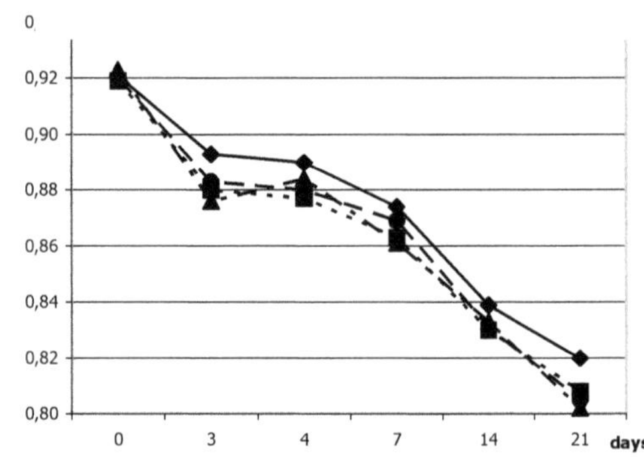

Figure 4: Changes in a_w values during production of "sremska" sausages.

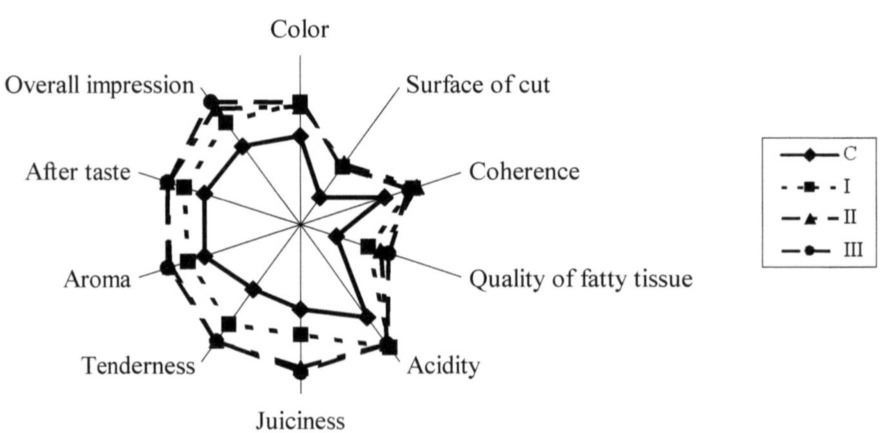

Figure 5: Evaluation of sensory characteristics of "sremska" sausages.

EFFICACY OF SIMPLE METHODS OF CLEANING FOR RED MEAT ABATTOIR LAIRAGES

A. Small*[1,3], C. James[2], G. Purnell[2], P. Losito[1], S. James[2] and S. Buncic[1]

[1.] *Division of Farm Animal Science, University of Bristol, Langford House, Bristol BS40 5DU, UK.* [2.] *Food Refrigeration and Process Engineering Research Centre (FRPERC), University of Bristol, Churchill Building, Langford House, Bristol BS40 5DU, UK.* [3.] *Current address: Food Science Australia, PO Box 3312, Tingalpa DC, QLD 4173, Australia. Email:alison.small@csiro.au*

Keywords: abattoir, lairage, cleaning, washing

Introduction

In the abattoir lairage, many animals from numerous sources are gathered in close proximity (Jarvis *et al.*, 1996), potentially carrying foodborne pathogens on their coats and in their intestines (Paiba and Gibbens, 2000; Reid *et al.*, 2002). Such organisms may be shed into the environment during the lairaging period, and can remain on the surfaces within the lairage for a number of days, particularly in the presence of faecal material (Gibson, 1961; Small *et al.*, 2003), and these organisms may contaminate the animals passing through the facility (Collis *et al.*, 2004). Abattoir operators are recommended to keep lairages clean, but routine cleaning procedures are often insufficient to remove significant pathogens from the environment (Swanenburg *et al.*, 2001, Schmidt *et al.*, 2004). This study was carried out to evaluate some simple methods of lairage cleansing found to be in use at commercial abattoirs in the UK (Small *et al.*, submitted for publication).

Materials and Methods

A mechanical rig was designed to effect consistent repeatable application of cleaning treatments, and this was used to clean concrete slabs designated as visually dirty (painted with bovine faeces inoculated with field strains of *E. coli* and *Salmonella* Kedougou) or visually clean (painted with a broth culture of the two organisms). Each of 30 visually clean and 30 visually dirty concrete slabs were cleaned using the following methods: *(i)* Plain hose (mains pressure); *(ii)* Pressure washer *(iii)* Pressure washer with Janitol branded sanitiser (DEB Limited); *(iv)* Steam under pressure. From the results gained, a further 30 visually dirty slabs of concrete were cleaned with each of a combination of *(v)* plain hose followed by steam or *(vi)* pressure wash followed by steam.

Samples were taken from each concrete slab immediately prior to the onset of cleaning, immediately after cleaning and after a one-hour drying period, using a wet/dry swab technique over a templated area of $100cm^2$. Samples were taken into a peptone salt solution in all cases except for those samples taken after the use of Janitol sanitiser, in which case the samples were taken using a proprietary neutralising medium. Samples were processed using standard methods for the enumeration of *Enterobacteriaceae* and the results were analysed by ANOVA using MINITAB software.

Results and Discussion

There was no significant difference in overall reduction in *Enterobacteriaceae* count between plain hose and pressure washer on visually clean surfaces, although pressure washing gave a greater reduction in *Enterobacteriaceae* count immediately after cleaning, and neither was there a significant difference in overall reduction in *Enterobacteriaceae* count between Janitol sanitiser or steam under pressure on visually clean surfaces. There was no significant difference between the overall reductions in *Enterobacteriaceae* achieved by plain hose and pressure wash on dirty concrete when compared with clean concrete, nor between the immediate reductions obtained using pressure wash. However, plain hose gave a greater immediate reduction in *Enterobacteriaceae* count on dirty concrete (2.1 log) than on clean concrete (1.7 log)($P<0.01$), while Janitol sanitiser gave a greater immediate reduction on clean (5.2 log) than on dirty (4.4 log), but less overall (5.2 log versus 5.7 log)($P<0.01$). The use of steam under pressure gave good reductions in *Enterobacteriaceae* count (3.7 log immediate, 5.5 log overall) on a clean surface, this overall reductions being statistically similar to that achieved using Janitol sanitiser. However, in the presence of faecal material, steam under pressure gave the poorest reduction in *Enterobacteriaceae* count (0.9 log immediate, 1.8 log overall). Of the single treatments, Janitol sanitiser gave the greatest immediate reduction in *Enterobacteriaceae* (5.2 log on clean and 4.4 log on dirty surfaces) as a result of the cleaning process, but there was little further effect of drying.

Using a combination of pressure wash followed by steam on a visually dirty surface gave overall reductions in *Enterobacteriaceae* (5.8 log) comparable with those achieved using sanitiser (5.7 log)($P<0.01$), but there was a greater component of drying where the combination cleanse was used. This combination also gave reductions comparable with those seen using steam alone on a visually clean surface (5.5 log), but a combination of plain hose and steam was less effective in cleansing a visually dirty surface. This combination gave results comparable with those achieved using a pressure wash alone (4.1 log and 3.9 log)($P>0.01$). It is possible that allowing a drying phase between the two phases of the pressure and steam combination may give greater reductions in *Enterobacteriaceae*.

The use of a chemical cleaning agent has been reported to be an important step in reducing microbial numbers on stainless steel for the dairy industry (Dunsmore, 1981), but the efficacy of chemical disinfectants or sanitisers is often much reduced in the presence of organic material (Sprenger, 1997), or by usage with water at temperatures below 25°C

(Gelinas *et al.*, 1984). The current study found that where a concrete surface is visually clean, the use of a proprietary sanitiser at maximum recommended concentration, or the application of steam under pressure gave the greatest reductions in microbial contamination. Plain hose or pressure washing gave similar results to one another, findings similar to those reported in the 1970s comparing hot water at low pressure to cold water at high pressure (Dempster, 1977), and were only slightly less effective than steam or sanitiser. Where the concrete surface was visually contaminated with the faecal material, the use of a pressure wash followed by immediate steam application gave reductions in microbial contamination comparable with the use of a proprietary sanitiser at maximum recommended concentration. The use of a pressure wash alone, or plain hose followed by immediate steam application would rank second in effectiveness, both giving similar reductions in microbial contamination, and the use of plain hose alone would rank third. The reduced effect observed in the presence of faecal contamination could be accounted for by the organic material forming a protective layer containing the organisms, and becoming firmly adherent to the concrete surface during the post-deposition period. The use of steam alone on a visually dirty surface was not an effective means of reducing microbial contamination, and was not even sufficient to remove visual faecal contamination.

Conclusions

When cleaning a faecally soiled area such as a holding pen floor, pressure washing immediately followed by steam under pressure gives comparable microbiological cleaning to use of a proprietary sanitiser at maximum recommended concentration. Pressure washing is not as good, but better than plain hose with steam, which in turn is better than plain hose alone. Drying of the surface following cleaning is important to maximize reductions in microbial load. Further work is required to explore the effects of parameters such as water temperature or pressure and angle of application.

References

Collis, V.J., C.-A. Reid, M. L. Hutchison, M. H. Davies, K. P. Wheeler, A. Small A. and S. Buncic (2004). Spread of marker bacteria from the hides of cattle in a simulated livestock market and at an abattoir. Journal of Food Protection, 67, 11, 2397-2402.

Dempster, J. F. (1977). Cold water, ultra-high pressure cleaning of abattoirs. Journal of Hygiene, Cambridge, 78, 11-16.

Dunsmore, D. G. (1981). Bacteriological control of food equipment surfaces by cleaning systems. I. Detergent effects. Journal of Food Protection, 44, 1, 15-20.

Gelinas, P., Goulet, J., Tastayre, G. M. and Picard, G. A. (1984). Effect of temperature and contact time on the activity of eight disinfectants - a classification. Journal of Food Protection 47, 11, 841-847.

Gibson, E. A. (1961). Salmonellosis in calves. The Veterinary Record, 73, 1284.

Jarvis, A. M., Harrington, D. W. J. and Cockram, M. S. (1996). Effect of source and lairage on some behavioural and biochemical measurements of feed restriction and dehydration in cattle at a slaughterhouse. Applied Animal Behaviour Science, 50: 83-94.

Paiba, G.A. and Gibbens, J.C. (2000). Prevalence of faecal carriage of VTEC O157 and other foodborne pathogens by cattle and sheep at slaughter in Great Britain, VLA/MHS Collaborative Project.

Reid, C.-A., Small, A., Avery, S. M. and Buncic S. (2002). Presence of foodborne pathogens on cattle hides. Food Control, 13, 411-415.

Schmidt, P. L., O'Connor, A. M., McKean J. D. and Hurd H. S. (2004). The association between cleaning and disinfection of lairage pens and the prevalence of *Salmonella enterica* in swine at harvest. Journal of Food Protection 67, 7, 1384-1388.

Small, A., Reid, C.-A. and Buncic, S. (2003). Conditions in lairages at abattoir for ruminants in Southwest England and *in vitro* survival of *Escherichia coli* O157, *Salmonella* kedougou and *Campylobacter jejuni* on lairage-related substrates. Journal of Food Protection, 66, 9, 1570-1575.

Sprenger, R. A. (1997). Hygiene for Management. London, Highfield Publications.

Swanenburg, M., Urlings, H. A. P., Keuzenkamp, D. A. and Snijders, J. M. A. (2001). *Salmonella* in the lairage of pig slaughterhouses. Journal of Food Protection, 64, 1, 12-16.

Acknowledgements

This project was funded by the Food Standards Agency, UK.

ASSESSMENT OF THE USABILITY OF CHEMICAL INDICATORS ASSOCIATED WITH GROWTH OF LACTIC ACID BACTERIA IN A PROCESSED MEAT PRODUCT

J.J. Leisner*[1], B.G. Laursen[1] and D.V. Byrne[2]

[1] *Department of Veterinary Pathobiology (IVP), Royal Veterinary and Agricultural University (KVL), Grønnegårdsvej 15, DK-1870 Frederiksberg C., Copenhagen, Denmark. E-mail: jjl@kvl.dk* [2] *Department of Food Science (IFV), Sensory Science, Royal Veterinary and Agricultural University (KVL) Rolighedsvej 30, 5., DK-1958 Frederiksberg C., Copenhagen, Denmark.*

Keywords: processed pork meat, microbiology, *Carnobacterium*, *Lactobacillus*, *Leuconostoc*

Introduction

It has previously proven extremely difficult to apply conventional total or selective microbial counts for the prediction/indexing of 'shelf-life' as defined by sensory/consumer assessment for many types of food products (Anonymous 2000, McMullen and Stiles 1989). Particularly, in the case of processed meat products such as the sliced Danish pork meat product, "rullepølse", the use of such microbial counts has proven highly ineffective in the prediction of the limit of sensory defined shelf-life. This is largely because lactic acid bacteria (LAB) which are not prominent spoilage organisms, constitute the predominant micro flora in vacuum or modified atmosphere packaged products. Rullepølse (pronounced in English as rule-purls) consist of a large flat piece of boned pork belly with skin removed, spread with herbs and seasoning, and then rolled up. It is cooked and shaped into a rectangle, cooled and thinly sliced as a cold cut to be placed on an open-faced sandwich (Danish: smørrebrød) frequently garnished with raw onion rings. This product will on the one hand exhibit a comparatively long shelf life but on the other hand the exact sensory shelf life is very difficult to predict as the numbers of bacteria tend to plateau and ultimately decrease over storage time (Figure 1). Thus, a correlation to consistently decreasing sensory quality is not data analytically viable.

There is a pressing need for the development and evaluation of new methods for the prediction and indexing of decreasing consumer/sensory based shelf-life quality in processed meat products. The development of such an index will ultimately enable the shelf-life of these products to be controlled efficiently and ultimately, we postulate, may allow shelf-life extension from a sensory perspective, via more efficient control of these products during production. This research project proposal is aimed at developing a methodology whereby sensory deterioration to a 'rational rejection point' (a products 'shelf-life') can be established through monitoring chemically the changes in concentrations of bacterial metabolites by the natural micro flora (see Figure 1). Thus, a quality indexing method can be established giving a sound chemical methodology for the prediction of 'sensory based product shelf-life'. This will enable improvement in product quality, potentially enhance designated shelf-life, and certainly reduce product waste. Table 1 gives an overview over potential LAB chemical indicators in rullepølse.

Table 1: Potential lactic acid bacteria chemical indicators of rullepølse quality included in this study.

Indicator	Producer organisms	Present in sterile meat	Usability
D-lactic acid	*Leuconostoc*, some *Lactobacillus*	-	++[a]
L-lactic acid	All lactic acid bacteria	+	+[b]
Acetic acid	*Carnobacterium*, *Leuconostoc*, some *Lactobacillus*	-	+++
Tyramine	*Carnobacterium*	-	++[a]
Arginine/Ornithine	*Carnobacterium*, some *Lactobacillus*	-	+++[a]

[a]Producer organism is not consistently present in products which limit the application
[b]The fact that the compound is also a result of post-mortem glycolysis complicates its use as an indicator

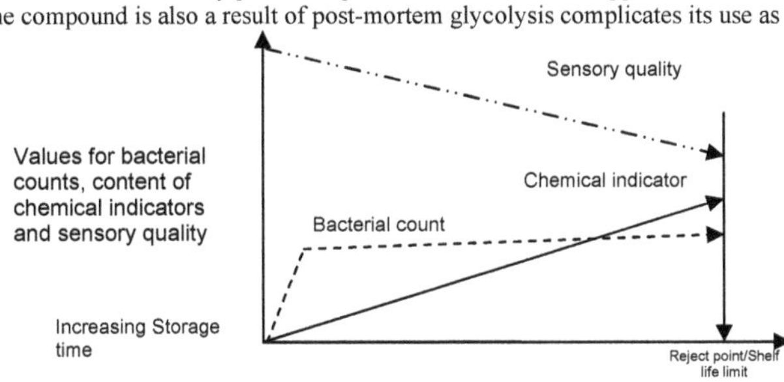

Figure 1: Chemical indicators as a quality index for processed meat products.

Materials and Methods

Bacteriological standard methods included PCA for aerobic counts and NAP agar for LAB counts. Tyramine, arginine and ornithine contents were analyzed by use of a HPLC post column derivatization method (Leisner *et al.* 1994). Total contents of lactic and acetic acid were analysed by a standard HPLC procedure and contents of D and L-isomers of lactic acid were determined enzymically. A meat extract experiment was conducted by using sterile-filtered meat juice inoculated with pools of LAB strains previously isolated from rullepølse products. The same raw material was used for meat juice as is being used for manufacturing rullepølse

Results and Discussion

A preliminary survey examined the content of aerobic mesophilic bacteria, LAB, pH and contents of biogenic amines and the amino acids arginine and ornithine in a range of different rullepølse products (Table 2). Among the LAB only a few different species, *Carnobacterium divergens*, *Lactobacillus sakei* and *Leuconostoc carnosum* were encountered and then only in a selection of the products. Only *C. divergens* produces tyramine and tyramine producers were therefore unevenly distributed in Rullepølse products. This limits the application of this chemical indicator (Table 1). It is therefore of interest to note that both carnobacteria and *Lb. sakei* were able to convert the arginine into ornithine (results not shown). Thus, the arginine/ornithine contents may offer an alternative to tyramine for establishing a rational rejection point.

With this point in mind we conducted a meat juice experiment. Both carnobacteria and *Leuc. carnosum* grew rapidly in this medium whereas *Lb. sakei* showed biphasic growth probably due to exhaustion of a substrate, *e.g.* glucose. Neither *Lb. sakei* nor *Leuc. carnosum* produced any tyramine whereas *C. divergens*_produced tyramine starting at the middle exponential phase and continuing into the stationary phase (results not shown). These results confirms that tyramine production by carnobacteria, if present, appear to be a candidate as an indicator of "the microbial age" of a product.

We also analysed the metabolism of arginine into ornithine and both *C. divergens* and *Lb. sakei* but not *Leuc. carnosum* were able to conduct this reaction (results not shown). Finally, the content of organic acids was analysed by use of HPLC and for some samples the contents of the D and L-isomers of lactic acid were determined. The implications of these findings will be discussed.

Table 2: Characterisation of rullepølse products (MAP or VP[a], pH 5.7-6.6) by aerobic counts (PCA), Lactic Acid Bacteria (LAB) counts, identification (ID) of microflora and analyses of tyramine, arginine and ornithine.

Product	PCA Log (cfu/g)	LAB Log (cfu/g)	ID (NP)[b]	ID (PCA)[b]	Tyramine (µg/ml)	Arginine (µg/ml)	Ornithine (µg/ml)
A	5.4 ± 0.2	5.5 ± 0.1	*Ls*	*Ls*:	<0.5		
B	7.3 ± 0.1	7.3 ± 0.0	*Ls*#, *Cd*§	*Ls*#:, *Cd*§	<0.5		
C	3.9± 0.1	4.1 ± 0.0	*Cd*, *Lc*	*Cd*, *Lc*	0.4	20.5	1.7
D	7.0 ± 0.2	6.8 ± 0.1	*Ls*#, *Lc*¤	*Ls*#, *Lc*¤	<0.5		
E	4.7 ± 0.1	4.7 ± 0.2	*Ls*, *Lc*	*Ls*, *Lc*	0.3	28.9	3.1
F	4.2 ± 0.1	4.2 ± 0.1	*Lc*¤, *Ls*#	*Lc*¤, *Ls*#, NI	<0.5		
G	8.2 ± 0.1	8.2 ± 0.1	*Lc*¤, NI	*Lc*¤, NI	1.9	10.5	9.5

[a]MAP=Modified Atmosphere Packaged, VP=Vacuum Packaged

[b]*Cd*: *Carnobacterium divergens* §or other species, *Lc*: *Leuconostoc carnosum* ¤or other species, *Ls*: *Lactobacillus sakei* #or other species. NI=LAB not identified to species level.

Conclusions

Tyramine producers (*i.e. Carnobacterium* spp.) are unevenly distributed in various products and batches of Rullepølse. This may limit the application of this chemical indicator. An alternative is to monitor the content of free arginine and ornithine as carnobacteria and lactobacilli isolates from rullepølse are able to convert arginine into ornithine by the arginine deiminase pathway. We are currently investigating this issue further.

References

Anon. (2000). ATV-rapport (In Danish): Micro organisms in our food – A focussed approach for higher quality. Final report from the ATV-comittee on microbiological criteria for foods. Academy for technological sciences, ATV.

Leisner, J.J., Millan, J.C., Huss, H.H., Larsen, L.M. (1994). Production of histamine and tyramine by lactic acid bacteria isolated from vacuum-packed sugar-salted fish. Journal of Applied Bacteriology, 76:417-423.

McMullen, L.M., Stiles, M.E. (1989). Storage life of selected meat sandwiches at 4°C in modified gas atmosphere. Journal of Food Protection, 52:792-798

MASS SPECTROMETRY-BASED SENSOR ANALYSIS OF RAW MEAT FOR THE AUTHENTICATION OF ANIMAL FEEDING

V. Vasta*[1,2], J. Ratel[1] and E. Engel*[1]

[1.] *Institut National de la Recherche Agronomique, UR Qualité des Produits Animaux, Equipe Typicité Aromatique et Authentification, 63122 Saint-Genès-Champanelle, France.* [2.] *Dottorato di Ricerca in Produzioni Foraggere Mediterranee, Università di Palermo, Viale delle Scienze 13, 90128 Palermo, Italy. Email: vvasta@unict.it*

Keywords: volatile compounds, raw meat, dynamic headspace-GC-MS, traceability, animal feeding system

Introduction

The study of volatile compounds (VC) in ruminant meat is of great interest because of the role that volatiles have in meat flavour (Rousset-Akrim, *et al.*, 1997). Most of the studies on meat VC have been performed on heated meat fat (Young *et al.*, 1997, Priolo *et al.*, 2004) or on cooked meat comprising muscle external fat (Elmore *et al.*, 1997) in order to reproduce consumers' eating conditions. Apart from the sensorial aspect, consumers demand healthier food and they prefer meat from pasture-fed than meat from stall-fed animals for its "green" image (Grunert *et al.*, 2004). It has been shown that VC in ruminant fat can be used as markers of the animal feeding system, as their occurrence is affected by diet (Vasta and Priolo, 2006). Although the muscle is the portion usually eaten by consumers, to date no studies are available on the use of muscle VC as tracers of animal feeding background probably because the low percentage of the intramuscular fat in which volatiles are accumulated and the presence of water which reduces VC extraction yield (Canac-Artega *et al.*, 2000). The aim of the present study was to investigate raw muscle volatile profile for the authentication of animal feeding system by DHSE-Mass Spectrometry (MS). Firstly, extraction operative conditions were set up by using a DHSE-Gas Chromatography (GC)-MS device: different extraction temperatures (trial 1) and sample sizes (trial 2) were tested in order to reduce the formation of heat-induced compounds and to minimise the release of water during the analysis. Once the best analytical conditions were determined, a third trial was undertaken to assess the application of the DHSE-MS analysis of raw muscle to the authentication of animal diet (pasture *vs* concentrate) (trial 3).

Materials and Methods

Trials 1 and 2. The muscle *semimembranosus* was excised from the carcass of an adult cattle after slaughter and was stored at 4°C for 6 days. Meat was trimmed of any visible fat, cut into small pieces, wrapped in aluminium foil, vacuum packed and stored at -80°C. Meat was defrosted overnight at 4°C and finely ground. For trial 1 fifteen tablets of ground meat were made, each of them weighing 50g (± 2.5g), while for trial 2 ground meat was shaped into four series of five tablets, weighing 50g (± 2.5g), 25g (± 1.25g), 12.5g (± 0.6g) and 6.25g (± 0.3g) respectively. Tablets were immersed into liquid nitrogen for 2 min, wrapped in aluminium foil, vacuum packed and stored at -80°C. Twenty-four hours before analysis, samples were stored at 4°C. *Trial 3.* The right *semimembranosus* muscles were excised from the carcasses of eighteen adult lambs: nine lambs were allowed to graze for 85 days (H= herbage group), while the other nine lambs were confined in stall being offered commercial concentrate for the same period (C= concentrate group). Muscles were handled as described above and from each muscle tablets weighing 6.25g were prepared. *Extraction and Analysis of Volatile Compounds.* Raw meat volatile compounds were extracted and analysed by DHSE-GC-MS (GC: Agilent 5890; MS: 5973N). After defrosting, tablets were spread and cut in small squares to increase the surface of meat samples. Meat was put into a glass cartridge partially filled with glass wool. The meat samples were purged with He as carrier gas (60mL min^{-1}) for 30 min. In trial 1 three different purge temperatures were tested (35°C, 60°C and 90°C) while for trial 2 and 3 purge temperature was set at 35°C. After the purge, a dry-purge step was held for 5 min. Volatiles were trapped using a Tenax A column (Tekmar 3100, Cincinnati, USA); Operative parameters were set as follows: preheat: 15 min at purge temperature; desorption: 10 min at 180°C; cyofocusing at -150°C; desorption into the GC column (Supelco, DB5, 60mx0.32mmx1μm) at 225°C for 2 min; splitless injection; oven temperature: 40°C for 5 min, increasing up to 230°C at 3°C min^{-1}, 230°C for 5 min; GC-MS transfer line temperature: 280°C; electronic impact ionization energy: 70 eV; scanned mass range: 33-230 amu. The identification of volatiles was achieved by comparing mass spectral data with those of the 270 Wiley library (Hewlett Packard Co.) and by comparing the experimental retention indices (KI) with those reported in published databases (Kondjoyan and Berdagué, 1996). The area of each peak was integrated by using ChemStation software (Hewlett Packard Co.). For trial 3, the global spectral were obtained from the sum of the mass spectra generated by the DHSE-GC-MS device. DHSE-GC-MS data were analysed by one way ANOVA of peak area according to the model : peak area = extraction temperature (trial 1), or sample size (trial 2). Similarly, one way ANOVA was processed on the abundance of the DHSE-MS fingerprint fragment according to the model : abundance = feeding system (trial 3). When significant differences were observed, ($P<0.05$), mean values were compared by Newman-Keuls multiple comparison tests. Principal component analysis (PCA) was processed on the DHSE-MS fingerprint to assess the potential of meat VC to discriminate between animal diet .

Results and Discussion

Trial 1: A number of the 222, 214 and 213 peaks respectively were quantified using the extraction temperatures of 90°C, 60°C and 35°C. Among these compounds, 140 were affected ($p < 0.05$) by the extraction temperature. Eighty-eight of them were identified, including 19 alcohols, 16 ketones, 12 aldheydes, 8 aromatic hydrocarbons, 11 alkanes, 7 heterocyclic compounds, 3 sulphuric compounds, 7 halogens and 5 terpenes. Eighty-four identified VC were present at significantly higher amounts in the samples extracted at 90°C compared to those purged at 35 °C or 60 °C ($p < 0.05$) including some potential diet tracers like the 1-alcohols, the 1-alken-3-ols, 2- and 3-ketones or 3-aldehydes. However, the main qualitative information remained exploitable at 35°C and, according to confidence interval values, the quantification of VC was more reproducible. Consequently, the purge temperature was set at 35°C for the following trials. ***Trial 2***: Among the 80 VC that were quantified, 18 were detected at a higher amount ($p < 0.05$) in the samples weighing 6.25g compared to the 50g samples. Probably, the small sample size has decreased the amount of water released from the matrix reducing its interference on VC absorption on the Tenax trap. The analysis of the 6.25g meat samples has the additional advantage of being less destructive and this sample size was selected for trial 3. ***Trial 3***: Thiry-four mass fragments were found to discriminate meat samples according to animal diet ($p < 0.05$). The PCA first map of the DHSE-MS fingerprint (Figure 1) was useful in resolving animals fed different regimes (H or C) demonstrating that muscle volatile compounds allow us to discriminate animal feeding background. Additionally, data of stall-fed lambs showed greater dispersion compared to pasture-fed lambs.

Figure 1: PCA of DHSE-MS fingerprint. Red and blue squares refer to grass- and concentrate-fed lambs.

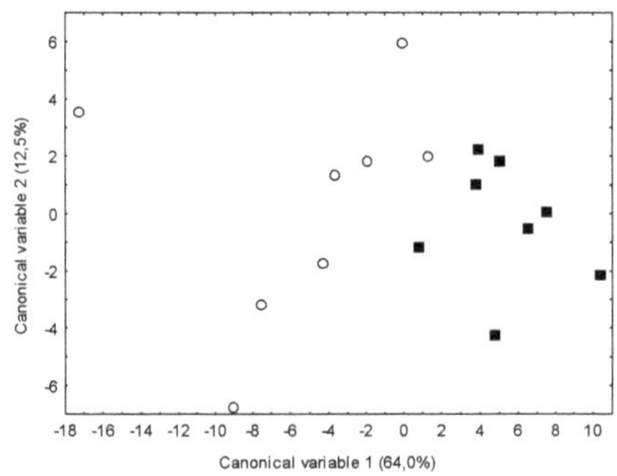

Conclusions

A soft headspace method was set up to extract the volatile compounds released from raw lean muscle. It was found that animal diet (grass or concentrate feeding) affects the appearance of volatiles in lamb muscle. The analysis of the fingerprint of muscle VC allows us to discriminate between meat samples of animals raised at pasture or in stall, thus representing a useful and rapid tool for the authentication of the animal feeding system. The analysis of muscle VC by DHSE-GC-MS could also allow the study of the metabolic origin of molecular markers of animal feeding system.

References

Canac-Artega, D., Viallon, C. and Berdagué, J.L. (2000). Analytical artifacts caused by the presence of water vapour in the headspace of food products. Analysis, 28: 550-556.

Elmore, J.S., Mottram, D.S., Enser, M. and Wood, J.D. (1997). Novel thiazoles and 3-thiazolines in cooked beef aroma. Journal of Agricultural and Food Chemistry, 45, 3603-3607.

Grunert, K.G., Bredahl, L. and Brunsø, K. (2004). Consumer perception of meat quality and implications for product development in the meat sector-a review. Meat Science, 66, 259-272.

Kondjoyan, N. and Berdagué, J. L. (1996). A compilation of relative retention indices for the analysis of aromatic compounds. Publisher: Laboratoire Flaveur, Clermont Ferrand, France.

Priolo, A., Cornu, A., Prache, S., Krogmann, M., Kondjoyan, N., Micol, D. and Berdagué, J.L. (2004). Fat volatiles tracers of grass feeding in sheep. Meat Science, 66: 475-481.

Rousset-Akrim, S., Young, O.A. and Berdagué, J.L. (1997). Diet and growth effects in panel assessment of sheapmeat odour and flavour. Meat Science, 45, 169-181.

Young, O.A., Berdagué, J. L., Viallon, C., Rousset-Akrim, S. and Theriez, M. (1997). Fat-borne volatiles and sheepmeat odour. Meat Science, 45, 183-200.

Vasta, V. and Priolo, A. (2006). Ruminant fat volatiles as affected by diet. A review. Meat Science, 73, 218-228.

MYCOLOGICAL AND MYCOTOXICOLOGICAL ANALYSIS OF SAGE USED IN THE MEAT INDUSTRY

L. Petrović*[1], V. Jankovic[2], M. Škrinjar[1], J. Vukojević[3] and M. Radmili[2]

[1] University of Novi Sad,Faculty of technology, Bulevar cara Lazara 1, Novi Sad, Serbia [2] Institute of meat hygiene and technology, Kaćanskog 13, Belgrade, Serbia [3] University of Belgrade, Faculty of biology, Takovska 43, Beograd, Serbia Email:vessnaantares@yahoo.com

Keywords: moulds, mycotoxins, contamination of spices

Introduction

The production of spices and constantly growing dependance of the food industry as a major consumer of spices, especially the meat industry, is affected by problems related to the presence of microorganisms. Spices are never completely sterile. They contain predominantly sporogenous bacteria and moulds. Control of raw materials and finished products for the presence of moulds and their toxic metabolites – mycotoxins – is exceptionally important in the drive to manufacture products which are safe and good for the human health. Major groups of toxins include: aflatoxins (produced by Aspergillus species; the most important is aflatoxin $B_1(AB_1)$ which is classified in Group 1 of human carcinogens) (Scimeca, 1995), ochratoxins (toxic products of storage moulds of the genera Aspergillus and Penicillium; ochratoxins appear in several forms, the most widespread and also most toxic being ochratoxin A(OA), which is responsible for the Balkan endemic nephropathy) (Mašić *et al.*, 2000) and zearalenon (ZEA) (produced by Fusarium species, primarily by F. graminearum) (Bočarov,1996).

Materials and Methods

Mycological and mycotoxicological research

The mycological research encompassed the determination of the total number of moulds in 1 g of the tested sage (five different samples) and their identification according to the standard laboratory procedure (Škrinjar, 2000), (Raper and Fenel, 1965), (Samson and van Reenen-Hoekstra, 1988), (Ainsworth *et al.*, 1973), (Gerlach and Nirenberg, 1982). Two types of selective culture media were used: Sabouraud-maltose agar (SMA) and maltose yeast extract agar with 50% glucose. The presence of Aflatoxin B_1, ochratoxin A and zearalenon were determined by the ELISA testing method.

Results and Discussion

Table 1: Total viable count of moulds per 1g.

sample	total no. of moulds / g SMA	MY 50 G
1	$4.1 \cdot 10^2$	$7.7 \cdot 10^2$
2	$4.9 \cdot 10^3$	$1.2 \cdot 10^2$
3	$1.5 \cdot 10^2$	$5.4 \cdot 10^2$
4	$3.0 \cdot 10^2$	$2.5 \cdot 10^2$
5	$4.0 \cdot 10^2$	$2.4 \cdot 10^2$

Table 2: Types of moulds isolated in the samples.

genus	species	percentage of total (%)
Aspergillus	*A. flavus*	66.60
	A. niger	
	A. rubrum	
	A. fumigatus	
Fusarium	*F. moniliforme*	16.60
Rhizopus	*R. stolonifer*	16.60

The Total viable count on the two different mediums is shown in Table 1 while Table 2 shows that most of the species found were from the genus *Aspergillus*; they made up 66.6 % of all the isolated species. As shown in Figure 1, storage moulds, most of them toxicogenic, were dominant in the samples. *Aspergillus* accounted for 97.8 % of the overall sage contamination myco-population. Figure 2. shows that *A. flavus was* predominant in all samples, especially in samples 2 (69.76%) and 3 (70%). The frequency of *A. rubrum* in sage's mycopopulations is shown in Figure 3. While Figure4 presents the frequency of *Aspergillus* spp. in sage mycopopulation, with the highest presence of *A .flavus.* AB1, OA and zearalenon found in all samples at a high concentration (Table3).

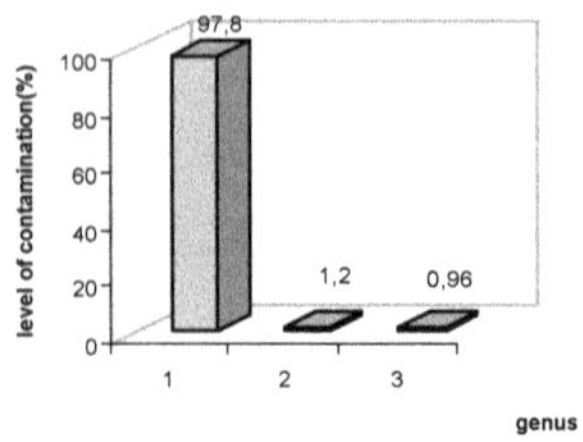

Figure 1: Frequency of genera in the sage myco-population (1-*Aspergillus*, 2-*Fusarium* 3-*Rhizopus)*.

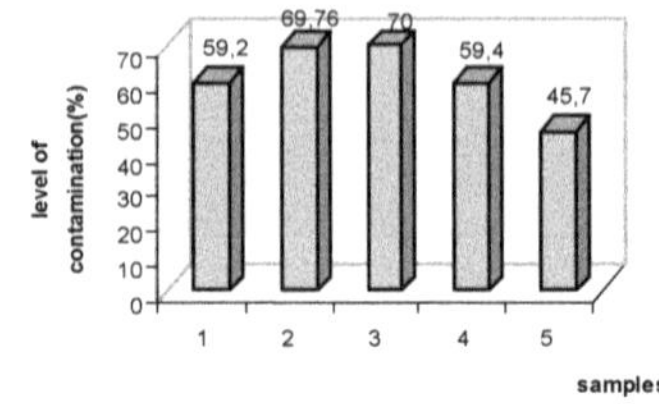

Figure 2: *Aspergillus flavus* frequency in tested samples.

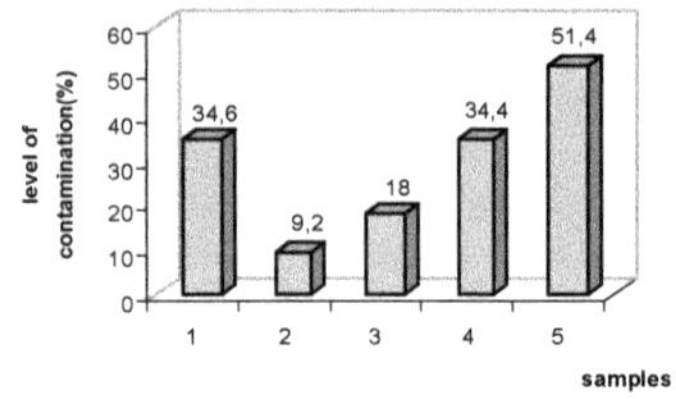

Figure 3: Frequency of *A. rubrum* in tested samples.

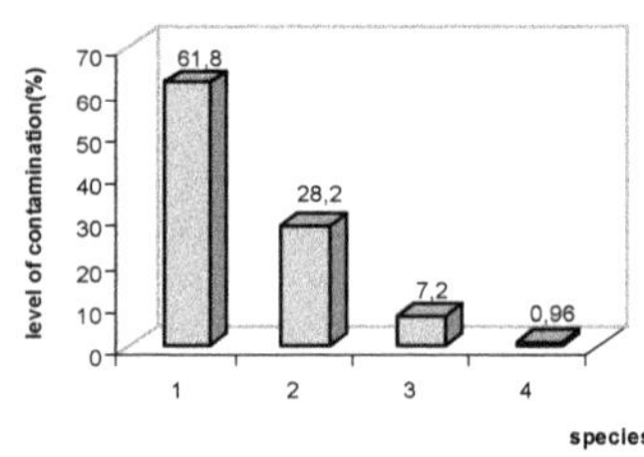

Figure 4: Frequency of *Aspergillus* species in sage (1-*A. flavus,* 2-*A. rubrum*, 3- *A.niger*, 4-*A. fumigatus)*.

Table 3: Mycotoxins in five different samples ($\mu g \cdot kg^{-1}$).

sample	AB1	OA	ZEA
1	7,5	63	20,0
2	7,0	60	25,0
3	12,0	90	25,0
4	10,0	78	12,5
5	8,0	42	19,0

Conclusion

According to the FAO more than 25% of the world's agricultural production is contaminated with mycotoxins. Most countries have adopted regulations to limit exposure to mycotoxins, having strong impact on food and animal crop trade.The presence of mycotoxins is unavoidable and therefore testing of raw materials and products is required to keep our food and feed safe.

References

Ainsworth, G. C., Sparrow, F.K., Sussman, A.S. (1973). The fungi. Academic Press, New York.

Bočarov-Stančić, A. (1996b). Uticaj ekoloških i drugih faktora na rasprostranjenje plesni i mikotoksina u žitaricama imogućnost njihove dekontaminacije. Doktorska disertacija. Tehnološki fakultet, Univerzitet u Novom Sadu.

Gerlach, W., Nirenberg (1982). The genus *Fusarium*. A pictorial atlas mitt boil. Bundesanst, Berlin-Dahlem.

Mašić, Z., Kljajić, R., Bočarov-Stančić, A., Škrinjar, M. (2000). Mikotoksini u stočnoj hrani kao factor poremećaja zdravlja životinja. Zbornik radova I kretkih sadržaja XII savetovanja veterinara Srbije, Vrnjačka Banja, 64-74.

Raper,K.B., Fenell, D.I. (1965). The genus *Aspergillus*.Williams and Wilkins Co., Baltimore.

Samson,R.A.,van Reenen-Hoekstra, E.S. (1988).Introduction to food borne fungi. Centraalbureau voor Schimmelcultures, Baam-Delft, The Netherlands.

Scimeca, J.A. (1995). Naturraly occurring orally active dietary carcinogens. In: Handbook of human toxicology, Massaro, E., J., CRC Press, 435-437.

Škrinjar, M. (2000). Metodi mikrobiološke kontrole životnih namirnica, Tehnološki fakultet, Univerzitet u Novom Sadu.

WHO-PCB-TEQ OF HERBS FROM DIFFERENT ORIGINS

S. Kleinhenz[1], K.-H. Schwind[1], and K.-O. Honikel[1]*

[1]*Federal Research Centre for Nutrition and Food, Location Kulmbach, Institute for Chemistry and Physics, E.-C.-Baumann-Straße 20, 95326 Kulmbach, Germany, karl-otto.honikel@bfel.de*

Keywords: polychlorinated biphenyls, PCB, herbs, origin

Introduction

Polychlorinated biphenyls (PCBs) are a group of 209 congeners with different degrees of chlorination (up to ten chlorine atoms). They are man-made, very stable and persistent substances. Since 1929, PCBs have been industrially produced (1) for the utilisation among other purposes as transformer oil, hydraulic fluid and softening agent. PCBs have been found also as contaminants in several industrial chemicals. Since 1989 (2) the production, the import and export or the sale of PCBs is forbidden in Germany. Nowadays PCBs are still ubiquitous pollutants in the environment of most countries (3). This is the reason why it is very probable that also herbs are contaminated with PCBs. Herbs are among other spices ingredients of meat products and are able to contaminate them in detectable amounts if the PCB concentrations are high. In former studies we detected in some meat products considerable amounts of PCB even if below the maximum residue limits (MRL).

Materials and Methods

Material, reagents and samples

$^{13}C_{12}$ -labelled PCB standards and the recovery standards (deuterised or $^{13}C_{12}$ –labelled) were purchased from Promochem (Wesel, Germany). Internal standard (50µl of each standard solution) was added before sample extraction. The analysed herbs were harvested from private and commercial cultivation. The samples were supplemented by herbs from different regions of the world through a spice company.

Extraction

After lyophilisating of the fresh plant material, extraction cells were filled with sample material, drying substance (poly (acrylic acid) - partial sodium salt-graft-poly (ethylene oxide) from Aldrich (Missouri, USA)) and sea sand. Then the extractions were performed by accelerated solvent extraction (ASE 200 from Dionex). For extraction n-hexane was used at a pressure of 100bar and a temperature of 100°C. The samples were extracted in two cycles with a static time of ten minutes each.

Clean up

The evaporated extracts were separated from unwanted substances on a gel permeation chromatography (GPC) column with an ID of 25 mm that was filled with 60 g Bio-Beads S-X3 (200-400 mesh), a cross-linked divinylbenzene-styrene copolymer, from Bio-Rad Laboratories (München, Germany). The GPC (Abimed Gilson, Langenfeld, Germany) was carried out with a solvent mixture of cyclohexane-ethyl acetate (1:1; v/v) and a flow rate of 5 ml/min. The extracts which were evaporated again were eluted from Florisil columns (8 ml-SPE columns (ID: 12 mm) filled with 3g Florisil deactivated with 4 % water (w/w)) with 60 ml of toluene. After evaporation of the solvent nearly to dryness the samples were dissolved in 1 ml of n-hexane. 8 ml-SPE cartridges with an ID of 12 mm were filled with NaOH-silica and H_2SO_4-silica. The sample extracts were eluted with 60 ml of n-hexane. To divide the extract (concentrated to a volume of 1 ml) into two fractions, a chromatography on activated charcoal followed. In the first fraction (20 ml n-hexane/toluene; 99:1; v/v) the di- and mono-ortho-substituted PCBs and in the second fraction (20 ml n-hexane/toluene; 75:25; v/v) the non-ortho-substituted PCBs were collected. The first fractions were concentrated to about 100µl and the second fractions were concentrated about 15µl. After adding recovery standard the samples were analysed by GC-MS.

Instrument and method

The GC-HRMS system used was a Hewlett-Packard 5890 series II gas chromatograph (Hewlett-Packard, Germany) with an injection in split/splitless mode and combined with a VG Autospec Micromass (Manchester, UK) high resolution mass spectrometer. A ZB-5ms column (ID = 0.25 mm, film thickness = 0.25µm, 5% polysilarylene, 95% polydimethylsiloxane) from Phenomenex (Torrance, USA) with a length of 60m was used. The GC program for the PCB measurement was as follows: 70°C (2 min) to 180°C (0 min) at 30°C/min, to 290°C (0 min) at 5°C/min and subsequently at 20°C/min to 320 °C, then maintained for 10 min at 320°C. PCBs were measured working in the electron impact (EI) positive ion mode with an electron energy of 35eV, the source temperature was set at 250°C. Data was recorded in SIR mode.

Results and Discussion

The analysed herbs were harvested from five different private and commercial places in Germany and obtained from a company. Their sources of origin were Germany, France, Peru, China and Egypt. Among other varieties following species could be analysed: parsley, chive, lovage, thyme, oregano and marjoram. More than hundred samples were analysed.

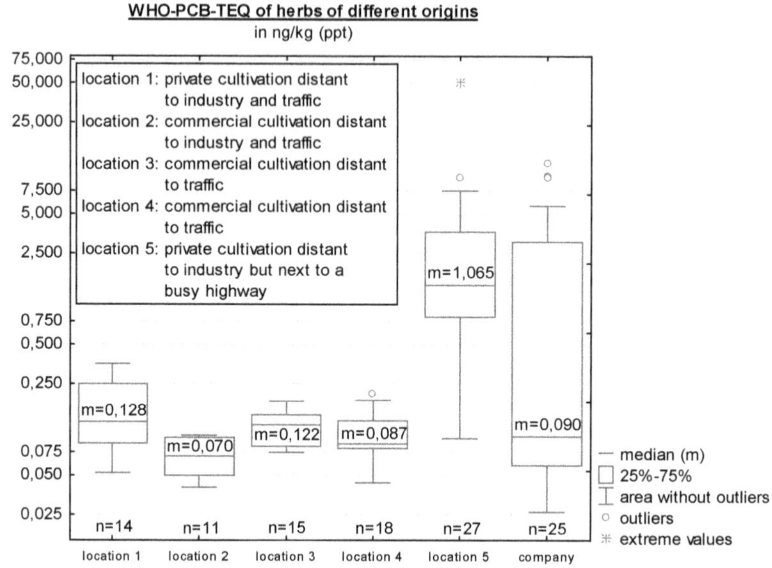

Figure 1: Comparison of WHO-PCB-TEQs in herbs from different cultivation locations in ppt (logarithmic representation) presented as box-whisker-plots.

Comparing the WHO-PCB-TEQs of the herbs in relation (Figure 1) to the different cultivation regions it is important to know the position of the cultivation field. The first four locations including one private and three commercial cultivations of herbs are areas with approximately no contact to any exhaust fumes from industry or traffic. This is the likely reason for the low medians and the small variation of the WHO-PCB-TEQs of these herbs (between 0.070 and 0.128ppt). In the case of location 5 the field for cultivating of the herb are very near to a busy highway. Despite a high distance to industrial areas the contamination of the herbs with PCBs on this location is very high. A median of 1,065ppt (WHO-PCB-TEQ) was measured. With this knowledge it is not surprising that herbs from the company with spices from many countries show a big variation in the measured WHO-PCB-TEQs but with a median of 0.090ppt. Combining all results for all analysed herbs a median of 0,123ppt arises for the WHO-PCB-TEQs.

Conclusions

18 PCBs with a toxicity equivalence factor were analysed for more than hundred herb samples. Most of the herbs have a low contamination with PCBs. Despite the long ban of PCB production (since 1989) traffic near the herb field enhances the PCB contamination by a factor of about ten.

References

Deutsche Forschungsgemeinschaft, Polychlorierte Biphenyle – Bestandsaufnahme über Analytik, Vorkommen, Kinetik und Toxikologie, Mitteilung XIII der Senatskommission zur Prüfung von Rückständen in Lebensmitteln, VCH Verlagsgesellschaft mbH, Weinheim (1988).

Verordnung zum Verbot von polychlorierten Biphenylen, polychlorierten Terphenylen und zur Beschränkung von Vinylchlorid (PCB-, PCT-, VC-Verbotsverordnung), BGBl.I vom 18. Juli 1989, S. 1482.

Kimbrough, R.D., (1985). Laboratory and human studies on polychlorinated biphenyls (PCBs) and related compounds, Environ. Health Persp., 59: 99-106.

Kleinhenz, S., Jira, W., Schwind, K.-H. (2006). Dioxin and polychlorinated biphenyl analysis: Automation and improvement of clean-up established by example of spices. Molecular Nutrition and Food Research, 50 (4-5): 362-367.

STUDY OF THE CONTENT OF CADMIUM AND LEAD IN MEAT OF BROILERS GROWN IN THE NORTH-CAUCASUS REGION OF THE RUSSIAN FEDERATION

I.M. Tchernoukha*, Yu.E. Kalinova and O.N. Orlova

The V.M.Gorbatov All-Russian Meat Research Institute, Talalikhina 26, 109316, Moscow, Russia.
Email: imcher@ inbox. ru

Keywords: heavy metals, broilers' meat

Introduction

Consumer protection from contaminated foods and prevention of foodborne diseases are two main principles of the program to ensure safety of foods. The risk of raw materials and food contamination with potentially dangerous substances can be reduced only with the effective system of the control of foods safety on all stages of production and marketing. Therefore, an urgent problem is the improvement of methodology of assessment of raw materials, food quality and safety. A concept of production of safe meat products is being developed at GNU the V.M.Gorbatov All-Russian Meat Research Institute. It comprises a complex system of quality and safety assurance of protein products based, among other things, on the use of traceability system of safety and quality of the product in the whole food and technological chain of its production, transportation and marketing.

According to FAO/WHO, toxic elements of chemical origin have pronounced carcinogenic and mutagenic effect and take the first place in the list of chemical substances that are dangerous for the environment and human health. Among these substances, heavy metals, coming to the environment as a result of people's activities, take one of the leading positions. The waste and by-products of production processes, pesticides, wastewaters, slags, ash and gases, by-products of transport, enterprises of heavy industry, engineering, instrument engineering, heat and electric power stations, contain a large amount of heavy metals, among which cadmium and lead are the most toxic ones. Owing to their high migration capacity, a tendency to bioaccumulate, specific toxic effects, heavy metals in feeds and foods, a deterioration in sanitary quality, and, if higher than limit values, they are dangerous for the health of people and animals.

Therefore, to develop a traceability system a task was set to create a databank of the level of residual noxious substances present in farm animals from the environment.

Materials and Methods

The study of the level of heavy metals (cadmium, lead) was carried out in the samples of muscular tissue of broilers after 42 days of feeding, grown in agroindustrial complexes (APK) and small private chicken farms, located at different distances from the sources of ecological contamination in the North-Caucasus region of RF (conventionally considered as favourable and unfavourable regions).

The level of heavy metals was determined in the skin, fat, white meat (thoracic muscle) and red meat (leg muscle) of broilers, feed and water with the help of Atomic Absorption Spectrometer VARIAN SpectrAA 220 FS, with background correction (D_2-lamp). All measurements were carried out in triplicate.

Results and Discussion

The levels of cadmium and lead level in the tested samples are presented in Figs 1 and 2.

Analysis of the results of tests of water from APK and private chicken farms, situated in favourable and unfavourable regions shows that the level of both cadmium and lead is ten times lower than the limit values (not higher than 0.02 mg/kg for lead and 0.002 mg/kg for cadmium). The level of lead in all the samples of feeds is considerably lower than the limit values (not more than 2.0 mg/kg). The level of cadmium in the sample of feed also doesn't exceed the established norms (not more than 0.2 mg/kg), however, the sample of the feed from APK, situated in ecologically unfavourable region approached the limit value (0.157 mg/kg). At the same time, no accumulation of cadmium in muscle and fat tissue of broilers grown in APK was observed. The level of cadmium and lead in the white meat of chickens (thoracic muscle) is within the established norms. No significant differences in the level of cadmium were observed in the samples obtained from the farms with different forms of property. At the same time, a higher level of cadmium is observed in red meat (leg muscle 0.02 – 0.026 mg/kg) as compared to white meat (thoracic muscle 0.014-0.019 mg/kg).

There were no differences in lead level for different broiler muscle types. The level of lead in fat tissue samples of broilers, grown in the farms situated in favorable regions of the North-Caucasus region of RF approaches the limit values (not more than 0.1mg/kg), and for similar samples from broilers, grown in the farms situated in unfavorable regions is within the established norms. It should be noted that this pattern doesn't depend upon the type of property of the farms. The level of cadmium in the samples of fat tissue of broilers grown in the APK of both regions is significantly lower than the established norms (not more than 0.03 mg/kg). The level of cadmium in the tested samples of fat tissue of broilers from private farms is different. For the sample from ecologically unfavourable regions the measured values approach the limit values (nor more than 0.03 mg/kg), and for the sample from ecologically favourable regions one can observe practically a three-fold exceeding of cadmium level. The level of cadmium in the skin samples is significantly lower than the established norms (not more than 0.3 mg/kg), and there are no large differences between the samples from different farms. It is also the case with the skin samples of broilers, as far as the content of lead is concerned, however, here, the skin sample from the APK of unfavourable region the measured level of lead is considerably lower as compared to other samples.

Conclusions

- The investigations carried out have shown that the most intensive accumulation of cadmium and lead in the organism of broilers, grown in North-Caucasus region of RF was in the fat tissue.
- The analysis of the obtained data has shown that the division of regions into ecologically favourable and unfavourable is rather conventional, and the analysis of the raw materials of poultry factories should be carried out for the indices of safety everywhere in the North Caucasus region.

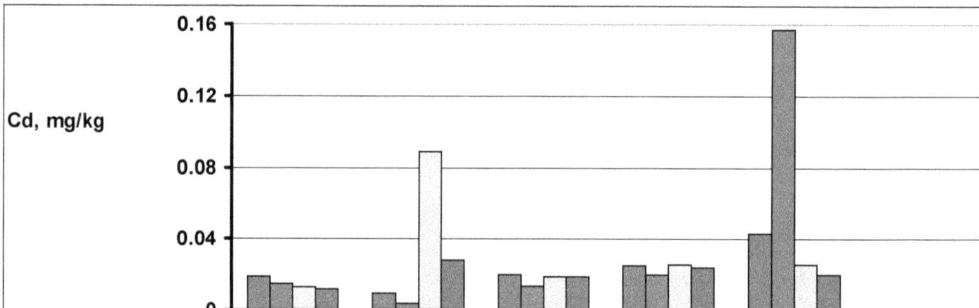

	skin	fat	thoracic muscle	leg muscle	feed	water
favorbl. reg. APK	0.019	0.01	0.02	0.025	0.043	0.0003
unfavrbl. reg. APK	0.015	0.004	0.014	0.02	0.157	0.0004
favorbl. reg. priv.	0.013	0.089	0.019	0.026	0.026	0.0002
unfavrbl. reg. priv.	0.012	0.028	0.019	0.024	0.02	0.0003

Figure 1: Content of cadmium in the samples of poultry meat, feed, water (APK and private farms of North Caucasus region of RF).

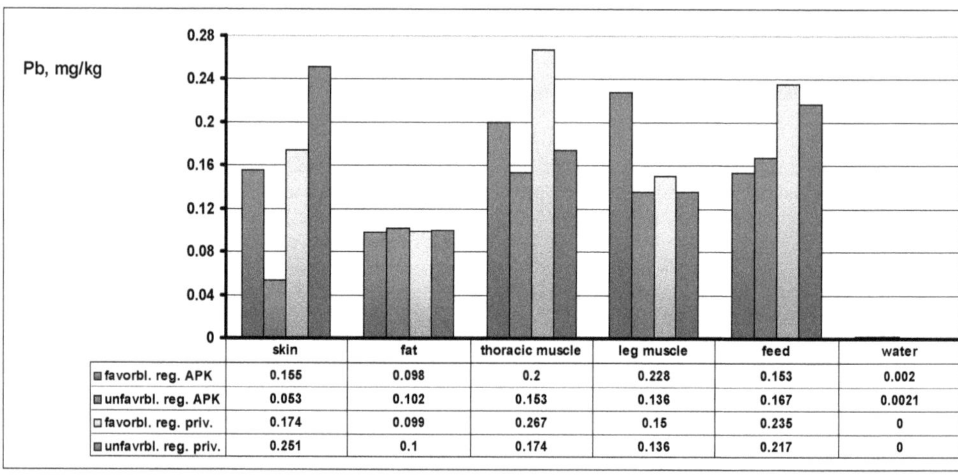

	skin	fat	thoracic muscle	leg muscle	feed	water
favorbl. reg. APK	0.155	0.098	0.2	0.228	0.153	0.002
unfavrbl. reg. APK	0.053	0.102	0.153	0.136	0.167	0.0021
favorbl. reg. priv.	0.174	0.099	0.267	0.15	0.235	0
unfavrbl. reg. priv.	0.251	0.1	0.174	0.136	0.217	0

Figure 2: Content of lead in the samples of poultry meat, feed, water (APK and private farms of North Caucasus region of RF).

HPLC METHOD FOR THE ANALYSIS OF DEXAMETHASONE IN FEED AND WATER FOR LIVESTOCK

M. Reig, L. Mora, M. Flores* and F. Toldrá

Instituto de Agroquímica y Tecnología de Alimentos, (CSIC), P.O. Box 73, Burjassot 46100, Valencia, Spain.
Email: mflores@iata.csic.es

Keywords: dexamethasone, corticoids, growth promoters, feed residues, water residues

Introduction

Dexamethasone (9-α-fluoro-16-α-methylprednisolone) is a synthetic glucocorticoid that is authorised for therapeutic use in veterinary medicine but its use as a growth promoting agent is banned in the European Union. Low concentrations of glucocorticosteroids are known to increase weight gain, to improve feed conversion and to have a synergetic effect with other molecules like beta-agonists or anabolic steroids. Due to these effects, dexamethasone has been illegally used to obtain an economic benefit through increased muscle development. A number of methods for the detection, determination and confirmation of dexamethasone in different biological matrices, like urine, faeces, liver, milk or feed, have been previously reported (Delahaut *et al.*, 1997; Creaser *et al.*, 1998; Stolker *et al.*, 2000; Draisci *et al.*, 2001; Cherlet *et al.*, 2004). Recently, the performance of methods and the criteria for the interpretation of test results of official control laboratories within the European Union has been regulated in the Decision 2002/657/EC (EC, 2002). The presented method, based on inmunoaffinity chromatography followed by Reverse phase HPLC with diode array detection at 242nm, has been validated according to this Decision.

Materials and Methods

Sample preparation: 2g of feed were weighed in a test tube and the internal standard (flumethasone) was added. After fortification and equilibration, the sample was extracted with 10mL of TBME (*tert*-butyl-methyl-ether), shaken for 20 min and centrifuged for 10 min at 2700rpm. The supernatant was collected in a clean glass tube, and the extraction was repeated. Both supernatants were loaded into an amino propyl (NH_2) cartridge (500mg) and eluted with 4 mL of methanol-water (80:20, v/v). The elutant was evaporated to dryness at 45°C under a stream of nitrogen.

Chromatography: The evaporated sample was resuspended and loaded into an immunoaffinity column, containing specific antibodies for dexamethasone. In the case of water samples (5mL), the internal standard (flumethasone) was added and then loaded in the immunoaffinity column. Corticosteroids were eluted with 4mL of ethanol:water (70:30 v/v), pH 5.0, and collected into a test tube that was evaporated to dryness at 45°C under nitrogen stream. Then, the evaporated sample was resuspended in 200 µL of mobile phase consisting of acetonitrile:mili-Q water (30:70), 20µL were injected into an Agilent series 1100 HPLC equipped with a diode-array detector. The column was a Synergi Max RP, 150mm x 4.6mm, from Phenomenex. The mobile phase, at a flow rate of 1mL per min, was a solution consisting in acetonitrile:mili-Q water (30:70) and the eluent was monitored at 242nm.

Results and Discussion

Stability: Dexamethasone solutions (10µg mL^{-1}) were kept under frozen storage (-20° C) up to 6 months and showed good stability for the full period of time.

Specificity: The method discriminated very well between the analyte (dexamethasone) and closely related substances, as can be appreciated in the chromatogram (see Figure 1) as they eluted at different retention times.

Recovery: The recovery was determined by experiments using a total of 24 fortified blank water samples. The recoveries, standard deviations and coefficients of variation (CV) were determined (see Table 1).

Repeatability: For feed, aliquots of the same sample were fortified at levels of 190, 285 and 380 ng mL^{-1} At each level, the analysis was performed with 6 replicates and the mean concentrations, standard deviations and coefficients of variation were determined (see Table 1).

Decision limit (CCα): 22 blank water/feed samples were analysed. The decision limit was set as 3 times the signal to noise ratio. This gives a CCα = 26 ng mL^{-1} for water and 190 ng mL^{-1} for feed.

Detection capability (CCβ): The value of the decision limit plus 1.64 times the standard deviation of the within-laboratory reproducibility of the measured content equals the detection capability. The obtained CCß was 30 ng mL^{-1} for water and 217 ng mL^{-1} for feed.

Within-laboratory reproducibility: Analysis were performed by different operators, days and reactives. The mean concentrations, standard deviations and coefficients of variation were determined (see table 1).

Table 1: Recovery, repeatability and within-laboratory reproducibility.

	Level (ppb)	Recovery %	Recovery SD	Recovery CV (%)	Repeatability Mean Conc. (ppb)	Repeatability SD	Repeatability CV (%)	Within-lab- reproducibility Mean Conc. (ppb)	Within-lab- reproducibility SD	Within-lab- reproducibility CV (%)
Water	26	105.10	1.73	6.33	27.90	2.05	7.35	26.96	1.69	6.27
	39	98.50	1.08	2.80	44.94	0.73	1.62	39.61	2.89	7.30
	52	94.50	1.09	2.22	59.25	2.86	4.83	54.65	4.24	7.76
Feed	190	108.91	15.31	7.40	205.06	14.45	7.05	196.15	17.35	8.85
	285	118.22	7.22	2.12	314.14	18.97	6.04	314.14	18.97	6.04
	380	108.68	16.81	4.11	430.91	34.66	8.04	430.91	34.66	8.04

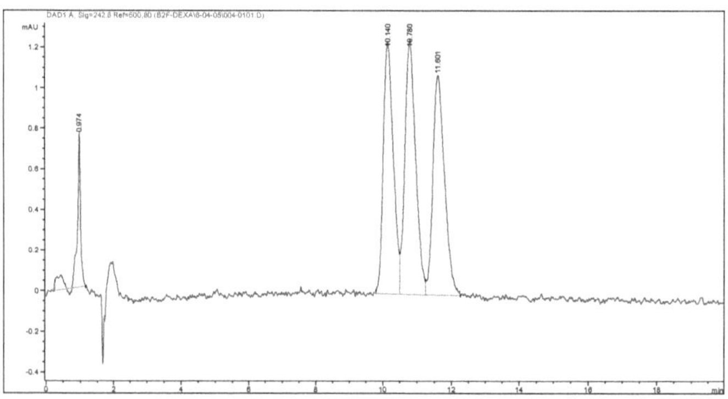

Figure 1: Chromatogram of dexamethasone and closely related substances for the specificity study. Retention times of 10.14, 10.78 and 11.60 min were for betamethasone, dexamethasone and flumethasone, respectively.

Conclusions

The method based on immunoaffinity chromatography followed by RP-HPLC for the analysis of dexamethasone in livestock drinking water and feed has been validated using water/feed fortified at levels up to 150ng mL^{-1} for water and 380 for feed. The main recovery is 99.4 ± 1.3%. The decision limit (CCα) is 26ng mL^{-1} for water and 190ng mL^{-1} for feed, detection capability (CCβ) is 30ng mL^{-1} and 217ng mL^{-1} for feed. Specificity, sensitivity and repeatability have also been validated using this protocol.

Acknowledgements

Work supported with a grant from Conselleria d´Agricultura, Pesca i Alimentació, Generalitat Valenciana (Spain).

References

Cherlet, M., De Baere, S., De Backer, P. (2004). *Journal of Chromatography B*, 805, 57-65.

Creaser, CS., Feely, SJ., Houghton, E., Seymour, M. (1998). *Journal of Chromatography A*, 794, 37-43.

Delahaut, Ph., Jacquemin, P., Colemonts, Y., Dubois, M., De Graeve, J., Deluyker, H. (1997). *Journal of Chromatography B*, 696, 203-215.

Draisci, R., Marchiafava, C., Palleschi, L., Cammarata, P., Cavalli, S. (2001). *Journal of Chromatography B*, 753, 217-223.

EC (2002). Commission Decision implementing Council Directive 96/23/EC concerning the performance of analytical methods and the interpretation of results. *Official Journal of the European Communities*, 17.8.2002.

Stolker, A.A.M., Schwillens, P..LW.J., Van Ginkel, L.A., Brinkman, U.A.Th. (2000). *Journal of Chromatography A*, 893, 55-67.

FORMATION OF BIOGENIC AMINES IN FRANKFURTER SAUSAGES TREATED WITH HIGH PRESSURE

C. Ruiz-Capillas*, J. Carballo and F. Jiménez Colmenero

Department of Meat and Fish Science and Technology, Instituto del Frío (CSIC). Ciudad Universitaria. 28040 Madrid, Spain. Email: claudia@if.csic.es.

Keywords: biogenic amines, microbiota, frankfurters, high pressure, chilled storage

Introduction

It is known that the consumption of products with high concentrations of biogenic amines may have a toxicological risk for the consumer (Halász *et al.*, 1994). The production of biogenic amines is conditioned by the microbial load in the product and the type of microbiota (Halász *et al.*, 19943, Ruiz-Capillas and Jiménez-Colmenero, 2004), which in turn will depend on several factors: characteristics of the raw material, incorporation of processing additives, storage conditions and the application of different storage technologies like high pressure. In recent years this technology has aroused great interest for its use in meat products (Cheftel and Culioli, 1997). In the case of frankfurter sausages there is a risk of recontamination in the different preparation stages. The application of high pressure (HP) after packaging in this process would lead to less contamination. The purpose of this study was to discover how the application of HP conditions the formation of biogenic amines throughout storage.

Materials and Methods

The frankfurter sausages were prepared from post-rigor pork meat, back fat, water and additives (sodium chloride, sodium tripolyphosphate and 150 ppm of sodium nitrite) according to Jiménez Colmenero *et al.* (1995). The frankfurters were heat processed in a force air oven until the temperature inside reached 70°C. After this the frankfurters were cooled, and the casing was removed before they were vacuum packed in plastic bags (Cryovac® BB4L). Then the bags were divided into three lots. The first lot was treated with pressurization processing in a high-pressure pilot unit ACB (France), using water as the pressurizing medium: 400 MPa/10 min/30°C (HP). The second lot was heated in a boiling water bath for 2 minutes (B). The third lot was the control and was not subjected to treatment (C). The three lots were stored in a chill room at 2°C during storage. The parameters assessed were pH, microbiological analysis (total viable count, lactic acid bacteria *Enterobacteriaceae*), and biogenic amines following the methodology of Ruiz-Capillas and Moral (2001). The results were analyzed statistically using the SPSS 13.0 statistical package.

Results and Discussion

The application of treatments (HP and B) to reduce the contamination produced during the packaging process of frankfurter sausages led to a decrease in microbiota and this was greater in the HP lot, where there was a decrease of over 4 log cfu/g in the total viable count (Table 1). However, these treatments did not produce any significant changes in the biogenic amine levels, except in the case of spermine in the HP lot. This amine exhibited very high levels, between 24.20 and 27.66 mg/kg, similar to those observed for agmatine (25.24-26.86 mg/kg).
An increase in total viable count and lactic bacteria microbiota was observed throughout chilled storage. Lot C was the one that exhibited the highest levels of lactic and total microorganisms of 6.01 log cfu/g at 48 days of storage (Table 1). Similar levels were found in lot B until 114 days of storage, while in the HP lot levels were less than 3 log cfu/g throughout storage (Table 1). These recounts may be related to the decreases observed in pH that reached values of 5.8, as described by other authors (Samelis *et al.*, 1998).

Table 1: Viable count and acid lactic bacteria (log cfu/g) in the different lots.

Microbiota	Samples	Days of storage				
		0	13	48	62	141
	C	4.63±0.08	4.81±0.13	6.01±0.05	6.24±0.02	
Total Viable	B	3.36±0.12	3.47±0.61	3.22±0.11	3.59±0.09	6.90±0.60
	HP	2.47±0.08	2.95±0.07	2.15±0.21	2.15±0.15	2.80±0.28
	C	4.27±0.02	4.66±0.08	6.01±0.04	5.57±0.07	
Acid lactic	B	3.15±0.15	3.00±0.09	2.15±0.21	3.59±0.02	6.29±0.02
bacteria	HP	2.19±0.06	2.50±0.71	2.00±0.03	2.15±0.15	2.50±0.71

Spermidine, spermine, histamine and agamatine levels remained constant ($p \geq 0.05$) throughout storage. An increase ($p \leq 0.05$) in tyramine was observed at 48 days of storage and levels increased to 11.54 mg/kg in lot C, whereas in the other lots changes were observed at the end of storage (141 days), the period when tyramine levels were higher in lot B than in the HP lot (Figure 1). Increases ($p \leq 0.05$) in putrescine and cadaverine were observed only at the end of storage, and the highest levels of cadaverine were observed in the sample heated with hot water (8.22 mg/kg) (Figure 1).

Generally, the increases observed in biogenic amines matched the increases observed in the main microbiota responsible for the formation of these amines (Edwards *et al.*, 1987, Halász *et al.*, 1994).

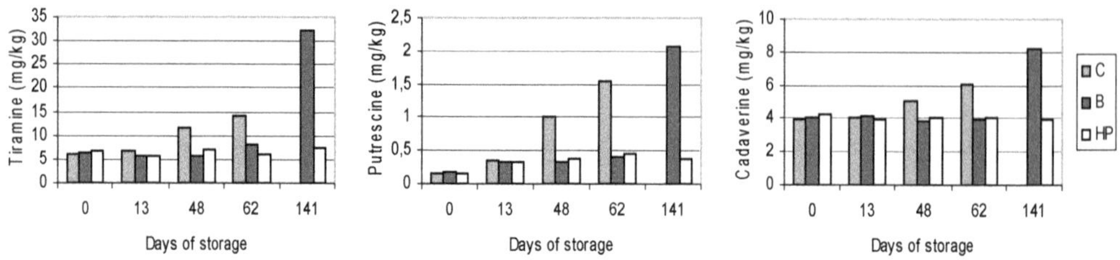

Figure 2: Evolution of biogenic amines in the different lots.

Conclusions

The application of high pressures can be used as an effective technology for reducing micobiota and limit the formation of biogenic amines that can grow and form respectively, during chilled storage of these products.

References

Cheftel, J-C., Culioli, J. (1997). Effects of high pressure on meat: a review. *Meat Science,* 46, 211-236.

Edwards, R.A., Dainty, R.H., Hibard, C.M., Ramantanis, S.V. (1987). Amines in fresh beef of normal pH and the role of bacteria in changes in concentration observed during storage in vacuum packs at chill temperatures. *Journal of Applied Bacteriology*, 63, 427-434.

Halász, A., Baráth, A., Simon-Sarkadi, L., Holzapfel, W. (1994). Biogenic amines and their production by microorganisms in food. T*rends in Food Science and Technology*, 5, 42-49.

Jimenez Colmenero, Carballo, J., Solas M.T. (1995). The effect of use of freeze-thawed pork on the properties of bologna sausages with two fat levels. *International Journal of Food Science and Technology*, 30, 335-345.

Patterson, M.F. (2005). Microbiology of pressure-treated foods. *Journal Applied Microbiology*, 98, 1400-1409.

Ruiz-Capillas, C., Jiménez-Colmenero, F. (2004). Biogenic amines in meat and meat products. *Critical Reviews in Food Science*, 44, 489-599.

Ruiz-Capillas, C., & Moral, A. (2001). Production of biogenic amines and their potential use as quality control indices for hake (*Merluccius merluccius, L.*) stored in ice. *Journal of Food Science*, 66, 1030-1032.

Samelis, J., Kakouri, A., Georgiadou, K.G., & Metaxopoulos, J. (1998). Evaluation of the extent and type of bacterial contamination at different stages of processing of cooked ham. *Journal of Applied Microbiology*, 84, 649-660.

Acknowledgments

This research was supported under project AGL2003-00454. Plan Nacional de Investigación Científica, Desarrollo e Innovación Tecnológica (I+D+I).

BEHAVIOUR OF THREE PATHOGENS IN PORTUGUESE CHOURIÇO INNOCULATED WITH SELECTED INDIGENOUS LACTIC ACID BACTERIA.

L. Patarata[1], M.J. Fraqueza*[2], M. Elias[3], M. Rua-Pinto[1], A. Esteves[1], A.S. Barreto [2], and M.C. Martins[1]

[1.] University of Trás-os Montes e Alto Douro, CECAV, 5000-911 Vila Real, Portugal, [2.] Facultulty of Veterinary Medicine, CIISA, UTL, Av. da Universidade Técnica, 1300-477, Lisbon, Portugal, [3.] University of Évora, ADCA, Apartado 94, 7002-554, Évora. Email:mjoaofraqueza@fmv.utl.pt

Keywords: fermented sausage, biological hazards, LAB, challenge test

Introduction

In Portugal there is a large variety of traditional dry fermented meat products. Among these, chouriço is made with pork and fat grossly grounded, with a seasoning varying according to the region, but salt, garlic, pimiento paste and wine are frequently used. The batter is stuffed into pork casings, and the product is smoked and dry-cured for a few weeks. In the manufacture of these non heated sausages, besides the effects of the low pH, the reduced water activity and the protective effect of nitrite and/or nitrate, there is strong evidence that the presence of lactic acid bacteria plays a role in the safety of these products more than the reduction of pH (Hugas, 1998). Even with centuries of experience manufacturing these products, occasionally foodborne outbreaks related to fermented sausages and other non-heated ready-to-eat meat products occur (Moore, 2004).

The aim of this work was to evaluate the effect of starter cultures prepared from indigenous microflora on the behaviour of three pathogens frequently considered hazards in the preparation of not heated ready-to-eat sausages: *Salmonella* spp., *S. aureus* and *L. monocytogenes,* under different experimental conditions.

Materials and Methods

Bacterial strains and preparation of inoculum: Three strains of *Salmonella* spp., *L. monocytogenes* and *S. aureus* were used. One reference and two strains isolated from meat products or its production environment. Freshly prepared suspensions of cells in NaCl 0.85% were prepared to achieve the level of inoculation required. *S. equorum*, *L. sakei*, *L. plantarum* were prepared basically as the pathogens.

Experiments on model sausages: After mincing the meat, model sausage were prepared (100 g each) in polyethylene bags, sealed and incubated at the regimen temperature.

A factorial design was performed for the three effects in study: Salt level (2%; 4%); Starter culture (control, St1 – *L. sakei* + *S. equorum*; St2 – *L. plantarum* + *S. equorum*); Temperature regimen (constant 22°C; oscillating 12 h at 10°C and 12 h at 18°C). For each one of the 12 possible experimental units, 3 replicates were prepared. Samples were contaminated with the pathogens and incubated for 2 h, 3 and 7 days of incubation.

Experiments on sausages: A challenge test was performed using sausages prepared with meat and pork fat, salt, pimiento paste, garlic paste and water. Meat and fat were chopped and other ingredients were added to the batter. After 10 min of mixture, bacteria were added. First the mix of pathogens to a level of ca. 2 log ufc/g (mixing during 5 min) and diveded in two portions: one control and one batch with a starter culture - (St1) to a level of inoculation of ca. 6 log cfu/g. After 24h at 4±2°C the batter was stuffed in casings and smoked overnight. Sausages were then dried at 15°C, 85% RH during the first week, and 80% the second week. Samples were drawn for analysis (in triplicate) at 4 h, 2, 7 and 14 days.

Microbial counting: Were performed in Compass *Listeria* (Biokar 06508), Compass *Salmonella* (Biokar 06608) and Baird Parker with RPF (Biokar 074). Lactic acid bacteria were enumerated in MRS (Cultimed 413785).

Results and Discussion

Experiments on model sausages: The results of the experiment performed in a model sausage are presented in Table 1. The main aspect that must be stressed is the loss of control of the growth of *Salmonella* spp. in every combination of conditions. However, as observed by the results of ANOVA, it was sensitive to all the effects under study. The worst situation observed for the growth of that Gram negative pathogen was, as expected, when the environmental condition were more favorable – low amount of salt and constant high temperature – when any starter culture was added. Different situation was observed for the gram positive pathogens. For the most of the combinations of factors, the worst situation was always when no starter culture was added. When starters were present, in 75% of the 16 possible cases (2 pathogens * 2 salt levels * 2 temperatures * 2 Starter cultures) the growth of the pathogens were sustained, as observed through the number of survivors lower than 100. When the means of bacterial counts were compared among the three possibilities of starter cultures (control, St1 and St2), it was observed that the counting of *S. aureus* was similar for both starters, but different from the control. For *L. monocytogenes*, the situation was different, with St2 similar to the control, and the count obtained with St1 significantly different ($p<0.05$).

Experiments on sausages: The microflora had a reduction along the procedure (Table 2). *Salmonella* spp. experienced a slight increase between the first and the second period of analysis, when starter culture was absent. This increase was favoured by the heating associated to the smoking procedure (ca. 25ºC). In both cases – control and with starter culture, *Salmonella* spp. was not detected in the final product. However, the reduction to below detection level was faster when starter culture was added. The differences in the counting of *Salmonella* spp. among the samples with or without starter culture were significant ($p<0.001$) at day 7. The behaviour observed with *L. monocytogenes* was not as favourable as that observed for *Salmonella* spp., once it was not possible to produce a reduction in the counting of the pathogen. Even so, the addition of the starter culture was advantageous because *L. monocytogenes* counts remained singnificantly lower at days 7 and 14. The effect of using the starter culture in the counting of *S. aureus* was interesting, once that starter microflora stopped the overgrowth of the pathogen during the smoking. The following reduction rendered a sausage with no detectable pathogen at the end of 7 days. The LAB count was always higher, significantly different ($p<0.001$) in samples with no starter culture added.

Table 1: Percentage of survivors of pathogens between 2h and 7 days of incubation in model sausage; results of the analysis of variance of the counting for the effects: salt levels; temperature; starter cultures (and one control) after 7 days of incubation.

Effects					
Salt level	Starter culture	Temperature regimen	*Salmonella*	*S. aureus*	*L. monocytogenes*
2%	St 1	K 22ºC			
2%	St 1	O 10-22ºC	115.2	97.9	90.3
2%	St 2	K 22ºC	122.9	92.8	89.1
2%	St 2	O 10-22ºC	131.0	107.7	91.8
2%	control	K 22ºC	220.4	138.0	104.2
2%	control	O 10-22ºC	130.2	103.5	101.3
4%	St 1	K 22ºC	122.7	96.8	94.8
4%	St 1	O 10-22ºC	119.0	93.8	95.6
4%	St 2	K 22ºC	120.5	101.5	98.2
4%	St 2	O 10-22ºC	128.8	111.1	101.7
4%	control	K 22ºC	126.8	153.0	105.2
4%	control	O 10-22ºC	132.8	107.5	94.4
Significance of the effects at day 7					
Salt level (NaCl)			***	***	ns
Starter culture (St)			***	***	***
Temperature (T)			**	**	ns

Table 2: Results of the challenge test. Counting of *Salmonella* spp., *S. aureus and L. monocytogenes* in control sausages and inoculated with starter cultures.

Time	*Salmonella* spp.			*S. aureus*			*L. monocytogenes*		
	control	Starter	Sig	control	Starter	Sig	control	Starter	Sig
4 h	2.30	2.62	ns	2.53	2.53	ns	2.26	2.16	ns
2 d	3.60	1.53	ns	3.11	1.33	ns	3.61	1.65	ns
7 d	2.10	0.00	***	1.43	0.00	**	2.95	2.10	**
14 d	0.00	0.00		0.00	0.00		3.27	2.26	***

Conclusions

It is possible to infer that the use of starter cultures in the Portuguese sausage studied results in safer products, due to its contribution in pathogen reduction. Independently of the use of starter cultures, the process of smoking and drying is responsible for an important reduction in *Salmonella* spp. and *S. aureus*. In order to avoid the overgrowth of *L. monocytogenes*, the use of the starter tested obtained from indigenous isolates is determinant.

Acknowledgments

This work was financially supported by EU programme QLK1-CT2002-02240.

References

Hugas, M. (1998) Bacterocinogenic lactic acid bacteria for the biopreservation of meat and meat products. Meat Science, 49:S139-S150

Moore, J. E.. (2004) Gastrointestinal outbreaks associated with fermented meats. Meat Science, 67:565-568.

SESSION 3: MEAT PROCESSING AND PACKAGING TECHNOLOGY

EVOLUTION OF INTRAMUSCULAR FATTY ACID COMPOSITION DURING LAMB MEAT STORAGE IN HIGH OXYGEN OR VACUUM PACKS

I. Álvarez*[1], J. De La Fuente[2] M.T. Díaz[3], S. Álvarez[1], C. Pérez[4], S. Lauzurica[2] and V. Cañeque[1]

[1.] Departamento de Tecnología de los Alimentos, INIA, 28040 Madrid, Spain. [2.] Departamento de Producción Animal, Facultad de Veterinaria, UCM, 28040 Madrid, Spain. [3.] Estación Tecnológica de la Carne, Instituto Tecnológico Agrario de Castilla y León, 37770 Guijuelo, Salamanca, Spain. [4.] Departamento de Fisiología Animal, Facultad de Veterinaria, UCM, 28040 Madrid, Spain. Email: acero@inia.es

Keywords: lamb meat, fatty acids, modified atmosphere packaging, vacuum packaging

Introduction

Vacuum (VP) and modified atmosphere packs containing high oxygen levels (MAP) are commonly used to maximise the shelf life of meat. The main advantage of MAP is that it imparts a bright-red colour that consumers find attractive. On the contrary, VP provides purple-red colours that consumers associate with loss of quality. For this reason, the use of VP in retail marketing is limited. However, VP provides longer microbiological shelf-life than MAP and it is probable that the absence of oxygen could minimise lipid oxidation of unsaturated fatty acid. From a nutritional point of view, lipid oxidation implies a loss of quality in meat not only by hydroperoxide accumulation but also changes in fatty acid composition during refrigerated storage. Previously lipid oxidation intensity has been only studied by Thiobarbituric Acid Reactive Substances (TBARS) accumulation (Lauzurica *et al.*, 2005), but evolution of fatty acid composition in lamb meat has not been studied yet. The aim of this study was to compare the evolution of intramuscular fatty acid (FA) composition during lamb meat storage in MAP and VP.

Materials and Methods

Nine weaned male Manchego breed lambs were fattening from an initial live weight of 13.2 ± 0.5 to a slaughter weight of 26.2 ± 0.3 kg, and then were slaughtered in a commercial abattoir. After a 24h chilling period (4ºC) the right *longissimus dorsi* muscle (LD) was dissected and cut in 4 slices which were randomly assigned to 0, 14, 21 and 28 days of MAP storage (70% O_2; 30% CO_2). Similarly, left LD was dissected and cut in 4 slices which were randomly assigned to 0, 14, 21 and 28 days of VP. Muscle slices were kept in darkness at 2 ± 1 ºC during storage. Intramuscular fat was extracted from LD at each time of storage (Hanson and Olley, 1963). FA methyl esters were formed according to Morrison and Smith (1964) and then analysis of samples by gas chromatography was performed. Data were statistically analysed using the MIXED procedure of the Statistical Analysis System package (SAS, 1996).

Results and Discussion

Table 1 shows the FA profile of MAP and VP samples at each time of storage.

Significant interaction was observed between packaging type and storage time for monounsaturated FA (MUFA), polyunsaturated FA (PUFA), PUFAn6 and PUFA/SFA ratio. MAP samples had the highest change in FA profile during storage, increasing SFA (10%) and MUFA (7%) after 28 days of storage respect to initial values whereas PUFA and PUFA n-6 decreased (40% and 38% respectively) in the same period. In VP samples, there was also a change in FA profile but moderate, because only an increase in SFA (5%) and a decrease on PUFA n-6 (10 %) was significant. Lipid oxidation is a process in which molecular oxygen reacts with FA to form hydroperoxides, attacking easily the double bonds in fatty acids. Thus, MAP containing high oxygen promoted lipid oxidation, principally in PUFA, whereas VP maintained profiles more constant. Cayuela (2003) concluded previously in pork that samples in high oxygen atmosphere showed higher levels of lipid oxidation, monitored by measuring TBARS, than vacuum packed ones during time of storage. From a nutritional point of view, due to a decrease in PUFA, PUFA/SFA ratio decreased significantly in MAP from 0.43, which came closest to the nutritional recommendation of 0.45 (Department of Health, 1994), to 0.23, whereas VP samples did not modify their ratio during refrigerated storage. Not only did higher loss of nutritional quality of MAP samples occur during storage, but also a reduction in the safety of the MAP samples occurred because higher lipid oxidation implies higher hydroperoxides accumulation. These results could suggest the need for studying the use of antioxidants when lamb meat is stored in MAP.

Conclusions

Modified atmosphere packaging containing high oxygen levels promoted fatty acid oxidation in lamb meat during storage, which implied changes in fatty acid profile, decreasing PUFA percentage and increasing SFA and MUFA percentages. Vacuum packing was shown to be the best method for lamb fatty acid profile preservation during 28 days of storage.

Table 1: Fatty acid profile (mg of FA per 100 mg of total FA) of MAP and VP samples during time of storage.

	Time	Packaging (P)							Sig.		
	(T)	MAP	SE		VP	SE		S_p	P	T	P*T
SFA (%)	**0**	41.67	0.63	z	41.20	0.66		ns	ns	***	ns
	14	42.56	0.65	y	42.11	0.79		ns			
	21	43.53	0.69	xy	41.76	0.65		ns			
	28	45.99	0.73	x	43.47	0.73		*			
S_t		***			*						
MUFA (%)	**0**	40.61	0.87	y	41.77	0.87		ns	ns	***	**
	14	43.77	0.89	y	42.09	0.98		ns			
	21	44.61	0.91	y	42.19	0.89		ns			
	28	43.41	0.94	x	41.35	0.91		ns			
S_t		***			ns						
PUFA (%)	**0**	17.72	1.09	x	17.94	1.09		ns	*	***	**
	14	13.68	1.12	y	16.33	1.28		ns			
	21	11.89	1.16	yz	15.80	1.12		*			
	28	10.56	1.21	bz	16.58	1.16	a	**			
S_t		***			ns						
$PUFA_{n6}$ (%)	**0**	14.34	0.95	x	15.70	0.95		ns	*	***	*
	14	11.68	0.98	y	13.88	1.13		ns			
	21	10.27	1.02	yz	13.25	0.98		*			
	28	8.94	1.07	bz	14.19	1.02	a	**			
St		***			*						
$PUFA_{n3}$ (%)	**0**	1.41	0.25		1.87	0.25		ns	*	ns	ns
	14	1.60	0.19		2.09	0.22		ns			
	21	1.35	0.17		2.16	0.17		*			
	28	1.01	0.19	b	2.00	0.18	a	**			
St		*			ns						
PUFA/SFA	**0**	0.43	0.03	x	0.42	0.03		ns	ns	***	**
	14	0.32	0.03	y	0.38	0.03		ns			
	21	0.27	0.03	yz	0.38	0.03		*			
	28	0.23	0.03	bz	0.36	0.03	a	**			
St		***			ns						

SFA: C10:0, C12:0, C14:0, C15:0, C16:0, C17:0, C18:0, C20:0; MUFA: C17:1, C16:1, C18:1; PUFA: C18:2n6, C18:2 *Cis-9 Trans-11*, C18:3n3, C20,3n6, C204n6, C20:5n3,C225n3, C226n3. SE: Standard error; Sig.: [1]Significance of the Model $y_{ijk}= \mu + P_i + T_j + P_i x T_j + \varepsilon_{ij(k)}$; S_p: Significance of dietary supplementation of VE within each storage period; S_t: Significance of storage period within each dietary supplementation VE level; ns: no significant; * P < 0.05; ** P < 0.01; *** P < 0.001; a.b Means in the same row with different superscripts differ significantly (P < 0.05).; x.y.z Means in the same column with different superscripts differ significantly (P < 0.05).

References

Cayuela, J.M. (2003). Efecto de la suplementación con acetato de α-tocopherol y del envasado en atmósferas protectoras en la oxidación de la carne de cerdo. Tesis doctoral Universidad de Murcia. Facultad de Veterinaria.

Cayuela, J.M., Garrido, M.D., Bañón, S. J. and Ros, J.S. (2003). Simultaneous HPLC Analysis of α-tocopherol and Cholesterol in fresh pig meat. Journal of Agricultural Food and Chemistry, 51, 1120-1124.

Department of Health. (1994). Report on health and social subjects. No. 46. Nutritional aspects of cardiovascular disease. London: HMSO.

Hanson, S.W.F. and Olley, J. (1963). Application of the Bligh and Dyer method of lipid extraction to tissue homogenates. Biochemistry Journal, 89: 101-102.

Lauzurica, S., De la Fuente, J., Díaz, M.T., Álvarez, I., Pérez, C. and Cañeque, V. (2005). Effect of dietary supplementation of vitamin E on characteristics of lamb meat packed under modified atmosphere. Meat Science, 70, 639-646.

Morrison, W.R. and Smith, L.M. (1964). Preparation of fatty acid methyl esters and dimethyl acetals from lipids with boron fluoride-methanol. Journal of Lipids Research, 5: 600-608.

SAS/STAT, (1996). SAS User´s Guide, Version 8. SAS Institute, Cary, NC.

COLOUR OF MINCED MEAT UNDER CARBON MONOXIDE

P. Pipek*[1], M. Izumimoto[2], O. Smékal[1] and E. Milatová[1]

[1] *Institute of Chemical technology, Technická 3, 166 28 Praha, Czech Republic. E-mail: petr.pipek@vscht.cz*
[2] *Okayama University, Okayama, Japan*

Keywords: colour, carbon monoxide, oxidation, heme pigments, MAP, minced meat

Introduction

The stability of packaged fresh meat colour is a problem, especially in the case of minced meat. Due to oxidation the red myoglobin and/or oxymyoglobin convert to brownish metmyoglobin. This reaction is accelerated due to interaction with lipid radicals and hydro peroxides. Usually a modified atmosphere (MA) is used in packaging of retail meat. MA with a high level of oxygen increases the formation of oxymyoglobin that inhibits for some time the oxidation. However the lipid oxidisation is eased by this high oxygen concentration. The growth of aerobic micro organisms is facilitated also (Martinéz, *et al.*, 2005; John, *et al.*, 2005). Carbon monoxide produces very stable carboxy derivatives with heme pigments. Although carbon monoxide is a toxic gas, the amount in packages is regarded as safe. The objection against its use is that it can mask the spoilage of meat (Sørheim, *et al.*, 1997; Stenzel and Feldhusen, 2004). A question is if the only spoilage criterion is colour or if it is advantageous to add oxygen in MA and to risk oxidisation and microbial growth.

Materials and Methods

The use of carbon monoxide in MA for packaged minced meat was studied under industrial conditions. Minced beef and mixture of beef and pork meat were packed in three different modified atmospheres: 1: 80% O_2+20% CO_2; 2: 25% CO_2+74% N_2+1% CO and 3: 25% CO_2+25% O_2+49% N_2+1% CO. All samples were stored at 4°C for 24 days. The colour of meat on the surface, internally and on the bottom of the meat surface was measured by reflectance spectrophotometry and video image analysis.

Reflectance spectrophotometry. The reflectance spectra were measured using a Chroma Meter Minolta CM-2600d spectrophotometer. The light source D_{65} and the standard observer angle 10° were used. The measured data were calculated using software Spectra Magic Ver. 3.3 (Minolta 2001, Japan) and the results expressed in terms of lightness (L*), redness (a*), yellowness (b*), chromacity (C*) and hue (h).

Calculation of metmyoglobin ratio. In the case of beef samples the myoglobin form (met, oxy and deoxy) ratios were calculated from the reflectance spectra by the Izumimoto and Ozawa method (Izumimoto and Ozawa, 1993).

Video image analysis was performed using software LUCIA 3.52. The results were expressed in RGB system as mean red (R), mean green (G) and mean blue (B), and as brightness (MB) and saturation (MS). From these values the ratios for red (r), green (g) and blue (b) were calculated as following ratios: $r = R/(R+G+B)$; g; b.

Lipid oxidation was evaluated by TBA test (absorbance at 538 nm).

Results and Discussion

The use of carbon monoxide and the elimination of oxygen suppressed oxidation of heme pigments and lipids and stabilized the colour of minced meat. From the comparison of the three different mixtures of MA it was found that the samples of minced meat under atmospheres containing CO and no oxygen are more stable then these ones under common oxygen containing MA. Results were similar for both beef and for a mixture of pork/beef.

The red colour ratio measured using VIA (Figure 1) on the beginning of the storage period was nearly identical for both MA. But although the oxygen atmosphere stabilized the colour forming oxymyoglobin, during the storage this value decreased due to the heme pigment oxidation. The oxidation of heme pigment was confirmed using calculations after [5] (see Figure 4). In the deeper layers these changes were slower.

Very similar results were achieved also by reflectance spectrophotometry; most apparent are changes of redness a* (see Figure 2, 3). Although the red colour seems to be stable at the samples packaged under oxygen atmosphere (80% O_2) on the beginning of the storage, the reflectance spectrophotometry uncovered that the redness a* and r ratio decreased since the beginning and that conversion of oxymyoglobin to metmyo-globin starts immediately. The change at this point, while detectable instrumentally, was not apparent visually, i.e. these objectively measured differences were below the detection level visually. Similar observation is described also by [Nicolade, *et al.*, 2005). Opposite to this, at the samples, where the modified atmosphere contained carbon monoxide and was free of oxygen, the colour remained stable for all storage times as documented by both methods used (Figure 1-3). Under MA3 (1% CO and 25% oxygen) the measured values were inconsistent; next measurement are necessary using different oxygen content. In the literature a positive effect of this is reported (Luňo, *et al.*, 1998).

The oxidation of lipids in the same samples was measured. It is evident that whereas under anaerobic atmosphere with CO almost no oxidation was obvious in the oxygen atmosphere a steep increase of TBA number occurred (Figure5 and 6). It can be assumed that this oxidation is delayed under CO atmosphere not only by the absence of oxygen, but also by the direct effect of the formation of carboxymyoglobin that is not oxidised and thus less Fe^{III+} is available as catalyst for lipid oxidation.

Conclusions

Carbon monoxide can stabilize the red colour of minced meat. In comparison with oxygen atmosphere, the lipid oxidation is lowered. Video image analysis proved to be a suitable simple method to enable monitoring of colour changes during storage.

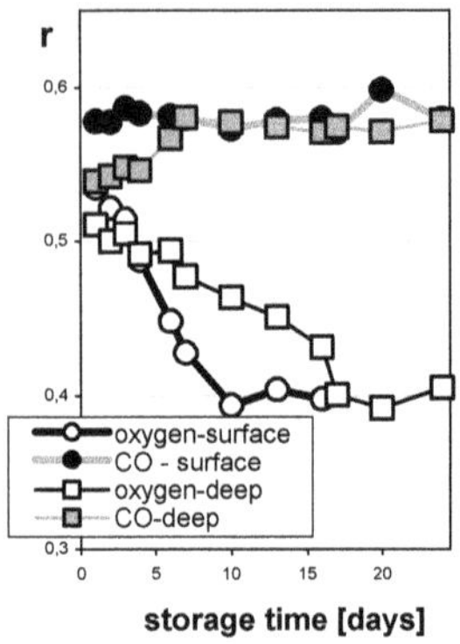

Figure 1: r-ratio changes during storage.

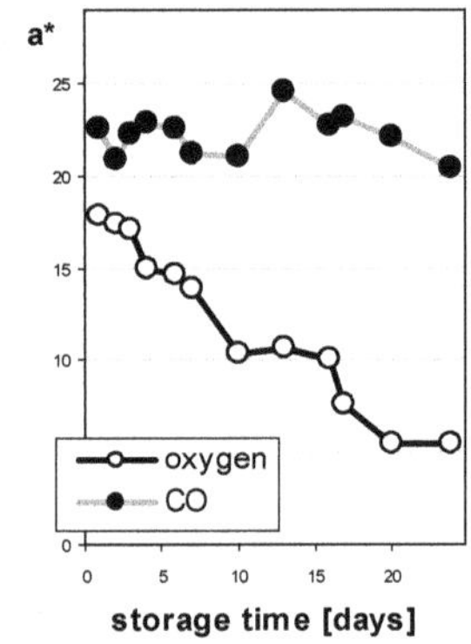

Figure 2: redness a* during storage of beef.

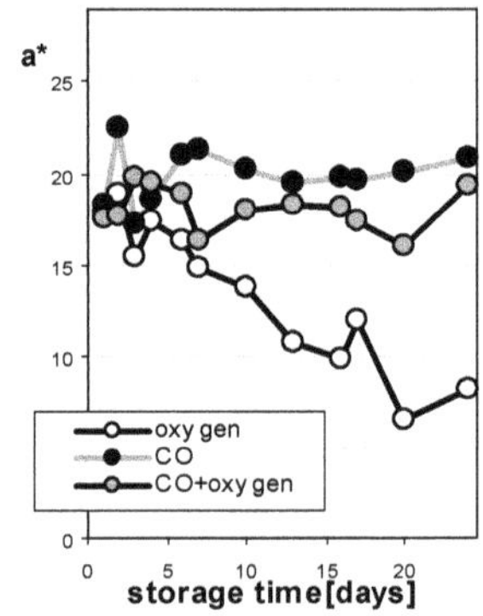

Figure 3: redness a* during storage of mixture of beef and pork.

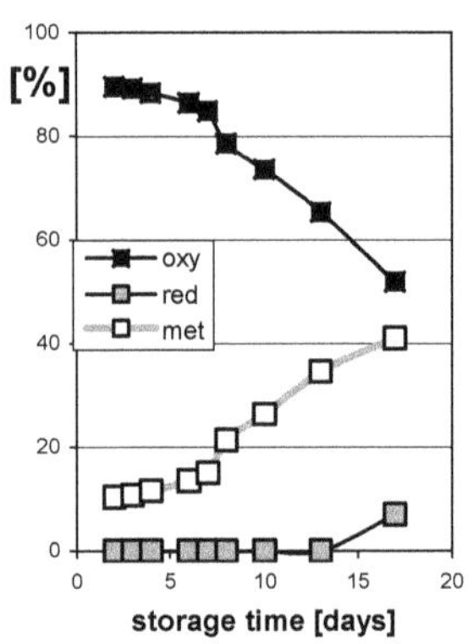

Figure 4: Oxidation of heme pigments during storage.

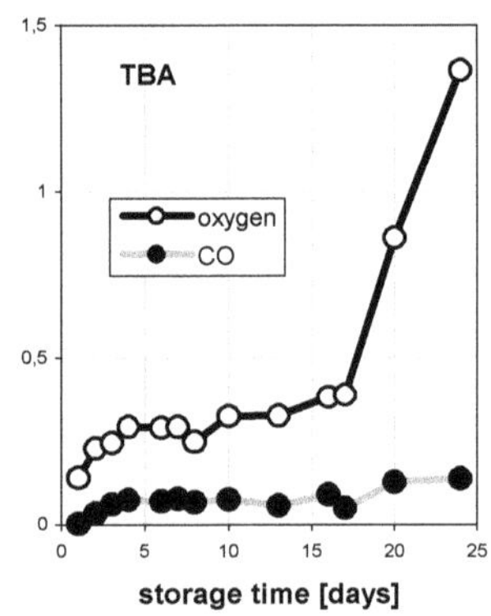

Figure 5: TBA during storage of beef.

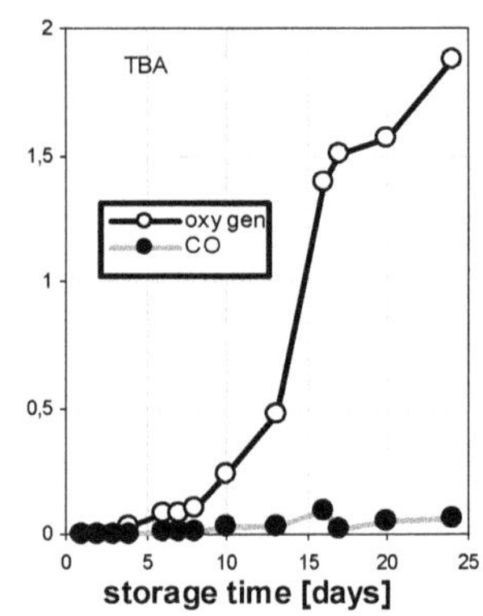

Figure 6: TBA during storage of mixture of pork and beef.

References

Izumimoto,M. and Ozawa, S. (1993) Reflectance spectrophotometric methodology on the myoglobin in intact beef. Japanese Journal of Dairy and Food Science, 42, 157-169.

John, L. *et al.*, (2005) Colour and thiobarbituric acid values of cooked top sirloin steaks packaged in modified atmospheres of 80 % oxygen, or 0.4 % carbon monoxide or vacuum. Meat Science 69 (3) 441-449.

Luňo, M., Beltran, J. A., Roncales, P. (1998) Shelf-life extension and colour stabilisation of beef packaged in a low O2 atmosphere containing CO: loin steaks and ground meat. Meat-Science, 48 (1/2) 75 - 84.

Martinéz, L. *et al.*, (2005) Effect of different concentrations of carbon dioxide and low concentration of CO on the shelf-life of fresh pork sausages packaged in modified atmosphere. Meat Science, 71 (3) 563-570

Nicolade, C. Stetzer, A.J. Tucker, E.M., cKeith,F.L. and Brewer,M.S. (2005) Development of a Model System to Mimic Beef Bone Discoloration. J.Fd.Sci. 70 (9) C575-C579.

Sørheim,O., Aune,T., and Nesbakken,T. (1997) Technological, hygienic and toxicological aspect of CO used in modified-atmosphere packaging of meat. Trends in food science and technology 8, 307 – 312

Stenzel, W.-R., Feldhusen, F. (2004) Aspekte der Verwendung von Kohlenmonoxid bei Fleisch- und Fischerzeugnissen. Fleischwirtschaft, 94 (9) 131 - 136.

DISPLAY LIFE OF FRESH PORK SAUSAGE: A COMPARISON BETWEEN TWO DIFFERENT PACKAGING SYSTEMS

E. Chiavaro[1], E. Zanardi* [2], B. Bottari[3], and A. Ianieri[2]

[1]*Dip. di Ingegneria Industriale, Università degli Studi di Parma, viale Usberti 181/A, 43100 Parma, Italy* [2]*Dip. di Produzioni Animali,Biotecnologie Veterinarie, Qualità e Sicurezza degli Alimenti, Università degli Studi di Parma, via del Taglio 8, 43100 Parma, Italy* [3]*Dip. di Genetica, Università degli Studi di Parma, viale Usberti 11/A, 43100 Parma, Italy. (E-mail: emanuela.zanardi@unipr.it)*

Keywords: fresh pork sausage, oxygen-permeable film, MA, colour, TBARS, microbiological counts

Introduction

Meat industries have shown growing interest in development of suitable techniques to extend shelf life and improve consumer acceptance while maintaining nutritional quality and ensuring safety. The combination of light and oxygen may cause discolouration during illuminated retail display life both in air and modified atmosphere packages. Recently, Schivazappa *et al.*, (2004) evaluated lipid oxidation and browning of fresh pork sausages packed under MA finding that the application of oxygen enriched atmospheres (60-80%) was advantageous up to 7 days. Tremonte *et al.*, (2005) reported that the application of 60% O_2 and 40% CO_2 atmosphere on fresh sausages resulted in an inhibitory effect of bacterial growth positively affecting colour and other physical attributes. The aim of this work was to evaluate shelf life of fresh pork sausage packed under an over-wrapping O_2-permeable film or a MA (65% O_2 and 35% CO_2) packaging system and stored under a refrigerated display cabinet. Changes of pH, water activity (a_w), lipid oxidation, colour and microbiological counts of pathogenic and spoilage bacteria were evaluated.

Materials and Methods

A total of 150 fresh pork sausages were purchased from a local manufacturer that also performed their packaging on trays containing 5 sausages each. The packaging treatments were: O_2Pfilm = packaging in a solid PS tray and over-wrapping with PVC O_2-permeable film; MAP= packaging in a solid PP tray coated by a PA-PP film under modified atmosphere (65% O_2 and 35% CO_2). Sausages were stored in a refrigerated display cabinet (5 ± 0.3°C), exposed for 24hr per day at a light intensity of 1000 lux for the following times after packaging: 3, 6 (commercial term) and 9 days for O_2Pfilm; 6, 12 (commercial term) and 15 days for MAP. For each packaging x storage temperature x time, 3 packages were used (one for microbiological analyses and two for the other analytical determinations). All the determinations were carried out immediately after the opening of the trays and removal of the casing. pH was measured directly on the fresh sausages with a penetration electrode 52-32 (Crison Strumenti, Carpi, I). a_w was measured on fresh sausages with a AQUALAB SERIES 3TE water activity meter (Decagon Devices, Inc. Washington USA). Lipid oxidation was evaluated by Thiobarbituric Acid Reactive Substances (TBARS, mgMDA/kg) test (Novelli *et al.*, 1998). Colour determination of a* (redness) was carried out by means of a digital imaging method developed by M. Riva (2006), scanning external surface of sausages with a flatbed scanner (Model Scanjet 8200, HP, Cupertino, USA) together with 12 standard coloured charts (Pantone® Inc. Carlstadt, USA) that covered the colour range of the sausages. The images were calibrated and quantitatively analyzed with Adobe Photoshop standard software (Version 6.0, Adobe Systems Inc., San Jose, USA). Microbiological analyses were performed on serial decimal solution prepared in Ringer's solution by 20 g of samples, after homogenization, on Plate count agar (PCA, Liofilchem) for total viable counts; de Man, Rogosa and Sharpe agar (MRS, Liofilchem) for lactic acid bacteria; Mannitol salt agar (MSA, Oxoid, Basingtoke, UK) for Micrococcaceae; Yeast Glucose Chloramphenicol agar (YGC, Merck, Darmstadt, G) for moulds and yeasts.

Results and Discussion

The pH of fresh samples was 5.70 and significantly decreased up to the commercial term to 5.30 for O_2Pfilm over-wrapped samples and to 5.23 for MAP remaining substantially unvaried extending shelf life. pH decrease was probably related to the growth of lactic acid bacteria (LAB), but it is not possible to exclude the influence of CO_2 adsorption for the explanation of the larger pH decrease observed for MAP sausages in comparison to O_2Pfilm samples. a_w of fresh samples was 0.974 and did not significantly change under all storage conditions for MAP while a slight decrease to 0.965 was observed during storage for O_2Pfilm sausages, as expected by the low gas barrier property of the film. TBARS values significantly increased in all samples during the first days of storage compared to fresh product, as shown in Figure 1, remaining substantially unvaried for O_2Pfilm sausages after 3 day storage. On the contrary, MAP samples showed a significant increase in lipid oxidation, resulting from a great light exposure influence on the rate of lipid oxidation for storage under O_2/CO_2 enriched atmosphere. MAP sausages also showed higher TBARS values than O_2Pfilm samples, as expected by higher O_2 percentage of MAP in comparison to air. a* values significantly decreased for O_2Pfilm sausages after day 3 (Figure 2). A significant decrease in a* was noticed for MAP, already after 6 days of storage, too. Schivazappa *et al.*, (2004) also noticed a discolouration of sausage surface after 7 days of storage under MAP (O_2 percentage higher than 20%) related to the metmyoglobin formation. The authors measured the change of MA composition during storage, reporting a marked O_2 percentage decrement (about 20%), starting from 7 days storage at 4

°C, that had probably reduced the rate of oxymyoglobin formation. On the contrary, no significant differences were found by Tremonte *et al.*, (2005) for a* values of sausages stored under 60% O_2 and 40% CO_2 in comparison to air.

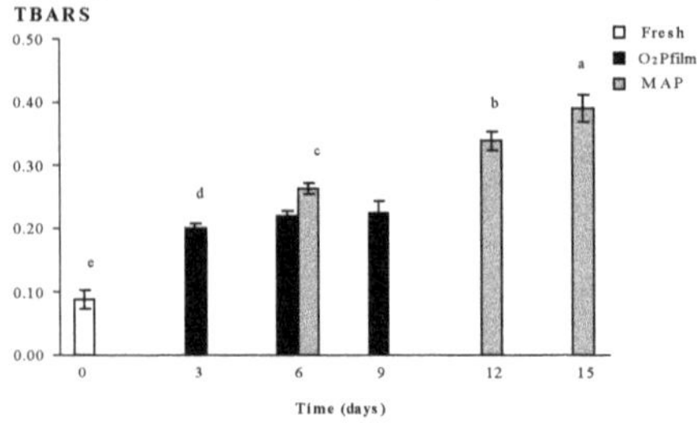

Figure 1: TBARS values during storage

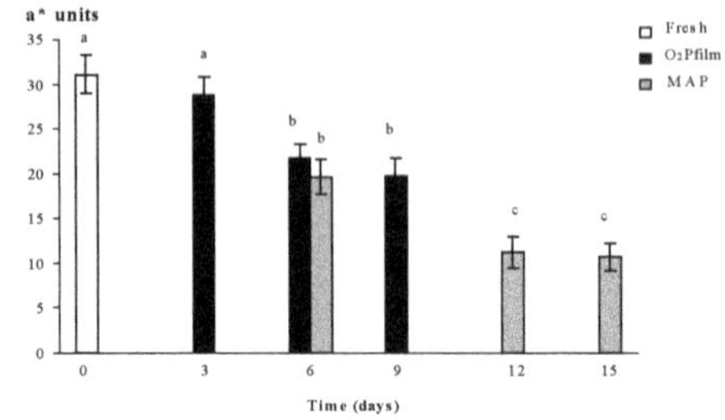

Figure 2: a* values during storage

Microbiological counts were evaluated on fresh sausages and stored samples both at the end and after the commercial term. Figures 3 and 4 show the PCA, MRS, MSA and YGC counts of O_2Pfilm and MAP, respectively.

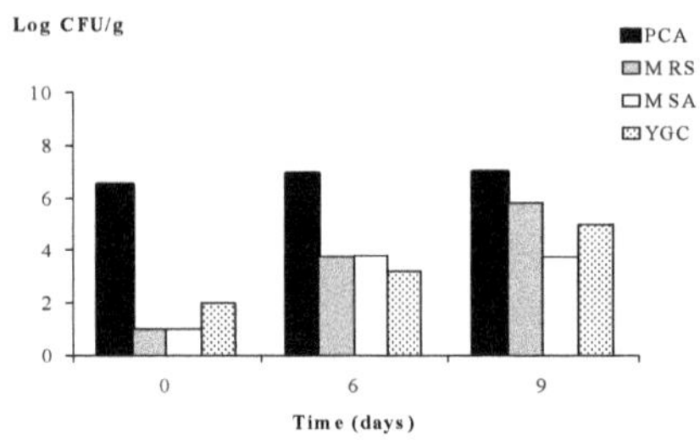

Figure 3: Microbiological counts for O_2Pfilm

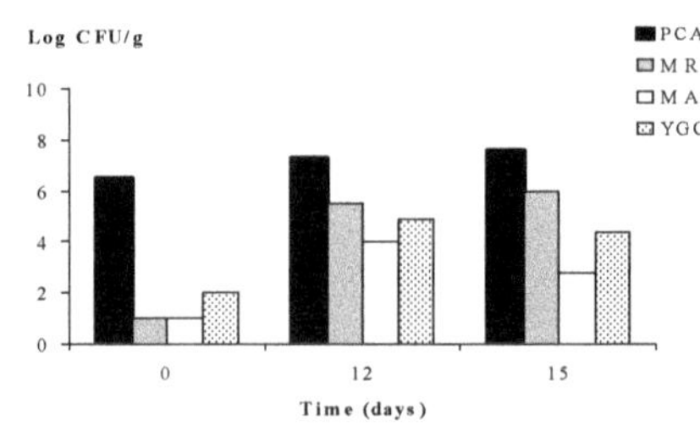

Figure 4: Microbiological counts for MAP

Total viable counts (PCA) did not significantly vary for O_2Pfilm and MAP samples during storage. Mesophilic lactic acid bacteria (MRS) increased in O_2Pfilm sausages and in MAP samples also after 12 days when a decrease in the oxygen level probably occurred in the modified atmosphere composition, showing to be tolerant to high CO_2 percentage. *Micrococcaceae* (MSA) were enabled to grow in both packaging systems until the commercial term while the subsequent count decrease of MAP samples could be related to the diminution of pH values due to LAB growth. Moulds and yeasts (YGC) showed a similar trend, increasing up to the commercial term to decrease until the end of the storage experiment probably because of the lower growing ability under diminishing oxygen only in MAP samples.

Conclusions

An extension of shelf life to 9 days could be probably applied on sausages over-wrapped by PVC film because of the substantial maintenance of the physico-chemical and microbiological properties, after the initial modification, in comparison to fresh product. Contradictory results were shown by MAP practice where a discolouration was noticeable after day 6 of storage, showing that high O_2 percentage is not able to control metmyoglobin formation for long time. In addition, TBARS values increased more under MAP, although the values were relatively lower than other literature data probably due to the presence of ascorbate in the formulation. In conclusion, MAP application on fresh pork sausage should be reconsidered paying particular attention to the advantages derived from reduction in distribution costs (*i.e.* fewer deliveries over longer distances) and the disadvantages not only derived by added costs (*i.e.* increase pack volume leading to increased retail display and transport costs) but also by the difficult maintenance of product quality for such long time.

References

Novelli, E., Zanardi E., Ghiretti G.P., Campanini G., Dazzi, G., Madarena, G., Chizzolini, R. (1998). *Meat Science*, 48, 29-40.

Riva, M. (2006). Available from http://users.unimi.it/~distam/info/Standard_colore.pdf.

Schivazappa, C., Virgili, R., Bovis, N., Pedrelli, T. (2004). *Industria Conserve*, 79, 441-451.

Tremonte, P., Sorrentino, E., Succi, M., Reale, A., Maiorano, G., Coppola, R. (2005). *Journal of Food Protection*, 68, 2686-2692.

HIGHER COLOUR STABILITY IN STEAKS OF BEEF LOIN AGED IN VACUUM COMPARED WITH HIGH-OXYGEN MODIFIED ATMOSPHERE

G. Lindahl, A.-C. Enfält, Å. Lagerstedt and K. Lundström*

Department of Food Science, Swedish University of Agricultural Science, P.O.Box 7051, SE-750 07 Uppsala, Sweden. Email: kerstin.lundstrom@lmv.slu.se

Keywords: beef, colour stability, vacuum storage, high oxygen modified atmosphere, *longissimus dorsi*

Introduction

Meat colour is an important quality attribute for the consumer. Beef is often pre-packaged and displayed for several days either in air or in high-oxygen modified atmosphere (MAP). However, poor colour stability during retail display has become an increasing problem, which might be related to large variations in the age of the meat and storage atmosphere before retail packaging. The colour stability of beef, i.e. the ability to withstand accumulation of metmyoglobin (MetMb) at the meat surface, depends on respiratory and reducing enzyme activities in the muscle (Mancini and Hunt, 2005) as well as on environmental factors such as oxygen partial pressure and temperature. This study focuses on colour stability of beef loin as an effect of ageing time in vacuum or in MAP preceded by vacuum.

Materials and Methods

Ten steers from the same herd were slaughtered according to standard commercial procedures in a Swedish abattoir. At 48 hours *post mortem*, *M. longissimus dorsi* (LD) from both sides of the carcass was removed and divided into 6 parts randomly assigned to packaging in vacuum (Vac) or vacuum followed by modified atmosphere with 80% O_2 + 20% CO_2 (MAP) for ageing according to Table 1. After each ageing time, steaks (2 cm thick) were cut and the newly cut surfaces were wrapped with oxygen-permeable plastic film and displayed in air at 4°C for five days. The colour stability during display was assessed by colour measurements after 1 and 3 hours and 1, 2, 3, 4, and 5 days of display using a Minolta CR-2500d spectrophotometer (Minolta Co, Ltd, Osaka, Japan) with specular reflectance excluded, 8 mm measuring aperture, illuminant D65 and 10° Standard Observer. The measuring aperture was covered with a glass plate, and the instrument was calibrated against a white tile. The plastic film was kept on the meat during measurements. The average of 4 measurements across the surface was used. The relative contents of deoxymyoglobin (Mb), oxymyoglobin (MbO_2) and MetMb were calculated from the reflectance curve according to Krzywicki (1979). Reflectance values at wavelengths not given by the instrument (473, 525 and 572 nm) were calculated using linear interpolation. Statistical analysis was carried out with SAS ver. 9.1 using the mixed procedure with degrees of freedom estimated by the Satterthwaite method. The model included fixed effects of ageing system, display time and their interaction and random effect of animal.

Table 1: Ageing time in vacuum (Vac) and modified atmosphere with 80% O_2 + 20% CO_2 (MAP).

Ageing system/ Ageing days	Vac 0 days	Vac 5 days	Vac 15 days	Vac 25 days	Vac 5 days + MAP 10 days	Vac 15 days + MAP 10 days
Vac	0	5	15	25	5	15
MAP	0	0	0	0	10	10
Total	0	5	15	25	15	25

Results and Discussion

Ageing time significantly affected the relative contents of MetMb (P=0.001), MbO_2 (P=0.001) and Mb (P=0.001) during display in air. The most striking effect was that of LD displayed in air at 48 hours *pm* (= Vac 0 days) compared with LD aged in vacuum for 5, 15 or 25 days before display (Figure 1). More MetMb was formed during the whole display time and more MbO_2 and less Mb, i.e. more blooming, was found during the first 2 days of display. Furthermore, increasing ageing time in vacuum resulted in more MetMb and MbO_2 and less Mb. Ageing in MAP for 10 days, preceded by vacuum ageing for 5 or 15 days, further increased the content of MetMb and decreased the content of MbO_2 during display compared with ageing only in vacuum during the same time period (Figure 2). It is known that enzyme activities of the muscle related to colour changes are depleted post-mortem (Mancini and Hunt, 2005), which explains the effect of ageing on colour stability. However, the more detrimental effect of ageing in high-oxygen MAP on colour stability is more difficult to explain, since a new meat surface from the interior of the whole loin that had not been directly exposed to the high oxygen atmosphere was exposed during the display. No other studies on this effect of high-oxygen MAP was found in the literature.

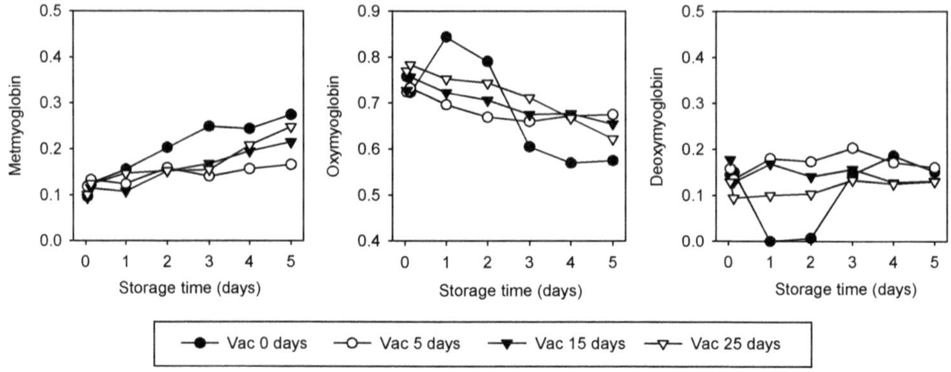

Figure 1: Changes in the relative contents of metmyoglobin (MetMb), oxymyoglobin (MbO_2) and deoxymyoglobin (Mb) during display of *M. Longissimus dorsi* in air at 4°C preceded by ageing in vacuum (Vac) for 0, 5, 15 and 25 days, respectively, at 4°C. Standard errors: 0.01 for MetMb, 0.02 for MbO_2 and 0.01 for Mb.

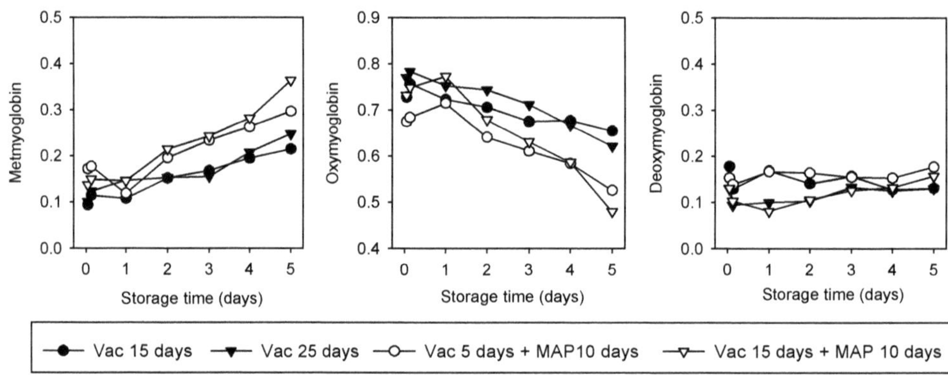

Figure 2: Changes in the relative contents of metmyoglobin (MetMb), oxymyoglobin (MbO_2) and deoxymyoglobin (Mb) during display of *M. Longissimus dorsi* in air at 4°C preceded by ageing only in vacuum (Vac) for 15 and 25 days, respectively, or in vacuum for 5 and 15 days, respectively, followed by high-oxygen modified atmosphere (MAP) for 10 days at 4°C. Standard errors: 0.01 for MetMb, 0.02 for MbO_2 and 0.01 for Mb.

Conclusions

Higher colour stability in beef loin steaks occurs

- after ageing for 5, 15 or 25 days in vacuum compared with no ageing time
- after ageing for 5 days in vacuum compared with 25 days in vacuum
- after ageing in vacuum compared with vacuum followed by high-oxygen modified atmosphere during the same total time period.

References

Krzywicki, K. (1979). Assessment of relative content of myoglobin, oxymyoglobin and metmyoglobin at the surface of beef. Meat Science, 3: 1-10.

Mancini, R.A. and Hunt, M.C. (2005). Current research in meat colour. Meat Science, 71: 100-121.

POSSIBLE APPLICATIONS OF OVOTRANSFERRIN-INCORPORATED K-CARRAGEENAN-BASED FILM AS PACKAGING FOR FRESH CHICKEN FILLET DURING COLD STORAGE

K.H. Seol, D.K. Lim and M. Lee*

Lab. of Animal Origin Food Science, Department of Animal and Food Biotechnology, Seoul National University, San 56-1, Sillim-dong, Gwanak-gu, Seoul, 151-742, Korea. Email:moohalee@snu.ac.kr

Keywords: ovotransferrin, κ-carrageenan-based film, chicken fillet

Introduction

Ovotransferrin (OTf) is the major component of the avian egg's antimicrobial defense system (Tranter and Board, 1982) as it renders iron unavailable to microbial growth within the albumin; this is the so called nutritional immunity of albumin (Ibrahim, 2000). Antimicrobial food packaging is one of the most promising concepts of active packaging. Cha *et al.*, (2002) investigated antimicrobial films based on Na-alginate and κ-carrageenan with some antimicrobials such as lysozyme, nisin, grapefruit seed extract and chelating agent (EDTA) against food spoilage bacteria and pathogens.However, the possibility of using OTf as an antimicrobial agent and the effect of antimicrobial films on shelf-life of fresh meat has not been thoroughly researched. Thus, we have investigated the effects of κ-carrageenan-based film incorporating OTf on the shelf-life of fresh chicken fillet.

Materials and Methods

κ-carrageenan-based films were prepared by modification of the methods used by Pavlath *et al.* (1999) and Cha *et al.* (2002). The compositions of κ-carrageenan-based films tested are listed in Table 1. Typical meat product bacterial contaminants used were *S. aureus* (ATCC 12692), *E. coli* (ATCC 25922), *S. typhimurium* (ATCC 14028), and the mould *C. albicans* (ATCC 10231). Chicken fillets were purchased from market to investigate the effect of films on shelf-life of fresh chicken fillet during cold storage (4~5°C). Antimicrobial activity of films was measured by a modified agar diffusion method according to Chen *et al.*(1996). Effects of the prepared films on chicken fillet were evaluated by modification of the method used by Ouattara *et al.* (2002);. experiments were replicated five times. One-way ANOVA with Duncan's multiple range test was carried out to analyse the significant differences between treatments ($P<0.05$) using the SAS program for Windows (V9.1; SAS Institute, Cary, NC, USA).

Results and Discussion

The effect of films on inhibition zone area against indicator microorganisms is given in Table 2 and Figure 1. In all indicator microorganisms, control (C), OTf (T1), and potassium sorbate-incorporated film (T3) revealed little inhibitory effect (indicated by minimal growth underneath film pieces) while EDTA-incorporated films (T2, T4 and T5) did reveal an inhibitory effect. The film incorporating OTf and EDTA (T2) revealed a strong inhibitory effect against *E. coli* and *C. albicans*. The EDTA may induce antimicrobial agents to kill microorganisms by destabilizing cell membranes and limiting the availability of cations (Shelef and Seiter, 1993). Total microbial counts on chicken fillets increased during storage in all treatments and after 7 days of storage those of C were >7.0 log CFU/g (Figure 2). Total microbial counts on chicken fillets covered by T1 and T3 showed no significant difference with C during the entire storage period ($p>0.05$). However, the fillets covered with T2 had significantly lower total microbial counts throughout the storage period ($p<0.05$) and those of fillets covered with T4 and T5 were significantly lower than those of the control after 7 days storage ($p<0.05$). The number of *E. coli* on chicken covered with T2 did not increase significantly during storage ($p>0.05$) and it was significantly lower than any other treatments ($p<0.05$). Those of fillets covered with all treated films (from T1 to T5) were significantly lower than C during the storage period ($p<0.05$). These results agree with the result that T2, T4 and T5 had inhibitory effects on microbial growth and show the possibility of using OTf and chelators in combination as an additive for antimicrobial films.

Table 1: Compositions of κ-carrageenan-based films.

Code	Film Treatments
Control (C)	κ-carrageenan (2%) + plasticiser[a] (1.5%)
T1	Control + Ovotransferrin (OTf, 25mg/g κ-carrageenan)
T2	Control + Ovotransferrin (OTf, 25mg/g κ-carrageenan) + EDTA (5mmol)
T3	Control + Potassium sorbate (PS, 10mg/g κ-carrageenan)
T4	Control + Potassium sorbate (PS, 10mg/g κ-carrageenan) + EDTA (5mmol)
T5	Control + EDTA (5mmol)

[a]Plasticiser : 50% PEG (polyethylene glycol) + 50% glycerol.

Table 2: Effect of κ-carrageenan-based films on inhibition zone area against indicator microorganisms.

Film type	*S. aureus*	*E. coli*	*S. typhimurium*	*C. albicans*
C	nd	nd	nd	nd
T1	nd	nd	nd	nd
T2	+	++	+	++
T3	nd	nd	nd	nd
T4	+	+	+	++
T5	+	+	+	+

nd : not detected, **+** : Inhibition is weak (diameter of inhibition zone is < 2㎜), **++** : Sensitive (diameter of inhibition zone is 2~5㎜)
C: κ-carrageenan + plasticiser; **T1**: C + OTf; **T2**: C + OTf + EDTA; **T3**: C + PS; **T4**: C + PS + EDTA; **T5**: C + EDTA

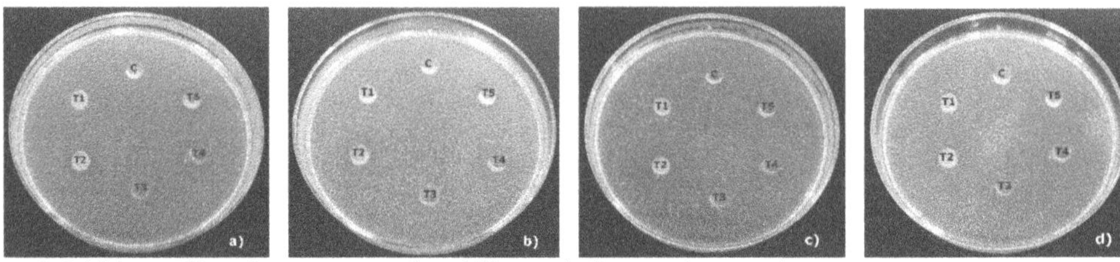

Figure 1: Inhibitory effect of κ-carrageenan-based films on indicator microorganisms.
a) *S. aureus*, **b)** *E. coli*, **c)** *S. typhimurium*, **d)** *C. albicans*
C: κ-carrageenan + plasticiser; **T1**: C + OTf; **T2**: C + OTf + EDTA; **T3**: C + PS; **T4**: C + PS + EDTA; **T5**: C + EDTA

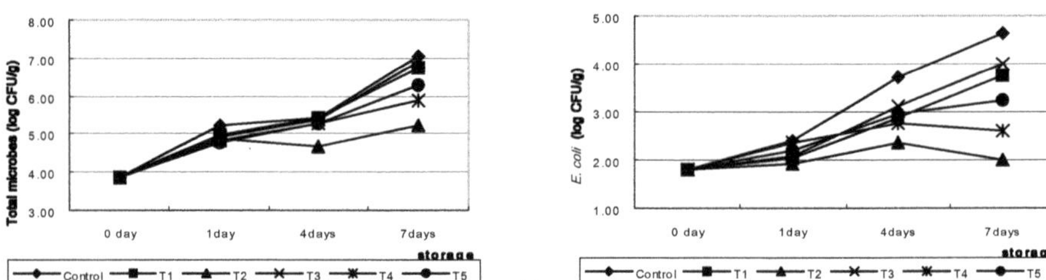

a) b)

Figure 2: Changes in the number of total microbes and *E. coli* on chicken fillet covered with κ-carrageenan-based films during storage. **a)** total microbes, **b)** *E. coli* **Control**: κ-carrageenan + plasticiser; **T1**: C + OTf; **T2**: C + OTf + EDTA; **T3**: C + PS; **T4**: C + PS + EDTA; **T5**: C + EDTA

Conclusions

OTf combined with EDTA showed a strong inhibitory effect against various microorganisms and has possibilities for application in active food packaging. Further studies concerning the applicability of OTf antimicrobial peptide are required.

References

Tranter, H. S. and Board, R. G. (1982). The antimicrobial defence of avian eggs: biological perspective and chemical basis. Journal of Applied Biochemistry, 4, 295-338.

Ibrahim, H. R. (2000). Ovotransferrin. In: Natural food antimicrobial systems, ed. A. S. Naidu, pp. 211-226. Florida: CRC Press, Inc.

Cha, D. S., Choi, J. H., Chinnan, M. S.. Park, H. J. (2002). Antimicrobial films based on Na-alginate and κ-carrageenan. Lebensmittel-Wissenschaft und Technologie , 35, 715-719.

Pavlath, A. E., Gossett, C., Camirand, W., Rovertson, G. H. (1999). Ionomeric films of alginic acid. Journal of Food Science, 4(1), 61-63.

Chen, M. C., Yeh, G. H. C., Chiang, B. H. (1996). Antimicrobial and physicochemical properties of methylcellulose and chitosan films containing a preservative. Journal of Food Processing and Preservation, 20, 279-390.

Ouattara, B. Simard, R. E., Piette, G., Bégin, A., Holley, R. A. (2000). Inhibition of surface spoilage bacteria in processed meats by application of antimicrobial films prepared with chitosan. International Journal of Food Microbiology, 62, 139-148.

Shelef, L. and Seiter, J. (1993). Indirect antimicrobials. In: Antimicrobials in Food, ed. P.M. Davidson & A. L. Braner, 2nd ed., pp. 544-555. New York: Marcel Dekker.

SHELF LIFE OF PORK IN A MODIFIED ATMOSPHERE PACKAGING WITH HUMIDITY ADSORBER SYSTEM

O. Almeida, M.J. Fraqueza and A. Barreto*

Faculty of Veterinary Medicine CIISA, Technical University of Lisbon. Av.Universidade Técnica, Pólo Universitário. Alto da Ajuda. 1300-477 Lisbon. Portugal(Email:mjoaofraqueza@fmv.utl.pt)

Keywords: sliced pork, MAP, shelf life, active package

Introduction

Modified atmosphere packaging (MAP) involves the removal of air from the food pack and its replacement with a single gas or a mixture of gases using one of two different techniques: gas flushing or compensated vacuum. In the first case the efficiency of the system is limited, since replacement of the air is accomplished by dilution (Blakistone, 1999). It is known that several factors affect MAP meat products' shelf life: temperature control, hygiene control, raw material quality, time prior to packaging, gas-to-product ratio, gas composition, and residual gas composition (dependent on the MAP technique used), gas purity, package film permeability; pack design, adsorption of gases to product surfaces, $CO_2/O_2/N_2$ permeability ratio. Presently, commercial systems with desiccant capacity are available in adsorbent sheets. These materials are basically a super-absorbent polymer placed between two polyolefin layers (López-Rubio *et al.*, 2004). Use in meat packaging avoids the accumulation of water surrounding the meat, thus reducing the micro flora growth rate.

The aim of this work was to establish the shelf life of sliced meat pork, packaged in a MAP (O_2/CO_2), with a humidity adsorber system, in accordance with the normal work conditions of a flow pack machine industrial slaughter processing room.

Materials and Methods

The samples were collected in an industrial slaughter processing room in accordance with real work conditions. Sliced samples of approximately 200 g were taken from a pork leg. They were packaged on a MAP flow pack machine, (Speed up to 40m/min, gas mixture: 70%O_2 / 30% CO_2 - ULMA-Alaska HRX-300, Spain) on a polystyrene tray with a humidity adsorber (Linpac, Terplus, England) using a polyolefin, coextruded, multilayer shrink film (Cryovac BDF 8050, France) with a permeability of -18 cm^3/m^2 for O_2 and 52 cm3/m^2 for CO_2. Control pork meat samples were also packaged under aerobic conditions using the same polystyrene tray with the humidity adsorber and a extensible PVC film (NOVUM, Spain) with a permeability of - 25000cm^3/m^2 for O_2 and 130000cm^3/m^2 CO_2. All the samples were stored at 2±1 °C in the dark. Microbial and chemical analyses were performed at days 1, 4, 8, 10 and 14. Measurements of O_2% and CO_2% on MAP Packaging Headspace were performed. Five trials in the conditions previously referred to, were made.

The following analyses were carried out, Microbial analysis: Mesophylic aerobic total counts in Plate count agar (Sharlau, Spain) at 30 °C for 2 days; *Enterobacteriaceae* counts in violet red bile agar (Merck Germany) at 37 °C for 2 days*; Escherichia coli* counts in Tergitol-BCIG (Biokar Diagnostics, Spain) at 44.5 °C for 24 hours; Detection of *Listeria monocytogenes* according to ISO/DIS 11290-1, 1995; Detection of *Salmonella* spp. according to ISO/DIS 6579, 2002; Detection of *Staphylococcus aureus* (NP-4400-2, 2002). Chemical analysis: TVBN (total volatile basic nitrogen) (NP-1848, 1987). Packaging Headspace Gas Analyze: pack headspace evaluation was carried out in a gas analyzer PBI-Dasensor A/S model CheckMate 9900. Statistical Analysis: the mean values were compared using the paired samples t Student test for time factor and independent samples for package conditions, according to SPSS 11.5 for Windows.

Results and Discussion

Initial samples of sliced pork meat samples show values of approximately 5 log_{10} CFU/g for mesophylic aerobic total counts and 4 log_{10} CFU/g for *Enterobacteriacea* counts (Figure 1). Low *E. coli* counts showed that all samples were packaged according to good hygienic practice. Meat samples packaged under work conditions of a flow pack machine presented approximately 20% of atmospheric air in the gas mixture due to a dilution factor (Figure 2). The high O_2% and approximately 25% of CO_2 in the MAP sliced pork did not to inhibit the growth of *Enterobacteriaceae* (Figure 1B). However, an inhibition occurred regarding the mesophylic aerobic total counts promoted by this gas mixture in packaged pork samples. After 8 storage days there is a significant difference (($p<0.05$, 1 to 2 log_{10} CFU/g) between total flora in MAP sliced pork and samples control. This can be explained by the CO_2 effect on main spoilage meat flora such as *Pseudomonas* (Gill, 1986; Erkmen, 2000). TVBN evolution (Figure 3) agrees with that as observed for mesophylic aerobic total counts, revealing an increase on meat spoilage promoted by the microbial flora utilization of amino acids, apart the enzymatic action. The increase of TVBN values over than 14 mg NH_3/100g and a total bacterial count higher than 7 log_{10} CFU/g would not allow us to extend the shelf life of pork meat under MAP with a humidity adsorber stored at 2 °C, more than 10 days. Concerning the pathogenic micro flora (Table 1) *Listeria monocytogenes* and *Salmonella* spp were not detected in pork under MAP storage conditions. *Staphylococcus aureus* was present in few samples but no increased prevalence was observed either in control or MAP meat.

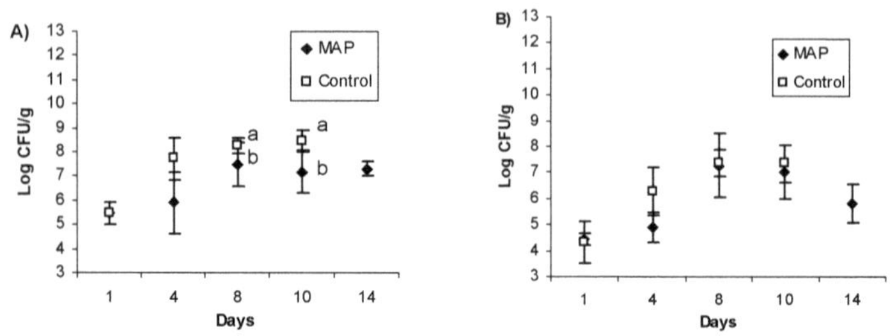

Figure 1: Mesophylic aerobic total (A) and *Enterobacteriaceae* (B) counts evolution on sliced pork packaged under MAP (CO_2/O_2) and Control conditions, stored at 2° C.

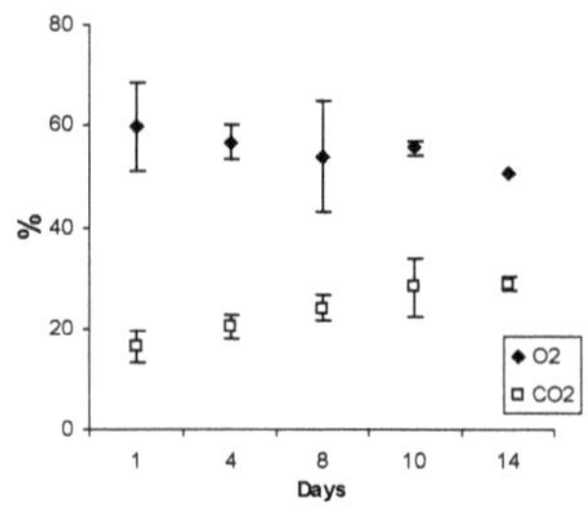

Figure 2: Packaging Headspace O_2/CO_2 (%) evolution in MAP studied conditions.

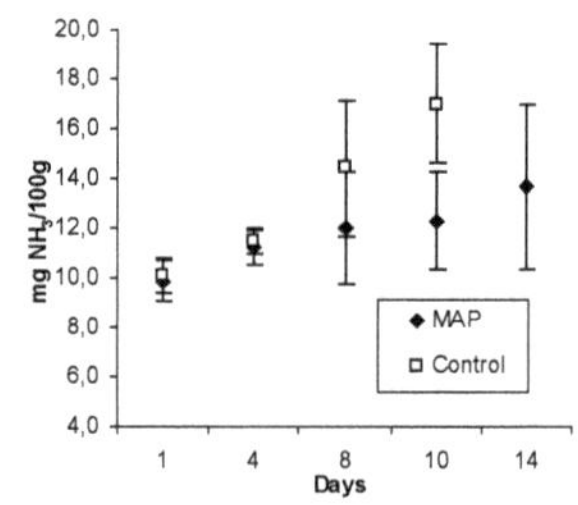

Figure 3: Total volatile basic nitrogen evolution on sliced pork packaged under MAP (CO_2/O_2) and Control conditions, stored at 2° C.

Table 1- Frequency (%) of *Listeria monocytogenes*, *Staphylococcus aureus* and *Salmonella* spp detection on sliced pork packaged under MAP (CO_2/O_2) and Control conditions, on different days of storage at 2° C (n=5).

Day	1		4		8		10		14	
Package	MAP	Control	MAP	Control	MAP	Control	MAP	Control	MAP	Control
Listeria monocytogenes	0	0	0	0	0	20	0	20	0	-
Staphylococcus aureus	0	40	40	0	20	0	40	20	0	-
Salmonella spp.	0	0	0	0	0	0	0	0	0	-

Conclusions

The increase of TVBN values greater than 14 mg NH_3/100g and a total bacterial count higher than 7 log CFU/g does not allowed the extension of the shelf life of pork meat greater than 10 days with the use of the MAP technique with a humidity adsorber, stored at 2 °C, no more than 10 days.

References

Blakistone, B.A.(1999).Principles and Applications of modified atmosphere packaging of foods. Maryland. Aspen Publishers, Inc. 2nd edition.

Erkmen, O.(2000). Antimicrobial effects of pressurized carbon dioxide on *Brochotrix thermofacta* in broth and foods. Journal of the Science of Food and Agriculture, 80, 1365-1370.

Gill, C.O.(1986). The control of microbial spoilage in fresh meat. *In*: Advances in Meat Science. Vol 2 Meat and poultry microbiology. Eds. A.M. Pearson & T.R. Dutson, AVI publishing Co. Inc. Wetsport, Connecticut. Pp. 49-88.

Lopéz-Rubio, A.(2004). Overview of Active Polymer-Based Packaging Technologies for Food Applications.Food Reviews International, Vol. 20, No. 4, pp. 367.

SHELF LIFE OF VACUUM PACKAGED COOKED HAM SLICED ON HYPERMARKET SHOP CONDITIONS

M.I. Teixeira[1], A.S. Barreto[1]*, J. Barbosa[2] and M.J. Fraqueza[1]

[1.] Faculty of Veterinary Medicine. CIISA..Technical Lisbon University. Av.da Universidade Técnica. Polo Universitário. Alto da Ajuda. 1300-477 Lisbon, Portugal. Email:mjoaofraqueza@fmv.utl.pt, [2.] Companhia Portuguesa de Hipermercados. Travessa Teixeira Junior nº1, 1300-553 Lisbon, Portugal.

Keywords: shelf life, cooked ham, vacuum, package, hypermarket

Introduction

Some hypermarkets, for benefit of consumer's time, in order to avoid long waiting while shopping, make sliced ham aerobic and vacuum packages which are placed into open refrigerated exposure equipment at temperatures between 0°C and 5°C. In former legislation, revoked by 2000/13/CE Directive, the pre-packaged foodstuffs for immediate sale were not allowed to be exposed on the day after package. However, the referred Directive (article 14) gives autonomy to each member-state to decide the labelling indications including the durability period in those particular cases, being necessary to determine the shelf life of pre-packaged ham in the point of sale conditions. The types of package, the environmental hygiene where product is sliced, and temperature control at which the ham is exposed, are fundamental factors influencing its shelf life. The aim of this work was to make product aging studies in order to determinate shelf life of vacuum packaged cooked ham, sliced on hypermarket shop conditions, to evidence a durability period longer then 24 hours.

Materials and Methods

The sampling of sliced ham took place in the hypermarket shop itself. The cooked ham was sliced (1.25mm thick) in a ham cutter (Bizerba VS 12 D, Portugal) which had previously been sanitized, and was immediately divided into portions which were weighted (approximately 150g) and packaged in aerobiosis on polystyrene platters (Hutamaki, United States) covered with extensible film (Lin-Wrap Film PVC, Linpac, Ovarpack, Portugal) and under vacuum inside polyethylene and polyamide (PE, PA) impermeable bags (Rovac A Bag 90 µm, CasFil-Indústrias Plástico SA., Portugal) with O_2 permeability of 45 ml/m^2/day/bar, sealed in a vacuum machine (Audionvac VMS 153, Holland). All these operations occurred in the hypermarket, under normal working conditions. The packages were rapidly transported under refrigeration to the laboratory, and were stored at 4°C in the dark, to simulate the temperature shop conditions, due to the impossibility of storing test samples with commercially packaged cooked available ham.

Microbial analysis were performed in the aerobic packaged sliced ham after storage for 0, 3, 5, 7, 9 and 11 days, whereas those samples under vacuum packaged were analysed at days 0, 5, 7, 9, 11, 15 and 21. At least a minimum of three replicates were made in different days.

Microbial analysis: Total Aerobic Mesophylic Counts (Tryptone Glucose Extract Agar, TGE, Scharlau) - incubation for 72 hours at 30°C; *Escherichia coli* (Tergitol BCIG Agar, Biokar Diagnostics) - incubation for 24 hours at 44.5°C; *Enterobacteriaceae* (Violet Red Bile Dextrose Agar, VRBD, Scharlau) - incubation for 48 hours at 37°C; *Pseudomonas* (Pseudomonas Agar Base, Oxoid and Pseudomonas C-F-C Supplement-SR103 E, Oxoid) –incubation for 48 hours at 25°C; *Brochothrix thermosphacta* (Streptomycin Thallous Acetate Actidione Agar, STAA Agar Base, Oxoid and STAA Selective Supplement SR151 E, Oxoid) – incubation for 48 hours at 25°C; Lactic Acid Bacteria, LAB (Man, Rogosa, Sharpe, MRS Agar, Scharlau) – anaerobic incubation for 72 hours at 30°C. Counts were expressed in log cfu/g. The detection of *Listeria monocytogenes* in 25g of sample was made according to international standard ISO 11290-1. Statistical analysis of data was done performing T-Student Test with SPSS 11.5 for Windows.

Results and Discussion

Sliced cooked ham samples initially presented 3.17 log cfu/g mesophylic aerobic counts. The *Enterobacteriaceae* initial counts were 1.28 log cfu/g and *Escherichia coli* count was above analytical method sensitivity, denoting good hygiene practices during slicing (Figure 1) (Holley *et al.*, 1996).

The results obtained show that until the day 5 of pre-packaged cooked ham storage there was no significant increase in bacterial counts levels (Figure 1), either for aerobic (A) or vacuum (B) packaged. The different microbial groups stayed in lag phase, due to refrigeration temperatures inhibition. The spoilage flora was mainly LAB and *B. thermosphacta.* Both had a significant growth increase from the 5th day of storage in aerobic packaged sliced ham ($p<0.05$), and from the 7th day in samples under vacuum condition ($p<0.05$). LAB counts tended to increase in both type of packages not being inhibited by package conditions, while *B. thermosphacta* had significant increases in aerobically packaged samples stored at 4°C (Samelis *et al.*, 2000). Although, we can consider this bacteria as the main responsibility for cooked sliced ham spoilage when vacuum packaged (Figure 1B). Only after the 11th storage day, significant counts differences were observed between samples aerobic and vacuum packaged, for total aerobic mesophylic ($p< 0,001$) and *B. thermosphacta* ($p<0,01$) (Figure 2A and B). Considering microbiological criteria for cooked sliced fine grocer's products as 3 x 10^5 cfu/g total aerobic mesophylic counts (Journaux Officiels, 1998), the counts of this bacterial group in aerobic packaged sliced ham were below that limit until the day 11 of storage, whereas the vacuum samples reach the limit only on the 15th storage day (Figure 2A). From the beginning to the end of storage, safety was assured due to absence of *Listeria monocytogenes* in 25g of sample for all analysis done (accomplishing 2073/2005/CE Regulation). The *E. coli* count was bellow 0 log cfu/g for all samples analysed.

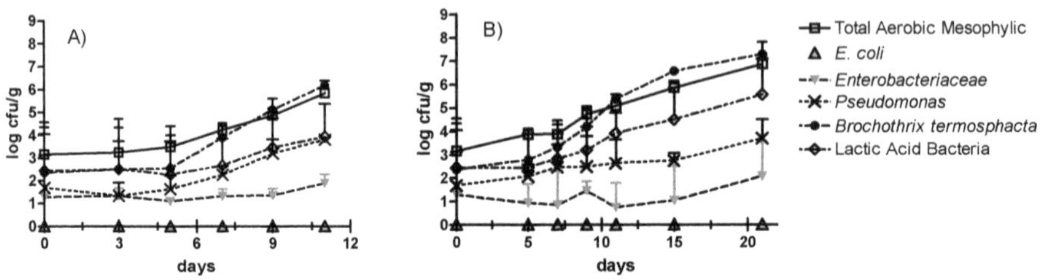

Figure 1: Aerobic (A) and vacuum (B) packaged sliced ham microbiological counts evolution.

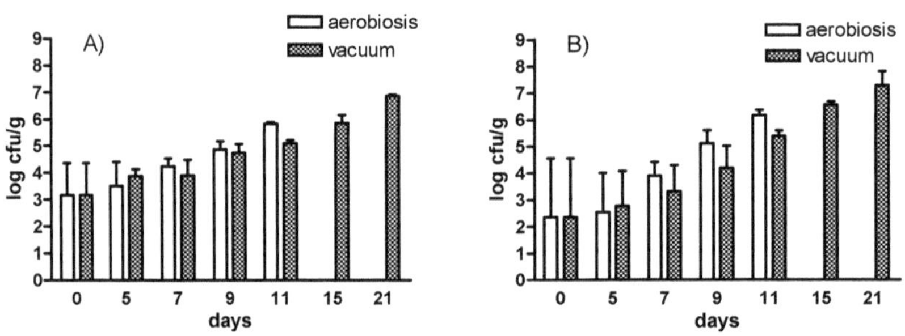

Figure 2: Total mesophylic aerobic (A) and *Brochothrix thermosphacta* (B) counts evolution comparing aerobic and vacuum packaged sliced ham.

Conclusions

When good hygienic practices are respected during product slicing procedures in sale point shops, and storing refrigeration temperature is regarded, the sliced ham aerobically packaged (attending the referred microbiological limit criteria) has a hygienic shelf life of 11 days, whereas the sliced ham vacuum packaged has 15 days. The retail sliced ham in this study conditions was safe for consumption due to the absence of *Listeria monocytogenes.*

Acknowledgments

The authors thank Auchan hypermarket Jumbo, Almada, Portugal, for providing sliced cooked ham and processing facilities.

References

Directive 2000/13/CE. (2000) Official Journal of the European Communities. L 109. European Parliament and Council of 20 March 2000. pp. 29-42.

Holley, R. A., Doyon, G., Rodrigue, N. & Carbonneau, M. (1996). Post-process, packaging-induced fermentation of delicatessen meats. Food Research International, Vol. 29, N°1., pp. 35-48.

Journaux Officiels, (1998). Higiène alimentaire. Textes généraux. Législation et Réglementation. Journaux Officiels n° 1488-I Paris. France. 308 p.

Regulation (CE) N° 2073/2005. Official Journal of the European Communities. L338. Commission of European Communities of 15 November (2005). pp 1-25.

Samelis, J., Kakouri, A., Rementzis, J. (2000). Selective effect of the product type and the packaging conditions on the species of lactic acid bacteria dominating the spoilage microbial association of cooked meats at 4°C. Food Microbiology, 17, 329-340.

INFLUENCE OF CARBON MONOXIDE/HIGH CARBON DIOXIDE ATMOSPHERE PACKAGING ON TURKEY MEAT SHELF LIFE AND THE SUBSEQUENT DEVELOPMENT OF A PINK COLOUR AFTER COOKING

M.J. Fraqueza*, M.C. Ferreira and A.S. Barreto

Faculty of Veterinary Medicine. CIISA.Technical University of Lisbon. Av. da Universidade Técnica. Polo Universitário. Alto da Ajuda. 1300-477 Lisbon. Portugal. Email: mjoaofraqueza@fmv.utl.pt

Keywords: MAP, shelf life, carbon monoxide, turkey, colour

Introduction

The use of gas mixtures oxygen free/high CO_2 gas mixtures benefits meat shelf life and promotes the inhibition of pathogens (Farber *et al.*, 1996; Jensen, 2002). However, the meat colour is unattractive to the consumer due to the myoglobin being in the deoxymyoglobin form. In spite of being forbidden in the EU, gas mixtures containing carbon monoxide (CO) are used in the USA for beef packaging to stabilise colour during storage (Sørheim, 2003; FDA, 2004). One of the problems of the adoption of CO in poultry meat, is the development of a pink colour after cooking or roasting turkey meat attributed to the presence of CO and nitric oxide (Sørheim *et al.*, 1997 and 2001). The aim of this work was to evaluate the effect of 0.5% CO in mixtures of gases containing 50 and 80% CO_2 on sliced turkey meat. The shelf life and the development of a pink colour after cooking was investigated. N_2 was used to package the sliced turkey meat.

Materials and Methods

Breast muscles from turkey carcasses were collected in a deboning room and sliced according to commercial practices. The muscles were transported under refrigeration to the laboratory in less than one hour. The sliced meat were individually packaged in an aerobic atmosphere (polystyrene trays wrapped in an oxygen permeable polyvinyl film) and in four different modified atmospheres (50% N_2 and 50% CO_2; MAP-1, 49.5% N_2, 50% CO_2, 0.5% CO; MAP-2, 19.5% N_2, 80% CO_2, 0.5% CO, MAP-3 and 100%N_2, MAP-4) in "HBX-070" bags (a multilayer film EVOH-based) sealed with a packaging machine (EVT-7-CD, Tecnoprip, Barcelona). The aerobic and modified atmosphere packaged meat have been immediately stored (0±1°C in the dark) for 12 and 25 days respectively. At least five replicates were performed for each study condition. On days 0, 5, 12, 19 and 25 the following microbiological analyses were carried out: Total mesophylic aerobic counts (Plate Count Agar, Sharlau, Spain) at 30°C for 2 days, *Pseudomonas* spp. counts (CFC agar base, Oxoid, UK) after incubation at 30°C for 2 days, lactic acid bacteria (LAB) counts [Man Rogosa Sharpe Agar (Oxoid, UK)] incubated at 30°C for 3 days and *Brochothrix thermosphacta* counts [streptomycin, actidione, thallous acetate agar (STAA, Oxoid, UK)] incubated for 2 days at 30°C. Counts were expressed as log cfu/g. Objective measurement of colour was performed over the surface of raw sliced samples which were aerobically exposed for 30 min. and on cooked samples (80°C, 15 min) with a Minolta Colorimeter CR-300 (Minolta, Osaka, Japan) using the **L**, **a**, **b** coordinates (CIELAB colour system). Statistical analysis of data was performed using SPSS 11.5 for Windows.

Results and Discussion

The evolution of total mesophylic aerobic, *Pseudomonas* spp., lactic acid bacteria and *Brochothrix thermosphacta* counts on sliced turkey meat packaged in aerobic and different modified atmospheres (MAP-1, MAP-2, MAP-3 and MAP-4) during storage at 0°C are presented on Figures 1 and 2. *Pseudomonas* spp., the main spoilage flora of turkey meat, was inhibited by 50% CO_2 in gas mixtures (Figure 1B). The increase of CO_2 concentration from 50% to 80% did not have any observable effect on mean spoilage microbial flora (Figures 1B and 2). *Brochothrix thermosphacta* was not inhibited by CO_2 but by the anoxic conditions created by MAP (Figure 2B; MAP-1 and MAP-4). The introduction of 0.5% CO in the mixtures (MAP-2 and MAP-3) did not had any effect on *Pseudomonas* spp. counts which agrees with Luno *et al.* (1998) who found that beef package with gas mixture containing 1% CO did not inhibit psychotrophic aerobic flora. However, the synergic effect of a higher concentration of CO_2 (80%) with 0.5% CO was responsible for the significantly lower total mesophylic aerobic counts ($p<0.001$). In fact, the introduction of 0.5% CO in MAP-2 and MAP-3 promoted a significant inhibition of *Brochothrix thermosphacta* (Figure 2B), which can benefit meat sensorial characteristics retarding the development of spoiled smell and taste (Jensen, 2002) and extending meat shelf life to 25 days. Sliced turkey meat under MAP-2 and MAP-3 had significantly higher a values ($p<0.001$), 13.25 and 13.73 respectively, due to

carboxymyoglobin formation (Sørheim *et al.*, 2001). However, cooked turkey meat after being exposed to CO during a storage period of 25 days had no significant differences in l, a, and b values (Figure 3). The meat exposure to 0.5% CO during long period of storage and the formation of carboxymyoglobin did not induced a subsequent pink colour defect on cooked turkey meat.

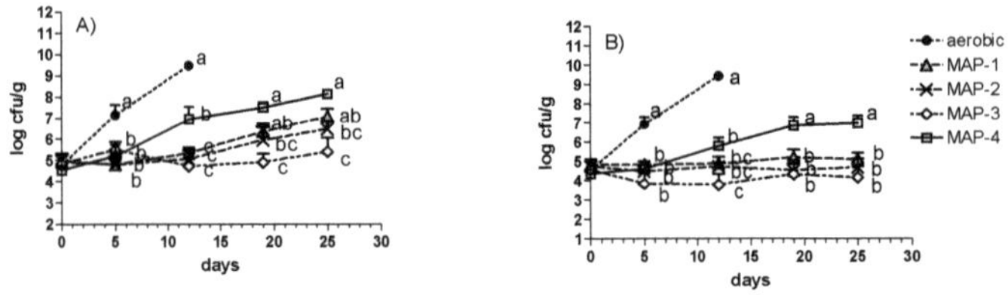

Figure 1: Total mesophylic aerobic (A) and *Pseudomonas* spp. (B) counts evolution on sliced turkey meat packaged in aerobic and MAP during storage at 0ºC ([abc] means with different letters for the same day are significantly different).

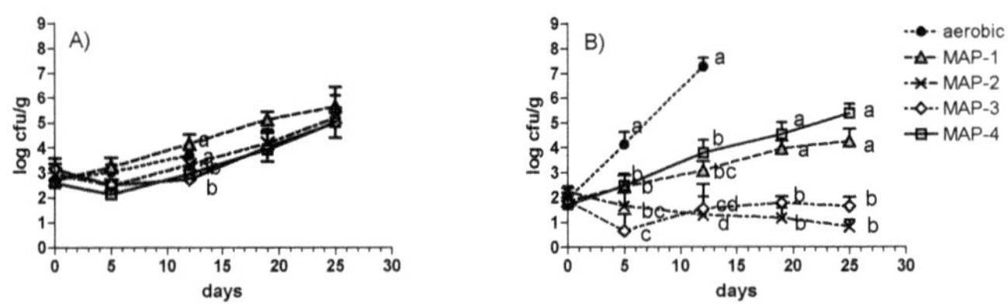

Figure 2: Lactic acid bacteria (A) and *Brochothrix thermosphacta* (B) counts evolution on sliced turkey meat packaged in aerobic and MAP during storage at 0ºC ([abc] means with different letters for the same day are significantly different).

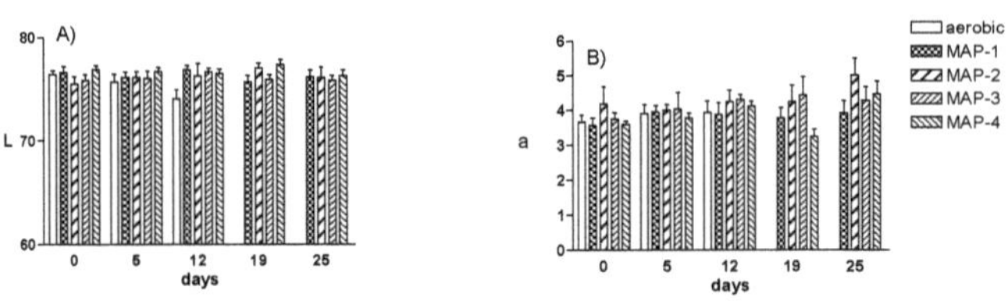

Figure 3: **L** (A) and **a** (B) values on sliced turkey meat cooked after being storage under aerobic and MAP packages.

Conclusions

The introduction of 0.5% CO in gas mixtures with higher CO_2 concentration promoted a significant inhibition of *Brochothrix thermosphacta* which can benefit meat sensorial characteristics, extending meat shelf life to 25 days. The meat exposure to 0.5% CO during long period of storage and the formation of carboxymyoglobin did not induced a subsequent pink colour defect on cooked turkey meat.

References

Farber, J.M., Cai, Y., Ross, W.H. (1996). Predictive modeling of the growth of *Listeria monocytogenes* in CO_2 environments. Int. J. Food Microbiology, 32, 133-144.

FDA, 2004. Agency Response Letter GRAS Notice No. GRN 000143. Center of Food Safety and Applied Nutrition, Office of Food Additive Safety, July 29. http://www.cfsan.fda.gov/~rdb/opa-g143.html

Jensen, J.S. (2002). Critical packaging parameters in MAP of meat products and their optimisation. Nordic Foodpack Seminar, 4-6 September. 6p.

Luño, M., Beltrán, J.A., Roncalés, P. (1998). Shelf-life extension and colour stabilization of beef packaged in a low O_2 atmosphere containing CO: Loin Steaks and ground meat. Meat Science, 48, 75-84.

Sørheim, O. 2003. Request for an amendment inserting CO as a packaging gas in annex 1 of Directive 95/2 EC. 2p.

Sørheim, O., Erlandsen, T., Nissen, P., Lea, P., Høyem, T. (1997). Effects of modified atmosphere storage on colour and microbiological shelf life of normal and pale, soft and exudative pork. Meat Science, 47, 147-155.

Sørheim, O., Lea, P., Nissen, H., Nesbakken, T. (2001). Effects of high CO_2/low CO atmosphere on colour and yield of cooked ground beef patties. Proceedings of 47th International Congress of Meat Science and Technology, ICoMST. Kraków, Poland. pp. 74-75.

CARBON DIOXIDE INJECTION GRINDING AND HIGH OXYGEN MODIFIED ATMOSPHERE PACKAGING EFFECTS ON GROUND TURKEY MEAT QUALITY

P.L. Dawson, J.C. Acton*, I.Y. Han and R. Dhananjayan

Department of Food Science and Human Nutrition, 224 Poole Agricultural Center, Clemson University, Clemson, South Carolina 29634-0316, U.S.A. Email: jcacton@clemson.edu

Keywords: carbon dioxide, grinding, packaging, ground turkey meat, quality

Introduction

Microbial outgrowth and loss of surface redness are major causes of quality changes in meat products, particularly in fresh ground meat. Modified atmosphere packaging (MAP) is widely practiced as an industry method to extend the marketing shelf life of "case ready" meats. In MAP systems involving either O_2 exclusion or high O_2 content, CO_2 is utilized in the atmosphere to slow microbial outgrowth. In high O_2 systems, the typical atmosphere utilized is 60-85% O_2, 15-40% CO_2 and sufficient N_2 to maintain package shape since CO_2 absorbs and dissolves in the meat's aqueous phase (Farber, 1991; Hotchkiss, 1994). The high oxygen level serves to increase the depth of the surface oxymyoglobin layer while the CO_2 concentration slows microbial growth. Colour stability and time to microbial spoilage for meat packaged in high O_2 modified atmospheres are approximately doubled when compared to simple aerobic over-wrapped product (Gill, 1990). Typical grinding involves meat from a mixer being fed into an auger, exiting the cutting assembly, being formed or shaped for packaging and then packaged with the selected modified atmosphere. All steps are conducted in an aerobic environment until the meat encounters the modified atmosphere. The depth of the antimicrobial effect of CO_2 is thus limited by the depth of penetration at the surrounding atmosphere-to-meat surface. This research involves comparison of typical grinding with grinding in a modified CO_2-injection grinder and the effects on surface colour and microbial outgrowth within layers for MAP ground turkey meat.

Materials and Methods

Diced, fresh turkey breast meat was ground by a food processor (Rival Manufacturing, Kansas City, MO) modified with 4 injection holes drilled into the grinding chamber and a copper fitting placed in each hole. A 4.8 mm diameter flexible manifold of 4 flexible tubes from a main CO_2 line was attached to each of the copper fittings of the grinder barrel. The inlet CO_2 pressure was maintained at 35 psi during grinding. The meat was thoroughly mixed and ground with CO_2 by flushing CO_2 at positive pressure into the chamber during grinding. The ground meat was then held in a barrier bag flushed 3 times with CO_2. The bag was heat sealed and stored at 4°C for 1 hr to ensure that there would be no loss of gas until packaging. For non-CO_2 ground samples, the turkey meat was ground as in typical grinding (in air contact) and also held in a heat sealed bag for 1 hr at 4°C. Control (0 day) samples of meat from the bags were analyzed as described below.

Following the brief holding period, meat patties of 220 g each were prepared and then packaged in expanded polystyrene barrier trays sealed with barrier lidstock film (Cryovac – Sealed Air Corporation, Duncan, SC) at an approximate 3:1 headspace-to-meat volume ratio. Trays were packaged using high O_2 (>90% O_2) on a Ross Preformed Tray Packaging Machine (Reiser, Canton, MA) and displayed in light at 4°C ± 2°C until analysis on days 2, 4, and 6. One patty from each treatment was removed from refrigeration and sampled for package gas headspace, meat colour and microbiological counts. For analysis each patty was divided into three layers: top, middle and bottom in contact with the tray.

Headspace analysis was conducted on 3 separate meat samples using a Gow-Mac Series 580 gas chromatograph (Gow-Mac Instrument Company, Bethlehem, PA) equipped with a CTR column operated at 30°C. Surface colour (CIE L*a*b*) of patties was determined using a Minolta Chroma Meter CR-300 (Minolta Corporation, Ramsey, NJ) with an 8-mm diameter orifice and illuminant C. Total aerobic plate counts were determined using pour plates with Plate Count Agar (Difco Laboratories, Detroit, MI) and incubation at 37°C for 48 hr. Aerobic counts were expressed as $\log_{10}$ colony forming units (CFU) per g meat.

Results and Discussion

There was no change ($P \geq 0.05$) in O_2 or CO_2 concentration in the package headspace of ground turkey meat from either treatment during the 6 days of display with the amount of O_2 ranging from 92-96%. The CO_2 ranged from 6% at day 0 to 3-4% at days 2 – 6. The slight shift was likely due to remaining CO2 dissolution into the meat.

Turkey meat ground in the CO_2 environment had lower ($P<0.05$) overall average total aerobic counts within each layer compared to the meat ground in the typical aerobic environment (Table 1). This is likely due to the enhanced effect of CO_2 dissolving in the meat surface during grinding.

Table 1: Grinding treatment average log $_{10}$ CFU/gm of total aerobic bacterial counts in the three layers of the turkey meat patty packaged in high O_2 and refrigerated at 4°C ± 2°C.

Layer of Meat Patty	Ground in CO_2	Ground in Air
Top	4.16 ^{b}	4.84 ^{a}
Middle	4.32 ^{b}	4.98^{a}
Bottom	4.30 ^{b}	4.98^{a}

Means with the same superscript did not significantly differ (P>0.05), n= 36.

In the first 2 days of storage (day 0 to day 2), there was an increase (P>0.05) in total aerobic bacteria counts for both the meat ground in CO_2 and non-CO_2 ground meat although the increase was less with CO_2 present (Table 2). Also, from day 0 through day 6, bacterial counts increased in all layers (top, middle and bottom) for meat ground in air at a faster rate than meat ground in CO_2. No layer effect was found within each grinding treatment. Although CO_2 was not used at recommended headspace levels (>15%) when packaging the meat, it is apparent that CO_2 injection during grinding has a positive microbial benefit by slowing outgrowth.

Table 2: Days of display and grinding treatment effects on log $_{10}$ CFU/gm of total aerobic bacterial counts in the three layers of the ground turkey meat patties packaged in high O_2 and refrigerated at 4°C ± 2°C.

Days of Package	Ground in CO_2			Ground in Air		
Display	Top	Middle	Bottom	Top	Middle	Bottom
0		3.25 ^{e}			3.36 ^{de}	
2	3.65 ^{d}	3.82 ^{d}	3.75 ^{d}	4.45 ^{c}	4.53 ^{c}	4.61 ^{c}
6	5.34 ^{b}	5.55 ^{b}	5.62 ^{b}	6.44 ^{a}	6.51 ^{a}	6.37 ^{a}

Means with the same superscript did not significantly differ (P>0.05), n= 6.

Surface colour measured as CIE +a* (redness) remained more stable overall in the patties from turkey meat ground in CO_2 compared to patties prepared with meat ground in air (Table 3). The degree of redness for turkey meat patties is much less of that expected for ground beef or poultry thigh meat due to the lower concentration of myoglobin.

Table 3: Surface CIE +a* (redness) for ground turkey meat patties packaged in high O_2 for 6 days under refrigeration (4°C ± 2°C).

Treatment	Days of Storage			
	0	2	4	6
Ground in CO2	5.96a	4.33b	4.33b	2.55c
Ground in Air	4.98b	5.01b	2.94c	1.47d

$^{a-d}$ Means with the same superscript did not significantly differ (P>0.05), n=36.

Oxymyoglobin formed in the high O_2 packaging condition was retained better during the 6 days of display in patties from CO_2 ground breast meat compared to meat ground in air. Maintaining the higher initial CIE +a* value obtained via dissolved CO_2 during grinding and use of the high O_2 concentration in packaging retarded the oxidation of oxymyoglobin to metmyoglobin, the undesirable brown surface pigment. also observed that CO_2 maintained low microbial counts and aided in maintaining ground turkey meat color.

Conclusions

Grinder modification for direct CO_2 injection during turkey meat grinding aided in retarding aerobic bacteria outgrowth in formed meat patties. When combined with packaging in a high O_2 atmosphere, surface redness retention was also improved. This technique of CO_2 enhancement to the meat may also aid in preventing package collapse that occurs with high CO_2 use in MAP for meat products.

References

Farber, J.M. (1991). Microbiological aspects of modified-atmosphere packaging technology – a review. J. Food Prot. 54:58-70.

Gill, C.O. (1990). Controlled atmosphere packaging of chilled meat. Food Control 1:74-78.

Hotchkiss, J.H. (1994). Packaging muscle foods. In: Muscle Foods (Kinsman, D.M., Kotula, A.W. and Breidenstein, eds.) pp. 475-496, Chapman and Hall, New York.

SURFACE PIGMENT SHIFTS DURING LIGHT DISPLAY FOR GROUND BEEF IN HIGH OXYGEN MODIFIED ATMOSPHERE PACKAGING

J. Buchanan and J.C. Acton*

Department of Food Science and Human Nutrition, 224 Poole Agricultural Center, Clemson University, Clemson, South Carolina 29634-0316, U.S.A. Email: jcacton@clemson.edu

Keywords: ground beef, oxymyoglobin, metmyoglobin, colour, high oxygen packaging

Introduction

Modified atmosphere packaging (MAP) with headspace gas concentrations of 70-80% O_2 and 20% CO_2 is known to extend the colour life of fresh red meats. Jakobsen and Bertelsen (2000) found that mixtures with 55% to 80% O_2 in MAP maintained fresh beef colour better than lower concentrations. The high O_2 concentration oxygenates myoglobin (Mb) well below the meat surface and thereby aids in maintaining the bright red colour as oxymyoglobin (OMb) for a longer period of time than is possible with simple aerobic film over-wraps. MAP systems for fresh red meats are termed "case ready" and use is growing rapidly. By 2010, case-ready packaging for meats is expected to account for over 70% of retail meat packages in the U.S. (Makely, 2003).

Light catalyzes OMb oxidation in meats to the brown pigment metmyoglobin (MMb) (Bertelsen and Skibsted, 1987) and even small amounts of brown surface discoloration likely means product will be rejected by consumers although it may still be microbiologically safe. Daly and Acton (2004) reported that for packages of high O_2 MAP ground beef stored in the dark for up to 6 days at 0°C, Hunter +a values did not decline on the first day of display. Product stored in the dark for 2, 4, or 6 days prior to display had a colour life of approximately 2 days. This research was conducted to examine the shift in surface spectra, OMb and MMb, and redness via CIE +a* and visual panel ratings for ground beef packaged in high O_2 MA and displayed in lighting for 3 days.

Materials and Methods

Fresh ground beef (93-96% lean) in high-oxygen (80% O_2 target) MAP was obtained on the same day of grinding and packaging from a local USDA-inspected plant. Packaging was conducted with a Ross Reiser Model 3320 Packaging Machine (Reiser & Co., Inc., Canton, Massachusetts). Each package contained approximately 0.454 kg of ground beef placed on a pink barrier foam tray sealed with lid-stock film (Lid 550™) (Cryovac Division, Sealed Air Corporation, Duncan, South Carolina). The barrier foam trays had an oxygen transmission rate (OTR) of ≤0.1 cc/tray/24hr (@ 22°C, 0%RH, 1atm) and the top transparent, flexible multi-layer film had an OTR of ≤6cc/m^2/24hr (@22°C, 0%RH, 1atm). Product was held at approximately 3-4°C in the dark for 3 days to simulate shipping prior to light display.

Day 0 was defined in this study as the day packages were removed from dark storage and placed in lighted display. A surface light intensity of 120 foot-candles (ft-c) was used as measured by Extech Light Meters (Extech Instruments, Tampa, Florida). In the display, the light source was GE F20T12-CW fluorescent tubes. A temperature of 3.7°C was used and packages were continuously displayed for 3 days.

To measure surface colour of the MAP ground beef, a package was removed from display, inverted, and placed on a rigid surface. The bottom portion of the tray was cut with a scalpel and removed. A rigid rectangular plastic plate was placed against the bottom of the ground beef and used to press the original surface against the lid film. Sufficient pressure to ensure full contact between meat surface and lid film was applied and care was taken to avoid expressing pigment from the meat. The package was then held under slight pressure and placed at the 2.54 cm diameter aperture on a HunterLab UltraScan XE spectrophotometer (Hunter Associates Laboratory Inc., Reston, Virginia). Initially (day 0) and on each full day (day 1-3) of display, surface colour characteristics were measured at 8 different loci of each package. Two packages were measured for each of 3 study replications.

For measures in the CIE L*a*b* colour space, the UltraScan XE was operated with Illuminant C, 10° observer and specular reflectance excluded. Reference myoglobin pigment forms were determined by the method of AMSA (1991). Percent reflectance between 360-750 nm was recorded simultaneously with each colour measurement and reflectance values were converted to K/S. Percent OMb and MMb were calculated using the methods of Mancini *et al.*, 2003). On day 0 at the time of placing packages in the lighted display, 2 packages were also placed on a white background under 52 ft-c of fluorescent lighting for panel evaluation of surface color. At day 3, 2 additional packages were removed from display and evaluated under the same viewing conditions. A panel of 6 evaluators consisting of faculty members and graduate students with experience in meat science was used for subjective evaluation of ground beef lean colour. The evaluation scale used was as follows: 8 = very bright cherry red, 7 = bright cherry red, 6 = cherry red, 5 = slightly cherry red, 4 = slightly dark red, 3 = dark red, 2 = very dark red, 1 = grey.

Data were analyzed using PROC GLM in SAS® (SAS Institute, Cary, NC). Mean comparisons were made using least significant difference (LSD) with $\alpha = 0.05$.

Results and Discussion

To confirm the high O_2 MA conditions, packages were analysed for headspace gases using a GOW-MAC Series 580 gas chromatograph (GOW-MAC Instrument Company, Bethlehem, Pennsylvania) equipped with a Hewlett Packard 3395 integrator (Hewlett Packard, China). From initial display (0 day) through day 3, the MA consisted of the following: 79.6-78.5% O_2, 14.7-14.3% CO_2, and 6.3% N_2 thus confirming the high O_2 MAP system.

K/S spectra from intial product at day 0 through day 3 are shown in Figure 1 for wavelengths appropriate for OMb and MMb. A characteristic double peak for OMb occurred at 540-542 and 578-580 nm and a single peak for MMb occurred at 630-634 nm. The percent surface OMb decreased ($P<0.001$) each day of display and was 83% on day 3 whereas the percent MMb increased ($P<0.001$) each day to 15% by day 3.

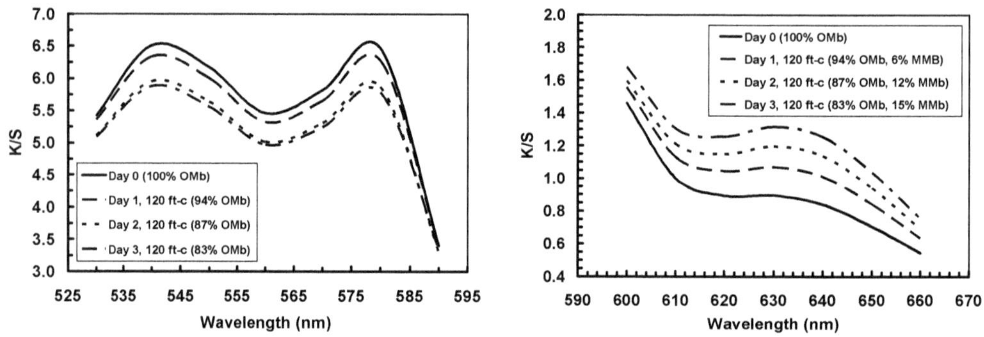

Figure 1: K/S spectra of the ground beef surface as packaged in a high O_2 MA. The spectra for surface oxymyoglobin (OMb, left) and metmyoglobin (MMb, right) from initial display (0 day) through day 3 in 120 ft-c of fluorescent lighting at 3.7°C confirms decreasing OMb with increasing MMb concentrations.

CIE +a* (redness) and panel scores (Table 1) for visual colour confirmed the decrease in redness due to the decline in OMb. There was a daily decrease ($P<0.05$) of +a* with no significant change in surface lightness (CIE L*). There was a decrease ($P<0.05$) for yellowness (CIE +b*) from day 0 (+4.0) to day 3 (+3.6). The visual colour change ($P<0.05$) determined by the panelists was a shift from near "bright cherry red" to "slightly cherry red".

Table 1: Colour parameters CIE +a* indicating a "redness" attribute and visual colour scores by panelists during the period of display for ground beef packaged in high O_2 MA at 3.7°C in 120 ft-c of fluorescent lighting.

Colour Parameter	Days of Light Display			
	0	1	2	3
CIE +a*	21.4 ^{a}	19.3 ^{b}	17.6 ^{c}	16.5 ^{d}
Colour Score	6.75 ^{a}	--	--	5.25 ^{b}

Means in each row having a different superscript letter are different at $P<0.05$.

Conclusions

Increase of product display time in lighting resulted in K/S surface spectra shifts denoting increased loss of redness for high O_2 MAP ground beef. Panelists scored visual colour as changing from bright cherry red at 0 day to slightly cherry red at the end of day 3 in display.

The findings suggest that even with the shifts that were measured, 2 days and possibly 3 days of shelf life were attained after product was placed in lighted display.

References

Daly, M.P. and Acton, J.C. (2004). Effects of dark storage time and UV-filtered fluorescent lighting during display on color stability of high-oxygen MAP ground beef. Journal of Muscle Foods, 15:1-22.

Jakobsen, M. and Bertelsen, G. (2000). Colour stability and lipid oxidation of fresh beef. Development of response surface model for predicting the effects of temperature, storage time and modified atmosphere composition. Meat Science, 54:49-57.

Makely, W. (2003). Packaging beefs up case-ready meat. Food and Drug Packaging, 67:26-35.

Mancini, R.A., Hunt, M.C. and Kropf, D.H. (2003). Reflectance at 610 nanometers estimates oxymyoglobin content on the surface of ground beef. Meat Science, 64:157-162.

EFFECT OF DIFFERENT ANTIOXIDANTS (VEGETABLE EXTRACTS) ON MEAT COLOUR STABILITY AND LIPID OXIDATION OF MINCED BEEF

E. Thomas[1], V. Scislowski[1], and M. Renerre * [2]

[1] *ADIV, Rue Chappe, 63039 Clermont-Ferrand Cedex 2, France* [2] *INRA, Unité Quapa, Theix, 63122 St Genès-Champanelle, France. Email : renerre@clermont.inra.fr*

Keywords: beef, antioxidant, colour, lipid oxidation, modified atmosphere packaging

Introduction

One of the most important quality traits of fresh beef in the decision to purchase by the consumer is the appearance of the meat and more particularly its colour. The colour of meat depends on many factors and particularly of the haeminic pigment level and the chemical state of these pigments (purple-red myoglobin, bright-red oxymyoglobin and gray-brown metmyoglobin) (Renerre, 1999). During meat storage, lipid oxidation is another cause of loss of quality with, particularly after cooking, the production of off-flavours. Lipid oxidation is a chain reaction which involves the production of free radicals, particularly from polyunsaturated fatty acids, and affects the nutritional quality of meat. These oxidative processes are more important with minced meat due to its loss of structure and high quantity of lipids. In France, minced beef is an important product which represents nearly 30% of slaughtered animals. To prolong the meat shelf-life, and preserve its red colour, the use of modified atmosphere (with O_2 and CO_2) was developed; however, O_2 also increases oxidation. The combined use of antioxidants and modified atmosphere packaging represents an attractive strategy to increase the shelf life of fresh meat and, perhaps, nutritional quality. Some years ago, many antioxidants from herbs and spices were discovered and tested in many foods and some of them, such as particularly rosemary extracts, are always tested. To overcome the problems of odour, colour and taste, a more refined rosemary extract can be developed and used. The phenolic diterpenes, carnosol and carnosic acid, make up 90% of its antioxidant activity (Offord *et al.*, 1997).The aim of this work was to evaluate the effectiveness of different vegetable extracts to delay oxidative reactions and thus extend the shelf life of minced beef packaged in a modified atmosphere.

Materials and Methods

Minced beef, with 15% lipids, was used for this experiment and was divided into 13 batches with either no addition (control) or many rosemary extracts (*Rosmarinus officinalis*), as liquid or powder, green tea (*Camellia sinensis L.*) extract, grape seed extract or onion extract as powder. For some of them, extraction techniques were different. Twelve extracts were tested and only seven were kept with respect to their origin, form, odour, concentration and solubility. The concentration of each antioxidant product was chosen with respect to industrial recommendations and varied between 0.4 and 1.5 g / kg meat.

Minced meat for each treatment was placed in modified-atmosphere packaging (MAP) composed of 70% O_2 / 30% CO_2, under artificial lighting during 16h / 24h, and stored at 4°C for a maximum of 9 days. Lipid oxidation was measured by the 2-thiobarbituric acid method of Lynch & Frei (1993) by measuring absorbance at 535 nm. The analyses were performed on fresh samples stored 3, 6 or 9 days. After 6 days storage, some steaks were cooked at 55°C during 2min 30s and analysed for TBARS. The results were expressed as equivalent mg MDA / kg sample. Colour measurement was performed at the surface of meat using a reflectance spectrophotometer equipped with an integrating sphere. Colour coordinates were calculated in the CIELAB (1976) system. The results were expressed as lightness (L*), redness (a*), yellowness (b*), hue (h*) and chroma (C*). The rates of meat discoloration were determined using R_{630}-R_{580} (Renerre, 2000).

Results and Discussion

The final choice of different antioxidants was done after sensorial analysis; if the meat developed pleasant tastes (results not shown), the vegetable extracts were kept and their antioxidant activity was tested. Only seven extracts were chosen in relationship with their good flavour.

Lipid oxidation: Figure 1 showed the changes in TBARS values for all samples throughout the storage, indicative of lipid oxidation. For the majority of the samples, storage time had a low effect on TBARS values particularly for control. Whatever the time of storage, control samples had the highest values throughout ($P < 0.0001$) while all steaks with added vegetable extracts had the lowest ones. At day 3 (Figure 1), extract n° 4 (grape seed) was the most efficient with a mean inhibition of lipid oxidation near of 56%, while this inhibition was 47% for extracts n° 7 (rosemary) and n° 20 (onion) and only 36% for extract n° 3 (rosemary). If the most efficient sample seemed to be n° 4 (grape seed), it was not statistically different from rosemary extracts (n°s 7, 9, 11) and onion extract (n° 20). After 6 days storage, no difference in TBARS values was noted between extracts but all were significantly lower than control sample. At day 9, the majority of the extracts showed an important inhibition of TBARS values close to 60%. After meat cooking at day 6 (55°C in core), it was observed that antioxidant extracts were all efficient to significantly lower the TBARS values compared to control (Figure 2) with extracts n°s 4, 7 and 9 as the most efficient ($P < 0.0001$). These results are in agreement with those of Sanchez *et al.* (2001), Ahn *et al.* (2002), Djenane *et al.* (2003) and Guiavarc'h *et al.* (2004).

Colour measurement: During a 9 days storage, a low loss of redness (a*) was observed but it differed between samples (Figure 3). The best results to stabilise red colour were obtained with samples n°3 (rosemary), better than control (P<0.0001), and n°9 (rosemary). Direct addition to minced packaged beef of rosemary gave good results for colour stability and better than those obtained with a dietary supplementation of animals (Maher *et al.*, 2002). Whatever the storage time, the worst results were obtained with the extract n°20 (onion) and particularly at days 6 and 9 of storage (P <0.001) with a loss of redness and the oxidation of oxymyoglobin (Renerre, 2000). Similarly, by using R_{630}-R_{580} ratio (Figure 4), the best results were obtained with extracts n°4 (grape seed) at day 3 ($P< 0.05$), extracts n° 3 and n° 9 (rosemary) at day 6 and extracts n° 9 and n° 11 at day 9. The worst results were observed with extracts n° 17 (not communicated) and n° 20 (onion). Other works have shown the positive effect of added rosemary (as essential oil, powder extract…) on colour stability of beef patties (Mc Carthy *et al.*, 2001, Sanchez *et al.* 2001; Ahn *et al.*, 2002;).

Conclusion

Direct addition of rosemary and grape seed extracts improve lipid stability, and to a lesser extent, colour of minced beef during a 9 days storage in MAP; further research is needed to know the exact levels and different forms of vegetable extracts to use for a better enhancement of beef quality during storage. The phenolic antioxidants found in rosemary can be used in the food industry as stabilizing agents but also with beneficial effects on human health.

Acknowledgements: Funded by Interbev and Ofival in 2005.

References

Ahn, J., Grun, I.U., Fernando, L.N., 2002. Food Chemistry and Toxicology, 67, 1364-1369.

Djenane,D., Sanchez-Escalante, A., Beltran, J.A., Roncalès, P. (2002). Meat Science, 64, 417-426.

Estevez, M., Ventanas, S., Cava, R. (2005).Journal of Food Science, 70, C427-C432.

Guiavarc'h E., Thomas, E., Peyron, A., Renerre, M., Gatellier, P., Gladine, C., Bauchard, D., Durand, D. (2004), 10èmes Journées des Sciences du Muscle et de la Technologie de la Viande, Clermont-Ferrand.

Lynch, S.M., Frei, B. (1993). Journal of Lipid Research, 34, 1745-1751.

Maher, M., O'Grady, M.N., Buckley, D.J., Troy, D., Moloney, A.P., Kerry, J.P. (2002). 48th IcOMST, 2, 500-503.

McCarthy, T.L., Kerry, J.P., Kerry, J.F., Lynch, P.B., Buckley, D.J. (2001). Meat Science, 57, 45-52.

Mercier, Y., Gatellier, P., Viau, M., Remignon, H., Renerre, M. (1998). Meat Science, 48, 301-318.

Offord, E., Guillot, F., Aeschbach, R., Loliger, J., Pfeifer, A.(1997) In: *Natural Antioxidants*, AOCS Press, Champaign, p.88.

Renerre, M.(1999). Review, Proceedings of the 45th ICoMST, Yokohama, 344-352.

Renerre, M.(2000). In : *Antioxidants in Muscle Foods*, John Wiley, New-York, p.113.

Sanchez-esacalante, A., Djenane, D., Torrescano, G., Beltran, J.A., Roncalès, P. (2001). Meat Science, 58, 421-429.

Figure1

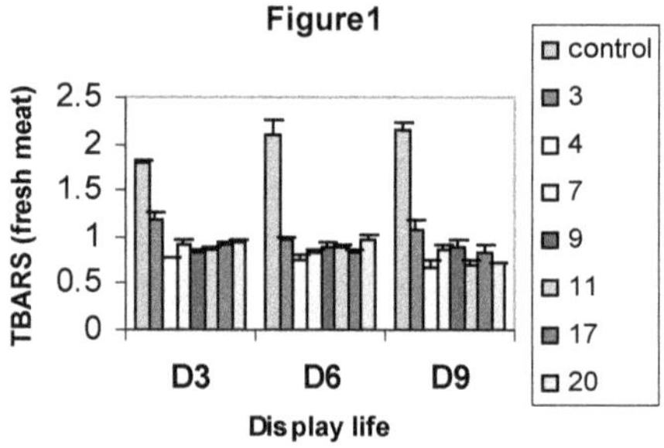

Figure 2

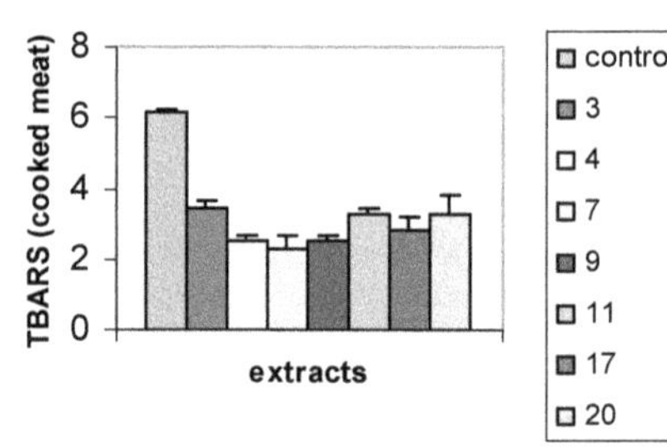

Figure 3

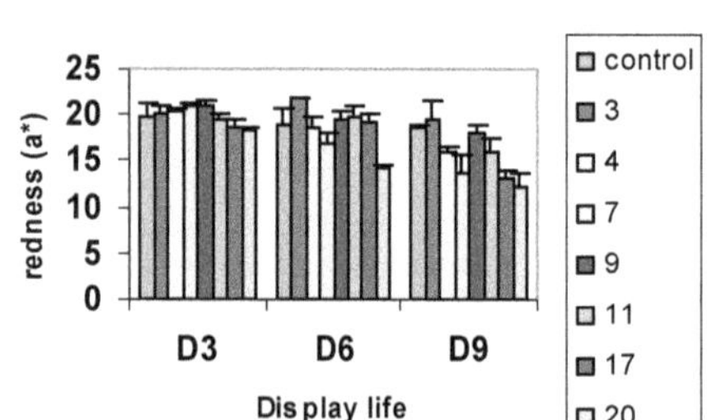

Figure 4

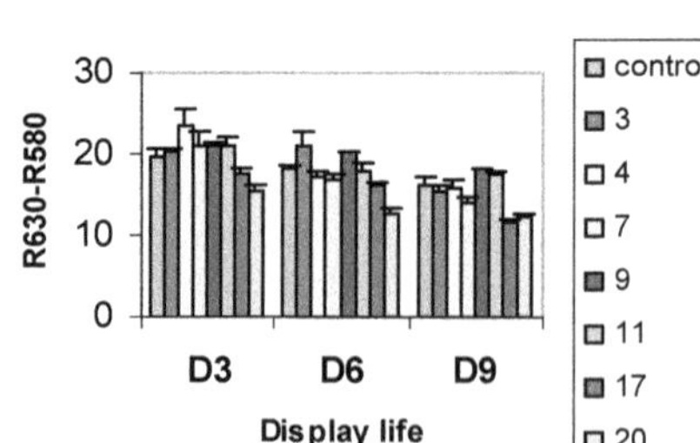

EFFECT OF RESIDUAL OXYGEN AND LIGHTING MODE ON COLOUR AND LIPID OXIDATION OF TURKEY COOKED-HAM STORED IN MAP

E. Lemoine[1], E. Roussel [2], D. Riff [3], P. Gatellier [4] and M. Renerre* [4]

[1] ADIV, 2 Rue Chappe, 63039 Clermont-Ferrand Cedex 2, France [2] Société Standa, 68 rue Robert Kaskoreff, 14050 Caen Cedex 4, France [3] Société Madranges, Feytiat, BP 138, 87004 Limoges Cedex, France [4] INRA, Unité Quapa, Theix, 63122 St Genès-Champanelle, France. Email : renerre@clermont.inra.fr

Keywords: cured cooked-ham, colour, lipid oxidation, light, oxygen

Introduction

Colour is one of the most important quality parameters at the time of purchase of meat and meat products (Renerre, 2000). For cured meat, the pink, pink-red colour is viewed as an indicator of product freshness. Colour stability is dependent on many factors such as product characteristics, pH, nature of additives, storage temperature, wavelength and intensity of light and packaging mode (Andersen *et al.* 1989; Moller *et al.*, 2002). In the presence of oxygen, nitrosylmyoglobin, the nitrite-cured meat pigment, is transformed after heating in nitrosylmyochrome which is particularly sensible to light and gives denatured metmyoglobin. The change in colour from bright red to grey and brown is unacceptable by the consumer. To avoid the light induced discoloration, which is photo-oxidation, the level of residual oxygen in the package must be very low. Lights used by the retailers can vary between incandescent, fluorescent and metal halide lamps but fluorescent lamps are the most used. In the meat industry, some lights are recommended by commercial firms but often without serious scientific basis. With modified atmosphere packaging (MAP), the critical level of residual oxygen in fresh meat during chill storage must be below 0.1% to avoid discolouration (Rousset and Renerre, 1991). In MAP pasteurised ham, only Moller *et al.*, (2000) have clearly shown that, among other factors, the critical level of residual oxygen must be 0.1% or less to avoid light discolouration. The aim of this work was to study the effect of residual oxygen in MAP of turkey cooked-ham, combined or not with light exposure used in practice, which can cause photo-oxidation of the pigment and greyness at the surface.

Materials and Methods

In the ADIV laboratory, thawed turkey fillets (Madranges) were roughly minced and meat was injected with brine containing nitrited salt (18g/kg final product), dextrose (10g/kg), erythorbate (0.5g/kg). Then the fillets were packed in cellulose pouches and cooked under vacuum until a core temperature of 69°C. After rapid cooling at 0°C, meat was stored for 3 days in a chill cabinet at 4°C. Turkey fillets were then sliced and packed in modified atmosphere (70% N_2 / 30% CO_2) (MAP) with a laminate film which OTR was 8 cc/m^2/atm./24^h at 23°C (Pechiney-Soplaril R1385). Two levels of residual oxygen were used for the packaging atmosphere: 0% and 1%. Conditioning with about 1% O_2 was obtained by using a vacuum pressure equal to -700 mbars before gas reinjection. To obtain a percentage closest to 0% oxygen, after vacuum packing and during MAP, an oxygen scavenger of great capacity (Atco, Standa) was added to the package; the rate of residual oxygen was measured by a gas analyser (Abiss-Tom 12). The packages were placed in forced air chilled cabinets at 0-2°C. Four light sources were tested: a "warm" source (OSRAM, *Lumilux de luxe, Blanc chaud)*, **S4**, a "cold" source (OSRAM, *Lumilux de luxe, Biolux*, **S3**), and two light sources recommended by the manufacturers: Philips TLD *Food Pro* (**S2**) and OSRAM *Natura de luxe* (**S1**). Control samples were also kept in darkness (**S5**). At days 3, 10 and 17, packages were opened and tristimulus colour parameters measured (Renerre, 2000) using a spectrocolorimeter. Oxidative rancidity was measured by the 2-thiobarbituric acid (TBARS) index (Lynch and Frei, 1993).

Results and Discussion

The fluorescent lamps used showed a spectrum containing from 5 to 10 intensity peaks (not shown) in the visible range, with similarities to those showed by Djenane *et al.* (2001) with Mazda fluorescent lights. For Moller *et al.* (2002), the reaction between oxymyoglobin and nitric oxide yields peroxynitrite ($OONO^-$) which may initiate lipid oxidation in the meat product and were significantly influenced by light and oxygen level.

Colour measurements: Firstly, whatever the treatments, neither the residual oxygen nor the lighting mode had influence on lightness (L*) during meat storage (not shown). For redness, with 0% oxygen in MAP, during the 17 days storage (Figure 1), no important decrease in a* was observed whatever the treatment. Conversely, with 1% oxygen (Figure 2), lighting of samples induced an important decrease in redness (4> a*> 2) compared to samples stored in darkness (a* > 6) (*P< 0.001*) except with S1 at day 10. Among the tested lights, S4 source gave the worst results with the lowest a*. These differences in redness index are well correlated to sensorial analysis according to the results of Djenane *et al.* (2001). By calculation of hue (results not shown), in the presence of 1% oxygen, hue value was near 40 in darkness and near 100 for illuminated samples indicative of an important oxidation. As shown respectively in Figures 3 and 4, the accumulation of metmyoglobin was near 30% whatever the lighting conditions in 0% oxygen packs. With 1% oxygen, the metmyoglobin percentage was near 30% in the darkness, but from 50 to more than 60% (between day 3 and day 17 of storage) when meat, of a grey colour, was displayed under varying wavelengths. It was found that 40%

metmyoglobin on meat surface was synonymous with meat rejection by the consumers (Renerre, 2000). For Andersen *et al.* (1989), photooxidation of nitrosomyoglobin showed little wavelength dependence.

Lipid oxidation: Firstly, it was observed that with 0% oxygen in the package, the TBARS of turkey fillets were very low whatever the treatments (0.5 < TBARS < 0.8) but less than 0.5 for meat stored in darkness even after 17 days in MAP (Figure 5). In the presence of 1% residual oxygen and in darkness, TBARS values remained equal to about 0.5-0.8 even after 17 days storage (Figure 6). Conversely, with different lighting conditions, TBARS increased significantly ($P<0.05$) between day 3 and day 17 of storage. Mean TBARS of all samples showed values near to 0.8 at day 3 but equal to about 1.2 at day 10 and comprised between 1.2 and > 2, for some of them, after a 17 days storage indicating the appearance of off-odours and off-flavours.

Conclusion: it can be concluded that lighting and quantity of residual oxygen on slices of turkey cooked–ham will influence the colour and lipid stability after MAP. In the absence of residual oxygen, the lighting mode has no detrimental influence on colour, compared to darkness, and TBARS values remain largely under 1 (acceptable limit) until day 10 indeed day 17. With residual oxygen (near to 1%), myoglobin oxidation is important, colour is grey (sometimes even at day 3), and TBARS value can be three fold higher than the control in darkness, with development of off-flavours. Globally, sources S1 and S2 are not better than the other two.

References

Andersen, H.J., Berthelsen, G. and Skibsted, L.H. (1989). Meat Science, 25, 155-159.

Djenane, D., Sanchez-Escalante, A., Beltran, J.A. and Roncalès, P. (2001). Journal of Food Science, 66, 181-186.

Lynch , S.M. and Frei, B. (1993). Journal of Lipid Research, 34, 1745-1751.

Moller, J.K.S., Berthelsen, G. and Skibsted, L.H. (2002). Meat Science, 60, 421-425.

Moller, J.K.S., Jensen, J.S., Olsen, M.B., Skibsted, L.H. and Berthelsen, G. (2000). Meat Science, 54, 399-405.

Renerre, M. (2000). In: *Antioxidants in Muscle Foods*, John Wiley, New-York, p.113.

Rousset, S. and Renerre, M. (1991). International Journal of Food Science and Technology, 26, 641-652.

Acknowledgements: funded by ACTIA in 2005.

Figure 1.

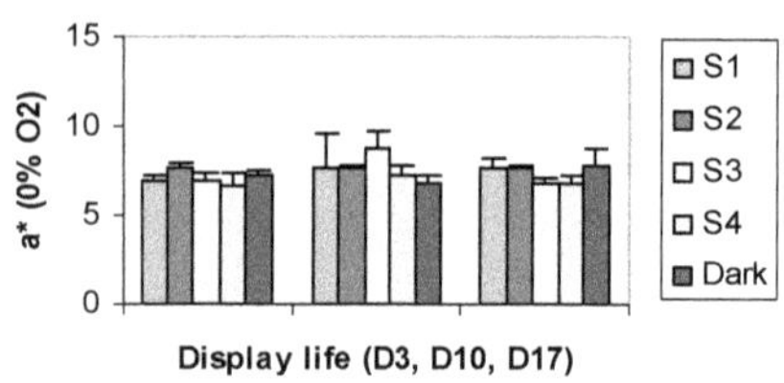

Figure 2.

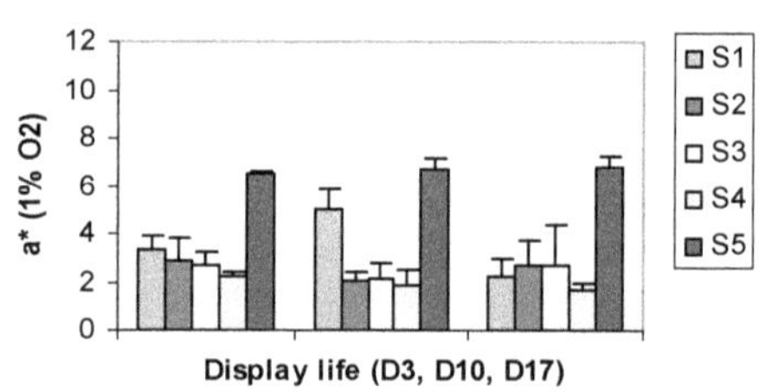

Figure 3.

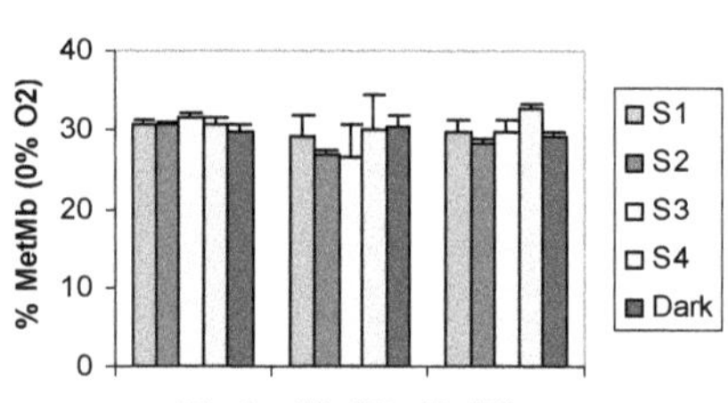

Figure 4.

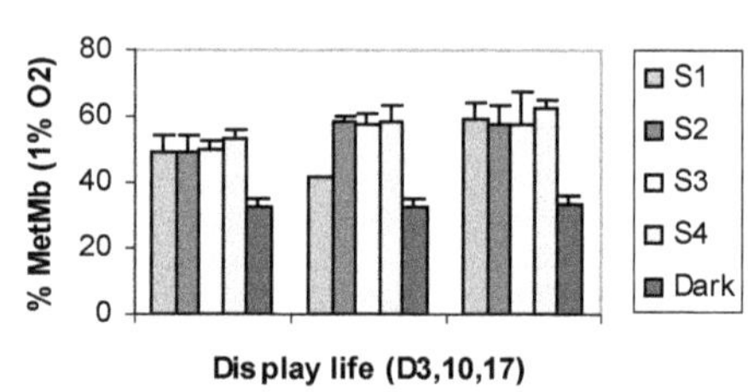

Figure 5.

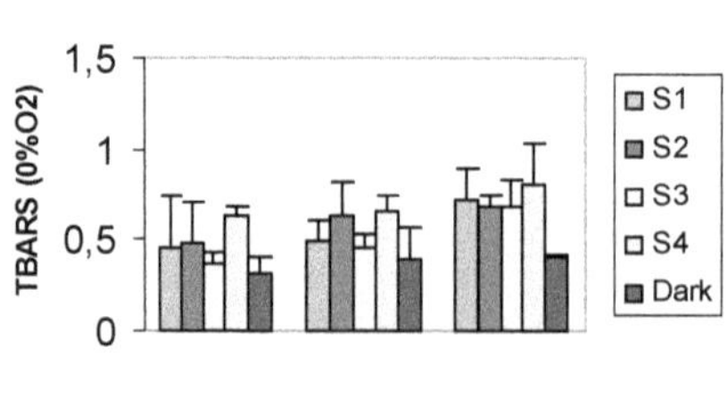

Figure 6.

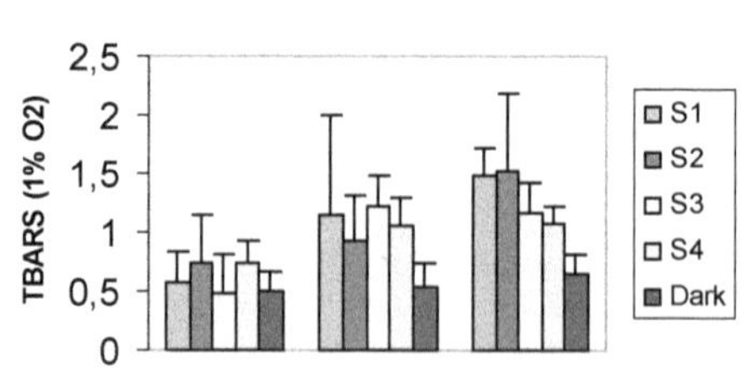

EVOLUTION OF MICROBIOLOGICAL AND SENSORY QUALITY ALONG STORAGE OF PACKAGED "SALCHICHÓN" ENRICHED IN MUFA AND PUFA

B. Rubio[1], B. Martínez[1], M.J. Sánchez[1], M.D. García-Cachán[1], J. Rovira[2] and I. Jaime[2]*

[1.] *Estación Tecnológica de la Carne. Instituto Tecnológico Agrario de Castilla y León. C/ Filiberto Villalobos s/n. 37770 Guijuelo (Salamanca). Spain,* [2.] *Departamento de Biotecnología y Ciencia de los Alimentos. Facultad de Ciencias. Universidad de Burgos. Plaza Misael Bañuelos s/n. 09001 Burgos. Spain.*
Email: ijaime@ubu.es

Keywords: modified atmosphere packaging, salchichon, shelf life

Introduction

Salchichon is one of the most important dry-cured sausages prepared in Spain. Nowadays, due to the recent changes in shopping and consumption habits, packaged sliced meat products are an important retail selling method. Furthermore, nutritionists have persistently recommended changes in the composition of fat associated with meat, towards a reduction of saturated acids to the advantage of mono and polyunsaturated fatty acids, for their positive effects. For this, the problem of safe preservation in the meat industry has become more complex as today's products require longer shelf life and greater assurance of protection from microbial spoilage. Therefore, industrial trends try to extend the shelf life of this type of products by storage practices such as modified atmosphere packaging (vacuum and gas mixtures packaging). The aim of this work was to evaluate the shelf life of salchichon manufactured with raw meat with different fatty acid composition, packaged under two kinds of conditions: vacuum and 20/80% CO_2/N_2 currently used by the meat industry. The assessment of the shelf life was based on the microbiological counts, pH and a_w measurements as well as on the sensory evaluation.

Materials and Methods

Samples: The sausages were manufactured according to a traditional formulation with lean and backfat obtained from pigs fed with three different diets: 1) control (CO), 2) enriched in monounsaturated fatty acids (HO) and 3) enriched in polyunsaturated fatty acids (HL). Two sausages, randomly selected for each batch, were sliced at 1 mm thickness and 100 g of slices were placed in polystyrene trays.

Packaging and storage: The trays were packed under: (a) Vacuum: trays were introduced in plastic bags (polyamide/polyethylene) which were subjected to vacuum and sealed using a packer (mod. EVT-7-TD Tecnotrip, Barcelona, Spain), or (b) Gas mixture: trays were evacuated and flushed with the selected gas mixture, 20/80% CO2/N2, (Carburos Metálicos S.A., Barcelona, Spain). Then, they were closed with a high barrier film by heat-sealing with a packer (mod. TSB-100 Tecnotrip, Barcelona, Spain). Packages had a headspace volume ratio of 1:1. All samples were stored at 6°C until their analysis. Samples for all treatments were analysed at 0, 15, 30, 60, 90, 150 and 210 days of storage. The whole experiment was replicated twice.

Microbiological analysis: The samples were analysed for aerobic mesophilic bacteria (3M Petrifilm Aerobic Count Plate (Bioser, Barcelona, Spain) at 30°C for 48 h), psychrotrophic bacteria (Plate Count Agar (Scharlau, Spain) at 7°C for 10 days), anaerobic bacteria (Schaedler Agar (Scharlau, Spain) at 37°C for 48 h), *Enterobacteriaceae* (3M Petrifilm Enterobacteriaceae Count Plate (Bioser, Barcelona, Spain) at 37°C for 24h), *Enterococcus* (Slanetz Bartley Agar (Scharlau, Spain) at 37°C for 24 h), *Pseudomonas* (*Pseudomonas* Agar (Oxoid, Spain) supplemented with Cetrimide, Fucidine and Cephaloridine (CFC, Oxoid, Spain) at 30°C for 48 h), lactic acid bacteria (LAB) (MRS Agar (Scharlau, Spain) at 30°C 72 h), *Micrococcaceaes* (MSA (Scharlau, Spain) at 37°C for 48 h), yeasts and molds (3M Petrifilm Yeast and Mold Count Plate (Bioser, Barcelona, Spain) at 25°C for 5 days).

Physicochemical analysis: Water activity (a_w) was measured by CX2 AQUA LAB equipment. The pH values were determined by blending with a pH meter equipped with a glass electrode.

Sensory evaluation: This was carried out on "salchichon" slices after each storage time by an experienced 8-member sensory panel. The sensory attributes (colour, odour, taste, hardness, juiciness and overall acceptability) were scored using 5-point scales (5 = excellent, 4 = good, 3 = acceptable, 2 = fair and 1 = unacceptable).

Statistical analysis: Data were statistically analysed using one-way analysis of variance (ANOVA) in order to determine the effect of storage time for each batch and for each type of storage. Furthermore, a two-way ANOVA test was carried out in order to determine significant differences among sausages depending on the type of formulations and type of storage. Means were separated by Tukey honest significant difference test at 5% level. Data analyses were conducted using the statistical package STATISTICA 7.0.

Results and Discussion

The results of the microbiological analysis showed that the lactic acid bacteria constituted the major microflora of the sausages (7.7-8.5 log cfu/g). Because of the good adaptation of lactic acid bacteria to the meat environment and their faster growth rates during fermentation and ripening of sausages, they become the dominant microflora (Drosinos *et al.*, 2005). In general, microorganims in either vacuum or gas mixture packed salchichon, remained nearly constant for up to 210 days (data not shown). The stability in the counts during storage time, could be explained by taking into account that packaging in anoxic environments retards microbial growth and delays spoilage due to slow proliferation of bacteria capable of tolerating anaerobic conditions (Martínez *et al.*, 2006). However, enterobacteria and pseudomonad counts were subjected to a significant inhibition when the salchichon was packaged under vacuum and with 20/80% CO_2/N_2 and values under detection limit were found after 15 days of storage, probably due to the strong competitive effect of lactic acid bacteria on the rest of the endogenous flora. Lactic acid bacteria suppress the growth of Gram-negative bacteria by producing organic acids and various antibacterial metabolic products (Holzapfel *et al.*, 1995). Regarding packaging method, no significant differences ($p>0.05$) were found in counts for flora studied between both types of packaging for the same salchichon group. Therefore, vacuum and gas mixtures (20/80% CO_2/N_2) packaging did not show differences with respect to the shelf life of salchichon.

Storage time had a significant effect on pH values of different sausages and during storage the pH values were significantly reduced in all groups (from 5.05-5.10 to 4.77-4.99). The decrease in pH values during the refrigerated storage of vacuum and gas mixture packed samples could be attributed to the activity of lactobacilli and to the dissolution of CO_2 into meat product. The results of water activity (a_w) showed a_w remained constant during storage under modified atmospheres packaging and no differences were found between both types of packaging. Mean values obtained were 0.862, 0.877 and 0.869 for CO, HO and HL, respectively.

Storage time had significant effects ($p<0.05$) on sensory parameters (Figure 1) considered and the sensory parameters values decreased linearly aver time in vacuum and 20/80% CO_2/N_2 packed samples of sausages. In general, a loss of spice odour and taste and a higher acid taste were detected in samples by the judges. In addition, they perceived a slight increase of hardness. Taken together, these results indicate that packaging under both modified atmospheres (vacuum and 20/80% CO_2/N_2) increases the shelf life of the product, though without entirely preventing the maturational changes that occur in dry fermented sausages. These results agree with those provided by Fernández-Fernández *et al.* (2002) for Galician chorizo sausage.

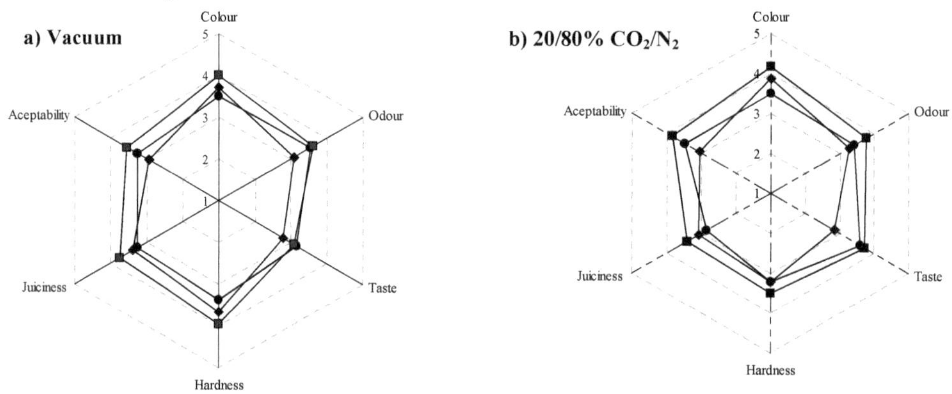

Figure 1: Sensory parameters evaluated on salchichon: (•) CO control (■) HO high oleic (♦) HL high linoleic at 210 days of storage under vacuum (a) and under 20/80% CO_2/N_2 (b).

Conclusions

The results of this study indicate that packaging under both modified atmospheres (vacuum and 20/80% CO_2/N_2) increases the shelf life of the CO, HO and HL salchichon. In terms of overall acceptability, the shelf life of HO and CO salchichon packed under vacuum and with 20/80% CO_2/N_2 were 210 days, and 150 days for HL salchichon.

References

Drosinos, E.H., Mataragas, M., Xiraphi, N., Moschonas, G., Gaitas, F. and Metaxopoulos, J. (2005). Characterization of the microbial flora from a traditional Greek fermented sausage. Meat Science, 69: 307-317.

Fernández-Fernández, E., Vázquez-Odériz, M.L. and Romero-Rodríguez, M.A. (2002). Sensory characteristics of Galician chorizo sausage packed under vacuum and modified atmospheres. Meat Science, 62: 67-71.

Holzapfel, W.H., Guisen, R. and Schillinger, U. (1995). Biological preservation of foods with reference to protective cultures, bacteriocins and food-grade enzymes. International Journal of Food Microbiology, 24: 343-362.

Martínez, L., Djenane, D., Cilla, I., Beltrán, J.A. and Roncalés, P. (2006). Effect of varying oxygen concentrations on the shelf-life of fresh pork sausages packaged in modified atmosphere. Food Chemistry, 94: 219-225.

STORAGE OF PACKAGED SLICED "SALCHICHÓN" ENRICHED IN MUFA OR PUFA: EFFECT ON COLOUR AND OXIDATION STABILITY

B. Rubio [1], B. Martínez [1], C. Molinero [1]*, M.D. García-Cachán [1], J. Rovira [2] and I. Jaime [2]

[1.] *Estación Tecnológica de la Carne. Instituto Tecnológico Agrario de Castilla y León. C/ Filiberto Villalobos s/n. 37770 Guijuelo (Salamanca). Spain.,* [2.] *Departamento de Biotecnología y Ciencia de los Alimentos. Facultad de Ciencias. Universidad de Burgos. Plaza Misael Bañuelos s/n. 09001 Burgos. Spain. Email:molsascr@itajcyl.es*

Keywords: modified atmosphere packaging, fermented sausage, fatty acid composition, TBA, colour

Introduction

From the nutritional point of view, fermented sausages are an important source of proteins of high biological value (Beriain *et al.*, 2000); however, these traditional meat products show some negative aspects as a consequence of their high animal fat content. Nowadays, the meat industry tries to offer the consumer healthy products obtained by different strategies, improving the nutritional quality of food by modification of the lipid fraction: increasing the monounsaturated fatty acids (MUFA) and polyunsaturated fatty acids (PUFA). Moreover, a high percentage of MUFA and PUFA makes these products highly sensitive to oxidation. In this sense, packaging methods in modified atmospheres (vacuum and gas packaging) allow a longer product shelf life as well as variety in presentation method. The aim of this work was to investigate the effect of different storage conditions (vacuum and gas mixture packaging) on oxidation and colour stability of sliced salchichon, manufactured with raw meat enriched in MUFA and PUFA, and stored for a long period time (210 days) at 6 °C.

Materials and Methods

Samples. The sausages were manufactured according to a traditional formulation with lean and backfat obtained from pigs fed with three different diets: 1) control (CO), 2) enriched in monounsaturated fatty acids (HO) and 3) enriched in polyunsaturated fatty acids (HL). Two sausages, randomly selected for each batch, were sliced at 1mm thickness and 100g slices were placed in polystyrene trays.

Packaging and storage. The trays were packed under: (a) Vacuum: trays were introduced in plastic bags (polyamide/polyethylene) which were subjected to vacuum and sealed using a packer (mod. EVT-7-TD Tecnotrip, Barcelona, Spain), or (b) Gas mixture: trays were evacuated and flushed with the selected gas mixture, 20/80% CO_2/N_2, (Carburos Metálicos S.A., Barcelona, Spain). Then, were closed with a high barrier film by heat-sealing with a packer (mod. TSB-100 Tecnotrip, Barcelona, Spain). Packages had a headspace volume ratio of 1:1. The gas content of each pack and the residual O_2 was controlled using a gas analyzer1450 B3 Servomex (Aries, Madrid, Spain).

All samples were stored at 6ºC until their analysis. Samples for all treatments were analysed at 0, 15, 30, 60, 90, 150 and 210 days of storage. The whole experiment was replicate twice.

Determination of fatty acids. Fatty acids were determined in the lipid extracted from ripened sausages for each packaging method and after each storage time. The Bligh and Dyer (1959) method was used for the lipid extraction. Fatty acids composition was determined by gas chromatography. Boron trifluoride/methanol was used for the preparation of fatty acid methyl esters (Morrison and Smith, 1964). Fatty acids methyl esters were identified by comparison with standards run previously alone or together with samples and were quantified as percentage of total methyl ester percentage.

TBA (2-Thiobarbituric acid) analysis. TBA measurements were made using the Maraschiello *et al.* (1999) method. The TBA values are expressed as micrograms of malonaldehyde per gram of sausage.

Instrumental colour measurement. Objective measurement of colour was performed at the surface of the salchichon using a reflectance spectrophotometer (Minolta CM-2002; Osaka, Japan). Colour coordinates were determined in the CIE-LAB system and the results were expressed as lightness (L*), redness (a*) and yellowness (b*).

Statistical analysis. Data were statistically analysed using one-way analysis of variance (ANOVA) in order to determine the effect of storage time for each batch and for each type of packaging. In addition, a two-way ANOVA test was carried out in order to determine significant differences among sausages depending on the type of formulations and type of storage. Means were separated by Tukey honest significant difference test at 5% level. Data analyses were conducted using the statistical package STATISTICA 7.0.

Results and Discussion

Regarding fatty acids composition, significant differences were observed between the group CO and those with modified fat composition. An increase (2.67%) of oleic acid was observed for the HO group compared to CO group and a greater increase in linoleic and linolenic acids (48.08 and 57.27% respectively) was obtained for the HL group compared to the CO group. With these fatty acids profiles, modified products (HO and HL) showed better ratios than control products (CO) from a nutritional point of view. Higher MUFA+PUFA/SFA (1.63 and 1.51) and PUFA/SFA (0.40 and 0.47) in HO and HL products were found than in control ones (1.36 and 0.32 respectively), which is considered beneficial in relation to blood lipids. In Figure 1, the evolution of TBA values can be observed for vacuum packaged salchichon (Figure 1a) and for 20/80% CO_2/N_2 packaged salchichon (Figure 1b). Higher values were observed in the HL salchichon, followed by HO and CO samples. This may be due to the fact that SFA and MUFA fatty acids are less susceptible to oxidation than PUFA. With respect to the lipid oxidation evolution, for all treatments, TBA values increased ($p<0.05$) at 150 days of vacuum storage and then began to decrease ($p<0.05$). For the samples stored under 20/80% CO_2/N_2 no changes ($p>0.05$) were observed up to 150 days and then, TBA values decreased ($p<0.05$) at the end of storage period. Janero (1990) pointed out that a decrease in TBA values during storage could be attributed to reaction of MDA with amino acids, sugars and nitrite in complex formulations. With respect to storage conditions, gas packaged samples showed lower TBA values ($p<0.05$) than those found in vacuum packaged samples after 90 days of storage.

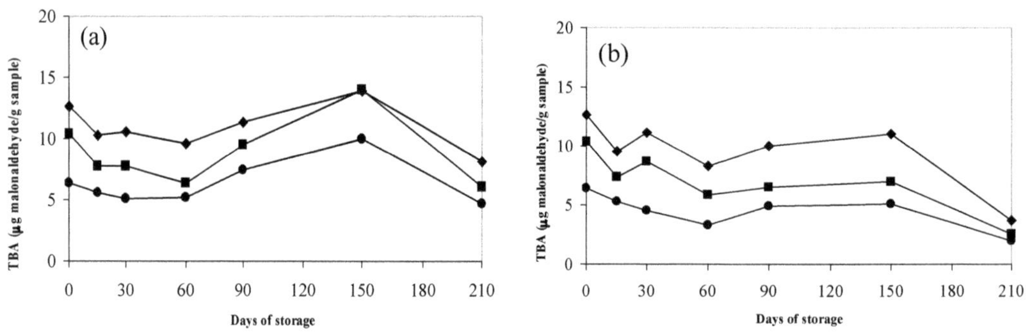

Figure 1: Evolution of TBA values on ripened salchichon with three formulations: (•) CO control (■) HO high oleic (♦) HL high linoleic along 210 days of storage under vacuum (a) and under 20/80% CO_2/N_2 (b).

Regarding colour parameters of salchichon packaged under vacuum and with 20/80% CO_2/N_2, no differences were found for L* values along storage, and no differences ($p>0.05$) were obtained for this parameter between packaging methods at each storage time. Besides, no differences ($p>0.05$) were found in L* values between salchichon types, except at day 0 of storage, where significant differences were found between CO salchichon (49.21) and HO group (42.96). With regard to redness value (a*), which is used as an indicator of colour stability in meat and meat products, did not show differences ($p>0.05$) during the storage period, and values at the end of storage were similar to those at the packaging moment (7.23-11.93). As expected, a* values remained constant throughout all storage period because nitrite was added, which acts as a colour stabiliser (Harms *et al.,* 2003). The b* value is a colour parameter concerning with lipid oxidation. In this work, storage time did not determine changes ($p>0.05$) on parameter b* in vacuum and gas mixtures packaged samples, and no differences ($p>0.05$) were obtained for this parameter between packaging methods at each storage time. However, statistical differences ($p<0.05$) were observed for this parameter between sausages with different fat composition when sausages were packed under 20/80% CO_2/N_2, although these differences did not follow a clear trend.

Conclusions

On the basis of the results obtained in this work, it can be concluded that it is possible to store sliced dry fermented sausages enriched in MUFA or PUFA, (desirable from a nutritional point of view) for a long time period without a significant increase on the colour parameters and on oxidation if a suitable packaging method is used.

References

Beriain, M.J.; Chasco, J. and Lizaso G. 2000. Relationship between biochemical and sensory quality characteristics of different commercial brands of salchichon. Food Control, 11: 231-237.

Bligh, E.G. and Dyer, W.J. 1959. A rapid method of total lipid extraction and purification. Can. 42: 371-386.

Janero, D.R. 1990. Malonaldehyde and thiobarbituric acid reactivity as diagnostics indices of lipid peroxidation and peroxidative tissue injury. Free Radicals and Biological Medicine, 9: 515-540.

Maraschiello, C., Sárraga, C. and García-Regueiro, J.A. 1999. Glutathione peroxidase activity, TBARS and □-tocopherol in meat from chickens fed different diets. J. Agric. Food Chem. 47: 867-872.

Morrison, W.R. and Smith, L. M. 1964. Preparation of fatty acid methyl esters and dimethylacetals from lipid with boron fluoride-methanol. J. Lipid Res, 5:600-608.

Harms, C., Fuhrmann, H., Nowak, B., Wenzel, S. and Sallmann, H.P. 2003. Effect of dietary vitamin E supplementation on the shelf life of cured pork sausage. Meat Science, 63: 101-105.

EFFECT OF PACKAGING AND ENDPOINT TEMPERATURE ON INTERNAL COOKED COLOUR OF PORK CHOPS MEASURED BY VIDEOMETERLAB 2

C. Borggaard, M. Hviid and T. Fihl

Danish Meat Research Institute, Maglegaardsvej 2, DK-4000 Roskilde. E-mail: clb@danishmeat.dk

Keywords: packaging technique, reflection spectra, internal cooked colour, pork

Introduction

It is well documented that the chemical properties of myoglobin in meat affects the colour of cooked meat (Hunt *et al.*, 1999). Warren *et al.*, (1996) were the first to conclude, that the internal cooked colour can be related to the oxidative state of myoglobin present in meat prior to cooking. The oxidative state of myoglobin can be affected by the packaging procedure, e.g. when using modified atmosphere packaging containing oxygen nearly all myoglobin will be oxygenated forming oxymyoglobin. By adding carbon monoxide to the packaging atmosphere myoglobin will turn into carboxymyoglobin. Different pigment forms have different stability and denaturate at different temperatures. Metmyoglobin denatures at a lower temperature than oxymyoglobin, and deoxymyoglobin is the most heat stable myoglobin form of the three (Machlik, 1965). John *et al.*, (2004) found that meat packaged with carbon monoxide remains red until a temperature of 79°C, temperatures above the expected denaturation temperature for myoglobin. The objective of this study was to examine the effect of the packaging technique and endpoint temperature on the internal cooked colour of pork chops measured by a sensory panel and with VideometerLab2, a picture generating reflectance spectrophotometer.

Materials and Methods

Paired loins from six pigs with normal pH value slaughtered at a commercial slaughterhouse were selected. The loins were deboned, trimmed, sliced to a thickness of 2.5 cm, and packaged by three different packaging techniques. In each packaging the chops were randomly packaged with two neighbouring chops in a skin pack imitation, an oxygen MAP (70%O_2/30%CO_2), and a carbon monoxide MAP (70%N_2/30%CO_2/0.4%CO). The skin pack imitation was made by putting the pork chops in shrinking bags under vacuum. Afterwards the bags were lowered in almost boiling water (approximately 85°C) and elevated after a few seconds. After packaging, the chops were placed in cold storage (4°C) for 7 days with retail display simulation for 12 hours the last day of the storage period. After storage, the pork chops were cooked to endpoint temperatures of 50, 55, 60, 65, 70, 75, and 80°C, turned every second minute on a 155°C hot frying plate. The internal cooked colour of the chops was evaluated by a trained 8 member sensory panel by determining the doneness on a 15 cm unstructured scale and afterwards measured with VideometerLab2 (www.videometer.com). The spectra from VideometerLab2, which are made of measurements at 18 discrete wavelengths, were evaluated in the Unscrambler software package (www.camo.com), version 9.2. Sensory data was evaluated in SAS (www.sas.com), version 9.1.

Results and Discussion

Results from the sensory evaluation are shown in Figure 1.

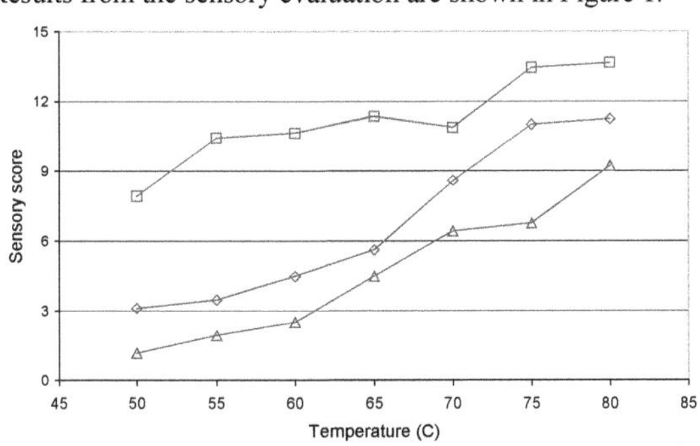

Figure 1: Least square mean values of the internal cooked colour of pork chops packaged using three different packaging techniques cooked to seven different end point temperatures. Packaging technique: ◊ skin pack; □ oxygen MAP; Δ CO MAP.

Pork chops packaged in oxygen MAP were judged higher in doneness than the pork chops packaged in skin pack and CO MAP at all endpoint temperatures. Furthermore, the skin packaged pork chops were rated higher than the pork chops packaged in CO MAP at all endpoint temperatures. As expected, increasing endpoint temperature reduces the red appearance of the interior of the pork chop. Pork chops packaged in oxygen MAP had a done appearance when cooked to an endpoint temperature of 50-55°C. Pork chops packaged in CO MAP and skin pack had to be cooked to an endpoint temperature of 75-80°C before appearing slightly done. These findings suggest that pork chops packaged in oxygen MAP have a tendency to brown prematurely, which is in agreement with the findings of Hunt *et al.*, (1999).

Furthermore, it is noticed that the pork chops packaged in CO MAP barely reaches a done appearance when cooked to 80°C, which are in agreement with the findings of John *et al.,* (2004).

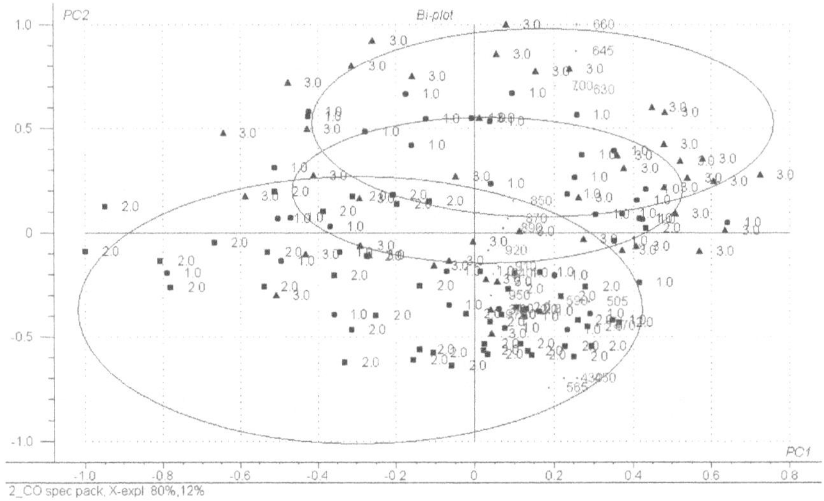

Figure 2: The Unscrambler bi-plot grouped according to packaging. No. 1: Skin pack; 2: O_2 MAP; 3: CO MAP. 3-figure numbers correspond to the measured wavelengths.

The bi-plot in Figure 2 is made by the Unscrambler and shows the spectra grouped according to packaging method. Basically all spectra from pork chops packaged in oxygen MAP (No.2) are located at the bottom of the plot (lower circle) together with the blue and green wavelengths (430-600 nm). The spectra from pork chops packaged in skin pack (No.1) (middle circle) and CO MAP (No.3) are located at the top of the plot (upper circle) together with the red wavelengths (600-700 nm). This indicates that generally the skin packaged and CO MAP packaged pork chops have a more red internal cooked colour than pork chops packaged in high oxygen MAP. Furthermore, the spectra from pork chops packaged in CO MAP are tilted in favour of the red wavelengths compared to the skin packaged pork chops indicating that they are more red in the interior than skin packaged pork chops after cooking.

An Unscrambler PLS analysis (partial least squares regression) shows a correlation between VideometerLab2 spectra with the sensory evaluation of doneness, which gives an r-value of 0.94.

Conclusions

VideometerLab2 is able to distinguish between pork chops packaged in three different packagings and cooked to seven different endpoint temperatures. Additionally there is good correlation between obtained spectra and the evaluation of doneness. It is noticed that the chemical state of the myoglobin form in the interior of the meat is of importance to the internal cooked colour, as concluded by Warren *et al.,* (1996) and furthermore, the chemical state of myoglobin can be influenced by packaging.

References

Hunt, M.C., Sorheim, O. and Slinde, E. (1999). Color and heat denaturation of myoglobin forms in ground beef. *Journal of Food Science*, 64:847-851.

John, L., Cornforth, C.E., Sorheim, O., Pettee, B.C. and Whittier, D.R. (2004). Comparison of color and thiobarbituric acid values of cooked hamburger patties after storage of fresh beef chubs in modified atmospheres. *Journal of Food Science*, 69:608-614.

Machlik, S.M. (1965). The effect of heat on bovine myoglobin derivatives in model systems and in beef semitendinosus muscle. Ph.D. dissertation, Purdue University, West Lafayette, IN, USA.

Warren, K.E., Hunt, M.C., Kropf, D.H., Hague, M.A., Walder, C.L., Stroda, S.L. and Kastner, C.L. (1996). Chemical properties of ground beef patties exhibiting normal and premature brown internal cooked color. *Journal of Muscle Foods*, 7:303-314.

www.camo.com (2006-02-07).

www.sas.com (2006-04-28).

www.videometer.com (2006-01-20).

EFFECT OF TWO TYPES OF PACKAGING ON QUALITY OF PREVIOUSLY FROZEN LAMB MEAT

M. Mota[1], N. Soto[1], M.F.J. Peñúñuri[2], C.J. Velásquez[3], G. Torrescano[1], A. Sánchez-Escalante[1] and E. Díaz*

[1]Department of Meat and Seafood Products. Meat Science Laboratory. Centro de Investigación en Alimentacióny Desarrollo, A.C. PO Box 1735. Hermosillo, Sonora, 83000, México.[2] INIFAP-Carbó. Blvd. Bosque #7. Colonia Valle Verde, Hermosillo, Sonora, México. [3] UNISON-DAG, Blvd. Luis Encinas y Rosales, Col. Centro, Hermosillo Sonora, México. Email: armida-sanchez@cascabel.ciad.mx

Keywords: pelibuey, modified atmosphere, colour, lipid oxidation, meat quality

Introduction

The consumption of ovine meat is mainly focussed on lamb and adult animals. The consumption per capita in Mexico of this meat is 0.4 kg (SAGARPA, 2004), mainly prepared at informal gatherings such as barbecues and mixotes, a typical Mexican food. The production of lamb meat in Mexico does not meet the demand that exists and for that reason, it is necessary to import it from other countries, in cuts generally frozen.

The Pelibuey breed, well-known like Tabasco, is one alternative that can contribute to satisfying the demands of meat in our population. The interest and demand for Pelibuey sheep in Sonora has been increasing. The evidence of this trend is the increase in the production of this species and the incorporation of new producers. The Pelibuey is the most widespread breed of sheep in Cuba and is an animal very tolerant to extreme environmental conditions. Different packaging methods exist, including, modified atmosphere (MAP) and over-wrapped.

The objective of this investigation was to evaluate the quality of the lamb meat after frozen over wrapped during six months, and the effect of two types of package (modified atmosphere and over-wrapped) in order to observe if the quality of the product is altered over the course of time at refrigerated storage.

Materials and Methods

Raw material. Back legs from Pelibuey lambs at age 200 days were used; legs of 48 hours *post-mortem* were frozen for six months. *Semimembranosus* muscles from each leg were obtained and 6 steaks of 1.27cm thickness were packaged. Packaging and storage. Two different methods of packaging were used: 1) modified atmosphere packaging ($80\%O_2/20\%CO_2$) and 2) over-wrapped. All treatments were stored either for 18 days at 2-4°C, or under darkness. Sampling and analyses. The sampling and analyses were carried out on days 0, 4, 7, 11, 14 and 18. Percentage of moisture and fat were determined at day 0. Other quality analyses were colour (L*, a*, b*), lipid oxidation (TBARS), haeminic pigments, water holding capacity, pH and weight loss by purge. Experimental designs. A completely randomised design in duplicate with three experimental repetitions was used. The data were analysed using an ANOVA. The statistical analysis was carried out with software NCSS (2000).

Results and Discussion

Moisture and fat. No significant differences were showed between treatments (data not shown).

Purge. The averages of weight between treatments were stable during the time of storage (Figure 1).

pH. The pH values of treatments (Figure 2) did not show significant difference (P>0.05). These values are lower than those reported by Tschirhart (2003) with values of 5.90, but similar to those reported by Koohmaraie and Wheeler (1994).

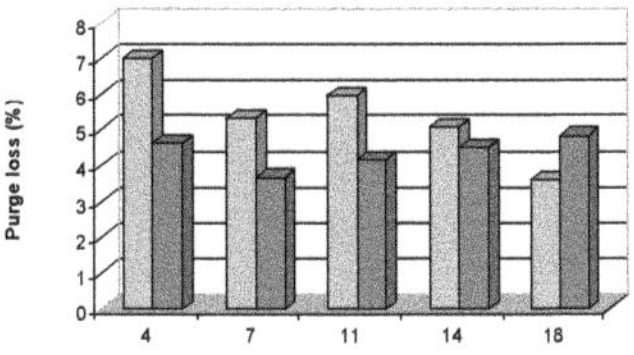

Figure 1: Loss of weight by purge in lamb meat.

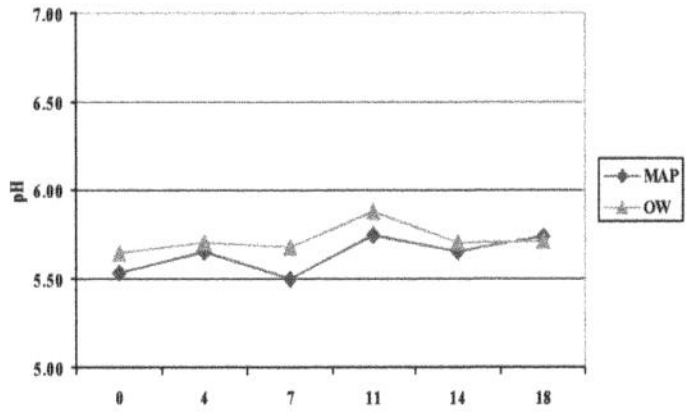

Figure 2: Evolution of pH in lamb meat.

Water holding capacity (WHC). The values associated with WHC, expressed as loss of free water are shown in Figure 3. No statistically significant difference existed between treatments (P>0.05), but higher values were found compared to those reported by Tschirhart (2003) in the *Semimembranosus* muscle of lamb.

Haeminic pigments. Results (Figure 4) indicate a reduction in the pigment content based on the time for both treatments. Larzul (1997) reports values of 1.91 ± 0.03 and 1.98 ± 0.03 in fresh tissue pig, in female and male reproducers, respectively. On the other hand, Cano Expósito *et al.* (2003) showed similar values to day 0 in *L. dorsi* and *T. brachii* muscles with 3.57 and 3.47 mg Mb/g, respectively.

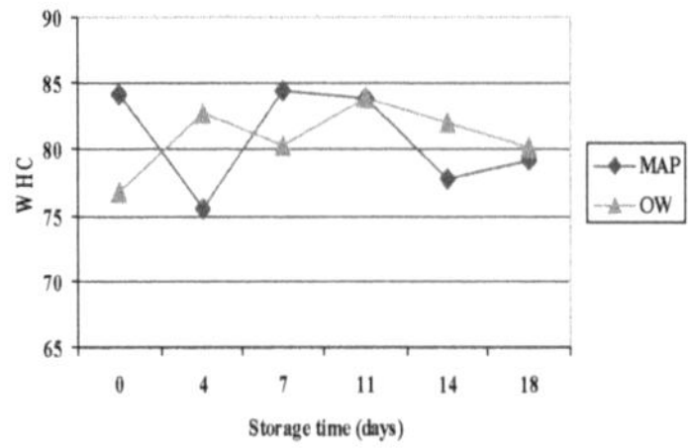

Figure 3: Water holding capacity in lamb meat.

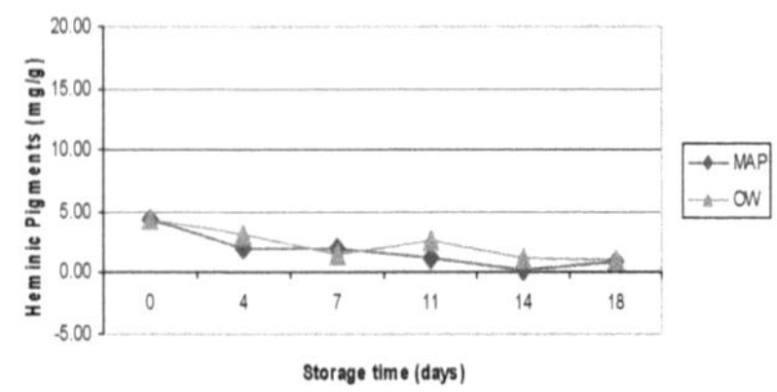

Figure 4: Haeminic pigments (mg/g) in lamb meat.

Lipid oxidation. In both treatments, lipid oxidation (TBA) (Figure 5) revealed similar behaviours between treatments. However, initial TBA values were high, indicating that the lamb meat was oxidised before beginning the experiment; this was probably due to the over-wrapped packaging used during freezing of the meat.

Colour. The values of a* and Hue (Figure 6) indicate that no differences exist between treatments (P>0.05). Hue values were in the range of 48.12±5.02 and 50.72±9.01, for over-wrapped and modified atmospheres respectively. The a* values decreased for both treatments during the first four days, indicating loss of red colour and corresponding to increased %MetMb, however, it remained constant during storage.

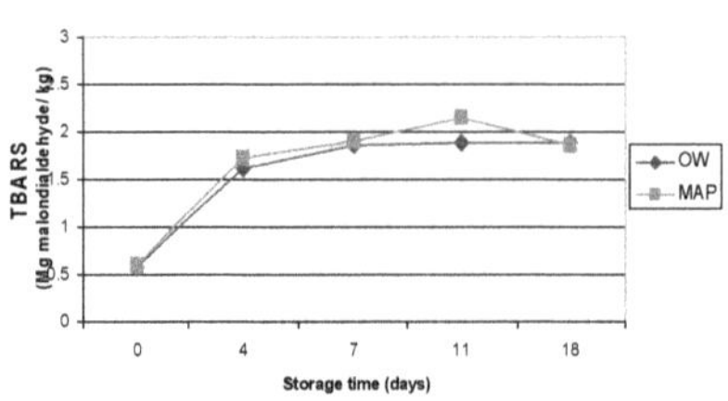

Figure 5: Lipid oxidation (TBA) in lamb meat.

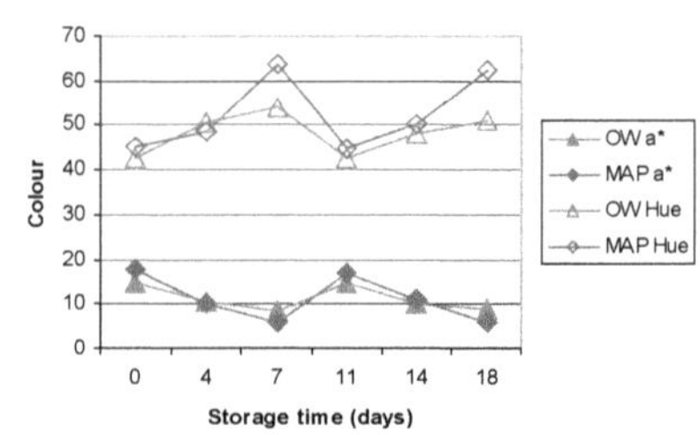

Figure 6: Colour in lamb meat.

Conclusions

The types of packaging used for *Semimembranosus* muscle of lamb do not show significant differences (P>0.05) during the time of refrigerated storage, indicating that type of packaging used during freezing was an important factor that influenced the quality of meat.

References

Cano Expósito, T., F. Peña Blanco, J. Martos Peinado, V. Domenech García, M.J. Alcalde Aldea, A. García Martínez, M. Herrera García, E. Rodero Serrano and R. Acero de la Cruz. (2003). Calidad de la canal y de la carne en corderos ligeros de raza Segureña. *Arch. Zootec.* 52: 315-326.

Koohmaraie, M., Kent, M. P., Shackelford, S. D., Veiseth, E., and Wheeler, T. L. (2002). Meat tenderness and muscle growth: is there any relationship. *Meat Science, 62*, 345-352.

Larzul C., Lefaucheur L., Ecolan P. Gogué J., Talmant A., Sellier P., Le Roy P., and. Manin G. (1997). Phenotypic and Genetic Parameters for Longissimus Muscle Fiber Characteristics in Relation to Growth, Carcass, and Meat Quality Traits in Large White Pigs. *Journal of Animal Science*, 75,3126-3137.

SAGARPA. Coordinación General de Ganadería. Centro de estadística Agropecuaria. (2004). Tomado de su dirección electrónica (www.sagarpa.gob.mx).

Tschirhart, T.E. (2003). Histological, Physical, and Chemical Factors of Various Lamb Muscles. manuscript in progress. http://www.tffc.state.tx.us/tamusheep/progress/2002-2003/sheepute3.pdf.

SENSORY EVALUATION OF BEEF STEAKS STORED IN MA COMBINATIONS OF CO, CO_2, N_2 AND O_2

N.T. Madsen* and I. Clausen

Danish Meat Research Institute, Maglegaardsvej 2, 4000 Roskilde. Email:ntm@danishmeat.dk

Keywords: MAP, beef, sensory evaluation, colour, CO, CO_2, O_2

Introduction

Studies have demonstrated that in steaks of beef loin, sensory properties like tenderness and juiciness decrease during storage in a modified atmosphere containing high oxygen and that warmed over flavour (WOF) may increase (Seideman *et al.*, 1979, Tørngren, 2003; Clausen, 2004; Sørheim *et al.*, 2004). Case ready meat packaging is a fast growing segment, where it is important to use packing that preserves quality during storage. MAP (modified atmosphere packaging) with high O_2 also increases the amount of oxymyoglobin, which gives rise to a well-done appearance at lower temperatures (Hunt *et al.*, 1999). An alternative to high O_2 MAP is a combination of CO_2, N_2 and CO. CO_2 is added to the gas mixture because of its antimicrobiological properties (Jakobsen and Bertelsen, 2002) and increases the shelf life. CO_2 has, however, shown adverse effect on beef steak quality, where more than 20% increase the development of pores and fissures after cooking caused by rapid release of CO_2 from the meat (Bruce *et al.*, 1996; Penny, 1999). Increased CO_2 may also lead to snug down (low pressure) if CO_2 is absorbed in the meat. N_2 is an inert gas, but beef meat will appear purple (deoxymyoglobin), if O_2 is completely excluded from the gas mixture, or brown (metmyoglobin), if a little O_2 (½-1%) is left in the pack. Small amounts of CO (0.4%) gives the meat the desirable shiny red colour corresponding to meat exposed to O_2. Adding 0.3-0.5% CO in a gas mixture is estimated not to constitute any health risk by the European Commission (2001) but it is presently not on the list of accepted gas combinations as it is in the USA. In this study, the effect of using CO in a packing gas on the sensory quality of steaks is compared to several other packing alternatives, to obtain a full picture of the available options for retail packing of beef.

Materials and Methods

Source of meat: 24 M. *longissimus dorsi* (LD) from twelve dairy cows and heifers (27-54 months; 232-300 kg; low voltage electrically stimulated and chilled at 10°C the first 12 hours after stunning, stored at 2°C for 3 days; pH 5.5-5.6) were cut in three and vacuum packed. After 14 days, the six samples from the same animal were randomised with respect to packaging gas: ① 60% CO_2 / 40% N_2, ② 60% CO_2 / 39.6% N_2 / 0.4% CO, ③ 30% CO_2 / 70% N_2, ④ 30% CO_2 / 69.6% N_2 / 0.4% CO, ⑤ 30% CO_2 / 70% O_2 and ⑥ Vacuum skinpackaging. The LD was sliced into 22mm thick steaks, weighed and packed and stored at 2°C for six days. After storage the steaks were weighed again to calculate drip loss during storage.

Packaging materials: Vacuum skinpacking (an upper co-extruded film and a bottom semi-rigid film in the form of a tray): Max. O_2 permeability (upper film and bottom): 2 $cm^3/m^2/$ x d x bar (Cryovac, Sealed Air Corporation). Vacuum: 5-10 mbar (Multivac packing machine). MAP: Tray (13x18x4 cm) covered with transparent film (TOPSEAL™PP MAP AF 57), O_2 permeability: <100 cm^3/m^2/d, bar.

Colour: Measured immediately after the removal of the wrap (Minolta (L*, a*, b*).

Cooking and sensory evaluation: Steaks were equilibrated at room temperature (approx. 20°C) to an internal temperature of max. 15°C prior to cooking on a preheated frying pan (155°C), turned every 2 minutes until an internal temperature of 62±1°C had been reached. Steaks were weighed before and after cooking to calculate cooking loss. Steaks were cut and served in pieces of 2½ x 3cm. Samples were evaluated by 9 trained assessors using a 15-point non-structured line (0=slight and 15= intense). The attributes comprised tenderness, juiciness, Warmed Over Flavour (WOF), meat-flavour and doneness (internal colour).

Statistics: Data were analysed in an analysis of variance model (mixed procedure, SAS version 8.2). Fixed effects in the model were main effects.

Results and Discussion

Results of of the study are shown in Table 1.

Meat Colour: CO packed steaks were higher in a* (redness) and lower in b* (yellowness) than O_2 packed steaks, but we consider the differences to be small and of no concern to the consumer. MA packed steaks (without O_2 and CO) were lower in a* and higher in b* and L* (lightness) than vacuum-skin packed. The explanation is probably development of metmyoglobin despite the low O_2 in the package (<0.1%), but O_2 permeability was rather high.

Drip loss: Drip loss was higher for vacuum-skinpacked steaks (1%), than MAP (irrespective of gas combination; 0.6-0.7%).

Eating quality: Addition of CO did not affect eating quality or the internal colour (doneness) of the cooked steaks. Clausen and Madsen (2005) found that CO did not affect eating quality of beef patties but the internal colour (after cooking to 76°C) appeared less done. Beef steaks packed with high CO_2 (60%) were less juicy than steaks packed with low CO_2 (30%). Clausen and Madsen (2005) have also shown that high CO_2 decreases juiciness. High CO_2 increased doneness after cooking compared to packs with low CO_2. Steaks packed in O_2 were less tender than vacuum skinpacked steaks. They had less meat flavour, developed extreme WOF and looked well done although the internal temperature was only 62°C (premature browning). Several investigations have shown that O_2 is detrimental to the eating quality and causes premature browning (Tørngren, 2003, Clausen, 2004, Sørheim 2004).

Cooking loss: Cooking loss was lower for O_2 MAP than high CO_2 MAP.

Table 1: Mean drip loss, colour (Minolta) and sensory score*

Packaging atmosphere (measured **)	Drip loss	Colour L*	Colour a*	Colour b*	Tender-ness	Juici-ness	WOF	Meat flavour	Done-ness	Cook-ing loss
31% CO_2 / 69% N_2	0.6^{b}	36.4^{b}	12.6^{d}	6.3^{c}	9.0	9.1^{a}	1.5^{b}	7.8^{a}	5.6^{c}	12.8^{ab}
28% CO_2 / 71.6% N_2 / 0.4% CO	0.6^{b}	37.7^{a}	24.9^{a}	8.5^{b}	8.9	9.0^{a}	1.2^{b}	7.6^{a}	5.7^{c}	14.4^{ab}
54% CO_2 / 46% N_2	0.7^{b}	36.1^{b}	12.0^{d}	5.3^{d}	8.8	8.2^{ab}	1.5^{b}	7.8^{a}	6.9^{bc}	15.6^{a}
58% CO_2 / 41.6% N_2 / 0.4% CO	0.6^{b}	37.9^{a}	24.7^{a}	8.2^{b}	8.7	7.7^{b}	2.1^{b}	7.1^{a}	7.2^{b}	14.8^{a}
30% CO_2 / 70% O_2	0.7^{b}	38.6^{a}	23.7^{b}	11.8^{a}	8.5^{b}	8.5^{ab}	11.9^{a}	4.5^{b}	12.3^{a}	12.6^{b}
Vacuum-skinpackaging	1.0^{a}	33.9^{c}	17.2^{c}	3.5^{e}	9.6^{a}	9.0^{a}	1.2^{b}	7.8^{a}	6.3^{bc}	14.1^{ab}
Significance level	$P<0{,}001$	$P<0{,}001$	$P<0{,}001$	$P<0{,}001$		$P<0{,}05$	$P<0{,}001$	$P<0{,}001$	$P<0{,}001$	$P<0{,}05$

*(a non-structured line scale, anchored to the extremes; 0=slight, 15=intense) steaks of beef loin stored in 6 different atmospheres (n=12)

** the delivered gas combinations was not exactly as specified

Conclusions

CO packed steaks appeared shiny red, and differed only sligthly in a* and b* values from O_2 packed steaks. CO_2/N_2 packed steaks differed in L*, a* and b* compared to vacuum packed steaks, probably due to metmyoglobin development. CO did not affect the eating quality of beef steaks. High CO_2 (60%) packaging resulted in steaks that were less juicy than low CO_2 packed. O_2 packed steaks were less tender and scored lower in meat flavour and considerably higher in WOF and well done appearance than vacuum-skin packed steaks.

References

Bruce, H. L., Wolfe, F. H., Jones, S. D. M. and Price, M. A. (1996). Porosity in cooked beef from controlled atmosphere packaging is caused by rapid CO_2 gas evolution. Food Research International. 29: 189-193.

Clausen, I. (2004). Sensory evaluation of beef loin steaks stored in different atmospheres. 50[th]International Congress of Meat Science and Technology, Finland. August.

Clausen, I. and Madsen, N. T. (2005). Sensory evaluation of ground beef stored in different atmospheres. 51[th]International Congress of Meat Science and Technology, USA. August.

European Commission. Health and Consumer Protection Directorate-General. (2001). Opinion of the Scientific Committee on Food on the use of carbon monoxide as component of packaging gases in modified atmosphere packaging for fresh meat. B-1049 Bruxelles – Belgium. pp.1-9.

Hunt, M. C., Soerheim, O. and Slinde, E. (1999): Colour and heat denaturation of myoglobin forms in ground beef. Journal of Food Science. 64:847-851.

Jakobsen, M. and Bertelsen, G. (2002): The use of CO_2 in packaging of fresh red meats and its effect on chemical quality changes in the meat. A review. Journal of Muscle Foods. 13. 143-168.

Penny, N., (1999) Influence of Carbon Dioxide in Meat Packaging Atmospheres on Spoilage, Microflora Development, Drip Loss and Cooked Meat Appearance. Meat New Zealand. Technical Report. MIRINZ. ISSN 0465-4390. 1-5.

Seideman, S.C., Carpenter, Z. L., Smith, G. C., Dill, C. W. and Vanderzant , C. (1979). Physical and sensory characteristics of beef packaged in modified atmospheres. *Journal of Food Protection. 42*. 233-239.

Sørheim, O., Wahlgren, M., Nielsen, B. N. and Lea, P. (2004): Effects of high oxygen packaging on tenderness and quality characteristic of beef *Longissimus* muscles. 50th International Congress of Meat Science and Technology, Finland. August.

Tørngren, M. A. (2003): Effect of packing method on colour and eating quality of beef loin steaks. International Congress of Meat Science and Technology. Brazil, September. 495-496.

ANTIOXIDANT ACTIVE PACKAGING FOR BEEF USING OREGANO EXTRACTS

J. Camo*, T. Pérez, J.A. Beltrán and P. Roncalés

Departamento de Producción Animal y Ciencia de los Alimentos. Universidad de Zaragoza, Miguel Server 177, 50013 Zaragoza, Spain. Email: jcamo@unizar.es

Keywords: beef, active packaging, antioxidant, oregano

Introduction

The three most important factors that limit the shelf life of fresh meat are colour, microbial growth and lipid oxidation. Lipid oxidation is perhaps a critical point for red meat; nevertheless, lipid oxidation is not the main limitation for storage for meat packed under aerobic conditions because it occurs at a slower rate than discolouration and microbial growth (Jakobsen and Bertelsen, 2000).

To retard or minimise oxidative deterioration, effective antioxidants could be added in such products. Synthetic antioxidants have long been used, but their use has recently come into dispute due to a suspected carcinogenic potential (Chen *et al.*, 1992 and Imaida *et al.*, 1983) and the general rejection of synthetic food additives by consumers. There is, therefore, a growing interest in the identification of new, natural antioxidants that would serve as alternatives to the synthetic compounds.

Oregano, a characteristic spice of Mediterranean cuisine obtained by drying leaves and flowers of *Origanum vulgare* subsp. *hirtum* plants, is well known for its antioxidative activity (Economou *et al.*, 1991).

The combined use of antioxidants and modified atmosphere packaging for meat represents a realistic and attractive strategy to increase the shelf life of fresh meat (Giese, 1996). The interest in the application of naturally occuring antioxidants has increased over recent years. Besides this, the development of active packaging is currently attracting the attention of food technologists. We are now focusing on applying the antioxidant potential of oregano to the design of a new active packaging system.

The aim of this work was to investigate the effect of treatment with oregano extract and antioxidant active packaging on the display life of beef steaks packaged in modified atmospheres.

Materials and Methods

Whole loins were obtained from the abattoir 48h post-slaughter. Thick steaks 1.5 cm were prepared: (1-3) without any addition, (4) sprayed on the surface with an oregano extract. Batches 1 and 4 were packaged in a modified atmosphere of 70% O_2, 20% CO_2, 10% N_2; batches 2 and 3 were packaged in the same atmosphere with active films containing 1 and 2% oregano extract, respectively. The samples were stored in the dark at 1±1 °C. Samples were taken at selected times (5, 10, 19, 22, 25, 28 and 32 days) for subsequent analysis.

Meat colour was measured at the surface of beef steaks using a reflectance spectrophotometer (Minolta CM-2002; Osaka, Japan). The metmyoglobin (MetMb) percentage of the total myoglobin perceptible at the steak surface was estimated spectrophotometrically, according to Stewart *et al.*, (1965), by measuring steak surface reflectance at 525 and 572 nm (Minolta CM-2002; Osaka, Japan). Lipid oxidation was assessed in duplicate by the 2-thiobarbituric acid (TBA) method of Pfalzgraf *et al.*, (1995). Meat samples were evaluated by a six-member expert panel. All three attributes were scored using a 5-point scale. For 'Red colour', 1 denoted extremely high and 5 denoted extremely low. Scores for 'Discolouration' referred to percentage of discoloured surface, according to Djenane *et al.*, (2001): 1=none, 2=0–10%, 3=11–20%, 4=21–60%, and 5=61–100%. Scores for 'Fresh Meat Odour' were: 1=excellent, not different from fresh meat; 2=good, but slightly poorer than fresh meat; 3=acceptable, but obviously poorer than fresh meat; 4=hardly acceptable as fresh meat; and 5=non acceptable.

Results and Discussion

Values of CIE a^* (redness) are depicted in Figure 1. Treatment with antioxidant led to significant differences ($P<0.05$) with the control from day 8 of storage onwards. At the end of the storage period, untreated sample had very low a^* values, below 5, while samples sprayed with oregano extract and oregano film showed a^* values above 12, representative of a bright red colour.

Figure 2 shows the results of metmyoglobin, expressed as percentage of metmyoglobin of total surface myoglobin. Results demonstrated that the antioxidant extract sprayed effectively delayed metmyoglobin formation. Packs with oregano film were significantly ($P<0.05$) less effective than the sprayed sample.

Figure 3 shows the results of TBA reactive substances (TBARS) throughout the storage of treated and untreated steaks. All of the treated samples showed a highly significant ($P<0.05$) inhibitory effect on the formation of TBARS, with only small differences between surface treatment with the oregano extract and packaging with the active films. Differences were significant ($P<0.05$) from day 9 of storage onwards.

Results of sensory analysis of meat samples, including evaluation of red colour, discolouration and fresh meat odour, are summarised in Table 1.

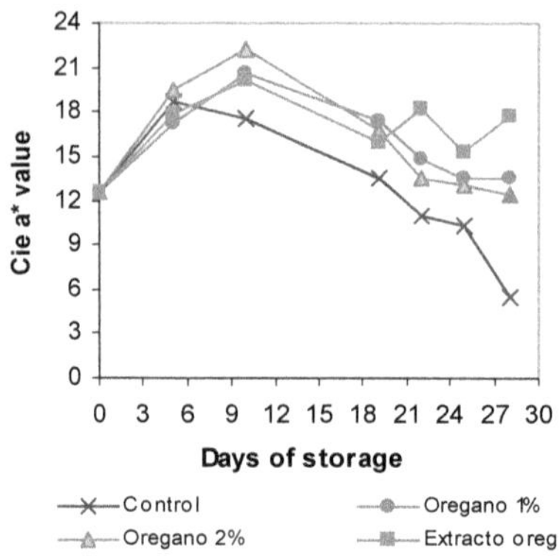

Figure 1: Values of CIE a^* (redness).

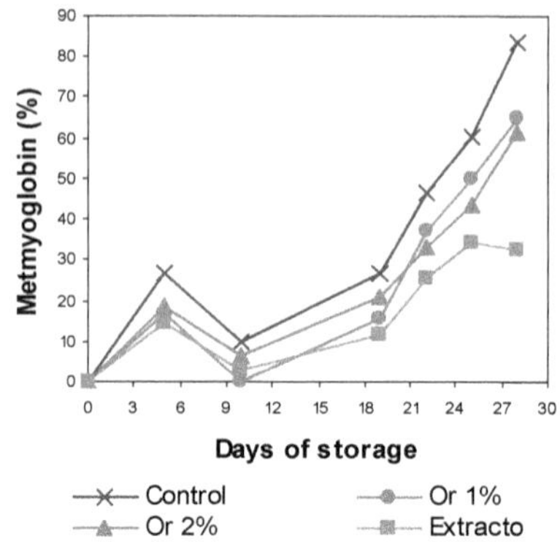

Figure 2: Metmyoglobin, (expressed as percentage of metmyoglobin of total surface myoglobin).

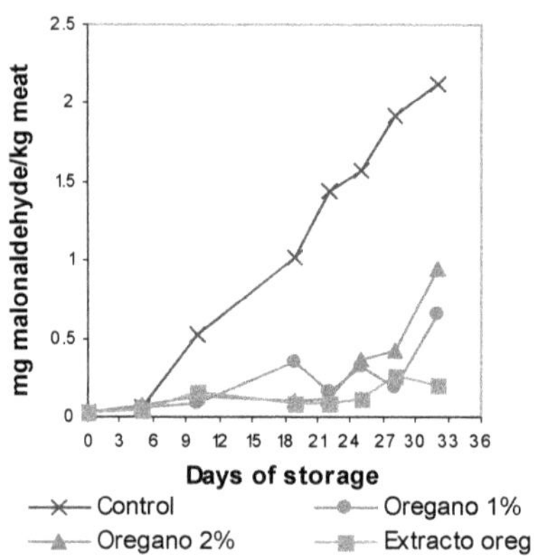

Figure 3: TBA reactive substances (TBARS).

		Days of storage						
Parameter	Sample	0	5	10	19	22	25	28
Red colour	Control	1	1	1	2	2	4	4
	Oregano 1%	1	1	1	1	2	2	3
	Oregano 2%	1	1	1	2	2	2	4
	Extract	1	1	1	1	1	1	2
Discolouration	Control	1	1	1	2	1	4	5
	Oregano 1%	1	1	1	1	1	2	2
	Oregano 2%	1	1	1	2	2	3	3
	Extract	1	1	1	1	2	1	2
Off odour	Control	1	1	2	2	2	3	4
	Oregano 1%	1	1	1	1	2	2	3
	Oregano 2%	1	1	1	2	1	2	3
	Extract	1	1	1	1	1	1	2

Table 1: Sensory Analysis of meat samples.

Conclusions

Both the addition of the oregano extract and packaging with an oregano active film resulted in enhanced oxidative stability of beef steaks packaged in modified atmosphere at 1°C, although to different extent. Oxymioglobin and lipids were protected against oxidation during storage, resulting in a better colour and odour stability than the control, extending the self life for at least 3 days.

References

Jakobsen M. and Bertelsen G., (2000). Colour stability and lipid oxidation of fresh beef. Development of a response surface model for predicting the effects of temperature, storage time, and modified atmosphere composition. *Meat Science* 54, 49-57.

Chen C.H., Pearson A.M. and Gray J.I., (1992). Effects of synthetic antioxidants (BHA, BHT and PG) on the mutagenicity of IQ-like compounds. *Food Chemistry,* 45, 177-183.

Imaida K., Fukishima S., Shirai T., Ohtami M., Nakamishi K. and Ito N., (1983). Promoting activities of butylated hydroxyanisole and butylated hydroxytoluene on 2-stage urinary bladder carcinogenesis and inhibition of gamma-glutamyl trans-peptide-positive for development in the liver of rats. *Carcinogenesis,* 4, 895-899.

Economou, Oreopoulou, V., and Thomopoulos, C.D., (1991). Antioxidant properties of some plant extracts of the Labiatae family. *Journal of the American Oil Chemical Society,* 68, 109-113.

Giese, J., (1996). Antioxidants: tools for preventing lipid oxidation. *Food Technology,* 50 11, 73-80.

Stewart, M.R., Zipser, M.W. and Watts, B.M., (1965). The use of reflectance spectrophotometry for the assay of raw meat pigments. *Journal of Food Science,* 30, 464-469

Pfalzgraf, A., Frigg, M. and Steinhart, H., (1995). Alpha Tocopherol contents and lipid oxidation in pork muscle and adipose tissue during storage. Journal of Agricultural and *Food Chemistry,* 43, 1339-1342.

Djenane, D., Sanchez-Escalante, A., Beltrán, J.A. and Roncalés, P., (2001). Extension of the Retail Display Life of Fresh Beef Packaged in Modified Atmosphere by Varying Lighting Conditions. *Journal of Food Science,* 66, 181-186

MODIFIED EDIBLE COATINGS FOR MEAT PRODUCTS

A.B. Lisitsyn * E.A. Evstafyeva and E.I. Ukrainskaya

The V.M.Gorbartov All-Russian Meat Research Institute, Talalikhina 26,109316 Moscow, Russia
Email: vniimp @ inbox.ru

Keywords: chitosan, collagen, coating, composition

Introduction

Many kinds of meat products without protective coatings which are manufactured according to traditional technology are intended for long-term storage. In this case, the changes in moisture content affect the quality and lead to weight losses, as well as fat oxidation and changes in the colour of muscle tissue. Under the influence of air oxygen, the surface becomes contaminated, and the development of moulds and bacteria are inevitable. Therefore, creating protective coatings would be a great advantage. Protective coatings are the most effective methods of packaging, because they can be formed directly on food products without any contact with oxygen from the air, and they are formed with simplified packaging technology.

In the last few years, several kinds of coatings and films were investigated in an attempt to maintain the quality of fresh and frozen meat, poultry and seafood products. There are good prospects for edible coatings and water-soluble coatings (US Patent 5824751, 1998). There is a film-forming composition, containing starch, monoglycerides of edible animal fats, sorbic acid, water or meat broth, for coating meat and meat products (European patent 829503, 1998), which has a drawback that there are relatively high meat losses. A composition of a protective coating containing distilled acetylated monoglycerides and distilled monoglycerides for smoked sausages and pork products has been patented, (Japanese patent №27446525, 1998), but the limitation of this composition is formation of rigid coating on the product, that can result in loss of the mass of a meat product during chilling and freezing. Nevertheless, edible protective coatings on pre-cooked meat products have not been used widely up to the present time.

The purpose of the study was the creation of the compositions on the basis of polysaccharides of different surface activity, and of water-soluble high molecular protein components, possessing biological activity, to protect prepared meat products from oxidative and microbial spoilage, and impart them specific properties, that allows us to extend the range of the manufactured products.

Materials and Methods

Collagen and chitosan were chosen as the main raw materials for the coating. The presence of active end groups in the molecule of collagen, a complicated spatial conformation with the trend to form fibrillar structures provides the possibility of its binding with various low- and high molecular compounds. Chitosan, a polysaccharide, obtained from shells of crustacea is such a compound. A polycation fibrillar structure of chitosan provides the effective adhesion affinity to structural proteins, to collagen also. This allows the creation of various film-forming complexes on the basis of collagen and chitosan and introduce into them active materials, absorbing or excreting specific substances.

Comparative investigations of the developed compositions and experimental film structures have been carried out to choose the optimum composition. In the investigations the following parameters were determined: structural viscosity of the composition on the rotary Viscosimeter Reotest-2; strength and elongation during stretching of the model films, vapor permeability of the model films, microbiological and organoleptical characteristics of meat products in edible coatings; acid and peroxide number of meat products in edible coatings during storage.

The protective edible coating with optimum parameters was formed on the surface of smoked-cooked pork and beef products by their immersion in a 2% composition mixture with its subsequent coagulation in a setting bath, containing a 25% solution of common salt, and drying during 36 hours at the air relative humidity 65-75% and temperature 20-25^{0}C.

Results and Discussion

Several versions of film-forming composition mixtures have been developed. They contained protein and polysaccharide fractions in different relationships and additives with functional effects, such as acetic and lactic acid, and essential oil of ginger.

At the first stage, the influence of lactic acid on the viscosity of composition mixtures was investigated. It was shown that the composition mixture, obtained with the use of the mix of 0.5 M lactic and acetic acid in the ratio 1 : 3 as a solvent, had a viscosity 48.26 Pa*s which is 1.2 times lower than the viscosity of the composition mixture, obtained by the solution in 0.5 M acetic acid. This is connected with the plastifying effect of lactic acid through the reduction of inter-molecular interaction and the complex formation of the latter with polar groups of collagen structure.

Further increase in the amount of incorporated lactic acid reduced the viscosity of the composition mixture, but led to stickiness of the films. It is connected with a significantly larger amount of the introduced lactic acid, than the forces of the inter-molecular interaction could join, and it is in unbound condition in inter-molecular space.

At the second stage, essential oil of ginger was incorporated into the composition mixture at 0.3; 1.0 and 2.0% in relation to the mass of its dry matter. The composition mixture contained the solution of collagen (SC) and chitosan obtained by solving in the mix of 0.5 M of lactic and acetic acids in the ratio 1: 3. It was found that the essential oil of ginger actually had no influence on viscosity, strength characteristics and vapor permeability of films obtained from these composition mixtures. However, an increase in the concentration of the essential oil to 2% led to a very sharp smell of model films, that could affect the flavour characteristics of the final product. Therefore, in subsequent experiments the composition was used, containing SC-chitosan, solved in the mixture of 0.5 M of lactic and acetic acids in the ratio 1: 3 added with 1% essential oil of ginger with the parameters, as follows: viscosity of the composition – 50 Pa* with the deformation velocity 1 c^{-1} ; pH value – 3.2 –4.0; thickness of the coating on the product – 0.06 ± 0.005 mm; vapor permeability – not more than 400 g/m^2 for 24 hours; strength – not less than 9 MPa.

The composition with such parameters was applied to meat products from pork and beef. Analysis of the obtained results has shown that during storage the indices of total plate count of the samples having protective edible coating did not exceed the norm, while in the control sample these indices were by one order higher - 1* 10^3 CFU/g.

Sensory investigations have shown that taste and flavour typical for smoked-cooked products were preserved longer in the samples in edible coating, than in the controls.

Along with microbiological investigations, the acid and peroxide numbers of the samples of meat products in edible coatings during storage were determined. It was found that these indices for all the meat products with coatings were lower than in the control samples. When the weight losses of the samples in edible protective coatings and in the control ones were investigated during storage, it was found that in the coated samples they were lower by 6%, than in the samples without coatings.

Conclusions

The investigations carried out have shown that a film-forming composition mixture developed on the basis of the solutions of collagen and chitosan. provided the formation of protective edible coating on the surface of smoked and cooked meat products. This edible coating has a number of optimum characteristics with regard to thermal stability, vapour-, gas permeability, and providing preservation of quality, reduction of weight losses and increase in storage life of ready meat products.

References

US Patent 5824751, 1998
European Patent 829503, 1998
Japan Patent 2746525, 1998

COLLAGEN PASTE OBTAINED FROM PIGSKIN DEFATTED WITH SUPERCRITICAL CARBON DIOXIDE

S. Beltrán, E.M. Vaquero, M.T. Sanz and I. Jaime*

Área de Ingeniería Química, Facultad de Ciencias, Universidad de Burgos, Plaza Misael Bañuelos s/n, 09001 Burgos, Spain. Email: ijaime@ubu.es

Keywords: pigskin, collagen paste, casing meat products, food industry

Introduction

There are several types of collagen pastes commonly used in the food, pharmaceutical and related industries. A collagen paste is a white suspension with a jelly aspect, obtained from denaturalisation of collagen. One of its main applications is as casing of sausages. Most of the procedures reported to obtain such collagen pastes use different mammal's skin as raw material (Checa) due to their high collagen content. In fact, collagen is the most abundant protein in mammals making up over 75% of the dry fat-free weight of the skin (Cleary, 1996). Cattle skin is most widely used to obtain collagen paste. In some cases it is combined with skin from other mammals with good results, but there is usually a strong dependence on cattle skin for this process with around 60% of cattle skin required. In order to provide an alternative to cattle skin, we propose in this work the use of pigskin, without combining it with other collagen sources, as raw material to obtain collagen paste. The main inconvenience of using pigskin for this process is its high fat content which makes it difficult to obtain consistent collagen pastes. Pigskin is only used if mixed with other sources of collagen with a lower fat content. Although there are several methods to eliminate or reduce the fat content of animal skin, we have used a mild method, supercritical fluid extraction. The solvent used was supercritical carbon dioxide (sc-CO_2). Carbon dioxide is the most frequently used supercritical solvent. The reasons are its good properties including non flammability, low critical temperature (that allows carrying out the extraction under mild operation conditions, which is very important when collagen is involved in the process), and phase equilibrium properties relatively favourable for fat extraction and later separation.

In this work, the pigskin used was obtained from experiments carried out in a pilot plant using supercritical fluids for extraction, to optimise the process to defat pigskin (Vaquero, 2004). The results showed the possibility of obtaining collagen paste from this raw material, although further studies need to optimise the yield and final operation parameters of the process, and to evaluate the consistency of the collagen paste.

Materials and Methods

Pigskin with different fat contents, ranging from 5 to 20 g of fat per gram of protein was used as source of collagen. The procedure followed to obtain the collagen paste of interest was the one patented by Checa (1) with some modifications, since the raw material used in this work did not require fat elimination that had been already done by supercritical fluid extraction (Vaquero, 2004).

This procedure consists of three consecutive stages: Firstly, around 10g of defatted pigskin, cut into pieces of around 2 cm wide and long and 5 mm thick, were immersed in a basic solution for 24 hours. The vessel was continuously stirred in a platform shaker (New Brunswick Scientific – Innova 2000). The basic aqueous solution was achieved either with NaOH or Ca $(OH)_2$ in the concentration necessary to reach a pH around 12. This treatment aims to eliminate hair residuals in the skin including the hair root.

After being washed with distilled water, the pigskin was immersed in acid solutions. In this second stage, two acid solutions were used consecutively. In the first one, acidity was achieved with concentrated acetic acid to reach a pH of 3.5 or 4. This solution produces a swelling of the pigskin samples. After being washed, the pigskin was immersed in a lactic acid solution. HCl was used to adjust the pH to between 1 or 2. This solution produces a partial denaturalisation of the collagen contained in the pigskin. Finally, the pigskin was washed and ground to obtain a homogenous paste. The consistency of the collagen paste was evaluated by extending it in a glass that was then immersed in an aqueous solution of NaCl. The film obtained was visually tested for consistency.

Results and Discussion

The evolution of the pigskin after being immersed in the different solutions seems to be important to characterise the final product.

The consistency of the samples treated with NaOH was partially lost and the solution surrounding the samples became fairly viscous while no visual changes were observed in the samples treated with $Ca(OH)_2$.

After being ground, the samples treated with NaOH formed a homogeneous white paste that could provide a fairly consistent film (Figure 1).

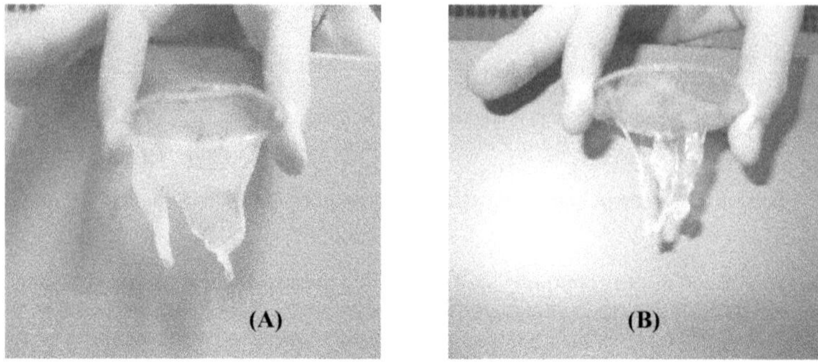

Figure 1: Collagen paste film obtained when (a) an aqueous Na(OH) solution was used and (b) when $Ca(OH)_2$ solution was used.

On the contrary, the samples treated with $Ca(OH)_2$ were difficult to grind, probably because swelling of those samples was not visually observed. After the basic and acid treatments, the collagen paste obtained from these samples also consisted of a white paste, but it was not as homogenous as the one obtained from the samples with NaOH. The effect of the fat content in the pigskin samples on the final consistency of the collagen paste was also clearly observed. The lower the initial fat content of the pigskin samples, the stronger the consistency of the film that was finally obtained from the collagen paste produced.

Conclusions

The main conclusion from the work presented, is that pigskin can be used as a unique source of collagen as long as it is defatted to a low fat content.

References

Checa, M.A. Procedimiento de obtención de una pasta de colágeno. Patente española ES 2 017 564.

Cleary EG. (1996) Skin. In: Comper WD, editor. Extracelular Matrix, Volume 1, Harwood Academic Publishers GmBH, Amsterdam, p. 77-109.

Vaquero, E.M.; Beltran, S and Sanz, M.T. (2004) Extraction of fat from pig skin with supercritical carbon dioxide. Journal of supercritical fluid, (37) 142-150.

EFFECTS OF PH AND CORN OIL ADDITION ON MECHANICAL AND PERMEABILITY PROPERTIES OF GELATIN-BASED FILMS

L.Z. Wang*[1], S.A. Hogan[1], L. Liu[1], J. Holmes[2], J.F. Kerry[1] and J.P. Kerry[1]

[1] *Department of Food & Nutritional Sciences, University College Cork, Cork, Ireland,* [2] *Department of Chemistry, University College Cork, Cork, Ireland. E-mail address: lzwang2001@yahoo.com*

Keywords: biopolymer film, response surface methodology, gelatin

Introduction

Biopolymer films and coatings produced from renewable substances have attracted much research interest in recent years. Gelatin is an important by-product of the beef processing industry and offers considerable potential, due to its abundance, low-cost and functionality, for use in biodegradable food packaging applications (Arvanitoyannis, 2002). The sensitivity of gelatin-based films to moisture, however, limits their potential as a potential food packaging material. In contrast, films made from lipids have suitable water vapour barrier properties but poor mechanical properties (Kamper and Fennema, 1984). The production of composite films incorporating both gelatin and lipid components may enhance the suitability of gelatin-based edible films. This study determines the effects of pH and corn oil addition on the mechanical and permeability properties of gelatin films and uses response surface methodology (RSM) in attempting to optimise film formulation.

Materials and Methods

Glycerol (Gly) was added to 4% (w/w) gelatin (G) solutions to give a G to Gly ratio of 2:1 (w/w). Corn oil (CO) was added at concentrations ranging from 7.5 to 55.18% (of G powder weight) and the pH was adjusted to range from 3.46 to 10.54. Solutions were heated to 80°C and homogenised at 480 bar. Films were cast by pouring solutions onto level Teflon-coated perspex plates and dried for 24h at 50 ± 5% RH and 23h ± 2°C. Dried films were subsequently peeled from the casting plates and held as above for a further 12h prior to testing. Film thickness was measured using a 25mm micrometer screw gauge. Mechanical properties (tensile strength (TS), elongation at break point (E), and puncture strength (PT)) were evaluated according to the ASTM-D882 (ASTM, 1985) standard test methodologies. Water vapour permeability (WVP) was determined according to McHugh *et al.* (1993). Oxygen permeability (OP) was conducted at 23 ± 2°C, 50 ± 3 % RH according to Papkovsky *et al.* (2000). Film morphology was determined by scanning electron microscopy (SEM) (JSM-5510, JEOL Ltd. Tokyo, Japan) at 10kV. The experimental design and statistical analyses were performed using Minitab 14 for Windows (Minitab Inc., PA, USA).

Results and Discussion

TIK, TS and PT increased with increasing addition of CO but were not significantly affected by pH (Figures 1, 2 and 4). The highest TS (15.30MPa) and PT values (28.30N) were observed from films produced using a CO content of 47% at pH 9.5. Such effects may be attributed to the degree of crystallinity and orientation of the crystalline protein network in which CO is actively entangled by gelatin molecules. Composite films showed a heterogeneous structure (Figure 7), composed of a continuous matrix with CO incorporation. Increasing CO content of films resulted in increased opacity. Elasticity was greatest in films prepared at extreme alkali values (10.54) and CO of 27.25% (Figure 3). It is probable that protein denaturation and charge repulsion effects at such pH values contributed to a stretching of polymer chains. CO addition had a greater effect on WVP than pH adjustment. The lowest WVP values (50.55 $g{\cdot}mm/kPa{\cdot}d{\cdot}m^2$) were observed in films produced with a combination of pH 10 and CO addition of 55.18% (Figure 5). It was evidenced by SEM that CO addition led to a decrease in free volume within the film matrix and that more dense films were less permeable to WVP (Figure 7). The lower WVP values obtained with higher CO addition may also be explained by formation of an interconnecting lipid network within the film matrix (Krochta, 1997). OP was highest at neutral pH (6.95) and increased linearly with increasing CO content (Figure 6). This may be due to the increasingly hydrophobic nature of high CO films and their associated facility for oxygen transfer.

Conclusion

Results showed that pH had a greater effect on E, WVP and OP than that exerted by CO addition. CO content of gelatin-based films showed a linear relationship with TIK, TS, PT and quadratic effects with both WVP and OP. The optimum formulation for gelatin-based test films, determined by RSM, was for 4% (w/w) gelatin solutions containing 55.18% CO adjusted to a pH of 10.54. Such films show considerable potential as alternative (biodegradable) materials for food packaging and increase the possible end uses for collagen.

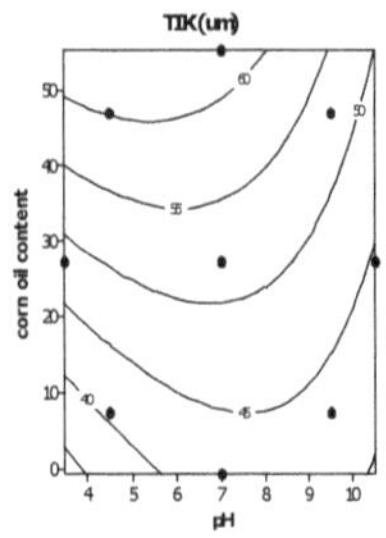

Figure 1.

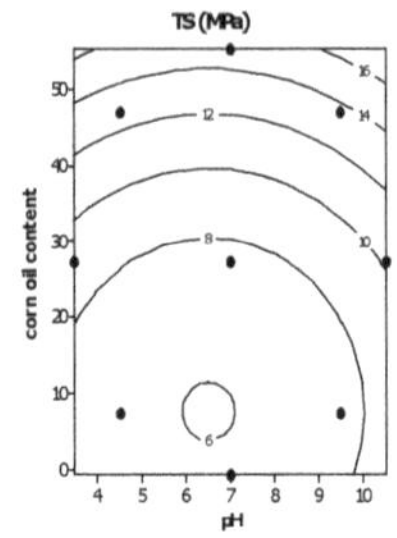

Figure 2.

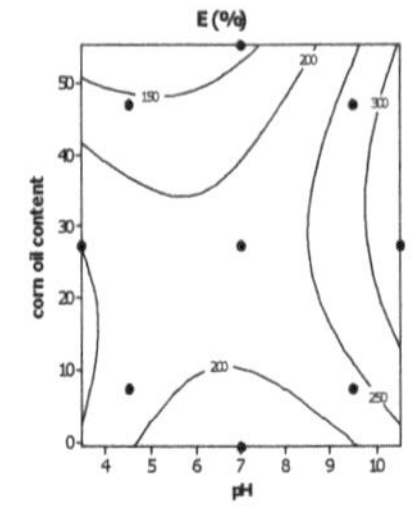

Figure 3.

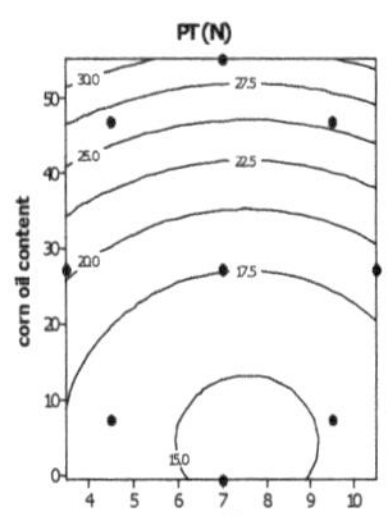

Figure 4.

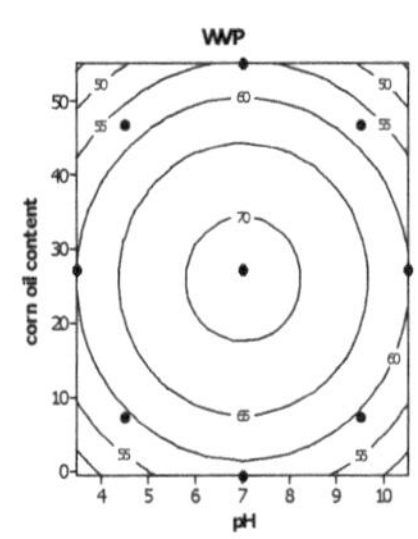

Figure 5.

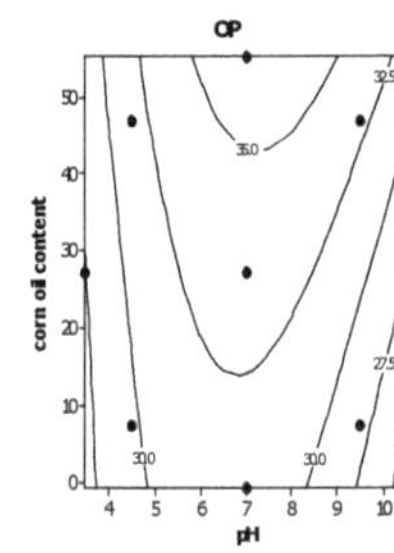

Figure 6.

Figure 1, 2, 3, 4, 5, 6 are counter plots of the effects of pH and CO on TIK, TS, E, WVP and OP respectively.

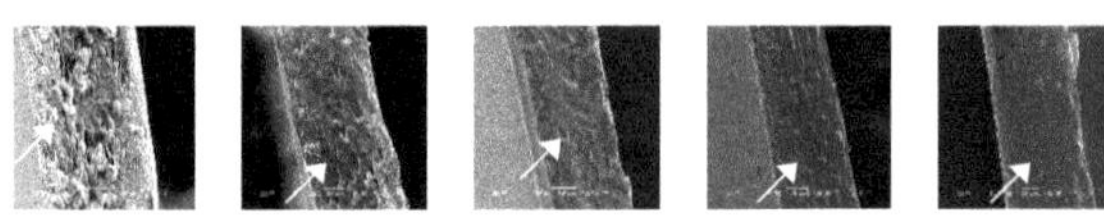

0 % 7.5% 27.5% 47 % 55.18 %

Figure 7: Scanning electron micrograph cross-sections of G films formed from solutions with varying CO contents (arrows indicate film cross-section).

References

Arvanitoyannis, I. S. (2002). Formation and properties of collagen and gelatin films and coatings. In Gennadios, A. (Ed.), Protein-based films and coatings. Boca Raton: CRC Press, pp: 275-304.

ASTM. (1985). Standard Test Methods for Tensile properties of thin plastic sheeting. Standard designation: D882. In Annual Book of American Standards Testing Methods, Philadelphia, pp. 182-188.

Kamper, S.L. and Fennema, O. (1984). Water vapour permeability of an edible fatty acid bilayer film. Journal of Food Science, 49, 1482-1485.

Krochta, J. M. (1997). Edible composite moisture-barrier films. In 'Packaging Yearbook: 1997', Ed., B. Blakistone. National Food Processors Association. Washing, DC: pp. 38 - 51.

McHugh, T.H., Avena-Bustillos, R., Krochta and J.M. (1993). Hydrophilic edible films: modified procedure for water vapour permeability and explanation of thickness effects. Journal of Food Science, 58, 899-903.

Papkovsky, D.P., Papkovskaia, N., Smyth, A., Kerry, J. and Ogurtsov, V.O. (2000). Phosphorescent sensor approach for non-destructive measurement of oxygen in packaged foods: optimisation of disposable oxygen sensors and their characterization over a wide temperature range. Analytical Letters, 33, 1755-1777.

HISTOLOGICAL CHARACTERISATION OF THE EFFECT OF ADDED SALT IN FRESH BEEF SAUSAGE

R. Labas*[1], T. Astruc[1], R. Taylor[2], J.L. Martin[3] and J.L. Vendeuvre[3]

[1.] *INRA, Unité Qualité des Produits Animaux, Theix 63122 St Genès Champanelle, France* [2.] *EWOS Innovation 4335 Dirdal, Norway* [3.] *CTSCCV 94704 Maison Alfort, France. Email : Roland.Labas@clermont.inra.fr*

Keywords: beef, minced meat, sodium chloride, histology

Introduction

The EU regulation (EC) N° 853 /2004 on the hygiene of products of animal origin which applies from January 1st, 2006 assigns meat preparations to meat if the product has undergone a process sufficient to modify the internal muscle fibre structure of the meat. Therefore histology was used in this study to characterise changes in the muscle fibre structure of fresh beef sausages due to the addition of salt and to try to assign products to the EU-definitions.

Materials and Methods

Minced meat with 3 levels of sodium chloride: 0, 0.8 and 1.6% was prepared from various cuts of beef forequarters. Three samples of each mince (1x1x1cm^3) were frozen in isopentane chilled by liquid nitrogen (-160°c). Frozen sections (10 µm thick) were stained using Hematoxylin Eosin Safran (HES) coloration to visualise general structure, and picro-Sirius red coloration (Flint and Pickering, 1984) which reveals the collagen of perimysium and endomysium. Immunohistology was done using antibodies against myosin and laminin. Histological sections were studied with a Polyvar Reichert microscope, Sony CCD video camera, computer with a Matrox image acquisition card and image analysis software (Noesis, France).

To quantify muscle structures, sections were analysed with an optic microscope using a 25x lens, by systematic vertical then horizontal scanning. One random field of every four was photographed. Fifty images were captured per section with triplicate repeats by level of salt (150 images per treatment group). Each image was visually assessed and assigned/ ranked to a class as: 0 class (all fibres of the field are intact, Figure 1A), 25 class (1 to 25% of the field is damaged), 50 class (26 to 50% of the field damaged), 75 class (51 to 75 % of the field damaged, Figure 1B), 100 class (76 to 100% of the field damaged, Figure 1C).

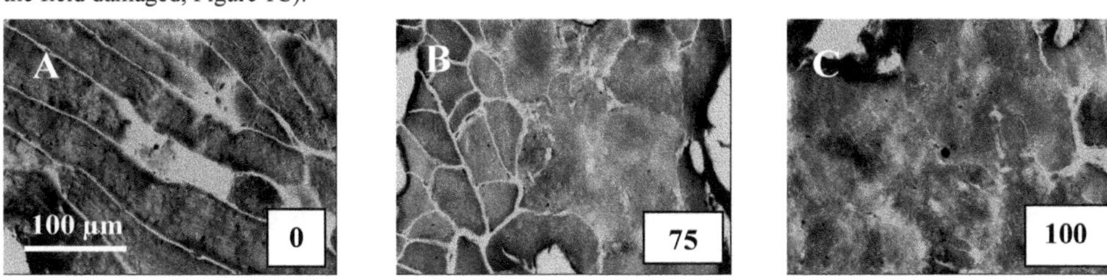

Figure 1: Distribution in classes in relation with % damaged fibres (1A 0%; 1B 75%; 1C 100%).

Statistical comparison of values of % damage for the three levels of salt was done using ANOVA and comparison of means with the "Linear Model" procedure (Statistical Analysis System).

Results and Discussion

As a first approach, suggested by a previous study (Ripoche *et al.,* 2001), to evaluate the effect of salt on muscle structure, in response to an EU request allowing characterization of minced meat, we decided to evaluate the sarcomeric A band as a potential indicator structure. Sarcomeric striations were evident in sections colored with HES or immunohistology (Figure 2, staining of myosin to show sarcomeres). Visualization depends on the sarcomeres being chosen from longitudinal cuts to the fibres. In fact sarcomere orientation was too random in minced meat to allow quantification. Therefore we next evaluated sarcolemma to sarcomere attachment as an index of salt and grinding. We also used a scoring of general fibre structure as an index of meat integrity.

Using the criteria of general fibre integrity and fibre to fibre attachment as our damaged fibre index it is apparent that salt added to mince at both 0.8 and 1.6% causes extensive loss of normal fibre structure. It is evident in these sections, as reported by Offer and Knight (1988), that the perimysium is a barrier of salt diffusion in minced meat, as shown by the fibre damage being limited by the perimysium (Figure 3). The endomysium was not a damage barrier. Quantification of fibre damage (Figure 4) shows that virtually all fibres in control samples were classified as intact, whereas 0.8 and 1.6% salt resulted in 49% and 18% of fibres remaining intact respectively. 1.6% salt was necessary to cause extensive fibre damage (42% of fibres) with 0.8% salt causing intermediate damage. Statistical analysis indicated a significant difference ($P<0.05$) between three levels of salt.

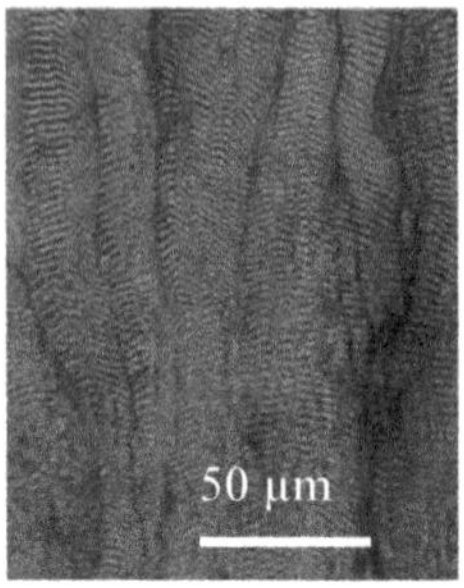

Figure 2 : Antibody staining of myosin.

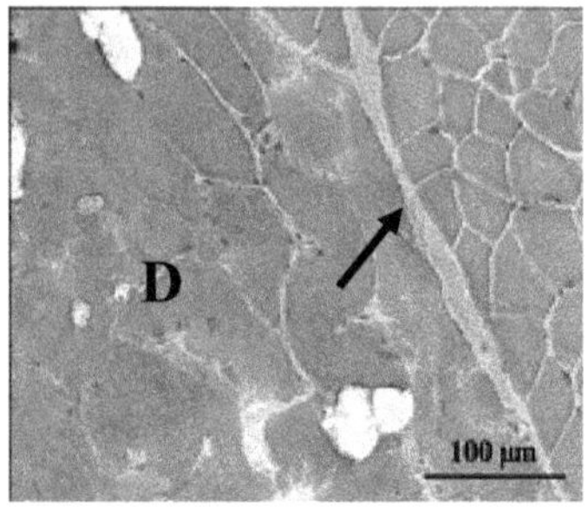

Figure 3: Damaged fibres (D) HES coloration Perimysium: arrow.

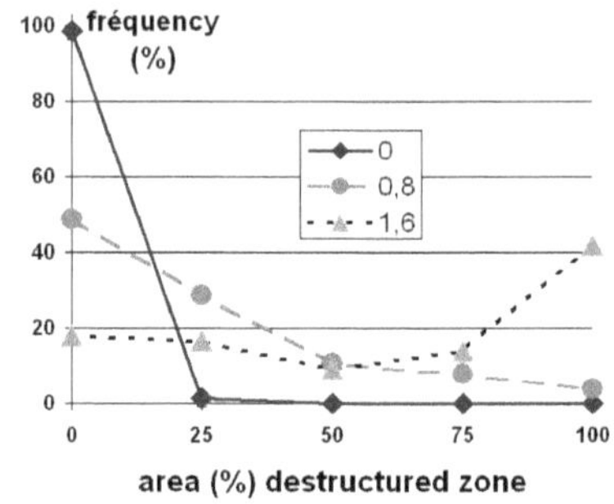

Figure 4 : Evolution of % fibre damage in relation to salt concentration.

We also evaluated other staining methods as tools to characterise fibre damage. Antibody staining of laminin, basal lamina's constituent, was used. As shown in Figure 5A, red sirius staining of connective tissue, also tested, highlights the endomysium sufficiently to see that it is continuous around intact fibres but not at damaged fibres. Image analysis was applied to determine if automatic quantification was possible.

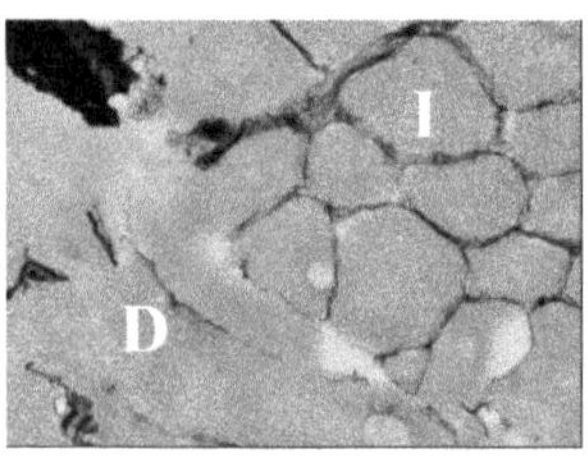

Figure 5A: Red sirius staining of endomysium.

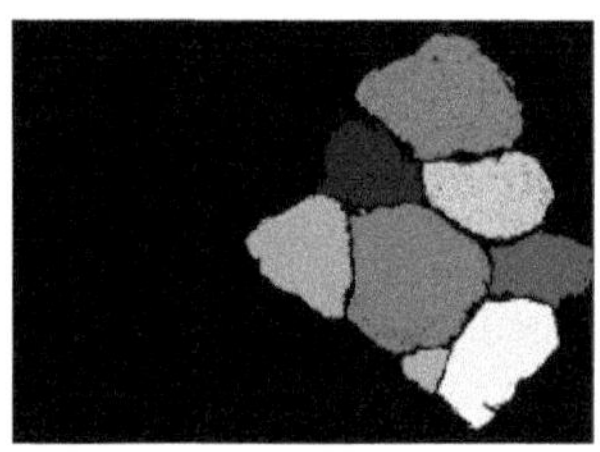

Figure 5B: Image analysis.

As shown in Figures 5A and 5B, intact fibres (I) can be identified and the boundaries determined but for damaged regions (D) we were not able to mark boundaries which by visual inspection corresponded to these regions. However, since this test should be applied to determine industry standards, an automated method is desirable.

Conclusions

These results demonstrate that meat structure standards can be developed using standard microscopy techniques. HES staining of frozen sections coupled with visual classification of fibre destruction clearly distinguishes intact control samples from altered muscle fibres in ground sausages with added salt. However, due to the variability of fibre orientation and damage, automated image analysis with and without specific antibody stains cannot by applied routinely. The perimysium is the barrier limiting salt penetration in meat samples, so further investigation on the relationship of meat beef standards should consider connective tissue content as an important variable.

References

Flint, F. and Pickering, K. (1984) Demonstration of collagen in meat products by an improved picro-sirius red polarisation method. Analyst, 109: 1505-1506.

Offer, G. and Knight, P. (1988). The structural basis of water holding in meat. In Developments in meat science, vol 4. Lawrie R, editor, pp 63-173 New York. Elsevier Applied Science.

Ripoche A. *et al.*, (2001) Sausage structure analysis. Journal of Food Science, 66, 5: 670-674

Acknowledgements

This study was supported by OFIVAL (French Interprofessional Meat, Breeding and Poultry Agency) and Interbev (French Interprofessional Cattle and Meat Association).

SALT EFFECTS ON BEEF FRESH SAUSAGE ULTRASTRUCTURE

T. Astruc*[1], R. Labas[1], B. Gaillard-Martinie[2], R. Taylor[3] J.L. Martin[4] and J.L. Vendeuvre[4]

[1.] Unité Qualité des Produits Animaux, INRA de Theix 63122 St Genés Champanelle, France. [2.] Plateau Technique Microscopie Electronique INRA de Theix. [3.] EWOS Innovation, 4335 Dirdal, Norway. [4.] CTSCCV, 94700 Maison Alfort, France. Email : astruc@clermont.inra.fr

Keywords: beef, minced meat, salt, ultrastructure

Introduction

The aim of this study was to characterise the salt and process effects on meat ultrastructure in beef fresh sausage and check if that level of magnification might help assessing it in meat preparations or in meat products according to the definitions laid down by EU-regulation (EC) N° 853/2004 Annex I- under 115 and 71 whether the process undergone was insufficient or not to modify the internal muscle fibre structure.

Materials and Methods

1) Meat sausage processing

Minced meat was prepared from beef chuck (fattening mark = 3) with a 3.2 to 3.5mm diameter mincing grid at 0°C temperature. Fat in the mixture was less than 20% and the ratio of collagen/protein was 15. Three salt concentrations were tested in the mixture: the first mixture without NaCl addition (control), the second including 8g of NaCl per kg of minced meat (0.8 % salt) and the last with 16g of NaCl per kg of minced meat (1.6 % salt). Beef minced meat was stored for 24 h at 0°C before examination.

2) Sample preparation

Five sausage samples of about 5mm^3 were taken for each preparation and fixed overnight at 4°C by immersion in 2.5 % glutaraldehyde in 0.1M sodium cacodylate buffer, pH 7.2. Small blocks (1 to 3mm^3) were post-fixed in 1% osmium tetroxide in the same sodium cacodylate buffer for 1 hour at ambient temperature. The blocks were dehydrated through an ethanol gradient and embedded in epoxy resin (TAAB, Eurobio France). Semi-thin sections (1.5µm) were stained with toluidine blue and observed with an optical microscope (Reichert Jung). An area presenting muscle fibres in a longitudinal state was selected for ultrastructure observations. Ultra-thin sections (90nm) were stained with uranyl acetate and lead citrate, and observed with a Morgagni electron microscope using a 90KV acceleration voltage. Micrographs were made using a numeric camera system coupled with the microscope.

Results and Discussion

1) Mincing effects on fibre and myofibril morphology

Mincing had an effect on fibre structure as shown in Figure 1A where changes of orientation of the fibres are seen. At the ultrastructural level, myofibrils can keep their normal alignment (1B) or show undulation or torsion of the sarcomeres (1C). At low magnification (1A), torsion due to mincing is evident, that at higher magnification (1C) is seen as loss of normal A and I band definition.

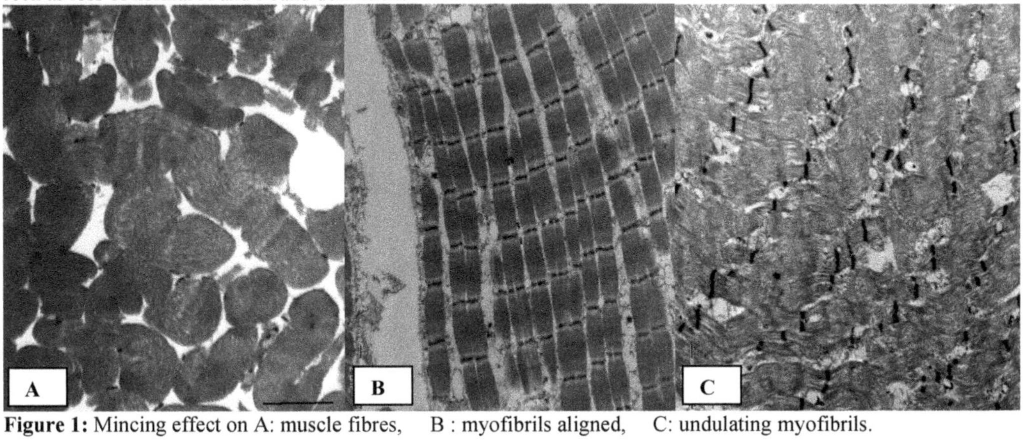

Figure 1: Mincing effect on A: muscle fibres, B : myofibrils aligned, C: undulating myofibrils.

2) Salt effects on ultrastructure

A 0.8% salt concentration didn't modify the general structure of the fibres (D, 3 fibres are seen) and myofibrils (D and E) but had some effect on the sarcomeres, espescially along the Z line as shown by arrows on the micrograph E.

A 1.6 % salt concentration led generally to heterogenous changes. In micrograph F, the general morphology of the myofibrils was preserved with Z and M lines still visible (about 50% of the oberved fields). However, the Z line width diminished and the "alignment organisation" of the myofilaments is less evident. In the last micrograph (E), the ultrastructure of the myofibrils was destroyed with an important loss of the integrity of the sarcomere and of the Z lines (50% of the observed fields).

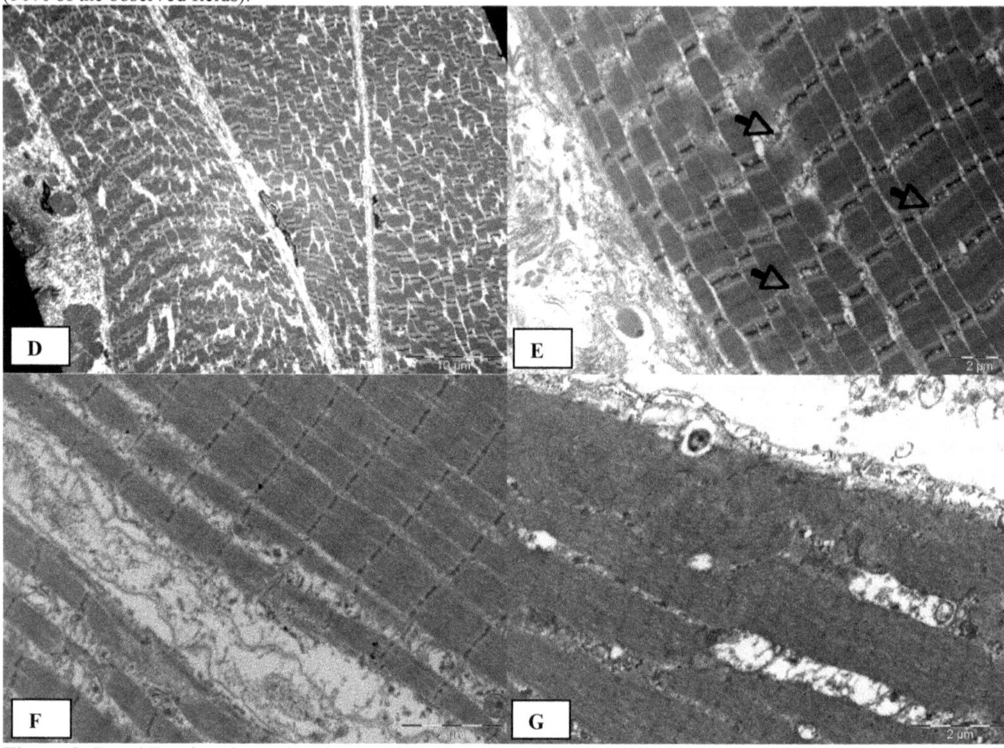

Figure 2: D and E: minced muscle mixed with 0.8% salt F and G : minced muscle mixed with 1.6% salt.

According to Offer and Knight (1988) and Ripoche *et al.* (2001), the main results indicate that salt modification of sarcomere ultrastructure depends on salt concentration. Theses changes, which seem to begin by a solubilisation of the Z line, are probably due to the interaction between the increase of the ionic strength and the mincing effect. Important differences in preservation seen in the same preparation (micrographs F and G) suggests a heterogeneity in salt concentration in the muscle. Ripoche *et al.* (2001) observed that the sarcomeres and Z lines of sausage meat with 1.3% salt are totally destroyed. However, the preparation was based on pork muscle. These results indicate a different behaviour of myofibrils in the presence of salt depending on the muscle species used in the preparation.

Conclusion

A previous classification system of the extent of muscle alteration during processing of pork cannot be applied directly to beef treated by similar mincing and salt. Beef shows better preservation of myofibrils even with higher salt concentration.

References

Ripoche A., Le Guern L., Martin J.L., Taylor R.G. and Vendeuvre J.L. (2001). Sausage structure analysis. Journal of Food Science, 66, 670-674.

Offer G., Knight P. (1988). The structural basis of water holding in meat. In Developments in meat science, vol 4. Lawrie R, editor, pp 63-173, New York. Elsevier Applied Science.

Acknowledgements

This study was realised with the support of OFIVAL (French Interprofessional Meat, Breeding and Poultry Agency) and INTERBEV (French Interprofessional Cattle and Meat Association).

INFLUENCE OF ONION ON PHYSICO-CHEMICAL CHARACTERISTICS OF "MORCILLA DE BURGOS"

I. Jaime, L. González-Arnáiz*, A.M. Diez and J. Rovira

Department of Biotechnology and Food Science. University of Burgos.
Pza Misael Bañuelos s/n 09001. Burgos. Spain.
Email: ijaime@ubu.es

Keywords: Morcilla de Burgos, physico-chemical analysis, horcal onion

Introduction

"Morcilla de Burgos" is a popular cooked blood sausage produced in Burgos, in the north of Spain, consisting in a mixture of chopped onion, rice, animal fat, blood, salt and different spices such as black pepper, paprika and oregano. According with the elaboration process, the kind and amount of spices, and the proportion of ingredients different types of Morcilla de Burgos can be differentiated.

One of the more characteristic ingredients, primordial in the elaboration of Morcilla the Burgos, is the kind of onion used. The most traditional onion is a regional variety known as Horcal. This onion is white, with elliptical shape and bigger size compared with other types of onions. The horcal onion is a seasonal crop (harvest between September and November) and a short period of storage, but it gives to Morcilla de Burgos a better sensory quality. According with the first draft of the regulation to obtain the quality label "Protected Geographical Indication" (PGI) the amount of Horcal onion included in formulation will be one of the main factors to classify the different types of Morcilla de Burgos. The project is to establish three types of morcilla, the first one called "Morcilla de Burgos" is a product with at least 35% of Horcal onion. The second type ("Morcilla de Burgos Tradicional") must include at least 45% of Horcal onion and the third one ("Morcilla de Burgos Matancera") includes 50% of Horcal onion.

The purpose of this study is to determine the influence of Horcal onion on the physic-chemical characteristics of Morcilla de Burgos, in order to determine if the percentage of onion is suitable as a main factor to classify morcilla and besides to find physico-chemical parameters of control for PGI of this product.

Materials and Methods

Sample preparation: Three batches of 60 Morcillas were made with the same formulation, with the only difference being the percentage onion added. One batch was made with 35% of Horcal onion, because this is the minimum amount for the morcilla to be called "Morcilla the Burgos". The second batch was made with 20% of Horcal onion, amount that is not included in the regulation for the PGI, and the third batch was made with 47% of Horcal onion, because this is the middle amount between the percentages proposed in PGI for the other two categories.

After removing casings, four morcillas from each batch, randomly chosen, were homogenised in a lab blender and used for physicochemical analysis.

Physicochemical analyses: pH, water activity (a_w), moisture content, ether-extractable fat, protein, starch, total sugar, ash and total dietary fibre were determined by sixtuplicate for each morcilla type.

Statistical analysis: Data was statistically analysed using one-way analysis of variance (ANOVA) in order to determine the effect of onion percentage on the physicochemical properties of morcilla. Besides, Principal Component Analysis (PCA) was also applied to obtain the parameters of higher weight on variability of product characteristics. Data analyses were conducted using the statistical package Statgraphics Plus for Windows ver. 5.1.

Results and Discussion

Table 1 shows the results of the physicochemical parameters of morcilla. The statistical analysis denotes that the physicochemical parameters that vary with the onion percentage are pH, total sugar and dietary fibre. These parameters differs significantly between the three samples with different percentage of onion with a 0.05 level of significance ($P<0.05$).

The pH increased when the percentage of horcal onion added to "morcilla de Burgos" decreased, however, total sugar and dietary fibre decreased. This linear correlation between amount of onion and content of sugar and total fibre on the composition of the product could allow using both parameters in PGI control. Moreover, the ratio between the parameters total sugar and dietary fibre is 1 ± 0.3, irrespectively of percentage of onion. This ratio is an important additional control parameter to avoid a source of total sugar and dietary fibre different from onion, which is not allowed in PGI. If only total sugar and dietary fibre were individually controlled a high value in one of these parameters could be due to the addition of sugars or rice with a high content of fibre, respectively and not because of the onion added to "Morcilla de Burgos", but in this case the ratio will be outside the normal interval. The rest of the composition parameters do not differ between the three batches or present punctual differences, as was expected since they are not related with the onion components. Concerning to the parameters moisture content and ash statistically significant differences were found, but they do not seem to be related with the different percent of horcal onion added to the three experimental batches of "Morcilla de Burgos".

Table 1: Means of physicochemical parameters of the "morcilla" elaborated with different onion percentages (n = 6), expressed on dry matter basis.

% Onion	47	35	20
pH	$6{,}27^{a}$	$6{,}443^{b}$	$6{,}58^{c}$
aw	$0{,}982^{a}$	$0{,}981^{a}$	$0{,}984^{b}$
Moisture content	$55{,}15^{b}$	$53{,}33^{a}$	$58{,}09^{c}$
Fat	$35{,}25^{b}$	$35{,}10^{b}$	$34{,}02^{a}$
Protein	$12{,}45^{a}$	$11{,}31^{a}$	$12{,}64^{a}$
Total sugar	$4{,}20^{c}$	$3{,}54^{b}$	$2{,}79^{a}$
Starch	$34{,}81^{a}$	$33{,}99^{a}$	$32{,}68^{a}$
Ash	$3{,}59^{b}$	$3{,}35^{a}$	$3{,}97^{c}$
Dietary fibre	$4{,}23^{c}$	$2{,}74^{b}$	$2{,}17^{a}$

Means in the same row with different letters (a-c) are significantly different ($P<0.05$).

A Principal Component Analysis was applied to data of all physical-chemical analysis to obtain which parameters permit discrimination of the three types of morcilla. Two components with eigenvalues greater or equal to 1 extracted. These components accounted for 79.97% of the variability in the original data. The first and second principal component explained 61.78% and 18.19% of the overall variation, respectively. The PCA clearly separated morcilla samples into three groups that match with the three types of "morcillas" manufactured, and they are especially defined by the parameters total sugar and dietary fibre linearly related with onion percentage.

Conclusions

The horcal onion has a great influence in the content of total sugar and dietary fibre expressed on a dry matter basis and there is a linear correlation between these two parameters and the percentage of horcal onion in the formulation of the product. So, the percentage of horcal onion used in the elaboration of "Morcilla de Burgos" is a very important factor to classify different types of "Morcilla" in order to obtain a Protected Geographical Indication. Besides, total sugar content and dietary fibre, together with the ratio between both parameters, could be suitable parameters to control the established types of morcilla satisfied the requirements of PGI.

References

AOAC (1997) "Official Methods of Analysis", 16th edition, Association of Official Analytical Chemists, Inc., Arlington, Virginia.

Belitz-Grosch, Food Chemistry (2004) Second edition, Springer, Chapter 22: Spices, Salt and Vinegar. 905-919.

Borrador del reglamento de la Indicación Geográfica Protegida "Morcilla de Burgos".

Cuevas, L.I. (2005). Proyecto fin de carrera, Universidad de Burgos.

Hernández, T., Hernández, A. and Martinez, C. (1995). "Fibra alimentaria. Concepto, propiedades y métodos de análisis."Alimentaria, Abril 95:19-30.

Santos, E.M. (2001). "Caracterización, tipificación y conservación de la Morcilla de Burgos". Tesis doctoral, Burgos.

Santos, E.M., González-Fernández, C., Jaime, I. and Rovira, J. (2002). "Elaboración y caracterización fisicoquímica de la morcilla de Burgos". Eurocarne 103, Enero-Febrero 02.

Santos, E.M., González-Fernández, C., Jaime, I. and Rovira, J. (2003). "Physicochemical and sensory characterization of Morcilla de Burgos, a traditional Spanish blood sausage". Meat Science, 65, 893-898.

Acknowledgements

We want to express our thanks to Excma. Diputación Provincial de Burgos for supporting with a grant this project. We also want to express to gratitude to the Junta de Castilla y León project BU 28/04 and to Mercadona S.A for supporting this research project.

RELATIONSHIP BETWEEN TOTAL DIETARY FIBRE AND THE AMOUNT OF ONION USED IN MANUFACTURE OF "MORCILLA OF BURGOS".

M. Gomez-Recio* L. Gonzalez-Arnaiz., L.I. Cuevas, I. Jaime and J. Rovira

Deparment of Biotechnology and Food Science. University of Burgos. Pza Misael Bañuelos s/n 09001. Burgos. Spain. Email: jrovira@ubu.es

Keywords: dietary fibre, onion, sausages

Introduction

"Morcilla de Burgos" is a popular, cooked blood sausage produced in the province of Burgos in the North of Spain. It consists of a mixture of chopped onion, rice, animal fat, blood, salt and different spices such as black pepper, paprika and oregano. Previous studies have pointed out the sensory importance of the amount of onion in "morcilla" (Santos *et al.*, 2003). Onion is a major source of water, sugars and fibre. Water is released from the onion during cooking and is absorbed by the rice starch, while most of the sugars and fibre remain in the final product. EU regulation 2081/92 established two regulations to protect traditional foodstuffs in Europe; these are the protected designation of origin (PDO) and the protected geographical indication (PGI). "Morcilla de Burgos" producers want to apply for a PGI designation. The draft regulation considers three different types of "morcilla" based on the amount of onion, with a minimum value of 35% w/w. For this reason, the amount of fibre found in the "morcilla" could be of interest as an indicator of the amount of onion used in its manufacture. The aim of this study is to evaluate the feasibility of using total dietary fibre (TDF) content as indicator of the amount of onion in a "morcilla" sausage.

Materials and Methods

Although raw fibre content has previously been determined in "morcilla" (Santos *et al.*, 2003), in this study, total dietary fibre (TDF) has been considered. This is because raw fibre only includes in its definition hemicelluloses, celluloses and lignins while TDF includes the above plus pectins, gums, nondigestible oligosaccharides and waxes. Total dietary fibre has been analysed by the gravimetric enzymatic method described by the AOAC (1997). This consists of weighing out 1.0000 ± 0.0005 g in duplicate of freeze-dried, ground sample. Each sample is suspended in Mes-Tris buffer (40ml) and three different enzymes (α-amylase, protease and amyloglucosidase) are added to hydrolyse starch, protein and amylose respectively. Next, ethanol (78%) is added to precipitate soluble dietary fibre. The remaining material is filtered and washed with ethanol (96.5 %) and acetone after which it is dried and weighed. One of the duplicates is used to measure residual protein and the other for ash determination. Total dietary fibre content results are calculated as the difference between the weight of the remainder and the residual protein plus ash.To show the influence of fibre content in "morcilla de Burgos", two experiments were performed: (1) Two batches of three different "morcilla de Burgos" sausages elaborated with 20%, 35% (minimum to be considered "morcilla de Burgos") and 47% w/w onion were analysed for TDF; data obtained from those analyses were statistically analysed (Statgraphics plus v.5.1). Different statistical models (polynomial, lineal regression and inverse of Y) were investigated to correlate the amount of fibre present in the sample with the amount of onion added by producers. (2) Eleven "morcilla" sausages from different geographical areas of the Burgos province were analysed for TDF content and, the statistical models developed previously were used to predict the amount of onion in each sample.

Results and Discussion

The best predictive model uses the "inverse of Y" and produced the regression equation - $Y= 1/(0.635539-0.0082758X)$, [X being the percentage of onion] with a correlation coefficient R= 0.9791 (Figure 1). The other models studied produced lower correlation coefficients.

Santos *et al.*, (2001) established that it was possible to distinguish between three different types of "morcilla" in the Burgos region according to their manufacturing process, level and type of spices and percentage of onion employed. , These different types were located in different geographical parts of the region; in the North, less onion is used (around 20-25%), in the South the usage level is around 35-40% and in the Centre more than 45% w/w. Table 1 shows the percentage of TDF (expressed on a dry matter basis) found in different "morcilla" products. Using only natural components, the amount of fibre found in this kind of products is remarkable. Among the eleven products analysed, 6 belonged to the Centre of the region, 2 to the South (products 10 and 23), and 2 to the North (18 and 24). In the same table, the estimated percentage of onion content in the different products is alos shown. Taking into account the minimum percentage of TDF detectable by the model (2.87% w/w), all produtrs have higher amounts of fibre except number 24 (2.367% w/w).

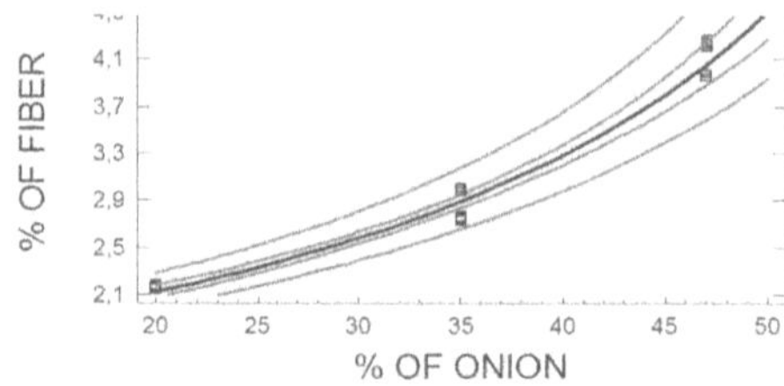

Figure 1: Inverse of Y correlation between percentage onion content and percentage fibre.

When a cluster analysis is applied to all fibre data obtained (see Figure 2), it is possible to distinguish four different groups. In group 1 two products from the Southern part of the region are included; in group 2, which is split in two subgroups, all producers from the Centre of the region are concentrated. All members of group 3 are products located in the same village, also located in the Centre, and finally group 4, which is different from the others, comprises products from the Northern part of the region. When fibre data from this latter group is analysed (Table 1), a paradox arises because this group contains products with higher and lower amounts of fibre in the "morcilla". These data suggest that this group is formed by outliers. However, the amount of onion predicted for product 24 is in the range of that described by Santos *et al.* in their work (2001). In the case of product 18, it is possible to suggest that there it contains sources of fibre other than those derived from onion.

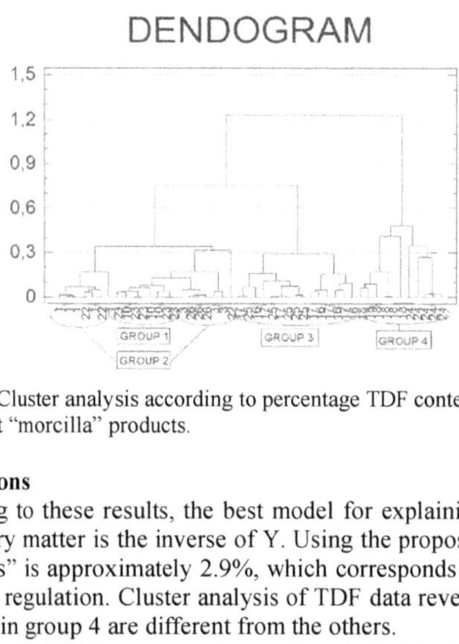

Figure 2: Cluster analysis according to percentage TDF content in different "morcilla" products.

Table 1: Prediction of onion content from analysed fibre content (dry matter basis).

PRODUCT	% FIBRE	% ONION
1	6,080^{d}	52,811
3	3,542^{bc}	39,969
10	4,265^{b}	45,406
23	4,489^{b}	46,742
22	4,336^{b}	45,621
25	6,251^{d}	53,863
16	5,630^{cd}	51,724
17	5,876^{d}	52,668
18	7,542^{e}	56,887
24	2,367^{a}	24,128
26	4,176^{a}	44,866

Conclusions

According to these results, the best model for explaining the correlation between percentage onion and percentage of TDF in dry matter is the inverse of Y. Using the proposed model, the minimum detectable amount of TDF in "morcilla de Burgos" is approximately 2.9%, which corresponds to the minimum amount of onion (35% w/w) established in the draft PGI regulation. Cluster analysis of TDF data reveals that products of groups 1 to 3 are very similar to each other, but those in group 4 are different from the others.

The content of TDF alone is not a good predictor for the amount of onion employed in the manufacture of "morcilla de Burgos". This is because other sources of fibre, arising for example from higher inclusion rates of spices or the use of rice with higher bran content, can increase its value.

References

AOAC (1997). Official Methods of Analysis, 16th edition. Association of Official Analytical Chemists Inc., Arlington, Virginia, USA.

Cuevas, L.I. (2005). Proyecto de fin de carrera, Universidad de Burgos.

Hernández T., Hernández A., Martínez C. (1995). Fibra alimentaria. Concepto, propiedades y métodos de análisis. Alimentaría, 95, 19-30.

Santos E.M., González-Fernández, C., Jaime, I., Rovira, J. (2003). Physicochemical and sensory characterisation of "morcilla de Burgos", a traditional Spanish Blood sausage. Meat Science, 65, 893-898.

Santos E.M. (2001). Caracterización, tipificación y conservación de la "morcilla" de Burgos. PhD Thesis. University of Burgos, Spain.

This study was financed by Junta de Castilla y León project BU28/04 and the retail company Mercadona, S.A.

COMPOSITION AND PHYSICOCHEMICAL CHARACTERISTICS OF 'CHORIZO' FROM THE HIDALGENSE HUASTECA REGION IN MEXICO

V. Austria[2], S. Soto[2], I. Caro[2], N.B. Fonseca*[1], Güemes[2] and J. Mateo[1]

[1.] *Departamento de Higiene y Tecnología de los Alimentos, Universidad de León, 24071 León, Spain.* [2.] *Cuerpo Académico de Alimentos de Origen Animal, Universidad Autónoma del Estado de Hidalgo, 43000 Tulancingo, Hgo., Mexico.. Email: dhtjmo@unileon.es*

Keywords: chorizo, sausages, composition, characteristics, Huasteca

Introduction

'Chorizos', in Spanish, or 'Chouriços', in Portuguese, are popular raw-meat sausages in Iberian and most of Latin American countries, showing regional differences regarding recipe and processing. Main ingredients of Mexican-style 'chorizo' including 'Huasteco chorizo', are minced pork and fat, salt, dry chilies and paprika (both *Capsicum* spp.), vinegar, and a mix of several spices, namely, cumin, pepper, and others (Personal communication, 2006); sometimes curing salts, sugars and other vegetable matter are also added to the mix. In general, the resulting sausage is bright red with a distinctive aroma. Usually, it is not intentionally subjected to ripening. However, a spontaneous ripening may occur if the product is subjected to long enough storage. This storage would take place with the sausage hung for periods of hours to weeks at room temperature which largely depends on local weather conditions, and involves fermentation and dehydration (Escartin *et al.*, 1999).

The 'Hidalguense Huasteca' is a region located amidst a mountain range with semi-tropical wet climate and abundant vegetation (Schryer, 1986). 'Huasteco chorizo' is typically produced at small plants or butchers and its manufacture is differentiated from other 'chorizos', i.e. after stuffing in natural casings sausages are tied into short segments (5-10 cm long) by means of small strips made from dry husks (special leaves that protect the ear of the corn plant), then sausages are dried a few days at room temperature, and finally at the moment of purchasing in the market sausages are wrapped with banana leaves, instead or paper or plastic (Figure 1).

Figure 1: Photographs of 'Huasteco chorizo' as presented in the market.

Considering the social and commercial interests of regional food characterization (Tregear *et al.*, 1999), the facts that 'Huasteco chorizo' is a typical regional product, and that to our knowledge it has not been studied before, this study was aimed at contributing to the characterization of this typical sausage by acquiring knowledge about its chemical composition and other basic physicochemical properties.

Materials and Methods

Ten samples of 'Huasteco chorizo' were purchased in the local retail market and then transported to the lab where an amount c.a. 150 g of each was homogenized. Subsequently homogenized samples were wrapped in aluminum foil and frozen (-40°C) until aliquots were taken for analysis. Proximate composition was determined according to official methods of analysis and the pH, a_w and colour parameters with a pHmeter, aw-meter device (Decagon Aqualab, model CX-2), and colorimeter (Minolta, model CM-508d), respectively. Salt (NaCl) and lactic and acetic acid contents were analyzed on an aqueous extract of the sample by HPLC with a chromatograph (Waters, model 2690) equipped with a cation-exchange column (Bio Rad, Aminex HPX-87H) using aqueous H_2SO_4-solutions as mobile phases, and with a differential refractometer detector (Waters, model 410) for NaCl and a photodiode array detector (Waters, model 996) for the organic acids, following Van Riel and Olieman (1986) and Bruna *et al.*, (2003) methodologies, respectively.

Results and Discussion

Compositional and physicochemical parameters of 'Huasteco chorizo' are shown in table 1a) and b), respectively. Protein and fat contents, expressed as percentage of dry matter, were similar to those found in European ripened sausages (Moretti *et al.*, 2004; Osorio *et al.*, 2004), which imply that fat levels of the initial mix (before stuffing) of 'Huasteco chorizo' were c.a. 25-30% (fresh matter). In dry fermented sausages, moisture content changes from 55 to 65% (initial) to lower final values which depend on sausage properties, time and ripening conditions; normally, after one month drying moisture decreases on average to 25-35%; the decrease being more sharp during the first few days (Stiebing and Rödel, 1988; Pérez-Álvarez, 1999; Zanardi *et al.*, 2002; Moretti *et al.*, 2004). In this sense, considering moisture content of 'Huasteco chorizo', and a_w too, a few days drying must have occurred. NaCl content was in the highest part of the range for initial mixes of fermented sausages content was usually around 2-3% common salt (Incze, 1992). Lactic acid content of 'Huasteco chorizo' implies that fermentation of sugars took place; sugars will come mainly from chillies and paprika (Aguirrezábal *et al.*, 1998). To reach pH<5.3 and aw<0.95 values in a short time appears to be critical objectives for high acid fermented sausage stability (Incze, 1992); this seems easy to fulfil in the case of 'Huasteco chorizo' by basic control. With regard to colour, differences between chorizos are expected; these could be mainly due to amount and type of *Capsicum* spp added and, in general, sausage composition (Pérez-Álvarez *et al.*, 1999).

Table 1: Characteristics of Huasteco chorizo. a) Composition expressed as weight percentage of fresh matter # or of dry matter*; b) Physicochemical parameters of 'chorizo' (n=10).

a)	Protein*	Fat*	Moisture	NaCl#	Lactic acid#	Acetic acid#
Mean	37.1	51.6	51.8	3.23	1.95	0.13
Minimum	31.6	48.2	45.1	2.52	1.33	0.02
Maximum	40.3	56.9	56.4	4.33	2.67	0.40
D.E.	3.0	3.0	3.9	0.59	0.48	0.13

b)	pH	aw	L*	a*	b*
Mean	4.82	0.955	40.8	16.1	18.5
Minimum	4.67	0.942	36.5	13.6	14.4
Maximum	5.07	0.969	44.1	19.9	23.1
D.E.	0.14	0.008	2.2	2.0	2.5

Conclusions

Proximate composition, pH, aw and colour of 'Huasteco chorizo' have been determined. This product can be classified as a high acid sausage. Controls should be done to guarantee a quick pH and aw (under 0.95) decrease.

References

Aguirrezábal, M.M., Mateo, J., Domínguez, M.C., Zumalacárregui, J.M. (1998). Spanish paprika and garlic as sources of compounds of technological interest for the production of dry fermented sausages. Sciences des Aliments, 18: 409-414.

Bruna, J.M., Hierro, E.M., De La Hoz, L., Mottram, D.S., Fernández, M., Ordoñez, J.A. (2003). Changes in selected biochemical and sensory parameters as affected by the superficial inoculation of *Penicillium camemberti* on dry fermented sausages. International Journal of Food Microbiology, 85: 111-125.

Escartin, E. F., Castillo, A., Hinojosa-Puga, A., Saldaña-Lozano, J. (1999). Prevalence of Salmonella in chorizo and its survival under different storage temperatures Food Microbiology, 16: 479-486

Incze, K. (1992). Raw fermented and dried meat products. Fleischwirtschaft, 72:58-62.

Moretti, V.M., Madonia, G., Diaferia, C., Mentasti, T., Paleari, M.A., Panseri, S., Pirone, G., Candini, G. (2004). Chemical and microbiological parameters and sensory attributes of typical Sicilian salami ripened in different conditions. Meat Science, 66: 845-854.

Osorio, M.T., Cabeza, E.A., Zumalacárregui, J.M., De Castro, S., Mateo, J. (2004). Aportaciones a la caracterización del chorizo elaborado en la provincia de Zamora. Eurocarne, 125: 151-162.

Pérez-Álvarez, J.A., Sayas-Barberá, E., Fernández-López, J., Aranda-Catalá, V. (1999). Physicochemical characteristics of Spanish-type dry-cured sausage. Food Research Internacional, 32: 599-607.

Schryer, F. J. (1986) Peasants and the Law: A History of Land Tenure and Conflict in the Huasteca**.** Journal of Latin American Studies, 18: 283-311.

Stiebing, A. and Rödel, W. (1988). Influence of relative humidity on the ripening of dry sausage. Fleischwirtschaft, 68: 1284-1291.

Tregear, A., Kuznesof, S., Moxey, A. (1998). Policy initiatives for regional foods: some insights from consumer research. Food Policy, 23: 383–394.

Van Riel, J.A.M., Olieman, C. (1986). High-performance liquid chromatography of sugars on a mixed cation-exchange resin column. Journal of Chromatography A, 362: 235-242.

Zanardi, E., Dorigoni, V., Badiani, A., Chizzolini, R. (2002). Lipid and colour stability of Milano-type sausages: effect of packing conditions. Meat Science, 61: 7-14.

QUALITY OF HALVES AND MEAT OF PIGS OBTAINED IN DIFFERENT MODELS OF CROSSBREEDING WITH LARGE YORKSHIRE

Lj. Petrović*[1], N. Džinić[1], V. Tomović[1], S. Timanović[2], P. Ikonić[1] and T. Tasić[1]

[1]*Faculty of Technology, University of Novi Sad, Bulevar cara Lazara 1, 21000 Novi Sad* [2] *Industrija mesa "Neoplanta" Novi Sad, Serbia and Montenegro. E-mail: ljiljapet@uns.ns.ac.yu*

Keywords: pigs, cross-breeding, quality of halves, meat quality

Introduction

The use of different selection and crossbreeding models in modern production of pork results in a bigger production range and higher yield of meat in halves. Fattening hybrids are obtained by permanent selection in pure races and their two-, three- and four-race cross-breeding *e.g.* maximal use of heterozys and complementarity. The characteristics of these hybrids are high growth and good quality of halves *e.g.* best characteristic of each cross-bred race. The quantitative values of the above characteristics are considered to have reached their maximum, however, other research (Fischer *et al.*, 2000; Oliver *et al.*, 1999), and our previous experience (Džinić *et al.*, 2001; Tomović *et al.*, 2003) point to the fact that the higher the meat yield in halves, the lower the technological meat quality. The high incidence of extremely changed quality – PSE meat – is unacceptable. The aim of this work was to find how the set model of selection and cross-breeding of pigs with Large Yorkshire as the basic race, affects the meat yield in halves, *e.g.* the quality of *M. semimembranosus.*

Materials and Methods

The investigations included 7 genotypes of pigs obtained as the result of selection and crossbreeding on a pig farm, (»Čenej« IM »Neoplanta«) in Novi Sad, Serbia and Montenegro: Large Yorkshire (LY=29); Large Yorkshire♂ x Svedish Landrace♀ (LY x SL=36); Large Yorkshire♂x(Large Yorkshire♂xSvedish Landrace♀)(LYx(LYxSL))=26; Duroc♂x(Large Yorkshire ♂ xSvedish Landrace ♀)(Dx(LYxSL))=25; Hampshire ♂x(Large Yorkshire♂xSvedish Landrace♀)(Hx(LYxSL))=18);(Duroc♂xPietren♀)♂x(LargeYorkshire♂xSvedishLandrace♀)♂ ((DxP)x(LYxSL)= 20);(Hampshire♂ x Pietren♀) ♂ x (Large Yorkshire♂x Svedish Landrace♀)♂ (HxP)x(LYxSL)=10. The pigs were fed standard feed during fattening, and pigs of approximately uniform age and mass were transported to the abattoir (*ca* 15km). After a night's rest, the pigs were stunned, slaughtered and processed by the standard technological procedure.
The pH_i was determined in right halves 45 min p.m., and pH_u 24h *post mortem*., in caudo-medial part *of M. semimembranosus* (MS), using the portable pH-meter ULTRA X. Quality and class (SEUROP) of meat were determined by the partial dissection method of cooled left halves (Commission Regulation (EC) No 3127/94., 1994; Walstra and Merkus, 1996). Samples (200-300g) taken from the caudio-cranial part (MS) were used for the determination of $colour_u$ and water holding capacity (WHC_u). Colour was determined using the MOM Colour 100 device and the colour charateristics were expressed in CIEL*a*b* system (Robertson, 1977). WHC_u was determined by compression method and expressed as % of bound water (Grau and Hamm, 1953). On the basis of parameters and criteria for MS quality (Petrović, 2000; Džinić, 2005) (PSE: pH_i<5.8, pH_u<6.2, WHC_u<50%, L*>50; RSE: pH_i<5.8, pH_u<6.2, WHC_u<50%, L*=43-50; RFN: pH_i>5.8, pH_u<6.2, WHC_u>50%, L*=43-50; PFN: pH_i>5.8, pH_u<6.2, WHC_u>50%, L*>50; DFD: pH_i>5,8, pH_u>6.2, WHC_u>70%, L*<43), the incidence of different MS quality in the investigated hybrids was determined.

Results and Discussion

The highest average relative meat yield in halves was found in four-race hybrids (H×P)×(LY×SL) – 58.35% (class E), folowed by four-race hybrids (D×P)×(LY×SLe). with average relative meat yield 57.85% (class E), reciprocal hybrid LY×(LY×SL) – 57.39% (class E), LY rac – 56.75% (class E), three-race hybrids H×(LY×SL) – 56.21% (class E), three-race hybrids D×(LY×SL) – 54.52% (class U), while the two-race hybrids LY×SL had the lowest average relative meat yield – 53.9% (class U). The difference of mean values of meat yield between LY race and two-race hybrids LY×SL is statistically significant ($P > 0.05$). Statistically significant difference of mean values of meat yield, determined by partial dissection was also found between two-race hybrids LY×SL and hybrids LY×(LY×SL), (D×P)×(LY×SL), (H×P)×(LY×SL); as well as between hybrids D×(LY×SL) and (D×P)×(LY×SL) with 95 probability. The analysis of meat yield values, obtained by partial dissection of pig halves, showed that the applied selection and crossbreeding model resulted in the effect of complementarity (Cartwright, 1970). The produced lines of commercial hybrides (H×P)×(LY×SL) and (D×P)×(LY×L) are characterized by high quality of halves, however, further work on quality improvement of sow LY×SL, showed that quality can be higher. The results presented in Figure 1 show that the incidence of DFD muscles (SM) in two-race hybrids LY×SL is 11.11%, of reciprocal three-race LY×(L×SL) 3.85%, three-race D×(LY×SL) and four-race (H×P)×(LY×SL) 10%. The highest incidence of PFN muscles was found in halves of two-race hybrids LY×SL – 52.78%. The incidence of RFN muscles (SM) in halves of pure race LY was 44.83%, lower in four-race hybrids (H×P)×(LY×SL) – 30%, and very low in other investigated halves, in the range from 4%, for three-race hybrids D×(LY×SL) to 22.22% in two-race hybrids LY×SL. In SM muscles from halves of four-race hybrids, the incidence of RSE was the highest – 40%. The incidence of PSE muscles was very high; the highest in three-race hybrids H×/LY×SL) – 66.66%, in multi-race hybrids D×(LY×SL), LY×(LY×SL), (H×P)×(LY×SL) and (D×P)×(LY×SL) the incidence was 52%, 50%, 40% and 30%, respectively.

Table 1: Effect of genotype on meat yield (%) class of halves.

Genotype	Prinos mesa (%)	Klasa polutki
VJ	56,75[a] ± 4,27	E
VJ×ŠL	53,90[b] ± 3,85	U
VJ×(VJ×ŠL)	57,39[a] ± 5,68	E
D×(VJ×ŠL)	54,52[a] ± 4,01	U
H×(VJ×ŠL)	56.21[a] ± 2,45	E
(D×P)×(VJ×ŠL)	57,85[a] ± 4,67	E
(H×P)×(VJ×ŠL)	58,35[a] ± 4,45	E

Figure 1: Effect of genotype on incidence of different meat and quality of *M. semimembranosus.*

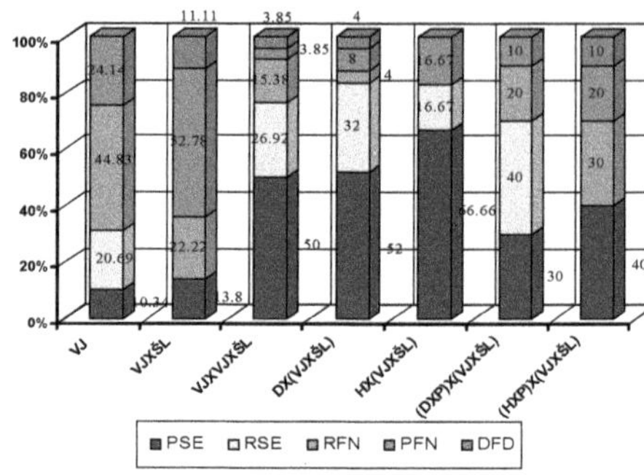

Conclusions

The set selection and crossbreeding model resulted in lines of hybrids with average meat yield in halves from 53.9% (LY×SL) to 58.35% (H×P)×(LY×SL). The higher the level of crossbreeding, the higher the meat yield in halves; however, the increase of meat yield is followed by decrease of technological meat quality.

Acknowledgments

This paper presents the results of work on project BTN 351008, supported by Serbian Ministry of Science and Environmental Protection.

References

Džinić Natalija, Petrović Ljiljana, Manojlović Danica, Tomović, V., Timanović, S., Trišić-Ilić Svetlana Kurjalov Nada, (2001). Carcas and quality of purebred and four–race hybrids. 47th ICoMST,Krakow, Poland,146.

Fisher P., F. D. Mellett, L. C. Hofman, (2000). Halothane genotype and pork quality, Carcass and meat quality characterirtich of three halothane genotypes. Maet Science, pp 96-105.

Grau, R. und Hamm, R. (1953). Eine einfache Methode zur Bestimmung der Wasserbindung im Muskel. Naturwissenschaften, 40, 29 – 30.

Oliver, M. A., Gou, P., Gispert, M., Diestre, A., Arnau, J., Nouguera, J. L., Blasko, A. (1994). Comparison of five types of pig crosses. II French meat quality and sensory characteristic of dry cured ham. Livestock Productions Science, 40:2, 179-185.

Petrović Ljiljana, 2000. S.4.28. 50.0037. NOVOTEH 2000, Katalog rezultata, Beograd.

Tomović, V., Ljiljana Petrović, Natalija Džinić, Danica Manojlović, S. Timanović, Dragica Vidarić, Nada Kurjakov i Svetlana Trišić-Ilić. (2004). Učestalost pojavljivanja različitog kvaliteta M. Semimembranosus sa polutki svinja četvororasnih hibrida, Simpozijum: "Veterinarstvo i stočarstvo u proizvodnji zdravstveno bezbedne hrane", Zbornik kratkih sadržaja, Herceg Novi, 23.

Džinić Natalija (2005). Uticaj endogenih i egzogenih faktora na kvalitet mesa svinja, doktorska disertacija, Univerziet u Banjoj Luci, Tehnološki fakultet, Banja Luka.

Walstra, P. and Merkus, G. S. M. (1996). Procedure for assessment of the lean meat percentage as a consequence of the new EU reference dissection metod in pig carcass classification. DLO – Research Institute for Animal Science and Health (ID – DLO), Research Branch Zeist, The Netherlands.

Robertson, A.R. (1977). The CIE 1976 Color-Difference formulae, Color Research Aplied, 2,7.

PRODUCT QUALITY OF LOIN HAMS MANUFACTURED WITH VARIOUS SALT LEVELS

H.C. Lee*, J.S. Hwang and K.B. Chin

Department of Animal Science, Chonnam National University, Gwangju, 500-600 Korea. Email: harmony0318@naver.com

Keywords: loin hams, product quality, salt levels

Introduction

More recently, consumers are selecting healthier foods for the benefit of their health. In the manufacture of healthier meat and meat products, fat and salt levels should be reduced in the formulations. Healthier foods have been consumed to prevent or treat certain disease as well as to provide nutrition (Jimenez-Colmenero *et al.*, 2001). In the west, people prefer low-fat or low-salt meat products as recommended by the World Health Organization (WHO) (Resurreccion, 2003). However, most people consumed sodium more than recommended (AHA, 2000). Following a similar trend, Koreans consumed more sodium than western people (KHIDI, 2001). Since fat and salt contributed to flavour and texture, these ingredients are normally added in the manufacture of meat products. In particular, salt contributes to reduction of cooking loss (Girard *et al.*, 1990) and extension of shelf-life in meat products (Brewer *et al.*, 1995). However, research on the effect of salt on the product quality of whole muscle ham has not been extensively studied. Thus, the objective of this study was to select the optimum salt level without quality defects for the manufacture of loin hams containing various salt levels (0~2.0%).

Materials and Methods

The loin hams were manufactured with various salt levels (0, 0.5, 1, 1.5, and 2%). The brine solution was injected into the loin hams using multiple injectors at about 15% of original weight. The pork loins were then injected with brine solutions and then cooked and smoked in a smoke chamber until the internal temperature reached 71.7°C. After cooking, the loin hams were chilled, packaged and stored in a refrigerator (4°C) until analysed. Proximate analyses were measured according to AOAC (1995) procedure. Colour measurements were performed using a colour meter and expressed as L (brightness), a (redness) and b (yellowness) values. Water holding capacity (WHC, %) and cooking loss (CL, %) were measured to determine the functional properties. WHC was measured according to the modified method of Jauregui *et al.* (1981) and CL was evaluated by a weigh difference of cooking before and after. Warner Bratzler and Allo-Kramer shear values were measured using an Instron Universal Testing Machine. Microbial counts included total bacteria, lactic acid bacteria and *Enterobacteriaceae* were determined. Sensory evaluation performed by a 9-membered sensory panel which evaluated colour, flavour, texture and overall acceptance using an 8-point hedonic test. Statistical analyses were performed using one-way analysis of variance (ANOVA) using the SPSS 12.0 (2003) program, and then the significant differences among treatments were evaluated by Student-Newman-Keuls multiple range test ($p<0.05$). Dunnett's-T test was used to compare each treatment to the control.

Results and Discussion

Table 1 shows the physicochemical, functional and textural properties, and sensory characteristics of loin hams with various salt levels. pH values, moisture (%), fat (%) and protein (%) content of loin hams were 5.99~6.11, 62.9~68.6, 2.15~5.62 and 20.2~26.6, respectively. The salt level of loin hams did not affect physicochemical properties and functional properties, whereas cooking loss (CL, %) was significantly affected by salt level ($p<0.05$). Loin hams containing higher than 1% salt level had lower CL than those containing lower 0.5% salt ($p<0.05$). These results were confirmed with previous results reported by Girard *et al.* (1990). However, loin hams with 1% salt level had similar CL to the control ($p>0.05$). In textural properties, no differences in Warner-Bratzler and Allo-Kramer shear values were observed among treatments including the control ($p>0.05$). The total bacterial counts were lower than 10^2cells/g, and lactic acid bacteria and *Enterobacteriaceae* were not detected in all treatments (data not shown). On the other hand, most sensory characteristics were affected by reduced salt level (Figure 1). The loin hams containing less than 1% salt received lower preference scores in sensory evaluation, for attributes such as texture, juiciness, colour and overall acceptance. These results also indicated that loin hams containing 1.0% or higher salt might be acceptable for consumers because they had better sensory characteristics. Thus, the optimum salt level without quality defect was at least 1%.

Table 1: Product qualities of loin hams manufactured with various salt levels.

Treatments	Various salt levels				
	0%	0.5%	1.0%	1.5%	2%
pH	6.11	6.04	6.04	5.99	5.99
Moisture (%)	62.9	65.5	67.4	68.6	68.0
Fat (%)	5.62	4.57	3.89	3.72	3.57
Protein (%)	26.6	21.3	20.2	20.9	22.2
Hunter L	73.0	69.2	68.8	70.6	69.6
Hunter a	11.2	12.1	9.11	9.66	9.74
Hunter b	4.99	4.84	2.82	3.75	3.11
Cooking Loss (%)	35.0 ^{a*}	28.6 ^{b*}	21.3 ^{c}	19.4 ^{c}	18.2 ^{c}
Expressible Moisture (%)	17.5	15.4	22.9	20.9	21.1
Allo-Kramer (kgf/g)	6.11	5.47	4.63	5.03	4.86
Warner-Bratzler (kgf)	2.89	2.25	2.46	2.45	2.31

$^{a-c}$ Means having same superscript within same column are not different ($p>0.05$);
* Paired comparisons (CTL 2.0 vs. treatments) significant at the $p<0.05$ level using Dunnett's-T test.

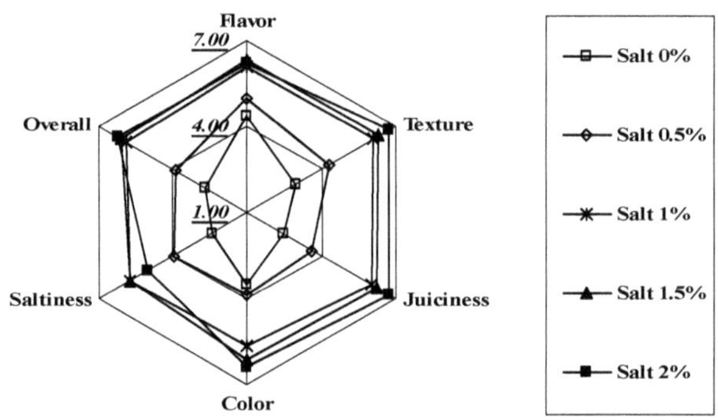

Figure 1: A spider web graphic representation of sensory data from the loin hams containing various salt levels.

Conclusions

Loin hams containing 1% salt had similar product qualities to the control in physicochemical, functional, textural and sensory properties ($p>0.05$). Increasing salt level up to 2% improved sensory characteristics and reduced cooking loss. Thus, the optimum salt level in the manufacture of low-salt loin hams without quality defects was at least 1%.

References

AHA (2000). American Heart Association Guidelines: Revision 2000. Circulation, 102(18): 2284-2299.

AOAC (1995). Association of Official Analytical Chemists. (15th ed). Washington, DC.

Brewer, M.S., Rostogi, B.K., Argoudelis, L., and Sprouls, G.R. (1995). Sodium lactate/sodium chloride effects on aerobic plate counts and color of aerobically packaged pork. Journal of Food Science, 60: 58-62.

Girard, J.P., Culioli, J., Maillard, T., Denoyer, C., and Touraille, C. (1990). Influence of Technological Parameters on the Structure of the Batter and the Texture of frankfurter type Sausages. Meat Science, 27: 13-28.

Jauregui, C.A., Regenstein, J.N., and Baker, R.C. (1981). A simple centrifugal method for measuring expressible moisture, a water-binding property of muscle foods. Journal of Food Science, 46: 1271-1273.

Jimenez-Colmenero, F., Carballo, J., and Cofrades, S. (2001). Healthier meat and meat products: their role as functional foods. Meat Science, 59: 5-13.

KHIDI (2001). Korean national health and nutrition survey 2001. Chapter 3. Research results: pp. 511-512, table 7-47, 8. Korea Health Industry Development Institute.

Resurreccion, A.V.A. (2003). Sensory aspects of consumer choices for meat and meat products. Meat Science, 66: 11-20.

THE EFFECT OF RADIO FREQUENCY POWER ON TEMPERING TIMES AND TEMPERATURES IN BEEF BLENDS OF VARYING COMPOSITION

K. Farag*, D.J. Morgan, D.A. Cronin and J.G. Lyng

School of Agriculture, Food Science and Veterinary Medicine, College of Life Sciences, University College Dublin, Dublin 4, Ireland Email: james.lyng@ucd.ie

Keywords: tempering, meat, radio frequency heating, temperature

Introduction

Radio frequency (RF) is an emerging electroheating technique which has potential to reduce cooking times (Rowley, 2001). When used for heating solids it differs from conventional heating where heat is transferred through the product from the outside to the centre primarily by conduction, resulting in a temperature differential between the edge and centre. In contrast, RF heating involves the application of a high voltage alternating electromagnetic field to a product placed between two electrodes where volumetric heating takes place, primarily by an ionic depolarisation mechanism (Zhang *et al.*, 2004). Since the late 1980s RF heating has achieved some commercial success in the area of post bake drying of cookies and snack foods (Mermelstein, 1998). There have also been a number of recent publications in the area of RF pasteurisation of meat products (Zhang *et al.*, 2004; Brunton *et al.*, 2005). Research in the area of RF defrosting of meat has focused on thawing and not tempering and has been relatively limited in comparison to research in other areas. Early RF thawing applications were proposed for eggs, fruits, vegetables, fish and meat (Cathcart *et al*, 1947). Jason and Sanders (1962) thawed fish using an RF unit at a frequency range from 36 to 40 MHz, and their results showed a significant reduction in thawing time and quality improvement as less drip was observed. Sanders (1966) and Bengtsson (1963) also reported reduced thawing times with the former author also claiming lower drip loss. The objective of the present study was to develop an RF tempering protocol and to evaluate temperature distribution in a range of beef blends during conventional and RF tempering. Specifically, the purpose of the tempering was to heat the product from the storage temperature of -18°C to a temperature range of -2 to -5°C where it can be processed mechanically.

Materials and Methods

Meat handling and preparation

Beef lean (95% visual lean) and fat was purchased from a local supplier. The lean and fat were ground through a 10 mm plate using a mechanical mincer and subsequently bowl chopped for 3 minutes. A 50:50 mixture was prepared by bowl chopping an equal amount of lean and fat for 3 min. Thirty boxes (20×20×10 cm) were prepared using freezer grade cardboard (0.3 cm thick) and equal numbers were filled with lean, fat or 50:50 mixture. Care was taken to exclude air inclusion during box filling and then the meat surface was covered with a cellophane film. Boxes were frozen for 48 hour at -20°C and subsequently predrilled at specified locations to facilitate temperature measurement.

RF and conventional tempering

An RF oven consisting of a low power RF generator (600W) with a complimentary automatic impedance matching network and controller at 27.12 MHz was used to temper the blocks of meat. The box was placed at the centre of the lower electrode and tempered using the time/power combinations summarised in Table 1. Conventional tempering procedures were developed using a refrigerated incubator, programmed to mimic industrial methods. The target temperature range was -2 to -5°C.

Temperature measurement following RF and conventional tempering

A time temperature logger was used to record post tempering temperatures at fifteen specified points within the product using a thermocouple jig fitted with Type K thermocouples.

Results and Discussion

There was a large reduction in tempering time using RF compared to the conventional method. With the latter the average time to reach the target temperature range was 335, 348 and 391 min for lean, fat and 50:50 mixture respectively. By contrast in the case of lean beef, RF achieved target temperatures in 11-12 min (400W), 10-11 min (500W) and 9-10 min (600W). Corresponding times for the mixture were 13, 12 and 11 min for 400, 500 and 600 W respectively. However, with fat alone the target range could not be achieved without runaway heating developing within the sample at any power level.

The overall mean temperatures at the end of RF (using 600 W as an example) and conventional tempering are compared in Figure 1. As seen in the Figure there was a greater temperature range within RF heated products with a high fat content making the greatest contribution to the observed effect.

Table 1: RF time/power combinations used.

Meat type	Power (W)	T1 (min)	T2 (min)	T3 (min)	T4 (min)
Lean	400	9	10	11	12
	500	8	9	10	11
	600	7	8	9	10
50:50 Mixture	400	10	11	12	13
	500	9	10	11	12
	600	8	9	10	11
Fat	200	14	18	-	-
	400	10	11	-	-
	500	8	9	10	-
	600	8	-	-	-

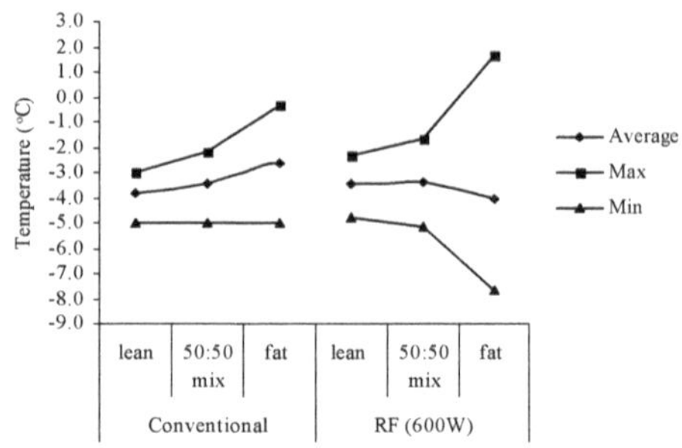

Figure 1: Post tempering temperatures[1].

As expected for conventional tempering the coldest spot was located at the geometric centre of the samples. In contrast, for RF tempering the geometric centre was the warmest spot for both lean and 50:50 mixture, though for fat the top surface was the warmest area.

With lean meat tempered by RF, the temperature at the power settings used ranged between -2 to -5°C while for the 50:50 mixture the range at similar power settings (-1.8 to -5.5°C) was slightly outside the target. However, for fat the range was much wider (*e.g.* at 600W for 8 min, temperatures ranged from +1.7 to –7.7 °C (Figure 1))

It is clear that for the RF oven configuration used, high fat levels had a major negative effect on tempering resulting in increased treatment time and less uniform temperature distribution.

Conclusion

The current study shows that the RF tempering protocol developed was successful for tempering lean beef and a 50:50 lean/fat mixture though the latter was on the temperature limits of the target range chosen. With the current setup the very large (~35 fold) reduction in tempering time compared to conventional treatments implies that for beef of a normal fat content, RF heating may have considerable potential in facilitating the rapid preparation of frozen meat for processing. However, further modifications to the application method will be required to optimise the system for tempering high fat meats.

References

Bengtsson, N. E. (1963). Electronic defrosting of meat and fish at 35 and 2450 MHz-a laboratory comparison. Food Technology, 17 ,10, 97–100.

Brunton N. P., Lyng, J. G., Li W., Cronin, D.A., Morgan D. and McKenna, B. (2005). Effect of Radio Frequency (RF) heating on the texture, colour and sensory properties of a comminuted pork meat product. Meat Science 38, 337-344.

Cathcart, W. H., Parker, J. J. and Beattie, H.G. (1947). The treatment of packaged bread with high frequency heat. Food Technology 1: 174-177.

Jason, A. C. and Sanders, H.R. (1962). Dielectric thawing of fish. Experiments with frozen herrings. Experiments with frozen white fish. Food Technology, 16(6):101-112.

Mermelstein, N.H. (1998). Microwave & radio frequency drying. Food Technology, 52(11), 84– 86.

Rowley, A.T., (2001). Radio frequency heating. In: Richardson, P. (Ed.), Thermal Technologies in Food Processing. Woodhead Publishing Ltd., Cambridge, UK, pp. 163–177.

Sanders, H. R. (1966). Dielectric thawing of meat and meat products. Journal of Food Technology, 1,183-192.

Zhang, L., Lyng, J. G., Brunton, N. P., Morgan D. and McKenna, B. (2004). Dielectric and thermophysical properties of meat batters over a temperature range of 5–85 °C. Meat Science, 68:173-184.

[1] Power level of 600 W at 10, 11 and 8 min for lean, mixture and fat respectively.

EFFECTS OF VARIOUS TREATMENTS ON THE TEXTURE SOFTENING OF POST-BREEDING MATURE COWS MEAT

A.M. Ahhmed*[1], N. Matsumoto[1], S. Kawahara[1], K. Ohta[1], R. Kuroda[2], T. Okayama[3], K. Nakade[4], M. Numata[4], T. Nakamura[5] and M. Muguruma[1]

[1.] Faculty of Agriculture, University of Miyazaki, Miyazaki 889-2192, Japan, [2.] Junwa Medical Center for Rehabilitations, Miyazaki 880-2112, Japan, [3.] Faculty of Agriculture, Kobe University, Kobe 657-8501, Japan, [4.] Central Research Institute, Ito Ham Food, Inc. Moriya 302-0104, Japan, [5.] R&D Center, Unicafe Inc, Tokyo105-0003, Japan. Email: abdo772201@yahoo.co.uk

Keywords: beef, tenderisation, softening, vacuum packaging

Introduction

The processing of post-breeding mature cows meat has become a primary development theme in meat industries. The use of new materials sometimes can help to meet the challenges faced in meat manufacturing. In recent decades, many meat suppliers started to create some sort of soft meat, through the utilization of several techniques, such as mechanical tenderisation. The physical disruption of the muscle structure reduces the density of cooked meat fibres and therefore improves its tenderness (Hokins, 2004). The technologists need to pay particular attention to those variables that would directly impact the texture quality of meats that are tough, thereby difficult to chew, (e.g. mechanical processes, additives, heat treatments, packaging type). The use of new materials that originally come from leguminous and fruity sources has become increasingly used in meat tenderization. Products treated by organic, or plant-derived additives can contribute to promote the consumers to use those products as daily goods. This study was aimed to soften tough mature meat through the use mechanical tenderisations and the use of softening additives at two different heat treatments.

Materials and Methods

The meat used in this study was Top Loin Steak (Japanese Black Cows). The samples were designed as slices (4mm thick) and intentionally mechanically tenderised by tenderizer (OHMICHI, OMTR-270, Tokyo, Japan) to create clefts in the flesh. Additives, such as a fermented apple solution (FAS), fermented paste of beans (miso) and non-sterilized soy sauce were used to tenderise the meat. Samples used in the present study were divided to three groups. The first group was treated with 4% FAS, and set in a solution for 3 hours. The proportion of meat to the solution was 2:1. The second group was treated with miso. The miso was pasted on both sides of the sliced meat for one day. The proportion of the meat to miso was 5:1. The third sample set was treated with diluted (two times in distilled water before being used) non-sterilised soy sauce. Meat was set in the soy sauce for 3 hours; the proportion of meat to soy sauce was 2:1. The samples were subjected to two different heat treatments, the first of which involved being packaged and boiled at 75°C for 15 min, while the second condition involved meat that was grilled in an oven at 170°C for 5min, on each side of the sliced meat. The water-holding capacity was measured by checking the weight of the samples before and after the heat treatments. Another part of this research investigated the effect of vacuum packaging on the beef softening after adding FAS only. FAS was added to beef slices of the same thickness (4mm), and then packaged in clear plastic bags. The samples were subjected to the breaking strength, determination by using creep meter (Yamaden RE2-33005S, Tokyo, Japan).

Results and Discussion

Fracturing the fibre of the flesh facilitated the absorption of the tissues and allowed for greater penetration with the additives. It was clear that the breaking strength of the samples treated with FAS, miso and soy sauce, respectively, were reduced. Moreover, the mechanically tenderised and boiled samples showed a decrease in the breaking strength, greater than that found in non-mechanical tenderised samples. Fig 1 shows the effects of mechanical tenderisation and all the additives on the breaking strength of boiled samples. These data indicate that the breaking strength was significantly reduced. Moreover, the breaking strength was decreased in the grilled samples and those processed by the same treatments (Fig 2). In a comparison between the untreated samples and samples processed by FAS and vacuum packaged, the breaking strength of the treated samples was drastically reduced (Fig 3 and 4). These data suggest that the vacuum packaging provides the opportunity for the additive solution to penetrate into the meat tissues. The total remaining weight examination was conducted to determine whether the various treatments had an effect on the water-holding capacity and total remaining weight of treated samples. Table 1 presents the values of the total remaining weight after the samples were treated at two different temperatures. The best recorded values were found in the samples treated by FAS and miso. Each of the samples was also subjected to sensory trait test. A total of 15 participants of different ages were engaged in this test. The taste of the samples treated with miso and soy sauce differed from the original quite a bit and the flavour of those samples was dominated by miso and soy sauce taste (data not shown). The taste of the samples treated by both the miso and the soy sauce were reportedly strongly influenced by those additions.

The meat in each case took on the flavour of the solution. However, the taste of samples treated with FAS was good. In addition, the vacuumed packaging with FAS contributed to the tenderization of post-breeding mature cow's meat. Data suggest that the use of mechanical tenderization and other additives had a positive affection the meat texture of mature beef.

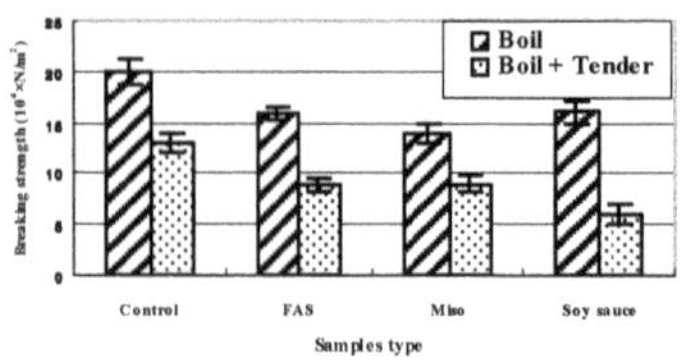

Figure 1: Breaking strength of tenderised and boiled samples.

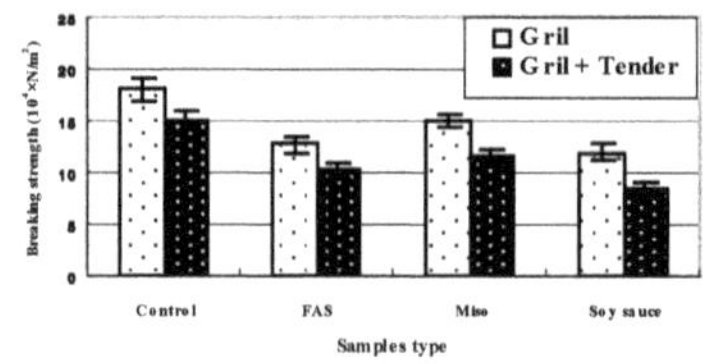

Figure 2: Breaking strength of tenderised and grilled samples.

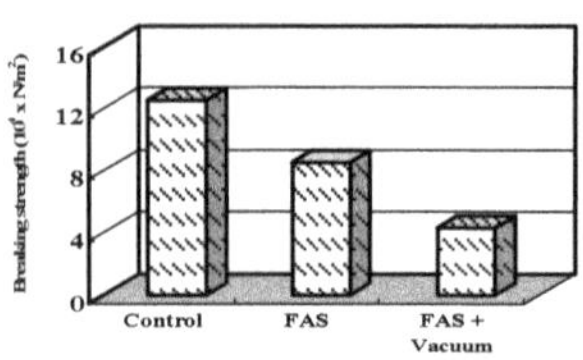

Figure 3: Breaking strength of samples boiled after treated with FAS and vacuum package for 3hours.

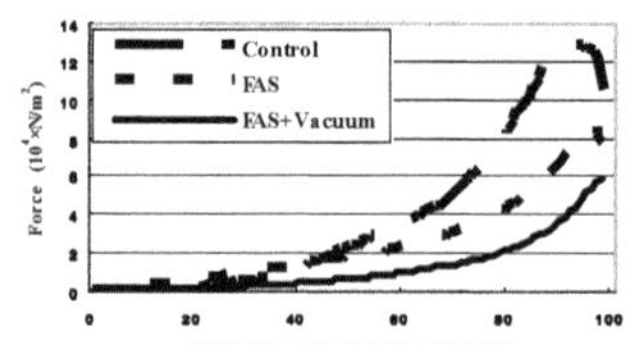

Figure 4: Force of a creep meter knife to samples treated by FAS and vacuum packaged.

Table 1: The total remaining weight of treated samples with plant-derived additives after subjected to two different heat treatments.

Grilled group		Boiled group	
Samples type	Samples weight (%)	Samples type	Samples weight (%)
Control	62.07	Control	70.81
FAS	68.03	FAS	75.32
Miso	66.17	Miso	76.27
Soy sauce	62.6	Soy sauce	69.72

Conclusions

The study was engaged using additives in order to soften the texture of mature beef. The feasibility of this study purposes to address a significant problem, in many countries, mature beef is considered to be an inconsumable meat. This study demonstrates that additives can play a role in solving such a problem. The use of FAS, miso and soy sauce, along with mechanical tenderization treatment can contribute to the softening of the meat of post-breeding mature cows. Furthermore, the breaking strength of the boiled samples was also significantly reduced. However, the taste of the samples treated with miso and soy sauce was largely influenced by the taste of those additives themselves, making them less palatable to the participants in this study. In conclusion, the samples treated with the combination of meat tenderizing and addition of FAS along with vacuum packaging proved to be effective in beef softening. The vacuum packaged samples gave the lowest value of breaking strength, which was found to be 60% less than the breaking strength in the control samples. This study provides evidence that may contribute to our understanding of the advantages of using plant-derived additives, along with meat tenderization, in order to present a consumable meat from mature beef that would otherwise be considered inedible by consumers.

Reference

Hokins, L. D. Tenderizing mechanisms/mechanical. In Jensen, W., Devine C. and Dikeman M. (2004) (Eds), *Encyclopedia of Meat Science.* (Vol. 3) (pp. 1355-1369) Oxford: Elsevier Ltd.

RESIDUAL NITRITE LEVELS IN DIFFERENT TYPES OF SAUSAGES

S. Saicic*, D. Vranic and D. Trbovic

Institute of Meat Hygiene and Technology, 11000 Belgrade, Serbia and Montenegro.
Email: meatinst@beotel.yu

Keywords: nitrite, sausages, permitted level

Introduction

Nitrite is widely used as additive in cured meat products for the reason of several beneficial effects such as redding (stabilizing the colour component of the products and obtaining characteristic flavour of cured meat), as preservative and as antioxidant. Nitrite in meat greatly delays development of botulinum toxin (botulism) and destroys the other pathogenic microorganisms (Christiansen *et al.,* 1975). In spite of their positive influence on ripening, colour and stability of meat products, nitrite also has mutagenic and toxic properties that are leading to formation of carcinogenic N-nitrosoamines and N-nitrosoamides (Sugimura *et al.,* 1986; Hotchkiss *et al.,* 1992). The known acute and chronic toxic effect of nitrite is the production of methaemoglobin (MetHg) by reacting with haemoglobin, which does not possess the ability to carry oxygen. This intoxication is named methaemoglobinemia (Kennedy *et al.,* 1997). Therefore, usage of nitrites that are quite effective on human health, have to be done with utmost care and kept under control. Hence, this study was undertaken to determine residual nitrite levels in different kinds of sausages.

The main objective of this investigation was to determine nitrite levels in fermented dry sausages, boiled and cooked sausages according to the Regulations of Food Additives quality . (Off. Journal of SCG, No 56/2003).

Materials and Methods

The 78 fermented dry sausages, 89 minced boiled sausages, 83 grounded boiled sausages, 33 boiled sausages with meat pieces and 86 cooked sausages were used for determination of nitrite content. All samples of sausages originated from Serbian producers. For the determination of nitrite content, the standard method was used (JUS ISO 2918/1999). The principle of the method is reaction between dinitronium salt and aromatic amine (N-(1-Naphthyl)ethylenediamine dihydrochloride) to give a red coloured aminonitro compound, whose absorbance is measured spectrophotometrically at 538 nm.

Results and Discussion

Nitrite content (minimum, maximum and mean value), maximum permitted values and number of samples which ara not in compliance with the Regulations are presented in Table1.

Table 1: Nitrite content in different types of sausages, mg/kg.

Type of sausages	Number of samples	Nitrite content, mg/kg min	max	Mean value mg/kg	Maximum permmitted values	N*	N*,%
Fermented dry sausages	78	0.00	59.19	4.15	50.00	1	1.28
Minced boiled sausages	89	0.00	102.36	38.33	100.00	1	1.12
Grounded boiled sausages	83	0.76	93.94	39.30	100.00	0	0
Boiled sausages with meat pieces	33	2.39	74.84	45.66	100.00	0	0
Cooked sausages	86	0.00	100.00	16.99	100.00	0	0

N* - Number of samples which is not in compliance with the Regulations

The average nitrite concentrations in fermented dry sausages were lower than the concentrations found in other types of sausages, 4.15 mg/kg (0.00-59.19 mg/kg), because of the fact that residual nitrite contents may be reduced due to storage time and also microorganism activities, when starter cultures are used. In Serbia and Montenegro, residual quantities of nitrite in the final products are limited by the Regulations of Food Additives quality (Official Journal of SCG, No 56/2003). Maximum permitted level in fermented dry sausages is 50 mg/kg and in the other types of products

is 100 mg/kg. The nitrite level from only one sample (1.28%) of the investigated fermented dry sausages was higher than the level permitted by the Regulations.
The average nitrite levels in group of boiled sausages were: 38.33 mg/kg (0.00-102.36 mg/kg) in minced boiled sausages, 39.30 mg/kg (0.76-93.94 mg/kg) in ground boiled sausages and 45.66 mg/kg (2.39-74.84 mg/kg) in boiled sausages with meat pieces. Only one sample (1.12%) of minced boiled sausages was not in compliance with the Regulations.
The average nitrite content in cooked boiled sausages was between fermented dry sausage and boiled sausage levels, 16.99 mg/kg (0.00-100.00 mg/kg).

Conclusions

On the basis of the research results which were done on different kinds of sausages, the following conclusions should be emphasised:
- In total, 369 sausages were investigated, only two samples (0.54%) were not in compliance with the Regulations of Food Additives quality (Official Journal of SCG, No 56/2003).
- These results also showed that in the meat industry, curing salt mixture are used according to the Official Journal rules.

References

Christiansen, L.N., Tompkin, R.B., Shaparis, A.B., Johnston, R.W. and Kautter, D.A. (1975). Effect of sodium nitrite and nitrate on Clostridium botulinum growth and production in summer style sausage. Journal of Food Science, 488-490.

Hotchkiss, J.H., Helser, M.A., Maragos, C.M and Weng, Y.M. (1992) Nitrate, nitrite and N-nitroso compounds in Food Safety Assesment (Eds J.W.Finley, S. Kobinson and D.J. Armstrong). American Chemical Society, Washington DC, 400-418.

Kennedy, N., Smith, C.P., Mc Whinney, P. (1997) Faulty sausage production causing methaemoglobinaemia. Archives of Disease in Childhood, 76: 367-368.

Sugimura, T., Sato, S., Ohgaki, H., Takayama, S. Nago, M. and Wakabayashi, K. (1986) Mutagens and carcinogens in cooked food. Genetic Toxicology of the Diet (Ed. J. Knudsen), Alan R. Liss, New York, 85-107.

SALT-REDUCED LYONERS WITHOUT INGREDIENTS DECLARED BY E-NUMBERS

R. Hadorn[1,]*, M. Suter[1,2], and J. Javor Qvortrup[2]

[1]Agroscope Liebefeld-Posieux (ALP), Swiss Federal Research Station for Animal Production and Dairy Products, Berne, Switzerland. [2] Department of Food Technology, University of Applied Sciences Waedenswil (HSW), Wädenswil, Switzerland. Email: ruedi.hadorn@alp.admin.ch

Keywords: boiled sausage, lyoner, curing, E-Number, salt reduction

Introduction

In Swîtzerland, the per capita consumption per year of boiled sausages is about 6 kg, which represents approximately 2/3 of the whole sausage consumption (Proviande, 2005). All additives used during boiled-sausage processing have either to be declared by their names or by E-numbers. This is often badly received in the market place because a lot of consumers have a negative association with E-numbers. Except nitrite curing salt and ascorbic acid/ascorbate, food additives are also banned from organic meat products. Therefore, a trial was designed to test whether processing of boiled sausages without E-number ingredients is possible. Due to its large calibre in relation to the curing process, lyoners were chosen as experimental boiled sausage. The following two aspects were tested in combination: 1. adaptation of processing procedures in order to produce without additives (stabilisators, flavour enhancers) and 2. the use of alternative ingredients which contain some technological important substances naturally and therefore have not to be declared by E-numbers (nitrate from vegetable mixtures, ascorbic acid from acerola cherries). The level to which extent the salt content of lyoners may be reduced without having any adverse effects on technological and sensorial attributes as described by Martin (2002).

Materials and Methods

In two series, 2×12 different lyoners (caliber: 90 mm, weight: 1.5 kg) based on the same ground recipe (31% pork III, 15% veal II, 12% neck backfat, 10% sausage backfat, 10% processed calf rind, 22% ice-water, 6 g per kg of a conventional spice mixture), different ingredient mixtures and different conditions during curing (between stuffing and boiling) were produced (Table 1).

The sausages were analysed for physical [peeling, sliceability, jelly formation, colour (by CIE L*a*b*-system), pH], chemical [dry matter (DM), crude protein (CP), crude ash (CA), crude fat (CF), total sugars (SUG), chloride (Cl), sodium (Na), calcium (Ca), magnesium (Mg), potassium (K), nitrate (NO_3^-), nitrite (NO_2^-)] and microbiological parameters [total viable counts, Enterobacteriaceae, *Clostridium perfringens*, Salmonella spp.: all bacteria were determined 1, 14, 28 and 42 days after processing]. Furthermore, the lyoners were assessed for their sensory profile on different odour, flavour and texture criteria on a 10-point intensity scale by a panel of thirteen trained internal panelists. Also a visual evaluation was performed for all treatments.

Table 1: Experimental design of the two series.

Series 1 (Curing conditions: 1440' and 2°C)	Salt supplementation		
	2.0%	1.5%	1.0%
Table salt, acerola[1]	1_01	1_02	1_03
Sea salt, acerola	1_04	1_05	1_06
Table salt, acerola, vegetable I[2], starter I[3]	1_07	1_08	1_09
Nitrite salt, ascorbic acid[4], Na-ascorbate[5]	1_10	1_11	1_12

Series 2 (Salt supplementation 2.0%)	Curing condition		
	90' / 45°C	240' / 19°C	1440' / 2°C
Sea salt, acerola[1]	2_01	2_02	2_03
Table salt, acerola, vegetable I[2], starter I[3]	2_04	2_05	2_06
Table salt, acerola, vegetable II[6], starter II[7]	2_07	2_08	2_09
Nitrite salt, ascorbic acid[4], Na-ascorbate[5]	2_10	2_11	2_12

[1] *Lyophilisate of acerola cherries: 1 g per kg*
[2] *Lyophilisate of a vegetable mixture: 8 g per kg*
[3] *Lyophilisated ferments of Staphylococcus carnosus (producer A): 0.2 g per kg*
[4] *Ascorbic acid: 0.25 g per kg*
[5] *Na-ascorbate: 0.25 g per kg*
[6] *Lyophilisate of a vegetable mixture: 5 g per kg*
[7] *Lyophilisated ferments of Staphylococcus carnosus (producer B): 0.2 g per kg*

Results and Discussion

From the physical analyses, the choice of the nitrate/nitrite-source had an influence only on colour, mainly the a*-value. a*-values for the treatments with table salt and sea salt were between 5.2 and 6.3 and with vegetable mixture I between 5.4 and 6.8. Highest a* values (9.5 to 10.4) were determined for nitric-salt treatments. Salt reduction lowered water-binding capacity and jelly was formed between stuffing and sausage casing. Salt supplementation of 2.0, 1.5 and 1.0% led to significantly different jelly percentages of 1.6, 7.6 and 18.3%, respectively. Therefore peeling force was reduced from 1.46 N (2.0% salt) to 0.66 N (1.0% salt) and sliceability values increased from 712 mJ (2.0% salt) to 818 mJ (1.0% salt) due to the higher DM content of the remaining sausage mass ($p \leq 0.05$). Curing at 19°C doubled jelly formation compared to the ones close to the freezing point or to the point of sarcoplasma protein denaturation at about 45°C, which caused sealed surfaces.

Na-, Cl- and Ca-contents were influenced by salt supplemention and therefore jelly formation. The latter caused a slight dilution of SUG due to its water solubility as well as concentration of other nutrients as CF, CP and DM in the stuffings ($p \leq 0.05$). Ca-, K- and Mg-contents were independent from the degree of salt supplementation. Curing conditions had no influence on nutrient and mineral contents, because jelly formation did hardly differ between the different curing conditions. Due to the source of nitrate/nitrite, SUG-, Ca- and K-contents were slightly increased for the treatments supplemented with vegetable powder being rich in carbohydrates and minerals. Vegetable mixtures brought in important amounts of Na-nitrate (vegetable mixture I: 38 mg/kg stuffing; vegetable mixture II: 28 mg/kg stuffing) into the stuffing. This nitrate was only partly reduced by the starter cultures to nitrite, which could be seen in non-cured areas during visual evaluation.

With the exception of the lyoners processed with vegetable mixture I and cured for 240 minutes and 19°C or 1440 minutes and 2°C, respectively, total viable counts were between 10^2 und 10^3 colony-forming units per gram (CFU/g). The corresponding values for the mentioned exceptions were between 10^4 and 10^5 CFU/g. Neither the differences during conservation nor the ones between the treatments were significant. Counts for *Clostridium perfringens* and Enterobacteriaceae were below the detection limits of 10 CFU/g for all treatments. Also no salmonellae could be detected in the 25-gram samples of all lyoners.

The degree of salt supplementation was important for sensorial evaluation ($p \leq 0.05$): 1. directly by tasting the salt in the attribute "salty", 2. indirectly by its effect on enhancing taste and spice effects in the attributes "acid" and "spicy" and 3. by influencing jelly formation which was followed by texture changes seen in the attributes "cracking" and "hard". Due to the choice of nitrate/nitrite-sources as well as the one of curing conditions, only minor effects on sensory traits were seen.

Visual evaluation of the sausages showed that colour of lyoners processed with table salt, sea salt and vegetables I and II never reached the one of the ones processed with nitrite salt. By cutting the sausages of treatment 2_04 (vegetable I, 90 minutes, 45°C), some spots could be observed where curing had been completed. For lyoners at the lowest salt level, jelly deposition occurred not only between stuffing and skin, there was also some jelly formation in the stuffing itself. This influenced the impression of the cross section negatively.

Conclusions

Replacing nitrite salt by vegetable mixtures rich in nitrate combined with the use of a nitrate-reducing starter culture and the right choice of curing conditions can lead to complete curing of lyoners. In order to achieve complete curing, at least 40 mg $NaNO_3$ per kg stuffing, which are necessary for aroma and colour formation (Jira, 2003), have to be provided by the vegetable mixtures. Depending on the starter culture, curing should be performed during 2 to 4 hours at a temperature of 45°C before the boiling process starts.

High salt supplementations are desirable from the technological point of view. Due to sensorial as well as health aspects, 1.6 to 1.8 g NaCl per kg stuffing is recommended. A further reduction of NaCl supplementation without addition of other salt-ions or phosphates has adverse technological effects (jelly formation, stuffing smoothness).

Also referring to practical experiences, it can be stated that stuffing production for boiled sausages is possible without using phosphates or modified starches, if not too much ice-water is added and slightly increased weight losses are considered.

Replacing ascorbic acid and Na-ascorbate by the acerola lyophilisate rich in ascorbic acid did not cause any problems. This is due to its desired effects on biochemical processes as well as having no adverse influence on sensorial traits.

References

Jira, W. (2003). In: Chemische Vorgänge beim Pökeln und Räuchern. Chemie des Lebensmittels Fleisch, Kulmbacher Reihe, Band 18, p. 119-150.

Martin, M. (2002). Meat Curing Technology. In: Hui, Y.H., Hip, W.K., Rogers, R.W., Young, O.A. (Hrsg.): Meat Science and Applications. Marcel Dekker Inc., New York.

Proviande (2005). Konsumzahlen Proviande 2004. Marktforschungsinstitut IHA-GFK, Hergiswil (personal communication).

ANTIOXIDANT EFFECTS OF MALAYSIAN PLANT EXTRACTS IN MECHANICALLY DEBONED CHICKEN MEAT SAUSAGE

A.S. Babji*[2], A. Noriham[1] and A. Aminah[2]

[1]Department of Food Technology, Faculty of Applied Sciences, Universiti Teknologi MARA, 40450 Shah Alam, Malaysia [2]School of Chemical Sciences and Food Technology, Faculty of Science and Technology, Universiti Kebangsaan Malaysia, 43600 Bangi, Malaysia Email: daging@pkrisc.cc.ukm.my

Keywords: antioxidant, Malaysian plants, mechanically deboned chicken meat sausage

Introduction

Consumption of poultry meat and poultry meat products is growing all over the world. For several reasons people prefer this kind of meat to beef and pork. The use of mechanically deboned chicken meat (MDCM) in frankfurters, various loaf products, fermented sausages and restructured products has been well documented (Froning, 1976; Lai *et al.*, 1991). A major problem with products manufactured with MDCM is the rapid onset of oxidative rancidity which results in off-flavours and odours (Mac Neil *et al.*, 1973).

It has long been known that herbs and spices possess antioxidative and antimicrobial effects, and since ancient times they have been used to prolong shelf-life and improve the taste of meat products. In various studies rosemary belonging to the mint family (*Labiateae*), has shown to be the most potent natural antioxidant among the common species (Hermann, 1981; Madsen and Bertelsen, 1995).

Thus, this study was carried out to investigate the effectiveness of two types of local plants water extracts, *kesum* and *tenggek burung* incorporated in the refrigerated MDCM sausage stored for 6 months in comparison with rosemary extracts, BHA/BHT and a control treatment without antioxidant.

Materials and Methods

Preparation of MDCM sausage

MDCM was purchased from Dinding Poultry Sdn. Bhd., Sitiawan, Perak, Malaysia and kept frozen prior to use. Five MDCM sausages with and without plant extracts were prepared. They were F1: Control; F2: with 200 ppm BHA/BHT; F3: with 200 ppm *kesum* water extract; F4: with 400 ppm *tenggek burung* water extract and F5: with 400 ppm commercial rosemary extract.

Plant material and extraction

Plant materials were extracted according to Duh (1998).

Peroxide value

Peroxide value (PV) was determined using AOAC methods (1984).

TBA analysis

Thiobarbituric acid reactive substances (TBARS) assay was performed as described by Buege and Aust (1978).

Statistical Analysis

The statistical analysis was performed using SAS 6.12 package (SAS Institute, 1995).

Results and Discussion

Antioxidant activity

In general all formulations showed an increased in POV after a period of time after which it started to decline, after 3 months storage for control and rosemary treated sausage while after 4 months for samples treated with BHA/BHT, *kesum* and *tenggek burung* water extracts (Fig 1). The increase in oxidation as reflected by POV increase was expected even though the sausages were kept at -18°C. Igene *et al.*, (1979) reported that the triglycerides and phospholipids played significant roles in the oxidation process of MDCM. Crapiste *et al.*, (1999) reported maximum POV will be attained and started to decline after the decomposition of hydroperoxides to other products. However POV of treated samples were significantly lower than the control throughout the 6 months storage. This indicated that the plant extracts and the synthetic antioxidant can lower the initial lipid oxidation rate.

TBARS value is routinely used as an index of lipid oxidation in meat products in store (Raharjo & Sofos 1993) and is shown in Figure 2. Malonaldehydes react specifically with thiobarbituric acid showing an increase in TBARS values after 6 months of storage in all samples. TBARS values of model systems that had been treated with *kesum* and *tenggek burung* extracts were much lower than the control thereby indicating protection of the meat constituents of the extract against autoxidation; complementary research supports the view that the compounds providing antioxidant activity are polyphenolics. In this study, samples treated with *kesum* extract were found to be significantly better than the synthetic antioxidant BHA/BHT after 6 months storage.

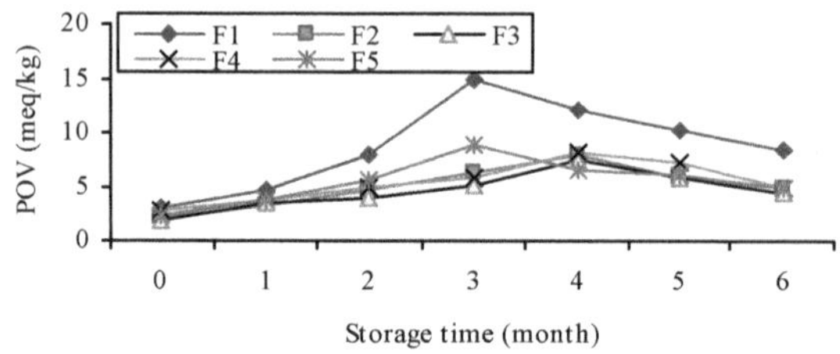

Figure 1: Effects of plant extracts and BHA/BHT on the POV in frozen MDCM sausage.

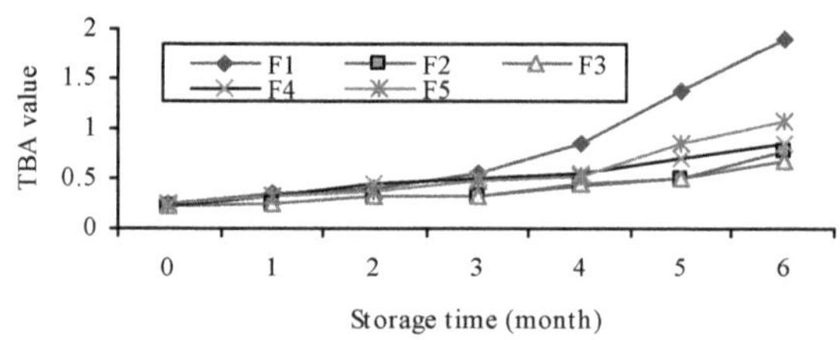

Figure 2: Effects of plant extracts and BHA/BHT on 2-thiobarbituric acid (TBA) value in frozen MDCM sausage.

Conclusions

This study concluded that *kesum* and *tenggek burung* water extracts provide antioxidant benefits to MDCM sausage during frozen storage. Of the two extracts studied *kesum* demonstrated the most potent effect in terms of antioxidative activity which is comparable to that of BHA/BHT.

References

A.O.A.C. (1984). *Official Methods*. 14th ed. USA:.Association of Official Analytical Chemists, Inc.

Buege, J.A. and Aust, S.D. (1978). Microsomal lipid peroxidation. *Methods in Enzymology,* 52: 306

Crapiste, G.H., Breveden, M.I.V. and Carelli, A.A. (1999). Oxidation of sunflower oil during storage. *Journal of the American Oil Chemists Society,* 76:1437-1442.

Froning, G.W. (1976). Mechanically-deboned poultry meat. *Food Technology,* 30(9): 50-63.

Hermann, K. (1981). The antioxidative action of spices. *Deutshe Lebensmittel Rundshau,* 77:134-139.

Igene, J.O., Pearson, A.M., Merkel, R.A. and Coleman, T.H. (1979). Effect of frozen storage time, cooking and holding temperature upon extractable lipids and TBA values of beef and chicken. *Journal of Animal Science*, 49:701-707.

Madsen, H.L. and Bertelsen, G. (1995). Spices as antioxidants. *Trends in Food Science and Technology,* 6:271-277.

Raharjo, S. and Sofos, J.N. (1993). Methodology for measuring malonaldehyde as a product of lipid peroxidation in muscle tissues: A review. *Meat Science*, 35:145-169.

SAS (1995). SAS/STAT user's guide, Version 6.21. SAS Institute.

EFFECTS OF ADDING CLOVE BUDS AND CINNAMON BARK EXTRACTS ON THE OXIDATIVE STABILITY OF CHICKEN MEATBALLS

K.W. Chan and A.S. Babji*

Department of Food Science, School of Chemical Sciences and Food Technology, Faculty of Science and Technology, University Kebangsaan Malaysia, 43600 Bangi, Malaysia. Email:chankw_antioxidant@yahoo.com

Keywords: antioxidant, clove buds extract, cinnamon bark extract, chicken meatball

Introduction

Lipid oxidation is one of the major deteriorations in meat products, thus the prevention of lipid oxidation in meat products during processing and storage is vital to maintain its quality and safety (Buckley *et al.*, 1995). In order to retard lipid oxidation, synthetic antioxidants such as butylated hydroxytoluene (BHT) and butylated hydroxyanisole (BHA) are widely used (Verhagen *et al.*, 1990). Somehow, the usage of these chemicals is always related to health risk (Shahidi and Wanasundra 1992). Thus, the interest in using natural antioxidants in meat products has increased in recent years. Clove and cinnamon are important ingredients in Malaysian curry powder. Yet, the bioavaibility and functionality of the polar/water soluble compound/oleoresins are seldom studied since their essential oils can quote a higher price in the market. The objective of this study was to evaluate the antioxidant activities of clove buds (*Eugenia caryophylata)* and cinnamon bark (*cinnamomum burmanii*) water extracts on the oxidative stability of chicken meatballs.

Materials and Methods

Clove buds (*Eugenia caryophylata)* and cinnamon bark (*cinnamomum burmanii*) extracts were obtained through hot water extraction. Generally, 30 g of spice powder was stirred with 600ml of freshly boiled distilled water for 15 minutes. The mixture was then filtered through filter paper Whatman No 1. and the filtrate was dried at 70°C with a rotary evaporator (Buchi Rotovapor R-114) until a constant weight was attained. Extracts were stored in the dark and frozen (- 18°C) prior to usage. A set of 6 treatment samples differing only in the extract/antioxidant added were prepared in this study (C: control meatball; T1:meatball + 200 ppm clove buds extract; T2: meatball + 200 ppm cinnamon bark extract; T3: meatball + 100 ppm clove buds extract + 100 ppm cinnamon extract ;T4: meatball + 200 ppm ascorbic acid; T5: meatball + 100 ppm BHA +100 ppm BHT). Meatballs were prepared according to a conventional formula: 65% chicken breast meat, 20% palm fat, 6.5% water, 6% potato starch, 1.5% sunflower oil and 1% salt. Firstly, chicken meat, salt, water and potato starch were homogenized in a bowl mixer for 1.5 minutes, followed by addition of palm fat and sunflower oil (3.5 minutes). Then, spice extract/ antioxidant was mixed into the emulsion for 5 minutes. Meatballs were formed manually (15g, 20-25mm) and flash fried in palm oil at 190°C for 30s. Subsequently, the meatballs were cooked in a forced draught oven at 250 °C for 4 minutes. The cooked meatballs were then placed in a chiller (2-5°C) immediately to reach a product temperature below 12 °C. After reaching the packing temperature, samples were placed into plastic containers and sealed with a layer of semi-permeable PVC film. A 12 day shelf-life study was conducted on the meatballs which were stored at 8±1 °C in darkness to follow an accelerated shelf-life study determination protocol (IFST, 1993). Sampling and storage conditions records from each treatment took place at 1,3,6,9 and 12 days (storage time) and every sample was analysed promptly. Peroxide values (PV) of samples were determined using AOAC methods (1984). Fat extraction was carried out using Kinsella *et al.*, (1977) extraction method with slight modification. Thiobarbituric acid reactive substances (TBARS) analysis was performed as described by Buege and Aust (1978).The statistical analysis (ANOVA and Duncan grouping) was performed using SAS 6.12 package (SAS Institute, 1995) to identify the significant difference ($P<0.05$) between samples. All experiments in this study were conducted in replication.

Results and Discussion

In this study, ascorbic acid and the combination of butylated hydroxytoluene (BHT) and butylated hydroxyanisole (BHA) (in ratio of 1:1) were used as reference antioxidants due to its efficacy and solubility/polarity. Ascorbic acid is a water soluble natural identical antioxidant whereas BHT and BHA are fat soluble synthetic antioxidant. Figure 1 shows the peroxide value (POV) changes in respective chicken meatball sample (8±1 °C) during 12 days of chilled storage. POV is the most commonly used method and a good indicator for primary lipid autoxidation, which will express its value in milliequivalent active oxygen or peroxides per kilogram sample. Generally, all the samples with antioxidants/extracts did not showed significant increse ($P>0.05$) in POV throughout the shelf–life. Yet, the control sample (C) exhibited significant POV change ($P<0.05$) after every six days of storage duration. This indicated that the studied spice extracts (T1, T2 and T3) were able to lengthen the induction period (IP) of lipid oxidation and subsequently delay the occurrence of lipid deterioration. Samples with spice extracts and reference antioxidants (T1, T2, T3, T4 and T5) were found to have lower POV ($p<0.05$) as compared to the control sample after 6 days of storage. No significant differences in POV were observed between samples with extracts and antioxidants (T1, T2, T3, T4 and T5). The result indicated that clove buds extract, cinnamon bark extract and their combination (1:1) were good primary antioxidants able to inactivate free radicals that contribute to the formation of hydroperoxides even in low concentration (200 ppm).

Thiobarbituric acid reactive substances (TBARS) analysis is a most widely used analysis to quantify the content of malonaldehyde (a secondary lipid oxidation product) in meat products. Secondary lipid oxidation products like malonaldehyde (MDA) are always related to the development of rancidity in food. Figure 2 shows the TBA value changes in respective chicken meatball sample (8±1 °C) during 12 days of chilled storage. MDA content in all studied samples remained stable (P>0.05) throughout the shelf life study. Somehow, the control sample exhibited a higher TBA value (P<0.05) compared to other samples (T1,T2, T3 T4 and T5) throughout the storage. No significant differences in TBA values were observed between samples with extracts and antioxidants (T1, T2, T3, T4 and T5). The TBA analysis result indicated that clove buds extract, cinnamon bark extract and their combination (1:1) were efficient secondary antioxidants that were comparable to strong natural identical antioxidant (ascorbic acid) and synthetic antioxidant (BHA and BHT) at the same concentration in meat products. The major function of a secondary antioxidant is to prevent the decomposition of hydroperoxides in fat.

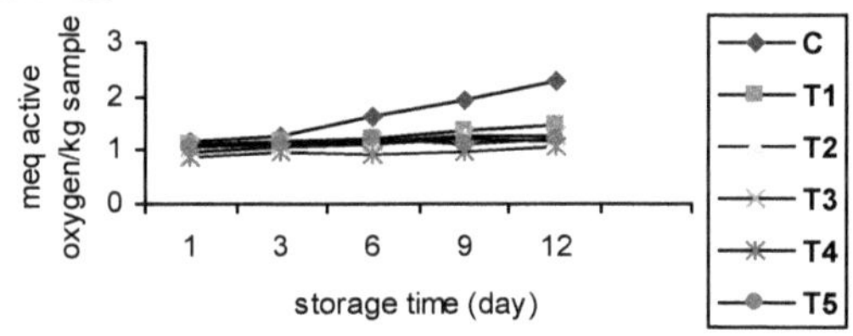

Figure 1: Effects of spice extracts and reference antioxidants on the peroxide value (POV) in chicken meatballs (8±1 °C). C: control meatball; T1: meatball + 200 ppm clove buds extract; T2: meatball + 200 ppm cinnamon bark extract; T3: meatball + 100 ppm clove buds extract + 100 ppm cinnamon extract ;T4: meatball + 200 ppm ascorbic acid;T5: meatball + 100 ppm BHA+100 ppm BHT.

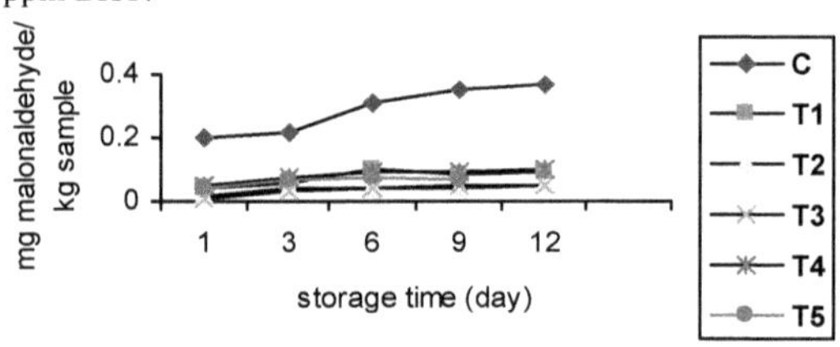

Figure 2: Effects of spice extracts and reference antioxidants on 2-thiobarbituric acid (TBA) value in chicken meatballs (8±1 °C). C: control meatball; T1:meatball + 200 ppm clove buds extract; T2: meatball + 200 ppm cinnamon bark extract; T3: meatball + 100 ppm clove buds extract + 100 ppm cinnamon extract ;T4: meatball + 200 ppm ascorbic acid;T5: meatball + 100 ppm BHA+100 ppm BHT.

Conclusions

This study concluded that clove buds and cinnamon bark water extracts were good primary and secondary antioxidants, which were comparative to ascorbic acid and synthetic antioxidants (BHA-BHT) at the same concentration. Yet, no significant synergistic antioxidant effect was observed in a mixture of both extracts (T3). The addition of 200 ppm of clove buds extract, cinnamon bark extracts or their combination (1:1) were able to improve the oxidative stability of chicken meatballs.

References

A.O.A.C. (1984). *Official Methods*. 14th ed. USA:.Association of Official Analytical Chemists, Inc.

Buckley, D.J., Morrissey, P.A. and Gray, J. I. (1995). Evalaution of warmed-over flavour during chill storage of cooked broiler breast, thigh and skin by chemical, instrumental and sensory methods. *J. Food Sci.* 55:644-648, 673

Buege, J.A. and Aust, S.D. (1978). Microsomal lipid peroxidation. *Methods in Enzymology*. 52: 306

IFST (1993). Shelf-life of foods- Guidelines for its determination and prediction, London:Institute of Food Science and Technology

Kinsella, J.E., Shimp, J.L., Mai, J. and Weihrauch, J. (1977). Fatty acid content and composition of freshwater in fin fish. *J. Am. Oil Chem. Soc.* 54:424-429.

Shahidi, F. and Wanasundra, P.K.J.P.D (1992). Phenolic antioxidant. CRC Critical Rev. Food Sci. Nutr. 32: 67-103

SAS (1995). *Statistical analysis system: User's guide statistics*. Version 6.21. Cary, NC:SAS Institute, Inc.

Verhagen, H., Deerenberg, I., Marx, A., Ten Hoor, F., Henderson, P.T. & Kleinjans, J.C.S. (1990). Estimation of daily dietary intake of butylated hydroxyanisole and butylated hydroxytoluene in Netherlands. *Food Chem. Tox.* 28: 181-183.

STABILITY OF ENDOGENOUS ANTIOXIDANT ENZYMES IN CHINESE SEMI-DRY SAUSAGE AS AFFECTED BY DRY-TEMPERATURE AND STORED TIME

T.-T. Chang and K.-J. Lin*

Department of Animal Science, National Chiayi University, 300 University Road, Chiayi 600, Taiwan, R.O.C. E-mail: kjlin@mail.ncyu.edu.tw

Keywords: Chinese semi-dry sausage, superoxide dismutase, glutathione peroxidase, catalase

Introduction

The muscle tissue contains various endogenous antioxidant compounds, which includes enzymic antioxidants, superoxide dismutase, glutathione peroxidase, catalase, which function as free radical scavenging to reach the oxidative stability of meat. The purpose of this study was to evaluate the stability of endogenous antioxidant enzymes in Chinese semi-dry sausage as affected by various drying temperatures (55, 50, 45 and 10°C) and when stored for 8 wks at 0-5°C.

Materials and Methods

(1) Raw materials: The chilled shoulder picnic (85/15) 85% and backfat (0/100) 15%, was purchased from Shang-Li Food Industry Company Ltd., Nantou, Taiwan, R.O.C., in a 100 kg meat block.
(2) Formula ingredients: The ingredients used are based on a 100 kg meat block, sugar 8%, sorbitol 3%, salt 1.5%, phosphate 0.25%, sodium nitrite 0.015%, Na-erythorbate 0.05%, glucono-lactone 0.1%, ethyl maltol 0.1%, nucleotide 0.05%, white pepper 0.2%, cinnamon 0.05%, garlic 0.05%, allspice 0.05%.
(3) Manufacturing procedures: The chilled shoulder picnic was ground through a 9mm plate, and the backfat cut into size 0.3 cm3. All the ingredients were mixed in a mixer for 5 min, stored at 0-5°C and cured for 3 days. The cured mixes were stuffed into size 30-32 mm hog casing, linked at 10 cm intervals, then dried by 4 various drying temperatures until the moisture content of the lean meat of sausage reached 45~48%; treatment 1 55°C hot-air drying for 3 hours; treatment 2 50°C hot-air drying 4 for hours; treatment 3 45°C hot-air drying for 5 hours; treatment 4 10°C, 80-85% relative humidity cold-air drying for 2 days. After cooling, the sausage was vacuum-packaged (vacuum degree -0.92bar), stored at 0-5°C cold storage, and sampled for determination at 0, 1, 2, 3, 4, 5, 6, 7, 8 weeks.
(4)Antioxidant enzymes activity determination: The endogenous antioxidant enzyme assay for all samples were performed as described by De Vore and Greene (1982); Günzler and Flohé (1985).
a. Superoxide dismutase (SOD) activity was performed by the riboflavin/nitro blue tetrazolium (NBT) photochemistry analysis method as described by Beauchamp and Fridovich (1971).
b. Glutathione peroxidase (GSH-Px) activity analysis was performed as described by De Vore and Greene (1982).
c. Catalase (CAT) activity analysis was performed as described by Aebi (1983) and Mei *et al.*, (1994).

Results and Discussion

Figure 1 shows the changes in absorbance variation of SOD of Chinese semi-dry sausage using various temperatures and stored for 8 wks at 0-5°C. The result showed that SOD activity of products dried using 10°C cold air were significantly higher than the others ($P<0.05$), and SOD activity of various drying temperature were significantly ($P<0.05$) decreased with the increased stored time. Figure 2 shows the changes in absorbance variation of GSH-Px of Chinese semi-dry sausage dried by various temperatures and stored for 8 wks at 0-5°C. The results show that GSH-Px activity of products by the cold air drying group at 0 wk being significantly ($P <0.05$) higher than the other hot air drying groups, and the GSH-Px activity of all treatments samples were significantly ($P<0.05$) decreased with the increased storage time, and GSH-Px activity by hot air drying groups was completely inactivated at 5 wks. Figure 3 shows the changes in absorbance variation of CAT of Chinese semi-dry sausage dried by various temperatures and stored for 8 wks at 0-5°C. The result showed that the CAT activity of all treatments samples had the same tendency, CAT activity of products dried by 10°C cold air was significantly ($P<0.05$) higher than the others, and CAT activity of all samples was significantly ($P<0.05$) decreased with the increased storage time. These results partly confirm previous observations by Mei *et al.*, (1994) who found that cooking ground pork and beef to an internal temperature of 70°C resulted in total inactivation of CAT and partial inactivation of GSH-Px , as GSH-Px ,which differed from CAT, was capable of reacting with both lipid and H_2O_2. On the other hand, during the later storage periods, Chinese semi-dry sausage may have started an auto-oxidative reaction which could produce more H_2O_2 than the Fenton reaction to cause the activity of GSH-Px of all tested samples to be completely inactivated at 5 wks.

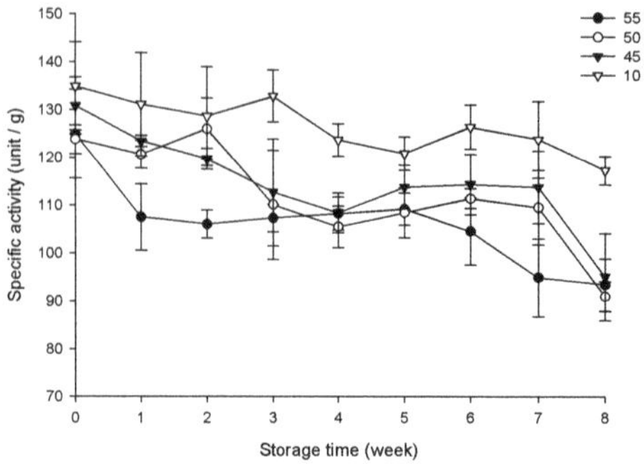

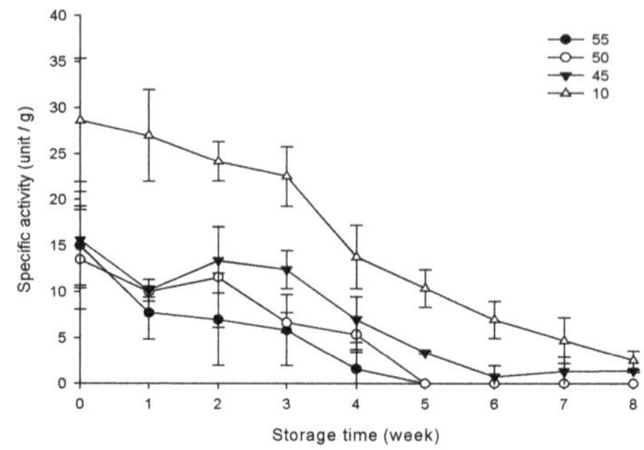

Figure 1 (Left): Changes in absorbance variation of superoxide dismutase from Chinese style semi-dry sausage dried by different temperatures and stored for 8 wks at 0-5°C.

Figure 2 (Right): Changes in absorbance variation of glutathione peroxidase from Chinese style semi-dry sausage dried by different temperatures and stored for 8 wks at 0-5°C.

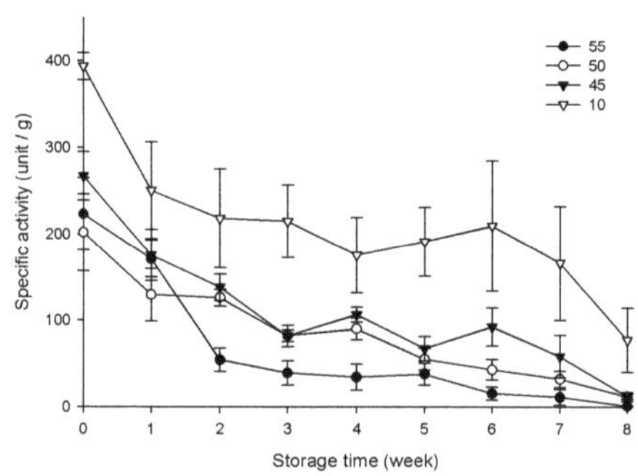

Figure 3: Changes in absorbance variation of catalase (CAT) from Chinese style semi-dry sausage dried by different temperatures and stored for 8 wks at 0-5°C.

Conclusions

The results show that the endogenous antioxidant enzyme activity in Chinese semi-dry sausages dried by 10°C cold air and 80-85% relative humidity were significantly ($P<0.05$) higher than the others, and GSH-Px of each group was completely inactivated following 5 weeks storage time. Although the antioxidant enzyme activity of all samples were significantly ($P<0.05$) decreased with the increased stored time, the stability of SOD is the highest among all tested samples.

References

Aebi, H. E. (1983). Catalase. In H. U. Bergmeyer (Eds.), Methods of enzymatic analysis, Vol. 3, pp. 273-286. Weinheim, Germany: Verlarg Chemie.

Beauchamp, C., and I. Fridovich. (1971). Superoxide dismutase : Improved assays and an assay applicable to acrylamide gel. Anal. Biochem. 44: 276-287.

De Vore, V. R., and Greene, B. E. (1982) Glutathione peroxidase in post-rigor bovine semitendinosus muscle. Journal of Food Science. 47: 1406-1409.

Gűnzler, A., and L. Flohé. (1985). Glutathione peroxidase. In Greenwald, R. A. (Eds.), CRC handbook of methods for oxygen radical research. Vol. 1, pp. 285-290. CRC Press Inc., Boca Raton, FL.

Mei, L., A. D. Curm, and E. A. Decker (1994). Development of lipid oxidation and inactivation of antioxidant enzymes in cooked pork and beef. J. Food Lipid. 1: 273-283.

EFFECT OF GARLIC ON LACTIC ACID AND BACTERIOCINS PRODUCTION OF BACTERIOCIN-PRODUCING LACTIC ACID BACTERIA ASSOCIATED IN NHAM (THAI FERMENTED MEAT) DURING FERMENTATION

A. Swetwiwathana*[1], N. Lotong[2], J. Nakayama[3] and K. Sonomoto[3]

[1.] *Department of Agro-industry, Faculty of Agricultural Technology, King Mongkut's Institute of Technology Ladkrabang (KMITL) Bangkok, 10520 Thailand.* [2.] *Departmnet of Microbiology, Faculty of Science, Kasetsart University, Bangkok, 10900 Thailand.* [3.] *Laboratory of Microbial Technology, Department of Bioscience and Biotechnology, Faculty of Agricultural Graduate School, Kyushu University. 6-10-1 Hakozaki, Higashi-ku, Fukuoka 812-8581, Japan. Email: adisorns@hotmail.com*

Keywords: garlic, nham, bacteriocin, pediocin PA-1, nisin Z

Introduction

Garlic, a flavouring food plant, is used as ingredient in various meat products, especially in Nham. The amount of garlic used in Nham is about 3-5%. Many reports are published on the effects of garlic on the growth of lactic acid bacteria (LAB) starter cultures and their lactic acid production in various kinds of fermented meat products including Nham (Swetwiwathana *et al.*, 1999a). Moreover, many publications implied some inhibitory effects of garlic on various food poisoning bacteria (Dababneh and Al-Delaimy, 1984; Swetwiwathana *et al.*, 1999b). Thus, we realized that garlic must also present some effects on lactic acid and even bacteriocins production of bacteriocin-producing LAB associated in Nham such as pediocin PA-1 producer (*Pediococcus pentosaceus* TISTR 536) and nisin Z producers (*Lactococcus lactis* subsp. *lactis* N100 and N190) (Swetwiwathana *et al.*, 2003). Hence, this paper is to report the effect of garlic on lactic acid and bacteriocins production from Nham-associated pediocin PA-1 and nisin Z producers during Nham fermentation.

Materials and Methods

Microorganisms: A strain of *P. pentosaceus* TISTR 536 (Bangkok, MIRCENS) and 2 strains of *Lc. lactis* subsp. *lactis* N100 and N190, which were isolated from Nham and were confirmed to produce pediocin PA-1 and nisin Z respectively (Swetwiwathana *et al.*, 2003), were used for this study. *Listeria innocua* LTH 3096 was used as the indicator for bacteriocin activity determination.

Medium: MRS broth was used as a cultivation medium for *P. pentosaceus* TISTR 536, *Lc. lactis* subsp. *lactis* N100 and N190. The incubation condition for all cultures was 30°C for 24 h. Trypticase soy broth + 0.6 % yeast extract (TSBYE) was used as a cultivation medium for *L. innocua* LTH 3096. The incubation condition for all cultures was 30°C for 24 h. Trypticase soy agar + 0.6 % yeast extract (TSAYE) was used as an agar base and Lactobacilli agar AOAC (LAA, Difco) was used for bacteriocin activity determination.

Study on the production of lactic acid and bacteriocin by *P. pentosaceus* TISTR 536, *L. lactis* subsp. *lactis* N100 and N190 during Nham fermentation:

Preparation of Nham was carried out in 1 kg batches per each strain of isolated bacteriocin-producing LAB from Nham as starter culture. The samples without starter culture (control batch) were also prepared for natural fermentation. The recipe of Nham was described in Swetwiwathana *et al.*, (1999 a, b)

The mixed ingredients of Nham samples (with and without garlic), except the control sample, were inoculated with either strain of *P. pentosaceus* TISTR 536, *L. lactis* subsp. *lactis* N100 or N190 at a level of 10^6 cfu/g and mixed with a KitchenAid model K45SS for 1 min. Approximately 25 g of both Nham control batch mixture without starter culture and Nham with starter cultures were stuffed into 10x14 cm plastic bags. Then, the mixture was squeezed to the closed bag end and tightly wrapped with a rubber band. All samples were fermented at 30°C for 3 days. Two bags per sample were examined every 12 h of fermentation for pH (Swetwiwathana *et al.*, 1999a), lactic acid production (Lotong and Swetwiwathana, 1990) and bacteriocin production.

Determination of the concentration of antimicrobial product during Nham fermentation:

Two bags per each studied condition of Nham samples were sampling for 10 g. The ratio of 1:1 dilution of Nham sample homogenates (Nham 10 g: distilled water 10 ml) were prepared. After centrifugation of each Nham sample homogenate (2,700 x g for 10 min), the supernatant was adjusted to pH 6 with 5 N NaOH and then filter-sterilized (0.20 µm pore-size polysulfone). Bacteriocins activity of each cell-free filtrate was determined by using spot-on-lawn technique as described by Mayr-Harting *et al.* (Mayr-Harting *et al.*, 1972).

Results and Discussion

Among the use of the three LAB strains (*P. pentosaceus* TISTR536, *L lactis* subsp. *lactis* N100 and N190) for Nham fermentation, pediocin PA-1 producer (*P. pentosaceus* TISTR536) was found to be the best strain as a starter culture of Nham production. The strain exhibited good results in decreasing pH, higher lactic acid production (Figure 1) and more stability of its pediocin PA-1 produced (Table 1), while nisin Z producer of both N100 and N190 showed less decrease in pH and lactic acid production including unstable bacteriocin production during Nham fermentation. This might be explained by the fact that the unstablity of nisin during Nham fermentation is due to a specific reaction with a compound of glutathione (GSH) in raw meat, which is the main material for Nham production. It has been proven that nisin was inactivated when bound to GSH which presents normally in animal tissues (Rose *et al.,* 1999). GSH is generally high with 156 to 627 nmol/g wet weight in beef, chicken and pork (Wierzbicka *et al.,* 1989). There is essentially no GSH in dairy products which would explain why nisin is a more effective antimicrobial agent in dairy products than in animal source of foods (Rose *et al.,* 1999; Wierzbicka *et al.,* 1989).

(a) **(b)** **(c)** **(d)**

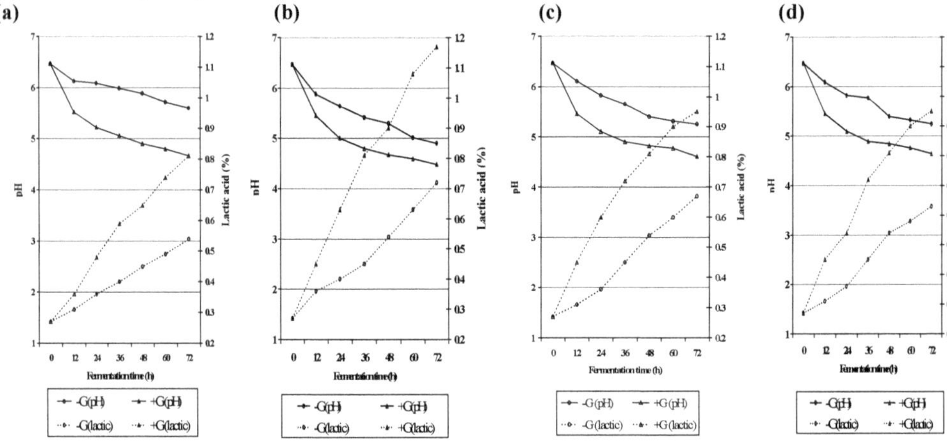

-G = no garlic added, + G = 5% fresh garlic was added

Figure 1: Effect of fresh garlic on the lactic acid production (%) and pH reduction during Nham fermentation using *P. pentosaceus* TISTR536 (b), *L. lactis* subsp. *lactis* N100 (c) and N190 (d) as starter cultures compared to control without starter culture (a).

Table 1: The production of pediocin PA-1 by *P. pentosaceus* TISTR 536 and nisin Z by *L. lactis* subsp. *lactis* N 100 and N 190 in Nham with and without garlic

Fermentation time (h)	Bacteriocin activity (AU/ml)*					
	Pediocin PA-1		Nisin Z (N100)		Nisin Z (N190)	
	Nham–G	Nham+G	Nham–G	Nham+G	Nham–G	Nham+G
0	0	0	0	0	0	0
12	0	100	0	200	0	200
24	100	400	100	800	200	800
36	200	1,600	400	1,600	800	1,600
48	400	3,200	1,600	800	1,600	800
60	800	3,200	400	400	400	200
72	1,600	3,200	100	200	0	100

* = *L. innocua* LTH 3096 was used as indicator strain , + G = with 5% fresh garlic , - G = without garlic

Conclusions

The study implies that the pediocin PA-1 producer of *P. pentosaceus* TISTR 536 showed more possibility to be used as starter culture for microbiological quality improvement of Nham than nisin Z producers of *L. lactis* subsp. *lactis* N100 and N190. Pediocin PA-1 was more stable after production during Nham fermentation than nisin Z.

References

Dababneh, B.F.A., and Al-Delaimy, K.S. 1984. Lebensm.-Wiss. U. –Technol. 17 : 29-31. 4.

Swetwiwathana, A., Leutz, U., and Fischer, A. 1999a. Fleischw. Int. 1/99 : 26-29.

Swetwiwathana, A., Leutz, U., Lotong, N., and Fischer, A. 1999b. Fleischw. 79 : 124-128.

Swetwiwathana, A. 2005. Microbiological Quality Enhancement of Thai Fermented Meat Product (Nham) Using Nham Associated Pediocin-producing Lactic Acid Bacteria (*P. pentosaceus* TISTR 536). Ph.D. Thesis of Agricultural Scienc, Kyushu University, Japan.

Lotong, N., and Swetwiwathana, A. 1990. Rep. ASEAN-Thailand Food Techn. Res. Dev. Proj. 1985-1990. p. 87-97.

Mayr-Harting, A., Hedges, A. J., and Berkeley, RCW. 1972. Meth. Microb. 7A : 315-422.

Rose, N.L., Sporns, P., Stiles, M.E., and McMullen, L.M. 1999. J. Food Sci. 64 : 759-762.

Wierzbicka, G.T., Hagen, T.M., and Jones, D.P. 1989. J. Food. Comp. Anal. 2 : 327-337.

PHYSICOCHEMICAL CHARACTERISTICS OF CHINESE MEATBALLS WITH MECHANICALLY DEBONED BULLFROG MEAT ADDED

F.J. Tan*[1], Y.J. Chen[1], J.M. Su[1], P.J. Shiu[1] and Y.C. Wu[2]

[1.] *Department of Animal Science, National University of Science and Technology, Pingtung, 912, Taiwan,* [2.] *Department of Animal Science, Tunghai University, Taichung, 407, Taiwan. Email: tanfj@mail.npust.edu.tw*

Keywords: mechanically deboned meat, *Rana catesbeiana*, bullfrog, Chinese meatball

Introduction

Mechanical deboning is removal of soft tissue from animal or poultry bone frames, which are meat processing by-products, with low commercial values, but still contain many useful and nutritional materials. The mechanical deboner produces a batter-type material which can be further utilized in food manufacturing. This increases the utilization and values of these byproducts (Hedrick *et al.*, 1994). Use of mechanically deboned meat in many products has been well documented (Thomsen and Zeuthen, 1988; Mielnik *et al.*, 2002; Daros *et al.*, 2005). Bullfrog (*Rana catesbeiana*) is a popular food material in Taiwan. Bullfrog legs are the only consumed parts; the other parts are dumped or used for animal feed. How to make good use of this animal byproduct and produce high quality Chinese meatballs, a very popular meat product in Taiwan and many other areas has become of interest to meat industries. The objective of this study was to evaluate the physicochemical qualities of Chinese meatballs containing various amounts of mechanically deboned bullfrog meat (MDBM) when the product was stored under refrigeration.

Materials and Methods

Raw delegged bullfrogs (*Rana catesbeiana*) were obtained from a local frog raising farm in Pingtung, Taiwan, mechanically deboned through a deboner to obtain mechanically deboned bullfrog meat (MDBM), and stored at -20°C prior to manufacturing meatballs. Fresh pork leg and pork back fat were ground utilizing a 3 mm plate. Meatball samples were formulated with 25% fat. For the samples with MDBM added, 7.5, 15, 22.5 or 30% MDBM were first mixed with ground pork, salt (1.8%) and phosphate (0.2%) for 3 min, then the fat (25%) and other non-meat ingredients including 3.5% sugar, 2.0% potato flour, 1.0% MSG, 0.3% white pepper powder, and 0.05% sodium erythorbate were added, and mixed thoroughly for approximately another 4 min. Then, the meat mixtures were manually formed into meatballs with approximate average weight of 15-25 g, cooked in a water bath at 75°C until the internal temperature of meatballs reached 70°C for 20 min, cooled in a ice bath for 10 min, dried at room temp for 10 min, packed in plastic bags, stored at 7°C for 14 days, and analyzed at day 0, 7 and 14. Moisture and ash contents were measured according to the AOAC (1980) method. Crude fat was measured using a fat extractor. Crude protein was measured using the Kjeldahl method. Water holding capacity was measured according to the Dagbjartsson and Solberg (1972) method. The L*, a*, and b* values, pH, cooking loss of samples were measured. TBA values of the samples were determined according to the methods described by Salih *et al.* (1987). Data was analyzed using the SAS software.

Results and Discussion

The results showed that meatballs that had more MDBM added tended to have higher moisture, and lower protein and fat contents (Table 1). Also, adding more MDBM resulted in a significantly lower L* colour values in the samples which indicated a darker colour while there were no significant differences in the a* and b* colour values among the samples. Adding more MDBM resulted in a significantly ($p < 0.05$) lower water holding capacity in the samples (Table 2) and adding more than 22.5% resulted in significantly ($p < 0.05$) lower water holding capacity. This reduction of water holding capacity of the product was probably because the MDBM texture was partially destroyed during mechanical separation and processing, thus resulting in less functionality of the raw materials. Adding more MDBM during manufacturing tended to increase the cooking loss of the meatballs (Table 2). For example, samples that had 30% MDBM added had significantly ($p < 0.05$) higher cooking loss of 7.11% when compared to samples that had 0 to 15% MDBM which resulted in losses between 4.27 to 4.70%. Higher cooking loss probably resulted from the lower water holding capacity which resulted in higher weight losses of the samples, and probably would result in less juiciness and tenderness of the products. This increase of cooking loss may also be due to the samples that had more MDBM added which resulted in more destroyed texture material, and lower water holding capacity. However, adding MDBM up to 15% did not significantly affect the cooking losses of the samples. The data showed that the pH values of Chinese meatballs decreased when stored under refrigeration for 14 days (Table 3). In addition, adding more MDBM during manufacturing resulted in a significantly ($p < 0.05$) higher pH value of the products during 14 days of refrigerated storage. As expected, the TBA values of Chinese meatball samples increased with storage time, indicating that lipid oxidation had occurred in the meatball samples during refrigerated storage. The data also indicated that the samples with no MDBM added had significantly ($p < 0.05$) lower TBA values than the samples with MDBM added. In addition, samples that had 7.5 and 15 % MDBM added had significantly ($p < 0.05$) lower TBA values than those with 22.5 or 30 % MDBM added during the refrigerated storage. This result indicated adding MDBM up to 22.5 and 30% might result in a higher lipid oxidation which was undesirable.

Table 1: Proximate composition and colour values of Chinese meatballs with various amounts of mechanically deboned bullfrog meat (MDBM) added.

Items	Mechanically deboned bullfrog meat added (%) 0	7.5	15.0	22.5	30.0
Moisture	59.95 ± 0.75^{w}	60.61 ± 0.74^{wx}	61.03 ± 0.80^{xy}	61.43 ± 0.69^{xy}	62.43 ± 1.24^{y}
Protein	24.07 ± 0.46^{w}	23.76 ± 0.35^{wx}	23.39 ± 0.69^{x}	22.56 ± 0.36^{y}	21.87 ± 0.20^{z}
Fat	17.22 ± 0.88^{w}	15.44 ± 0.29^{x}	15.04 ± 0.26^{x}	14.23 ± 0.14^{y}	14.16 ± 0.04^{y}
Ash	2.40 ± 0.45	2.03 ± 0.97	1.99 ± 0.46	1.93 ± 0.46	1.92 ± 0.27
L* value	63.66 ± 1.17^{w}	62.46 ± 2.08^{wx}	62.23 ± 1.28^{wx}	60.82 ± 0.76^{x}	59.14 ± 1.29^{y}
a* value	2.82 ± 0.78	2.35 ± 0.14	2.51 ± 0.11	2.50 ± 0.04	2.62 ± 0.05
b* value	11.54 ± 0.07	11.62 ± 0.46	11.64 ± 0.31	11.71 ± 0.48	11.79 ± 0.40

$^{w, x, y, z}$Means within a row that have different superscripts are significantly different ($p < 0.05$).

Table 2: Changes in water holding capacity and cooking loss of Chinese meatballs with various amounts of mechanically deboned bullfrog meat (MDBM) added during refrigerated storage.

MDBM added (%)	Water holding capacity	Cooking loss
0	37.30 ± 4.99^{a}	4.27 ± 0.71^{a}
7.5	30.21 ± 2.60^{b}	4.33 ± 0.96^{a}
15.0	27.71 ± 5.69^{bc}	4.70 ± 1.20^{a}
22.5	24.29 ± 3.99^{cd}	6.14 ± 2.35^{ab}
30.0	21.91 ± 3.65^{d}	7.11 ± 2.11^{b}

$^{a, b, c, d}$Means within a column that have different superscripts are significantly different ($p < 0.05$).

Table 3: Changes in pH and TBA values of Chinese meatball with various amounts of mechanically deboned bullfrog meat (MDBM) added during refrigerated storage.

MDBM	Storage time (day)			Storage time (day)		
added (%)	0	7	14	0	7	14
		pH values		TBA (mg malonaldehyde/kg sample)		
0	6.37 ± 0.06^{az}	6.11 ± 0.04^{ay}	5.81 ± 0.05^{ax}	1.59 ± 0.64^{dx}	1.70 ± 0.14^{dx}	1.89 ± 0.18^{dx}
7.5	6.43 ± 0.02^{bz}	6.17 ± 0.00^{by}	5.87 ± 0.02^{bx}	3.85 ± 0.65^{cy}	3.93 ± 0.14^{cy}	4.17 ± 0.11^{cxy}
15.0	6.47 ± 0.0^{bcz}	6.22 ± 0.01^{cy}	5.99 ± 0.01^{cx}	4.09 ± 0.73^{cy}	4.19 ± 0.12^{cy}	4.38 ± 0.06^{cxy}
22.5	6.51 ± 0.01^{cdz}	6.29 ± 0.02^{dy}	6.02 ± 0.01^{cx}	5.20 ± 0.87^{bx}	5.40 ± 0.37^{bx}	5.60 ± 0.46^{bx}
30.0	6.54 ± 0.01^{dz}	6.37 ± 0.02^{ey}	6.08 ± 0.01^{dx}	6.50 ± 1.22^{ay}	6.58 ± 0.80^{ay}	6.77 ± 1.01^{axy}

$^{a, b, c, d}$Means within a column in the same test that have different superscripts are significantly different ($p < 0.05$).
$^{x, y, z}$Means within a row in the same test that have different superscripts are significantly different ($p < 0.05$).

Conclusions

In conclusion, Chinese meatballs that had 7.5% of mechanically deboned *Rana catesbeiana* meat added during manufacturing was acceptable based on the physic-chemical qualities.

References

Daros, F.G., Masson, M.L. Amico, S.C. (2005). The influence of the addition of mechanically deboned poultry meat on the rheological properties of sausage. Journal of Food Engineering, 68: 185-189.

Hedrick, H.B., Aberle, E.D., Forrest, J.C., Judge, M.D., Merkel, R.A. (1994). Principles of meat science, 3 ed. Kendall/Hunt Publishing Company, Dubuque, Iowa.

Mielnik, M.B., Aaby, K., Rolfsen, K., Ellekjr, M.R., Nilsson, A. (2002). Quality of comminuted sausages formulated form mechanically deboned poultry meat. Meat Science, 61: 73-84.

Thomsen, H.H., Zeuthen, P. (1988). The influence of mechanically deboned meat and pH on the water-holding capacity and texture of emulsion type meat products. Meat Science, 22: 189-201.

EVALUATION OF *TOONA SINENSIS* WATER EXTRACT ON PHYSICO-CHEMICAL CHARACTERISTICS OF PORK PATTIES DURING STORAGE

F.J. Tan* and S.Y. Lu

Department of Animal Science, National University of Science and Technology, Pingtung, 912, Taiwan.
Email: tanfj@mail.npust.edu.tw

Keywords: *Toona sinensis*, physico-chemical, pork, patty

Introduction

Recently, antimicrobial and antioxidant properties of plant and spice extract have been widely reported. *Toona sinensis* (TS) is a perennial tree that has become widely grown in Taiwan and China (Edmonds and Staniforth, 1998). Its tender buds have a special aroma and are often consumed in Taiwan. In addition, TS has also been reported to have some antimicrobial activity (Shi, 2003). The objective of this study was to evaluate the qualities of pork patties which had *Toona sinensis* water extract added in the formula and then were stored at refrigeration temperatures.

Materials and Methods

Lean tissue from fresh pork legs and pork back fat were ground twice through a 6 mm plate. The formula included 80% fresh legs, 20% pork back fat, 2.4% sugar, 2.0 % corn starch, 2.0% yam starch, 1.2% salt, 1.0% monosodium glutamate, 0.2% garlic powder, 0.2% onion powder, 0.2% dried celery powder, 0.2% black pepper powder, and 0.1% polyphosphate. Lean tissue was first mixed with salt and phosphate for 5 min. After adding fat and other ingredients, it was mixed for another 6 min. For those samples containing TS, water extract was added at the rate of, 0.4, 0.6, or 0.8% (v/w) and mixed with the meat mixture. Then, the mixtures were manually formed into patties that had approximate thickness of 1.5 cm and a diameter of 8 cm and an average weight of 90±5g. Pork patties were packaged in polyethylene film, and stored at 4°C for 14 days, and analyzed at day 0, 7, and 14. Fresh tender buds of TS were washed, dried in an oven at 60°C for 8 hrs, and ground into powder. Ten grams of TS powders were mixed with 200 ml distilled water, heated in a water bath at 50°C for 24 hrs and then filtered. The filtered residues of water extract were mixed with another 200 ml distilled water, heated, filtered, collected and added to the meat mix. Moisture and ash contents were measured according to the AOAC (1990) method. Crude fat was measured using a fat extractor. Crude protein was measured using the Kjeldahl method. The Hunter L*, a*, and b* values, and pH values of the samples were measured. TBA values of the samples were determined according to the methods described by Salih *et al.*, (1987). Volatile basic nitrogen was determined by the Conway micropipette diffusion method. The data was analyzed using the SAS software.

Results and Discussion

Adding TS water extract in the formula up to 0.8% did not significantly affect the proximate composition of the pork patties (Table 1). The pH values of the products decreased significantly and reached approximate 5.7 after 14 days of refrigerated storage (Table 2). Adding TS water extract up to 0.8% did not significantly affect the pH values of the pork patties. During refrigerated storage, the L* colour values remained stable or slightly increased but with only small changes in values (Table 3). Adding more TS water extract, especially adding 0.8% extract, tended to decrease the L* colour values of the products. After refrigerated storage for 7 days, the samples had significantly lower a* colour values which indicated less red colour when compared with the samples at day 0. Samples that had more TS water extract tended to have lower L* and higher a* colour values, which indicated samples had a darker and a redder color. The TBA values significantly increased as the storage time increased which indicated more fat rancidity occurred as expected (Table 4). Adding more TS water extract tended to decrease the TBA values of the products. This decrease of TBA values was more obvious when adding more than 0.6% extract and stored under refrigeration for 14 days. Hsieh *et al.* (2004) indicated that methyl gallate which was extracted from *Toona sinensis* was an effective antioxidant compound. The VBN values increased significantly especially after storage for 14 days (Table 4). Adding more TS water extract tended to decrease the VBN values of the products especially when 0.8% extract was added.

Conclusions

In conclusion, adding *Toona sinensis* water extract up to 0.8% did not significantly affect the proximate composition and pH values of pork patties. Samples with more extract had darker and redder colors. Adding more extract tended to decrease the TBA and VBN of the products.

Acknowledgements

This work was supported by grant NSC 93-2313-B-020-028 from the National Science Council, Taiwan, ROC.

Table 1: Proximate composition of pork patties containing *Toona sinensis* leaf water extract.

	Extract added (v/w)			
Parameter	0%	0.4%	0.6%	0.8%
Moisture (%)	58.88 ± 1.11	59.02 ± 1.42	58.57 ± 0.80	59.85 ± 1.18
Ash (%)	3.33 ± 0.30	3.32 ± 0.20	3.10 ± 0.40	3.40 ± 0.32
Crude fat (%)	11.26 ± 1.06	12.02 ± 1.66	10.18 ± 0.24	10.98 ± 0.46
Crude protein (%)	21.06 ± 0.39	20.79 ± 0.33	21.29 ± 0.58	21.04 ± 1.03

No significant differences ($P > 0.05$) were found due to added extract

Table 2: Changes in pH values of pork patties containing *Toona sinensis* leaf water extract during refrigerated storage.

	Extract added (v/w)			
Storage time (day)	0%	0.4%	0.6%	0.8%
0	6.06 ± 0.26^{a}	6.09 ± 0.24^{a}	6.13 ± 0.19^{a}	6.12 ± 0.18^{a}
7	6.11 ± 0.19^{a}	6.11 ± 0.13^{a}	6.18 ± 0.06^{a}	6.19 ± 0.02^{a}
14	5.67 ± 0.20^{b}	5.66 ± 0.26^{b}	5.67 ± 0.28^{b}	5.67 ± 0.30^{b}

$^{a\text{-}b}$ values within a column that have different superscripts are significantly different ($P < 0.05$).

Table 3: Changes in colour values of pork patties containing *Toona sinensis* leaf water extract during refrigerated storage.

Storage time	Extract added (v/w)			
(day)	$0\%^{a,b}$	0.4%	0.6%	0.8%
	L* values			
0	56.69 ± 1.49^{bx}	56.68 ± 1.33^{x}	56.03 ± 2.08^{x}	53.69 ± 2.51^{by}
7	58.95 ± 1.01^{ax}	57.41 ± 1.98^{xy}	57.12 ± 2.70^{xy}	55.58 ± 2.45^{aby}
14	59.32 ± 1.65^{a}	56.93 ± 1.60	56.48 ± 1.59	55.88 ± 2.12^{a}
	a* values			
0	7.88 ± 0.22^{ax}	7.66 ± 0.45^{ay}	7.56 ± 0.42^{axy}	8.49 ± 0.22^{ax}
7	5.60 ± 0.16^{by}	6.16 ± 0.15^{bx}	6.10 ± 0.21^{cx}	5.83 ± 0.04^{cx}
14	5.48 ± 0.91^{by}	6.40 ± 0.28^{bx}	6.59 ± 0.87^{bx}	6.34 ± 0.15^{bx}
	b* values			
0	13.52 ± 0.16^{ax}	13.28 ± 0.12^{ay}	13.33 ± 0.19^{axy}	12.15 ± 0.34^{ay}
7	12.87 ± 0.11^{bx}	12.90 ± 0.20^{bx}	12.85 ± 0.20^{bx}	12.49 ± 0.44^{by}
14	12.82 ± 0.31^{b}	12.68 ± 0.18^{c}	12.72 ± 0.09^{b}	12.66 ± 0.14^{b}

$^{a\text{-}c}$ values within a column in the same test that have different superscripts are significantly different ($P < 0.05$).
$^{x\text{-}y}$ values within a row in the same test that have different superscripts are significantly different ($P < 0.05$).

Table 4: Changes in TBA and VBN values of pork patties containing *Toona sinensis* leaf water extract during refrigerated storage

Storage time (day)	Extract added (v/w)			
	0%	0.4%	0.6%	0.8%
	TBA (mg malonaldehyde / kg)			
0	0.89 ± 0.63^{c}	0.67 ± 0.49^{c}	0.61 ± 0.45^{c}	0.58 ± 0.43^{c}
7	2.11 ± 0.75^{b}	1.81 ± 0.58^{b}	1.62 ± 0.45^{b}	1.55 ± 0.44^{b}
14	3.25 ± 0.22^{ax}	2.89 ± 0.05^{ax}	2.79 ± 0.02^{ay}	2.26 ± 0.44^{az}
	VBN (mg / 100g)			
0	6.58 ± 0.70^{bx}	6.24 ± 0.80^{bx}	3.81 ± 1.33^{by}	3.81 ± 1.33^{y}
7	6.58 ± 0.13^{bx}	6.53 ± 0.70^{bx}	6.24 ± 0.80^{abx}	4.51 ± 0.69^{y}
14	9.01 ± 1.39^{ax}	8.67 ± 0.69^{ax}	9.01 ± 0.80^{ax}	5.54 ± 1.13^{y}

$^{a\text{-}c}$ values within a column in the same test that have different superscripts are significantly different ($P < 0.05$).
$^{x\text{-}y}$ values within a row in the same test that have different superscripts are significantly different ($P < 0.05$).

References

Edmonds, J.M., Staniforth, M. (1998). *Toona sinensis*. Curtis's Botanical Magazine, 15: 186-193.

Hsieh, T.J., Liu, T.Z. Chia, Y.C., Chern, C.L., Lu, F. J. Chuang, M.C., Mau, S.Y., Chen, S.H., Syu, Y.H., Chen, C.H. (2004). Protective effect of methyl gallate from *Toona sinensis* (Meliaceae) against hydrogen peroxide-induced oxidative stress and DNA damage in MDCK cells. Food and Chemical Toxicology, 42: 843-850.

Shi, Y.F. (2003). Herb knowledge guide. Suncolor Culture Co., Ltd., Taipei, Taiwan.

SENSORY AND COLOUR CHARACTERISTICS OF BULGARIAN DRY-FERMENTED SAUSAGES WITH ADDITION OF NATURAL ANTIOXIDANTS

S. Dragoev* and D. Balev

University of food technologies – Plovdiv, Department "Meat and Fish Technology", 26 Maritza blvd., Plovdiv, 4002, Bulgaria, E-mail: stefan_dragoev@yahoo.co.uk

Keywords: dry fermented sausage, antioxidant, sensory characteristics, colour

Introduction

The typical taste and flavour of dry-fermented sausages is due to the production materials, derivatives from carbohydrate fermentation, lipid hydrolysis and oxidation, proteolysis and spices and salts, used as preservatives (Demeyer *et al.*, 1986). The protein dissociation in the dry-fermented sausages has partially microbial and endogen origin (Igene and Pearson, 1979). The production of free amino acids and short-chain peptides is significant for dry fermented sausage flavour improvement, however their role in the flavour formation seems to be limited in comparison with lipid hydrolysis and oxidation (Berdague *et al.*, 1991).

The objective of this study was to estimate the efficiency of natural antioxidants, on the sensory and colour characteristics of Bulgarian dry-fermented "lukanka" sausages.

Materials and Methods

The experiments were carried out using "Monastery's lukanka". The model system "lukanka" was prepared from:

a) Product produced from meat raw materials chilled at 0 – 4°C: 1) beef type CL 95 stored 72 h *post mortem* and frozen to minus 5°C 6 h prior to mincing; 2) pork - sort 50/50, stored 72 h *post mortem* and frozen to minus 10°C 6 h prior to cutting; 3) and back bacon stored for 24 h *post mortem* frozen to -10°C, prior to use. Parallel samples were produced: **Control sample** –without antioxidants; **Sample No 1** – with the addition of 1,124 g/kg natural antioxidants composition (Liquid form No 1), liquid rosemary extract in combination with 35.48% rutine and 16.13 % sodium erythrobate. **Sample No 2** – with the addition of 1,112 g/kg natural antioxidants composition (Dry form No 2), dry rosemary extract in combination with 78.57% dry extract of Japanese acacia (*Sophora japonica*) flower bud (containing 53.33% quercetin – aglicon of the natural glycoside rutine) and 17.85% sodium erythrobate.

b) Product produced from meat raw materials stored for three months at minus 18°C: 1) beef type CL 95; 2) pork – sort 50/50 and back bacon. Parallel samples were produced: **Control sample** – without antioxidants; **Sample No 1** – with the addition of 1,124 g/kg natural antioxidants composition (Liquid form No 1), liquid rosemary extract in combination with 35,48 % rutine and 16,13 % sodium erythrobate. **Sample No 2** – with the addition of 1,112 g/kg natural antioxidants composition (Dry form No 2), dry rosemary extract in combination with 78,57 % dry extract of Japanese acacia (*Sophora japonica*) flower bud (containing 53,33 % quercetin – quercetin – aglicon of the natural glycoside rutine) and 17,85 % sodium erythrobate.

The sensory characteristics of the "lukanka" samples were assessed using the nine point examination scale developed by Higher Scientific Research Institute of Meat Processing Industry, Moscow, Russia and adapted to traditional Bulgarian meat products.

Instrumental colour properties of the "lukanka" cross-section surface were estimated according to the CYELAB system using a PAY UNICAM spectrophotometer (Model 8800, Perkin Elmer, UK). Brightness - L*, red colour component - a*, yellow colour component - b*, colour tone - C and colour saturation – H were recorded on days 0, 15 and 30 after production.

Results and Discussion

The sensory characteristics of the experimental samples were rated higher than the control samples (table. 1). Statistically significant differences were recorded between sensory texture of the control sample and sample No 2. Sensory evaluation scores of "lukanka" produced from chilled meat raw materials, with the addition of blend of antioxidants in liquid and dry form were not significantly different. In the control samples the appearance and colour, cross-section surface colour, taste and flavour were all rated lower than experimental samples. In samples containing liquid antioxidants uncharacteristic flavour was detected. In samples prepared using frozen meat raw materials a slight flavour and precursor of rancidity was detected. This is referred to as "WOF" (Mottram and Whitfield, 1995). The taste of samples produced from chilled raw materials and dry blend antioxidants scored the highest. The texture of the product was not influenced by the addition of antioxidants ($P > 0,05$).

The appearance of the experimental samples scored higher than the control sample ($P < 0,05$), which possessed darker colour surface and had a greater presence of "white noble mould". This result indicates that the addition of natural antioxidant blends, irrespective of the raw material state (frozen or chilled) or form of the blend (dry or liquid), improves the product appearance.

Surface section colour rated higher in samples produced from chilled meat raw materials. Therefore, it can be concluded that the addition of antioxidants improves surface section colour as evaluated by sensory panellists.

Table 1: Sensory characteristics of "lukanka".

Sample	Appearance and colour	Surface section colour	Taste	Flavour	Consistence
		Frozen meat is used as starting raw material			
Control sample	6,84 ± 0,11	6,91 ± 0,15	6,45 ± 0,11	6,52 ± 0,15	7,61 ± 0,08
Sample No 1	7,41 ± 0,14	7,32 ± 0,11	7,12 ± 0,10	7,08 ± 0,12	7,74 ± 0,12
Sample No 2	7,20 ± 0,14	7,48 ± 0,12	7,62 ± 0,11	7,71 ± 0,14	7,65 ± 0,10
		Chilled meat is used as starting raw material			
Control sample	7,95 ± 0,11	7,62 ± 0,14	7,48 ± 0,11	7,81 ± 0,12	8,04 ± 0,10
Sample No 1	8,81 ± 0,14	8,31 ± 0,15	8,12 ± 0,10	8,05 ± 0,11	8,14 ± 0,16
Sample No 2	8,76 ± 0,11	8,40 ± 0,11	8,61 ± 0,10	8,64 ± 0,13	8,18 ± 0,12

Table 2: Instrumental colour properties of "lukanka" cross-section surface.

Sample	Day d	Brightness (L*)	Red colour component (a*)	Yellow colour component (b*)	Colour tone (C)	Colour saturation (H)
			Frozen meat is used as starting raw material			
Control sample	0	38,93 ± 0,25	9,61 ± 0,12	12,46 ± 0,11	19,20 ± 0,32	40,46 ± 0,38
	15	37,71 ± 0,22	10,52 ± 0,37	9,64 ± 0,11	14,96 ± 0,29	39,71 ± 0,34
	30	33,84 ± 0,17	11,54 ± 0,41	9,12 ± 0,25	14,05 ± 0,31	38,72 ± 0,56
Sample No 1	0	38,93 ± 0,25	9,61 ± 0,12	12,46 ± 0,11	19,20 ± 0,32	40,46 ± 0,38
	15	37,98 ± 0,22	12,86 ± 0,16	11,47 ± 0,19	17,23 [b] ± 0,21	39,75 [c] ± 0,27
	30	35,12 ± 0,11	13,30 ± 0,21	10,87 ± 0,14	17,17 [b] ± 0,13	39,27 [c] ± 0,22
Sample No 2	0	38,93 ± 0,25	9,61 ± 0,12	12,46 ± 0,11	19,20 ± 0,32	40,46 ± 0,38
	15	37,28 ± 0,16	13,00 [a] ± 0,20	16,23 ± 0,21	18,80 ± 0,19	39,31 [d] ± 0,27
	30	33,47 ± 0,24	13,76 [a] ± 0,17	11,15 ± 0,20	17,70 ± 0,24	39,02 [d] ± 0,31
			Chilled meat is used as starting raw material			
Control sample	0	48,21 ± 0,19	10,54 ± 0,14	13,65 ± 0,21	18,56 ± 0,22	51,21 ± 0,29
	15	47,31 ± 0,25	12,31 ± 0,26	10,23 ± 0,18	15,21 ± 0,19	40,95 ± 0,23
	30	43,57 ± 0,27	13,78 ± 0,21	11,04 ± 0,22	14,98 ± 0,23	32,14 ± 0,20
Sample No 1	0	48,21 ± 0,19	10,54 ± 0,14	13,65 ± 0,21	18,56 [b] ± 0,22	51,21 ± 0,29
	15	47,92 ± 0,22	11,95 ± 0,19	10,89 ± 0,17	18,25 [b] ± 0,21	42,84 ± 0,17
	30	46,12 ± 0,11	13,65 ± 0,19	11,24 ± 0,24	17,41 ± 0,20	38,51 ± 0,26
Sample No 2	0	48,21 ± 0,19	10,54 ± 0,14	13,65 ± 0,21	18,56 ± 0,22	51,21 ± 0,29
	15	47,01 ± 0,26	12,85 ± 0,21	11,65 [a] ± 0,24	17,21 [c] ± 0,19	46,28 ± 0,21
	30	45,51 ± 0,24	13,21 ± 0,18	11,24 [a] ± 0,20	17,02 [c] ± 0,26	39,51 ± 0,21

Statistically significant differences are estimated between the samples, marked with letter index.

The L* value or brightness of the samples produced from frozen meat raw materials differed significantly ($P < 0.05$) from samples produced using chilled meat raw materials. A trend towards colour brightness reduction was detected irrespective of the type of the blend used (table.2). Similar trends occurred in values recorded for yellow colour component b*, colour tone C and colour saturation H. The red colour a* of products increased, and this increase was statistically significant in samples produced from chilled raw materials. The same trend in colour changes was observed for all experimental samples whether chilled or frozen meat raw material was used.

Conclusion

The addition of natural antioxidant blends, in both dry and liquid form led to more acceptable taste and flavour development and preserved the preferred colour properties of the dry-fermented sausages.

References

Berdague, J. L., Denoyer C., Le Quere J.L. and Semon E. (1991). Volatile components of dry-cured ham, J. Agric. Food Chem., 39 (12): 1257 - 1261.

Demeyer, D., Verplaetse A. and Gistelinck M. (1986). Fermentation of meat: un integrated process, Proc. 32-th EMMRW, Gent, Belgium, pp. 241 - 247.

Igene, J.O. and Pearson A.M. (1979). Role of phospholipids and triglycerides in warmed over flavour development in meat model systems, J. Food Sci., 44 (6): 1285 - 1290.

Mottram D.S., Whitfield, F.B. (1995). Volatile compounds from the reaction of cycteine, ribose, and phospholipids in low-moisture systems, J. Sci. Food Agric., 43 (4): 984 – 988.

PHEFFECT OF NATURAL ANTIOXIDANTS ON LIPOLYSIS AND PH OF DRY-FERMENTED SAUSAGES "LUKANKA" TYPE

S. Dragoev* and D. Balev

University of food technologies – Plovdiv, Department "Meat and Fish Technology", 26 Maritza blvd., Plovdiv, 4002, Bulgaria, E-mail: stefan_dragoev@yahoo.co.uk

Keywords: dry fermented sausage, antioxidant, lipolysis, pH

Introduction

Recently, scientists are directing their efforts towards clarification of the questions: which features and taste properties of dry-fermented sausages are formed as a result of the lactic-acid bacteria activity; which are due to the protein and lipids hydrolysis and to what extent are these factors mutually dependent? To a great extent the lipolysis in the dry-fermented sausages is passing during the drying and maturing processes and is a result most of all of the endogen lipases found in muscles and fats (Johansson *et al.,* 2000). In order to produce the desired taste and flavour it is necessary for the free fatty and free amino acids to go through consecutive chemical transformations (Diaz *et al.,* 2005). The aim of this study was to establish the efficiency of a blend of natural antioxidants, on the lipolysis depth and the pH changes in the Bulgarian dry-fermented "lukanka" sausage.

Materials and Methods

The experiments were carried out with "Monastery's lukanka". The model system "lukanka" was prepared from:
a) Product produced from meat raw materials chilled at 0 - 4°C: 1) beef type CL 95 stored 72 h *post mortem* and frozen to -5°C 6 h prior to mincing; 2) pork - sort 50/50, stored for 72h *post mortem* and frozen to -10°C 6h prior to cutting; 3) back bacon stored for 24 h *post mortem* frozen to -10°C, prior to use. Parallel samples were produced: **Control sample** – without antioxidants; **Sample No 1** – with the addition of 1,124 g/kg blend of natural antioxidants (Liquid form No 1), liquid rosemary extract in combination with 35.48% rutine and 16.13% sodium erythrobate. **Sample No 2** – with the addition of 1,112 g/kg blend of natural antioxidants (Dry form No 2), dry rosemary extract in combination with 78.57% dry extract of Japanese acacia (*Sophora japonica*) flower bud (containing 53.33% quercetin – aglicon of the natural glycoside rutine) and 17.85% sodium erythrobate.
b) Product produced from meat raw materials stored three months at minus 18°C: 1) beef type CL 95; 2) pork – sort 50/50 and back bacon. The same three parallel samples were produced.
For the laboratory analysis the following reagents were purchased: chloroform and n-hexane (Sigma Chemical Company Ltd. St. Louis, MO 63178, USA, affiliate Deisenhofen, Germany), silver nitrate (Fluka Chemie AG Buchs CH-9470, Switzerland). The remaining chemicals were AR or GPL quality and were supplied by an affiliate of "Aldrich Chemical" Co. (Gillingham, Dorset SP8 4JL UK in Germany D-89555 Steinheim, Germany). The total lipids from the sausages were extracted by the method of Bligh and Dyer (1959). The degree of lipolysis in the extracted lipids was determined by the acid value (AOAC, 1980), and pH was measured using a pH-meter (Microsyst MS 2004, Microsyst, Plovdiv), equipped with combined pH electrode (Sensorex Combination Recorder 450 CD, pH Electrode Station, CA 90680, USA).

Results and Discussion

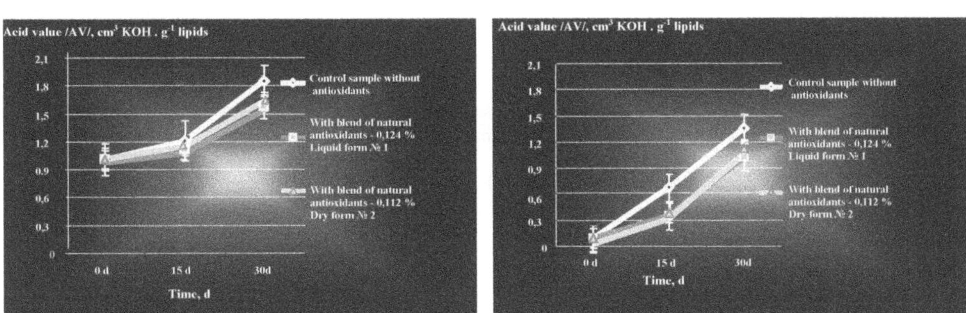

Frozen meat raw materials — Chilled meat raw materials

Figure 1: Acid value of the lipids extracted from "lukanka" with the addition of blend of natural antioxidants.

The application of natural antioxidants blend suppresses the lipolytic processes during the "Monastery's lukanka" production (Figure 1). In the experiments with frozen meat raw materials, the acid value of the test samples was statistically significantly lower ($P < 0.05$) as late as on day 30 of the experiment, whilst in the tests with chilled meat raw

materials statistically significant differences ($P < 0.05$) were estimated no more than the day 15 of the experiment. According to this indicator, both variances of the composition show similar properties (Figure 1).
Irrespective of the significant reduction in free fatty acids levels, the acid value of the final product "lukanka" produced from frozen meat raw materials is higher than that critical for maximum product quality - 1 cm^3 NaOH/g lipids (Dominguez Fernandez and Zumalacarregui Rodrigez, 1991). In the "lukanka" produced from chilled meat with the addition of antioxidants, similar composition critical values are estimated in the final product on day 30 of the experiment (Motilva *et al.*, 1993).
The addition of blend of natural antioxidants causes certain pH reduction of "Monastery's lukanka" during the drying and maturing processes when the final product is produced from frozen meat raw materials (Fig 2). As the meat is in the process of autolysis (maturing) pH of the filling mass is comparatively high 6.3.

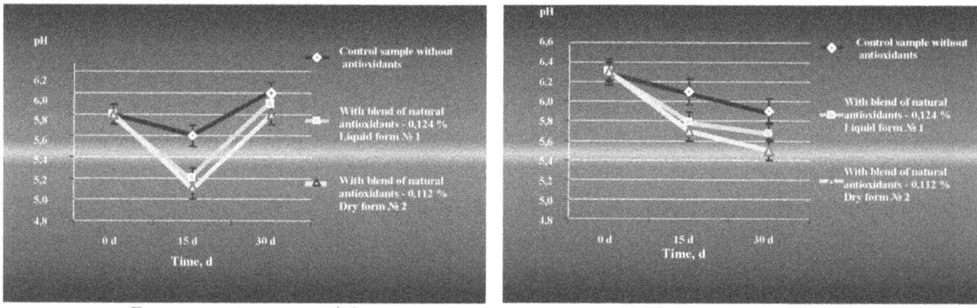

Frozen meat raw materials | Chilled meat raw materials

Figure 2: pH of "lukanka" with the addition of blend of natural antioxidants.

During the drying, pH decreases and in the final product (day 30) it reaches approximately 5.9 in the control sample. The pH of the experimental samples decreases faster and on the day 15 in Sample No 1 (with the addition of antioxidants composition) is approximately 5.8, and 5.7 in sample No 2. This trend goes deeper in the final product (day 30). The addition of natural antioxidant blend causes insignificant pH reduction of the "Monastery's lukanka" during the drying and maturing processes, when the product is produced from chilled meat raw materials (Figure 2). As the meat changes its condition from *rigor mortis* to initial autolysis, pH of the filling mass is higher than the samples produced from frozen meat – 5.8. During the first 15 days of the processing the pH of both experimental samples decreases statistically significantly ($P < 0.05$) and varies around levels of 5.1–5.2; then starts increasing and in the final product the values are near to the levels in the control sample 5.8–5.9. A little lower than the rest are pH values in sample No 2.

Conclusion

The blend of natural antioxidants provides evidence of antilipolytic activity and leads to pH reduction probably due to chemical substances obtained from the plant extracts.

References

Diaz, E., Meza C., Cervantez G., Torrescano G. and Sanchez-Escalante A. (2005). Effect of display conditions and ascorbic acid on color, lipid oxidation and sensory properties of fresh pork lion chop, In: Proceedings 51-th ICoMST, Baltimore, Maryland, USA, 7 – 12 August 2005, Section I, T 44, pp. 43.

Dominguez Fernandez, M.C. and Zumalacarregui Rodrigez J. M. (1991). Lipolysis and oxidative changes in "Choriso" during ripening, Meat Science, 29 (2): 99 - 107.

Johansson, M.A.E., I. Nilson, and G.K. Johansson (2000). Antioxidative Maillard reaction products in heated pork meat juice from two phenotypes (RN- and rn+rn+) from Hampshire crossbred pigs, Czech J. Food. Sci.,UZPI, 18 (Spec. №): 106 - 107.

Motilva, M.J., Toldra F., Nieto P., Flores J. (1993). Muscle lipolysis phenomena in the processing of dry-cured ham, Food Chem., 48 (2): 121 - 125.

EVALUATION OF ANTIOXIDATIVE EFFICIENCY OF CHINESE MAHOGANY (*TOONA SINENSIS ROEMOR*) IN FRESH CHICKEN SAUSAGES DURING STORAGE AT 4°C

D.C. Liu*[1], R.T. Tasu[1], Y.C. Lin[1] and F.J. Tan[2]

[1] *Department of Animal Science, National Chung Hsing University, Taichung, 402, Taiwan.* [2] *Departament of Animal Science, National University of Science and Technology, Pingtung, 912, Taiwan. Email: dcliu@dragon.nchu.edu.tw*

Keywords: Chinese mahogany (*Toona sinensis Roemor)*, antioxidative efficiency, fresh chicken sausage

Introduction

During storage, a rancid odour easily develops in fresh chicken sausages due to a higher amount of unsaturated fatty acid, which results also in rejection by consumers. For this reason, antioxidants such as BHA, BHT or natural herbs were added to sausages or meat products (Mielnik *et al.*, 2003; Nussue *et al.*, 2003). Chinese mahogany (*Toona sinensis Roemor*) is a traditional herb used in meal in Taiwan due to its special flavour and health properties. Recently, several studies mentioned that higher anti-oxidative ability existed in the shoot (leaves) of Chinese mahogany (Park *et al.*, 1996; Hsieh *et al.*, 2004). The objective of this study was to estimate the antioxidative efficiency of Chinese mahogany (*Toona sinensis Roemor*) in fresh chicken sausages stored at 4°C for 14 days.

Materials and Methods

Fresh shoot of Chinese mahogany (*Toona sinensis Roemor*) was obtained from a local herb store and dried at 60-65°C for 8hr then blended. Breast meat and skin of broilers were purchased from a local poultry plant. Dry blended shoot of Chinese mahogany (*Toona sinensis Roemor*) were divided into 3 groups (T5=500ppm; T10=1000ppm; T15=1500ppm) by phenol compounds then mixed into raw meat and stuffed into pork casing. All products were packed with film and carton then stored at 4°C for 14 days. In this study, pH value, TBARS, and sensory analysis were determined to evaluate the anti-oxidative efficiency of Chinese mahogany (*Toona sinensis Roemor*) in fresh chicken sausage during storage at 4°C. Data was analyzed using the SAS software (SAS, 2003).

Results and Discussion

The phenol compounds of the fresh and dried style were 55mg/g and 395.3 mg/g, respectively. The dry style Chinese mahogany was used in this experiment due to its higher phenol compound content and convenience. Higher pH values were obtained in the fresh chicken sausages with different amounts of Chinese mahogany when compared with that of the control. During storage, with increasing time the pH value of all fresh chicken sausages decreased but those with Chinese mahogany still had a higher pH value than the control (Figure 1). The TBARS value of the control had significantly higher value when compared with that of fresh chicken sausage with different amounts of Chinese mahogany during storage for 14 days (Figure 2). This result also indicated that Chinese mahogany actually had the better anti-oxidative effect in poultry meat products. The sensory score of fresh chicken sausage with different concentrations of Chinese mahogany at the initial time was showed in Table 1. The data analyses indicated that addition of Chinese mahogany did not effect the colour and aroma of fresh chicken sausages. Although a slight herb odour existed in fresh chicken sausage when the amount of Chinese mahogany was 500 ppm and above, the best flavour of fresh chicken sausages with 500 ppm was still evaluated by the panelists in this study. In addition, most of the panelists thought overall acceptance of these products were depended on the flavour and colour. Thus, in this study the best overall acceptance was the fresh chicken sausage with 500 ppm Chinese mahogany.

Conclusions

In conclusion, Chinese mahogany held higher total phenol compounds and also showed that a higher anti oxidative efficiency with amount when this herb was applied in fresh chicken sausage. According to the results of the sensory panel, the optimum amount of Chinese mahogany in fresh chicken sausage should not be above 500 ppm.

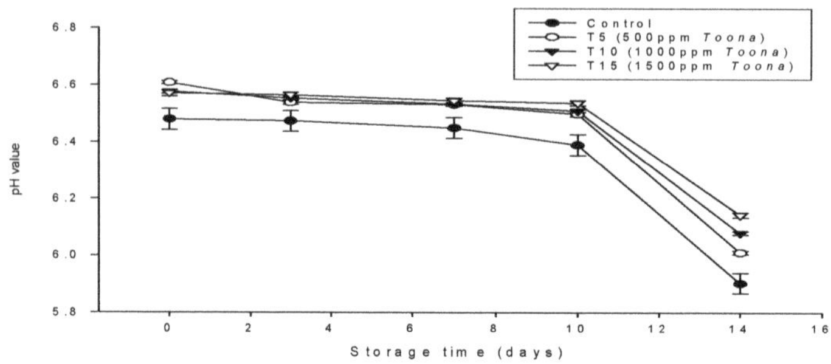

Figure 1: The change of pH value of fresh chicken sausage with different concentrations of Chinese mahogany *(Toona sinensis Roemor)* during storage at 4°C.

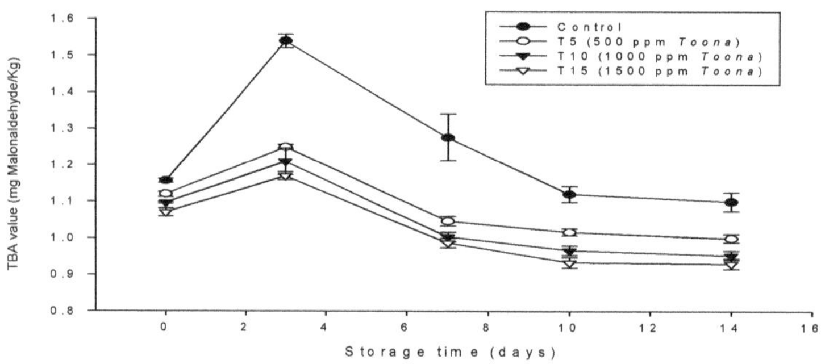

Figure 2: The change of TBAR value of fresh chicken sausage with different concentrations of Chinese mahogany *(Toona sinensis Roemor)* during storage at 4°C.

Table 1: Sensory score of fresh chicken sausage with different concentrations of Chinese mahogany (*Toona sinensis Roemor*) at the initial during storage at 4°C.

Item \ Group[1]	C	T5	T10	T15
Color	4.42±0.23^{ab}	4.75±0.30^{a}	4.50±0.15^{ab}	4.33±0.14^{b}
Aroma	4.33±0.26^{a}	4.67±0.40^{a}	4.58±0.26^{a}	4.50±0.29^{a}
Odour	1.92±0.29^{b}	2.08±0.29^{ab}	2.58±0.34^{a}	2.58±0.54^{a}
Flavour	4.12±0.31^{b}	5.08±0.26^{a}	4.83±0.30^{a}	5.00±0.25^{a}
Overall Acceptance	4.83±0.24^{b}	5.25±0.25^{a}	5.00±0.21^{ab}	5.08±0.19^{ab}

mean ± S.D., n=12.[1].C=without any spices ; T5=500ppm ; T10=1000ppm ; T15=1500ppm。

^{a,b} : Means within the same row without the same superscript are significantly different ($p < 0.05$).

References

Mielnik, M. B., K. Aaby and G. Skrede. 2003. Commercial antioxidants control lipid oxidation in mechanically deboned turkey meat. Meat Science, 65:1147-1155.

Nassu, R. T., L. A. G.. Goncalves, M. A. A. P. Silva and F. J. Beserra. 2003. Oxidative stability of fermented goat meat sausage with different levels of natural antioxidant. Meat Science. 63:43-49.

Park J. C., Y. Y. Yu, J. H. Lee, J. S. Chol and K. D. Ok. 1996. Phenolic Compounds form the Rachis of Cedrela sinensis. Kor J Pharmacogn. 27 (3):219-223.

Hsieh T.J., T.Z. Liu, Y. C. Chia, C. L. Chern, Lu FJ, Chuang MC, Mau SY,S. H. Chen, Y. H. Syu and C. H. Chen. 2004. Protective effect of methyl gallate from *Toona sinensis* (Meliaceae) against hydrogen peroxide-induced oxidative stress and DNA damage in MDCK cells. Food Chem. Toxicol. 42 (5):843-50.

POSSIBILITY OF PREVIOUSLY BLANCHED AND CURED BEEF TO BE USED FOR CORNED BEEF PRODUCTION

L. Turubatović* and P. Radetić

Institute of Meat Hygiene and Technology, 11000 Belgrade, SCG, e-mail: meatinst@beotel.yu

Keywords: corned beef, colour, nitrite, curing, blanching

Introduction

The aim of this task was to determine the extent of the chemical reaction between sodium nitrite *i.e.* nitrogen monoxide and denaturated myoglobin in thermally treated / blanched meat at 95-100°C for 15 minutes (Trial 2). Myoglobin was denaturated at 60°C (Hamm, 1996). Corned beef was produced by adding sodium nitrite to the blanched meat. The colour of this product was compared to the colour of a corned beef product produced with the same amount of sodium nitrite that was cured for 48 hours. Composition of raw material and chemical composition of the final product is presented in Tables 1 and 2. Furthermore, amounts of total pigments and nitrosylmyoglobin (NOMb) content were determined (Table 4) and the results of instrumental determination of colour in the final product are presented in Table 3.

Materials and Methods

Water content, proteins, ash and hydroxyproline content were determined according to ISO 1442/1998, ISO 937/1992, ISO 1444/1998, ISO 936/1999 and ISO 3496/2001 methods respectively. Product colour was determined on a Chroma meter CR-400 instrument (Minolta). On the basis of CIE Lab (Robertson, 1977), the following parameters were determined: psychrometric tone-a (level of red colour), psychrometric chroma-b (level of yellow colour) and psychrometric lightness (L). Total pigments content (TP) and nitrosylmyoglobin (NOMb) content were determined by the Möhler modification of the Hornseye's method (1958). In Trial 1, meat of the 1st and 3rd grade was cured with 2% of nitrite salt for 48 hours at +4°C and then blanched. In the second trial, 2.4% of the nitrite salt was added after blanching to the non-cured meat.

Results and Discussion

In Table 1, corned beef composition of both trials is shown. Table 2 (Trials 1 and 2) shows the results of chemical analysis of corned beef. Results indicate that there are no significant differences between maximal and minimal values for any of the examined properties in both Trial 1 and Trial 2. However, these parameters clearly indicate that in both trials high quality products with high nutritional values were obtained. Equal chemical composition of the products in both trials was provided by previous chemical analysis of product ingredients. Therefore, standardised quality has been ensured, not only considering sensory properties, but also the chemical composition. There were no differences in chemical composition and sensory properties between corned beef from meat cured before blanching and corned beef from meat cured immediately after blanching. Table 3 shows the results of instrumental determination of colour in corned beef samples from Trial 1 and Trial 2. Values of psychrometric light indicate that products made from meat that was cured after blanching were of brighter red and exhibited weaker intensity of yellow. However, the parameters clearly indicate that in both cases, the colour of the product was characteristic for cured, thermally treated meat. Most auditors found these properties very acceptable and desirable. Total pigments (TP) content and NOMb content (Table 4.) were significantly higher in corned beef made from meat cured 48 hours before blanching (Trial 1) compared with products cured after blanching *i.e.* before stuffing, formation and filling into the cans (Trial 2). These results confirm the previous observations that the darker red colour achieved in Trial 1 is a consequence of the conversion of higher myoglobin amount into nitrosyl form. However, products from Trial 2 also had a higher percentage of total pigments converted into nitrosylmyoglobin. This provided stable colour, characteristic for cured beef although of slightly brighter hue.

Table 1: Corned Beef Ingredients

Trial 1		Trial 2	
Ingredients	%	Ingredients	%
Cured/blanched beef 1st grade	54.96	Blanched beef 1st grade	53.85
Cured/blanched beef 3rd grade	9.80	Blanched beef 3rd grade	9.61
Blanched tallow	14.72	Blanched tallow	14.42
Cooked beef tendons	11.78	Cooked beef tendons	11.54
Broth	7.85	Broth	7.70
Gelatine	0.49	Gelatine	0.48
Nitrite salt	0.40	Nitrite salt	2.40
Total	**100,00**	**Total**	**100,00**

Table 2: Chemical Composition of Corned Beef (Trial 1 and Trial 2)

	Trial 1			Trial 2		
	minimum	maximum	$\bar{x}$	minimum	maximum	$\bar{x}$
Proteins, %	23.14	23.50	23.34	23.11	23.60	23.35
Fat, %	15.80	16.19	15.99	16.30	17.10	16.70
Water, %	57.63	58.28	57.88	57.50	57.70	57.60
Ash, %	2.61	2.78	2.66	2.50	2.60	2.55
Hydroxyproline, %	0.310	0.326	0.320	0.315	0.326	0.320
Connective tissue, %	2.48	2.61	2.57	2.52	2.61	2.56
Connective tissue-proportion in total proteins, %	10.72	11.11	10.91	10.90	11.06	10.96
Lean meat	104.23	105.86	105.13	104.09	106.31	105.20
Total meat	120.03	122.03	121.12	120.40	123.41	121.90

Table 3: Results of instrumental colour determination in Corned Beef-u (Trials 1 and 2)

	Trial 1			Trial 2		
	L*	a*	b*	L*	a*	b*
I sample	49.59	19.67	13.76	53.16	17.35	12.59
II sample	49.53	18.80	13.39	55.90	15.47	12.33
III sample	50.04	20.84	14.57	51.72	18.55	13.32
Average value	49.72	19.77	13.91	53.59	17.12	12.75
SD	0.28	1.02	0.60	2.12	1.55	0.51

L* - psychrometric light – samples with higher L* values are brighter
a* - psychrometric tone: positive a* values correspond to red.
Higher a* values correspond to higher intensity of red
b* - psychrometric chrome: positive b* values correspond to yellow.

Table 4: Content (μg/g) of total pigments (TP) and nitrosylmyoglobin (NOMb) in Corned Beef

	TP	NOMb
Trial 1	297.84	164.72
Trial 2	203.32	118.03

The results indicate that in blanched meat (thermally treated at 95-100°C), some of the pigments remain and react with nitrites creating thestable light red colour characteristic of cooked cured beef. The fact remains that thermally treated non-cured beef contains less pigment, however, the level of nitrosylmyoglobin is higher in this type of product. This relation is important for formation of desirable colour and other properties significant from a sensory point of view.

Conclusions

Beef, thermally treated in water at 95-100°C, achieved desirable and stable colour by the reaction of pigments with curing (nitrite) salt.

Technology of blanching the non-cured meat and subsequent addition of nitrite salt in the production of corned beef considerably shortens the production process and provides energy savings.

Acknowledgments

Data published in this paper was acquired from the Project № BTN-351001B financed by the Ministry of Science and Environmental Protection of Republic of Serbia.

References

Hamm, R., (1966). Compendium from the book: The physiology and biochemistry of muscle as a food. Univers. of Wisconsin;

Möhler, K., (1958). Zur Bestimmung der Pökelfarbstoffes, Mitteilung aus der Deutschen Forschungsanstalt für Lebensmittelchemie, München;

Robertson, A. R., (1977). Colour Difference Formular, Colour Research Applied, 2,7.

INFLUENCE OF DIFFERENT CUTTER KNIVES ON BATTER QUALITY AND PROPERTIES OF COOKED SAUSAGES

W. Schnäckel*, I. Micklisch, J. Krickmeier and D. Schnäckel

Faculty of Agriculture, Nutrition and Landscape Architecture, Department of Food Technology, Anhalt University of Applied Sciences, 06 406 Bernburg, Germany. Email: schnaeckel@loel.hs-anhalt.de

Keywords: cutter knives, cutting process, energy consumption, batters quality, cooked sausages

Introduction

The consumption of cooked sausages is on a high level in many countries and classical cutter machines are the most important equipment for the production of cooked sausage batter.

For the technical construction of cutter machines the following aspects have high importance:

- technical construction of the cutting mechanism (shape and volume of the cutting room, form of bowl)
- different chopper knives, sharpness and mode of grinding
- number and arrangement of the knives in the cutting machine
- knife speed and speed of the bowl in the cutting machine.

Limited new information about cuttting (chopper) knives can be found and it is published only by a handful of authors (Hammer *et al.*, 2003), (Schnäckel *et al.*, 2004), (Micklisch *et al.*, 2004). Meantime the competition between meat producing factories expands to the equipment suppliers, especially producers of chopper knives. Therefore possibilities to reduce the cutting time and energy consumption for cutting, especially for cooked sausage batter, are in demand. On empirical way producers of chopper knives try to change some construction aspects of the knives. Alternatives are hole- and hack-hole-knives or block knives (3 knives combined as one) (Haack *et al.* 2006).

The aim of this work is to test different forms of cutter knives for the production of cooked sausage batter and to find out the most effective one. The authors would like to investigate the influence of knife surface area and grinding mode on batters quality, energy consumption and properties of sausages regarding the best knife form.

Materials and Methods

In the first test series the authors want to investigate if the 5 different cutter knives (Figure 1) lead to different batters and sausage qualities. Secondly the influence of cutter knife area and grinding mode were analysed. For the experiments minced meat (3 mm) was used. The pouring media was ice. The composition consisted of 45 % pork, 35 % fat tissue and 20 % ice.

The pre-minced meat materials together with all other additives (broken ice, nitrite curing salt, ascorbic acid, spices) were cut in a one step procedure. The cutting conditions were as follows: knife revolutions: 2500 1/min, bowl revolutions: a) first 30 sec. 7 1/min b) till the end 14 1/min. The process was stopped at 12 °C batters temperature. The lean minced meat of each batch was filled into sausage covers (∅ 60 mm) and treated in a smoking-cooking procedure.

The following parameters were measured and recorded on-line: electrical power parameters of the machine, temperature, viscosity of the lean minced meat and electrical conductivity of batters. In steps of 30 sec emulsion stability, penetration hardness and colour of the lean minced meat were detected. The final products were examined for their texture, colour and sensoric value.

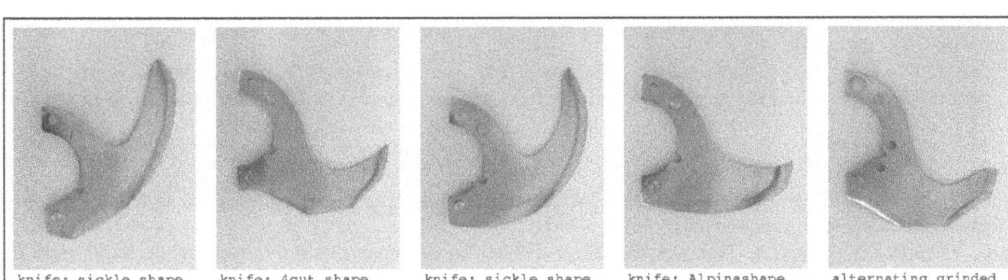

Figure 1: Types of knives used.

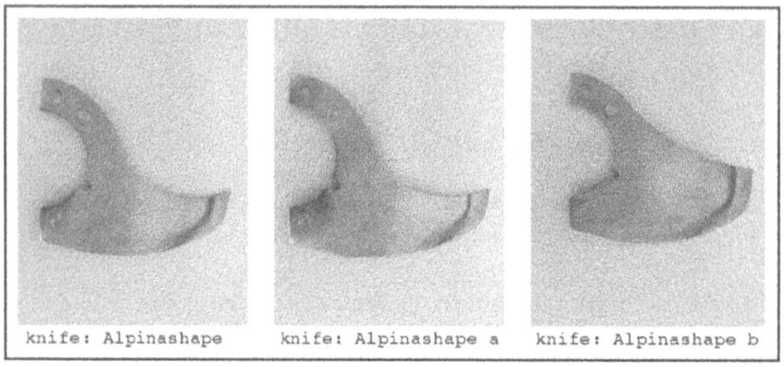

Figure 2: Knives with the same basic shape (M4), varying surface and grinding mode.

Results and Discussion

The first experimental series showed that knives M2, M4 and M5 lead to less temperature increase and thus to a longer total cutting time in comparison to knives M1 and M3 (Figure 3).The engine power of the main drive using knives M2, M4, M5 was significantly lower than using M1 and M3, while the total energy consumption stayed relatively constant according to the longer cutting time. A longer cutting time corresponds to a higher total number of knife revolutions and hence a more homogeneous batter and a better sensorically evaluated final product can be achieved. The development of emulsion stability and penetration hardness of the lean minced meat during the cutting time shows no significant differences between the tested knives.

The comparison of standard knives (sample M4) and knives with a double side grinded blade (M4a; equal surface according to M4) show that the changed grinding leads to an accelerated temperature increase and thus to a shorter cutting time (Figure 4). A minor total energy consumption should be quite relevant. No significant differences in the quality of the final product, especially in the texture, can be found between the samples.

Knives of sample M4b (larger surface according to M4) also lead to an accelerated increase in temperature and to shorter cutting time. The average electrical engine power is relatively constant. Therefore an increased surface reduces energy consumption. The extended knife surface however causes a softer texture of the final products.

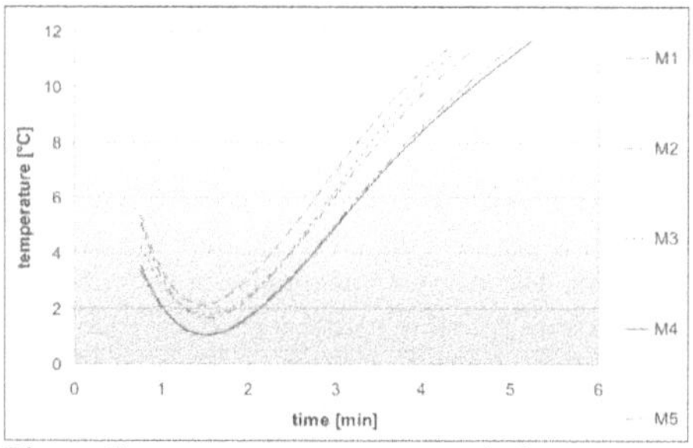

Figure 3: progress temperatures main forms.

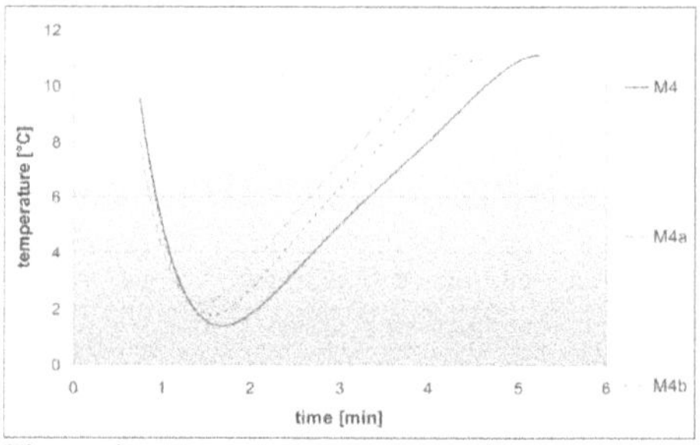

Figure 4: progress temperatures series M4.

Conclusions

In the production of cooked sausages the application of steep knives with short cutting edge is superior to the use of even knives with long cutting edge.

Knives with a double side grinded blade reduce cutting time and energy consumption without any influence on the quality of the final products. The extension of the knife surface area saves energy but the final products show a softer texture.

References

Haack, E., E. Ehrle, D. Kallweit (2006). There are a lot of unknown influences – meat processing in cutter machines with high capacity knives. Fleischwirtschaft, 86: 1, 47- 51.

Hammer, G.,G. Haller, S. Höpfl, H. Jankowitsch (2003). Effects of different chopper knives for the production of cooked sausage batter. Bundesanstalt für Ernährung und Lebensmittel, Standort Kulmbach, Institut für Technologie, Jahresbericht.

Micklisch, I., W. Schnäckel, J.Krickmeier, E. Haack, E. Ehrle (2004). Using different forms of cutter knives. Fleischwirtschaft, 84: 12, 100-105.

Schnäckel,W., E. Ehrle, O. Haack (2004). Perforated chopper knives reduce stress. Fleischwirtschaft, 84: 1, 51- 56.

EFFECT OF RADIATION PROCESSING OF PORK PRODUCTS ON ITS LIPIDS

E. Zanardi* [1], A. Battaglia[1], S. Ghidini[1], M. Conter[1], A Badiani[2] and A. Ianieri[1]

[1]*Dip. di Produzioni Animali, Biotecnologie Veterinarie, Qualità e Sicurezza degli Alimenti, Università degli Studi di Parma, via del Taglio 8 – 43100 Parma, Italy* [2]*DIMORFIPA, Alma Mater Studiorum - Università di Bologna, via Tolara di Sopra 50 - 40060 Ozzano Emilia (Bologna), Italy. E-mail: emanuela.zanardi@unipr.it*

Keywords: irradiation, pork, lipid, oxidation, 2-alkylcyclobutanones

Introduction

Radiation processing is one of the emerging technologies to ensure the microbiological safety of meat. This is one of the most important issues of the meat industry, especially for producers exporting to USA and Japan where the absence of some pathogenic bacteria is required. Although the effectiveness against most pathogens is well known, radiation brings about biochemical changes that could affect the nutritional adequacy of food. Lipids are one class of nutrients affected by irradiation treatment. Irradiation-induced lipid oxidation products are dose-dependent and their formation is influenced by packaging, storage and other processing conditions before and after radiation processing. In addition, the irradiation of lipids results in the formation of 2-alchylcyclobutanones, unique radiolytic products from precursor fatty acids. These molecules, proposed as a marker of irradiation of fat-containing food, have been recently reported as genotoxic (Knoll *et al.*, 2006).

The present study attempted to establish whether radiation processing of three different Italian pork products resulted in changes to the lipid oxidation status and 2-alchylcyclobutanone content during storage under vacuum and refrigerated conditions.

Materials and Methods

Salame Milano, coppa and pancetta were the three traditional Italian processed pork products investigated. Salame Milano is a typical minced, dry cured and fermented pork product; coppa is made up by the entire neck muscle, deboned, cured and afterwards matured in natural casing; pancetta is made up of the muscular and adipose tissues of the ventral region of the pork, cured and seasoned. The meat products at the end of maturing were packed under vacuum, placed in cardboard boxes and electron irradiated under refrigerated condition at 2, 5 and 8 kGy absorbed dose by using a linear accelerator. The three doses are within the range of 10 kGy which has been set by the World Health Organisation as a level presenting non-toxicological hazard. The degree of lipid oxidation and the content of 2-alchylcyclobutanones were determined after 0 (T_0) and 60 days (T_1) of vacuum storage in refrigerated conditions. Lipid oxidation was determined by the thiobarbituric acid reactive substances (TBARS) method as proposed by Novelli *et al.*, (1998). 2-alchylcyclobutanones (2-ACB) were determined according to the European Standard EN 1785 (2003) based on GC/MS analysis. 2-dodecylcyclobutanone (2-DCB), 2-tetradecylcyclobutanone (2-TDCB) and 2-(5'-tetradecenyl)cyclobutanone (2-TdecCB) were the three molecules considered because they are derived from the major fatty acid precursors (respectively palmitic, stearic and oleic acid) found in the processed pork products investigated.

Results and Discussion

The TBARS values are presented in Table 1. The pork products irradiated up to 5 kGy radiation dose did not differ significantly from the non-irradiated ones. However, the radiation dose of 8 kGy was able to accelerate the lipid oxidation process. Among the three pork products, pancetta showed the highest TBARS values. This could be attributed to its higher fat content (33.9 ± 3.9 %) compared to that of salame Milano (31.2 ± 1.5 %) and coppa (20.0 ± 5.1 %). Vacuum packaging was very effective for inhibiting irradiation-induced lipid oxidation during storage, in agreement with Jo *et al.*, (2002). TBARS values measured just after the irradiation treatment (T_0) did not differ significantly from those measured after 60 days of storage (T_1), with the exception of radiation dose 5 kGy for all the three pork products and 2 kGy for pancetta.

Table 2 shows the 2-alchylcyclobutanone content. These molecules were absent in the non-irradiated pork products. Similar to TBARS values, pancetta showed the highest content of 2-ACB. The concentration of 2-ACB is dose dependent. At 2 kGy radiation dose they were found in a few samples at concentrations very near to the limit of quantification. At 5 and 8 kGy they were present in all the samples, except 2-TdecCB, the presence of which was only detected at 8 kGy radiation dose. Vacuum packaging was not sufficient to limit the formation of 2-ACB. In most cases the content of the 2-ACB measured after 60 days of vacuum storage (T_1) was significantly higher compared to the amounts detected just after the treatment (T_0). 2-DCB and 2-TDCB are the two molecules most frequently detected. Their concentration was always lower than 0.60 μg/g. When it was present, 2-TdecCB reached highest levels, which ranged between 0.99 and 3.81 μg/g. This result is in agreement with the findings of Kim *et al.*, (2004) who detected the 2-ACB in dried shrimps at a radiation dose of 0.5 kGy with 2-TdecCB found in the highest concentration.

Table 1: TBARS values (mg MDA/kg) in the irradiated processed pork products (a,b, in a column, and z,y, in a line stand for significant differences, P≤0.05, Scheffè's test).

Pork product	Irradiation dose (kGy)	T_0	T_1
Salame Milano	0	0.108±0.019 ^{b,z}	0.104±0.010 ^{a,z}
	2	0.135±0.021 ^{ab,z}	0.233±0.135 ^{a,z}
	5	0.124±0.007 ^{ab,y}	0.199±0.037 ^{a,z}
	8	0.148±0.024 ^{a,z}	0.202±0.154 ^{a,z}
Coppa	0	0.085±0.009 ^{b,z}	0.075±0.036 ^{b,z}
	2	0.088±0.014 ^{b,z}	0.091±0.025 ^{b,z}
	5	0.083±0.020 ^{b,y}	0.204±0.079 ^{a,z}
	8	0.163±0.045 ^{a,z}	0.146±0.056 ^{ab,z}
Pancetta	0	0.187±0.080 ^{ab,z}	0.179±0.066 ^{b,z}
	2	0.227±0.042 ^{ab,y}	0.394±0.098 ^{a,z}
	5	0.130±0.027 ^{b,y}	0.273±0.030 ^{b,z}
	8	0.260±0.024 ^{a,z}	0.245±0.049 ^{b,z}

Table 2: 2-alchylcyclobutanone content (μg/g) in the processed pork products irradiated ([1] not detectable; [2] detected in less than 3 samples out of 5; a,b, in a column, and z,y, in a line, stand for significant differences, P≤0.05, Scheffè's test).

2-ACB	Irradiation dose (kGy)	Salame Milano		Coppa		Pancetta	
		T_0	T_1	T_0	T_1	T_0	T_1
2-DCB	0	n.d.[1]	n.d.	n.d.	n.d.	n.d.	n.d.
	2	traces [2]	traces	0.04±0.01	n.d.	0.05±0.01^{c}	traces
	5	0.03±0.01^{b,y}	0.07±0.02^{b,z}	0.05±0.02^{b,z}	0.07±0.01^{b,z}	0.12±0.03^{b,z}	0.11±0.03^{b,z}
	8	0.13±0.03^{a,y}	0.35±0.13^{a,z}	0.13±0.06^{a,y}	0.26±0.05^{a,z}	0.26±0.07^{a,y}	0.57±0.20^{a,z}
2-TdecCB	0	n.d.	n.d.	n.d.	n.d.	n.d.	n.d.
	2	n.d.	n.d.	n.d.	n.d.	n.d.	traces
	5	traces	traces	n.d.	traces	1.09±0.38^{b,z}	0.72±0.10^{b,z}
	8	0.99±0.10^{y}	2.34±0.77^{z}	0.73±0.38^{y}	1.73±0.41^{z}	2.52±0.80^{a,z}	3.81±1.28^{a,z}
2-TDCB	0	n.d.	n.d.	n.d.	n.d.	n.d.	n.d.
	2	traces	traces	0.04±0.01^{b}	n.d.	traces	traces
	5	0.05±0.02^{b,y}	0.09±0.03^{b,z}	0.06±0.03^{ab,z}	0.08±0.03^{b,z}	0.10±0.02^{b,z}	0.09±0.03^{b,z}
	8	0.13±0.01^{a,y}	0.33±0.11^{a,z}	0.11±0.05^{a,y}	0.31±0.09^{a,z}	0.24±0.08^{a,z}	0.42±0.14^{a,z}

Conclusions

The three Italian processed pork products investigated (salame Milamo, coppa and pancetta) submitted to radiation processing at 2 and 5 kGy did not present a significant increase in lipid oxidation products either just after the treatment or after 60 days of vacuum storage. However, a marked increase was observed at 8 kGy radiation dose. The content of 2-alchylcyclobutanones increased with radiation dose. Their presence has been recorded occasionally at 2 kGy and constantly at higher radiation doses (5 and 8 kGy). A significant increase of 2-alchylcyclobutanones was observed during vacuum storage of irradiated pork products.

References

EN 1785 (2003) "Detection of irradiated food containing fat–Gas-chromatographic/mass spectrometric analysis of 2-alkylcyclobutanones".

Jo C., Ahn D.U., Byun M.W. (2002) Irradiation-induced oxidative changes and production of volatile compounds in sausages prepared with vitamin E-enriched commercial soybean oil. *Food Chemistry*, 76, 299-305.

Kanatt S.R., Chander R., Sharma A. (2006) Effect of radiation processing of lamb meat on its lipids. *Food Chemistry*, 97, 80-86.

Kim Kyong-Su, Seo Hye-Young, Lee Jeong-Min, Park Eun-Ryong, Kim Jun-Hyoung, Hong Cheul-Hee, Byun Myung-Woo (2004) Analysis of radiation-induced hydrocarbons and 2-alkylcyclobutanones from dried shrimps. Journal of Food Protection, 67, 142-147.

Knoll N., Weise A., Claussen U., Sendt W., Marian B., Glei M., Pool-Zobel B.L. (2006) 2-Dodecylcyclobutanone, a radiolytic product of palmitic acid, is genotoxic in primary human colon cells and in cells from preneoplastic lesions. *Mutation Research*, 594, 10-19.

Novelli E., Zanardi E., Ghiretti G.P., Campanini G., Dazzi G., Madarena G., Chizzolini R. (1998) Lipid and cholesterol oxidation in frozen stored pork, salame Milano and mortadella. *Meat Science*, 48, 29-40.

DEVELOPMENT OF A MODIFIED DRY CURING PROCESS FOR PORK

J.E. Hayes*, T. Kenny, P. Ward and R. Kennedy

Ashtown Food Research Centre, Teagasc, Ashtown, Dublín 15, Ireland. Email: jenny.hayes@teagasc.ie

Keywords: pork, dry curing, tumbling, vacuum pulsing

Introduction

Bacon is one of the most widely produced traditional dry cured food items. Such products have a high variety of flavours and textures as compared to the corresponding wet-cured items. In keeping with the trend towards enriching our sensory perceptions, many consumers are getting more interested in dry-cured meat products (Toldrá, 2002). However, modifications to the traditional dry curing process are desirable in order to accelerate the process, improve the yield and increase the uniformity of the product. Tumbling causes disruption of the meat fibres (Lawlis *et al.*, 1992), which together with the curing additives allows the meat to bind more water (Chow *et al.*, 1986). Vacuum also increases cure absorption as well as extraction of the salt-soluble proteins (Sharma *et al.*, 2002). Tumbling without vacuum has been found to improve colour, sliceability, taste, aroma and yields of hams (Krause *et al.*, 1978). For this reason, various different vacuum tumbling and pulsing treatments were tested on pork M. *longissimus thoracis et lumborum* to further accelerate the curing process without having a detrimental effect on the product quality and safety.

Materials and Methods

Whole pork loins (*M. longissimus thoracis et lumborum)* were assigned to treatments at 72 hours post-mortem (pH 5.6-5.9). The following treatments were applied: 1) Control (conventional static dry cure) 2) Vacuum pulsed tumble (6 hour tumble, 4 revolutions per minute, vacuum at 90% (900mBar) for 20 minutes followed by 0% for 3 minutes) 3) Vacuum pulse (no tumble, vacuum at 90% for 20 minutes followed by 0% for 3 minutes), 4) Vacuum tumble (6 hour tumble, 4 revolutions per minute, vacuum at 90% with no pulsing), 5) Tumble only (6 hour tumble, 4 revolutions per minute, no vacuum or pulsing). The treatments were compared to a control conventional (static) dry cure method. The dry cure mixture was prepared in bulk and contained 66.5%w/w fine sea salt 32.0% brown sugar, 0.5% sodium nitrite and 1.0% sodium nitrate. It was applied over the surface of the pork at a rate of 6.0% by weight. The meat was then placed in the tumbler (Ruhle Ltd. Turbovac, Type SB 415, 200 liter) and the required programme selected. The test curing treatments were applied to the whole loins in order to simulate industry practice, but analysis of the products was confined to the eye muscle component. It had been observed in preliminary trials that conventional dry cured loins appeared to be cured at day 7 hence all other treatments were compared at day 7 of curing. At day 7 of curing, samples form each treatment were taken for chemical analysis and separated into core and outer regions. Salt content was determined by titration using the Mohr method (Kirk and Sawyer, 1991). Samples were analysed for moisture, fat and protein content (Bostian *et al.*, 1985; Sweeney and Rexford, 1987). Colour was measured using the HunterLab system. An 8 member trained panel was employed to evaluate sensory quality of sample treatments (AMSA, 1995). Results were analysed using a two-way analysis of variance (ANOVA).

Results and Discussion

Salt content differed ($p<0.001$) between the core and outer regions for all treatments therefore results show that salt is not uniformly distributed. Comparing the cores only, moisture levels were similar for 4 of the treatments and lower ($p<0.01$) for the vacuum pulse treatment, corresponding also to the salt being highest for the latter. Results for Hunter L values showed that the vacuum pulse and vacuum tumble treatments were found to be lighter ($p<0.01$) than the control. The vacuum tumble treatment was paler ($p<0.001$) in cured colour in comparison to the control. All treatments had 'good' to 'very good' colour uniformity. The treatments had no effect on juiciness, tenderness overall flavour and overall acceptability. The cellular alteration caused by tumbling is also believed to influence juiciness (Dzudie *et al.*, 2000), but again sensory results vary between studies (Chow *et al.*, 1986). Cassidy *et al.*, (1978) reported differences in the ultra structure after tumbling and suggested that an increased disruption contributed to a superior tenderness; however, sensory results deviate from this theory (Lawlis *et al.*, 1992). The control and vacuum pulse tumble treatments were rated by the panel as having an 'ideal' salt level compared to the remaining three treatments which were found to be 'slightly' salty. It should be noted here that the control and vacuum pulse tumble treatments had 2% salt in the core compared to the other treatments which had a 3% salt level, and 3% and 4% respectively for the outer values. There was a difference ($p<0.001$) between the core and outer for all of the treatments, therefore results show that salt is not uniformly distributed.

Conclusions

The treatments had no effect on juiciness, tenderness overall flavour and overall acceptability. These findings indicated that the combinations of tumbling, pulsing and vacuuming did not significantly accelerate the diffusion of curing ingredients in the pork loins or improve the overall quality and acceptability in comparison to the static conventional dry cure method. Further research on these techniques is required which will also include a detailed shelf-life study.

Table 1: Sensory analysis results on cured bacon.

Treatment Type	Cured Colour	Colour Uniformity	Juiciness	Tenderness	Saltiness	O/F	O/A
Control	5.5^{ac}	4.7^{ab}	5.0^{a}	5.6^{a}	3.2^{a}	4.2^{a}	4.2^{a}
Vacuum PT	5.7^{c}	5.1^{a}	4.7^{a}	5.5^{a}	3.3^{a}	4.0^{a}	4.2^{a}
Tumble	4.9^{ab}	4.9^{ab}	5.0^{a}	5.2^{a}	4.1^{b}	3.9^{a}	4.0^{a}
Vacuum tumble	4.6^{b}	5.0^{a}	4.7^{a}	5.3^{a}	4.0^{b}	3.6^{a}	3.9^{a}
Vacuum pulse	5.0^{ab}	4.3^{b}	4.7^{a}	5.5^{a}	4.1^{b}	4.0^{a}	3.9^{a}

[a, b] Means in the same column with unlike superscripts are different ($p<0.001$)

Vacuum PT = vacuum pulse tumble; O/F= Overall Flavour; O/A= Overall Acceptability

Cured colour, colour uniformity, saltiness, O/F and O/A were evaluated by means of a six-point hedonic scale (6 = very pa excellent/100% uniformity, very salty, extremely good flavour, extremely acceptable respectively; 1= very dark, very poor, very mild, ve poor, not acceptable respectively). Juiciness and tenderness were evaluated by means of a eight-point hedonic scale (8 = Extremely jui and extremely tender respectively; 1= Extremely dry and extremely tough respectively)

Table 2: Comparison of salt and moisture results on the core and outer samples of each treatment type of cured bacon.

Treatment Type		% Salt	% Moisture
Control	Core	2.1^{a}	71.9^{a}
Control	Outer	4.3^{b}	70.3^{b}
Vacuum pulse tumble	Core	2.0^{a}	71.9^{a}
Vacuum pulse tumble	Outer	3.2^{b}	71.1^{b}
Tumble only	Core	2.9^{a}	70.9^{a}
Tumble only	Outer	3.8^{b}	70.5^{a}
Vacuum tumble	Core	2.7^{a}	71.2^{a}
Vacuum tumble	Outer	3.9^{b}	70.7^{a}
Vacuum pulse	Core	3.4^{a}	71.2^{a}
Vacuum pulse	Outer	4.2^{b}	70.7^{b}

[a, b] Means in the same column with unlike superscripts are different ($p<0.001$).

References

AMSA (1995). Research guidelines for cooking, sensory evaluation and instrumental tenderness measurements of fresh me American Meat Science Association, 10-11.

Bostian, M. L., Fish, D. L., Webb, N. B. and Arey, J. J. (1985). Automated methods for determination of fat and moisture meat and poultry products: collaborative study. *Journal of the Association of Official Analytical Chemists* 68, 876-880.

Cassidy, R.D., Ockerman, H.W., Krol, B., Van Roon, P.S., Plimpton, R.F. Jr and Cahill, V.R. (1978). Effect of tumbli method, phosphate level on final cook temperature on histological characteristics of tumbled porcine muscle tiss *Journal of Food Science*, 43, 1514-1518.

Chow, H.M., Ockerman, H.W., Cahill, V.R. and Parrett, N.A. (1986) Evaluation of cured, canned pork shoulder tiss produced by electrical stimulation, hot processing and tumbling. *Journal of Food Science*, 51, 288-291.

Dzudie, T., Ndjouenkeu, R and Okubanjo, A. (2000). Effect of cooking methods and rigor state on the composition ar tenderness and eating quality of cured goat loins. *Journal of Food Engineering* 44, 149-153.

Fidel Toldrá (2002). *In:* Dry cured meat products. Blackwell Publishing. 1 – 4.

Kirk, R.S. and Sawyer, R. (1991). *In:* Pearson's Composition and Analysis of Foods. Longman Scientific & Technical. 14-

Krause, R.J., Ockerman, H.W., Krol, B. Moerman, P.C., and Plimpton, R.F. Jr. (1978). Influence of tumbling, tumbling tin trim and sodium tripolyphosphate on quality and yield of cured hams. *Journal of Food Science, 43*: 853.

Lawlis, T.L., Plimpton, R.F., Ockerman, H.W. and Parrett, N.A (1992). Electrical stimulation and tumbling affect pre-rig cured, sectioned and formed ham roasts. *Journal of Food Science*, 57, 564-568, 616.

Sweeny, R. A. and Rexroad, P. R. (1987). Comparison of LECO FP-228 "Nitrogen Determinator" with OAC copper catal Kjeldahl method for crude protein. *Journal of the Association of Official Analytical Chemists*, 70, 1028-1030.

Sharma, B.D., Kumar, S., Nanda, P.K and Kumar, S (2000). Optimisation of short term tumbling schedule for the processi of cured and restructured buffalo meat blocks. *Indian Journal of Animal Science*, 72, 684-688.

MODIFICATION OF A DRY CURING PROCESS FOR BEEF

J.E. Hayes* and T. Kenny

Ashtown Food Research Centre, Teagasc, Ashtown, Dublin 15, Ireland. Email: jenny.hayes@teagasc.ie

Keywords: beef, dry curing, tumbling, vacuum pulsing

Introduction

Vacuum tumbling is a mechanical method used in industry to improve brine distribution. The application of vacuum tumbling has been found to produce more extractable protein in beef than non-vacuum condition (Ghavimi *et al.*, 1986). Better binding ability and water absorption has been found in vacuum tumbled meat (Wiebe and Schmidt, 1982). Deumier *et al.*, (2003) found that pulsed vacuum increased salt gain and decreased water loss. Products obtained by pulsed vacuum brining had more uniform water and salt distribution from the product surface to the core. Consumers are demanding convenience foods, but are increasingly concerned about issues of food safety, quality and health. This is reflected in growing niche markets for products perceived as healthier and of higher quality than mainstream products. Dry-cured meat products fit this description. This project aims to develop a modified dry-curing process using a number of novel accelerated curing techniques such as vacuuming, tumbling and vacuum pulsing which would increase the relatively slow through-put associated with traditional curing allowing for more uniform and consistent products. For this reason, beef *M. supraspinatus* was applied various different vacuum tumbling and pulsing treatments to further accelerate the curing process without having a detrimental effect on the product quality and safety.

Materials and Methods

The *M. supraspinatus* were obtained from a local meat plant at 72 hours post-mortem. The pH of the *M. supraspinatus* was checked to ensure pH was within the normal range of 5.4 to 5.6. Six replicate loins chosen at random were used for each dry curing treatment. The following treatments were applied: 1) Control (conventional static dry cure); 2) Vacuum pulsed tumble (6 hour tumble, 4 revolutions per minute, vacuum at 90% (900mBar) for 20 minutes followed by 0% for 3 minutes); 3) Vacuum pulse (no tumble, vacuum at 90% for 20 minutes followed by 0% for 3 minutes); 4) Vacuum tumble (6 hour tumble, 4 revolutions per minute, vacuum at 90% with no pulsing); 5) Tumble only (6 hour tumble, 4 revolutions per minute, no vacuum or pulsing). The treatments were compared to a control conventional (static) dry cure method. The dry cure mixture was prepared in bulk and contained 66.5% w/w fine sea salt 32.0% brown sugar, 0.5% sodium nitrite and 1.0% sodium nitrate. It was applied by rubbing uniformly over the surface of the pork at a rate of 6.0% by weight. For the conventional control method 6% by weight of dry cure mix was rubbed onto the loins which were then placed in the chill at 2°C. It had been observed in preliminary trials that conventional dry cured loins appeared to be cured at day 7 hence all other treatments were compared at day 7 of curing. At day 7 of curing, samples for each treatment were taken for chemical analysis. The sample section for chemical analysis was separated into core and outer regions. Salt content was determined by titration using the Mohr method (Kirk and Sawyer, 1991). Samples were analysed for moisture, fat and protein content (Bostian *et al.*, 1985; Sweeney and Rexford, 1987). The remainder of the muscles were then cooked in the oven at 85°C to an internal temperature of 72°C. When cooked, the beef muscles were refrigerated at 4°C. An 8 member trained panel was employed to evaluate sensory quality of sample treatments (AMSA, 1995). Colour was measured using the HunterLab system. One-way analysis of variance (ANOVA) was used to compare means and to identify significant differences ($p<0.05$) between treatments.

Results and Discussion

Comparing core regions only the salt content for vacuum tumble and tumble only treatments were found to be higher ($p<0.001$) then the control. This result indicated the potential of vacuum tumbling and tumbling only treatments in accelerating the curing processes i.e. rate of diffusion of curing ingredients to the core of the muscle. All of the dry cure treatments were found decrease ($p<0.001$) cook loss compared to the control, hence increasing cook yield (Table 1). The control and vacuum pulse treatments were rated 'slightly darker' in cured colour in comparison to the remaining three treatments which were 'slightly pale' in colour (Table 2). For colour uniformity the control and vacuum pulse treatments had only 'fair' colour uniformity with the presence of brown uncured areas in the centre of the slice, while the other three treatments had 'good' colour uniformity ($p<0.001$). Juiciness results found that both the control and vacuum pulse treatments were 'slightly dry', whereas the other three treatments were found to be 'slightly juicy' according to the panellists. It should be noted that both the control and vacuum pulse treatments also had lower moisture content. The control and vacuum pulse treatment were perceived by the panel to be 'slightly tough' compared to the other treatments, which were rated as 'slightly' to 'moderately' tender. Saltiness results showed that the control and vacuum pulse treatments were 'ideal' to 'slightly' salty whereas all other treatments were 'slightly' salty. It should be noted that control and vacuum pulse treatments had the lowest average of 2% core salt whereas the other treatments had an average of c. 3% core salt. Both the control and vacuum pulse treatment also had an average outer salt level of c. 3% compared to the other three treatments, which had an average 4% salt level. Vacuum pulse tumble and vacuum pulse treatments were observed as similar overall acceptability to the control while vacuum tumble and tumble only treatments were found to be higher ($p<0.05$). Williams (2004) found that vacuum-tumbling had minimal and inconsistent effects on quality attributes in M. *Supraspinatus* steaks. No significant effect on colour was found between the outer and core samples of each of the treatments. However, visual differences were apparent between the different treatments, with the control and vacuum pulse treatments showing visible colour differences between the core and outer regions of the beef sample. These different regions indicate that the beef was not fully cured hence uneven distribution of curing ingredients.

Table 1: Effect of treatment on cook loss, salt and moisture content of core and outer regions at day 7.

Treatment	% Cook loss	Core region		Outer region	
		% Salt	% Moisture	% Salt	% Moisture
Control	26.5^{a}	2.1^{a}	72.3^{a}	3.8^{a}	68.5^{a}
Vacuum pulse tumble	22.4^{b}	2.4^{a}	71.2^{a}	3.6^{a}	69.8^{b}
Tumble only	20.8^{b}	2.0^{a}	71.9^{a}	3.3^{a}	69.8^{b}
Vacuum tumble	21.5^{b}	3.3^{b}	69.5^{b}	4.0^{a}	70.1^{b}
Vacuum pulse	20.1^{b}	3.1^{b}	71.0^{a}	3.6^{a}	70.1^{b}

$^{a, b}$ Means in the same column with unlike superscripts are different ($p<0.001$).
SED: standard error of the difference of the means

Table 2: Sensory analysis results of cured beef.

Treatment Type	Cured Colour	Colour Uniformity	Juiciness	Tenderness	Saltiness	O/F	O/A
Control	3.4^{ab}	2.8^{a}	4.0^{a}	4.5^{a}	3.6^{a}	3.2^{a}	2.9^{a}
Vacuum PT	3.6^{ab}	4.3^{b}	4.9^{b}	5.3^{bc}	4.5^{b}	3.3^{a}	3.4^{ab}
Vacuum pulse	2.9^{a}	3.8^{b}	4.0^{a}	4.5^{ab}	3.5^{a}	3.3^{a}	3.2^{ab}
Vacuum tumble	3.6^{b}	4.2^{b}	4.9^{b}	5.6^{c}	4.1^{ab}	3.7^{a}	3.7^{b}
Tumble only	4.6^{c}	4.3^{b}	4.8^{b}	5.8^{c}	3.8^{ab}	3.7^{a}	3.7^{b}

Vacuum PT = vacuum pulse tumble, O/F= Overall Flavour, O/A= Overall Acceptability,
$^{a, b}$ Means in the same column with unlike superscripts are different ($p<0.001$). SED: standard error of the difference of the means
Cured colour, colour uniformity, saltiness, O/F and O/A were evaluated by means of a six-point hedonic scale (6 very pale, excellent/100%uniformity, very salty, extremely good flavour, extremely acceptable respectively; 1= very dark, very poor, very mild, very poor, not acceptable respectively). Juiciness and tenderness were evaluated by means of a eight-point hedonic scale (8 = Extremely juicy and extremely tender respectively; 1= Extremely dry and extremely tough respectively)

Conclusions

This research has shown the benefits of vacuum tumbling technique. Further research will include a detailed shelf-life study, to be carried out on the vacuum tumble treatment, which was chosen as the optimum treatment for dry curing of beef. This will be compared against the static conventional methods in the development of cured products with enhanced organoleptic quality.

References

AMSA (1995). Research guidelines for cooking, sensory evaluation and instrumental tenderness measurements of fresh meat. American Meat Science Association, 10-11.

Bostian, M.L., Fish, D.L., Webb, N.B., and Arey, J.J. (1985). Automated methods for determination of fat and moisture in meat and poultry products: collaborative study. *Journal Association of Official Analytical Chemistry, 68, 876.*

Deumier, F., Bohuon, P., Tystram, G., Saber, N. and Collignan, A. (2003). Pulsed vacuum brining of poultry meat: experimental study on the impact of vacuum cycles on mass transfer. *Journal of Food Engineering,* 58, 75-83.

Ghavimi, B., Rogers, R.W., Althen, T.G. and Ammerman, G.R. (1986). Effects if non-vacuum, vacuum and nitrogen back-flushed tumbling on various characteristics of restructured cured beef. *Journal of Food Science*, 51, 116-1168.

Kirk, R.S. and Sawyer, R. (1991). *In:* Pearson's Composition and Analysis of Foods. Longman Scientific & Technical. 14-15.

Sweeny, R. A. and Rexroad, P. R. (1987). Comparison of LECO FP-228 "Nitrogen Determinator" with OAC copper catalyst Kjeldahl method for crude protein. *Journal of the Association of Official Analytical Chemists*, 70, 1028-1030.

Wiebe Jr, W.R. and Schmidt, G.R. (1982). Effects of vacuum mixing and precooking on restructured steaks. *Journal of Food Science,* 47:386-387, 396.

Williams, T.A (2004). Effects of beef enhancement with non-meat ingredients, blade tenderization and vacuum tumbling on quality attributes of 4 beef cuts stored in high oxygen. MSc Thesis Texas A & M University p20-51.

THE MECHANISMS CONTROLLING HEAT AND MASS TRANSFER ON FRYING OF BEEF BURGERS

B. Kovácsné Oroszvári, E. Bayod, I. Sjöholm and E. Tornberg*

Department of Food Technology, Engineering and Nutrition, Lund Institute of Technology, P.O. Box 124, S-22100 Lund, Sweden. Email: eva.tornberg@food.lth.se

Keywords: mass and heat transfer, beef burgers, frying, contraction and permeability

Introduction

During thermal processing, such as frying, minced meat products lose weight as a result of heat denaturation and contraction of the meat proteins. There is a need to understand the mechanisms controlling heat and mass transfer during frying in order to minimise frying losses and frying time, while at the same time achieving a tender and juicy product and ensuring a safe product, having a core temperature of 72°C.

Cooking loss can be divided into two main components, namely, moisture and fat loss. Moisture can be lost either as vapour through the crust-core interface or as drip, while fat leaves the patty only in the form of drip, since at normal frying temperatures fat does not vaporise. In this study an overall approach has been taken, considering both water and fat transport, heat transfer, shrinkage and porosity development during double-sided pan frying of beef burgers to gain better understanding of the main mechanisms governing these phenomena.

Special emphasis has been put on investigating the fraction of water loss achieved by drip and vaporisation, respectively. The effects of different factors such as the chemical composition of the meat raw material, the pan temperature and the meat patty diameter on the characteristics of heat and mass transfer were studied. The evolution of the characteristics of heat and mass transfer was investigated as a function of frying time. Ultimately, a technique was developed to measure the permeability of the porous meat patty. This property and the related mass transfer were observed at different cooking temperatures up to 80°C.

Materials and Methods

Material: Different beef meat material (Shank, Lean Brisket, Rib and Fat Brisket) have been used to obtain beef burgers of low, medium and high fat content.

Frying procedure: A double-sided pan fryer with heating surfaces of 100, 150 and 175°C was used. In all the frying experiments the beef burgers were fried from -20°C until a temperature of 72°C at the centre of the beef burger was reached.

Heat transfer: For the temperature measurements two K-type thermocouples (T/T-40-K) were used, one at the centre and one 2mm below the surface.

Mass transfer: Water and fat loss were calculated as a percentage related to the weight of the raw beefburger. The flux of water and fat was calculated, by dividing the water and fat loss by the original total and lateral area of the beefburger and by the total frying time

Porosity and permeability: The air-filled porosity [%] was calculated from the volume derived from the difference between the calculated meat patty volume related to fat and water losses, and the volume calculated from the diameter shrinkage. The intrinsic permeability of the heat-treated beefburgers (K) was calculated by Darcy's law : $K = \frac{\bar{u} \cdot \mu \cdot \Delta x}{\Delta P}$ where u is the flux of water and fat [$m \cdot s^{-1}$], μ is the viscosity of water or fat [$kg \cdot m^{-1} \cdot s^{-1}$], Δx is the height of the beefburger [m], ΔP is the pressure gradient [Pa].

Results and Discussion

The most determinant factor for the water flux is the temperature gradient, and for the fat flux, the fat content.

The water flux during the first 90 seconds of frying (thawing period) decreases. However, most of the water is lost during that period. After 90 seconds crust formation starts and the water flux increases, probably due to the enhancement of water loss by evaporation through the crust-core interface. The fat content governs both the amount and the initial time of the fat release during frying. The higher the rate of heat penetration, by using beef burgers of smaller diameters and higher cooking temperature, the higher the rate of water losses induced and the faster the crust is formed, which in turn results in less shrinkage and higher porosity of the heat processed meat. The highest permeability value was found in beef burgers prepared from the meat cooked at 60°C, where after the highest contraction between 60 and 70°C caused a much denser beef burger. The main mechanism for mass transfer during frying was suggested to be a flow of fat and water, created due to the denaturation and contraction of the meat proteins, and moving firstly inwards and there after in the radial direction and ultimately migrating out through the circumference of the burger. Therefore the pressure-driven water loss, i.e. drip, is the main mechanism governing the water loss in the frying of beefburgers and not the evaporation of water.

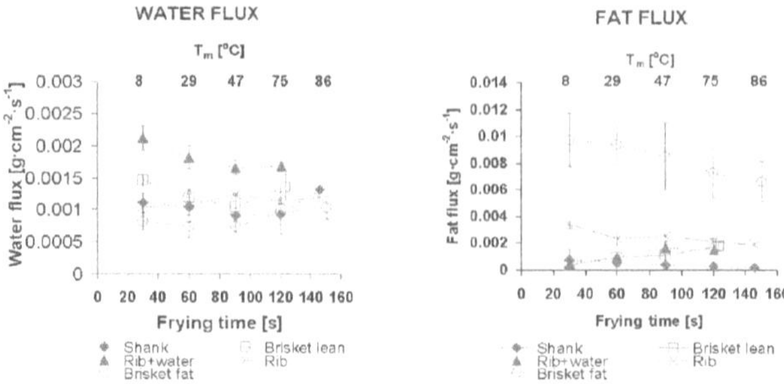

Figure 1: Water and fat flux as a function of the frying time and the average temperature for beefburgers made of five different meat raw materials.

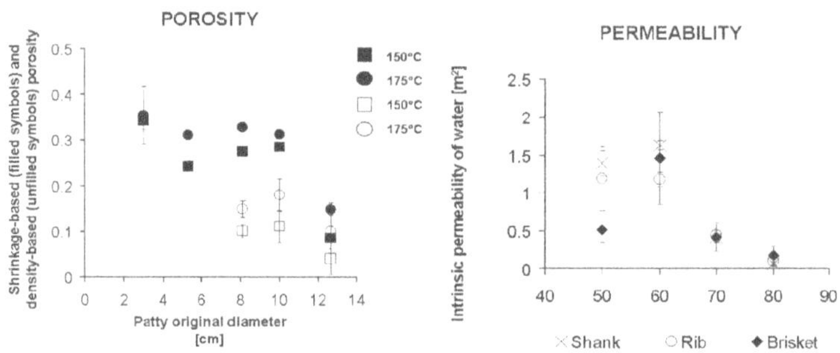

Figure 2: Porosity as a function of original patty diameter at 150°C and 175°C pan temperature and permeability of beefburgers, using different meat raw materials, as a function of endpoint-temperature.

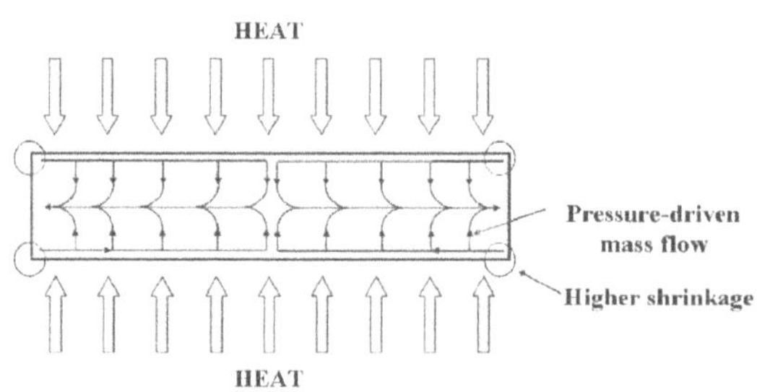

Figure 3: Schematic drawing of heat and mass transfer in the beefburger during double-sided pan frying.

Conclusions

As a result of the thermal denaturation and contraction of the meat proteins during pan-frying of beefburgers, a flow of water and fat is created moving towards the centre of the patty, which in turn speeds up the heat transfer. The water and fat drip leaves the patty through the circumference of the burger.

References

Kovácsné Oroszvári, B., Sjöholm, I. and Tornberg, E. (2004). The mechanisms controlling heat and mass transfer on frying of beefburgers. I. The influence of the composition and comminution of meat raw material. Journal of Food Engineering, 67, 4, 499-506.

Kovácsné Oroszvári,B.,Bayod, E., Sjöholm, I. and Tornberg, E. (2005).The mechanisms controlling heat and mass transfer on frying of beefburgers. II. The influence of the pan temperature and patty diameter. Journal of Food Engineering, 71, 18-27.

Kovácsné Oroszvári, B., Bayod, E., Sjöholm, I. and Tornberg, E. (2006).The mechanisms controlling heat and mass transfer on frying of beefburgers. III. Mass transfer evolution during frying. Journal of Food Engineering 76, 169-178.

Kovácsné Oroszvári,B., Rocha, C.S., Sjöholm, I. and Tornberg, E. (2006). Permeability and mass transfer as a function of the cooking temperature during the frying of beefburgers. Journal of Food Engineering, 74, 1-12.

SIMULATION OF A DESIGN OF A LAB SCALE RADIO FREQUENCY HEATING UNIT OF HAMBURGERS

M. Rayner and E. Tornberg*

Department of Food Technology, Engineering and Nutrition, Lund Institute of Technology, P.O. Box 124, S-22100 Lund, Sweden. Email: eva.tornberg@food.lth.se

Keywords: radio frequency heating, beef burgers, simulation, FEMLAB, resistance

Introduction

With regard to efficient food processing and cost effective production lines, Radio Frequency (RF) heating is a prime example. Thermal processing of food has gained an ever growing importance as food manufactures strive after increasing the number and quality of value added products on the market. Solid food, such as meat and vegetables, is more difficult to heat efficiently since it cannot be pumped or stirred. This is an area where RF cooking can provide significant benefits due to the volumetric nature of dielectric heating. Since heat is generated internally in the product, by inducing an electric field through the food, which acts as a dielectric material, and does not rely on conduction, convection or radiation, heating takes place evenly in the product without temperature gradients. RF is also superior to micro-wave heating in this respect as the wave can penetrate food items up to several meters in diameter whereas the penetration is limited to a few centimeters at micro-wave frequencies. This means that there are no hot-spots, cold-spots or overcooking of surfaces and the total time is significantly reduced and losses are minimised.

In order to fully take advantage of RF heating there is a need for improved equipment flexibility and design methods. The main reason for this is the high level of investment costs for this equipment. However, this existing RF technology has a well-know limitation: it is generally difficult to modify specialized equipment in order to do something even slightly different with it (Marchand and Meunier, 1990). So, for many new or non-standard applications, because of these design and starting-up difficulties, the investment costs remain too high. Trial and error is not an optimal design method since in RF heating efficiency and process outcomes are highly dependent on the product characteristics rather than just the choice process parameters. Therefore, there is an undeniable need to develop new methods and techniques that can be applied rapidly and efficiently in the quest for better optimized industrial systems. According to Marchand and Meunier (1990) to overcome this barrier different compromises between standardization and versatility must be found and the recent developments in measuring and computing techniques must be taken full advantage of. Metaxas (2000) asserts that with the advent of powerful computers many simulation packages have been developed, which pave the way for better technology transfer for RF heating techniques in the medium-term future.

There are also several challenges directly related to the physics of RF heating which need to be overcome, two major ones being arching and thermal run away heating (Zhao *et al.* 2000). Thermal run away heating or "hot-spots" in a heterogeneous medium is caused by the fact that in some materials dielectric loss factors often exhibit a strong positive correlation to temperature. *i.e.* the hotter something is the faster it heats (Zhao *et al.* 2000). Arching (or dielectric breakdown) is caused if the electric field strength across the sample is too high.

This paper describes the modeling work carried out in conjunction with the preliminary design of a lab scale RF-heating apparatus to overcome these problems. The adaptation of 50 Ω technology means that RF applicator can be designed for optimum performance with regard to electrode shape and spacing, since independent of, rather than being a part the tuning system. The placement and design of electrodes affect the resistance in the circuit and the distribution of current in the material being heated. The higher the resistance the greater the degree of RF heating. The more uniform the current density, the more uniform the heating is.

Materials and Methods

The purpose with this simulation is to design a lab scale RF-heating apparatus being able to cook hamburger patties with a dimension of 10mm thick and 30mm in diameter. This particular size was chosen in order to be able to compare results with previous work from our department using double sided frying (Oroszvári *et al.*, 2006). This work does not represent a final design, but rather tries to identify which process and design parameters are of greatest importance, *i.e.* recognising and quantifying what we need to know in order to make informed decisions while constructing the apparatus.

Results and Discussion

Resistances and Currents are significantly affected by applicator geometry. Conductivity is an intrinsic material property and constant at a given frequency and temperature, but the geometry can be changed to increase the resistance. Dielectric data for beef at 27.12MHz: $\varepsilon'=70.5$ $\varepsilon'' = 419$, electrical conductivity $\sigma=2\pi f\varepsilon''\varepsilon_o = 0.63$ [S/m].

If we consider the simplified 2 dimensional case, the hamburger can be heated in two possible directions. Either by applying the field in the axial direction between 2 plates above and below and shown in Figure 1, Case A (left) or in the radial direction from plates positioned on the sides, shown in case B (right).

Case A

$$R \propto \frac{d}{L\sigma} \propto \frac{1}{3\sigma}$$

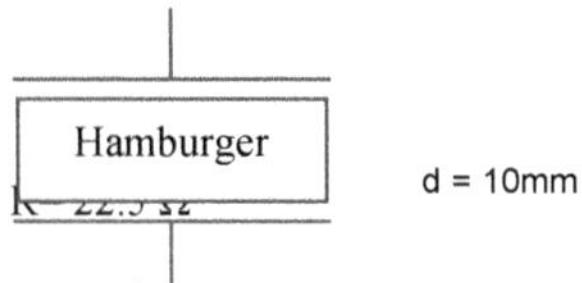

Case B

$$R \propto \frac{L}{d\sigma} \propto \frac{3}{\sigma}$$

R= 2[illegible]Ω

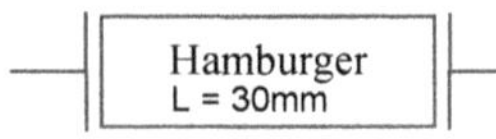

3D FEMLAB model $\sigma \nabla^2 V = 0$, where V is volts, the resistance was calculated from the potential divided by the current. The current was found by integrating the current density over grounded boundary.

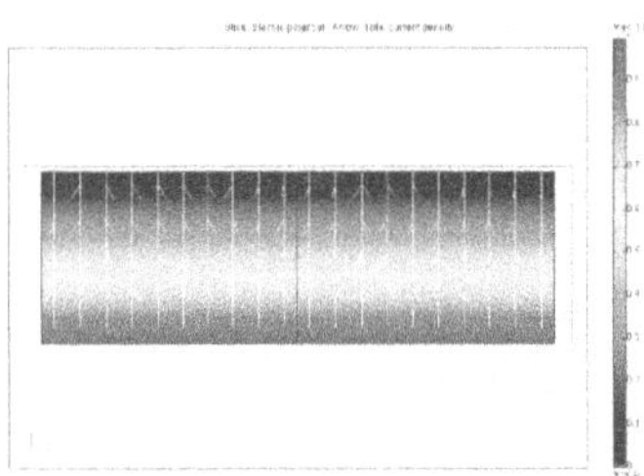

R = 22.5 Ω, same as analytical

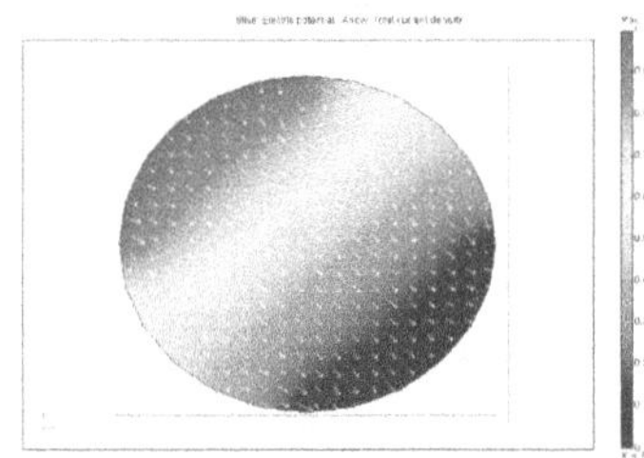

R=176 Ω, smaller than analytical because current will take the shortest path. This caused the current density vary through the material and can cause uneven heating.

Beyond the plane the EM field is applied, the distance between the burger and the electrodes was also found to have a significant effect on the heating pattern. The figures below show the effect of placing the electrodes at a distance of 60% larger than the burger (left) and 8% larger than the burger (right).

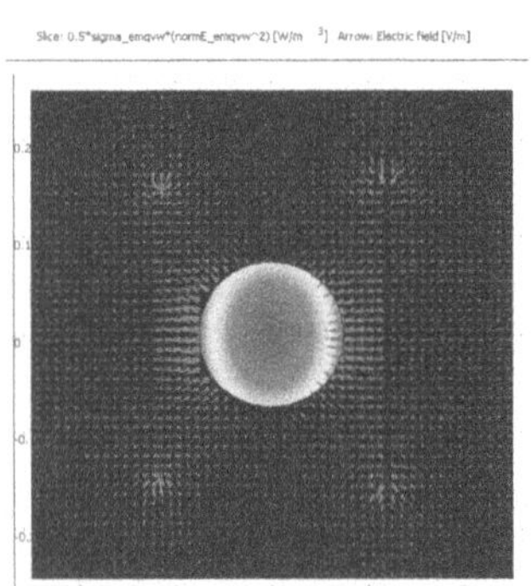

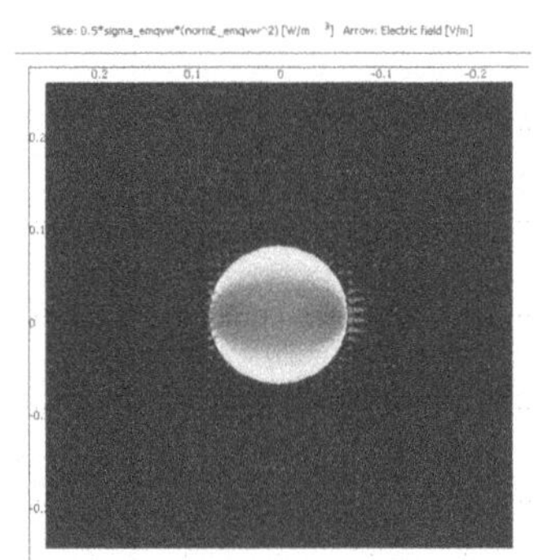

Conclusions

By using FEMLAB, a computers simulation package, the 3D resistance in hamburger patties with a dimension of 10mm thick and 30mm in diameter, using Radio Frequency (RF) heating was simulated. By applying the RF field in the axial direction between 2 plates above and below or in the radial direction from plates positioned on the sides it was shown that the latter configuration led to approximately 8 times higher resistance in the hamburger patty compared to the former and thereby a more efficient heating. Electrode placement has an important effect on heating distribution and therefore needs to be considered in applicator design.

References

Marchand, C. and Meunier,T. (1990). Journal of Microwave Power and Electromagnetic Energy 25, 1, 39-46.

Metaxas, R. (2000). Power Engineering Journal April, 51.

Zhao Y.Y., Flugstad, B., Kolbe, E., Park, J.W. and Wells, J.H. (2000). Journal of Food Process Engineering, 23, 1, 25-55.

Kovácsné Oroszvári, B., Bayod, E., Sjöholm, I. and Tornberg, E. (2006). The mechanisms controlling heat and mass transfer on frying of beefburgers. III. Mass transfer evolution during frying. Journal of Food Engineering 76, 169-178.

CHEMICAL CHARACTERISTICS OF DESALTED AND NON-DESALTED ISTRIAN DRY-CURED HAMS

M. Krvavica[1], S. Vidaček*[2], H. Medić[2] , K. Botka-Petrak[3] and T. Petrak[2]

[1] Marko Marulić Polytechnics in Knin, Petra Krešimira IV 30, 22300 Knin, Croatia, [2] Faculty of Food Technology and Biotechnology, University of Zagreb, Pierottijeva 6, 10000 Zagreb, Croatia, [3]Faculty of Veterinary Medicine, University of Zagreb, Heinzelova 55, 10000 Zagreb, Croatia. Email:svidacek@pbf.hr

Keywords: dry cured Istrian ham, desalting, ham quality

Introduction

The traditional production of dry-cured ham in the coastal areas of Croatia (Istria and Dalmatia) is likely to be as old as found in other Mediterranean countries. By definition, Istrian ham is a lasting cured meat product, processed in the Istrian manner with or without the aitch bones, salted with sea salt and natural herbs, and cured in the air without smoke. Considering that the desalting process contributes to a more uniform distribution of salt in the dry-cured ham and impacts the coarse of lipolysis and proteolysis, the objective of this work was to determine the effect of salt concentration on the most important chemical parameters of Istrian dry-cured ham (Andres *et al.*, 2005; Martin *et al.*, 1998).

Materials and Methods

21 Swedish Landrace breed hogs underwent an extended fattening process to a final body mass of 190-200 kg. Hogs were slaughtered, cut up, cooled and the raw hams trimmed. Legs were processed in the traditional Istrian manner, medially and laterally without rind and the subcutaneous adipose tissue, with the aitch bones. The continued technological process was conducted in Istria according to traditional Istrian technology. *Dry salting* of hams was conducted using a mixture of course and fine sea salt, with an addition of herbs. *Pressing* was carried out by burdening the hams with a mass of 200 kg/m². Half the hams were *desalted* by soakage in pure, cold water. After desalting, hams were subjected to *drying* and *ripening* processes. The process of production of Istrian dry cured ham lasted for 14 months. To carry out the chemical analysis a section was made by a longitudinal cross-section of the muscle on the caudal side of the leg, from the *tuber ishiadicum* to the *tuber calcanei*, and the following muscles were cross-sectioned: *m. semimembranosus*, *m. semitendinosus*, *m. gracilis*, *m. biceps femoris*, Chemical analysis included determination of moisture content and dry matter, total protein content, intramuscular fat content, ash content, mineral content, amino acid content and fatty acid content.

Results and Discussion

The technological process of desalting did not significantly influence the chemical composition of the dry-cured hams, with the exception of salt content (Figure 1). Statistical analysis of the macro and microelement content showed a significant difference for the mean content of the macroelement calcium (desalted: 0.26 g/kg; non-desalted: 0.23 g/kg). The most represented free amino acids in the experimental dry-cured hams are shown in Figure 2, expressed in g/100 g of sample. These amino acids were also found in Iberian and Serrano hams (Ruiz *et al.*, 1999; Toldrá, 1998). In the majority of cases, the non-desalted hams had a greater content of individual amino acids than the desalted hams, however the differences in the arithmetic means were not statistically significant, with the exception of methionine, which had a content statistically significant higher than in non-desalted hams (P=0.013). Several studies have shown a positive effect of salt on lipolysis (Motilva and Toldrá, 1993), while other papers have not shown this correlation (Coutron-Gambotti and Gandemer, 1999). Statistical analysis of the presence of free fatty acids in muscle tissue of dry-cured ham in this study showed that desalting had a statistically significant impact on the content of certain fatty acids in the intramuscular fat of dry-cured ham (Figure 3).

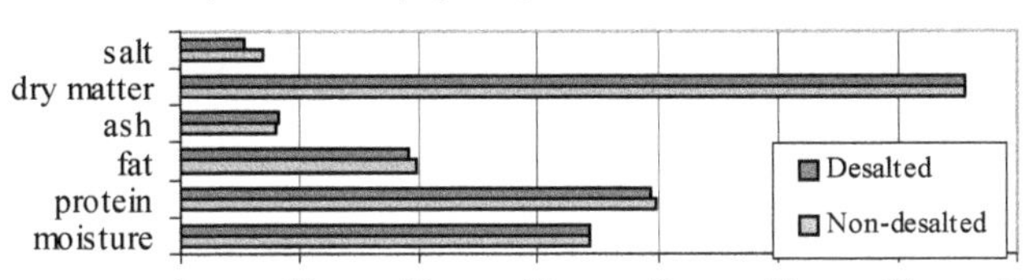

Figure 1: Chemical composition of desalted and non-desalted Istrian dry-cured hams (a statistically significant difference was found for the salt content (non-desalted : 6.83%; desalted: 5.33%; p<0.001)).

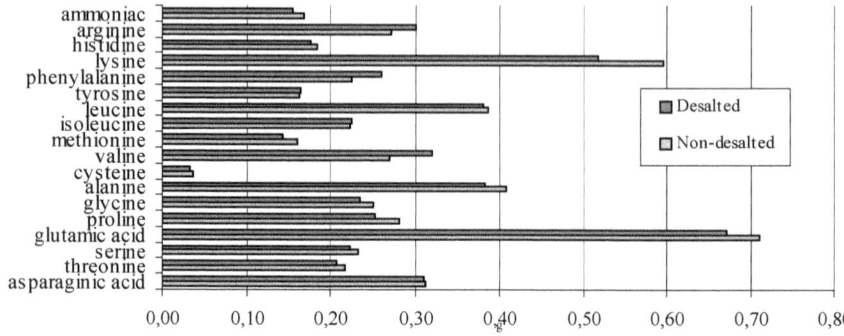

Figure 2: Amino acid composition of desalted and non-desalted Istrian dry-cured hams (g/100 g sample) with a statistically significant difference for the methionine content (P=0.013).

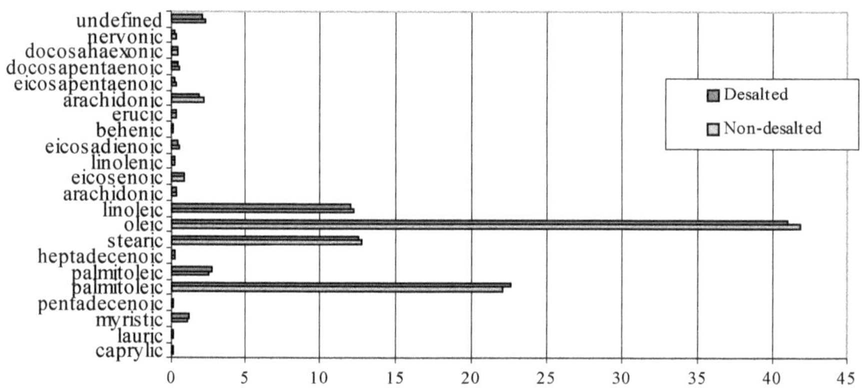

Figure 3: Free fatty acids in Istrian dry-cured ham muscle tissue (%).

A statistically significant difference (P<0.05) was found for the content of eikosapentaenoic acid (C20:5ω-3), which had a higher content in desalted dry-cured ham (P=0.019), while a statistically highly significant difference (P<0.01) was found for the content of heptadecanoic (C17:0) and docosapentaenoic (C22:5ω-3) acids; the content of heptadecanoic was higher in desalted (P=0.003) and docosapentaenoic acids in non-desalted dry-cured ham (P=0.002))

Conclusion

It is assumed that varying salt concentrations in dry-cured ham tissues during processing impacts the course of biochemical processes in the dry-cured ham, leading to a statistically significant difference in the chemical composition in non-desalted and desalted hams.

References

Andrés, A.I., Cava, R., Martin, D., Ventanas, J. and Ruiz, J. (2005) Lipolysis in dry-cured ham: Influence of salt content and processing conditions. Food Chemistry 90, 523-533.

Coutron-Gamboti, C.and Gandemer, G. (1999) Lipolysis and oxidation in subcutaneous adipose tissue during dry-cured ham processing. Food Chemistry 64, 95-101.

Martín, L., Córdoba, J.J., Antequera T., Timón, M.L. and Ventanas, J. (1998). Effects of salt and temperature on proteolysis during of Iberian ham. Meat Science, 49, 145-153.

Motilva, M-J. and Toldrá, F. (1993) Effect of curing agents and water activity on pork muscle and adipose subcutaneous tissue lipolytic activity. Zeitschrift fur Lebensmittel Untersuchung und Forschung 196, 228-231.

Ruiz, J., García, C., Díaz, M.C., Cava, R., Tejeda, J.F.and Ventanas, J. (1999) Dry cured Iberian ham non-volatile components affected by the length of the curing process. Food Research International, 32, 643-651.

Toldrá, F. (1998) Proteolysis and lipolysis in flavour development of dry-cured meat products. Meat Science, 49, 101-110.

EFFECT OF THE FREEZING PROCESS ON RAW MATERIAL DURING THE MANUFACTURE PROCESS OF "CECINA DE LEÓN"

C. Molinero*, B. Martínez, B. Rubio and M.D. García-Cachán

Estación Tecnológica de la Carne. Instituto Tecnológico Agrario de Castilla y León.
C/ Filiberto Villalobos s/n. 37770 Guijuelo (Salamanca). Spain. Email: molsascr@itacyl.es

Keywords: freezing, dry cured beef, microbiology, sensorial, texture

Introduction

Dry-cured beef "Cecina de León" is a traditional intermediate meat product, manufactured in the province of León (Northwest of Spain). The processing is similar to that followed in the production of dry-cured ham. During the ripening of this meat product, proteins undergo various modifications, particularly proteolysis. As a consequence of these changes peptides, free amino acids and other small nitrogenized compounds are generated. Nowadays, due to the continuous increase in "Cecina de León" consumption it is necessary alternatives for increasing "Cecina de León" production without affecting its quality. Frozen raw material is an appropriate preservation technique to enable production independently of season. However, it must be taken into account that, some biochemical mechanisms, like proteolysis, can be affected. The aim of this work was to evaluate the microbiological and the sensorial evolution in "Cecina de León" manufactured with both frozen and refrigerated raw material, looking for possible differences between them.

Materials and Methods

Twenty knuckles (composed by *M. rectus femoris* and *vastus lateralis*, *vastus. medialis* and *vastus intermedius*), weighing 5 ± 1 kg were procured from a slaughterhouse. The raw pieces were divided in two batches: (batch 1) the raw material was refrigerated (2-3 ºC) and (batch 2) the raw material was frozen (7 days, -20ºC) and thawed (4 days, 4ºC). The manufacture process was according to the specifications of Protected Geographical Indication "Cecina de León" (BOCyL, 1994). Microbiological analysis (mesophilic aerobic bacteria, lactic acid bacteria and *Micrococaceae*) and instrumental texture profile analysis (TPA) were carried out at the pre-salting stage (day 0), at the middle of the post-salting stage (day 37), at the end of the smoking stage (day 55), during the drying phase (days 90, 145 and 180) and at the end of the process (day 210). In order to detected sensory differences between batches, a triangular test was performed at 210 days with an experienced 8-members panel. Besides, the panellists carried out a descriptive test using a 5-point hedonic scale for to evaluate the main sensory attributes of "Cecina de León". Statistical analysis of data was carried out by one-way analysis of variance, and means were separated by Tukey Honest Significant Difference test using at 5% level (Statistica software package).

Results and Discussion

Results obtained for microbiological analysis are shown in table 1. Mesophilic aerobic bacteria counts increased from day 0 to day 37 day and then, these counts remained constant until 187 days. At the end of manufacture process (210 days) a decrease was observed for mesophilic aerobic bacteria numbers and the values obtained at this point were similar to those found at day 0. *Micrococcaceae* counts increased until day 55 for batch 1 and these count remained constant until day 187 to decrease at the end of the process. On the other hand, the counts in batch 2 increased until 90 day and a decrease was observed at the end of the process. Regarding lactic acid bacteria counts did not shown a clear trend along manufactured process. In general, *Micrococcaceae* and lactic acid bacteria constituted the predominant flora of "Cecina de León". This fact could be explained because coccus and lactic acid bacteria are salt tolerant and are not inhibited by osmotic pressure or partial desiccation (Cornejo *et al.*, 1992; Carrascosa and Cornejo, 1991). No significant differences ($P>0.05$) were found with either mesophilic aerobic bacteria or *Micrococcaceae* between batches. These results agree with data reported by other authors for "Cecina de León" (Garcia *et al.*, 1995) and for dry-cured Iberian ham (Rodriguez *et al.*, 1995). However, significant differences ($P<0.05$) between batches were found in lactic acid bacteria but these differences did not allow establishing clear differences using refrigerated raw and frozen/thawed raw material. Table 2 shows the changes in the instrumental parameters of texture during the manufacture process of "Cecina León". There were evident textural changes, particularly regarding hardness, which increased significantly ($p<0.05$). Besides, cohesiveness and springiness decreased during the manufacture process ($p<0.05$). Changes in these texture parameters could be attributed to both loss water content and protein changes (Serra *et al.*, 2005). The chewiness, as expected, showed an analogous tendency that hardness. On the other hand, no difference ($p>0.05$) were found between batches respect to the parameters of instrumental texture. Regarding results obtained with the sensorial panel, the triangular test showed differences between batch 1 and batch 2 ($P<0.05$). Besides, the sensory scores obtained in descriptive test (table 3) showed significant differences ($P<0.05$) in colour intensity, marbling and chewiness between batches, probably due to the rupture of tissue membranes during freezing.

Table 1: Evolution of aerobic mesophilic bacteria, lactic acid bacteria and *Micrococacceae* (mean values) (log cfu/g) along manufacture process of "Cecina de León" made using refrigerated raw material (batch 1) and frozen/thawed raw material (batch 2).

	Batch	*Days*						
		0	37	55	90	145	187	210
Mesophilic aerobic	1	$^{A}5.52_{a}$	$^{A}6.50_{cb}$	$^{A}6.59_{c}$	$^{A}6.69_{c}$	$^{A}6.3_{cba}$	$^{A}6.12_{cba}$	$^{A}5.16_{a}$
bacteria	2	$^{A}5.18_{a}$	$^{A}6.54_{cb}$	$^{A}6.60_{cb}$	$^{A}6.84_{c}$	$^{A}6.15_{b}$	$^{A}5.47_{a}$	$^{A}5.50_{a}$
Micrococacceae	1	$^{A}3.13_{a}$	$^{A}4.74_{b}$	$^{B}6.38_{c}$	$^{A}6.18_{c}$	$^{A}6.11_{c}$	$^{A}6.06_{c}$	$^{A}5.06_{b}$
	2	$^{A}3.15_{a}$	$^{A}4.73_{b}$	$^{A}5.15_{dcb}$	$^{B}6.90_{e}$	$^{A}6.06_{ed}$	$^{A}5.69_{dc}$	$^{A}5.00_{cb}$
Lactic acid bacteria	1	$^{B}5.35_{bc}$	$^{A}4.33_{a}$	$^{A}4.47_{a}$	$^{B}6.07_{d}$	$^{B}6.39_{d}$	$^{B}5.93_{cd}$	$^{A}4.95_{ab}$
	2	$^{A}4.23_{a}$	$^{B}5.12_{b}$	$^{A}4.39_{a}$	$^{A}5.74_{c}$	$^{A}5.39_{c}$	$^{A}4.30_{a}$	$^{A}5.30_{b}$

a, b, c, d, e Means with different letters in the same row indicate significant differences during the processing ($p<0.05$).
A,B Means with different letters in the same column indicate significant differences ($p<0.05$) between batches.

Table 2: Evolution of instrumental texture parameters (mean values) along manufacture process of "Cecina de León" made using refrigerated raw material (batch 1) and frozen/thawed raw material (batch 2).

	Batch	*Days*				
		55	90	145	187	210
Hardness (g)	1	$^{A}5133.0_{a}$	$^{A}9253.5_{ab}$	$^{A}12452.0_{b}$	$^{A}12039.1_{b}$	$^{A}9364.9_{ab}$
	2	$^{A}3486.0_{a}$	$^{A}6885.5_{a}$	$^{A}12101.0_{b}$	$^{A}14330.8_{b}$	$^{B}11963.1_{b}$
Cohesivenes	1	$^{A}0.49_{b}$	$^{A}0.48_{b}$	$^{A}0.49_{b}$	$^{A}0.33_{a}$	$^{B}0.39_{a}$
	2	$^{A}0.45_{b}$	$^{A}0.43_{b}$	$^{A}0.48_{b}$	$^{A}0.34_{a}$	$^{A}0.34_{a}$
Springiness	1	$^{B}0.76_{b}$	$^{A}0.62_{b}$	$^{A}0.60_{a}$	$^{A}0.50_{a}$	$^{A}0.51_{a}$
	2	$^{A}0.60_{b}$	$^{A}0.60_{b}$	$^{A}0.58_{ba}$	$^{A}0.50_{a}$	$^{A}0.51_{ab}$
Chewiness (g)	1	$^{A}1683.6_{a}$	$^{A}2690.8_{ab}$	$^{A}3821.8_{b}$	$^{A}2002.6_{a}$	$^{A}1939.3_{a}$
	2	$^{A}933.7_{a}$	$^{A}1948.4_{ab}$	$^{A}3470.9_{c}$	$^{A}2358.5_{bc}$	$^{A}2107.1_{c}$

a, b, c Means with different letters in the same row indicate significant differences during the processing ($p<0.05$).
A,B Means with different letters in the same column indicate significant differences ($p<0.05$) between batches.

Table 3: Results of sensorial parameters (mean ± S.D.) evaluated on dry-cured "Cecina de León" made using refrigerated raw material (batch 1) and frozen/thawed raw material (batch 2).

	Batch 1	Batch 2
Colour homogeneity	$^{A}3.50\pm1.06$	$^{A}3.92\pm0.53$
Colour intensity	$^{A}4.29\pm0.49$	$^{B}3.69\pm0.63$
Marbling	$^{A}1.50\pm0.64$	$^{B}2.61\pm0.98$
Yellowness	$^{A}1.75\pm0.84$	$^{A}1.80\pm0.90$
Intermuscular fat	$^{A}1.33\pm0.61$	$^{A}1.77\pm0.81$
Odour intensity	$^{A}2.70\pm0.96$	$^{A}2.77\pm1.07$
Chewiness	$^{A}2.77\pm1.07$	$^{B}3.20\pm0.78$
Juiciness	$^{A}2.21\pm1.25$	$^{A}2.11\pm1.29$
Pastiness	$^{A}1.21\pm0.39$	$^{A}1.23\pm0.38$
Flavour intensity	$^{A}3.00\pm1.18$	$^{A}2.88\pm1.32$

A,B Means with different letters indicate significant differences between batches ($p<0.05$).
Values rates on a 5 point scale (1=lowest values,-5=highest values).

Conclusions

The freezing of the raw material had hardly any influence on the microbial and texture parameters evolution along manufacture process of "Cecina de León". However, the freezing of the raw material had an effect on sensorial characteristics of dry-cured "Cecina de León" according to panellists.

References

BOCyL (Boletín Oficial de Castilla y León) (1994). Orden de 17 de enero de 1994, por la que se aprueba el Reglamento de la Denominación Específica "Cecina de León" y su Consejo Regulador. Boletín Oficial de Castilla y León N° 14, de 21 de enero de 1994, 444-450.

Cornejo, I., Carrascosa, A. V., Marin, M.E. and Martin, P.J. (1992). Consideration about the origin of microorganisms that grow on the deep muscle tissues of dry-cured Spanish hams during processing. *Fleischwirtsch.* 72, 1405-1407.

Carrascosa, A. V. and Cornejo, I. (1991). Characterization of Micrococcaceae strains selected as potential starter cultures to Spanish dry cured ham processes. 2. Slow process. *Fleischwirtsch.* 71, 1187-1188.

García, I., Zumalacárregui, J.M. and Díez, V. (1995). Microbial succession and identification on Micrococcaceae in dried beef "Cecina", and intermediate moisture meat product. Food Microbiol. 12, 309-315.

Rodríguez-Nuñez., E., Aristoy, M.C. y Toldrá, F. (1995). Peptide generation in the processing of dry-cured ham. Food Chem. 53, 187-190.

Serra, X., Ruiz-Ramírez, J., Arnau, J. and Gou, P. (2005).Texture parameters of dry-cured ham *m. bíceps femoris* samples dried at different levels as a function of water activity and water content. Meat Science 69, 249-254.

Acknowledgments The authors thank the PGI "Cecina de León for providing the samples used in this work. This research was supported by the"Instituto Tecnologico Agrario de Castilla y Leon" project (2003/01001) and Fondos Feder.

INFLUENCE OF NEW PRESERVATION TECHNIQUES ON SENSORY CHARACTERISTICS OF "MORCILLA DE BURGOS"

A.M. Diez*[1], L. González-Arnaiz[1], E.M Santos[2], I. Jaime[1] and J. Rovira[1]

[1]*Departamento de Biotecnología y Ciencia de los Alimentos, Universidad de Burgos, Burgos, SPAIN.* [2]*Centro de investigaciones Químicas, Universidad Autónoma del Estado de Hidalgo, Unidad Universitaria, Pachuca, Hidalgo, MEXICO. Email: ijaime@ubu.es*

Keywords: preservation, shelf life, vacuum packaging, organic acids

Introduction

"Morcilla de Burgos" is a traditional cooked blood sausage produced in the Burgos region, in the north of Spain. This product is commercialised as a fresh product with a shelf-life of around 10 days. However, vacuum packaging of this product has increased in order to extend the shelf-life and to increase market impact. The product is typically consumed by deep frying in vegetable oils, roasted or boiled as part of other traditional dishes. Cooked meat products, including "Morcilla de Burgos", are liable to contamination during post-cooking handling, particularly during the chilling step, just before vacuum packaging. In order to extend the shelf-life and safety of the product high pressure treatments and addition of organic acids have been applied. The aim of this work was to study the sensory evolution of vacuum packaged "morcilla" during a 50 day period, following the application of different preservation methods (HPP and addition of organic acids).

Materials and Methods

Samples: Four batches of 50 samples were treated (C: control; A: 3% of a mixture of lactate-Na and lactate-K (Purasal LITE. Purac.) (the Na content was adjusted in the product); H: 10 min at 600 MPa (these conditions were chosen due to previous studies with "morcilla" (Borek *et al.*, 2002; Diez *et al.*, 2005) which were shown to be optimal conditions), and AH: 3% lactate and 10 min at 600 MPa).

High Pressure equipment: comprised plant equipment with Wave 6000/120 with horizontal vessel of 300 mm diameter and 2400 mm of length, with a total capacity of 120 L developed by NC Hiperbaric (Nicolas Correa Group. Burgos. Spain).

Sensory analysis: two types of sensory analysis were conducted: a discriminative test and a sensory profile.

Difference test: a paired comparison between different treatments was done by untrained panellists. "Morcilla" was cooked in a microwave until the temperature reached 70-75°C in the core and served to the panellist in slices of 1 cm thickness. This test was completed for each batch including the control.

Sensory profile analysis: Quantitative descriptive analysis was carried out by a trained panel scoring over a five point scale, where 1 correspond to absence and 5 to maximum intensity of each parameter, (Santos *et al.*, 2005), in five sessions, on days 8, 15, 22, 29 and 36. The sensory profile consisted of 7 descriptors grouped in two blocks. The first block related to visual attributes including gas production, presence of spots, and presence of slime. The second block included the evaluation of texture and olfactory attributes including smell, flavour and taste. Samples were evaluated at room temperature for the first two blocks while for the texture and flavour evaluation, ''morcillas'' were cut in 1cm thick slices and were cooked in a domestic microwave to a core temperature of 70–75 °C.

Results and Discussion

Table 1 presents the results for the difference tests. No significant difference was found between treatment C (control) and A (lactate). However, for the rest of the comparisons, significant differences between treatments were found. Most of the differences perceived by the panellists were related to taste and texture of the products, although these differences were not always negative for the product, because the control was not always the preferred, as it is shown in table 1. Previous studies on "morcilla" (Diez *et al*, 2005), showed similar results for high pressure treatment, and other authors (Mor-Mur and Yuste, 2003), obtained significant difference between control and HPP samples, while consumers illustrated a preference for samples that had been subjected to the highest pressure treatment.

Table 1: Results of paired comparison test analysis.

Pairs	Panellists	Correct replies	Significant difference	p	Panellist differentiate by:					Pref.	Pref. %
					A	C	O	T	S		
C vs. H	45	33	Yes	p<0,01	10	8	6	17	28	C	54.54
C vs. A	24	15	No	-	-	-	-	-	-	-	-
C vs. AH	22	16	Yes	p<0,05	4	4	5	8	15	AH	62.50
AH vs. H	44	29	Yes	p<0,05	6	5	12	20	21	AH	62.07
AH vs. A	44	30	Yes	p<0,05	4	9	4	17	20	AH	56,66

Control (C); lactate-Na and lactate-K (A); HPP (H); HPP+ lactate-Na and lactate-K (AH).
p: signification level; A: appearance; C: colour; O: smell; T: texture; S: taste; Pref.: preferred sample;
Pref. %: percentage of panellists that prefer the sample.

The treatments had significant effects (P <0.05) on sensory properties. Scores on all sensory parameters dramatically increased for the control and to a lesser extent with the application of lactate (A) during storage (Table 2), while significant differences did not exist (P < 0.05) for the treatments H and AH during the same period. High pressures treatments were more effective than lactate to extend the sensory shelf-life of "morcilla de Burgos". The same results were observed in previous studies in "morcilla" (Borek *et al.*, 2002; Diez *et al.*, 2005, Santos *et al.*, 2005). When the score was greater than 3, the product was considered unfit for consumption. Products treated with HPP (H and AH) never exceeded this level for the parameters studied. The lactate treatment was less effective than HPP and only extended shelf-life until day 29.

Table 2: Results of sensory profile analysis.

Day	Spots	Slime	Gas	Texture modifications	Acid smells	Off-flavours	Off- tastes
Treatment C							
8	$^{a}1,30_{A}$	$^{a}1,10_{A}$	$^{a}1,00_{A}$	$^{a}1,20_{A}$	$^{a}1,10_{A}$	$^{a}1,10_{A}$	$^{a}1,10_{A}$
15	$^{ab}1,90_{A}$	$^{ab}1,50_{A}$	$^{a}1,20_{A}$	$^{a}1,80_{A}$	$^{b}2,40_{B}$	$^{a}1,60_{A}$	$^{a}1,40_{A}$
22	$^{abc}2,14_{A}$	$^{bc}2,43_{A}$	$^{b}2,43_{B}$	$^{b}2,88_{A}$	$^{bc}3,57_{B}$	$^{bc}2,71_{A}$	$^{b}3,14_{A}$
29	$^{bc}2,88_{A}$	$^{cd}3,38_{B}$	$^{bc}3,25_{B}$	$^{b}3,00_{A}$	$^{c}4,38_{B}$	$^{bc}2,50_{A}$	$^{b}3,75_{B}$
36	$^{c}3,25_{B}$	$^{d}3,88_{C}$	$^{c}4,00_{B}$	$^{b}3,00_{B}$	$^{c}4,50_{C}$	$^{c}3,63_{A}$	$^{b}4,00_{B}$
Treatment A							
8	$^{a}1,20_{A}$	$^{a}1,00_{A}$	$^{a}1,00_{A}$	$^{a}1,30_{A}$	$^{a}1,10_{A}$	$^{a}1,20_{A}$	$^{a}1,10_{A}$
15	$^{a}1,20_{A}$	$^{a}1,20_{A}$	$^{a}1,20_{A}$	$^{a}1,10_{A}$	$^{a}1,20_{A}$	$^{a}1,20_{A}$	$^{a}1,40_{A}$
22	$^{ab}1,71_{A}$	$^{ab}1,71_{A}$	$^{ab}1,43_{AB}$	$^{a}1,72_{A}$	$^{a}2,14_{A}$	$^{a}1,57_{A}$	$^{a}2,00_{A}$
29	$^{ab}1,75_{A}$	$^{ab}1,75_{AB}$	$^{ab}1,38_{A}$	$^{ab}2,00_{A}$	$^{b}3,50_{AB}$	$^{ab}2,50_{A}$	$^{b}4,00_{B}$
36	$^{b}2,38_{AB}$	$^{b}2,38_{B}$	$^{b}2,12_{A}$	$^{b}3,00_{B}$	$^{b}3,38_{BC}$	$^{b}3,50_{A}$	$^{b}3,50_{B}$
Treatment H							
8	$^{a}1,00_{A}$	$^{a}1,00_{A}$	$^{a}1,10_{A}$	$^{a}1,30_{A}$	$^{a}1,20_{A}$	$^{a}1,10_{A}$	$^{a}1,10_{A}$
15	$^{a}1,30_{A}$	$^{a}1,30_{A}$	$^{a}1,20_{A}$	$^{a}1,40_{A}$	$^{a}1,30_{A}$	$^{a}1,20_{A}$	$^{a}1,80_{A}$
22	$^{a}1,43_{A}$	$^{a}1,43_{A}$	$^{a}1,43_{AB}$	$^{a}2,14_{A}$	$^{ab}2,58_{A}$	$^{ab}1,86_{A}$	$^{a}2,00_{A}$
29	$^{a}1,75_{A}$	$^{a}1,86_{AB}$	$^{a}1,63_{A}$	$^{a}2,13_{A}$	$^{bc}2,50_{A}$	$^{ab}1,88_{A}$	$^{a}1,75_{A}$
36	$^{a}2,00_{A}$	$^{a}2,00_{AB}$	$^{a}1,63_{A}$	$^{a}2,25_{AB}$	$^{c}2,75_{AB}$	$^{b}2,75_{A}$	$^{a}1,88_{A}$
Treatment AH							
8	$^{a}1,00_{A}$	$^{a}1,00_{A}$	$^{a}1,00_{A}$	$^{a}1,50_{A}$	$^{a}1,30_{A}$	$^{a}1,10_{A}$	$^{a}1,20_{A}$
15	$^{a}1,30_{A}$	$^{a}1,10_{A}$	$^{a}1,10_{A}$	$^{a}1,50_{A}$	$^{a}1,30_{A}$	$^{a}1,30_{A}$	$^{a}1,60_{A}$
22	$^{a}1,57_{A}$	$^{a}1,43_{A}$	$^{a}1,14_{A}$	$^{a}2,00_{A}$	$^{a}1,58_{A}$	$^{ab}1,43_{A}$	$^{a}2,00_{A}$
29	$^{a}1,38_{A}$	$^{a}1,38_{A}$	$^{a}1,25_{A}$	$^{a}1,75_{A}$	$^{a}2,00_{A}$	$^{ab}1,88_{A}$	$^{a}2,12_{A}$
36	$^{a}1,5_{A}$	$^{a}1,38_{A}$	$^{b}1,86_{A}$	$^{a}1,80_{A}$	$^{a}2,00_{A}$	$^{b}2,50_{A}$	$^{a}2,13_{A}$

Control (C); lactate-Na and lactate-K (A); HPP (H); HPP+ lactate-Na and lactate-K (AH). Not Detected (N.D.).
a–f Averages with different letter in the same column are significantly different (P < 0.05), across time.
A–C Averages with different letter in the same column are significantly different (P < 0.05), for each treatment.

Conclusions

"Morcilla" treated with HPP shows slight differences in taste and texture compared to the control. However, these differences did not affect the preferences of panellists. Furthermore, the treatment extended the sensory shelf-life of the product by at least 15 days more than the control. Lactate treatments also increased the shelf life of the "Morcilla" product by 7 days.

References

Borek, S., Molinero, C., Santos, E.M., Jaime, I., Rovira, J. (2002). Application of high pressure to improve shelf-life of the typical blood sausage. In: 48th International Congress of Meat Science and Technology, Roma. Vol 1, 172-173.

Diez, A.M., Santos, E.M., Jaime, I., Rovira, J. (2005). Influencia de las altas presiones en la conservación de la morcilla de Burgos. III Congreso Nacional de Ciencia y Tecnología de los Alimentos, Burgos, España. 189.

Mor-Mur M., Yuste J. (2003). High pressure processing applied to cooked sausage manufacture: physical properties and sensory analysis. Meat Science 65, 1187-1191.

Santos, E.M., Diez, A.M., González-Fernández, C., Jaime, I., Rovira, J. (2005). Microbiological and sensory changes in "Morcilla de Burgos" preserved in air, vacuum and modified atmosphere packaging. Meat Science 71, 2, 249-255.

EFFECT OF ASPARTIC PROTEASE FROM *ASPERGILLUS ORYZAE* ON THE TENDERNESS OF BEEF

Z. Pietrasik* and P.J. Shand

Department of Applied Microbiology and Food Science, University of Saskatchewan, Saskatoon, SK, S7N 5A8, Canada. Email: zeb.pietrasik@usask.ca

Keywords: beef, enzymatic tenderisation, tenderness, texture

Introduction

Tenderness, the single most important palatability factor affecting consumer satisfaction of beef, is mainly associated with the structural integrity of myofibrillar and connective tissue proteins. One approach to increasing beef tenderness is to significantly reduce the amount of detectable connective tissues without causing extensive degradation of muscle fibres. This purpose may be achieved by controlled proteolysis of targeted proteins by addition of exogenous enzymes. Although enzyme tenderisation of meat has been used for many years, most commonly using enzymes of plant origin such as papain, bromelain, and ficin, previous reports on the effects of enzymatic tenderisation on tenderness of cooked meat have been unsatisfactory. Due to their broad substrate specificity, these enzymes tend to indiscriminately break down the major muscle proteins which often results in an extensive degradation of the meat structure and undesirable flavour and texture (McKeith *et al.*, 1994; Stefanek *et al.*, 2002).

Current technology in the area of enzyme engineering has revealed other novel sources of proteolytic enzymes, mainly of fungal origin. There has been however, little research on using novel proteases obtained from alternative sources (Ashie *et al.*, 2002; Stefanek *et al.*, 2002). The application of enzyme technology may provide a useful means of meeting consumer expectations for product quality and consistency.

The current study was conducted to determine the range of acceptable aqueous concentrations of proteinase from *Aspergillus oryzae* injected at 105% of raw weight and to determine if the enzyme preparation was active at refrigeration temperature following injection. In the second study we examined effectiveness of refined levels of the protease during moist and dry cooking.

Materials and Methods

Semimembranosus (SM) muscles from young Canada Grade A carcasses were used for this study. The major variables investigated in experiment one were level of proteinase from *Aspergillus oryzae* (Amano Enzyme USA Co., Ltd., Elgin, IL) (to deliver 0.001%, 0.0025%, 0.005% in injected product) and post-injection storage time (1, 7, 14 days). Each SM muscle was cut into four roasts (700g). The roasts designated for enzyme treatment were injected to 105% over the raw meat weight with enzyme solution and then each section was cut into three 2.5 cm steaks which were individually vacuum packaged and refrigerated (4°C) for 1, 7 and 14 days. Steaks were cooked in a conventional oven at 163°C to a final temperature of 71°C.

In experiment two, paired SM muscles were divided into two equal portions to yield four sections. One section was kept as a non-injected control, while the other three were injected to 105% over the raw meat weight with enzyme solution to give 0.0005%, 0.001% and 0.0015% of proteinase from *Aspergillus oryzae* in the final product. After injection, the muscle samples were vacuum packaged and stored overnight at 4°C. The following day each section was cut into four equal-sized roasts that were used for texture measurements after dry- and moist-heat cooking to 71 or 79°C. Roasts for moist-heat cooking were wrapped in aluminium foil before cooking.

Warner Bratzler shear force (WBSF) of 1.27x1.27x2.54cm core samples sheared perpendicular to the fibre direction was determined in experiment one. The variables measured on each roast in experiment two included: cook yield (% of green weight), expressible moisture (EM), WBSF, and instrumental texture profile analysis (TPA).

Results and Discussion

Experiment 1: Enzyme-treated meat showed a gradual reduction in shear force with an increase of enzyme concentration. The lowest concentration (0.001%) of enzyme reduced WBSF by about 25% as compared to the control while maintaining acceptable appearance and texture. However, steaks injected with the highest (0.005%) concentration of proteinase had localized spots with unacceptable mushy and creamy texture. Our results indicated no obvious difference amongst shear values following the different lengths of refrigerated storage (1, 7, 14 days) after injection. This suggested that this enzyme was relatively inactive at refrigerated temperatures and the tenderising effect occurs mainly during cooking. It appeared that the protease functioned quite effectively once cooking commenced and substrate temperature increased.

Experiment 2: Regardless of the cooking method, roasts cooked to a higher internal temperature appeared very dry/grainy and had lower cook yield compared to those cooked to 71°C (Table 1). This observation was well supported by the values obtained for EM with roasts cooked to 79°C having significantly ($p<0.001$) lower EM, thereby indicating lower amounts of free moisture due to loss during cooking. Enzyme-injected muscles had lower cook yield than non-

injected control samples. Enzyme-treated roasts had also significantly lower EM than roast beef made from non-injected meat indicating that there was less free moisture in the product as a result of larger cook loss.
Roasts cooked to 71°C exhibited lower TPA hardness, chewiness and springiness, but were more cohesive than those processed to 79°C. Roasts processed by the moist cooking regime had lower shear force values than those cooked by the dry method. However, the moist cook regime appeared to be more effective in the enzyme treated roasts. Enzyme injected roasts cooked by moist method were more tender (lower shear values) in comparison to non-injected control; while with the exception of the highest (0.0015%) enzyme treatment, there were no differences in shear force among samples baked in the dry environment. In addition, a significant reduction of WBSF values of moist cooked roasts had already been achieved at 0.0005% enzyme level; whereas dry cooked roasts required 0.0015% enzyme. Regardless of the endpoint temperature, increased level of protease resulted in a gradual decrease in hardness of moist cooked roasts, but its effect was insignificant for roasts cooked by the dry regimen.
The significant decrease of hardness and shear force values due to enzyme treatment in moist cooked roasts compared to roasts processed by the dry method might have been due to the initially slower heating rate of the moist method and approx. 15 min longer period of time in the temperature range for increased activity of injected proteinase (20°C to 50°C).

Table 1: Processing and textural characteristics of cooked beef *semimembranosus* roasts.

		Cook yield (%)	Expressible moisture (%)	WBSF (N)	Hardness (N)	Cohesiveness (-)	Springiness (cm)	Chewiness (N*cm)
Endpoint temperature								
	71	66.45a	10.46a	67.8	202b	0.401a	0.24b	20.8b
	79	59.77b	6.89	70.7	226a	0.386b	0.27a	25.9a
Cooking method x Enzyme level								
Dry	0	65.9a	9.15a	86.4a	238.8ab	0.405a	0.27ab	26.1ab
Dry	0.0005	62.6b	9.84a	83.3a	216.4abc	0.407a	0.26abc	22.7bc
Dry	0.0010	62.0b	7.71b	78.8a	217.9abc	0.404a	0.25bc	22.5bc
Dry	0.0015	61.1b	7.46b	59.3c	208.6bc	0.375c	0.25bc	19.9c
Moist	0	65.2a	10.14a	82.6a	250.9a	0.407a	0.28a	28.4a
Moist	0.0005	63.6ab	9.55a	69.5b	229.6ab	0.399ab	0.28a	25.9ab
Moist	0.0010	62.8b	7.88b	59.3c	188.7cd	0.386bc	0.25bc	18.3cd
Moist	0.0015	61.8b	7.65b	53.8c	162.5d	0.379c	0.24c	15.0d

a-c, Means with different letters in the same column are significantly different ($P<0.05$).

Conclusions

Exploratory testing indicated that a negligible amount of proteolysis occurred at refrigeration temperature, suggesting that enzyme-treated meat could therefore be stored without any adverse enzymatic changes in product characteristics. This result shows the superiority of the proteinase from *Aspergillus oryzae* to currently used enzymes such as papain which retains its activity at refrigerated temperature even after cooking thus increasing the risk of both texture deterioration and flavour defects.
Generally, the positive effect of increased enzyme concentration on tenderness of beef was more pronounced in roasts cooked by a moist heat method. Owing to a longer time in the temperature range from 20°C to 50°C, the moist cooking regime yielded products with lower shear force values than those cooked by the dry method.

References

Ashie, I.N.A., Sorensen, T.L. and Nielsen, P.M. (2002). Effects of papain and microbial enzyme on meat proteins and beef tenderness. Journal of Food Science, 67: 2138-2142.

McKeith, F.K., Brewer, M.S. and Bruggen, K.A. (1994). Effects of enzyme applications on sensory, chemical and processing characteristics of beef steaks and roasts. J. Muscle Foods. 5: 149-164.

Stefanek, J.L., Scanga, J.A., Belk, K. E. and Smith, G.C. (2002). Effects of enzymes on beef tenderness and palatability traits. [On-line] Available: http://ansci.colostate.edu/ran/beef/2002/pdf/jls02 [3 April 2006].

EFFECT OF SALT TREATMENT ON BEEF MEAT PROTEIN THERMAL BEHAVIOUR

D. Pighin[1,3], A. Pazos*[1,3] and C. Gonzalez [1,2,4]

[1]*Instituto Tecnología de Alimentos, CIA, INTA CC77 B1708WAB, Morón, Argentina,* [2]*Consejo Nacional de Investigaciones Científicas y Técnicas (CONICET),* [3]*Facultad de Agronomía y Ciencias Agroalimentarias, Universidad de Morón,* [4]*Universidad Nacional de San Martín. E-mail: apazos@cnia.inta.gov.ar*

Keywords: *Semitendinosus* muscle, sodium chloride, differential scanning calorimetry, sodium tripolyphosphate

Introduction

Differential scanning calorimetry (DSC) has become a useful technique to study the thermal behaviour of proteins (Stabursvik and Martens, 1980). Analysis performed in different muscles established the presence of three distinctive peaks, 54-58°C, 67°C and 71-83°C (Parson and Paterson, 1986, Mietsch *et al.*, 1994), mainly related to the thermal denaturation of myosin heads; myosin tails, sarcoplasmic proteins and collagen; and actin respectively. Studies of thermal transitions have involved protein conformation changes induced by environmental factors, such as pH and ionic strength (Findlay and Barbut, 1990) or by the addition of different additives (Farkas and Mohácsi-Farkas, 1996). These changes are responsible for the variation of meat protein stability (Ensor *et al.*, 1991). Previous research of the group was focused on the increase of water holding capacity (WHC) of beef products containing added sodium chloride (SC) and sodium tripolyphosphate (STPP) and cooked by the *sous vide* system (Vaudagna *et al.*, 2005). The purpose of the present study was to investigate the thermal behaviour of muscle proteins, including isolated myofibrillar proteins, when SC and/or STPP were injected in beef *Semitendinosus* muscle.

Materials and Methods

Ten *Semitendinosus* muscles were obtained 72 h *postmortem* from a commercial processor. Salt solutions were injected into the muscles (10% w/w) with an automatic multineedle injector to give the final concentrations (g/100 g injected muscle) indicated in Table 1.

Table 1.

Salt Treatment	SC	STPP
1	0.7	0.25
2	1.2	0.25
3	0.7	-
4	-	0.25
5	non injected	

After the injection, muscles were first vacuum packaged, then continually tumbled at 5 rpm for 8h (1.5±0.5°C) in a Lance Industries tumbler, and stored at -80°C until analyses were performed. All salt treatments were assay in duplicate. Myofibrillar proteins (MP) isolation was conducted according to Culler *et al.*, (1978). Final protein content was determined by estimating the concentration of total N by Kjeldahl method. MP patterns were monitored using discontinuous 3-12% SDS-PAGE according to Laemmli (1970).

A Perkin-Elmer Pyris-1 DSC was used to study the thermal denaturation of the whole muscle proteins and of the isolated MP. Samples (5-15 mg) were weighed into the aluminium pan, then hermetically sealed, allow to equilibrate at 25°C for 2 min., and finally heated from 25 to 100°C at a rate of 10°C/min. An empty pan was used as reference. Heat capacity profiles (endotherm: C_p vs T), calorimetric enthalpy (ΔH) and endothermic peak (T_m) were obtained using the software Pyris-7 (Perkin-Elmer). Finally, samples were rescanned to corroborate the irreversibility of the denaturation process.

All experiments were replicated twice with duplicate measurements of each one. Analysis of variance (ANOVA), means, and standard errors were determined using SAS (1994). Least significant differences ($p<0.05$) were used to establish differences among salt treatments.

Results and Discussion

Electrophoretic pattern of MP showed non significant differences among treatments, suggesting the absence of structural changes related to protein mobility, induced by the presence of salts.

Thermal analysis of non injected whole muscle is shown in Figure 1 as an example. Data show three endothermic peaks at 56.4±0.5°C (I); 68.4±0.5°C (II) and 79.4±0.4°C (III), corresponding to myosin heavy chain (MHC), sarcoplasmic proteins/collagen and actin, respectively. ΔH parameter of these peaks were 1.80±0.13 J/g protein; 0.56± 0.10 J/g protein and 1.71± 0.90 J/g protein, respectively.

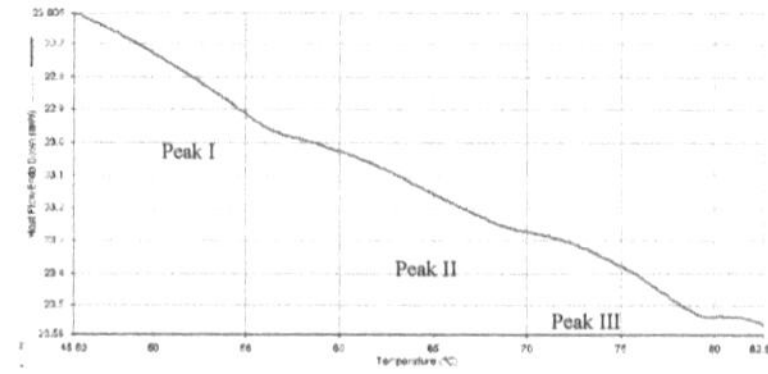

Figure 1: Thermal analysis of non injected Semitendinosus muscles.

When 0.7%SC was injected, actin T_m significantly decreased, denoting the protein thermal sensitivity increment. This effect is probably due to the perturbation of protein hydration induced by SC. In contrast, myosin and sarcoplasmic proteins/collagen T_m showed non significant reduction. Similar results were published by Benjakul *et al.*, (2000), who reported that actin T_m decreased when a threadfin bream muscle was injected with 2.5%SC. On the other hand, injection of 0.25%STPP did not lead to significant changes in T_m or ΔH, with the exception of sarcoplasmic proteins/collagen ΔH, which was non significantly reduced. When both salts were injected in combination, 0.7%SC and 0.25%STPP, a significant decrease in actin T_m and sarcoplasmic proteins/collagen ΔH was observed. This thermal behaviour, increase in actin instability, is similar to the one caused by 0.7%SC injection. Muscles injected with higher SC concentration (1.20%SC + 0.25%STPP) showed a significant decrease in actin T_m, and a small decrease in the T_m of the other proteins. These results are in accordance with Shand *et al.*, (1994), who had reported that in S*emimembranosus* muscles injected with 0.35%STPP and 1%SC, the first transition peak (myosin T_m) was not affected, and actin transition became smaller and its T_m shifted to a lower temperature.

MP thermal analysis of non-injected muscles showed the presence of only two of major thermal transitions, peak A at 59.4°C (4.29 J/g protein) and peak B at 72.9°C (2.28 J/g protein) corresponding to myosin and actin respectively. It is interesting to denote that in the isolated myofibrils there is a shift to lower transition temperatures for actin. This modification could probably be due to the effect of pH and ionic strength variation induced by the isolation conditions (Ma and Harwalkar, 1991; Murphy *et al.*, 1998). When 0.7%SC was injected, both myosin and actin ΔH decreased. The reason for the difference in myosin instability increment obtained between the whole muscle and the isolated proteins is the higher exposure of the extracted proteins to the environmental conditions. Similarly to the whole muscle, non significant differences were found when 0.25% STPP was injected. When salts were injected together (0.7%SC + 0.25%STPP), both myosin and actin ΔH decreased. Even more, when SC was increased (1.20%SC + 0.25%STPP), both thermal parameters (T_m and ΔH) of myosin and actin decreased. The present results corroborate previous ones, suggesting the important effect of SC on protein thermal stability. Again, in the case of isolated proteins (MP) greater protein instability was demonstrated by the reduction of both thermal parameters, involving probably the mentioned effect of the isolation conditions. The thermal behaviour of the isolated MP corroborates and supports the protein thermal profile of whole muscle.

Conclusions

Present results demonstrate that MP are effectively involved in the thermal behaviour depicted by the whole muscle. SC has an important destabilising effect on muscular myofibrillar proteins in a salt concentration dependent manner. The effect of SSTP was not clear and further studies are needed.

References

Benjakul, S, Visessanguan, W. and Kookkeaw, P. (2000). *Songklanakarin Journal of Science Technology*, 22(3), 329-338.
Culler, R.D., Parrish, F.C.(Jr.), Smith, G.C. and Cross, H.R. (1978). *Journal of Food Science,* 43, 1177.
Ensor, S.A., Sofos, J.N. and Schmidt, G.R. (1991). *Journal of Food Science*, 56, 175-179.
Farkas, J. and Mohácsi-Farkas, C. (1996). *Journal of Thermal Analysis and Calorimetry*, 47(6), 1787-1803.
Findlay, C.J. and Barbut, S. (1990). In: VR Harwalkar, CY Ma, (Eds), Thermal analysis of Foods. Elsevier. New York.
Laemmli, U.K. (1970). *Nature*, 227, 680-685.
Ma, C.-Y. and Harwalkar, V. R. (1991). In Advances in food and nutrition research, Academic Press, Inc., New York.
Mietsch, F., Halász, A. and Farkas, J. (1994) Die Nahrung, 38 (1), 47-52.
Murphy, R. Y., Marks, B. P. and Marcy, J. A. (1998). *Journal of Food Science*, 63(1), 88-91.
Parsons, S.E. and Patterson, R.L.S. (1986). *J. Food Technol.*, 21(2):123-131.
SAS (1994) Cary, N.C.: SAS Institute.
Shand, P.J., Sofos, J.N. and Schmidt, G.R. (1994). *Journal of Food Science*, 59(4), 711-715.
Stabursvik, E. and Martens, H. (1980). *Journal of Science and Food Agriculture*, 31, 1034-1042.
Vaudagna, S. R., Lasta, J. A. and Sánchez, G. (2005). In Y. H. Hui, I. Guerrero Legarreta & M. R. Rosmini (Eds.), Ciencia y tecnologías de carnes. México D.F.: Limusa S.A.

EFFECT OF WHEY PROTEIN CONCENTRATE, SODIUM CHLORIDE AND COOKING TEMPERATURE ON PHYSICOCHEMICAL AND TECHNOLOGICAL PARAMETERS OF *SOUS VIDE* COOKED BEEF MUSCLES

N. Szerman[1,2], A.M. Sancho[1], C.B. Gonzalez[1,2], A.A. Pazos[1]* and S.R. Vaudagna[1,2]

[1.] *Instituto Tecnología de Alimentos, CIA, INTA CC77 B1708WAB, Morón, Argentina,* [2.] *Consejo Nacional de Investigaciones Científicas y Técnicas (CONICET) E-mail: apazos@cnia.inta.gov.ar*

Keywords: beef, whey protein, yield, *sous vide* system, response surface methodology

Introduction

Application of *sous vide* system to beef appears to be an interesting alternative to diversify the consumption of ready-to eat beef products. The main drawback of this approach is the retention of liquid released during cooking inside the package, which affects product appearance. In order to increase water-holding capacity (WHC), and hence reduce cooking weight loss, the addition of whey protein concentrate (WPC) seems to be a novel alternative to replace or reduce conventional ingredients [sodium chloride (SC) and alkaline phosphates]. Thus, the aim of this work was to evaluate the effect of WPC and SC concentrations and cooking temperature on physicochemical and technological parameters of *sous vide* cooked beef muscles using a response surface methodology.

Materials and Methods

Response surface methodology (Doehlert design; Araujo and Brereton, 1996) was used to study the simultaneous effect of WPC concentrations (0, 0.9, 1.7, 2.6, 3.5% on raw injected muscle basis), SC concentrations (0, 0.6, 1.2, 1.9, 2.5% on raw injected muscle basis) and cooking-pasteurisation temperatures at the slowest heating point (SHP) of the muscles (65°C, 70°C, 75°C). The cooking-pasteurisation time applied at the muscle SHP for each temperature was the thermal death time suggested by FAIR (1999) to achieve 6D Log reduction of *Listeria monocytogenes* (9 min-65°C, 2 min-70°C, 26s-75°C).

Forty-five *Semitendinosus* beef muscles (48h post-slaughter, average weight: 1708.73±165.62g and average pH: 5.57±0.07) were used. The muscles were vacuum packaged and continually tumbled at 8.5rpm for 0.5h (1.5±0.5°C). Then, they were injected at 130% (over green weight) using an automatic multi-needle injector. Brines, prepared with WPC (Lacprodan80, Arla Food Ingredients S.A) and SC, were formulated to give the concentrations of ingredients defined by the experimental design. After the injection, muscles were continually tumbled at 8.5rpm for 5h (1.5±0.5°C), and then submitted to *sous vide* cooking in a water cascading retort (Microflow Barriquand). After cooking, samples were stored at 1.5±0.5°C for 18h until further testing.

The pH of raw and cooked muscles was measured on the slurries. Post-injection tumbling (PostIT, P_1) and cooking (P_2) weight loss percentages were determined using the following relationship: $P_i = (m_i - m_f)*100/$ trimmed raw muscle weight, where for P_1: m_i is the weight of the injected muscle and m_f is the weight of the muscle after PostIT treatment, for P_2: m_i is the weight of the muscle after PostIT treatment and m_f the one after thermal process. Total yield was calculated as TY= cooked muscle weight*100/raw non-injected muscle weight. Expressible moisture (EM) was calculated as the percentage of liquid lost after centrifugation (20min, 4°C, 4800xg) related to the initial sample weight. Statistical analysis was performed using SAS 8e for Windows (1991-2001).

Results and Discussion

The pH of cooked beef muscles was significantly ($P<0.05$) affected by WPC concentration and cooking temperature applied (Table 1). The pH increased proportionally to the increment of WPC concentration, probably induced by the pH of the injected brines (range 6.64-7.08). Cooking temperature had linear and quadratic effect. However, their estimated coefficients were lower than the one obtained for WPC, suggesting a minor influence on the pH parameter.

In the case of P_1 parameter, it was significantly affected ($P<0.05$) by SC concentration (linear and quadratic effects). The lowest P_1 value was obtained with the highest SC level (Table 1). Instead, WPC addition did not affect significantly this parameter (Table 1). The increment of SC content inside the tissue would induce fibre swelling, protein hydration properties modification, promoting an increase in WHC as a result. This increase can be evidenced by the observed reduction of PostIT weight loss.

Regarding P_2 parameter, at a fixed cooking temperature P_2 diminished linearly following the increment of SC and WPC concentrations (Table 1, Figure 1). This result could be a consequence of the augmented concentration of tissue proteins because of the addition of whey proteins, and/or a higher myofibrillar protein extraction exerted by the presence of the salt. The mechanisms involved in the water retention are difficult to describe, probably solubilised proteins are able to form a gel that entrap and retain water inside, and/or due to the interaction between water and added proteins. The P_2 parameter increased linearly and quadratically with cooking temperature. This result could be associated with the effect of heating upon tissue structure, such as the damage of cell membranes, transversal and longitudinal shrinkage of muscle fibres, and connective tissue contraction, which reduce in turn meat WHC (Tornberg, 2005).

The effect of ingredient concentrations and cooking temperature on TY were almost the same as those observed for P_2 (Table 1). In the regression model for TY, the interactions SC^2 and WPCxSC were significant ($p<0.05$). The WPCxSC interaction implies that WPC concentration effect was highly dependent on the SC concentration. Up to a SC concentration of 1.6%, the incorporation of WPC had a positive influence on TY (Figure 2). Beyond this concentration no effect was observed. This result let us assume that a similar TY can be obtained replacing part of the SC added by WPC.

Expressible moisture is a measured of the content of free water in the cooked product. This parameter may be used as an indicator of the purge expected in the packaged product (Boles and Shand, 2001). The highest EM value was observed in muscles processed at the lowest cooking temperature and the highest SC concentration (Figure 3). As expected, the EM parameter had a significant ($P<0.05$) positive correlation (R^2=0.789) with TY.

Conclusions

There is a positive effect of SC concentration on tissue WHC. Thus, the increased SC concentration was capable of reducing weight losses in the different stages of the process. In addition, the WPC increment produces an enhancement of total yield at low SC concentrations. Similar results were observed at the different cooking temperature studied, however the maximum yield was obtained at 65°C.

Table 1: Regression coefficients and analysis of variance of the regression model for cooked beef pH, post-injection tumbling (P_1) and cooking (P_2) weight loss percentages, total yield (TY), and expressible moisture (EM).

		pH	P_1	P_2	TY	EM
Constant		6.04*	7.4*	25.8*	96.2*	32.78*
Lineal	WPC	0.11*	-0.3	-5.6*	4.7*	-0.07
	SC	0.02	-8.6*	-8.4*	18.6*	3.26*
	T	-0.07*	-	8.7*	-9.0*	-2.46*
Quadratic	WPC^2	0.03	1.2	-0.1	-0.3	-0.67
	SC^2	0.08	1.9*	0.8	-4.6*	-1.90*
	T^2	0.01*	-	5.9*	-3.6*	-2.58*
Interactions	WPCxSC	0.03	2.5	1.9	-5.0*	-3.50*
	SCxT	-0.03	-	-0.5	2.7	-0.25
	WPCxT	0.00	-	-1.1	2.1	-3.10*
R^2		0.57	0.91	0.95	0.96	0.79
p-value		*<0.05*	*<0.05*	*<0.05*	*<0.05*	*<0.05*

*significant p<0.05. Regression coefficients calculated for coded values.

References

Araujo, P.W. and Brereton, R.G. (1996). Experimental design II. Optimization. *Trends in analytical chemistry, 15(2)*, 63-70.

Boles J.A. and Shand P.J. (2001). Meat cut and injection level affects the tenderness and cook yield of processed roast beef. *Meat Science*, 59, 259-265.

FAIR CT96-1020 (1999). Harmonization of Safety Criteria for Minimally Processed Foods. Rational and Harmonization Report, FAIR Concerted Action FAIR CT96-1020.

Tornberg, E. (2005). Effects of heat on meat proteins – implications on structure and quality of meat products. A review. *Meat Science, 70*, 493-508.

Figure 1: Effect of whey protein concentrate (WPC) and sodium chloride (SC) on cooking weight loss (cooking temperature fixed at 70°C).

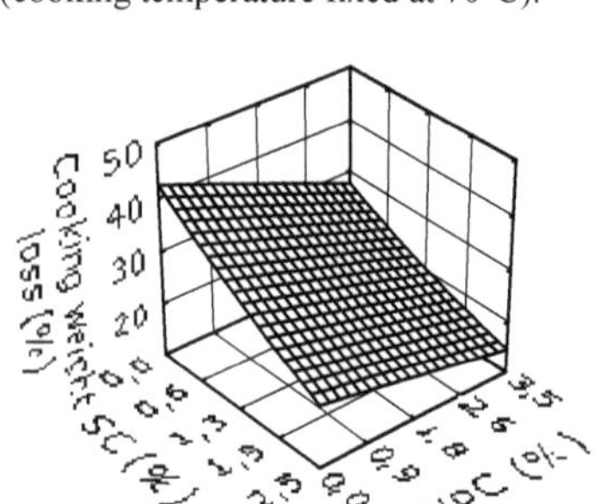

Figure 2: Effect of whey protein concentrate (WPC) and sodium chloride (SC) on total yield (cooking temperature fixed at 70°C).

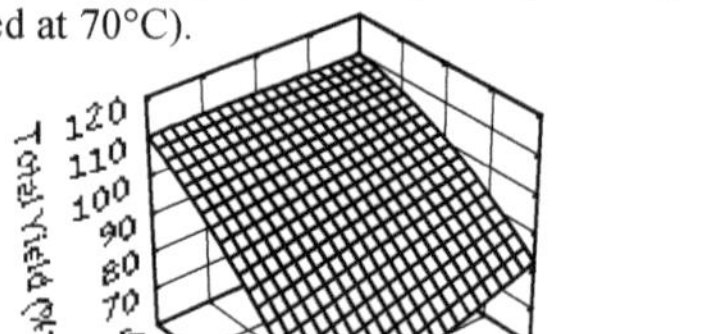

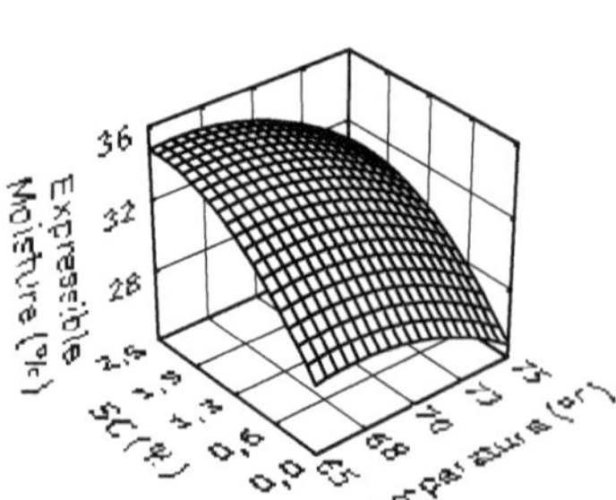

Figure 3: Effect of sodium chloride (SC) and cooking temperature on expressible moisture (whey protein concentrate concentration fixed at 1.25%).

ANTIOXIDANT ACTIVITY OF GRAPE SEED EXTRACT IN GROUND PORK MEAT

A.I. Mandić

University of Novi Sad, Faculty of Technology, Bul. Cara Lazara 1, 21 000 Novi Sad, Serbia and Montenegro.
Email:anam@tehnol.ns.ac.yu

Keywords: pork, TBARS, ground meat, antioxidant, grape seed extract

Introduction

Lipid oxidation is one of the major factors in the deterioration of meat and meat products during storage and processing, causing changes in the functional and sensory characteristics, and decreasing the shelf life. Synthetic antioxidants, such as BHT, BHA, and propyl gallate, have been utilised to extend the shelf life of a product by retarding the development of rancidity. However, the use of synthetic antioxidants in food has been under scrutiny for toxicological reasons (Madhavi *et al.*, 1995), and thus the interest in natural antioxidants has been steadily increasing. The antioxidant and radical scavenging activities of a large number of polyphenolic compounds, isolated from plants, have been studied by various authors (Bravo, 1998; Robards, 2003; Moure *et al.*, 2001; Pokorny *et al.*, 2001). As the grape seed extract (GSE) contains a lot of polyphenols, it is considered as an excellent natural antioxidant and is used as a health promoting product.

GSE has been evaluated for its antioxidant effect on a few meat types. The inhibition of the development of thiobarbituric acid reactive substances (TBARS) in dark poultry meat with pre- and post-mortem use of GSE has been published (Lau and King, 2003). Mielnik *et al.*, (2006) have supplemented ground turkey meat with GSE before processing and it has been proven that the addition of GSE improved the lipid stability in the cooked turkey meat stored cold. The aim of this research was to determine the possible protective effect of GSE, as a natural antioxidant, on ground pork during refrigerated storage. The development of the oxidation process during storage was monitored by the TBARS values.

Materials and Methods

The pork meat, obtained 24 h *post mortem* from a commercial processing plant, was minced in an industrial meat grinder and transported to the laboratory in plastic bags in a hand-held refrigerator.

GSE was characterised using spectrophotometric methods for its total phenolic and flavan-3-ols content and the radical scavenging activity on DPPH˙. The amount of total soluble polyphenols in the GSE was determined spectrophotometrically according to the Folin-Ciocalteau method (Singleton *et al.*, 1999). Gallic acid was employed as a calibration standard. The amount of total flavan-3-ols was assayed colorimetrically by the vanillin method using catechin as a standard (Sun *et al.*, 1998; Nakamura *et al.*, 2003). The antioxidant activity of the GSE was determined as the measure of radical scavenging using DPPH˙, measured by the spectrophotometric method. The value of the efficient concentration (EC_{50}), defined as the amount of the antioxidant necessary to decrease the initial free radical concentration by 50 %, was calculated for the GSE (Espin *et al.*, 2000).

The GSE was dissolved in sterile water and the following final concentrations were prepared: 0.01; 0.1; 0.5 and 1.0 % (w GSE / w of pork meat). The ground pork was divided into portions and each portion was mixed with the 5 % (w/w) antioxidant solution using a hand-held utensil, according to the following formulation: control (no antioxidant, 5 %, w/w, sterile water was added); 0.01 % GSE; 0.1 % GSE; 0.5 % GSE; 1.0 % GSE and 0.01 % BHA (5 %, w/w, solution of 0.01 % BHA was prepared in oil).

The patties (50 g each) formed from each portion of either water, or BHA or GSE supplemented ground pork were placed in polyethylene bags and stored in a refrigerator at +2° ± 0.5° C. The experiment lasted for 11 days. The determination of TBARS was performed according to the method of Botsoglou *et al.*, (1994). TBARS values were determined on the 1st, 4th, 6th, 7th, 9th, 10th and 11th day of storage. All determinations were made in triplicate and average values were reported.

Results and Discussion

The GSE used in this experiment contains 95.2 % total soluble polyphenols determined according to the Folin-Ciocalteau method. The total flavan-3-ols, assayed by the vanillin method, showed that 99.5 % of GSE contained flavan-3-ols. The investigation showed that the antioxidant activity increased while increasing the concentration of GSE. It was confirmed that the GSE influenced DPPH˙ depending on the concentration of the given extract. The value of EC_{50} obtained for the DPPH˙, determined spectrophotometrically, was 0.8627 mg GSE/mg DPPH˙.

The ground pork used in this experiment contained 5.32 % of total fats. The TBARS values represent the content of the secondary lipid oxidation products, mainly aldehydes, which contribute to off-flavours in the oxidised meat. Malondyaldehyde (MDA), a major degradation product of lipid peroxides, was used as a marker for assessing the extent of lipid peroxidation. The TBARS value is expressed as MDA content (ppb). The effect of the BHA and GSE antioxidants on TBARS value of the ground pork patties over 11 days of the refrigerated storage is shown in Figure 1.

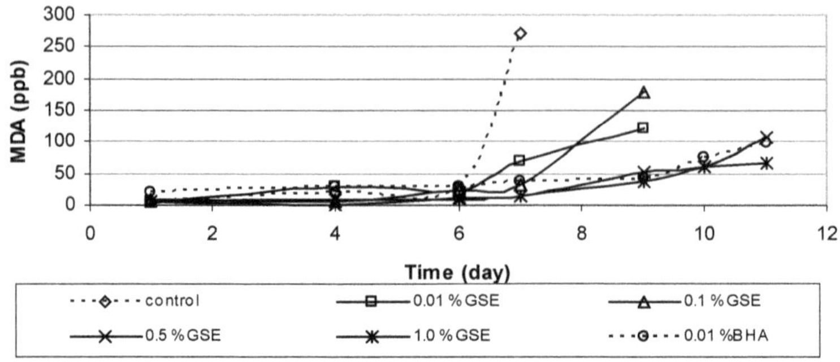

Figure 1: Evolution of TBARS values in ground pork samples.

During the first 6 days of the experiment MDA content in all samples was low and showed poor increase. On the 7th day of the experiment MDA content of the control sample reached 270 ppb, while MDA content of all supplemented samples maintained the moderate increase, except for the sample containing 0.01 % GSE (69 ppb). On the 9th and 10th day of the experiment the MDA content in the samples with 0.01 % and 0.1 % GSE were much higher than in the samples with 0.5 %, 1.0 % GSE and 0.01 % BHA. On the 11th day of the experiment the samples supplemented with 0.5 % GSE and 0.01 % BHA reached 108 and 99 ppb of MDA, respectively, while the sample with 1.0 % GSE contained 67 ppb of MDA.

Conclusions

The ability of the antioxidants examined in retarding lipid oxidation in the ground pork patties throughout the refrigerated storage was in the following order: 1.0 % GSE > 0.5% GSE = 0.01 % BHA > 0.1 % GSE > 0.01% GSE > control. It is evident that the addition of GSE exhibits antioxidant properties and extends the shelf life of ground pork.

References

Botsoglou N.A., Fletouris D.J., Papageorgiou G.E., Vassilopoulos V.N., Mantis A.J. and Trakatellis A.G. (1994). Rapid, Sensitive, and Specific Thiobarbituric Acid Method for Measuring Lipid Peroxidation in Animal Tissue, Food, and Feedstuff Samples. *J. Agric. Food Chem*, 42: 1931-1937.

Bravo L. (1998). Polyphenols: Chemistry, Dietary Sources, Metabolism, and Nutritional Significance. Nutrition Reviews, 56: 317-333.

Espin J.C., Soler-Rivas C. and Wichers H.J. (2000). Characterization of the Total Free Radical Scavenger Capacity of Vegetable Oils and Oil Fractions Using 2,2-Diphenyl-1-picrylhydrazyl Radical. *J Agric Food Chem*, 48: 648-656.

Lau D.W. and King A.J. (2003). Pre- and Post-Mortem Use of Grape Seed Extract in Dark Poultry Meat to Inhibit Development of Thiobarbituric Acid Reactive Substances. *J. Agric. Food Chem*, 51: 1602-1607.

Mielnik M.B., Olsen E., Vogt G., Adeline D. and Skrede G. (2006). Grape Seed Extract as Antioxidant in Cooked, Cold Stored Turkey Meat. LWT - *Food Science and Technology*, 39: 191-198.

Food Antioxidants: Technological, Toxicological, and Health Perspectives (1995). edited by D. L. Madhavi, S. S. Deshpande and D. K. Salunkhe, New York: Marcel Dekker Inc.

Moure A., Cruz J.M., Franco D., Domínguez J.M., Sineiro J., Domínguez H., Núñez M.J. and Parajó J.C. (2001). Natural Antioxidants from Residual Sources. *Food Chem*, 72:145-171.

Nakamura Y., Tsuji S. and Tonogai Y. (2003). Analysis of Proanthocyanidins in Grape Seed Extracts, Health Foods and Grape Seed Oils. *J Health Sci*, 49: 45-54.

Antioxidants in Food (2001). ed: J Pokorny, N. Yanishlieva and M. Gordon. Woodhead Publishing Ltd and CRC Press LLC.

Robards K. (2003). Strategies for the Determination of Bioactive Phenols in Plants, Fruit and Vegetables. *J Chromatogr A*, 1000:657-691.

Singleton V.L., Orthofer R. and Lamuela-Raventos R.M. (1999). Analysis of Total Phenols and other Oxidation Substrates by Means of Folin-Ciocalteu Reagent. *Methods Enzymol*, 299:152-178.

Sun B., Ricardo-da-Silva J.M. and Spranger I. (1998). Critical Factors of Vanillin Assay for Catechins and Proanthocyanidins. *J Agric Food Chem*, 46: 4267-4274.

EFFECTS OF REDUCING THE AMOUNT OF NITRITE IN ORGANIC MEAT PRODUCTS

T.J. Verkleij*[1], F.K. Stekelenburg[1] and D. Stegeman[2]

[1]TNO Quality of Life, P.O. Box 360, 3700 AJ Zeist, the Netherlands, [2]WUR Agrotechnology and Food Sciences Group, P.O. Box 17, 6700 AA Wageningen, the Netherlands. Email: verkleij@voeding.tno.nl

Keywords: organic meat, reduced nitrite, cured meat products, shelf life, colour stability

Introduction

Organic meat, especially pork, and products thereof is a small but steadily growing segment in the Netherlands and the EU. With the growing demands for organic production of meat, there is also more interest in organic meat products. The organic production chain of pork is well controlled and only a small list of additives (mentioned in EU Directive 2092/91) may be used. For years profound discussion is going on in the EU on the use of nitrite (E250) as an additive in meat products. Nitrite is used in meat products for formation and stabilising the red, cured meat color. Nitrite does also have antimicrobial and anti-oxidative properties. Colour and colour stability of cured meat products is of great importance also for consumers of organic products. Therefore NGO's and producers of organic products, organised by Biologica, decided to work out to which extent nitrite could be avoided or produced while still obtaining a stable and safe product. The objective of the study was to determine the amount of nitrite salt sufficient to obtain colour stable and safe organic meat products.

Materials and Methods

- Recent literature survey about sodium nitrite and alternatives in meat products.
- Practical experiments with organic meat products including shelf life monitoring.
- Visual and instrumental (CIElab) colour comparison.
- Challenge testing with *Clostridium botulinum.*

Results and Discussion

Based on literature information, organic cured ham and luncheon meat with three levels (0, ca 50 and ca 85 mg/kg) ingoing sodium nitrite were prepared. A generally applied product flow of 5 weeks at 4°C before slicing, MAP packaging of sliced product, and retail storage for 5 weeks at 7°C with light source of 880 lux (12 hours on a day) only the last week was simulated. At several stages, analyses were performed.

Table 1: Analyses of nitrite content in organic luncheon meat and cured ham during storage.

Nitrite content [mg/kg]	Luncheon meat			Cured ham		
Ingoing nitrite	0	41	82	0	59	94
After 1 week storage 4°C	0	6.1	6.4	0	16.7	24.7
At slicing, after 5 weeks storage at 4°C	0	4.3	3.5	0	12.1	16.3
4 Weeks after slicing and MAP packaging 7°C	0	2	2.3	0	4.9	7.1
5th Week after slicing, last week at 880 lux	0	2.5	5.6	0	5.7	4

The residual nitrite analysis showed that nitrite concentrations declined sharply after production for both type of products to about 15–25% of the ingoing level and steadily declined further during storage time to a final amount of 2.5– 6ppm. The visual evaluations of the colour of both products with nitrite were still acceptable compared to the one without nitrite. Colour measurements during bulk storage, slicing and presentation in a chilled cabinet were performed by using the CIElab method.

The a* value of both products without nitrite was significantly lower (Figure 1 and 2). During storage, slicing and presentation in the cabinet, the a* value of products without nitrite decreased further.

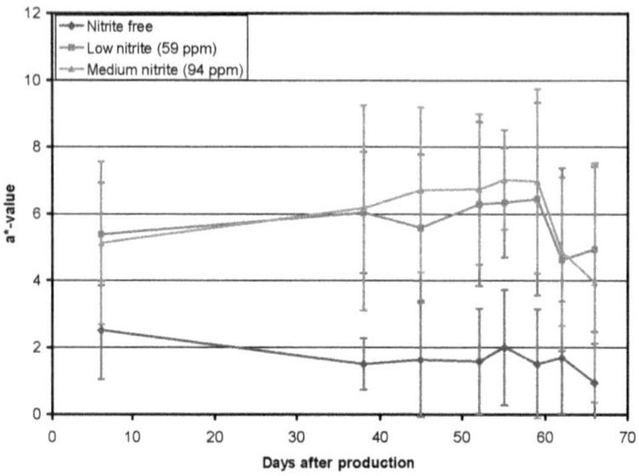

Figure 1: Red colour change of luncheon meat during storage.

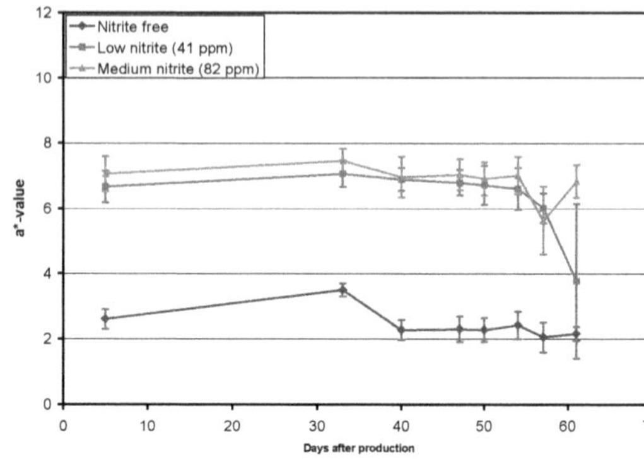

Figure 2: Red colour change of cured ham during storage.

Challenge tests with C.*botulinum* (100 spores/g) were performed with a lean (10% fat) and fat (30% fat) pasteurized (P_{70} = 55 min) luncheon meat type product with 0, 54 and 108 mg/kg ingoing nitrite respectively. The a_w-values of the end products varied between 0,968 and 0.975. The products were stored at 7°C, 10°C and 15 °C over 12 weeks.

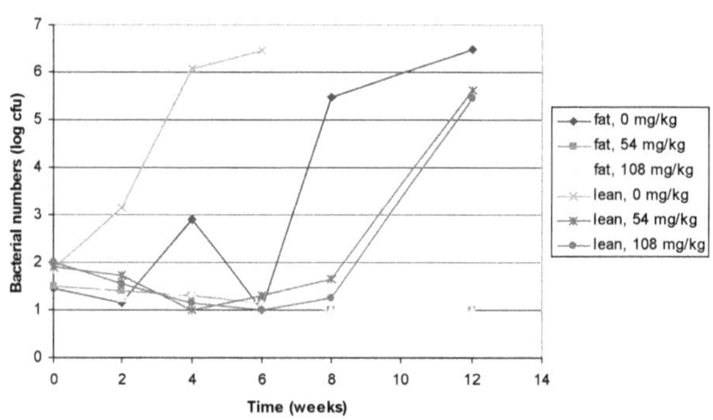

In products with or without nitrite growth of C. *botulinum* was not detected during the whole storage period at 7°C and 10°C. Botuline toxin was not determined.

At 15°C growth of C. *botulinum* was detected earlier in products without nitrite (Figure 3). Botuline toxin was detected in lean luncheon meat without nitrite after 6 weeks, in fat luncheon meat without nitrite after 8 to 12 weeks and in lean luncheon meat with 54 mg/kg nitrite after 12 weeks.

Figure 3: Growth of *Clostridium botulinum* in luncheon meat, stored at 15°C.

Conclusion

Complete abandoning of nitrite in the processing of organic meat products is not advisable. It would result in serious colour deficiency of the product compared to conventional products and from this point of view the product would not look as consumers expect. Furthermore, the risk of botulin toxin formation will increase at higher storage temperatures. By a reduction of the amount of nitrite to approximately 80 mg/kg ingoing, the colour forming and stability will be sufficient to reach a shelf life of 65 days. During production, heating, cooling and storage, the level of residual nitrite will be far less than the 50 mg/kg specified in the EU Directive 2092/91.

EFFECTS OF POTASSIUM LACTATE AND DISPLAY LIGHTING ON SHELF-LIFE OF FRESH-PORK-SAUSAGE PATTIES

S.M. Ryan*[1], M. Seyfert[1], M.C. Hunt[1], J.P. Grobbel[1], D.E. Johnson[2] and R.A. Monderen[3]

[1.] Department of Animal Sciences, Kansas State University, Weber Hall, Manhattan, KS 66506 U.S.A., [2.] Department of Statistics, Kansas State University, Dickens Hall, Manhattan, KS 66506 U.S.A. 3. PURAC America, Inc., 111 Barclay Boulevard, Lincolnshire, IL 60069 U.S.A. Email: smr6666@ksu.edu

Keywords: colour, display lighting, fresh pork sausage, lipid oxidation, potassium lactate

Introduction

Potassium lactate (KL) is used as an antimicrobial in various meat products. Research demonstrates its effectiveness in promoting colour stability in enhanced pork (Jensen *et al.*, 2003) and enhanced beef (Kim *et al.*, 2005). A potential mechanism for this colour stabilizing effect has been presented (Kim *et al.*, 2005). Inclusion of KL in fresh-pork sausage decreased microbial counts, but has inconsistent effects on lipid oxidation and colour stability (Bradford *et al.*, 1993ab). However, those studies focused on refrigerated storage and not on lighted retail display. Our objective was to examine the effects of KL on aerobic plate counts, colour stability, and lipid oxidation of fresh-pork-sausage patties in lighted and unlighted display.

Materials and Methods

Fresh-pork-sausage patties were formulated with 0% or 3% KL, placed on Styrofoam trays, over wrapped with polyvinyl chloride film, frozen at -4 °C for 18 d, and were thawed in the dark at 4 °C for 24 h prior to display. Fat and pH were determined after thawing. Patties were displayed at 4 °C for 10 d in lighted display or for 7 d in unlighted display (covered by foil) followed by 3 d in lighted display. Instrumental (CIE L*a*b*, Illuminant A) and visual colour were evaluated on d 0, 1, 2, 4, 7, and 10 (lighted display) and d 0, 7, and 10 (unlighted display). Visual colour was evaluated as 1 = bright red, 2 = red, 3 = dull red, 4 = slightly dark red, 5 = reddish tan, 6 = tannish red, and 7 = tan/brown. Aerobic plate counts (d 0 and 10) and thiobarbituric acid reactive substances (TBARS; d 0, 7, and 10) were determined on patties in both lighted and unlighted display. The experiment (n = 3 replications) was analyzed using the mixed procedure of SAS.

Results and Discussion

Patties with 0% KL had 34.3% fat and pH of 5.9 while patties with 3% KL had 32.9% fat and pH of 6.4. Initial aerobic plate counts were similar ($P > 0.05$) for both KL levels, but on d 10, patties without KL had greater ($P < 0.05$) counts (7.8 logs cfu/g) than the patties with KL (5.9 logs cfu/g).

For all treatment combinations, redness (a*) decreased ($P < 0.05$) each day (Figure 1; data not shown for d 1 and 2). Patties displayed in the dark had greater a* ($P < 0.05$) than patties in the light on d 7 and 10. Patties with KL on d 0, 4, 7, and 10 were more red (greater a*; $P < 0.05$) than patties without KL. On d 10, patties with KL exposed to light for 10 d were similar in a* ($P > 0.05$) to patties displayed in the dark without KL. For all treatments, visual-colour scores increased ($P < 0.05$) during display indicating patties became less red (Figure 1; data not shown for d 1 and 2). Treatments with KL were more red (lower visual scores; $P < 0.05$) than patties without KL after d 0. On d 7 and 10, patties displayed in the light for 10 d were less red (lower visual scores; $P < 0.05$) than patties displayed in the dark for 7 d followed by 3 d in the light. When exposed to light, more pigment oxidation occurred, but this research shows KL can help avert myoglobin oxidation in fresh-pork sausage during lighted display.

Values for TBARS increased ($P < 0.05$) for all treatment combinations during display except for patties with 3% KL that were in the dark for the first 7 d ($P > 0.05$; Figure 2). Adding KL decreased ($P < 0.05$) TBARS values on d 7 for patties that had been in the light, and for patties with both KL levels on d 10 ($P < 0.05$). After exposure to light for 10 d, patties had greatly increased ($P < 0.05$) TBARS values, but KL limited the increase over 3-fold in patties displayed 10 d. In previous studies, 2% or 3% KL (Bradford *et al.*, 1993ab) did not influence lipid oxidation of fresh-pork sausage in refrigerated storage; however, they did not expose sausage to light which increases lipid oxidation.

Conclusions

Addition of 3% potassium lactate will prolong the shelf life of fresh-pork-sausage patties by lowering aerobic plate counts, promoting colour stability, and limiting lipid oxidation. Inclusion of potassium lactate can help protect against lighting's negative influence on myoglobin and lipid oxidation.

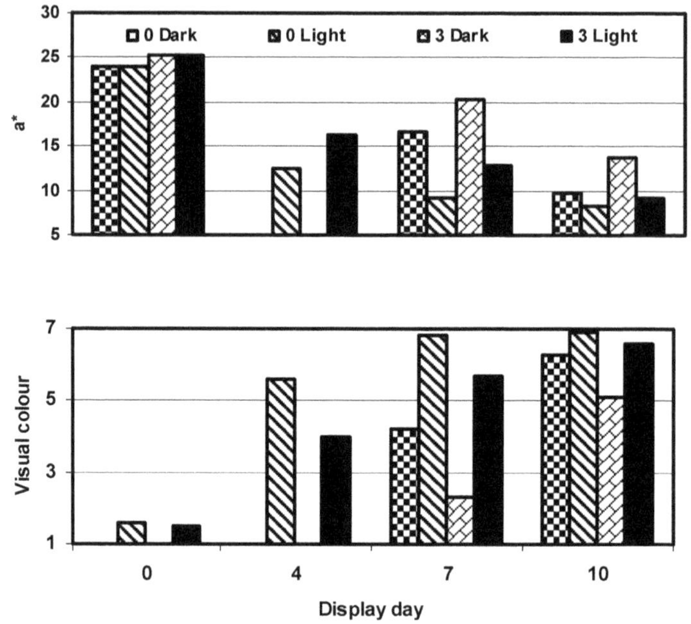

Figure 1: LS Means for a* and visual colour of fresh-pork sausage patties with 0% or 3% potassium lactate during lighted and unlighted retail display.

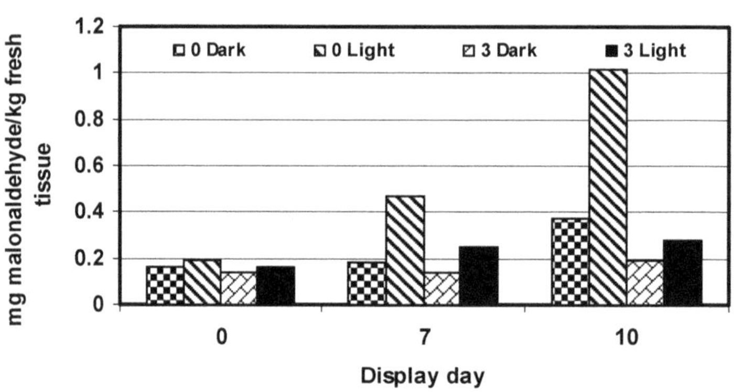

Figure 2: LS Means for thiobarbituric acid reactive substances of fresh-pork sausage patties with 0% or 3% potassium lactate during lighted and unlighted retail display.

References

Bradford, D.D., Huffman, D.L., Egbert, W.R., Mikel, W.B. (1993a). Low-fat fresh pork sausage patty stability in refrigerated storage with potassium lactate. Journal of Food Science, 488-491.

Bradford, D.D., Huffman, D.L., Egbert, W.R., Mikel, W.B. (1993b). Potassium lactate effects on low-fat fresh pork sausage chubs during simulated retail distribution. Journal of Food Science, 58, 1245-1253.

Jensen, J.M., Robbins, K.L., Ryan, K.J., Homco-Ryan, C., McKeith, F.K., Brewer, M.S. (2003). Effects of lactic and acetic acid salts on quality characteristic of enhanced pork during retail display. Meat Science 63(4):501-508.

Kim, Y.H., Hunt, M.C., Mancini, R.A., Kropf, D.H., Smith, J.S. (2005). Metmyoglobin reduction through lactate-NAD-LDH system *in vivo* and *in vitro*. *In* Proceedings 51st International Congress of Meat Science Technology, 7-12 August 2005, Baltimore, Maryland, U.S.A.

Acknowledgements: Contribution No. 06-301-A from the Kansas Agricultural Experiment Station, Manhattan, KS.

COMBINED TREATMENT OF PORCINE PLASMA WITH MICROBIAL TRANSGLUTAMINASE AND CYSTEINE-EFFECTS ON THE FOAMING AND EMULSIFYING PROPERTIES

N. Fort*, P. Salvador, D. Parés, M. Toldrà, C. Carretero and E. Saguer

Institut de Tecnologia Agroalimentària (INTEA). Universitat de Girona Av. Lluís Santaló s/n. 17071. Girona
E-mail address: nuri.fort@udg.es

Keywords: plasma, microbial transglutaminase

Introduction

Different studies focused on the improvements of heat-induced gel properties of plasma proteins at acidic pH using microbial transglutaminase (*MTGase*) have carried out, the most recent one that combining the enzymatic treatment with cysteine (Fort *el al.*, 2006). The results of this work indicate that cysteine does not enhance the enzyme activity, but both improve gel texture separately. However, the effects of these treatments on other functional properties of plasma proteins have been not elucidated. Plasma is frequently used in comminuted meat products for its emulsifying properties and it is also being considered a possible alternative to egg proteins as foaming agent. So, the aim of this study is to determine how the treatment with *MTGase* and/or cysteine modifies the surface properties of plasma.

Materials and Methods

Three porcine plasma samples came from an industrial abattoir. Cysteine (1.02838) was purchased from Merck & Co., Inc. (New Jersey, USA), and *MTGase* commercial preparation (ACTIVA®, Ajinomoto Co. Inc., Japan) was supplied as a mixture containing 99 % maltodextrin and 1% *MTGase*. The enzyme concentration is reported as the commercial product concentration and referred as % MTG.

Experimental design: Each sample was separated in 5 aliquots, each one being submitted to one of the next treatments: a) native solution (control); b) *MTGase* (MTG) treatment; c) cysteine (Cys) treatment; d) simultaneous *MTGase* and cysteine (MTG+Cys) treatment; and e) *MTGase* first and adding cysteine afterwards (MTG//Cys) treatment. The enzymatic treatment was carried out with 3 % MTG (w/v) for 3 h at 30 °C and pH 7, and the cysteine concentration applied was in every case 0.25 % (w/v).

Foaming properties: A 200-mL sample of 0.25 % (w/v) plasma protein solution adjusted to pH 5.5 was blended in a double beater Multimix M700 (Braun, Kronberg, Germany) for 10 min. The foaming capacity (FC) was determined as foam volume recorded 2 min after blending. To measure foam stability (FS), a discontinuous gravimetric method was used placing a known foam quantity in a dry stainless sieve and monitoring the percentage of remaining foam as a function of time after. The relative foam stability (RFS) was calculated as time (min) taken for 50 % foam weight loss by fitting data to an exponential function.

Emulsifying properties: A 150 mL sample of 0.35 % (w/v) plasma protein solution adjusted to pH 5.5 was mixed with 50 mL of corn oil and homogenized in a HC-5000 Microfluidizer homogenizer (Microfluidics, Newton, MA, USA). Immediately, 20 mL of the emulsion were diluted with 50 mL of 0.1 % SDS solution. The absorbance of this solution was measured at 500 nm. EAI ($m^2 \cdot g^{-1}$ of protein) was calculated as follows: EAI=$(2T/\phi C)\cdot$dilution factor, where T= turbidity, ϕ= oil volume fraction, C= weight of protein per volume unit in the aqueous fraction before emulsification, and the dilution factor= 2500. To determine the emulsion stability index (ESI), a new sample of diluted emulsion was taken 10 min after emulsification.

Statistical Analysis: Data were subjected to Proc GLM by considering sampling days as blocks in a randomized complete block design. The Tukey test was used to compare means. The significance level for all tests was α= 0.05.

Results and Discussion

A significant (P<0.05) reduction in foaming capacity (FC) of the control samples was observed when the plasma was treated only with *MTGase* (Table 1). This was in all probability due to a reduction in protein flexibility as a result of an increase in the degree of protein cross-linking, although contradictory results have been reported with respect to foaming properties (Babiker, 2000). The presence of cysteine seemed to partially diminish the foaming effect. The tested treatments did not significantly (P>0.05) affect the relative foam stability (RFS), although in *MTGase* treatments (both with and without cysteine) the mean values tended to be lower.

Table 1. Foaming properties of porcine plasma; effect of treatment with *MTGase* (MTG) and/or cysteine (Cys). Mean ± sd (n=3). Different letters indicate significantly differences between treatments.

Treatment	FC (mL)	RFS (min)
Control	785 ± 14 a	32 ± 12
MTG	708 ± 2 b	26 ± 3
Cys	742 ± 34 ab	34 ± 12
MTG + Cys	728 ± 28 ab	26 ± 4
MTG // Cys	732 ± 30 ab	22 ± 5

These results indicate that the viscoelastic properties of film surrounding the gas bubbles were not greatly affected by the applied treatments. In addition, the majority of foam was lost during the first forty minutes post foaming (Figure 1).

A significant ($P<0.05$) decrease in the emulsifying activity index (EAI) of plasma proteins was detected post treatment with *MTGase* and/or cysteine (Figure 2), thus indicating that their efficiency as emulsifiers was always reduced. For treatments with *MTGase* only, similar results have been reported by other authors working on different proteins, which have been attributed to a decrease in protein solubility (Siu *et al.*, 2002). However, emulsions were stable (no change in absorbance values) for 10 min.

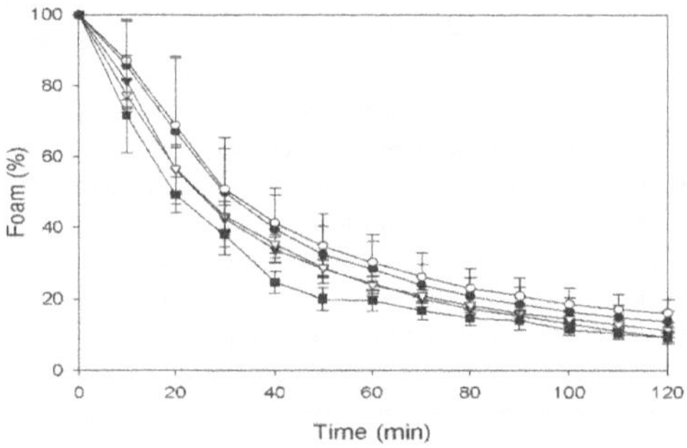

Figure 1: Foam stability of porcine plasma. Effect of different treatments: control (●), MTG (▼), Cys(○), MTG+Cys(▽), and MTG//Cys (■). Mean ± sd (n=3).

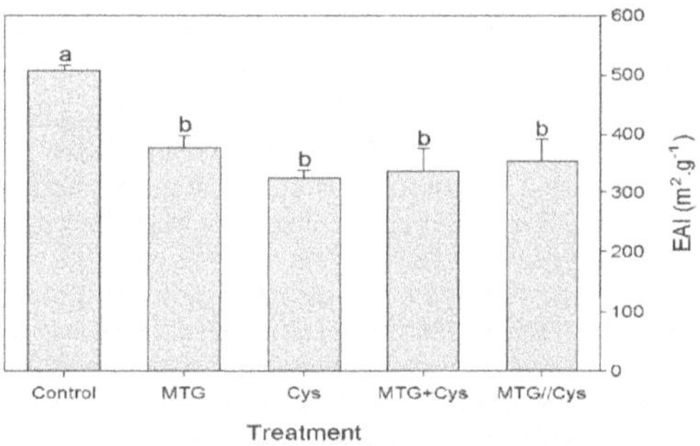

Figure 2: Emulsifying activity index (EAI) of porcine plasma; effect of treatment with cysteine (Cys) and/or *MTGase* (MTG). Mean ± sd (n=3). Different letters indicate significantly differences between treatments

The results indicate that the treatment of plasma proteins with *MTGase* alone or in combination with cysteine supposes an important reduction in their surface properties.

References

Babiker, E.E. (2000). *Food Chem.*, 70: 139-145.

Fort, N.; Carretero, C.; Parés, D.; Toldrà, M.; Saguer, E. (2006). *Food Hydrocolloid.* (In revision)

Siu, N.C.; Ma, C.-Y.; Mock,W.-Y.; Mine, Y. (2002). *J. Agric. Food Chem.*, 50: 2666-2672

QUALITY OF TURKEY HAMBURGERS PROCESSED WITH GROUND *SCUTELLARIA BAICALENSIS* GEORGI ROOTS

M. Korzeniowska[1]*, A. Wojdylo[2], M. Sobota[1], A. Sujecka[1] and W. Kopec[1]

[1]Department of Animal Products Technology and Quality Management, [2]Department of Fruit and Vegetables Processing, Agricultural University, 25 Norwida Str., 50375 Wroclaw, Poland. E-mail: gosiak2000@yahoo.com

Keywords: turkey, hamburger quality, *Scuttelaria baicalensis* Georgi, colour and sensory evaluation

Introduction

Convenient foodstuffs, including meat products, have become more and more popular, especially in rapidly-growing and well-educated populations. Apart from the fact that people are looking for easy-to-prepare meals, they very often search for products containing natural substances. *Scute*llaria baicalensis Georgi is known for its high antioxidant activity and its incorporation in food products in order to slow down lipid oxidation and prolong shelf-life is now being considered. Lipid oxidation is one of the main limiting factors for the quality and acceptability of meat products. Synthetic and natural antioxidants can inhibit this process (McCarthy *et al.*, 2001, Fernandez-Lopez *et al.*, 2005, Wojdylo *et al.*, 2005). The aim of this study was to evaluate quality parameters of hamburgers made from turkey meat with the addition of *Scutellaria baicalensis* Georgi roots.

Materials and Methods

Ground, dry roots of Skullcap (*Scutellaria baicalensis* Georgi) were analysed by high perfrormance liquid chromatography (HPLC) to determine flavone content (Oszmianski and Wojdylo, 2005). The main study was carried out on turkey hamburgers prepared from mechanically-deboned turkey meat, pork fat, breadcrumbs, potato flour and spices. The addition of 1000 ppm, 3000 ppm and 5000 ppm of Scullcap was tested. All ingredients were carefully mixed with ice and shaped into burgers of 6 cm diameter and 1.5 cm height. Finally they were deep fried in rapeseed oil until a temperature of 72°C was reached in the geometric centre of the product, chilled quickly, vacuum-packed and stored at 4°C for 15 days. Samples were taken at day 0, 5, 10 and 15. Lipid oxidation (TBARS) was determined by the method of Pikul *et al.*, (1989). Sensory evaluation of hamburgers was performed by a trained panel using a 5-point scale of acceptance. Objective measurement of colour (L*, a*, b*) was performed at the surface and in a cross-section of hamburgers. Hardness of the product was analysed on Zwick. All results were statistically analysed using STATISTICA (version 6.0) at a significance level of $P=0.05$.

Results and Discussion

HPLC analysis of Scullcap powder revealed it to contain baicalin (12.78% w/w), wogonin 7-glucoronoid (3.94% w/w), baicalein (0.04% w/w) and other unidentified compounds (3.24% w/w) (Figure 1). Scullcap used in turkey hamburgers significantly reduced lipid oxidation (Figure 2) as expressed by TBARS value. Sensory evaluation of the products did not reveal any significant differences between the control and hamburgers processed with Scullcap (Table 1). The highest lightness value (L*) of the product was analysed for control samples either on the surface or on the cross-sections (Table 2). However, colour measured on cross-sections during storage of hamburgers processed with Scullcap was more stable. Similarly to the sensory evaluation, instrumental texture profile analysis did not reveal any significant differences between hardness of experimental hamburgers (Figure 3). Moreover, the longer the storage time, the higher the hardness of all analysed products despite the level of added Scullcap roots.

Conclusion

Turkey hamburgers processed with Scullcap roots were characterised by similar to control samples in terms of values for quality parameters tested; however, the range of lipid oxidation was lower than in control samples. Roots of Scullcap can be utilised in hamburger processing.

References

Fernandez-Lopez J., Zhi N., Aleson-Carbonell L., Perez-Alvarez J.A., Kuri V. (2005). Antioxidant and antibacterial activities of natural extracts: application in beef meatballs. Meat Science, 69, 371-380.

McCarthy T.L., Kerry J.P., Kerry J.F., Lynch P.B., Buckley D.J. (2001). Evaluation of the antioxidant potential of natural food/plant extracts as compared with synthetic antioxidants and vitamin E in raw and cooked pork patties. Meat Science, 57, 45-52.

Oszmianski, J., Wojdylo, A. (2005). *Aronia melanocarpa* phenolics and their antioxidant activity. European Food Research and Technology, 221, 809-813.

Pikul, J., Leszczynski, D.E., Kummerow, F.A. (1989). Evaluation of three modified TBA methods for measuring lipid oxidation in chicken meat. Journal of Agricultural and Food Chemistry, 37, 1309-1313.

Wojdylo, A., Oszmiański, J., Sokol-Letowska, A., Korzeniowska, M. (2005). Butter stabilisation by plant phenolic antioxidants. Polish Journal of Food and Nutrition Sciences, 14/55 SI1, 121-128.

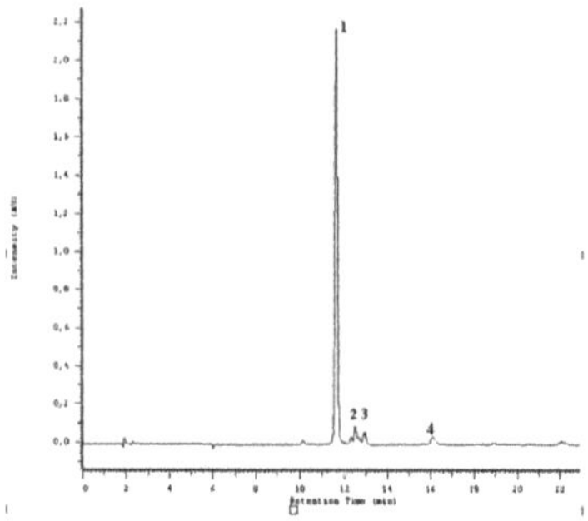

Figure 1. HPLC chromatograms of flavones of skullcap detected at 280 nm: 1-baicalin; 2, 3,- flavones unidentified; 4- baicalein.

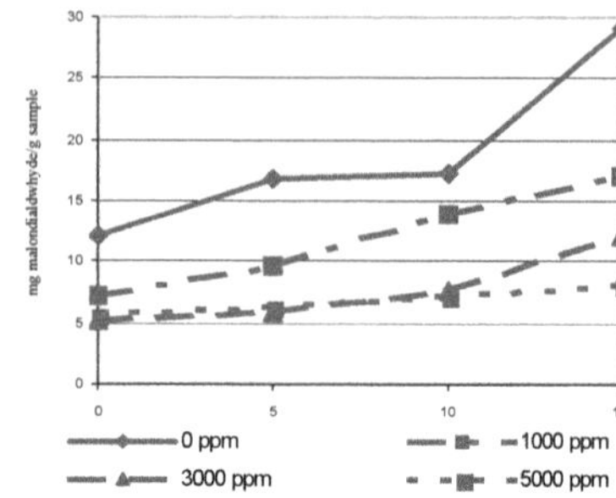

Figure 2. Effect of skullcap root addition on TBARS values of turkey hamburgers stored at 4°C for 15 days.

Table 1: Sensory evaluation of turkey hamburgers processed with roots of *Scutellaria baicalensis* Georgi and stored at 4°C for 15 days (sensory evaluation scale:1-not acceptable, 5-very acceptable).

0	Colour				Flavour				Taste				Consistency				Overall acceptability			
Storage time [days]	0	5	10	15	0	5	10	15	0	5	10	15	0	5	10	15	0	5	10	15
control	4.49	4.28	4.17	3.99	4.37	4.20	4.25	4.05	4.07	4.02	4.21	4.07	4.24	4.15	3.97	4.14	4.33	4.30	4.19	4.15
1000 ppm	4.27	4.09	4.09	3.73	4.22	4.28	4.06	3.77	4.10	3.87	4.23	3.96	4.30	4.10	3.61	3.94	4.19	4.15	4.12	3.77
3000 ppm	3.97	4.06	3.80	3.69	4.29	4.18	4.01	4.03	4.26	4.36	4.14	3.99	4.18	4.13	3.95	4.04	4.29	4.17	4.03	3.89
5000 ppm	3.96	3.54	3.53	3.40	4.35	4.20	4.05	3.81	4.13	4.08	4.01	3.90	4.20	4.19	3.98	3.81	4.29	4.08	3.84	3.89

Table 2: Colour measurements taken on the surface and cross-sections of turkey hamburgers processed with roots of *Scutellaria baicalensis* Georgi and stored at 4°C for 15 days (L*a*b* scale).

	Colour of the hamburger surface				Colour of the hamburger cross-sections			
Storage time [days]	0	5	10	15	0	5	10	15
			*L parameter					
control	51.68	53.89	53.82	54.64	57.78	58.72	58.72	58.88
1000 ppm	49.56	52.10	53.46	52.79	56.42	56.93	56.61	56.80
3000 ppm	46.40	50.62	50.29	50.42	55.47	55.75	55.49	55.86
5000 ppm	45.76	49.70	48.63	50.93	54.29	55.46	53.84	54.61
			a* parameter					
control	5.56	4.83	5.54	4.74	3.78	4.35	3.81	3.97
1000 ppm	4.72	4.23	3.70	3.62	2.88	3.21	3.03	3.06
3000 ppm	4.88	3.25	3.54	3.93	2.59	3.11	2.68	2.55
5000 ppm	4.53	3.80	4.19	3.18	2.76	3.02	2.66	2.90
			b* parameter					
control	16.39	16.49	17.07	16.81	12.77	11.76	12.14	11.68
1000 ppm	15.30	15.28	14.17	14.26	11.89	11.30	10.83	10.88
3000 ppm	15.70	14.56	13.40	12.87	12.38	11.19	10.75	10.71
5000 ppm	14.61	14.93	13.32	12.00	12.67	11.79	11.03	10.62

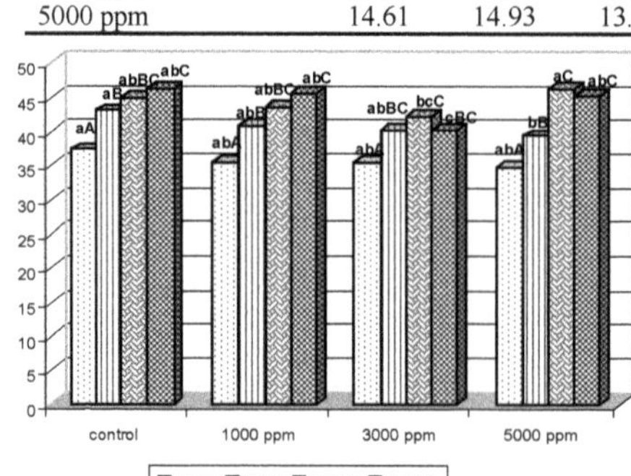

Figure 3: Effect of skullcap root addition on hardness of turkey hamburgers stored at 4°C for 15 days measured on Zwick equipment. a,b,c - means with different superscripts are significantly ($p \leq 0.05$) different when analysed for Scullcap addition, ABC - means with different superscripts are significantly ($p \leq 0.05$) different when analysed for storage time.

Acknowledgements: Research was supported by Polish Scientific Committee grant no. 2P06T04428

QUALITY OF CURED AND SMOKED BEEF TENDERISED BY VARIOUS CHEMICAL SUBSTANCES

T. Szmanko, M. Korzeniowska* and M. Jezlowicka

Department of Animal Products Technology and Quality Management, Faculty of Food Science, Agricultural University of Wroclaw, 25 Norwida Str, 50375 Wroclaw, Poland. E-mail: gosiak2000@yahoo.com

Keywords: beef, tenderisation, curing, quality

Introduction

Tenderness is one of the most important quality parameters of beef (Carmack *et al.,* 1994). Beef tenderness can be modified by different methods such as genetic (Cundiff, 1992), enzymatic (Katsaras *et al.*, 1984), physical (Miller *et al.*, 1998) and chemical (Berge *et al.*, 1998). However, the results of application of some chemical substances for improving beef quality are sometimes controversial mainly due to a negative correlation between tenderness and other quality parameters of beef such as colour, flavour or production yield (Seuss and Martin, 1991). This method is still of interest of meat researchers. The majority of literature data concerns the quality of culinary beef, whilst there is still a lack of knowledge about improving tenderness of beef cuts, which are further used in meat technology. The objective of this study was to create the most suitable chemical method for tenderisation of beef which is to undergo further processing.

Materials and Methods

The study was carried out on round beef muscle *musculus semitendinosus* (pH_{24}<5.75) cut off from 9 carcasses of Polish black and white young bulls of 500 kg average weight.. Each muscle was divided into four parts and randomly assigned as **C** – control muscle cured with standard brine solution; **CC** – muscle injected with brine containing 0.3 M calcium chloride; **LA** – muscle soaked in 1.2% of lactic acid (20 h) before standard curing procedure; **BS** – muscle soaked in 30% of sodium bicarbonate (8 h) before standard curing. Standard curing solution contained curing salt, polyphosphates, sucrose, carrageenan, sodium isoascorbate and monosodium glutamate. Beef samples were injected with brine solution up to the level of 40% and tumbled for 17 h at 4°C. Then, meat from all experimental groups was smoked at 50°C and scalded at 80°C until the temperature at the geometric centre of the product reached 72°C. The products were then chilled at 4°C for 24h.

Reaction (pH) was measured in raw meat 24 h after slaughter as well as in semi-final (after curing) and final products. Sensory analysis especially evaluation of tenderness and overall acceptability was performed by a trained panel using a 6-point hedonic scale (1= not acceptable and 6 = very acceptable). Colour measurements (L*, a*, b* system) were performed using Minolta equipment. Moreover, colour stability (ΔE6, ΔE12, ΔE24) was calculated on the basis of L*, a*, b* data measured respectively after 6 h, 12 h and 24 h (cross-sections of the products were exposed on white fluorescent light with an intensity of 250 lux). Free amine group content was determined according to Kuchroo *et al.*, 1983) and free nitrates(III) content on the basis on the colour reaction of $NaNO_2$ with Griess' reagent. All the results were then statistically analysed using STATISTICA (version 6.0) at a significance level of P=0.05.

Results and Discussion

Reaction (pH) of beef measured 24 h after slaughter was quite consistent and equaled 5.48 average. pH values increased after curing processes by about 0.3 for control sample, 0.4 for beef cured with brine containing $CaCl_2$, and 1.8 for meat treated with $NaHC0_3$. In final products from control and CC groups, pH values were similar (average 6.00) while in **LA** samples it was statistically lower (4.8), resulting in a reduction of sensory acceptance. Final beef products treated with sodium bicarbonate were characterised by high pH (8.05), which consequently diminished the durability of analysed meat products. The highest increase in yield after curing was noticed for beef from **LA** and **BS** experimental groups (33% and 34% respectively). Moreover, those products were characterised by extremely different pH values i.e. 5.24 for **LA** and 7.28 for **BS**. An increase in free amine group content was the result of protein degradation caused mainly by tenderising substances used in the study. The highest amount of free amine group was found in the most tender beef products from **LA** and **BS** blocks (Figure 1); 7853 µg Gly/g protein and 7640 µg Gly/g protein, respectively. Moreover, beef treated with brine containing sodium bicarbonate or lactic acid were distinguished by extremely different sensory properties, especially in overall acceptability, colour and flavour (Table 2, Figure 2). Lower sensory scores obtained for **LA** beef arose fom large variations in colour. There was a darker layer, about 1 cm wide, which was clearly seen on the circumference of the cross-section of the final product. Moreover, products made from beef previously cured with brine containing lactic acid had a slightly acidic taste. Beef treated with **LA** or **BS** were characterised by lower values of L* than the control group, while $CaCl_2$ solution used as a tenderising agent lightened the products. The participation of red colour (a*) in the whole hue spectrum of the control beef was comparable with products from the **BS** group and higher than measured for **CC** and **LA** products. However, beef cured with brine containing **LA** and **BS** was characterised by lower participation of yellow colour (b*) in the spectrum than the beef analysed for **C** and **CC** groups. Moreover, the colour of beef treated with **BS** was the most stable, which was explained by low values of Δ6, Δ12 and Δ24 (Table 1). Free nitrite(III) content in final experimental beef products was not effected by tenderising agent used. The highest amount (but still considered to be safe) of free nitrite (92.38 ppm) was found in beef cured with sodium bicarbonate. The final products from this experimental block were also characterised by the most acceptable sensory and physicochemical properties.

Conclusion

From technological, physicochemical and sensory points of view, sodium bicarbonate emerged as the most effective substance for beef tenderisation using a curing process. Tenderness of culinary beef can also be improved by application of lactic acid.

References

Berge, P., Ertbjerg, P., Melchior-Larsen, L., Astruc, T., Moller, A. (1998). Tenderisation of beef by lactic acid treatment applied at different times post-mortem. Proceedings of the 44th ICoMST, 710-711.

Carmack, C.F., Kastner, C.L., Dikeman, M.E., Schwenke, J.R., Garcia Zepeda, C.M. (1994). Sensory evaluation of beef-flavor-intensity, tenderness, and juiciness among major muscles. Meat Science 39, 143-142.

Cundiff, L.C. (1992). Genetic selection to improve the quality and composition of beef carcasses. Proceed. Recip. Meat Conference, 45, 123-131.

Katsaras, K., Tezlaff, G., Budras, K.D. (1984). Morphologische Veranderungen der Rindermuskulatur wahrend der Reifung. Fleischwirtschaft, 64, 1252-1256.

Kuchroo, C.N., Rahilly, J., Fox, P.F. (1983). Assessment of proteolysis in cheese by reaction with trinitrobenzene sulphonic acid. Irish Journal of Food Science and Technology, 7, 129-133.

Miller, M.F., Ramsey, C.B., Hoover, L.C., Carr, M.A., Crockett, K.L. (1998). Consumer thresholds for establishing the value of beef tenderness. AMSA Recip. Meat Conference Proceedings, 51, 4-9.

Seuss, I., Martin, M. (1991). Einfluss der Marinierung mit Genusssauren auf Zusammensetzung und sensoriche Eigenschaften von Rindfleisch. Fleischwirtschaft, 71, 1269-1272.

Table 1. The dynamics of colour stability changes in final beef products after 6, 12 and 24 h exposure to light (ΔE_6, ΔE_{12}, ΔE_{24}) (n=9)

Experimental groups		ΔE_6	ΔE_{12}	ΔE_{24}
C	x	2.27^b	2.67^b	3.55^b
	sd	1.19	1.15	1.94
CC	x	2.96^b	2.98^b	3.82^b
	sd	3.20	1.07	1.40
LA	x	2.00^b	2.98^b	3.93^b
	sd	0.74	1.19	2.06
BS	x	1.45^a	2.00^a	2.42^a
	sd	0.84	0.86	0.91

Table 2. The sensory evaluation of beef products: overall acceptability and colour (n=9)

Experimental groups		Overall acceptability	Colour
	x	4.57^b	4.28^b
C	sd	0.76	0.58
CC	x	4.40^b	4.37^b
	sd	0.62	0.61
LA	x	2.30^a	1.80^a
	sd	0.88	0.69
BS	x	5.28^b	5.10^c
	sd	0.82	0.86

a,b,c… - means within a column with different superscripts are significantly ($p \leq 0,05$) different.

Figure 1. Tenderness of final beef products evaluated by a sensory panel (n=9)

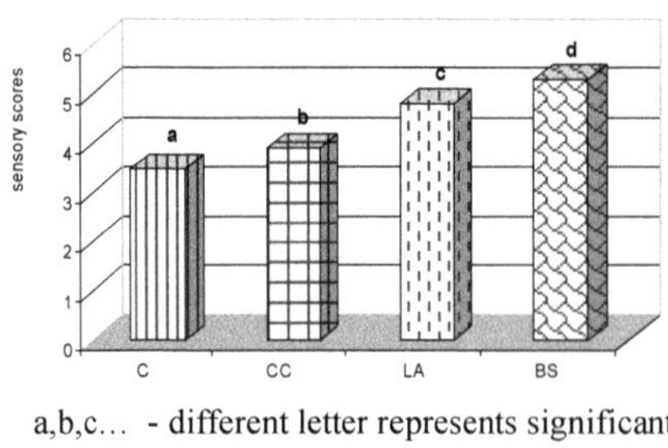

Figure 2. Flavour of final beef products evaluated by a sensory panel (n=9)

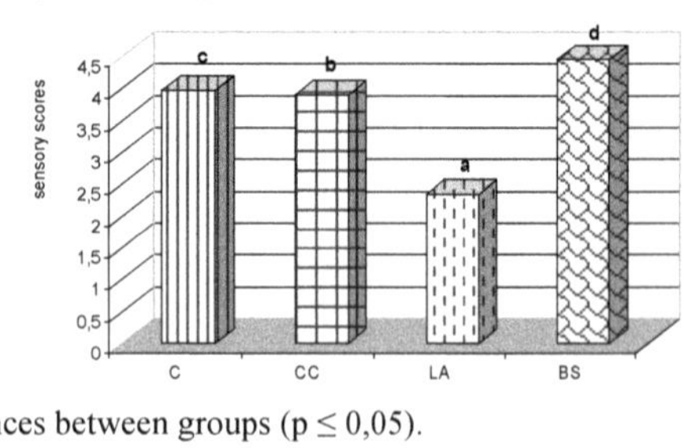

a,b,c… - different letter represents significant differences between groups ($p \leq 0,05$).

ACTIN FRAGMENTS GENERATED DURING DRY-CURED HAM PROCESSING

M.A. Sentandreu*[1], M. Armenteros[1], J.J. Calvete[2], M.C. Aristoy[1] and F. Toldrá[1]

[1.] *Instituto de Agroquímica y Tecnología de Alimentos (CSIC). P. O. Box 73, 46100. Burjassot (Valencia), Spain.* [2.] *Instituto de Biomedicina de Valencia (CSIC). Jaime Roig 11, 46010. Valencia, Spain. Email: ciesen@iata.csic.es*

Keywords: dry cured ham, proteolysis, actin, peptide sequencing, proteomics

Introduction

Dry-cured ham is a traditional food requiring a long processing period for development of its appreciated texture and flavour characteristics. During this time, there is an intense degradation of muscle proteins due to the action of endogenous proteolytic enzymes. This gives rise to an important generation of free amino acids and oligopeptides that contribute directly or indirectly to flavour characteristics of the final product. The nature and properties of free amino acids has been well established but, on the contrary, little is known about the identity of the peptide fraction at the end of curing. The main objective of the present work was to gather more information on this peptide fraction, and to relate the results to flavour development in dry-cured meat products.

Materials and Methods

Twenty-five grams of muscle *semimembranosus* were taken from a Spanish Serrano dry-cured ham in order to obtain a deproteinised ham extract. Five ml of this extract were fractionated on a Sephadex G-25 size exclusion column, pooling fractions eluting in the range 215-230 ml. These fractions were concentrated to a final volume of 2 ml that was subsequently injected on a SymmetryPrep C18 (7.8 x 300 mm) reverse phase column. Peptide separation was made by a linear gradient from 0 to 50 acetonitrile in 55 min at a flow rate of 3 ml/min. Separation was monitored at λ=214 nm and prominent peaks were collected manually, dried and stored at –20 °C. The selected fraction was redissolved in 6 µl of 50 % acetonitrile containing 1% TFA, then spotting 1 µl on a MALDI plate and mixed with an equal volume of a α-cyano-4-hydroxycynnamic acid solution. Molecular mass determinations were performed in an Applied Biosystems MALDI-TOF mass spectrometer, operated in delayed extraction and positive reflector mode.

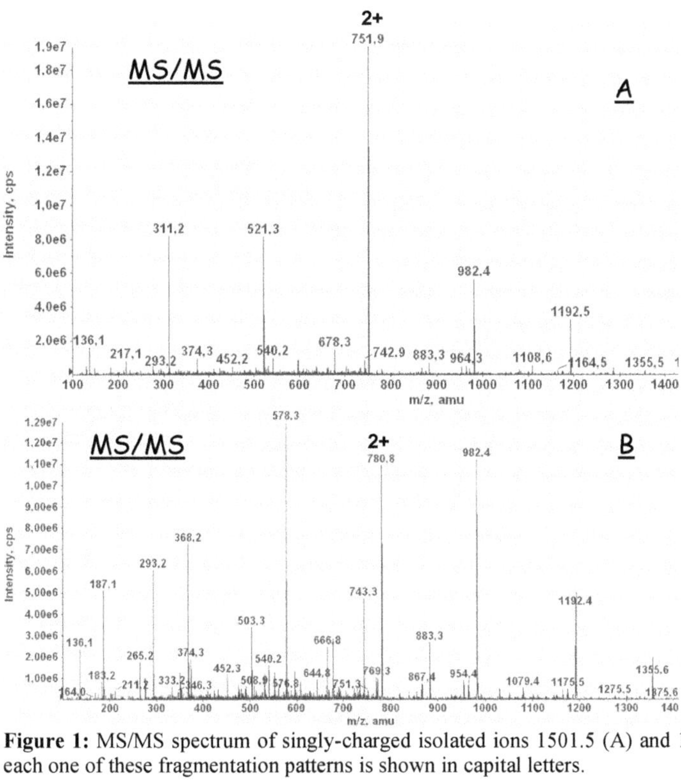

Figure 1: MS/MS spectrum of singly-charged isolated ions 1501.5 (A) and 1559.5 (B). Peptide sequence matching to each one of these fragmentation patterns is shown in capital letters.

Determination of peptide sequence. Relevant peptide ions obtained from MALDI-TOF were subjected to ESI mass spectrometry analysis using a Qtrap mass spectrometer (Applied Biosystems). Doubly-charged ions were fragmented using the Enhanced Product Ion with Q_0 trapping option. Enhanced resolution was performed at 250 amu/s across the entire mass range.

Results and Discussion

Dry-cured ham extract was fractionated on Sephadex G-25, taking fractions corresponding to 215-230 ml elution volume. This corresponded to a molecular mass of around 1530 Da according to the curve made with standards of known molecular weight. Pooled fractions were injected on a Symmetry C18 reverse phase column and a peak eluting at a concentration of 14 % acetonitrile was collected and selected for molecular mass determination. MALDI-TOF MS of this fraction yielded two singly-charged ions having a molecular mass of 1501.5 and 1559.5 Da, respectively, which was in accordance to the estimated peptide size obtained with Sephadex G-25. MS/MS spectra of these ions are shown in Figure 1. Doubly-charged forms of these peptides are clearly appreciated at m/z 751.9 (Figure 1A) and 780.8 (Figure 1B), respectively. Interpretation of mass spectra was performed by using MASCOT software (www.matrixscience.com). The fragmentation pattern shown in Figure 1A matched the sequence DSGDGVTHNVPIYE (peptide A), whereas data of Figure 2B gave as unequivocal result the sequence DSGDGVTHNVPIYEG (peptide B). The only difference between the two peptides was a C-terminal Gly present in peptide B but not in peptide A, being the responsible for the difference in their molecular masses. Sequence homology analysis using BLAST search engine (www.bork.embl-heidelberg.de) revealed a 100 % identity of these peptides with an internal region of muscle actin, as illustrated in Figure 2. The present results give clear evidence of the intense actin proteolysis occurring during dry-cured ham processing. Previous works reported the progressive actin degradation during postmortem muscle proteolysis either during meat ageing (Morzel *et al.* 2004) but specially during dry-curing where most myofibrillar proteins are extensively degraded (di Luccia *et al.* 2005). Other authors reported a notable increase of the peptide fraction during dry-cured ham processing, indicative of an intense proteolysis that is closely related to flavour development (Toldrá, 2002). However, this is the first time that small peptides coming from actin are isolated and fully identified from dry-cured ham. The fact that two fragments with and without Gly^{170} in C-terminal position have been identified may be indicative of any carboxypeptidase action during the curing period. According to previous observations, degradation of actin and other myofibrillar proteins during postmortem muscle proteolysis could be due to the action of lysosomal cathepsins (Delbarre-Ladrat *et al.*, 2004), a hypothesis that would be supported by their relatively good stability observed during great part of the curing period (Toldrá 2002).

MCDEDETTAL VCDNGSGLVK AGFAGDDAPR^{30} AVFPSIVGRP RHQGVMVGMG QKDSYVGDEA^{60}
QSKRGILTLK YPIEHGIITN WDDMEKIWHH^{90} TFYNELRVAP EEHPTLLTEA PLNPKANREK^{120}
MTQIMFETFN VPAMYVAIQA VLSLYASGRT^{150}TGIVLDSGDG VTHNVPIYEG YALPHAIMRL^{180}
DLAGRDLTDY LMKILTERGY SFVTTAEREI^{210}VRDIKEKLCY VALDFENEMA TAASSSSLEK^{240}
SYELPDGQVI TIGNERFRCP ETLFQPSFIG^{270}MESAGIHETT YNSIMKCDID IRKDLYANNV^{300}
MSGGTTMYPG IADRMQKEIT ALAPSTMKIK^{330}IIAPPERKYS VWIGGSILAS LSTFQQMWIT^{360}
KQEYDEAGPS IVHRKCF^{377}

Figure 2: Primary structure of porcine skeletal muscle actin, indicating the position of the fragments isolated and identified in the present work from a dry-cured ham extract at the end of processing. Underlined amino acids are the common sequence of the two peptides, whereas squared Gly^{170} was present only in peptide B.

Conclusion

Two actin fragments of small size (1501.5 and 1559.5 Da) have been isolated and identified for the first time in a dry-cured ham extract at the end of processing, confirming the extensive proteolysis of this protein during curing. These findings contribute to meat research by revealing new information about the complex mechanisms taking place in postmortem muscle and also the enzyme groups that are mainly implicated in the proteolytic processes.

References

Delbarre-Ladrat, C., Verrez-Bagnis, V., Noel, J. and Fleurence, J. (2004). Relative contribution of calpain and cathepsins to protein degradation in muscle of sea bass (Dicentrarchus labrax L.). *Food Chemistry*, 88 (3): 389-395.

di Luccia, A., Picariello, G., Cacace, G., Scaloni, A., Faccia, M., Liuzzi, V., Alviti, G. and Spagna Musso, S. S. (2005). Proteomic analysis of water soluble and myofibrillar protein changes occurring in dry-cured hams. *Meat Science*, 69 (3): 479-491.

Morzel, M., Hamelin, M., Santq-Lhoutellier, V., Sayd, T., Monin, G. and Chambon, C. (2004). Proteome changes during pork meat ageing following use of two different pre-slaughter handling procedures. *Meat Science*, 67 (4): 689-696.

Toldrá, F. (2002). *Dry cured meat products* Food and Nutrition Press, Trumbull, Conn.

THE EFFECT OF MUSCLE TYPE AND STORAGE ON BINDING STRENGTH AND COOKING LOSS OF RESTRUCTURED PORCINE MEAT

B.R. Madsen and L. Kristensen*

*Department of Food Science, Royal Veterinary and Agricultural University, Rolighedsvej 30, DK-1958 Frederiksberg C, Denmark * E-mail: lak@kvl.dk.*

Keywords: binding strength, cooking loss, restructured meat, muscle type, storage

Introduction

The functional properties of meat in the production of meat products is mainly determined by the amount and properties of the salt soluble myofibrillar protein. The solubility of myofibrillar proteins is influenced by pH, ionic strength, animal species, post mortem history and muscle fibre type (Asghar *et al.*, 1985; Samejima *et al.*, 1992). It has also been shown that storage time increases the WHC of meat (Kristensen and Purslow, 2001).

In restructured meat products the salt soluble proteins (SSP) are extracted with brine during the tumbling process. The SSP will afterwards cover the surface as a sticky substance of all the meat pieces which have been tumbled. When the meat pieces are lightly compressed and heat treated, the SSP denature and bind the pieces together, i.e. SSP act as a glue. In the production of the restructured meat product cooked ham, different muscles from the ham are used. These muscles vary in fibre type composition, which probably influence the functional properties. The muscles also vary in pH, connective tissue characteristics and sarcomere length, which also might influence the functional properties.

The aim of the work was to investigate if muscle type and storage time affects the functional properties of meat. The functional properties were determined by measuring the binding strength and cooking loss of a model of a restructured cooked ham produced from a single muscle type.

Materials and Methods

Samples of *Semitendinosus* (ST), *Psoas Major* (PM) and *Vastus Intermedius* (VI) from 8 different animals 24 h post mortem were cut into 3 pieces, which were stored at 5°C for either 1, 3 or 6 days post mortem. All samples were afterwards stored at -18°C.

From each sample two 0.5cm thick slices were cut across the fibre direction. These were placed in a vacuum bag with 20% brine (2 M NaCl and 0.01M sodium pyrophosphate, pH 9.12). The bags were vacuum packed at 0.5 bar and placed in a vacuum tumbler. The samples were tumbled for 3h using a cycle of 5min tumbling and 5min rest. Then the slices were placed on top of each other and vacuum packed. Each sample was placed between two acryl plates and cooked at 75°C for 1h in a water bath. The samples were cooled in ice-water and left overnight at 5°C. Three pieces (1cm x 1cm) from each sample were glued between two aluminium blocks. The aluminium blocks were then mounted on an Instron Texture Analyser and the binding strength was determined by pulling the blocks apart. Only samples that separated between the 2 slices were used, i.e. samples separating between the aluminium block and the meat, were discharged.

The following measurements were made: pH prior (pH 1) and post (pH 2) storage and after tumbling (pH 3), drip loss post freezing, sarcomere length post freezing (only day 1 samples), weight gain during the tumbling process, total protein concentration in the tumbling exudates, cooking loss and binding strength. Data were analysed with SAS 8.2 using the Proc Mixed procedure.

Results and Discussion

Muscle type had a significant influence on both binding strength and cooking loss (Table 1Fout! Verwijzingsbron niet gevonden.). There was no difference in binding strength between VI and PM ($p = 0.14$), however, there was a clear difference between ST and the two other muscles ($p < 0.0001$).

Muscle and the interaction between muscle and storage had an effect on cooking loss (Table 1). PM has the lowest cooking loss followed by ST and VI. This is the opposite of what was expected from the pH of the meat since PM has the lowest pH (5.45) followed by ST (5.52) and VI (5.68). A higher pH normally results in higher water holding capacity.

Table 1: Least squares means of binding strength and cooking loss for cooked ham produced of *Semitendinosus* (ST), *Psoas Major* (PM) and *Vastus Intermedius* (VI) as affected by muscle, storage time and their interaction.

	Muscle			*P* - values		
	ST	PM	VI	Muscle	Storage	M × S
Binding strength, N	3.7	1.8	2.4	***	-	-
Cooking loss, %	34	28	38	***	-	***

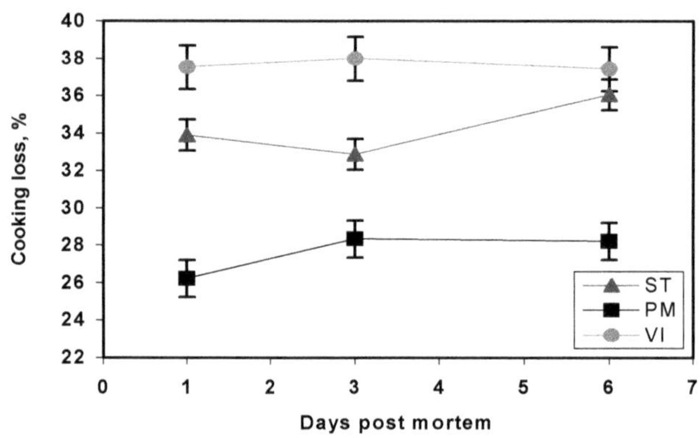

Figure 1: Effect of storage time on the cooking loss of cooked ham produced of *Semitendinosus* (ST), *Psoas Major* (PM) and *Vastus Intermedius* (VI). Error bars represent SEM.

Storage has no effect on cooking loss but there was an interaction with muscle (p = 0.0012). Cooking loss increased with storage for ST and PM, however was stable for VI (Figure). The value of pH 3 also increased with storage for ST and PM indicating loss of buffer capacity of the muscles. These effects may result from proteolytic activity during storage.

Storage time did not have any effect on either binding strength or cooking loss (p > 0,.05). There was no random effect of animal and also, there were no effects (p > 0.05) of any of the following covariates pH 1, pH 2, drip loss, weight gain, and total protein.

Conclusions

The results show a clear effect of muscle type on the functional properties of meat when producing restructured cooked ham. No main effect of storage time was observed, however the cooking loss of ham produced from ST and PM seems to increase with storage time of the meat before processing. The results suggest that muscle fibre type composition of muscles has a large influence on the functional properties of meat.

References

Asghar, A., Samejima, K. and Yasui, T. (1985). "Functionality of Muscle Proteins in Gelation Mechanisms of Structured Meat-Products." *Crc Critical Reviews in Food Science and Nutrition.* 22:27-106.

Samejima, K., Lee, N.H., Ishioroshi, M. and Asghar, A. (1992). "Protein Extractability and Thermal Gel Formability of Myofibrils Isolated from Skeletal and Cardiac Muscles at Different Postmortem Periods." *Journal of the Science of Food and Agriculture.* 58:385-393.

THE EFFECTS OF RIPENING TEMPERATURE ON SOME PROPERTIES OF TURKISH DRY-FERMENTED SAUSAGE (SUCUK)

G. Yıldız-Turp, I. Eren, M. Serdaroğlu and F. Kaymak-Ertekin

Ege University, engineering Faculty, Food Engineering Department 35100 Bornova İzmir Türkiye e-mail:meltem.serdaroglu@ege.edu.tr

Keywords: Sucuk, dry sausage, fermentation, beef, drying

Introduction

Turkish dry-fermented sausage, which is called sucuk, is one of the most popular traditional meat products in Turkey. To obtain sucuk dough, minced beef and fat are mixed with curing ingredients and spices, then sucuk dough is filled into natural or artificial casings. Sucuk is produced in two stages; the first stage is the fermentation either by added starter culture or natural flora (Bozkurt and Erkmen, 2002). The second stage is drying of the sucuks under controlled conditions. The combination of fermentation and drying stages is known as the ripening period. Ripening conditions should be strictly controlled to prepare good quality product (Bozkurt and Erkmen, 2004; Bozkurt and Bayram, 2006). Gökalp (1986) concluded that starter culture addition and ripening at high temperatures resulted in faster drying. The objective of this study was to examine the effects of different ripening temperatures on some physical and chemical properties of sucuk.

Materials and Methods

Fresh boneless beef cuts (80%) and beef fat (20%) were used as a raw material. The following ingredients were added per kg of meat mixture, 27.5g spices, 10g garlic, 28g curing ingredients and 0.5g lyophilised starter culture mixture (*Staphylococcus xylosus+Pediococcus acidilactici*). Fat and meat are separately ground then mixed with the ingredients. Sucuk dough was stuffed into collagen casings. Cylindrical rod shaped sucuks which had a length of 40cm and a diameter of 4cm were ripened in a laboratory scale incubator at a constant air velocity (0.5 m/s) and at four different air temperatures (15 (S15), 20 (S20) 25 (S25) and 30°C (S30) for 7 days. Average RH of the air inside the drying chamber varied and was between 65% and 85%. At 0, 1, 2, 3, 4, 5, 6 and 7 days, moisture content (AOAC,2000), pH (Landvogt, 1991) and colour (CIE, 1986) was measured.

Results and Discussion

Changes in the pH values of sucuk samples during ripening are given in Figure 1. The initial pH of sucuk samples ranged from 5.5 to 5.6 and the final pH varied from 5.4 to 4.4. The pH values of sucuk samples were significantly affected by ripening temperature. Similar results in sucuks were reported by Soyer *et al.* (2005). There was a rapid decrease in pH at 25 and 30°C. A sharp decrease was observed in samples ripened at 20, 25 and 30° C on the second day of ripening. These results were in agreement with the literature that pH values of sucuk samples decreased sharply during the first days of ripening due to the production of lactic acid (Kayaardı and Gök, 2003, Bozkurt and Bayram, 2006).

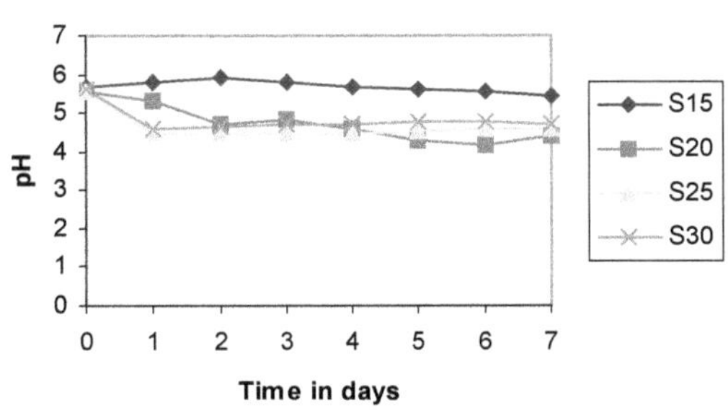

Figure 1: Effect of ripening temperature on pH values of sucuks.

Figure 2 shows the moisture changes of sucuk samples. During the drying period, the moisture content decreased as a result of the moisture loss at high ripening temperatures. At day 0, moisture contents of S15, S20, S25 and S30 samples were 65.8, 59.5, 63.2 and 63.2 respectively. Due to slow dehydration at a lower temperature, sucuk samples ripened at 15°C had higher moisture content than those ripened at 20, 25 and 30°C by day 7. Faster pH decline could be the reason for rapid moisture movement in sucuk samples. Gökalp (1986) concluded that ripening at higher temperatures resulted in faster drying. Similar results were reported by Soyer *et al.* (2004) who indicated that moisture decreased slowly at low ripening temperatures.

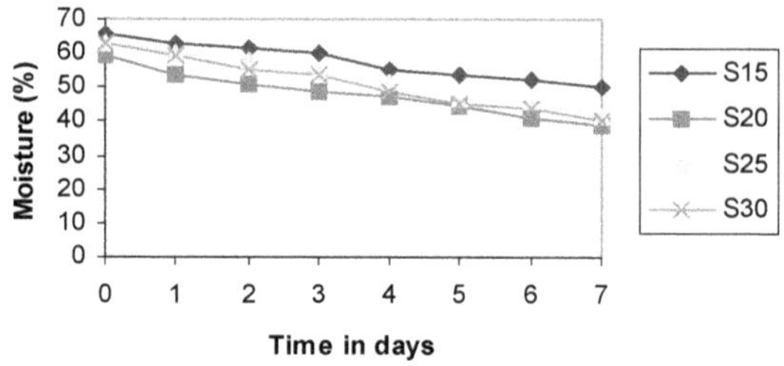

Figure 2: Effect of ripening temperature on moisture changes in sucuk samples.

Table 1: Colour values (L,a,b) of sucuk samples at day 0 and day 7.

	*L**	*a**	*b**
0 Day			
S15	41.3	15.2	19.6
S20	42.2	17.2	20.5
S25	41.0	16.8	21.7
S30	41.6	15.1	19.0
P	NS	NS	NS
7th Day			
S15	38.1a	13.0	15.2b
S20	46.3c	14.3	16.1b
S25	45.4bc	13.9	16.1b
S30	42.8ab	12.0	13.6a
P	0.018	NS	0.032

(a-c)Different letters in the same column indicate significant differences. NS: Non significant,P: Significance

The L*a* and b* values of treatment groups were not significantly different from each other at day 0 of ripening (Table 1). On the 7th day of ripening there was an increase in L* values for the treatment groups except for the S15 group. S15 samples had lower L* values than other treatment groups on the 7th day of ripening. Similarly Bozkurt and Bayram (2006) found that L* values of sucuk samples increased in the early ripening period. No differences were found in a* values of sucuk samples. S30 group had lower b* values (less yellow colour) than other treatment groups. The possible reason for lower b* values in S30 samples might be a partial denaturation of nitrosomyoglobin because of the higher levels of lactic acid in those samples.

Conclusion

Ripening of Turkish dry-fermented sausage (sucuk) at high ripening temperatures resulted in rapid pH decline and rapid drying. Further research should focus on investigating the effects of high ripening temperatures on the sensory properties of sucuk.

References

Bozkurt, H. And Erkmen,O. (2002). Effect of starter cultures and additives on the quality of Turkish style sausage (sucuk) Meat Science, 61, 149-156.

Bozkurt,H. and Erkmen,O. (2004). Effect of nitrite/nitrate on the quality of sausage (sucuk) during ripening and storage. Journal of Food Science and Agriculture, 84, 279-286.

Bozkurt, H. And Bayram, O. (2006). Effects of some commercial additives on the quality of sucuk (Turkish dry-fermented sausage). Meat Science. In Press.

Gökalp, H.Y. (1986). Turkish style fermented sausage (soudjouk) manufactured by adding different starter cultures and using different ripening temperatures. Fleischwirtschaft, 66,573-575.

Horwitz, W. (Ed) (2000). Official Methods Of Analysis Of AOAC International, 17th Edn. AOAC International, Gaithersburg, MD USA.

Kayaardı, S. and Gök, V. (2003). Effect of replacing beef fat with olive oil and quality characteristics of Turkish dry fermented sausages. Meat Science, 66, 249-257.

Landvogt, A. (1991). In: Proceedings of 37th International Congress of Meat Science and Technology, 8, 10, 1159-1162.

CIE (1986). Colorimetry, 2nd Edition. Commission Internationale De l'Eclairage, Vienna.

Soyer, A., Ertaş, S.H. and Üzümcüoğlu,Ü. (2005). Effect of processing conditions on the quality of naturally fermented Turkish sausages (sucuks). Meat Science, 69,135-141.

EFFECTS OF DRIED APRICOT PULP ON FUNCTIONAL PROPERTIES OF BEEF SAUSAGES

Ç. Purma and M Serdaroglu *

Food Engineering Department, Ege University, 35100, Bornova, İzmir, Turkey.
Email: meltem.serdaroglu@ege.edu.tr

Keywords: apricot pulp powder, beef, sausages, dietary fibre, restructured meat

Introduction

In recent years, consumers are aware of the importance of diet for human health. Therefore, demand for healthy foods (functional foods) is increasing day by day. For the production of healthy foods, functional ingredients are used in the food industry. Residues of fruit juice production can be used as a source of functional ingredients. Dried pulp and molasses, fibre pectin, cold-pressed oils, essences, *D*-Limonene, juice pulps and pulp wash, ethanol, seed oil, pectin, ascorbic acid, limonoids and flavonoids are found in residues of citrus juice production (Fernández-López *et al.*, 2004). Fruit juice by-products are commonly used as animal feed or fertiliser in Turkey. Dietary fibres are not only desirable for their nutritional properties but also for their functional and technological properties (Fernández-Ginés *et al.*, 2003). Fibre is suitable for meat products and it has previously been used in cooked meat products to increase cooking yield due to its water and fat binding properties and to improve texture (Fernández-Ginés *et al.*, 2003; Cofrades *et al.*, 2000). The use of peach dietary fibre as a food substitute could be a good alternative in the preparation of both low fat and high dietary fibre food products. Also, peach dietary fibre has a high water binding capacity (Grigelmo-Miguel *et al.*, 1999). Our objective is to evaluate the effects of adding different levels of dried apricot pulp on water holding capacity, cook loss, production yield and gel and fat separation of beef sausages and compare with control samples.

Materials and Methods

Apricot pulp (residual of fruit juice production) was dried to 9% moisture in a drum dryer and milled to a powder. Water absorption index. (Anderson *et al.*, 1969), Water solubility index (Anderson *et al.*,1969), swollen volume (Gould *et al.*, 1989), total and invert sugar analysis (Egan, 1981) and pH analyses of dried apricot pulp were carried out. Lean beef was divided into 4 batches. The first batch was used as the control with no added dried apricot pulp. Dried apricot pulp was added to the other three batches at the levels of 5, 10, and 15%. Four different formulations were obtained with different dried apricot pulp. For sausage preparation, lean beef was ground through a 3mm plate and mixed with curing ingredients, and half of the ice/water and chopped for 1 min in a silent cutter. Than beef fat, seasonings, cornstarch, dried apricot pulp together with the remaining ice/ water were added and the batter was chopped at high speed for 4 min. Each formulation was stuffed into natural sausage casings and sausages were placed in a smoking chamber (drying at 40°C for 1 hr, smoking at 40°C for 30 min) and heat processed in water (85°C) to an internal temperature of 73°C. WHC of sausages were determined using the procedure of Hughes *et al.,* (1997) with a modification. Weight of sausages before and after cooking was recorded and the cooking loss was expressed as a percentage difference between the raw and cooked weights. Before heating, the process weight of sausage batters in casing were recorded and after heating and smoking, sausages were held in a refrigerator at 2°C for 12h. Then the samples were weighed and production yield was expressed as a percentage difference between the first and the second weights. Water and fat seperation analysis was determined using the procedure of Bloukas and Honikel (1992). The trial was performed twice and the data was evaluated by one-way analysis of variance (ANOVA). Significant difference was defined as $p < 0.05$.

Results and Discussion

Functional and chemical properties of dried apricot pulp are shown seen in Table 1.

Table 1: Functional and chemical properties of sausages.

pH	4.91- 4.90
Moisture (%)	9
Water Absorption Index (%)	11.54
Swollen volume	34 ml/100g
Total sugar (%)	17.38
Invert sugar (%)	11.52

WHC, cook loss, production yield and gel and fat separation of sausages were given in Table 2. Addition of dried apricot pulp had significant effects on all investigated parameters. Increasing dried apricot pulp level in sausage formulation resulted in lower WHC. This result is possibly due to the decrease of protein ratio in sausage formulation. Yılmaz (2004), pointed out that a suitable amount of rye bran to be added to low-fat meatballs was no more than 10%.

Table 2: WHC, cook loss, production yield and gel and fat separation values of beef sausages.

Treatment	WHC (%)	Cook Loss (%)	Production Yield (%)	Gel and Fat Separation
Control	$83.11^{d}\pm0.90^{A}$	$10.29^{c}\pm1.04$	$90.78^{a}\pm0.52$	$9.71^{a}\pm0.01$
5%	$81.34^{c}\pm0.63$	$9.98^{bc}\pm0.78$	$90.83^{a}\pm0.50$	$9.75^{a}\pm0.01$
10%	$80.04^{b}\pm0.80$	$9.14^{ab}\pm0.49$	$92.77^{b}\pm0.25$	$9.73^{a}\pm0.03$
15%	$78.21^{a}\pm0.77$	$8.41^{a}\pm0.22$	$93.74^{c}\pm0.50$	$11.46^{b}\pm0.11$
p < 0,05 (significant)	0.000	0.011	0.000	0.000

[A] Standard Deviation. Different letters in the same column indicate significant differences ($p < 0.05$).

The same effect was seen for cooking loss values of sausages. Cooking loss occurs by loss of fat and evaporation of moisture. With higher levels of powder addition, lower moisture content in sausages was obtained and the cook losses were decreased while cooking. Similar results were obtained by other researchers (Mansour and Khalil, 1997, Anderson and Berry, 2001, Yılmaz, 2003), For production yield there was a reverse effect; addition of 10 and 15% dried apricot pulp increased product yield. Addition of dried apricot pulp had no effect on gel and fat separation values except at the 15% level.

Conclusions

Dried apricot pulp can be used in emulsion-type sausages up to 10% without adversely affecting the functional properties. Further research should focus on investigate the effects of dried apricot pulp in low fat high added water sausage formulations.

References

Anderson, E. T., Berry, B. W. (2001). Effects of inner pea fibre on fat retention and cooking yield in high fat ground beef. *Food Research International*, 34: 689-694.

Anderson, R. A., Conway, H. F., Pfeifer, V. F., Griffin Jr., E. L. (1969). Roll and extrusion-cooking of grain sorghum grits. *Cereal Science Today*, 14: 372-375, 381.

Bloukas, J.G. and Honikel, K.O. (1992). The influence of additives on the oxidation of pork backfat and its effect on the water and fat binding in finely comminuted batters. *Meat Science*, 32: 31-43.

Cofrades, S., Guerra, M.A., Carballo, J., Fernández-Martín, F. and Jiménez-Colmenero, F. (2000). Plasma protein and soy fibre content effect on bologna sausage properties influenced by fat level. *Journal of Food Science*. 65: 281-287.

Egan, H., Kirk, R.S. and Sawyer, R. (1981). Pearson's Chemical Analysis of Foods, 8th Edition, Churchill Living Stone, London, 591.

Fernandez-Gines, J.M., Fernandez-Lopez, J., Sayas- Barbera, E., Sendra, E. and Perez-Alvarez, J.A. (2003). Lemon albedo as a new source of dietary fibre : Application to bologna sausages. *Meat Science*, 67: 7-13.

Fernandez-Lopez, J., Fernandez-Gines, J.M., Aleson-Carbonell, L., Sendra,E., Sayas-Barbera, E. and Perez-Alvarez, J.A. (2004). Application of functional citrus by-products to meat products. *Trends in Food Science and Technology*. 15: 176-185.

Gould, M.J., Jasberg, B.K., Dexter, L.B., Hsu, J.T., Lewis, S.M. and Fahey, G.C. (1989). High-fibre, noncaloric flour substitute for baked foods. Properties of alkaline peroxide treated lignocelulose. *Cereal Chemistry*, 66: 201-205.

Grigelma-Miguel, N. and Martin-Belloso, O. (1997). Peach dietary fibre as a food ingredient. IFT Annual Meeting Book of Abstracts. 13C-19: 36.

Grigelmo-Miguel, N., Abadias-Seros M. I. and Martin-Belloso O. (1999). Characterization of low-fat high- dietary fibre frankfurters. *Meat Sci*. 5:, 247-256.

Hughes, E., Cofrades, S. and Troy, D. J. (1997). Effect of fat level, oat fibre and carragenan on frankfurters formulated with 5, 12 and 30% fat. *Meat Science*, 45(3): 273-281.

Mansour, E. H., and Khalil, A. H. (1997). Characteristics of low-fat beefburger as influenced by various types of wheat fibres. *Food Research International*, 30, 199-205.

Yılmaz, I. (2003). Effects of rye bran addition on fatty acid composition and quality of low-fat meat-balls. *Meat Science*, 67: 245-249.

Yılmaz,I (2004). Effects of rye bran addition on fatty acid composition and quality characteristics of low fat meatballs. *Meat Science*, 67:245-249.

ANTIOXIDATIVE EFFECT OF SODIUM ASCORBATE, GARLIC AND ONION IN FRESH PORK BELLY AND LOIN DURING REFRIGERATED STORAGE

S.Y. Park* and K.B. Chin

Department of Animal Science, Chonnam National University, Gwangju, 500-600 Korea.
Email:bishace3114@hanmail.net

Keywords: antioxidation, antimicrobial activity, sodium ascorbate, garlic, onion

Introduction

The palatability of food for consumers is affected by sensorial factors, such as appearance, taste, colour and flavour. In particular, flavour released from meat products has a particular mouthfeel which allowdistinguishes them from other foods. However, oxidative products containing undesirable rancidity are developed due to the oxidation of lipids with long-period storage, and it is one of the reasons for reduced quality and safety of meat and meat products. Thus, natural or synthetic antioxidants for inhibition of oxidative deterioration have been widely studied. The objectives of this study were to evaluate antioxidant and antimicrobial activity of sodium ascorbate (SA), garlic and onion in fresh pork belly and loin, to find volatile compounds to induce rancidity, and to inhibit the production of these compounds.

Materials and Methods

Fresh pork loin and belly slaughtered one day previously were purchased from a wholesale market. Belly and loin cuts were pumped to 110% of their original weight with solution containing 0.5% antioxidants (sodium ascorbate, SA; garlic; onion), 0.5% salt and 0.4% sodium tripolyphosphate for comparison with belly and loin without antioxidants as a control. Thiobarbituric acid reactive substance (TBARS) values were measured by the method of Witte *et al.* (1970). Lipid was extracted by method of Folch *et al.* (1957). Free fatty acid (FFA) and peroxide values (POV) were determined by method of the AOCS (1987) and Lea (1952), respectively. Volatile compounds were extracted by the simultaneous distillation extraction (SDE) method of Heath and Reineccius (1986). Extracted volatile compounds were quantitatively and qualitatively analysed by GC and GC–MS. Total plate count (TPC) agar and violet red bile (VRB) agar were used to determine total bacterial counts and *Enterobacteriaceae*, respectively. The experiment was carried out in triplicate, and data were analysed using two-way analysis of variance in SPSS 12.0 software, with treatment and storage time as factors. Means were separated by the Duncan's multiple range test.

Results and Discussion

Belly containing 0.5% onion had higher TBARS values than control and garlic bellies, and 0.5% garlic belly had lowest TBARS value (Table 1). However, TBARS values were increased with increased storage time in loin, and those of garlic loin were higher than those of the control, SA and onion loins (Table 1). pH values were affected by antioxidants in both belly and loin. These results were due to the different pH of the injection solution (control 7.53, SA 7.39, garlic 7.12, onion 7.02). As shown Table 2, FFA values were increased with increased storage time in both loin and belly, regardless of the addition of antioxidants except for onion loin. However, FFA values of garlic and onion bellies were lower than the control and SA bellies at later storage times, and were decreased by 53 and 31%, respectively, after 28 days of storage, as compared to the control. PO values of control belly and belly containing SA, garlic and onion reached the maximum after 7 and 3, 28 and 28 days of storage, respectively. Therefore, production rates of PO in belly with antioxidants were reduced except for belly containing SA. Approximately 22 volatile compounds were identified from belly and loin, regardless of antioxidants. Among them, C-16 compounds containing hexadecanal were predominant compounds. 2-Pentyl-furan and aldehydes containing 4-pentanal and trans-2-octenal were not detected in SA, onion and garlic loins (Table 3). It took 7 days to reach the total plate counts (TPC) of 10^7 log cfu/g in control and SA bellies, and 7 to 14 days in garlic and onion bellies, respectively. To reach 10^7 log cfu/g of TPC, control, garlic and onion loins took 7-14 days, whereas SA loin took 14 days of storage. Microbial counts of *Enterobacteriaceae* were also similar to those of TPC.

Table 1: Changes of thiobarbituric acid reactive substance (TBARS), pH, Hunter colour values of pork belly and loin as affected by antioxidants during storage at 8°C.

		Treatments				Storage time (day)					
Cut	Parameters	CTL[2]	SA[3]	Garlic	Onion	0	3	7	14	21	28
Belly	TBARS (MDA mg/kg)[1]	0.27^{bc}	0.45^{ab}	0.23^{c}	0.47^{a}	0.27	0.24	0.23	0.36	0.57	0.44
	PH	6.30^{a}	6.32^{a}	6.24^{ab}	6.14^{b}	6.32^{a}	6.24^{ab}	6.34^{a}	6.16^{b}	6.22^{ab}	6.24^{ab}
Loin	TBARS (MDA mg/kg)	0.04^{b}	0.05^{b}	0.10^{a}	0.03^{b}	0.02^{b}	0.02^{b}	0.05^{ab}	0.07^{ab}	0.09^{a}	0.09^{a}
	PH	5.96^{a}	5.80^{b}	5.83^{b}	5.83^{b}	5.79	5.85	5.87	5.83	5.89	5.90

[1]TBARS = thiobarbituric acid reactive substance; [2]CTL = control; [3]SA = sodium ascorbate; [a-c]Means with a same superscript within a row are not significantly different (P>0.05).

Table 2: Changes in free fatty acid (FFA) and peroxide value (POV) of pork belly and loin as affected by various antioxidants during storage at 8°C.

Cuts	Parameters	Treatment	Storage time (day) 0	3	7	14	21	28
Belly	FFA[1] (%)	CTL[3]	1.03^{bA}	1.18^{b}	1.29^{abA}	1.50^{abA}	1.90^{abA}	2.22^{aA}
		SA[4]	0.95^{bA}	0.83^{b}	0.98^{bAB}	1.39^{abA}	1.90^{aA}	1.92^{aAB}
		Garlic	0.58^{cB}	0.66^{bc}	0.69^{bcB}	0.87^{abcB}	0.92^{abB}	1.05^{aB}
		Onion	0.61^{cB}	0.76^{bc}	0.90^{bcAB}	0.93^{bcB}	1.18^{abB}	1.54^{aAB}
	POV[2] (meq/kg)	CTL	1.00^{bA}	0.67^{bB}	2.90^{aA}	1.73^{ab}	0.56^{bB}	1.00^{bB}
		SA	0.70^{bAB}	2.41^{aA}	1.10^{bB}	1.33^{b}	0.79^{bB}	1.43^{bB}
		Garlic	0.45^{bB}	0.71^{bB}	1.19^{abB}	1.58^{a}	0.68^{bB}	1.79^{aB}
		Onion	0.86^{cA}	1.81^{bcA}	0.89^{cB}	1.27^{c}	3.00^{bA}	5.88^{aA}
Loin	FFA (%)	CTL	3.49^{cAB}	3.76^{c}	4.82^{bc}	5.87^{ab}	6.57^{a}	5.81^{ab}
		SA	2.71^{bB}	2.67^{b}	3.42^{b}	3.98^{b}	4.53^{b}	6.44^{a}
		Garlic	3.42^{bAB}	3.66^{b}	3.70^{b}	5.06^{a}	5.19^{a}	5.33^{a}
		Onion	4.54^{A}	4.48	4.20	5.45	4.51	6.36
	POV (meq/kg)	CTL	0.80^{d}	3.18^{cdB}	7.41^{abA}	8.17^{aAB}	4.27^{bcdB}	4.67^{bcB}
		SA	1.18^{c}	2.20^{bcB}	2.59^{bcC}	8.95^{aAB}	4.19^{bB}	4.32^{bB}
		Grlic	0.46^{d}	1.55^{dB}	4.90^{cdB}	12.7^{abA}	8.76^{bcA}	15.5^{aA}
		Onion	0.45^{d}	6.01^{abA}	8.60^{aA}	5.74^{abB}	2.12^{cdB}	4.25^{bcB}

[1]FFA = free fatty acid; [2]POV = peroxide value; [3]CTL = control; [4]SA = sodium ascorbate; [a-c]Means with a same superscript within a row are not significantly different (P>0.05). [A-C]Means with a same superscript within a column are not significantly different (P>0.05).

Table 3: Changes of volatile compounds of loin as affected by various antioxidants during storage at 8°C

Compounds	Treatment	Storage time (days) 0	3	7	14	21	28
2-pentyl-furan	CTL[1]	5.42^{A}	1.59^{A}	2.11^{A}	2.32^{A}	1.69^{A}	1.57^{A}
	SA[2], Garlic, Onion	t^{B}	t^{B}	t^{B}	t^{B}	t^{B}	t^{B}
4-pentenal	CTL	16.9^{A}	1.66^{A}	5.63^{A}	6.52^{A}	1.73^{A}	4.55^{A}
	SA, Garlic, Onion	t^{B}	t^{B}	t^{B}	t^{B}	t^{B}	t^{B}
trans-2-octenal	CTL	18.1^{A}	1.57^{A}	3.18^{A}	4.79^{A}	4.26^{A}	1.98
	SA, Garlic, Onion	t^{B}	t^{B}	t^{B}	t^{B}	t^{B}	t

[1]CTL = control; [2]SA = sodium ascorbate; [a-c]Means with a same superscript within a row are not significantly different (P>0.05). [A-C]Means with a same superscript within a column are not significantly different (P>0.05). t : trace amount.

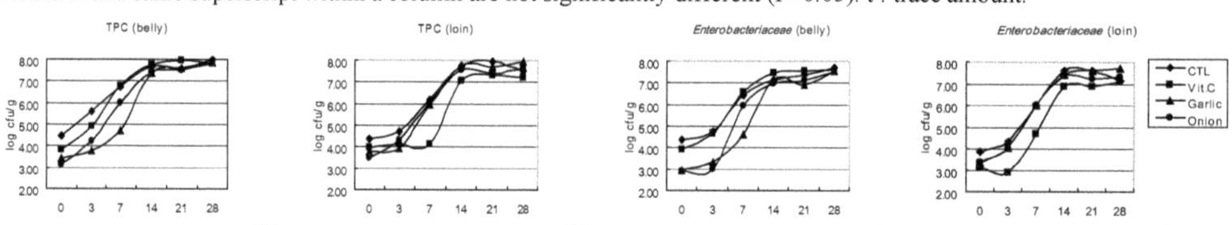

Figure 1: Change of microbiological of pork belly and loin as affected by antioxidants and storage time at 8°C (TPC = total plate count; CTL = control) .

Conclusions

Belly was spoiled faster than loin due to the high amounts of fat. The antimicrobial effect of natural ingredients was affected by the muscle cut. Garlic and onion bellies showed antimicrobial effect, resulting in increases in shelf-life. Natural ingredients used in this study showed antioxidant and antimicrobial effect, garlic, in particular, of the natural ingredients may be considered as the best natural antioxidant to control lipid oxidation of food system.

References

AOCS. (1987). Official and Tentative Methods of the American Oil Chemists' Society, BS 684, 1.5.

Folch, J., Lees, M., Stanley and G.H.S. (1957). A simple method for the isolation and purification of total lipids from animal tissues. Journal Biology Chemistry, 226: 497-509.

Heath, H.B. and Reineccius, G. (1986). Isolation of food flavours. In flavour chemistry and technology, eds. H. B. Heath and G. Reineccius. AVI, New York. pp. 6.

Lea, C.H. (1952). Methods for determination peroxide in lipid. Journal of the Science of Food and Agriculture, 3, 586.

Witte, V.C., Krause, G.F. and Bailey, M.E. (1970). A new extraction method for determining 2-thiobarbituric acid values of pork and beef during storage. Journal of Food Science, 35, 582-58.

EVALUATION OF THE EFFECT OF DIFFERENT COOKING TREATMENTS ON THE TENDERNESS AND COOK LOSS OF INJECTED BOVINE FOREQUARTER MUSCLES

H. Walsh [1] *, S. Martins[1], E.E. O'Neill[1], J.P. Kerry[1], T. Kenny[2] and P. Ward[2]

[1]*Department of Food and Nutritional Sciences, University College, Cork, Ireland.*
[2]*Ashtown Food Research Centre, Teagasc, Ashtown, Dublin 15, Ireland.* **Email: Helena.Walsh@ucc.ie*

Keywords: beef, tenderness, cooking, forequarter, muscle

Introduction

Recently, numerous studies have shown that injecting bovine muscles with brine enhances tenderness (Molina *et al.*, 2005; Lennon *et al.*, 2006). Cooking has a major influence on the tenderness of meat. However, there is little information available on the effect of different cooking regimes on the eating quality of injected bovine muscles. Collagen shrinks during cooking. However, with extended heating in the presence of moisture, collagen is solubilised and therefore meat becomes more tender (Sims and Bailey, 1992). The aim of this study is to investigate the effects of extended cooking regimes on the tenderness and cooking losses of brine injected forequarter bovine muscles.

Materials and Methods

Forequarter muscles were excised from O grade steer carcasses at 48 hours (h) *post-mortem*, namely, *pectoralis profundus* (PP), *infraspinatus* (IF), *supraspinatus* (SP) and *triceps brachi caput longum* (TB). PP, TB and IF muscles were cut into 400g pieces. SP was not sectioned. Individual muscles/muscle sections were selected randomly from each muscle group and assigned to one of the treatments below, such that a total of twelve representatives of each muscle were used for each treatment. Each muscle/muscle section was injected with brine at an injection rate of 10% to give a residual 0.5% NaCl content. The muscles or sections of muscles were equilibrated at 1°C overnight and then vacuum packed in cook-in shrink bags before cooking. Five cooking regimes were evaluated for IF, SP and TB as follows: (a) Cooking to an internal core temperature (ICT) of 72°C using an oven temperature of 80°C (72°C); (b) Cooking to an ICT of 72°C using a delta (Δ) 10 cook procedure such that a constant differential of 10°C is maintained between the product core temperature and oven temperature (Δ72C°); (c) Cooking to an ICT of 69°C and holding for 4h (69°C+4h); (d) Cooking to an ICT of 69°C using a delta 10 cooking procedure and holding for 4h (Δ69°C+4h); (e) Cooking to an ICT of 69°C and holding for 12h (69°C+12h)

Sections of PP were (a) Cooked to an internal core temperature of 69°C using a delta 10 cooking procedure and held for 4, 6, 8, 10 and 12h (Δ69°C+4, 6, 8, 10, 12h) or (b) Cooked to an ICT of 69°C and held for 12h (69°C+12h). Warner Bratzler shear force (WBSF) values were determined on 1.27 cm diameter cores (n=6), cut parallel to muscle fibre orientation, on a texture analyser (Stable Micro Systems, UK). Cook loss was determined by expressing the weight loss during cooking as a percentage of the before cooking weight. Sensory analysis was performed using an 8-member in-house trained panel.

Results and Discussion

PP was the toughest of the four muscles investigated as indicated by WBSF values (Tables 1 and 2), which may be attributed to its high collagen content (Torrescano *et al.*, 2003). Slow Δ 10 cooking did not significantly enhance the tenderness of brine injected SP compared with conventional cooking when meat was cooked to an end point cooking temperature (EPCT) of 72°C or to an ICT of 69°C and held at 69°C for 4h as indicated by WBSF values and sensory scores. In contrast, Δ 10 cooking significantly reduced the WBSF values of brine injected IF and TB heated to 69°C or 72°C compared with conventional cooking. While the sensory tenderness ratings followed the same trend, the differences were not significant. For brine injected IF, TB and SP, Δ 10 cooking did not significantly affect % cook loss and overall there was no significant difference in the % cook losses when meat was cooked to an EPCT of 72°C or to an ICT of 69°C and held at 69°C for 4h. For brine-injected IF and TB, increasing the holding time at 69°C from 4 to 12h significantly increased % cook loss, while no significant effect was observed for brine injected SP. Increasing the cooking time at 69°C from 4 to 12h significantly increased the tenderness scores and decreased the WBSF values of brine injected IF, TB and SP.

Table 1: Mean ± SD Warner Bratzler shear force values (WBSF) (N), sensory analysis tenderness scores (Tend) and cook loss (%CL) for IF, SP and TB muscles.

	M	Δ69°C+4h	69°C +4h	Δ72°C	72°C	69°C +12h
	IF					
WBSF		26.0 ± 3.0[a]	31.2 ± 3.0[b]	25.6 ± 2.4[a]	32.6 ± 3.6[b]	16.1 ± 2.0[c]
Tend		6.5 ± 0.7[a]	5.8 ± 0.9[a]	6.4 ± 1.1[a]	5.3 ± 0.8[a]	7.4 ± 1.0[b]
% CL		36.7 ± 1.6[ac]	34.2 ± 1.6[c]	37.5 ± 1.8[a]	36.1.9[ac]	41.7 ± 1.7[b]
	TB					
WB		26.2 ± 3.0[a]	34.3 ± 3.8[b]	27.2 ± 3.2[a]	33.8 ± 3.9[b]	17.8 ± 2.1[c]
Tend		5.8 ± 0.7[a]	5.1 ± 0.7[a]	5.6 ± 0.8[a]	5.3 ± 1.1[a]	7.3 ± 0.9[b]
% CL		33.9 ± 1.8[a]	33.3 ± 1.9[a]	32.5 ± 1.2[a]	34.8 ± 1.7[a]	41.7 ± 1.6[b]
	SP					
WB		44.2 ± 3.6[a]	44.3 ± 6.5[a]	41.7 ± 6.1[a]	45.7 ± 6.3[a]	31.4 ± 5.9[b]
Tend		3.6 ± 0.8[a]	3.5 ± 0.7[a]	3.0 ± 1.1[a]	3.6 ± 0.9[a]	6.0 ± 0.9[b]
% CL		37.0 ± 0.9[a]	39.6 ± 0.7[b]	39.5 ± 1.3[b]	38.9 ± 1.6[b]	40.2 ± 1.8[b]

[abc] Values with a different superscript across the line denotes significance ($p<0.05$).

When brine injected PP was cooked to 69°C using a Δ 10 heating programme and held at 69°C for 4h, the meat was rated as being unacceptably tough by the sensory panel. Increasing the cooking time to 8h significantly decreased the WBSF values. There was no further significant decrease in WBSF values when the holding time was extended to 10 and 12h. Brine injected PP held at 69°C for 12h was significantly more tender than that held at 69°C for 4h as indicated by the sensory tenderness scores and WBSF values. Increasing the holding time from 4 to 8h did not significantly increase % cook loss of brine injected PP. However the % cook loss at 10 and 12h was greater than that at 4h. When a holding time of 12h was used, there was no significant difference observed in the tenderness or % cook loss of conventionally cooked PP and those cooked using a Δ 10 programme.

Table 2: Mean ± SD Warner Bratzler shear force values (WBSF) (N), sensory analysis tenderness scores (Tend) and cook loss (%CL) values for PP muscles Δ cooked to 69°C and held for 4, 6, 8, 10 and 12h and conventionally cooked to 69°C and held for 12h.

	Δ 69°C+4h	Δ 69°C +6h	Δ 69°C +8h	Δ 69°C +10h	Δ 69°C +12h	69°C +12h
WBSF	50.8 ± 3.3[a]	47.8 ± 5.1[ac]	43.2 ± 4.2[bc]	43.0 ± 3.9[bc]	37.2 ± 1.9[b]	35.9 ± 3.1[b]
Tend	2.8 ± 0.9[a]	n/d	n/d	n/d	4.9 ± 1.4[b]	4.5 ± 1.3[b]
% CL	34.6 ± 2.9[a]	37.0 ± 2.3[ab]	35.5 ± 2.0[ab]	37.6 ± 1.4[b]	37.2 ± 1.5[b]	37.6 ± 1.7[b]

[abc] Values with different superscript across the line denotes significance ($p<0.05$); n/d not determined

Conclusions

Extended cooking times increased tenderness of brine injected IF, TB, PP and SP. However, there was also a significant increase in % cook loss for 3 of the 4 muscles examined. Slow delta cooking programmes decreased the WBSF values of IF and TB but had no significant effect on sensory tenderness scores.

References

Lennon, A.M., Moon, S.S., Ward, P., O'Neill, E.E. and Kenny, T. (2006). Effects of enhancement procedures on whole and reformed beef forequarter muscles. Meat Science, 72, 513-517.

Molina, M.E., Johnson, D.D., West, R.L and Gwartney, B.L. (2005). Enhancing palatability traits in beef chuck muscles. Meat Science, 71, 52-61.

Sims, T.J. and Bailey, A.J. (1992). Structural aspects of cooked meat. In: The chemistry of muscle based foods, Ledward, D.A., Johnston, D.E and Knight, M.K. (Eds.). Royal Society of Chemistry, Thomas Graham House, Science Park, Cambridge, pp 106-127.

Torrescano, G., Sanchez-Escalante, A., Gimenez, B., Roncales, P and Beltran, J.A. (2003). Shear values of raw samples of 14 bovine muscles and their relation to muscle collagen characteristics. Meat Science, 64, 85-91.

Acknowledgements:

This project was funded by the Irish Government under the National Development Plan 2000-2006.

CHARACTERISATION OF ONION *ALLIUM CEPA* L. VAR. HORCAL VOLATILE COMPOUNDS IN "MORCILLA DE BURGOS"

L. González-Arnáiz, A.M. Diez, I. Jaime and J. Rovira

Department of Biotechnology and Food Science. University of Burgos. Pza Misael Bañuelos s/n 09001. Burgos. Spain. Email: luciagarnaiz@yahho.es.

Keywords: Morcilla de Burgos, volatile, GC-MS, Solid Phase Dynamic Extraction

Introduction

"Morcilla de Burgos" is a popular cooked blood sausage produced in Burgos, in the north of Spain, consisting of a mixture of chopped onion, rice, animal fat, blood, salt and different spices such as black pepper, paprika and oregano. According to the method, kind and amount of spices, and the proportion of ingredients, it is possible to classify Morcilla de Burgos in three different types.

One of the ingredients more characteristic and important in the elaboration of "Morcilla de Burgos" is the kind and percentage of onion used. The onion should be from a regional variety known as Horcal. This onion is white, with a seasonal crop and a short period of storage, but it gives "Morcilla de Burgos" a better sensory quality, also the horcal onion has a flat and elliptical shape and bigger size compared with other types of onions.

Depending on the amount of horcal onion it is possible to classify the different types of "Morcilla de Burgos" in three categories according with the first draft of the regulation to obtain the EU quality label "Protected Geographical Indication", (PGI).The first type is a morcilla with at least 35% of horcal onion. The second one has to have at least 45% of horcal onion and the third one has 50% of horcal onion.

The purpose of this study is the determination of the influence of different horcal onion percentages in the elaboration of "Morcilla de Burgos" in the total volatile profile.

Materials and Methods

Sample preparation: Three batches of 60 morcillas were made with the same formulation with the only difference being the percent of onion added according to the first draft of the regulation for the "Protected Geographical Indication". One batch was made with 35% of horcal onion, because this is the minimum amount for the morcilla to be called "Morcilla the Burgos". The second batch was made with 20% of horcal onion, an amount that is not included in the regulation for the PGI, and the third batch was made with 47% of horcal onion, because it is intermediate between the other two categories.

After peeling four morcillas randomly chosen, they were homogenised in a lab blender, and 2g of this homogenate were introduced into a 20 ml headspace vial.

HS-SPDE sampling: All Solid Phase Dynamic Extraction (SPDE) (Chromtech, Idstein, Germany) sampling steps were automatically controlled by the CTC-Combi PAL software. The SPDE needle coated inside is a PDMS-AC. The vials were all equilibrated at 70°C for 1 min, 50 Strokes, the plunger speed for extraction: 40µl/s; helium volume for desorption: 0.5ml; plunger speed for desorption: 15µl/s; pre-desorption time in the GC injection port: 30 s; desorption temperature: 250°C. Each analysis was performed three times for each sampling.

GC-MS Conditions: Gas chromatographic analyses were performed with an Agilent Technologist 6890N Series GC System (Agilent Technologist, Palo Alto. CA. USA) coupled to a Agilent Technologist 5973i mass spectrometer (Agilent Technologist, Palo Alto. CA. USA). The SPDE syringe was injected automatically into the GC injection port, and thermally desorbed at 250°C. Compounds were separated on a HP5 capillary column (50m length x 0.32mm I.D fused silica capillary column coated with 1.05µm film thickness (Quadrex Corporation. New Haven. USA)). The temperature of the column was programmed starting at 40°C, after injection for 5 min, afterwards temperature was increased at a rate of 3°C/min from 40 to 240°C. Helium gas with a flow of 1ml/min was used as the GC carrier gas. The temperature of the transfer line was 280 °C. During desorption of the SPDE syringe the injector split valve was closed. The effluent from the capillary column went directly into the mass spectrometer, operated in the electron impact (EI) mode with an ionisation voltage of 70eV.

Compounds were identified by comparing their mass spectra with a NIST and Wiley spectrum libraries. These compounds were related to the different ingredients used in the elaboration of "Morcilla de Burgos", by comparing them with previous published spectra obtained from different *Allium cepa* L. varieties, and spices, and also from samples analysed, in the same conditions mentioned above.

Results and Discussion

The results obtained show 117 different volatile compounds found in "Morcilla de Burgos", taking into account all compounds that appear in all the samples analysed. Eighty-three of these compounds have been tentatively identified using the NIST and WILEY libraries. Three of this 83 compounds were acids, 7 were alcohols, 17 were aldehydes, 9 were sulphur compounds, 42 cyclic compounds, 4 hydrocarbons and 1 ketone.

As it was expected, all three samples of morcilla presented the same volatile profile except for the compounds related to the presence of onion. These compounds were the 9 sulphur compounds and the aldehyde 2-Methyl 2-Pentenal.
The number and intensity of these chromatographic peaks decreased with the percent of onion in the formulation of the "Morcilla de Burgos". In this way, "Morcilla de Burgos" showed 10, 9 and 6 chromatographic peaks for morcilla with 47, 35, and 20% of onion in its formulation.
Table 1 shows seven out of ten volatile compounds related with horcal onion and three volatile compounds related with the more important species for the samples analysed.

Table 1: Volatile compounds extracted in the headspace with PoliDiMethylSiloxane-Active Charcoal-SPDE (expressed en AAU x 10 $^{-7}$) in each one of the samples analysed.

% Onion		47	35	20	
Peak Number	Retention Time	Area	Area	Area	Compound
10	18.78	63.41^{c}	44.59^{b}	34.12^{a}	2-Methyl 2-pentenal
11	21.60	1.17^{c}	0.56^{b}	nd	2,4-Dimethyl- thiophene
13	23.19	7.39^{c}	3.85^{b}	2.09^{a}	3,4-Dimetil- thiophene
15	24.74	16.49^{c}	11.73^{b}	4.83^{a}	Methyl propyl disulphide
37	34.64	14.35^{c}	10.52^{b}	6.18^{a}	Dipropyl disulphide
38	35.07	3.25^{c}	1.98^{b}	0.76^{a}	Trans-Propenyl propyl disulphide
43	45.98	0.53^{b}	nd	nd	Dipropyl trisulphide
21	27.61	43.78^{a}	39.32^{a}	44.41^{a}	β-Pinene
25	29.36	155.53^{a}	136.94^{a}	158.73^{a}	Δ 3-Carene
29	30.33	118.66^{a}	107.12^{a}	129.77^{a}	dl-Limonene

Means in the same row with different letters (a-c) are significantly different ($P< 0.05$).
nd: not detected.
AAU: Arbitrary Area Units.

Conclusions

According to the results obtained in this study, it can be conclude that the differences in the volatile profile of the samples analysed are due to the amount of Horcal onion used in the elaboration of the morcilla. These differences are shown in the number of peaks related with the onion percent in the composition of morcilla and also in the different intensity of the chromatographic peaks, in terms of the percent of onion.

Acknowledgements

We want to express our thanks to Excma. Diputación Provincial de Burgos for supporting this project with a grant. We also want to express our gratitude to the Junta de Castilla y León project BU 28/04 and to Mercadona S.A for supporting this research project.

References

Bicchi, C., Cordero, C., Liberto, E., Rubiolo, P. and Sgorbini, B. (2004). Automated headspace solid phase dynamic extraction to analyse the volatile fraction of food matrices. Journal of Chromatography A, 1024, 217-226.
Borrador del reglamento de la Indicación Geográfica Protegida "Morcilla de Burgos"; 2004.
González, L., Diez, A. M., Ena, J. M., Jaime, I. and Rovira, J. (2004). Volatiles characterization of the blood sausage "morcilla" de Burgos. In: Proceedings of the 50th International Congress of Meat Science and Technology, Helsinki, Finland.
Järvenpää, E., Zhang, Z., Huopalahti, R. and King, J. (1998). Determination of fresh onion (*Allium cepa L.*) volatiles by solid phase microextraction combined with gas chromatography-mass spectrometry. Z Lebensm Unter Forsch A, 207:39-43.

EFFECT OF TRANSGLUTAMINASE ON THE TEXTURAL PROPERTIES OF MEAT EMULSIONS

M.J. Beriain*, M.J. Arrizubieta, D. Pascual, M.V. Sarries, P.M. Diéguez[1], G. Indurain, M. San Roman, A. Igea and K. Insausti

E.T.S. Ingenieros Agrónomos, [1]E.T.S. Ingenieros Industriales y de Telecomunicación Universidad Pública de Navarra, Campus Arrosadia s/n 31006 Pamplona, Navarra, Spain. Email: mjberiannavarra.es

Keywords: transglutaminase, meat, enzyme

Introduction

Transglutaminases (TGase) are enzymes of great interest to food technologists due to their ability to catalyse the formation (and or strengthening) of gel-type structures through protein cross-linking (Motoki and Seguro, 1998, Kuraishi *et al.*, 2001). The major utility of TGase in meat products is its hability to change the functional properties of the meat products, such as gelling capacity, viscosity, or the water holding capacity and also they contribute to stabilise the network of the proteins (Motoki and Seguro, 1998). The commercial TGase used in this study comes from the microorganism *Streptomyces mobaraense*. There are many works which have studied the effect of this enzyme in the properties of the gel and the emulsions in models of procesing foods (Sakamoto *et al.*, 1994; Dickinson, 1997, Kuraishi *et al.*, 2001) but not at different cooking temperatures and times of fabrication. In addition to the gelification effect this enzyme can stabilize emulsions where there is protein in the interfase as it happens in meat emulsions (Dickinson, 1997).

In this study the time course of the effect of the microbial TGase on the textural properties of a meat emulsion was analysed from the moment of fabrication when the enzyme is added to the moment of cooking at different temperatures.

Materials and Methods

Two types of meat emulsions were fabricated with 44% turkey, 15% olive oil, 36% water, 1% fibre, 1.5% soya and 2% salt. One of the emulsions was added 60ppm of TGase Active (Ajinomoto). The emulsions were refrigerated at 5°C during 30, 60, 120, 180, 240 minutes and overnight. In order to determine the TGase activity, the viscosity was determined at these time points. A VistoTester 7R (Brookifield Method) (HAAKE) was used to measure the viscosity at 30 rpm. Then, each emulsion was cooked at 45, 60 and 72°C and then they were cut in cubes of 10x10x10mm (Klettner, 1989) to be analyzed with a texture analyzer (Stable Micro System Model TA-XT2i) using the TPA compression test. Data were analysed statistically with a SPSS 12 program.

Results and Discussion

Figure 1 shows the apparent viscosity values obtained for both types of meat emulsions at different time points after production. In general, a higher apparent viscosity is measured for the sample with added TGase than for the control, and the effect of the enzyme on viscosity is time dependent, increasing with longer incubation periods. Figure 2 compares the apparent viscosity values obtained with the viscosimeter working at 12 rpm. The values obtained for the control emulsion during the first 24 hours oscilate around an average value of about 150 Pa. In contrast, the apparent viscosity of the emulsion with added TGase increases progressively with incubation time. At two hours the apparent viscosity of the sample with added enzyme is significantly higher than the values obtained for the control emulsion ($p<0.001$). By four hours the apparent viscosity of the sample doubles that of the control, and this proportion still increases during further incubation. It has to be pointed out that the apparent viscosity of the sample could not be measured at 24 hours of incubation because the value was above the range detectable under this experimental conditions. Thus, the cross-linking action of the enzyme starts having a detectable effect on the viscosity of the emulsion two hours after production. However, the action of the enzyme is not complete after four hours of incubation, since a longer incubation time leads to higher apparent viscosity values.

There was different texturising activity as a result of the cooking treatment. In this regard, the compression test of the emulsion with TGase at 45°C and 60°C showed higher hardness and higher chewiness than the emulsion without TGase ($p\leq0.05$). However, the most important significant differences were shown when the emulsions were cooked at 72°C. At this temperature the emulsion with TGase showed lower elasticity ($p\leq0.05$) and higher hardness ($p\leq0.01$), cohesiviness and chewiness ($p\leq0.001$) (Table 1).

Conclusion

These results show that at 5°C and at this enzyme concentration, the meat emulsion requires a long incubation period (longer than four hours) before the full effect of the enzyme is reached (in agreement with the Ajinomoto). In addition, the effect of the enzyme on the textural properties of the meat emulsion is apparent even after cooking, and the sample with added enzyme presents higher breaking strength than the control.

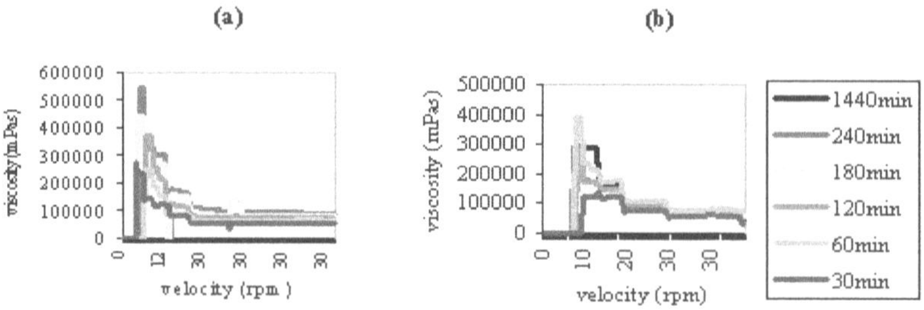

Figure 1: Viscosity evolution of the emulsion with TGase (a) and without TGase (b), over time.

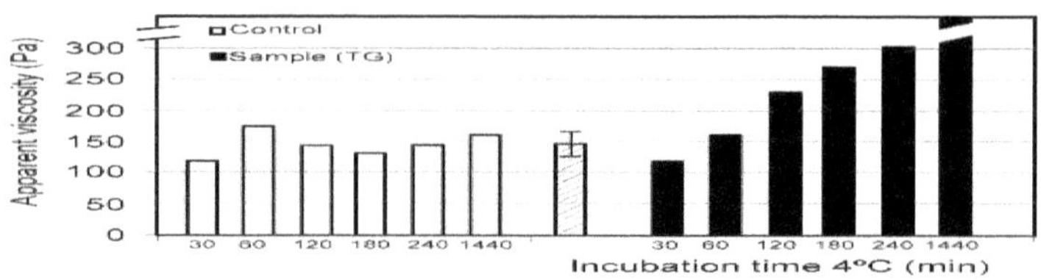

Figure 2: Time course of the apparent viscosity values obtained for the control (without added TGase) and sample (with added TGase) meat batters during the first 24 hours after fabrication.
* Hatched bar corresponds to the average of the apparent viscosity values (standard deviation represented) obtained for the control at the various time points.

Table 1: Means, standard error and significant differences between the control emulsion and the emulsion with Tgase cooked at 45°C, 60°C and at 72°C measured with the TPA compression test.

	Hardness			Chewiness			Elasticity	Cohesiviness
	45°C	60°C	72°C	45°C	60°C	72°C	72°C	72°C
Control	0.36(0.05)	0.48(0.06)	0.36(0.06)	1.51(0.22)	1.49(0.28)	0.88(0.21)	6.19(0.18)	0.46(0.05)
TGase	0.49(0.05)	0.63(0.06)	0.74(0.07)	2.20(0.22)	2.27(0.28)	2.52(0.22)	5.44(0.19)	0.61(0.06)
Significance	*	*	**	*	*	***	*	*

*$p \leq 0.05$; **$p \leq 0.01$; ***$p \leq 0.001$.

References

Dickinson, E. (1997). Enzymic crosslinking as a tool for food colloid rheology control and interfacial stabilization. Trends in *Food Science and Technology,* 8, 334-339.

Klettner, P.G. (1989) "Comminution techniques for frankfurters-type sausage". *Fleischwirthsachaft International,* 1: 32-37.

Kuraishi, C., Yamazaki, K. and Susa, Y. (2001). Transglutaminase: its utilization in the food industry. *Food Reviews International,* 17: 221-246.

Motoki, M. and Seguro, K. (1998). Trasnsglutaminase and its use for food processing. *Trends in Food Science and Technology,* 9: 204-210.

Sakamoto, H., Kumazawa Y., Motoki, M. (1994). Strength of protein gels prepared with microbial transglutaminase as related to reaction conditions. *Journal of Food Science,* 59: 866-71.

EVALUATION OF THE POSSIBILITY OF APPLICATION OF PORK HAMS WITH PSE SYNDROME TO PRODUCTION OF DRY CURED HAMS

M. Olkiewicz*, P. Moch and H. Makała

Meat and Fat Research Institute, 4, Jubilerska St., 04-190 Warszawa, Poland, E-mail: mol@ipmt.waw.pl

Keywords: dry cured ham, PSE syndrome, ham quality

Introduction

It is the common view that the production of dry cured hams, which mature (ripen) in the raw state, needs specific raw material, without PSE syndrome, with a high level of marbling (fat content > 3.5% and fat acid components with high resistance to oxidation, available mainly in the Mediterranean region from traditional, extensive breeding). In Poland, conditions exist to produce such raw material using traditional breeds like puławska, wielka biała, złotnicka and their cross-breeding with the duroc race, but the most widely available is the raw material from no race (race-less) breeding, often with PSE syndrome, which results in its lower processing possibilities, not to mention hams, which mature (ripen) in the raw state. In our institute an attempt has been made to evaluate if such raw material is suitable for the production of hams, which mature (ripen) in the raw state.

Materials and Methods

In the trial, the model production of hams, which mature (ripen) in the raw state, made of meat from fatteners (baconers) of three groups has been carried out: 1 – with deep PSE syndrome (pH_{45} – av. 5.6) code PSEM, 2 – with mild PSE syndrome (pH_{45} – av. 5.9) code PSES, 3 – of good quality (pH_{45} – av. 6.5) code NORM. The number of hams in each group was 10. The production of smoked hams, which mature (ripen) in the raw state, has been carried out by the traditional "dry" method consisting of rubbing raw hams with a mix of salt with nitrite and nitrate curing mix, with the additions of starter cultures and a mix of sugar and spices. The production process lasted about 52 days and the finished product was obtained in about 75% yield. In the finished product the following elements were tested: weight after the production process (yield), pH, colour parameters (L, a*, b*), chemical composition (water content, proteins, fat, NaCl, $NaNO_2$, sum of $NaNO_2$ and $NaNO_3$), rheological characteristics using the CASRA method (plasticity – P, elasticity – E, fluidity – F, minimum deformation – D_{min} and plastic deformation - D_{plast}) as well as the sensorial profile analysis were carried out and the internal concentration of brine and the Feder number have been calculated. The obtained results were subjected to statistical tests, including analysis of variance.

Results and Discussion

It has been stated that the samples of the PSEM hams had substantially higher finished product weights, a substantially lower pH measured in the semimembranous muscle and a substantially higher clearness parameter "L", compared to the PSES and NORM hams (Table 1). The majority of chemical parameters did not show substantial differentiation (variety) for the tested trial variants ($P \leq 0.05$).

Table 1: Physical and colour parameters of dry cured hams.

Variant	Mass [kg]	pH 1	pH 2	pH 3	L	a*	b*
PSEM	3.22 ^{b}	5.47 ^{a}	5.57	5.50 ^{a}	44.28 ^{b}	7.84	8.74
PSES	2.90 ^{a}	5.59 ^{b}	5.66	5.56 ^{ab}	41.76 ^{a}	7.66	8.06
NORM	2.77 ^{a}	5.62 ^{b}	5.66	5.64 ^{b}	42.64 ^{ab}	7.03	8.75
LSD	*0.29*	*0.10*	*0.15*	*0.12*	*2.38*	*1.39*	*0.70*

P≤0.05, PSEM – hams with deep PSE syndrome (mean pH_{45} -5.6), PSES – hams with mild PSE syndrome (mean pH_{45} - 5.9), NORM – hams of good quality (mean pH_{45} - 6.5), pH 1 – *m.semimembranosus,* pH 2 – *m.semitendinosus*, pH 3 – *m.biceps femoris*

Only the concentration of NaCl and – consequently –brine concentration were substantially lower for the samples of the PSEM hams with very high PSE (Table 2). The rheological parameters of the tested samples due to the type of raw material were on the same level as well, except for the substantially lower D_{plast} of the PSES variant. However, rheological parameters of the tested samples differed according to the muscle type used. The semitendinous muscle (*m. semitendinosus*), in all the tested variants of hams, was characterised by substantially lower plasticity and substantially higher elasticity, liquidity, minimum deformation D_{min} and plastic deformation D_{plast} compared to the semimembranous muscle (*m. semimembranosus*) and biceps (*m. biceps femoris*) (Table 3). All the above testifies to the fact that the semitendinous muscle is more delicate which, most probably, is connected with the higher degree of marble-like structure evaluated visually (higher fat and collagen contents).

Table 2: Chemical characteristics of tested dry cured hams.

Variant	Water content [%]	Protein content [%]	Fat content [%]	Feder's number	NaCl content [%]	Brine Conc. [%]	$NaNO_2$ [g/kg]	$NaNO_2$+ $NaNO_3$ [g/kg]
PSEM	61.48	28.84	5.54	2.13	4.07 ^{a}	6.23 ^{a}	27.88	108.66
PSES	60.08	27.90	6.62	2.16	4.76 ^{b}	7.35 ^{b}	35.74	105.36
NORM	59.78	28.56	6.18	2.10	4.45 ^{ab}	6.93 ^{ab}	12.16	118.30
LSD	*1.84*	*2.01*	*3.47*	*0.13*	*0.64*	*1.05*	*47.70*	*24.26*

$P \leq 0.05$, PSEM – hams with deep PSE syndrome (mean pH_{45} -5.6), PSES – hams with mild PSE syndrome (mean pH_{45} - 5.9), NORM – hams of good quality (mean pH_{45} - 6.5).

Table 3: Rheological characteristics of dry cured hams.

Factor	Variant	Plasticity [x10^5 Nm^{-2}]	Elasticity [x10^{-6} m^2N^{-1}]	Fluidity [x10^{-8} $m^2N^{-1}s^{-1}$]	D_{min} [%]	D_{plast} [%]
Type of hams	PSEM	17.10	1.11	3.16	11.36	85.90 ^{b}
	PSES	17.73	1.16	3.08	10.83	83.37 ^{a}
	NORM	16.07	1.21	3.38	11.15	87.43 ^{b}
	LSD	*1.45*	*0.13*	*0.28*	*1.56*	*2.45*
Type of musculus	*m.semimembranosus*	17.01 ^{b}	1.13 ^{b}	2.97 ^{b}	9.44 ^{a}	83.41 ^{a}
	m. semitendinosus	13.00 ^{a}	1.44 ^{c}	4.26 ^{c}	13.72 ^{b}	90.61 ^{b}
	m. biceps femoris	20.89 ^{c}	0.91 ^{a}	2.40 ^{a}	10.17 ^{a}	82.68 ^{a}
	LSD	*1.45*	*0.13*	*0.28*	*1.56*	*2.45*

$P \leq 0.05$, PSEM – hams with deep PSE syndrome (mean pH_{45} -5.6), PSES – hams with mild PSE syndrome (mean pH_{45} - 5.9), NORM – hams of good quality (mean pH_{45} - 6.5)

The results need to be confirmed in future research work. The sensorial evaluation has shown that the majority of the evaluated features for all the variants were on the same level. The only substantial differences have been stated in colour intensity and in the desirability of taste (hedonic rating). The PSEM hams were characterised by a substantially higher colour intensity and substantially lower desirability of taste (hedonic rating) compared to other variants. The overall hedonic rating > 6.5 confirms the high quality of tested trial variants of dry cured hams. The highest overall hedonic rating was that of the PSES variant with mild PSE syndrome (Table 4).

Table 4: Sensory analysis of tested dry cured hams.

Variant	marble-like structure [c.u.]	Odour intensity [c.u.]	Odour desirability [c.u.]	Colour intensity [c.u.]	Colour desirability [c.u.]	Taste intensity [c.u.]	Saltiness [c.u.]	Acidicity [c.u.]	Consistency intensity [c.u.]	Consistency desirability [c.u.]	Overall hedonic rating [c.u.]
PSEM	2.38	6.81	6.41	6.38 ^{b}	6.53	6.08 ^{a}	4.71	3.77	4.57	6.42	6.67
PSES	2.59	6.86	6.57	6.08 ^{ab}	6.63	6.53 ^{b}	4.93	3.85	4.73	6.60	6.93
NORM	2.63	6.88	6.16	5.77 ^{a}	6.26	6.33 ^{ab}	4.96	3.92	4.52	6.45	6.69
LSD	*0.57*	*0.36*	*0.51*	*0.48*	*0.44*	*0.34*	*0.62*	*0.62*	*0.65*	*0.35*	*0.38*

$P \leq 0.05$, PSEM – hams with deep PSE syndrome (mean pH_{45} -5.6), PSES – hams with mild PSE syndrome (mean pH_{45} - 5.9), NORM – hams of good quality (mean pH_{45} - 6.5)

Conclusions

From the presented test results we may conclude that, contrary to common opinion, hams, of sufficient consistency and good sensorial quality, which mature (ripen) in the raw state, may be produced from raw material with PSE present. The above observation should be confirmed in future research work and better documented. Confirmation of the obtained results may provide a valuable guideline for the industry.

THE INFLUENCE OF PROCESSING PARAMETERS ON LEUCINE CATABOLISM BY A STRAIN OF *STAPHYLOCOCCUS SIMULANS*

H.Y. Wang, P.L. Li and C.W. Ma*

College of Food Science & Nutrition-Engineering, China Agricultural University, Beijing, 100083, China. Email: chwma@cau.edu.cn

Keywords: *Staphylococcus simulans*, leucine, processing parameters, 3-methylbutanal

Introduction

In the production of fermented sausages, various *Staphylococcus* spp. are widely used as starter cultures and participate in the development of typical flavour. Numerous flavour compounds have been identified in previous studies (Berdagué *et al.*, 1993; Stahnke, 1995). Among them, more attention has been paid to the formation of the branched – chain aldehyde, especially 3-methylbutanal, which is derived from the microbial catabolism of L-leucine and has a strong impact on sausage aroma (Stahnke, 1995). *Staphylococcus xylosus* and *S. carnosus* are popular species used in commercial starter cultures, and the volatiles originating from branched-chain amino acid degradation by these different stains have been studied (Søndergaard and Stahnke, 2002; Olesen and Stahnke, 2003). However, other species such as *S. simulans* were also proposed, as starters (Coppola *et al.*, 1997). As it is generally accepted that the ability for leucine catabolism is highly strain dependent, and processing conditions may have a great influence on the growth of strains and on their aroma formation. Therefore, it is necessary to select strains with high aroma-producing potential as starter culture to improve the sensory quality of fermented meat products. A strain referred to as S52 was screened from Chinese dry-cured bacon and identified as *S. simulans* on the basis of its biochemical characteristics and 16S rRNA analysis (data unpublished). It was catalase positive, coagulase negative, DNase negative and urease positive. It possessed nitrate reductase activity and produced acetoin. Moreover, it had moderate lipolytic and proteolytic activities, and did not produce pigment and haemolytic ring. Therefore, it was thought to be safe after primary tests and was chosen as a potential strain as a starter. The objective of the present study was to investigate the effect of various processing parameters on the capacity of *S. simulans* S52 to catabolize leucine into volatile compounds.

Materials and Methods

S. simulans S52 was grown in MC medium described by Talon *et al.*, (1999). After overnight incubation at 30°C under shaking (100 rpm), cells were harvested by centrifugation (10,000×g, 10 min, 4°C) and washed twice with sterile saline solution (0.9% NaCl), then resuspended in sterile saline solution. The optical density of the cell suspension was measured at 600 nm and adjusted to $OD_{600} \approx 1.0$. The reaction mixture was composed of leucine (2 mM), pyridoxal-5 phosphate (2 mM), α-ketoglutaric acid (10 mM), and resting cells (0.5 ml). It was prepared with sterile phosphate buffer KH_2PO_4-Na_2HPO_4 (0.067 M, pH 6.5) and the reagents were sterilized by a filter. Fifteen-ml screw-top clear vials, hole cap with PTFE/Silicone septa (Supelco, Bellefonte, PA, USA) containing 3.5 ml of liquid were incubated for 3d and 9d, respectively, at 30°C under static conditions. The control was prepared with 0.5 ml sterile saline solution instead of resting cells. All the practices were performed under aseptic conditions. To study the effects of different factors on the production of 3-methybutanal, a factorial experi-mental design with four factors was set up. The factors were temperature, pH, the concentration of NaCl and nitrite, respectively. The experimental data were analyzed in triplicate by variance analysis with SPSS 11.5 software. The volatile compounds of the reaction mixtures were extracted and analyzed according to Song *et al.*, (2005). Standard solutions of 3-methylbutanal and 3-methlybutanoic acid were extracted and analysed under the same conditions.

Results and Discussion

The result showed the metabolites of leucine produced by *S. simulans* S52 during the incubation of nine days. 3-methylbuntanal was the only degradation product after three days, and reached the highest quantity at the fifth day when 3-methylbutanoic acid was also produced. After a week, the amount of 3-methylbuantanal decreased, meanwhile, the 3-methylbutanoic acid increased. On the ninth day, there was only 3-methylbutanoic acid. The sequence of aldehyde and acid production revealed that *S. simulans* S52 first metabolised leucine into branched-chain aldehyde, which was then oxidised into acid. The ability of *S. simulans* S52 to catabolize leucine was strongly affected by changing the processing factors ($P<0.01$). Increasing the temperature and pH resulted in an increased generation of 3-methylbutanal, however, high level of NaCl and $NaNO_2$ inhibited its production. The optimal temperature and pH for the production of 3-methylbutanal were 40°C and 7.0, without NaCl and $NaNO_2$. Our work in the future aims to find out the prevailing factor influencing the production of 3-methylbutanal as well as the interaction of different factors.

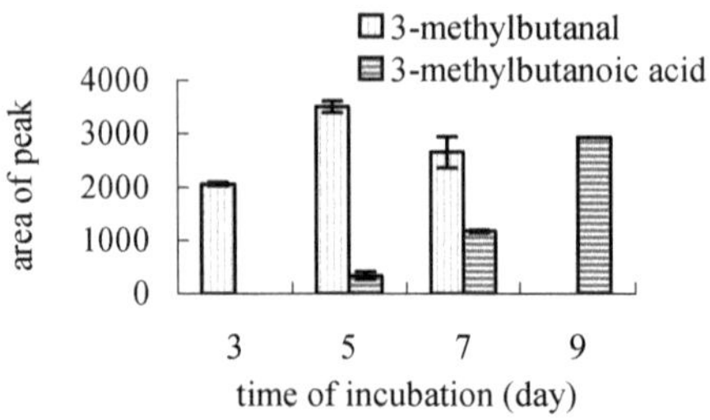

Figure 1: Mean peak area of methyl-branched aldehyde and acid produced by *S. simulans* S52 during the incubation.

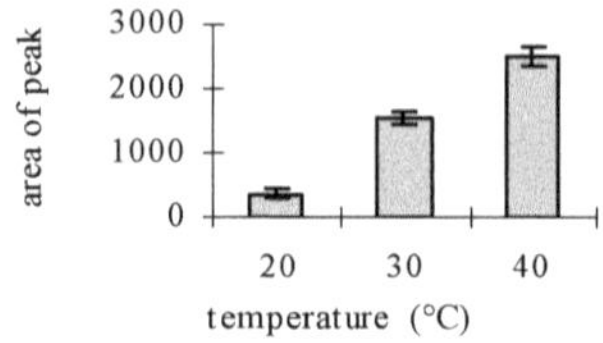

Figure 2: Effect of temperature on the production of 3-methylbutanal by *S. simulans* S52 incubated for 3 days.

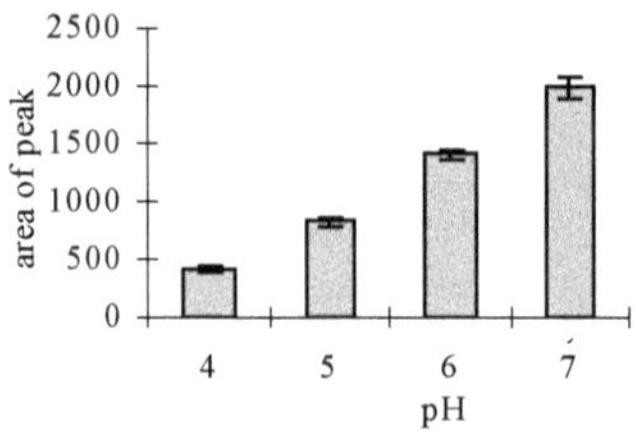

Figure 3: Effect of pH on the production of 3-methylbutanal by *S. simulans* S52 incubated for 3 days.

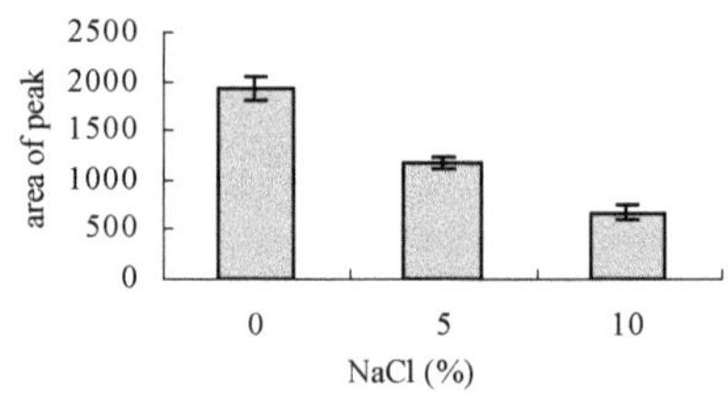

Figure 4: Effect of NaCl on the production of 3-methylbutanal by *S. simulans* S52 incubated for 3 days.

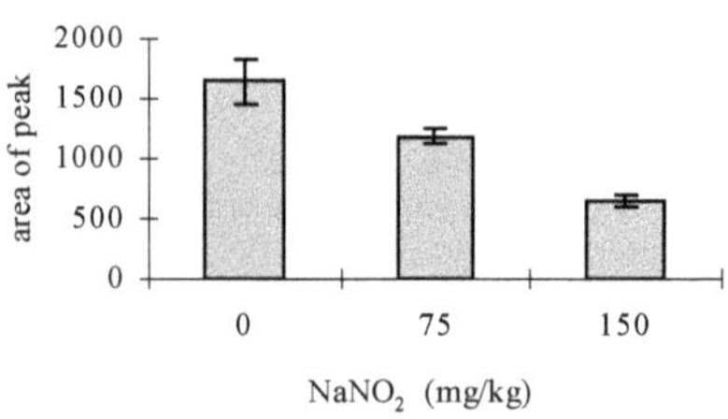

Figure 5: Effect of nitrite on the production of 3-methylbutanal by *S. simulans* S52 incubated for 3 days.

Conclusions

S. simulans S52 could catabolize L-leucine into 3-methylbutanal and 3-methylbutanoic acid, and processing parameters such as temperature, pH, NaCl and $NaNo_2$ influenced differently its capability.

References

Berdagué, J. L., Monteil, P., Montel, M. C., Talon, R. (1993). Effects of starter cultures on the formation of flavour compounds in dry sausage. *Meat Science*, 35: 275-287.

Coppola, R., Iorizzo, M.m Saotta, R., Sorrentino, and E., Grazia, L. (1997). Characterization of micrococci and staphylococci isolated from Sapperssata molisana, a southern Italy fermented sausage. *Food Microbiology*, 14: 47-53.

Olesen, P. T. and Stahnke, L. H. The influence of precultivation parameters on the catabolism of branched-chain amino acids by *Staphylococcus xylosus* and *Staphylococcus carnosus*. *Food Microbiology*, 2003, 20, 621-629.

Søndergaard, A. K. and Stahnke, L. H. (2002). Growth and aroma production by *Staphylococcus xylosus*, *S. carnosus* and *S. equorum*-a comparative study in model systems. *International Journal of Food Microbiology*, 75: 99-109.

Song, Y., Bi, H. X., Yang, H. J., Ma, C. W., Du, M. (2005). SPME-GC/MS analysis of key volatile compounds generated from unheated amino acid and reducing sugar mixtures. International Congress of Meat Science and Technology, 51st, Baltimore, USA.

Stahnke, L. H. (1995). Dried Sausages Fermented with *Staphylococcus xylosus* at Different Temperatures and with Different Ingredient Levels - Part II. Volatile Components. *Meat Science*, 41: 193-209.

Talon, R., Walter, D., Chartier, S., Barrière, C., Montel, M.C. (1999). Effect of nitrate and incubation conditions on the production of catalase and nitrate reductase by staphylococci. *International Journal of Food Microbiology*, 52: 47-56.

OHMIC COOKING OF NON COMMINUTED BEEF *BICEPS FEMORIS*

M. Zell*, J.G. Lyng, D.A. Cronin and D.J. Morgan

UCD Agriculture and Food Science Centre, School of Agriculture, Food Science and Veterinary Medicine College of Life Sciences, University College Dublin, Belfield, Dublin 4, Ireland. Email:markus.zell@ucd.ie

Keywords: ohmic heating, *biceps femoris*, cooking values

Introduction

Ohmic heating is a thermal process which occurs when an electrical current is passed directly through a food and the resistance imposed by the food leads to the generation of heat within the product. The basic principles as well as the factors influencing ohmic cooking have been explained by Sastry (1994) and Ruan (2004). Sastry and Palaniappan (1992) and Fryer *et al.*, (1993) reported that ohmic heating can be used in a continuous flow mode to cook and sterilize liquid food and solid-liquid mixtures. Nowadays ohmic heating is viewed as an alternative heating system for pumpable foods and there are currently a number of commercial scale processing plants in various countries (UK, Italy, Mexico) producing fruit and/or vegetables in sauces and also pasteurised orange juice and liquid egg. While a number of the early patents in ohmic heating of meat were in the area of meat processing the amount of in-depth research conducted to date has been quite limited in spite of the fact that ohmic heating has the potential to cook meat in a much shorter time than the conventional cooking procedures. Shirsat *et al.*, (2004) and Piette *et al.*, (2004) showed that it is possible to cook comminuted meat emulsions ohmically to a comparable quality to conventional cooked samples. However, these products represent a relatively small proportion of total cooked meats and no results have yet been presented on the quality of ohmically cooked non comminuted meats. Lyng and McKenna (2006) discussed the differences between non comminuted and comminuted meats in the context of ohmic heating. The ohmic method requires uniform conductivity values within the meat which means that a perfectly even distribution of injected salt or brine solutions must be achieved in the case of non-comminuted meats. The aim of the current research is to compare selected quality attributes of conventionally and ohmically cooked entire non-comminuted whole muscle meat cooked to similar cook values.

Materials and Methods

Muscle: High quality single muscle beef *biceps femoris* was purchased from a local supplier (Kepak, Clonee, Co. Dublin, Ireland). This muscle was selected primarily because of its uniform myofibrillar structure.

Meat preparation: Six uniform pieces of *biceps femoris* (146 g) were prepared to incorporate a specified amount of a 3% salt solution. After injection the meat was tumbled at 4°C in a vacuum tumbler.

Cooking equipment: For ohmic cooking the tumbled meat was filled into a casing which was placed in a cylindrical stainless steel cell (length 0.075 m and internal diameter 0.073 m). The cell was lined with Teflon tape and insulated. The ends of the cell were fitted with plastic caps within which platinum-coated titanium electrodes (0.069 m diameter) were housed. The casing was left open at the ends to ensure direct contact between the electrodes and the meat. Temperatures were measured with a multi thermocouple probe (with one thermocouple located in the geometric centre and the other near to the surface) and monitored at 1 s intervals with a data logger. The meat pieces were cooked with a 3.5 kW (15 A, 0-240 V, 50 Hz) batch ohmic heater at a constant voltage-density of 12.5 V/cm. The electrodes were spring-loaded to ensure optimal contact between the meat and the electrode surface. Reference meat samples were cooked conventionally in a thermostatically controlled steam oven set at 80°C. Internal product temperatures were recorded at 30 s intervals using a Type K thermocouple (placed at the geometric centre of the product) and a data logger. The product was cooked until an endpoint temperature (coldest spot) of 72°C was reached and this was followed by a 2 min holding time. Cook values (C_s) were subsequently calculated (Lyng and McKenna, 2006) from this time temperature data. In order to reach similar C_s values between ohmic and steam cooking, preliminary experiments showed that a combined pre heat and hold interval amounting to a maximum of 12 min was required. Pasteurisation units were also calculated using equations given in Lyng and McKenna (2006). For the cooling process all samples were cooled for 1 h in a circulating water bath (8°C) followed by 24 h cooling in a refrigerator (4°C).

Quality assessment: The cook loss was determined by measuring the difference between the fresh and cooked weights and dividing by the fresh weight. Total salt content was analyzed using the method of Fox (1963). Hunter L*a*b* values were measured using a Minolta colorimeter and Hue angle (H) and Chroma (S) were calculated using equations given in Shirsat *et al.*, (2004). Colour readings were taken at six different points in the centre area of the cooked meats. For Warner Bratzler (WB) shear force measurements 10 strips 3 cm long and 1×1 cm in section were prepared by cutting the meat in the plane perpendicular to the direction of the fibre bundles. Maximum shear force (N) was recorded by an Instron universal testing machine following meat pre-equilibration at 25°C with the blade cutting the slices at right angles to the fibre direction.

Statistical analysis: Warner Bratzler shear force, salt content, cook loss and colour were analysed by ANOVA.

Results and Discussion

Significantly shorter cooking times were required in ohmic vs. conventional heating to produce similar C_s values (12 vs. 36 min respectively). With the ohmic process meat was heated to its final temperature in 65 s. The main problem that appeared was the heat loss which occurred in regions of the product adjacent to the outer surfaces. So a cook time that ensured these regions were heated to the target temperature (72-74°C) for times that gave similar C_s values to steam-cooked meat led to core temperatures of 80-84°C. From steam cooked products the range from core to the outside of the product was approximately 6°C. Cook values of 3.82 and 3.99 were reached for the ohmic and conventional cold spots respectively though a value of 6.63 was reached in the hottest region of the ohmically heated samples. This data is shown in Figure 1. Pasteurisation units calculated from the time temperature profiles showed values of 2800 for ohmic vs. 1480 for steam cold spots indicating the ohmic process would have induced approximately double the amount of kill of the reference microorganism used in pasteurisation unit calculation. Cook loss was lower in ohmic samples which also had significantly higher L* and lower a* values ($p<0.05$) indicating a lighter, less red meat and this difference was also visible to the naked eye. In relation to tenderness ohmic heated samples were tougher ($P<0.05$) though the difference was only 5.08 N (Table 1) which is unlikely to be detected by a sensory panel. Furthermore, the centre regions of the ohmic samples were heated to higher temperatures to compensate for the heat loss from the outer regions and therefore definite conclusions on texture cannot be made until this surface cooling problem can be resolved.

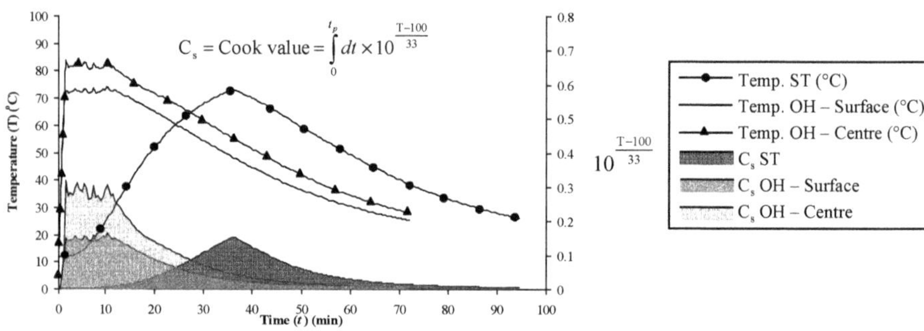

Figure 1: Time temperature and time cook value (C_s) profiles for ohmically (OH) and steam (ST) cooked beef *biceps femoris.*

Table 1: Quality parameters of ohmic and steam cooked whole beef *biceps femoris.*

	Salt (%)	WB (N)	Cook loss (%)	L*	a*	b*	H	S
Ohmic	1.16	52.48[a]	27.7[b]	59.62[a]	2.37[b]	10.09	76.61[a]	10.38
Steam	1.15	47.40[b]	37.9[a]	55.80[b]	2.79[a]	10.07	74.47[b]	10.46

[a,b] means in a column with different superscript letters are significantly different ($P<0.05$)

Conclusions

The results show that it is possible to ohmically cook injected whole muscle (*biceps femoris*) beef to produce lower cook losses and levels of quality which are comparable to those achieved by conventional methods. Additional research is needed to eliminate local temperature variations caused by heat loss.

References

Fox, P.F. (1963). Potentiometric determination of salt in cheese. Journal of Dairy Science, *46*, 744-745.

Fryer, P.J., De Alwis, A.A.P., Koury, E., Stapley, A.G.F. and Zhang L. (1993). Ohmic Processing of Solid-Liquid Mixtures: Heat generation and Convection effects. Journal of Food Engineering, *18*, 101-125.

Lyng, J.G. and McKenna, B.M. (2006). Ohmic pasteurisation of meat and meat products. In *Food Machinery Handbook* (W. A. Myer Kutz, ed.), in press.

Piette, G., Buteau, M.L., De Halleux, D., Chiu, L., Raymond, Y., Ramaswamy, H.S. and Dostie, M. (2004). Ohmic Cooking of Processed Meats and its Effects on Product Quality. Journal of Food Science, *69*(2), FEP71-FEP78.

Ruan, R., Ye, X., Chen, P., Doona, C.J., and Taub, I. (2004). Ohmic heating. In: *Thermal technologies in food processing*, (P. Richardson, ed.) pp. 241-265, Woodhead Publishing Limited, Cambridge.

Sastry, S.K. (1994). Ohmic heating. In: *Minimal processing of Food and process optimization*, (R.R. Singh and F.A.R. Oliveira, eds.) pp. 17-33, CRC Press, London.

Sastry, S.K. and Palaniappan, S. (1992). Ohmic heating of liquid-particle mixtures. Food Technol. *46*(12), 64-67.

Shirsat, N., Brunton, N.P., Lyng, J.G., McKenna, B.M. and Scannell, A. (2004). Texture, colour and sensory evaluation of conventionally and ohmically cooked meat emulsion batter. Journal of Science Food andAgriculture, *84*, 1861-1870.

COMPARISON OF MINKE WHALE (*BALAENOPTERA ACUTOROSTRATA*) FERMENTED SAUSAGES WITH MORE COMMON SALAMI PRODUCTS FROM PORK AND BEEF

B. Egelandsdal[1], T.C. Johannesen[2], D. Ekeberg[1], T. Bruun[1] and E. Slinde[3]

[1]Department of Chemistry, Biotechnology and Food Science, University of Life Science, P.O. Box 5003, N-1430-Ås, Norway, [2]Matforsk, Oslovn.1, N-1430-Ås, Norway, [3]Institute of Marine Research, P.O. Box 1870, Nordnes, N-5817 Bergen, Norway. Email: bjorg.egelandsdal@umb.no

Keywords: Minke whale, beef, pork, salami, fermentation kinetics, flavour

Introduction

Whaling has a long tradition in some countries. Whale meat is still consumed in certain countries and regarded as a special and traditional treat. Minke whale hunting is today recommended by Norwegian Authorities due to its high abundance. Most whale recipes describe simple processing (http://www.80dager.no/hval) involving frying the meat as a steak. In Norway processing of Minke whale meat into fermented sausage has not been widely commercially exploited. No publication exists on the quality of this product compared to products produced in accordance with the "Norwegian Salami Technology". The aim of this study was to compare Minke whale salami with more common Norwegian fermented sausages produced from pork and beef-pork.

Materials and Methods

Minke whale meat was collected late June at the end of the catching season. The meat was vacuum-packed and frozen until the fermented sausages were produced in mid-September. Frozen pork and beef meat were used, and salami minces were produced according to standard production routines (Skjelkvåle *et al.*, 1974). A standard recipe (for pork) is given in Table 1. Smoking was carried out on the 3rd day after production. All recipes had the same initial protein, fat and water content. The starter culture was Biocarna ferment CXK (Danisco, Denmark) containing *Staphylococcus carnosus, Staphylococcus xylosus* and *Lactobacillus sakei.* The fermentation kinetics were recorded with continuous pH loggers (L, Table 2) inserted in the raw batters at 20°C. The fermentation is thus accelerated, since the temperature was not reduced. The acid production was calculated from logger and titration data. Lag time, maximum production rate of lactic acid as well as ripening due to base production ($pH= ae^{bt}$, where b is a measure of base production and a is a constant) were calculated. The sausages were withdrawn at intervals during the fermentation and their pH and moisture loss determined. The fermented sausages were produced three times with three different batches of raw materials. The sausages were removed from the chamber after approx. 30% weight loss, vacuum-packed and kept at 4°C until analysed approx. 3 weeks later

(-40°C and 20 weeks for GC-MS). **Water activity,** a_w, was measured at 25°C (± 1°C) after about 2-hrs equilibrium (AquaLab model CX-2, Decagon Devices Inc., Pullman, Washington, USA). **Organic acids** were analysed by HPLC as described by Skeie *et al.* (1997). **Accelerated rancidity** was induced in sausage slices, covered by a transparent film, using fluorescent light for 17-20 days before thiobarbituric acids were analysed in accordance with Tarlagdis *et al.*, (1960). **GC-MS**: Static headspace was measured on samples (headspace to sample 1:2) preheated to 50°C. GC was performed with an Agilent Technologies 6890 N, in combination with an Autospec Ultima (Micromass, Manchester,UK) mass spectrometer with EBE geometry, operated in EI mode at 70eV. The column (CP-WAX 52 CB, 30 m, ID 0.32 mm and DF 0.50µm) was run at 200°C. The results are reported as number of compounds identified (in % of total numbers). **Statistics:** Minitab version 14 (Minitab Inc., Pennsylvania, USA) was used for one-way ANOVA, for testing for equal variances (Bartlett's test) and for descriptive statistics.

Results and Discussion

The recipes (Table 1) were all with pork back fat as the major fat source. The lean meat originated from three different species. The general pattern of pH change involves first a lag phase, thereafter a decline in pH to a minimum value followed by an increase in pH. No arrest in base production during the ripening phase was observed, *i.e.* during logging pH increased exponentially post-acidification. The average pH drop was 1 unit using 0.7% sugar (Table 2). Whale meat tended towards the lowest drop ($pH_{min} - pH_{init}$). This suggests that whale meat has a higher buffer capacity than beef-pork or pork. The evaporation of water tended to be highest for pork meat. This appears due to pork's lower ability to bind water, as reflected in the moisture loss. Most common lag times were 20-27 hrs. Longer lag times were observed for one of three batches of whale and beef-pork. This is evident from the significantly larger standard deviation for lag time of the sausages from these two species (Table 2). This might be due to difficulties in establishing the lactic acid bacteria as the dominant bacteria. Once established, the acid production stopped at a final value of about 11 - 12 gram of lactic acid *per* kg of the initial wet weight (Table 2). The acid production rate was highest in the pork meat. The glucose level dropped to 0.1-0.3ppm in the final products (not shown). Only the whale sausage contained a substantial, but variable among batches, amount of acetic acid. With respect to the other organic acids analysed *i.e.* citric, orotic, pyruvic, succinic, formic, uric and propionic, only succinic acid was present above 0.5 g/kg in the raw batter, but was later metabolised by the bacteria (Table 2). The slowest production of base equivalents took place in the beef-pork sausages. This increase in pH during ripening is generally regarded as important for flavour development (Toldra, 2002). The final pork sausages were all accepted according to the Norwegian salami standard and were not regarded as oxidised (not shown). However, the accelerated test for oxidation induced extensive oxidation in all products. Most pronounced oxidation was present in Minke whale sausage due to its high myoglobin content. The gas chromatographic run differentiated between the three samples (Table 2). The most frequent aldehyde in the headspace of the whale sausage was decanal that originates from lipid degradation.

Conclusion

Salami sausage characteristics of Norwegian salami and Minke whale sausages had largely the same fermentation characteristics. The Minke whale sausage would benefit from slightly more sugar in order to obtain the same pH drop as in the more traditional salamis. The largest variation in lag time and amount of acetic acid produced was observed for the whale material. Minke whale has a high base production capacity that might be utilized to optimise flavour development. The shelf life of the sausage is limited due to its high amount of the pro-oxidant myoglobin even when pork back fat is used as fat source.

Table 1: A standard pork sausage recipe.

Raw materials/ingredients	Amount (g)
Pork 6% fat	7,570
Pork back fat	2,000
Ascorbic acid	5
NaCl	100
Salt (0.6% $NaNO_2$)	220
Spices incl. garlic	30
Starter culture	10
Glucose	70
Sum	10,000

Table 2: Kinetic and flavour characteristics of the sausages (S=sausage; Raw batter=RB; L=logger data obtained on RB). The standard deviations given are the standard deviations of different productions. Significantly different mean values ($p<0.05$) are indicated with letters a, b and c. The letters x, y reflect data with unequal variances ($p<0.05$).

	Whale	Pork	Pork-beef
Initial pH-RB	5.68±0.14	5.60±0.15	5.56±0.04
Minimum pH	4.71±0.13	4.51±0.11	4.44±0.10
a_{w}-S	0.905±0.005	0.891±0.014	0.904±0.008
Moisture loss(%)-S	30.9±1.7	31.6±2.5	30.6±1.7
Lag phase(hrs)-L	38^{x} ±12	23^{y} ±4	27^{x} ±9
Acid production-doubling time (hrs)-L	$7.1^{a,b}$ ± 1.0	5.3^{a} ±1.1	8.0^{b} ±0.7
Base eqv.*(1/hr)-L	0.012^{a} ± 0.002	0.012^{a} ± 0.002	0.0072^{b}±0.002
Lactic acid(g/kg) RB⇒S	8.4⇒11.0±2.1	5.0 ⇒11.4±0.9	5.4 ⇒12.1±0.5
Acetic acid (g/kg) RB⇒S	0.16⇒1.2^{x}±0.7	0.0⇒0.19^{y}±0.03	0.0⇒0.22^{y}±0.03
Succinic acid (g/kg) RB⇒S	0.66^{a}⇒0.09±0.03	0.60^{a}⇒0.14±0.05	0.96⇒0.12±0.04
TBARS* (mg/kg) RB⇒S	0.8⇒22.8^{a}±2.1	0.9⇒6.7^{b}±0.8	0.3⇒11.4^{c}±0.2
GC-MS**-% acids	64	35	47
GC-MS-% aldehydes	27	18	Not identified
GC-MS-% alcohols	Not identified	18	24

References

Skeie, S., Narvhus, J.A., Ardö, Y., Thorvaldsen, K. and Abrahamsen, R.K. (1997). The effect of reduced salt content on the function of liposome-encapsulated neutrase and heat-treated lactobacilli in rindless low-fat cheese. *Le Lait*, 77: 575-585.

Skjelkvåle, R., Tjaberg, T.B. and Valland, M. (1974). Comparison of salami sausage produced with and without addition of sodium nitrite and sodium nitrate. *Journal of Food Science*, 39, 520-524.

Tarladgis, B.G., Watts, B.M. and Younathan. M.T. (1960). A distillation method for the quantitative determination of malonaldehyde in rancid foods. *The Journal of the American Oil Chemists' Society*, 37,44 -48.

Toldra, F. (2002). Dry-cured meat products. Food and Nutrition Press, Inc., Chap. 5, page 108.

Acknowledgements

Karin Solgaard and Kari Olsen are thanked for their excellent assistance. A special thank to the Graduate students of Muscle Food Processing Course (years 2003- 2005) from University of Life Science for carrying out parts of this work.

EFFECT OF IONISING RADIATION ON THE COLOUR AND OXIDATIVE RANCIDITY OF STORED MEAT PUREES CONTAINING ANTIOXIDANTS

E.M. Stewart, W.D. Graham and B. Moss*

Agriculture, Food & Environmental Science Division, Agri-Food & Biosciences Institute (AFBI), Newforge Lane, Belfast BT9 5PX, UK. Email: bruce.moss@afbini.gov.uk

Keywords: ionising radiation, antioxidant, colour, oxidative rancidity

Introduction

Irradiation of meat has been shown to result in a number of colour changes, with chicken breast meat becoming 'pinker' (Miller *et al.*, 1995) and beef becoming 'browner' (Millar *et al.*, 2000). A number of pigment forms have been proposed to account for these observed changes in appearance following irradiation. Moss *et al.* (2000) suggested that the differences in spectra of irradiated and unirradiated pork could be due the formation of carboxymyglobin. The production of free radicals during irradiation results in degradation of a number of chemicals and increased oxidation of fatty acids leading to rancidity and off-odours (Nam *et al.*, 2003). One way of overcoming this problem is to include antioxidants in the foodstuff and given consumer concerns about food additives, the use of natural antioxidants such as flavonoids could be an option. The following study was undertaken to study the effect of inclusion of natural antioxidants on the free radical induced oxidation in meat following irradiation with the effects of their addition of the colour also being examined.

Materials and Methods

Fresh minced beef (8% lipid) was purchased locally in Belfast and mixed to provide a homogenous bulk sample. A total of 100 patties (50 g weight) were prepared. The antioxidants α-tocopherol, epicatechin, quercetin and resveratrol were included at concentrations of 25, 50, 100, 250, 500 and 700 μmol/kg whilst other patties were prepared with no antioxidant included. The samples were packed in plastic containers, double wrapped with cling film and treated with an irradiation dose of 2.5 kGy or left unirradiated. They were then stored for 1 or 14 days at 4 ± 1°C. Samples were analysed for oxidative rancidity by measuring their thiobarbituric acid number (TBA) according to the method of Pikul *et al.* (1989) but using trichloroacetic acid in place of perchloric acid. The reflectance spectra of the patties were measured immediately after opening using a Monolight spectrophotometer and CIELAB values calculated (Moss *et al.*, 2000). The experiment was repeated on three separate occasions.

Results and Discussion

Irradiation at 2.5 kGy resulted in significantly higher a* values and significantly lower a* and b* values (Table1) than control samples. Storage for 14 days in overwrap packs resulted in significantly lower a* and b* values compared to fresh samples. There was no statistically significant effect of either type of antioxidant added or its level on any of the colour parameters studied (Table 1). There was a statistically significant interaction between Irradiation and storage time which showed that a* and b* values decreased less with storage for irradiated samples than unirradiated samples.

TBA values were significantly higher ($p<0.001$) after irradiation at 2.5 kGy and storage for 14 days (Table 1). Inclusion of antioxidants in the patties significantly reduced ($p<0.001$) oxidative rancidity. Resveratrol proved to be the most effective antioxidant with TBA values being half those of the control samples. Quercetin and epicatechin produced similar effects with oxidation levels being significantly lower than those of the controls but higher than samples containing resveratrol. α-Tocopherol was the least effective antioxidant. Antioxidant concentration had a significant effect ($p<0.05$) on the TBA values. The decrease in TBA values with increasing concentration of resveratrol did not appear to be significant as similar values were obtained at each concentration. Thus, it would appear that inclusion levels of as low as 25 μmol/kg would effectively reduce oxidation in meat purees containing resveratrol as an antioxidant. There were statistically significant interactions fro TBA values between; irradiation and type of antioxidant; irradiation and storage; and storage with type of antioxidant (Table1).

Conclusions

Resveratrol is a highly effective antioxidant for reducing oxidation in meat puree samples, with quercetin and epicatechin also proving to be effective although to a lesser extent. Although, α-tocopherol did reduce oxidation of the meat it was significantly less effective than the other three antioxidants employed. Irradiation of beef purees resulted in samples which were less red and less yellow, however the colour of irradiated samples showed less change during storage.

Table 1: Effect of ionising radiation and storage on the colour (CIEBLAB values) and TBA values (mg malonaldehyde/kg muscle) of overwrapped meat purees containing antioxidants.

	L*	a*	b*	TBA
Effect of irradiation (kGy)				
0	46.03	18.32	17.09	0.261
2.5	47.35	17.39	15.85	0.386
Significance of effect	**	*	***	***
SEM (n = 150)	0.317	0.259	0.161	0.0110
Effect of storage (days at <3°C)				
1	47.01	21.66	17.13	0.211
14	46.36	14.06	15.80	0.436
Significance of effect	n.s.	***	***	***
SEM (n = 150)	0.317	0.259	0.161	0.0110
Effect of antioxidant (AO)				
Control (n = 12)	42.97	17.88	15.92	0.480
α-Tocopherol (n = 72)	46.77	17.53	16.23	0.410
Epicatechin (n = 72)	46.97	17.57	16.38	0.311
Quercetin (n = 72)	46.93	17.81	16.56	0.306
Resveratrol (n = 72)	46.70	18.51	16.79	0.242
Significance of effect	n.s.	n.s.	n.s.	***
SEM (max)	0.457	0.374	0.233	0.0159
(min)	1.120	0.915	0.570	0.0389
Effect of AO concentration				
0 (n = 12)	42.97	17.88	15.92	0.480
25 (n = 48)	46.09	17.52	16.22	0.361
50 (n = 48)	46.38	17.94	16.41	0.345
100 (n = 48)	47.12	18.04	16.43	0.320
250 (n = 48)	46.66	18.23	16.67	0.295
500 (n = 48)	47.41	17.64	16.49	0.290
700 (n = 48)	47.39	17.77	16.73	0.294
Significance of effect	n.s.	n.s.	n.s.	*
SEM (max)	0.560	0.457	0.285	0.0195
(min)	1.120	0.915	0.570	0.0389
Significance of interaction				
A x C	n.s.	n.s.	n.s.	n.s.
I x A	n.s.	n.s.	n.s.	***
I x C	n.s.	n.s.	n.s.	n.s.
I x S	***	***	**	**
S x A	n.s.	n.s.	n.s.	***
S x C	n.s.	n.s.	n.s.	n.s.

where: I = irradiation treatment; S = storage; A = antioxidant; C = antioxidant concentration; SEM = standard error of the mean; ***=p<0.001; **=p<0.01; *=p<0.05; n.s.=p>0.05 (not significant)

References

Millar, S.J., Moss, B.W., McDougall, D.B. and Stevenson, M.H. (1995). The effect of ionising radiation on the CIELAB colour coordinates of chicken breast as measured by different instruments. International Journal of Food Science and Technology, 30, 663-674.

Millar, S.J., Moss, B.W. and Stevenson, M.H. (2000). The effect of ionising radiation on the colour of beef, pork and lamb. Meat Science, 55, 349-360.

Moss, B.W., Millar, S.J. and Kilpatrick, D.J. (2000). The use of discriminant analysis for the interpretation of the reflectance spectra of irradiated porcine *M. longissimus dorsi*. Meat Science, 55, 337-348.

Nam, K.C., Min, B.R., Yan, H., Lee, E.J., Mendonca, A., Wesley, I. and Ahn, D.U. (2003). Effect of dietary vitamin E and irradiation on lipid oxidation, color, and volatiles of fresh and previously frozen turkey breast patties. Meat Science, 65, 513-521.

Pikul, J., Leszczynski, D.E. and Kummerow, F.A. (1998). Evaluation of three modified TBA methods for measuring lipid oxidation in chicken meat. Journal of Agricultural and Food Chemistry, 37, 1309-1313.

THE EFFECT OF STARTER CULTURES AND FERMENTATION TIMES ON THE PROCESSING AND SENSORY PROPERTIES OF CURED BEEF PREPARED FROM FOREQUARTER MUSCLES

D. Keenan*[1], T. Kenny[1] and J.P. Kerry[2]

[1]Ashtown Food Research Centre, Teagasc, Ashtown, Dublin 15, Ireland.
[2]Department of Food and Nutritional Sciences, Food Technology and Nutrition, University College Cork – National University of Ireland, Cork, Ireland. Email: derek.keenan@teagasc.ie.

Keywords: beef, curing, starter cultures, flavour, *biceps femoris*

Introduction

The flavour of cured meat products is governed by curing ingredients, such as salt, nitrite and spices but also, by the processes to which they are subjected e.g. smoking. Additionally, the microbial decomposition of carbohydrates, lipids and proteins contributes to several flavour components. The fermentation process, i.e. the anaerobic breakdown of carbohydrates by non-pathogenic micro-organisms to yield various alcohols and organic acids can preserve or enhance the organoleptic characteristics of meat products (Campbell-Platt & Cook, 1995). The addition of starter cultures is common practice in the meat fermentation industry, so as to increase the uniformity of batches. The most commonly used bacterial species are Lactic acid bacteria (LAB) and *Micrococcaceae* (Houde *et al.*, 1998). These are designed to give products a novel flavour profile by acidification and related changes in the meat. The objective of this study is to determine the effects of two different starter cultures and two different fermentation times (4 hours and 10 hours) in enhancing flavour development in a full-moisture cured beef product.

Materials and Methods

Silverside flat cuts (M. *biceps femoris*) were sourced from a local meat producer. The starter cultures used were a commercial single strain culture containing *Staphylococcus carnosus* and a mixed culture containing *Staphylococcus carnosus, Staphylococcus xylosus* and *Lactobacillus sakei,* obtained from Christian Hansen GmbH. Brine solutions containing the single strain culture (SC1) and the mixed strain (SC2), were prepared. The boneless cuts were pumped to 115% of their green weight with brine and tumbled for 2 hours at 10 rpm. Tumbled muscles were then enclosed in elastic netting and vacuum packed. The fermentation stage was carried out in a steam-air oven using at 30°C and 95% relative humidity (RH), for 4 or 10 hours after which time the products were cooked to a core temperature of 72°C. Cooked samples were analysed for moisture, fat and protein content (Bostian *et al.*, 1985; Sweeney and Rexford, 1987). Cook loss and yield were calculated. Texture Profile Analysis (TPA) was carried out using Instron model 5543. An 8-member trained panel was employed to evaluate sensory quality of sample treatments (AMSA, 1995). Results were analysed using a two-way analysis of variance (ANOVA).

Results and Discussion

Brine composition had a small effect ($P \leq 0.05$) on moisture, with cured beef produced from brine containing SC1 yielding a drier product than the control. This result corresponds with yield measurements that showed products made from brines containing SC1 having greater cook losses ($P \leq 0.05$) and therefore, poorer final yields ($P \leq 0.01$). The SC1 product also had a lower ($P \leq 0.05$) salt content. However, these differences were not reflected in sensory panel scoring, which showed no effect on juiciness and saltiness ($P \geq 0.05$). Taste panellists preferred ($P \leq 0.05$) SC1 samples to those produced with SC2, but neither was rated as different to the control (Table 1).

Table 1: Effect of brine formula on overall flavour and preference in cured beef.

Treatment	Overall flavour	Preference
Starter culture 1	3.6[a]	0.3[a]
Starter culture 2	3.3[a]	0.1[b]
Control	3.6[a]	0.3[a]

[a-b] Means in the same column with different letters are different ($P \leq 0.05$)

Brine composition affected many of the TPA attributes, with the product from SC1 being harder ($P \leq 0.01$), more cohesive ($P \leq 0.05$), gummier ($P \leq 0.001$) and chewier ($P \leq 0.01$) than the control. Despite this apparent detrimental effect on texture, the sensory panel score for overall acceptability of the product was unaffected ($P \geq 0.05$). Colour and pH values were not affected ($P \geq 0.05$) by brine formula or fermentation time. Unlike SC1 product, chemical analysis of the SC2 product showed that brine formula and fermentation time had no effect ($P \geq 0.05$) on moisture. However, fermentation time affected cook loss ($P \leq 0.01$) and yield ($P \leq 0.001$) with the 10 hour fermentation period resulting in less cook loss and better yields. Furthermore, sensory analysis showed no effects ($P \geq 0.05$) for brine formula and fermentation time on overall flavour, juiciness, saltiness and overall acceptability (Figure 1). Texture and colour were unaffected ($P \geq 0.05$) by brine formula and fermentation time.

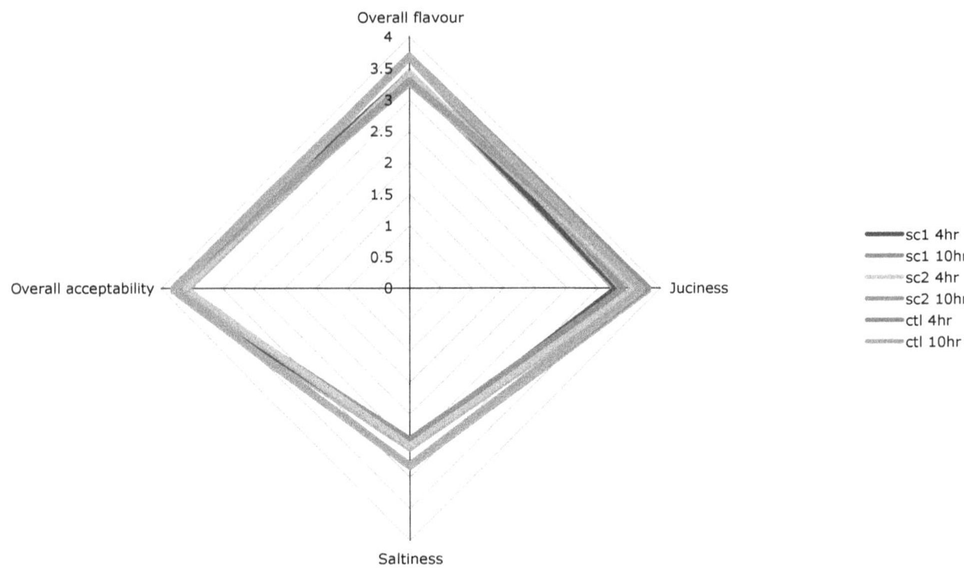

Figure 1: Effect of brine formula and fermentation time on the sensory aspects of cured beef products.

Conclusions

The results indicate that the addition of starter cultures, single or mixed, did not give an improvement in flavour in the processed products. Although the addition of the single starter culture did not have detrimental effect on flavour, the control product was preferred by taste panellists to that made with the mixed culture. Further research would be required to investigate the application of starter cultures in full moisture products and to determine a flavour profile that would specifically suit the Irish consumer.

References

AMSA (1995). Research guidelines for cooking, sensory evaluation and instrumental tenderness measurements of fresh meat. American Meat Science Association, 10-11.

Bostian, M. L., Fish, D. L., Webb, N. B. and Arey, J. J. (1985). Automated methods for determination of fat and moisture in meat and poultry products: collaborative study. Journal of the Association of Official Analytical Chemists, 68, 876-880.

Campbell-Platt, G., and Cook, P. E. (1995). Starter cultures for meat fermentations. *In:* Fermented Meats (1st Edition). Blackie Academic and Professional (Chapman and Hall Ltd.). 130-159.

Houde, A., Dussault, F., van Beek, S., and Bard, L. (1998). Characterisation of lactic acid bacteria used in commercial meat starter cultures: strain identification and exocellular polysaccharide reduction. *In.* Proceedings 44th International Congress of Meat Science and Technology (ICoMST). C5, 828-829.

Sweeny, R. A. and Rexroad, P. R. (1987). Comparison of LECO FP-228 "Nitrogen Determinator" with AOAC copper catalyst Kjeldahl method for crude protein. Journal of the Association of Official Analytical Chemists, 70, 1028-1030.

Acknowledgements

This project is funded by the Irish National Development Plan, under the Food Institutional Research Measure (FIRM) and administered by The Department of Agriculture, Food and Rural Development. We are grateful to Christian Hansen GmbH. for providing technical advice and samples of the starter cultures.

THE EFFECT OF HOT BONING AND ELEVATED BRINE TEMPERATURE ON THE PROCESSING AND SENSORY PROPERTIES OF CURED BEEF PREPARED FROM SILVERSIDE AND FOREQUARTER MUSCLES

D. Keenan*[1], J. Hayes[1], T. Kenny[1] and J.P. Kerry[2]

[1]Teagasc, Ashtown Food Research Centre, Ashtown, Dublin 15, Ireland. [2]Department of Food and Nutritional Sciences, Food Technology and Nutrition, University College Cork – National University of Ireland, Cork, Ireland. Email: derek.keenan@teagasc.ie

Keywords: beef, hot boning, elevated brine temperature, curing, *biceps femoris, pectoralis profundus*

Introduction

Hot boning has not been adopted by meat plants in Ireland, despite offering significant savings in space, energy, labour, materials and supplies as well as improvements in functional properties of the product. Concern has centred on issues, such as re-training of staff, hygiene regimes (as meat is handled while warm) and resultant added costs.). The main concern is the possible loss of tenderness associated with hot boning. Conventionally boned muscles are held in a stretched state by the carcass but unrestrained, hot-boned muscles are more susceptible to contraction and toughening in the absence of this skeletal support at low chilling temperatures (Locker, 1960). Therefore, processing directly after hot boning may offset some of its effects. The rate of curing is controlled by a number of temperature-dependant diffusion processes and chemical reactions. By letting the process take place at a higher temperature than the normal 0-5°C, curing could be accelerated (Pearson and Gillett, 1996). This would be a significant advantage, in particular, for dry cured products as the cure ingredients are rubbed into the meat by hand and therefore require a prolonged time period for cure distribution and aging (Ranken, 2000). The rate of migration of cure ingredients will be examined by determining salt and nitrite distribution in samples at different time intervals.

Materials and Methods

Two muscles, the M. *biceps femoris* (silverside) and M. *pectoralis profundus* (brisket in forequarter), were tested in accordance with a randomised experimental design. Beef cuts were hot-boned within 1.5 hours post mortem. Cold boned carcasses were chilled at 0-2° C before excision at 24 hours post mortem. Brine solutions at normal (1-4 °C) and elevated (12-16 °C) temperatures, were employed, resulting in a total of 4 treatments per muscle. The boneless cuts were pumped to 115% of their green weight with brine and tumbled for 2 hours at 10 r.p.m. Tumbled muscles were enclosed in elastic netting, vacuum packed and steam cooked to a core temperature of 72 °C. Cooked samples were analysed for moisture, fat and protein content (Bostian *et al.*, 1985; Sweeney and Rexford, 1987). Cook loss and yield were calculated. Texture Profile Analysis (TPA), Warner-Bratzler Shear Force (WBSF) and Kramer Shear Force (KSF) measurements were made using Instron 4464 and 5543 models. Mohr's method was used to determine salt content. Nitrite was determined by a purple azo dye colourimetric titration (Ruzicka and Hansen, 1981) using a Foss FIAstar™ 5000 flow-rate injection analyser (Foss Tecator AB, Sweden). An 8- member trained panel was employed to evaluate sensory quality of samples (AMSA, 1995). The effects on storage life of the combined hot boning and warm brine treatments were assessed by storing MA-packed samples for up to 21 days at 4°C in a chill display cabinet. Results were analysed using a two-way analysis of variance (ANOVA).

Results and Discussion

Hot-boned silverside beef had lower ($P \leq 0.001$) salt content than the control beef after tumbling but not after cooking. Table 1 showed that cooked silverside samples exhibited lower nitrite levels ($P \leq 0.05$) than raw meat after tumbling. This expected reduction is a result of nitrite being broken down to nitric oxide, which is fundamental to the formation of the stable cured colour pigment (Ranken, 2000). Cook losses and total yields were not affected ($P \geq 0.05$) by boning method or brine temperature for both silverside and forequarter samples. West (1983) surmised that advantages of moisture retention in hot-boned products may be maintained in the early stages of processing but are often offset in later processing e.g. cooking. Tenderness of silverside was not affected ($P \geq 0.05$) by boning method and only slightly affected ($P \leq 0.05$) by the higher brine temperature, with samples from the latter being rated tougher on day 21. In the case of the forequarter (brisket) beef, the higher brine temperature resulted in higher toughness rating on all days of the shelf-life study, with the effect being more marked ($P \leq 0.001$) on day 14 than day 21 ($P \leq 0.05$). Juiciness was rated as less in samples from the higher brine temperature on day 7 and day 21 ($P \leq 0.05$) for the silverside and on day 7 ($P \leq 0.001$) for the forequarter. Although tenderness was indicated to be higher for hot-boned silverside beef in terms of TPA hardness ($P \leq 0.05$), KSF values ($P \leq 0.01$) and WBSF values ($P \leq 0.001$) than conventional boning, however, this was not supported by sensory analysis data or literature evidence (Jerimiah *et al.* 1985, Bowles-Axe *et al.*, 1983). Hot-boned silverside beef was darker ($P \leq 0.05$) than cold-boned. This agrees with sensory data and the work of authors such as Rees *et al.* (2002b) and Claus *et al.* (1981) who found that hot boning produces a darker coloured meat, attributed to higher pH, increased water holding capacity and a more densely-packed myofibrillar protein matrix.

Combinations of hot boning and higher brine temperatures led to expected higher bacterial numbers. The retail shelf-life of products (estimated as time to reach 10^7 cfu /g) was maintained up until day 21 under retail display conditions (2-4° C) in MA packaging for both silverside and forequarter beef (Table 2).

Table 1: Effect of brine diffusion on nitrite and % salt in cured silverside and forequarter beef.

Treatment	Nitrite		% Salt	
	BF	PP	BF	PP
After tumbling	56.5[a]	67.2[a]	2.1[a]	2.5[a]
After cooking	20.5[b]	14.8[b]	1.7[b]	2.6[a]

[a-b] Means in the same column with different letters are different ($P \leq 0.05$).

Table 2: Effect of shelf-life on microbiology of cured silverside and forequarter beef.

Treatment	Hot-boned		Cold-boned	
	(0 - 4° C)	(12 - 16° C)	(0 - 4° C)	(12 - 16° C)
BF				
TVC D.14	1.9×10^5	1.1×10^6	5.6×10^4	1.4×10^5
TVC D.21	4.7×10^6	1.4×10^6	2.9×10^5	4.1×10^5
PP				
TVC D.14	5.0×10^5	2.1×10^5	3.7×10^4	2.0×10^5
TVC D.21	3.7×10^6	7.0×10^5	1.2×10^5	7.7×10^5

Conclusions

Hot boning did not give the expected increase in yield in the cured products. Although processing directly after boning appeared to prevent the adverse effects of hot boning, such as cold shortening- related toughness, sensory data did not support the instrumental measurements in the silverside product. PP products did not exhibit additional toughness as a result of hot boning. While the use of higher brine temperatures did not compromise retail shelf-life guidelines for bacterial counts, it had detrimental effects on juiciness and tenderness as perceived by taste panellists especially in the product from brisket. Incorporation of additional functional ingredients such as may be required to reduce the apparent loss of moisture of the meat as well as novel heating system in the hope of producing a more rapid curing system.

References

AMSA (1995). Research guidelines for cooking, sensory evaluation and instrumental tenderness measurements of fresh meat. American Meat Science Association, 10-11.

Bostian, M. L., Fish, D. L., Webb, N. B. and Arey, J. J. (1985). Automated methods for determination of fat and moisture in meat and poultry products: collaborative study. Journal of the Association of Official Analytical Chemists 68, 876-880.

Bowles-Axe, J. E., Kastner, C. L., Dikeman, M. E., Hunt, M. C., Kropf, D. H., and Milliken, G. A. (1983). Effects of beef carcass electrical stimulation, hot boning, and ageing on unfrozen and frozen longissimus dorsi and semimembranosus steaks. *Journal of Food Science*. 48, 332-336.

Claus J. R., Kropf, D. H., Hunt, M. C., Kastner, C. L., and Dikeman M. E. (1981). Muscle colour display life as influenced by electrical stimulation and hot boning. *Journal of Animal Science*. 53 (S1), 210.

Irish Farmers Journal Interactive – Irish agricultural publication (2003). 7c per kg extra for beef. http://www.farmersjournal.ie/2003/0301/agribusiness/companycoop/analysis.htm, Accessed 12/12/2005.

Jeremiah, L. E., Martin, A. H., and Murray, A. C. (1985). The effects of various post-mortem treatments on certain physical and sensory properties of three different bovine muscles. *Journal of Meat Science*. 12, 155-176.

Locker R. H. (1960) Degree of muscular contraction as a factor in the tenderness of beef. *Food Research*. 25, 304-311.

Pearson, A. M. and Gillett, T. A. (1996). *In*: Processed Meats (3rd Edition). Chapman and Hall. 53-78.

Ranken, M. D., (2000). Processing principles and Curing. *In*. Handbook of: Meat Product Technology (1st Edition). Blackwell Science Ltd. 25-58.

Rees, M. P., Trout, G. R. and Warner, R. D. (2002b). Tenderness, ageing rate and meat quality of pork m. longissimus dorsi et lumborum after accelerated boning. Part 2. *Journal of Meat Science*. 61 (2), 215-224

Rizuicka, J. and Hansen E. H. (1981). Flow injection analysis. J. Wiley and Sons, New York.

Sweeny, R. A. and Rexroad, P. R. (1987). Comparison of LECO FP-228 "Nitrogen Determinator" with AOAC copper catalyst Kjeldahl method for crude protein. Journal of the Association of Official Analytical Chemists, 70, 1028-1030.

West, R. L. (1983). Functional characteristics of hot-boned meat. *Food Technology*, 57-66.

Acknowledgements

This project is funded by the Irish National Development Plan, under the Food Institutional Research Measure (FIRM) and administered by The Department of Agriculture, Food and Rural Development.

RELATIONSHIP BETWEEN FAT LEVEL AND QUALITY CHARACTERISTICS OF GROUND PORK PATTIES COOKED BY MICROWAVE ENERGY

J.H. Choi[1], J.Y. Jeong[2], Y.S. Choi[1], E.S. Lee[3], J.M. Kim[1], and C.J. Kim*[1]

[1.] Department of Food Science and Biotechnology of Animal Resources, Konkuk University, Seoul 143-701, South Korea. [2.] Animal Resources Research Center, Konkuk University, Seoul 143-701, South Korea. [3.] Department of Applied Microbiology and Food Science, University of Saskatchewan, Saskatoon, Saskatchewan S7N 5A8, Canada.
Email: kimcj@konkuk.ac.kr

Keywords: microwave cooking, fat level, pork patties, cooking properties, cooking loss

Introduction

Microwave cooking has provided several benefits in terms of speed and convenience, but its full potential has not been realised because of disadvantages, such as non-uniformity of heating, edge overheating, soggy texture, and lack of browning (Datta *et al.*, 2005; Ryynanen *et al.*, 2004). Many factors, such as composition, cooking methods, and physical properties influence the product quality in ground meat patties during cooking (Ikediala *et al.*, 1996; Troutt *et al.*, 1992). In particular, it is well documented that the amount of fat plays an important role in the quality of ground meat products. Numerous studies have evaluated the effect of fat levels on the sensory and physicochemical properties of ground meat products (Colmenero, 2000; Hoelscher *et al.*, 1987). Fat levels and cooking methods of patties are the most important factors, influencing the quality of cooked patties (Cannell *et al.*, 1989). Depending on cooking methods, patties are often overcooked, leading to a deterioration in textural quality. Microwave cooking has been shown to produce patties with the least amount of fat and calories compared to other cooking methods and reduced fat retention in beef patties (Berry and Leddy, 1984). Jeong *et al.* (2004) found that as fat levels of microwaved ground pork patties increased, cooking loss increased, while conversely, shear force decreased. It is therefore important to understand the interrelationships between these factors on fat content because of their relationship to the ultimate quality of the cooked patties. Therefore, this study was carried out to elucidate the relationship between different levels of fat and the quality characteristics of ground pork patties cooked to reach 75°C in a microwave oven.

Materials and Methods

Fresh pork hams were purchased from a local market at 48h post mortem. Pork back fat was also collected. Lean materials were initially ground through a 13 mm plate and the pork back fat was ground through an 8 mm plate. The fat was added to the lean meat to create patties with fat levels of 5%, 10%, 15%, 20%, and 25%. The mixtures from each batch were mixed by hand for 3 min and twice ground to through a 3mm plate. The ground mixtures were hand-mixed and then formed into patties (90g each) using sterile 15 × 90mm Petri dishes. Patties were then packaged with Nylon/PE film, frozen, and stored at -25°C until testing.

Samples were cooked in a household-type microwave oven (Model RE-M400, Samsung Electronics Co. Ltd., Suwon, Korea) with full power (700W). Patties from all fat levels were held at 1°C for 24-36h prior to cooking. Each patty was placed in microwave-safe plastic containers, located in the centre of the oven, and cooked until the centre of the patty reached the designated end-point temperature (75°C). The container was rotated inside the microwave chamber during the cooking period. The internal temperature of the cooked samples was obtained by inserting an iron-constantan thermocouple probe into the geometric centre of the patty immediately after removal from the oven. Preliminary time-temperature trials were conducted to determine the length of cooking time needed to reach the designated internal temperature. The heating times necessary for each type of patty were: 5% fat = 103 sec; 10% fat = 100 sec; 15% fat = 96 sec; 20% fat = 93 sec; 25% fat = 86 sec. The oven was turned off before an internal temperature of 75°C was reached, and a short period was allowed for the temperature to rise. After the temperature was achieved, the container including the patty and then the patty alone were immediately weighed to determine cooking loss. Cooking rates were calculated as differences in initial and final internal temperature divided by cooking time. Total cooking loss was determined for twenty patties per fat level by calculating the weight differences of patties before and after cooking. Each patty was cut with a knife into 2.5 cm-wide sections and the sections were sheared in two separate locations with a Warner-Bratzler blade set attached to a texture analyser (TA-XT2*i*, Stable Micro System Ltd., Surrey, UK). Data were analyzed in a one-way analysis of variance through the analysis of variance (ANOVA) procedure of the SAS statistical package (SAS Institute Inc., USA). The procedure CORR of the SAS package was used to calculate correlations between fat level and quality characteristics of ground pork patties cooked by microwave energy.

Results and Discussion

The relationship between fat level and cooking rate of ground pork patties cooked by microwave energy was highly significant ($p<0.001$) and showed a non-linear regression. The correlation coefficient between the two measurements had a high positive value ($r=0.72$). The relationship between fat level and total cooking loss of ground pork patties cooked by microwave energy showed a non-linear regression, and the determination coefficient was $R^2=0.9178$. The correlation coefficient between the two measurements was a very high positive value (*r*=0.89). Total cooking loss increased with increasing fat levels in the patties (p<0.001). As fat level of ground pork patties increased, shear force after cooking decreased. The correlation coefficient between the two measurements was a very high negative value (*r*=-0.86). The relationship between fat level and shear force of ground pork patties cooked by microwave was highly significant (p<0.001) and showed a non-linear regression (R^2=0.954). The cooking rate was positively correlated with total cooking loss (*r*=0.64, p<0.001), but negatively correlated with shear force (*r*=-0.06, p<0.001). Total cooking loss correlated negatively with shear force (*r*=-0.93, p<0.001).

Table 1: Coefficients of correlation among measurements of quality in ground pork patties with different added fat levels cooked by microwave energy.

Measurements	Fat level	Cooking rate	Total cooking loss	Shear force
Fat level	1	0.72^{***}	0.89^{***}	-0.86^{***}
Cooking rate		1	0.64^{***}	-0.60^{***}
Total cooking loss			1	-0.93^{***}
Shear force				1

* p<0.05; ** p<0.01; *** p<0.001.

Conclusions

Fat level was significantly and highly correlated with cooking properties of ground pork patties cooked by microwave. Fat level correlated positively with cooking rate, total cooking loss, and cooking drip loss. On the other hand, fat level correlated negatively with evaporation loss and shear force. To improve the quality characteristics of ground pork patties cooked by microwave energy, more studies of the fat level, and the addition of other ingredients in ground meat patties are necessary.

References

Berry, B.W. and Leddy, K.F. (1984). Effects of fat level and cooking method on sensory and textural properties of ground beef patties. Journal of Food Science, 49, 870-875.

Cannell, L.E., Savell, J.W., Smith, S.B., Cross, H.R. and John, L.C.S.T. (1989). Fatty acid composition and caloric value of ground beef containing low levels of fat. Journal of Food Science, 54, 1163-1168.

Colmenero, F.J. (2000). Relevant factors in strategies for fat reduction in meat products. Trends Food Science Technology, 11, 56-66.

Datta, A.K., Geedipalli, S.S.R., Almeida, M.F. (2005). Microwave combination heating. Food Technology, 59, 1,36-40.

Hoelscher, L.M., Savell, J.W., Harris, J.M., Cross, H.R. and Rhee, K.S. (1987). Effect of initial fat level and cooking method, cholesterol content and caloric value of ground beef patties. Journal of Food Science, 52: 883-885.

Ikediala, J.N., Correia, L.R., Fenton, G.A. and Ben-Abdallah, N. (1996). Finite element modeling of heat transfer in meat patties during single-sided pan-frying. Journal of Food Science, 61: 796-802.

Jeong, J.Y., Lee, E.S., Paik, H.D., Choi, J.H. and Kim, C.J. (2004). Microwave cooking properties of ground pork patties as affected by various fat levels. Journal of Food Science, 69, C708-C712.

Ryynanen, S., Risman, P.O. and Ohlsson, T. (2004) Hamburger composition and microwave heating uniformity. Journal of Food Science, 69, M187-M196.

SAS Institute (1999). SAS User's Guide. Statistical Analysis Systems Institute, Cary, NC, USA.

Troutt, E.S., Hunt, M.C., Johnson, D.E., Claus, J.T., Kostner, C.L., Kropt, D.H., Stroda, S. (1992). Chemical, physical and sensory characterization of ground beef containing 5 to 30 percent fat. Journal of Food Science, 57, 25-29.

COLOUR STABILITY OF BEEF STORED UNDER VACUUM PACKAGING AND ITS RELATIONSHIP TO CARCASS FATNESS

K. Insausti, G. Indurain, M.V. Goñi, V. Sarriés, B. Hernández, and M.J. Beriain*

Escuela Técnica Superior de Ingenieros Agrónomos, Universidad Pública de Navarra, Campus Arrosadía sn, 31006 Pamplona, Spain. Email: mjberiain@unavarra.es

Keywords: beef, colour stability, ageing, display

Introduction

Beef colour and its relationship with beef palatability are important quality factors for the industry (Faustman *et al.*, 1992). Beef colour depends on the muscle structure and the concentration and chemical state of myoglobin in the muscle surface which varies with time and packing conditions (Renerre and Labadie, 1993). Myoglobin can be present in the muscle surface as reduced myoglobin (deoxymyoglobin, Mb), oxygenated myoglobin (oxymyogobin, MbO_2) and oxidised myoglobin (metmyoglobin, MMb) (Rosset and Roussel-Ciquard, 1978). Half of consumers would refuse beef when MMb percentage in the muscle surface achieves 20% (Renerre and Mazuel, 1985). A brown colour is evident with 50% of MMb and this beef is unacceptable in all cases (Van den Oord and Wesdorp, 1971). Vacuum packaging allows increasing time of storage of beef (Jeremiah and Gibson, 2001) and it can be associated with an improvement of the organoleptic quality of beef and a higher quality assurance. However, chemical reactions, oxidation in particular, are responsible for changes in colour and often determine the practical storage life of meat products. Besides, it has also been shown that shelf-life of retail cuts on display decreases when stored for long periods (Greer and Jones, 1991). Fatness parameters, related to lipid oxidation, have been related to meat pigment oxidation (Mercier *et al.*, 1995; Insausti *et al.*, 1999).

The objectives of the current research were to study beef colour parameters after ageing under vacuum packaging and display and its relationship with carcass colour and fatness.

Materials and Methods

Sixty yearling bulls were slaughtered and carcasses were classified according to the European classification scales for conformation and fatness. Conformation and fatness scales were subdivided in 15 points. After 24 h *post-mortem,* carcasses were sliced at the 6th rib and colour coordinates were measured on *longissimus dorsi* muscle (LM) after the muscle was oxygenated at 4°C for 60 minutes (LM60). Colour measurements were taken using a spectrocolorimeter (Minolta CM2002, Japan) (λ: 400-700 nm; $\Delta\lambda$: 10 nm; D65; 10°). The representation scheme was based on the CIE (1976) L*a*b* colour coordinates, hue (H*) and chroma (c*). Five readings were taken for each sample at different locations on the muscle. Fat thickness was measured and marbling was measured on the sliced surface of LM by computer image analysis software (Optimas 6.5, Media Cybernetics Inc., USA) (Table 1). For each carcass a steak (1.5 cm thick) was collected at the 8th rib, vacuum packaged and stored 7 days (4 C°). After storage, samples were removed from the vacuum bags and placed in plastic foam trays and overwrapped with a gas permeable film. Colour coordinates and reflectance between 400 and 700 nm were measured with the spectrocolorimeter after 1 h blooming (7days + 60min) and 48 h display (7 + 2 days) at 4°C. Mb, MbO_2 and MMb percentages were calculated (Mancini et al., 1991). A t-student test was used to compare the pigment oxidation state of beef after display. Pearson correlation coefficients between %MMb of beef aged 7 days and carcass conformation and fatness, marbling, fat thickness, and early colour measurements were calculated.

Results and Discussion

The 48 h display led to the increase in MMb percentage, in beef aged 7 days, and a reduction in percentage of the other two chemical states of myoglobin (Table 2). The accumulation of MMb on the meat surface after display could be related to a decrease in reducing activity in meat during storage (Faustman and Cassens, 1990). Beef aged 7 days did not show MMb percentages higher than 20% after 1 h blooming. However, after 48h display, 29% of samples showed %MMb higher than 20% and 5% of samples higher than 40%. As lipid oxidation is a promoter of myoglobin oxidation during aerobic storage, (Mercier *et al.*, 1995; Insausti *et al.*, 1999), in the present study, the correlation between marbling and %MMb was studied (Table 3). Beef colour stability could be explained by the differences in intramuscular fat content, but this difference is not reflected by the current carcass grading system. In an attempt to predict colour stability by early measurements on-line, colour coordinates recorded on the LM after cutting the carcasses and 1 h blooming were correlated to %MMb after 7 days of ageing. Beef samples with higher %MMb showed higher a*, b* and C* but lower L* coordinate values. Haem pigment is related to a* coordinate, so that beef with higher pigment content showed lower colour stability after ageing 7 days and less acceptable colour scores (Renerre et al., 1996).

Table 1: Carcass characteristics for conformation, fatness and *longissimus dorsi* colour After 24 h *post-mortem,* and oxygenation at 4°C for 60 minutes. Mean and standard deviation (SD).

	Mean	SD
Conformation score	10.90 (U-)	0.77
Fatness score	4.97 (2-)	1.16
Marbling	0.03	0.01
Fat thickness	0.32	0.17
L* LM60	38.39	3.54
a* LM60	14.78	2.12
b* LM60	10.23	2.99
C* LM60	18.09	3.04
H* LM60	34.00	7.08

Table 2: Mean and standard error of myoglobin states of *longissimus dorsi* muscle from yearling bulls aged 7 days plus 1 hour blooming or 48 hours display.

	7 days + 60min	7 days + 2 days display	
	Mean ± SE	Mean ±SE	
%Mb	5.98 ± 1.14	2.52 ± 1.08	***
%MMb	2.33 ± 1.40	14.37 ± 1.33	***
%MbO_2	97.45 ± 1.35	88.47 ± 1.28	***

Table 3: Correlation coefficients between %MMb of beef aged 7 days and blooming 1 hour or display 2 days, with carcass conformation and fatness, marbling, fat thickness and early colour measures at *longissimus dorsi* muscle.

	7 days + 60min	7 days + 2 days display
Conformation score	0.09	-0.08
Fatness score	0.02	0.12
marbling	0.31(*)	0.34(*)
Fat thickness	-0.04	0.00
L*LM 60'	-0.31(*)	-0.23
a*LM60'	0.30(*)	0.29(*)
b*LM60'	0.34(*)	0.43(**)
C*LM60'	0.35(*)	0.38(**)
H *LM60'	0.22	0.40(**)

Conclusions

It can be stated that after 48h display a high percentage of beef samples aged 7 days under vacuum packaging might be rejected by consumers. There was a relationship between colour stability and fatness, but not with carcass fatness score. Besides, there was observed a relationship between early instrumental colour measurements on the carcass (*longissimus dorsi* muscles) and colour stability of beef after 7 days of ageing.

References

Faustman, C. and Cassens, R. G. (1990). Journal of Muscle Food, 55, 1278-1279.

Faustman, C., Specht, S. M., Malkus, L. A. and Kinsman, D. M. (1992). Meat Science, 31, 351-359.

Greer, G. G. and Jones, S, D. M. (1991). Canadian Institute Food Science and Technology Journal, 24, 181-185.

Insausti, K., Hernández, B., Alberti, P., Gorraiz, C., Alzueta, M. J., Beriain, M. J. and Purroy, A. (1999). Proceedings of the 1st International Congress on Pigments in Food Technology (289-293), Seville, Spain.

Mancini, R. A., Hunt, M. C. and Kropf, D. H. (2003). Meat Science, 64, 157-162

Jeremiah, L. E. and Gibson, L. L. (2001). Food Research International, 34, 815-826.

Renerre, M., Dumont, F. and Gatallier, P. (1996). Meat Science, 43, 111-121.

Renerre, M. and Labadie, J. (1993). In: Proceedings of the 39th ICOMST (pp. 361-387), Calgary, Canada.

Renerre, M., Mazuel, (1985). Sciences des Aliments, 5, 541-557.

Rosset, R. and Roussel-Ciquard, N. (1978). Industries Alimentaraires et Agricoles, 357-369.

Van den Oord, A. H. and Wesdorp, J. J. (2001). Journal of Food Technology, 6, 1-8.

SESSION 4: CONSUMER TOPICS

CRITICAL POINTS IN THE TRANSPORT OF LAMBS TO SLAUGHTER IN SPAIN THAT MAY COMPROMISE THE ANIMALS' WELFARE

J. Escos, G.A. María, J. López, S. Alierta, S. García-Belenguer and G. Liste*

Departamento de Producción Animal y Ciencia de Alimentos. [1] Departamento de Patología Animal. Facultad de Veterinaria, Universidad de Zaragoza. Miguel Servet 177, 50013. Zaragoza, Spain. Email: mgliste@unizar.es

Keywords: transport, lamb, critical points, animal welfare

Introduction

The meat production chain ends with the transport and slaughter of the animals in the slaughterhouse. This very important stage of the system can put at risk all of the work carried out for months by the cattle dealers. It is advisable that the sector pays attention to ending this process well, offering a good image to the consumer and assuring benefits to the sector. There is little information available on transport problems in the countries of the Mediterranean area. Project OVITRANS analyses the effect of transport and a delay to the slaughter, on the well-being of the animals and the quality of the meat. In the present communication we present the results of module 1 of the project the objective of which is to describe the transport situation of ovines in relation to prior handling up to transport, the transport itself, the delay in the slaughterhouse and the slaughter, with the purpose of identifying the strategically important points of the transport process that could affect the well-being of the animal and the quality of the meat.

Materials and Methods

The procedures for loading, transporting and unloading lambs, and slaughterhouse practices in Spain were surveyed from November to March 2006, and involved 60 farms, 30 truck drivers, 20 slaughterhouses and 11 centres of classification that receive lambs from more than 1000 livestock farms. Questionnaires were completed either face-to-face, by telephone or by mail. The original questionnaire is available at http://wzar.unizar.es/catra. The same questionnaire was used for each interviewee and farmers were asked to provide specific information about their loading and handling facilities, in addition to the farm's general housing and husbandry practices.

Data management was performed with ACCESS and descriptive statistics were calculated for each of the relevant questions. All the averages presented are mean ± standard deviation (number of data).

Results and Discussion

Farms

We analysed surveys from 60 farms growing Rasa Aragonesa. Before the animals are transported they are separated in 32 out of 39 cases (82.05%) for 52.82 ± 79.64 (34) hours. The animals are led in groups (88.89%). Preloading pens are present in only 14 out of 37 cases (37.84%). Feedlots are mixed 13 out of 36 times (36.11%). The animals are driven using humane aids together with ramps and hydraulic elevators. Loading angles range from 15° to 20°. Climatic and environmental control is not present (94.74%). Veterinary control is generally not required (97.44%). Average unloading time is 16.36±7.84 (39) minutes.

The travelling distance from the farms to the slaughterhouse ranges from 190 to 2 km distance.

Before transportation, the animals are separated from their original pen for an average of 21.86±33.67 (7) hours before they were transported in 90.91% of the cases. The animals are normally fed before transportation. The loading is performed by both transporters and farmers. Only in 27.27 % of the cases the farmer is the legally responsible person present, while in most cases (72.73%), it is both farmer and transporters. Seven out of nine (77.78%) of the feedlots had preloading pens.

Unloading at the farm was typically short and similar to the loading. No use of sticks or electric elements was recorded in any case. Normally, both the driver and farmer employees unloaded the truck, but the driver unloaded by himself at 18.18% of the feedlots, and the remainder were unloaded by both. The legal person responsible for unloading is eight out of eleven, and only 2 out of eleven cases it was the transporters. The animals were unloaded in groups.

The average number of loading animals was 319 with 55 animals per compartment and an average density of 0.36 $m^2 sheep^{-1}$.

Transport

Transport of livestock in Spain is performed by drivers with averaged experience of 20.71 ± 10.60 (14) years. Of this group, 65.38% are the owners and 19.23 % are employees. From twenty seven surveys analysed, the drivers had special permission in 81.48% of the cases, 33.33 % passed special examinations and 48.15 % received training.

Most of the drivers interviewed owned small lorries with two to three axles (64.52%). Most of the trucks are rigid (58.06%), 25.81% are trailers, and only 16.13% are trucks with trailers. Most of them had pneumatic suspension (70.97%). Truck transports a single species per trip (93.55%), but they can transport different species in different trips (63.33%).

The truck box is always separated from the truck (100%) and may be open (36%) or closed (48%). The average height of the box-tack is 3.48 ± 0.66 (22) meter. Box aeration can be either natural (65.38%), acclimatised (7.69 %), forced (11.54 %). They present drinking facilities in 46.15% of the cases. Average driving time is 7.82 ± 2.67 (20) hours. The number of compartments per levels varies between 2 (13.79%) to 12 (6.90%) evenly distributed among those values.

Slaughter

The mean number of animals slaughtered each day at each slaughterhouse was 659.74 ± 909.49 (20); 3032.37 ± 4544.92 (19) per week. Most of the lambs are slaughtered are Lechal and Ternasco. Slaughters are performed evenly in spring-summer (30.43%) and in autumn-winter (43.48%). During unloading in the slaughterhouse the animals were kept in groups (77.27%) unmixed (91.93%).

Most of the slaughterhouses had a waiting place after unloading (94.74%), covered with a roof (95.65%) and climatic control (55%). The average density of lambs per square meter was 2.57 ± 0.53 (7). The animals waiting before unloading varied between 19.13 ± 18.96 (15) and 4.42 ± 8.64 (12) minutes. Average time of unloading oscillated between 22 ± 18.69 (15) and 5 ± 4.71 (14) minutes.

CORRELATION BETWEEN SENSORY AND PHYSICO-CHEMICAL PARAMETERS USED IN EVALUATION OF SUCKLING LAMB MEAT QUALITY

I. Revilla* and A.M. Vivar-Quintana

*Area de Tecnología de Alimentos. Universidad de Salamanca, E.P.S de Zamora, Avda Requejo 33, 49022 Zamora, Spain. *Email: irevilla@usal.es*

Keywords: instrumental measurement, taste panel, meat quality, colour, texture

Introduction

Meat is a complex product and its quality evaluation is not always easy. Many studies on meat and meat products have been designed to develop instrumental methods to replace the sensory tests, mainly because sensory methodologies have been seen as subjective and expensive (Risvik, 1994). However, although instrumental methods have the advantages of being objective and cheaper, they only provide partial explanations of the complex set of interactions that occur when cooked meat is consumed. Karlsson (1992) proposed that when a large number of measures were used to assess meat quality and they were correlated, they could be replaced by fewer measures without a significant loss of information. This technique has been already used to assess relationships between carcass characteristics (Hernández *et al.*, 2000, Sarti *et al.*, 1992), between meat characteristics (Karlson, 1992) and between carcass and meat characteristics (Okeudo and Moss, 2005) however comprehensive reports on correlations between sensory and physico-chemical parameters in sheep are relatively scarce (Cañeque *et al.*, 2004) and the sensory parameters can differ from one panel test to the other. Such information is important because it provides a broad picture on the probable effects of a change in one quality factor on the vast range of other quality factors, including consumer acceptance. The aim of this work was to study the relationship between several meat quality traits (physico-chemical measurements) and sensory parameters used to asses the quality of suckling lambs.

Materials and Methods

Nineteen suckling lambs (animals that were raised exclusively on maternal milk from birth to slaughter at one month of age), from three breeds: 30 Churra lambs, 30 Castellana lambs and 30 Assaf lambs were slaughtered according to standard commercial procedures in Magnus abattoir (Arcenillas, Zamora, Spain). At 72h post-mortem, carcasses were cut into two halves and the *M. Longissimus dorsi* from the left half of carcasses was taken to the laboratory. pH measurements were recorded up to 24hrs and the rest of physico-chemical parameters up to 72 hours post-mortem. Meat colour was measured on fat free surface of the *M. Longissimus thoracis* (between 8th and 9th ribs) after 1 h blooming at 4ºC and fat colour was determined on fat-cover of the left loin. Lightness (L*), redness (a*) and yellowness (b*) were recorded with a D65 illuminant at a 10ºC standard observer in the CIELab space. Other physico-chemical parameters included fat, moisture (IR method, Lurueña *et al.*, 2004) and expressible juice (Grau and Ham, 1953). Warner-Bratzler shear force was determined on *M. Longissimus thoracis* (9th-12th rib level). The muscle was wrapped in aluminium foil and cooked on a pre-heated double hot plat grill at 200ºC until the internal temperature reached 70ºC. Then six rectangular parallelepipeds, 1 x 1 cm-thick and 2-3 cm long, were cut parallel to the muscle fibres using a TX-T2i*plus* (Stable Micro Systems, Surrey, England).

Sensory analysis was developed by a fifteen member trained panel using a quantitative descriptive analysis on 2 cm-thick slices of *M. Longissimus lumborum* wrapped in aluminium foil and cooked on a pre-heated double hot plat grill at 200ºC until the internal temperature reached 70ºC. Then, they were cut into portions, each one uniform and free from obvious connective tissue and fat streaks. The sub-samples were immediately wrapped in aluminium foil, codified and kept at 60ºC. Meat was served following a randomised design for order and carry over effects. Each attribute was rated on a nine-point structured scale with score 1 equivalent to the lowest intensity and score 9 to the highest intensity of the attribute.

Statistical analysis was performed using the Statgraphic Plus program for Windows (Manugistics Inc. 1995). Outliers were detected by means of Box and Wisker plot afterwards, multivariate analysis was used to construct a linear correlation matrix and significance of the correlation coefficient (r) was established using the *t* test.

Results and Discussion

Results (Table 1) showed there were several significant correlations between sensory and instrumental parameters differing from other studies where few significant correlations were found (Cañeque *et al.*, 2004). Regarding colour parameters there was significant negative correlation between brown (external meat colour after cooking) and L* but this correlation was positive with yellow (b*) showing that raw darker meats (lower value of L*) were also perceived as darker by the panellists. This parameter was also correlated with fat content, because the higher the slaughtering weight of lambs was, the higher the fat content was and this fat have more intense yellow colour (b* value). It was possible to observe a positive correlation between pink (internal meat colour after cooking) and L* (both of meat and fat) and this sensory parameters was positively correlated with the red colour of meat. This indicated that pale pink raw meat kept this colour after cooking inside the meat.

Regarding texture parameters, hardness did not showed significant correlation with WBSF revealing this instrumental method was not a good predictor for sensory hardness. On the other hand, significant correlation between red colour and hardness was observed and also a negative correlation between moisture and hardness. This negative correlation was also observed for fibrous sensation revealing that a low moisture content produce that muscular fibres were less hydrated and were more difficult to cut (hardness) and to chew (fibrous sensation).

The juiciness was positively correlated with intramuscular fat but not with moisture or expressible juice. These results pointed out that fat content is more important than moisture for juiciness because it produces a sustained juiciness in the mouth. A positive correlation between % fat and a* have been previously observed (Okeudo and Moss, 2005) that may explain the positive correlation observed between juiciness and a*. Expressible juice was significant correlated with springiness and fat because the higher de water holding capacity, the fibres were more turgent and recovered easily the initial form.

Finally regarding flavour and odour determinations, the most important correlation founded was the positive correlation between odour intensity and fat percentage, because fatty acids are important precursors of aroma compounds in meat and meat products. This correlation was not found for taste intensity, revealing these two parameters were independents.

Table 1: Correlation coefficients between sensory parameters and meat quality parameters.

	Brown	Pink	Hardness	Juiciness	Fibrous	Spring	Taste int.	Odour int.	Liver odour
WBSF								-0.250	
L^*_{meat}	-0.392**	0.263	0.259						
a^*_{meat}		0.401**	0.208	0.363*		0.286*			
b^*_{meat}	0.221								
L^*_{fat}		0.490**							
a^*_{fat}		-0.328**	-0.282*				-0.246	0.221	-0.291*
b^*_{fat}	-0.200		-0.213			0.249			
Expr. juice						0.280*			
Moisture%			-0.272*		-0.384				
Fat%	0.297*			0.200		0.204		0.345**	
pH									

* Significant correlation at α=0.1 level; ** Significant correlation at α=0.05 level.

Conclusions

Results revealed that an increase in moisture and fat produced that suckling lamb meat was more tender and juicy, with higher odour intensity and lower fibrous sensation. However, pH was not correlated with any sensory parameter and WBSF opposite as expected did not showed correlation with hardness. Finally, CIELab parameters were shown to be good predictors of sensory colour.

Acknowledgments: Authors are grateful for financial support from the Asovino Cooperative.

References

Cañeque, V., Pérez, C., Velasco, S., Díaz, M.T., Lauzurica, S., Alavarez, I., Ruiz de Huidobro, F., Onega, E. and De la Fuente, J. (2004). Carcass and meat quality of light lambs using principal component analysis. Meat Science, 67: 595-605.

Grau, R. and Hamm, R. (1953). Eine einfache Methode zur bestimmung der Wasserbindung in Muskel. Naturwissenshaften, 40: 29-30.

Hernández, P., Pla, M., Oliver, M.A. and Blasco, A. (2000). Relationships between meat quality measurements in rabbits fed with three diets of different fat types and content. Meat Science, 55, 379-384.

Karlsson, A. (1992). The use of principal component analysis (PCA) for evaluation results from pig meat quality measurements. Meat Science, 31, 423-433.

Lurueña-Martínez, M.A., Vivar-Quintana, A.M. and Revilla, I. (2004). Effect of locust bean/xanthan gum addition and replacement of pork fat with olive oil on the quality characteristics of low-fat frankfurters. Meat Science, 68: 383-389.

Okeudo, N.J. and Moss, B.W. (2005). Interrelationships amongst carcass and meat quality characteristics of sheep. Meat Science, 69: 1-8.

Risvik, E. (1994). Sensory properties and preferences. Meat Science, 36: 67-77.

Sarti, D., Morbidino, L., Panella, F., Cavalletti., C. and Lavvaf, A. (1992). Qualitative characteristics of the carcass and meat of Italian lamb. In: Proceedings of the 43rd Reunión Anual de Federación Europea de Zootecnia, Madrid, Spain.

INTER AND INTRAMUSCULAR VARIATION ON SENSORY CHARACTERISTICS OF BEEF STEAKS

S. Velasco*, M.T. Silva, F. Ruiz de Huidobro and E. Miguel

IMIDRA. Departamento de Agroalimentación, Comunidad de Madrid, 28800 Alcalá de Henares, Madrid, Spain. Email: susana.velasco@madrid.org

Keywords: beef, texture, sensory analysis, collagen, *longissimus dorsi*

Introduction

Meat tenderness is the most important quality trait for consumer acceptance and the variability in this concept has become an increasing problem for the retailers. Tenderness is extremely variable between carcasses, between muscles and even within the same muscle, and in part can be due to total collagen content and its solubility, influencing the so-called "background" toughness of meat (Denoyelle and Lebihan, 2003, Torrescano *et al.*, 2003). Sensory analysis in meat determines the quality attributes that consumers are going to value. In this sense, Shackelford *et al.*, (1995) compared sensory and instrumental values of hardness in ten different bovine muscles and found that sensory values in the loin were very variable. The objective of this study was to observe inter and intramuscular variability on sensory characteristics and quality parameters of three muscles of different regions of the beef carcass included in the label from the PGI "Carne de la Sierra de Guadarrama".

Materials and Methods

Fifteen male yearlings from Limousin, Charollais or Brown Swiss bulls, and Avileña-Negra Ibérica dams (a bovine breed from the Central area of Spain) were studied. All animals were the quality label from the PGI "Carne de la Sierra de Guadarrama" (a Protected Geographical Indication of Madrid). Animals employed in this study were entire males 13-15 months old, 391 kg HCW and a carcass classification value of R3 (European Union, 1991). They were slaughtered in a commercial abattoir and processed according to the rules. M. *longissimus dorsi* (thoracic portion) (LD), M. *semitendinosus* (ST) and M. *supraspinatus* (SS) were removed at 7 days of ageing postmortem and cut in steaks (2 cm thick) of three homogeneous regions (1, 2 and 3). The regions in LD were cranial, middle and caudal, while that in ST and SS muscles were dorsal, middle and ventral regions, respectively. Attributes measured included pH, water holding capacity, collagen composition, and intramuscular fat content. Also, a compression textural measurement by TPA (texture profile analysis) method was studied (Velasco *et al.*, 2005). One cm x 1 cm-strips were made from each steak for texture assessment on raw meat. The texture analysis was made with a cylindrical 10 mm-diameter probe of ebonite and with a deformation force of 75%. Sensory analysis was performed on samples from three muscles from each of fifteen animals, grill-cooked till 80°C (internal temperature). Samples were randomly served to a trained twelve member sensory panel. Sensory analysis evaluated hardness, springiness and juiciness, completed with the parameters of fat sensation (in mouth), flavour intensity, number of chewings until swallowing and overall analysis. Scales used to assess intensity of every parameter were interval scales of 10 cm long. Statistical analyses were carried out using the analysis of variance of the GLM procedure of the Statgraphics Plus (1994). Differences between means were compared using Duncan test.

Results and Discussion

Effects of different muscles and regions inside these muscles, on pH, water-holding capacity, and intramuscular fat and collagen contents are shown in Table 1.

Table 1: Arithmetic means and mean squares of the error of pH, WHC, intramuscular fat and collagen parameters of muscles and regions studied.

	Muscles (M)				Regions (R)					
	LD	ST	SS	Sig	1	2	3	Sig	MxR	MSE
pH	5.53^a	5.54^a	5.65^b	*	5.59	5.54	5.59	NS	NS	0.015
WHC (%)	15.59^a	18.21^b	18.90^b	*	18.51	17.24	16.95	NS	NS	9.16
Intramuscular fat (%)	2.04^{ab}	1.42^a	2.50^b	**	2.07	2.00	1.90	NS	NS	0.82
TC (mg/g muscle)	6.04^a	11.54^b	10.39^b	***	9.19	9.49	9.28	NS	NS	5.48
IC (mg/g muscle)	4.77^a	9.11^b	7.53^b	***	7.17	7.03	7.21	NS	NS	3.60
Solubility (% total)	13.07	14.84	12.90	NS	13.67	12.79	14.35	NS	NS	23.90

Means in the same row not followed by a common letter differ significantly. Sig: Significance; NS (non significance), * ($p<0.05$), ** ($p<0.01$), *** ($p<0.001$). MSE: mean square of the error. LD= M. *longissimus dorsi*, ST= M. *semitendinosus*, SS= M. *supraspinatus*; Regions: 1 (cranial in LD, and dorsal in ST and SS), 2 (middle), 3 (caudal in LD, and ventral in ST and SS). WHC (water-holding capacity, expressed as percentage of expelled liquid), TC (total collagen), IC (insoluble collagen).

Significant intermuscular variation was detected at 7 days postmortem, with higher pH value in SS than in LD and ST muscles. Intramuscular fat was higher in LD and SS than in ST, while a higher quantity of exudate was seen in ST and SS than in LD muscle. Thus it could be due to the fact that the greater the pH the greater the WHC (Purchas, 1990). Total and insoluble collagen contents were higher in ST and SS than in LD as Torrescano *et al.*, (2003) noted. No difference within regions in the three muscles studied was observed for these parameters.

Table 2: Arithmetic means and mean squares of the error of textural and sensorial parameters of muscles and regions studied.

	Muscles (M)				Regions (R)					
	LD	ST	SS	Sig	1	2	3	Sig	MxR	MSE
TPA Hardness (N)	206.33^{a}	369.96^{b}	302.72^{b}	***	262.19^{a}	251.61^{a}	365.20^{b}	**	NS	8476
TPA Springiness (g)	74.60^{a}	78.38^{b}	76.80^{ab}	*	74.63	77.14	78.01	NS	NS	20.55
TPA Chewiness	4211.9^{a}	11966.6^{b}	8527.5^{c}	***	6879.2^{a}	7044.2^{a}	10782.7^{b}	*	NS	1.5 10^{7}
Hardness	4.47^{a}	5.08^{b}	5.06^{b}	*	4.92	4.66	5.03	NS	NS	0.21
Springiness	4.42	4.50	4.88	NS	4.80	4.42	4.59	NS	NS	0.22
Juiciness	2.90^{a}	1.97^{b}	2.42^{ab}	*	2.40	2.37	2.53	NS	NS	0.52
Fat sensation	2.56	2.06	2.35	NS	2.23	2.34	2.41	NS	NS	0.24
Flavour intensity	4.46	3.95	4.17	NS	4.23	4.22	4.13	NS	NS	0.24
Chewings (number)	24.24^{a}	27.50^{b}	27.19^{b}	*	25.52^{a}	24.37^{a}	29.04^{b}	**	NS	6.23
Overall	4.62^{a}	3.58^{b}	4.42^{a}	**	4.28^{a}	4.20^{a}	4.15^{b}	*	NS	0.32

Means in the same row not followed by a common letter differ significantly. Sig: Significance; NS (non significance), * ($p<0.05$), ** ($p<0.01$), *** ($p<0.001$). MSE: mean square of the error. LD= M. *longissimus dorsi*, ST= M. *semitendinosus*, SS= M. *supraspinatus*; Regions: 1 (cranial in LD, and dorsal in ST and SS), 2 (middle), 3 (caudal in LD, and ventral in ST and SS).

Effects of different muscles and regions inside these muscles on textural and sensorial parameters are displayed in Table 2. According to compression measurements by TPA method, each muscle presented a same gradient of hardness and chewiness. The region 3 showed a higher hardness ($p<0.01$) and chewiness ($p<0.05$), although only for SS and ST muscles (ventral region), while the LD muscle did not vary within the region in this study. These differences corresponded with the most number of chewings until swallowing ($p<0.01$) and a less overall ($p<0.05$) by sensorial analysis. There were no differences in sensorial hardness between the three muscles studied, similar to that found by Searls *et al.*, (2005) who did not observe tenderness differences among steak locations in four beef muscles.
The results based on sensorial analysis showed a significant difference between the muscles studied. LD was more tender ($p<0.01$) and with less springiness ($p<0.05$) than ST and SS muscles. Also LD showed less chewiness by TPA ($p<0.001$) than ST and SS muscles. Low compression rates, in raw meat, are clearly influenced by ageing in the development of meat tenderness and high compression rates demonstrate the influence of connective tissue composition (Campo *et al.*, 2000). ST muscle displayed a smaller juiciness ($p<0.01$) and overall ($p<0.01$) than LD and SS muscles, principally due to a smaller percentage of intramuscular fat and a higher content of total collagen. Springiness, fat sensation and flavour intensity were no different between muscles studied.

Conclusions

Cranial region in LD and dorsal in ST and SS showed the most overall. Between muscles we observed that ST was sensory depreciated principally due to its lower fat content, higher total collagen value and greater hardness and chewiness by TPA method than others, which was a good indicator of sensory hardness.

References

Campo M.M., Sañudo C., Panea B., Alberti P. and Santolaria P. (1999). Breed type and ageing time effects on sensory characteristics of beef strip loin steaks. Meat Science, 51: 383-390.

Denoyelle C. and Lebihan E. (2003). Intramuscular variation in beef tenderness. Meat Science, 66: 241-247.

Purchas R.W. (1990). An assessment of the role of pH differences in determining the relative tenderness of meat from bulls and steers. Meat Science, 27: 120-140.

Searls G.A., Maddock R.J. and Wulf D.M. (2005). Intramuscular tenderness variation within four muscles of the beef chuck. Journal of Animal Science, 83: 2835-2842.

Shackelford S.D., Wheeler T.L. and Koohmaraie M. (1997). Relationship between shear force and trained sensory panel tenderness ratings of 10 major muscles from *Bos indicus* and *Bos taurus* cattle. Journal of Animal Science, 73: 3333-3340.

Torrescano G., Sánchez-Escalante A., Jiménez B., Roncalés P. and Beltrán J.A. (2003). Shear values of raw samples of 14 bovine muscles and their relation to muscle collagen characteristics. Meat Science, 64: 85-91.

Velasco S., Huidobro F., Miguel E. and Blázquez B. (2005). Prediction of sensory analysis of bovine meat from two methods of instrumental texture. EFFoST 2005. Congress Proceedings. Innovation in Traditional Foods, vol I, (Ed. P. Fito and F. Toldrá), pp: 169-172., Valencia, Spain.

SURVEY OF US RETAIL BEEF SHEAR FORCE

K.L. Voges*[1], C.L. Adams[1], J.C. Brooks[2], R.J. Delmore[3], D.B. Griffin[1], D.S. Hale[1], W.R. Henning[4], D.D. Johnson[5], C.L. Lorenzen[6], R.J. Maddock[7], R.K. Miller[1], J.B. Morgan[8] and J.W. Savell[1]

[1.] Dept. of Anim. Sci., Texas A&M University, TAMU 2471, College Station, TX, USA 77843 [2.] Dept. of Anim. Sci., Texas Tech University, PO Box 42141, Lubbock, TX, USA 79409,[3.] Dept. of Anim. Sci., California Poly Technical University, Anim. Sci. Bldg 24, San Luis Obispo, CA, USA 93407, [4.] Dept. of Anim. Sci., Pennsylvania State University, 0307 AgSci and Ind Bldg, University Park, PA, USA 16802, [5.] Dept. of Animal Science, University of Florida, PO Box 110910, Gainesville, FL, USA 32611, [6.] Dept. of Food Science, University of Missouri, 256 William C. Stringer Wing, Columbia, MO, USA 65211, [7.] Dept. of Animal Science, South Dakota State University, 120 AnSci Complex, Brookings, SD, USA 57007, [8.] Dept. of Animal Science, Oklahoma State University, 10413 Animal Science, Stillwater, OK, USA 74078 Email: kristinvoges@tamu.edu

Keywords: beef, tenderness, sensory, consumer panel, market survey

Introduction

The first National Beef Tenderness Survey (Morgan *et al.*, 1991) determined beef tenderness of beef in US retail cases. Neely *et al.*, (1999) showed that tenderness is a major and contributing factor to consumers' perception of taste. With tenderness problems found in beef and the elapsed time since that survey, it became clear that it was important to update the information. Morgan *et al.*, (1991) focused solely on the retail sector; however, with the increasing prevalence of foodservice, the 1998 Tenderness Survey included a foodservice portion (Brooks *et al.*, 2000). The 1998 Tenderness Survey found that retail cuts from the round still required more attention in processing and preparation to ensure acceptable tenderness, while chuck cuts improved in tenderness. Providing a benchmark for beef palatability allows the industry to identify where improvements have been made and where tenderness issues may still exist. Thus, a new survey was conducted to determine the tenderness of beef from the retail case based on Warner-Bratzler shear (WBS) force.

Materials and Methods

Collaborators sampled retail stores in eleven U. S. cities: Seattle, WA; Los Angeles, CA; San Francisco, CA; Denver, CO; Houston, TX; Chicago, IL; Kansas City, MO; Atlanta, GA; Tampa, FL; Philadelphia, PA; and New York City, NY. Retail cut types were taken and shipped via overnight delivery in insulated containers containing commercial ice packs and processed under refrigerated conditions (2 to 4°C). Each steak was individually identified, individually vacuum packaged, and frozen (-23°C). Steaks were thawed in a 4°C cooler for 48 h before cooking. Grated, non-stick electric grills were preheated for 15 min to a temperature of 177°C. During cooking, all steaks were turned after reaching 35°C and cooked to a final internal temperature of 70°C. Cooked steaks were removed from the grill and allowed to cool for approx. 4 hours. At least six cores (1.3 cm in diameter) were removed parallel to the muscle fibre orientation and sheared once, perpendicular to the muscle fibres, on a Warner-Bratzler shear machine (United 5STM-500, Huntington Beach, CA) using an 11.3 kg load cell. Means were separated using the p-diff option of PROC GLM of SAS. The percentages of steaks stratified into tenderness categories were analyzed using PROC FREQ of SAS.

Results and Discussion

The WBS values for retail cuts are presented in Table 1. Clod steaks had lower WBS values than bottom and eye of round steaks ($P < 0.05$). Brooks *et al.* (2000) found WBS values for the clod, bottom round, and eye of round to be 3.01, 5.09, 4.19 kg, respectively. Morgan *et al.* (1991) reported WBS values for the clod, bottom round, and eye of round to be 4.01, 4.38, and 4.67 kg, respectively. However, these workers braised the steaks to an internal temperature of 85°C, compared to 70°C in our study and the 1998 tenderness study. The bone-in strip steak, bone-in ribeye steak, and t-bone steak had the lowest WBS values found in our study. Brooks *et al.* (2000) reported the t-bone and porterhouse steaks to have the lowest WBS values.

Tenderness categories developed by Belew *et al.*, (2003) and Shackelford *et al.* (1991) are based on WBS values and were used to determine percentages of retail cuts that fell into each group (Table 2). Top round, bottom round, and eye round steaks were the only cuts shown to have WBS values over 4.6 kg. These percentages are much lower than those found by Brooks *et al.* (2000) and Morgan *et al.*, (1991). Our study had lower percentages for all cuts exceeding 3.9 kg WBS values. Consistent cooking methods allowed for the determination of tenderness between all cuts sampled in this study and the previous 1998 tenderness study. However, the single cooking method did not allow for the use of other cooking methods that may optimize the palatability of cuts that contain higher connective tissue levels (Brooks *et al.*, 1998).

Table 1: Least squares means and standard errors for WBS values.

Retail Cut	n	Mean, kg	Standard Error
Clod Steak	23	2.83^{b}	0.12
Ribeye Steak	81	2.37^{bcd}	0.06
BI Ribeye Steak	19	2.16^{bcde}	0.13
Strip Steak	8	2.34^{bcd}	0.20
BI Strip Steak	15	2.14^{bcde}	0.14
T-Bone Steak	48	2.27^{bcde}	0.08
Porterhouse Steak	32	2.32^{bcd}	0.10
Top Sirloin Steak	70	2.51^{bc}	0.07
Top Round Steak	39	3.02^{b}	0.09
Bottom Round Steak	27	3.67^{a}	0.11
Eye Round Steak	29	3.38^{a}	0.10

^{a,b,c,d,e} Means lacking a common letter differ ($P < 0.05$).

Table 2: Percentage of retail cuts stratified into tenderness categories.

Retail Cut	Very Tender (WBS <3.2 kg)	Tender (3.2 kg > WBS < 3.9 kg	Intermediate 3.9 kg > WBS < 4.6 kg	Tough WBS <4.6 kg
Clod Steak	69.57	30.43		
Ribeye Steak	95.06	4.94		
BI Ribeye Steak	100.00			
Strip Steak	87.50	12.50		
BI Strip Steak	100.00			
T-Bone Steak	97.02		2.08	
Porterhouse Steak	93.75	6.25		
Top Sirloin Steak	87.14	12.86		
Top Round Steak	61.54	25.64	10.26	2.56
Bottom Round Steak	22.22	48.15	18.52	11.11
Eye Round Steak	34.48	55.17	6.90	3.45

Conclusions

Comparing WBS values from this survey to those reported by Morgan *et al.* (1991) and Brooks *et al.* (2000) show that tenderness of US beef retail cuts appear to be improving over time. For the most part, cuts sampled from the chuck, rib, and loin had very desirable WBS values and only some round cuts had WBS values that were categorized as "intermediate" and "tough."

References

Belew, J. B., Brooks, J. C., McKenna, D. R. and Savell, J. W. (2003). Warner-Bratzler shear evaluations of 40 bovine muscles. Meat Science. 64:507-512.

Brooks, J. C., Belew, J. B., Griffin, D. B., Gwartney, B. L., Hale, D. S., Henning, W. R., Johnson, D. D., Morgan, J. B., Parrish, F. C., Reagan, Jr., J. O. and J. W. Savell. (2000). National Beef Tenderness Survey-1998. Journal of Animal Science. 78:1852-1860.

Morgan, J. B., J. W. Savell, D. S. Hale, R. K. Miller, D. B. Griffin, H.R. Cross, and S. D. Shackelford. (1991). National Beef Tenderness Survey. Journal of Animal Science. 69:3274-3283.

Neely, T. R., C. L. Lorenzen, R. K. Miller, J. D. Tatum, J. W. Wise, J. F. Taylor, M. J. Buyck, J. O. Reagan, and J. W. Savell. (1998). Beef Customer Satisfaction: Role of cut, USDA quality grade, and city on in-home consumer ratings. Journal of Animal Science 76:1027-1033.

Shackelford, S. D., J. B. Morgan, J. W. Savell, and H. R. Cross. (1991). Identification of threshold levels for Warner-Bratzler shear force in top loin steaks. Journal of Muscle Foods 2:289-296.

MULTIVARIATE PREDICTION OF SENSORY TENDERNESS/HARDNESS FROM WARNER BRATZLER SHEAR PRESS CURVES OF BEEF

K.I. Hildrum* and B. Narum

MATFORSK, Norwegian Food Research Institute, Osloveien 1, N-1430 Aas, Norway
Email: kjell.ivar.hildrum@matforsk.no

Keywords: beef, WB shear press, sensory hardness/tenderness, chemometrics

Introduction

Inconsistent meat tenderness has been identified as the major cause of consumer dissatisfaction with beef worldwide (Thompson, 2002). The tenderness problem has resulted in both increased interest in new efficient methods for tenderness assessment and improving existing methods. With the most used method, Warner-Bratzler (WB) shear force, the reported correlations with sensory analysis of tenderness have often been found to be less than satisfactory. The maximum value of the WB curve (in kg/cm^2 or N/cm^2) is usually used as estimator of tenderness, while the potential information in the remaining WB curve is neglected. The purpose of this study was to clarify if extraction of information by multivariate methods from the full WB shear force curve would contribute to improved and consistent correlations between WB shear force and sensory tenderness.

Materials and Methods

Longissimus thoracis (LT) samples from 27 young bulls were excised approx. 60min after stunning and four 15cm long pieces from each muscle were randomly assigned to different treatments and aged at 4°C for an additional 1 and 6 days (Hildrum *et al.*, 1999). Meat slices of 3.5cm thickness were vacuum-packed, heated at 70°C for 50 min in a water bath, cooled in ice water for 45min and, frozen at - 40°C, stored for 1-3 weeks, thawed for 18h at 4°C and slices of 1cm thickness cut along the fibre direction of the muscles. The second cut was also performed in the fibre direction to give samples cross-section dimensions of 1cm x1cm. Ten subsamples (replicates) were cut perpendicular to the fibre direction with the WB triangular device in an Instron Materials Testing Machine. 126 shear press values at equidistance along the WB curve of the replicates were used. Sensory analysis was performed on duplicate samples tempered at 20°C for 2 hours and served to 11 trained assessors (ISO 6564-1983). The variables assessed were hardness (first bite; across the fibre direction) and tenderness (whole chewing process) with a continuous intensity scale from tough (1) to tender (9) for tenderness and from soft (1) to hard (9) for hardness. The results were calculated by PLS regression on the Unscrambler®, version 9.2 and presented as the regression coefficient (r) and root mean square error of cross validation (RMSECV):

$$RMSECV = \sqrt{I^{-1} * \sum_{i=1}^{I} \left(y - \hat{y} \right)^2}$$

where the sample numbers are i [1,2,3,..,I], and y and ŷ are sensory method values and WB the predicted values, respectively. Jack-knifing was used to identify spectral regions that contributed to the models.

Results and Discussion

As both hardness and tenderness were assessed on the same samples, a very high univariate correlation was found between these variables (r = - 0.994) (Figure 1). Also reasonably high correlations were observed between sensory hardness/tenderness when only the WB maximum values were used (Figure 2, Table 1). PLS prediction of hardness from the whole WB curves (Figure 3) yielded a modest increase in the correlation coefficient, while the prediction error was reduced. Jack-knifing showed that information extracted was from almost the entire curve (Figure 4). However, performing predictions with only 3 values from the WB curve (early, max and late) yielded prediction results on a similar level. This means that making good baseline corrections for the curves are important for optimal correlations.

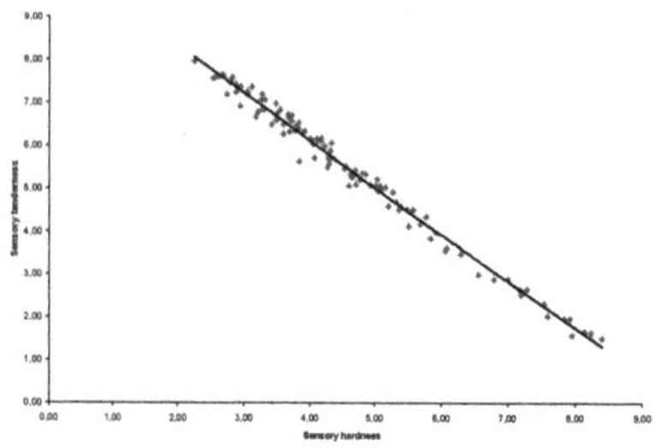

Figure 1: Correlation between sensory hardness and tenderness (108 samples).

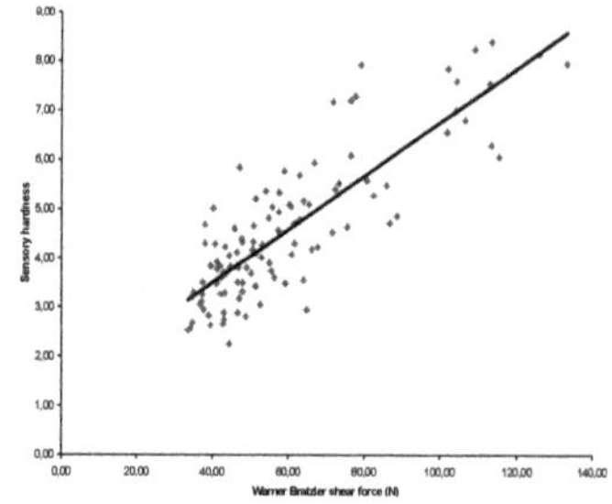

Figure 2: Correlation between sensory hardness and WB shear press (108 samples).

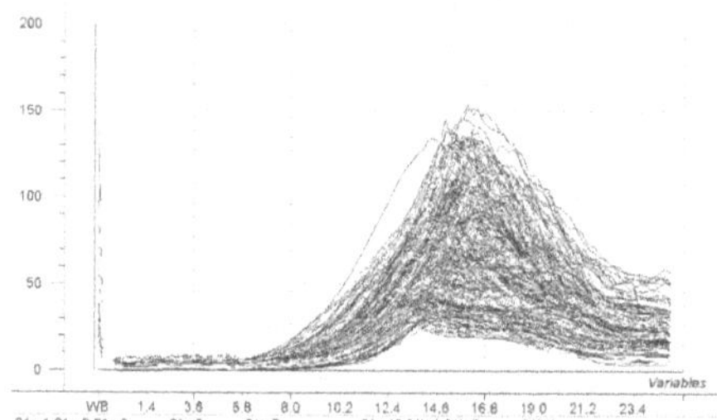

Figure 3: All replicate WB curves (1080) in the analysis.

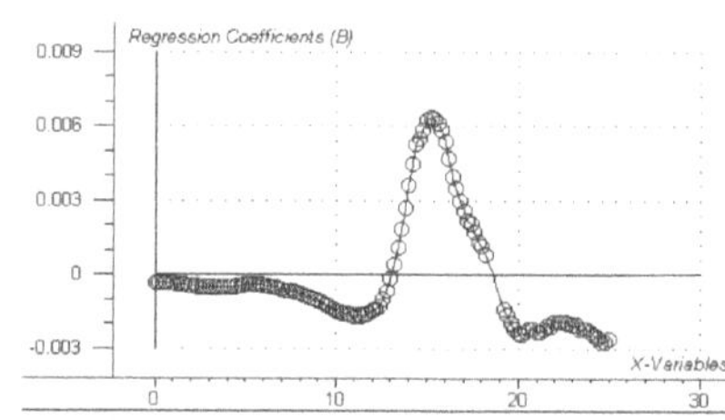

Figure 4: Jack-knifing showing the curve parts used in PLS prediction (y=regression coef.).

Table 1: Prediction of sensory variables from the WB shear press curve.

		PC's in model	r	RMSECV
Sensory tenderness				
	From WB max value	1	0.833	0.900
Sensory hardness				
	From WB max value	1	0.831	0.819
	From whole WB curve	3	0.849	0.776
	Best 3 WB values	2	0.847	0.783

Conclusion

Using PLS to extract information from the full WB shear force curves gave improved results in predicting sensory hardness from WB shear force. A corresponding improvement was attained by using 3 selected WB values (start, max, end) to secure a good baseline correction for the curves. Besides improving the prediction results, the use of multivariate techniques in extracting data from the WB curves presents the opportunity for efficient surveillance and quality control of the data.

Acknowledgements

The project was supported by a grant from The Norwegian Research Council and the Fund for the Research Levy on Agricultural Products.

References

Hildrum, K.I., Solvang, M., Nilsen, B.N., Frøystein, T. and Berg, J. (1999). Combined effects of chilling rate, low voltage electrical stimulation and freezing on sensory properties of bovine M. longissimus dorsi. Meat Science 5:1-7.

Thompson, J. (2002) Managing meat tenderness. Meat Science, 62: 295-308.

DEVELOPMENT OF A NEW SOFT CHICKEN LOAF PARTICULARLY FOR ELDERLY PERSONS AND INFANTS

M. Muguruma*[1], A.M. Ahhmed[1], Y. Namba[1], S. Kawahara[1], K. Ohta[1], H. Tanabe[2], R. Kuroda[3], K. Nakade[4], M. Numata[4] and T. Nakamura[5]

[1.] *Faculty of Agriculture, University of Miyazaki, Miyazaki 889-2192, Japan,* [2.] *Seitoku University, Mutsudo 271-8555, Japan,* [3.] *Junwa Medical Center for Rehabilitations, Miyazaki 880-2112, Japan,* [4.] *Central Research Institute, Ito Ham Food, Inc. Moriya 302-0104, Japan,* [5.] *R&D Center, Unicafe Inc, Tokyo 105-0003, Japan. Email: muguruma@cc.miyazaki-u.ac.jp*

Keywords: chicken, loaves, softening, geriatric food, gelatin

Introduction

The development and manufacturing of food products that meet the special needs of consumers is a current focus in the field of meat science (Hollingsworth group, 2003). Some people require eating meat and meat products that are soft because they are facing difficulty to bite and chew meats that are stiff. Babies and elderly people, in particular, fall within the general scope of people with these needs. Losing the ability to chew and swallow food has become a significant problem for many. Consequently, that fewer elderly people are eating meat and meat products. In addition to their most fundamental food (milk), babies at around 8 months of age begin to add solid, nutritional foods to their diet. However, it is difficult for them to process meat products at this age because their teeth are not yet strong enough or children at that age might not have many teeth. The soft chicken loaf (SCL) can provide the opportunity for these people to eat meat. Generally, SCLs are composed of chicken meat, egg white and some vegetables, which results in a complete, balanced, nutritional food product. The purpose of this study is to develop and present a new SCL including some vegetable ingredients and egg white, the aim also was intended to commercialize this product particularly for elderly persons and infants.

Materials and Methods

The main meat that was used in this research was frozen chicken legs with egg white and fat of beef. Gelatin and some vegetable ingredients were used such as fresh ginger, potato starch powder, onion, herbals (labiate) and miso (soybean paste). Moreover, some additives were utilized such as salt, fresh cream and fish powder (dried bonito), which was used to add flavour.

The frozen chicken was thawed at a refrigerated temperature; the meat was cut into cube-shaped pieces. The chicken (450 g) were mixed with 12.5 g of salt, 10 g of ginger, and 30 g of potato starch powder, 150 g fat of beef and 30 ml of water. The mentioned ingredients, including the chicken meat, were placed in a food processor for 5 min. Afterward, 50 ml of water was added again and mixed for an additional min in the same processor. As a final step in the process, 12 g of miso, 25 g of fresh cream, 2 g of fish powder and 2 g of herbal additive were added to the previously described mixture, and then all the additives were combined in the food processor for 3 min, which created a chicken paste. In addition, gelatin was added at two different ratios, 1% and 2%, to the chicken paste. In order to rid the chicken paste of the fibres or any blotches it was filtrated by a stainless steel strainer. Then, the whipped egg whites were added to the chicken paste and mixed at 90 rpm for 1 min. Each portion (40 g of the paste) was placed in a plastic bowl. These portions were steamed in a steam convection oven (Rational CM101G, Germany): 100% humidity, 80°C, for 20 min. Eventually, the fundamental measurements related to the aim of this study (determining the breaking strength and cohesiveness of the meat product) were carried out on this type of loaf by using a creep meter (Yamaden RE2-33005S, Tokyo, Japan). A sensory test analysis was conducted in a geriatric ward of a local hospital (Average age; 82 years-old). The participants included 21 people, 7 men and 14 women. The participants had a variety of teething conditions, 4 had their own teeth, 14 had artificial teeth, and 3 had not teeth. The latter group uses their gums to chew their food.

Results and Discussion

A new SCL containing vegetable ingredients, egg whites, fat of beef, fish powder and fresh whipped cream was developed. Before the essential preparations of chicken loaves came to an end, the chicken paste was divided into two groups, differentiated by the percentage of gelatin added, 1% and 2% respectively. The paste was then formed into different loaves, both of which were steamed at 80°C for 20 min, 100 % humidity. Each group (1% and 2%) had two types of samples, a control sample and a treated sample (filtered by a strainer). Additionally, another loaf, without gelatin, was ordinary soft food group used as a reference for breaking strength measurements. A sensory test analysis was conducted in geriatric ward (Average age; 82 years-old). The force resistance of loaves to the plunger sink is shown in Fig 1. The best value was of SCL filtered by a strainer and containing 1% gelatin. The breaking strength value in the samples filtered by a strainer in both groups (1% and 2% gelatin) is roughly reduced by a rate of 50% when compared to non-filtered samples. Moreover, the breaking strength of the filtered samples was significantly less than in the samples of soft ordinary food (Fig. 2). The tenacity of foods is a very important factor for easily eating and swallowing meat products. The loaves containing 1% gelatin kept their texture cohesiveness better than soft ordinary food samples, despite having been filtered by a strainer (Fig. 3). Fig 4 depicts the ease with which a participant (female, 88 years-old) was eating and swallowing the SCL, despite the fact that she was chewing with her own gums. Moreover, it was the first meal for her after using a feeding tube for several weeks.

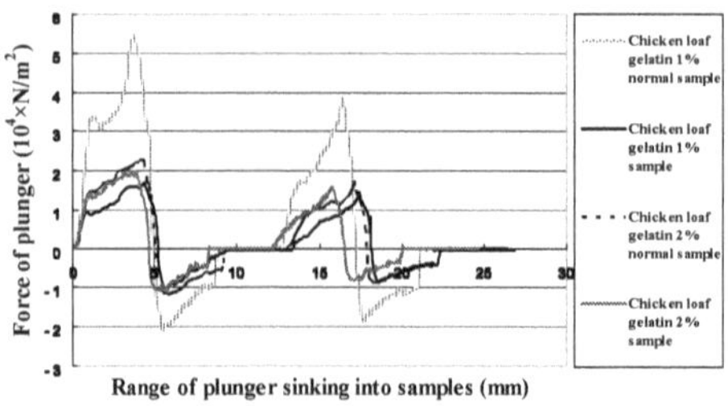

Figure 1: Chart shows the force resistance of loaves to the gelatin and various treatments such as filtration.

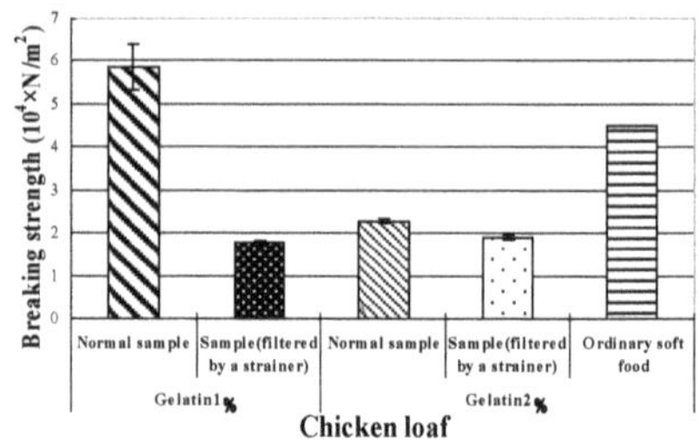

Figure 2: Breaking strength of chicken loaves with different concentrations of gelatin and ordinary soft food.

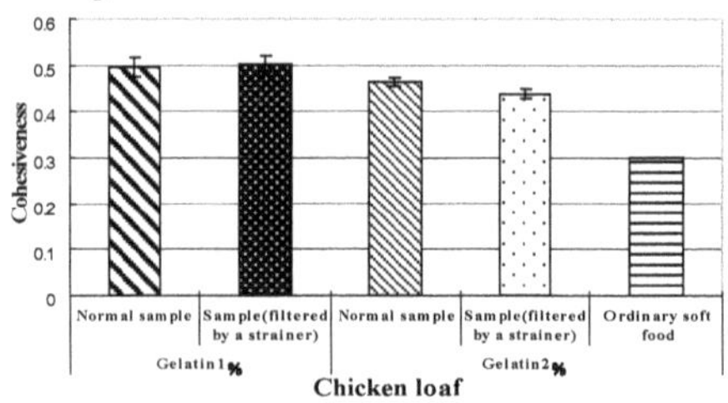

Figure 3: Cohesiveness of SCL samples with different levels of gelatin and of ordinary soft food.

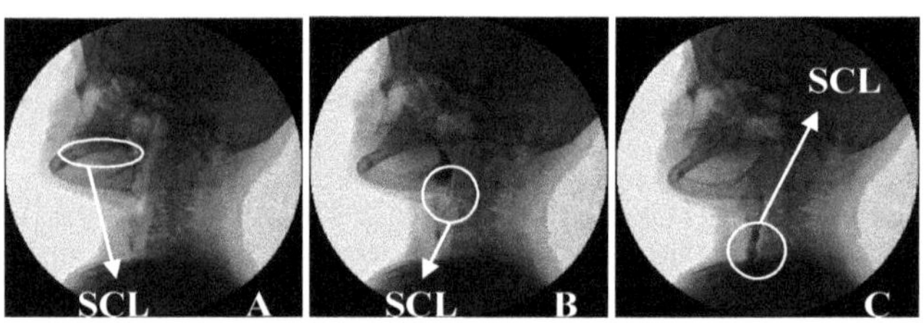

Figure 4: Profiles of video-fluorographic examination show 3 major steps of eating SCL by geriatric participant (Female, 88 years-old). (A) Without teeth able to chew and bite SCL, (B) Swallowing the chewed SCL, (C) SCL is passing through the esophagus of a geriatric participant.

Conclusions

The data suggest that the loaf was tender and the taste was acceptable. The samples containing 1% gelatin and filtered by a strainer gave the lowest value on the plungers force resistance test, which indicates 1.7×10^4N/m^2. Moreover, the SCL was smooth and easily to swallow by elderly participants (Average age = 82). The breaking strength value of filtered samples (1% gelatin) was less than the breaking strength of the normal samples and ordinary soft food as well. The cohesiveness of this type of loaf is improved by the addition of gelatin in comparison with to normal food, even though it was filtered by a strainer (data not showed). This new technique is perhaps the first of its kind available for chicken processing, in order to present and commercialize a new SCL particularly for elderly persons and infants. The results of this study suggested that the new chicken loaves were accepted to humans with dysphagia (3rd grade of soft food local classification table).

Reference

President of the Hollingsworth group: Food and the aging consumer, *Food Technology,* 57(7), 28-30 (2003).

PRODUCTION OF BEEF CARCASS ACCORDING TO CONSUMERS DEMANDS

S. Aleksic*, M.M. Petrovic, B. Miscevic, L. Sretenovic, V. Pantelic and D. Tomaševic

Institute for Animal Husbandry, 11080 Beograd-Zemun, Serbia and Montenegro, E-mail: saleksic@mail.com

Keywords: beef, crossing, baby beef, carcass, Milanese cut

Introduction

The future of meat export in Serbia lies in the production of meat of such a quality that meets the demands of a target market. In this regard, on the basis of our previous experience and production of beef for already known customers as well as according to their demands, the markets could be divided into two of the following groups: Italian and Greek market. A prerequisite for any kind of beef export is: identification and registration of all heads of cattle in the territory of the Republic of Serbia, application of international standards of control and monitoring of the production process and application of EU regulations relating to animal welfare (Official Journal C 340, 1997). Besides these mentioned prerequisites it is also necessary to provide a quality fattening material, that is, female calves for the production of the meat type called "baby beef". Baby beef fattening means that the duration of fattening is 11.5 months and during that time realised body weights of females should be 455kg, and of males 530 kg. For the Greek market, both females and males are produced, and as regards the way of processing a standard medial dissection is demanded. For Italian market, especially, the meat from female cattle not older than 11.5 months is required. This same market requires the cutting of the carcass to be performed by the so called Milanese cut. Contrary to increasing demands for animals intended for fattening, there is a decline in the number of calves and young cattle (Republic Bureau of Statistics, 2005). One way to ensure fast and efficient production of quality calves for fattening is the application of the method of industrial crossing of Domestic Spotted breed of lower production traits (Aleksic *et al.*, 1995).

Materials and Methods

In Serbia, there are several big companies engaged in production and export of beef. Technology of production is the same in all of them. Production of fattening calves is organised on farms of their subcontractors. After calving, at the age of 50–70 days, the calves are bought from subcontractors and placed into fattening facilities. Young cattle are fattened in the free housing system with a capacity of 500–1500 animals per cycle. Feeding and health care of fattening young cattle is under the constant supervision of experts of the companies engaged in meat export. For the Greek market, both male and female animals are used and as regards the way of processing a standard medial dissection is demanded. Cutting of the carcass along the spine is a classical, long-standing internationally accepted method of carcass cutting into two halves (Codex Committee on Meat and Meat Products, 1965). The halves are beheaded and spinal cord removed. As regards quality, the carcass of good conformation and low content of fat tissue is in demand. Especially for the Italian market, the meat of young cattle from female animals not older than 11.5 months is required as well as the cutting of carcass by the so-called Milanese cut. The cutting of the carcass "baby beef Milanese" is performed in two phases. The first phase is standard cutting of beef carcass along the spinal column into two halves. The second phase of cutting is the production of a commercial unit of "baby beef Milanese". For the production of "baby beef Milanese" conventionally obtained halves of beef carcasses are used. A hind part of carcass half is separated along the 5th or 6th rib at a right angle onto the spine, then cutting parallel to the spine, the belly of the carcass is separated without the hind shank. Regarding meat quality, light red colour, equal marbling, fat tissue of white to light yellow colour and good covering of the carcass by fat tissue is required. A trial included 60 heads of cattle, 30 heifers and 30 young bulls, slaughtered at the age of 11.5 months. A carcass yield was calculated in relation to animal live weight. The yield of Milanese and front quarter was calculated in relation to animal live weight.

Results and Discussion

Purchasing of the cattle was organised in a way that the prices per kg of live weight were variable and depended on market supply and demand. The prices of fattened young cattle and carcasses were relatively stable.

Table 1 shows the values of studied traits of Simmental male cattle. On the basis of economic parameters presented in the table it can be seen that market value of live weight prior to slaughtering is EUR 954,00. After slaughtering market value of carcass processed by standard cut increases by 1,5%. The transport and slaughter costs are covered by the income realised from the skin and giblets. Due to the small difference (1.5%) the state supports the export by 20% per kg of carcass.

Table 1: Average values and prices per kg for male cattle (n=30).

Male cattle	Average values	Price per kg (EUR)	Total (EUR)
Live weight on 65 day	147,5	2,95	435,13
Live weight on 337 day	530,00	1,80	954,00
Daily gain during fattening, g	1406,00	-	
Warm carcass weight, kg	307,40	3,15	968,31
Carcass yield, %	58,00	-	

In Table 2 we can see that the market value of cattle prior to slaughter of EUR 773,50 increased after slaughtering by 2,98%. By further processing of the same carcass into "baby beef Milanese", this value increased by 5,43%. The state supports the export of Milanese by 20%.

Table 2: Average values and prices per kg for female cattle (n=30).

Female cattle	Average value	Price per kg (EUR)	Total (EUR)
Live weight on 65 day, kg	130,50	2,61	340,61
Live weight on 337 day, kg	455,00	1,70	773,50
Daily gain during fattening, g	1171,00	-	
Warm carcass weight, kg	252,80	3,15	796,32
Carcass yield, %	55,47		
Weight of Milanese, kg	96,62	4,40	425,13
Yield of Milanese, %	21,20		
Weight of front quarter, kg	156,16	2,50	390,40
Yield of front quarter, %	34,27		
Milanese +front quarter, kg	252,78		815,53

Due to increasing pressure on the producers to produce the meat from heifers called "baby beef" farmers are forced to fatten quality female heads instead of using them in further reproduction aimed to improve the genetic potential of the Simmental breed. Because of the reduced supply of fattening calves, as well as the lower quality, there is a need for new ways and methods for improvement of the production of beef. Based on obtained research results we can recommend the method of industrial crossing with French fattening breeds which can include 20-25% breeding females of total population of Domestic Spotted breed (Aleksic *et al.,* 1997). In the same time, the quality of meat of crossbreds of F1 generation can fully meet all demands of consumers, especially those (Italy and Greece) who are traditionally the importers of our beef (Miscevic *et al.,* 2000).

Conclusion

Taking into consideration the demands of developed countries for the import of quality beef meat, the Republic of Serbia must in the future, in a fast and efficient way, provide sufficient quantities of meat of high quality. In Serbia there are about 750,000 cows and heifers. Domestic Spotted cattle of Simmental type makes up 50%, Domestic Spotted of lower production traits makes up about 40% while the remaining 10% are other breeds and crossbreeds. One of the methods for fast and efficient production of quality calves for fattening is the application of industrial crossing of Domestic Spotted cattle of lower production traits (40% of total population) with French fattening breeds.

References

Aleksic S. R. Lazarevic, B. Miscevic, Petrovic M.M. and Josipovic S. (1995) Proizvodni i ekonomski aspekti ukrštanja domaće šarene rase sa francuskim tovnim rasama. Tehnologija mesa 2-3, god. XXXVI, p.77-79.

Aleksic S. R. Lazarevic, B. Miscević, Petrovic M.M. and Josipovic S. (1997) Nove tehnologije za povećanje prinosa i kvaliteta junećeg mesa domaće šarene rase namenjenog izvozu. Biotehnologija u stočarstvu, posebna edicija, p. 17-24.

Miscević B., R. Lazarevic, S. Aleksic, M.M.Petrovic, and S. Josipovic (2000) Evaluation of the genetics parameters of fattening traits of young cattle of different genotypes. Biotehnology in Animal Husbandry Vol. 16, (1-2), p.19-24.

Offical Journal C 340, 10/11/1997 Protocol on protection and welfare of animals

Republički Zavod za statistiku (2005) br.70 LIV.

SEGMENTATION OF JAPANESE CONSUMERS' BEEF CHOICE ACCORDING TO RESULTS OF CONJOINT ANALYSIS

K. Sasaki*[1], H. Aizaki[2], M. Motoyama[1] and M. Mitsumoto[1]

[1.] *National Institute of Livestock and Grassland Science, 2 Ikenodai, Tsukuba, Ibaraki 305-0901 and* [2.] *National Institute of Rural Engineering, 2-1-6 Kannondai, Tsukuba, Ibaraki 305-8609, Japan. Email: ksuk@affrc.go.jp*

Keywords: beef, appearance, labelling, Japanese consumer, conjoint analysis

Introduction

In Japan, marbling score in *longissimus* muscle has been the most important factor for breeding, production, classification, and consumption of beef cattle. On the other hand, another important trend in recent years is the increasing diversification of consumers' requirements for food quality. In a previous study, we classified Japanese consumers into four groups by their requirements for beef quality (Sasaki and Mitsumoto, 2004). These groups included a 'Japanese brand high-marbled beef' oriented and a 'lean, fresh and safety beef' oriented group. The diversification is observed not only in marbling but also in price, shelf life, safety, and other characteristics of beef. But the diversification was not based on the quantitative impact of individual beef quality and labelling information in Japanese consumers.

In the present study, we segmented Japanese consumers according to the quantitative importance and utilities of beef appearance and labelling information obtained from conjoint analysis, which is a popular marketing research technique to determine the impact and utility of quality characteristics of products.

Materials and Methods

Preparation of profiles

Eight profiles of beef package were prepared by using four attributes presented in Table 1, according to 4^5 orthogonal arrays. These profiles were presented to consumers by profile cards as presented in Figure 1.

Questionnaire study

Respondents were recruited among the general public attending the 'open-day' of the National Institute of Livestock and Grassland Science on October 2003. We presented eight profile cards to participants, and asked them to rank profiles from most preferred to least preferred. In the present study, 247 guests responded completely.

Statistical analysis

The rank of profiles in each answer was applied to conjoint analysis carried out by TRANSREG procedure of SAS (Version 6.12, SAS Institute, Cary, NC). For the diversification of consumers, participants were segmented by cluster analysis carried out by SAS FASTCLUS procedure according to the importance of attributes and utilities of levels in each participant. Differences in importance of attributes among consumer segments were analyzed by SAS GLM procedure. Differences in characteristics among consumer groups were analyzed by chi-square analysis carried out by SPSS (Version 12.0J, SPSS Inc., Chicago, IL).

Table 1: Attributes and levels for profile preparation of beef.

Attributes	Marbling	Colour	Price (yen / 100g)	'Tenderness' guarantee
Levels	Low (BMS No.2)	Dark	140	Yes
	High (BMS No.4)	Blight	200	No
			260	
			320	

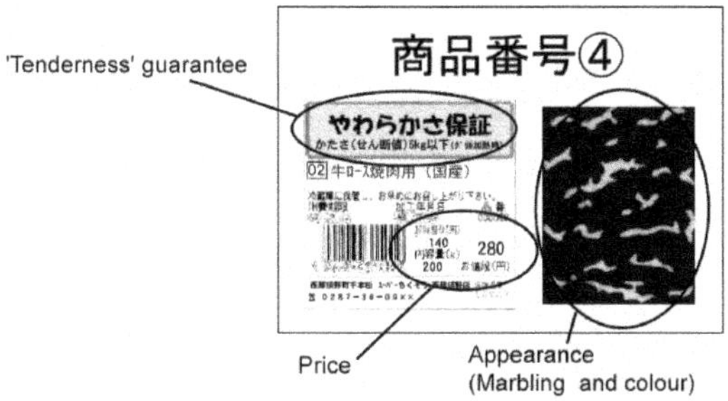

Figure 1: Appearance of profile cards used in the present questionnaire study: An example.

Results and Discussion

We divided participants into four groups by cluster analysis. The number of respondents and the importance of attributes in each group were presented in Table 2. Importance of marbling was significantly (P<.05) higher in groups 1 and 2 than in group 3 and 4, Group 3 placed significantly (P<.05) higher importance in 'Tenderness' guarantee and price than the other groups, respectively.

Utilities of levels in each group were indicated in Figure 2. Group 1 preferred low-marbled beef and group 2 accepted high-marbled beef. Group 1 and 2 had opposite utilities for marbling, although both groups 1 and 2 had high importance in marbling. The utility of each level in colour and 'Tenderness' guarantee was similar in four groups. Utilities of price levels were different among the groups. In group 4, which had highest importance in price among the groups, utility decreased with price increasing. But in groups 1 and 3, highest utilities were observed in high price and middle price, respectively. In group 2, utilities were unchanged in all price levels.

Chi-square analysis indicated that the age characteristics were different among the four groups. In groups 1 and 2, number of high-aged participants (60s and/or over) were higher than the expected value (P<.05). Group 4 had higher number of panelists in 20s and/or under and 40s than expected values (P<.05) (data not shown).

From these analysis, groups 1 to 4 were characterised as follows:

Group 1: 'Lean meat' and high price oriented and higher-aged group.
Group 2: 'Marbling' and 'Tenderness' guarantee oriented and higher-aged group.
Group 3: Middle price, 'Tenderness', and 'Marbling' oriented group.
Group 4: Low price and 'Tenderness' oriented and younger-aged group.

Table 2: Importance of attributes in groups of answerers.

Group	n	(%)	Importance (%)			
			Marbling	Colour	Price	'Tenderness' guarantee
1	30	(12.1)	36.9 ± 2.7 a	12.0 ± 2.3 ab	34.6 ± 3.2 b	16.5 ± 2.8 b
2	64	(25.9)	42.8 ± 1.8 a	18.0 ± 1.6 a	21.8 ± 2.2 c	17.4 ± 1.9 b
3	50	(20.2)	14.6 ± 2.1 b	11.6 ± 1.8 b	42.8 ± 2.5 b	30.9 ± 2.2 a
4	103	(41.7)	15.2 ± 1.4 b	10.9 ± 1.3 b	55.9 ± 1.7 a	18.0 ± 1.5 b

Values with different superscript in each attribute mean differ among groups significantly (P<.05).

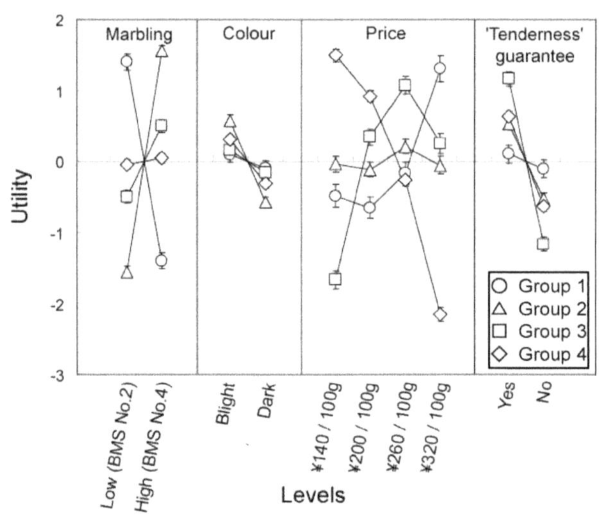

Figure 2: Utility of each level in four participant groups. Values are expressed as least square means ± SE.

Conclusions

Japanese consumers were segmented into four groups according to quantitative impact of each characteristics of beef obtained from conjoint analysis. The groups were characterized by their importance and utilities. It was also indicated that age characteristics were different among the four groups.

References

Sasaki K. and Mitsumoto M. (2004). Questionnaire-based study on consumer requirements for beef quality in Japan. Animal Science Journal, 75: 369-376.

DOES FREEZING POSITIVELY OR NEGATIVELY AFFECT SHEAR FORCE AND SENSORY QUALITY IN BEEF?

Å. Lagerstedt*, M. Marcazzani, L. Enfält, L. Johansson and K. Lundström

Department of Food Science, Swedish University of Agricultural Sciences, Box 7051, SE-750 07, Uppsala, Sweden. E-mail: asa.lagerstedt@lmv.slu.se

Keywords: beef, freezing, tenderisation, cooking loss, sensory quality

Introduction

The eating satisfaction of beef is due to a combination of tenderness, juiciness and flavour (Koohmaraie, 2002), yet the most important quality aspect for many consumers is tenderness (Koohmaraie, 1994). Consumers prefer meat that has not been frozen. This means that the industry and the retailers supply the market with fresh meat, which has not first been frozen and then thawed. However, most research studies evaluating meat tenderness and shear force are done on aged meat that has, until heat treatment, been frozen. When meat is frozen the cell membranes are damaged, which results in a lower water holding capacity and a higher cooking loss, and consequently, less juicy meat. It has been found that the freezing can positively affect shear force for beef *longissimus dorsi* (LD) (Enfält *et al.*, 2004). Meat that was tenderized for 7 days and then frozen and thawed had the same shear force values as LD that was tenderized for 21 days. Similar results have been found by Shanks *et al.*, (2002). The aim of this study was to find whether shear force values, cooking losses and sensory qualities differ between fresh and frozen meat judged by both consumers and a sensory panel.

Materials and Methods

Nine young bulls of the Swedish Holstein breed, 13-17 months of age (average 15.5 months), were slaughtered at a commercial slaughter plant. All animals came from the same farm and were fed the same feed. They were graded as O to O- (average O) with a fat content of 2+ to 3 (average 2+) according to the EUROP scale and had a slaughter weight of 289.9-311.6 (average 302.6 kg). Both LD muscles from each animal were used. The pH was measured 48 h post mortem. Both LD muscles were cut in 4 pieces (8 samples per animal) for shear force and sensory analysis. The pieces were weighed, randomized within animal and vacuum packed and stored at 4°C for 2, 7 or 14 days, day 0 being the day of slaughter. At each of those times one sample from each animal was either cooked to measure shear force or frozen at -20°C for cooking later. The frozen pieces were thawed for 24 h at 4°C before heat treatment. All samples were heat treated in a 70°C water bath for 2 h. After heat treatment the pieces were cooled in a water bath with running cold water for 30 min and then stored at 4°C until next day before being analyzed. Shear force was measured with a Warner-Bratzler shear force instrument (WB). Sensory analyses were performed with a sensory panel and a consumer test. The meat for the sensory testing was wrapped in aluminium foil and heat treated in an oven at 150°C, until a final temperature of 69°C. The meat was stored at 4°C until next day and cut in 3-mm-thick slices. The meat had room temperature at testing. Parameters judged by the 8 panellists were tenderness, juiciness and meat taste. The scale used for the panel test was 1-10 (1 very low, 10 very high). For the consumer test 140 consumers had to choose whether they preferred the fresh or the frozen sample (from the same animal). Cooking and purge or thawing losses were measured on samples after 7 days of ageing. The combined losses (purge, thaw and cooking) were measured for samples day 2, 7 and 14. Data were analyzed using the mixed procedure in Statistical Analysis System (version 9.1, SAS institute Inc., Cary, NC, USA). The model used for shear force and losses included the fixed effects of aging time, treatment (fresh or frozen), their interaction, and animal as a random effect. For the sensory results the model included treatment as a fixed effect, and animal and panellist as random effects.

Results and Discussion

Both ageing and freezing influenced shear force values. As expected shear force values declined with longer ageing time ($p < 0.0001$) in agreement with results by e.g. Monsón *et al.*, (2004). The frozen samples had significantly ($p=0.005$) lower shear force values compared to the fresh samples (LSM 49.1 and 56.4, respectively). Meat aged for 2 days and then frozen had the same shear force value as unfrozen meat that had been aged for 7 days (Table 1), which agrees with results by Shanks *et al.*, (2002) and Enfält *et al.*, (2004). The fresh meat had improved sensory qualities, as shown by higher meat taste, juiciness and tenderness. The consumer test showed no significant differences between the fresh and frozen meat ($p=0.25$). Water loss was higher in the frozen meat on all testing occasions except at 14 days (Table 2).

Table 1: Shear force (N), (LSM) in fresh and frozen meat after 2, 7 or 14 days of ageing.

	Treatment		Levels of significance		
	Fresh	Frozen	Treatm.	Day	T*D
Day					
2	68[a]	58[b]			
7	58[bc]	51[cd]			
14	43[d]	38[e]			
Overall	56	49	0.005	0.001	0.634

Different letters indicate significant differences ($p<0.05$) between values.

Table 2: Water Loss (LSM) in fresh and frozen meat after 2, 7 or 14 days of ageing.

	Treatment	
	Fresh	Frozen
Sensory test, 7 days		
Thawing/purge loss	2.5[a]	4.2[b]
Cooking loss	27.3[a]	32.2[b]
Combined loss, WB		
day 2	16.7[a]	22.1[b]
day 7	23.9[a]	29.4[b]
day 14	20.0	23.8

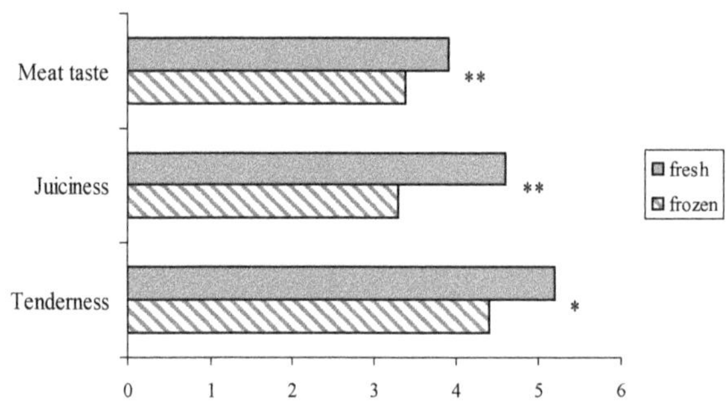

Figure 1: Sensory quality (LSM) in fresh and frozen LD; *($p<0.05$), **($p<0.01$), ***($p<0.001$).

Conclusions

Meat that was frozen and then thawed is more tender according to shear force measurements but is not perceived as more tender by the sensory panel. On the contrary, they experience the fresh meat to be more tender, juicy and having more meat flavour. It might be that meat flavour and juiciness influence the overall sensory impression, leading to the perception of a higher tenderness in fresh meat in comparison with frozen. The panellist's perception that the fresh meat was juicier concurred with the water loss results.

References

Enfält, A.C., Lundesjö Ahnström, M., Svensson, K., Hansson, I. and Lundström, K. (2004). Tenderness in m. *longissimus dorsi* from cows – effect of pelvic suspension and ageing time. ICoMST 2004, Helsinki, Finland: 81.

Koohmaraie, M. (2002). Meat tenderness and muscle growth: is there any relationship. Meat Science, 62: 345-352.

Koohmaraie, M. (1994). Muscle proteinases and meat ageing, Meat Science, 36: 94.

Monsón, F., Sañundo, C. and Sierra, I. (2004). Influence of cattle breed and aging time on textural meat quality. Meat Science, 68: 592-602.

Shanks, B.C., Wulf, D.M. and Maddock, R.J. (2002). Technical note: The effect of freezing on Warner-Bratzler shear force values of beef *longissimus* steaks across several post-mortem aging periods. Journal of Animal Science, 80: 2122-2125.

LABEL INFLUENCE ON KID'S MEAT ACCEPTABILITY BY SPANISH CONSUMERS

M.M. Campo[1]*, C. Sañudo[1], I. Cilla[1], J.L. Olleta[1], M.R. Jiménez-Badillo[1], E. Muela[1], R. Delfa[2], A. Horcada[3] and M.J. Alcalde[3]

[1] *Dept. Animal Production and Food Science, University of Zaragoza, 50013-Zaragoza, Spain* [2]*Unit of Animal Production, Agricultural Technology and Research Centre, 50080-Zaragoza, Spain* [3] *Dept. Agricultural and Forestry Science, University of Sevilla, 41013-Sevilla, Spain-Email: marimar@unizar.es*

Keywords: goat, hedonic, consumer perception, labelling

Introduction

In Spain there are 5860 goat farms for a total of 2.5 million heads, but only 14.4 % have more than 50 heads in the exploitation (INE, 2002). From this percentage, most of them rear dairy goats, making Spain are the third largest country in Europe in goat's milk production, behind France and Greece, contributing to a 3.7% of the total world goat's milk (FAO, 2006). Such a small farming size supports the status of social and traditional husbandry, keeping human population in difficult areas since it prevents emigration, with an ecological role due to the cleaning of forest against fire, aspects mainly related with meat breeds. The production of most of these farms is a suckling kid that is sold around Christmas, competing in the market with a suckling lamb and other products, all of them traditionally included in family's menus. The suckling lamb is mainly produced in the centre of the country and has a quality label (Lechazo de Castilla y León, PGI), coming mainly from a dairy sheep breed (Sañudo *et al.*, 1997). However, there has not been attempts to produce a quality label in kid's meat even when the production, although locally, is extended to most of the country. The objective of this work was to assess the perception that the Spanish consumer has about suckling kid in relation to its most direct competitor in the market.

Table 1: Consumer's characteristics, n=120.

		%
% consumers	≤ 29 years old	20.0
	30-44 years old	33.3
	45-59 years old	31.7
	≥ 60 years old	15.0
	Female	50.0
	Male	50.0

Figure 1: Percentage of consumers who would buy with certainty kid's meat when its retail price is higher, lower or equal to the selling price of lamb.

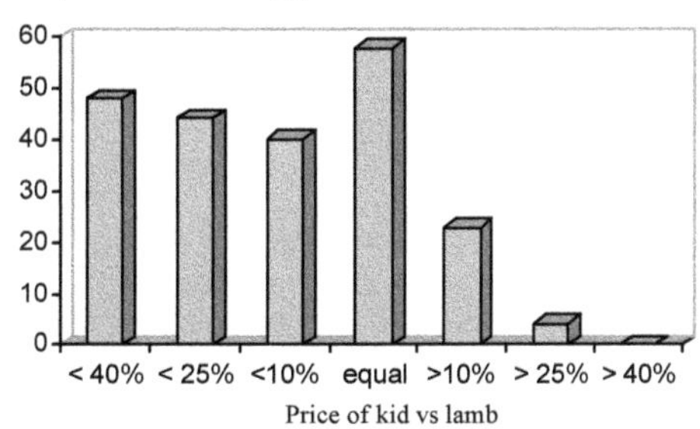

Materials and Methods

A total of 60 suckling male animals, comprising of five breeds of goats (Moncaína, (MO), Blanca Celtibérica, (BC), Negra Serrana, (NS) and Pirenaica, (PI), meat purpose, and Murciano-Granadina, (MG), dairy purpose) and one of lamb Churra, (CH) were used in this study. All animals were reared in the farms of origin throughout the country (MO and PI in Aragón; BC in Castilla-La Mancha and Andalucía; NS and MG in Andalucía, and CH in Castilla-León) using common husbandry practices, slaughtered at less than 30 days of age in local licensed abattoirs and sent in refrigerated transport to the laboratory 24 hours post mortem. The M. *longissimus dorsi* from both sides was dissected, cut in 7-cm thick pieces, vacuum packaged and aged for 3 days at 4ºC. Samples were then frozen and kept at -18ºC.

Samples were thawed at 4ºC for 24h prior to the analysis at the facilities of the University of Zaragoza. Meat was grilled until reaching an internal temperature of 70ºC. Then, three 2-cm thick pieces were cut from each loin, wrapped in aluminium foil and codified as suckling kid, suckling lamb or without label. All three pieces were tasted by the same consumer in a given order to avoid first-over and carry-over effects (Macfie *et al.*, 1989) in individual boots under red light to mask colour differences. Half of the consumers tasted only kid's meat and half of them only lamb meat, although they did not have this information since they thought they tasted what was writing on the label (Dransfield *et al.*, 2005). A total of 120 consumers (Table 1) assessed overall acceptability on an 8-point scale: like extremely, like very much, like moderately, like slightly, dislike slightly, dislike moderately, dislike very much and dislike extremely.

Table 2: Percentage of consumers (n=120) whose acceptability is positively (+) or negatively (-) influenced by tasting meat labelled as either suckling kid or suckling lamb, in comparison with a blank sample.

		Suckling kid			Suckling lamb		
		-	0 *	+	-	0 **	+
	Female	26.7	33.3	40.0	23.3	35.0	41.7
	Male	23.2	35.0	41.7	23.3	46.7	30.0
Female	≤ 29 years old	45.5	18.2	36.3	27.3	27.3	45.4
	30-44 years old	10.0	50.0	40.0	15.0	30.0	55.0
	45-59 years old	26.3	36.8	36.9	15.8	47.4	36.8
	≥ 60 years old	40.0	30.0	30.0	50.0	30.0	20.0
Male	≤ 29 years old	7.7	46.2	46.1	23.1	46.2	30.7
	30-44 years old	25.0	40.0	35.0	35.0	45.0	20.0
	45-59 years old	26.3	26.3	47.4	10.5	52.6	36.9
	≥ 60 years old	37.5	25.0	37.5	25.0	62.5	12.5

* % of consumers who scored the same value to suckling kid than to a blank sample

** % of consumers who scored the same value to suckling lamb than to a blank sample

Results and Discussion

Only 15.8% of consumers were not influenced by the label and scored the three samples with the same value. This segment of the population was the only one to follow organoleptic characteristics as the exclusive criteria to value the meat, confirming that sensory perceptions can change depending on which non-sensory information is made available to the consumer (Stefani *et al.*, 2006). The majority of consumers changed their score depending on the information given in the label, increasing or decreasing their liking for samples labelled as suckling kid and/or lamb in comparison with the same sample labelled as blank (Table 2). Nevertheless, there were more people positively influenced, matching the sample the image that they had about suckling animals. In general, men changed their opinion less than women, with a higher number of uninfluenced opinions for both kid and lamb. However, men were more positively influenced by the suckling kid label than by the suckling lamb, the negative scores remaining in the same percentage.

Nevertheless, market segments were observed in the behaviour of gender in different age groups. As age increased, men were more negatively influenced by the label suckling kid and less influenced, positive or negatively, by the label suckling lamb. Women on the range between 30-44 years old were the least influenced by suckling kid (50% of consumers in this group did not differ their score from the blank) but were positively influenced by the label of suckling lamb. Since suckling lamb is well known in the region due to the PGI label, expectations of this group of consumers may have been higher for lamb than for kid.

Information in the label can increase the price that the consumer is willing to pay for a product by 5% (Dransfield *et al.*, 2005). However, 23% of consumers claimed that they would buy with certainty kid's meat even when it was 10% more expensive than lamb (Figure 1). This percentage decreased down to 4.2 % when kid was 25% more expensive than lamb. As expected, the intention of consumption decreased with price. Nevertheless, the percentage of consumers who would buy kid when it was cheaper than lamb was lower than when the price was the same between the two of them, showing a certain lack of confidence of consumers towards this product.

Conclusions

Consumers can be influenced in their acceptability of meat by the information that they perceive in the label. Meat from suckling animals is well appreciated in Spain. There is a potential market for commercialising suckling kid in the same way as there is an existing market for suckling lamb, although certain segmentation has been observed.

References

Dransfield, E., Ngapo, T.M., Nielsen, N.A., Bredahl, L., Sjöden, P.O., Magnusson, M., Campo, M.M. and Nute, G.R. (2005). Consumer choice and suggested price for pork as influenced by its appearance, taste and information concerning country of origin and organic pig production. *Meat Science*, 69, 61-70.

FAO. (2006). http://faostat.fao.org

INE. (2002). www.ine.es

Macfie, H. J., Bratchell, N., Greenhoff, K. and Vallis, L. V. (1989). Designs to balance the effect of order presentation and first-order and carryover effects in hall tests. *Journal of Sensory Studies*, 4, 129–148.

Sañudo, C., Campo, M.M., Sierra, I., María, G.A., Olleta, J.L. and Santolaria, P. (1997) Breed effect on carcass and meat quality of suckling lambs. *Meat Science* 46, 357-365.

Stefani, G., Romano, D. and Cavicchi, A. (2006). Consumer expectations, liking and willingness to pay for speciality foods: Do sensory characteristics tell the whole story? *Food Quality and Preferences*, 17, 53-62.

Acknowledgements

Authors are grateful to Goat's Breeders Associations and CICYT for financial support.

THE RELATIONSHIP BETWEEN SUBJECTIVE RANKING OF STEAKS IN A NATIONAL COMPETITION WITH OBJECTIVE QUALITY MEASUREMENTS

R. Bickerstaffe*[1], A.E. Bekhit[1], K. Gately[1] and F. Greig[2]

[1.] *Agriculture and Life Sciences, Lincoln University, Canterbury, New Zealand.* [2.] *Beef and Lamb Marketing Bureau, Auckland, New Zealand. Email: bickerst@lincoln.ac.nz*

Keywords: steak competition, evaluation, supply chain, specifications, consumer

Introduction

It is recognised that consumer preferences ultimately determine the economic viability of the meat industry. The meat supply chain, however, consists of the consumer, retailer, processor and producer. All are linked and inter-dependent but research usually concentrates on one component of the supply chain. In this study we have utilised the logistics of collecting steaks to identify the best national eating quality steaks in four categories: Best of European; Best of British; Best of Crossbreeds and Best of Retail Branded. We have obtained additional information on the supply chain and analysed the steaks for specific quality attributes. Integration of this information will contribute to the discussion 'Can producers or retailers claim that their branded steaks match the required preferences of consumers'.

Lincoln University, in conjunction with a large supermarket chain, currently operate a supermarket guaranteed meat quality endorsement programme whereby the branded meat matches the consumer's expectations in regards to tenderness. This is based on our published research on the ability of consumers to differentiate meat with varying levels of tenderness (Bickerstaffe *et al.*, 2001).

The objectives of this research is to test the hypothesis 'Will specific breeds, processing plants, environment or management influence the final eating quality experience of steaks to enable an additional premium price to be claimed by the producer and/or retailer.

Materials and Methods

Striploins (*M. longissimus dorsi*) originating from a known breed and producer, were processed at a given date at a specific processing plant and subsequently forwarded to Lincoln University. All striploins were coded and aged under vacuum packed conditions at 3°C for 14-18 days. After ageing, four 2.5 cm thick steaks were removed from the centre of the striploin for analysis and eating quality evaluation.

One steak was analysed for its pH, colour characteristics, intramuscular fat content, tenderness and cooking loss characteristics using standard techniques within the laboratory. From this analysis, the top four steaks in each of the categories; Best of European, British, Crossbreed and Retail Brand were selected based on tenderness values and pH <5.8. Tenderness was determined on steaks from the middle of the striploins.

The sixteen steaks identified as finalists were vacuum packed and frozen prior to being subsequently evaluated by an expert panel of four judges consisting of two experienced chefs and two beef eating celebrities. Prior to judging the steaks were thawed at 4°C. A separate group of sixteen steaks were also evaluated, on an individual basis, by the general public.

Results and Discussion

Results have been obtained from three national competitions in 2003, 2004 and 2005. The 2006 competition is currently in progress. A complete set of analytical and consumer results are not available for each year due to the variations in organisation of the competition and requirements varying slightly across the years. Data available across the years is breed, processor, ageing period, tenderness values, intramuscular fat content, cooking loss, objective ranking of steaks within a class judged by the panel of experts and overall winner.

In the 2004 and 2005 competitions there were a total of 102 and 195 competitors. In the Best of European there were 10 and 25 competitors, Best of British 42 and 105 competitors, Best of Crossbreed 21 and 20 competitors. In each year, the most competitive class with the largest number of entrants was the Best of British. Each year this class of animals produced the highest number of very tender steaks (<5.0 kg F) and the highest number of steaks which were light to moderate marbled (>39% intramuscular fat). Some steaks which did not reach the final for the 'Best of British' class would have reached the final of the other three classes if they had been judged in that class on tenderness.

In all the three years there was an agreement between the selection of the finalist within a class and the meat quality ranking based on tenderness. This reflects the ability of the judges to accurately assess tenderness. However, there was always a difference in ranking the second, third and fourth in each class between the 'judges and 'machine assessment'. This is probably due to the judges taking into account subtle differences in intramuscular fat, juiciness, taste and appearance of the steaks. Such factors have not been incorporated into the machine assessment' but a computer model programme, with additional data from the 2006 competition will be developed in 2006. In all the three years, the grand finalist was from the 'Best of British' class and, interesting, in all the years the steak was lightly marbled containing between 4-6% intra-muscular fat. Other research has also examined the consumer sensory acceptance of steaks (Sitz *et al.*, 2005).

The results show that all breeds can produce tender meat in New Zealand independently of the processor and environmental region, reflecting the high level of expertise within the domestic supply chains with the ability to produce high quality meat. British breeds, however, produced consistently a higher percentage of steaks perceived by judges to be high quality steaks. Such steaks were characterised by being very tender (<5.0 kg F) and light-moderately marbled (4-6% intra-muscular fat). It is evident that there is a wide variation in tenderness and marbling within any breed and that in the future the selection of steaks, based on eating quality characteristics, needs to take precedence over selecting on productive characteristics if the industry is to move to a supply chain which provides products matching consumer expectations.

Conclusion

The competition illustrates that the industry can supply high quality branded products (Nelson *et al.*, 2004) which should capture a premium price providing the supply chain is integrated from producer to retailer and the supply chain is based on specifications provided from research data as outlined.

References

Bickerstaffe, R., Bekhit, A.E.C., Robertson, L.J., Roberts, N, Geesink, G.H. (2001). Impact of introducing specifications on the tenderness of retail meat. Meat Science, 59: 303-315.

Sitz, B.M., Calkins, C.R., Fenz, D.M., Umberger, W.J., Estridge, K.M. (2005). Consumer sensory acceptance and value of domestic Canadian and Australian grass fed beef steaks. Journal of Animal Science, 83: 2863-2868.

Nelson, J.L., Dolezal, H.G., Ray, F.K., Morgan, J.B. (2004). Characterisation of certified Angus Beef steaks from round, loin and chuck. Journal of Animal Science, 82: 1437-1444.

INTEGRATION OF SENSORY AND CONSUMER DRIVERS IN QUALITY CONTROL TO OPTIMISE PRODUCT PRODUCTION AND DEVELOPMENT IN THE MEAT INDUSTRY ACRONYM: THE CON-SENSE APPROACH

D.V. Byrne

The Royal Veterinary and Agricultural University, Department of Food Science, Sensory Science, 1958 Frederiksberg C, Denmark, Email: dby@kvl.dk

Keywords: sensory, consumer, quality control, multivariate analysis

Introduction

To realise the importance of sensory and consumer evaluation in quality control, one must ask: what are consumers really buying when purchasing food products? They may note nutrition, convenience, and image, but ultimately consumers are buying sensory properties (Muñoz, 2002). Many new products fail in the consumer realm because food production and development does not focus systematically on consumer preferences and perceptions of sensory properties. Thus, there is a need for strategies and methodologies that allow producers and developers in the meat industry to introduce key aspects of consumer expectations and demands into the quality control cycle of the product (Pecore and Kellen, 2002; Weller and Stanton, 2002). The ultimate aim of this project is to develop and implement a quality control strategy in the meat industry for production and development, based on consumer and expert sensory perception of product properties.

Materials and Methods

Thus, we present a holistic sensory/consumer based quality control strategy for food production and development (Figure 1). The strategy involves three distinct phases, where initially, specific products with sensory quality variation issues from an in-house sensory, expert sensory and consumer perspective are identified and analysed, secondly sensory based changes are implemented in product production and their effectiveness in improving quality from a sensory and marketplace consumer perspective is explored, and thirdly, an implementation of the strategy in a product development context is presented, the aim being 'preemptive problem prevention'.

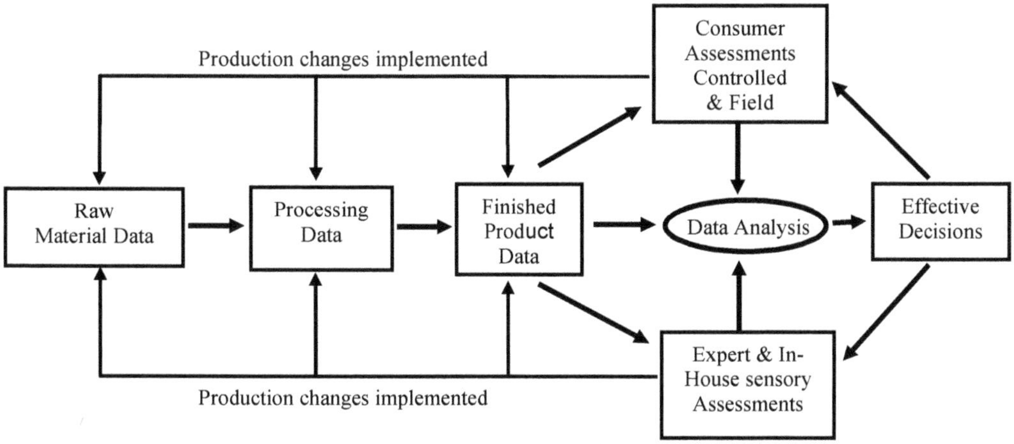

Figure 1: A schematic of the constituent elements and their interaction in the novel integration of sensory and consumer science in quality control to optimise food production and development in the meat industry.

Overall, the approach involves multivariate data analytical modeling of measurements from critical stages in the production chain, and their causal and predictive relationships to the sensory properties of processed, competitive, high quality meat products. Specifically, interactions within and between; raw material storage, processing steps and finished product treatment in relation to sensory measurements are investigated (see Figure 1). The strategy allows for acquisition of information on how qualities that inherently exist in the raw material can be transferred, utilised and preserved in products such that they are consistently of superior quality from a sensory perspective. Critically the approach involves the end-user, such that, when products are produced, manipulated and developed the effects on consumer perception and acceptability are used as key criteria in the implementation of changes in the production process.

Results and Discussion

The constituent elements will add much needed fundamental knowledge in relation to the interaction of the most important consumer quality assessment criteria in food, and allow incorporation of this knowledge into food production and development such that the consumer gains direct benefit through consistent quality improvement.

- The development and implementation of the presented quality strategy is on-going and is being conducted in close collaboration with the Danish meat industry
- The final strategy will lead to much needed fundamental knowledge in relation to the interaction of the most important consumer quality assessment criteria in foods, and will allow incorporation of this information into food production and development such that the consumer gains direct benefit through quality improvement.
- The approach and its determined consumer/sensory critical control points and methodological procedures will be published as guidelines that can be applied to increasing quality in the food production industry per se.

Acknowledgements

This research (Project Acronyms: Sense-Index and ConSense) was funded by The Danish Ministry of Food, Agriculture and Fisheries under the research programmes "Food technology, safety and quality", and "The Innovations Law".

References

Muñoz, A. M. (2002) Sensory evaluation in quality control: an overview, new developments and future opportunities. Food Quality and Preference, 13, 329-339.

Pecore, S., and Kellen, L. (2002). A consumer-focused QC/sensory program in the food industry. Food Quality and Preference, 13, 369-374.

Weller, J.N., and Stanton, K.J. (2002). The establishment and use of a QC analytical/descriptive/consumer measurement model for the routine evaluation of products at manufacturing facilities. Food Quality and Preference, 13, 375-383.

FATTY ACID PROFILE AND CONSUMER TASTE EVALUATION OF PORK AS INFLUENCED BY DIETARY LINSEED OR FISH OIL SUPPLEMENTATION

L. Haak*[1], K. Van Walleghem[2], S. De Smet[1], D. Fremaut[2] and K. Raes[1]

[1.] *LANUPRO, FBE, Ghent University, 9000 Ghent, Belgium;* [2.] *Technical University Ghent, 9000 Ghent, Belgium.*
Email: Lindsey.Haak@UGent.be

Keywords: pork, fatty acid profile, consumer, sensory, omega-3 fatty acids

Introduction

Higher intakes of n-3 polyunsaturated fatty acids (PUFA) are recommended from a health point of view. Pork, as part of the diet, can contribute to this higher intake when its fatty acid (FA) profile is steered by supplementing the feed with n-3 FA sources. However, these changes may influence the shelf-life and palatability of the meat and meat products. PUFA are more prone to oxidation causing rancid, fishy and other abnormal flavours. In addition, the FA profile may affect the formation of volatile compounds during preparation.

In this experiment, dietary supplementation with linseed or fish oil, being a rich linolenic and n-3 long chain PUFA source (LCPUFA) respectively, was compared to standard feeding practice. Intramuscular FA composition was determined. Meat was grilled and three types of meat products (fermented sausage, cooked and dry cured ham) were manufactured and tested by an untrained consumer panel at a fair.

Materials and Methods

Experimental setup and material: Crossbred pigs (n=66; Topigs40xPietrain) at a mean (sd) live weight of 36.4 (4.5) kg were randomly allocated to one of three feeding groups. Each group was housed in 2 pens of 11 animals, and consisted of an equal number of barrows and gilts. All diets (S, L and F) were grain-soybean meal based and were formulated for an equal energy supply (2225 kcal). In group L and F, 1.2% of the dietary fat was supplied by linseed (group L; α-linolenic acid (α-LNA) supply) or by fish oil (group F; eicosapentaenoic acid (EPA) and docosahexaenoic acid (DHA) supply) respectively. These 2 diets and the standard diet had an equal fat content of 4%, adjusted by the addition of animal fat. The trial lasted for 14–17 weeks. Animals were slaughtered in 2 groups with a 3 week interval in a commercial slaughterhouse (Westvlees, Belgium). From each feeding group, 12 animals were selected on average live weight. Mean (sd) live weight at slaughter and cold carcass weight was 97.1 (6.1) kg and 78.6 (5.1) kg respectively. The *M. longissimus thoracis* (LT; starting from the 7th rib, left carcass side) was sampled at 24 h *post mortem* (day 0) and sliced into 2.5 cm thick chops which were then immediately vacuum packed and stored at –18°C. In addition, 4 animals from each feeding group were randomly selected for the preparation of cooked cured ham, dry cured ham and fermented sausage according to standard recipes and manufacturing procedures at a butchery school (KTA, Diksmuide, Belgium). The fermented sausage was prepared by using 1/3 meat and 1/3 backfat from the experimental animals, and 1/3 beef. A taste panel was organised using these meat products on the occasion of the agricultural fair Agriflanders (Ghent Expo, Belgium, 2005).

Fatty acid analysis: Fatty acids were extracted according to Folch *et al.,* (1957), methylated and analysed by gas chromatography as described by Raes *et al.,* (2001).

Taste panel: Because of the impossibility of grilling meat at the fair, a smaller taste panel session was organised at the laboratory to have taste panel data which included grilled pork. Visitors at the fair were invited to taste one of the three meat products and were given a brief explanation about the aim and setup of the experiment. The participants were asked to score meat pieces for taste preference, taste intensity and overall preference on a 1 to 5-scale (1 = dislike, weak; 5 = like, strong). Overall preference was defined as the combination of the other two traits and the mouth-feel impression. Meat samples were served under red light in order to mask possible colour differences. The participants received 4 pieces per meat product (1.5*1.5*1.5 cm), with one piece from each feeding group and a 4th piece as replicate of one of the other 3 pieces. This was done to test for the consistency of scoring. Data from panelists not scoring consistently or not having completed the questionnaire thoroughly, were excluded from the statistical analysis.

Statistics: For the LT fatty acid data, a General Linear Model with fixed factors feeding group and gender was used. The interaction term was not significant. Effect of feeding group on the taste panel data was tested by one-way Anova for each type of meat product separately. The feeding group means were further compared with the Bonferroni *post hoc* test. The analyses were performed using S-Plus (version 6.0).

Results and Discussion

(*:P<0.05; **:P<0.01; ***: P<0.001; ns: not significant; ^{a,b,c}: means with different superscripts differ significantly at P<0.05)

Fatty acid composition:

Table 1: Fatty acid profile (g/100g FA) of the LT muscle.

Fatty acid	S	L	F	P
C18:2n-6	11.23^{a}	10.44^{ab}	8.84^{b}	*
C20:4n-6	2.90^{a}	2.38^{b}	1.65^{c}	***
C22:4n-6	0.39^{a}	0.29^{a}	0.15^{b}	***
C18:3n-3	0.55^{b}	1.24^{a}	0.47^{b}	***
C20:5n-3	0.22^{c}	0.54^{b}	1.37^{a}	***
C22:5n-3	0.47^{b}	0.75^{a}	0.82^{a}	***
C22:6n-3	0.14^{b}	0.18^{b}	1.02^{a}	***
Σn-3 PUFA	1.38^{c}	2.72^{b}	3.68^{a}	***
Σn-6 PUFA	15.09^{a}	13.66^{a}	11.12^{b}	**
n-6/n-3	11.02^{a}	5.23^{b}	3.03^{c}	***
SFA	34.48^{ab}	34.03^{b}	35.48^{a}	**
MUFA	44.59	44.48	45.62	ns
PUFA	16.81	16.66	15.07	ns
Total FA (g/100g)	1.38^{b}	1.49^{b}	1.79^{a}	*

Total MUFA and PUFA proportion did not significantly differ between feeding groups. Although the total n-3 PUFA supply by the feed was higher in group L compared to group F, the n-3 PUFA proportion was significantly higher for group F > group L > group S. As expected, the n-3 LCPUFA proportion was highest in group F, whereas the α-LNA content was highest in group L. The supply of the n-3 LCPUFA precursor α-LNA in group L yielded a significant increase in the proportion of EPA and DPA compared to group S, but did not result in an increase in DHA. Hence, this study supports other literature findings that specific dietary supply of DHA is needed in order to increase the DHA content of pork. The n-6 PUFA proportion was significantly higher in both the S and L group compared to the F group, as a result of the lower feed LA supply in the F group. The proportion of LA was not significantly different between the L and S group. However, the highest proportion of arachidonic acid (C20:4n-6; AA), the major desaturation/elongation product of LA, was found in group S, probably because of the higher proportion of α-LNA in this group which can be in competition with LA for the enzymes of desaturation/elongation.

Taste panel:

Table 2: Taste parameter scores.

	S	L	F	P
cooked cured ham				
taste preference	3.65^{a}	3.30^{b}	2.98^{c}	***
taste intensity	3.62^{a}	3.14^{b}	2.94^{c}	***
overall preference	3.58^{a}	3.22^{b}	2.99^{c}	***
fermented sausage				
taste preference	3.29^{a}	3.13^{a}	2.88^{b}	***
taste intensity	3.38^{a}	3.20^{ab}	3.14^{b}	**
overall preference	3.35^{a}	3.09^{b}	2.88^{c}	***
dry cured ham				
taste preference	3.23	3.29	3.26	ns
taste intensity	3.31	3.31	3.31	ns
overall preference	3.15	3.24	3.22	ns
grilled meat				
taste preference	3.75^{a}	3.34^{ab}	3.17^{b}	*
taste intensity	3.58^{a}	3.10^{ab}	3.06^{b}	*
overall preference	3.73^{a}	3.31^{ab}	3.06^{b}	**

For the dry cured ham, the cooked cured ham and the fermented sausage, data from 210, 208 and 255 people respectively were retained for the statistics. For the grilled meat, 30 people were retained. For the cooked cured ham, the scores for all parameters were significantly higher for group S > group L > group F. For the taste intensity and taste preference of the fermented sausage, group S again scored significantly higher compared to group F, whereas the score for the L group was intermediate. Overall preference for the fermented sausage was significantly different between the groups and scored highest for group S > group L > group F. For the dry cured ham, the taste parameters did not reveal significant differences between the groups. For the grilled meat, scores for all taste parameters were significantly higher for the S group compared to the F group, whereas results for the L group were intermediate. In general, for all meat products, differences between feeding groups were small and the general appreciation was on average 3 or higher (on a scale of 1 to 5). However, the meat products of the fish oil fed animals were scored lower.

Conclusions

From this study, it seems that meat products originating from animals fed on a relatively low level of dietary fish oil were generally less appreciated by an untrained consumer panel compared to meat products originating from animals fed on a standard diet or a diet containing linseed. However, the differences, although significant, were not pronounced. On the other hand, the level of supplementation did induce significant increases in the proportion of beneficial n-3 long chain PUFA.

References

Folch, J., Lees, M., Stanley and S.G.H. (1957). A simple method for the isolation and purification of total lipids from animal tissues. *Journal of Biological Chemistry*, 226: 497-509.

Raes, K., De Smet, S. and Demeyer, D. (2001). Effect of double-muscling in Belgian Blue young bulls on the intramuscular fatty acid composition with emphasis on conjugated linoleic acid and polyunsaturated fatty acids. *Animal Science*, 73: 253-260.

EFFECTS OF ELECTRICAL STIMULATION ON SENSORY QUALITY OF REINDEER MEAT

L. Johansson*[1], E. Wiklund[2], G. Aguiar[2], P.J. Bechtel[3] and G. Finstad[2]

[1]Fjärdhundragatan 32, Uppsala, Sweden, [2]University of Alaska Fairbanks, Reindeer Research Program, P.O. Box 757200, Fairbanks, AK 99775-7200, USA, [3]USDA-ARS, Subarctic Agricultural Research Unit, Fairbanks, AK 99775-7220, USA. Email: lisbjoha@yahoo.se

Keywords: reindeer meat, venison, tenderness, trained panel, consumer test

Introduction

The majority of field slaughtering of reindeer in Alaska is conducted during the winter months on snow to reduce the risk of contamination and to eliminate the need for refrigeration and freezing facilities. Ambient air temperatures during field slaughtering are usually very low and immediate chilling and freezing of the carcass occurs. This un-inspected reindeer meat can be sold within the state of Alaska if the carcasses are field dressed in the winter and the carcasses are kept frozen until they reach the consumer. Rapid chilling and freezing of carcasses can result in meat tenderness problems due to cold shortening of the muscles. Electrical stimulation accelerates post-mortem glycolysis and rigor onset, so that rapid cooling or freezing of carcasses may be carried out soon after slaughter, reducing the risk of muscles cold shortening. Electrical stimulation has been widely adopted as a meat tenderisation method for beef, lamb and goat carcasses, and also for deer carcasses in New Zealand.

In this study the effect of electrical stimulation on meat tenderness of reindeer loin (*M. longissimus*) and shoulder muscles (*Mm. triceps brachii, supra spinatus* and *infra spinatus*) was evaluated by a trained sensory panel and consumer preference tests. In addition, the study included a comparison of carcasses and meat from an uninspected field-slaughter situation and slaughter at an inspected processing facility.

Materials and Methods

A total of 26 castrated reindeer bulls (>3 years old) were used to investigate the effects of low voltage stimulation on sensory attributes of the meat. Reindeer were gathered and herded just outside Nome in the Seward Peninsula, Alaska, and shot in the field. After bleeding carcasses were randomly allocated to either an electrical stimulation treatment (ES) or a no electrical stimulation treatment (NES). Electrical stimulation was applied immediately, followed by removal of the viscera and hide. Shoulder meat was boned out from the warm carcasses in the field, packed in plastic-lined cardboard boxes and left to freeze at ambient temperature (to simulate the quality of un-inspected meat). The remaining carcasses were transported by truck to a meat processing facility in Nome were they were hung in the chilling room (+3°C). At 2 days post slaughter (simulating an inspected slaughter) carcasses were boned and both the right and left loins (*M. longissimus*) were removed, loins from each side were randomly allotted for either proximate analysis or sensory evaluation. All samples were vacuum-packaged and frozen at –20°C until analysis.

The frozen and boxed shoulder meat was shipped to SFOS Fishery Industrial Technology Center located at Kodiak Island, Alaska. The frozen meat was removed from the boxes and either ground, sliced or cubed while still frozen. The meat samples were then vacuum-packaged and transported to Fairbanks and kept frozen until analysis. The cubed meat was cooked in the vacuum-bags in a water bath to an end temperature of 74°C. The slices and ground meat were roasted on an oven tray until well done (74°C).Three paired-comparison consumer tests (cubed meat, sliced meat and ground meat from ES vs. NES carcasses) were performed in collaboration with the Cooperative Extension Service (CES) at UAF in Fairbanks using a total of 212 consumers to evaluate the cubed, sliced and ground meat.

At the CES Research Kitchen the loin samples were thawed overnight in a refrigerator, thermocouples inserted and cooked in a conventional oven at 150°C to a core temperature of 70°C. A selected and trained sensory panel consisting of 7 members performed a descriptive test on the reindeer loin samples. All assessments were done in a sensory laboratory with separate booths and under standard white light. At every session, the panel members were served meat samples from 7-9 animals at the same time, each sample consisting of one slice of meat. Samples were placed in plastic cups coded with three-digit numbers and were served to the panel members in randomised order, at room temperature and in two replicates. The following attributes were selected and unanimously agreed upon during panel training; tenderness, juiciness, bloody flavour, gamey flavour and sweet flavour. An unstructured continuous line scale from 0 (low intensity) to 10 (high intensity) was used.

Results and Discussion

When consumers tasted the cubed and sliced meat from the shoulder they judged the meat from the electrically stimulated carcasses to be more tender for both products ($P \leq 0.05$ and $P \leq 0.001$ respectively). No significant difference ($P \geq 0.05$) was found between the two treatments for ground meat. However, there was a trend ($p=0.01$) towards increased tenderness in meat from electrically stimulated carcasses. While primal cuts like tenderloin (*M. psoas major*), strip loin (*M. longissimus dorsi*) and topside (*M. semimembranosus*) are easily marketed, less valuable cuts like meat from the shoulders have a reduced market share because of a high percentage of connective tissue (lack of tenderness). Hot-boning of less valuable carcass cuts in the field combined with the treatment of electrical stimulation could be a way for value-added processing. Frozen, boneless meat can be further processed at regional processing facilities to increase the value of a typically low quality product.

The trained panel did not find significant differences in any of the measured attributes when comparing loin samples from the electrically stimulated and non-stimulated carcasses that had been conditioned in a cooler for 48 hours prior to removing and freezing the loin samples. These results agree with previously reported Warner-Bratzler tenderness measurements for the same reindeer loin samples and with other studies on tenderness in reindeer meat, where it was concluded that reindeer meat from high quality meat cuts were very tender, especially in comparison with beef (Wiklund *et al.*, 1997; Barnier *et al.*, 1999;Wiklund *et al.*, 2004). The loins evaluated by the trained panel simulated the treatment of reindeer carcasses and meat from an inspected slaughter where the carcass was chilled to + 3°C before boning. In this situation the electrical stimulation did not have any effect on tenderness. However, the effect of electrical stimulation on tenderness in other less valuable cuts than the loin needs to be determined under the conditions of an inspected slaughter.

Conclusions

Results from this study demonstrate that ES increased tenderness in the cubed and sliced products made from field slaughtered reindeer shoulder meat. Loins from carcasses conditioned in a cooler for 48 hours prior to removing and freezing the loin samples were very tender regardless of treatment. The ES techniques can be used in field slaughter systems for reindeer to significantly increase the quality and potential value of meat. It is possible ES will also have a role in enhancing reindeer meat quality in conventional USDA slaughter operations; however, this remains to be further investigated.

References

Barnier, V. M. H., Wiklund, E., van Dijk, A., Smulders, F. J. M. and Malmfors, G. (1999). Proteolytic enzyme and inhibitor levels in reindeer (*Rangifer tarandus tarandus L*) vs. bovine longissimus muscle, as they relate to ageing rate and response. Rangifer, 19, 13-18.

Wiklund, E. Finstad, G. and Bechtel, P. (2004). Effects of electrical stimulation on the technological quality of reindeer *M. longissimus*. Proceedings: 13th Nordic Conference on Reindeer Research, 23-25 August, Røros, Norway.

Wiklund, E., Barnier, V. M. H., Smulders, F. J. M., Lundström, K. and Malmfors, G. (1997). Proteolysis and tenderisation in reindeer (*Rangifer tarandus tarandus* L) bull longissimus thoracis muscle of various ultimate pH. Meat Science, 46, 33-43.

CHARACTERISTICS OF BEEF AND PORK CONSUMPTION IN MEXICO: A CASE STUDY

L.M. Sagarnaga Villegas*, J.M. Salas González, J.L. Chávez Barraza and F. Cervantes Escoto

Departamento de Zootecnia, Universidad Autónoma Chapingo, Km 38.5 Carretera México-Texcoco. Chapingo, México. CP 56230. México. E-mail: myriamsv@correo.chapingo.mx; myriamsv@prodigy.net.mx

Keywords: consumption, meat, beef, pork, Mexico

Introduction

Meat consumption in Mexico is very sensitive to the economic environment. In 2005 the average consumption was a total of 48.8 kg of meat, of which 11.8 kg was beef, and 9.6% pork. These estimates are the result of the efforts of the Mexican government and national producers to stimulate meat consumption. The amount of meat consumed per capita and the accessibility of this product to the less economically favoured has been a reason for the government's favouring the importation of meat, despite strong controversy. Regardless of the reasons that have caused its growth, consumption is presently 43% above that in the nineties. Nevertheless, it is important to mention that it is lower than in the eighties when 20kg of pork alone was consumed per capita yearly. It is a fact that the Mexican consumer eats more meat everyday and is more critical with regards to its characteristics. People are more concerned with the quality of their lifestyle and take better care of their health as a result of this attitude. This new lifestyle imposes new demands on the market. It no longer suffices to produce healthy animals which in turn yield healthy products. The industry is under pressure to evolve and development better production techniques, processing, distribution, packing and presentation of meat products.

The state of Guanajuato is an important pork producer in Mexico, though its beef production is not notable, to the point of being deficient. The state also has an important hog slaughtering industry. The pork industry in Guanajuato is third in importance, generating 9% of the national production; on the other hand, it only generates 2% of the national beef production (Sagarnaga, 2005).

The producers and state government are presently undertaking a project to modernise the state meat industry, principally those segments that include the raising of calves up to their slaughtering, wholesale distribution, and primary cuts. A lack of information oriented towards the decision making process is one of the main limitations that need to be overcome.

Materials and Methods

The objective of this paper is to evaluate the consumption characteristics and habits, and preferences of meat consumers. The analysis is based on the information collected through a questionnaire applied to 154 consumers in the state of Guanajuato. The results are considered to be exploratory.

Results and Discussion

The people who buy meat are middle aged (40-49 years of age), females (92%), with small families (3-5 members), and a monthly income of under 6,000 Mexican pesos (less than 400 euros). The consumers interviewed consumed an average of 40.2kg yearly, of which 23.3 is beef, and 16.9kg pork. This is above the national average. The size of the family determines the amount of meat consumed. In families with fewer than 2 members, the per capita beef consumption is 880g weekly, and of pork, 420g. This amount decreases to 140 grams of beef and only 60 grams of pork weekly in families with more than 10 members.

The amount of meat consumed is also affected by family income. The demand for beef increases according to the income level. The weekly consumption of low income families is 1.6kg, while the meat consumption of high income families is 2.30kg weekly. On the other hand, pork consumption decreases as the income level increases. Low income families acquire 1.2kg of this type of meat, while high income families only buy 0.7kg of pork weekly.

The majority of the consumers interviewed (54%), eat meat from 3 to 4 days a week; an important part (31%), eat meat only once or twice a week. Beef is bought 8 times a month, which implies that consumers customarily buy meat twice a week. Pork is bought 5 times a month, on an average of about once a week. Meat products are mainly bought at butcher shops found in public markets; given that 55% and 57% of the beef and pork products, respectively, are bought in this type of establishments. The second most important place of purchase is neighbourhood meat markets, 39% and 40% of those interviewed buy beef and pork products in this type of establishment, respectively.

The main reason for customarily buying meat products in the aforementioned places For 34% of the consumers interviewed is the freshness of the products offered; an equal number of those interviewed mentioned hygiene as another of the main reasons; 33% mention their closeness; and 31% the quality of the service. Other reasons of lesser influence in the selection of where to buy meat products are: convenience, price, variety, and the sale of other products. The type of meat most sought by the retail consumers is hot meat (51.3%), that is, meat from animals that have been recently slaughtered. Consumers in Guanajuato prefer regionally produced meat. Those interviewed said that they preferred meat that had been produced locally (51%); for an important percentage of those interviewed, (37%), the source of origin was not a determining factor when buying meat. Of those interviewed, 81% preferred meat that had not been previously packaged or cut, that is, they preferred meat in whole sections so that they might select the part they wanted when they were served. The cuts preferred were 86% beefsteak, while pork cuts had a greater variety. Those preferences were: ribs (35%), loin (32%), and jams (31%). The majority of those interviewed (62%), indicated their conformity to paying a higher price for meat if hygiene and quality were guaranteed. The average surcharge that the consumers are willing to pay is 9%. The price is flexible depending on the income level. The demand for beef in the case of high and medium income consumers is less flexible with regards to price; a large number of consumers (80% and 49%, respectively), maintain their consumption level independent of the changes in price. Low income consumers are more sensitive to price changes; a larger number of them (41%), significantly reduce their consumption when there are price changes. Something similar occurs with regards to pork; high and low income consumers maintain their consumption despite price increases. In the case of low income consumers (68%) reduce their pork consumption as the price increases.

Conclusions

It can be generally stated that the consumer in Guanajuato retains traditional buying and consumption habits which should be considered when modernizing the state meat industry. Such behavior is different form the average Mexican consumers. Along the same line, this population consumes more meat than the average Mexican, which is an opportunity to this market.

References

Sagarnaga, V.M. *et al.,* (2005). La ganadería en el estado de Guanajuato en sistemas de producción de no rumiantes y especies menores. Resultados de investigación 2004. Universidad Autónoma Chapingo. México.

PERCEPTIONS OF BEEF EATING QUALITY IN NORTHERN IRELAND

L.J. Farmer*[1,2], A.W. Gordon[1], T.D.J. Hagan[1], J. Hanna [2], E.L.C. Tolland[1], I.J. Tollerton[1], D.J. Devlin[1], B.W. Moss[1,2] and N.F.S. Gault[1,2]

[1] *Agri-Food and Biosciences Institute, Newforge Lane, Belfast BT9 5PX, UK.* [2] *Department of Food Science, Queen's University Belfast, Newforge Lane, Belfast BT9 5PX, UK. Email: linda.farmer@afbini.gov.uk*

Keywords: beef, consumer, demographic

Introduction

Irish consumers have been found to associate several types of 'risk' with the purchase of beef (McCarthy and Henson, 2005); the greatest of these is financial risk ('wasting money'), performance ('the product will not meet taste expectations') and safety. 'Quality' has been shown to be a key issue for consumer behaviour, and this term encompasses tenderness, juiciness and anticipated taste as well as overall appearance, nutritional value, perceived safety and provenance (Krystallis and Arvanitoyannis, 2006). Eating quality attributes are very difficult to predict by the consumer. Focus groups have indicated that cut, colour, visible fat, price, origin and production information are 'cues' used by consumers or purchasers, together with place of purchase (Grunert *et al.*, 2004). However, few of these influence eating quality. It is, therefore, not surprising that consumers are reported to have difficulty predicting eating quality at time of purchase (Grunert *et al.*, 2004).

In Northern Ireland, the consumption of 'carcase meat' is a little higher than elsewhere in the UK (Anonymous, 2003); beef consumption is about 150g per person per week out of over 1000g meat and meat products per person per week (DARD, 2002). The aim of this work was to determine whether this pattern reflected satisfaction with the quality of beef purchased and whether Northern Ireland's purchasers felt able to predict eating quality of the beef they buy.

Materials and Methods

Trial 1. A total of 175 customers were questioned in five supermarkets in and around Belfast. Surveys were conducted both during the week and at weekends and in mornings, afternoons and evenings to provide a distribution of ages and socioeconomic backgrounds. Customers who were seen to select beef were approached on a 'next come, next asked' basis. The country of origin of the beef they had chosen was not recorded and will have included beef from both Northern Ireland and elsewhere.

Purchasers were asked about their choice of beef, how frequently they were satisfied with selected attributes of beef and whether they were able to predict the eating quality of a certain cut of beef by the price. In addition, consumers were asked socio-economic questions (Bailey, 1978; De Vaus, 1996). Although the number of beef purchasers sampled was small, it showed a very similar distribution of household income to that reported for Northern Ireland (NISRA, 2001). Purchasers included a higher proportion of women (136) than men (40) and more older people than younger: <25 (11), 25-34 (17), 35-44 (35), 45-59 (68), 60+ (45). The first three age groups were combined for statistical analyses.

Trial 2. As part of a larger study, consumers of beef were asked to taste a range of cuts, which were grilled or roasted (360 consumers each) under controlled conditions. Consumers were untrained members of the public, who attended once and were selected for these panels on the basis that they ate beef on a regular basis. Panels were conducted at a central location. Consumers were asked to score seven pieces of beef for satisfaction on a category scale (unsatisfactory / satisfactory everyday quality / better than everyday quality / premium) and were also asked to complete a questionnaire covering a number of demographic questions.

Statistics. Associations between answers to selected questions were assessed by means of chi-squared probabilities calculated from contingency tables using the method of Pearson.

Results and Discussion

In Trial 1, each purchaser was asked 'Are you satisfied with the beef that you buy in terms of its tenderness / flavour / juiciness / fat content / gristle content?' Overall the proportion of purchasers who claimed to be 'usually' or 'always' satisfied with the tenderness was 84%, flavour, 63%, juiciness, 63%, fat content, 63% and gristle content, 65%. Thus, while most purchasers are content with the quality of beef that they buy, up to 37% of purchasers were 'never', 'rarely' or only 'sometimes' satisfied. When asked: 'Do you feel you can predict the eating quality [*of the cut they had selected*] by the price?', the answers were: never (16.5%), rarely (34%), sometimes (16.5%), usually (22%), always (11%). Thus, half of purchasers felt that price 'never' or 'rarely' predicted the final eating quality.

Age had a significant effect on purchasers' satisfaction with tenderness ($P<0.05$), flavour ($P<0.01$), juiciness ($P<0.001$), fat content ($P<0.01$) and gristle content ($P<0.05$). In all cases, purchasers younger than 45 were less satisfied than older consumers, as illustrated in Figure 1(a) and (b) for juiciness and fat content. In an apparent contradiction, older people felt significantly less ($P<0.01$) able to predict eating quality by price, with 57% of those 45 and older feeling that price never or rarely predicts eating quality compared with only 34% of those younger than 45. This may reflect the greater experience of the >45 age group and, if younger people expect to be able to predict eating quality by price, this may also contribute to their dissatisfaction.

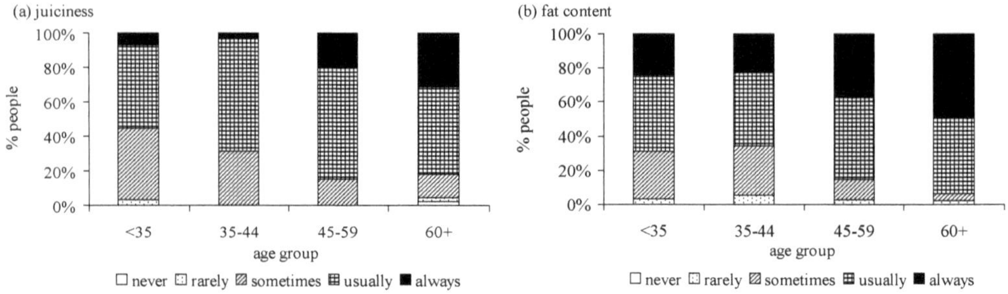

Figure 1: Effect of age group on satisfaction with beef eating quality in terms of (a) juiciness and (b) fat content.

Neither gender, income group nor frequency of consumption influenced purchasers' satisfaction with eating quality, but household composition had some effect. Purchasers from households with children were more likely to respond that they were 'usually' satisfied with tenderness, flavour and juiciness (all $P<0.05$), while those from adults only households gave a greater spread of responses, with a greater proportion 'always satisfied'. This may be due to children being more discerning (or hard to please), or because the person questioned was responding for the whole family and not just for themselves.

Trial 2 evaluated the effect of age, sex, income group and household composition on the actual satisfaction scores given by consumers to a range of beef muscles; only age had any effect ($P<0.001$). For a given range of roast topside joints, the '60+' age group found 45% to be 'better than everyday quality' or 'premium', compared with only 30% for the younger age groups. The '60+' age group found only 17% of the sample to be 'unsatisfactory' compared with 30% younger assessors. The same differences were not observed for sirloin steak, perhaps due to its generally better eating quality scores.

A similar effect of age on satisfaction with beef has also been recorded in Australia (Cox *et al.*, 1997) and Greece (Krystallis and Arvanitoyannis, 2006).

Conclusions

Most Northern Ireland purchasers are 'usually' satisfied with the quality of the beef they purchase. However, a high level of satisfaction with eating quality and fat content is a characteristic of older purchasers, which is not shared by those under 35. This effect is reflected, in part, when consumers are presented with beef to assess for eating quality. This suggests that, as has been observed in other countries, the younger generation of beef purchasers is more discriminating and demanding with regard to eating quality than their parents.

Acknowledgements

The assistance of supermarkets in and around Belfast is gratefully acknowledged and we also thank the Agricultural Economics Branch of the Agri-Food and Biosciences Institute for advice.

References

Anonymous (2003). Family food in 2001-02. A National Statistics Publication, UK.

Bailey K.D. (1978). Methods of Social Research. Free Press: Collier Macmillan, London; New York.

Cox, R.J., Thompson, J.M., Cunial, C.M., Winter, S. and Gordon, A.J. (1997). The effect of degree of doneness of beef steaks on consumer acceptability of meals in restaurants. Meat Science, 45, 75-85.

DARD (2002). The Agricultural Census in Northern Ireland. Results for June 2002. Department of Agriculture and Rural Development - Economics and Statistics Division, Northern Ireland, UK.

De Vaus, D.A. (1996). Surveys in social research. UCL Press, London.

Grunert, K.G., Bredahl, L. and Brunso, K. (2004). Consumer perception of meat quality and implications for product development in the meat sector - a review. Meat Science 66: 259-272.

Krystallis, A., Arvanitoyannis, I.S. (2006). Investigating the concept of meat quality from the consumers' perspective: The case of Greece. Meat Science 72: 164-176.

McCarthy, M., Henson, S. (2005). Perceived risk and risk reduction strategies in the choice of beef by Irish consumers. Food Quality and Preference 16: 435-445.

NISRA (2001). Family Resources Survey. Northern Ireland 2000-01. Northern Ireland Statistics and Research Agency, Department for Social Development, UK.

KEY FLAVOUR PRECURSORS IN CHICKEN

L.J. Farmer*[1,2] and M. Aliani[2,3]

[1] *Agri-Food and Biosciences Institute, Newforge Lane, Belfast BT9 5PX, UK.* [2] *Department of Food Science, Queen's University Belfast, Newforge Lane, Belfast BT9 5PX, UK.* [3] *Faculty of Science, Kingston University of London, Kingston Upon Thames, Surrey, KT1 2EE, UK Email: linda.farmer@afbini.gov.uk*

Keywords: chicken, flavour, precursors

Introduction

Flavour and taste are the 'chemical senses' by which most creatures assess the chemistry of the world around them and the wholesomeness of food. The volatile compounds contributing to meat flavours and odours are formed during cooking by numerous reactions, which occur between the components of raw meat. Many volatile odour compounds of importance to chicken flavour have been identified, and the chemical pathways important for their formation studied. A previous paper identified chickens with differing scores for chicken aroma and sweet aroma and investigated the differences in volatile odour compounds (Farmer *et al.*, 2000). The results suggested that the observed differences were largely due to differences in sulphur-containing and other odour compounds. These compounds may be derived from the Maillard reaction between amino acids or proteins and sugars or the sugar-containing nucleotides, or from the thermal breakdown of thiamine. However, the relative importance of each of these pathways for chicken flavour has not been fully elucidated. This paper reports an investigation of the role of selected precursors for odour and flavour formation in cooked chicken.

Materials and Methods

Sugars, sugar phosphates, thiamine, amino acids, nucleotides, inosine and hypoxanthine were determined in chicken breast fillets (*M. pectoralis major*) as described (Aliani and Farmer, 2002; Aliani and Farmer, 2005b). Post-slaughter changes in ATP breakdown products were monitored (Aliani and Farmer, 2005a) in breast meat at intervals post-slaughter, by the same methods (Aliani and Farmer, 2002; Aliani and Farmer, 2005b).

Sensory evaluation of cooked chicken to which precursors had been added was conducted using profiling and paired comparison tests as described previously (Aliani and Farmer 2005c). Precursors were dissolved in water at the required concentration and homogenised with the raw meat. Breast meat from twelve individual chickens that had been analysed for ribose and ribose phosphate was subjected to paired comparison tests between breasts with naturally high concentrations of ribose and those with lower concentrations (Aliani and Farmer, 2005c).

Results and Discussion

Figure 1 shows the odour profile for cooked chicken obtained following addition of selected precursors (Aliani and Farmer, 2005c). Added thiamine, ribose, cysteine and cysteine + ribose gave significant increases odours compared to the control. Glucose (180 mg/100g), glucose-6-phosphate (300mg/100g) and inosine 5'-monophosphate (IMP, 400mg/100g) did not give significantly different odour scores to the control.

Table 1: The natural variation of selected flavour precursors in chicken meat.

Precursor	No. chickens	Mean conc. [a]	CV% [b]	Range[c]
IMP	30	84	45	2
Inosine	30	36	30	2-3
Hypoxanthine	30	12	43	3-4
Ribose	30	25	34	3-4
Glucose	24	40	44	3-4
Ribose phosphate	24	14	50	3-4
Glucose phosphate	24	17	80	3-4
Thiamine	6	0.22	26	1.5
Cysteine	6	trace	trace	-

a: CV = Coefficient of variation; b: mg/100g wet weight; c Range = Max/min values

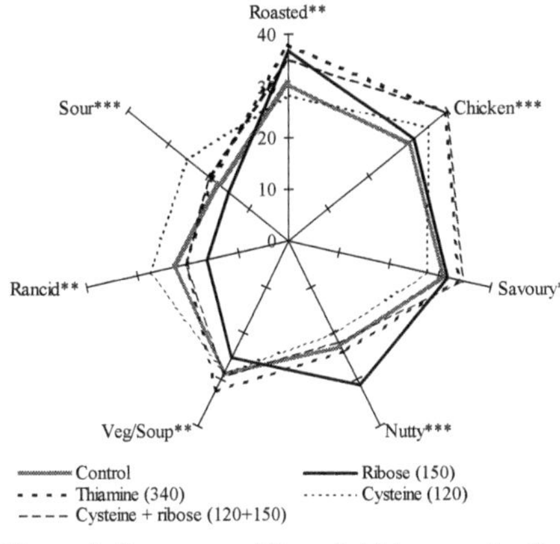

Figure 1: Sensory profiling of chicken meat with precursors added (mg/100g) prior to cooking.

This experiment confirms that many precursors have the potential to contribute to meaty odour and flavour. However, the concentrations added in Figure 1 are, in most cases, based on literature reports which were often conflicting. Analysis of chickens from commercial sources showed that the concentrations of these precursors (Table 1) were often considerably lower than expected. Considerable natural variation between birds (Table 1) and between commercial sources (Aliani and Farmer, 2002; Aliani and Farmer, 2005b) was demonstrated; this variation could be up to 4 fold between the maximum and minimum concentrations determined

On the basis of these data, further sensory studies were conducted to determine whether such differences would be large enough to explain the difference between well-flavoured and bland-flavoured chicken. Sensory studies, using added concentrations of thiamine, IMP and ribose-5-phosphate at low multiples of the natural concentration, did not alter 'chicken' or 'roasted' odour (Aliani and Farmer, 2005c). However, one reducing sugar, ribose, significantly ($P<0.05$) and consistently increased 'roasted' or 'chicken' odour when added at only 2-4 fold the natural concentration, which is close to the natural variation of ribose in meat (Table 1). Similar experiments showed that added ribose could also increase "roasted flavour" (Aliani and Farmer, 2005c).

It is impossible to add precursors into the same biochemical environment that they would naturally occupy in raw meat. Therefore, a sensory comparison was made of chicken breasts with naturally high concentrations of ribose (23-32 mg 100 g^{-1}) and those with lower ribose concentrations (7-12 mg 100 g^{-1}). This 2-3 fold natural difference in ribose concentration was sufficient to significantly increase 'roasted chicken flavour' (Aliani and Farmer, 2005c).

These results confirm that small changes in ribose concentration can affect chicken flavour.

The origin of ribose in meat is generally proposed to be via the breakdown post-slaughter of ATP to ADP, AMP and IMP, then to inosine and further into hypoxanthine and ribose (Lawrie, 1985). Analyses showed that the concentration of ribose gradually increases after slaughter and during chilled storage (Figure 3). However, this reaction does not go to completion and most of the ribose present in meat is present as a component of IMP and inosine, even after 200 hours post-slaughter (Aliani and Farmer, 2005a).

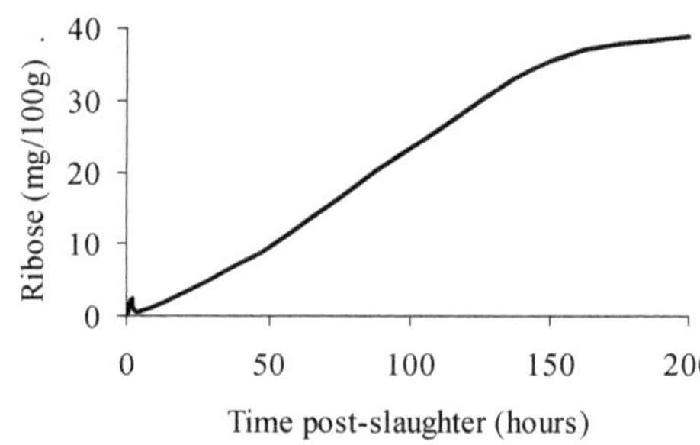

Figure 2: Effect of time post-slaughter on ribose concentration in chicken breast meat.

Conclusions

Sensory and analytical studies have provided strong evidence to suggest that ribose is a key precursor of flavour in cooked chicken meat. Differences in ribose concentration as small as 2-4 fold can cause perceptible differences in odour and flavour of cooked chicken. Concentrations of ribose increase post-slaughter until beyond the 'sell-by date'. Even at this time, a considerable proportion of potential ribose remains 'locked up' as IMP and inosine. Further studies are needed to determine the factors important for the formation of ribose in meat.

Acknowledgements

We gratefully acknowledge funding and support from the Department of Education and Learning and from Moy Park Ltd, both from Northern Ireland.

References

Aliani M. and Farmer L.J. (2002) Postcolumn derivatization method for determination of reducing and phosphorylated sugars in chicken by high performance liquid chromatography. Journal of Agricultural and Food Chemistry, 50, 2760-2766.

Aliani M. and Farmer L.J. (2005a). Identification of key flavour precursors in chicken meat: In: Proceedings of the 11th Weurman Symposium, Roskilde, Denmark, June 2005 (in press).

Aliani M. and Farmer L.J. (2005b). Precursors of chicken flavor. I. Determination of some flavor precursors in chicken muscle. Journal of Agricultural and Food Chemistry 53, 6067-6072.

Aliani M. and Farmer L.J. (2005c). Precursors of chicken flavor. II. Identification of key flavor precursors using sensory methods. Journal of Agricultural and Food Chemistry 53, 6455-6462.

Farmer L.J., Nolan M. and McAlinden E. (2000). Studies on the origin of chicken flavour. In: Schieberle, P., Engel, K.-H. (eds), Frontiers of Flavour Science. Deutsche Forschungsanstalt fur Lebensmittelchemie, Garching, Germany, p 21-28.

Lawrie RA (1985). Meat Science. Pergamon Press, Oxford.

BIOGENIC AMINES AS A QUALITY INDEX FOR LIGHTLY PRESERVED MEAT PRODUCTS: ACRONYM (BIAMINDEX)

D.V. Byrne*[1] and J.J. Leisner[2]

[1] Department of Food Science (IFV), Sensory Science, Royal Veterinary and Agricultural University (KVL) Rolighedsvej 30, 5., DK-1958 Frederiksberg C., Copenhagen, Denmark. [2] Department of Veterinary Pathobiology (IVP), Royal Veterinary and Agricultural University (KVL), Grønnegårdsvej 15, DK-1870 Frederiksberg C., Copenhagen, Denmark. E-mail: dby@kvl.dk

Keywords: sensory, consumer, shelf life, multivariate analysis

Introduction

The overall aim of the present project proposal is to develop a methodology whereby sensory deterioration to a 'rational sensory rejection point' (a products 'shelf-life') can be established through monitoring chemically the production of biogenic amines by the natural microflora in 'lightly preserved meat products'. Thus, a quality indexing method can be established giving a sound chemical methodology for the prediction of 'sensory based product shelf-life' (Figure 1). This will enable improvement in product quality, potentially enhance designated shelf-life, and certainly reduce product waste.

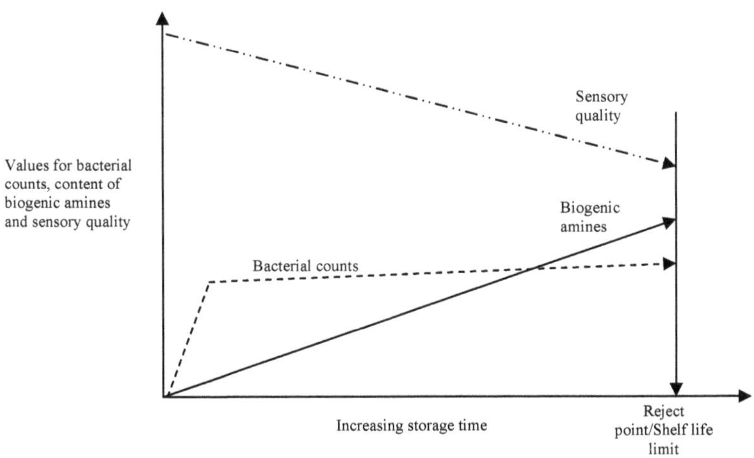

Figure 1: Schematic of the project: Biogenic amines as a quality index for lightly preserved meat products.

It has previously proven extremely difficult to apply conventional total or selective microbial counts for the prediction/indexing of 'shelf-life' as defined by sensory/consumer assessment for many types of food products (McMullen and Stiles 1989). Particularly, in the case of 'lightly preserved meat products', the use of such microbial counts has proven highly ineffective in the prediction of the limit of sensory defined shelf-life. These products, predominated by lactic acid bacteria (LAB), will on the one hand exhibit a comparatively long shelf life but on the other hand the exact sensory shelf life is difficult to predict as the numbers of bacteria tend to plateau and ultimately decrease over storage time. Thus, a correlation with the consistently decreasing sensory quality has not been data analytically viable.

Materials and Methods

In the present project we will monitor and utilise the production of biogenic amines produced by LAB, as well as Enterobacteriaceae, in lightly preserved meat products while stored vacuum packed in modified atmosphere packaging and under aerobic storage. As the biogenic amines produced by these bacteria increase consistently with storage (Jørgensen *et al.*, 2000), they may be used in combination with measurements of sensory deterioration over storage time to predict and derive a quality index, allowing the designation of a 'rational reject point' which will designate the products 'point of sale' shelf-life (Figure 1). Additionally, measurements of the opened product will allow the indexing of the products shelf-life in the consumer context (Muñoz and Chambers, 1993). The composition of the bacterial flora in this product as well as the ability of representative bacterial isolates to produce amines will be examined. The effect of parameters such as pH, content of salt, storage temperature and atmosphere composition on bacterial production of amines will be analysed by use of pure and mixed bacterial cultures growing in laboratory media as well as in meat extracts of 'lightly preserved' meat products.

Subsequently, the project will examine the sensory quality and shelf life of products on the basis of expert and consumer sensory analyses of products in close collaboration with the meat industry. The sensory data will be compared with other product data including traditional total microbial counts, selective counts for lactic acid bacteria as well as indicative counts for amine producing bacteria and chemical data, specifically the content of biogenic amines, pH and salt content in addition to processing data including time/temperature relationships during processing and distribution of products.

A multivariate data analytical strategy utilizing regression based predictive/causal analysis (e.g. Partial Least Squares Regression or PLSR) will be utilized in the linkage of the various data sets (Martens and Martens, 2001), leading ultimately to the development of a 'biogenic amine based sensory quality index', BiAmIndex in acronym (Figure 2).

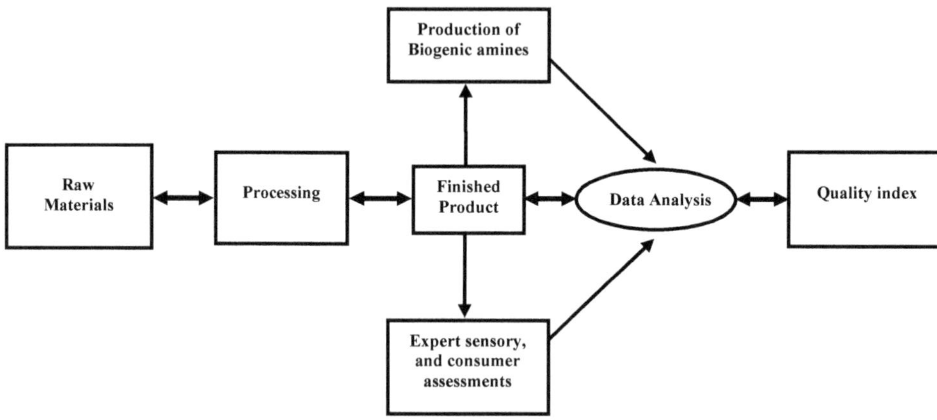

Figure 2: Schematic of the use of biogenic amines and sensory/consumer measurements in the product production chain as a quality index for 'lightly preserved meat products'.

Results and Discussion

Overall, there is a pressing need for the development and evaluation of new methods for the prediction and indexing of decreasing consumer/sensory based shelf-life quality in 'lightly preserved' meat products. The development of such an index will enable the shelf-life of these products to be controlled efficiently and ultimately, we postulate, may allow shelf-life extension from a sensory perspective, via more efficient quality control of these products during production.

Acknowledgements

This research is funded by the Danish Meat Association under the Basic Research Funding Programme III and Danish Ministry of Food, Agriculture and Fisheries under the research programmes "Food technology, safety and quality", and "The Innovations Law".

References

Jørgensen, L.V., Dalgaard, P. and Huss, H.H. (2000). Multiple compound quality index for cold-smoked salmon (*Salmo salar*) developed by multivariate regression of biogenic amines and pH. Journal *of Agriculture and Food Chemistry*, 48, 2448-2453.

McMullen, L.M. and Stiles, M.E. (1989). Storage life of selected meat sandwiches at 4°C in modified gas atmosphere. *Journal of Food Protection*, 52, 792-798.

Muñoz, A. M. and Chambers IV, E. (1993). Relating measurements of sensory properties to consumer acceptance of meat products. *Food Technology*, 47, 128-131, 134.

Martens, H. and Martens, M. (2001). Multivariate Analysis of Quality. An Introduction. United Kingdom: John Wiley & Sons.

LIPID OXIDATION AND SENSORIAL ACCEPTANCE OF GAS-STUNNED SUCKLING LAMB MEAT

M.B. Linares[1,2], R. Bórnez[1,2], M.I. Berruga[1,2] and H. Vergara *[1,2]

[1.] Departamento de Ciencia y Tecnología Agroforestal, Universidad de Castilla-La Mancha, 02071, Albacete, Spain, [2.] Sección Calidad Alimentaria, Instituto de Desarrollo Regional, Universidad de Castilla-La Mancha, 02071, Albacete, Spain. Email: herminia.vergara@uclm.es

Keywords: lamb, rancidity, acceptance, carbon monoxide

Introduction

Lamb meat is highly valued by consumers in Spain, even though it is considered as a luxury item in the daily diet due to its organoleptic characteristics and high price (Vergara and Gallego, 2001). However, the increased price factor is of secondary importance upon purchase, since consumers are more influenced by such factors as image and quality (Bernabeu and Tendero, 2005). Some processes such as lipid oxidation of muscle could contribute to a change in sensorial characteristics, mainly colour and odour (Liu *et al.*, 1995). Consumers in many countries complain of the rancid taste in meat packed in the habitual modified atmospheres rich in oxygen (Knut and Guy, 2006). It is therefore necessary to maintain an attractive visual colour of meat without off-odour even up to three weeks post-packing (Krause *et al.*, 2003) in order to reduce loss of freshness and to offer greater possibilities of commercialization.

Materials and Methods

Ten suckling lambs of the Manchego Spanish breed were slaughtered at 12.8 kg live weight by commercial procedure after gas stunning. The *Longissimus dorsi* (LD) muscle from both sides was removed from each chilled carcass (at 24 h post-mortem) and cut into 9 portions of similar size, giving a total of 90 samples. The samples were preserved in modified atmosphere (MA). Three types of gas mixtures were used, Type A: 30%CO_2+70%O_2; Type B: 30%CO_2+69.3%N_2+0.7%CO; Type C: 40%CO_2+60%N_2. Ten samples were analysed per atmosphere type and per sampling time (at 7, 14 and 21 days of storage at 2°C). Sample rancidity was determined in duplicate from 2 g of the LD muscle as described by Botsoglou *et al.*, (1994) by determining 2-thiobarbituric acid-reactive substances (TBARS). A panel of five experts assessed sensorial colour and odour. Colour was categorised in closed packs, and odour immediately after opening the pack. For both colour and odour, samples were scored as 1= not acceptable, 2= acceptable or 3= very acceptable. Data were analysed using analysis of variance to determine the effect of atmosphere type on rancidity. When the differences among types of modified atmospheres were significant ($p < 0.05$), Tukey's test was carried out to check the differences between pairs of groups.

Results and Discussion

In general, acceptability of packs upon visual and odour evaluation was higher in Type B atmosphere (with CO) and lower in Type A (with oxygen) at all times of storage (Table 1). MA type C was intermediate.

Table 1: Percentage of acceptable packs according to sensorial colour[1] and odour[2].

	Time post-packaging	Type A	Type B	Type C
Colour	7 days	100	100	96
	14 days	36	100	83
	21 days	25	100	96
Odour	7 days	82	100	79
	14 days	25	93	79
	21 days	14	46	50

[1]Closed packs, [2]When packs were opened; Modified Atmosphere: Type A:30%CO_2+70%O_2, Type B:30%CO_2+69.3%N_2+0.7%CO, Type C:40%CO_2+60%N_2. Samples were categorised (for both colour and odour) as: 1 = not acceptable (strong off-odour/colour), 2 = acceptable (slight off-odour/colour) or 3 = very acceptable (no off-odour/colour).

Samples packed in Type B showed the lowest TBARS values and highest oxidative stability throughout time without significant differences at 7, 14 and 21 days post-packaging (Figure 1). According to Krause *et al.*, (2003) the use of a low concentration of CO could ensure the inhibition of the lipid oxidation while improving the visual appearance (colour) and odour of meat.
On the other hand samples from the MA type A (high O_2 concentration) obtained the highest rancidity levels. The increase in the TBARS level coincided with the slight off-odours detected by the sensory panellist at 14 days of storage.

The high oxygen concentration in the packs favours oxidative processes, while the products of lipid oxidation have been associated with off-odours, thus agreeing with Jeremiah (2001).
Samples packed with Type C (without O_2) raised the intermediate TBARS values and an adequate acceptance by panellists, although in general lower than Type B. The presence of a certain residual level of oxygen inside the packs could be enough to produce lipid oxidation and a slight but not excessive alteration of sensorial meat quality parameters such as colour and odour (Smiddy *et al.,* 2002).

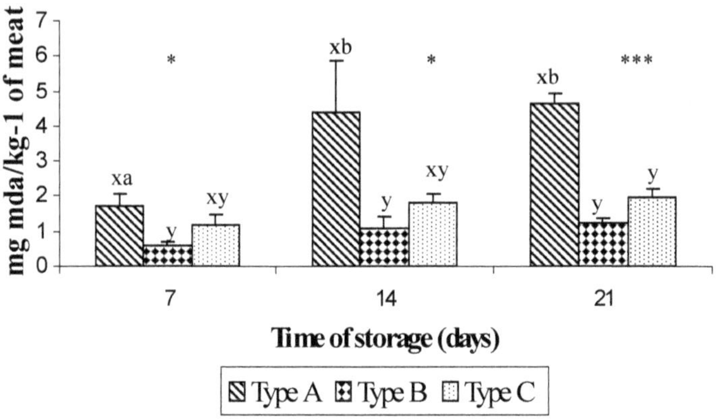

Figure 1: [a,b] indicate significant differences ($p < 0.05$) for the same type of modified atmosphere (Type A:70%O_2+30%CO_2; Type B:69.3%N_2+30%CO_2+0.7%CO; Type C: 60%N_2+40%CO_2) at different times of storage. [x, y] indicate significant differences ($p < 0.05$) between different types of modified atmosphere for the same period of storage. *, *** indicate significant levels at 0.05 and 0.001 respectively.

Conclusions

A low concentration of carbon monoxide could be a possible alternative to the packaging of fresh lamb meat since it can guarantee suitable colour and odour, along with minimum lipid oxidation.

References

Bernabéu, R.,Tendero, A. (2005). Preference structure for lamb meat consumers. A Spanish case study. Meat Science, 71: 464-470.

Bostoglou, N. A., Fletouris, D. J., Papageorgiu, G. E., Vassilopoulus, V. N., Mantis, A. J., Trakatellis, A. G. (1994). Rapid, sensitive and specific thiobarbituric acid method for measuring lipid peroxidation in animal tissue, food and feedstuff samples. Journal of Agricultural and Food Chemistry, 42: 1931-1937.

Jeremiah, L. E. (2001). Packaging alternatives to deliver fresh meats using short- or long-term distribution. Food Research International, 34:749-772.

Knu, Ph., Guy, D. (2006). Envasado con CO: una nueva tecnología de envasado sin oxígeno para la industria cárnica de la UniónEuropea. Eurocarne, 143: 195-199.

Krause, T. R., Sebranek, J. G., Rust, R. E., Honeyman, M. S. (2003). Use of carbon monoxide packaging for improving the shelf life of pork. Journal of Food Science, 68(8): 2596-2603.

Liu, Q., Lanari, M. C., Schaefer, D. M. (1995). A review of dietary vitamin E supplementation for improvement of beef quality. Journal of Animal Science, 73: 3131-3140.

Smiddy, M., Papkovskaia, N., Papkovsky, D. B., Kerry, J. P. (2002). Use of oxygen sensors for the non-destructive measurement of the oxygen content in modified atmosphere and vacuum packs of cooked chicken patties: impact of oxygen content on lipid oxidation. Food Research International, 35:577-584.

Vergara, H., Gallego, L. (2001). Effect of gas composition in modified atmosphere packaging on meat quality of Spanish Manchega lamb. Journal of the Science and Agriculture, 81: 1353-1357.

THE USE OF TASTE AND ODOUR SENSORS TO EVALUATE THE FLAVOUR OF BEEF DURING CONDITIONING

T. Okayama*[1], M. Iwasaki[1], H. Kamisoyama[1], M. Muguruma[2], M. Kaya[3], K. Nodake[3], M. Numata[3], M. Okada[4], J. Kita[4], E. Tokuyama[5] and T. Uchida[5]

[1.] *Faculty of Agriculture Kobe University, Kobe 657-8501, Japan,* [2.] *Faculty of Agriculture, University of Miyazaki, Miyazaki 889-2192, Japan,* [3.] *Central Research Institute, Ito Ham Food , Inc. Moriya 302-010, Japan,* [4.] *Analytical and Measuring Instruments Division, SHIMADZU Corp. Kyoto 604-8511, Japan,* [5.] *School of Pharmaceutical Sciences, Mukogawa Women's University, Nishinomiya 663-8179, Japan. Email:okayama@kobe-u.ac.jp*

Keywords: beef, conditioning, flavour, taste sensor, odour sensor

Introduction

In general, it is well known that beef flavour is improved by conditioning. The flavour of beef has become one of the most significant factors in consumer decision processes (Bruce *et al.*, 2005). Flavour results from compounds are divided into two broad classes: Those responsible for taste and those responsible for odours, the latter often designated as aroma substances (Belitz *et al.,* 2000). Compounds responsible for taste are generally nonvolatile at room temperature. The five important basic taste perceptions are provided by: sour, sweet, bitter, salty and umami compounds. Especially, glutamic acid, IMP and peptides in beef contribute to the delicious taste (umami). Aroma substances are volatile compounds which are perceived by the odour receptor sites of the smell organ. There are many odour substances which contribute to the aroma of boiled beef (Coppock and MacLeod, 1977). Matsuishi *et al.,* (2004) reported that aroma is the most important contributor to the identification of the animal spices of meat, and the contribution of taste appears much smaller than that of aroma and texture. On the other hand, the objective sensory evaluation of meat flavour is made so difficult by the conditions of selection of panelists, sample procurement and preparation, cooking and sensory presentation, establishing beef flavour profiles etc. Recently, taste and odour sensors technology imitating human's tongue and nose is steadily improving. Already, Taste Sensing System SA402B equipped with taste sensors and Fragrance and Flavour Analyzer FF-2A equipped with odour sensors have been developed by Japanese companies. The purpose of this study was to examine the possibility of the use of taste and odour sensors to evaluate the flavour of beef during conditioning.

Materials and Methods

1. Materials: *Semimembranosus* muscles from five Japanese Black Cattles approximately 29 month of age were used. 2. Methods of conditioning: Samples are conditioned at 2-4°C in refrigerator for 7, 14, 21 and 28 days after slaughter. 3. Types of sample for sensor measurements: Raw, boiled at 75-80°C for 30 min and cooked at 160-200°C for 2.3 min on a hot plate (griddle). Samples were subjected to various sensor measurements. 4. Sample preparation: For taste sensor measurement, raw, boiled and cooked samples were homogenized in deionized water at 10,000 rpm for 2 min. The homogenate was centrifuged at 10,000 x g for 30 min. The upper layer filtered through an Advantec 5A filter paper. The filtrate was frozen in a deep freezer at -80°C in preparation for a taste sensor measurement. Sample for odour sensor was thin sliced and frozen in deep freezer at -80°C. Each sample was thawed out in refrigerator and used to the measurement. 5. Sensory evaluation: Taste and odour sensory evaluation was conducted by paired comparison method. Panelists were our laboratory staff and students. 6. Sensor instruments: Taste Sensory System SA402B (Intelligent Sensor Technology, Inc. Japan) was used for taste sensor imitating human's tongue. This system, the first "Equipment carrying out scientific knowledge of the taste" in the world, has been developed in Japan. This instrument constructed to detect tastes such as bitterness, sourness, astringency, sweetness, umami etc., a robot arm and computer. The detecting sensor part of the equipment consists of eight electrodes composed of lipid/polymer membranes. SHIMADZU Fragrance and Flavour Analyzer FF-2A was used for odour sensor imitating human's nose. This electronic nose resembles a human nose, with the receptor proteins replaced by multiple sensor elements having different properties. The signals generated by these sensor elements in response an odour are sent as vectors to a computer, which plays the role of the human brain. This system can display the odour and smell in absolute value and can reject the humidity interferences by using the trap tube.

Results and Discussion

From the measurement of beef taste conditioned days 7, 14, 21 and 28 using taste sensors, saltiness and sustainable umami intensity were shown to have a gradual higher tendency during conditioning of beef (Figure 1). Figure 2 shows that changes in umami derived from amino acids, nucleotides and sustainable umami intensity during conditioning. Umami and sustainable umami intensity were measured by a separate taste sensor with deferent characteristics. Umami by amino acids and nucleotides was not changed during conditioning, but sustainable umami intensity in the beef conditioned for 14, 21 and 28 days was higher than in the beef conditioned for 7 days, except for a single sample conditioned for 21 days. It is interesting that the sustainable umami intensity evaluated by taste sensor was strengthened during conditioning of beef.

From the radar chart of cooked steaks odour conditioned days 7, 14, 21 and 28 using odour sensors, the intensity of some odour types appeared to have a slightly higher tendency during conditioning of beef (Figure 3). Figure 4 shows the amines intensity was rapidly enhanced from days 14 to days 28, and hydrogen sulfide intensity of steaks was promoted from 14 days to 21 days of conditioning. Further investigations are needed to ascertain the relationship between sensory evaluation using panelists and the output of taste and odour sensor systems.

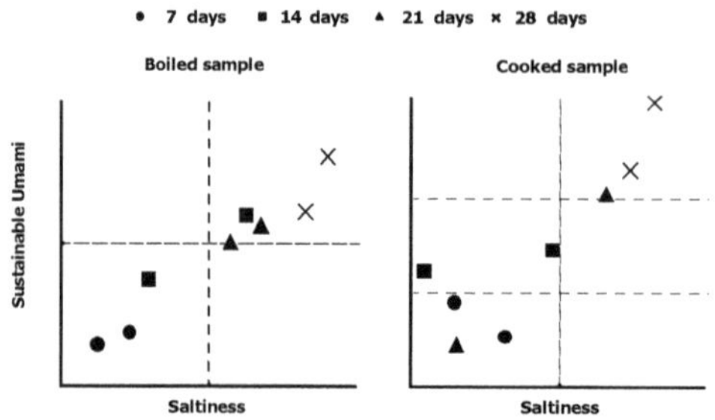

Figure 1: Changes in sustainable umami and saltiness during conditioning of beef measured by taste sensors (n=2).

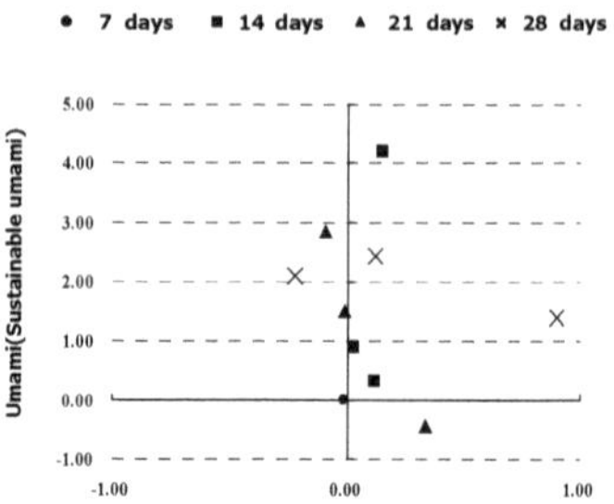

Figure 2: Changes in sustainable umami and umami during conditioning of beef measured by taste sensors (n=3). Based on the calculated data, value of sample conditioned 7 days was denoted in zero point.

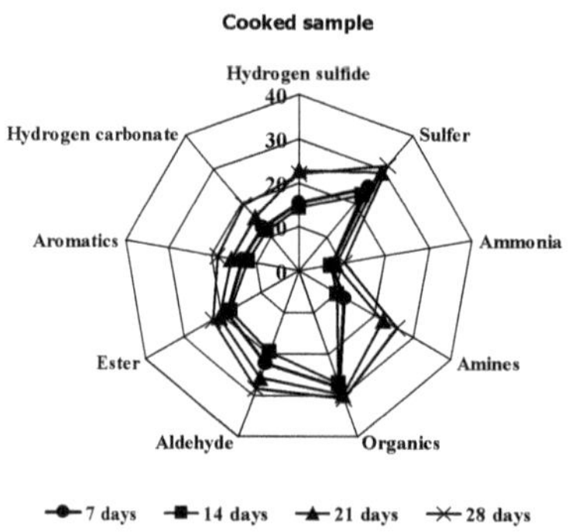

Figure 3: Changes in odour intensity during conditioning of beef measured by odour sensors.

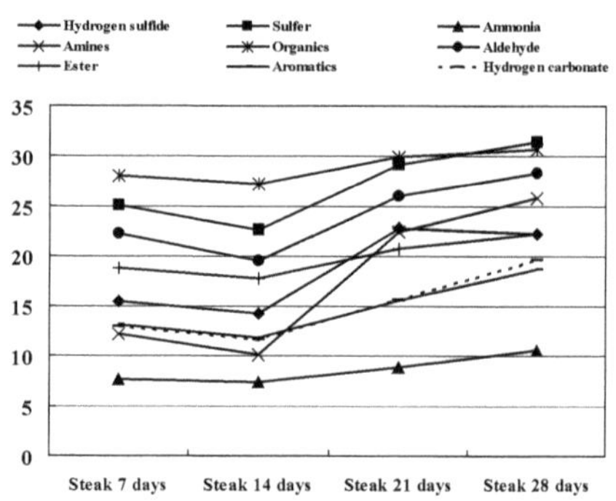

Figure 4: Changes in odour quality during conditioning of beef measured by odour sensors.

Conclusions

This study was aimed to examine the possibility of the use of taste and odour sensors to evaluate the flavour of beef during conditioning. From the results of beef taste conditioned at days 7, 14, 21 and 28 using taste sensors, sustainable umami intensity was increased by conditioning of beef. The results of radar chart and the intensity of some quality of odour of steaks using odour sensors exhibited the change of aroma of beef by conditioning. It is suggested that the taste and odour sensors from these results can be utilized as useful and powerful support tools of sensory evaluation.

References

Belitz,H.-D., Grosch.W., Schieberle,P. (2000). Food Chemistry, 3rd Eds. pp.342. Springer-Verlag.

Bruce,H.L., Beilken, S.L., Leppard,P. (2005). Variation in flavor and textural descriptions of cooked steaks from bovine *m.longissimus thoracis et lumborum* from different production and aging regimes. Journal of Food Science, 70:S309-S316.

Coppock,B.M., MacLeod,G. (1977). The effect of ageing on the sensory and chemical properties of boiled beef aroma. Journal of Science and Food Agriculture, 28: 206-214.

Matsuishi,M., Igeta,M,. Takeda,S., Okitani,A. (2004). Sensory factors contributing to the identification of the animal speices of meat. Journal of Food Science, 69: S218-S220.

RELATIONSHIP BETWEEN INTERNAL TEMPERATURE AND PARAMETERS MEASURED BY AN ELECTRONIC NOSE DURING MEAT COOKING

P. Cornale[1], S. Barbera*[1], G. Sala[2], and G. Masoero[2]

[1]*Dipartimento di Scienze Zootecniche, Università di Torino, via L. da Vinci 44, 10095 Grugliasco (TO), Italy* [2]*C.R.A. - Istituto Sperimentale per la Zootecnia, via Pianezza 115, 10151 Torino, Italy Email: salvatore.barbera@unito.it*

Keywords: electronic nose, cooked meat, temperature

Introduction

Food contains several volatile compounds that are responsible for, among other properties, odour (Maul, 1998). The literature relating to the volatile compounds found in meat suggests that over 1000 chemicals have been identified (Mottram, 1998). Raw meat has only a blood-like taste but during cooking, mainly due to the Maillard reaction and thermal degradation of lipids, a wide range of volatile compounds is developed. Aroma and flavour evaluation are of great importance in food quality assessment. These analyses are usually carried out by trained panels and/or gas chromatography/mass spectrometry (GC/MS). However, the human sensory panel has a number of inherent disadvantages including subjectivity, repeatability, set-up time and costs. Also, instrumental techniques like GC/MS require much time and have a high operating cost. So researchers are constantly investigating alternative analytical solutions. One response to this problem is the electronic nose. From Persaud and Dodd (1982) this new technology is used in the agricultural sector to evaluate odours of products and food like meat, grains, coffee, beer, cheese, sugar, mushrooms etc (Schaller *et al.*, 1998). This experiment was to evaluate the relationship between the increasing internal temperature during meat cooking and the parameters measured by an electronic nose with the aim of defining the odour profile of meats.

Materials and Methods

Eight beef and 8 pork *Longissimus dorsi* muscle portions refrigerated at 4°C were used as meat sources. From each steak, a 1 cm thick circular sample (Steak) was obtained according to the MCS method proposed by Barbera (2006). A 1 cm thick circular hamburger sample (Minced) weighing 30 g was produced using the minced meat of the same steak. The internal sample temperature was measured before cooking (0 min) which was performed at 165°C for 15 min in an electric forced-air convection oven. By forcing air into the oven, after passing through an active charcoal filter, regular cooking odour out-flow was guaranteed. PEN 2 (AIRSENSE Analytics GmbH, Hagenower, Germany) is a portable electronic nose (EN) with 10 metal oxide sensors (MOS) that change their resistance in the presence of oxidising and reducing gaseous compounds (Kohl, 1992). The ten PEN2 sensors analyse 10 classes of chemicals: 2 sensors for aromatic (w1c and w3c), broad range (w5s), hydrogen (w6s), aromatic-aliphatic (w5c), broad-methane (w1s), sulphur-organic (w1w), broad-alcohol (w2s), sulphur-chloride (w2w) and methane-aliphatic (w3s). This instrument was utilised in the continuous monitoring of meat volatile compounds produced during 15 minutes cooking. The temperature probe was put into the sample at the start of cooking and the internal temperature was recorded every 30 s. In all, 32 steak and minced samples were used. Relationships between internal temperatures and sensor parameters were detected and analysed. Statistical analysis was performed by SAS (2001).

Results and Discussion

Several average internal temperatures are presented in Table 1. The internal temperature shows an obvious increasing trend during cooking but after the 11th minute, the sample preparation (steak or minced) causes a significant difference. The inclusion of air into minced meat can account for the observed delay in temperature increase. The Pearson correlation coefficients between internal temperature during cooking and the ten sensor parameters show a strong relationship ($P<0.0001$) for both steak and minced treatments.

Table 1: Means ± SD at different cooking times.

Time (min)	Temperature °C		
	Minced	Steak	Pr
0	17.2±1.34	17.7±1.83	NS
5	43.0±3.44	40.8±3.48	NS
10	68.6±3.01	69.9±4.51	NS
15	84.1±4.66	91.3±4.05	<0.0001

As shown in Figure 1, using the standardised data, the correlation is not always linear but polynomial. In general, sensors show an initial, linear phase: an increase in temperature causes a corresponding signal increase. The following phase shows a lack of correlation as the sensor parameter does not rise with the temperature (steak). Finally the sensor response starts rising again with the increase in temperature. It is possible to note the different trend between minced and steak preparations.

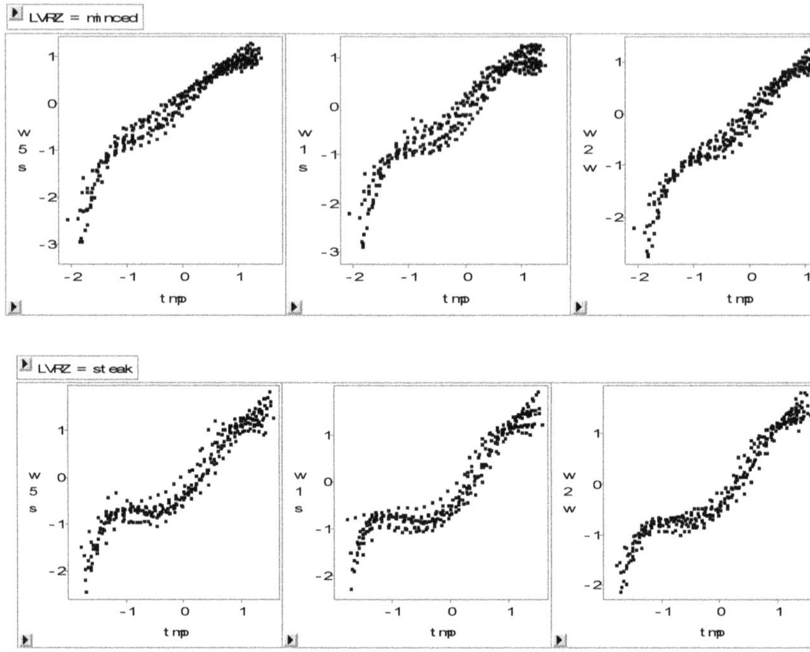

Figure 1: The relationships in minced and steak samples (lvrz) between the internal temperature (tmp) during cooking and the standardised output of three sensors (w5s, w1s, w2w) measured by an electronic nose.

Conclusions

The electronic nose appears to be useful for the investigation of global aroma characteristics in cooking meat. It emerges from this experiment that the chemicals produced during cooking are dependent in different ways on the internal temperature and the sampling mode. A minced preparation is considered to be a more representative sampling in comparison to steak. It releases volatiles in more homogeneous patterns and is less dependant on human error. Further experiments, to better define the correlation patterns, are necessary.

References

Barbera, S. (2006). Meat Cooking Shrinkage: measurement of a new meat quality parameter. Meat Science (in press).

Kohl, D. (1992). Oxidic semiconductor gas sensors. In: Sberveglieri, G. (Ed), Gas sensors. Dordrecht: Kluwer Academic Publishers, pp. 43-88.

Maul, F. (1998). Application of an "electronic nose" technology for quality assessment of food. In: Developing premium-quality programs for fresh fruits and vegetables. Florida Postharvest Horticulture Institute and Industry Tour. University of Florida and the Florida Fruit and Vegetables Association.

Mottram, D.S. (1998). Flavour formation in meat and meat products: a review. Food Chemistry, 62, 415-424.

Persaud, K., Dodd, G.H. (1982). Analysis of discrimination mechanisms in the mammalian olfactory system using a model nose. Nature, 299, 352-355.

SAS (2001). SAS User's Guide. Carey NC: SAS Institute.

Schaller, E., Bosset, J.O., Escher, F. (1998). Review: 'electronic nose' and their application to food. Lebensmittel Wissenschaft und Technologie, 31, 305-316.

MULTIPLE PROTEIN PHASES APPROACH FOR BETTER UNDERSTANDING MEAT PRODUCT QUALITY - UNRAVELLING KEY INTERACTIONS BY USING A MULTIPLE PHASE AGGREGATION MODEL

C. Don, R.P. Happe, T.J. Verkleij* and G. Wijngaards

TNO Quality of Life, P.O. Box 360, 3700 AJ Zeist, the Netherlands. Email: verkleij@voeding.tno.nl.

Keywords: meat product quality factors, sarcoplasmic protein, myofibrillar protein, collagen, multi phase aggregation

Introduction

There is a general consensus that factors affecting meat quality and the quality of meat-based products are related to proteins. Over the years it has become an accepted notion that myofibrillar proteins are crucial in meat texture. (Acto)myosin in particular, is regarded as a protein that governs important textural characteristics of meat products, like stiffness and elasticity. Clearly, muscle proteins like myosin and actin are important estimators of meat quality. A cooking experiment with two types of deboned chicken meat with similar protein content reveals that stiffness and elasticity parallels myofibrillar protein content. Furthermore, cooking loss increases with decrease of myofibrillar protein. In this case it was clear that the textural quality of the cooked meat must be related to myofibrillar proteins. Systematic studies using meat models that focus on muscle protein provide useable insight, but they are single-protein-phase oriented in nature. Thereby important data on interactions between meat proteins, like salt-soluble myofibrillar (*incl.* myosins and actins), (water and salt insoluble) collagens and water-soluble sarcoplasmic proteins, are left out in the meat quality specification. In this paper we will discuss the importance of looking at the various protein phases besides myosin.

Materials and Methods

In this study 4 different types of chicken meat and 4 industrial meat composite samples were used.

Table 1: Samples and protein content.

sample	% total protein
A	10.92
B	8.42
C	19.87
D	23.16
E	14.34
F	15.53
G	16.25
H	16.85

Samples A – D, are various chicken derived meat fraction: from skin-rich fraction to chicken filet. Samples E- H are various deboned chicken meat samples.

The total protein and the protein content of the distinguished protein fractions: myosin, collagen and sarcoplasmic were determined. With each type of meat, a heat-induced gel was prepared by emulsifying the protein in a suspension of water with 2% NaCl. Cooking loss was determined. This resulted in a model-type meat sausage. Of these meat sausages the consistency was measured with a texture analyser. The textural properties were compared with various protein factors. For the content of specific fractions the cooking loss was corrected for, resulting in protein effectively present in the cooked meat gel.

Results and Discussion

When specifying meat quality, protein content of the uncooked starting material is often related to end product properties. In Figures 1and 2, it is demonstrated that protein content has no relation to texture.

Next to myosin there are also sarcoplasmic proteins present in meat products, however, this class of proteins also does not reveal an encouraging correlation with final textural properties (Fig 3). The result in figure 4 reveals an interesting correlation. The correlation revealed here indicates that the collagen phase and the myosin phase act together in determining final textural properties. Total protein, or myosin protein correlations do not reveal a good relationship with technological quality. Hence we propose a multiphase approach. This multiphase approach is discussed in the following section.

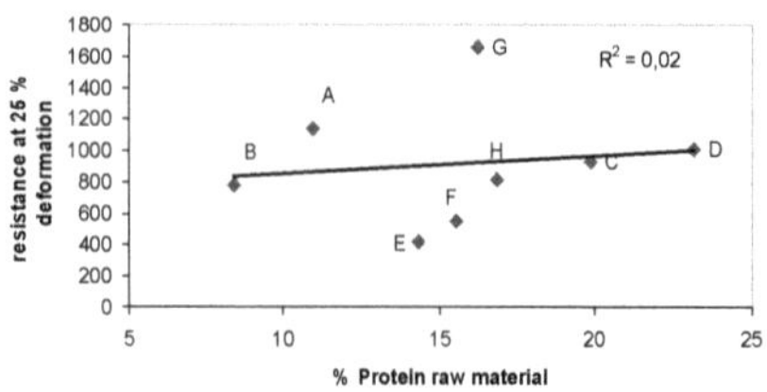

Figure 1.

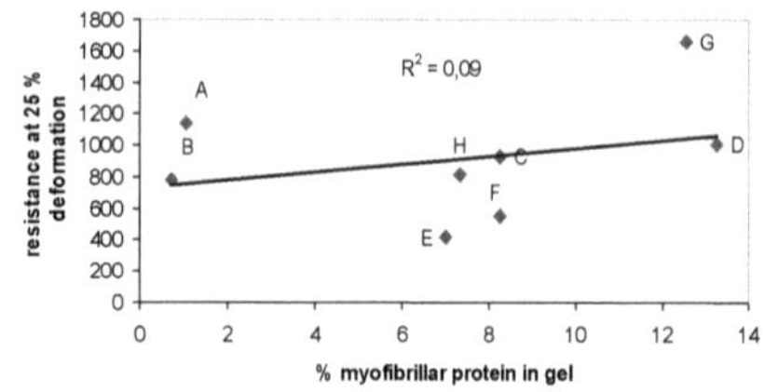

Figure 2.

Another factor often looked at, is the content of muscle proteins: myosin. This was determined for the gelled protein sausages but also here a poor correlation is revealed (Figure 2).

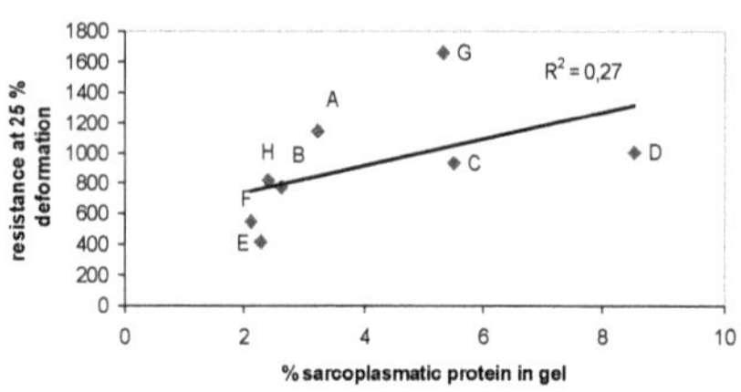

Figure 3.

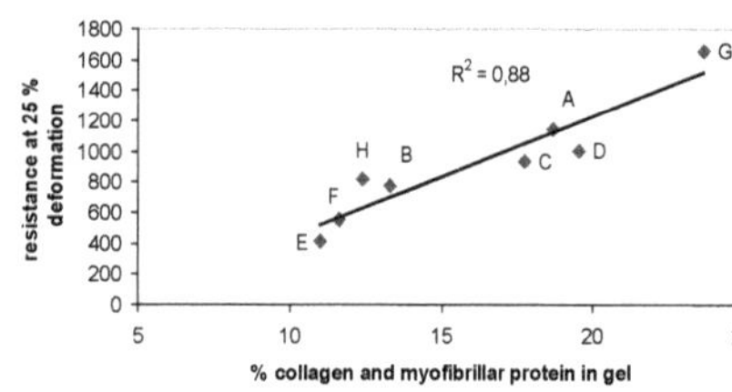

Figure 4.

In order to improve meat and meat product quality assessment, a multiple-protein-phases aggregation model is proposed. Viewing the various meat protein constituents as being of separate phase behaviour is for one justified by the fact that isolation of each protein requires specific conditions for solubility, in this case, extraction. These differences in solubility criteria indicate also a different behaviour during cooking. Sarcoplasmic proteins readily dissolve in water, during cooking these proteins denature and coagulate at 40-60°C. Myofibrillar proteins are not readily soluble in water, but when excess salt is added solubility improves. Myofibrillar proteins change structure at 60- 80°C. Collagens require strong base (NaOH) or SDS - urea to dissolve. During heat-up, collagens shrink at 50-60°C, only after prolonged heating collagens may soften and form a more gelatine-like phase. All these changes take place during the preparation of meat products.

By taking these typical differences into account the multiple-protein-phases aggregation model provides a framework to better understand the network formation phenomena in meat products. On the macro-level, heat-induced meat protein network can be viewed as a particle network. For the meso-level 3 types of protein-network phases are hypothesised:

1) a large aggregate network phase 'skeleton structure'
2) colloidal particle network sub-phase
3) polymeric gel sub-phase.

The water caught in the network formed has constituents that are molecularly dissolved; this is defined as liquid phase.

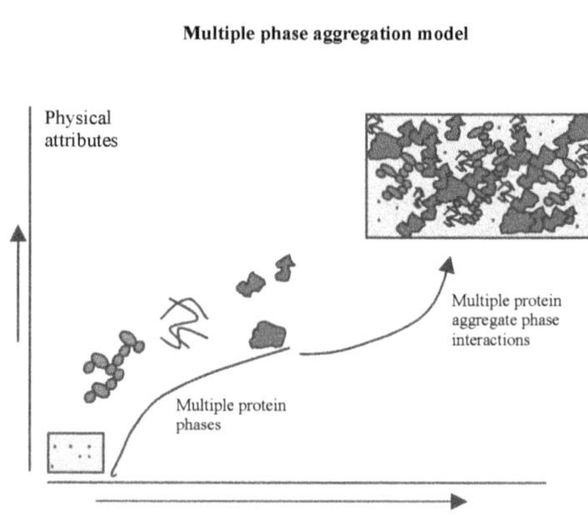

Conclusion

The framework of the multiple protein phase aggregation model helps to improve our understanding of final meat product quality factors. This will support the industry to select more effectively among various meat sources and the vast amount of additives intended to improve final product properties. Our latest results will be addressed during the ICoMST meeting.

THE INFLUENCE OF END-POINT COOKING TEMPERATURE ON SHEAR FORCE MEASUREMENTS

C.E. Devine[1]*, R.W. Wells[1] and M. North[2]

[1] *Bioengineering and Biosensors, HortResearch, PB 3123 Hamilton, New Zealand,* [2] *AgResearch MIRINZ Centre, PB 3123 Hamilton, New Zealand. Email: cdevine@hortresearch.co.nz*

Keywords: beef, cooking temperature, tenderisation, *longissimus dorsi*

Introduction

The way meat is cooked influences its palatability, hence for objective testing and to allow comparisons, we need to standardise the method of cooking. Tenderness (a consumer derived measurement) is determined by feeding cooked meat to consumers, whereas shear force is an objective measurement on a cooked piece of meat obtained from an instrument. The methods for preparing meat for shear force assessment have been widely varied; it may be cooked from fresh (AMSA,1995) or from the frozen state (Graafhuis *et al.*, 1991; Duckett *et al.*, 1998) or thawed for an indeterminate time (often overnight) before cooking (Locker and Daines, 1973; Jeremiah *et al.*, 1991; Ferguson *et al.*, 2000). In addition, there are variations in the end-point temperature to which meat is cooked, ranging from 65-80°C, with significant differences in tenderness between the lowest and highest temperatures (Graafhuis *et al.*, 1991; Bejerholm and Aaslyng, 2004), or meat may be cooked for a defined period of time at a given temperature (Ferguson *et al.*, 2000). Even if meat samples are cooked to the same end-point temperatures, using a different heating temperature (e.g. 160°C or 100°C) will result in vastly different temperature profiles from outside to the interior (Bejerholm and Aaslyng, 2004) creating layers at different degrees of doneness. Adding to the problem further is that many different cooking methods have also been used including contact frying, hot air roasting, radioactive grilling, liquid immersion and even microwaves.

As meat ages, the cytoskeletal proteins degrade and the contribution of these compounds diminishes with none being present in fully aged meat. Thus it would be expected that cooking temperature effects may differ for aged and non-aged meat. Actin degradation in the range 70-80°C is another influencing factor and hence end-point temperatures should be kept as consistent as possible and under or overshooting should be minimised.

The present experiments investigate the effect of cooking meat to four end-point temperatures from 65-80°C over a three-day period of ageing at 15°C. The chosen temperature range covered the likely end-point temperatures encountered, especially if the meat is not removed from the heating medium before there is an overshoot.

Materials and Methods

Eight *m. longissimus lumborum* from bulls that were electrically stunned and not electrically stimulated were obtained from an abattoir undertaking hot boning. The muscles were immediately transported to a laboratory and within 90min from slaughter were wrapped tightly in polyethylene film and held at 15°C to go into *rigor mortis* (pH values at *rigor mortis* ranged from 5.5-5.7). For convenience, the muscles were taken to a 4°C room and held overnight to reduce the rate of ageing. The next day the muscles were placed in a 15°C room to equilibrate and age at this temperature.

After 12, 24 and 48 hours of ageing, samples were taken from each muscle and were cooked in the following manner: From each muscle, four 30 mm thick slices (~250 g) were obtained and thermocouples placed into the centre of the muscles. The four samples were placed in weighted plastic bags and placed in an 85°C water bath. The temperatures were closely monitored so that when the first piece of meat reached 65°C, it was immediately removed and placed in ice water; this was repeated as the remaining meat samples reached 70, 75 and 80°C respectively. The cooking procedure was then repeated for each of the eight muscles at each of the ageing times. Every cooked sample was then cut along the muscle fibre axis using scalpel blades to produce at least six sub samples (termed bites) with a 1cm x 1cm cross section. Each bite was sheared using a MIRINZ tenderometer with a wedged shaped tooth (as described by Graafhuis *et al.*, 1991) and the peak shear force value was obtained. The entire experimental procedure gave shear force values at four cooking temperatures with eight *m. longissimus lumborum* samples aged for 3 different times.

Results and Discussion

It took approximately 13 minutes for the meat to reach 65°C and 20 minutes for the meat to reach the 80°C end-point temperature. In order to avoid an overshoot, the samples were removed from the water bath early so that the measured end-point temperature was as close as possible to the nominal value. Most measured end-point temperatures were within ± 1°C of the nominal value.

There was a wide variation in shear force values at all end-point temperatures, as shown by the standard error bars in Figures 1A and 1B. End-point temperature had the biggest impact on shear force values from meat aged for 12 and 24 hours, with much less impact after 48 hours of ageing (Figure 1A). Figure 1B showed significant differences between meat cooked to 65, 75 and 80°C ($p < 0.01$) after 12 hours of ageing, but no significant differences between the different end-point temperatures after 24 and 48 hours of ageing.

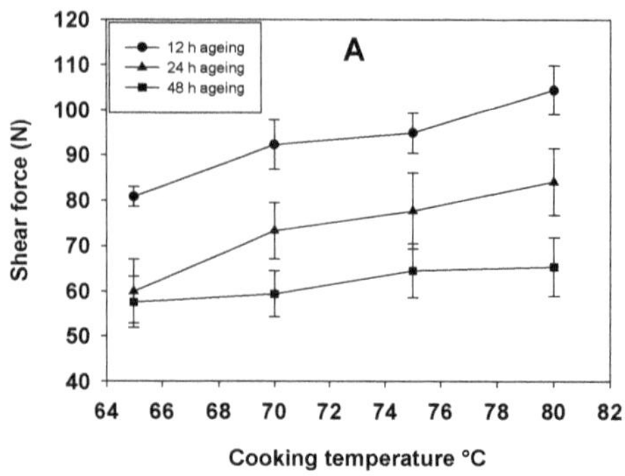

Figure 1A: The effect of ageing duration on the shear force values of meat cooked to various end-point temperatures.

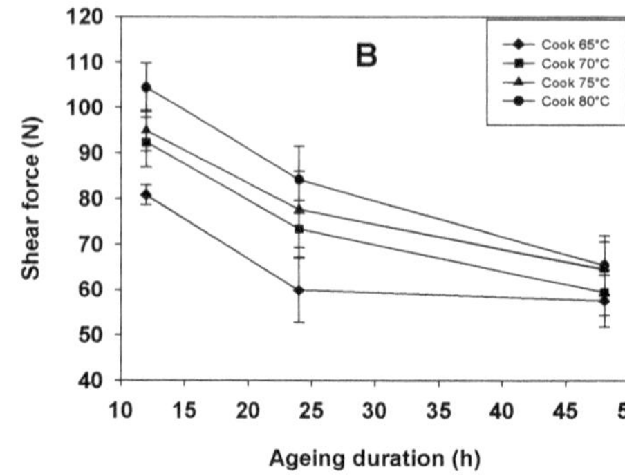

Figure 1B: The effect of end-point cooking temperature on the shear force values at three different ageing durations.

Analysis of the mean values showed that there was an average increase of 1.6N per degree for the range 65 to 80°C at 12 hours ageing, which reduces to 0.7N per degree for 24 hours ageing and 0.2N per degree for 48 hours ageing. Hence, at a nominal end-point temperature, a 5°C overshoot will result in mean shear force values that differ by about 8N for meat aged at 15°C for 12 hours. The differences in N in this instance are greater than the standard deviation of the bites of each meat sample where the range at 12h ageing is 24N.

Conclusions

Although an increase in the mean shear force values was observed for higher end-point cooking temperatures, increases for individual readings were generally obscured by the variation in shear force measurement. The end-point temperature only had a significant impact at the shortest ageing duration of 12 hours at 15°C. After longer ageing periods, the end-point temperature had no significant impact on shear force values.

References

American Meat Science Association (AMSA), (1995). Research guidelines for cookery, sensory evaluation and instrumental tenderness measurements of fresh meat, Chicago, IL.

Bejerholm, C. and Aaslyng, M.D. (2004). Cooking. In: Encyclopedia of Meat Science (Eds) Jensen, W.J., Devine, C.E. and Dikeman, M., 343-349, Academic Press, Oxford.

Cross, H.R., Stansfield, M.S. and Koch, E.J. (1976). Beef palatability as affected by cooking rate and final internal temperature. Journal of Animal Science, 43: 114-121.

Duckett, S.K., Klein, T.A., Dodson, M.V. and Snowder, G.D. (1998). Tenderness of normal and Callipyge lamb aged fresh or after freezing. Meat Science, 49: 19-26.

Ferguson, D.M., Jiang, S.T., Hearnshaw, H., Rymill, S.R. and Thompson, J.M. (2000). Effect of electrical stimulation on protease activity and tenderness of M. longissimus from cattle with different proportions of Bos indicus content, Meat Science, 55: 265-272.

Graafhuis, A.E., Honikel, K.O., Devine, C.E. and Chrystall, B.B. (1991). Meat tenderness of different muscles cooked to different temperatures and assessed by different methods. In: Proceedings 37th International Congress of Meat Science and Technology, 365-368, Kulmbach, Germany.

Jeremiah, L.E., Tong, A.K.W. and Gibson L.L. (1991). The usefulness of muscle color and pH for segregating beef carcasses into tenderness groups. Meat Science, 30: 97-114.

Locker, R.H. and Daines, G.J. (1973). The effect of repeated freeze-thaw cycles on tenderness and cooking loss in beef, Journal of the Science of Food and Agriculture, 24: 1273-1275.

VARIATION OF MEAT SHEAR FORCE MEASUREMENTS – A CHARACTERISTIC OF MEAT

C.E. Devine[1,*], R.W. Wells[1] and M. North[2]

[1.] *Bioengineering and Biosensors, HortResearch, PB 3123 Hamilton, New Zealand,* [2.] *AgResearch MIRINZ Centre, PB 3123 Hamilton, New Zealand. Email: cdevine@hortresearch.co.nz*

Keywords: lamb, shear force, sample variation, *longissimus dorsi*

Introduction

An objective measurement of tenderness is obtained by determining the force required to shear a cooked standardised piece of meat using a mechanical method of measurement (AMSA, 1995). However, there is wide variation in individual shear force values obtained. The mean values do not always reflect this variation and the variation itself does not follow a normal distribution. These issues have implications when developing non-destructive methods of measuring meat tenderness, such as near infrared spectroscopy (McGlone *et al.*, 2005). Such methods characteristically measure only a portion of the meat and it is important to know whether such measurements are representative or whether a larger series of measurements needs to be obtained to obtain a true representation of the meat being evaluated. The present study investigates the shear force variation from *rigor mortis* until meat is fully aged.

Materials and Methods

Frozen samples of *m. longissimus thoracis et lumborum* (LTL), aged for varying periods, were obtained from 462 sheep. The samples were cooked in a water bath at 85°C until reaching an internal temperature of 75°C and then chilled in ice. Each cooked sample was then cut along the muscle fibre axis using scalpel blades to produce six sub samples (termed bites) with a 1cm x 1cm cross section. All the bites were sheared with a MIRINZ tenderometer with a wedged shaped tooth, as described by Graafhuis *et al.* (1991).

Results and Discussion

To examine the variation between individual readings and mean readings, the first three bites from each sample were plotted against the second three bites from the same sample (Figure 1) and the mean values from each group of three were then plotted against each other (Figure 2). Figure 1 shows that there was a wide scatter between individual readings ($r^2 = 0.7883$) from the same sample. Unsurprisingly, Figure 2 shows that when the means of the first three bites were compared with the means of the second three, the scatter was reduced ($r^2 = 0.9162$). These results suggest that there is a wide variation of individual shear values for each meat sample, but this is obscured when the means are obtained.

The extraordinarily wide variation in shear values for the same piece of meat has not been generally addressed in previous studies, although it is known to occur as the same pattern is indicated with the standard deviations of the mean of each bite. It is known that there is a similar variation between different types of instruments, even on the same pieces of meat (Bekhit *et al.*, 2003) but this variation is also present when the same instrument is used, as in this study.

None of these factors are generally incorporated into tenderness evaluations, as most interpretations are not only focused on mean values, but means of several muscles. For any given piece of meat therefore, the individual shear value variation is masked when the means are compared, but a consumer would note this variation while eating the meat, and therefore it will influence the final perception of the meat quality. Consumers eat larger pieces of meat than are used for tenderometer measurements and make more than one bite on each portion of meat.

The variation may be caused by the measurement device, but his appears unlikely, because the variation reduces at lower shear force values, suggesting it is most likely dominated by the inherent variation of meat, which decreases as ageing progresses. This large variation will have a significant effect on the development of correlations to use near infrared spectroscopy to predict meat tenderness (McGlone *et al.*, 2005).

Conclusions

The results indicate that there is a large range of shear force values with each meat sample that is significantly reduced when these values are averaged. When meat has fully aged the variation is least. While mean values give a good representation of meat that has been fully aged, they do not accurately represent the characteristics of meat that has been insufficiently aged. The importance of this variation with regard to consumer acceptance and when it is used as a basis for the development of new measurement methods should not be underestimated.

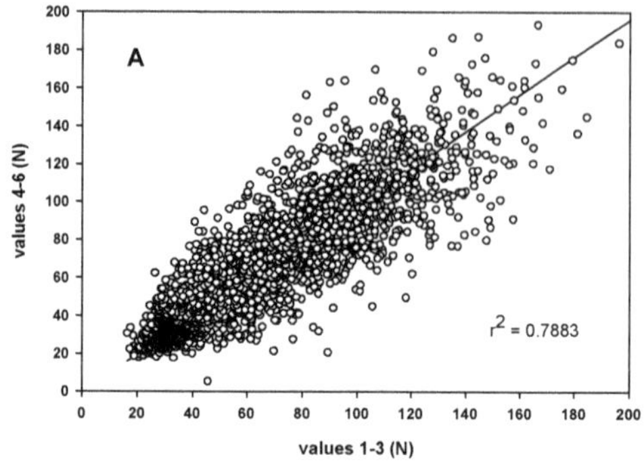

Figure 1: Scatter plot of 2772 shear force values from sheep LTL - Individual readings from the same samples (bites 1-3 vs. bites 4-6).

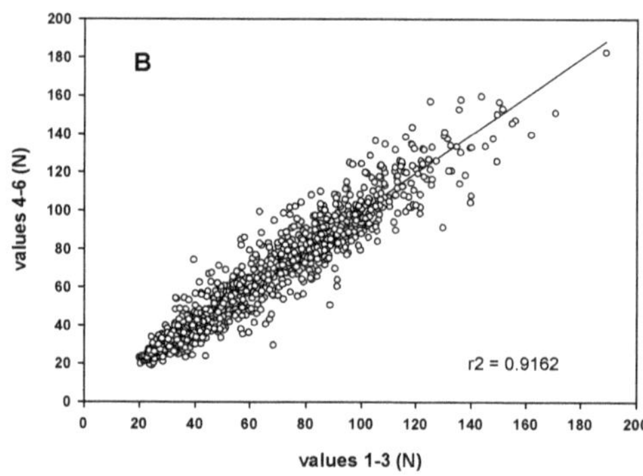

Figure 2: Scatter plot of 924 mean shear force values from sheep LTL - Mean readings from the same samples (mean of bites 1-3 vs. mean of bites 4-6).

References

American Meat Science Association (AMSA) (1995). Research guidelines for cookery, sensory evaluation and instrumental tenderness measurements of fresh meat, Chicago, IL:

Bekhit, A.E.D., Devine, C.E, Morton, J.D. and Bickerstaffe, R. (2003). Towards unifying meat shear force measurement systems to determine meat tenderness. In: Proceedings 49th International Congress of Meat Science and Technology, 221-222, Buenos Aries, Brazil.

Graafhuis, A.E., Honikel, K.O., Devine, C.E. and Chrystall, B.B. (1991). Meat tenderness of different muscles cooked to different temperatures and assessed by different methods. In: Proceedings 37th International Congress of Meat Science and Technology, 365-368, Kulmbach, Germany.

McGlone V.A, Devine, C.E. and Wells, R.W. (2005). Detection of tenderness, post rigor age and water status changes in sheep using near infrared spectroscopy. Journal of Near Infrared Spectroscopy, 13: 277-285.

INTERACTION BETWEEN PAN TEMPERATURE AND END POINT TEMPERATURE ON THE EATING QUALITY OF PORK WITH VARIOUS RAW MEAT QUALITIES

S. Støier[1]*, M.D. Aaslyng[1], L. Lauridsen[1,2], W.L.P. Bredie[2], and A.S. Jørgensen[2]

[1]*Danish Meat Research Institute, Maglegaardsvej 2, DK-4000 Roskilde,* [2]*The Royal Veterinary and Agricultural University, Rolighedsvej 30, DK-1958 Frederiksberg C, Email: sst@danishmeat.dk*

Keywords: pork, cooking, eating quality, meat quality

Introduction

It is well known that different cooking techniques and end point temperatures result in different eating qualities (Bejerholm *et al.*, 2003) just as variations in raw meat quality can result in variation in the eating quality (Candek-Potokar et al., 1998; Candek-Potokar *et al.*, 1999). Furthermore, recent studies have shown that an interaction exist between the cooking procedure and the raw meat quality, where the effect of the cooking technique depends on the raw meat quality (Bejerholm *et al.*, 2003). Frying meat on a pan is a common cooking technique. The effect of end point temperatures (Wood *et al.*, 1993) as well as pan temperatures (Lauridsen *et al.*, 2006) has been investigated. The interaction between pan temperature and end point temperature on the eating quality of various raw meat qualities has, however, not been investigated.

The aim of the current study was to investigate the interaction between pan temperature (PT) (150°C versus 220°C) and end point temperature (EPT) (65-68°C versus 85-88°C) depending on the raw meat quality (MQ) (normal versus low carbohydrate content in the meat, heavy versus light carcass weight and castrates versus females).

Materials and Methods

Shoulder clod and blade roll were excised from 80 DLY slaughter pigs the day after slaughter, vacuum-packed and frozen at -18°C. The 80 pigs were divided equally between castrates and females, heavy (120 kg live weight) and light (90 kg live weight), and fed to a normal carbohydrate content or strategically fed to a low carbohydrate content (Rosenvold *et al.*, 2001) combined with 48 hours fasting before slaughter, resulting in 10 pigs per group. The meat was defrosted and minced through a 3mm hole size. The meat from 10 pigs of the same group was mixed, portioned and frozen again (-18°C).

Before sensory analysis, the meat was defrosted overnight and formed into patties of 95 ± 0.3g each, 9.1cm in diameter and 1.4cm high. The patties were fried at either 220°C or 150°C PT measured by a surface thermometer (Digitron, 2038T, Sifram Instruments Limited, Torquay, England) to either 65-68°C or 85-88°C EPT measured by a handheld probe (Testo 926, Testoterm, Buhl and Boendsoe, Virum, Denmark). A small amount of grape seed oil was used on the pans before frying. The patties were first fried 30 seconds on each side and then turned every second minute until the aimed EPT was reached. The sensory quality was evaluated by 8 trained assessors. The following attributes were assessed on an unstructured 15cm scale going from nothing (0) to very high intensity (15): Odour: Fried pork (surface), burnt/bitter (surface), piggy (internal), sour (internal). Appearance: Uneven fried surface, degree of fried surface, wet surface, internal colour (from pink to grey). Texture: Firmness when pressing in the middle of the patty, crispiness of the crust when cutting, juiciness after 5 chews. Flavour/Taste: Fried pork, burnt/bitter, sweet, sour, piggy. After taste: Sour, burnt/bitter. An APLS regression was performed; using the cooking methods and the 8 MQs as a design matrix (X-matrix) and the sensory score as Y-matrix. The MQ was regarded as 8 treatments in the first statistical analysis of variance.

Results and Discussion

The main variation in the data set was between the 4 cooking methods (Figure 1, 55% Y-variation explained). Cooking to 85°C EPT resulted in higher firmness and less juiciness compared to 65°C EPT, but also in a tendency to a more fried pork flavour and crispy crust. A PT of 220°C gave more burnt/bitter taste and after taste, but also more fried pork odour and flavour compared to the more intensive piggy odour and flavour than that cooked at 150°C. A high PT seemed to have increased the Maillard reaction to a higher degree than the high EPT, while the high EPT especially decreased the juiciness. In this case, the high PT was so high, that the fried flavour changed to a more burnt/bitter flavour. Therefore, an intermediate PT might be preferred. The MQs were described by PC3 explaining only 2% of the Y-variance. The main distinction was between the two feedings. The normal feeding had a more intensive burnt/bitter odour, but also a more intense piggy odour and flavour, while the low carbohydrate feeding had a more intensive sour flavour and after taste. The expected higher carbohydrate content in normal fed pigs therefore seems to have resulted in a higher degree of Maillard reaction, leading to more burnt/bitter flavours.

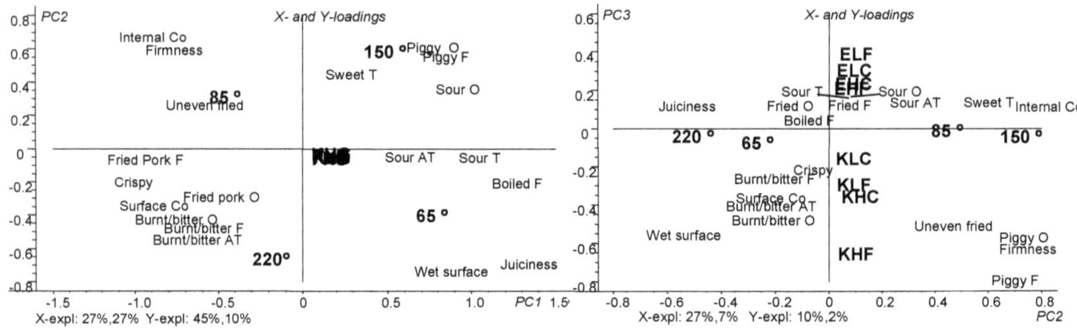

Figure 1: APLSR of the sensory profile. E: Experimental feeding, K: Control feeding, L: Light, H: Heavy, C: Castrates, F: Female, O: Odour, F: Flavour, AT: After taste, Co: Colour

The analysis of variance showed a significant effect of PT, EPT or MQ on all sensory attributes. A significant interaction between PT and EPT, PT and MQ or EPT and MQ was furthermore seen for all odours, after taste and appearance attributes as well as for burnt/bitter taste, piggy flavour, crispiness of the crust and juiciness.

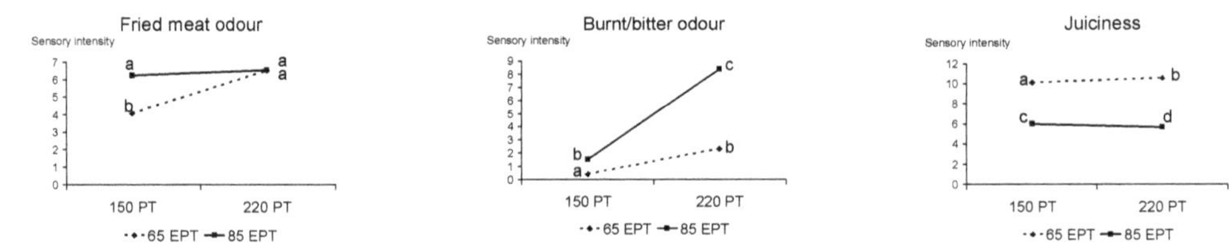

Figure 2: Sensory intensity of fried meat odour, burnt/bitter odour and juiciness.

When the patties were fried at 150°C PT, 85°C EPT resulted in a higher intensity of fried odour than at 65°C. When fried at 220°C PT no difference between EPTs was seen. Burnt/bitter odour showed the opposite pattern - the difference between EPTs was largest at 220°C PT. There was no difference between 65°C EPT/220°C PT and 85°C EPT/150°C PT, showing that the burnt/bitter odour at the high PT could be reduced by choosing a low EPT while the EPT did not mean so much for the burnt/bitter odour at the low PT. The main difference in juiciness was between EPTs, 65°C being the juiciest.

Conclusions

Feeding influenced the flavour of pork, but to a lesser degree than cooking conditions. PT had a large effect on the fried and burnt/bitter odours, whereas juiciness was more influenced by EPT. At low EPT, a high fried odour can be gained when using high PT, but be careful not to develop burnt flavours.

References

Bejerholm, C. and Aaslyng, M. D. (2003). The influence of cooking technique and core temperature on the results of sensory analysis of pork. Food Quality and Preference, 15, 19-30.

Candek-Potokar, M., Zlender, B., and Bonneau, M. (1999). Effects of breed and slaughter weight on longissimus muscle biochemical traits and sensory quality in pigs. Ann.Zootech., 47, 3-16.

Candek-Potokar, M., Zlender, B., Lefaucheur, L., and Bonneau, M. (1998). Effects of age and/or weight at slaughter on longissimus dorsi muscle: Biochemical traits and sensory quality in pigs. Meat Science, 48, 287-300.

Lauridsen, L., Andersen, L. T., Bredie, W. L. P., Bjergegaard, C., and Aaslyng, M. D. (2006). Chemical and sensory characterisation of pan-fried pork flavour: Interactions between raw meat quality, ageing and frying temperature. Meat Science, submitted.

Rosenvold, K., Petersen, J. S., Lærke, H. N., Jensen, S. K., Therkildsen, M., Karlsson, A., Møller, H. S., and Andersen, H. (2001). Muscle glycogen stores and meat quality as affected by strategic finishing feeding of slaughter pigs. Journal of Animal Science, 79, 382-391.

Wood, J. D., Nute, G. R., and Cuthbertson, A. (1993). Optimum cooking conditions for pork. Meat Focus International, October: 453-455.

POSSIBILITY OF NITRITE, PHOSPHATE AND PORK FAT SUBSTITUTION IN POULTRY FRANKFURTERS PRODUCTION

Z. Pejkovski [1], M. Kratovalieva[1], T. Polak[2], A. Rajar[2], L. Gašperlin[2] and B. Žlender*[2]

[1.] University Ss Cyril and Methodius – Skopje, Faculty for Agriculture and Food – Skopjebld. Aleksandar Makedonski b.b., P.O. box 297, 1000 Skopje, Republic of Macedonia, [2.] University of Ljubljana, Biotechnical Faculty, Jamnikarjeva 101, SI-1111 Ljubljana, Slovenia. Email: bozidar.zlender@bf.uni-lj.si

Keywords: poultry frankfurters, nitrite, phosphate, fat, colour

Introduction

Red sausages (frankfurters) are popular among consumers and sold in large quantities in many countries. By choosing the proper combination of available raw materials and ingredients, a high-quality and healthy product may be created. Sensory attributes such as colour, taste and texture are important for consumer acceptance. Colour of the chicken frankfurters is limited by the ratio between white and dark meat and use of various colour ingredients. Texture (firmness) is traditionally restricted by biochemical composition (protein, fat, type of fat, use of polyphosphate, etc.) (Dingstad *et al.*, 2005; Tan *et al.*, 2006). Sunflower and palmoil in ratio 6:4 were incorporated into frankfurters in order for pork fat to be substituted. Also, various additives (cochineal, paprika extract, no colorant) were tested in order for the colour function of nitrite to be copied. Phosphates were not applied. An attempt to make frankfurters for safety nutrition was made. The effects of used oils and colour additives on some technological, sensorial and nutritional (i.e. fatty acid composition) quality were studied.

Materials and Methods

The experiment was conducted on four kinds of model chicken frankfurters, each group prepared in three repetitions according to recipes described in Table 1. Meat emulsions were prepared by Stephan UMC 5 electronic bowl chopper, at 80% vacuum until 12°C was reached. Additives were solubilized in cold water before the preparations of each meat batter. Emulsion was stuffed in artificial casings Naturin (ϕ=23 mm) and thermal treated by combined Fessmann Turbomat oven (T_{core}= 72°C, t=120 min). Sensorial analysis of frankfurters was performed by a panel of four qualified panelists. The assessment was based on analytical descriptive tests whereas sensorial attributes were evaluated on a scale made on the basis of a preliminary test. The attributes are scored on a scale from 1 to 7 where 1 point means that the attribute is either not sufficiently expressed or it is completely unacceptable, whereas 7 points means that the attribute is strongly expressed or it is regarded as excellent.

Table 1: Ingredients present in the recipes.

g/kg of mass	Nitrite and phosphate	Cochineal, no phosphate	Cochineal+ paprika,no phosphate	No colorant and no phosphate
Leg:breast (1:1)	420	500	500	500
Sunflower oil	0	120	120	120
Palm-oil	0	80	80	80
Pork fat	250	0	0	0
Ice	310	280	280	280
Starch	0	20	20	20
Caseinate	0	10	10	10
Lactate	21	21	21	21
Vitamin C	0.5	0.5	0.5	0.5
Citrate	0	0.2	0.2	0.2
Salt	0	11	11	11
Spices	2	2	2	2
Cochineal (E120)	0	0.03	0.03	0
Paprika extract (E160c, 60000 CU)	0	0	0.3	0
Phosphate	2	0	0	0
Nitrite	11	0	0	0
Supro	20	0	0	0

Some attributes were scored by system 1 - 4 - 7 where 1 point means too low expressed attribute, 4 points mean an optimally expressed attribute and 7 points are given to too much expressed attribute. Sensorial attributes are listed in Table 2.

Four instrumental measurements of CIE $L^*a^*b^*$ values were made on products surface and on the freshly cut slice of frankfurters also. A Minolta CR 200b colorimeter (Illuminant C, 0° viewing angle) was used to determine the CIE L^* (lightness), a^* (+/-, red to green), and b^* (+/-, yellow to blue) values. A white ceramic tile with the specification of Y = 93.8, x = 0.3134, and y = 0.3208 was used to standardise the colorimeter.

Warner-Bratzler Shear Force (WBSF) was measured seven times and expressed in N. Cylinders (d=8 mm, h=25 mm) were removed parallel to the longitudinal orientation of frankfurters. Each cylinder was sheared once at the centre with a Warner-Bratzler shear attachment using a TA.TX plus texture analyser (Stable Micro Systems). The crosshead speed was $3.3\ 10^{-3}$ m/s. The data were analysed by the method of the least squares using the GLM procedure (SAS, 1990).

Results and Discussion

Generally, sensorially evaluated quality of three experimental groups of frankfurters was comparable to standard one (nitrite and phosphate). Frankfurters prepared with vegetable oils, cochineal and red pepper had even better surface and cross-section colour, texture profile and slightly worse smell and flavour then standard frankfurters. Frankfurters prepared without colorant and phosphate had an unacceptable colour, appropriate emulsion stability and texture but relatively poor flavour (Table 2). All samples had the same saltiness (optimal) and slightly expressed acidity (data not presented). Colour of standard frankfurters after 1h at room temperature (20 °C) became unacceptable. Decrease of a^* values (Table 3) confirmed the findings. Slices of other frankfurters under the same conditions remaited the colour stability. One month of storage in a refrigerator (vacuum packed, dark, 4°C) did not affect the colour. Significantly lower WBSF in frankfurters with cochineal + paprika extract and frankfurters without nitrite and phosphate than in standard one was found. This statement is in slight disagreement with sensorial analysis of texture.

Table 2: Sensorial quality of chicken frankfurters.

Attribute (point) / group		Nitrite and phosphate	Cochineal, no phosphate	Cochineal + paprika, no phosphate	No colorant and no phosphate	Sign.
Surface colour:	Characteristic (1-7)	$5.4±0.5^b$	$6.1±0.2^a$	$6.3±0.3^a$	$5.3±0.2^b$	***
	Intensity (1-7)	$5.4±0.6^c$	$5.8±0.3^b$	$6.2±0.5^a$	$5.3±0.3^c$	***
Cross-section colour:	characteristic (1-7)	$4.8±0.2^c$	$5.9±0.2^a$	$5.6±0.2^b$	$2.5±0.0^d$	***
	Characteristic 1 h/20 °C (1-7)	$3.5±0.4^c$	$6.0±0.0^a$	$5.3±0.3^b$	$2.5±0.0^d$	***
	Intensity (1-7)	$3.4±0.2^c$	$5.0±0.1^b$	$5.8±0.4^a$	$1.0±0.0^d$	***
	Uniformity (1-7)	$4.8±0.3^d$	$6.0±0.0^b$	$5.8±0.3^c$	$6.5±0.0^a$	***
Other attributes:	Emulsion stability (1-7)	$5.9±0.2^{ab}$	$6.2±0.5^a$	$6.1±0.4^{ab}$	$5.8±0.5^b$	Ns
	Structure (1-7)	$5.7±0.4^b$	$6.0±0.3^a$	$6.2±0.4^a$	$5.4±0.5^b$	***
	Texture (1-4-7)	$3.8±0.3^b$	$4.3±0.3^a$	$4.3±0.3^a$	$3.9±0.6^b$	**
	Mouth feeling (1-7)	$5.8±0.3^a$	$5.8±0.3^a$	$5.7±0.4^a$	$5.6±0.4^a$	Ns
	Juiciness (1-7)	$5.9±0.2^a$	$5.8±0.3^a$	$5.8±0.3^{ab}$	$5.5±0.3^b$	*
	Fattiness (1-7)	$1.6±0.2^a$	$1.5±0.1^{ab}$	$1.5±0.3^{ab}$	$1.4±0.3^b$	Ns
	Smell (1-7)	$5.9±0.3^a$	$5.8±0.3^{ab}$	$5.8±0.3^a$	$5.5±0.3^b$	*
	Flavour (1-7)	$6.0±0.3^a$	$5.7±0.3^{ab}$	$5.7±0.4^{ab}$	$5.5±0.4^b$	*

Mean values ± standard deviation. Sign. – Statistically not significant: Ns – P>0.05; statistically significant: * P≤0.05 and ** P≤0.01; highly statistically significant: *** P≤0.001; a,b,c,d means with a different superscript row differ significantly (P≤0.05).

Table 3: Difference in instrumentally measured cross-section colour and texture of chicken frankfurters.

Parameter	Group Storing	Nitrite and phosphate	Cochineal, no phosphate	Cochineal + paprika, no phosphate	No colorant and no phosphate	Sign.
L^*	1	$80.8±0.8^{dx}$	$84.1±0.4^{bxy}$	$81.8±0.7^{cxy}$	$86.7±1.0^{ax}$	
	1 A	$80.7±0.6^{bx}$	$83.3±0.4^{az}$	$81.2±0.7^{by}$	$84.0±3.0^{ay}$	P_C***
	30	$80.5±0.7^{dx}$	$84.4±0.6^{bx}$	$82.4±1.7^{cx}$	$87.2±1.0^{ax}$	P_S***
	30 A	$80.5±0.5^{dx}$	$83.9±0.4^{by}$	$81.8±0.7^{cxy}$	$85.3±2.9^{axy}$	$P_{C×S}$**
a^*	1	$3.5±0.4^{cx}$	$7.0±0.5^{by}$	$7.6±0.7^{axy}$	$0.1±0.8^{dy}$	
	1 A	$2.1±0.4^{by}$	$7.4±0.5^{ax}$	$7.9±0.8^{ax}$	$0.5±1.3^{cy}$	P_C***
	30	$3.4±0.2^{cx}$	$7.1±0.3^{bxy}$	$7.6±0.5^{axy}$	$1.3±0.4^{dx}$	P_S***
	30 A	$2.2±0.4^{cy}$	$6.2±0.3^{bz}$	$7.1±0.4^{ay}$	$0.3±0.3^{dy}$	$P_{C×S}$***
b^*	1	$13.6±1.1^{bz}$	$10.6±0.2^{cy}$	$15.2±0.4^{ay}$	$13.6±0.4^{by}$	
	1 A	$15.5±0.7^{ay}$	$11.6±0.9^{cx}$	$15.7±0.2^{ax}$	$14.8±1.1^{bx}$	P_C***
	30	$15.7±1.0^{ay}$	$10.9±0.3^{dy}$	$15.0±0.7^{by}$	$13.5±0.6^{cy}$	P_S***
	30 A	$17.3±1.1^{ax}$	$11.5±0.3^{dx}$	$15.8±0.3^{bx}$	$14.6±0.9^{cx}$	$P_{C×S}$***
WBSF	1	$4.1±0.8^{ab}$	$4.2±0.5^a$	$3.6±0.5^{bc}$	$3.3±0.7^c$	P_C**

Mean values ± standard deviation.
1 – 1st day.
30 – 30th day.
A – 1 h/20 °C.
Sign. – statistically significant: ** P≤0.01; highly statistically significant: *** P≤0.001; a,b,c,d means with a different superscript row differ significantly (P≤0.05) ; x,y,z,w means with a different superscript column differ significantly (P≤0.05).

Conclusions

Sensorially evaluated quality of three experimental groups of frankfurters (Cochineal, no phosphate; Cochineal + paprika, no phosphate; No colorant and no phosphate) was comparable to standard one (nitrite and phosphate). Colour of frankfurters without colorants and without phosphate is unacceptable. Texture of all groups, despite instrumentally demonstrated differences is suitable to standard.

References

Dingstad G.I., Kubberød E., Næs T., Egelandsdal B. (2005). Critical quality constraints of sensory attributes in frankfurter-type sausages, to be applied in optimization models. LWT - Food Science and Technology, 38: 665-676.

Tan S.S., Aminah A., Zhang X.G., Abdul S.B. (2006). Optimizing palm oil and palm stearin utilization for sensory and textural properties of chicken frankfurters. Meat Science, 72: 387-397.

SAS Software. Version 8.01. (1999). Cary, SAS Institute Inc.

Acknowledgements: This research was financed by the NATO PROGRAMME and Slovenian and Macedonian government (NATO CBP.EAP.CLG 981493 and BI-MK/04-06/004 projects).

RELATIONSHIP BETWEEN CONSUMER SCORES AND OXIDATIVE STATUS OF BEEF

V. Sierra[1], B. Caballero[2], M. Mocha[1], M.J. Martínez[1], K. Osoro1, D. Tolivia[2], M.J. Rodríguez-Colunga[2], A. Coto-Montes[2]*† and M. Oliván [1]†

[1.] *S. E. R. I. D. A,. Apdo. 13, 33300 Villaviciosa, Asturias, Spain.* [2.] *Departamento de Morfología y Biología Celular, Universidad de Oviedo, 33006 Oviedo, Asturias, Spain, (†) Joint senior authorship: M. Olivan and A. Coto-Montes.*

Keywords: beef, tenderisation, muscular hypertrophy, oxidative damage, sensory quality

Introduction

The most important quality traits required by consumers in beef are tenderness, juiciness and flavour. Those parameters are affected not only by pre-slaughtering management but also by processes of proteolysis and lipid oxidation occurring during meat *post-mortem* storage. Lipid oxidation produces a deterioration of the sensory quality of beef due to the development of undesirable flavour and odour notes. In muscle, lipid oxidation initiates at the membrane level in the phospholipid fractions as a free-radical autocatalytic chain mechanism. However, there is a group of enzymes that can act as antioxidants absorbing energy from reactive species of oxygen. The main groups are Superoxide Dismutase (SOD), Catalase (CAT) and Glutathione Reductase (GR). The aim of this work was to study the relationship between consumer perception of flavour and acceptability of beef after long *post-mortem* storage (3 to 21 days), the oxidative status of meat evaluated by the amount of malonaldehyde (TBARS) and the antioxidant activity of enzymes SOD, CAT and GR.

Materials and Methods

Twenty yearling bulls of two local breeds from northern Spain, Asturiana de los Valles (AV) and Asturiana de la Montaña (AM) were studied. Animals of the AV breed were homozygous (*mh*/*mh*), heterozygous (*mh*/+) or normal (+/+) for muscular hypertrophy gene. Bulls were fattened by feeding concentrate meal and barley straw *ad libitum* and were slaughtered at a live weight of 512.4 ± 37.2 kg. At 24 h *post-mortem* the *Longissimus dorsi* muscle was sliced, aged at 4°C and stored at -20°C for further analysis. Samples for TBARS and sensory analysis were aged under vacuum conditions, while samples for antioxidant enzymes analysis were kept in bags permeable to O_2. All meat samples were aged 7, 14 and 21 days (except animals of double-muscled AV bulls which were aged 3, 7 and 14 days, as faster maturation was supposed). Thiobarbituric acid reactive substances (TBARS) were analyzed by the method of Botsoglou *et al.* (1994) and expressed as mg of malonaldehyde per kg of lean muscle. The activity of antioxidant enzymes SOD, CAT and GR were analysed as described by Caballero *et al.*, (2006). Sensory analysis was conducted using 120 consumers. Consumers evaluated flavour quality and overall acceptability of meat using a nine point hedonic scale (1= very low, 9= very high). Linear regressions were assessed between the sensory traits and the chemical variables (SPSS 11.5, 2002).

Results and Discussion

There was a significant correlation ($p<0.001$) between consumer scores for flavour and acceptability and TBARS values (Figure 1), indicating a clear inverse relationship between the consumer perception of meat quality and the increase in lipid oxidation. This agrees with findings of Campo *et al.*, (2006) who described that oxidation produces a deterioration of beef flavour that can be detected by TBARS measurements. When looking to the evolution of TBARS values along meat aged separately by genetic groups (Figure 2), meat of AV bulls homozygous (*mh*/*mh*) for muscular hypertrophy showed higher values of TBARS in the early *post-mortem* period (3 to 14 days). The rest of genotypes showed an increase of TBARS that is, of lipid oxidation at later ageing times (14 to 21 days). This could be due to the different fat quantities and composition in the loin of these genotypes (Aldai *et al.*, 2006), with lower fat content but higher PUFA proportion in the intramuscular fat of double-muscled (*mh*/*mh*) AV animals with respect to the others. This could make them more sensitive to early oxidation. From these data it appears there are parallels between the evolution of lipid oxidation and the evolution of meat tenderization in meat of the different genotypes. Oliván *et al.*, (2004) described that meat of AV bulls homozygous (*mh*/*mh*) for muscular hypertrophy showed higher cathepsin activities at early ageing periods. Also a higher tenderization rate than normal AV or AM bulls for which the end of the tenderisation process should be expected at longer ageing times than the figures detailed in this study (21 days).

Another parallel was observed between antioxidant enzymes and the cathepsins. It was found that, under refrigerated conditions (4°C), antioxidant enzymes increased activitie earlier for AV (*mh*/*mh*) than for other genotypes. The period of ageing studied was too short for showing high antioxidant activity in meat of AM (+/+) (Caballero *et al.*, 2006), that is, the genotype with slower tenderisation.

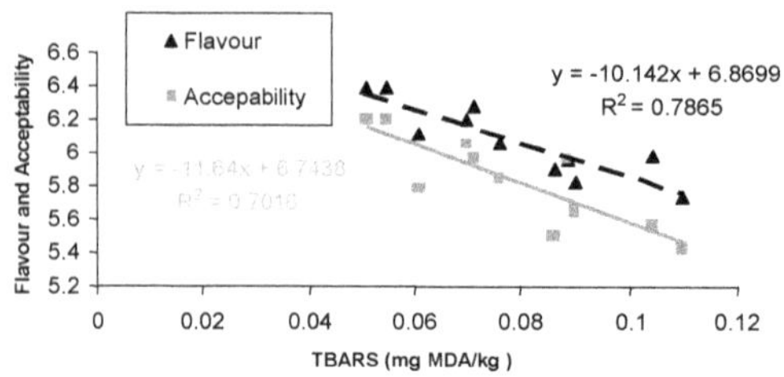

Figure 1: Linear regressions between TBARS values and sensory attributes flavour and acceptability.

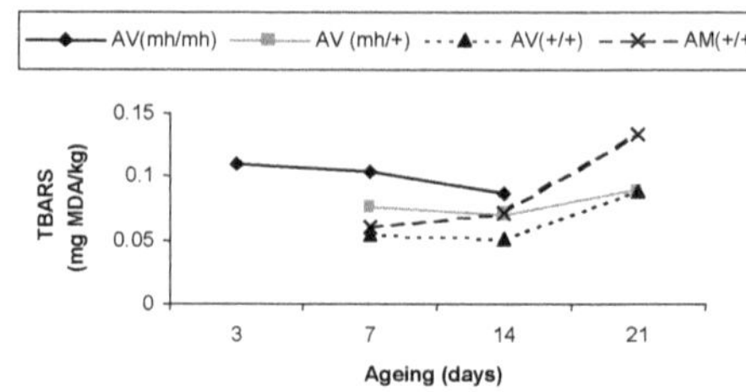

Figure 2: TBARS evolution along ageing for the four genotypes.

It is known that the activity of antioxidant enzymes decreases free radical damage by converting free radicals or derivates that would affect negatively proteins, lipids and other components of meat into harmless products for tissues. This data supports the hypothesis that antioxidant enzymes have a synergic action with proteolytic enzymes applied in the process of meat tenderization by avoiding free radical inactivation of proteolytic enzymes, since both enzymatic groups showed similar evolution along the maturation in all studied genotypes. On basis of exposure (detailed above), a certain role in the tenderisation pathway can be hypothesised to be antioxidant enzymes. From the different antioxidant enzymes studied, GR showed stronger relationship with sensory quality of meat detected by consumers, especially with global meat acceptability (R^2=0.82). This correlation data showed a relationship between sensory quality and oxidative processes along with meat ageing. However, this implication requires further study.

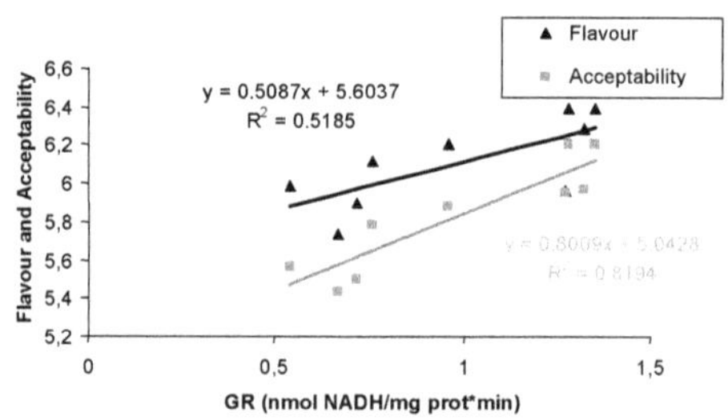

Figure 3: Linear regressions between the activity of GR and sensory attributes flavour and acceptability.

Conclusions

Consumer scores for acceptability and flavour quality were highly correlated with the oxidative status of meat measured as TBARS and the activity of some antioxidant enzymes (mainly GR). TBARS follow the same pattern along ageing than proteolytic enzymes, which are related to the different tenderization pattern of studied genotypes along maturation.

References

Aldai N., Murray E.M., Oliván M., Martínez A., Troy D.J., Osoro K. and Nájera A.I. (2006) The influence of breed and mh-genotype on carcass conformation, meat physico-chemical characteristics, and the fatty acid profile of muscle from yearling bulls. Meat Science 72, 486-495.

Botsoglou N.A., Fletouris D.J., Papageorgiu G.E., Vassilopoulos V.N., Mantis A.J. and Trakatellis A.G. (1994) Rapid sensitive and specific Thiobarbituric Acid Method for Measuring Lipid Peroxidation in Animal Tissue, Food and Feedstuff Samples. Journal of Agriculture and Food Chemistry, 42, 1931-1937.

Caballero B., Sierra V., Vega-Naredo I., Tomás-Zapico C., Rodríguez-Colunga M.J., Tolivia D., Oliván M. and Coto-Montes A. (2006) Evolución de los enzimas antioxidantes en la maduración de carne procedente de dos cabañas autóctonas asturianas: Asturiana de los Valles y Asturiana de la Montaña. ITEA. Accepted.

Campo M.M., Nute G.R., Hughes S.I., Enser M., Wood J.D. and Richardson R.I. (2006) Flavour perception of oxidation in beef. Meat Science, 72, 303-311.

Oliván M., Coto-Montes A, Caballero B., Sierra V., Aldai N., Rodríguez-Colunga M.J. and Osoro K. (2004) Meat tenderization by acidic lysosomal proteinases associated with genotype in beef. In: Proceedings of ICoMST 2004, Helsinki.

NON-VOLATILE TASTE COMPOUNDS OF DIFFERENTLY PROCESSED DUCK

Y. Liu, X.L. Xu and G.H. Zhou*

College of Food Science and Technology, Nanjing Agricultural University, Nanjing 210095, PR. China. Email: ghzhou@njau.edu.cn

Keywords: duck, processing, free amino acid, peptides, nucleotides

Introduction

Flavour is a very important component of the eating quality of meat and much research has been aimed at determining those factors during the production and processing of meat which influence flavour quality (Mottram, 1998), among which was the study of the non-volatile substances that may stimulate the taste buds-taste compounds.
Cooked duck products are popular food in China and other parts of world. However, to our knowledge, no study on taste regarding duck meat products has so far been reported. This research focuses on the taste compounds of three processed ducks including Nanjing water-boiled salted duck (NJWSD), roasted duck (ROD) and water-boiled duck (WBD). Our work sought to determine the quantities of the majority of the taste compounds, *i.e.* free amino acid, peptides and nucleotides in processed duck, and to study the effects of different processes on the taste compounds in duck to provide a theory bases for the quality improvement of duck meat products.

Materials and Methods

Eighteen lean-type Cherry Valley ducks were taken on the day after slaughter, each of which was about 1.5kg. Six ducks were processed to NJWSD by dry curing, brining, roasting and boiling. Each duck was dry-cured using 100g stir-fried salt with *Illicium verum* Hook. f. for 2h. As for brining, the brine contained not only excessive salt, but also some spices components such as *Allium fistulosum* L. etc. The brining process was 4 h. Roasting process was 1 h at 90°C. A low boiling temperature from 85°C to 90°C was used for the tender taste, and the boiling time was 40 minutes. The WBD was processed by boiling at 100°C for 40 min (Six ducks). Another six ducks were roasted at 180°C for 40 min, then were roasted at 240°C for 10 min.
Free amino acids (FAAs) were analyzed with an 835-50 amino acid auto-analyser (Hitachi co., Japan). Peptides were analysed following the procedure described by Martín *et al.* (1998). Nucleotides were analysed following the procedure described by Flores *et al.* (1999).

Results and Discussion

Free amino acids. Total contents of FAAs varied and ranged from 504.67 to 722.78 mg 100 g^{-1} dry matter (DM) (Figure 1). NJWSD contained the highest amount of total FAAs whereas the others contained a lower amount. As for most of FAAs, no difference were found between NJWSD and ROD, which were higher than that in WBD. Glu, Ala and Thr were found to be the three major FAAs, which were generally up to 44 mg 100g-1 of DM, which was consistent with the report in Iberian ham and Parma ham (Martín *et al.*, 2001). While Cys, Phe and His were the lowest, usually less than 19 $100g^{-1}$ of dry matter (DM). Compared the threshold values and taste sensory characteristics, umami amino acids Asp and Glu with the sweety amino acid Ala could contribute to the sensory taste of processed duck. As for the contents of the above three amino acids, these were higher in NJWSD and ROD than in water-boiled salted duck. Furthermore, the Asp content in NJWSD was higher than that in ROD.
Peptides. A total of 14 peptide peaks corresponding to peptides were detected using reverse-phase HPLC (Figure 2). The method used to analyse the peptides of the perchloric acid soluble fractions did not allow the determination of their amino acid composition (Martín *et al.*, 2001). The results of small peptides were similar to that of free amino acids. No differences were found in most small peptides between NJWSD and ROD, whereas significant differences were observed between ROD and WBD. The small peptides 4, 10, 11, and the total small peptides were found in the ROD in the largest levels, which indicated more proteolytic breakdown took place leading to the formation of peptides in duck meat. The importance of peptides to the sensory perception of food has been recognised for some time. It has been demonstrated that peptides contribute to the improvement in flavour of meats during refrigerated storage (Nishimura *et al.*, 1988). The generation of peptides as they contribute to the specific flavour of duck product should be further investigated.

Nucleotides. Flavour 5'-nucleotides contents were significantly different among differently processed duck, and in the order: ROD> NJWSD> WBD. The contents of flavour nucleotides in ROD and NJWSD were as high as that in some mushrooms such as king oyster mushrooms, shiitake, era mushrooms and black poplar mushrooms (Yang *et al.*, 2001), whereas flavour nucleotides decrease to undetectable levels during the long time curing process in sausages and in Parma hams (Mateo *et al.*, 1996). As we know, among the ATP derivatives, IMP is predominant in meat extract 24h after slaughter. This compound is gradually transformed into inosine and hypoxanthine in the meat flesh. As for our study, the duration of the full process was less than one day and the concentrations of flavour nucleotides especially for 5'-IMP found in NJWSD were highly beyond their umami taste threshold 35ppm. All these facts could imply that nucleotides would be very relevant to the taste of processed duck.

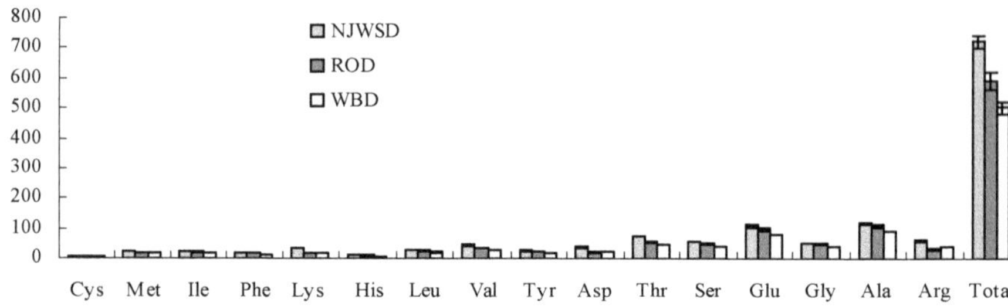

Figure 1: Free amino acids concentrations in breast meat of differently processed duck. Contents of free amino acids were in mg $100g^{-1}$ on the basis of duck meat dry matter and expressed as mean ± standard error (n=6).

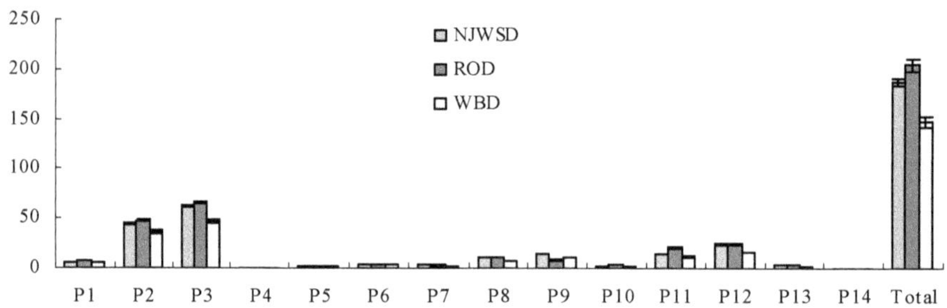

Figure 2: Peptides concentrations in breast meat of different processed duck. Contents of peptides were in the region of$100g^{-1}$ on the basis of dry matter and expressed as mean ± standard error multiplying 10^4 (n=6).

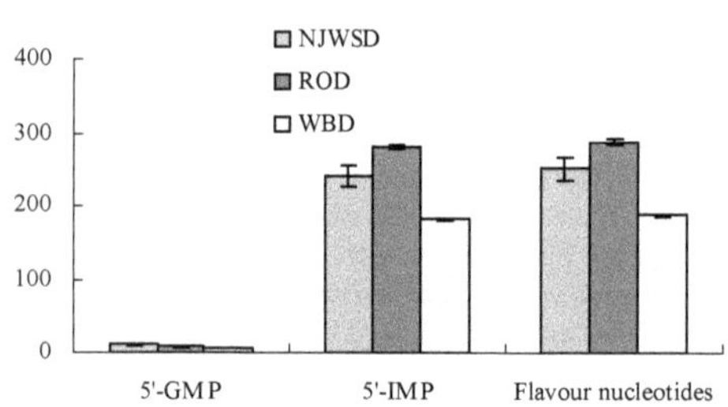

Figure 3: Nucleotide concentrations in breast meat of different processed duck[1] Contents of nucleotide were in mg 100g-1 on the basis of dry matter and expressed as mean ± standard error (n=6). 5'-IMP, 5'-Inosinic acid; 5'-GMP, 5'-Guanosine monophosphate; Flavour nucleotides: 5'-IMP + 5'-GMP.

Conclusions

Traditional NJWSD and ROD contained high levels of taste compounds. The delicate processing of NJWSD and roasting promote the proteolytic breakdown, and the latter retained more flavour nucleotides.

References

Flores, M., Armero, E., Aristoy, M. C. and Toldra, F. (1999). Sensory characteristics of cooked pork loin as affected by nucleotide content and post-mortem meat quality. Meat Science, 51, 53-59.

Martín, L., Antequera, T., Córdoba, J.J., Timón, M. L., and Ventanas, J. (1998). Formation of non-volatile flavour compounds in Iberian dry-cured ham during processing. In: Proceedings of the 44th ICoMST: 1008-1009. Barcelona.

Martín, L., Antequera, T., Ventanas, J., Benítez-Donoso, R., and Córdoba, J.J.. (2001). Free amino acids and other non-volatile compounds formed during processing of Iberian ham. Meat Science, 59: 363-368.

Mateo, J., Domínguez, M. C., Aguirrezábal, M. M. and Zumalacárregui, J. M. (1996). Taste compounds in chorizo and their changes during ripening. Meat science, 44: 245-254.

Mottram, D.S. (1998). Flavour formation in meat and meat products: a review. Food chemistry, 62: 415-424.

Nishimura, T., Rhue, M. R., Okitani, A. and Kato, H. (1988). Components contributing to the improvement of meat taste during storage. Agric. Biol. Chem., 52: 2323-2330.

Yang, J. H., Lin, H. C. and Mau, J. L. (2001). Non-volatile taste components of several commercial mushrooms. Food Chemistry, 72, 465-471.

GENERATION OF MEATY FLAVOURS DURING THE PROCESSING OF DRY FERMENTED SAUSAGES

A. Marco, J.L. Navarro and M. Flores*

Instituto de Agroquímica y Tecnología de Alimentos (CSIC), P. Box 73, 46100 Burjassot (Valencia), Spain.
E-mail: mflores@iata.csic.es

Keywords: aroma, curing, dry fermented sausage, meaty flavour

Introduction

Hundreds of flavour compounds have been identified in dry fermented sausages using different techniques (Berdagué *et al.*, 1993; Stahnke, 1994), but many of them do not have an important aromatic impact due to their high detection thresholds. Very few attempts have been made to apply olfactometric techniques to asses the aromatic impact of these volatile compounds (Blank *et al.*, 2001). Of the aromatic compounds which have been identified, esters are often found which provide fruity notes, and aldehydes providing herbal aromas (Stahnke, 1994). There are no compounds identified as meaty or cured flavours. Some authors, such as Wirth (1991), suggest that fermented sausages made with the addition of nitrate, rather than nitrite, have a better taste. In a previous work, Marco *et al.*, (2006) confirmed that the different use of nitrite or nitrate affects the volatile compound profiles of dry fermented sausages.
The aim of this study was to determine the generation of meaty aroma compounds in a fermentative process under the addition of different curing agents.

Materials and Methods

Preparation of dry fermented sausages and sampling. Two different batches containing nitrate (NO_2^-) or nitrite (NO_3^-) were manufactured as described by Marco *et al.*, (2006). Four sausages were collected at days 0, 14, 31, 45 (finished sausage) and 105 (vacuum stored) for the different analyses.
Volatile compounds GC-MS analyses. Volatile compounds were extracted, analyzed and quantified as described by Marco *et al.*, (2004).
Volatile compounds GC-O analyses. Volatile compounds were extracted using solid phase micro-extraction (SPME) using the same procedure cited for GC-MS analyses. The compounds adsorbed by the fibre were desorbed in a gas chromatograph (GC 8000 Top, CE Instruments, Milan, Italy) injection port for 6 min at 240°C in split-less mode, the split valve was opened after 1 min. The compounds were separated using a DB-624 capillary column (J&W Scientific, 60 m, 0.32 mm i.d., film thickness 1.8 μm). Helium was used as carrier gas with a linear velocity of 35.14 cm/s. The capillary column was split (2:1) into deactivated and uncoated capillaries connected with the sniffing port and FID, respectively. Six trained assessors evaluated the odours from the GC-effluent by smelling and recording the odour descriptors. For each assessment, evaluation of the odour took place over two different time intervals (0-35 and 35-70 min) in order to avoid olfactory fatigue for the assessors.

Results and Discussion

Among the 105 volatile compounds identified in the headspace of dry fermented sausages in this work, sixty different aromatic notes were described. Nine of these were described by the assessors as meat related flavours (fig. 1).

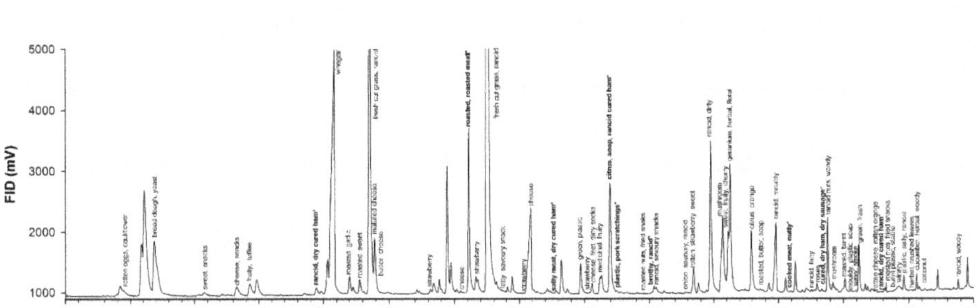

Figure 1: Aromatic regions within the GC-FID chromatogram of dry fermented sausages.

In order of elution, the volatile compounds responsible for these aromas were: 3-methyl-butanal (rancid, dry-cured ham), 1-pentanol (roasted, roasted meat), 2-hexenal (salty meat, dry-cured ham), heptanal (citrus, soap, rancid cured ham), 2-heptanol (plastic, pork scratchings), methional (brothy, rancid), 2,4-heptadienal (cooked meat, nutty), unknown (cured, dry ham, dry sausage), and heptanoic acid (rancid, dry-cured ham). Of these, 2-heptanol, the unknown compound and heptanoic acid were found only in trace amounts. 1-pentanol, 2-hexenal, heptanal and methional showed an increase during the drying stage but their amounts began to decrease during the vacuum storage. By contrast, 3-methyl-butanal and 2,4-heptadienal increased during storage (Figure 2). Only two compounds, methional and 2-hexanal were detected in significantly higher amounts in the samples with added nitrates, whilst no differences were found in the rest. All the meaty aroma compounds identified are derived from the lipid oxidation process except 3-methyl-butanal which comes from amino acid degradation (Marco *et al.*, 2006).

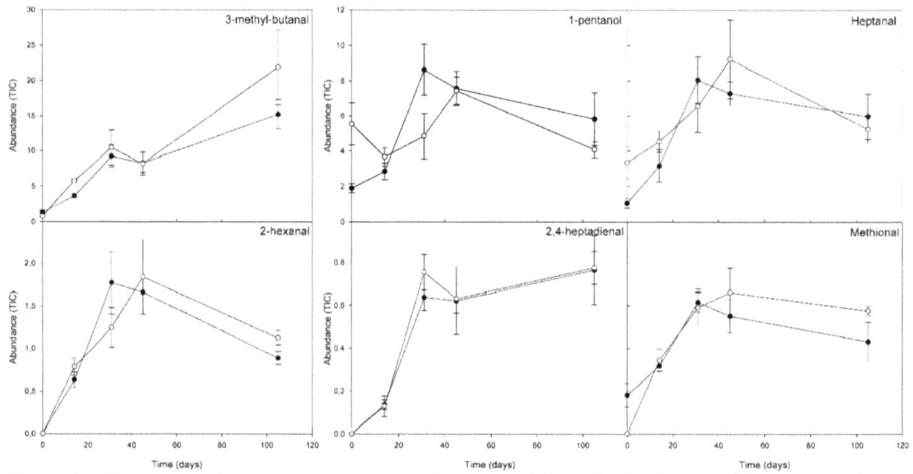

Figure 2: Generation of meaty aroma compounds extracted from the headspace of dry fermented sausages, quantified as AU×10^{-6} per g of dry matter. NO_2^- (●) and NO_3^- (○) batches.

Conclusions

Meaty aromas in dry fermented sausages are primarily generated during the drying process. However, during vacuum storage these compounds mainly decrease as they are derived from the lipid oxidation process. It is interesting to note the effect of vacuum storage on the generation of the volatile compounds responsible for the meaty aroma in dry fermented sausages.

Acknowledgements

The scholarship from CSIC/MEC to A. Marco and financial support from grants AGL2005-00713 (MEC, Madrid, Spain) and GV05/067 (Generalitat Valenciana, Spain) are fully acknowledged.

References

Berdagué, J. L., Monteil, P., Montel, C. and Talón, R. (1993). Effects of starter cultures on the formation of flavour compounds in dry sausage. Meat Science, 35: 275-287.

Blank, I., Devaud, S., Fay, L. B., Cerny, C., Steiner, M. and Zurbriggen, B. (2001). Odour-Active compounds of dry-cured meat: Italian-type salami and Parma ham. In: G. R. Takeoka, M. Güntert, and K. H. Engel (Eds.) Aroma-active compounds in foods. Chemistry and sensory properties. pp. 9-20. Washington, D.C.: American Chemical Society.

Marco, A., Navarro, J. L. and Flores, M. (2004). Volatile compounds of dry-fermented sausages as affected by solid-phase microextraction (SPME). Food Chemistry, 84: 633-641.

Marco, A., Navarro, J. L. and Flores, M. (2006). The influence of nitrite and nitrate on microbial, chemical and sensory parameters of slow dry fermented sausage. Meat Science, (In press).

Stahnke, L. H. (1994). Aroma components from dried sausages fermented with *Staphylococcus xylosus*. Meat Science, 38: 39-53.

Wirth, F. (1991). Restricting and dispensing with curing agents in meat products. Fleischwirtschaft, 71: 1051-1054.

FATTY ACIDS AND MINERALS AFFECT THE LIVER-LIKE OFF-FLAVOUR IN COOKED BEEF

B.E. Jenschke, J.M. Hodgen and C.R. Calkins*

Department of Animal Science, University of Nebraska, P.O. Box 830908, Lincoln, Nebraska, USA
Email: ccalkins@unlnotes.unl.edu

Keywords: beef, flavour, liver-like, fatty acids, minerals

Introduction

Flavour is an important organoleptic trait that consumers use in the determination of acceptability. Meat flavour is a result of reducing sugars and amino acids, differences in fatty acid composition that are responsible for species-specific flavour, off-flavour development due to lipid oxidation, microbial by-products, or other degradative mechanisms (Miller, 2001). Current research in our laboratory has specifically concentrated on the liver-like off-flavour. Miller (2001) indicated that the causes of the livery off-flavour in beef are lipid oxidation, heme iron content and elevated degrees of doneness. James and Calkins (2006) reported that slow cooking (36-40min) and subsequent holding for 1 h of specific muscles from the chuck and round prior to serving reduced off-flavour intensity while Meisinger (2005) found that heme-iron concentration and pH significantly (P=0.0004) accounted for 55% of the variation in the liver-like off-flavour of the *M. rectus femoris* (knuckle center). Although heme-iron content, lipid oxidation, and pH have been implicated as contributing to the liver-like off-flavour, these relationships are weak. Therefore, the objective of this research was to identify factors affecting the liver-like off-flavour in knuckles.

Materials and Methods

Samples were obtained using the screening procedure previously described by Jenschke *et al.,* 2006. Briefly, two trained sensory panelists tasted a 10g piece of the knuckle centre. Knuckles identified as having an off-flavour (n=30) and knuckles not having an off-flavour were vaccum-packaged and shipped to the Loeffel Meat Laboratory at the University of Nebraska. Following a 7d aging period at 1°C, the *M. rectus femoris* was isolated and cut into 2.54cm thick steaks, freezer wrapped and frozen (-16°C) until sensory analysis was conducted. Steaks were cooked to an internal temperature of 70°C on an electric broiler (FSR200, Farberware Inc., Prospect, IL). Internal temperature was monitored with a digital thermometer (Omega Engineering, model 450-ATT, Stamford,CT) with a type T thermocouple (Omega Engineering, Stamford, CT). Once the internal temperature reached 35°C, the steak was turned once until the final temperature was reached. The steak was then cut into 1.27cm x 1.27cm x 2.54cm cubes and served warm to the panelists, approximately 5 minutes post cooking. Panelists for this study were selected and trained according to the guidelines and procedures outlined by Meilgaard *et al.,* (1991). Six samples, identified using three-digit codes, were served on each day. Eight-point descriptive attribute scales (Muscle fibre tenderness: 1=extremely tough, 8=extremely tender; Connective tissue: 1=abundant, 8=none; Juiciness: 1=extremely dry, 8=extremely juicy; Off-flavour intensity: 1=extreme off-flavour, 8=no off-flavour) were used. Off-flavours were rated using a 15-point intensity scale (0=extremely bland; 15=extremely intense). Heme iron was determined using the acidified acetone extraction procedure while pH was determined using a penetrating pH probe. Moisture and ash (expressed as percentages) were quantified using a LECO Thermogravimetric Analyzer-601 while percent fat was determined using an ether extraction method. Minerals were isolated and quantified using an inductively-coupled argon plasma spectrometer. Fatty acids were extracted using a 2:1 chloroform:methanol solution and methylated using boron fluoride-methanol. Data were analysed using the REG procedure of SAS (Version 9.1.3), and the stepwise option was used to determine the final variables to be included in the model while the CORR procedure was used to generate correlation coefficients.

Results and Discussion

Table 1 includes the minerals and fatty acids tested in this study. The final model for predicting liver-like off-flavour is: Liver-like off-flavour rating=-0.21+ 0.0008 (Na) + 0.13 ((20:2 (n-6))- 0.005 (16:1) – 0.002 ((18:1 (n-7)) – 0.033 (20:3) which explained 46% of the variation in the liver-like off-flavour. It should also be noted that non liver-like samples had 3.5 times more vaccenic acid (18:1 n-7) when compared to liver-like samples. Simple correlations from our data indicated that vaccenic (r=-0.32; P=0.02), *cis*-11,14-eicosadienoic (20:2 n-6) (r=0.34; P=0.02) and eicosapentaenoic acid (20:5) (r=0.28; P=0.05) significantly affected the liver-like off-flavour in this study. Camfield *et al.,* (1997) reported simple correlation coefficients of fatty acids and their effect on the livery off-flavour. Vaccenic acid (r= -0.32; P<0.05) and *cis*-11,14-eicosadienoic acid (r=0.38; P<0.05) individually accounted for a significant amount of variation in the livery off-flavour which is in agreement with our data. Yancey (2002) reported in the *M. Gluteus medius* that palmitoleic (16:1), heptadecenoic (17:1), and elaidic (18:1 n-9t) acids were negatively correlated to the livery off-flavour and individually accounted for about 20% of the variation in the livery off-flavour. Although not significant, our results indicate that palmitoleic acid may also play a role in the development of the liver-like off-flavour (r=-0.25;

P=0.08). Meisinger (2005) indicated that heme iron might be a cause of the livery off-flavour in beef, but our data suggests that neither heme iron nor total iron play a role in the development of the liver-like off-flavour.

Table 1: Minerals and fatty acids quantified.

	Fatty Acids		Minerals
14:0	18:1 (n-7)	20:4	Zn
14:1	18:2	20:5	Fe
15:0	18:3	22:5	P
16:0	20:0	22:6	Mn
16:1	20:1	24:0	Mg
17:0	20:2		Ca
18:0	20:2(n-6)		Cu
18:1	20:3		Na

Conclusions

Data from this study indicate individual fatty acids and minerals play a significant role in the development of the liver-like off-flavour. Future studies to manipulate the fatty acid and mineral profiles of muscle might prove beneficial in lowering the incidence of the liver-like off-flavour in beef.

References

Camfield, P.K., A.H. Brown. Jr., P.K. Lewis, L.Y. Rakes, and Z.B. Johnson. (1997). Effects of frame size and time-on-feed on carcass characteristics, sensory attributes, and fatty acid profiles of steers. Journal of Animal Science, 75: 1837-1844.

James, J.M. and C.R. Calkins. (2006). The influence of cooking rate and holding time on beef flavor. Nebraksa Beef Report, 112-114.

Jenschke, B, A. Hamling, J. Hodgen, J. Meisinger, D. Moss, M. Lundesjo Ahnstrom, and C. Calkins. (2006). A novel procedure to identify off-flavored steaks. http://www.asas.org/midwest/2006/2006abstracts.pdf Accessed: May 1, 2006.

Meilgaard, M., G.V. Civille, and B. T. Carr. (1991). Sensory Evaluation Techniques, 2nd ed. CRC Press, Inc., Boca Raton, FL.

Meisinger, J.L. (2005). Flavor relationships among muscles from the beef chuck and round. M.S. Thesis, University of Nebraska, Lincoln, NE.

Miller, R.K. (2001). Beef Flavor: A White Paper. A paper prepared for the National Cattlemen's Beef Association, Centennial, CO.

Yancey, E. J. (2002). Determination of factors causing livery flavor in steaks from the beef loin and chuck. Ph.D. Diss., Kansas State University, Manhattan, KS.

VOLATILE AROMA COMPOUNDS OF COOKED SUCKLING LAMB MEAT

M.T. Osorio, J.M. Zumalacárregui, B. Fonseca* and J. Mateo

Departamento de Higiene y Tecnología de los Alimentos, Universidad de León, Campus Vegazana s/n, 24071 León, Spain. Email: dhtjmo@unileon.es

Keywords: lamb, suckling lamb, volatile, meat aroma

Introduction

In the Mediterranean Europe Area, sheep/dairy farms are frequent and, in addition to milk, meat from suckling lambs aged between 25 and 45 days and with a carcass weight lower than 7 kg is largely produced on these farms (Sañudo *et al.*, 1998). Suckling lamb meat is a product in demand because of its high eating quality, which could be partly due to its mild flavour compared to meat from older animals (Martínez-Cerezo *et al.*, 2005).

Volatile compounds in meat have been widely studied for their effects on meat flavour and their potential as tracers of animal feeding system (Vasta and Priolo, 2006). According to the origin, volatile compounds of cooked meat can be divided into three groups, those coming (directly or formed by animal metabolism) from feed ingested by animals and those formed during heating of meat by means of lipid oxidation and the Maillard reaction (Mottram, 1998).

To our knowledge there are no studies dealing with aroma compounds of meat suckling lamb. These studies could be useful to gain a better understanding of the flavour of this distinctive product and for traceability of the feeding diet of suckling lambs (Prache *et al.*, 2005). Considering these facts, the purpose of the present study was to develop an initial approach to determining the volatile compounds of suckling lamb meat.

Materials and Methods

Ten carcasses of Churra breed suckling lambs were randomly sampled in a local slaughterhouse. All lambs had been reared by ewes' milk exclusively. The right-hand pelvic limb of each carcass was boned, then the meat, thus obtained, was homogenised and a portion of 100 g was taken for volatile extraction. A Likens-Nickerson simultaneous distillation-extraction apparatus was used; diethyl-ether was the extraction solvent and a dispersion of the meat in water at boiling temperature was the source of volatiles. The extraction process lasted 4 hours. Then, ether solution with the extracted volatiles was concentrated by distilling in a Kuderna-Danish concentrator (until 1 ml solution) in a 50 °C water bath. Separation and identification of volatiles were carried out by GC-MS (Hewlett Packard-6890 Series GC system - Hewlett Packard-5973 Inert MSD Mass Selective Detector) equipped with a HP-5MS column (30 m x 0.25mm x 0.25 µm) and He as carrier gas. Four microlitres of the solution were injected (injector temperature was 230°C, and split mode 50:1). Initial oven temperature was 50°C which was increased to 95°C at a rate of 10 °C min^{-1} and then to 270°C at a rate of 10°C min^{-1}. The MS detector was activated after 6 min of injection. Compounds were identified by comparing the mass spectra with those contained in the Willey 275 database and then comparing their Kovacs' Index with those in the literature, if available. Quantification was carried out by area percentage.

Results and Discussion

A total of 68 volatile compounds were detected in suckling lamb meat using this methodology. The most abundant chemical families were carboxylic acids and esters (16 detected, 35% of total area), and aldehydes (16 detected and 30% of total area) (Table 1). The large majority of compounds were previously observed in cooked meats, and an important part of them came from lipids as was also expected for cooked lamb meat (Mottram, 1998; Elmoore *et al.*, 2000; Vasta and Priolo, 2006). The most abundant compounds were fatty acids (*e.g.* tetradecanoic, hexadecanoic, dodecanoic, or octadecenoic acids), feed-ingested fat-soluble volatiles such as 3,7,11,15-tetramethyl-hexadecene (phytene), or lipid-derived compounds originated by thermal degradation (oxidation products, e.g. hexadecanal, octadecenal, and octadecanal).

By comparing our results with those of other studies dealing with lamb volatiles (Elmoore et al., 2000; Vasta and Priolo, 2006), the present data showed elevated presence of medium and long chain acids and esters, higher relative abundance of phytane, and no o lower levels of low molecular weight compounds (e.g. hexanal, short-chain branched fatty acids, propanol, etc). Extraction methodology could account for the observed differences; the Likens-Nickerson simultaneous distillation-extraction apparatus is effective in collecting meat flavour volatiles of the mid- to high-molecular weight range.

With regard to phytene, it seems to be a grass-derived compound (Elmoore et al., 2004) which will be ingested with the ewes' milk by suckling lambs and then retained in the lambs' fat tissues. Based on this fact, it would be beneficial to study the potential use of this compound, as a plant biomarker, as suggested by Prache *et al.* (2005), for traceability of age at slaughter and type of rearing (ewes' milk vs milk-substitute).

Table 1: Volatile compounds detected in cooked suckling lamb (only the most abundant are shown in the table).

Compound *(number of compounds)*		Relative abundance (area %)[&]	Samples[*]	Method of Identification[#]	Observed KI[#]
Hydrocarbons (3)	15.4				
3,7,11,15-tetramethyl-hexadecene		15.2	10	MS+KI	1846
Aldehydes (16)	*30.0*				
Nonanal		0.4	10	MS+KI	1104
Tetradecanal		0.4	9	MS+KI	1613
Hexadecanal		22.8	10	MS+KI	1818
Octadecenal		3.8	5	MS+KI	1919
Octadecanal		2.7	6	MS+KI	2022
Ketones (4)	4.5				
2-Tridecanone		0.9	6	MS+KI	1498
2-Pentadecanone		2.1	10	MS+KI	1696
2-Heptadecanone		1.4	10	MS+KI	1902
Alcohols and hydroxyketones (5)	1.2				
1-Hexadecanol		0.7	9	MS+KI	1882
Carboxilic acids and esters (16)	35.2				
Decanoic acid		1.2	10	MS+KI	1372
Dodecanoic acid		4.3	10	MS+KI	1572
Tetradecanoic acid		15.3	10	MS+KI	1764
Hexadecanoato metil éster		1.0	8	MS+KI	1926
Hexadecanoic acid		8.5	10	MS+KI	1963
Methyl-octadecenate		0.6	6	MS	2104
Methy-octadecanoate		0.4	7	MS	2129
Octadecenoic acid		3.1	9	MS+KI	2145
Octadecanoic acid		1.3	10	MS	2163
Furan and furanoids (6)	2.0				
δ-Decalactone		0.8	10	MS+KI	1714
γ-Undecalactone		0.8	9	MS+KI	1932
Sulphur containing (8)	6.7				
A sulphur containing compound		1.0	10	MS	1196
A sulphur containing compound		2.6	10	MS	1687
A sulphur containing compound		1.5	10	MS	1747
A sulphur containing compound		1.4	10	MS	1945
Others and unknown (10)	4.1				
Isobuthyl-phtalate		0.7	9	MS	1875
Dibuthyl-phtalate		0.9	6	MS	1969

[&]: Percentage of the area of the peak with respect to the total sum of areas. [*]: Number of samples in which the compound appeared. [#]: MS, coincidence with the mass spectrum of the library; KI, coincidence with Kovats' Index (KI) obtained in literature.

Conclusion

Volatile compounds of suckling lamb meat have been determined. Among them, phytene could be considered for further studies as a potential tracer of feeding system or age.

References

Elmoore, J.S., Mottram, D.S. and Hierro, E. (2000). Two-fibre solid-phase microextraction combined with gas chromatography–mass spectrometry for the analysis of volatile aroma compounds in cooked pork Journal of Chromatography A, 905: 233–240.

Elmoore, J.S., Warren, H.E., Mottram, D.S., Scollan, N.D., Enser, M., Richardson, R.I. and Wood, J.D. (2004). A comparison of the aroma volatiles and fatty acid compositions of grilled beef muscle from Aberdeen Angus and Holstein-Friesian steers fed diets based on silage or concentrates. Meat Science, 68: 27-33.

Mottram, D. S. (1998). Some aspects of the chemistry of meat flavour. In: The flavour of meat and meat products F. Shahidi (Ed.). Chapman and Hall. London. (pp. 210-230).

Prache, S., Cornu, A., Berdagué, J.L. and Priolo, A. (2005). Traceability of animal feeding diet in the meat and milk or small ruminants. Small Ruminant Research, 59: 157-168.

Sañudo, C., Sánche, A. and Alfonso, M. (1998). Small ruminant production systems and factors affecting lamb meat quality. Meat Science, 49, S29-S54.

Vasta, V. and Priolo, A. (2006) Ruminant fat volatiles as affected by diet. A review. Meat Science. In press.

SESSION 5: MEAT TECHNOLOGY

- HOT TOPICS

INFLUENCE OF PELVIC SUSPENSION AND RN⁻ GENOTYPE ON SHEAR FORCE AND SENSORY QUALITY IN PORK LOIN

L. Enfält*, I. Hansson, L. Johansson and K. Lundström

Department of Food Science, Box 7051, SLU, SE-750 07 Uppsala, Sweden. Email: lotta.enfalt@lmv.slu.se

Keywords: pelvic suspension, pork, RN genotype, sensory quality, shear force

Introduction

Meat quality is complex and consists of several attributes, depending on where in the production line one is: producer, processor, retailer or consumer. Whereas carcass composition and weight are important traits for the producer, meat tenderness and juiciness are the most important for the consumer. Several treatments along the whole production chain also influence the different quality attributes. The dominant RN^- allele observed in Hampshire and Hampshire crosses influences several quality traits, resulting in more lean meat and higher glycogen content and lower protein content in the live animal. In meat, the RN^- allele gives lower ultimate pH, water holding capacity, processing yield and higher internal reflectance (Lundström & Andersson, 2001). The effect of the RN^- allele on sensory evaluated tenderness has been shown to be positive (Jonsäll *et al*., 2000), negative (Le Roy *et al*., 2000) and of low importance (Josell *et al*., 2003a), whereas the juiciness is mostly reported to be higher in meat from RN^- carriers (Jonsäll *et al*., 2000; Josell *et al*., 2003a). Along the slaughter line, various treatments can be used to influence the final meat quality, such as electrical stimulation, pelvic suspension and chilling conditions. Pelvic suspension has been shown to give lower shear force values in pork (Møller & Vestergaard, 1986) and higher tenderness with sensory evaluation (Tyler *et al*., 1995). Josell *et al*. (2003b) studied the combined effects of RN genotype, chilling rate and suspension method and found interactions between all three effects for tenderness. Highest tenderness was found in pelvic-suspended and slowly chilled RN^- carriers, whereas lowest tenderness was found in rapid chilled non-carriers. Greatest effect of the post-slaughter treatments was found in the non-carriers. The aim of this study was to further evaluate the effect of pelvic suspension on shear force and sensory quality in pork loins from carriers and non-carriers of the RN^- allele.

Materials and Methods

Female pigs (n=36) with a carcass weight between 80 to 90 kg were used in this study. pH (pH-meter WTW pH 340, Germany) was registered in *M. longissimus dorsi* (LD) at 45 min *pm*, to avoid carcasses with PSE (pH_{45}< 5.8). At 1 h *pm*, left carcass side was rehanged in the pelvic bone (*obturator foramen)*, while the right side was kept hanging in the Achilles tendon. At 24 h *pm*, pH was measured to avoid carcasses with DFD (pH_u>5.8) and samples were taken from the LD muscle on both sides of each carcass. Samples for both shear force and sensory analysis were weighed, vacuum packed, and aged for another 4 days and then stored in -20°C until analysis. Samples for shear force measurements were thawed overnight (+4°C) and cooked at 70°C for 90 min. Maximal shear force value was registered on 12 strips per animal 10 x 10 x 40 mm cut parallel with the fibre direction, using a Stable Micro Systems Texture Analyzer HD 100 (Godalming, UK), equipped with a Warner-Bratzler metal blade (1mm thick; speed 50 mm/min). Samples for sensory analysis were thawed overnight (+4 °C) and prepared wrapped in foil (Al) in an oven (150°C) to a final internal temperature of 70°C. The meat was served room tempered in 4-mm slices to 6 panellists, who judged tenderness, juiciness, meat taste, acidity and crumbliness on a scale from 0 (low intensity) to 100 (high intensity). RN genotypes were identified according to Milan *et al*. (2000). The model used for the statistical analysis (version 8e, SAS Institute Inc., Cary, NC, USA) contained the fixed effects of RN genotype, suspension method and their interaction, and the random effect of animal within RN genotype. For the sensory traits, the random effect of assessor was also included in the model.

Results and Discussion

Pelvic suspension gave lower shear force values within both RN genotypes (p-value for hanging= 0.001). RN genotypes differed significantly (p=0.031), with lower values for the RN^- carriers. Both the Achilles- and pelvic-suspended rn^+rn^+ loins reached similar shear force values as the Achilles-suspended loins from RN^- carriers. Lowest shear force values were observed in the pelvic-suspended RN^- carriers (Figure 1). This positive effect of pelvic suspension on shear force agrees with results by Møller & Vestergaard (1986), but differs with other studies, which found no significant differences (Tyler *et al*., 1995; Josell *et al*., 2003). Meat from RN^- carriers had significantly higher tenderness, juiciness, meat taste and acidity, and lower crumbliness (Table 1). Pelvic suspension gave higher tenderness, juiciness and meat taste, and lower crumbliness. Pelvic-suspended RN^- carriers had highest tenderness and Achilles-suspended non-carriers had lowest tenderness. In between these groups, pelvic-suspended non-carriers achieved the same tenderness as Achilles-suspended RN^- carriers. Juiciness was most affected by RN genotype, whereas meat taste and crumbliness were affected by pelvic suspension only within RN^- carriers. Tyler *et al*. (1995) also found higher tenderness in pelvic-suspended loins, but no effect on juiciness in contrast to our results. We found larger effects on both shear force and sensory attributes within both RN genotypes compared with Josell *et al*. (2003), who found the largest effect of suspension method within the non-carriers, and also an interaction between RN genotype and chilling system.

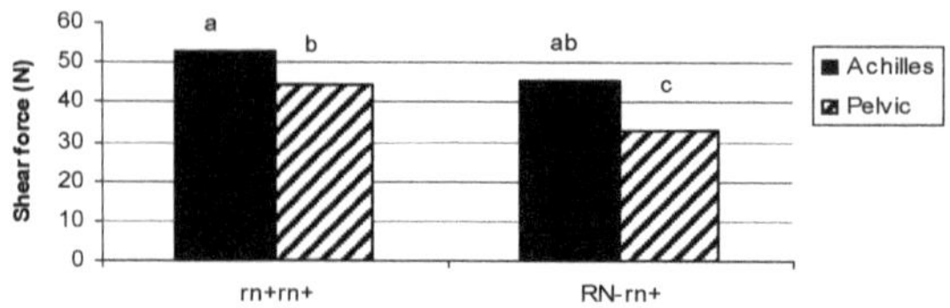

Figure 1: Shear force values (LSM) in LD from pelvic and Achilles suspended carcasses from non-carriers (rn^+rn^+) and carriers (RN^-rn+) of the RN^- allele. Bars with the same letter do not differ significantly ($p>0.05$).

Table 1: Sensory quality in pelvic and Achilles suspended LD from non-carriers and carriers of the RN^- allele (LSM).

	Genotype/ hanging				Levels of significance		
	rn+rn+		RN^-rn+		Genotype	Suspension	
Sensory trait	Achilles	Pelvic	Achilles	Pelvic	(G)	(S)	GxS
Tenderness	44.2^a	52.7^b	57.6^b	67.7^c	0.002	0.001	0.496
Juiciness	46.6^a	48.6^a	56.0^b	58.9^b	0.001	0.040	0.707
Meat taste	50.2^a	51.0^a	53.4^b	56.1^c	0.001	0.019	0.180
Acidity	54.5^a	55.7^{ab}	58.0^b	57.4^b	0.014	0.738	0.242
Crumbliness	43.6^a	40.2^a	32.3^b	25.6^c	0.001	0.001	0.252

Conclusion

Pelvic suspension of pork carcasses improved tenderness in loins of both non-carriers and carriers of the RN^- allele. RN^- carriers had higher tenderness than non-carriers, but with pelvic suspension it is possible to achieve the same tenderness for non-carriers as for Achilles suspended RN^- carriers.

References

Jonsäll, A., Johansson, L. and Lundström, K. (2000). Effects of red clover silage and RN genotype on sensory quality of prolonged frozen stored pork (*M. longissimus dorsi*). Food quality and preference, 11:371-376.

Josell., Å., Enfält, A.-C., von Seth, G., Lindahl, G., Hedebro-Velander, I., Andersson, L. and Lundström, K. (2003a). The influence of RN genotype, including the new V199I allele, on the eating quality of pork. Meat Science, 65:1341.

Josell, Å., von Seth, G. and Tornberg, E. (2003b). Sensory and meat quality traits of pork in relation to post-slaughter treatment and *RN* genotype. Meat Science, 66:113-124.

Le Roy, P., Elsen, J-M., Caritez, J-C., Talmant, A., Juin, H., Sellier, P. and Monin, G. (2000). Comparison between the three porcine RN genotypes for growth, carcass composition and meat quality traits. Genetics Selection Evolution, 32:165-186.

Lundström, K. and Andersson, L. (2001). The effect of the RN^- allele on meat quality and how the gene was discovered. In Proc. 54th Annual Reciprocal meat Conference pp. 143-147. Publ. American Meat Science Association, 1111 North Dunlap Avenue, Savoy, IL 61864, USA.

Milan, D., Jeon, J-T., Looft, C., Amarger, V., Robic, A. *et al.* (2000). A mutation in *PRKAG3* associated with excess glycogen content in pig skeletal muscle. Science, 288:1248.

Møller, A.J. and Vestergaard, T. (1986). Effects of altered carcass suspension during rigor mortis on tenderness of pork loin. Meat Science, 18: 77-87.

Tyler, A. A, Perry, A.M. and Warkup, C.C. (1995). Improving pork quality by electrical stimulation or pelvic suspension on carcasses. Meat Science, 39: 327-337.

A PROCEDURE TO ACCELERATE THE CHILLING RATE OF PORCINE SEMIMEMBRANOSUS MUSCLE

L. Voutila*[1], N. Linqvist[2], M. Ruusunen[1] and E. Puolanne[1]

[1.] *Department of Food Technology, University of Helsinki, P.O. Box 66, 00014 University of Helsinki, Finland,* [2.] *Oy Snellman Ab, Amerikankatu 11, 68600 Pietarsaari, Finland. Email: liisa.voutila@helsinki.fi*

Keywords: chilling rate, porcine, *semimembranosus*

Introduction

Pale zones in the deep part of porcine *M. semimembranosus* (SM) have been shown to be very similar to the classical PSE meat (Laville *et al.*, (2005). According to Offer and Knight (1988) inherited susceptibility to stress and severe stress at slaughter lead to a high rate of *post mortem* glycolysis. As a result of a high rate of *post mortem* glycolysis, temperature in PSE carcasses may rise by 3°C soon after slaughter. Lawrie (1998) noticed that increasing environmental temperature has also been known to accelerate *post mortem* glycolysis. According to Ruusunen and Puolanne (2004), as a light muscle, SM utilizes glycogen to a greater extent than dark muscles. This may lead to lactate accumulation in the muscle already in live animal (Pösö and Puolanne 2005). As a consequence, the initial *post mortem* pH can be quite low in SM. Chilling in commercial abattoir starts usually at 45 min *post mortem*. According to Offer and Knight (1988) the rate of cooling of muscle fibres is dependent on their depth from the meat surface and so is also the rate of glycolysis: higher temperature and lower pH could be expected in deep parts of muscle. Seyfert *et al.*, (2004) reported that colour in bovine *M. biceps femoris*, (BF) *M. recturs femoris* (RF) and *M. vastus lateralis* (VL), which were partly detached from the hind leg before chilling, was darker than in muscles of an intact hind leg. As the paleness in deep porcine SM seems to be of PSE origin, we set up a trial where the distal part of SM was partly detached from the body before chilling. The aim of this project was to study if it was possible to influence the rate of temperature and pH fall in deep SM by partly detaching the muscle before chilling.

Materials and Methods

A total of 19 pig carcasses weighing 79–110.5 kg (mean 94.3 kg) and of meat percentage 56.0–65.0 (mean 59.8) were selected in a standardised abattoir. The distal part of SM in the left hind leg of each carcass was detached from the body by knife at 45 min *post mortem*. The proximal part of the SM was left attached to the hind leg so the distal part of SM hung away from the body. The right hind leg was kept intact during chilling. During the two phase chilling, the carcasses went first through a blast chilling room (air temperature -15°C) during the period from 45 min to 2 h *post mortem*. After that, the carcasses were kept in a chilling room at a constant temperature of 2°C until 24 h *post mortem*. The temperature of the proximal (deep) part of SM was measured at the time of partial detaching at 45 min and 2 h, 2.5 h, 4 h and 24 h *post mortem* on both sides of the carcass. The pH value was measured at 45 min, 2.5 h, 4 h and 24 h *post mortem*. The sample for pH determination was taken from the deep SM with a stainless steel biopsy probe. The muscle sample of 0.3 g was mixed immediately in 3 ml Na-I-acetate solution (5 mmol Na-I-acetate + 150 mmol KCl). The pH value was measured at 20 °C with a pH meter Knick Portamess 911 (Knick Electronische Me Gerate Gmb & Co, Berlin, Germany) connected to an electrode Mettler Inlab 427 (Mettler-Toledo Process Analytical Inc, USA). At 24 h *post mortem* each SM was split and after a blooming time of 20 min colour (L* a* b*) was measured with Minolta CR-400 on the bloomed surface of each muscle. Drip loss was determined according to Honikel (1998): 80–100g of SM from the part next to the end of *M. adductor* was weighed and kept in a plastic bag for 2 d at 4°C. After that the piece was weighed again and the drip loss was expressed as a percentage of the original weight. The statistical analyses between the partly detached SM and the intact SM muscles were carried out using the Independent Samples T-test with SPSS 12.0.1 (2004) for Windows.

Results and Discussion

The pH of the deep SM was 6.49 and temperature 40.5°C at the time of partial detaching. We then observed a major difference in temperature decline during chilling: At 2 h, 2.5 h and 4 h *post mortem* the temperature in partly detached SM was 10–13°C lower than in intact SM ($p<0.001$) (Fig. 1.). At 24h *post mortem* the partly detached SM was only slightly colder than the intact SM ($p<0.05$). We also observed a significat change in the rate of pH fall in the partly detached SM compared to the intact SM. At 2.5 h and 4 h *post mortem* the pH of partly detached SM was 0.3–0.4 units higher than that of intact SM ($P<0.001$) (Fig. 2). At 24 h no difference was observed in the pH. It can be concluded that the *post mortem* glycolysis was slower but glycogen was utilized to the same extent as no difference in pH at 24 h *post mortem* was observed. Our pH and temperature data at 45 min *post mortem* are close to observations of Tikk *et al.*, (2006) who reported that the temperature of porcine deep SM was 40–40.5°C and the pH 6.25–6.5 at 45 min *post mortem*. In a study on beef Seyfert *et al.*, (2004) reported a faster temperature decline in BF, RF and VL when they had partly detached the *quadriceps* muscles (the beef knuckle). They also observed that the deep SM beside the partly detached beef knuckle was approximately 10°C colder than the SM beside the intact beef knuckle during the first seven hours *post mortem*. Seyfert *et al.*, (2004) recorded the rate of pH fall only in SM and found that the increased rate of chilling was not enough to slow the rate of pH fall in SM which is contrary to the results in the present study. We did not observe any difference in colour between the partly detached SM and the intact SM (P>0.1) which is contrary to the BF, RF and VL colour results of Seyfert *et al.*, (2004).

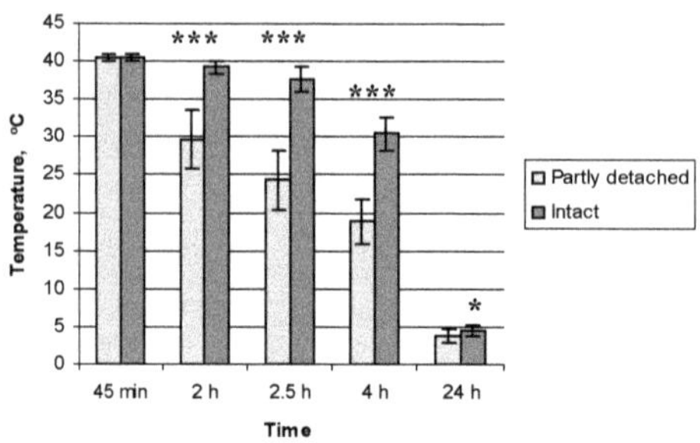

Figure 1: The temperature decline (± s.d.) of partly detached and intact SM. Difference between adjacent columns: *** p<0.001; *p<0.05.

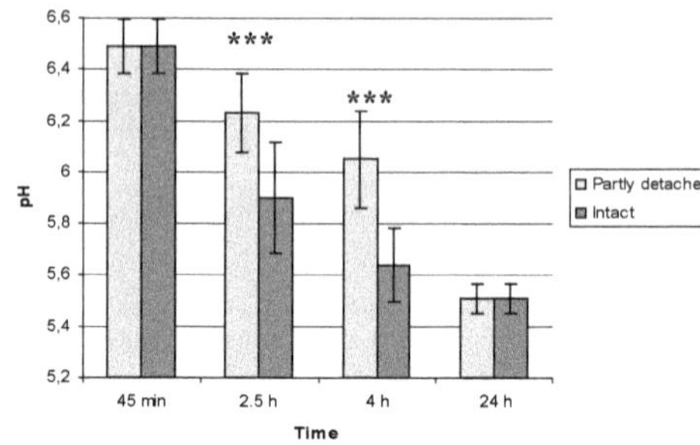

Figure 2: The pH decline (± s.d.) of partly detached and intact SM. Difference between adjacent columns: *** p<0.001.

A lowered rate of pH decline would be expected to improve the water retention both in raw meat and cooked meat products. In the present study the drip loss in the partly detached SM was lower than in intact SM (3.5% vs. 4.7%; p<0.01). More important from the industry point of view is, however, the total weight loss and total cost of the process. According to Daudin and Kuitche (1996) most of the weight loss is generated during the first hours after slaughter while the carcass is still warm. A new surface for water evaporation in a partly detached SM would be critical, and the total weight loss of the hind quarter would be an important parameter to measure. Further studies are needed to clarify the total effect of the partial detaching of SM on the costs and profits in the meat industry.

Conclusions

Partly detaching the SM from the carcass before chilling accelerates the chilling rate of deep SM, slows tremendously the rate of pH fall and reduces the drip loss in the deep SM. It does not influence the pH value at 24 h.

References

Daudin, J. D. and Kuitche, A. (1996). Modelling of temperature and weight loss kinetics during meat chilling for the time variable conditions using and analytical based method – III. Calculations versus measurements on pork carcass hindquarters. Journal of Food Engineering, 29: 39–62.

Honikel, K. O. (1998). Reference methods for the assessment of physical characteristics of meat. Meat Science, 49: 447–457.

Laville, E., Sayd, T., Santé-Lhoutellier, V., Morzel, M., Labas, R., Franck, M., Chambon, C. and Monin, G. (2005). Characterisation of PSE zones in semimembranosus pig muscle. Meat Science, 70: 167-172.

Lawrie, R. (1998). Chemical and biochemical constitution of muscle, Chapter 4. In: Lawrie, Editor, *Lawrie's Meat Science*, Sixth edition, Woodhead Publishing Ltd. Cambridge, England, pp. 58–95.

Offer, G. and Knight, P. (1988) The structural basis of water-holding capacity in meat. Part 2: drip losses. In: R. Lawrie, Editor, *Developments in Meat Science 4*, Elsevier Applied Science, London, pp. 172–243.

Pösö, A. R. and Puolanne, E. (2005). Carbohydrate metabolism in meat animals. Meat Science 70: 423–434.

Ruusunen, M., and Puolanne, E. (2004). Histochemical properties of fibre types in muscles of wild and domestic pigs and the effect of growth rate on muscle fibre properties. Meat Science, 67: 533–539

Seyfert, M., Hunt, M. C., Mancini, R. A., Hachmeister, K. A. Kropf, D. H. and Unruh, J. A. (2004). Accelerated chilling and modified atmosphere packaging affect colour and colour stability of injection-enhanced beef round muscles. Meat Science, 68: 209–219.

Tikk, K., Tikk, M., Karlsson, A. and Andersen, H. J. (2006). The effect of a muscle-glycogen-reducing finishing diet on porcine meat and fat colour. Meat Science 73: 378–385.

THE EFFECT OF ACCELERATED CHILLING OF CARCASSES ON PORK SEMIMEMBRANOSUS MUSCLE COLOUR

V. Tomović*, Lj. Petrović, N. Džinić, T. Tasić and P. Ikonić

Faculty of Technology, University of Novi Sad, Bulevar cara Lazara 1, 21000 Novi Sad, Serbia and Montenegro.
E-mail: tomovic@uns.ns.ac.yu

Keywords: pork, accelerated chilling, colour, *M. semimembranosus*

Introduction

Accelerated chilling of carcasses, e.g. rapid reducing intramuscular temperature as soon as possible after exsanguination can be an effective method to reduce the incidence of PSE pork (Savell *et al.*, 2001, review; Petrović *et al.*, 1996). Grandin (1994) believes that chilling process accounts for 20 to 40% of the variation in PSE meat. Kerth et al. (2001) reported a reduced incidence of PSE meat in loins and hams from stress sensitive pigs after accelerated chilling, however, not in stress resistent pigs. Some studies have shown, however, that rapid chilling, compared to conventional process, can affect the colour of pork (Jones et al., 1993; Milligan *et al.*, 1998).

Unlike those studies, Eilert (1997) and Hambrecht et al. (2004) stated that more rapid chilling procedures will not prevent PSE, e.g. have no effect on colour (Crenwelge et al., 1984a), since the changes in meat occured before the chilling process. The aim of this work was to determine the effects of two different rapid chilling regimes on colour of *M. semimembranosus*, in the scope of a broader program of decreasing the PSE meat incidence in Serbia.

Materials and Methods

Commercial pigs, slaughtered according to the standard procedure used in abattoirs in Serbia were investigated. At the end of the slaughterline, pH (pH_{30min}) in the medial part of *M. semimembranosus* (MS) and temperature (T_{30min}) at the center of the ham near to *femur* were measured. 40 carcasses with normal flow of glycolysis ($pH_{30min} \geq 5.8$) were chosen (n = 20 – Experiment 1; n = 20 – Experiment 2). Left sides (LS) were air chilled at –31 °C (AC) for 150 min. (Experiment 1) e.g. for 180 min. (Experiment 2), than placed in the conventional chiller at 2 – 4 °C until 24 h *post mortem*. The right sides (RS) were also chilled according to the conventional chilling (CC) regime at 2 – 4 °C until 24 h *post mortem*. Colour of MS was determined after measurement of pH_{24h} and T_{24h}. The procedure described by Honikel (1988) was used for colour measurements, in the CIE$L^*a^*b^*$ system using a Minolta Chroma Meter CR-400 device. On every MS sample, the colour was measured four times. On the basis of total number of instrumental colour measurements (L^*) (n = 160 – Experiment 1; n = 160 – Experiment 2) and according to criteria given by Joo et al. (1999) for colour, e.g. quality of pork (Pale: $L^* > 50$; Reddish-pink: $L^* = 43 – 50$; Dark: $L^* < 43$) the incidence of pale, reddish-pink and dark colour was determined. A five member panel assessed the MS colour according to the following criteria: 1 – Pale-pinkish-gray, 2 – Grayish-pink, 3 – Reddish-pink, 4 – Purplish-red, 5 – Dark-purplish-red.

Table 1: The effect of accelerated chilling on temperature and pH decline and colour of MS.

Trait	Experiment 1		Experiment 2		*P*–value (paired *t*-test)	
	LS – AC	RS – CC	LS – AC	RS – CC	Experiment	
	Mean ± SD	*Mean ± SD*	*Mean ± SD*	*Mean ± SD*	1	2
T_{30min}	41.36 ± 0.42	41.47 ± 0.30	41.73 ± 0.50	41.73 ± 0.54	0.533	1.000
T_{24h}	4.14 ± 0.39	5.52 ± 0.36	3.80 ± 0.53	5.16 ± 0.26	<<0.001	<<0.001
pH_{30min}	6.23 ± 0.22	6.25 ± 0.23	6.24 ± 0.26	6.19 ± 0.22	0.798	0.386
pH_{24h}	5.71 ± 0.16	5.66 ± 0.16	5.77 ± 0.22	5.65 ± 0.23	0.013	0.010
CIEL^* (lightness)	47.92 ± 2.65	48.15 ± 3.15	45.74 ± 4.50	48.24 ± 4.69	0.757	0.006
CIEa^* (redness)	8.79 ± 1.38	9.22 ± 1.60	9.32 ± 1.37	9.25 ± 1.60	0.401	0.883
CIEb^* (yellowness)	4.21 ± 0.76	4.47 ± 0.98	4.62 ± 1.27	4.81 ± 1.10	0.103	0.339
Colour – Sensory	2.90 ± 0.21	2.75 ± 0.42	2.90 ± 0.52	2.60 ± 0.64	0.343	0.132

Results and Discussion

At the beginning of chilling, 30 min. *post mortem* (Table 1), no significant difference ($P > 0.05$, Experiment 1, 2), was estimated between the average temperatures measured in the ham of left and right sides. However, at the end of the chilling process, 24 h *post mortem*, in accelerated chilled sides (LS – AC) a statistically significantly lower average temperature in the middle of the ham was found ($P << 0.001$; Experiment 1, 2), compared to conventionally chilled sides (RS – CC). The accelerated chilling e.g. the rapid decrease of intermuscular temperature, resulted in slower biochemical processes in MS. In rapid chilled sides (LS – AC) statistically significantly higher ($P = 0.013$; $P = 0.010$; Experiment 1, 2) average pH_{24h} values were found. The obtained results, which point to slower decrease of pH, are in accordance with the results obtained at similar conditions of rapid air chilling, but no significant difference between the ultimate pH values. In the first experiment the average lightness (L^*) of MS was not significantly affected ($P = 0.575$) by accelerated air chilling. However, in the second experiment significantly ($P = 0.006$) darker colour of MS was found in left sides rapid chilled for 180 min. (L^* = 45.74), compared to conventionally chilled sides (L^* = 48.24). The obtained results are in accordance with the results of other authors who also found significant effect of rapid chilling on lightness (L^*) in most cases. However, the accelerated air chilling had no effect on other colour characteristics (a^*, b^*, sensory, $P > 0.05$, Experiment 1, 2).

On the basis of lightness measurement (L^*) of MS (Figure 1), reduced incidence of pale colour was found in both rapid air chilling regimes; in the first experiment up to 9.1%, with highest share of reddish-pink colour – 59.1% (LS – AC), e.g. reduced incidence of pale colour of 11.5% in the second experiment with highest share of dark colour (LS – AC) – 35.7% compared to conventional chilling where a significant high share of pale colour was found – 40.9% and 42.5%, in experiments 1, and 2, respectively.

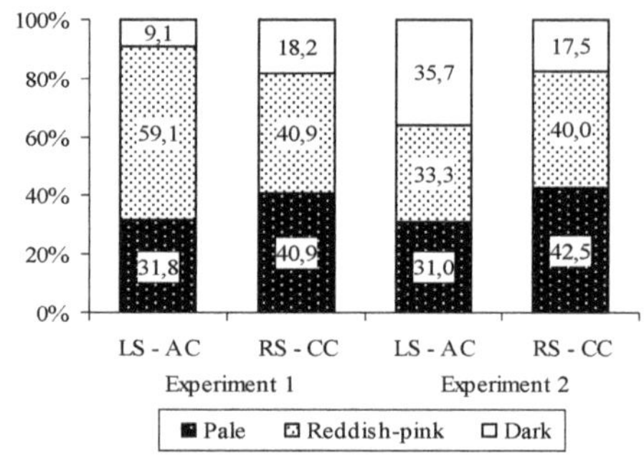

Figure 1: The effect of accelerated chilling on incidence of pale, reddish-pink and dark colour of MS.

Conclusion

Statistically significantly slower decline of pH was found in both experiments when pork was cooled using accelerated chilling. A siginificant effect on lightness (L^*) was found only in the second experiment, confirming once more that rapid air chilling can affect colour e.g. the colour is darker the longer the rapid air chilling.

Acknowledgments

This paper presents the results of work on project BTN 351008, supported by Serbian Ministry of Science and Environmental Protection.

References

Crenwelge, D.D., Terell, R.N., Duston, T.R., Smith, G.C., Carpenter, Z.L. (1984a). Effects of chilling method and electrical stimulation on pork quality. Journal of Animal Science, 59: 697-704.

Eilert, S.J. (1997). What quality controls are working in the plant? In: Proc. Pork Quality Summit. July 8-9, NPPC, Des Moines, IA, 59-63.

Grandin, T. (1994). Methods to reduce PSE and bloodsplash. Proc. Allen D. Leman Swine Conference, College of Veterinary Medicine, University of Minnesota, 21: 206-209.

Hambrecht, E., Eissen, J.J., de Klein, W.J.H., Ducro, B.J., Smits, C.H.M., Verstegen, M.W.A., den Hartog, L.A. (2004). Rapid chilling cannot prevent inferior pork quality caused by high preslaughter stress. Journal of Animal Science, 82: 551-556.

Honikel, K.O. (1998). Reference methods for the assessment of phisical characteristics of meat. Meat Science, 49: 447-457.

Jones, S.D.M., Jeremiah, L.E., Robertson, W.M. (1993). The effects of spray and blast-chilling on carcass shrinkage and pork muscle quality. Meat Science, 34: 351-362.

Joo, S.T., Kauffman, R.G., Kim, B.C., Park, G.B. (1999). The relationship of sarcoplasmic and myofibrillar protein solubility to colour and water-holding capacity in porcine longissimus muscle. Meat Science, 52: 291-297.

Kerth, C.R., Carr, M.A., Ramsey, C.B., Brooks, J.C. Johnson, R.C., Cannon, J.E., Miller, M.F. (2001). Vitamin-mineral supplementation and accelerated chilling effects on quality of pork from pigs that are monomutant or noncarriers of the halothane gene. Journal of Animal Science, 79: 2346-2355.

Milligan, S.D., Ramsey, C.B., Miller, M.F., Kaster, C.S., Thompson, L.D. (1998). Resting of pigs and hot-fat trimming and accelerated chilling of carcasses to improve pork quality. Journal of Animal Science, 76: 74-86.

Petrović Lj., Okanović, Đ., Rede., R. (1996). Possibility of cooked ham production from ham deboned early post mortem, 1. Influence of chilling rate on the properties of muscles deboned different time post mortem, International Fieischwirtschaft, 3, 27-32.

Savell, J.W., Mueller, S.L., Baird, B.E. (2005). The chilling of carcasses. Meat Science, 70: 449-459.

DEVELOPMENT AND EVALUATION OF A PILOT SCALE ATMOSPHERIC STEAM TREATMENT CABINET

G. Purnell*[1], S. James[1], J.E.L. Corry[2] and C. James[1]

[1]. *Food Refrigeration and Process Engineering Research Centre (FRPERC), University of Bristol, Churchill Building, Langford, Bristol BS40 5DU, UK,* [2]. *Division of Farm Animal Science, University of Bristol, Langford, Bristol, BS40 5DU, UK. Email: graham.purnell@bristol.ac.uk*

Keywords: steam, decontamination, equipment, temperature, surface

Introduction

Large scale steam "pasteurisation" systems are being employed commercially in the USA for treating red meat carcasses. However, there is still much that is unknown regarding the effectiveness and optimisation of steam treatments and there is also interest in using such technologies for treating much smaller meat carcasses, such as poultry. This paper describes the development and evaluation of a pilot atmospheric steam cabinet for treating meat carcasses.

Materials and Methods

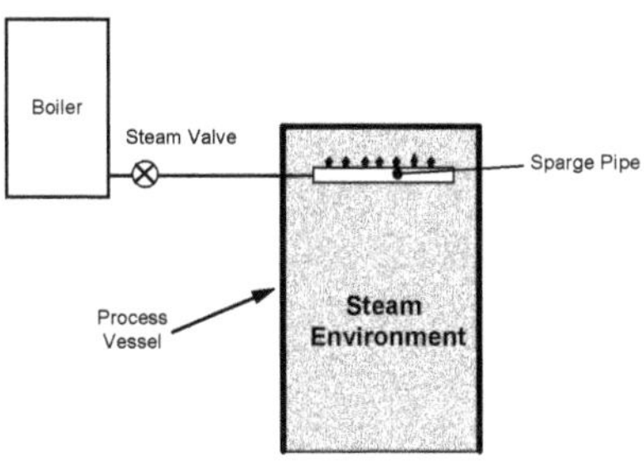

Figure 1: Schematic of initial atmospheric steam cabinet.

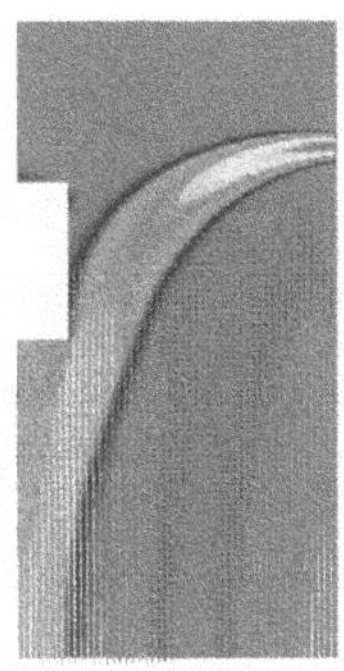

Figure 2: CFD predictions of a steam jet entering the cabinet.

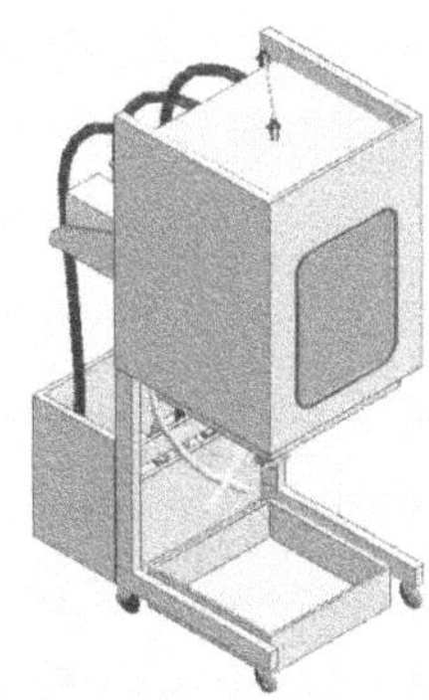

Figure 3: 3D CAD of final atmospheric steam cabinet.

The basic principle behind the use of atmospheric steam for decontamination is to utilise a vessel with an open base, filled with steam. The steam remains in the vessel because it is less dense than the surrounding air. To treat the product, it is raised into the steam environment for the required duration.

Initial studies were carried out using a very simple prototype rig (Figure 1) consisting of a large, food grade, plastic barrel with the base removed mounted on three metal legs. In the prototype, steam was supplied from a boiler with the steam line entering near the top of the barrel and was fed into an upward pointing sparge pipe to distribute the steam. The steam flow was controlled by opening the solenoid valve from the boiler and manually adjusting an isolation valve to achieve the desired flow rate. An initial temperature test was conducted using a mock food product with thermocouples attached to the surface. The measured temperature in the vessel was less than that expected for steam at atmospheric pressure (c.100°C). Investigations revealed this to be due to air entrapped by the turbulence of the steam coming out of the sparge pipe at a high velocity. Thus the method of steam introduction was altered to create a more stable flow. The steam inlet was moved to the top of the vessel where it exhausted into a plenum chamber. From the plenum chamber, steam entered the body of the vessel through a perforated baffle plate. This improvement produced a much more stable flow within the vessel, with excess steam gently spilling out from the base of the vessel. No air was entrapped and a much higher temperature was produced within the vessel. Tests were then carried out to determine the uniformity of steam distribution and the effect on the distribution of inserting samples. CFD modelling was carried out to see the effect of air entrainment caused by the product moving from the air into the steam (Figure 2).

Using the experience gained from the initial pilot plants and the CFD studies, a more user and food friendly atmospheric pilot plant has been designed and constructed in order to conduct investigation in commercial plants. The final experimental atmospheric steam treatment equipment (Figure 3) consists of a double-skinned stainless steel treatment chamber supported by two legs. The front of the vessel houses a double glazed unit to allow the product to be viewed during treatment. The treatment sample is raised into the chamber by a pneumatic cylinder controlled by an adjustable electronic timer. Steam is provided by three 2.8kW boilers operating in parallel. Steam enters a plenum chamber at the top of the vessel and is evenly distributed into the main chamber through a baffle. A lip around the bottom of the chamber collects condensate and drains it to the back of the unit, preventing hot water falling onto experimenters as the system is loaded.

Results and Discussion

The effects of various atmospheric steam treatments on the appearance, shelf-life and microbiological quality of chicken have been investigated. Initial experiments (James *et al.*, 2000a) showed that a 10s steam treatment on naturally-contaminated chicken breast portions resulted in a 1.65 $\log_{10}$ CFU (colony-forming units)/cm^2 reduction in aerobic plate count (APC). However, in comparison with untreated controls, this treatment did not extend the shelf-life. Overall, results indicated that significant reductions in microbial counts could be achieved for chicken meat using steam. Further work has looked at the effectiveness of steam against *Campylobacter* spp., under laboratory and commercial conditions (James *et al.*, 2005). In experimental studies whole chicken carcasses, inoculated with ca. 6 $\log_{10}$ CFU/cm^2 *Campylobacter jejuni* and *Escherichia coli* K12, were treated with steam at atmospheric pressure for up to 20s. Numbers of *C. jejuni* were reduced by ca. 1.8 $\log_{10}$ CFU/cm^2 in 10s and 3.3 $\log_{10}$ CFU/cm^2 in 20s. Corresponding reductions in numbers of *E. coli* K12 were 1.7 and 2.8 $\log_{10}$ CFU/cm^2. However, the 20s treatments caused the skin to shrink and change colour. The optimum steam treatment for maximum effect on *C. jejuni* and *E. coli*, least skin shrinkage and change of colour was concluded to be 10-12s. Additional trials in a commercial poultry plant using naturally contaminated carcasses were hampered by low initial levels of *Campylobacter* spp. (~1 $\log_{10}$ CFU/cm^2) but variable reductions of about 2 log cycles were obtained for pseudomonads and Enterobacteriaceae. Numbers of campylobacters were reduced, but not eliminated. It was considered that changes to appearance of skin-on carcasses or portions would be acceptable to many consumers. Unacceptable carcasses produced (or more severe treatments with greater kill potentials) could be used for production of 'skin-off' portions. Additional work has been carried out on pork, lamb (James *et al.*, 2000b), and beef (and also fish, fruits and vegetables).

The surface temperature is a critical parameter in the destruction of pathogenic and meat spoilage microorganisms in steam systems. However, measuring the temperature of the surface of the meat is a difficult procedure. There is a marked temperature profile from the surface into the environment and from the surface into the meat. Thermal modelling can be used to predict the surface temperature but reliable data on the surface heat transfer coefficient is required. An experimental protocol based on that of Creed and James (1985) has been used to determine the surface heat transfer coefficients in the atmospheric steam cabinet to enable further optimisation of the system. The overall effective surface heat transfer coefficient determined for the steam cabinet were 1215.8 W/m^2/K at 100°C, much higher than those used in previous studies on modelling steam decontamination systems (Hoke *et al.*, 2003). This should allow much better prediction of meat surface temperatures and subsequently aid the design and optimisation of similar steam intervention systems.

Conclusions

A pilot scale atmospheric steam cabinet has been developed and successfully used in many experimental trials involving poultry, pork, lamb, and beef (and also fish, fruits and vegetables) both in a laboratory setting and alongside commercial processing lines in industry to demonstrate the effectiveness of steam interventions in lowering microbial loads.

Acknowledgements

This project was funded by the Food Standards Agency, UK.

References

Creed, P. G. and James, S. J. (1985) Heat transfer during the freezing of liver in a plate freezer. Journal of Food Science. 50: 285-288.

Hoke, K., Housska, M., Kyyhos, K., Landfeld, A. and Pipek, P. (2003) Modelling of beef surface temperatures during steam decontamination. Journal of Food Engineering. 58: 95-102.

James, C., Gŏksoy, E. O., Corry, J. E. L. and James, S. J. (2000a) Surface pasteurisation of poultry meat using steam at atmospheric pressure. Journal of Food Engineering. 45: 111-117.

James, C., Thornton, J. A., Ketteringham, L. and James, S. J. (2000b) Effect of steam condensation, hot water or chlorinated hot water immersion on bacterial numbers and quality of lamb carcasses. Journal of Food Engineering. 43: 219-225.

James, C., Purnell, G., Hannay, N., James, S. J., Allen, V. M., Barbedo-Pinto, C., Howell, M. and Corry, J. E. L. (2005) Surface pasteurisation of chicken carcasses using hot water or steam at atmospheric pressure in order to reduce levels of Campylobacter spp. and other bacteria. Proceedings of 13th International Workshop on Campylobacter, Helicobacter and Related Organisms (CHRO 2005), Gold Coast, Queensland, Australia, 4-8th September, 2005.

THE TENDERISING EFFECT OF MARINATION WITH CITRUS JUICES ON BEEF WITH HIGH CONNECTIVE TISSUE CONTENT

G. Mutaf, E. Eyiler*, A. Yarali and A. Oztan

Hacettepe University, Department of Food Engineering, 06800 Beytepe-Ankara-Türkiye.
Email: eseneyiler@gmail.com

Keywords: tenderisation, beef, citrus juice marinades, Seville orange juice

Introduction

Many methods of altering meat tenderness including the use of marinades have been evaluated (Gault, 1985). Acidic marination involves the immersion of meat in an acidic solution of vinegar, wine or fruit juice (Lewis and Purslow, 1991; Stanton and Light, 1990). Marinades with a tenderizing capacity are particularly important in applications involving muscles rich in connective tissue. These muscles often make up the cheaper carcass cuts and the tenderizing effect of marinating offers a commercially important means of upgrading them (Gault, 1991; Lewis and Purslow, 1991). Burke and Monahan found (2002) that when shin beef strips were immersed in the citrus juice marinade (31% orange juice, 31% lemon juice, 38% distilled water), then the pH value decreased from 5.7 to 3.1 and the Warner-Bratzler shear force value decreased from 178 to 44 N cm^{-2} following marination. It has been also found out that the sensory analysis scores for tenderness and juiciness increased following marination. The purpose of this study is to investigate the potential marination of citrus juice on tenderizing the beefs with high connective tissue content. Previous researches have been performed for orange and lemon juice marinades. This study also contains the data for the marination of Seville orange juice.

Materials and Methods

Beef cuts (*M. semitendinosus*) used in this study were obtained from a local Turkish abattoir, 12 samples for 6 different marinades and control group were tested during the investigation. 4% salt and 0.5% sugar was added to deionized water in order to make a control group solution.

Table 1: The content of marinates.

Marinades	Juice (%)	Solution (%)
Control		100
1	Seville orange 31	69
2	Seville orange 62	38
3	Lemon 31	69
4	Lemon 62	38
5	Orange 31	69
6	Orange 62	38

The beef samples were immersed in these solutions. Six of the samples from each marinade were analyzed at the end of 24 hours, whereas 6 of the 12 samples were analyzed at the end of 48 hours. Each side of the samples was grilled for 4 min. The pH measurements, surface colour measurements and the texture analysis were done for the 24 hour-marinated and the 48-hour-marinated samples after cooking. Colour measurements were carried out at the surfaces of samples by using a Minolta chromometer (CM 3600 d). Texture analyses were performed by using Ta Plus Texture Analyser (Amatek Lloyd Instruments Ltd., UK). Correlations between the changes in pH, the a/b value; "the index of consumer testing (ICT)" and the hardness values were calculated.

Results and Discussion

The average pH values of raw meat samples were 5.26 and pure Seville orange juice, lemon juice and orange juice were 2.68, 2.80 and 3.43 respectively. The addition of salt and sugar to deionized water increased the pH of the solution. A decrease in the pH value of the samples was observed at the end of 24th and 48th hours according to the marination process. The pH value of the control group was the highest when compared with the marinated samples. Samples marinated with 62% lemon juice had the lowest pH value (a decrease of 1.51), while the samples marinated with 31 % orange juice had a higher pH value.

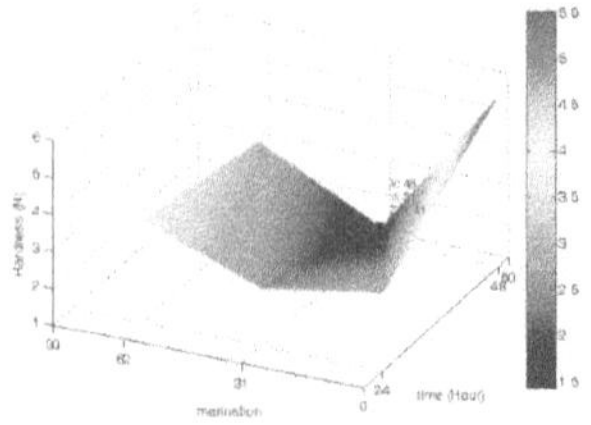

Figure 1:
The effect of Seville orange juice on the hardness of cooked beefs

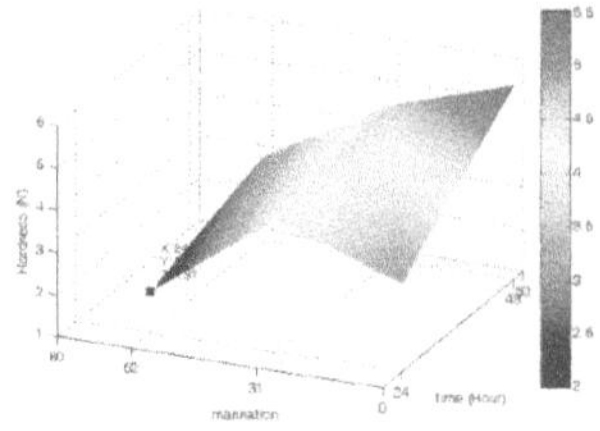

Figure 2:
The effect of lemon juice on the hardness of cooked beefs

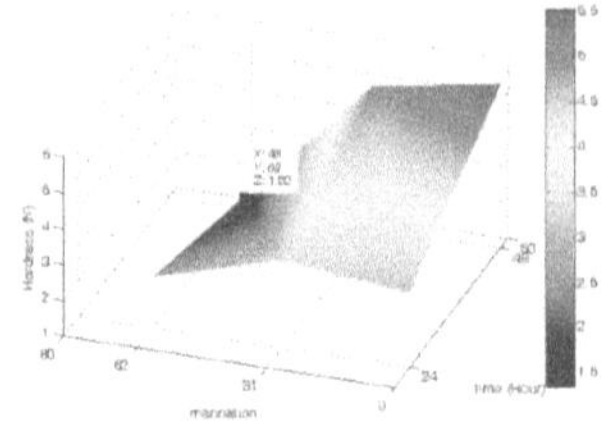

Figure 3:
The effect of orange juice on the hardness of cooked beefs

The hardness of the unmarinated samples increased during storage but marination process reduced this increase. The tenderising effect of marination increased, as the concentration increased except for the Seville orange juice. Although the cooked samples at 48th hour, which were marinated with 62 % content Seville orange juice, had high hardness value (3.08 N); cooked samples marinated with 31 % content Seville orange juice had a lower value (1.41 N) (Figure 1). 62% of lemon juice marination resulted in a rapid tenderising effect (1.97N) at the end of 24 hours compared with Seville orange and orange juices which were effective at the end of 48 hours (Figure 1, 2 and 3). As seen in Figure 3, 62% orange juice marinade gave the most tenderising character to the meat. The hardness values of the 31% Seville orange marinade and 62% orange marinade at the end of 48 hours were similar, which were 1.41 N and 1.32 N respectively. The pH results showed that orange juice have the lowest acidity. It gives the most tendering character at the end of the marination process.

According to the texture analysis, the control group samples had the highest shear force values. Therefore, the marination process can be recommended for these types of cuts.

The samples, which were marinated with 62 % content lemon juice, had the highest and the same ICT values at the end of 24 and 48 hours (0.62). It was calculated that the ICT and the hardness values had a negative correlation (r= - 0.33*). It was also found out that there was a negative correlation (r= - 0.34*) between pH and the ICT.

Conclusions

These results showed that the marination process provided tenderisation of meat after cooking. Using natural ingredients, the products can be perceived as healthier than those products with synthetic additives. Furthermore, Seville orange and lemons are found easily and are cheaper than other additives so it is much more economical to use the juices of these fruits as marinade ingredients. In this study, it was shown that other than orange and lemon juices; Seville orange juice also has tenderising effect on beef cuts. If it is needed to have a rapid tenderising effect it is recommended to use 62% lemon juice because the effect is observed at the end of 24 hours. However the tenderising effect of orange juice is the highest at the end of 48 hours.

When ICT results are considered, acceptability of the samples marinated with lemon juices are the highest ones.

Acknowledgments

The authors thank to Apikoglu Etsan for raw materials.

References

Gault, N. F. S. (1985). The relationship between water-holding capacity and cooked meat tenderness in some beef muscles as influenced by acidic conditions below the ultimate pH. *Meat Science, 15*, 15-30.

Lewis, G. J., and Purslow, P. P. (1991). The effect of marination and cooking on the mechanical properties of intramuscular connective tissue. *Journal of Muscle Foods, 2*, 177-195.

Stanton, C., and Light, N. (1990). The effects of conditioning on meat collagen: Part 4-the use of pre-rigor lactic acid injection to accelerate conditioning in bovine meat. *Meat Science, 27*, 141-159.

Gault, N. F. S. (1991). Marinated meat. In R. Lawrie (Ed.), *Developments in meat science-5* (pp. 191-245). London: Elsevier Science.

Burke R. M. and Monahan F. J. (2002). The tenderization of shin beef using a citrus juice marinade. *Meat Science, 63*, 161-168.

SYSTEMATIC DECOMPOSITION AND CONCEPTUAL MODEL OF THE COMPLEX INDEX OF EXTRUDATES QUALITY

A.B. Lisitsyn*, V.B. Krylova and A.O. Blinov

The V.M.Gorbatov All-Russian Meat Research Institute, Talalikhina 26, 109316 Moscow, Russia. Email: vniimp@inbox.ru

Keywords: system analysis, extrudates, quality, model, decomposition

Introduction

The problem of increasing the quality of extrudates is connected with six main aspects: preservation of and increase in food value (FV) improving structural-mechanical (SMI), functional-technological (FTI), organoleptical (OI) and safety indices (SI), and of the microstructure (M) of the products.

The model of the complex index of the extrudates quality can be presented as follows:

$K = <K_{FV}\ K_{SMI}\ K_{FTI}\ K_{OI}\ K_{SI}\ K_M>$, where $K_{FV}\ K_{SMI}\ K_{FTI}\ K_{OI}\ K_{SI}\ K_M$ are the models of *FV, SMI, FTI, OI, SI and M,* respectively.

To describe the problem of increasing quality of extrudates, allowing us to single out system components and determine the general system characteristics, the authors used the model of the purpose-oriented functioning of the system:

<purpose > ↔ <strategy> ↔ <resources>

The purpose determines the necessity of creation and development of the system. The purpose can be defined as a set $W_{(x)}$ of states $X_{(t)}$ of the system S, corresponding to the desired level of the complex index of quality (X(t) Ђ W(x)), and desired by the person, taking a decision (PTD), where t is time.
The description of the set W (x) consists of the formation of limit values on the vector X(t) (quantitative and qualitative) and in assigning to the vector E (W) the properties of the purpose (the vector of efficiency) which can be considered preferable (or which the PTD considers to be the preferred ones).
As the purpose states (X(t) Ђ W(x) can be achieved in three different ways (strategies), then in the general case there is a set Ω (Y) of strategies Y (t), (Y) (t) € Ω (Y), realizing the purpose W(x). The set Ω (Y) determines the function of the system and can be assessed by the vector of evaluation E (Ω).

Materials and Methods

The resources {d(t)} are the properties (material, power, finance, intellectual, *etc.*) that are necessary and sufficient for realisation of the function of the system S. The limited number of resources d(t) determines some field D(d) (d(t) Ђ D(d), the efficiency of which is designated through the vector of criteria E(D). The components of this vector can be: sufficient number of resources, their mobility, intensity of use, *etc.* It should be noted that the resources can be independent and interchangeable. As resources we understand all the means (initial raw materials, method and degree of their comminution, degree of their moistening or drying, uniformity of mixing of the components, technological parameters of the extrusion, form and diameter of matrix holes, type of auger elements and sequence of their assembly, that are necessary to increase the complex index of extrudates quality. In this case, the complex index of extrudates quality can be obtained by determinging optimum conditions, and the task of a research worker is to find this state with the desired resources vector from a set of possible states of the complex index of extrudates quality. The presence of a set of purposes for increasing the complex index of extrudates quality, the set of resources that are necessary and sufficient to achieve it, determines the problems of selection in these sets.

Results and Discussion

It should be noted that the choice of the optimum strategy to increase the complex index of extrudates quality to a large degree depends upon the subjective qualities of the research worker, and also on the initial conditions of each applied task, therefore, in general, it is not possible to make these strategies concrete, *i.e.* to propose the method of solution. But it is possible to indicate the way in which the choice can be made much easier. One of the methods for choosing the optimum strategy to increase the complex index of extrudates quality can be introduction of priorities for purposes and resources. It is known (Novoseltsev, 2006) that priorities should be formed prior to direct distribution of resources or during solution of the problem, when the obtained information about importance of one or another purpose, or about the efficiency of use of every existing resource is analysed.
Thus, the task of increasing the complex quality index is multi-purpose in nature, possesses many criteria for evaluation and alternatives for solution. This task requires a system approach, the main principles of which are the singularity of

functional-purposeful and cause-effect relationships on all the steps and phases of increasing the complex index of extrudates quality and on all the levels of its management.
The procedure for the synthesis of the system S is a complex process, connected with a large number of different problems. The end product of the process will be the optimum vector of the resources, agreed with the objectives of the task and possibilities of the objective of the investigations.
To formalise the next step of the investigations it is necessary to solve the problem of structuring the set of resources. Then the method of increasing of the complex index of the quality of extrudates can be described as a system:

CQI = < R, **RRR, d, C, E, V** CQI,R >, where R is the name of the resource that is used to increase the complex quality index of extrudates, RRR – the region of realisation of the resource.

Each resource (d) requires specific conditions for its use or realisation (C) with the objective of increasing the complex quality index of extrudates (E) by the method of realisation by the effect (E) and includes the set of the required values of the complex index of quality (CQI) of the extrudates and the set of requirements to the complex index of quality of extrudates (R), *i.e.*

CQI = {food value, structural-mechanical, functional- technological, organoleptical indices and safety indices, including microbiological ones, and microstructure of the products}.

It should be noted that each quality index of extrudates has its own special properties: division into groups, interrelationship, possibility of combination, absolute or relative values, the broadness of the list of indices, *etc.* We believe that if any two components of the system or phenomena have at least one property in common, then there is a connection between them by common nature of properties. If a specific functional relationship between different properties of two different components exists, then a connection between these components exists by mutual dependence of the properties. The connections existing between the components of the complex quality index of the extrudates are presented as a graph of the dependencies of the complex quality index components (Figure 1).

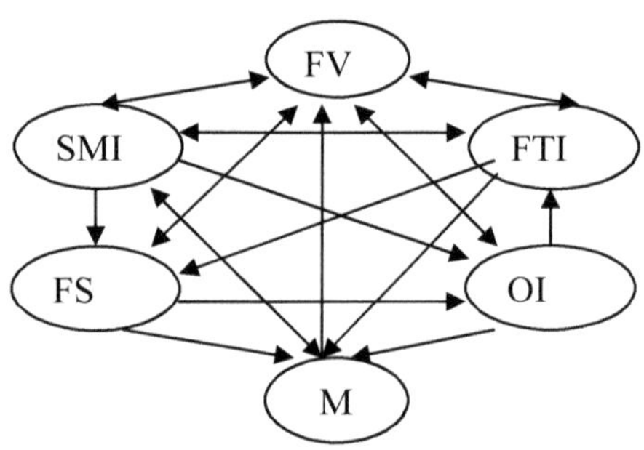

Figure 1: Graph of relationships of the extrudates quality components complex index

Conclusions
The analysis of relationships and interactions of extrudates quality complex index components, presented in Figure 1, allows us to judge the structural characteristic of the problem. This characteristic is one of the most common characteristics of the system, reflecting the set of steady relations, connections and interactions between the levels and components of the system under the conditions of external and internal effects.
In accordance with the classification of typologies of structures (Novoseltsev, 2006) as used in the system analysis, the present system problem can be related: by spacious topology - to the volumetric, having three-dimensional structure; by the nature of development – to extensive, for which the growth in the amount of links is typical; and by the kind of interactions – to the mixed, for which combination of material, energetic and informational types of structures is typical; by the nature of connection – to totally connected systems possessing high informative value.

Reference
Novoseltsev, V.I., Tarasov, B.V., Golikov V.K. and Demin B.E. (2006), Theoretical bases of system analysis. Edited by V.I.Novoseltsev – M.: Major, 592 pp.

EFFECT OF RELATIVE HUMIDITY OF DRYING AIR DURING THE RESTING PERIOD ON THE APPEARANCE, TEXTURE AND FLAVOUR OF DRY-CURED HAMS

J. Arnau*, M.D. Guàrdia, L. Guerrero and P. Gou

Institut de Recerca i Tecnologia Agroalimentàries. Unitat de Tecnologia i Enginyeria de Processos, Centre de Tecnologia de la Carn. 17121 Monells (Girona), Spain. E-mail: jacint.arnau@irta.es

Keywords: dry cured ham, relative humidity, resting, sensory property

Introduction

In dry-cured hams, a relative humidity (RH) lower than 75% produces a marked decrease in the equilibrium water content in salted fresh ham muscles (Comaposada *et al.*, 2000) and encourages oil drip on salted subcutaneous fat (Arnau and Gou, 2001). Arnau *et al.* (2003) observed that hams stored at 52±3 % RH during resting have a white appearance on some parts of the rind and have a larger surface area with oil drip. Hams stored at 78±3 % RH show no white rind at the end of resting, and at the end of the process the surface covered with oil drip is slightly smaller than the surface of those hams stored at 52%. Storage of hams at 85±3 % RH increases surface pH, phosphate crystals, NPN/N_T at the end of the resting period and it also increases superficial pH, decreases a_{ws} and the surface covered with oil drip at the end of the process. Moreover, the application of liquid subcutaneous pork fat on ham surface increases the aged flavour (Sanchez-Molinero and Arnau, 2005).

The aim of this study was to evaluate the effect of three different percentages of relative humidity during resting on the appearance, texture and flavour of dry-cured ham at the end of the process.

Materials and Methods

Raw material selection and processing

Two batches of 18 carcasses were selected at a commercial slaughterhouse. The pH in *Semimembranosus* muscle (SM) at 24 hours was lower than 6.2. The ham rind was partially removed to obtain the typical 'V' cut found on some Spanish Serrano hams. One ham per carcass was manually rubbed with 30 g NaCl, 0.5 g of sodium nitrite, 0.5 of potassium nitrate and 0.5 sodium ascorbate per kg of ham. After 7 days the hams were salted again with 30 g of NaCl per kg of green ham. Hams remained 18 days at 1-3 °C and 80-90 % RH. Afterwards, they were rinsed in cold water and randomly introduced into three resting chambers (70-75 %, 75-80 % and 80-85 % RH) at 2-4 °C until day 104. Then hams were hung together in a drying room for 12 months. The RH ranged from 60 to 70% and the temperature from 12 to 30 °C.

Sensory analysis

The external appearance of the hams was evaluated in one session by six experts in sensory analysis of dry-cured ham. The proportion of area covered with mould (mouldiness) and oil (oilness) was evaluated on the external part of ham; colour, marbling, brightness, red rings, white film and phosphate crystals on the cut surface and sweetness, saltiness, piquantness, bitterness, metallic, aged flavour, adhesivity, hardness, crumbliness, pastiness and stringiness on 2 mm thick slices in SM and *Biceps femoris* (BF) muscles by using a 10 points scoring scale where 0=absence and 10=maximum intensity. The assessors were selected and trained following ASTM standards (1981). The panelists had a minimum experience of 10 years in descriptive analysis of a wide range of foods.

Statistical analysis

The data were analysed by ANOVA using the GLM procedure of SAS (2001). The differences between the three different RH treatments were tested by Tukey test. A Chi Square test was applied to compare the distribution of the frequencies of aged flavour according to the intervals established.

Results and Discussion

No significant differences ($P>0.05$) were found in weight losses between treatments, which ranged from 37.3% to 38.9%. Table 1 shows the mean values for those attributes where significant differences ($P<0.05$) were observed. White film was higher in 70-75% RH treatment. The lean and subcutaneous fat surface covered with oil drip at the end of the process was lower when the RH during resting period was above 80%. This could be due to the higher pH (Arnau *et al.*, 2003) or to the higher microbial growth during the resting period at higher RH. The surface covered with mould increased significantly according to the RH on the lean surface, but on the fat surface there was a significant increase in fat mouldiness but only when the RH was above 80%. Hams with slight and moderate aged flavour scores were significantly higher in the 70-75% treatment and lower in 80-85% treatment in BF muscle, but the differences were not significant in SM muscle (Table 2). This could be due to the higher oil drip, which coincides with the higher aged flavour obtained when liquid subcutaneous pork fat is applied to the lean surface during ageing (Sanchez-Molinero and Arnau, 2005). The oil drip is oxydated more easily, and this could increase some aged flavour nuances, which are appreciated by some consumers of dry-cured ham. Piquantness was significantly higher in 80-85% RH than in 75-80%

RH. Bitterness was significantly higher at 80-85% with respect to the other two treatments and sweetness tended to decrease as the RH increased (p>0.05).

Table 1: Mean values for each batch regarding the significant attributes.

Attribute	p	Relative Humidity			RMSE
		70-75	75-80	80-85	
White film[1]	0.0361	6.7^{a}	4.8^{b}	4.6^{b}	1.9266
Fat Oilness[2]	<0.0001	6.5^{a}	4.8^{a}	1.9^{b}	2.2917
Fat Mouldiness[2]	<0.0001	1.0^{b}	1.5^{b}	5.4^{a}	2.9704
Lean Oilness[2]	<0.0001	4.8^{a}	2.9^{b}	1.0^{c}	1.5107
Lean Mouldiness[2]	<0.0001	2.3^{c}	4.6^{b}	7.7^{a}	1.8668
Piquantness[3]	0.0489	2.2^{ab}	2.0^{b}	2.7^{a}	0.6183
Bitterness[3]	0.0031	1.8^{b}	1.7^{b}	2.8^{a}	0.7547

[1]: White film on cut surface; [2]: on ham; [3]: on BF muscle; [a,b,c]: in a row mean values with different superscripts were significantly different (p<0.05); RMSE: Root mean standard error.

Table 2: Frequency of each category of aged intensity attribute evaluated on BF and SM muscles.

Aged flavour intensity[1]	*Biceps femoris*			*Semimembranosus*		
	70-75	75-80	80-85	70-75	75-80	80-85
Low	3	7	11	6	7	9
Slight	2	1	1	1	1	1
Moderate	7	4	0	5	4	2

Aged flavour intensity[1]: Low <0.5; Slight = [0.5, 1.5]; Moderate = (1.5, 4.0].

Conclusions

A RH higher than 80% during the resting period decreases oil drip during the ageing process and also the aged flavour and increases the bitterness of partially skinned dry-cured hams at the end of the process.

Acknowledgements

This work was supported with funds from the Spanish Ministry of Agriculture, Fishing and Feeding (INIA project RTA01-074).

References

Arnau, J. and Gou, P. (2001). Effect of air relative humidity on ham rind and subcutaneous salted fat during the resting period. *Meat Science*, 58: 65-68.

Arnau, J., Gou, P., and Comaposada, J. (2003). Effect of the relative humidity of drying air during the resting period on the composition and appearance of dry-cured ham surface. *Meat Science*, 65: 1275-1280.

ASTM. (1981). Guidelines for the selection and training of sensory panel members (ASTM STP 758). Philadelphia, USA: American Society for Testing and Materials.

Comaposada, J., Gou, P. and Arnau, J. (2000). The effect of sodium chloride content and temperature on pork meat isotherms. *Meat Science*, 55: 291-295.

Sánchez, F. and Arnau, J. (2005). Efectos de la fusión de la grasa ("sudado") del jamón. *Eurocarne* 135: 61-68.

SAS (Release 8.2, 2001). SAS User's guide, Cary NC: SAS Institute Inc.

ELECTRON BEAM TREATMENT AND STORAGE ON COLOUR AND LIPID OXIDATION OF DRY-CURED IBERIAN HAM SLICES

R. Cava[1], L. Ladero[1], D. Morcuende[1], M.R. Ramírez[1] and A. Carrasco[2]

[1] *Food Technology. Faculty of Veterinary Science. University of Extremadura. Cáceres 10071. Spain.* [2]*Department of Engineering. Instituto del Frío. CSIC. Madrid. Spain. E-mail: rcava@unex.es*

Keywords: dry cured Iberian ham, irradiation, colour, TBA-RS, hexanal, storage

Introduction

Irradiation is the most effective technology in eliminating pathogens in meat and meat based products. Significant progress has been made in this respect by using irradiation doses lower than 10 kGy to control the growth of pathogenic and spoilage bacteria on meat and meat products, such as *Listeria monocytogenes* and *Salmonella typhimurium*, *Escherichia coli* O157:H7 and *Yersinia enterocolitica* (Fu *et al.*, 1995; Shenoy *et al.*, 1998; Olson, 1998). One of the major concerns in irradiating meat and meat products, however, is its effect on meat and meat product quality, mainly because of free radical reactions resulting in the possibility of colour change, lipid oxidation and odour generation, and consumer responses to these quality changes are quite negative (Patterson and Stevenson, 1995; Ahn *et al.*, 2000; Jo and Ahn, 2000; Du *et al.*, 2002). Dry-cured Iberian ham is a meat product with a high consumer acceptance due to its particular sensory characteristics derived from both raw meat composition (high intramuscular fat and oleic acid contents) and the physical, chemical and biochemical changes that take place during the ripening period (Cava *et al.*, 1997; *Cava et al.*, 1999). Irradiation of dry-cured ham could have important effects on overall quality of dry-cured Iberian ham, characterized by a high content of intramuscular fat, heme pigments and unsaturated fatty acids and a desirable and characteristic flavour. The quality characteristics of irradiated dry-cured hams would be very important for the acceptance of the irradiation technology. The objective of the present study was to investigate the effect of using different ionization doses (0 kGy, 5 kGy and 10 kGy) and storage time on instrumental colour, lipid oxidation and hexanal changes of vacuum-packaged slices of dry-cured Iberian hams.

Materials and Methods

Dry-cured ham samples

Dry-cured Iberian hams from pigs free-range reared and feeding on acorns and grass were used for the experiment. *m. Biceps femoris* was dissected and sliced using a slicing machine obtaining 2mm thickness slices.

Vacuum-packaging, irradiation and refrigerated storage.

Dry-cured ham slices were vacuum-packaged in nylon/polyethylene bags (9.3ml $O_2/m^2/24h$ a 0ºC) containing 3 slices/pack. Dry-cured ham slices (3 slice/pack/treatment/sampling) were irradiated at a 0, 5, or 10 kGy dose using an Electron Beam irradiator (Rhodotron TT200, IBA, Louvain La Neuve, Bélgica). The irradiation process was conducted at room temperature with single layer display and single-sided dosage. To confirm the target dose, 2 alanine dosimeters per cart were attached to the top and bottom surfaces of the sample. The final doses were 5.21± 0.32 kGy for 5 kGy treatment and 10.47±0.65 for 10 kGy treatment. Irradiated samples were stored in a 4ºC refrigerator in the dark for 0, 30, 60 and 90 days.

Physico-chemical analysis

The CIELAB colour space was studied in accordance with Cassens *et al.* (1995). The colour parameters were determined using a Minolta CR-300 colorimeter (Minolta Camera Co., Osaka, Japan). The measurement was repeated on eight randomly selected locations on each loin slice and averaged for statistical analysis. Lipid oxidation was measured using a complete dry-cured Iberian ham slice and homogenised using a kitchen blender. The extent of lipid oxidation was estimated as TBA-RS using the method by Salih *et al.* (1987). TBA-RS were measured on one slice from each pack and were expressed as mg malondialdehyde (MDA)/kg meat. A Solid Phase MicroExtraction, SPME (Supelco Co., Bellefonte, PA) fibre (10 mm length) coated with DVB/Carboxen/PDMS (50/30 µm thickness) was used for the determination of hexanal content of the samples according to Carrasco *et al.*, (2005). Hexanal was tentatively identified by comparing its mass spectra with that contained in the NIST/EPA/NIH and Wiley libraries and by comparison of Kovats index with that reported in the literature. The quantification of the hexanal content was performed using an external standard.

Statistical analysis

Samples were completely and randomly assigned to the treatments. There were five replicates per dose. The effect of dose and time of storage were analyzed using the Analysis of Variance procedure of SPSS, version 10.0 (SPSS, 1999). Tukey's test was used to compare the mean values of the treatments. Mean values and standard error of the means (SEM) were reported ($P<0.05$).

Results and Discussion

After treatment, 10 kGy treatment increased significantly TBARS values of sliced dry cured Iberian ham when comparing to the 0 kGy dose level (Table 1). With respect to control, TBARS increased in ~12% and ~44% at 5 kGy and 10 kGy, respectively. After 30 days of refrigerated storage, treated samples at 5kGy and 10kGy showed higher TBARS than the control. After 60 and 90 days of storage, no differences in TBARS were found between treated and control samples. At day 0, the 10kGy dose significantly increased hexanal content in vacuum-packed slices of dry-cured Iberian ham (Table 2). However, 5kGy dose significantly reduced hexanal content when compared to control. At the end of storage, control samples (0 kGy) showed lower hexanal content than samples treated at 5 kGy ($p<0.05$) and 10 kGy ($p>0.05$).

Table 1: TBARS (mg MDA/kg meat) of vacuum-packaged slices o dry-cured Iberian ham stored for different period of time after treatment[a,b].

	Dose			
Days of storage	0 kGy	5 kGy	10 kGy	SEM [c]
Day 0	1.21 b xy	1.36 b	1.75 a	0.09
Day 30	0.98 b y	1.28 a	1.43 a	0.06
Day 60	1.18 xy	1.37	1.77	0.11
Day 90	1.35 x	1.48	1.54	0.10
SEM [d]	0.1	0.1	0.1	

Table 2: Hexanal contents (ng hexanal/g sample) of vacuum-packaged slices of dry-cured ham stored for different period of time after treatment [a,b].

	Dose			
Days of storage	0 kGy	5 kGy	10 kGy	SEM [c]
Day 0	21.3 b xy	17.9 c y	34.7 a x	2.3
Day 30	25.7 x	20.3 xy	25.0 y	1.4
Day 60	20.6 xy	24.0 x	22.3 y	1.0
Day 90	17.3 b y	24.8 a x	20.2 ab y	1.2
SEM [d]	1.1	1.2	1.5	

a Means with a different letter (a,b,c) within a row of the same day of storage are different ($p<0.05$). b Means with a different letter (x,y,z) within a column of the same irradiation dose are different ($p<0.05$). c SEM: Standard error of the mean among the same irradiation dose (n=15). d SEM: Standard error of the mean among the same storage day (n=20).

Table 3: The surface colour of vacuum-packaged slices of dry-cured Iberian ham stored for different period of time after treatment [a,b].

	Dosage			
	0 kGy	5 kGy	10 kGy	SEM [c]
CIE L*-value				
Day 0	37.56 xy	38.49 x	39.44 x	0.37
Day 30	38.59 a x	39.05 a x	35.40 b y	0.52
Day 60	36.95 xy	36.31 y	35.00 y	0.37
Day 90	35.44 y	36.97 xy	35.72 y	0.39
SEM [d]	0.37	0.35	0.50	
CIE a*-value				
Day 0	24.07 c x	27.11 b x	30.47 a x	0.77
Day 30	22.96 a x	21.93 a y	19.80 b y	0.49
Day 60	20.29 y	20.17 y	20.52 y	0.29
Day 90	19.02 y	17.90 z	19.22 y	0.50
SEM	0.55	0.82	1.13	
CIE b*-value				
Day 0	10.97 xy	11.51 x	12.08 x	0.21
Day 30	11.31 a x	10.64 a x	9.38 b y	0.26
Day 60	9.88 a yz	8.83 b y	8.46 b y	0.23
Day 90	8.92 z	8.73 y	9.17 y	0.14
SEM	0.26	0.29	0.34	

Regarding instrumental colour, treatment at 5kGy and 10kGy did not affect CIE L*-value (lightness) at day 0, 60 and 90 of storage. Lightness tended to decrease with time of storage being darker. At day 0, CIE a*-value (redness) was significantly affected by treatment, being linearly increased with dose. Storage significantly affected CIE a*-value, decreasing redness with time of storage. After 30 days of storage, 0kGy and 5kGy samples showed significantly higher CIE a*-values than those treated at 10kGy dose. No differences in redness were found at day 60 and 90 of storage.

Conclusion

Electron beam treatment induced changes were only of importance after treatment. During storage, decolouration and oxidative changes were similar in treated and non-treated samples, storage induced changes being of a greater importance than those induced by electron beam treatment. The absence of great differences in +30 days stored products treated with respect to non-treated products shows that application of 5kGy or 10kGy dose could not compromise their sensory characteristics, with the exception of hexanal content and thus their flavour.

References

Carrasco, A, Tárrega, R, Ramírez, MR, Mingoarranz, FJ. And Cava, R (2005). Colour and lipid oxidation changes in dry-cured loins from free-range reared and intensively reared pigs as affected by ionizing radiation dose level. *Meat Science,* 69, 609-615.

Cassens, R. G., Demeyer. D., Eikelenboom, G., Honikel, K. O., Johansson, G., Nielsen, T., Renerre, M., Richardson, I., and Sakata, R. (1995). Recommendations of reference methods for assessment of meat colour. Proc. 41st Int. Congr. Meat Sci. Technol. (vol. 2, pp.410-411). San Antonio.

Salih, A.M., Smith, D.M., Price, J.F. and Dawson, L.E. (1987). Modified extraction 2-thiobarbituric acid method for measuring lipid oxidation in poultry. *Poultry Science,* 66, 1483-1489.

SPSS (1999). SPSS Base 10.0. User manual: application guide. Republic of Ireland.

TENDER PORK THROUGH STEPWISE CHILLING

K. Rosenvold[1]*, U. Borup[1] and M. Therkildsen[2]

[1] *Danish Meat Research Institute, 4000 Roskilde, Denmark,* [2] *Department of Food Science, Danish Institute of Agricultural Sciences, Post box 50, 8830 Tjele, Denmark. Email: kro@danishmeat.dk*

Keywords: chilling, *longissimus dorsi*, pork, tenderness, WHC

Introduction

Implementation of groupwise treatment of slaugther pigs during transport, lairage and stunning has resulted in a reduced temperature in the carcass at the time of sticking as well as an improved water-holding capacity (WHC) (Støier *et al.*, 2001). However, the chilling process has not been optimised to the lower carcass temperature, and trials have indicated that pork is less tender today than before.

It is assumed that, the chilling process can be optimised to improve pork tenderness without compromising WHC. The assumption is based on the following:

1. Carcass temperature at 1min pm ($T_{1\ min}$) is essential for WHC as shown in a study where $T_{1\ min}$ and $pH_{2\ h}$ explained 89% of the variation in drip loss (range: 2 to 13%) (Schaëfer *et al.*, 2002). Implementation of groupwise transport, lairage and stunning reduce $T_{1\ min}$ (Køltoft, 2005).
2. Pre-rigor carcass temperature is crucial for tenderness in beef, lamb and pork (Tornberg, 1996; Rees *et al.*, 2003). In beef, pre-rigor temperatures in the interval of 10 to 15°C results in more tender beef compared to temperatures above and below this interval (Tornberg, 1996).
3. The *post mortem* pH-fall is a result of lactic acid formation from glycogen breakdown. Glycogen is broken down by glycogen phosphorylase and glycogen debranching enzyme (GDE) (Brown and Brown, 1996). Kylä-Puhju *et al.*, (2005) has shown that the activity of GDE is temperature dependent and that its activity at 15°C is identical to that at 4°C. *Post mortem* pH-fall may therefore be identical in the temperature interval of 4°C to 15°C.

Thus, combining the facts that 1) $T_{1\ min}$ is reduced because of groupwise transport, lairage and stunning, 2) the temperature interval 10 to 15°C results in maximal tenderness and 3) glycogen breakdown at 15°C is identical to that at 4°C, we hypothesised that a stepwise chilling method, composed of a rapid temperature reduction to 10 or 15°C (in chilling tunnel), a 6 h holding period at 10 or 15°C followed by rapid post chilling (in chilling tunnel) and finally equalization, would improve tenderness without compromising the waterholding capacity of the meat. This hypothesis was tested in the present study.

Materials and Methods

A total of 42 pigs, 21 female and 21 castrate crossbreed pigs (D(LY)) were slaughtered at the research facility at The Danish Institute of Agricultural Sciences. The pigs were exposed to 2 stepwise chilling methods. One half of the carcasses was chilled stepwise (F10 and F15), while the other half was the control (C10 and C15). The chilling methods were imposed 40 minutes post mortem: F10 was placed in the chilling tunnel (-22°C/3 ms^{-1}) for 69 minutes, and then placed in a chilling room at 10°C for 6 h, returned to the chilling tunnel for 12 minutes and finally placed in a chilling room at 4°C for equilization until the next day. F15 was placed in the chilling tunnel for 47 minutes, and then placed in a chilling room at 15°C for 6 h, returned to the chilling tunnel for 24 minutes and finally placed in a chilling room at 4°C until the next day. C10 and C15 were placed in the chilling tunnel for 75 minutes and then placed in a chilling room at 4°C until the next day. The temperature was measured continuously and pH was measured in *Longissimus dorsi* (LD) after 1 minute, 35 minutes, when the carcasses were moved to and from the chilling tunnel, 4 h, 6 h and 24 h *post mortem*. WHC was determined as drip loss in LD using the EZ-drip method. Sensory tenderness was determined in LD 3 days post mortem without prior freezing. Cooking and sensory method were as described by Hansen *et al.* (2004).

Results and Discussion

The mean $T_{1\ min}$ measured in LD in the 84 half carcasses was 39.5 ± 0.4°C. The maximal $T_{1\ min}$ was measured to 40.6°C. $T_{1\ min}$ was low – one of the arguments that the hypothesis tested was based upon. The temperature profiles in LD are shown in Figure 1. Clear temperature differences were obtained between the treated and the control carcasses. Furthermore, at the end of the 6 h holding period the temperatures were 10.4°C and 13.7°C in F15, hence close to the desired pre-rigor temperatures of 10°C and 15°C, respectively. $pH_{6\ h}$, $pH_{8\ h}$ and $pH_{24\ h}$ was sligthly (< 0.15 pH unit) yet significantly lower in F10 as well as F15 compared to C10 and C15 (data not shown). But this pH difference did not influence WHC, which was found to be low and identical in the four groups (Table 1). Chilling loss was increased in F15 compared to C15, but identical in F10 and C10.

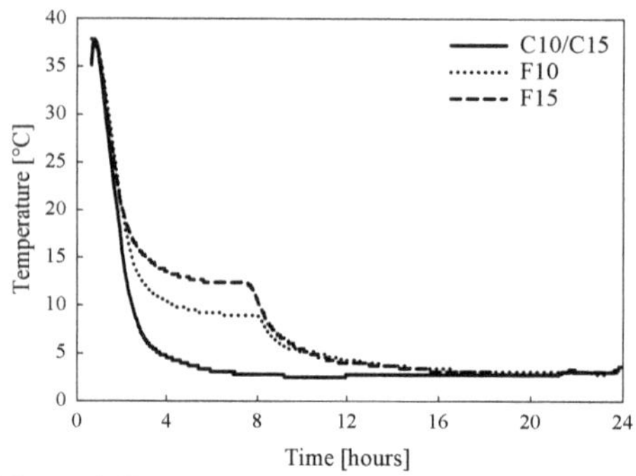

Figure 1: Temperature measured in the LD from 35 mins. to 24 h post mortem in **F10**, **F15** and **C10/C15.**

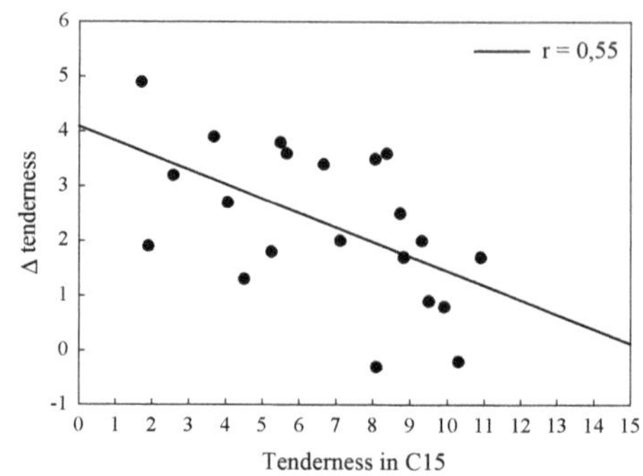

Figure 2: Δ tenderness between **F15** and **C15** plotted against tenderness in **C15**.

Table 1: Waterholding capacity (WHC) and sensory tenderness in LD from **F10**, **C10**, **F15** and **C15.**

	F10	C10	p-value	F15	C15	p-value
WHC [%]	1.5	1.9	0.023	1.4	1.7	0.79
Tenderness[1]	8.4	6.4	< 0.0001	9.0	6.7	< 0.0001

[1]sensory scale from 0 (not tender) to 15 (very tender).

Three days *post mortem*, sensory tenderness of LD was evaluated. Both F10 and F15 had improved the sensory tenderness of the meat by 2 sensoric units (on at 15 point scale) compared to C10 and C15 (Table 1).

The pig functioned as its own control, one carcass side being the control and the other being either F10 or F15. Hence, the specific effect of the stepwise chilling method could be found for each pig. In Figure 2, the difference in tenderness between F15 and C15 is plotted against tenderness in C15, indicating that the effect of stepwise chilling on tenderness was greater in pigs with tough meat whereas the effect is small in pigs with per se tender meat despite the rapid chilling process.

Conclusions

This study shows that stepwise chilling of carcasses can improve pork tenderness significantly without compromising WHC. The effect was greatest in meat that potentially would have been tough.

References

Brown, D.H. and Brown. B.I. (1966). In S.P. Colowic, and No.O. Kaplan (Eds.). Methods in enzymology (Vol. 8, pp. 525-524). New York: Academic Press.

Hansen, S., Hansen, T. Aaslyng, M.D. and Byrne, D.V. (2004). Sensory and instrumental analysis of longitudinal and transverse textural variation in pork. *Meat Science*, 68: 661-629.

Kylä-Puhju M., Ruusunen, M. and Puolanne, E. (2005). Activity of porcine muscle glycogen debranching enzyme in relation to pH and temperature. *Meat Science*, 69, 143-149.

Køltoft, P. (2005). Undersøgelse af kølemetodens betydning for kødkvaliteten. Meat trainee report, KVL, pp 1-12.

Rees, M. P., G. R. Trout, and R. D. Warner. (2003). The influence of the rate of pH decline on the rate of ageing for pork. II: Interaction with chilling temperature. *Meat Science*, 65:805-818.

Schäfer, A., Rosenvold, K., Purslow, P. P., Andersen, H. J. and Henckel, P. (2002). Physiological and structural events post mortem of importance for drip loss in pork. *Meat Science*, 61(4), 355-366.

Støier, S., Aaslyng, M. D., Olsen, E. V. and Henckel, P. (2001). The effect of stress during lairage and stunning on muscle metabolism and drip loss in Danish pork. *Meat Science*, 59, 127-131.

Tornberg, E. 1996. Biophysical Aspects of Meat Tenderness. *Meat Science*, 43(S), S175-S191.

MECHANICAL TENDERISATION AND ENHANCEMENT IMPROVE EATING QUALITY OF BEEF FROM DAIRY COWS

K. Rosenvold*, I. Clausen and N.T. Madsen

Danish Meat Research Institute, 4000 Roskilde, Denmark. Email: kro@danishmeat.dk

Keywords: beef, mechanical tenderisation, enhanced beef, tenderness, eating quality

Introduction

Denmark has a significant production of beef from calves and young bulls in addition to a large production of beef from dairy cows because of the large milk production. Beef from young bulls is primarily exported to Southern Europe, while veal and beef from the dairy cows mainly is sold on the Danish market. Beef from dairy cows has a larger variation in eating quality compared to beef from heifers and steers. Both mechanical tenderising and enhancing beef have been demonstrated to improve the eating quality of beef from beef cattle (Jeremiah *et al*, 1999; Baublits *et al.*, 2006), however the potential of applying the methods to beef from dairy cows have not been investigated. In this study the potential of applying mechanical tenderisation and enhancement to produce steaks from the loin and knuckle from dairy cows was investigated.

Materials and Methods

The study included 90 loins (*M. Longissimus dorsi*) and 90 knuckles from 45 dairy cows - mainly Holstein Friesian - slaughtered at the age of 54 months (± 10) in a commercial slaughter plant. After deboning the samples were vacuum packed and stored at 2°C. On day seven post mortem the loins were divided in two (the cranial and cardial end) and the knuckles were divided into *M. Vastus lateralis* and *M. Rectus femoris*. The samples had a pH mean of 5.6 and a standard deviation of 0.13 and were distributed for four treatments: 1) control (Control), 2) mechanical tenderisation using a Ross Tenderiser TC 700M (Mec), 3) enhancement by multi needle injection of 6% brine (0.6% salt and 0.3% sugar in the final product; Brine 1), and 4) enhancement by multi needle injection of 8% brine (0.6 salt, 0.4% sugar and 0.25% sodium-bicarbonate in the final product; Brine 2). The samples were cut into 2 cm steaks and packaged two and two in consumer packages. Three different packaging methods were used: 1) MA-packing in 80% O_2 and 20% N_2, 2) MA-packing in 70% N_2 and 30% CO_2 and 3) vacuum packing. The packages were stored at 2°C for seven days before the steaks were pan fried to an internal temperature of 65°C, followed by a sensory profiling of tenderness, juiciness, fried and boiled beef flavour and warmed-over flavour (WOF). Cooking method and sensory profiling as described by Clausen (2004). PROC MIXED in SAS was used to calculate least square means (LSM).

Results and Discussion

Mechanical tenderisation improved the tenderness of steaks from loin and knuckle compared to control steaks (Figure 1). The mechanically tenderised beef was slightly less juicy compared to control steaks, while the flavour was unaffected (Table 1).

Enhancement using flavour-neutral brines also improved the tenderness of steaks from loin and knuckle, and the enhanced beef was significantly more tender than the mechanically tenderised beef. Furthermore, steaks enhanced

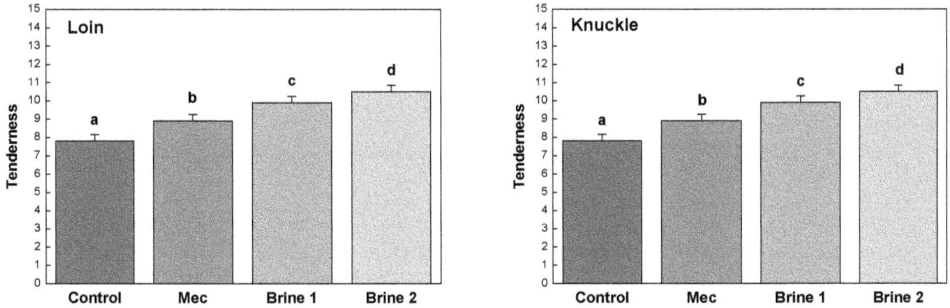

Figure 1: Tenderness in steaks of loin and knuckle, which have been mechanically tenderised (Mec), or enhanced with Brine 1 or Brine 2 and control steaks. Sensory scale from 0 (= nothing) to 15 (= very intense). LSM are shown and bars with different letters are significantly different.

Table 1: Juiciness, fried and boiled beef flavour and warmed-over flavour (WOF) in steaks of loin and knuckle, which had been mechanically tenderised or enhanced with brine. LSM are shown and rows with different letters are significantly different.

	Loin				Knuckle			
Treatment	Control	Mec	Brine 1	Brine 2	Control	Mec	Brine 1	Brine 2
Juiciness	9.4^{a}	8.8^{b}	9.9^{a}	9.9^{a}	9.4^{a}	8.3^{b}	9.9^{c}	9.1^{a}
Fried beef flavour	5.9^{a}	5.8^{a}	5.1^{b}	4.9^{b}	4.8^{ab}	5.0^{b}	4.5^{a}	4.5^{a}
Boiled beef flavour	1.5^{a}	1.6^{a}	2.3^{b}	2.2^{b}	2.0^{a}	2.2^{a}	2.4^{ab}	2.8^{b}
WOF	3.3^{a}	3.7^{ab}	4.5^{c}	3.9^{b}	3.7^{a}	3.6^{a}	4.8^{b}	4.5^{b}

* For abbreviations see Figure 1.

with Brine 2 were more tender than the steaks enhanced with Brine 1 (Figure 1). Marinating reduced the fried beef flavour, while boiled beef flavour and WOF were slightly more intense compared with the control steaks (Table 1). The changes in flavour were relatively small compared with the large positive effect of enhancement on tenderness.

Steaks from loins were generally more tender than steaks from knuckle. However, the effect of both mechanical tenderisation and enhancement was larger in steaks from knuckle compared with steaks from loin, as the tenderness improvement was 0.7 sensory units higher in steaks from knuckle compared with steaks from loin irrespectively of whether they was tenderised or enhanced.

The knuckle consists of several muscles. Prior to processing the cut was divided into *M. Vastus lateralis* and *M. Rectus femoris*, and these were distributed evenly on the four treatments. The results showed that there was a large difference in tenderness of the two muscles. The overall average tenderness in *M. Vastus lateralis* was 5.1 while it was 8.0 in *M. Rectus femoris.*

It is important to keep this difference in mind when producing steaks from the knuckle. When enhancing *M. Rectus femoris* using Brine 2, one can obtain steaks that are almost two sensory units more tender than control steaks from the loin. In contrast, steaks from *M. Vastus lateralis* enhanced with Brine 2 would still be less tender than control steaks from the loin (Figure 2).

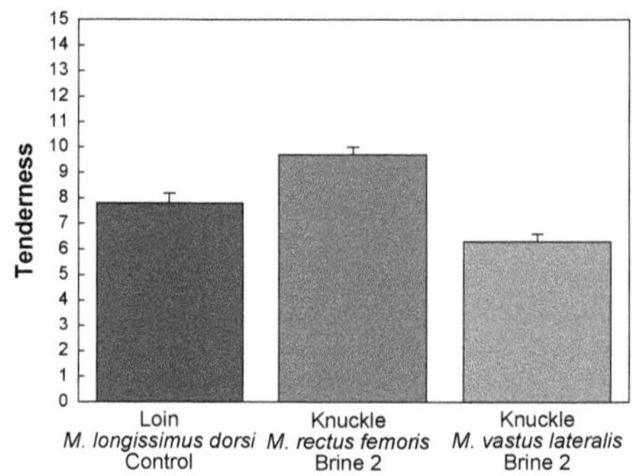

Figure 2: Tenderness in control steaks of loin and in steaks of knuckle enhanced with Brine 2 from either *M. Rectus femoris* or *M. Vastus lateralis*. Sensory scale from 0 (= nothing) to 15 (= very intense).

The effect of packaging methods will not be reported here. However, there were no interaction between treatments and packaging methods.

Conclusions

This study showed that mechanical tenderisation improved the tenderness of steaks from both loin and knuckle from dairy cows by more than one sensory unit, without affecting the flavour of the meat. Flavour-neutral enhancement by 6-8% improved tenderness to an even larger extent (> 2 sensory units), but also reduced fried beef flavour and enhanced boiled beef flavour and WOF. However, the flavour changes were small. The study also demonstrated, that it is possible to produce steaks from enhanced *M. Rectus femoris*, that are more tender than steaks from untreated loin.

Acknowledgements

This study was funded by the Danish Innovation Law, The Danish Livestock and Meat Board, Swedish Meats and Gilde Norsk Kjött.

References

Baulits, R.T., Pohlman, F.W., Brown Jr., A.H., Yancey, E.J. and Johnson, Z.B. (2006). Impact of muscle type and sodium chloride concentration on the quality, sensory, and instrumental colour characteristics of solution enhanced whole-muscle beef. *Meat Science*, 72: 704-712.

Clausen, I. (2004). Sensory evaluation of beef loin steaks stored in different atmospheres. In: Proceedings of 50th ICoMST, Helsinki, Finland, pp 377-380.

Jeremiah, L.E., Gibson, L.L. and Cunningham, B. (1999). The influence of mechanical tenderisation on the palatability of certain bovine muscles. *Food Research International*, 32: 585-591.

CONSUMER EVALUATION OF BEEF MUSCLES FOR USE AS "FAJITAS" AFTER MECHANICAL AND/OR ENZYMATIC TENDERISATION

L. Garcia*[1], D. Huerta-Montauti, R.K. Miller, C.E. Schuehle Pfeiffer, K.D. Pfeiffer, K.L. Voges, W.N. Osburn and J.W. Savell

Department of Animal Science, Texas A&M University, 2471 TAMU, College Station, TX 77843-2471.
E:mail:Lydarraider@tamu.edu

Keywords: beef fajita, consumer panel, papain, blade tenderisation

Introduction

Beef fajitas, a popular Mexican food dish in the U.S., were initially prepared using the outside beef skirt (diaphragm muscle). Today, the inside and outside skirts and flank muscle are commonly used as fajitas, and with increasing demand for these muscles, other sources are of interest to meet the growing demand for the raw materials for this product.

If other muscles are to be used for beef fajitas, their sensory characteristics will have to be comparable with that expected by consumers of this dish. Some muscles may have to be tenderised, either mechanically or with enzymes, to make them more suitable for this unique preparation. We evaluated seven muscles and different combinations of mechanical and enzymatic tenderisation treatments to see if we could find alternative muscles and processing procedures for beef fajita applications.

Materials and Methods

Sixty USDA Choice beef chucks, plates, outside skirts, inside skirts, and flaps were purchased and shipped to the Rosenthal Center at Texas A&M University. *M. rhomboideus* and *M. trapezius* were removed from beef chucks and *M. latissimus* and *M. serratus ventralis* were removed from beef plates. Four different treatments — papain (P), blade tenderisation (B), papain + blade tenderisation (P+B), and control (C) — were applied to all muscles. Muscles were cut into steaks and frozen at -10ºC for subsequent consumer evaluation. Steaks were thawed 48 hours and cooked to an internal temperature of 70°C. Randomly selected consumer panelists (n=81) evaluated overall like of flavour and tenderness using 9-point scales (9=like extremely; 1=dislike extremely). Steaks were cut into fajita strips (4.5 cm × 1 cm), and were presented to consumers rolled in flour tortillas (12 cm × 4.5 cm).

Statistical analyses were performed using SAS PROC MIXED (v. 6.12). For consumer data, Box-Cox transformations were used to produce normally distributed errors. Least squares means were generated and tested for significance ($P < 0.05$) using PDIFF.

Results and Discussion

Treatments

Of muscles and treatments combinations, consumers preferred the diaphragm, *M. transversus abdominis*, and *M. obliquus abdominis int.* muscles that were treated with papain. In contrast, the *M. trapezius* B and C treatment combinations were the least favored. Addition of papain to the *M. latissimus* greatly increased the consumer preference, which was similar to the P treated *M. transversus abdominis*.

Flavour

Consumers preferred the flavour of the diaphragm and *M. obliquus abdominis int.* treated with papain. The flavour of *M. trapezius* (C and B) and *M. latissimus* (C) was the least liked by consumers. The P and P+B treatments improved ($P < 0.05$) consumer flavour of inside skirt, flap, *M. rhomboideus*, *M. latissimus*, and *M. trapezius* fajitas.

Tenderness

Consumers preferred the tenderness of the diaphragm (P and P+B), *M. transversus abdominis* (P+B), and *M. obliquus abdominis int.* (P and P+B). However, consumers disliked the tenderness of the C and B of the *M. trapezius*. The *M. serratus ventralis* and *M. latissimus* with the P treatment can be compared in tenderness to the C steaks of the diaphragm. Consumer tenderness ratings were lower for P and P+B fajitas except in the *M. serratus ventralis* where fajitas from all treatments were tender. Inside skirt, *M. serratus ventralis,* and *M. latissimus* had similar ($P > 0.05$) tenderness when the P treatment was applied.

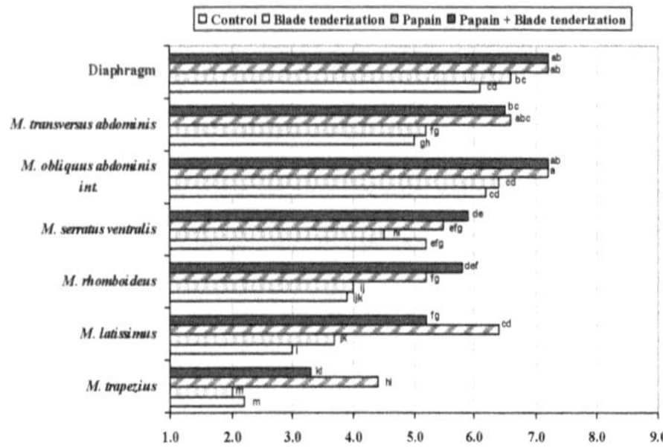

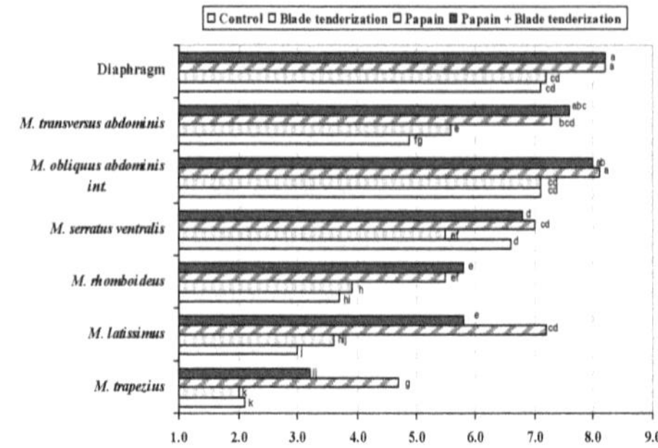

Conclusion

Consumers preferred the flavour and tenderness of beef fajita strips that were treated with P and P+B. Papain and blade tenderisation increased tenderness to a sufficient degree for consumers to have a positive eating experience. Regardless of the initial tenderness state of beef muscles, papain alone or the combination of papain and blade tenderisation improved tenderness. The addition of these two treatments to lower valued beef muscles may increase the supply of muscles that can be used for beef fajitas.

MEAT TENDERISATION OF BEEF MEAT BY DIFFERENT METHODS

S. Failla*, M. Contò, S. Ballico, F. Signorelli, A. Ficco and M. Mormile

C.R.A. Istituto Sperimentale per la Zootecnia. 00016 Monterotondo (Roma), Italy. Email betti.failla@isz.it

Keywords: tenderstretch, calcium infusion, very fast chilling, meat tenderisation

Introduction

Tenderness is the most important characteristic for meat acceptability and many studies have been carried out to understand meat tenderisation mechanisms, either structural or enzymatic processes, that are affected by production systems, biological factors (such as genotype, age, sex and so on) and technological processes (Dransfield, 1994).

To improve meat tenderness various physical and chemical techniques were checked. In this paper we analyse some principal ones: suspension of beef carcasses by the pelvic girdle (tenderstretch), as an alternative method of carcasses hanging, to produce during rigor a decrease in myofibrils shortening and a connective tissue structural change (Eikelenboom, 1998); very fast chilling (VFC), a fast cooling allows a strong muscle contraction with release of calcium into myofibrils and an outcoming activation of proteolytic enzymes (Joseph, 1996; Van Moeseke, et al. 2001); and *post mortem* $CaCl_2$ injection, that leads to an advanced and increased activity of Ca dependent enzymes (Gerelt *et al.*, 2002).

The aim of this work is to compare these tenderisation methods to establish the effect on meat quality properties, particularly on physical characteristics.

Materials and Methods

Six Friesian young bulls were slaughtered at 16 months of age. After slaughter, the left side was suspended from the aitch bone (pelvic suspension PS) for 24 hours, while the right side from the Achilles tendon (AT). From this latter, *Longissimus thoracis* (Lt) (between 7th and 13th rib) was soon taken and subdivided into three portions to carry out the three different experiments: VFC, obtained by putting a sample into a freezer at -80° C and stored until the core reached 1°C; $CaCl_2$, obtained by injecting 9% of meat weight (wt/wt) with 300mM calcium chloride solution; and the traditional ageing method (Control).

Both carcasses were aged for 8 days and stored at 2°C ±1°C. At dissection, two muscles were removed: *Longissimus thoracis* at the 6th rib (Lt) and *Gluteo biceps* (Gb), where physical analysis at 8 days only were determined, while on VFC, $CaCl_2$ and Control the same parameters but at three different times (24 hour, 5 and 8 days) were studied; in addition, temperature and pH values, with penetration through the meat, were monitored from 8 hours after slaughter with 1 hour steps.

On each sample, the following analysis for physical characteristics were performed:

-Drip loss with gravimetric method on the raw meat preserved at 4°C for 48h (Barton-Gade *et al.*, 1994)

-Cooking loss in a water bath at 75°C for 50 mins using vacuum-packed samples in a polyethylene bag

-Shear force on cooked meat samples (1x1cm cross section and 2cm long), using Warner Bratzler Shear apparatus on an Instron1011.

The statistical analysis of variance was performed using the GLM procedure of the SAS software (SAS 1985) using a bifactorial model (treatment and muscle for tenderstretch test, treatment and time for other tests)

Results and Discussion

The pH of meat samples treated with calcium felt slowly (Figure 1), at two hours from slaughtering already reaching the 5.8 values, the meat pH in the VFC experiment initially rose and then declined faster than the pH of the control group, at 8 hours after slaughtering; the pH of the $CaCl_2$ experiment was always significantly lower than the others until 5 days after slaughter, but at 8 days the differences were reduced obtaining a similar value among the three experiments (5.66). Shear force showed a decreasing trend with time for all experiments (from 10.4 to 8.2kg in average) even if the difference was significant only between 24 h and 5 days in the VFC group.

Fig.1 pH trend in the three different thesis

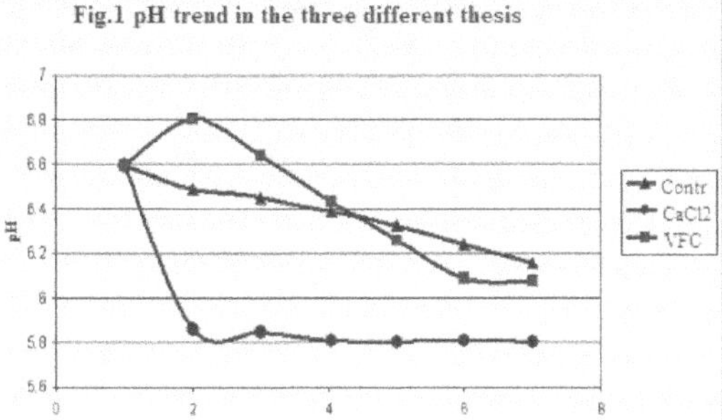

The $CaCl_2$ group showed a more tender meat already at 24h (about 50% less shear force than the control group). In fact, in the first days of ageing there was a greater activation of calcium-dependent enzymes (Gerelt *et al.*, 2002). The VFC experiment showed intermediate results and the greatest difference from the control group was highlighted at 5 days. The cold contraction increases free calcium ions content, that initially tends to increase toughness, due to shortening, but protease action increase is sufficiently fast and widespread to overlay the effect of contraction (Joseph, 1996).

Table 1: Shear force during aging of *longissimus thoracis* treated with three different methods (kg).

Experiment	Times			Significance	Thesis
	24h	5d	8d		means
Control	12.86	12.16	10.96	ns	11.99
$CaCl_2$	6.56	5.68	4.74	ns	5.60
VFC	11.92	9.20	8.80	*	9.97
Significance	***	***	***		***
Time means	10.44	8.88	8.24	*	RMSE 1.98

*ns = not significant; *= p≤0.05; **= p≤0.01; ***= p≤0.001.*

Water loss was higher in the $CaCl_2$ group at 24h, due to buffer injection, and in the VFC group where the muscular contraction, due to low temperature, led to flow of liquids from the muscle, but already at 5 days the effect was limited for both groups, at 8 days myofibrillar degradation caused a greater water loss. However drip loss increase with time was found for all experiments.

Table 2: Drip loss percentage during ageing of *longissimus thoracis* treated with three different methods.

Thesis	Times			Significance	Thesis
	24h	5d	8d		means
Control	0.83	1.36	2.05	**	1.41
$CaCl_2$	3.10	2.06	3.65	*	2.93
VFC	2.09	1.55	2.45	*	2.03
Significance	***	ns	**		***
Time means	2.00	1.65	2.71	**	RMSE 0.75

Table 3: Shear force of two different muscles aged with two different methods (kg).

Muscles	Thesis		Significance	Muscle
	AT	PS		means
Lt	6.90	4.47	**	5.70
Gb	4.82	3.80	ns	4.31
Significance	*	ns		*
Thesis means	5.86	4.13	*	RMSE 1.33

Cooked meat in tenderstretched carcasses was significantly more tender, mainly in *Longissimus* muscle; however GB muscle was less tough than LD, probably because the animals were young.
These techniques accelerate the tenderisation process, but differences decrease with meat ageing. In particular calcium injection showed significant effect on meat tenderisation but a lot of problems, particularly for acceptability of this treated product are hard to overcome.

References

Barton-Gade, P.A., Demayer, D., Honikel, K.O., Joseph, R.L., Poulanne, E., Severini, M., Smulders, F.S.M. and Tornberg, E. (1994). Final version of reference methods for water holding capacity in meat and meat products: procedures recommended by an OECD working group. In Proceeding of the 40st International Congress on Meat Science and Technology (ICoMST). The Hague, Nederlands 1994 S-V.05.

Dransfield, E. (1994). Optimisation of tenderisation, ageing and tenderness. Meat Science 36: 105-121.

Eikelenboom, G, Barnier, V.M.H., Hoving-Bolink, A.H., Smulders, F.J.M. and Culioli, J. (1998). Effect of pelvic Suspension and cooking Temperature on the Tenderness of electrically stimulated and aged beef, assessed with shear and compression test. Meat Science 49: 89-99.

Gerelt, B. Ikeuchi, Y. Nishiumi, T. and Suzuki, A. (2002). Meat tenderization by calcium chloride after osmotic dehydration. Meat Science 60,237-244

Joseph, R.L. (1996). Very fast chilling of beef and tenderness-a report from an EU concerted action. Meat Science, 43, 217-227.

SAS (1985). SAS user's guide statistics. Ed. Cary (NC) SAS Institute Inc. USA.

Van Moeseke, W., De Smet, S., Claeys, E. and Demeyer, D (2001).Very fast chilling of beef : effects on meat quality. Meat Science 59, 31-37.

NEW INDUSTRIAL METHOD OF MARBLING EVALUATION OF PORK MUSCLES

D. Lisiak*, A. Borys, K. Borzuta, E. Grześkowiak, and J. Strzelecki

Meat and Fat Research Institute in Warsaw, 04-190 Warszawa, ul. Jubilerska 4, Poland
Email:dsi.poznan@ipmt.waw.pl

Keywords: marbling, pork meat, model

Introduction

The content of intramuscular fat in muscles is considered as one of the most important factors influencing pork meat sensory quality (Egert *et al.*, 2002). Together with the increase in meatiness, deterioration of meat taste values can be observed. In addition, investigations revealed that the content of intramuscular fat correlates only slightly with the thickness of subcutaneous backfat (Blicharski and Pierzchała, 2005). Because of the considerable importance of marbling as an essential meat quality trait, attempts are made to find new methods of its assessment (Fernandez *et al.*, 1999). The most common methods include the analysis the ultrasound echo, computer image analysis and the application of the spectroscope (Altmann and Pliquunett, 2006). However, the above-mentioned methods have some defects. The ultrasound probe is very expensive and image analysis can only by carried out on cut muscles. On the other hand, endoscopy employing a special probe is an invasive method and, additionally, very slow. It can, therefore, be said that so far no simple and cheap method of marbling assessment in muscles has been proposed which could be commonly employed on an industrial scale in meat processing plants. Hence, the aim of the performed investigations was to develop a method which would be suitable for the application in industrial conditions.

Materials and Methods

The experimental material comprised 126 pig carcasses selected on slaughter lines of three meat processing plants. Using a 5-point scale, the degree of intramuscular fattening in the *m. gluteus medius* was evaluated on chilled half carcasses, where 1 point denotes the absence of fat traces and 5 points – very strong fattening. In order to determine the meat content in the carcass, carcasses were dissected according to the method of Walstra and Merkus (Walstra and Merkus 1996). Next, the following three muscles were prepared: *longissimus dorsi*, *semimembranosus* and *biceps femoris*. Using the same above-mentioned 5-point scale, the degree of intramuscular fattening was estimated. All the assessed muscles were photographed with the aim of elaborating fattening patterns of the *gluteus medius* muscle. The authors calculated correlation coefficients between the results of the marbling assessment of this muscle and the evaluation of the cross section of the other examined muscles.

Results and Discussion

Table 1 presents mean results from the dissection and marbling assessment of individual muscles.

Table 1: Mean results from the dissection and marbling assessment of individual muscles.

Trait	Mean	Minimum	Maximum	Standard deviation
Meatiness, %	51.01	35.72	68.40	6.85
LD, points	2.34	1.00	4.50	0.96
BF, points	2.76	1.50	4.50	0.74
SEM, points	2.42	1.00	4.50	0.75

It is evident from the performed investigations that the *longissimus dorsi* and *semimembranous* muscles were characterised by similar degrees of fattening, whereas the *biceps femoris* muscle was characterised, on average, by 0.3 – 0.4 points higher degree of marbling. That is why two separate regression equations were elaborated to estimate marbling of individual muscles; first – to assess marbling for the *longissimus dorsi* and *semimembranous* muscles and the second – for the *biceps femoris* muscle. The content of lean meat and the degree of marbling of the *gluteus medius* muscle were selected as independent variables because these traits showed high correlation coefficients (from r = 0.52 to r = 0.64) and they were easy to determine on the carcass.

The obtained equations assumed the following forms:

Equation 1 – for the *m. longissimus dorsi* and *m. semimembranosus*

$$Y = 13.98367 - 7.23056 \log x_1 + 0.27374 x_2 \qquad RSD = 0.68$$

Equation 2 – for the *m. biceps femoris*

$$Y = 9.69371 - 4.38471 \log x_1 + 0.21326 x_2 \qquad RSD = 0.59$$

where:

x_1 – meat content in the carcass, %
x_2 – evaluation of the marbling of the *gluteus medius* muscle, points

Mörlein *et al.*, (2005) obtained a smaller estimation error (RMSE = 0.36) but in their investigations they employed a very expensive ultrasound probe used in medicine.
The obtained equations were used to elaborate marbling patterns where carcass meatiness was placed in lines assuming the principle that one line corresponds to one percentage point in the range from 35% to 65%, while the points of assessment of the degree of fattening of the *gluteus medius* muscle were placed in columns assuming the principle that one column corresponds to 0.5 point in the range from 1 to 5 points. The estimated marbling value of the examined meat can be read at the intersection of the selected column and line. Because two regression equations were employed, also two marbling patterns were elaborated: one of them refers to the *longissimus dorsi* and *semimembranous* muscles and the second – to the *biceps femoris* muscle.
In the course of the performed investigations, special attention was paid to the fact that the marbling evaluation of the *gluteus medius* muscle must be carried out not earlier than three hours after slaughter, *i.e.* when this muscle is already partially chilled. This is due to the fact that warm intramuscular fat is transparent and, therefore, poorly seen with the naked eye. Once the muscle is cold, the fat becomes white and is clearly visible against the background of red meat and marbling can be evaluated.
The obtained photographs and tables allowed to develop a new pattern of marbling. On the face side of the page, we can see the photographs of the *gluteus medius* muscle in the 5-point scale of marbling, whereas on its reverse -the table with the marbling assessment of the appropriate muscle is presented. The adopted approach allows the grader to evaluate the degree of fattening of the largest muscles rapidly, cheaply, simply and in a non-invasive manner. It is enough to assess the *gluteus medius* muscle, which is always visible in the half carcass, using the adopted 1 - 5 point scale and then determine the meatiness of the carcass. The meat content in the carcass can be read from the print on the carcass skin or from the slaughter report. If it is impossible to determine carcass meatiness in the way mentioned above, it is still possible to use the allocated EUROP class adopting the mean meat content for a given class, *e.g.* E class – 57.5%; U class – 52.5%; R class – 47.5% etc. Knowing the above-mentioned values, the grader reads the required value of muscle marbling from the table placed on the back page of the pattern.

Conclusions
The authors developed a new method of evaluation of marbling of the *m. longissimus dorsi* and *semimembranous* muscle and the *biceps femoris* muscle. The method is based on the visual assessment of the degree of marbling of the *gluteus medius* muscle taking into account the meat content in the carcass. It is worth emphasising that the method is characterised by a small estimation error only slightly higher than that of the methods which employ expensive electronic devices.

References

Altmann M. and Pliquunett (2006). Prediction of intramuscular fat by impedance spectroscopy. Meat Science, 72, 666 – 671.
Blicharski T. and Pierzchała M. (2005). Tłuszcz śródmięśniowy u świń, Trzoda Chlewna 11, 62 – 64.
Eggert, J. M., Depreux, F. F. S., Schinckel, A. P., Grant A. L. and Gerrard D. E. (2002). Myosin heavy chain isoforms account for variation in pork quality. Meat Science, 53, 2, 117 – 126.
Fernandez, X., Monin, G., Talmant A., Mourot J., and Lebret B. (1999). Influence of intramuscular fat contentment on the quality of pig meat – 1. Composition of the lipid fraction and sensory characteristics of M. longissimus lumborum. Meat Science, 53,1, 67-72.
Mörlein D., Rosner F., Brand S., Janderka K. V. and Wicke M., (2005). Non – destructive estimation of the intramuscular fat content of the longissimus muscle of pig by means of spectral analysis of ultrasound echo signals. Meat Science, 69, 187 – 199.
Walstra P. and Merkus G.S.M. (1996). Procedure for assessment of the lean meat percentage as a consequence of the new UE reference dissection method in pig carcass classification - Raport ID-DLO 96.014, 1-22 - Inst. for Animal Sci. and Health Lelystud.

TESTING THE MEAT STANDARDS AUSTRALIA (MSA) MODEL ON IRISH BEEF

K. Brandon[1]*, P. Allen[1], R. Polkinghorne[2], A. Gee[3], J. Lau[4] and M. Drennan[5]

[1]*Ashtown Food Research Centre, Teagasc, Ashtown, Dublin 15, Ireland,* [2]*Marrinya, Agricultural Enterprises, Bairnsdale, 3875, Victoria, Australia,* [3]*Cosign Pty Ltd, 20 Eleventh Avenue, Sawtell, 2452, NSW, Australia,* [4]*Meat Livestock Australia, 23 Kyabra Street, Newstead Qld, PO box 2363, Fortitude Valley, Queensland, Australia,* [5]*Grange Research Centre, Teagasc, Grange, Co. Meath, Ireland.*
**Email: karen.brandon@teagasc.ie*

Keywords: beef grading, MSA, palatability, cooking method, muscle

Introduction

Meat and Livestock Australia has developed a meat grading scheme called Meat Standards Australia (MSA). The MSA grading scheme is based on the principles of Palatability Assurance at Critical Control Points (PACCP). The objective of PACCP is to identify and carefully control production and processing factors, which have the largest affect on palatability so that it is possible to accurately predict the quality of the final product (Polkinghorne *et al.*, 1999). The palatability of individual cuts is predicted by the model for a number of different cooking methods. In the MSA model, palatability is assessed by consumers as tenderness, juiciness, flavour and overall acceptability, with appropriate weightings to give a palatability score out of 100. Cut off points are used to give a quality rating, *e.g.* 3-star, 4-star and 5-star. The MSA model is currently being tested on Irish consumers using Irish and Australian beef to determine it's suitability as a grading system for the Irish beef industry. In order to do this Irish consumers are being asked to taste beef from both Irish and Australian heifers. The Australian and Irish heifers were chosen as similar homogenous groups, with the only notable difference being the country of origin. Consumer taste panels are being held at Ashtown Food Research Centre. The Australian beef is also being tasted by Australian consumers. The results will show whether the MSA model fits Irish beef and whether Irish consumers assess palatability in the same way as Australian consumers. Depending on the outcome, the model could be applied in Ireland or it may require some adaptation prior to adoption by the Irish industry.

The objective of this paper is to examine the predicted palatability scores for the Irish beef carcases for six muscles and two cooking methods using the MSA model and the grading data from the factory at time of slaughter.

Materials and Methods

Twenty crossbred (Charolais bulls x crossbred cows) heifers were selected from a homogeneous herd reared at Grange Beef Research Centre, Teagasc. The heifers were slaughtered together at a commercial abattoir (Meadow Meats, Rathdowney, Co. Laois). All heifers were stunned by captive bolt, shackled, hoisted up and bled within 30 seconds. The carcasses were dressed and centrally split into two sides, then hung by the aitchbone before chilling. The carcasses were graded using the Australian, American and European systems. The following muscles, M. *longissimus dorsi* (striploin), M. *psoas major* (tenderloin), M. *semimembransus* (topside), M. *gluteus medius* (rump), M. *biceps femoris* (outside flat) and M. *triceps brachii caput longum* (blade), were removed from the right side of the carcasses 4 days post slaughter, aged for 14 days and cut into steaks which were vacuum packed and frozen for use in consumer taste panels at Ashtown Food Research Centre. These muscles were chosen to give a range in palatability scores when combined with two cooking methods, grilling and Yakiniku (thin slice) cooking. This latter cooking method involves dry frying of thin slices on special domed hot plates.

The factory grading data was inputted into the MSA model (SP2004x04) and predicted palatability scores were recorded for the six muscles and two cooking methods.

Results and Discussion

Summary statistics for the 20 Irish heifers are shown in Table 1. Given that the Irish heifers were a relatively homogeneous group, (age at slaughter 559-646 days, carcass weight 257-334 kg) considerable variation was noted in rib fat depth (CV = 46%), US marbling score (CV = 36.7%), Australian marbling score (CV = 87%), conformation score (CV = 17.5%) and fat score (CV = 31.4%). Rib fat depth had a high correlation with US marbling score (r = 0.82), Australian marbling score (r = 0.69), and EUROP fat score (r = 0.67). EUROP fat score was also highly correlated with US marbling score (r = 0.80) and Australian marbling score (r = 0.73). Carcass weight was correlated with eye muscle area (r = 0.45) but not with age, US marbling score or Australian marbling score.

The predicted MSA palatability scores for the six muscles and two cooking methods are shown in Table 2 and the associated star ratings are shown in Table 3. Thin slice cooking has the effect of increasing the palatability score compared to grilling for all cuts except the tenderloin, for which there was a small reduction. The greatest increase in palatability score for the thin slice cooking was for the topside cut, which increased from 36 to 57 or from 'unsatisfactory' to 'good every day' quality. The range in mean palatability scores was greater for grilling (36 – 77) than for the thin slice method (58 – 74), indicating that this latter method is suitable for a wider range of cuts than

grilling. Of the six cuts included in the test, the striploin showed most variation between heifers in the predicted palatability score, whether grilled or cooked by the thin slice method. The heifer with the highest palatability scores across muscles and cooking methods also had the highest US marbling score and, conversely, the heifer with the lowest scores had the lowest marbling score, indicating the overriding importance of marbling score when other factors are relatively constant. All grilled blades and rumps were graded as 3-star (good everyday eating quality), while topsides were graded as 2-star (unsatisfactory) when grilled. Cooking method influenced the predicted eating quality of the muscles. The predicted eating quality of topsides increased from 2-star to 3-star when cooking method was changed from grill to thin slice. Grilled PM steaks had an average predicted star rating of 4.75 (better than every day quality to premium quality) while thin slice PM had a predicted star rating of 4.1 (better than every day quality).
When the results from the consumer trials are available it will be interesting to see how closely they match the predicted scores and whether any deviations are due to differences between Irish and Australian consumers or between Irish and Australian beef.

Conclusions
Predicted eating quality scores for a relatively homogeneous group of heifers showed little variation within muscle and cooking method, with the exception of the striploin. There was less variation in mean predicted eating quality score between cuts for the thin slice cooking method compared to grilling. In this relatively homogeneous group of heifers most of the variation in predicted palatability score was accounted for by variation in the marbling score.

References

Polkinghorne, R., Watson, R., Porter, M., Gee, A., Scott, J. and Thompson, J. (1999). Meat Standards Australia, A 'PACCP' based beef grading scheme for consumers. 1) The use of consumer scores to set grade standards. Presented at the 45th International Congress of Meat Science and Technology, Yokohama, Japan. 45:14-15.

Table 1: Summary statistics for the group of 20 heifers.

	Carc. wt (kg)	Age (d)	EMA (cm^2)	Rib fat (mm)	US marbling score	AUS marbling score	Conformation score (1-15)	Fat score (1-15)
Mean	303.5	609.1	104.2	4.1	320.5	0.83	8.7	7.6
SD	22.7	26.1	9.3	1.9	117.6	0.72	1.5	2.4
CV	7.5	4.3	8.9	46.1	36.7	86.9	17.5	31.4

Table 2: Mean predicted palatability scores for the six muscles using two different cooking methods.

	Striploin	Tenderloin	Blade	Rump	Topside	Outside Flat
Grill	56.4(50-66)	77.5(74-82)	54.1(52-58)	53.3(52-56)	35.9(33-41)	NA
Thin slice	60.3(54-70)	74.2(71-79)	59.7(57-64)	63.7(57-67)	57.8(55-62)	57.7(54-62)

Table 3: Mean predicted star ratings for the six muscles using two different cooking methods.

	Striploin	Tenderloin	Blade	Rump	Topside	Outside Flat
Grill	3	4	3	3	2	NA
Thin slice	3	4	3	3	3	3

(2 star = unsatisfactory; 3 star = good every day eating quality; 4 star = Better than every day eating quality; 5 star = premium quality)

EFFECTS OF ELECTRICAL STUNNING AND STIMULATION ON MEAT QUALITY UNDER COMMERCIAL CONDITIONS IN URUGUAYAN BEEF CATTLE

G. Brito*, C. Fiol, J.M. Soares de Lima and M. Del Campo

Programa Nacional de Carne y Lana, INIA Tacuarembó, 45000 Tacuarembó, Uruguay.
Email: gbrito@tb.inia.org.uy

Keywords: beef; electrical stunning; electrical stimulation; pH; tenderness

Introduction

There is a growing concern in the meat industry about the reliability of the stunning equipment and its advantages on meat quality. The non-penetrating and captive bolt guns might be encouraging the distribution of BSE prions from the brain to the edible products, making them an inappropriate alternative. The electrical stunning could solve this problem but one of its pitfalls is that it can cause blood splash. In commercial beef abattoirs pre-rigor rapid chilling of beef carcasses has been used to increase chiller capacity, minimize evaporative loss and improve microbiological conditions. The electrical current applied to the carcass affects *post-mortem* changes of muscle and meat quality: acceleration of *rigor mortis*, physical disruption of the myofibrillar matrix, acceleration of proteolysis (increasing intracellular calcium levels and activating calpain-calpastatin system) and meat appearance improvement (Morton *et al.*, 1999). The objective of this study was to evaluate the effects of electrical current (stunning and stimulation) on carcass pH decline, beef appearance and tenderness of the steaks at two ageing times.

Materials and Methods

Four hundred and forty three Hereford steers were randomly assigned to one of these treatments: T1: Electrical stunning (ES) followed by electrical stimulation (EE) (n=112), T2: ES (n=111), T3: Mechanical stunning (MS) followed by EE (n=105) and T4: MS (n=115). Animals were slaughtered in a commercial abattoir. Steers from T1 and T2 were electrically stunned using a voltage of 850 V (60 Hz and 1.5 A). The MS was done using a pneumatic air injection stunner. Carcasses of T1 and T3 were electrically stimulated using high voltage (400 V, 60Hz), 1,5 minutes after bleeding for 20 seconds in 3 times. The National Classification System (INAC, 1997) was registered in hot carcasses and the USDA Quality Grade System (USDA, 1997) in cold ones. Carcass pH and temperature were monitored using a portable pH meter (ORION, model 210A) with gel electrode and a thermometer (Barnant 115) with a type E thermocouple. The pH meter was calibrated a room temperature (25°C) before each measurement. Measurements were recorded at about 1, 3, 6, 12, 24 and 72 hours *post-mortem* in the M. *longissimus* of each left side at the location of 10th /11th rib interface. For each measurement the pH probe and the thermometer were inserted into carcass with a similar depth. A 7-cm thick steak was obtained from the M. *longissimus dorsi* at the mentioned location. All visible subcutaneous fat was removed and 3 steaks were prepared and vacuum packaged individually. Two of them (2.54-cm thickness) were aged for 7 and 14 days at 2-4 °C for tenderness determination. The bags were opened and the exudation on the surface of the steaks was removed with paper towel. The M. *longissimus* steaks were placed inside polyethylene bags and cooked in a water bath at 80 °C until an internal temperature of 70 °C was achieved. During cooking the internal temperature was tracked by the Barnant 115 thermometer with type E thermocouple. About six 1.27-cm diameter cores were removed from each steak parallel to the muscle fibre orientation. A single peak shear force measurement was obtained for each core using Warner Bratzler (model D2000) and an average value was calculated for each steak. Muscle colour was measured on the l*ongissimus dorsi* at 72 hours *post-mortem* at 10th/11th rib in the L*, a*, b* colour space using a colorimeter (Minolta Chroma meter C10) with an 8 mm diameter measurement area.

Results and Discussion

Neither hot carcass weight (HCW), fat thickness (Fat) or marbling (Marb) were affected by different treatments (P>0.05). The mean HCW values for each treatment were, T1: 260.4 kg; T2: 259.6kg; T3: 259.6kg and T4: 262.3 kg. For Fat were: T1 5.65mm, T2: 5.67mm, T3: 5.64mm and T4: 5.54 mm. Marb score average for all the groups was Slight (Sl^{50}). The age of the steers was estimated by dentition and there was no difference (P>0.05) among treatments ($\bar{x}$: 4 teeth). The shear force (SF) values were under the threshold to separate tender from tough steaks (4.5 kgF). The effect of the electrical current (T1 and T2) treatments was observed in the SF values at 7 (SF7) and 14 (SF14) days of aging (P<0.01; Table 1). The steers stunned electrically were more tender than the animals stunned mechanically and it seems that EE did not modify the levels of tenderness within each stunning method. Eilers *et al.* (1996) found that *longissimus* steaks from EE carcasses had lower SF values (P<0.05) and received slightly higher panel tenderness score (P<0.01) than did *longissimus* steaks from Control carcasses. Using objective colour measurements, the carcasses on T1 presented the highest values for L*, a* and b* colour parameters (P<0.01) compared with the other treatments (Table 1), showing the effect of the electrical current (ES+EE) on the meat appraisal (brightness and color). Roeber *et al.* (2000) found that *longissimus* muscles of EE treated sides were brighter, redder and less blue than the *longissimus* muscle of control side (P<0.05). There was a treatment effect on the rate of pH fall (Figure 1). The treatments with ES had a faster pH decline than those with MS. This difference was significant (P<0.01) at 1, 3, 6 and 12 h post-mortem comparing T1 and T2 vs. T3 and T4. Therefore, MS treatments showed differences between them. Carcasses in T3 had lower pH values at 1 ($P<0.05$), 3 and 6 h ($P<0.01$) *post-mortem* than those in T4. This suggests an EE effect on carcass pH decline, in agreement with previous studies (Hwang *et al.*, 2001; Hwang *et al.*, 2003; Strydom *et al.*, 2005). These authors reported differences on pH values were observed between carcasses that received EE and control ones for at least the first 4 to 6 h post-mortem.

Table 1: Mean values for L*, a*, b* parameters and shear force (kgF) at 7 and 14 days of aging for each treatment.

Treat	L*	a*	b*	SF7	SF14
1	37.2 a	21.7 a	13.8 a	3.60 b	3.34 b
2	36.7 b	21.0 b	13.4 b	3.64 b	3.35 b
3	36.5 b	20.8 b	13.2 b	3.96 a	3.67 a
4	36.5 b	20.2 b	13.1 b	4.02 a	3.59 ab

ab, Columns with different letter are statistically different $P<0.01$.

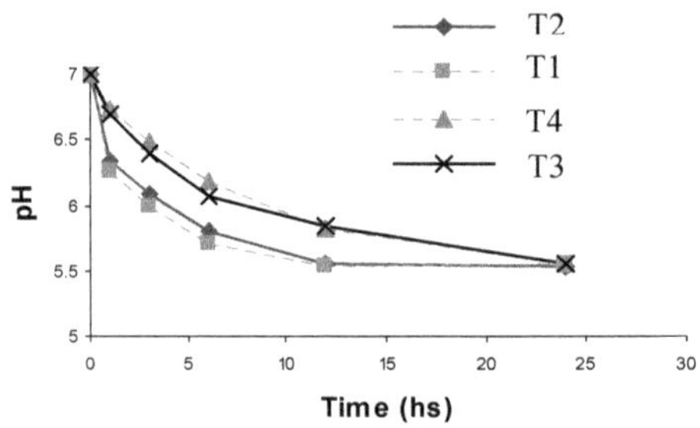

Figure 1: Rate of pH decline for each treatment (24 hs *postmortem* period).

Conclusions

There seems to be an effect by the ES on meat quality traits of Uruguayan Hereford steers in commercial conditions. In this population, the carcasses in which ES+EE was applied showed the highest muscle colour parameters, lowest pH values at 1, 3, 6 and 12 hs *postmortem* and most tender steaks. The use of ES also could have other advantages such as in food safety (i.e.: BSE). It was not clear if the EE had an effect on the measured meat quality variables, although in animals stunned mechanically, EE accelerated the pH fall in early *postmortem*. Further investigations should be conducted in Uruguayan beef cattle to study the effects of electrical stunning and stimulation on meat quality and food safety.

References

Morton, J.D., Bickerstaffe, R., Kent, M.P., Dransfield, E. and Keeley, G.M. (1999). Calpain-calpastatin and toughness in M. *longissimus* from electrically stimulated lamb and beef carcasses. *Meat Science*, 52, 71-79.

Eilers, J.D., Tatum, J.D., Morgan, J.B. and Smith, G.C. (1996). Modification of early postmortem muscle pH and use of *postmortem* aging to improve beef tenderness. *Journal of Animal Science*, 74, 790-798.

Strydom, P.E, Frylinck, L. and Smith, M.S. (2005). Should electrical stimulation be applied when cold shortening is not a risk? *Meat Science*, 70, 733-742.

Hwang, I.H. and Thompson, J.M. (2001). The effect of time and type of electrical stimulation on the calpain system and meat tenderness in beef *Longissimus Dorsi* muscle. *Meat Science*, 58, 135-144.

Roeber, D.L., Cannell, R.C., Belk, K.E., Tatum, J.D. and Smith, G.C. (2000). Effects of a unique application of electrical stimulation on tenderness, color, and quality attributes of the beef *longissimus* muscle. *Journal of Animal Science*, 78, 1504-1509.

Hwang, I.H. and Thompson, J.M. (2003) Effects of early postmortem pH on the quality of beef *longissimus*. *Asian-Australasian Journal of Animal Sciences,* 16(8), 1218-1223.

OPTIMISATION OF A NEW GENERATION PRE-DRESSING ELECTRICAL STIMULATION UNIT

E.S. Toohey*[1] and D.L Hopkins[2]

[1.] *NSW Department of Primary Industries Dubbo, PO Box 865 Dubbo NSW Australia,* [2.] *NSW Department of Primary Industries, Centre for Sheep Meat Development, PO Box 129, Cowra NSW 2794, Australia. Email: edwina.toohey@dpi.nsw.gov.au*

Keywords: lamb, mid-voltage, electrical stimulation

Introduction

The temperature at which a carcase enters rigor (pH 6.0) can be used to predict meat quality (Thompson *et al.*, 2005). If the carcase temperature falls too fast before the onset of rigor then cold shortening may result (Tornberg 1996) which can have adverse effects on meat tenderness. At the other end of the scale, slow rates of cooling can lead to heat toughening (Devine *et al.*, 1999) which may reduce ageing potential (Strydom *et al.*, 2005). Electrical stimulation has been identified in many studies as a useful processing technique to improve meat quality traits (e.g. Polidori *et al.* 1999) however few domestic abattoirs within Australia have adopted this method because of either the cost, safety considerations or a lack of space on the slaughter floor (Shaw *et al.*, 2005). Concerns about safety have been alleviated by development in Australia of new electrical stimulation technology that can be installed at either the start or end of a sheep chain and because current and pulse width are safe it requires minimal isolation from workers.

The Australian Sheep Meat Eating Quality (SMEQ) program identified various temperature ranges for optimal eating quality, depending on the market for the product. From this it was concluded that the target temperature range to achieve pH 6.0 should be 18-25°C for short aged meat destined for the domestic market (Thompson *et al.* 2005). However from a study by Toohey *et al.* (2006), which monitored three domestic Australian abattoirs, over 12 month, it was concluded that only 18.8 % of the 1197 carcases tested complied with the recommended pH-temperature window. Approximately 79 % of the carcases tested were at the cold end of the scale not reaching rigor (pH 6.0) by 18°C. This low compliance is also supported by data presented by Hopkins *et al* (2006) which show data from three other abattoirs which had a compliance rate from 0-16 %. This low compliance to the SMEQ window suggests that other processing techniques such as electrical stimulation need to be adopted.

The aim of the study was to optimise the current and pulse width of a new generation pre-dressing mid-voltage electrical stimulation unit, in order to maximise the number of carcases falling within the recommended SMEQ pH/temperature window (pH 6.0@18-25°C).

Materials and Methods

In total, 345 lambs from 14 different consignments were assessed. The lambs in the different consignments were of varying backgrounds and breed and represent lambs purchased by the company. Each consignment was exposed to different stimulation treatments having various levels of current and pulse width, using a pre-dressing mid voltage electrical stimulation unit. All treatments were applied for the same length of time (40 seconds). An outline of stimulation treatments, the number of consignments and the number of carcases sampled per treatment is shown in Table 1. Carcases were trimmed according to the specifications of AUS-MEAT. Hot carcase weights were recorded and the GR measured (total tissue depth over the 12th rib, 110 mm from the midline) using a GR knife. Carcase pH and temperature measurements were taken 30 min after death and then measured at every hour 5-7 times, depending on which day they were sampled. The carcases were stored in chillers with a mean temperature of 9.7°C and a range of 4-17.2°C. The pH and temperature measurements were taken in the left portion of the M. *longissimus thoracis et lumborum* (LL) muscle at the caudal end over the lumbar/sacral junction. A section of the subcutaneous fat and the M. *gluteus medius* was cut away to expose the LL and after each measurement the area was resealed with the overlaying tissue. Muscle pH was measured using a glass combination pH probe (potassium chloride) Ionode intermediate junction pH electrode, (TPS Pty Ltd., Brisbane, Queensland) attached to a data recording pH meter (TPS WP-80). Muscle temperature was measured using a stainless steel cylindrical probe attached to the same meter. The pH meter was calibrated before use and at regular intervals using buffers of pH 4 and pH 6.8 at room temperature.

Results and Discussion

There were some differences in both carcase weight and GR between treatments, where there was a recorded weight (Table 1). The predicted temperature at pH 6.0 was significantly different between treatments, but it should be noted that it was not possible to predict the temperature at pH 6.0 for all carcases. The greatest difference for predicted temperature at pH 6.0 was between the carcases in treatment 3 (14.9°C) and treatment 6 (24.5°C). Predicted pH at 25°C and 18°C were significantly different between treatment groups, with treatments 6 and 7 showing a lower pH for both. The percent of carcases that fell in the pH and temperature window varied between treatments. This was also the case for the proportion that had a pH less than 6.0 at 25°C and that had a pH greater than 6.0 at 18°C (Table 1).

Table 1: Predicted means (a.v. s.e.d.) of carcase weight (kg), GR (mm), predicted temperature at pH 6.0, predicted pH at 25°C,predicted temperature at 18°C, percent of carcases hitting the pH and temperature window, % with pH < 6.0 at 25°C and % with pH > 6.0 @ 18°C.

Treatment	1	2	3	4	5	6	7	8	9	av. s.e.d
Stim treat^	400x0.5	350x0.5	off	550x0.5	400x1.5	800x0.5	700x0.5	800x1.0	800x1.5	
Traits										
Animals	60	20	5	85	20	80	45	15	15	
Sampling days	3	1	1	5	1	5	3	1	1	
Consignments	3	1	1	9	1	11	7	3	3	
Weight	20.8ab	20.7ab	19.9a	20.4ab	n.a.	20.7ab	20.9ab	20.8ab	21.32b	0.65
GR*	9.8a	10.6abc	10ab	11.4abc	n.a.	12.2c	10.7abc	12.1bc	10.85abc	1.03
Predtemp@pH6#	18.7abc	20.5bcd	14.9a	21.6bcd	17.9ab	24.5d	23.7cd	17.7abc	17.32ab	2.66
Pred pH @ 25°C	6.26cd	6.20bc	6.38d	6.13abc	6.29d	6.00a	6.05a	6.07ab	6.12abc	0.072
Pred pH @ 18°C	6.00bcd	5.94abc	6.11d	5.92abc	6.04cd	5.82a	5.81a	5.88ab	5.94abc	0.079
%hit pH window	37	30	20	44	40	28	36	27	7	34 % +
%pH<6.0@ 25°C	3	0	0	28	5	54	47	20	20	28 % +
%pH>6.0@ 18°C	60	70	80	28	55	18	17	53	73	38 % +

Means followed by a different letter in a row (a,b,c,d) are significantly different ($P < 0.05$), ^stimulation treatment is the current (milliamps) by the pulse width (milliseconds), *Adjusted to a hot carcase weight of 20.7kg, # predicted values for 317 animals, + Average % across treatment groups, n.a. not available.

Conclusions

The results obtain from the 345 carcases in this study indicate that it is complicated to find the optimum setting across all animals killed. Application of a low current, such as 400 milliamps will result in a large percentage of carcases having a pH > 6.0 at 18 °C and application of a higher current, such as 700 or 800 milliamps, will result in a large percentage of carcases within a pH < 6.0 at 25°C. In this study an effective response was achieved with a current of 700-800 milliamps and a pulse width of 0.5 milliseconds with a moderate proportion of carcases hitting the window, but only a small proportion achieving a pH > 6.0 at 18 °C. Avoiding the cold end of the window must take precedence over avoiding the hot end as results from the SMEQ program showed that a detrimental effect on eating quality was more likely at the cold end (Shaw *et al.*, 2005). At the abattoir sampled in this study it is also likely that in the future a faster chill will be applied and for these reason the higher currents are recommended.

References

Devine C.E., Wahlgren N.M. and Tornberg E. (1999). Effect of rigor temperature on muscle shorting and tenderisation of restrained and unrestrained beef *m. longissimus thoracicus et lumborum. Meat Science* 51, 61-72.

Hopkins, D.L., Jacob, R.H., Toohey, E.S., Pearce, K.L., Pethick, D.W. and Richards, I. (2006). Electrical stimulation and hydration to optimise meat quality. *International Journal of Sheep and Wool Science* 54, No. 1. 42-47.

Polidori P., Lee S., Kauffman R.G., Marsh B.B. (1999). Low voltage electrical stimulation of lamb carcases: effects on meat quality. *Meat Science* 53, 179-182.

Shaw F.D., Baud S.R., Richards I., Pethick D.W., Walker P.J. and Thompson J.M. (2005). New electrical stimulation technologies for sheep carcases. *Australian Journal of Experimental Agriculture* 45, 575-583.

Strydom P.E., Frylinck L. and Smith M.F. (2005). Should electrical stimulation be applied when cold shortening is not a risk? *Meat Science* 70, 733-742.

Thompson J.M., Hopkins D.L., D'Souza D.N., Walker P.J., Baud S.R. and Pethick D.W. (2005). The impact of processing on sensory and objective measurements of sheep meat eating quality. *Australian Journal of Experimental Agriculture* 45, 561-573.

Toohey E.S., Hopkins D.L., McLeod B.M. and Nielsen S.G. (2006). Quantifying the rate of pH and temperature decline in lamb carcases at three NSW abattoirs. *Australian Journal of Experimental Agriculture* 46, (in press).

Tornberg E. (1996). Biophysical aspects of meat tenderness. *Meat Science* 43, 175-191.

Acknowledgements

This project was funded by Meat and Livestock Australia, the Australian Sheep Industry CRC and NSW DPI.

INCREASING THE STIMULATION RESPONSE FROM MEDIUM VOLTAGE ELECTRICAL STIMULATION

K.L. Pearce[1]*, D.L. Hopkins[2], D.W. Pethick [1], D. Gutzke[3], I. Richards[3], P.I. Fuller[1] and J.K. Phillips[1]

[1.] The Australian Sheep Industry Cooperative Research Centre, Murdoch University, Division of Veterinary and Biomedical Science, South Street, Murdoch, WA 6150, Australia, [2.] NSW Department of Primary Industries, Centre for Meat Sheep Development, PO Box 129, Cowra NSW 2794, Australia, [3.] Meat and Livestock Australia, 23 Kyabra St Newstead Qld 4006, Australia. Email: k.pearce@murdoch.edu.au

Keywords: electrical stimulation, lamb, pH decline

Introduction

Electrical stimulation is an important component of sheep meat processing because it reduces the variability in sheep meat eating quality (Hopkins and Toohey, 2006). Additionally the ability to meet a pH temperature window is increasingly becoming a requirement for the processing of lamb in Australia based on the results of the Sheep Meat Eating Quality (SMEQ) program. The SMEQ program has identified that for optimal eating quality the meat destined for the overseas (air freight) or domestic markets should reach pH 6 when the carcass temperature is between 18-25°C (Thomson *et al.*, 2005). Australian processors are installing electrical stimulation units given the low rate of compliance to the window (Toohey *et al.*, 2006). A new approach to stimulation has been developed in Australia based on medium voltages (MVS) (Shaw *et al.*, 2005) which are favoured over the traditional high voltage systems because the power levels and pulse widths comply with occupational health and safety regulations (Anon, 2002). The system is devised of 6 segmented electrodes which ensure that only one carcass contacts the electrodes at any one time. The current remains constant and the voltage is varied (peak 300V) by controlled electronics which determine the resistance of the carcass and this feed back system alters the voltage accordingly.

A preliminary validation study at an Australian abattoir into the effectiveness of the MVS post-dressing unit in terms of pH-temperature decline demonstrated that only 42% hit of carcasses met the SMEQ pH temperature guidelines (Pearce *et al.*, 2006). To reach the desired pH temperature window it appears necessary to increase the stimulation response to increase the pH drop. Alternative approaches to increase stimulation response were briefly touched on by Pearce *et al.* (2006) such as changing frequency and modulating the stimulation on/off times. The study by Pearce *et al.* (2006) indicated that the stimulation setting of 2.5ms pulse width, 1A current and 15hz was the best, resulting in 60% of the carcasses 'hitting' the pH temperature window. This study investigated the potential for the alternative settings to increase the stimulation response so to ensure a higher proportion of carcasses 'hit' the window.

Materials and Methods

The selections of treatments tested to increase stimulation response are listed in Table 1. Treatment 0 was the unstimulated control. Treatments 1 to 4 had a constant current, pulse width and frequency across the 6 electrode bars of the MVS unit which was hard wired into the abattoir. Treatments 5 to 10 had a constant current and pulse width but different frequencies across the 6 electrodes. The frequency of each electrode is specified in order i.e. treatment 5 had a frequency of 15Hz for electrodes 1 and 2, 20Hz for electrodes 3 and 4 and 25Hz for electrodes 5 and 6. Treatments 11 was a modulated treatment with constant current, pulse width and frequency across the 6 electrodes but the frequency of 25Hz was switched on and then off every 200 ms. The unstimulated control was tested over all 30 consignments, treatments 2, 3, 6 to 11 were tested over 8 consignments (n = 8x7 = 56), treatments 4 was tested over 13 consignments (n=91) and treatments 1 and 5 were tested over 23 consignments (n = 161). The total duration of stimulation was 34 s. The lambs in each consignment were of varying backgrounds and breed and selected from the abattoir daily kill sheets. Carcass pH and temperature measurements were taken at hourly intervals beginning approximately 25 mins post-slaughter. These carcasses were stored in chillers with a mean temperature of 2°C. The pH and temperature measurements were taken in the left portion of the *M. longissimus thoracis et lumborum* (LL) at the caudal end over the lumbar-sacral junction. Hot carcass weights were recorded and the GR tissue depth). Further details on sampling procedures and statistical analysis are outlined in Pearce *et al.* (2006).

Results and Discussion

A low initial pH, fast rate of decline, high temperature at pH 6.0 and high number of carcasses with a pH of 6.0 by 25°C were criterion used to define the treatment with the greatest stimulation response. Treatment 6 demonstrated the greatest stimulation response having the lowest initial pH, fastest rate of decline, highest temperature at pH 6 and highest number of carcasses with a pH of 6 by 25°C. The success of this treatment could be due to the frequency combination resulting in the greatest rate of muscle contraction, with a minimal degree of muscle fatigue which typically arises from stimulation with high frequencies (Takahashi *et al.*, 1986).

A commercial preference will be for a treatment that results in the high number of carcasses in the pH temp window, with a temperature at pH 6 between 18-25°C and a minimum number of carcasses not reaching pH 6 before 18°C. The treatments recommended to fit such criteria are 1 and 8. Treatment 8 had the lowest number below the pH temperature window, but had a temperature at pH 6 of above 25°C whereas treatment 1 had less in the window but an optimal temperature at pH 6 between 18-25°C. No difference was observed between treatments for hot carcass weight (HCW) and GR depth (P>0.1) (average HCW was 22.7 ± 0.7 kg and 11 ± 1 mm for GR depth).

Table 1: Initial pH, rate pH decline, temperature at pH 6, % of carcasses that reach pH 6 between 18-25°C above 25°C or below 18°C.

Tmt no	Pulse width (ms)	Current (A)	Frequency (Hz)	Count	Initial pH	Rate of pH decline (ph units/h)	Temp at pH6 (°C)	% at pH 6 18-25°C	% pH <6 at 25°C	% pH>6 at 18°C
0	No stimulation			210	6.73a[a]	-0.243a	16.4ab	18a	5a	77a
1	2.5	1	15	161	6.29d	-0.182b	22.3d	29de	35c	36e
2	5	1	15	56	6.15e	-0.108d	22.2d	19a	49d	32e
3	5	1	10	56	6.38c	-0.189b	25.5d	21a	62e	17g
4	5	1	18	91	6.26d	-0.162b	25.3d	15ab	61e	24f
5	2.5	1	15, 15, 20, 20, 25, 25	161	6.28d	-0.166b	22.6d	26d	32c	42d
6	2.5	1	10, 15, 25, 10, 15, 25	56	6.13e	-0.108d	28.7e	6c	83f	11gh
7	2.5	1	25, 15, 25, 15, 25, 15	56	6.21de	-0.133c	25.6d	33e	50d	17g
8	2.5	1	25, 25, 15, 15, 10, 10	56	6.25d	-0.142c	26.6d	28de	67e	6h
9	2.5	1	25, 25, 15, 15, 15, 25	56	6.16e	-0.116d	25.7d	20a	52d	28ef
10	2.5	1	35, 0, 35, 0, 35, 0	56	6.56b	-0.241a	14.9a	15ab	0a	85a
11	2.5	1	25hz on/off 200ms	56	6.51bc	-0.133c	15.2a	14ab	3a	83a
Ave					6.32[b] (6.34[c])	-0.159[b] (-0.164[c])	21.8 (21.5)	17.6 (17.7)	38.5 (36.6)	43.8 (45.7)
SED					0.04	0.06	2.2			

[a]Numbers down row not followed by same letters are significantly different (P<0.05), [b]Average of stimulated carcasses only, [c]Average of all treatments

Conclusions

Alternative frequency combinations such as treatment 6 resulted in the greatest stimulation response possibly due to a maximal rate of muscle contraction. This treatment may have application into abattoirs with installed MVS units that hot bone sheepmeat and require a fast pH declines to ensure minimal cold shortening of meat. A downside to this treatment is that under routine commercial practices the carcasses may be more predisposed to heat shortening (Thomson *et al.,* 2005). Therefore commercial application of treatment 1 and 8 appears more relevant to abattoirs that require an increased stimulation response due to the application of a fast chilling regime. Objective tenderness, drip loss and colour stability of treatments 0, 1, 5 and 8 will be assessed.

References

Anon (2002). 'Australian Standard AS/NZS 60479. Effects of current on human beings and livestock.' Standards Australia, Standards House, North Sydney, NSW.

Hopkins DL, Toohey ES (2006). Eating quality of conventionally chilled sheep meat. *Australian Journal of Experimental Agriculture In press.*

Pearce KL, Hopkins DL, Toohey E, Pethick DW, Richards I (2006). Quantifying the rate of pH and temperature decline in lamb carcasses using mid voltage electrical stimulation in an Australian abattoir. *Australian Journal of Experimental Agriculture In press.*

Shaw FD, Baud SR, Richards I, Pethick DW, Walker PJ, Thompson JM (2005). New electrical stimulation technologies for sheep carcasses. *Australian Journal of Experimental Agriculture* 45, 575-583.

Takahashi G, Wang SM, Lochert JV, Marsh BB (1986). Effects of 2-Hz and 60-Hz electrical stimulation on the microstructure of beef. *Meat Science* 19, 65-67.

Thompson JM, Hopkins DL, D'Souza DN, Walker PJ, Baud SR, Pethick DW (2005). The impact of processing on sensory and objective measurements of sheep meat eating quality. *Australian Journal of Experimental Agriculture* 45, 561-573.

Toohey ES, Hopkins DL, McLeod BM, Nielsen SG (2006). Quantifying the rate of pH and temperature decline in lamb carcases at three NSW abattoirs. *Australian Journal of Experimental Agriculture In press.*

THE IMPACT OF NEW GENERATION PRE-DRESSING MID-VOLTAGE ELECTRICAL STIMULATION ON COLOUR STABILITY

E.S. Toohey*[1] and D.L Hopkins[2]

[1.] NSW Department of Primary Industries Dubbo, PO Box 865 Dubbo 2830 NSW Australia, [2.] NSW Department of Primary Industries, Centre for Sheep Meat Development, PO Box 129, Cowra NSW 2794, Australia. Email: edwina.toohey@dpi.nsw.gov.au

Keywords: meat colour, lamb, mid-voltage, electrical stimulation

Introduction

Consumer decisions at the point of purchase are influenced by meat colour more than any other quality factor as consumers use the degree of discolouration as an indication of freshness (Mancini and Hunt 2005). For this reason it is economically important for retailers to ensure colour is desirable for consumers. A key ingredient in prolonged meat colour stability is the rate of metmyoglobin formation. The formation of metmyoglobin begins in a band between oxygenated and deoxygenated layers. The thickness of this band increases with time and will eventually extend all the way to the meat surface leaving a brownish colour (Channon *et al.,* 2005). New generation mid voltage electrical stimulation is currently being adopted by many sheep processors in Australia because of its minimal risk to workers and the versatility of installation. As outlined by Channon *et al.,* (2005) past studies which have examined the effects of high or low voltage stimulation on meat colour have produced contrasting results. The objective of this study was to determine the impact on meat quality of "optimal current and pulse width" settings for a new generation pre-dressing mid-voltage electrical stimulation unit with an emphasis on meat colour. The operation principles of the new unit have been outlined by Hopkins *et al.,* (2005).

Materials and Methods

In total, 80 lambs from 11 different consignments were assessed. The lambs were killed over two days with 40 killed on day one and 40 killed on day two. Of the 40 killed on each kill day there were two stimulation treatments (20 stimulated and 20 unstimulated). Each individual consignment had equal numbers of animals stimulated and unstimulated. The lambs in the different consignments were of varying backgrounds and breed and represented lambs purchased by the abattoir. In the stimulated treatment, carcases were exposed to a current of 800mA and pulse width of 0.5ms using a pre-dressing mid voltage electrical stimulation unit. The stimulation treatment was applied for approximately 60s. Carcases were trimmed according to the specifications of AUS-MEAT. Hot carcase weights were recorded and the GR measured (total tissue depth over the 12th rib, 110 mm from the midline) using a GR knife. The carcases were chilled at a mean temperature of 4.2°C. Carcase pH and temperature measurements were taken 30 mins after death and every hour after that to give 7 measurement times. A 28h pH and temperature measurement was also recorded. The pH and temperature measurements were taken in the left portion of the *M. longissimus thoracis et lumborum* muscle at the caudal end over the lumbar/sacral junction. From kill 1, 40 chumps and loins were removed from the left side of the carcase at 24h post-mortem. From each of the chumps and loins a sample for colour measurement (3 cm slice) and a 1g sample for final pH (5 day aged) were taken. Additionally in kill 2, 40 left legs were taken and gas flushed with a Secure Fresh – secure pack 15.2 gas flush machine using 100% CO_2 and then the legs were aged for 14 days. From these legs, the chumps and rounds were removed and a sample was taken for colour measurement (3cm slice) and a 1g sample for final pH (17 day aged) for each cut. In kill 2 the loin was also sampled the same as for kill 1. pH was determined using an iodoacetate method. The meat colour reflectance of the loin and chump from the first kill and the loin from the second kill were measured with an initial colour measured at 24h post-mortem followed by measurements taken twice a day (am/pm) for 5 days resulting in 10 measurements per sample. For the initial measurement a fresh surface was prepared by cutting in a transverse direction across the sample and after 30min a colour reading was measured. Samples were then positioned randomly on black plastic trays and over-wrapped with polyvinyl chloride clear film and placed under continuous lighting (1190 Lux) in a chiller at 4°C. A Hunter Lab MiniScan XE spectrophotometer was used to measure colour dimensions (Hunter Associates, Reston, VA, Model D45/0-s 6 mm port with 5mm area viewing set for L (lightness), a (redness) and b (yellowness) with a D65 at 10 degrees illuminate). The MiniScan was calibrated using both white and black tiles. Every slice of each cut was scored for colour acceptability by 2 people using the scoring system of Channon *et al.,* (2005). The loin and chump slices were scored at the start and end of the 5 day display period. Colour measurements were also taken from the chump and round which had been gas flushed and aged for 14 days. The initial measurement was taken on the 14th day in the afternoon and further measurements were taken twice a day (am/pm) for 3 days resulting in 7 measurements on these samples which were displayed at (1110 Lux) in a display cabinet at 4°C. For these cuts visual scores were given 4 times during the display period (0, 24, 46 and 72 hours) by 2 people. Traits were analysed using a residual maximum likelihood (REML) procedure (Genstat 7.1, 2004) which contained a fixed effect for treatment (stim or no stim), to estimate the means and standard errors of the differences with kill day and consignment as a random terms. The relationship between visual colour scores and objectively measured colour traits (ratio at 630nm/580nm, L and a values) was derived using linear and non-linear regression analysis. The ratio 630nm/580nm is a good indicator of the formation of metmyoglobin.

Results and Discussion

There was no difference between stimulated and unstimulated carcases for weight or GR. Differences were found between treatments for initial pH, rate of pH decline and predicted temperature at pH 6.0 where stimulated carcases had a lower initial pH, faster rate of pH decline and a higher predicted temperature at pH 6.0 (24.8°C as opposed to 13.9°C for unstimulated carcases). On average across both kills 67.5% of stimulated carcases hit pH 6.0 above 18°C opposed to 25% of unstimulated carcases. There was no significant difference ($P > 0.05$) between stimulated and non stimulated loin and chump meat colour after 5 days of display based on ratio, L, a or b values. However there was a difference ($P < 0.05$) between the initial a values of the loin (1 day post-mortem) with stimulated carcases having a higher value. The chump had higher initial values for the ratio 630nm/580nm compared with the loin and was redder than the loin. As can be seen for the loin there was no difference in the rate of formation of metmyoglobin based on the ratio values between treatments.

Table 1: Predicted means and standard error of difference (s.e.d) between stimulated and non stimulated carcases for kill 1 chump initial and final ratio, L, a, b values, kills 1 and 2 loin initial and final ratio, L, a, b values and gas flushed chump and round initial and final L, a, b and ratio values (kill 2).

	Initial (fresh)			Final				Initial (gas flushed)			Final (gas flushed)		
	Stim	No stim	*av. s.e.d*	Stim	No stim	*av. s.e.d*		Stim	Non stim	av. s.e.d	Stim	Non stim	*av. s.e.d*
Chump							Chump						
L	31.4a	31.7a	*1.25*	38.5a	38.3a	*1.26*	L	32.8a	32.7a	*0.81*	34.7a	35.0a	*0.78*
a	9.20a	8.81a	*0.56*	7.51a	7.94a	*0.37*	a	11.70a	11.99a	*0.54*	7.05a	6.45a	*0.50*
b	11.14a	10.97a	*0.49*	10.89a	11.40a	*0.61*	b	13.92a	14.11a	*0.48*	11.22a	10.47a	*0.47*
ratio	3.6a	3.5a	*0.19*	1.7a	1.9a	*0.09*	ratio	4.3a	4.4a	*0.17*	1.9a	1.7b	*0.10*
Loin							Round						
L	30.5a	31.3a	*0.93*	38.3a	39.4a	*0.91*	L	41.2a	40.4a	*0.78*	36.5a	35.8a	*0.88*
a	7.80a	7.11b	*0.32*	6.42a	6.13a	*0.25*	a	6.76a	7.20a	*0.32*	7.65a	7.25a	*0.49*
b	8.81a	8.39a	*0.38*	9.38a	9.04a	*0.31*	b	11.47a	11.98a	*0.36*	10.78a	10.66a	*0.50*
ratio	3.2a	3.0a	*0.14*	1.6a	1.6a	*0.04*	ratio	2.6a	2.8a	*0.08*	2.1a	2.0a	*0.09*

Means followed by a different letter in a row (a, b) are significantly different ($P < 0.05$).

There was no difference ($P > 0.05$) between stimulated and non stimulated chump and round colour values (L, a, b) at initial measurement or at final measurement after 3 days on display. The only difference for the ratio values was for the chump at the final measurement, where stimulated chumps had a higher ratio (less brown) than non-stimulated chumps ($P < 0.05$). There was an interaction between cut type and visual scores ($P < 0.001$) for ratio (630nm/580nm), L and a values, with the chump exhibiting a greater range in ratio and a values than the round. Visual colour scores and cut type explained 65% ($R^2 = 0.65$) of the variation in ratio values, whereas for L and a values only 33% and 36% of the variation could be explained based on linear modelling respectively. It is apparent that visual scores are more useful for detecting colour changes in the chump than the round and that visual scores are less useful for detecting changes in lightness (L) and redness (a) irrespective of the cut type. Based on the fitted models if chumps have measured ratio values less than 3.3 then they would be considered unacceptable (browny red) and for rounds the ratio value is 2.6.

Conclusions

From the results it can be concluded that using a new generation pre-dressing mid-voltage electrical stimulation unit set at 800 mA and a pulse width of 0.5 ms the rate of pH decline will increase and reduce the number of carcases entering rigor at cold temperatures. The use of new generation pre-dressing mid-voltage electrical stimulation had minimal effects on meat colour traits, with no indication of any detrimental affects.

References

Channon H.A, Baud S.R. and Walker P.J. (2005). Modified atmosphere packaging improves retail display life of lamb cuts with variation between loin and knuckle. *Australian Journal of Experimental Agriculture*, 45, 585-592.

Hopkins D.L., Shaw F.D., Baud S. and Walker P.J. (2005). Effects of level of current during lamb carcase electrical stimulation on post-mortem muscle changes and meat quality. Proceedings New Zealand Society of Animal Production 65, 247-251.

Mancini R.A. and Hunt M.C. (2005). Current research in meat colour. *Meat Science*, 71, 100-121.

Acknowledgements

This project was funded by Meat and Livestock Australia, the Australian Sheep Industry CRC and NSW DPI.

PHYSIOLOGICAL UNDERSTANDING OF ELECTRICAL STIMULATION

P.I. Fuller[1]*, K.L. Pearce[1], R. Taplin[2], K.E. Creed[1], D. Gutzke[3], D.W. Pethick[1] and J.K. Phillips[1]

[1.] Department of Veterinary and Biomedical Science, Division of Health Sciences; [2] Department of Engineering Science, Division of Science and Engineering, Murdoch University, W.A. 6150, Australia; [3] Meat and Livestock Australia, Newstead, Qld 4006, Australia. Email: p.fuller@murdoch.edu.au

Keywords: sheep, electrical stimulation, muscle contraction, electrophysiology

Introduction

It is well established in the meat processing industry that electrical stimulation enhances meat tenderness (Devine *et al.*, 2004), generally by hastening the process of rigor mortis. This results in an initial pH fall (ΔpH) followed by a change in the rate of pH decline. Within Australia, a new type of electrical stimulation system has been developed and is currently being implemented in sheep processing plants. These new mid-voltage stimulation (MVS) units are designed to impart the same type of response observed with traditional high voltage stimulation, but without the associated danger and cost (Devine *et al.*, 2004). This is achieved with shorter pulse widths and medium range voltages, delivered through segmented electrodes. Although there have been numerous studies conducted on optimising the effectiveness of stimulation by altering the electrical parameters, such as frequency, pulse width and current (Devine *et al.*, 2004), there have been few detailed studies on MVS units (Pearce *et al.*, 2006). Furthermore, few studies have clarified the physiological interaction and effects of stimulation on muscle contraction, nerve function and pH decline (Chrystall and Devine, 1978; Chrystall *et al.*, 1980; Devine and Chrystall, 1984), hence the effects on meat tenderness remain unclear. In many circumstances, the protocol and physiological effects are defined empirically, based on experience rather than an integrated understanding of the applied electrical parameters and the induced response in the muscle. Given the difficulties associated with studying the subtleties of electrical stimulation on muscle/nerve components in large animals, the use of an isolated nerve-muscle preparation circumvents these problems whilst allowing a detailed study of stimulation characteristics (Devine and Chrystall, 1984). Therefore, the emphasis of this preliminary study was to examine several parameters of MVS systems, particularly frequency and pulse width, and relate them to the mechanical response of muscle (i.e. amount of force generated) and the accompanying biochemical changes (ΔpH).

Materials and Methods

In the abattoir, muscle pH of the *M. semimembranosus* (SM) was measured using an Orion 250A pH meter with a glass body, spear-tipped probe (Orion Research Inc, MA), coupled with a temperature probe. pH and temperature measurements were taken within 1 min prior to stimulation and within 5 min post-stimulation. Table 1 outlines the treatments used for muscle pH. Thirteen consignments were tested, using 5 sheep per treatment (Tmt) and 8 treatments per consignment (Tmt 0 was the unstimulated control). Tmt 1 to 4 had constant current, pulse width and frequency across all 6 electrode bars of the MVS unit. Tmt 5 to 8 had a constant current and pulse width, but different frequencies across the electrodes, e.g. Tmt 5 had a frequency of 15Hz for the first 2 electrodes, 20 Hz for the next 2 electrodes, and 25Hz for the last 2 electrodes. The total duration of stimulation for all Tmts was 34 seconds. For the *in vitro* preparation, rats (male Wistars, aged 10–16 weeks) were deeply anaesthetized with carbon dioxide and decapitated using a guillotine. The SM muscles were immediately removed and immersed in Krebs solution at room temperature. Longitudinal strips (avg 22 x 3mm, 0.132g) of SM muscle were mounted vertically in organ baths containing Krebs solution at 34°C aerated with 5% CO_2 in 95% O_2 (pH 7.4; Chen & Creed, 2004). Following equilibration and before the use of the test parameters, the muscle was stimulated with 80Hz, 0.1ms and 60V to act as a control twitch for each strip, with subsequent responses analysed as a percentage of the control tension generated. Table 2 outlines the parameters used for the in vitro preparations to correlate the ΔpH with tension responses. These parameters were chosen to reflect those used in the abattoir. For muscle pH, data was analysed using an ANOVA and Tukey's Post Hoc test, with Tmt as a fixed variable. Analysis of the nerve/muscle preparation data (the percentage of tension generated, relative to the control twitch) used an ANOVA with treatment as fixed variable.

Results and Discussion

The results presented in Table 1 reveal that modulating the frequency (i.e. altering the frequencies during the stimulation period) produced the greatest influence on the drop in muscle pH, compared with other treatments that had a constant frequency. Furthermore, altering the pulse width had no effect on ΔpH. These results are in agreement with those of Chrystall and Devine (1978), who found that compared to other electrical parameters, pulse frequency had a greater effect on ΔpH and the amount of tension produced. However, results from our preliminary in vitro study (Table 2) found that modulating frequencies had no effect on tension generated, whereas varying pulse width did.

Table 1: Effect of electrical stimulation on muscle pH (mean ± S.E.).

Tmt No.	Frequency (Hz)	Pulse width (ms)	Current (A)	No. of Sheep	Initial pH	Post-stim pH
0		No stimulation		67	6.89 ± 0.03	6.77[a] ± 0.04
1	10	1	0.75	67	6.87 ± 0.04	6.54[b] ± 0.06
2	40	1	0.75	67	6.81 ± 0.04	6.52[b] ± 0.04
3	15	1	0.75	67	6.86 ± 0.04	6.47[ab] ± 0.03
4	15	5	0.75	67	6.86 ± 0.04	6.49[ab] ± 0.04
5	15, 15, 20, 20, 25, 25	1	0.75	67	6.91 ± 0.05	6.46[c] ± 0.03
6	25, 25, 20, 20, 15, 15	1	0.75	67	6.89 ± 0.03	6.77[a] ± 0.04
7	25, 25, 20, 20, 15, 15	3	0.75	67	6.87 ± 0.04	6.54[b] ± 0.06
8	25, 25, 20, 20, 15, 15	5	0.75	67	6.81 ± 0.04	6.52[b] ± 0.04
	Ave				6.86 ± 0.24	6.59 ± 0.28

* Treatments followed by different superscripts are significantly different (P<0.05)

Table 2: Effect of electrical stimulation on muscle tension.

Treatment	%tension
15hz 1ms 60v	23.58
15hz 5ms 60v	47.32
15hz 2.5ms 60v	40.98
15hz 1ms 100v	38.63
15hz 1ms 30v	17.85
i. 15hz 1ms 60v	19.93
ii. 20hz 1ms 60v	21.93
iii. 25hz 1ms 60v	25.53

Conclusion

We believed that a reasonable assumption would be that the biggest drop in muscle pH stems from the greatest peak contraction. However, this appears not to be the case. The magnitude of ΔpH is governed not only by stimulation characteristics (e.g. frequency and pulse width) but also muscle temperature and post-mortem delay of stimulation (Devine *et al.*, 2004). Further studies are required to investigate the influence that each of these elements have on ΔpH and the corresponding tension produced.

References

Chen X and Creed KE. (2004). Histochemical and contractile properties of striated muscles of urethra and levator ani of dogs and sheep. *Neurourology and Urodynamics* 23, 702-708.

Chrystall BB and Devine CE. (1978). Electrical stimulation, muscle tension and glycolysis in bovine sternomandibularis. *Meat Science* 2, 49-58.

Chrystall BB, Devine CE and Davey CL. (1980). Studies in Electrical Stimulation: Post-mortem decline in nervous response in lambs. *Meat Science* 4, 69-78.

Devine CE and Chrystall BB. (1984). Electrical stimulation of rats: Part 2 - The effect of electrical parameters on muscle tension and post-mortem glycolysis. *Meat Science* 10, 293-305.

Devine CE, Hopkins D, Hwang IH, Ferguson DM and Richards I. (2004). Electrical Stimulation. In *Encyclopedia of Meat Sciences*, ed. Jensen WK, Devine CE and Dikeman M, pp. 413-423. Elsevier Academic Press.

Pearce KL, Hopkins DL, Toohey E, Pethick DW and Richards I. (2006). Quantifying the rate of pH and temperature decline in lamb carcasses using mid voltage electrical stimulation in an Australian abattoir. *Australian Journal of Experimental Agriculture* In press.

EFFECTS OF LOW VOLTAGE ELECTRICAL STIMULATION ON BOVINE LONGISSIMUS AND ADDUCTOR MUSCLES

V. Høst[1]*, E. Veiseth[1,2], K. Hollung[1], T. Frøystein[3], X. Jia[1,4], L. Aass[5] and K.I. Hildrum[1]

[1.] *Matforsk AS, Norwegian Food Research Institute, Osloveien 1, N-1430 Ås,* [2.] *Department of Mathematical Sciences and Technology, Norwegian University of Life Sciences, PO Box 5003, N-1432 Ås,* [3.] *Norwegian Meat Research Centre, PO Box 396 Økern, N-0513 Oslo,* [4.] *Department of Chemistry, Biotechnology and Food Science, Norwegian University of Life Sciences, PO Box 5003, N-1432 Ås and* [5.] *Department of Animal and Aquacultural Sciences, Norwegian University of Life Sciences, PO Box 5003, N-1432 Ås .*
E-mail: vibeke.host@matforsk.no

Keywords: electrical stimulation, WB shear press, *post mortem,* pH decline, proteomics

Introduction

Electrical stimulation (ES) of beef carcasses is frequently used by industry to enhance meat tenderness. There is a general consensus that ES is beneficial in terms of quality, however reported results of the effectiveness of ES on improving meat tenderness show considerable variation (Hildrum *et al.*, 1999). Moreover, depending on their location on the carcass, it is expected that muscles will respond differently to ES. It is well known that ES will increase the biochemical activity in the muscle cells, leading to ATP and creatine phosphate disappearance, accumulation of lactate and accelerated pH decline. However, the biological and physiological changes induced by ES are poorly understood. Metabolic profiling (proteomics) should provide a promising tool to investigate these effects.
In the present on-going study, the effects of low voltage ES on meat quality parameters of two different bovine muscles (*Longissimus dorsi* (LD) and *Adductor* (AD)) are investigated. Proteomics is used in order to investigate the up- and down-regulation of metabolic and structural muscle proteins. Understanding the molecular changes stimulated by ES may lead to development of new strategies for optimisation of the ES processing parameters.

Materials and Methods

Young bulls of the Norwegian Red breed (n=7) from a performance test station (GENO) were slaughtered at a commercial abattoir, and carcasses were halved. One half received low voltage ES (90 V, 15 Hz, 32 sec) which was applied through muscles in the neck and leg approximately 40 min (± 4 min) after stunning, while the other half served as a non-stimulated control (NES). At approximately 1h *post mortem*, the LD and AD were removed from both halves of each carcass, further subdivided, packed in polyethylene bags and stored at 4°C for 8 days. Muscle pH was measured at 1, 3, 6, 10 and 24 h after ES treatment for both muscles, while sarcomere lengths were measured at 48 h *post mortem* for LD only, as described by Rødbotten *et al.* (2001). Following the 8-day aging period, 4-cm thick slices of the muscles were vacuum-packed in polyethylene bags, heated in a water bath at 70°C for 50 min and chilled in iced water for 50 min. The meat slices were cut into pieces of 1x1 cm^2 thickness and ten parallels were sheared perpendicular to the fibre direction with a Warner-Bratzler (WB) shear force device attached to an Instron Materials Testing Machine (Model 4202, Instron Engineering Corporation, High Wycombe, UK). The average maximum force for the ten parallels was used in the data analysis. For metabolic profiling, muscle samples were separated by 2-dimensional gel electrophoresis (2-DE) according to the method described by Jia *et al.* (2006).

Results and Discussion

The results in Figure 1 reveal an effect of ES on the initial pH decline in both muscles. The pH is lower in the AD than in the LD muscles the first 6 hours after ES. However the ultimate pH, attained after 24 hours, does not differ between treatment or muscle. The results in Figure 2 demonstrate that there is a positive effect of ES on tenderness in the LD. In contrast, there was no observed effect of ES on AD tenderness after 8 days storage. No effect of ES on sarcomere length was observed in the LD. Thus the changes observed in tenderness for the LD is not due to changes in sarcomere length. A preliminary investigation of metabolic proteins from the muscles is being conducted using 2-DE. A representative image of water-soluble proteins from LD is shown in Figure 3.

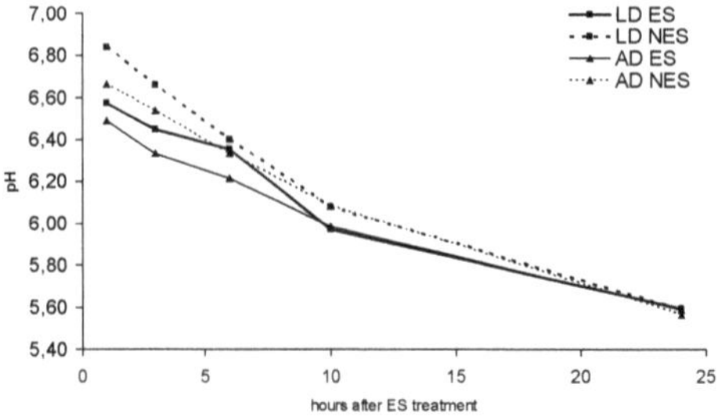

Figure 1: Mean pH measurements during the first 24 h *post mortem* in ES- and NES-treated LD and AD muscles from animals.

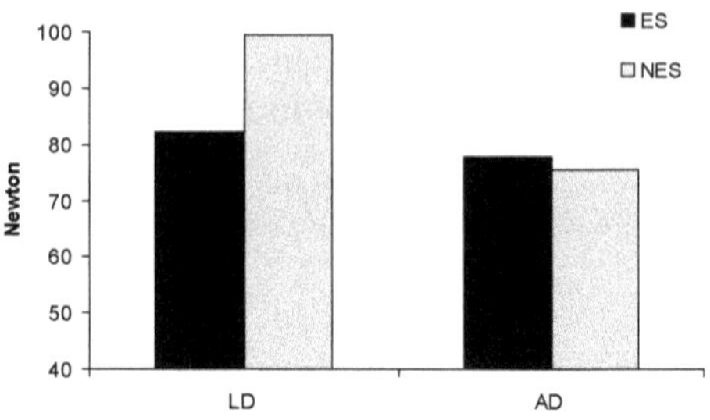

Figure 2: Mean WB shear force measured 8 days *post mortem* in ES- and NES-treated LD and AD muscles from 7 animals.

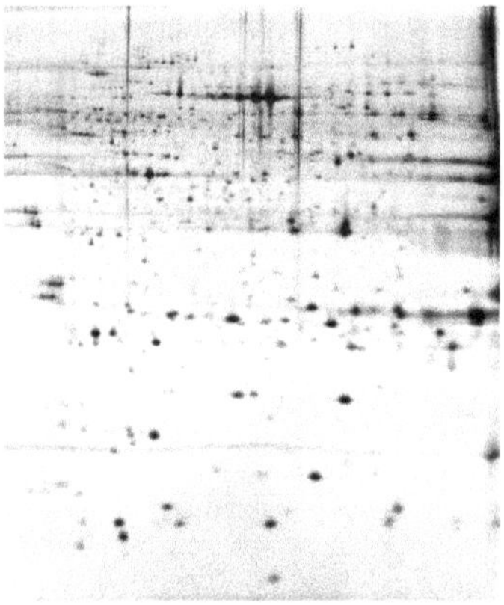

Figure 3: 2-DE image of water-soluble proteins from LD separated on a pH 4-7 gradient in the first dimension (left-right), followed by separation on 12% SDS-PAGE (top-down). Gels are silver stained.

Conclusions

We observed an effect of ES on the pH decline in LD and AD muscles of Norwegian Red cattle. However, an effect of ES on tenderness was only observed in the LD. Proteomics is a powerful tool to study the metabolic changes in muscle and will be used to unravel the metabolic mechanisms occurring during low voltage ES of bovine carcasses. Work is in progress to elucidate protein changes related to ES treatment, and up- and down-regulated proteins will be identified by MALDI-tof MS analysis.

Acknowledgements

The project was supported by a grant from The Norwegian Research Council and The Fund for the Research Levy on Agricultural Products.

References

Hildrum, K.I., Solvang, M., Nilsen, B.N., Frøystein, T. and J. Berg (1999). Combined effects of chilling rate, low voltage electrical stimulation and freezing on sensory properties of bovine M. *longissimus dorsi.* Meat Science, 52, 1-7.

Jia, X.; Hollung, K.; Therkildsen, M.; Hildrum, K. I.; Bendixen, E (2006). Proteome analysis of early post mortem changes in two bovine muscle types: M. *longissimus dorsi* and M. *semitendinosis*. Proteomics, 6: 936-944.

Rødbotten, R., Lea, P. and Hildrum, K.I. (2001). Relative influence of low voltage electrical stimulation, chilling rate and ageing time on tenderness of M. longissimus dorsi of Norwegian cattle. Acta Agriculturæ Scandinavica. Section A, Soil and Plant Science. 51, 184-191.

INTER-MUSCULAR RESPONSES TO EXTENDED LOW VOLTAGE ELECTRICAL STIMULATION OF LAMB

N.F.S. Gault*[1], A. Gordon[2] and E.L.C. Tolland[1]

[1.] *Agriculture, Food and Environmental Science Division[1], and Corporate Services Division[2], Agri-Food and Biosciences Institute, Belfast BT9 5PX, UK. Email: Norman.Gault@afbini.gov.uk*

Keywords: electrical stimulation, lamb, eating quality

Introduction

Traditional electrical stimulation (ES) increases the rate of post mortem glycolysis in red meat carcasses, hence facilitating rapid chilling pre-rigor and avoiding cold shortening (Chrystall and Devine 2000). However, most conventional ES procedures aim to decrease the pH of the *longissimus dorsi* muscle to no less than pH 6.3 1h post mortem to avoid the risk of heat shortening. Conversely, Gault *et al.*, (2000) showed that 300 pulses of 100v low voltage electrical stimulation (LVES) applied to chicken carcasses after bleeding could quickly bring the breast muscle pH to pH 6.0, inducing very early rigor onset, yet having no adverse effect on meat quality.

Unlike chicken breast muscle which comprises mainly αW (fast-twitch high glycolytic) muscle fibres, the muscles of red meat animals consist of varying proportions of αW, αR (fast-twitch high oxidative) and βR (slow-twitch intermediate/high oxidative) muscle fibre types (Monin 1981). Gault *et al.*, (2005) subsequently demonstrated that the responsiveness of lamb carcasses to 300 pulses of LVES reflected the fibre type composition of the muscles analysed. The greatest glycolytic response was seen with predominantly αW fibre-type muscles, and the least with predominantly βR fibre-types. Even though some muscles reached pH 6.0 within 1h of slaughter, the LVES treatments used had no detrimental effect on meat quality *per se*, and limited to some extent the susceptibility of muscles to shortening-induced toughness when de-boned early.

This study assessed the glycolytic response of extended LVES procedures, at different duty cycles, in lamb carcasses, and how this might impact on early de-boning of lamb muscles of different fibre-type composition.

Materials and Methods

Forty-four (24 male; 20 female) Texel x Greyface crossbred lambs were used; mean live weight 41.02 kg (range 38-47kg). Groups of 4 lambs were randomly allocated to one of 10 LVES treatments. Each carcass received either 600 or 1200 pulses of 100v 10ms pulse width LVES at a frequency of 5, 10, 20, 50 or 100Hz. LVES was applied 90s after stunning and bleeding. A control group had no LVES treatment. All carcasses were dressed and split within 10 minutes of bleeding.

The muscles selected for assessment were the *m. semitendinosus* (ST; αW fibre-type); *m. longissimus dorsi* (LD; αR fibre-type) and *m. supraspinatus* (SS; βR fibre-type). These were removed from alternate carcass sides within each treatment for analysis immediately after splitting (early de-boned), while the matching sides were Achilles hung until 24h post-mortem (late de-boned). Chilling was carried out at 9°C for the first 24h post-slaughter. Muscles were then stored aerobically for one week at 2°C when the remaining analyses were completed.

Muscle pH (Bendall, 1975) was monitored at 1h, 2h and 24h post mortem in the early de-boned muscles, and at 24h in the late de-boned muscles. Sarcomere lengths (SL) (Koolmees *et al.*,1986) were measured in samples of early and late de-boned muscles at 48h post-mortem and cooking loss (CL) and shear force (SF) (Gault *et al.*, 2000) assessed at 48h and 7 days post mortem. Data were analyzed by ANOVA (Genstat, 2003).

Results and Discussion

As found previously (Gault *et al.*, 2005), muscle fibre-type had a significant ($P<0.001$) effect on muscle responsiveness to LVES. The predominantly αW fibre-type ST gave the greatest response, the αR fibre-type LD gave an intermediate response, and the βR fibre-type SS gave the least response compared to the controls (Table 1). The lowest frequencies also gave significantly greater ($P<0.001$) glycolytic responses compared to the 100Hz treatments, which were ineffective. Likewise, there was no significant improvement ($P>0.05$) in glycolytic response in any muscle when the total number of pulses was increased to 1200, irrespective of frequency (Table 1). Although significant ($P<0.001$) intermuscular differences were found for each quality attribute (Table 2), no significant ($P>0.05$) differences due to LVES treatment were found other than for pH_{1h} and pH_{2h} values. Early de-boning resulted in significantly shorter ($P<0.001$) sarcomeres and higher ($P<0.001$) 48h and 7d shear force values in the ST and SS (Table 3), little difference being seen with the LD. In general terms, this would suggest that the intermediate glycolysing LD was unaffected by early de-boning. In contrast, the fast glycolysing ST, which showed characteristics of early rigor onset, and the slow glycolysing SS, were both prone to a degree of shortening and toughening when be-boned early. A more detailed analysis of specific LVES treatments may help provide an explanation for this anomaly.

Table 1: Effect of LVES (pulses x frequency (Hz)) on intermuscular pH_{1h} values.

		600 pulses					1200 pulses				
Muscle	Control	5	10	20	50	100	5	10	20	50	100
ST	6.51	6.01	6.12	6.07	6.20	6.47	5.98	6.05	6.04	6.15	6.41
LD	6.51	6.21	6.24	6.23	6.29	6.66	6.15	6.15	6.25	6.36	6.50
SS	6.71	6.47	6.54	6.62	6.63	6.61	6.36	6.39	6.46	6.54	6.71
mean	6.58	6.23	6.30	6.30	6.37	6.58	6.16	6.20	6.30	6.35	6.54
P	<0.001	<0.001	<0.001	<0.001	<0.001	<0.001	<0.001	<0.001	<0.001	<0.001	<0.001
n	4	4	4	4	4	4	4	4	4	4	4
lsd[a]	0.10	0.10	0.10	0.10	0.10	0.10	0.10	0.10	0.10	0.10	0.10
lsd[b]						0.16					

[a]least significant difference (P<0.05) within a column. [b]least significant difference (P<0.05) across columns.

Table 2: Intermuscular differences in the quality attributes of lamb for all treatments combined.

Muscle	pH_{1h}	pH_{2h}	pH_{24h}	SL_{48h} µm	CL_{48h} %	CL_{7d} %	SF_{48h} kg/cm^2	SF_{7d} kg/cm^2
ST	6.18	6.00	5.67	2.03	37.1	36.8	6.75	5.00
LD	6.32	6.15	5.56	1.88	31.7	31.6	4.94	3.95
SS	6.55	6.36	5.83	1.89	37.5	37.9	5.54	4.43
P	<0.001	<0.001	<0.001	<0.001	<0.001	<0.001	<0.001	<0.001
n	44	44	88	88	88	88	88	88
lsd P<0.05	0.03	0.04	0.02	0.03	0.49	0.48	0.25	0.19

Table 3: Effect of de-boning time on intermuscular quality attributes.

Muscle	pH_{24h} early	pH_{24h} late	SL_{48h} early	SL_{48h} late	CL_{48h} early	CL_{48h} late	CL_{7d} early	CL_{7d} late	SF_{48h} early	SF_{48h} late	SF_{7d} early	SF_{7d} late
ST	5.66	5.68	1.95	2.10	37.3	36.8	36.5	37.0	7.25	6.24	5.42	4.59
LD	5.56	5.55	1.88	1.89	31.3	32.2	31.3	32.0	4.88	5.00	3.93	3.98
SS	5.81	5.85	1.85	1.92	38.4	36.7	38.1	37.7	6.21	4.87	4.77	3.89
mean	5.68	5.70	1.89	1.97	35.7	35.2	35.3	35.6	6.11	5.37	4.70	4.15
P	<0.01		<0.001		<0.001		<0.05		<0.001		<0.001	
n	44	44	44	44	44	44	44	44	44	44	44	44
lsd[a]	0.02	0.02	0.04	0.04	0.70	0.70	0.68	0.68	0.35	0.35	0.27	0.27
lsd[b]	0.03		0.04		0.72		0.74		0.36		0.27	

[a]least significant difference (P<0.05) within a column. [b]least significant difference (P<0.05) across columns.

Conclusions

Enhanced LVES has limited potential in promoting rapid glycolysis and early rigor onset in all muscles of ruminant carcasses, due to the poor responsiveness of βR muscle fibres in multi-fibre-typed muscles. Likewise, more detailed investigations are needed to eliminate the potential for shortening defects when de-boning early.

References

Bendall J.R. (1975). Cold-contracture and ATP-turnover in the red and white musculature of the pig, post- mortem. Journal of the Science of Food and Agriculture, 26: 55-71.

Chrystall, B.B. and Devine, C.E. (2000). Meat and electrical stimulation. In: Encyclopedia of Food Science and Technology – 2nd edn. (F.J. Francis, ed). John Wiley & Sons, Inc., New York.

Gault, N.F.S., Burns, C., Tolland, E.L.C. and Kilpatrick, D.J. (2000). Effect of some electrical stimulation variables on wing flapping, post-mortem glycolysis and eating quality characteristics of broiler *pectoralis major* muscle. British Poultry Science, 41: 293-299.

Gault, N.F.S., Gordon, A. and Tolland, E.L.C. (2005). Inter-muscular differences in response to low voltage electrical stimulation of lamb. In: Proceedings of the 51st International Congress of Meat Science and Technology, 43.

Genstat 7 Committee. (2003). Genstat 7th edn., Lowes Agricultural Trust. VSL International Ltd.

Koolmees, P.A., Korteknie, F. and Smulders, F.J.M. (1986). Accuracy and utility of sarcomere length assessment by laser diffraction. Food Microstructure, 5: 71-76.

Li, Y., Siebenmorgen, T.J. and Griffis, C.L. (1993). Electrical stimulation in poultry: a review and evaluation. Poultry Science, 72: 7-22.

Monin,G. (1981). Muscle metabolic type and the DFD condition. In: The Problem of Dark-cutting in Beef. (D.E. Hood and P.V. Tarrant, eds.). pp 63-85. Martinus Nijhoff Publishers, The Hague.

TENDERNESS OF TYPICAL SOUTH AFRICAN CATTLE BREED CROSSES, THE EFFECTS OF FEED WITHDRAWAL PERIOD AND ELECTRICAL STIMULATION

L. Frylinck[1], P.E. Strydom*[1], E.C. Webb[2], J. Anderson[1] and J.D. Snyman[1]

[1.] *Agricultural Research Council of South Africa, Private Bag X2, Irene, South Africa, 0062.* [2.] *Department of Animal and Wildlife Sciences, University of Pretoria, 0002. E-mail: lorinda@arc.agric.za,*

Keywords: sarcomere length, calpain system, collagen, genotype, tenderness

Introduction

The concept of meat tenderness is very complex, as it is biologically dependent on a combination of many physiological factors such as: proteolytic degradation of cytoskeletal proteins during *post mortem* storage in the calpain system; changes in the nature of actin/myosin interaction - these changes are reflected in the sarcomere length; the amount of intra-muscular connective tissue and the nature of the intra-muscular connective tissue as measured by total collagen content and % collagen solubility (Purslow, 1999). Although it is reported that *Bos indicus* cattle generally produce tougher meat than *Bos taurus* cattle under specific pre-slaughter and slaughter conditions (Shackelford *et al.*, 1995), sometimes *Bos taurus* cattle can also produce tough meat (Frylinck *et al.*, 2003; De Bruyn, 1991). The calpain system explains most of these genetic differences. Due to the belief that all humped animals are *Indicus*, cattle breeds with humps are excluded from meat quality improvement schemes. Although certain indigenous South African breeds such as the Nguni, have humps, their genetic composition are predominantly *taurine* (Frylinck *et al.*, 2003) and they consistently produce tender meat. In the present study, we measured physiological factors to try to explain the tenderness characteristics of typical breed-crosses in the South African feedlots. Also, how the duration of stimulation and pre-slaughter feed withdrawal period would influence these meat tenderness characteristics.

Materials and Methods

Simmental cross, Brahman cross and Nguni cross (Sanga) bulls (n=180) were raised, treated and slaughtered as specified by Strydom and Frylinck (2005). Carcasses were treated and samples were taken as follows. *M. longissimus* aged for 1 or 14d were prepared and Warner Bratzler shear force (SF) was measured as the peak force (kg). Sarcomere lengths (SL) and myofibril fragment lengths (MFL) were measured using a Video Image Analyzer. Calpains and calpastatin were extracted and measured (Frylinck *et al.*, 2003). One unit of calpain activity was defined as an increase in absorbance at 366 nm of 1.0 per hour, at 25 °C. One unit of calpastatin activity was defined as the amount that inhibited one unit of m-calpain activity. Total collagen content and collagen solubility were extracted and measured (Frylinck *et al.*, 2003). Collagen content was expressed as hydroxy-proline nitrogen per total protein nitrogen. Collagen solubility was calculated by expressing hydroxy-proline in the filtrate, as a percentage of the total hydroxy-proline of the filtrate plus residue. The data were subjected to a three-way analysis of variance. Means for the main effects and their interactions were separated using Fisher's protected t-test least significant difference (LSD) at the 5% level.

Results and Discussion

Strydom and Frylinck (2005) reported that the Brahman and Nguni carcasses were similar in tenderness after 14 d *post mortem*, but they were both significantly more tender than that of the Simmental carcasses. Variation in SL did not explain that; Nguni cross SL measured at 1 d *post mortem* was the shortest on average, yet the meat was tender (shorter SL of Nguni could be considered a breed characteristic). The high SF values could be partly explained by the fact that the SL of the Simmental cross was less than 1.6μm (muscle shortening, not shown) when animals were on the long feed withdrawal period and not stimulated (NS). Improvement in SF due to 15 sec electrical stimulation was reflected by longer SL, but after 120 sec ES the SL corresponded with those of the not stimulated carcasses, probably as a result of heat shortening.

MFL of the Nguni at 7d and 14d *post mortem* was on average the shortest (indicating to the most effective *post mortem* proteolytic enzyme degradation for tenderness). MFL of the Simmental and the Brahman did not differ significantly from each other. These results correspond with that of the calpain system (Table1), where the Nguni had significantly more favourable μ-calpain levels and calpastatin/μ-calpain ratio than that of the Brahman and the Simmental (which had similar values in this respect). Pre-slaughter treatments (24h feed withdrawal vs. 3 h feed withdrawal) were found to significantly influence SF measured at 1d and 14d *post mortem* and SL measured at 1 d *post mortem* ($p<0.001$) (Table 1), but the calpain proteolytic system seemed to be influenced very slightly. Improvement in SF due to 15 sec ES was reflected by longer SL and shorter MFL measurements especially after 14 d *post mortem*, thus a simultaneous relaxation of the muscle fibres and an enhanced rate of proteolysis could have contributed to the tenderizing. It can be derived that all three cross breeds showed an improvement on tenderness with 15 sec ES, but that the toughest cross-breed *i.e.* the Simmental had the most benefit from this post-slaughter treatment. The percentage collagen solubility measured in the Nguni-X was the most favourable, and could explain some of the tenderness characteristics of the Nguni carcasses (Table 1).

Table 1: The effect of breed, feed withdrawal period and ES treatment on meat tenderness characteristics.

Breed	Brahman	Nguni	Simmental	SEM[1]	P-value
Shear force 1 d p.m. (kg)	6.9^{ab}	6.7^{a}	7.3^{b}	0.1627	0.027
Shear force 14 d p.m. (kg)	4.3^{a}	4.4^{a}	4.9^{b}	0.1324	0.005
Sarcomere length (µm)	1.72^{b}	1.65^{a}	1.70^{b}	0.0146	0.001
Total Collagen (Hypro N x 10^3/ Total N)	1.56	1.71	1.66	0.0588	0.201
% Collagen solubility	17.8^{b}	19.3^{a}	17.5^{b}	0.4190	0.005
MFL ˙14 d p.m. (µm)	26.5^{b}	24.7^{a}	26.3^{b}	0.510	0.026
µ-Calpain- (U/g meat)	1.71^{b}	2.03^{a}	1.86^{a}	0.064	0.002
Calpastatin/µ-calpain	1.74	1.52	1.66	0.060	0.028
Feed withdrawal	**20 hours**	**3 hours**		**SEM[2]**	**P-value**
Shear force 1 d p.m. (kg)	7.3^{a}	6.7^{b}		0.1329	0.004
Sarcomere length (µm)	1.67^{a}	1.72^{b}		0.0119	0.004
Calpastatin/µ-calpain	1.69	1.58		0.049	0.108
Electrical stimulation (ES)	**None**	**15 sec.**	**120 sec.**	**SEM[2]**	**P-value**
Shear force 1 d p.m. (kg)	8.7^{b}	6.3^{a}	6.0^{a}	0.1627	<0.001
Sarcomere length (µm)	1.67^{a}	1.74^{b}	1.65^{a}	0.0146	<0.001
MFL ˙14 d p.m. (µm)	25.43^{a}	24.56^{a}	27.55^{b}	0.510	<0.001

[a,b] Means within a row with different superscripts differ significantly ($P<0.05$),
[1] Standard error of means

Conclusions

Longer feed withdrawal coincided with higher shear force values in general, although it was more evident for NS carcasses. According to the Shackelford *et al.*, (1995) tenderness decreased as percentage *Bos indicus* inheritance increased, but this does not explain the Simmental's *(Bos Taurus)* low tenderness measured in this project and others (Frylinck *et al.*, 2003; De Bruyn, 1991). Variations in tenderness due to breed effects could be influenced by different combinations of pre-slaughter and slaughter practices.

References

De Bruyn, J.F. (1991) Production and product characteristics of different cattle genotypes under feedlot conditions. *D.Sc.-thesis*, University of Pretoria, Pretoria, South Africa.

Frylinck, L., Strydom, P.E., Smith, M.F., Naudé. R.T. and Heinze, P.H. (2003). Evaluation of meat tenderness of indigenous South African and other beef breeds. Proceedings of the 11th International Meat Symposium 29 and 30 January 2003, *Organised by ARC-Animal Nutrition and Animal Products Institute, Irene, South Africa.*

Purslow, P.P. (1999). The intramuscular connective tissue matrix and cell/matrix interactions in relation to meat toughness. *Proceedings of the 45th ICoMST.* Yokohama, Japan. 1, 210.

Shackelford, S. D., Wheeler, T. L. and Koohmaraie, M. (1995). Relationship between shear force and trained sensory panel tenderness ratings of 10 major muscles from *Bos indicus* and *Bos taurus* cattle. *Journal of Animal Science,* 73, 3333-3340.

Strydom, P.E. and Frylinck, L. (2005). The effect of genotype, duration of feed withdrawal and electrical stimulation on meat quality. 51st International Congress of Meat Science and Technology. August 7-12, 2005 – Baltimore, Maryland USA.

POLARIMETRIC OHMIC PROBES FOR THE ASSESSMENT OF MEAT AGEING

J.-L. Damez*, S. Clerjon, S. Abouelkaram and J. Lepetit

Unité Qualité des Produits Animaux (QuaPA), Institut National de la Recherche Agronomique (INRA), F-63122 Saint Genès-Champanelle, France. Email: damez@clermont.inra.fr

Keywords: sensor, meat ageing, tenderisation, quality, electrical properties, anisotropy

Introduction

A frequent criticism advanced by consumers is the lack of consistent meat tenderness and thereby it is one of the major problems to be resolved by the meat industry. It is well known that meat tenderness depends on the properties of the muscle fibres and the amount and type of connective tissue. Intramuscular connective tissue is rather immutable and contributes to a ''background toughness'' of the meat (Purslow, 2005). Meat tenderness is improved during refrigerated storage by factors of physico-chemical nature and of enzymatic nature but the ageing rate is tremendously variable (Zamora *et al.,* 1996). A possible way to optimise the duration of refrigerated storage is to assess the ageing state early in the *post-rigor* phase and to classify the meat according to rate of ageing (Lepetit and Hamel, 1998). The assessment of ageing state can be done by different biological or mechanical laboratory means. Biological methods are time consuming and mechanical ones are destructive. The development of rapid and non destructive methods which can give the same information on-site is needed. The structural modifications of the meat which occur during ageing affect not only the mechanical properties but also the electric and dielectric properties of the product (Swatland, 1980, Damez *et al.*, 2005) and particularly the anisotropy of these properties (Lepetit et al 2002). In this study, we attempted to demonstrate the feasibility of a sensor of ageing, useful in the meat industry, based on a measure of the electric anisotropy of the meat.

Materials and Methods

The sensor (under patent) that we designed is suitable to measure the anisotropy of the impedance properties with a single application, by quickly carrying out a multitude of measures in various directions within the muscle. Two probes with circular shape and different diameters were used in this study (Figure 1). The first one (P1) is 8 cm in diameter and has 10 pairs of stainless electrodes (1.65mm in diameter, 15mm in length). The second one (P2) is 4 cm in diameter with an external silver-plated crown and a central silver-plated contact to inject the current and 12 pairs of golden-plated needles, to measure the voltage, which are 2mm in length and 0.6mm in diameter. The impedance measurements were made with a HP 4194A Impedance/Gain-Phase analyzer at a frequency range of 1 KHz to 1.5 MHz and communication between the pairs of electrodes were carried out with an Analog Devices. Curves fitting with Matlab 7.0 were carried out using the equation:

$Z_\theta = Z_{\|}\cos^2\theta + Z_\perp \sin^2\theta$ where Z_θ is the impedance at θ angle with regard to the muscle fibres axis, $Z_{\|}$ is the minimum impedance (measured parallel to the direction of muscle fibres), $Z_\perp$ is the maximum impedance (measured perpendicularly to the direction of muscle fibres)(Figure 2). Statistical analyses to extract pertinent parameters were performed using SAS. The trials to evaluate ageing have been done on a population of 104 bovine muscles (28 *Rectus Abdominus* (RA) muscles, 48 *Semimembranosus* (SM) muscles, 28 *Semitendinosus* (ST) muscles). The electrical measurements have been put correlated with the mechanical stress at 20% of the compression assessed on raw meat samples using an Instron 5543 according to Lepetit and Buffiere, 1995.

Figure 1:
The polarimetric ohmic probes.

Circular probe Ø8cm 20electrodes (P1)

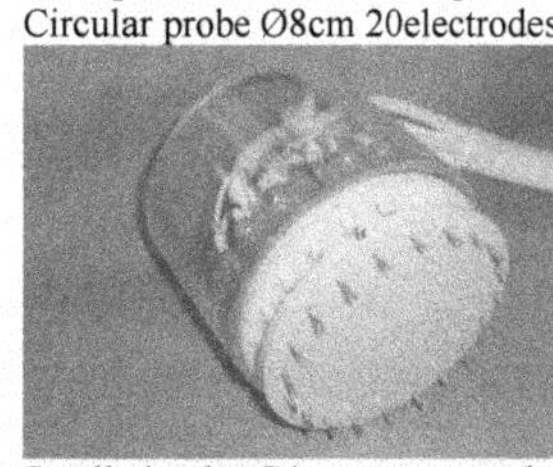

Small circular Ø4cm crown probe (P2)

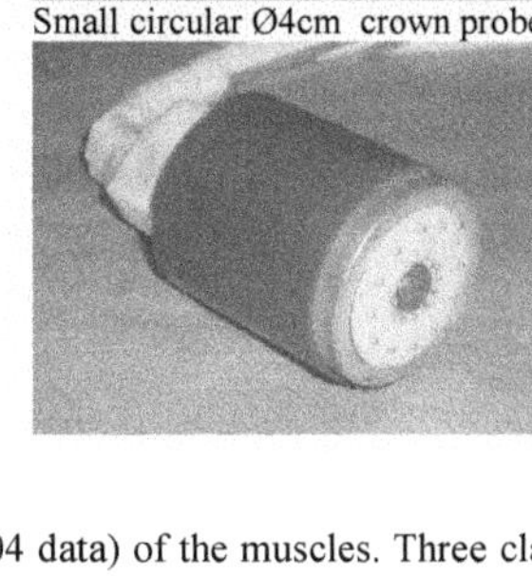

Results and Discussion

The validation test was carried out both in all (312 samples) and on the means (104 data) of the muscles. Three classes of ageing were arranged (aged < 10 N/cm², 10 N/cm² < mid-aged < 30 N/cm², not aged > 30 N/cm²). The database of 104 muscles comprises 51 aged muscles, 38 mid-aged muscles, 15 not aged muscles according to a classification carried out by the stress compression with 20% compression. In the database of 312 samples, 153 are aged, 114 are mid-aged and 45 are not aged. Good correlations were obtained between the results from the probes and the mechanical compression measurements, particularly those obtained with the P1 probe (Figure 3). The results of the discrimination using each polarimetric ohmic probe are presented (Table 1) from a set of 128 variables and a reduced whole brought back to 36 averaged variables showing that both sensors are performing in some cases a classification close to 90%.

Conclusions

The performances of the two probes presented here are good enough to carry out non-invasive on-line classification of bovine muscles, according to their ageing state, permitting the optimisation of refrigerated storage duration.

The integration in the sensors of electronics on the one hand to measure impedance and to perform communication between the electrodes, and on the other hand to carry out calculations in order to obtain a display of measurement, ought to allow in the near future the availability of equipment able to assess the state of ageing for an optimization of refrigerated storage duration and consistency of the meat tenderness.

References

Damez, J. L., Clerjon, S., and Abouelkaram, S. (2005). Mesostructure assessed by alternating current spectroscopy during meat ageing, 51st International Congress of Meat Science and Technology , 2005, Baltimore, August 7-12

Lepetit, J. and Buffiere, C. (1995). Meat ageing measurement: comparison between two mechanical methods. Fleischwirtschaft. 75(10):1220-1222.

Lepetit, J. and Hamel, C. (1998). Correlations between successive measurements of myofibrillar resistance of raw longissimus dorsi muscle during ageing, Meat Science; 49(2):249-254.

Lepetit, J.; Salé P.; Favier, R., and Dalle, R. (2002). Electrical impedance and tenderisation in bovine meat. Meat Science. 60:51-62.

Purslow, P. (2005). Intramuscular connective tissue and its role in meat quality, Meat Science; 70: 435–447

Swatland, H. J. (1980). Postmortem changes in electrical capacitance and resistivity of pork. Journal of Animal Science. 51(5):1108-1112.

Zamora, F., Debiton, E., Lepetit, J., Lebert, A., Dransfield, E., and Ouali, A. (1996). Predicting variability of ageing and toughness in beef *M. longissimus lumborum* et *thoracis*. Meat Science, 43, 321-333

Figure 2: Radial plot of impedances measured with the 8 cm in diameter probe (P1), at 10 KHz of a bovine *Rectus Abdominus* (RA) muscle aged for 2, 7, 15 and 23 days.

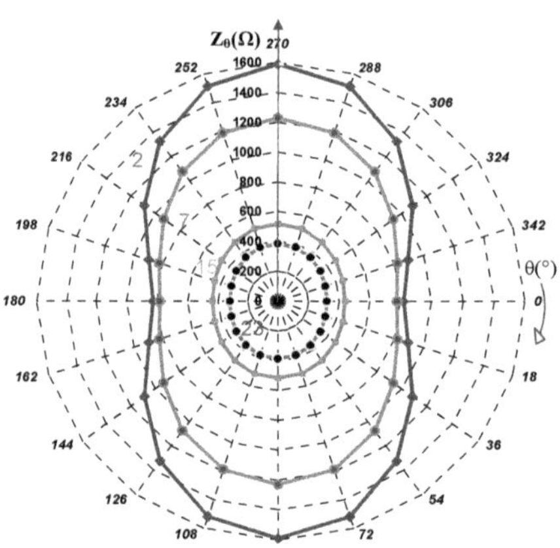

Figure 3: Correlations between the electrical and the mechanical measurements.

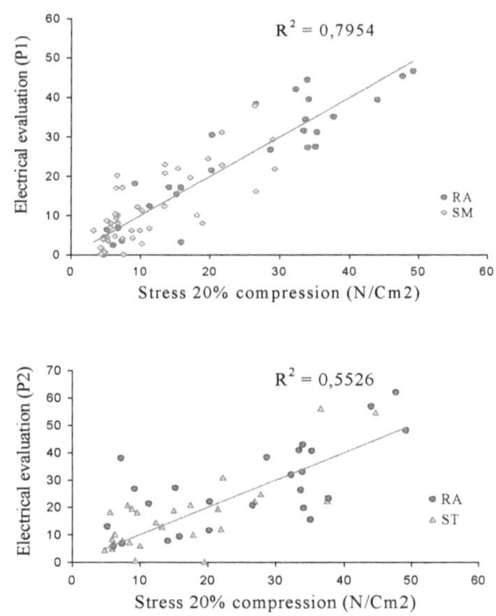

Table 1: Results of the discrimination analysis between 3 states of ageing using the two polarimetric ohmic probes.

	Nb observations	Nb parameters	Discrimination Aged	Mid-aged	Not aged
Circular probe 20 electrodes (P1)	312	36	115 (**75%**)	73 (**64%**)	34 (**76%**)
	312	128	135 (**88%**)	92 (**81%**)	40 (**89%**)
	104	36	45 (**88%**)	26 (**68%**)	13 (**87%**)
Small circular crown probe (P2)	312	36	104 (**68%**)	68 (**60%**)	31 (**69%**)
	312	128	124 (**81%**)	84 (**74%**)	37 (**82%**)
	104	36	37 (**73%**)	30 (**79%**)	13 (**87%**)

Acknowledgements

The authors wish to thank INTERBEV and OFIVAL for financial support.

FT-IR MICROSCOPY AS A TOOL TO INVESTIGATE PROTEIN SECONDARY STRUCTURAL CHANGES OF MUSCLE TISSUE IN INDUSTRIAL PROCESSES LIKE HEAT TREATMENT

R. Ofstad*[1], H.C. Bertram[2], U. Böcker[1], H.J. Andersen[2] and A. Kohler[1,3]

[1] *Centre for Biospectroscopy and Data Modelling, Norwegian Food Research Institute, Matforsk, 1430 ÅS, Norway,* [2] *Department of Food Science, Danish Institute of Agricultural Sciences, Research Centre Foulum, P.O. Box 50, DK-8830 Tjele, Denmark* [3] *Unité de Sensométrie et de Chimiométrie, ENITIAA, BP 82225, 44322 Nantes Cedex 3, France. Email: ragni.ofstad@matforsk.no*

Keywords: FT-IR micro-spectroscopy, pork meat, heating

Introduction

FT-IR spectroscopy has traditionally been used within analytical chemistry for structural analysis of organic compounds. Since the 1990ies, its application has been extended to the characterisation of biological samples, amongst others intact biological cells such as microorganisms or single mammalian cells, and complex tissues. FT-IR spectra of biological samples display absorption bands due to vibrations of functional groups of proteins, polysaccharides, lipids, and nucleic acids. The use of FT-IR spectroscopy is thus providing information about the overall biochemical composition of the sample in one single measurement. FT-IR spectra of biological samples often contain a number of overlapping bands that may be difficult to interpret, but the spectral features of proteins, carbohydrates and lipids are generally separated well. As the position of spectral bands is governed by the molecular structure, intra- and intermolecular interactions and the amounts of the present molecules, FT-IR spectra can be used to evaluate the structure of the components in biological samples. By the invention of FT-IR micro-spectroscopy, a combination of spectroscopy and microscopy, it became possible to acquire spatially resolved FT-IR spectra, from slices of intact tissue (Wetzel *et al.,* 1999). FT-IR spectroscopy has become a very important technique in the investigation of meat and fish muscle tissue, since it offers a high specificity regarding the characterisation of protein secondary structure. Furthermore, FT-IR microspectroscopy allows the investigation of single muscle fibres and the spatial distribution of secondary structural modes within one muscle cell. In this paper we show the excellence of FT-IR for studying protein structural changes in meat tissue during heating.

Materials and Methods

Pork meat (*M. longissimus dorsi)* with high (6.5) and normal (5.4) pH, that was heated at 10 different temperatures (28.5, 34.0, 40.0, 46.1, 52.0, 57.6, 63.0, 69.3, 75.5 and 81 °C), was used for this experiment. The spectra were acquired from cryo-sections obtained from small meat tissue blocks (5x5x2 mm), that were immediately frozen in liquid nitrogen and stored at –80 °C prior to sectioning. The 8 μm thick cryo-sections were thaw-mounted on infrared transparent (2 mm thick) CaF_2 slides for the FT-IR microscopic measurements (IRscope II coupled to an Equinox 55 FT-IR spectrometer, Bruker Optics, Germany). The IR spectra were collected from single myofibres in transmission mode from 4000 to 700 cm^{-1} with a spectral resolution of 6 cm^{-1}. For each spectrum 256 interferograms were co-added and averaged. A background spectrum of the CaF_2 substrate was recorded before each sample measurement in order to account for variations in water vapor and CO_2. Spectra were pre-processed and normalised using extended multiplicative signal correction (EMSC) according to Martens *et al.,* (2003).

Results and Discussion

Figure 1a shows the amide I region (1700 cm^{-1}-1600 cm^{-1}) of the 2nd derivative of the FT-IR spectra from the pork meat with normal pH. Prior to taking the 2nd derivative the spectra were pre-processed by EMSC as described above. The minima in the 2nd derivatives refer to maxima (bands) in the original spectra. In Fig. 1a we are able to identify 9 bands at the wavenumbers1695 cm^{-1}, 1682 cm^{-1}, 1668 cm^{-1}, 1660 cm^{-1}, 1652 cm^{-1}, 1639 cm^{-1}, 1628 cm^{-1}, 1619 cm^{-1}, and 1610 cm^{-1}. Table 1 presents a tentative assignment of these spectral bands.

In Fig. 1b shows a correlation loading plot obtained from an ANOVA-PLSR with the design variables (indicator variables) as *X* and the assigned FT-IR spectral bands as *Y*. The inner and outer circle in Fig. 1b refer to 50 and 100% explained variance, respectively. All bands have a very strong correlation to the temperature used for the heat-treatment. The correlation is also confirmed by considering plots of the single bands against temperature (not shown). The bands at 1695 cm^{-1}, 1668 cm^{-1}, 1628 cm^{-1} and 1619 cm^{-1} all have positive correlations towards temperature. The bands at 1682 cm^{-1}, 1660 cm^{-1}, 1652 cm^{-1}, 1639 cm^{-1} and 1610 cm^{-1} all have negative correlations towards temperature. PH showed no apparent effect on the protein structure. However, a potential effect of pH may be concealed by the influence of temperature.

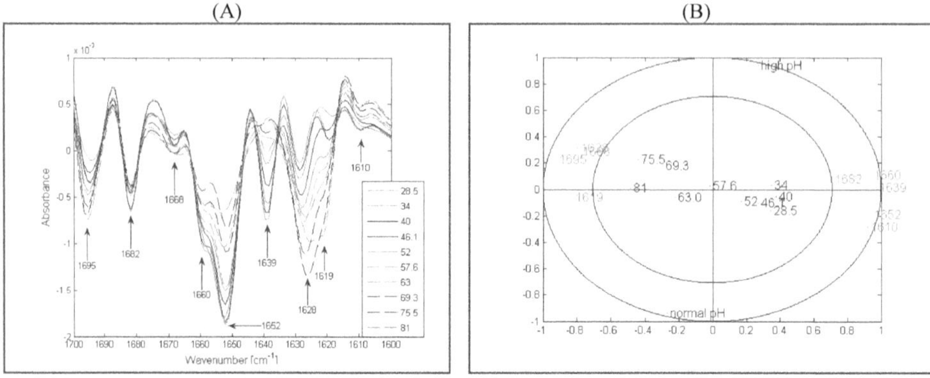

Figure 1: In (a) the 2nd derivative of the FT-IR spectra in the amide I region (1700 cm^{-1}-1600 cm^{-1}) of pork meat with normal pH (*M. longissimus dorsi)*at 10 different temperatures (28.5, 34.0, 40.0, 46.1, 52.0, 57.6, 63.0, 69.3, 75.5 and 81 °C). In (b) correlation of FT-IR spectral bands and design variables with the first and second PLS components of an ANOVA-PLSR (with design as X and selected FT-IR bands from amide I as Y) are shown.

Table 1: Band assignments (Jackson *et al.,* 1995, Fabian *et al.,* 2002).

1695 cm^{-1}	aggregated β-strands (intermolecular dipole coupling), high frequency
1682 cm^{-1}	native β-sheet structures (intramolecular), high frequency
1660 cm^{-1}	Possibly referring to native structures as loops in native structures
1652 cm^{-1}	α-helical structures
1639 cm^{-1}	native β-sheet structures (intramolecular), low frequency, O-H bending of water
1619/1628 cm^{-1}	aggregated β-strands (intermolecular dipole coupling), low frequency
1610 cm^{-1}	possibly the amino acid side chain tyrosine

The band positions shown above are very stable with respect to different animals and different treatments of the meat, though there are strong absorbance variations in the bands due to the different treatments that refer to protein structural changes. Our results show that FT-IR microspectroscopy can be used to measure structural changes in meat tissue proteins during heating. The type and extend of changes in secondary structure may have a potential as an indicator for product quality parameters of processed as well as aged meat.

References

Wetzel, D. L. and LeVine, S. M. (1999). Microspectroscopy: Imaging Molecular Chemistry with Infrared Microscopy. *Science*, 285, (5431), 1224-1225

Martens, H., Nielsen, J.P. and Engelsen, S.B. (2003). Light scattering and light absorbance separated by extended multiplicative signal correction. Application to near-infrared transmission analysis of powder mixtures. Analytical Chemistry, 75 (3), 394-404.

Fabian, H. and Mäntele, W. (2002). Infrared Spectroscopy of Proteins. In: Handbook of Vibrational Spectroscopy, J.M. Chalmers and P.R. Griffiths, Editors. John Wiley & Sons Ltd: Chichester. p. 3399-3425.

Jackson, M., Choo, L.P., Watson, P.H., Halliday, W.C. and Mantsch, H.H. (1995). Beware of connective tissue proteins: assignment and implications of collagen absorptions in infrared spectra of human tissues, Biochimica et Biophysica Acta, 1270, 1-6.

Bertram, H.C., Kohler, A., Böcker, U., Ofstad, R. and Andersen, H.J. (2006). Heat-induced changes in myofibrillar protein structures and myowater of two pork qualities. A Combined FT-IR spectroscopic and low-field NMR relaxometry study, Journal of Agricultural and Food Chemistry 2006, 54 (5), 1740-1746

THE MEASUREMENT OF PIG CARCASS LEAN MEAT PERCENTAGE WITH X-RAY COMPUTED TOMOGRAPHY

M. Judas*, R. Höreth, A. Dobrowolski and W. Branscheid

Institut für Fleischerzeugung und Vermarktung, BfEL Standort Kulmbach, E.-C.-Baumann-Str. 20, 95326 Kulmbach, Germany. Email: Michael.Judas@bfel.de

Keywords: pig carcass, grading, lean meat, dissection, PLS regression

Introduction

The physical dissection of pig carcasses is presently the official reference method for the determination of the real lean meat content of pig carcasses in the European Community. But it is a very expensive and time consuming procedure. The results of a joint European project (EUPIGCLASS) recommended the inclusion in the future of scanning by CT as an alternative to the physical dissection method (Dobrowolski *et al.*, 2004). Consequently, the German authorities provided a CT at the Federal Research Centre for Nutrition and Food, Location Kulmbach, with the objective of having the facilities for more regular tests of the grading systems in Germany. In this paper, we present the scanning protocol adopted, and the results of an evaluation of the accuracy of the CT method. In particular, we propose a spectral analysis of X-ray attenuation data by partial least squares (PLS) regression, refined by a smoothing of the PLS calibration model.

Materials and Methods

Carcasses were selected according to a sampling scheme stratified by fat depth, carcass weight, and sex (for details, see Judas *et al.,* 2005). Left half-carcasses were selected at the slaughter line for grading and dissection. The chilled carcasses were transported to the Institute of Meat Production and Market Research in Kulmbach, and were prepared according to the requirements of EC Council Regulation (EEC) № 3220/84. Dissection was carried out by expert technicians of the institute between one and four days *post mortem* and immediately after CT scanning. All cuts except the head were dissected. Lean meat weight (LMW) was determined with tendons and coarse fasciae removed.
A Siemens *Somatom Plus 4* spiral computer tomograph was used for scanning. Scanning parameters were 140 kV tube voltage, 146 mA tube current, a rotation time of 1 sec with pitch 1. Reconstruction algorithm AB50 was applied. Slice thickness was 10 mm with a 460 mm squared field of view at a resolution of 512x512 pixels. Scanning was performed in 2–3 spirals per carcass and lasted 15–30 min.
CT scanning produces a series of images that show tissues in shades of grey according to their density (Figure 1, left). Each pixel of an image represents a voxel with a constant volume, in this case 8.1 mm³. Each grey value is the visualization of the variation in X-ray attenuation measured in Hounsfield units (HU). The total volume per HU value has specific bi-modal distributions that differ between carcasses (Figure 1, right). Peaks around -70 HU represent adipose tissue, peaks around +70 HU represent muscle tissue.
A problem in the interpretation and analysis of the image information stems from mixed voxels which embody both adipose and muscle tissue, or muscle and bone tissue, or tissue and air. Therefore, a calibration with dissection LMW is required.

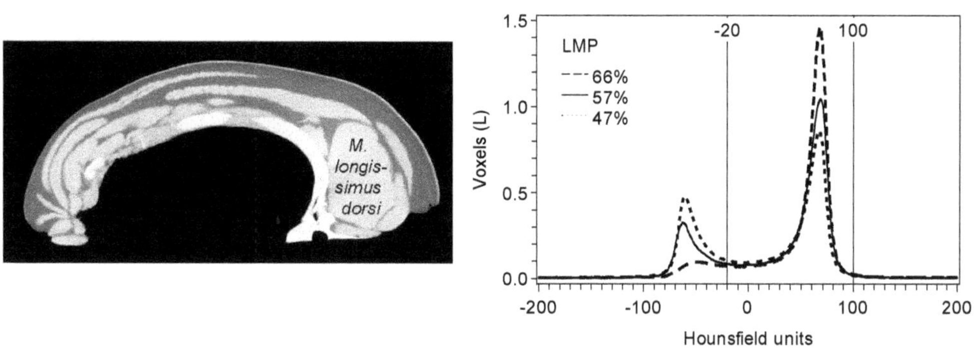

Figure 1: Left: Transversal grey-scale CT image of loin/belly. Right: Distribution of voxel-volume for discrete X-ray attenuation values (for three half-carcasses of 51 kg each, but with different LMP; the range -20 – 100 HU is optimal for the calibration of the PLS-model).

PLS is a regression method especially suited for the analysis of spectral data, and the distributions of voxel-volume *vs.* HU values can be treated as such. In this study, both cross-validation and test-set-validation have been used to identify an optimum model. For PLS, variables (i.e. volume per carcass per HU value) are centered and standardized. For the computation of estimated LMW from HU spectra, re-transformed unweighted regression coefficients have to be used (Figure 2, left).

Results and Discussion

The optimum fit was obtained when the range of HU values was restricted to -20 – 100 (Figure 2). Extending the range to lower or higher values increased the validation-RMSE determined by cross-validation. Values around 70 HU had the most influence on predicting LMW. But values of 85–90 HU had the highest raw correlation coefficients, which reflects their higher density. Negative regression coefficients at the extremes (<0 HU, >90 HU) stabilize the PLS predictions. This model proved also the best for an independent test data set, for which a sample of 19 carcasses was chosen (Figure 2, right). These test sample carcasses had been used for an initial evaluation of CT scanning parameters. They were selected from the same abattoir and with a similar stratification as the calibration data set.

The PLS regression coefficients display some erratic variation in the lower half and at the upper end of HU values (Figure 2, left). This is an expression of too little differentiation of HU spectra in these ranges, and of insufficient information from the sample to produce a continuous distribution that is to be expected theoretically. But this erratic result can easily be smoothed without any loss of prediction accuracy (Figure 2).

Lean meat percentage (LMP) was computed from LMW in relation to the weights of the dressed half-carcasses. There was a good fit between LMP from dissection and from CT analysis: the absolute RMSE was 0.5%-points, which corresponds to a relative RMSE of 0.9% of average dissection LMP (58.1%). Thus, the accuracy of determination of lean meat percentage with a CT is practically the same as with the reference dissection method (Nissen *et al.*, 2006).

Conclusions

High resolution CT scanning is suited to differentiation of carcass tissues with high precision. We conclude that CT scanning with a defined protocol, combined with well-calibrated PLS-regression, provides an applicable alternative to manual dissection. Smoothing of the PLS model allows the formulation of a general, robust relationship between Hounsfield spectra and lean meat content of carcasses.

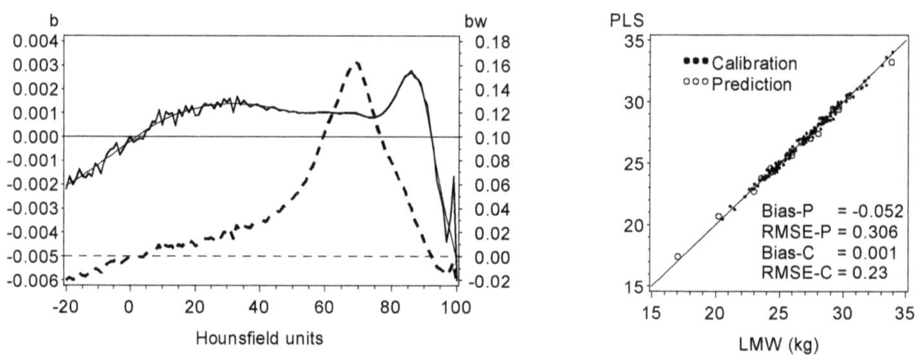

Figure 2: Left: PLS-model for -20 – 100 HU with 4 PC axes. The PLS model is optimized for centered and standardized data and produces weighted regression coefficients *bw* (dashed). The re-transformed raw regression coefficients *b* can be smoothed (solid curves). Right: Fit of the smoothed PLS model for the calibration data set (n=136) and the test data set (n=19); *LMW* — lean meat weight from dissection; diagonal — target line (identity).

References

Dobrowolski, A., Romvári, R., Allen, P., Branscheid, W., Horn, P. (2004). Schlachtkörperwertbestimmung beim Schwein — Röntgen-Computertomographie als mögliche Referenzmethode. Fleischwirtschaft, 84: 109–112.

Judas, M., Höreth, R., Dobrowolski, A. (2005). Computed tomography as an analytical method for pig carcasses. Mitteilungsblatt der Fleischforschung Kulmbach, 44, Nr. 169: 145-151.

Nissen, P.M., Busk, H., Oksama, M., Seynaeve, M., Gispert, M., Walstra, P., Hansson, I., Olsen, E. (2006). The estimated accuracy of the EU reference dissection method for pig carcass classification. Meat Science, 73: 22–28.

MID-INFRARED TRANSMISSION SPECTROSCOPY TO QUANTIFY INTRAMUSCULAR FAT, PROTEIN AND DRY MATTER CONTENT IN PIG MEAT

L. Enfält*[1], G. Zamaratskaia[1], V. Olsson[1], L.-O. Sjaunja[2] and K. Lundström[1]

[1]Dept of Food Science, Box 7051, SLU, SE-750 07 Uppsala, Sweden. [2]MIRIS AB, Uppsala, Sweden Email: lotta.enfalt@lmv.slu.se

Keywords: MIR spectroscopy, intramuscular fat, protein, pork

Introduction

Protein and intramuscular fat (IMF) contents are closely associated with meat quality and nutritional value and largely affects tenderness of meat. Pork with high content of IMF is expected to produce more desirable meat for consumption. Several methods have been developed to analyse IMF, the simplest of which is visual evaluation to obtain fat content. This method is often used to evaluate marbling (Stoller *et al.*, 2003). However, the method is subjective and the accuracy depends on the individual ability of the surveyor to describe the samples. In contrast, IMF determined by chemical methods is an objective measurement that quantifies the total fat content within the muscle. One important indicator of nutritional value of meat is its protein content, and total protein is widely determined by Kjeldahl digestion. Mid-infrared spectroscopy plays an important role in food analysis, and the applications of this technique have considerably increased (Wilson and Tapp, 1999). The objective of this study was to evaluate the capability of mid-infrared transmission (MIR) technique to analyse the contents of IMF, protein and dry matter.

Materials and Methods

Muscle samples from *M. longissimus dorsi* (LD) and *M. semimembranosus et aductor* (SMA) were obtained from female pigs at a live weight of approximately 107 kg. Muscle samples, approximately 10 g in duplicate, were diluted with LOSsolver MEAT (LOSAB, Uppsala, Sweden) and homogenized with LOSmixer (LOSAB, Uppsala, Sweden) for 3 min. Then, 500 µl of antifoam solution was added into each tube and mixed for another 2-3 sec. The homogenates were kept in a water bath at 45° C, for 60 min to remove foam followed by filtration through a 60-µm sieve. The samples were analyzed for fat, protein and dry matter with a mid-infrared instrument (FMA2001 Farm Milk analyzer). To evaluate extraction repeatability, two samples with low (0.82 ± 0.15) and high (1.53 ± 0.12) fat content were extracted 8 times, and analysed for fat, protein and dry matter. The MIR performance was evaluated by repeated measurement of the same extract.

Reference methods included the following. IMF was measured by an established reference method of Schmid, Bondzynski, and Ratzlaff (SBR), which uses hydrolysis with HCl and extraction with petroleum ether: diethyl ether (1:1 v/v). Nitrogen was determined by the Kjeldahl digestion by a Kjeltec Auto System (Tecator AB, Höganäs, Sweden) and a factor of 6.25 was used to convert the nitrogen content to total protein content. To determine dry matter content, the samples were dried in 105°C for 16 h and heated in an ash oven at 550°C for 4 h.

Results and Discussion

The repeatability of the extraction, calculated as the relative standard deviation (R.S.D.) of the repeated analysis of two samples, was sufficient for analysing fat, protein and dry matter contents (Table 1). The repeated measurements of the same extract demonstrated a good repeatability of the MIR technique. Variations (R.S.D.) in the measurements were 10.1 – 11.6% for IMF, 0.7 – 2.3% for protein, and 1.4 – 2.8% for dry matter contents.

Table 1: Intra-assay variations.

	Sample 1			Sample 2		
	Mean	SD	RSD, %	Mean	SD	RSD
Fat	0.82	0.15	18.4	1.53	0.12	7.76
Protein	21.92	0.29	1.33	22.80	0.33	1.44
Dry matter	24.43	0.55	2.27	25.47	0.59	2.30

The relationship between IMF contents obtained by MIR spectroscopy and reference method is presented in Figure 1a. Comparison of individual mean values demonstrated that MIR technique often underestimated the corresponding value detected by reference method. This should be taken into consideration if MIR is to be used for routine analysis of IMF content. The measurements of dry matter obtained using the two methods were moderately correlated (r=0.55). Equation of regression was y=0.334x + 16.89, where x is the dry matter content obtained by MIR.

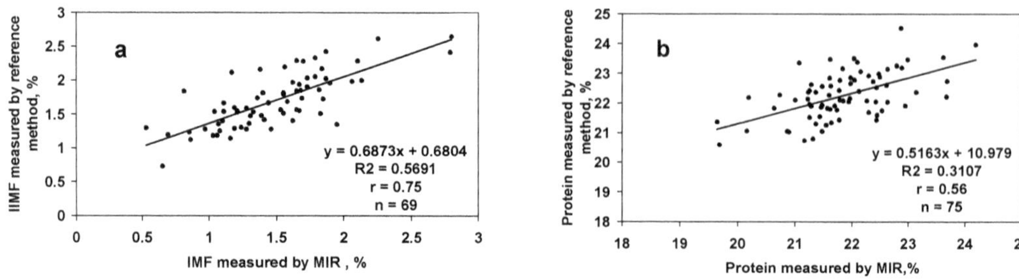

Figure 1: Comparison of IMF (a) and protein (b) content determined by MIR technique and reference method.

The relationship between protein content obtained by the two methods, was quite poor (Figure 1b). It should be noted, however, that in the present study the range of the measured protein was limited and therefore it was problematic to find a good statistical agreement between these two methods. Additionally, the Kjeldahl method is based on the measurements of nitrogen; such measurements likely contribute to uncertainty when converting the nitrogen content to total protein.

Various methods were developed to measure meat composition. Near-infrared spectroscopy (NIR) was shown to be an attractive technique to determine IMF with sufficient accuracy (Monin, 1998; Savenije *et al.*, 2006). Recently, Altmann and Pliquett (2006) described a method to predict IMF in beef and pork using impedance spectroscopy. The MIR method described here is a new attempt to develop a simple, rapid method to evaluate meat composition. Originally, the infrared instrument used in the present study (FMA2001 farm milk analyzer) was designed to analyse milk composition. Elvingson and Sjaunja (1992) successfully adapted a commercial milk analyzer (Milko-Scan, Multispec) to determine fat, protein and dry matter content in fish muscles. The present study demonstrated that this technique can also be used to evaluate those components in pig muscles. Compared with the reference methods, the MIR method has several advantages that make it attractive for routine purposes. Simplicity and rapidity of this method can considerably reduce the amount of time and labour needed to obtain the data. Three important components of muscle chemical composition are measured simultaneously, and the same extraction procedure is involved. Additionally, the procedure is comparatively cheap and does not involve the use of toxic extraction solvents. However, full validation of the proposed method is needed before it can be used for research purposes.

Conclusion

In the present study, the contents of IMF, protein and dry matter in pig muscle were analysed using MIR technique. This preliminary study demonstrated that the method is simple and repeatable, and can potentially be used to routinely analyse meat composition. Prior to this, more complete investigations are needed.

References

Altmann, M. and Pliquett, U. (2006). Prediction of intramuscular fat by impedance spectroscopy. Meat Science, 72: 666-671.

Elvingson, P. and Sjaunja, L.O. (1992). Determination of fat, protein and dry matter content of fish by mid-infrared transmission spectroscopy. Aquaculture and Fisheries Management, 23: 453-460.

Monin, G. (1998). Recent methods for predicting quality of whole meat, Meat Science, 48: S231–S243.

Savenije, B., Geesink, G.H., van der Palen, J.G.P. and Hemke, G. (2006). Prediction of pork quality using visible/near-infrared reflectance spectroscopy. Meat Science, 73: 181-184.

Stoller, G.M., Zerby, H.N., Moeller, S.J., Baas, T.J., Johnson, C. and Watkins, L.E. (2003). The effect of feeding ractopamine (Paylean) on muscle quality and sensory characteristics in three diverse genetic lines of swine. Journal of Animal Science, 81:1508-1516.

Wilson, R.H. and Tapp H. S. (1999). Mid-infrared spectroscopy for food analysis: recent new applications and relevant developments in sample presentation methods. TrAC Trends in Analytical Chemistry, 18: 85-93.

EFFECTS OF PH- AND TEMPERATURE DECLINE CURVES ON PORK QUALITY DETERMINED UNDER DIFFERENT COMMERCIAL CHILLING CONDITIONS

E. Kurt and R.E. Klont*

PIC Deutschland, Ratsteich 31, D-24837, Schleswig, Germany. Email:Ronald.Klont@pic.com

Keywords: pork, drip loss, carcass chilling, post mortem pH, temperature decline

Introduction

Pork quality is determined by the interaction between genetic and environmental factors. The genetic make-up of a pig, the entire system of live animal production and handling through initial chilling and finally storage and handling of the meat and case-ready products all play significant roles in influencing drip loss of pork (Andersen, 2000). Pork quality characteristics like drip loss and colour are determined by the interaction between postmortem pH and temperature in the muscle. High postmortem muscle temperature in combination with a low pH can cause denaturation of muscle proteins and a decrease in the electrostatic repulsion between myofilaments. The subsequent increase in light scattering properties and the extent of lateral shrinkage of the myofibrils provokes the meat to become pale, soft, and exudative (Offer, 1991). The rate of postmortem pH decline is determined by preslaughter stress conditions and the presence of the halothane gene (Cassens *et al.,* 1975; Klont, 1994). A rapid temperature decline in combination with a slow pH decline is essential to produce high quality pork. Rapid chilling of pig carcasses is a very important aspect of improving pork quality for of two reasons. First of all the direct effect on carcass temperature leads to less muscle denaturation. Secondly, quickly lowering the temperature postmortem will decrease the velocity of all biochemical reactions taking place in the muscle after slaughter, thereby also decreasing the rate of pH decline (Savell *et al.,* 2005). The objective of this study was to review the effects of carcass pH and temperature decline curves, in two commercial slaughterhouses with distinct differences in carcass chilling, on pork quality characteristics.

Materials and Methods

Two batches of 60 pigs of two different genetic backgrounds were delivered to slaughterhouse A and B during two different slaughter days. All pigs were offspring of Pietrain boars. The difference in genetic background was related to the percentage of Duroc (0 or 50%) in the sow. Each batch consisted of 30 female pigs and 30 castrates. Hot carcass weight for each individual carcass was collected. The pH at 45 minutes, 3, 6, and 24 hours after slaughter was measured in the loin of each carcass at the height of the 3rd and 4th rib. Temperature probes were inserted in loin muscles of six carcasses before the chilling process. The day after slaughter one loin per carcass was collected for measurements of ultimate pH, and Minolta colour values. A boneless loin sample of more than 100 grams was taken from the blade end of the loin, which was weighed in and stored in a case ready meat tray that contained a meat juice absorbing layer. The loin samples were weighed after 48 hours to determine the drip loss percentage. Data were analyzed using the GLM procedure of SAS. Fixed effects included in the model were: day of slaughter (day 1 or 2), genotype, sex and the interactions between slaughter day and genotype. Carcass weight was included as a covariate for all traits.

Results and Discussion

Slaughterhouse A exposed carcasses to a 25 minute cooling with a higher air flow before storage in the cooler at 2 to 4°C. The fastest chilling regime was applied by slaughterhouse B that started the chilling process after 45 to 50 minutes with a 2½ hour blast chilling of effectively -14°C after which the carcasses were kept at 2 to 3°C until cutting at 24 hours postmortem. In Figure 1 the pH decline curves are extrapolated based on the 4 postmortem pH measurements, and compared with the measured carcass cooling profiles. We arbitrarily defined a critical time zone, which is based on the time it takes to get the average pH below 5.8 after slaughter. Each graph has an arrow that shows the time after slaughter when the muscle temperature gets below 30°C. Our assumption is that the further this arrow reaches into the critical time zone the better the pork quality will be, because of the importance of the relationship between postmortem muscle temperature and pH. Low pre-slaughter stress levels and a fast chilling and cooling regime of slaughterhouse B created a relatively high starting pH at 45 minutes and a slow pH decline curve. The slower pH decline curve of slaughterhouse B compared with A will at least in part be caused by the more extreme chilling and cooling regime as described by Savell *et al.,* (2005). The consequences for the pork quality traits are shown in Tables 1 and 2. In general the best meat quality was produced at slaughterhouse B with the lowest drip loss % (2.13% ± 0.70, compared to 2.81% ± 1.12 for abattoir A) and darkest colour with an average L* value of 52.5 ± 4.0 (compared to 54.4 ± 3.9 for abattoir A). Bertram *et al.,* (2001) also found increased water holding capacity in pork with faster chilling rates.

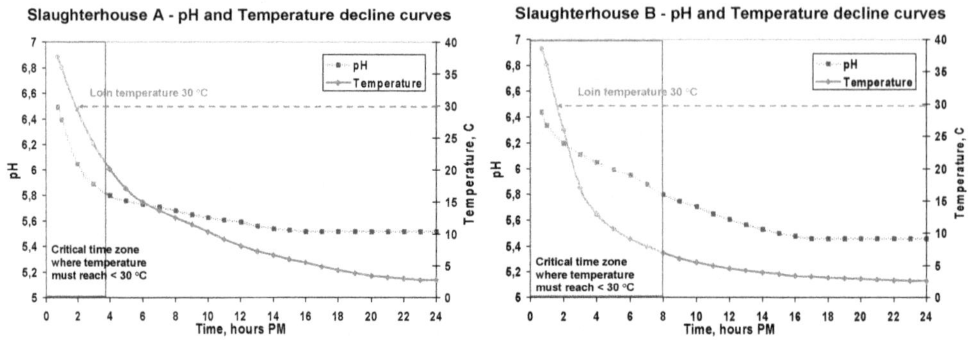

Figure 1: The pH and temperature profiles of pigs processed at 2 different slaughterhouses (A, B).

Table 1: Means and SD of meat quality measurements carried out at abattoir A with two different genotypes.

Variables	Day 1		Day 2		Significance level P		
	Genotype 1	Genotype 2	Genotype 1	Genotype 2			
	Means ± SD	Means ± SD	Means ± SD	Means ± SD	Breed	Day	Breed x Day
N =	56	58	60	59	P =	P =	P =
Carcass weight kg	91.4 ± 5.2	90.8 ± 5.7	93.4 ± 6.6	88.7 ± 6.1			< .01
L* value	55.6 ± 3.9	52.8 ± 3.9	54.4 ± 4.1	54.6 ± 3.0			<. 001
A* value	9.1 ± 1.4	9.9 ± 1.5	8.4 ± 1.8	9.3 ± 1.8	<. 0001	<. 01	
Drip loss %	2.30 ± 0.82	2.30 ± 0.70	2.94 ± 1.18	3.67 ± 1.13			< .01

Table 2: Means and SD of meat quality measurements carried out at abattoir B with two different genotypes.

Variables	Day 1		Day 2		Significance level P		
	Genotype 1	Genotype 2	Genotype 1	Genotype 2			
	Means ± SD	Means ± SD	Means ± SD	Means ± SD	Breed	Day	Breed x Day
N =	62	58	59	59	P =	P =	P =
Carcass weight kg	93.7 ± 5.6	89.2 ± 5.2	95.1 ± 6.5	88.9 ± 4.5	< .0001	n.s.	
L* value	53.9 ± 2.8	51.6 ± 3.3	54.3 ± 3.4	49.9 ± 8.1			< .05
A* value	6.8 ± 1.0	6.4 ± 1.4	7.6 ± 1.2	8.1 ± 1.4			< .01
Drip loss %	1.99 ± 0.64	2.32 ± 0.65	2.20 ± 0.75	2.02 ± 0.71			< .01

Conclusions

The results confirm the importance of fast chilling for good pork quality by both reducing carcass temperature and slowing down the rate of pH fall. Based on these results we recommend that as a general rule the temperature at 2 hours post mortem in the loin should be below 30 – 32 °C.

References

Andersen, H. J. (2000) What is pork quality? In: Quality of meat and fat in pigs as affected by genetics and nutrition. EAAP Publication No. 100, 15-26. Zurich, Switzerland.

Bertram, H.C., S. Donstru, A.H. Karlsson, H.J. Andersen, and H. Stodkilde-Jorgensen, (2001). Post mortem energy metabolism and pH development in porcine M. longissimus dorsi as affected by two different cooling regimes. A 31P-NMR spectroscopic study. Magnetic Resonance Imaging, *19*, 993-1000.

Cassens, R.G., D.N. Marple, and G. Eiekelenboom (1975). Animal physiology and meat quality. Adv. Food Res. *21*, 71-112.

Klont, R.E. (1994). Effects of preslaughter stress factors on muscle metabolism and meat quality. Studies on anaesthetized pigs of different halothane genotypes. PhD thesis, University of Utrecht, The Netherlands

Offer, G. (1991). Modeling of the formation of pale, soft, and exudative meat: effects of chilling regime and rate and extent of glycolysis. Meat Science, *30*, 157-171.

Savell, J.W., S.L. Mueller, and B.E. Baird (2005). The chilling of carcasses. Meat Science, *70*, 449-459.

QUALITY EVALUATION OF DIFFERENT TYPES OF DRY CURED HAM BY AN ELECTRONIC NOSE

J. González-González*, A. Camarero, I. Jaime and J. Rovira

Departamento de Ciencia y Tecnología de los Alimentos, Universidad de Burgos, 09001Burgos, Spain. Email: javigonzalez77@hotmail.com or pechuno@yahoo.es

Keywords: dry cured ham, electronic nose, Serrano ham, volatile compounds

Introduction

Nowadays, consumer trust is very important when you want to consolidate and situate a product into the market. For this reason, new techniques such as the Electronic Nose (e-nose) are available to safeguard the quality of food products. Dry cured ham is a product typically consumed in the Iberian Peninsula which has a high popular acceptability. For example, we understand "Serrano Ham" (a ham salted and dried for more than 210 days) can be displayed boned or deboned. In this type of product during storage rancidity and acid odours will increase significantly because of proteolysis and lipid oxidation (Monahan *et al.,* 1998). It is important to find easy, quick and cheap techniques to evaluate the quality of these products before consumer purchase.

In this paper an electronic nose is used to analyse volatile compounds of a wide variety of samples of "Serrano" Ham to assess if the odour profile provided by e-nose is a good approximation to the volatile sensation perceived by the nose, and if hams with different olfactory characteristics could be discriminated by e-nose.

Materials and Methods

Sample preparation:

Thirty batches of different Serrano ham with high quality and different origin were analysed. Samples of 1 g were placed in 10 ml headspace vials. The numbers of replicates were six for each ham.

It is important that the sampling was homogeneous and that the fat content of the samples was not too high, as samples with a high fat content inhibit volatile release into the headspace (Taylor *et al.,* 2000) and this can be an impediment for reproducibility.

Headspace sampling:

The analyses were performed by an electronic nose αFOX 4000 (AlfaMOS, Toulouse, France) with a sensor array of 18 metal oxide sensors. The vials are incubated in an oven at 50°C for generating the equilibrated headspace agitation cycles (5 s on and 2 s off) and 500 rpm of agitation speed were applied.

The temperature of the syringe during injection was 60°C and the injected volume was 1500 µl with a speed of 1500 µl/s and 120 s flushing time. The carrier gas was synthetic air with a flow of 150 ml/min.

Acquisition data:

The e-nose was controlled by AlphaSoft version 9.1 software and it took readings each 0.5s during 120s of acquisition time and 600s of acquisition delay. This software permit to execute a Principal Component Analysis (PCA). The sensor responses were performed with a statistical program STATGRAPHICS Plus to Windows ver.5.1.

Results and Discussion

In a sensory odour analysis of samples it was detected that some batches had an anomalous smell (acidity, rancidity and different types of taints).

Data from 172 samples obtained as sensor responses were statistically analysed. All analyses were performed with e-nose software and statistical program STATGRAPHICS Plus to Windows ver.5.1.

Principal Component Analysis of the results extracted by the e-nose separated three easily recognizable groups (Figure 1).

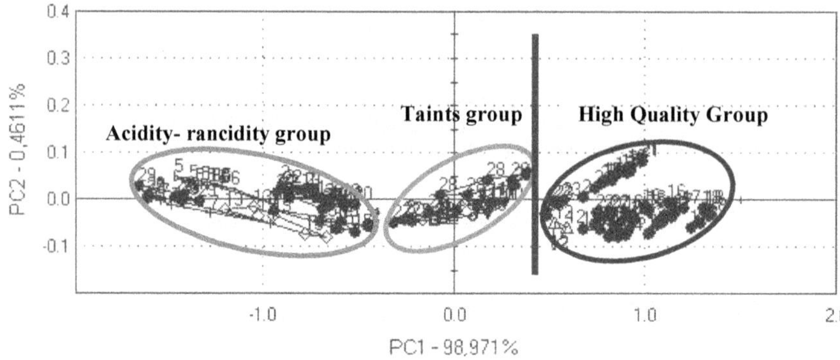

Figure 1: Distribution by PCA of 30 different samples of Serrano Ham with different quality.

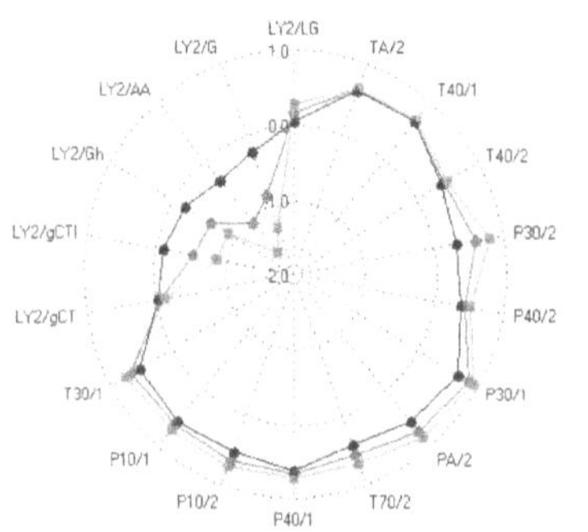

Figure 2: Average of sensor responses for each different group.

Analysis by STATGRAPHICS Plus indicated three groups (Figure 1). In this case the purpose was to obtain a small number of linear combinations of the 18 variables which account for most of the variability in the data. With this method, as shown in Figure 1, 98.97% of the variability in the original data is explained. The odour fingerprint of each detected group, resultant of the average for each sensor response group, is shown in Figure 2. It is remarkable that although most of sensors contribute to the discrimination of hams with defective and optimum odours, a small number of sensors seem to be more sensitive to the volatile compounds responsible of these odour differences (LY2/gCTI, LY2/Gh, LY2/AA, LY2/LG and P30/2)

In both figures the response for faulty hams with acidity and rancidity odours is pointed out in green; in red for faulty hams with different kind of taints; and finally, the hams with high quality sensorial properties are marked in black.

Conclusions

The electronic nose is a suitable tool to evaluate the quality and preservation of Spanish "Serrano" ham. It allows easy discrimination between hams with defective and optimum properties. Therefore, this instrument could be useful in deboned and sliced ham quality control. However, the sampling method must be optimized for application in boned hams.

References

Monahan, R.L., Brunton, N.P., Cronin, D.A. and Durcan, R. (1998). Determination of hexanal in cooked turkey using solid phase microextraction (SPME)/GC. 44th International Congress of Meat Science and Technology (ICoMST).Congress Proceedings, vol. I, 586-587.

Taylor, A.J., Linforth, R.S.T., Baek, I., Brauss, M., Davidson, J., Gray, D.A. (2000). Flavor release and flavor perception. Flavor Chemistry Industrial and Academic Research. Ed. Risch, S.J., Ho, C.T. ASC Symposium Series 756, Washington DC, 151-165.

PREDICTION OF BOVINE MEAT TENDERNESS USING IMAGE ANALYSIS TECHNIQUE

S. Abouelkaram[1], M. El Jabri*[2], J.-L. Damez[1], D. Roux[2] and B. Picard[1]

[1.] *INRA, Centre de Theix, F-63122 Saint Genès-Champanelle, France.* [2.] *Laboratoire de Mathématiques, Université Blaise Pascal.63000 Clermont-Fd, France. Email: Mohammed.Eljabri@math.univ-bpclermont.fr*

Keywords: image analysis, tenderness, connective tissue

Introduction

Tenderness is the most important meat quality attribute for consumers. Many studies have identified the muscle components which are responsible of tenderness while its inconstancy is still unexplained (Hedrick *et al.*, 1994). The muscle connective tissue plays a major role in meat tenderness and presents important variations in its constituent content and in its spatial organisation. It is mainly composed of collagen, whose content and distribution has an important effect on the background toughness of meat. Tools of meat quality prediction based on image analysis have been studied by several authors. Connective tissue information was obtained using image analysis to predict collagen content (Abouelkaram *et al.*, 2003). Lu *et al.*, (1998) have used colour and marbling image features in models of tenderness prediction which were found to be limited. Li *et al.*, (1999) used the same features and added image texture characteristics in their models to obtain R^2 values up to 0.70 for predicting beef tenderness.

The goal of our study was to develop prediction tools of meat tenderness using image analysis techniques. Our results show that tenderness is significantly related to connective tissue characteristics. Indeed we extracted from images of bovine muscle parameters related to connective tissue which were used in models of tenderness prediction.

Materials and Methods

Animals: the assays were performed on samples of semitendinosus (ST) muscle of young bulls from four French breeds: Aubrac (n=17), Charolaise (n=19), Limousine (n=18) and Salers (n=19). The animals were 15, 19 and 24 months old. The experimental design was described in Jurie et al. (2005).

Sensory analysis: sensorial qualities (tenderness, juiciness, flavour intensity) after 14 days of ageing were determined by experimented (QUERY; do you mean experienced?) panellists as described in Drandsfield et al. (2003).

Imaging material and method: Imaging bench is build with a black and white CCD video camera (Sony MACC77) remote-controlled by a PC, an UV and a polarised white lighting (Figure 1).

The images, of which size was 530 x 521 pixels, were processed following a threshold method for the objects extraction. A segmentation algorithm was used to obtain areas of the objects identified as belonging to the connective tissue network. A classification of these objects according their size allowed obtaining the image parameters used to characterise the meat samples.

The kept parameters hold the information of connective tissue organisation in term of size distribution of the connective network constituents (perimysium). The parameters were extracted from meat sample images acquired with white (20 variables) and UV (20 variables) lightings. The tenderness factor was analysed with data issued from a combination of the two types of variables (n=40).

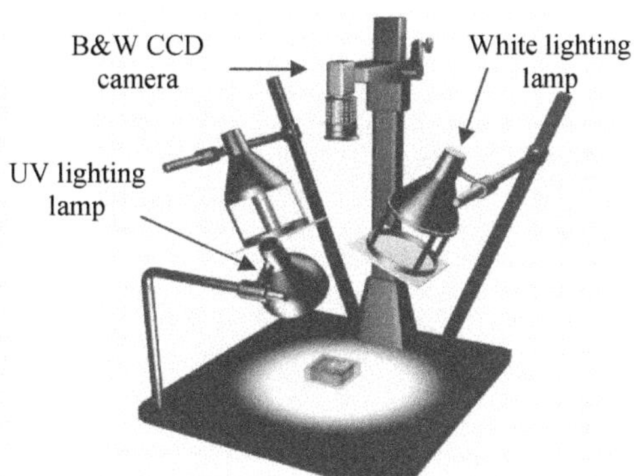

Figure 1: Imaging bench.

Statistical analysis: The whole data was analyzed using the SAS software. The prediction models of the tenderness, built with the image parameters, were determined by an approach based on multiple linear regressions (MLR). A step of parameter selection was performed in order to keep the most relevant variables for the MLR. The final model retained for tenderness prediction is determined according to a maximization criterion of R^2 ($p<0.05$).

Results and Discussion

In this study we compared the various results of statistical analysis carried out on the tenderness. This factor was explained according to the age of the animals.

The best results of regression were obtained by the combination of the two types of parameters. Table 1 gives the R^2 obtained by using the first six variables which maximize the R^2. The choice of the number of parameters was based on a cross validation criterion (Leave one out) in order to stop introducing additional variables in the regression model. Figure 2 gives charts of the values estimated by the models of regression according to the measured values, for the samples of the three animal ages.

Table 1: Coefficients of determination of the regression models for the three ages.

Results obtained with the combined parameters		
	R^2 (%)	R^2 (%)
Age (month)	Model	Validation
15	83.58	69.97
19	87.70	76.74
24	76.08	58.10

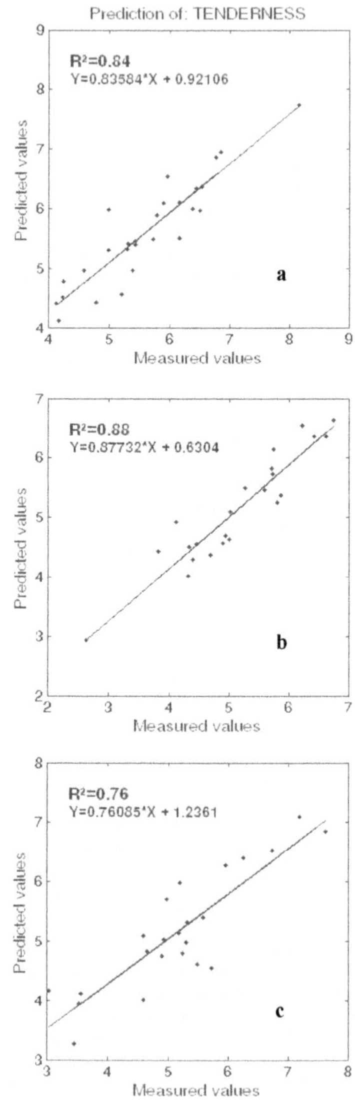

Figure 2: Predicted vs. measured values of tenderness of young bulls' meat. Curve 'a' corresponds to animals which were 15, 'b' 19 and 'c' 24 months old.

Conclusions

Results presented here show that a good estimate of the tenderness was achieved using these models. The image parameters which were used in these models give information on the distribution of the connective tissue. They depend partially on the method used for segmenting the images. Obviously these models can be improved by optimizing classification of object sizes and image segmentation method.

References

Abouelkaram S., Berge P., Hocquette J. F., Culioli J. and Listrat A. (2003). Image study analysis of the relationship between collagen content and distribution of the perimysial connective network in a bovine muscle. *Sciences des Aliments*, 231, 166-170.

Dransfield E., Martin J.F., Bauchart D., Abouelkaram S., Lepetit J., Culioli J., Jurie C., Picard B. (2003). Meat quality and composition of three muscles from French cull cows and young bulls. Animal Science, 76, 387-399.

Hedrick, H. B., Aberle, E.D., Forrest, J.C., Judge, M.D., and Merkel, R.A. (1994). *Principles of meat science.* Dubuque, Iowa: Kendal / Hunt publishing company

Jurie C., Martin J.F., Listrat A., Jailler R., Culioli J., Picard B. (2005). Effects of age and breed of beef bulls on growth parameters, carcass and muscle characteristics. Animal Science, 80, 257-263.

Li. J., Tan, J., Martz, F., and Heymann, H. (1999). Image texture features as indicators of beef tenderness. Journal of Meat science, 53, 17-22.

Lu, J., Tan, J., Gao, X., and Gerrard, G.E. (1998). USDA beef classification based on image processing. 1998 ASAE Mid-Central Conference (paper no. MC98131), St. Joseph, MI: ASAE.

THE USE OF NEAR INFRARED REFLECTANCE SPECTROSCOPY (NIRS) TO PREDICT BOVINE MARONESA BREED MEAT TENDERNESS

J.A. Silva[1], S. Andrés[2], A.L. Soares[3], C. Martins*[1] and A.M. Bruno-Soares[3]

[1]*Universidade de Trás-os-Montes e Alto Douro, (ICETA-UTAD) Departamento de Ciências Veterinária, Apartado 1013, 5000-911 Vila Real, Portugal;* [2]*Universidad de León, Departamento de Producción Animal I, E-24071, León, Spain;* [3] *Instituto Superior de Agronomia, Departamento de Produção Agricola e Animal, Tapada da Ajuda, 1399 Lisboa Codex, Portugal. Email: cmartins@utad.pt*

Keywords: Near Infrared Reflectance Spectroscopy, Maronesa beef meat, tenderness

Introduction

NIRS has the potential to predict different attributes of meat quality quickly and accurately. It allows rapid and frequent measurements, the sample preparation is fast and simple, is suitable for on-line use and for simultaneous determinations of different attributes (Prevolnik *et al.*, 2004).

Since tenderness is the single most important quality parameter of consumer importance (Rødbtten *et al.*, 2000, Leroy *et al.*, 2003), in the context of the application of NIRS to predict the quality of fresh meat, most attention has been focused on the prediction of this parameter (Geensink *et al*, 2003). Tenderness can be assessed as a measurement of meat mechanical resistance (Warner-Bratzler Shear Force, WBSF) which is a destructive and time-consuming method (Rødbtten *et al.*, 2000).

The aim of this study was to examine the accuracy of visible/NIR spectroscopy for the prediction beef quality characteristics such as pH post-mortem, WBSF and sarcomere length using the *longissimus thoracis* et *lumborum* muscle of young Maronesa bulls.

Materials and Methods

Young Maronesa bulls (n=30) aged between 9 and 11 months and with live weights ranging from 90 to 150 kg were slaughtered and selected according to the ultimate pH (pH measured at 24 h post-mortem) in the muscle *longissimus thoracis* et *lumborum*, to obtain an high range of pH values. This muscle was excised 24 hours post-mortem between the 8th rib and 2nd lumbar vertebra and divided into four parts. The first one was used for the laboratory procedures performed during the first day *post mortem* and the three remaining pieces were vacuum packed and aged at 4°C for 3, 7 and 14 days. At the end of each period, one slice (2.5 cm thick) was vacuum packaged and stored (–80°C) for NIRS analysis. Other slices for WBSF and other one for sarcomere length (SL) measurements were taken.

pH was measured directly in the muscle at 3 hours (pH_3) and 24 hours (pH_{24}) post-mortem. SL analyses were performed at day 7 *post mortem*; 4 g of meat was minced and homogenised at low speed (8000 rpm) in a chilled 0.25 M sucrose solution using an Ultra Turrax T25 mixer (Cross *et al.*, 1980-81). The length of 10 consecutive sarcomeres was measured (15 groups of 10 sarcomeres for each sample) using a phase microscope (40x objective). WBSF was measured (meat cooked at 70°C at the core) on 10 sub-samples of 1 cm^2 cross-section with fibres perpendicular to the direction of the blade attached to a Stevens QTS 25 texturometer (Silva *et al.*, 1999).

Prior to NIRS measurements, meat samples were thawed in a fridge for 24 hours, stored in a plastic bag and left at least one hour at ambient temperature. Each sample was trimmed to eliminate connective tissue, two pieces of intact meat of 35 mm diameter were cut parallel to the longitudinal orientation of the muscle fibres (Cozzolino and Murray, 2004) and put inside 35 mm diameter quartz cuvettes with aluminium foil backing and scanned in order to obtain a mean spectrum per sample.

Diffuse reflectance spectra were collected at 2 nm intervals from 400 to 2498 nm using a NIRSystems 6500 scanning spectrophotometer (FOSS NIRSystems, Silver Spring, MD, USA).

Calibration development and validation were performed using WinISI II version 1.02 (Infrasoft International, Port Matilda, PA, USA). Spectral data pre-treatments such as standard normal variate and detrending (SNV-D), multiplicative scatter correction (MSC) and first or second order derivatives were applied. Partial least squares regression (PLSR) with mathematical treatment (2, 12, 2, 2) was used. The accuracy of prediction is given by the standard error of cross validation (SE_{CV}).

Results and Discussion

Measurements of the parameters (Table 1) showed a wide range of variability as a consequence of the aging of meat and the different ultimate pH values of the samples. With regard to the mean pH values, as a consequence of the acidification during post-mortem aging process pH_{24} was slightly lower than pH_3.

Table 1: Range, mean, standard deviation (SD) and coefficient of variation (CV) of beef parameters.

Parameter	n	Mean	Range	SD	CV(%)
pH_3	26	6.64	6.17 – 6.95	0.203	3.06
pH_{24}	27	6.03	5.59 – 6.67	0.330	5.47
SL (μm)	99	1.710	1.51 – 1.84	0.083	4.85
WBSF (kg/cm)	112	10.44	3.85 – 19.88	3.889	37.26

n: number of samples

WBSF showed a mean value of 10.44 kg/cm^2 with a coefficient of variation equal to 37.26%. This variation is due mainly to aging and ultimate pH but the effect of sarcomere length may not be excluded.

Table 2: Prediction of beef parameters by near infrared reflectance spectroscopy.

Parameter	n	SEC	R^2	SE_{CV}	1-VR	RPD
pH_3	26	0.183	0.186	0.210	-0.054	0.97
pH_{24}	27	0.059	0.968	0.104	0.906	3.17
SL (μm)	107	1.87	0.608	2.05	0.532	0.04
WBSF (kg/cm)	112	2.30	0.650	2.67	0.532	1.46

n: number of samples, SEC: standard error of calibration; R^2: coefficient of determination for calibration; SE_{CV}: standard error of cross validation; 1-VR: coefficient of determination for validation; RPD; ratio performance deviation calculated as SD/SE_{CV}.

Currently, limited published information about the ability of NIRS to predict meat quality parameters is available. However, in the present study, absorbance data of meat samples aged during 24 hours showed good correlations with pH_{24} parameter. Therefore, this parameter could be accurately predict by NIRS (Table 2, R^2=0.97 and RPD=3.17). The good repeatability of the reference method in the present study could have contributed partially to the successful predictions of pH_{24} obtained by NIRS (SEL_{pH24}=0.102).
The prediction results for WBSF (R^2=0.650, RPD=1.46) shouldn't be ignored. These statistics are in agreement with that described by Rødbotten *et al.* (2000) showing R^2 values of 0.68.

Conclusions

It must be pointed out that although the beef samples studied in this work arose from a small number animals and samples, some conclusions can be drawn.
The results suggest that VIS/NIRS instruments may predict pH_{24} and have potential to provide information about WBSF.

Acknowledgements

This work was carried out with financial support from PROGRAM AGRO (Ministério da Agricultura, Desenvolvimento Rural e Pescas; União Europeia, Fundos Estruturais).

References

Cozzolino, D., Murray, I. (2004). Identification of animal meat muscles by visible and near infrared reflectance spectroscopy. Lebensmittel Wissenschaft-und Technolnologie. 37, 447-452.

Cross, H.R., West, R.L., Duston, T.R. (1981). Comparision of methods for measuring sarcomere length in beef *semitendinosus* muscle. Meat Science, 5, 261-266.

Geensink, G.H., Schreutelkamp, F.H., Frankhuizen, R., Vedder, H.W., Faber, N.M., Kranen, R.W., Gerritzen, M.A. (2003). Prediction of pork quality attributes from near infrared reflectance spectra. Meat Science, 65, 661-668.

Prevolnik, M., Candek-Potokar, M., Skorjanc, D. (2004). Ability of NIR spectroscopy to predict meat chemical composition and quality – a review. Czech Journal of Animal Science, 49, 500-510.

Rødbotten, R., Nilsen, B.N., Hildrum, K.I. (2000). Prediction of beef quality attributes for early *post mortem* near infrared reflectance spectra. Food Chemistry, 69, 427-436.

Silva, J.A., Patarata, L., Martins, C. (1999). Influence of ultimate pH on bovine meat tenderness during ageing. Meat Science, 53, 453-459.

USING IMAGE ANALYSIS OF TOMOGRAPHIC IMAGES FOR STUDYING CHANGES IN THE SUBCUTANE FAT LAYER DURING GROWTH OF YOUNG PIGS

C. Borggaard*[1], M.T. Madsen[2], F. McEvoy[3], E. Svalastoga[3] and H.H. Thodberg[4]

[1.] Danish Meat Research Institute, Maglegaardsvej 2, DK-4000 Roskilde. [2.] The Department for Nutrition and Reproduction, Danish Pig Production, Axeltorv 3, DK-1609 Copenhagen. [3.] Royal Veterinary and Agricultural University, DK-1870 Frederiksberg C. [4.] Visiana, Sollerødvej 57 C, DK-2840 Holte. Email: clb@danishmeat.dk

Keywords: pork, CT scans, X-Ray tomography, image analysis, subcutaneous fat, growth

Introduction

Computer assisted Tomography (CT) is becoming an important tool in carcass evaluation of pigs, and its potential within this field has been documented in the EUPIGCLASS project. In various studies (Romvari *et al.*, 2006) it has been demonstrated that CT scans of pig carcasses can be used as a less expensive and rapid alternative to total dissection for determining carcass composition. In this study we have used CT images for evaluating the growth pattern in the subcutaneous fat layer (in the following denoted the mantle) of live pigs during growth. Potentially, this will enable us to compare the effects of different feeding strategies on the growth rate of live animals and to determine the sites on the carcass providing optimal information for measuring fat and muscle thickness for on-line carcass classification.

Materials and Methods

A total of 27 live (Danish land race and Duroc) pigs were CT scanned. The mean body weights were 52, 95, and 112 kg. (here referred to as visit 1, 2 and 3). Scans were performed using a standard Siemens Somatom Emotion unit, designed for diagnostic imaging in human patients. When scanning the pigs in vivo, the animals were sedated so subject movement could be reduced during image acquisition. However, respiratory effects did show up as artifacts in the images, thus slightly complicating the following image analysis. The pigs were scanned using a helical scanning protocol, slice thickness 5 mm.

Results and Discussion

In order to demonstrate our methodology we have chosen to present data from a single pig. In figures 1a and 1b are shown tomographic slices at the same anatomical location (at the last rib, often denoted P2) through the same pig at visits 1 and 2. The images have been processed to locate the boundary between the fat mantle and the underlying muscle tissue, thus enabling us to calculate the thickness of the mantle around the full circumference of each slice.

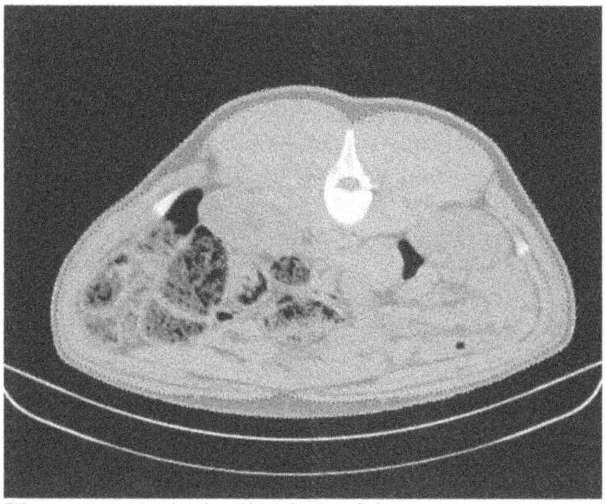

Figure 1A: A single slice from visit 1.

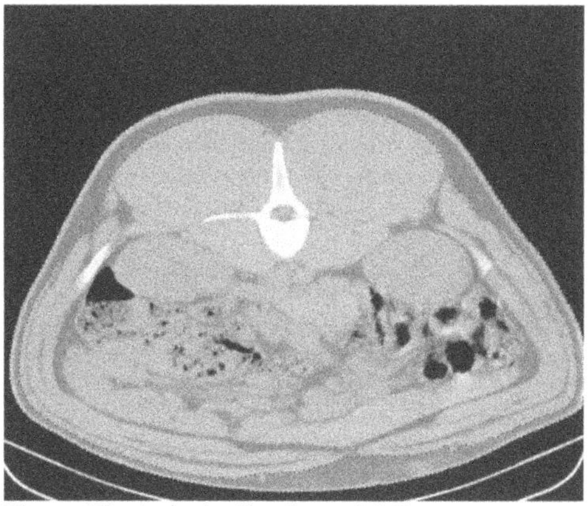

Figure 1B: A single slice from visit 2.

The entire fat mantle of the pig was then unfolded and is shown as a flat matrix representing the fat thickness at locations in both the sagital (longitudinal, parallel to the spine) and the transverse direction (running around the pig on its surface). This is illustrated in Figure 2 where the fat mantle of half the pig body is depicted. The resolution in the longitudinal direction is 5 mm and in the transverse direction 100 data points, having in average a 5 mm separation. The mantles are scaled in the longitudinal direction so that they match at three anatomical fixed points (lateral aspect of pelvis, P2, and 15 vertebrae away from the pelvis in the cranial direction) even though the pig has grown between scanning sessions. The imprint of the Longissimus Dorsi muscle can easily be seen in Figure 2.

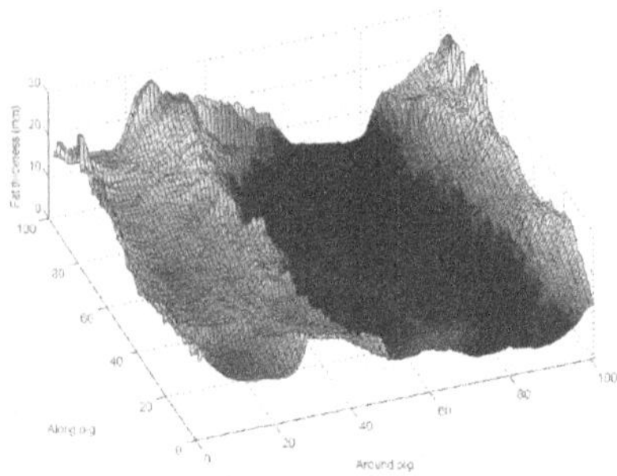

Figure 2: Fat mantle at visit 2.

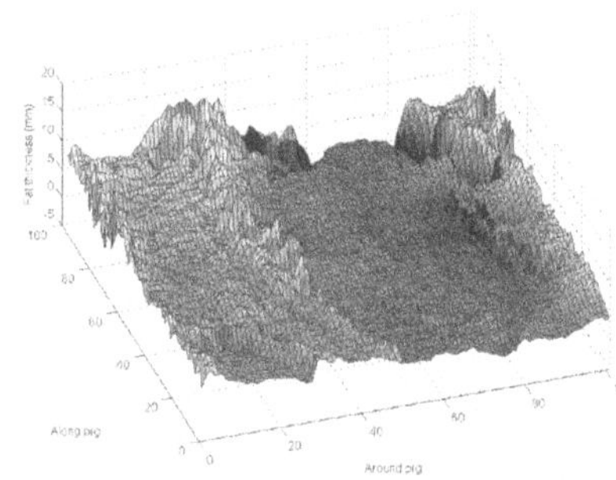

Figure 3: Mantle at visit 2 minus mantle at visit 1.

It is now possible to study in detail at which locations growth takes place in the fat mantle of pigs in the time between e.g. visit 1 and 2. This was done by simple subtraction of the fat mantles measured at these two times. The result is shown in Figure 3.

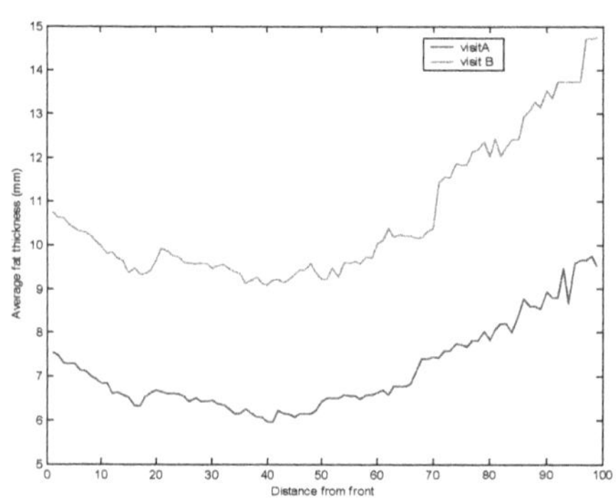

Figure 4.

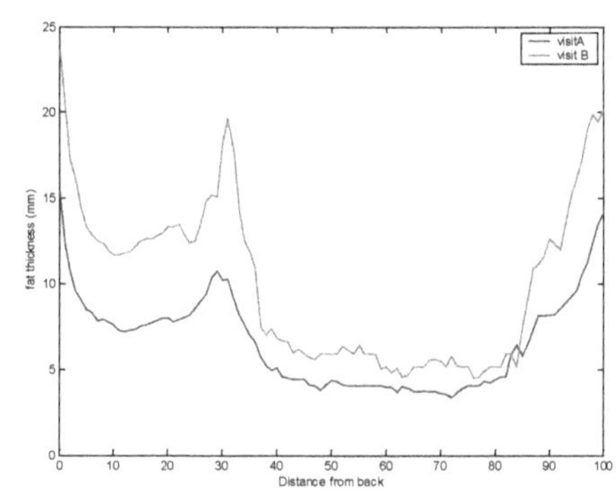

Figure 5.

The fat thickness of the mantle as seen by the CT scanner in visit 1 and 2 is also illustrated in Figure 4 and 5. The lower curve shows the fat thickness in mm of the pig at visit 1 (52 kg) and the upper curve the fat thickness at visit 2 (95 kg). In Figure 4 the x-axis denotes the distance from the fixed point in the front of the animal measured in slice numbers. The thickness is here calculated as an average mantle thickness of each slice. In Figure 5 the thickness of the fat mantle at the P2 slice is shown. The x-axis runs from mid-back to mid-belly. Again the imprint of the Longissimus Dorsi is easily recognized.

Conclusion

Using a single pig that has been scanned several times during growth, we have outlined a method which enables us to study in detail how growth takes place in the fat mantle of pigs.

References

EC Project: G6RD-CT-1999-00127 EUPIGCLASS of the Measurements and Testing activity (Competitive and Sustainable Growth Program).

Romvari, R., Dobrowolski, A., Repa, I., Allen, P., Olsen, E., Szabo, A., Horn, P. (2006). Development of a computed tomography calibration method for the determination of lean meat content in pig carcasses, Acta Veterinaria Hungarica 54(1), pp.1-10.

INVESTIGATING THE BEHAVIOURAL PROPERTIES OF ADIPOSE TISSUES USING CONFOCAL LASER SCANNING MICROSCOPY

D.A. Brüggemann and M.A. Lawson*

Department of Food Science and the LMC Centre for Advanced Food Imaging, The Royal Veterinary and Agricultural University, Frederiksberg C, Denmark. Email: mal@kvl.dk

Keywords: adipose tissue, microscopy, fatty acid composition

Introduction

The fatty acid (FA) composition of foodstuffs defines, in part, the healthiness of a food. Many muscle-based foods have long been criticised for having a high level of the undesirable saturated fatty acids in relation to the more desirable polyunsaturated fatty acids. In porcine systems, the ratios of different fatty acid components can be manipulated by varying feeding regimes but this manipulation can lead to adverse changes in back fat texture, taste and oxidation levels (Warnants *et al.*, 1996; Wood *et al*, 1999).

The composition of FAs in a tissue is the result of a complex biological process (Large *et al.*, 2004). Adipocytes are cells known to be the lipid storage depot for circulating fats in the body. The FAs contained in adipocytes are a mixture of molecules which are either imported into the cells from the blood and lymph or are newly-synthesised within the fat cell.

Using laser-scanning confocal microscopy, we show in this work that we can investigate not only tissue structure, but the behavior of different cells and different levels of maturity, resulting in a better understanding of the properties of adipose tissue.

Materials and Methods

Twenty-one female pigs were slaughtered at an average live weight of 112 kg (± 9 kg). Three days post mortem, samples for immuno-histochemical analysis of subcutaneous fat were taken at the P2 position and immediately frozen in liquid nitrogen. The samples were than stored at -21°C until use.

The actin cytoskeleton was visualized using Rhodamine-Phalloidin (Sigma-Aldrich Inc., St. Louis, USA). A primary monoclonal antibody against Perilipin was obtained from Progen Biotechnik, (Heidelberg; Germany) , and a monoclonal antibody to CD36 was obtained from Cell Science (Canton, MA, USA). Secondary Alexa-conjugated antibodies were purchased from Molecular Probes (Invitrogen, Carlsbad, CA, USA).

Serial sections (60 µm) were cut using a cryostat at -21 °C. Immunolabelling was performed following a standard protocol. Samples were viewed using a confocal laser microscope (SP, Leica Laser Technik GmbH, Heidelberg, Germany) equipped with an argon/krypton laser. During image acquisition, each frame was collected 20 times and averaged. The images shown are a projection of stacks of images in vertical directions at the maximum intensity mode. The resultant grey value of each pixel was the highest value present in the column of pixels in the stack with the same (x, y) position.

Results and Discussion

Existing studies on the activity of adipose tissue use PCR or biochemical methods for analysis. These methods give an inaccurate picture of the tissue as a whole since this adipose tissue is a mixture of cells at different developmental states. Figure 1 shows a typical area of tissue containing both immature brown fat cells and more mature white fat cells.

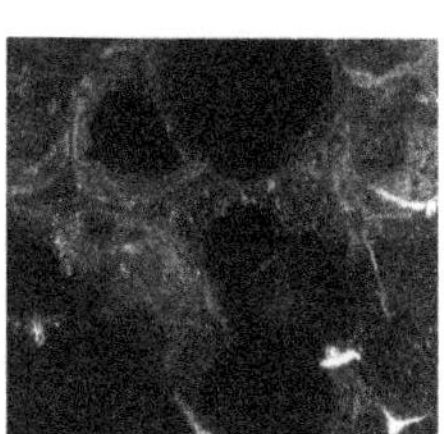

Figure 1: Co-existence of immature and mature adipocytes in porcine tissue at time of slaughter. Arrows – immature adipocyte.

CD36 is a fatty acid transporter in many cell types and is responsible for the uptake of free fatty acids. The level of CD36 on the adipocyte surface and the clustering of the protein on the cell membrane give information about the activity of the transporter. Figure 2 shows a section of porcine adipose tissue labeled for CD36. Punctate staining for this receptor can be seen surrounding a number of the adipocytes.

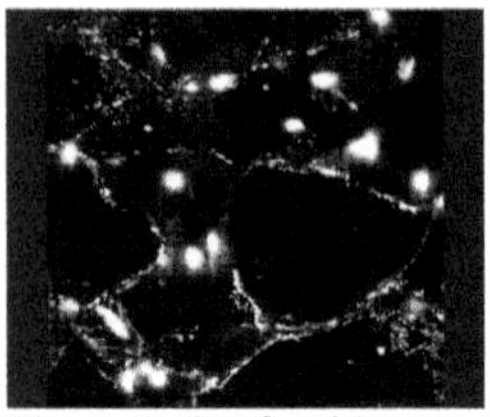

Figure 2: Confocal laser scanning micrograph of the fatty acid transporter CD36 in adipose tissue.

Perilipins are a family of phosphoproteins involved in the transport of lipids out of adipocytes. Perilipins interact with lipases that are responsible for the breakdown and release of triacylglycerides in fat droplets and allow activation of the lipase after hormonal activation. Figure 3 shows the localisation of perilipins on the surface of a white adipocye.

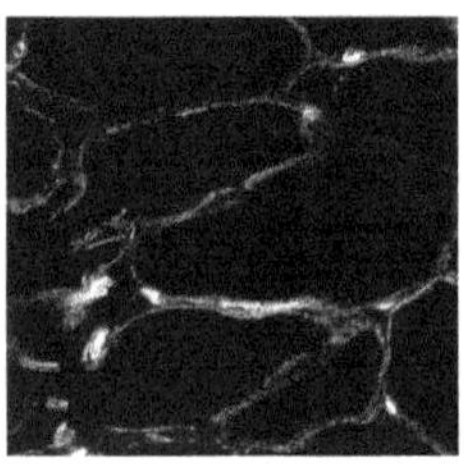

Figure 3: Confocal laser scanning micrograph of perilipins in adipose tissue.

Conclusions

Immunohistochemistry and laser-scanning confocal microscopy are powerful tools to resolve the function of different cells at different maturities in adipose tissue. These techniques can give valuable insights into the properties of adipose tissues with regard to meat eating and processing quality.

References

Large, V., Peroni, O., Letexier, D., Ray, H. and Beylot, M. (2004). Metabolism of lipids in human white adipocyte. Diabetes Metabo, 30, 294-309.

Warnants, N., Monique, J., van Oeckel, M.J. and Boucque, C.V. (1996). Incorporation of dietary unsaturated fatty acids into pork fatty tissues and its implications for the quality of the end products. Meat Science, 44, 125-144.

Wood, J.D., Enser, M., Fisher, A.V., Nute, G.R., Richardson, R.I. and Sheared, P.R. (1999). Manipulating meat quality and composition. Proceeinds of the Nutrition Society, 58,363-370.

NEAR INFRARED REFLECTANCE SPECTROSCOPY IN THE PREDICTION OF NUTRITIONAL AND RHEOLOGICAL QUALITY TRAITS OF *BAUERNSPECK*

S. Balzan*[1], F. Blaas[2], L. Serva[3], M. Mirisola[3], V. Giaccone[1], I. Andrighetto[3], S. Segato[3] and E. Novelli[1]

[1.] Department of Public Health, Comparative Pathology and Veterinary Hygiene, University of Padova, 35020 Legnaro (PD), Italy, [2.] Technical bearing for mountain agriculture, Autonomy Province of Bolzano, 39100 Bolzano (BZ), Italy, [3.] Department of Animal Science, University of Padova, 35020 Legnaro (PD), Italy. Email: stefania.balzan@unipd.it

Keywords: *bauernspeck*, dry cured meat, quality traits, Near Infrared Reflectance Spectroscopy

Introduction

Bauernspeck is a typical smoked and salted meat product of the South Tyroler Region (North Italy), made following the tradition and the custom of this mountain area. It is a pork product made of ham, shoulder, loin-belly or the entire half carcass, with rind and subcutaneous fat. After deboning, primal cuts are cured with salt and spices (frequently with sodium nitrite), smoked and ripened up to 4–8 months. At the end of processing it shows peculiar rheological and gustative characteristics. In order to improve quality control assessment, conventional determinations may be integrated with alternative analytical methods. Near infrared reflectance spectroscopy (NIRs) is a rapid, low cost and non-destructive method of analysis widely used to assess proximate composition of meat and fish (Cozzolino *et al.*, 2002). The aim of this study was to evaluate NIRs performance in the prediction of proximate composition and salt content of *Bauernspeck.* Moreover, for the muscular fraction (*Longissimus dorsi,* LD) the ability of NIRs to predict some ripening quality traits such as pH, water activity, proteolysis index, meat colour and firmness was evaluated.

Materials and Methods

The study examined 120 pigs reared in 2 farms of the province of Bolzano (Italy). Animals were equally divided between castrated and female and consisted of three different crossbreeds. The pigs were slaughtered at a live weight of 132.4±12.8kg. Primal cuts were randomly sent to six local manufacturers where they were submitted to a ripening productive process according to their own traditional method. After a period of 4 months of ripening on average, samples (n=120) of loin (*Longissimus dorsi*, rind and backfat included) were collected as 1.5cm thick whole slice (WS) and as lean cut (*Longissimus dorsi*, LD only). The WS and LD samples were analysed for proximate composition: moisture, crude protein (CP), ether extract and ash according to A.O.A.C. (2000). Salt content (% NaCl) was determined as reported by Meynier *et al.* (1999); in LD samples, non-protein nitrogen (NNP) (Meynier *et al.*, 1999) and instrumental colour (CIE L*a*b*) by means of Minolta RIO-TNE-005 (Minolta Camera Co. Ltd, Japan) were also measured. The pH was measured after homogenisation of sample in distilled water (1/10 w/v); water activity (A_w) determination was performed using an Aqualab CX-2 (Decagon). Firmness was measured with a Carcass Fat Hardness Meter 006. All samples were minced then scanned in duplicate by a FOSS NIRSystem 5000 (1100-2498 nm, gap 2nm). Spectra were recorded as log1/R averaging 32 scans of the sample after 16 scans of the white reference. Original spectra were interpolated every 2nm.WinISI 1.5 was used to acquire spectral data and to perform calibration adopting the modified PLS method of ISI. Spectra math treatments included scatter correction (standard normal variate and detrending), second derivative with gaps and smoothing between 5 data points. Evaluation of performances was based on the SEC standard error of calibration, SECV = standard error of cross validation, 1-VR = one minus variance ratio (Shenk and Westerhaus, 1993).

Results and Discussion

Chemical and rheological parameters considered in this study represented the main quality traits to assess the economic value of *Bauernspeck.* Moreover, A_w, moisture/crude protein ratio and NaCl/moisture ratio have been included in the voluntary regulation of production. Chemical composition of LD and WS are given in Table 1 and 2, respectively. *Bauernspeck* is a meat product characterised by a low moisture content and a fair variability in the fat percentage especially when the WS is considered. As reported in Table 1 for dry-cured LD the value of the R^2 of cross validation (1-VR) ranged from 0.34 to 0.99 suggesting an effective prediction for some of these parameters. The NIRs technique showed good accuracy in the prediction of proximate composition both for lean samples and fatter samples. These results were partially expected because of the low moisture percentage. In fact, high moisture levels (above 50%) influence the wavelength range over which the instrument can provide reliable analysis (Williams, 2002). Data reported in Table 2 confirmed the great accuracy in the prediction of proximate composition. The higher fat content of the WS was not detrimental as evidenced by 1-VR values which were still very high, except for NaCl content. The slight decrease in the prediction of this latter chemical parameter was probably due to the increasing value of fat content that caused the dilution of salt. A high fat content could also produce a lipidic film that probably limited the absorbance of incident light.

Table 1: Performance of calibration for the lean portion of *Bauernspeck* (only *Longissimus dorsi*). SD: standard deviation; SECV: standard error for cross validation; 1-VR: fraction of explained variance; CV(%): (SEVC/mean)x100.

(LD) samples	Mean ± SD	SECV	1-VR	CV(%)
Moisture, %	38.8 ± 6.3	0.1	0.99	0.3
Crude protein, %	48.9 ± 6.3	0.1	0.99	0.3
Ether extract, %	4.8 ± 1.8	0.1	0.99	0.5
Ash, %	7.4 ± 1.4	0.1	0.99	1.8
Moisture/crude protein	0.8 ± 0.2	0.1	0.99	1.5
NaCl, %	6.5 ± 1.2	0.4	0.89	6.5
NaCl/moisture	17.0 ± 4.2	1.1	0.93	6.7
Proteolysis index (NNP/CP)	9.00 ± 1.87	0.80	0.82	8.9
pH	6.13 ± 0.36	0.10	0.92	1.7
Water activity (A_w)	0.85 ± 0.05	0.01	0.98	0.8
Lightness (L*)	31.26 ± 4.97	3.66	0.47	11.7
Redness (a*)	5.02 ± 1.91	1.57	0.34	31.2
Yellowness (b*)	8.03 ± 2.16	2.01	0.14	25.0
Firmness, 0-1000 points	900 ± 90	49	0.72	5.4

Table 2: Performance of calibration for whole slice (WS) *Bauernspeck.*

(WS) samples	Mean ± SD	SECV	1-VR	CV(%)
Moisture, %	25.2 ± 5.8	0.4	0.99	1.6
Crude protein, %	25.8 ± 5.4	0.6	0.99	2.2
Ether extract, %	44.1 ± 11.3	0.8	0.99	1.8
Ash, %	4.4 ± 0.9	0.2	0.95	4.7
NaCl, %	4.4 ± 1.0	0.4	0.84	8.9

Good accuracy in the prediction of rheological parameters such as pH, A_w and proteolysis index was observed. Firmness was characterised by a medium performance magnitude, which probably could be enhanced by increasing the dataset. Poor performances were observed for colour instrumental data (1-VR: 0.14-0.47) attributable to the wavelength range (of the near infrared spectrum) which was unsuitable for this aim.

Conclusions

Results obtained from this research showed that proximate composition, physical and rheological traits of *Bauernspeck* can be successfully monitored by NIRs. Considering the importance of non protein nitrogen associated with moisture, salt, A_w and fat for the quality characterization of dry cured meat products (Schivazappa *et al.*, 2002), our results suggest that NIRs could be used as a rapid instrument to assess their ripening parameters and quality traits too. The determination of salt and fat content and proteolysis index are time consuming; thus the availability of NIRs as a rapid method for their measurement could improve the control of the productive process both reducing time and cost of analysis.

References

AOAC (2000). Official Methods of Analysis – Association of Official Analytical Chemists, (17th Ed.).

Cozzolino D., Murray I. and Scaife J.R. (2002). Near Infrared Reflectance Spectroscopy in the prediction f chemical characteristics of minced raw fish. Aquaculture nutrition, 8, 1-6.

Shenk J.S. and Westerhaus M.O. Analysis of Agriculture and Food Products by Near Infrared Reflectance Spectroscopy. Monograph. Infrasoft Int., Port Matilda, PA, USA. (1993).

Williams P.C. (2002). Implementation of Near-Infrared Technology. In: Near-Infrared Technology in the Agricultural and Food Industries, Chapter 8, 145-170. Edited by P. William and K. Norris.

Meynier, A., Novelli, E., Chizzolini, R., Zanardi, E., and Gandemer, G. (1999). Volatile compounds of commercial salami. Meat Science, 51, 175-183.

Schivazappa C., Degni M., Nanni Costa L., Russo V., Buttazzoni L. and Virgili R. (2002). Analysis of raw meat to predict proteolysis in Parma ham. Meat Science, 60, 77–83.

Acknowledgements

The research was supported by Federazione Zootecnica dell'Alto Adige – Bolzano.

3D MAPPING OF BEEF M. LONGISSIMUS USING NEAR INFRARED SPECTROSCOPY CHANGES AND MAHALONOBIS DISTANCE

P.W. Hansen*[1] L. Streeter[2,3], R. Burling-Claridge[2], M. North[2] and C.E. Devine[4]

[1.] EquatioNZ, Lyngby Hovedgade 11B, 2. th., DK-2800 Kongens Lyngby, Denmark, [2.] Meat Quality, AgResearch, PB 3123 Hamilton, New Zealand, [3.] University of Waikato, Private Bag 3105, Hamilton, New Zealand, [4.] Bioengineering and Biosensors, HortResearch, PB 3123 Hamilton, New Zealand. Email: PWH@equationz.dk

Keywords: beef, shear force, Mahalanobis distance, *longissimus dorsi*

Introduction

Ageing is the term given to the process of meat tenderisation. It occurs over the period following rigor mortis and is measured destructively by shear force measurements after cooking. We have shown that non-destructive measurements such as near infrared (NIR) spectroscopy can be correlated with tenderness although the correlation may not be sufficiently accurate and the question arises as to whether this is due to inherent variability of the meat itself, and if so, how is this variability characterised? (McGlone *et al.*, 2006). The theory of sampling (Petersen *et al.*, 2005) needs to be invoked in order to determine and understand the variability inherent in meat, and determine the appropriate measurement sites for calibration against NIR.

With meat it is clear that one cannot keep subdividing it into lots or fragments or consider the variation as it moves past a measurement device as the measurements are destructive. However it is possible to consider the variation of a muscle along its length and across its width, i.e. 3D mapping, by measuring NIR spectral changes using a combination of Principal Component Analysis (PCA) and Mahalanobis distance (D) (Mahalanobis, 1936). It is suggested to use D as a measure of how representative a given location inside a meat sample is for the whole meat sample. This study complements physical measurements on beef (Janz *et al.*, 2006) and pork (Hansen *et al.*, 2004) that show shear force gradients both across and along the muscles.

The Mahalanobis distance is a distance measure that takes the variability along the axis of the data set into account. In practical terms it fits an N-dimensional hyper ellipsoid around the data set and measures the distance using that as a ruler. This is different from the normal Euclidean distance where all directions have the same influence, independent of their variance. The advantage of D is that it allows even minor variations in the N-dimensional space to contribute to the distance. This is very useful when combining with PCA, which derives the principal components (PCs) according to their covariance.

Materials and Methods

Four pairs of beef *m. longissimus et lumborum* (LL) were removed 30 minutes after slaughter from steers shot with a captive bolt and not electrically stimulated. The LLs were wrapped in polyethylene cling film to prevent shortening (Devine *et al.*, 1999) and were held at 15°C to enter rigor mortis and aged at this temperature for 24 hours. Each of the wrapped muscles, with an approximate circular cross section, were positioned in a holder and a 14mm thick slice was removed from the rump end of the muscle and discarded. Forty-two NIR readings were then made on the face of the muscle in a grid arrangement with a spacing of 14mm. The readings at each grid position were taken using a fibre optic probe (with diameter of approximately 16mm) attached to a diode array spectrophotometer from KES Analysis (NY, USA). The instrument covered the spectral range from 400 to 1700 nm. At grid positions where there was no muscle, readings were omitted from the data analysis. After the forty-two readings on the face of the muscle were taken, another 14mm slice was removed from the rump end of the LL and the process was repeated. After approximately 20 slices, the whole LL was examined along its length. This process was carried out for all eight LLs.

A PCA model incorporating all good NIR spectral data was developed. Standard Normal Variate (SNV) pre-processing was applied to reduce scatter effects. The optimal number of PCs was defined as the point when the loadings started to look "noisy" and this happened after 11 PCs, which were finally used to calculate D. For each animal, each 2D position appears up to 40 times (20 slices x 2 sides), so the averaging of D was carried out over 40 data points.

Results and Discussion

The average D values (D mean) for four animals were made and the regions correspond to low values of D mean, i.e. positions that resemble the average meat sample most. The central positions that are dark in colour (Figure 1) showed least variation. On average, positions 17, 18, 19, and 26 appear to be the least variable on the face of the muscle. For one animal the inner portions showed greatest variation due to the meat being rolled the wrong way during wrapping so that the fat seam was at the centre and this muscle was therefore removed from calculations.

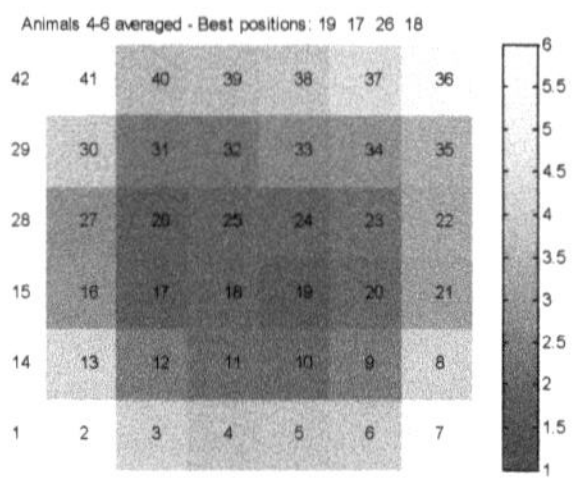

Figure 1A: A mean density pattern generated for the cross section

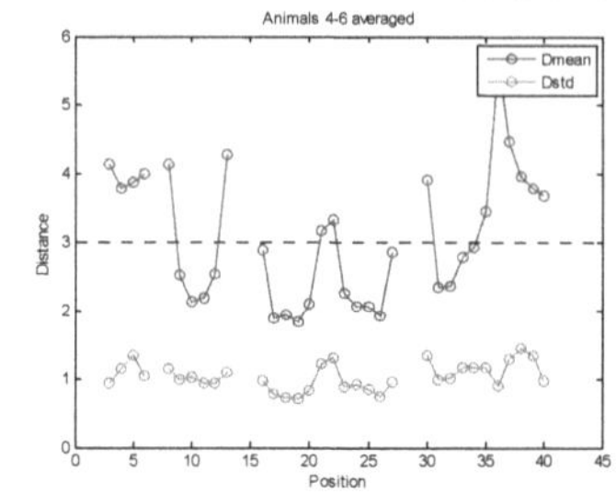

Figure 1B: D_{mean} (average of D) and D_{std} (standard deviation of D) over all slices from all three animals.The variation along the length of the muscle (i.e. the variability between slices) was determined with the resulting D_{mean} averaged over positions 17, 18, 19, and 26 for three animals shown in Figure 2A. The average and standard deviations for all three animals are shown in Figure 2B. Variation (D_{mean}) is lowest between slice numbers 8 and 20 (closer to the head end).

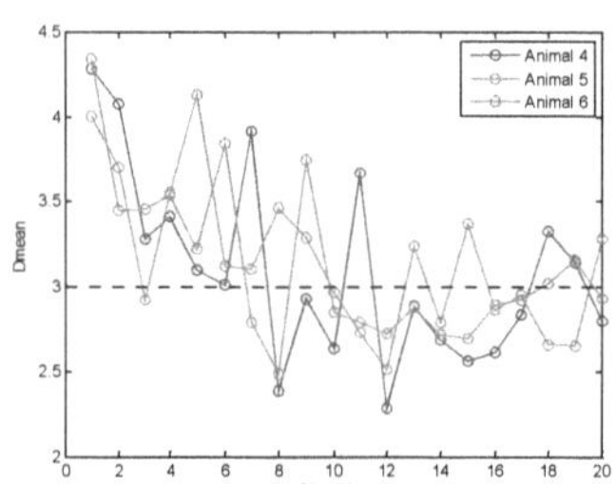

Figure 2A: Individual density values generated along the length for three animals.

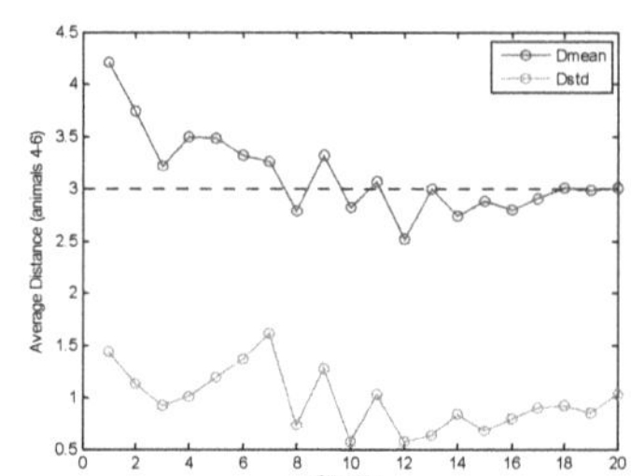

Figure 2B: The average and standard deviation for the three animals.

Conclusions

The results suggest that many shear force measurements are required to characterise a piece of meat and that the most representative results are obtained in the central regions of the LL, closer to the head end. These should be the best regions to use when developing NIR calibrations for meat tenderness.

References

Devine, C.E., Wahlgren, N.M. and Tornberg, E. (1999). Effect of rigor temperature on muscle shortening and tenderisation of restrained and unrestrained beef *m. longissimus thoracis et lumborum. Meat Science*, 51:61-72.

Hansen, S., Hansen, T., Aaslyng, M.D. and Byrne, D.V. (2004). Sensory and instrumental analysis of longitudinal and transverse textural variations in pork *longissimus dorsi. Meat Science*, 68:611-629.

Janz, J.A.M., Aalhus, J.L., Dugan, M.E.R. and Price, M. A. (2006). A mapping method for the description of Warner-Bratzler shear force gradients in beef *longissimus thoracis et lumborum* and *semitendinosus. Meat Science,* 72:79-90.

Mahalanobis, P. (1936). On the generalized distance in statistics. Procedings National Institute of Science, 12:49-55.

McGlone, V.A, Devine, C.E. and Wells, R.W. (2005). Detection of tenderness, post rigor age and water status changes in sheep using near infrared spectroscopy. *Journal of Near Infrared Spectroscopy*, 13:277-285.

Petersen, L., Minkkinen, P. and Esbensen, K. (2005). Representative sampling for reliable data analysis: Theory of sampling. *Chemometrics and Intelligent Laboratory Systems*, 77:261-277.

USE OF REFLECTANCE SPECTROSCOPY TO PREDICT ULTIMATE PH OF BEEF MEAT

S. Failla*, F. Signorelli, S. Ballico, M. Contò and F. Vincenti

C.R.A, Istituto Sperimentale per la Zootecnia. 00016 Monterotondo (Roma), Italy Email betti.failla@isz.it

Keywords: pH, visible spectroscopy, multivariate analysis

Introduction

Much research has investigated the relationship between muscle pH and temperature with some physical characteristics (Bendall, 1979; Devine and Graafhuis, 1995), since pH measurement is an important tool for optimising meat quality (Tompson, 2002). pH is normally measured using a glass electrode, but fat and protein material tends to contaminate it during use compromising the accuracy of the data, so recently some authors have been studying alternative methods to measure pH.

In particular, ultimate pH seems to affect the ageing rate of meat and of cytoskeleton protein degradation. This also produces a change in the capacity of meat to absorb and reflect light, causing modifications in the reflectance spectrum, so the analysis of visible spectrum alternative methods could give some information on ultimate pH in meat, which is a complex matrix (Swatland, 1989). The aim of this paper is to evaluate the possibility of estimating the ultimate pH and the rate of pH fall using the visible spectrum of reflectance.

Materials and Methods

The experiment was carried out on 62 entire males from 3 different Italian cattle breeds: 20 Chianini (CN), 21 Marchigiani (MG) and 21 Piemontesi (PD). At slaughter, carcasses were chilled at 4° C for 24 hours, then the *Longissimus thoracis* muscle (between 7th and 10th rib) was removed from the left side of the carcass and stored in a vacuum package at 2°C±1°C. At 24, 48, 72, 120, 192 hours the sample was extracted from the package and, after removing the external slice, cleaned by water and exposed to oxygen for 1 hour. pH values (obtained as the mean of 4 determinations) with electrode penetration through meat using a pH meter (Hanna Hi 98240) and visual reflectance spectra between 360-740 nm (in steps of 10nm), using a reflectance spectrophotometer (Minolta CM-2006d) were determined. The pH was also determined at 1 and 3 hours after slaughter to study the pH fall during ageing, using the negative exponential function as described by Bruce (2004): $pHt= pH_f+ ae^{-kt}$ where pHt is the muscle pH at different time, pH_f is final pH, "a" estimates the extent of pH decline and "k" is the exponential constant of pH decline.

The model was fitted to the pH data of each animal, using proc NLIN procedure (SAS institute, 1999) to obtain respective "a" and "k" coefficients.

The reflectance spectra obtained at different times were correlated with ultimate pH (pHu) to choose the best spectrum to use in multivariate analysis. PLS (Partial Least Square regression) multivariate statistical analysis (Martens and Naes, 1989) is used for creating calibration models for spectroscopic data. The wavelength selection procedure consists of finding the number of variables that give the minimum value of RMSEP (root mean square error of prediction), found by cross-validation. Before performing the PLS regression, spectral outliers were identified and eliminated. The multivariate data analysis was performed with a chemometric program the Unscrambler 9.1 (CAMO, Trondheim, Norway).

Results and Discussion

The corresponding Pearson correlation coefficients between the reference pH and visual spectra (360nm to 740nm) are shown in Figure 1; the maximum correlation coefficients were achieved at wavelength interval between 460nm and 640nm, Soret peak at 192h visual spectrum and between 450 and 580nm at 24h visual spectrum, while the other ones showed a pronounced decrease in a narrow spectral range around 420nm, coinciding with the Soret peak as reported by Andersen *et al.* (1999). The correlation coefficient between ultimate pH and the single wavelength highlights the possibility of estimating the ultimate pH from a spectrum performed at 24 hours, while the spectra obtained at 48 and 72 hours showed a higher variability. Particularly interesting was the possibility of associating spectrometrics with pH decline over time. PLS analysis (Table 1) evidenced an optimal RMSEP between spectrum data and coefficient of pH decay proposed by Bruce (2004); although a low value of RMSEP, the cross validation didn't show a good index of correlation and slope among predicted and measured pH values. Therefore, with these errors it is necessary to consider

Figure1 Correlation between single wavelength of different reflectance spectra and ultimate pH

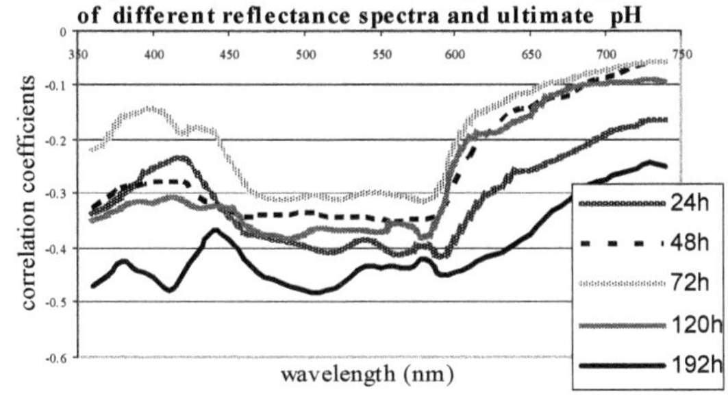

also the low capacity of the exponential equation to take account of anomalous courses of pH registered in some animals.
The cross validation between ultimate pH, coefficient "k" and spectrum data at 192 hours however resulted better then the other, in fact a good correlation index and slope was obtained. Moreover, for a good pH and decay prediction it was necessary to eliminate more sample outliers from the spectrum at 24 hours than the other one. The regression coefficients (Figure 2) evidenced principal points to predict the data, in particular the absorbances at the beginning and at the end of spectrum (360, 370nm and from 630 to 740nm), and at the absorption points of myoglobin (470, 580 nm), in particular about six factors of PLS were necessary to estimate the pHu in validation phase. Also if it was possible to estimate the ultimate pH from the visible spectrum of reflectance obtained both at 24 and at 192 hours. Chianini meat gave more problems: this in fact showed limited pH decay and some samples had maintained high pH also after eight days of ageing, in fact it had a greater frequency of outlier values (Table 2). Considering the relationship between predicted and measured pH, Chianina values were undervalued, while Marchigiana and Piemontese pH values were slightly overvalued.
The results obtained indicate that the use of a spectroscopic method for measuring pH may be possible, improving determination rapidity, while further studies are necessary to obtain from the spectrometric data information on the pH trend and an estimation of pH decline after rigor.

Table 1: Cross-validation results for prediction pH with visible spectrometer.

	Spectrum at 24 h		Spectrum at 192h	
	pHu	"k"	pHu	"k"
N outliers	8	15	4	9
PLS factors	7	6	6	6
RMSEP	0.0346	0.0503	0.0422	0.0417
Correlation index	0.79	0.64	0.81	0.74
Slope	0.694	0.56	0.741-	0.62

Figure 2. Estimated regression coefficient for predicting pHu by spectrum at 192 hours

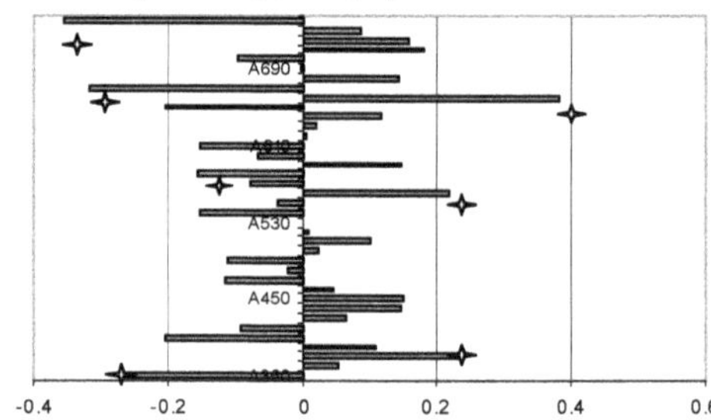

Fig.3 Predicted versus measured pH values

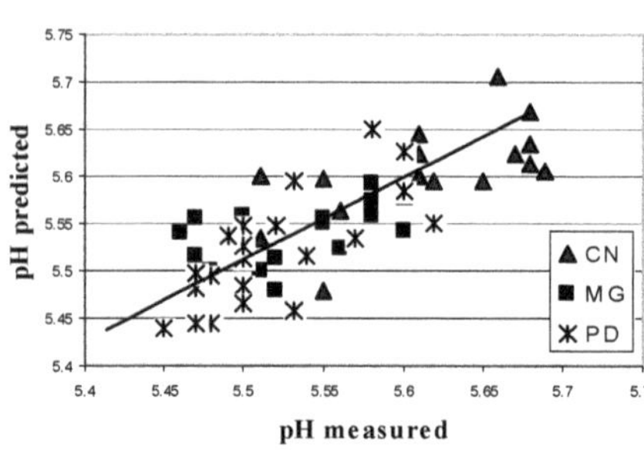

Table 2: pH and exponential decay constant "k" in the different breed.

	pH at 3hours	Ultimate pH	"k"
CN	6.20	5.62a	0.125
MG	6.24	5.53b	0.183
PD	6.21	5.51b	0.196
RMSE	0.073	0.051	0.092

different letters mean significant differences (p< 0.05),

References

Andersen J.R., Borggaard C., Rasmussen A.J., Houmoller L.P., (1999). Meat Science 53: 135-141.
Bendall, J. R. (1979). Meat Science, 3: 143–157.
Bruce, H.L., (2004) Meat Science 66: 507-512.
Devine C.E, and Graafhuis A.E., (1995). Meat Science 39: 285.
Martens, H., and Martens, M. (2001). London, UK: Wiley.
Swatland, H. J., (1989). Canadian Institute Food Science and Technology Journal, 22: 390–402.
Tompson J., (2002). Proceeding of the 48th International Congress of Meat Science and Technology, 1: 17-28.

DISCRIMINATION OF BEEF *VS.* PORK MEAT AND SAMPLE PREPARATIONS BY AN ELECTRONIC NOSE (EN): PRELIMINARY STATIC AND DYNAMIC RESULTS

S. Barbera*[1], P. Cornale[1], G. Sala[2] and G. Masoero[2]

[1.] *Dipartimento di Scienze Zootecniche, Università di Torino, via L. da Vinci 44, 10095 Grugliasco (TO), Italy salvatore.barbera@unito.it* [2.] *C.R.A. - Istituto Sperimentale per la Zootecnia, via Pianezza 115, 10151 Torino, Italy*

Keywords: electronic nose, beef, pork, partial least squares analysis, prediction

Introduction

Meat flavour is an aspect intrinsic to quality parameters. Electronic Nose (EN) devices are a promising family of instruments which operate rapidly and at low costs, both in qualitative mode, comparing patterns, and in quantitative mode by fitting specific volatile compounds. Several studies have shown the possibilities in the application of EN to meat products. Spanish researchers (Gonzales-Martin *et al.*, 2000; Otero *et al.*, 2003) attempted to classify special Iberian cured ham into very different commercial categories. Other researchers from Spain (Santos *et al.*, 2004) discriminated feeding regimen effects in pig meat and optimised some ripening time parameters and they also concluded that different types of Iberian ham can be discriminated and identified successfully by EN. The EN properties of raw meat materials were investigated by Hansen *et al.* (2005); they concluded that sensory quality of porcine meat loaf, based on measuring the volatiles in either the raw materials or the meat loaf produced from those raw material may be modelled in a predictive-causative multivariate analysis. The present work aims to preliminarily ascertain EN reliability in the investigation of the positive aspects of meat aromatic traits, induced by animals and technologies. Thus EN multichannel evaluation of samples of bovine and porcine meat, whole or minced, will be carried out when the examination occurs in static or dynamic modes. For the static mode, the raw and the cooked samples were utilised respectively, while in the dynamic mode gas flux from a furnace at 165°C was examined in continuous monitoring or in samples of air bags captured at 5 and 15 min.

Materials and Methods

Eight beef and 8 pork *Longissimus dorsi* muscle portions refrigerated at 4°C were used as the meat source. From each muscle a 1cm thick circular sample (steak, 5.5cm Ø) was obtained according to the MCS method proposed by Barbera (2006). A 1cm thick, hamburger sample (minced) weighing 30g was produced using minced meat of the same muscle. Cooking was performed at 165°C for 15 min in an electric forced-air convection oven. By forcing air into the oven, after passing through an active charcoal filter, regular cooking odour out flow was guaranteed to pass in the PEN 2 (AIRSENSE Analeptics GmbH, Hagenower, Germany) 10 metal oxide sensors (MOS) to carry out dynamic allotherm measurements. This instrument was utilised in the non-stop monitoring of volatile compounds of meat produced for 15 min cooking. The trial started performing a zero point trim to standardise the sensors conditions (oxidation-reduction). The continuous measurement included 120 second white reference (air in the bin), and 180 of raw sample aroma whiffing. Then the cooking trend was performed. The EN was connected to a spill duct of oven, thus in a same way a 120 second white reference of 165°C air of the oven and then 900 seconds of cooking sample aroma were carried out. After 1020 seconds, the cooked sample was taken into another smaller bin, where the trend of concentrate aroma after cooking was tested. At 420 and 720 seconds (5 and 10 minute of cooking) a Nalophan bag was connected with the inner oven and filled up with "cooking air". The bags were inflated under vacuum. The aim of this procedure was to try to understand if is possible to collect the aroma for subsequent or later analysis. The static isotherm, 100 sec long, EN analyses of bags were carried on within 6-15 hours directly by the EN aspirating for 90 sec at 400 ml/min flow. Two other static isotherm measurements of raw (70 sec) and cooked samples (150 sec) were, as previously reported, carried out under a glass bell (50 ml vol.) at 400 ml/min flow rate allowing input of pure filtered air. In all, 32 steak and minced samples were used. Chemometrics were performed by Modified Partial Least Squares method (ISI, 1994) using a cross-validation system to assess the optimal number of latent variables to be included in the equations. The design variables: Species (beef/pork) and preparation (steak/minced) were fitted to numbers (1/2). Binary discrimination between groups allowed one passage for elimination of few outliers (t>2; H>10) (Fearn, 1997). The points from sensors were adjacently vectored, then mathematically pre-treated as Standard Normal Deviate with Detrend and finally 1st derived and smoothed.

Results and Discussion

In the Table 1 the whole experiment is represented in term of R^2 values. In spite of the high values recovered in the calibration mode, the cross-validation process and the elimination of few outliers (on average 6% of the observations) resulted in some more reliable and robust causative models, thus special attention must be paid to the R^2_{cv} Coefficient. The Dynamic examination during the cooking process will account 81% of the variance observed in the cumulative MOS sensors due to the preparation of sample, because release of odour from the more homogeneous and aired minced samples was more linearly dependent on the raising of temperature during cooking (Cornale *et al.*, 2006) while only 23% occurred when species comparisons was attended. Better results for discrimination of beef vs pork meat by EN, than for the post-identification of sample preparation, emerged when EN examinations were carried out in Static mode. This was realised first in the pre-cooking raw muscle examination, R^2_{cv} 0.75 *vs.* 0.39 respectively; secondly the result was replicated when the examination was carried out immediately after the end of cooking on the warm meat (0.58 *vs.* 0.15). The examination of Nalophan bags did show significant differences between the two sampling points at 5 and 10 min (R^2_{cv} = 0.27) but when the data were pooled, the between species account of variability (0.48) was higher than those between preparations (0.37). The time delay between capture and examination of pure meat aroma may be not so critical for meat as it was for tobacco smoke, as investigated by van Harreveld (2003) who, in the frame of EN13725 "Air Quality-Determination of Odour Concentration by Dynamic Olfactometry" enhanced great decay of signals in bags after 12-30 hrs.

The discrimination of single specimens from beef or from pork pursued in this experiment represents only a preliminary challenge for the rapid analysis of studied factors in animal production experiment involving meat intrinsic quality. In this sense some R^2, in validation mode, of medium value (0.5-0.6) do not appear as accurate for individual prediction but it may be adequate for a preliminary and rapid evaluation of average effects. Partial Least Squares is a linear method and may not be so accurate as the non-linear one utilised by other researchers (Santos *et al.*, 2004) in the Iberian ham classification; however Neural Network Analysis becomes efficient in validation mode only with adequate and wide datasets. Finally, EN efficiency in the candidate meat products appears more suitable without sample heating: the freeze-drying process, which works with cold muscles, can capture volatile compounds and could be of major interest, but remains unexplored.

Table 1: Results of calibration and cross-validation chemometrics of the transformed olfactograms on the Species origin and on the sample Preparation.

Electronic Nose Analysis	Static Sample						Dynamic Sample	
	Raw		Cooked		Bags		Cooking	
Time measurement, sec	180		150		100		750	
R^2Coefficient	*RSQ*	R^2_{cv}	*RSQ*	R^2_{cv}	*RSQ*	R^2_{cv}	RSQ	R^2_{cv}
Species: Beef_1 vs Pork_2	*0.87*	0.75	*0.97*	0.58	*0.96*	0.48	*0.55*	0.23
Preparation: Steak_1 vs. Minced_2	*0.54*	0.39	*0.41*	0.15	*0.41*	0.37	*0.94*	0.81

Conclusions

The electronic nose was confirmed to be very useful in order to investigate global aroma characteristics in raw meat, and to a lesser extent in cooked meat or sampling gas in Nalophan bags. Dynamic monitoring of the cooking process in a wide oven appears not to be so reliable because it may interfere with sample preparation for cooking meat whole or ground.

References

Barbera, S. (2006). Meat Cooking Shrinkage: measurement of a new meat quality parameter. Meat Science (in press).

Cornale P., Barbera S., Sala G. and Masoero G. (2006). The relationship between the internal temperature and parameters measured by an electronic nose during meat cooking. In: Proceedings of the 52nd International Congress of Meat Science and Technology.

Fearn T. (1997). Qualitative analysis I. NIR news, 8, 4-5.

Gonzalez-Martin I., Perez-Pavon J. L., Gonzalez-Perez C., Hernandez-Mendez J. and Alvarez-Garcia N. (2000). Differentiation of products derived from Iberian breed swine by electronic olfactometry (electronic nose). Analytica Chimica Acta, 424, 279-287.

Hansen T., Byrne D.V. and Petersen M.A. (2005). Sensory based quality control utilising an electronic nose and GC-MS analyses to predict end-product quality from raw materials. Meat science, 69, 621-634.

ISI (1994). ISI - Infrasoft International, NIRS 2 Version 3.00. Routine operation and calibration software development for Near Infrared instruments. USDA NIRS Network,USDA, Washington D.C., USA: 270 pp.

Otero, L., Horrillo M.C., Garcia M., Sayago I., Aleixandre M., Fernandez M.J., Ares L. and Gutierrez J. (2003). Detection of Iberian ham aroma by a semiconductor multisensorial system. Meat Science, 65,1175-1185.

Santos J.P., Garcia M., Aleixandre M., Horrillo M.C., Gutierrez J., Sayago I., Fernandez M.J. and Ares L. (2004). Electronic nose for the identification of pig feeding and ripening time in Iberian hams. Meat science, 66, 727-732.

van Harreveld AP. (2003). Odour concentration decay and stability in gas sampling bags. J Air Waste Manag Assoc. 53:51-60.

DIFFERENTATION OF BEEF ACCORDING TO THE PRE-SLAUGHTER DIET OF CATTLE USING THE STABLE ISOTOPE RATIOS OF CARBON AND NITROGEN

A.P. Moloney*[1], B. Bahar[2,3], O. Schmidt[3], F.J. Monahan[2], P. O'Kiely[1] and C. Scrimgeour[4]

[1]Teagasc, Grange Research Centre, Dunsany, Co. Meath; [2]UCD School of Agriculture, Food Science and Veterinary Medicine, Belfield, Dublin 4, Ireland; [3]UCD School of Biology and Environmental Science, University College Dublin, Belfield, Dublin 4; [4]Scottish Crop Research Institute, Invergowrie, Dundee, DD2 5DA, UK; Email: aidan.moloney@teagasc.ie

Keywords: beef, stable isotopes, pre-slaughter diet

Introduction

Consumers demand reliable information about the food they buy. In particular, guarantees concerning the authenticity of meats are deemed fundamental to the assurance of food safety, quality and animal welfare (Verbeke and Viaene, 1999). Existing livestock traceability schemes depend ultimately on a paper trail and there is a need for scientific technologies for meat authentication to reassure consumers and to protect regional designations.

Stable isotope (SI) ratio analysis (SIRA) is one potentially useful technique for testing food authenticity. Since dietary C and N stable isotope (SI) compositions (expressed as $\delta^{13}C$ and $\delta^{15}N$) influence the isotope compositions of these elements in animal tissues, $\delta^{13}C$ and $\delta^{15}N$ of animal tissues are useful isotopic markers of diet. The $\delta^{13}C$ in livestock animal tissue primarily depends on the proportion of C_3 and C_4 photosynthetic plants consumed. The $\delta^{15}N$ in such animal tissue is usually less specific for particular dietary inputs but can reflect, among other husbandry practices, the presence of leguminous plants in the diet and also the intensity of agricultural land use in raising the feed crops. Recently, we have shown SIRA to have potential for the authentication of the geographic origin of beef and the proportion of C_3 and C_4 plant material in cattle diets and to distinguish between beef from organic or conventional production systems (Schmidt *et al.*, 2004; Bahar *et al.*, 2005). The objective of this study was to determine the potential of SIRA to distinguish between beef from cattle that consumed, during the finishing phase, a range of feedstuffs commonly available in Ireland.

Materials and Methods

In an indoor study, continental cross-bred beef steers (n=14/feedstuff) were offered grass silage, maize silage (cv. Benecia), fermented whole-crop wheat (cv. Soissons), alkalage whole-crop wheat (cv. Soissons) and *ad libitum* concentrates (83% rolled barley). The alkalage was ensiled with 45 kg Home 'N' Dry (Volac International Ltd.)/t dry matter. Forages were offered *ad libitum* through individual Calan gates and supplemented with 3kg concentrates/head/day. In a grazing study, continental cross-bred steers (n=14/feedstuff) were finished on either a conventionally managed sward i.e. the grass sward received approximately 200kg N ha^{-1}, or an optimally managed grass-clover sward that received 50kg N ha^{-1} in early spring. Animals were slaughtered after approximately 5 months of treatment. After cooling the carcasses for 24 h, *longissimus* muscle was sampled, vacuum packed and stored at -18°C until analysis. Natural abundance isotope ratios of carbon ($^{13}C/^{12}C$), and nitrogen ($^{15}N/^{14}N$) were measured on de-fatted muscle tissue by continuous flow isotope ratio mass spectrometry using a Europa Scientific ANCA-NT 20-20 Stable Isotope Analyser and are expressed in delta (δ) notation in parts per thousand. Data were analysed by (Multivariate) Analysis of Variance.

Results and Discussion

The stable isotope composition of the feeds is shown in Table 1. As expected, the $\delta^{13}C$ of maize silage was less negative than the other feeds. The $\delta^{15}N$ was highest for grass silage and lowest for the concentrate. Muscle SI composition is summarised in Table 2. Feed and muscle $\delta^{13}C$ were linearly related: Muscle = 0.41 (feed) – 14.02, se = 0.59, $P<0.05$, $R^2 = 0.95$. Beef from maize silage-fed cattle had the least negative $\delta^{13}C$ reflecting the SI composition of the feed, as previously observed (Bahar *et al.*, 2005). The $\delta^{13}C$ value distinguished ($P<0.05$) between beef from concentrate/wheat silage, grass silage, grazed grass and grazed grass/clover-fed cattle, but not between beef from alkalage and wheat silage-fed cattle or between beef from alkalage and concentrate-fed cattle. The relationship for $\delta^{15}N$ was: Muscle = 0.29 (feed) + 6.36, se = 0.61, P = 0.05, $R^2 = 0.47$. Using the $\delta^{15}N$ value resulted in a poorer discrimination of beef samples compared to using the $\delta^{13}C$ value.

A scatter plot of the individual data is shown in Figure 1. As before, beef from maize-silage fed cattle was clearly distinguished from other samples. The combined isotopic composition of carbon and nitrogen did not improve the discrimination of beef from wheat silage or concentrate-fed cattle beyond that achieved by considering only $\delta^{13}C$.

Table 1: Carbon and nitrogen stable isotope composition of feed ingredients.

Ration	Delta ^{13}C	sd	Delta^{15}N	sd
Alkalage	-28.20	0.10	4.37	0.73
Concentrate	-27.48	0.30	2.95	0.36
Grass silage	-30.65	0.11	9.31	0.73
Maize silage	-12.74	0.30	6.32	0.26
Wheat silage	-28.02	0.25	3.11	0.36
Grass	-30.37	0.56	5.88	1.14
Grass/clover	-30.24	0.64	4.88	0.97

Table 2: Carbon and nitrogen stable isotope composition of *Longissimus* muscle from cattle fed various diets pre-slaughter[1]

Ration	Delta ^{13}C	Delta ^{15}N
Alkalage	-25.36^{b,c}	6.93^{a}
Concentrate	-25.65^{c}	7.16^{a}
Grass silage	-26.18^{d}	8.82^{d}
Maize silage	-19.42^{a}	7.76^{b}
Wheat silage	-25.28^{b}	7.19^{a}
Grass	-27.51^{e}	9.05^{d}
Grass/clover	-27.19^{f}	8.17^{c}
Sed	0.145	0.167
Significance	P<0.001	P<0.001

[1]Within a column means with different superscripts differ (P<0.05)

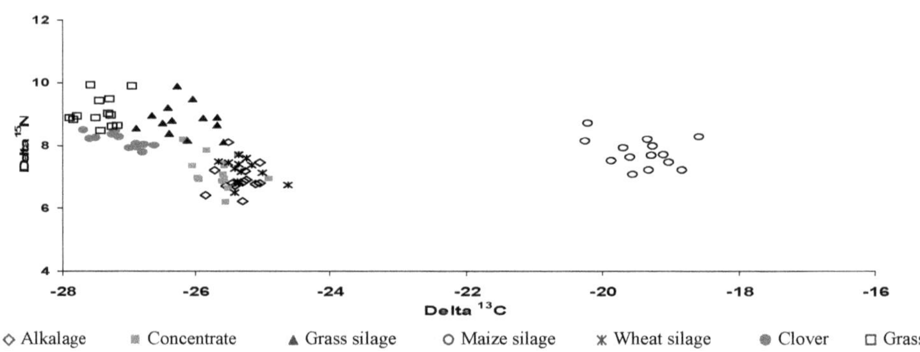

Figure 1: Carbon and nitrogen SI ratios in beef from cattle fed different diets.

Conclusions

Differences in the carbon and nitrogen stable isotope composition of the feeds examined were reflected in the muscle of cattle thereby allowing SIRA to be used as a component, at least, of a scheme to authenticate the dietary history of beef from cattle consuming these feeds.

References

Bahar, B., Monahan, F.J., Moloney, A.P., O'Kiely, P., Scrimgeour, C.M. and Schmidt, O. (2005). Alteration of the carbon and nitrogen stable isotope composition of beef by substitution of grass silage with maize silage. Rapid Communications in Mass Spectrometry, 19, 1937-1942.

Schmidt, O., Quilter, J.M., Bahar, B., Moloney, A.P., Scrimgeour, C.M., Begley, I.S. and Monahan, F.J. (2005). Inferring the origin and dietary history of beef from C, N and S stable isotope ratio analysis. *Food Chemistry,* 91, 545-549.

Verbeke, W. and Viaene, J. (1999). Consumer attitude to beef quality labelling and associations with beef quality labels. Journal of International Food and Agribusiness Marketing, 10, 45-65.

TEMPORAL CHANGE IN THE CARBON STABLE ISOTOPE RATIOS OF BEEF FOLLOWING A CHANGE IN RATION COMPOSITION

F.J. Monahan*[1], O. Schmidt[2], A.P. Moloney[3], P.G. Dunne[3], C.M. Scrimgeour[4] and I.S. Begley[5]

[1.] UCD School of Agriculture, Food Science and Veterinary Medicine, University College Dublin, Dublin 4, Ireland, [2.] UCD School of Biology and Environmental Science, University College Dublin, Dublin 4, Ireland, [3.] Teagasc, Grange Research Centre, Dunsany, Co. Meath, Ireland, [4.] Scottish Crop Research Institute, Invergowrie, Dundee, DD2 5DA, United Kingdom, [5.] Iso-Analytical Ltd., Unit 2, Zan Industrial Park, Sandbach, Cheshire CW11 4QD, United Kingdom.
Email: frank.monahan@ucd.ie

Keywords: beef, dietary history, stable isotopes

Introduction

Information concerning the production and origin of foods of animal origin is increasingly sought by consumers. In particular, guarantees of food authenticity and traceability, with appropriate labelling, are required. There is therefore a need to develop methodologies capable of providing information about the origin of animal-derived foods. On-going research aims to establish whether the isotopic composition of light elements (δ^{13}C, δ^{15}N, δ D, δ^{18}O and δ^{34}S) can be used as an intrinsic, biochemical marker tool for tracing and authenticating beef (Kelly et al., 2005). The specific objectives of the experiments conducted in Ireland were 1) to determine the temporal variability in the stable isotope composition of C (δ^{13}C) in feed materials over one growing season, and 2) to determine the speed at which temporal changes in the isotopic composition of beef muscle tissue occur following a change in diet.

Materials and Methods

From a group of 63 heifers, 15 were fed concentrates for 220 days and 36 were grazed for 97 days, then housed and offered in groups of 12 one of three diets differing in the proportion of concentrates, silage and grass (Figure 1). Heifers were slaughtered at approximately monthly intervals (n=3 per date per diet) and feed samples were collected regularly from June to November. Two control groups (n=6) were slaughtered at day 30 and day 101 (see Figure 1). Samples of the striploin (*M. longissimus dorsi*) were collected, freeze-dried and milled. The δ^{13}C of bulk muscle and bulk feed materials was determined by Isotope Ratio Mass Spectrometry.

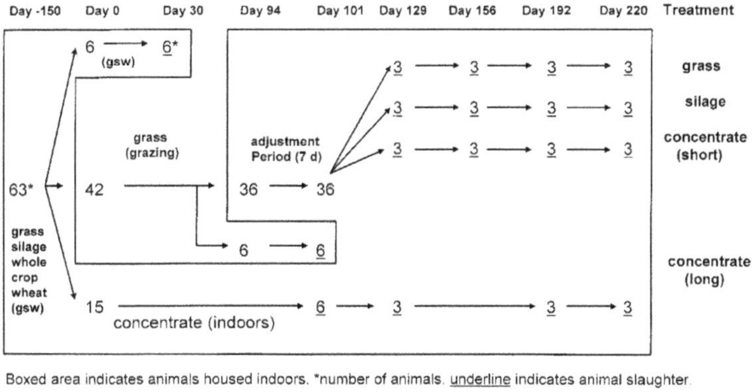

Figure 1: Experimental design.

Results and Discussion

The isotopic composition of the concentrate and grass silage showed little temporal variation (range <0.6‰ δ^{13}C) over a 5-month period. Grass, by contrast, showed marked temporal trends over this period with monthly values ranging from –29.6 to –31.0‰ in δ^{13}C. There was an isotopic difference (about 3‰ δ^{13}C) between concentrate and grass/grass silage feed materials (Figure 2).

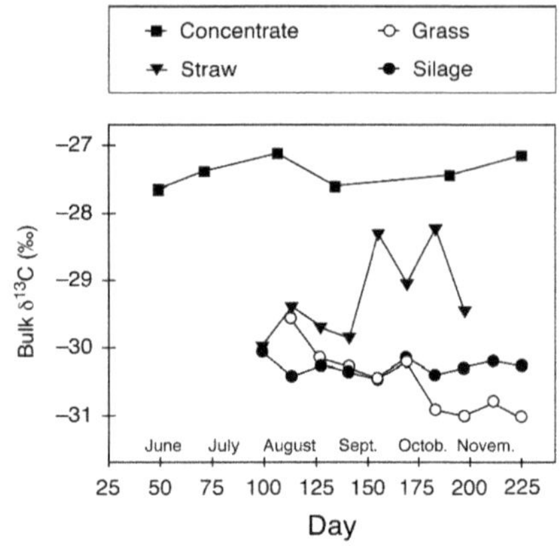

Figure 2: Isotopic composition of feed materials.

The isotopic composition of muscle tissue changed only slowly after diets were switched. However, even small isotopic differences between grass- and concentrate-based diets were consistently reflected in bulk muscle tissue (Figure 3). The final difference in muscle $\delta^{13}C$ between the grass and 'long concentrate' treatments was 1.9‰.

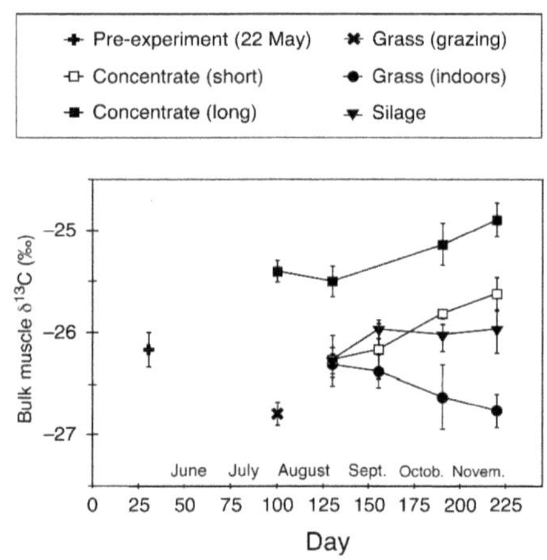

Figure 3: Isotopic composition of muscle tissue.

Conclusions

1) While $\delta^{13}C$ showed little temporal variation in concentrate and silage feed materials, it was more variable in fresh grass over one growing season.

2) Bulk muscle tissue reflected small isotopic differences between dietary components after about 100 days.

References

Kelly, S., Heaton, K and Hoogewerf, J. Tracing the geographical origin of food: the application of multielement and multi-isotope analysis. Trends in Food Science and Technology, 16, 555-567 (2005).

Schmidt, O., Quilter, J.M., Bahar, B., Moloney, A.P., Scrimgeour, C.M., Begley, I.S. and Monahan, F.J. Inferring the origin and dietary history of beef from C, N and S stable isotope ratio analysis. Food Chemistry, 91, 545-549. (2005).

MUSCLE STUDY WITH MULTISPECTRAL IMAGE ANALYSIS

S. Abouelkaram[1], S. Chauvet[1], M. El Jabri[2], P. Strydom[1], D. Bertrand[3] and J-L. Damez*[1]

[1] INRA, Centre de Clermont-Fd/Theix, 63122 Saint Genès Champanelle, France. [2] Laboratoire de Mathématiques, Université Blaise Pascal. 63000 Clermont-Ferrand, France. [3] Unité de Sensométrie et de Chimiométrie ENITIAA-INRA, ENITIAA, 44322 NANTES CEDEX 3, France. Email: damez@clermont.inra.fr

Keywords: multispectral imaging, muscle structure, meat quality

Introduction

A feasibility study of multispectral image analysis (MIA) was undertaken with the aim of relating meat components to sensorial and physical properties. The MIA technique has been used during last decade for the assessment of food products (Novales *et al.*, 1996) especially for the inspection of poultry carcasses in the Visible/NIR range (Park *et al.*, 1996). Contaminations of apples have also been studied using multispectral fluorescence imaging (Kim *et al.*, 2005). In this work we applied MIA to study meat component by targeting myofibres, collagen and lipids which play a major role in meat sensorial quality. In a previous study we established the specific spectral signals of muscle components (Skjervold *et al.*, 2003). This paper describes the multispectral imaging approach we have used in order to characterise muscle slices. A specific bench was built and settled to optimise the UV-Visible image of muscle samples according to specific spectral tissue information. A method of exploiting the resulting multiway image data was developed. It covers the different steps of data analysis *i.e.* from the processing of the raw image signal to statistical analysis of the data. The image segmentation was performed using a multiway method based on multispectral techniques (Novales *et al.*, 1996). This method allowed the capture of information on muscle composition and structure and particularly the distribution and quantification of the connective tissue. The analysis involved segmentation of the connective tissue making it possible to directly identify collagen and lipids, while myofibre information was obtained indirectly. Muscles of six animals were analysed in real conditions to demonstrate the feasibility of the multispectral technique for muscle characterisation.

Materials and Methods

Image analysis was performed on bovine muscles. Frozen *biceps femoris* muscles of six animals were cut into samples of 20 mm thickness. The set of images was produced with an optical bench using a UV sensitive digital camera (Sony XCD-SX900UV) (Figure1). Sample images of 5 cm X 4.5 cm were recorded. They were acquired at different wavelengths in the UV–Visible range. The light source was a 300W maximum power Xenon light used for pure white light or ultra violet light filtered at 320 and 380 nm. Each wavelength corresponded to one channel of the multispectral image.

In order to improve image quality both median and Gaussian low-pass filters were applied. The image segmentation is a chemometric technique based on pixel classification. It was used here to classify the three components of muscle tissue: myofibres, collagen and lipids. The chemometric technique consists of a Matlab toolbox developed by the authors. Image features were extracted from segmented images according to (Abouelkaram *et al.*, 2003). Parameters were selected in order to find the most pertinent ones to use in the prediction models. Data statistical studies, mainly based on multiple linear regression, made it possible to build specific statistical models which relate image features to meat properties.

Sensory parameters, including tenderness were assessed on cooked meat by a sensory panel while mechanical tests were performed on raw meat (Maunier-Sifre, 2005).

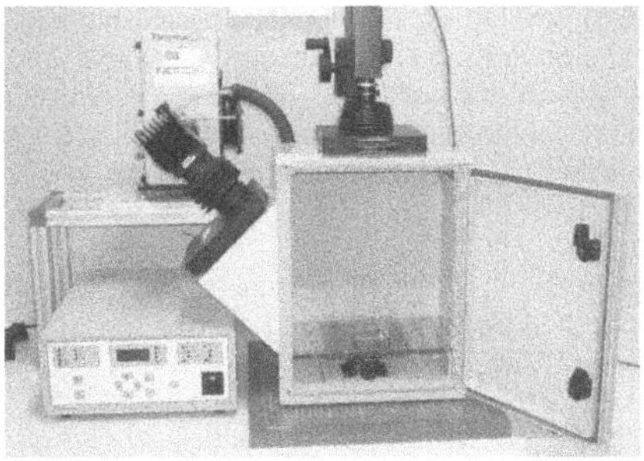

Figure 1: Multispectral imaging bench.

Results and Discussion

The multispectral imaging technique involves the following steps: image acquisition, image pre-processing, multiway segmentation, image features extraction and statistical analysis (Figure 2). The main default on the raw images was a non uniform lighting. We solved this problem by using a low pass Gaussian filtering (Figure 3). A typical three grey level segmented image resulting from the multiway method is given in Figure 4. It shows the discrimination between myofibrils (in black), collagen (in grey) and lipids (in white) allowing assessment of their quantities and distributions.

On resulting segmented images of muscle of six animals, image features were extracted and related to composition, mechanical and sensory measurements. Because of the low observations number, the predicting models were built with only the most relevant image parameter.

Results in predicting composition and sensory parameters are presented on Figure 5. The R^2 coefficients of the prediction models were satisfactory. However, due to a small number of observations, these results must be confirmed. The highest R^2 were obtained when predicting the proximate composition: collagen (R^2=0.88) and lipids (R^2=0.87). Tenderness gave the lowest R^2 (0.75) while R^2 associated with juiciness was equal to 0.84. Note that collagen and lipids could be assumed as primary parameters as they were visible on the image, whereas sensory characteristics were supposed to be second rank parameters. This can explain the score of tenderness which is also multifactorial.

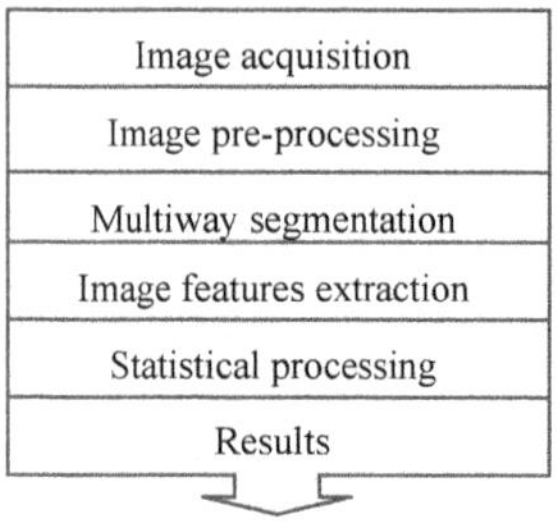

Figure 2: MIA steps.

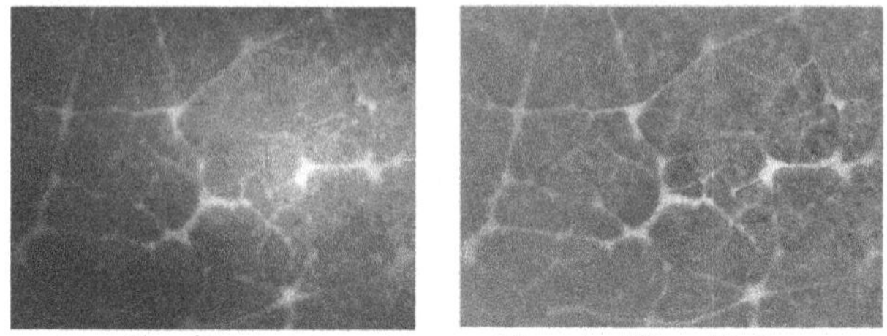

Raw image
Gaussian filtered image

Figure 3: Image pre-processing.

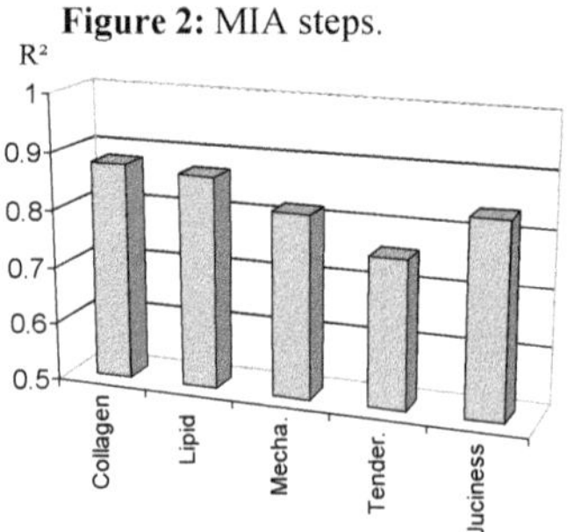

Figure 5: Meat quality prediction results.

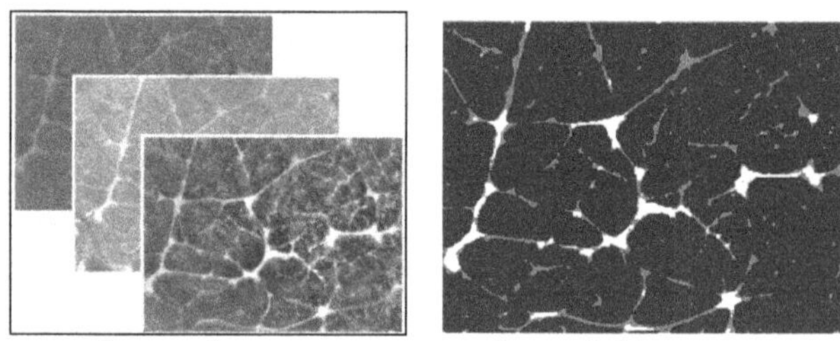

Image set
Segmented image

Figure 4: Multiway segmentation.

Conclusions

This work was focused on the application of a multispectral imaging technique to muscle characterisation. It was needed to develop specific equipment and method. The method covers the different steps of data analysis from the image pre-processing to the data statistical analysis. To complete this study, preliminary experiments were undertaken on muscles of six animals. The results obtained on this small sample population show the potential of MIA in estimating composition and structural organisation and predicting sensory parameters. These promising results have to be confirmed on a larger population.

References

Abouelkaram, S., Berge, P., Hocquette, J. F., Culioli, J. and Listrat A. (2003). Image study analysis of the relationship between collagen content and distribution of the perimysial connective network in a bovine muscle. *Sciences des Aliments*, 231 : 166-170.

Kim, MS, Lefcourt, A.M., Chen, Y.R., Tao (2005). T. Automated detection of fecal contamination of apples based on multispectral fluorescence image fusion. *Journal of Food Engineering*, 71 (1): 85-91.

Maunier-Sifre L., (2005). Organisation spatiale du tissu conjonctif intramusculaire : relation avec la texture de la viande bovine, Université d'Auvergne Université Blaise Pascal, N° d'ordre : 422, *PhD Thesis*.

Novales, B., Bertrand, D., Devaux, M.F., Robert, P. and Sire, A., (1996). Multispectral Fluorescence Imaging of Food Products. *Journal of the Science of Food and Agriculture*, 71: 376-382.

Park, B. Chen, Y. R., Huffman, R. W. (1996). Integration of Visible/NIR Spectroscopy and Multispectral Imaging for Poultry Carcass Inspection. *Journal of Food Engineering*, 30:197-207.

Skjervold, P. O., Taylor, R. G., Wold, J. P., Berge, P., Abouelkaram, S., Culioli, J., and Dufour, E. (2003). Development of intrinsic fluorescent multispectral imagery specific for fat, connective tissue and myofibres in meat. *Journal of Food Science*, 68: 1161-1168.

ESTIMATION OF VISIBLE FAT CONTENT IN BEEF *M. LONGISSIMUS* USING NEAR INFRARED SPECTROSCOPY

L. Streeter[1,2], P.W. Hansen*[3], R. Burling-Claridge[1], M.J. Cree[2], R. Kunnemeyer[2] and C.E. Devine[4]

[1.] AgResearch MIRINZ Centre, PB 3123 Hamilton, New Zealand, [2.] Department of Physics and Electronic Engineering, University of Waikato, Private Bag 3105, Hamilton, New Zealand, [3.] EquatioNZ, Lyngby Hovedgade 11B, 2. th., DK-2800 Kongens Lyngby, Denmark, [4.] Bioengineering and Biosensors, HortResearch, PB 3123 Hamilton, New Zealand. Main Author: PWH@equationz.dk

Keywords: beef, Near Infrared Reflectance Spectroscopy, fat, *longissimus dorsi*

Introduction

Near infrared (NIR) spectroscopy is a non-destructive measurement method that analyses the light reflected from a sample in order to infer information about the sample composition and structure. It has been shown that NIR spectroscopy can be correlated with the tenderness of lean meat (McGlone *et al.*, 2005). However, when assessing the NIR spectra of meat, it would be useful to objectively estimate the spectral contribution of substance that is not lean. Additionally, estimation of the visible fat content of meat may have potential for automated grading purposes.

Materials and Methods

Four pairs of beef *m. longissimus et lumborum* (LL) were removed 30 minutes after slaughter from steers shot with a captive bolt and not electrically stimulated. The LLs were wrapped in polyethylene cling film to prevent shortening (Devine *et al.*, 1999) and were held at 15°C to enter rigor mortis and aged at this temperature for 24 hours. Each of the wrapped muscles, with an approximate circular cross section, was positioned in a holder and a 14mm thick slice was removed from the rump end of the muscle and discarded.

To estimate the visible fat content across the face of the muscle an image was taken using a JAI machine vision colour camera. Then, forty-nine NIR readings were made on the face of the muscle in a grid arrangement with a 14 mm spacing. The NIR readings at each grid position were taken using a fibre optic probe (with diameter of approximately 16 mm) attached to a diode array spectrophotometer from KES Analysis (NY, USA). The instrument covered the spectral range from 400 to 1700 nm. At grid positions where there was no muscle, readings were omitted from the data analysis. After the forty-nine readings on the face of the muscle were taken, another 14 mm slice was removed from the rump end of the LL and the process was repeated. After approximately 20 slices, the whole LL was examined along its length using digital imaging and NIR measurements. This process was carried out for all eight LLs.

Digital colour imaging and image processing methods (Gonzalez and Woods, 2002) were used on the collected camera images to estimate the lean and fat content of meat at each grid position. To locate the grid positions, a simple colour difference technique was used. To analyse the visible fat content at each position, a colour-based principal component analysis (PCA) was employed to determine the maximum variation in colour, followed by a threshold to find candidate fat objects and finally an object filter to improve the specificity of fat object detection.

Data from three animals were used to analyse the relationship between the NIR spectra and the visible fat content in the camera images. Data from the fourth animal was kept exclusively for independent validation of the prediction method. To relate the NIR spectra to the visible fat content, Partial Least Squares (PLS) regression was used (de Jong, 1993). PLS works by finding and utilising the covariance between data and response. In the context of this work it obtained and utilised spectral information indicative of the interaction of NIR light with fat. Predominantly, this information was due to light interaction with water, hydrocarbon chains and the higher absorption of light by lean than by fat. The model was validated by cross-validation, *i.e.* a PLS model was built in turn on data from two animals and tested on the third (per animal hold out subset evaluation). Examination of spectra per animal increases independence of cross-validation test sets, simulating behaviour of a separate validation dataset (but does not reduce the importance of validation on an independent test data set).

Results and Discussion

Figure 1A shows two typical examples of NIR spectra from meat demarcated as being entirely lean and almost entirely fat by the image analysis. Due to the NIR probe window size no spectrum was entirely fat, so greater than 85% fat was chosen. Figure 1B shows the visible fat content from the image analysis against the PLS prediction result from the NIR spectra using per animal cross-validation training and testing with 95% confidence intervals computed via the root mean square error of prediction.

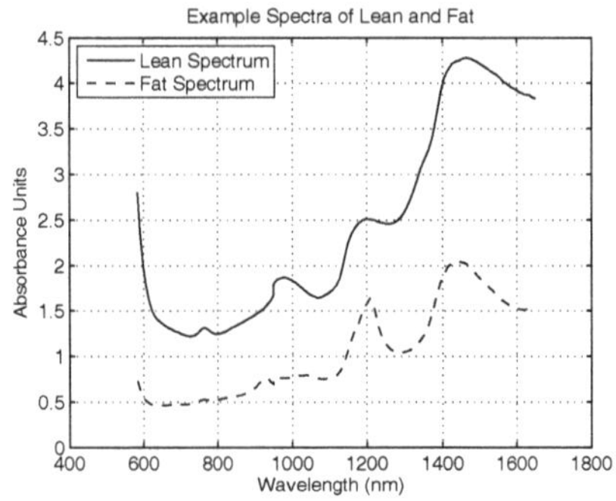

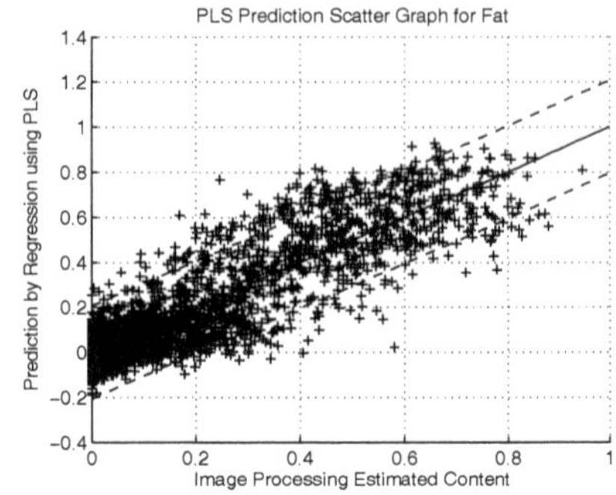

Figure 1A: Example NIR spectra of lean and fat samples.

Figure 1B: Scatter plot of visible fat predicted by NIR using PLS.
(95% prediction intervals are shown as dashed lines).

The spectral shapes of fat and lean in Figure 1A are very similar; however, several important differences are apparent. Most apparent is that the absorbance of light by fat is less than by lean. Examples of other useful variations are the overlapping water/hydrocarbon peaks around 1200nm and 1450nm and the peak in the lean spectrum at 970nm that is not in the fat spectrum. Figure 1B shows that an approximately linear relationship was found between the visible fat content and NIR measurements using PLS. The coefficient of correlation (R) between visible fat content and prediction of fat content using PLS on the NIR data was 0.90. The Root Mean Square Error of Prediction (RMSEP) was 0.103. Despite the strong correlation, the 95% confidence interval in prediction is 0.206, or approximately 21% of the maximum value.

Conclusions

Calibration of NIR data against visible fat content gave a correlation of 0.90 and an RMSEP of prediction of 0.103. Future work may include PLS and other standard NIR analysis of lean and connective tissue and investigations into non-linear mathematical techniques. In addition, consideration will be given to the problem of classifying a spectrum as lean or not lean.

References

de Jong, S. (1993). SIMPLS: An alternative approach to partial least squares regression. *Chemometrics and Intelligent Laboratory Systems* 18, 251-263.

Devine, C.E., Wahlgren, N.M., and Tornberg, E. (1999). Effect of rigor temperature on muscle shortening and tenderisation of restrained and unrestrained beef *m. longissimus thoracis et lumborum*. *Meat Science* 51, 61-72.

Gonzalez, R.C. and Woods, R.E. (2002). *Digital Image Processing*, 2nd ed., Prentice Hall, Upper Saddle River, NJ.

McGlone, V.A., Devine, C.E., and Wells, R.W. (2005). Detection of tenderness, post rigor age and water status changes in sheep using near infrared spectroscopy. *Journal of Near Infrared Spectroscopy* 13, 277-285.

POST MORTEM AGING AND FREEZING OF CAMEL MEAT (A COMPARATIVE STUDY)

R. Shariatmadari* and M. Kadivar

Department of Food Science and Technology, Isfahan University of Thechnology, Isfahan, 84156, Iran, r_shariatmadari@yahoo.com

Keywords: camel meat, physico-chemical properties, ageing, frozen storage

Introduction

Regarding the great impact of meat (as the major source of nutrients) in the human diet, high meat prices and its deficiency particularly in developing countries, introduction of cheaper sources of meat, such as that of camel, will serve consumers' interest. A fattened camel at an age of 7 years can produce a carcass of about 260 kg with a meat / bone ratio of 3.0 (Yousif, 1989). Elgasim *et al.*, (1987) concluded that the carcass characteristics were equal to those of other red meat animals. The meat of camels 3 years old or less is comparable in taste and texture to beef (Knoess, 1977). It is also nutritionally the same as common sources of meat (Elgasim and Alkanhal, 1992). Despite this potential, the quality of camel meat has received little attention. There is even a notion about its toughness and lower quality among consumers. However, in spite of this notion, no data have been published on the comparison of its physicochemical attributes with other common sources of meat. Moreover, aging and freezing which have profound effects on meat have not been widely studied in this regard. The objective of this study was to investigate the changes in various physicochemical properties of camel meat during aging and freezing compared with those of beef.

Materials and Methods

Three year old Iranian Zaboli camels (n=3) or hybrid Holstein cows (n=3), were randomly selected from two different slaughterhouses. Cows were stunned (150 to 180 mV) before slaughtering. Intact samples from leg (5 kg) were obtained within 45 min after slaughtering and stored at *ca* 15°C before subsampling. Each sample was then subsampled at 5h postmortem to determine the physicochemical properties of the fresh meat. The rest of each primary sample was subdivided in two lots. One lot which was used to determine the effect of postmortem aging on physicochemical properties, was refrigerated at 4 °C until further sampling at 24, 72 and 168 h postmortem. Analysis included measuring pH, tenderness (Smith *et al.*, 1978), water holding capacity [WHC] (Jauregui *et al.*,1981) and cooking loss [CL] at 5, 24, 72 and 168 h post mortem; tyrosine and TBA values (Strange and Benedict, 1977) and colour (L, a and b values) at 5, 72 and 168 h post mortem. The second lot of each sample was used to investigate the effects of freezing on the same characteristics. This lot was stored at -20°C for a maximum period of five months and subsampled at months 2 and 5 of storage. At each measuring time, samples were thawed for 24 h at 4°C. Data were compiled in split plot design format and analysed by analysis of variance and Duncan's multiple range test using SAS statistical analysis software. All measurements were taken in duplicate.

Results and Discussion

ANOVA indicated no significant difference between species for physicochemical properties except in pH (Table 1 and 2). Mean pH was higher in camel meat relative to beef ($p<0.05$). Also, for most of the factors, fresh camel meat and beef (5 h postmortem) were similar. However, differences in pH, tyrosine, TBA and "a" values were evidenced ($p<0.05$, Table 1). Aging caused significant differences: tenderness, CL, Tyrosine value, TBA value, colour L, a, and b values increased, whereas pH and WHC decreased (Table 1). As table 1 shows, during 72 h aging, tenderness, tyrosine value, colour L, a, and b values showed higher increases in camel meat relative to beef, so that camel meat was found to be more tender and brighter with higher "b" value at 72 h compared to the beef ($p<0.05$). This could be an indicator of a higher rate of proteolysis in camel meat during the initial stages of aging beef. Camel meat also showed lower TBA values than beef at different aging intervals. Frozen storage affected all parameters except tenderness and colour significantly (Table 2). WHC and pH decreased, whereas, CL, tyrosine and TBA values increased by freezing. In the frozen state, camel meat and beef were similar in all parameters except cooking loss and redness; although fresh camel meat had a smaller "a" value than beef, during frozen storage the redness of camel meat in contrast to the beef, showed a significant increase resulting in camel meat being found redder than beef after 5 months storage ($p<0.05$, Table 2). This can be attributed to the reduced oxygen consumption rate due to low freezing temperature leading to deeper oxygenated layers of meat and delayed oxidation. Camel meat showed higher amounts of cooking loss than beef during frozen storage ($p<0.05$).

Table 1: Physicochemical properties of meat as influenced by species and storage time during aging period.

	Aging time (h)	pH	Shear force (KgF)	WHC (expressible moisture)	CL (%)	Tyrosine value (mg/g)	TBA value (OD)	L	a	b
species(p<)		0.035	0.065	0.367	0.702	0.716	0.410	0.095	0.595	0.166
Aging(p<)		0.0001	0.0001	0.0001	0.0001	0.001	0.031	0.002	0.002	0.0001
Camel Meat	5	6.33^{a}	6.917^{a}	37.14^{d}	15.24^{d}	0.33^{d}	0.031^{c}	37.98^{bc}	9.43^{c}	7.18^{c}
	24	5.79^{cd}	5.317^{bc}	41.17^{cd}	13.40^{d}	-	-	-	-	-
	72	5.9^{bc}	5.093^{cd}	46.06^{ab}	20.32^{ab}	0.44^{abc}	0.033^{c}	43.20^{a}	12.70^{a}	12.04^{a}
	168	5.66^{dc}	3.857^{d}	49.67^{a}	22.22^{a}	0.48^{ab}	0.038^{b}	43.03^{a}	12.26^{ab}	12.51^{a}
Beef	5	6.02^{b}	7.173^{a}	37.26^{d}	15.91^{cd}	0.40^{c}	0.038^{ab}	35.33^{c}	11.25^{b}	8.06^{c}
	24	5.6^{e}	7.120^{a}	44.02^{bc}	18.47^{bc}	-	-	-	-	-
	72	5.63^{e}	6.390^{ab}	46.22^{ab}	18.50^{bc}	0.43^{bc}	0.040^{ab}	38.98^{b}	11.42^{ab}	10.25^{b}
	168	5.57^{e}	4.837^{cd}	49.91^{a}	21.24^{ab}	0.51^{a}	0.042^{a}	39.52^{b}	12.54^{ab}	11.14^{ab}

1- Means within a column with different superscripts differ significantly ($p<0.05$) according to Duncan's multiple range test

Table 2: Physicochemical properties of meat as influenced by species and storage time during frozen storage.

	Frozen time (m)	pH	Shear force (KgF)	WHC (expressible moisture)	CL (%)	Tyrosine value (mg/g)	TBA value (OD)	L	a	b
Species(p<)		0.046	0.284	0.628	0.245	0. 113	0.407	0.874	0.606	0.575
Frozen storage(p<)		0.0001	0.514	0.0001	0.0001	0.010	0.010	0.560	0.141	0.460
Camel meat	0	6.33^{a}	6.917	37.14^{b}	15.24^{d}	0.33^{c}	0.031^{b}	37.98	9.43^{c}	7.18
	2	5.71^{c}	7.373	45.72^{a}	29.92^{ab}	0.50^{ab}	0.108^{ab}	-	-	-
	5	5.50^{cd}	6.657	50.97^{a}	30.80^{a}	0.46^{abc}	0.160^{a}	34.01	12.45^{a}	8.17
Beef	0	6.02^{b}	7.173	37.26^{b}	15.91^{d}	0.40^{bc}	0.038^{b}	35.33	11.25^{ab}	8.06
	2	5.53^{cd}	7.770	46.17^{a}	24.75^{c}	0.53^{ab}	0.196^{a}	-	-	-
	5	5.42^{d}	6.757	46.96^{a}	27.00^{bc}	0.57^{a}	0.130^{ab}	37.76	9.84^{bc}	8.43

1- Means within a column with different superscripts differ significantly ($p<0.05$) according to Duncan's multiple range test

Conclusions

Meat quality was found to be mainly affected by aging and freezing rather than by species. Fresh camel meat and beef were similar in all parameters except pH, tyrosine, TBA and "a" values. Positive aging-related changes in tenderness and colour occurred earlier in camel meat compared to beef. It also showed less TBA values throughout aging period than beef. Considering the importance of tenderness, colour and lipid oxidation for meat consumers, it could suggest a possible advantage for camel meat.

References

Elgasim, E. A. and Alkanhal, M. A. (1992). Proximate composition, amino acids and inorganic mineral content of Arabian camel meat: Comparative study. Food Chem., 45: 1-4.

Elgasim, E. A., Elhag, G. A. and Elnawawi, F.A. (1987).Quality attributes of camel meat. Second Progress Report, the Scientific Council, King Faisal University, Alhasa, KSA.

Jauregui, C. A. and Regenstein, J. M. (1981). A simple centrifugal method for measuring expressible moisture, a water-binding property of muscle foods. J. Food Sci., 46: 1217, 1273.

Knoess, K. H. (1977). The camel as a meat and milk animal. World Anim. Rev., 22: 3-8.

Smith, G. C., Culp, G. R. and Carpenter, Z. L. (1978). Postmortem aging of beef carcass. J. Food Sci., 43: 823- 826.

Strange, E. D. and Benedict, R. C. (1977). Evaluation of rapid tests for monitoring alterations in meat quality during storage. J. Food Protect., 40: 843- 847.

Yousif, O. K. And Babiker, S. A. (1989). The desert camel as a meat animal. Meat Sci., 26: 245- 254.

FUNCTIONAL AND ULTRASTRUCTURAL PROPERTIES OF CAMEL MEAT

R. Shariatmadari* and M. Kadivar

Department of Food Science and Technology, Isfahan University of Thechnology, Isfahan, 84156, Iran, Email:r_shariatmadari@yahoo.com

Keywords: camel meat, functional properties, ultrastructure, freezing

Introduction

A few studies were carried out concerning the production of comminuted meat products from camel meat. Sadek (1966) reported that using camel meat in sausage manufacture losses its toughness. Heikal *et al.*, (1972) by studying the histological and technological aspects of smoked sausage from camel meat reported that a sausage with high quality could be prepared. Dawood (1995) reported that replacement of lean meat with 5 or 10% fat improved sensory attributes of camburger. However, with regard to the importance of the muscle proteins functionality in the final quality of comminuted meat products, no data has been published concerning functional properties of camel meat. Moreover, freezing can produce profound effects on the structure and functionality of meat whereas no information concerning such criteria has been reported in camel meat. The objective of this study was to investigate functional and ultrastructural changes in camel meat due to freezing, which in functionality compared with beef.

Materials and Methods

Four Iranian Zaboli camels (one approximately 12 months and the others 5 years old) and three hybrid Holstein cows (approximately 5 years old) were randomly selected from two different slaughterhouses. Cows were stunned in the range of 150 to 180 mv before slaughtering. Sample from *M. longissimus dorsi* at the 12th rib was obtained from 12-month old camel within 15-20 minutes after slaughtering and divided into two parts. One part was immediately fixed in glutaraldehyde 2.5% and used to study the ultrastructure of pre-rigor muscle. The other part was frozen stored at -20°C for 5 months and used to investigate the ultrastructural changes due to freezing. For electron microscopy muscle samples were subjected to several treatments (fixation, washing, dehydration, embedding and polymerization) to prepare ultrathin sections (70nm) which were examined with an electron microscope. In order to determine functional properties, intact samples from leg were obtained from 5-year old animals (n=6) within 45 minutes after slaughtering. Samples were refrigerated at 4°C until further sampling at 24h to determine the functional properties of fresh meat. The remaining lot of each sample was frozen stored at -20°C and subsampled after 2 and 5 months to investigate the effects of freezing. Samples were thawed for 24 h at 4°C. Analysis include pH, water holding capacity (WHC) (Jauregui *et al.*, 1981), emulsifying capacity (EC) and emulsion stability (ES) (Swift *et al.*, 1961), foaming capacity (FC) and foam stability (FS) (modification of method described by Lin et al., 1974) and buffering capacity (BC) over the ranges pHu-4.5 and pHu-7. Data was compiled in split plot design and analysed by analysis of variance and Duncan's multiple range test using SAS. Measurements were taken in duplicate.

Results and Discussion

Table 1: Functional properties of meat as influenced by species and frozen storage

	Frozen time (m)	pH	WHC (E.M.[f])	EC (ml/g)	ES (%)	FC (ml)	FS30[g] (%)	FS60[h] (%)	BC[i] (pHu-4.5)	BC (pHu-7)
species(p<)		0.064	0.821	0.018	0.0007	0. 552	0.469	0.305	0.063	0.853
storage(p<)		0.0001	0.514	0.0001	0.0003	0.709	0.019	0.014	0.0001	0.007
	0	5.81 ^{a}	39.80 ^{c}	40.69 ^{e}	68.16^{cd}	363.3	17.75 ^{c}	11.17 ^{c}	4.47 ^{bc}	3.81 ^{c}
camel meat	2	5.73 ^{ab}	46.42 ^{b}	46.75 ^{b}	74.63 ^{ab}	373.3	63.10 ^{ab}	49.63 ^{ab}	5.12 ^{b}	4.44^{abc}
	5	5.50 ^{c}	51.16 ^{a}	48.93 ^{a}	76.85 ^{a}	528.3	76.19 ^{a}	66.84 ^{a}	3.55^{d}	5.23 ^{a}
	0	5.61 ^{b}	41.03 ^{c}	42.57 ^{d}	61.99 ^{e}	355	23.67 ^{bc}	14.13 ^{bc}	5.00 ^{b}	4.02 ^{bc}
beef	2	5.55 ^{bc}	47.23 ^{ab}	44.11 ^{c}	65.13 ^{de}	490	34.72 ^{abc}	27.61 ^{bc}	6.56 ^{a}	4.62 ^{ab}
	5	5.44 ^{cd}	49.33 ^{ab}	44.81 ^{c}	71.55 ^{bc}	293.3	52.54 ^{abc}	37.55^{abc}	3.98 ^{cd}	4.64 ^{ab}

[a-e] Means within Columns with different superscripts differ significantly ($p<0.05$) according to Duncan's multiple range test
[f] Expressible Moisture [g] Foam stability after 30 min. [h] Foam stability after 60 min. [i]mmol H^{+}/pH*100g meat

With regard to the functional properties, camel meat and beef were similar in all factors except in emulsifying properties (Table 1). Camel meat showed higher EC (M=45.46) and ES (M=71.23) relative to beef (respectively, M=43.83 and M=66.22); ($p<0.05$). Freezing caused significant differences: EC, ES, FS and BC increased, whereas pH and WHC decreased due to freezing. Although fresh camel meat had lower EC than beef, it showed more EC after 2 and 5 months storage relative to beef ($P<0.05$).

With regard to the ultrastructure, the electron micrograph of camel LD in pre-rigor state (Figure 1), shows parallel myofibrils to be intact with characteristics transverse striations; the sarcomers are clearly bordered by Z-lines. A-, I-, and H-segments, also M-bands are visible. Large mitochondria can be seen between myofibrils. In frozen muscle, electron micrograph (Figure 2) shows compressed myofibrils to be attached parallel with each other. They are relatively thin and dense. These are consistent with the observations of Rahelic *et al.,* (1985) in the LD of beef frozen at -22 °C and Grujic *et al.,* (1993) at the slow freezing rates. This compression is attributed to the pressure of ice crystals. The structure of sarcomer is clearly seen in many places. Mean sarcomer length reduced from 2.02μm to 1.57μm after frozen storage. Fissures inside sarcomers particularly in I-segments and Z-lines are noticeable. In few parts these fissures elongated to A-segments or even M-bands. Rahelic *et al.,* (1985) observed differently shaped gaps in several fibrils, mostly at Z-lines and in the I-bands.

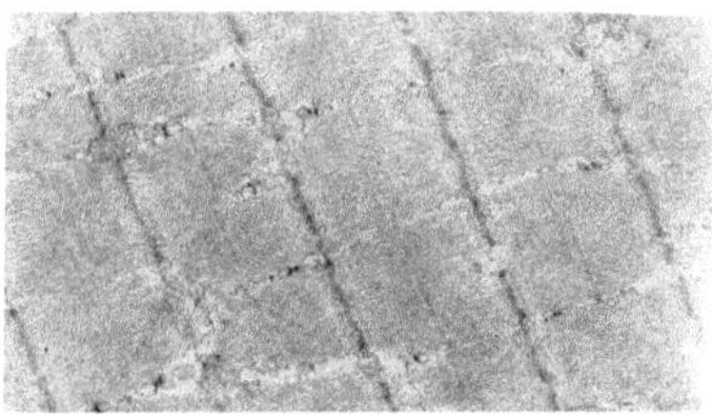

Figure1: Electron micrograph of Camel *Longissimus dorsi* in prerigor muscle, (x 22250).

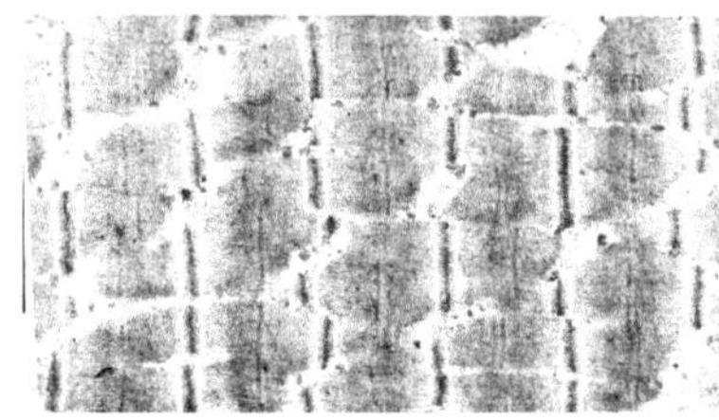

Figure 2: Electron micrograph of Camel *Longissimus dorsi* after 5 months frozen storage, (x 22250).

Conclusions

In general, it seems that functional properties of camel meat are similar to those of beef; freezing also affected functionality of both meats in a similar pattern. So, our findings suggest that camel meat can be used in the manufacture of comminuted meat products; as previous studies regarding technological and sensory aspects of comminuted meat products from camel meat (Sadek, 1966; Heikal *et al.,* 1972; Dawood, 1995) have shown. It also may have an edge for consumers as well as meat products producers over beef due to its lower price. With regard to the ultrastructure, camel LD shows normal structure of a striated muscle. Frozen storage at -20 °C affected the ultrastructure of camel LD similar to that reported by other researchers in beef.

References

Dawood, A. A. (1995). Physical and sensory characteristics of Najdi camel meat. Meat Science, 39: 59- 69.

Grujic, R. and petrovic, L. (1993). Definition of the optimum freezing rate.1- Investigation of structure and ultrastructure of beef M. Longissimus dorsi frozen at different freezing rates. Meat Science, 33: 301- 318.

Heikal, H. A. and El- Dashlouty, M. (1972). Biochemical, histological and technological changes occurring during the production of sausage from camel meat and beans. Agricultural Research Review, 50: 244-252.

Jauregui, C. A. and Regenstein, J. M. (1981). A simple centrifugal method for measuring expressible moisture, a water-binding property of muscle foods. Journal of Food Science, 46: 1217, 1273.

Lin, M. J. Y. and Humbert, E. S. (1974). Certain functional properties of sunflower meal products. Journal of Food Science, 39: 368- 372.

Rahelic, S. and Gawwad, A. H. (1985). Structure of beef longissimus dorsi muscle frozen at various temperatures: Part 2- Ultrastructure of muscle frozen at- 10, - 33, -23, -78 and- 115°c. Meat Science, 14: 73- 81.

Sadek, M. 1966, in: Yagil, R. (1982). Camels and Camel Milk. FAO, Rome, 31- 32.

Swift, C. E. and Sulzbacher, W. L. (1963). Comminuted meat emulsions: factors affecting meat proteins as emulsion stabilizer. Food Technology, 17: 224- 226.

BEEF CATTLE COMMERCIALISATION IN MEXICO: A CASE STUDY

J.M. Salas González*, L.M. Sagarnaga Villegas, J.L. Chávez Barraza and F. Cervantes Escoto

Departamento de Sociología Rural, Universidad Autónoma Chapingo, Carretera México-Texcoco Km 38.5, Chapingo, México, CP 56230, México. E-mail: jmsalas@correo.chapingo.mx

Keywords: commercialisation, cattle, cattle ranching, Mexico

Introduction

Cattle ranching is an important economic as well as social activity in Mexico. Mexico exported 1,374,737 live calves to the United States of America (the U.S.A.) in 2004[2], along with other bovine products which generated US$619,362,000, an equivalent of 72% of the total animal products exported. Calf exports have generated important incomes in those areas that have export permits because of their disease-free status. Nevertheless, the increased demand for calves for export has had a negative impact on the national beef industry. An important proportion of the fattening and slaughtering capacity is underutilised, while meat imports show accelerated growth. Imports in this sector reached US$1,109 million in 1994, consolidating this sector's position as a net importer.

In 1994, the state of Guanajuato produced 33,370 metric tons of slaughtered beef, only 2% of total Mexican production. Said production is lower than consumption in the state so it is necessary to bring cattle from other states in the country. The state is awaiting disease free certification which would permit access to the export market. The state government, along with the producers, undertook a project to modernise the state cattle industry in the hopes of better facing the impact of calf exports.

Materials and Methods

In order to evaluate the conditions under which buying and selling of live beef occurs in the state of Guanajuato, México (the producer-introducer segment in the commercial chain), an analysis, based on a questionnaire applied to 88 producers, and on semi structured interviews of opinion leaders in the sector, was made.

Results and Discussion

The majority, (87%), of the producers interviewed had herds of fewer than 50 productive cows. The principal production system, (53%), is raising calves, 38.6% are corral fattened, 3.4% are grazing fattened, 3.4% have dual purpose production (meat and milk), and the remaining 1.2% produce breeding cows.

According to the national evaluation of the Programa de Fomento Ganadero de la Alianza para el Campo (FAO-SAGARPA 2005), national cattle producers are typically middle aged or older and have a low level of education. The average age of the cattle ranchers interviewed in Guanajuato is 49; these have primary schooling of 6 to 7 years. This limits their positive disposition for change.

Only 35% of those interviewed are dedicated full time to cattle raising. The rest do other activities to supplement the family's income: agriculture, (65%), commerce (14%), and services (7%) are mentioned.

Of the cattle ranchers who are also farmers, 44% have irrigable land, at an average size of 21.7ha. Of these, 84% also have seasonal lands, on average 17.2ha. The principal crops of said lands are corn, (91%), sorghum, (40%), and beans (32%).

The cattlemen who have grasslands (44%) have an average of 117ha that has an average capacity of 145 animal units. According to the cattlemen, they are only exploited at 63.7% of their capacity. In order to separate the property and the pasture lands, 67% of those interviewed have a perimeter fence on their development and only 57% have dividing fences, which can be an indication of bad planning with regards to the operation/regulation of grasslands.

The available infrastructure on the ranches is limited: 72% have troughs, 66% have management corrals, 47% have mills, and 43% have tractors. The infrastructure related to the commercialisation of cattle is even more limited: 45% have management corrals, 41% have loading docks, 15% have a press, and only 11% have an animal livestock scales.

The animals are handled principally by the producer and his family (56.5%). Feeding is mainly based on grasslands and in a few ranches (36%) supplements are given. Corral fattening is based on the same diet during the entire period of production.

With regards to sanitary management related to meat hygiene, it was discovered that an important proportion of the producers used chemical products in handling the animals, such as: vaccinations (87%), medications and antiparasite medication (86%), and growth inducers (19%). The growth inducers mentioned are Ralgro and Zilmax, though no producer mentioned the use of Clenbuterol. Of those interviewed, 94% stated that these products were stopped 30 days before the animals were sent to market.

---

[2]The most recent official statistics available are from 2004.

The most common sale procedure is that when the animals are ready to be sold, the producer contacts the buyers (55%), or the buyers regularly pass by the development (35%). Only 6% of the producers have a delivery agreement for a set period of time.
According to the information provided by the producers, the criteria followed by the buyers to establish the price for the cattle are, in order of importance: that the animals be well formed (35%), that the animals look healthy (30%), the breed (29%), and the animal weight (24%). With regards to the last criterion, given that the ranchers generally do not have a scale, even when weight is a determining factor, the sale price is established as a whole.
Of the producers interviewed, 13% said that they receive a bonus payment on the sale of their livestock. The average bonus payment is 1.60 pesos per kilo (about 0.14 Euros), though it can range from 50 cents up to 3 pesos per kilo (€0.04 to €0.21).
Payment, according to the majority of those interviewed, (49%), is deferred. The waiting time for payment is lengthy, from one to two weeks, (31%), from two to three weeks, (12%), more than four weeks, (5%), while only 10% of the producers are paid in less than a week.
A significant group of producers stated that they had been cheated by the buyers because they lacked information about livestock prices on at least one occasion, (5%), and more than once (14%).
Twenty-two percent of the state's cattle herds are specialised in milk production. This production system generates an important number of calves that are fattened and commercialised within the same state. An unexpected finding is that the state's consumers preferred this type of cattle for its lower fat content. An analysis of this segment of livestock ranching and its impact on the meat industry should be studied further in later research.

It was concluded that this activity is considered as a complementary activity for family income. State livestock ranching is of a rustic family type. The level of technology could be greatly improved. The producer depends on middle-men, who determine the buying conditions, almost exclusively to commercialise their livestock.

Livestock ranching shows a minimal level of development in Guanajuato, Mexico which has to be improved: first of all, in order to satisfy the state's demand, and then in order to explore the export market.

References

FAO-SAGARPA, (2005). Informe de Evaluación Nacional Programa de Fomento Ganadero. Evaluación Alianza para el Campo 2004.

DEVELOPMENT OF A MEAT GOAT SELECTION, CARCASS EVALUATION AND FABRICATION GUIDE

K.W. McMillin*[1] and F. Pinkerton[2]

[1]Department of Animal Sciences, Louisiana State University Agricultural Center, Baton Rouge, Louisiana, USA 70803-4210 and [2]197 Thomas Drive, Martindale, Texas USA 78655 Email: kmcmillin@agctr.lsu.edu

Keywords: goat, conformation, carcass evaluation, fabrication, meat cuts

Introduction

Goats are an important food source, especially in technologically developing regions (Casey, 1992). Market value or price differentials are based on characteristics of quality (palatability) and quantity (cutability) through animal and carcass grading systems (Sims, 1985). Value of commodities or products can be increased by decreased costs or by higher relative final product values (McMillin and Brock, 2005). Recommended goat carcass evaluation procedures were proposed (Pinkerton and McMillin, 2000) to complement standard institutional goat meat cuts (USDA, 2001). The project objectives were to develop a printed manual and digital presentations for live goat and carcass conformation descriptions, meat goat carcass evaluation, and summarization of standardized fabrication of goat carcasses.

Materials and Methods

This project utilized data, pictures, and techniques from previous studies (McMillin et al., 1999). More than 600 electronic images and 700 slides (5.1 x 5.1 cm) of live meat goat front, rear, or side views were previewed to eliminate those that were blurry, depicted abnormal muscle tension, or did not allow clear distinction of the goat body. An expert panel of livestock judges familiar with different meat goat sizes, ages, and breeds reviewed the resulting 425 electronic images and 150 slides of kid meat goats against the three written selection classification descriptions (USDA, 2001) to reach a consensus selection class score for each image. Only images with the different views of an individual animal with selection scores within 20 points of one another were considered for inclusion in the manual. It was decided that images representative of the midpoints of the three selection classification groups would be desired for the manual and presentation materials. An expert panel of certified meat graders and meat scientists familiar with goat carcasses and cuts reviewed more than 120 electronic images of goat carcass conformation (rear and side views) and 40 images of flank colour score, kidney fat, and external fat cover scores that had been previously screened to eliminate blurry or incomplete images. A draft manual of the selected images and explanatory text and compact diskettes with presentations on the manual, goat and goat meat marketing, carcass fabrication, and value-added goat meat products were developed.

Results and Discussion

The manual was developed as a guide for the meat goat industry in evaluating live meat goats, measuring carcass traits, and standardizing cutting procedures for goat carcasses. There are no official USDA grades or standards for live meat goats or goat carcasses, but the terminology in this manual was been adapted from USDA Institutional Meat Purchase Specifications (IMPS) Selection Criteria (USDA, 2001). The manual can be revised as the meat goat industry produces kid goats with higher conformation and heavier muscling or needs additional or different cuts of goat meat. Live goat traits that influence meat yield are conformation, proportions of muscle to fat and bone, and relative body size. Selection 1 goats have a superior conformation that yields a high proportion of meat from the carcass, while Selection 2 goats have an average meat type conformation and Selection 3 goats have inferior conformation and a low ratio of meat to body components. External or subcutaneous fat is deposited on kid meat goats over the ribs and behind the shoulder, with minimal fat over the back muscle (*Longissimus dorsi*), except with high concentrate energy feeding. The distinctive hip and leg structure of goats changes during cooling of carcasses to give meat goat carcasses a more elongated and stretched appearance, even with heavy muscled carcasses, compared with other red meat species. The carcass evaluation system developed easily estimates the traits influencing lean meat yields and customer preferences with hot carcass weight; conformation (Selection 1, Selection 2, or Selection 3); external fat score (1= minimal/none, 2= fat over rib/shoulder, 3= excessive fat cover); kidney, heart, and pelvic internal fat as a percentage of hot carcass weight; and lean flank colour (A- pale red, B- dark red, C- very dark red). The five meat goat carcass fabrication styles (platter, roasting, barbeque, food service, hotel) for different weight and sizes of carcasses in the USDA Fresh Goat IMPS Series 11 (USDA, 2001) are described for shanks, neck, foresaddle, shoulder, outside shoulder, inside shoulder, rack, ribs, breast, back, loin, sirloin, and legs.

Conclusions

The goat manual integrated data and images from different goat, carcass, and goat meat projects with the detailed knowledge of goats and meat carcass evaluation of the expert panel. The manual provides visual images for use by goat producers, meat processors, and meat goat purchasers and standardizes communication among the different market channel segments.

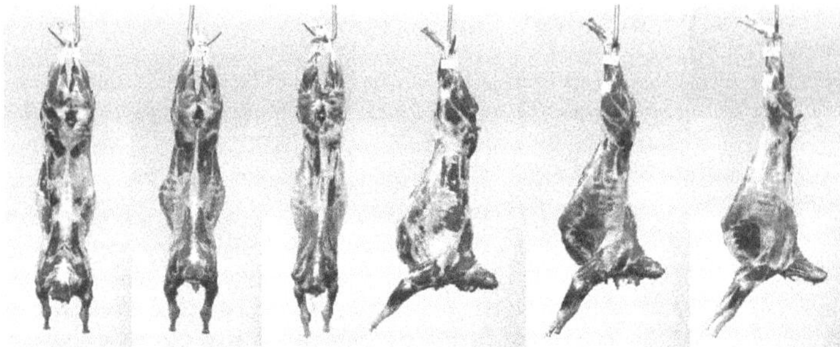

Figure 1: Comparisons of the three meat goat carcass selection classifications (1^{50}, 2^{50}, 3^{50}, left to right).

References

Casey, N.H. (1992). Goat meat in human nutrition. In:Proceedings of the V International Conference on Goats, March, 1992, New Delhi.

McMillin, K.W. and Brock, A.P. (2005). Production practices and processing for value-added goat meat. Journal of Animal Science, 83(E Suppl.): E57-E68.

McMillin, K.W., Olson, S., Phelps, O., Pinkerton, F. and Michel, M. (1999). Goat meat carcass evaluation and IMPS cuts. Reciprocal Meat Conference Proceedings, 52: 79.

Pinkerton, F. and McMillin, K.W. (2000). Evaluation of slaughter goat selection and goat meat classification systems. Goat Rancher, February: 7-10.

Sim, D.W. 1985. Grading and classifying meat: An international view. Reciprocal Meat Conference Proceedings, 38: 57-61.

USDA. (2001). Institutional Meat Purchase Specification for Fresh Goat Series 11. United States Department of Agriculture Agricultural Marketing Service Livestock and Seed Program, Washington, DC.

SESSION 6: MEAT AND HEALTH

HEALTHY BEEF – THE EFFECT OF DIET ON THE FATTY ACID COMPOSITION OF BEEF FROM HUNGARIAN GREY AND HOLSTEIN-FRIESIAN YOUNG BULLS

G. Holló*[1], J. Seregi[1], K. Ender[2], I. Repa[1] I. Holló[1] and K. Nuernberg[2]

[1.] *Institute of Diagnostic Imaging and Radiation Oncology, University of Kaposvár, Guba S. 40.Hungary*
[2.] *Research Institute for Biology of Farm Animals, Dummerstorf, Wilhelm-Stahl-Allee 2, Germany.*
Email: hollo.gabriella@sic.hu

Keywords: beef, fatty acid composition, human health

Introduction

According to current human dietary principles a reduction in the fat content of the diet and the modification of fatty acid composition of meat is recommended. The meat with "healthy" fatty acid composition contains a lower amount of saturated fatty acids; its maintenance of the myristic and palmitic acid content is low, the ratio of PUFA/SFA is higher than 0.45, the ratio of *n*-6/*n*-3 PUFA is lower than 5:1 and the fat contains a high amount of the conjugated linoleic acid (CLA) (Scollan *et al.*, 2001, Nuernberg *et al.*, 2005). It was demonstrated in cattle, that feeding concentrates enriched in *n*-3 fatty acids, increased the content of long chain *n*-3 PUFA in beef muscle (Nuernberg *et al.*, 2005). On the other hand, in several studies differences in fatty acid composition among cattle breeds under same feeding regimes were established (Nuernberg *et al.*, 1998). Consumers are not only concerned about meat quality and its healthiness, but also about its origin and how animals are reared. In the EU many local breeds are kept under traditional conditions and their meat is labelled for a quality guarantee. This gives the Hungarian old animal breeds (especially Hungarian Grey,) a greater opportunity. Knowledge of the meat quality of traditional animal breeds is needed for the effective product output. It is an important research target to examine and evaluate the nutritional value of meat for human health.
The aim of our fattening trial was to analyse the effect of extensive vs. intensive nutrition on beef quality and fatty acid composition from human nutritional point of view using Hungarian Grey (HG) cattle bulls in comparison to Holstein-Friesian (HF) bulls.

Materials and Methods

HG and HF growing-finishing bulls were fed rations consisting of either grass, grass silage and concentrate or maize silage and concentrate with and without linseed supplementation according to a 2 x 2 factorial experimental design in four groups. Ten animals were assigned per group into the above treatments. In the extensive and intensive groups days on feed lasted for 221 and 201, respectively. There were no significant differences among groups in the initial live weight. The nutrient content of different feedstuffs can be seen in Table *1*. Concentrates for extensive groups were supplemented 20 % linseed meal containing 44.77 % linolenic acid fed in the last month of growing-finishing period. Average final weights were actually identical in all groups (512.4±58.4 kg). The animals were slaughtered and the left half carcasses were dissected according to DLG method. After a 24hr chilling *longissimus* (LD) samples were taken from the left carcasses. The fatty acid composition of samples was determined according to method described by Nuernberg *et al.*, (2002). Data processing was performed using SAS program package.

Table 1: Chemical composition of feed.

%	intensive			extensive		
	Maize silage	Hay	Concentrate	Grass silage	Grass	Concentrate with linseed
Dry matter	36.4	91.4	88.0	23.3	23.6	90.8
Crude protein	3.2	7.0	16.4	2.3	2.8	18.0
Crude fat	0.9	1.7	2.0	0.9	0.6	4.3
Crude fibre	7.0	35.9	13.0	8.8	6.6	9.4

Results and Discussion

Higher final weights were recorded for the intensively fed bulls at the end of fattening (i:555.00±35.62 kg, e:469.75±44.03 kg), with differences between breeds not significant (HG:506.35±67.01 kg, HF:518.40±49.66 kg).
The extensive feeding resulted as less fat deposition and more lean into their carcass, on the other hand the HG bulls had more lean meat content in both feeding groups, than the HF (HG:69.41+2.26 %, HF:66.62+2.44 %, $P<0.001$).
Table 2 summarises the main fatty acids of LD. Rumen biohydrogenation in cattle limits the amount of long chain PUFA which can be transferred into ruminant muscle. As a result of this, the PUFA/SFA ratio for beef is typically very low. Findings reveal that in the case of the extensive fed HG the ratio of SFA/UFA of *longissimus* could be more favourably altered, than in case of the HF bulls. On the other hand the HG deposited more SFA in muscle tissues opposite to HF bulls. The process of biohydrogenisation also results in the production of CLA intermediates. It is recognised that CLA has numerous positive effects on the human health. The CLA content of *longissimus* muscle was

influenced by breed as well as by feeding. The CLA percentage of *longissimus* muscle in HG bulls was higher, than that of HF considering any groups. On the other hand, with extensive feeding CLA content in muscle can be two times higher.

In nutritional guidelines there are separate recommendations for *n*-6 and *n*-3 classes of PUFA. There are beneficial effects of *n*-3 PUFA in the prevention of cardiovascular disease and cancer in contrast to opposite effects indicated for *n*-6 PUFA. The amount of *n*-3 and *n*-6 fatty acids was significantly higher in extensive fed groups. In the *n*-6 fatty acid content of beef, there was a significant difference between breeds. HG had in both feeding groups a lower value, consequently their ratio of *n*-6/*n*-3 fatty acid is more advantageous for human nutrition.

Table 2: Fatty acid composition of *longissimus* muscle.

Fatty acids (%)	intensive				extensive				Significance
	HF		HG		HF		HG		P<0.05
	mean	SE	mean	SE	mean	SE	mean	SE	
C16:0	25.94	0.49	28.40	0.49	21.13	0.49	21.41	0.49	D,B,B*D
C16:1	2.52	0.13	2.82	0.13	1.46	0.13	1.57	0.13	D
C18:0	15.64	0.50	15.96	0.50	19.09	0.50	17.67	0.50	D
C18:1*trans*-11	1.88	0.27	0.89	0.27	4.46	0.27	4.78	0.27	D ,B*D
C18:1*cis*-9	32.86	0.82	35.21	0.82	25.88	0.82	25.97	0.82	D
C18:1*cis*-11	1.34	0.04	1.21	0.04	1.40	0.04	1.31	0.04	D ,B
C18:2*trans*	0.22	0.02	0.17	0.02	0.37	0.02	0.24	0.02	D ,B
C18:2 *n*-6	8.20	0.65	5.44	0.65	10.36	0.65	10.79	0.65	D,B*D
C18:3 *n*-3	0.75	0.18	0.61	0.18	2.73	0.18	3.21	0.18	D
*c*9,*t*11 CLA	0.27	0.03	0.36	0.03	0.58	0.03	0.81	0.03	D, B, B*D
C20:3 *n*-6	0.50	0.05	0.28	0.05	0.67	0.05	0.59	0.05	D,B
C20:4 *n*-6	2.36	0.26	1.18	0.26	3.55	0.26	2.87	0.26	D,B
C20:5 *n*-3	0.11	0.04	0.14	0.04	0.37	0.04	0.63	0.04	D,B,B*D
C22:4 *n*-6	0.30	0.03	0.13	0.03	0.41	0.03	0.22	0.03	D,B
C22:5 *n*-3	0.28	0.05	0.25	0.05	0.61	0.05	0.75	0.05	D
C22:6 *n*-3	0.04	0.01	0.05	0.01	0.10	0.01	0.13	0.01	D
SFA	45.80	0.68	49.20	0.68	44.58	0.68	43.80	0.68	D,B*D
UFA	54.20	0.68	50.80	0.68	55.42	0.68	56.20	0.68	D,B*D
PUFA	13.36	1.21	8.83	1.21	20.29	1.21	20.76	1.21	D,B*D
n-3 fatty acids	1.30	0.28	1.13	0.28	4.14	0.28	5.05	0.28	D
n-6 fatty acids	11.26	0.94	7.05	0.94	14.80	0.94	14.44	0.94	D,B,B*D
n-6/*n*-3 ratio	9.27	0.34	6.24	0.34	3.61	0.34	2.86	0.34	D,B,B*D

D- diet, B - breed

Conclusions

The fatty acid composition of intramuscular fat of beef can be modified favourably by feeding grass and concentrate supplementation which is rich in *n*-3 fatty acids, however the beef of HG has a more beneficial impact on human health (CLA content, *n*-6/*n*-3 ratio), compared with HF bulls.

References

Nuernberg, K., Dannenberger, D., Nuernberg, G., Ender, K., Voigt, J., Scollan, N., Wood, J., Nute, G., Richardson, I. (2005). Effect of a grass-based and a concentrate feeding system on meat quality characteristics and fatty acid composition of longissimus muscle in different cattle breeds. *Livestock Production Science*, 94 137-147.

Nuernberg, K., Wegner, J. Ender, K. (1998). Factors influencing fat composition in muscle and adipose tissue of farm animals. *Livestock Production Science,* 56 (3) 145-156.

Scollan, N.D., Dhanoa, M., S., Choi, N.J., Maeng, W.J., Enser, M., Wood, J.D. (2001). Biohidrogenisation and digestion of long chain fatty acids in steers fed on different sources of lipid. *J. Agric. Sci,* 136 345-355.

EVALUATION OF FERMENTED SAUSAGES MANUFACTURED WITH FUNCTIONAL STARTER CULTURES AND VARIOUS FAT LEVELS

Y.J. Kim*, H.C. Lee and K.B. Chin

Department of Animal Science, Chonnam National University, Gwangju, 500-600 Korea, Email: stress99@naver.com

Keywords: fermented sausage, fat level, functional lactic acid bacteria, physico-chemical properties

Introduction

Fermented meat products are manufactured with comminuted meat and fat, mixed with salt, curing agents, sugar, spices and inoculated lactic acid bacteria, and formed by changes that occur during the ripening. Since lactic acid bacteria produces lactic acid, pH values are decreased. As a result, undesirable microbial growth is inhibited, and these safety products have a sourish taste during ripening. However, consumers tend to avoid purchasing a fermented sausage if it contains a high salt content and excess fat. Dry fermented sausages normally have a high fat content. In the case of research by Wirth *et al.,* (1998), sausages with a 32% fat content in the initial mixture, had a fat content of approximately 40~50% after 4 months. Thus, overeating fat, which leads to obesity, is one of the main reasons for cardiovascular disease, such as high blood pressure and arteriosclerosis. Therefore, the objectives of this study were to assess the effect of a fat replacer on the product quality of low-fat fermented sausages, as compared to regular-fat counterparts and to determine the addition of inoculated functional lactic acid bacteria having anti-cholesterol and anti-hypertension activity on the functionality of the fermented sausages.

Materials and Methods

Sausage mixtures containing fresh pork ham, back fat, other ingredients and lactic acid bacteria were stuffed into fibrous casings (diameter 45mm) and placed in the ripening chamber. The fat replacer used in the low-fat fermented sausage was prehydrated soy protein isolate (1:4 ratio). Starter cultures containing ACE inhibitory and cholesterol assimilation activities were prepared and compared to the commercial one (LK-30 plus, Gemurzmueller Co., Germany). The mixed lactic acid bacteria, *Lactobacillus plantarum* L167 and L155 were added to the initial mixture of these sausages. Samples from each treatment were taken for analysis on day 0, 2, 3, 6, 14 and 21. Moisture, fat and protein were determined according to standard AOAC (1990) methods. Colour, pH, weight loss and shear force values were determined during the 21-day ripening time. In addition, microbiological analysis, total bacterial counts, lactic acid bacteria and *Enterobacteriaceae* were evaluated. Sensory evaluation of fermented sausages was performed after 21 days of ripening using a seven-member sensory panel. ACE inhibitory activity of sausage was measured by method of *Cushman et al.* (1971). The cholesterol assimilation activity of sausages was determined by the method of Searcy and Bergquist (1960). Data were analysed using two-way analysis of variance (ANOVA) in the SPSS program (12.0). Means were separated by the Duncan's multiple range test.

Results and Discussion

The fat content and M:P ratios of fermented sausages with various fat levels are shown in Table 1. Reaching a weight loss of 30%, RFC (regular-fat control, ~40% fat) and RFT (regular-fat with cholesterol treatment) took approximately 14 days of ripening, whereas MFC (medium-fat control, ~28%) and LFC (low-fat control, ~6%) and LFT (low-fat with anti-hypertension treatment) took 6~14 days, and 6 days of ripening, respectively (Table 2). The lower the fat level the higher the weight losses over the same processing time (Muguerza, 2002). The pH of fermented sausage was decreased during the fermentation period, and the lowest pH value was observed at the 3rd day of ripening (Table 3). Yellowness of the sausages decreased with increased ripening time (Table 3). However, shear force value and cholesterol content were increased with increasing ripening time. Hunter-L values of regular-fat sausages were higher than those of other fermented sausages. Hunter-b values of low-fat fermented sausages were higher than those of other sausages. Fermented sausage containing high fat had higher cholesterol content compared to the reduced-fat counterparts. Although ACE inhibitory activity was not affected by various starter cultures, it increased with increased ripening time, until 6 days of ripening. In sensory evaluation, colour and flavour in MFC, texture in RFC, saltiness and juiciness in RFT had the highest scores.

Conclusions

The ripening time of low-fat fermented sausages was shortened approximately two weeks, as compared to regular-fat counterparts. LFT containing functional lactic acid bacteria had similar ACE activity to LFC. Cholesterol content of the sausages decreased with decreased fat level. In sensory evaluation, MFC (~28%, fat) had the highest score for overall acceptance. Further studies will investigate the anti-microbial activity of these sausages with inoculated pathogens.

Table 1: Changes of fat level and M:P ratios of fermented sausages as affected by various starter cultures during ripening.

	TRT[1]	Ripening time (day)			
		0	6	14	21
Fat (%)	RFC	21.5^{bX}	-	36.5^{aW}	39.5^{aX}
	RFT	22.5^{bX}	-	33.0^{aX}	36.5^{aX}
	MFC	13.5^{cY}	-	23.0^{bY}	28.0^{aY}
	LFC	2.50^{aZ}	4.00^{b}	6.00^{cZ}	-
	LFA	2.00^{bZ}	3.50^{ab}	5.50^{aZ}	-
M:P ratio	RFC	3.82^{aXY}	-	1.49^{bX}	1.32^{b}
	RFT	3.81^{aXY}	-	1.39^{bXY}	1.21^{b}
	MFC	3.72^{aY}	-	1.40^{bXY}	1.09^{b}
	LFC	3.92^{aXY}	2.10^{b}	1.12^{cY}	-
	LFA	4.00^{aX}	1.98^{b}	1.08^{cY}	-

$^{a-c}$Means with same row having same superscript are not different (p>0.05).
$^{X-Z}$ Means with same column having same superscript are not different (p>0.05).
TRT[1] = treatment : RFC = regular fat control (LK30); RFT = regular fat treatment (L55); MFC = medium fat control (L55+L67+LK30); LFC = low fat control (LK30); LFA = low fat treatment (L67)

Table 2: Changes of weight loss (WL, %) in fermented sausages with low, medium and regular-fat fermented sausages as affected by various starter cultures during ripening.

	TRT[1]	Ripening time (day)					
		0	2	3	6	14	21
Weight loss (%)	RFC	-	6.78^{d}	9.79^{dY}	19.8^{cZ}	30.0^{bZ}	34.0^{aY}
	RFT	-	6.86^{c}	10.8^{cXY}	20.7^{bZ}	31.1^{aZ}	34.7^{aY}
	MFC	-	7.46^{e}	11.1^{dXY}	24.2^{cY}	36.9^{bY}	41.4^{aX}
	LFC	-	6.92^{d}	13.1^{cXY}	30.5^{bX}	49.6^{aX}	-
	LFA	-	8.90^{d}	14.9^{cX}	32.2^{bX}	51.1^{aX}	-

$^{a-e}$Means with same row having same superscript are not different (p>0.05).
$^{X-Z}$ Means with same column having same superscript are not different (p>0.05).
TRT[1] = treatment : RFC = regular fat control (LK30); RFT = regular fat treatment (L55); MFC = medium fat control (L55+L67+LK30); LFC = low fat control (LK30); LFA = low fat treatment (L67)

Table 3: Changes of pH, Hunter colour (L, a, b), bacterial counts (TPC, MRS, VRB), shear force (SF), cholesterol content, and ACE inhibitory activity in fermented sausages as affected by fat content and various starter cultures.

	Treatment[1]				
	RFC	RFT	MFC	LFC	LFA
pH	4.94^{ab}	4.84^{b}	4.93^{ab}	5.03^{a}	5.03^{a}
L	60.0^{ab}	62.3^{a}	57.2^{b}	56.8^{b}	57.8^{b}
a	14.4	14.8	15.4	14.2	13.6
b	5.95^{ab}	5.76^{ab}	4.83^{b}	6.52^{a}	6.79^{a}
TPC (log cfu/g)	7.87^{b}	7.81^{b}	7.94^{b}	8.52^{a}	8.53^{a}
MRS (log cfu/g)	7.65^{b}	7.74^{b}	7.83^{b}	8.33^{a}	8.41^{a}
VRB (log cfu/g)	$*^{b}$	$*^{b}$	$*^{b}$	2.15^{a}	2.11^{ab}
SF (kgf/g)	6.22^{b}	6.46^{b}	9.13^{a}	8.53^{a}	9.91^{a}
Cholesterol	73.9^{a}	73.3^{a}	58.6^{ab}	42.6^{b}	-
ACE inhibition(%)	-	-	-	23.7	23.1

^{a,b}Means with same row having same superscript are not different (p>0.05). *: not detected(< 10^2 cfu/g)
TRT[1] = treatment : RFC = regular fat control (LK30); RFT = regular fat treatment (L55); MFC = medium fat control (L55+L67+LK30); LFC = low fat control (LK30); LFA = low fat treatment (L67)

References

Cushman, D.W., and Cheung, H.S. (1971). Spectrophotometric assay and properties of the angiotensin-converting enzyme of rabbit lung. Biochemical Pharmacology, 20: 1627-1648

Rudel, L.L. and Morris, M.D. (1973). Determination of cholesterol using *o*-phthalaldehyde. Journal of Lipid Research, 14, 364.

Wirth, F. (1988). Technologies for making fat-reduce meat products. What possibilities are there? Fleischwirtschaft, 68, 9, 1153-1156.

Incze, K. (1992). Raw fermented and dried meat products. Fleischwirtschaft, 72,1, 58-62.

Muguerza, E., Fista, G., Ansorena, D., Astiasaran, I. and Bloukas, J.G. (2002). Effect of fat level and partial replacement of pork backfat with olive oil on processing and quality characteristics of fermented sausages. Meat Science, 61, 397-404.

Searcy, R.L. and Bergquist, L.M. (1960). A new color reaction for the quantitation of serum cholesterol. Clinica Chimica Acta, 5: 102.

DEVELOPMENT OF MEAT PRODUCTS WITH LUTEIN FOR EYE HEALTH

I. Csapo*, K. Incze, A. Kovacs, L. Zelenak and J. Zsigo

Hungarian Meat Research Institute, 1097 Budapest, Gubacsi ut 6/. Hungary. E-mail: csapo@ohki.hu

Keywords: lutein, meat products

Introduction

Lutein is an oxygenated derivative of carotenoids called xantophylls. It is a yellow pigment present in the macular region of the eye retina. Lutein is not produced in the human body, it is obtained from the diet. Patients with age related macular degeneration have lower levels of lutein in the macula compared to healthy individuals. Consumption of both lutein supplements and foods rich in lutein (green vegetables, fruits, egg yolk) are associated with lower risk of age related macular degeneration (Granado *et al.*, 2003).

The recommended daily allowance for lutein is 6mg. However the average lutein intake in Hungary is only 1.3 mg/day and even less in the elderly. The reason for this is the insufficient consumption of vegetables and fruits, especially those (spinach, sorrel, parsley, dill) high in lutein. The aim of the present study was 1) to measure the lutein content of Hungarian vegetables and meat raw materials; 2) to determine the effects of meat processing operations on lutein content of added vegetables and added lutein preparations; 3) to develop meat products with lutein content of at least one third of the RDA in one single portion, *i.e.* 900µg/50g. Meat products were chosen because: they are consumed every day or every two days also by old people; meat is a rich source of zinc and selenium which also protect eye health; and the fat in meat products enhances absorption of lutein.

Materials and Methods

Green vegetables (spinach, sorrel, mangold, parsley, Brussels sprout, broccoli, green of celery, dill, lettuce, cabbage) were obtained from the local market or from a stockyard garden. The edible parts of the vegetables were chopped by hand and analysed freshly or after frozen storage at -20°C for one or two months. Frozen and dried vegetables were purchased in supermarkets. Pork (loin, butt, leg), pork fat, chicken and turkey breast, pork and chicken liver were chopped and frozen at -20°C until use.

Comminuted pork products and liver pate were manufactured in the laboratory and in a meat processing plant with 0.5-1% dried parsley and various amounts of lutein preparation (FS20 Lutein, Roche).

For the extraction of lutein, chopped sample and solid magnesium carbonate were homogenised with extractant (hexane-acetone-methanol-toluene with BHT). The resulting suspension was filtered through Macherey-Nagel filter papers. The residue was re-extracted twice. The filtrate was transferred into a separating funnel. The lower phase was separated and washed twice with hexane. The upper layers were combined and dried for at least two hours over anhydrous sodium sulphate. The volume of lutein extract was measured. In case lutein-esters were present, the extract was saponified with methanolic potassium hydroxide solution at room temperature overnight followed by neutralisation with some drops of acetic acid. Duplicates of 1.5 ml extract were evaporated to dryness. The residues were dissolved in eluent (acetonitrile-methanol-dichloromethane-*n*-heptane), agitated and centrifuged. From the supernatant 20µl was injected into a reverse-phase HPLC (Shimadzu LC-10AD with 3PD-M10A diode array detector) employing a C_{18} column in isocratic mode.

Meat products were tested by a sensory panel. Colour was measured with Minolta CR-300.

Results and Discussion

Lutein content of fresh Hungarian vegetables (Table 1) is in the range presented in the literature (Granado *et al.*, 1992; Mangels *et al.*, 1993; Souci *et al.*, 1994). Of the vegetables dried leaves have the highest lutein content and seem most suitable for addition to meat products. No lutein was detected in meat and fat samples, while in pork liver 79µg/100g, in chicken liver 80.5µg/100g lutein were found.

Recovery of lutein from heat-treated meat-vegetable mixtures is above 100%. This finding was also reported by Updike and Schwartz (2003) and explained by increased extractability of lutein from a heat-denatured protein matrix. On the contrary, recovery of free lutein being sensitive to light and oxygen is only 35-42% (Table 2).

Preliminary sensory tests showed that the amount of dried parsley that can be added to meat products is 0.5-1.0%. This ratio is much higher than commonly used and it was necessary to strengthen the taste by adding triple the amount of dried onion and allspice. Addition of 0.5-1.0% dried parsley, however, does not result in the targeted lutein content in the meat products therefore lutein preparation also has to be added. Lutein preparation gave yellow colour to the product (Table 3) which was partly balanced with carmine acid. The yellow shade of the liver pate was more accepted by the sensory panel than that of the comminuted meat product.

Table 1: Lutein content of vegetables (µg/100g).

Vegetable	Fresh	Frozen	Dried
spinach	2674	682	-
parsley	11980	3929	18997
sorrel	5958	980	-
mangold	5543	968	-
light leaves	3768	-	-
dark leaves	7844	-	-
cabbage	26	-	-
Brussels sprouts	168	118	-
broccoli	290	30	-
lettuce	374	-	-
green of celery	6520	-	12190
dill	7156	-	23001

Table 2: Recovery of lutein from heat treated mixtures.

Mixture	Heat treatment	Recovery of lutein (%)
Pork butt+ free lutein	75°C, 1 hour	42.4
	75°C, 2 hour	38.4
	85°C, 1 hour	34.8
Pork liver+free lutein	75°C, 1 hour	37.2
	75°C, 2 hour	35.9
	85°C, 1 hour	37.4

Table 3: Instrumental colour of meat products with parsley and lutein.

Colour parameter	Comminuted meat product with parsley	Comminuted meat product with parsley and lutein	Liver pate with parsley	Liver pate with parsley and lutein
L*	63.2±9.0	60.4±1.8	56.4±1.5	5.2±1.0
a*	3.1±1.1	5.5±0.9	6.6±1.3	6.6±0.6
b*	11.5±0,5	22.0±1.0	12.9±0.1	21.0±0.4

Conclusions

Comminuted meat products and liver pate of good sensory characteristics and of targeted lutein content (900 µg/50g) can be produced by adding dried vegetables and lutein preparation. Introduction of meat products with lutein to the market calls for an intensive promotion to inform consumers of the advantages of lutein for eye health.

References

Granado, F., Olmedilla, B., Blanco, I. and Rojas-Hidalgo, E. (1992). Carotenoid composition in raw and cooked Spanish vegetables. Journal of Agricultural and Food Chemistry. 40:2135-2140.

Granado, F., Olmedilla, B. and Blanco, I. (2003). Nutritional and clinical relevance of lutein in human health. British Journal of Nutrition, 90:487-502.

Mangels, A.R., Holden, J.M., Beecher, G.R., Forman, M.R. and Lanza, E. (1993). Carotenoid content of fruits and vegetables: An evaluation of analytic data. Journal of American Dietetic Association. 93:284-296.

Souci, S.W., Fachman, W. and Kraut, H. (1994). Food Composition and Nutritional Tables, 5th ed. Medpharm Scientific Publishers, Stuttgart, Germany.

Updike, A.A. and Schwartz, S.,J. (2003). Thermal processing of vegetables increases isomers of lutein and zeaxanthin. Journal of Agricultural and Food Chemistry, 51:6184-6190.

EFFECT OF ENCAPSULATED CONJUGATED LINOLEIC ACID ON MEAT QUALITY, CARCASS AND FATTY ACID COMPOSITION OF BEEF STEERS

R.I. Richardson*[1], J.D. Wood[1], K. Hallett[1], A.V. Fisher[1] and K.J. Shingfield[2]

*[1]Division of Farm Animal Science, University of Bristol, Langford, Bristol, BS40 5DU, UK. [2]Department of Animal Science, The University of Reading, Reading, RG6 6AT, UK, present address, MTT, Agrifood Research Finland, Jokoinen, FIN-31600, Finland. * E-mail: ian.richardson@bristol.ac.uk*

Keywords: beef, CLA, carcass fatness, fatty acid composition, shelf life

Introduction

Dietary supplements of a mixture of CLA isomers have been shown to alter lipid metabolism and reduce body fat accretion in mice (West *et al.*, 1998) and pigs (Ostrowska *et al.*, 1999). Evidence from studies in lactating dairy cows (Baumgard *et al.*, 2000) have shown that the *trans*-10, *cis*-12 isomer is responsible for the anti-lipogenic effects. Shingfield *et al* (2004) demonstrated the potential of rumen protected supplements of a mixture of CLA isomers to reduce milk fat content and improve the energy status of dairy cows during early lactation. The efficacy of calcium salts of a mixture of CLA methyl esters containing *trans*-10, *cis*-12, on milk fat synthesis, is relatively low (Bernal-Santos *et al.*, 2003) which may explain the lack of effect on animal performance of heifers fed a total mixed ration containing (20g/kg) calcium salts of a four isomer CLA (Gillis *et al.*, 2004). The lack of performance responses may also be related to the short duration of CLA supplementation, during the last 32 or 60 days before slaughter, and the relatively low level of *trans*-10, *cis*-12 CLA in the diet (2.04g/kg). The current experiment used a lipid-encapsulated supplement containing equal amounts of *cis*-9, *trans*-11 and *trans*-10, *cis*-12 CLA methyl esters on the performance and carcass characteristics of Limousin steers during a 100d finishing period.

Materials and Methods

Forty Limousin steers at the University of Reading beef unit had *ad libitum* access to a basal diet (forage (equal parts grass and maize silage):concentrate ratio 60:40 on a dry matter basis). On reaching a live weight of 425kg, each steer was allocated at random to one of four experimental treatments in a 2 (grass or maize silage, forage:concentrate ratio 60:40 on a DM basis) x 2 (control or CLA lipid supplement) factorial arrangement according to a complete randomised block design. The lipid-encapsulated CLA supplement was prepared by binding methyl esters of CLA (Luta-CLA 60 supplied by BASF, Germany) to a silica matrix and coating this complex with hydrogenated soyabean oil by Balchem Encapsulates (New Hampton, New York). It was expected that the intake of *trans*-10, *cis*-12 CLA would be *circa* 75g of *trans*-10, *cis*-12 CLA, significantly higher than in earlier studies. Major fatty acids in the CLA mix (g/100 g total fatty acids) were: C16:0, 5.26; C18:0, 4.12; *cis*-9 C18:1, 25.66; C18:2*n*-6, 1.99; *cis*-9, *trans*-11 CLA, 29.32; *trans*-10, *cis*-12 CLA, 29.09.

After 100d on diet steers were transported to Bristol University for slaughter and subsequent analysis. At 48h post-mortem, samples of m. *longissimus thoracis* were removed and blast frozen for fatty acid and vitamin E analysis. Lipid was extracted using chloroform/methanol and separated into neutral and phospholipid. Fatty acid methyl esters were prepared by alkaline hydrolysis followed by methylation with diazomethane and analysed on a CP Sil 88, 100m x 0.25mm ID column (Chrompack, UK). An additional section of muscle was conditioned at 1°C for 12 days in vacuum pack and 4 steaks 20mm thick were cut, packed in modified atmosphere trays (O_2:CO_2, 75:25) and subjected to simulated retail display (700lux lighting for 16h a day, 4°C±1°C). Colour (L*a*b*) was measured on the surface of 2 steaks at three points, daily with a Minolta Chromameter. A further 2 steaks were taken at 10d of display and analysed for lipid oxidation as thiobarbituric acid reacting substances (TBARS).

Results and Discussion

Overall, the total amount of lipid in subcutaneous fat was reduced by the protected CLA supplement but only significantly so for the maize diet (not shown), a non-significant but similar trend being seen in muscle total fatty acids. Feeding CLA increased total tissue levels of *trans*-10, *cis*-12 CLA 10 to 16-fold in muscle (Table 1) and 11 to 12-fold in adipose tissue. Concentrations of *cis*-9, *trans*-11 CLA were increased up to 2-fold in muscle and 1.5-fold in adipose tissue. The former is much higher than achieved with diet alone in previous studies, but the latter is less than typical on high concentrate diets. For the MUFA C16:1*n*-5, C18:1*n*-9 and C18:1*n*-11 decreased whilst *trans*-C18:1 increased. As expected there was more C18:2 in the meat and adipose tissue of maize silage-fed animals and this decreased with CLA supplementation, whilst there was less C18:3*n*-3 in tissues of maize silage-fed animals which was less affected by CLA. The total CLA increased more with protected CLA feeding on grass-silage based than maize-silage based diets. There was no effect of grass versus maize silage or the addition of the protected CLA in the diet on carcass composition except for a small but significant reduction in the deposition of adipose fat around the kidney (kidney knob and clod fat) and, in a forelimb sample joint dissection, the total and % fat content was reduced non-significantly in animals fed CLA.

Table 1: Selected fatty acids in total lipid of m. *longissimus thoracis*, weight (mg/100g) and % composition, of steers fed a grass or maize silage diet with or without CLA supplementation and measures of meat stability.

Diet	Grass	Maize	Grass +CLA	Maize +CLA	sed	sig	Grass	Maize	Grass +CLA	Maize +CLA	sed	sig
weight							%					
14:0	81	89	121	72	15.3	*	2.4	2.5	3.2	2.6	0.18	**
16:0	903	978	1033	716	140.9	ns	27.2	27.3	26.9	25.7	0.69	ns
18:0	490	515	623	453	81.1	ns	14.8	14.3	16.4	16.2	0.60	**
t-18:1	53	64	110	89	12.4	***	1.6	1.8	2.9	3.2	0.18	***
9c18:1	1174	1287	1278	934	185	ns	35.1	36.0	33.3	33.8	0.72	**
11c18:1	45.7	48.8	45.6	36.8	5.3	ns	1.44	1.40	1.20	1.36	0.083	*
18:2n-6	119	124	93	94	6.7	***	3.4	3.7	2.6	3.6	0.41	**
18:3n-3	13.7	8.7	15.2	7.5	1.27	***	0.44	0.26	0.22	0.28	0.042	***
9c11tCLA	9.7	11.7	19.3	15.2	2.06	***	0.29	0.33	0.51	0.55	0.029	***
10t12cCLA	0.42	0.58	6.97	5.75	0.66	***	0.01	0.02	0.19	0.20	0.015	***
20:3n-6	11.5	12.8	8.9	7.5	0.8	***	0.39	0.39	0.24	0.29	0.049	**
n-6:n-3	5.8	8.0	4.3	6.9	0.63	***						
Chroma d7	21.8	21.1	22.0	20.6	1.1	ns						
Chromad11	17.3	17.4	20.6	17.8	1.5	ns						
TBARS*	1.9^{a}	3.5^{b}	1.5^{a}	2.2^{a}	0.46	***						
Vitamin E*	2.2^{b}	1.6^{a}	2.3^{b}	1.9^{a}	0.25	*						

* mg/kg muscle fresh weight.

No additional Vitamin E was fed in these trials and the concentration of vitamin E in the meat was lower than the 3-3.5mg/kg meat required for optimum stability. This was expected for the maize silage, but good quality grass silage usually results in higher concentrations of vitamin E than found here, although it was still significantly greater than for the maize silage. Feeding protected CLA did not affect vitamin E concentrations. Colour stability, as measured by saturation, was good after 7 and 11 days simulated retail display, despite the lower concentratrion of vitamin E, but TBARS were higher than in previous trials, being worst on the maize diet. Feeding protected CLA appeared to give some protection against lipid oxidation, but only significantly so for the combination of maize silage with CLA. It has been suggested that CLA is preferentially oxidised and may thus spare other fatty acids from oxidation.

Conclusions

The CLA was protected to a sufficient extent to produce high levels of incorporation, in particular of the *trans*-10, *cis*-12 isomer, into beef meat and adipose tissue. This had no significant effect on body fat accretion, but reduced total fatty acids in adipose tissue, mainly to the detriment of mono- and poly-unsaturated fatty acid content. There was some indication that CLA improved lipid stability.

References

Baumgard L.H., Corl, B.A., Dwyer, D.A. and Bauman D.E. (2000). Identification of the conjugated linoleic isomer that inhibits milk fat synthesis. American Journal of Physiology, 278: R179-R184.

Bernal-Santos, G., Perfield II, J.W., Barbano, D.M., Bauman, D.E. and Overton, T.R. (2003). Production responses of diary cows to dietary supplementation with conjugated linoleci acid (CLA) during the transition period and early lactation. Journal of Dairy Science, 86: 3218-3228.

Gillis M.H., Duckett, S.K., Sackmann, J.R., Realini, C.E. Keisler, D.H. and Pringle, T.D. (2004). Effects of supplemental rumen-protected conjugated linoleic acid or linoleic acid on feedlot performance, carcass quality, and leptin concentrations in beef cattle. Journal of Animal Science, 82: 851–859.

Ostrowska, E., Muralitharan, M., Cross, R.F., Bauman, D.E. and Dunshea, F.R. (1999). Dietary conjugated linoleic acids increase lean tissue and decrease fat deposition in growing pigs. Journal of Nutrition, 129: 2037-2042.

Shingfield, K.J., Beever, D.E., Reynolds, C.K., Gulati, S.K., Humphries, D.J., Lupoli, B., Hervas, G. and Griinari, J.M. (2004). Effect of rumen protected conjugated linoleic acid on energy metabolism of dairy cows during early to mid-lactation. Journal of Dairy Science, 87: (Suppl. 1) 307.

West, D.B., Delany, J.P., Camet, P.M., Blohm, F., Truett, A.A., Scimeca, J. (1998) Effects of conjugated linoleic acid on body fat and energy metabolism in the mouse. American Journal of Physiology 44: R667–R672.

EFFECT OF NUTRITION ON FATTY ACID PROFILE AND QUALITY OF PORK IN TWO GENOTYPES

G. Mas*[1], C.E. Realini[2], M. Llavall[1], M.A. Oliver[2], M. Gispert[2], I. Díaz[2], J. Tibau[2], R. Roca[3], and D. Coll[1]

[1.] *UPB España S.A., Ctra Viver, Km 6,3, Viver i Serrateix, Barcelona 08673, Spain,* [2.] *IRTA, Institute for Food and Agricultural Research and Technology, Granja Camps i Armet, Monells 17121 Spain,* [3.] *Grupo Omega de Nutrición Animal, SL. Email: carolina.realini@irta.es*

Keywords: pork, fatty acids, diet, genotype

Introduction

Modern society is concerned about the importance of meat in the diet, and the consequences of consuming pork fat in human health. High levels of fat consumption and particularly of saturated fatty acids have been associated with increased serum LDL-cholesterol, and increased risk of coronary heart disease in humans (Department of Health, 1994). The trend in human nutrition is towards a higher intake of polyunsaturated fatty acids, especially n-3 fatty acids. However, greater contents of highly unsaturated fatty acids in pork fat may cause reduced storage stability, rancidity, off-flavours, and poor processing properties. Miller *et al.* (1990) showed that diets with high levels of monounsaturated fatty acids increased the amount of oleic acid in adipose tissues without adversely affecting carcass quality. Vitamin C has been used in different ways to enhance lean colour stability of meat during display. The aim of this study was to evaluate the effect of feeding a diet rich in oleic acid and supplementation with antioxidants (Biocitro®) on the fatty acid profile of intramuscular and subcutaneous fat and pork quality in two genotypes.

Materials and Methods

One hundred sixty eight pigs from 2 sexes (barrows and gilts) and 2 genotypes were fed one of three finishing diets: 1) grain and soy diet (CONTROL); 2) grain and soy plus 3.83 % high oleic acid supplement (HO); and 3) grain and soy plus 3.83% HO plus 200 ppm Biocitro® (HO-ANTX). Animals (60 kg live weight) were fed CONTROL (n=53) and HO (n=115) diets until they reached 90 kg live weight. During the finishing period animals were fed the 3 diets (CONTROL n=53, HO n=55, and HO-ANTX n=60) until slaughter. Diets had similar nutrient content and diet composition is shown in Table 1. Genotypes were crosses of Landrace(LD)*Large White(LW) with Pietrain (negative for the halothane gene, NN) or Large White and were slaughtered at 115 (fresh pork market) and 130 kg (processed ham market), respectively. The high oleic acid supplement is a natural byproduct of the olive industry composed of a mixture of Ca-salts rich in oleic acid. Biocitro® is a natural byproduct of the citric juice industry and a rich source of vitamin C and flavonoids. Muscle samples of *longissimus thoracis* and subcutaneous fat from *semimembranosus* muscle were taken at 24 h postmortem. Lipids were extracted following the chloroform-methanol procedure of Folch *et al.* (1957), and converted to fatty acid methyl esters following the method of ISO 5509-1978 (E) and analyzed using GC (BP70-SGE, USA). Colour and lipid stability were evaluated on *longissimus* chops packaged in MAP (70%O_2, 30%CO_2) and displayed for 10 days at 4°C in a lighted cooler using a Minolta chromameter (CR-400, Osaka, Japan) and TBARS procedure (Kornbrust and Mavis, 1980), respectively. There were no interactions (P>0.05) between sex and diet, and results are presented as main effects.

Table 1: Diet composition.

Composition, %	CONTROL	HO	HO-ANTX
Corn	25.0	5.0	5.0
Barley	26.9	47.2	47.2
Wheat	20.0	20.0	20.0
Soy	22.3	20.3	20.3
Fat	2.90	1.39	1.39
High oleic acid supplement	-	3.83	3.83
Biocitro®	-	-	0.02
Lysine	0.24	0.27	0.27

Results and Discussion

There were no differences (P>0.05) in the fatty acid profile of intramuscular and subcutaneous fat between HO and HO-ANTX for Pietrain and Large White crosses. CONTROL muscle from Pietrain animals had lower percent of C16:0, C18:1 and MUFA, greater percent of C18:2 and PUFA, and higher PUFA:SFA and n-6:n-3 ratios compared with *longissimus* from HO and HO-ANTX fed animals. Intramuscular fatty acid profile from Large White animals showed similar pattern to Pietrain, but only C18:0 percent was lower and n-6:n-3 ratio was higher (P<0.05) for CONTROL compared with HO and HO-ANTX. Values for PUFA:SFA are above the recommended ratio of 0.45 (Department of Health, 1994). However, n6:n3 ratios are well above the recommended value of 4.0, being lower (P<0.05) for animals fed HO compared with CONTROL. Subcutaneous fat was more unsaturated than intramuscular fat for all diets and both genotypes. CONTROL subcutaneous fat from Pietrain pigs had lower percent of C18:1 and MUFA, greater percent of C18:0, C18:2, SFA, and higher n-6:n-3 ratio than HO and HO-ANTX fed pigs. Subcutaneous fat from Large White CONTROL animals had lower percent of C18:1 and MUFA, greater percent of C16:0, C18:2, SFA, PUFA and higher n-6:n-3 ratio than HO and HO-ANTX. Although the n-6:n-3 ratio is lower for subcutaneous compared with intramuscular fat, values are higher than 4.0 and lower for HO and HO-ANTX compared with CONTROL. Minolta and TBARS results (data not shown) indicated that animal supplementation with Biocitro® did not offer advantages in improving colour and lipid stability of pork chops displayed during 10 days at 4°C.

Table 2: Fatty acid composition of *longissimus thoracis* intramuscular fat and *semimembranosus* subcutaneous fat from CONTROL, HO and HO-ANTX in two genotypes.

Fatty Acid, %	(LD*LW) * Pietrain				(LD*LW)*Large White			
Intramuscular fat (*LT**)	CONTROL	HO	HO-ANTX	RMSE†	CONTROL	HO	HO-ANTX	RMSE†
16:0, *palmitic*	22.31^{a}	23.02^{ab}	23.55^{b}	1.283	22.37	22.75	23.09	2.118
18:0, *estearic*	12.88	12.79	12.92	0.841	12.63^{a}	12.93^{ab}	13.58^{b}	1.118
18:1, *n-9 oleic*	32.62^{a}	36.72^{b}	36.44^{b}	4.287	31.61	34.46	35.38	5.262
18:2, *n-6 linoleic*	16.97^{b}	13.29^{a}	13.35^{a}	4.091	17.32	14.38	13.99	5.615
18:3, *n-3 linolenic*	0.47	0.46	0.44	0.083	0.43	0.45	0.44	0.126
SFA	36.77^{a}	37.45^{ab}	38.13^{b}	1.958	36.46	37.20	38.10	3.174
MUFA	40.05^{a}	44.69^{b}	44.42^{b}	4.776	39.16	42.05	42.57	5.842
PUFA	23.23^{b}	17.85^{a}	17.50^{a}	5.977	24.38	20.70	19.33	8.735
PUFA:SFA	0.64^{b}	0.49^{a}	0.46^{a}	0.184	0.69	0.59	0.53	0.282
n-6:n-3	21.79^{b}	18.89^{a}	19.29^{a}	3.733	23.32^{b}	19.77^{a}	19.25^{a}	2.983
Subcutaneous fat (*SM**)								
16:0, *palmitic*	19.83	19.38	19.50	0.900	21.23^{b}	20.09^{a}	20.89	1.128
18:0, *estearic*	10.17^{b}	9.08^{a}	9.35^{a}	1.073	11.10	10.52	11.36	1.284
18:1, *n-9 oleic*	41.21^{a}	44.46^{c}	43.19^{b}	1.735	41.58^{a}	44.78^{b}	43.94^{b}	1.261
18:2, *n-6 linoleic*	16.93^{b}	15.12^{a}	16.04^{ab}	1.747	14.70^{b}	13.31^{a}	13.20^{a}	1.628
18:3, *n-3 linolenic*	1.30	1.31	1.36	0.197	1.04	1.10	1.08	0.163
SFA**	32.00^{b}	30.45^{a}	30.87^{a}	1.594	34.39^{b}	32.46^{a}	34.15	1.985
MUFA**	47.95^{a}	51.41^{c}	50.02^{b}	1.907	48.30^{a}	51.64^{b}	50.25^{b}	1.616
PUFA**	20.05^{b}	18.14^{a}	19.11^{ab}	2.045	17.38^{b}	15.90^{a}	15.60^{a}	1.884
PUFA:SFA	0.63	0.60	0.63	0.089	0.51	0.49	0.46	0.082
n-6:n-3	11.31^{c}	9.83^{a}	10.61^{b}	0.942	12.05^{b}	10.78^{a}	10.76^{a}	0.593

†RMSE: root mean-square error, *LT: *longissimus thoracis*, SM: *semimembranosus*; **SFA: saturated, MUFA: monounsaturated, PUFA: polyunsaturated fatty acids. Means within the same row with different letters differ $P<0.05$

Conclusions

Dietary supplementation with 3.83% of high oleic acid supplement was effective in increasing oleic acid and MUFA percent of intramuscular and subcutaneous fat in finishing Pietrain and Large White crossed pigs offering the consumer a healthier product. Treated animals showed a more favourable n-6:n-3 ratio than control, but higher than current nutritional recommendations. Supplementation with 200 ppm Biocitro® did not improve colour and lipid stability of fresh pork.

References

Department of Health. (1994). Report on health and social subjects. No.46. Nutritional aspects of cardiovascular disease. London: HMSO.

Kornbrust, D.J. and Mavis, R.D. (1980). Relative susceptibility of microsomes from lung, heart, liver, kindney, brain and testes to lipid peroxidation: correaltion with vitamin E content. Lipids, 15:315-322.

Miller, M. F, Shackelford, S.D., Hayden, K.D. and Reagan, J.O. (1990). Determination of the alteration in fatty acid profiles, sensory characteristics and carcass traits of swine fed elevated levels of monounsaturated fats in diet. Journal of Animal Science, 68:1624-1631.

PREVENTION OF LIPID OXIDATION AND HEME IRON ACCESSIBILITY BY ADDITION OF ROSEMARY (*ROSMARINUS OFFICINALIS L.*) AS AN INGREDIENT IN HAMBURGER

M. Lucarini*[1], M. Delfini[2], L. D'Evoli[1], P. Salvatore[1], S. Nicoli[1], P. Gabrielli[1], F. Piccioni[2], L. Marletta[1] and G. Lombardi-Boccia[1]

[1.] *National Institute for Food and Nutrition Research-Rome, Italy,* [2.] *Department of Chemistry-University of Rome "La Sapienza"-Rome, Italy. Email: lucarini@inran.it*

Keywords: beef meat, lipid oxidation, heme iron, metmyoglobin, rosemary

Introduction

Both natural and synthetic antioxidants are used to retard lipid oxidation in meat (Mielche and Bertelsen, 1994). The shelf-life of meat is limited because of the biochemical changes that involve muscle lipids (lipolysis, lipid oxidation) starting after slaughtering. Oxidation in muscle foods is a degradation process responsible for losses in the nutritional value of meat and for undesirable changes in appearance (flavour, colour) which affect the acceptance by consumers. The most frequent mechanisms of food deterioration are microbiological spoilage and oxidation. In fresh meat and meat product, colour is a strong indicator of quality. The colour of meat is due to a balance between oxymyoglobin oxidation and metmyoglobin reduction. Lipid oxidation may initiate the oxidation of myoglobin to metmyoglobin, and thus, change the colour from red to brown. Several authors have postulated that pigment and lipid oxidation are interrelated and that the rate of meat discolouration is related to the rate of myoglobin oxidation. Spices have been reported to contain components which may be effective like BHA or BHT as an antioxidant.

The aim of this study was to evaluate the effect of increasing levels of rosemary (0.33% and 1%) on lipid oxidation and on iron release from the heme moiety in beef hamburger during refrigerated storage. Studies are going on with other spices (oregano, sage and saffron) and on their respective olive oils extract (D'Evoli *et al.*, in press).

Materials and Methods

Beef muscle, obtained from industry, was minced twice and mixed into a homogeneous sample. Rosemary (*Rosmarinus officinalis L.*) was dried at 40°C in a ventilated oven and then ground; the herb was added to the meat sample at 0.33% and 1% (w/w). Rosemary 0.33% and rosemary 1% were added to 60g meat aliquots, formed into burgers and stored at 4°C for 9 days under fluorescent light. A control sample (a 60g hamburger) was stored in the same condition.

Thiobarbituric acid reactive substances (TBARS) assay was performed as described by Ulu (Ulu, 2004). 10g of raw meat was extracted with 5% TCA and the filtrate was applied to TBA reaction. TBA values were expressed as mg/kg sample.

Metmyoglobin (MMB) was extracted with cold 0.04 M phosphate buffer (pH 6.8) with a sample to buffer ratio of 1:10. After homogenization, samples were centrifuged, and the filtered supernatant was read at 525, 572, and 730nm and the percentage of MMB was determined as indicated by Krzywicki (Krzywicki, 1979).

Heme iron was determined using the acidified acetone extraction method of Lombardi-Boccia *et al.* (Lombardi-Boccia, 2002). The heme iron was determined spectrophotometrically at 640nm and was expressed in mg/100g of meat.

Results and Discussion

Lipid oxidation: Fig. 1 shows the changes in TBARS value. Rosemary 1% hamburgers showed the lowest TBARS values during refrigerated storage. On the other hand rosemary 0.33% didn't show significant differences in TBARS values with respect to the control sample during the storage.

Heme iron destruction: in the control sample the decrease in heme iron content was up to 49% after 8 days of refrigerated storage, by contrast in both the rosemary treated samples it was up to 28% (Fig. 2).

MMB formation: The metmyoglobin formation is showed in Fig. 3. Only the addition of 1% rosemary to hamburger showed to have inhibitory effect on MMB formation during the storage period (below 40%).

Conclusions

In hamburger the presence of rosemary at 1% was effective in inhibiting both lipids and myoglobin oxidation, thus increasing the shelf-life of meat. On the other hand, heme iron degradation was slowed by both the rosemary concentrations (1 and 0.33%); this result suggests that the mechanism which inhibits the heme iron accessibility is quite different from the inhibition mechanism involved in lipids and myoglobin oxidation.

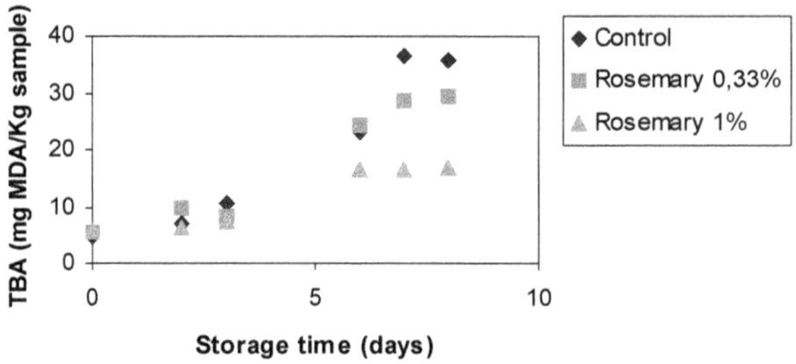

Figure 1: Effect of rosemary on the TBA values in hamburger beef during storage for 8 days at 4°C under fluorescent light.

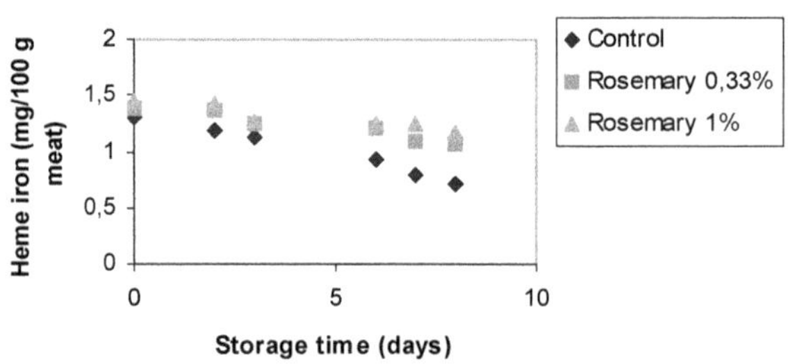

Figure 2: Effect of rosemary on heme iron content in hamburger beef during storage for 8 days at 4°C under fluorescent light.

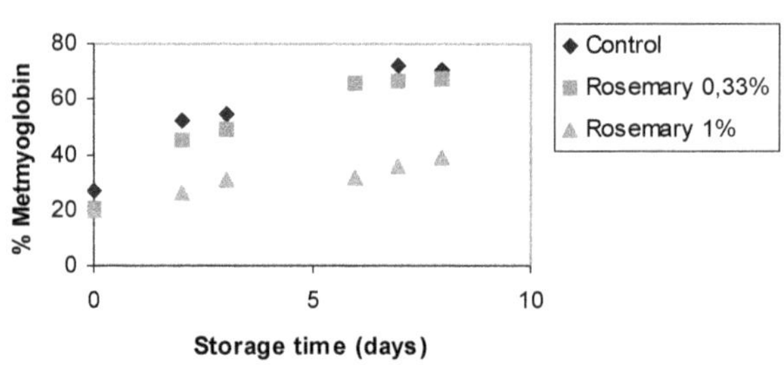

Figure 3: Effect of rosemary on metmyoglobin production in hamburger beef during storage for 8 days at 4°C under fluorescent light.

References

Mielche, M.M. and Bertelsen, G. (1994). Approaches to the prevention of warmed-over flavour. Trends in Food science and Technology, 5: 322-327.

D'Evoli, L., Huikko, L., Lampi, A.M., Lucarini, M., Lombardi-Boccia, G., Nicoli, S. and Piironen, V. Influence of rosemary (Rosmarinus officinalis, L.) on plant sterol oxidation in extra-virgin olive oil. (In press)

Lombardi-Boccia, G., Martinez-Dominguez, B. and Aguzzi, A. (2002). Optimization of heme iron analysis in raw and cooked raw meat. Food Chemistry, 78, (4): 505-510.

Ulu, H. (2004). Evaluation of three 2-thiobarbituric acid methods for the measurement of lipid oxidation in various meats and meat products. Meat Science, 67; 683-687.

Krzywicki, K. (1979). Assessment of relative content of myoglobin, oxymyoglobin, and metmyoglobin at the surface of beef. . Meat Science, 3: 1-10.

TERRINCHO-PDO LAMB MEAT SEEMS TO HAVE HIGHER CONTENTS OF TOTAL LIPID AND CHOLESTEROL THAN TRANSMONTANO-PDO KID MEAT

I. Trigo Rodrigues, M.A.G. Quaresma*, C.M. Alfaia and J.A.M. Prates

Faculdade de Medicina Veterinária-CIISA, Avenida da Universidade Técnica, Pólo Universitário do Alto da Ajuda, 1300-477 Lisboa, Portugal. Email: mquaresma@fmv.utl.pt.

Keywords: lamb meat, kid meat, lipid content, cholesterol content, *longissimus lumborum*

Introduction

The Northeast territory of Portugal, named as Trás-os-Montes, produces some of the most traditional and appreciated Portuguese products. Among them are the Terrincho-PDO milk lamb meat, produced by purebred flocks of "Churra da Terra Quente" ewes, and the Transmontano-PDO milk goat kid meat, obtained from purebred flocks of Serrana goats. Churra da Terra Quente ewe's breed and Serrana goat's breed are two important Portuguese autochthonous breeds. Both production systems are certified under the Protected Designation of Origin (PDO), which obliges producers to fulfil the traditional rearing methods based on extensive grazing, and restrict feeding supplementation with local agricultural by-products during the snowfall season. Globally milk lamb and kid meat production is considered a by-product of the small ruminant dairy sector. Light lamb or kid meat is traditionally consumed for Portuguese Christmas lunch. Demand and price for both meats peak over the Christmas period, which has led producers to synchronize most of their flocks' birth just before Christmas. The income of this "Christmas market" represents the most important part of the overall herdsman income and contributes significantly to the socio-economical wellbeing of the rural populations. World Health Organization (WHO), together with other international health organizations, strongly recommends that fat should provide between 15 and 30% of all diet energy and cholesterol intake should not exceed 300 mg per day (Chizzolini *et al.*, 1999). These dietary recommendations are health related concerns since higher fat and cholesterol intake are associated with obesity and hypercholesterolemia, conditions that predispose to several chronic diseases. The quantity of fat in meat and meat products is a general concern for consumers, while dieticians are not only concerned with the quantity but also with the quality of intramuscular fat. Cholesterol content in meat has long been identified as the single most important characteristic of overall meat quality. The aim of this study was the compare the levels of total lipid and cholesterol in Terrincho-PDO lamb meat and Transmontano-PDO kid meat.

Materials and Methods

Meat samples were taken from Terrincho-PDO suckling lambs (n=15; 6.3±1.0 kg of carcass weight and 47.3±10.9 days), and Transmontano-PDO suckling goat kids (n=15; 4.3±0.6 kg of carcass weight and 50.8±18.0 days) raised with their dams under an extensive production system based on natural pastures, as postulated by the PDO certification system. During the Christmas season in 2005, meat samples from the mid loin (*longissimus lumborum* from the 13th rib to the point of the hip bone) were collected 2-3 days after slaughter (+1°C). Total lipids were extracted from meat (dry matter) by ultrasonication, using methylene-chloride (4:1 v/v) (3×) and *n*-hexane (1×), as was described by Fritsche *et al.*, (2000). Total lipid contents of the meat samples were calculated in duplicate by weighting the residues obtained after solvent evaporation under a stream of nitrogen. Total cholesterol was extracted from meat, after saponification. Briefly, 0.75 g of homogenised meat sample was placed in a screw teflon-lined cap tube, in duplicate, to which 0.2 g L-ascorbic acid and 5.5 ml saponification solution were added. The saponification solution, freshly prepared each week, contained 11% w/v potassium hydroxide in a mixture of 55% v/v absolute ethanol and 45% v/v distilled water. The sample was then immediately vortexed in order to avoid meat agglomeration. After vortexing, the air was eliminated from the reaction, by displacement with nitrogen gas and the sample was further shaked until the ascorbic acid was completely dissolved. The saponification was carried out in a shaking water bath (200 rpm) at +80 °C for 15 min. Cholesterol was quantified by normal-phase HPLC, using an UV-visible photodiode array as described by Prates *et al.* (2006). The contents of total cholesterol in meat were calculated, in duplicate for each muscle sample (values accepted for CV <6%), based on the external standard technique, from a standard curve of peak area *vs.* concentration. The data were analysed by the Student's t–test with a significance level of 5% INSTAT for Windows (GraphPad Software, San Diego, CA) was used to perform the analysis and data was presented as the mean ± standard deviation of the mean.

Results and Discussion

Table 1: Total lipid and cholesterol contents (mg/g meat) and specific cholesterol content (mg/g lipids) in mid loin from Terrincho-PDO lamb meat and Transmontano-PDO kid meat.

	Transmontano-PDO kid meat	Terrincho-PDO lamb meat
Contents of total lipids	13.3±2.2^{a}	17.6±1.9^{b}
Contents of total cholesterol	0.67±0.04^{a}	0.73±0.8^{b}
Specific contents of total cholesterol	12.0±1.6^{c}	9.8±1.4^{d}

Means in the same row with different superscripts are significantly different: P<0.001 (between a and b) and P<0.05 (between c and d).

As observed in Table 1, Terrincho-PDO lamb meat has significantly higher contents of total lipids and total cholesterol than Transmontano-PDO goat kid meat (P<0.001). On the other hand, the specific content of total cholesterol was significantly higher in Transmontano-PDO kid meat (P<0.05). The lower lipid and cholesterol contents found in goat kid meat are characteristic of the species (Babiker *et al.*, 1990). Total lipid contents in Transmontano-PDO goat kid meat is in between the values found by Marichal *et al.*, (2003) in mid loin obtain from goat kids of 6 kg (9.6±4.5 mg/g meat) and 10 kg (16.4±11.2 mg/g meat) of live weight. The total lipid contents of Terrincho-PDO lamb meat are lower than the total lipid contents found in an Italian study (31.4-40.0 mg/g meat) (Salvatori *et al.*, 2004) and in a Brazilian study (68.5-107.9 mg/g meat) (Rowe *et al.*, 1999), both obtained from heavier lambs. The total cholesterol content in Terrincho-PDO lamb mid loin is slightly higher than those found by Rowe *et al*, (1999), on both drylot and grazing lambs (0.58 and 0.62 mg/g meat respectively) and those found by Salvatori *et al.*, (2004) (0.60-0.63 mg/g meat). The higher total cholesterol contents found in Terrincho-PDO lamb meat in comparison with previous studies may be explained by Arsenos *et al.*, (2000), which has shown that younger lambs have higher total cholesterol contents when compared with older animals. They demonstrated that total cholesterol contents in lambs meat is inversely related with maturity degree at slaughter time. A comparison made with our previous results obtained in beef and veal of Portuguese autochthonous breeds, PDO certified (Quaresma *et al.*, 2004), showed that lamb meat and kid meat are leaner meats but their higher contents of total cholesterol does not mean that they are healthier meats.

Conclusion

The evaluation performed indicated that Transmontano-PDO kid meat has lower contents of total lipid and cholesterol than Terrincho-PDO lamb meat, although both meats are lean and contain relatively high cholesterol contents.

Acknowledgments

Sampling assistance (Eng. Francisco from CAPRISERRA and Mr. Manuel Pires from OVITEQ) and financial support (grant CIISA/2005/76) are acknowledged.

References

Arsenos, G., Zygoyjannis, V., Kufidis, D., Katsaounis, N., Stamataaris, C. (2000). The effect of breed, slaughter, weight and nutritional management on cholesterol content of lamb carcass. Small Rum.t es., 36:275-283.

Babiker, SA., El Khider, I.A., Shatie, S.A. (1990). Chemical composition and quality attributes of goat meat and lamb. Meat Sci. 28, 273-277.

Chizzolini, R. Zanardi, E., Dorigoni, V., Ghidini, S. (1999). Calorific value and cholesterol content of normal and low fat meat and meat products. Trends in Food Science & Technology, 10, 119-128.

Fritsche, S., Solomon, M.B., Mossoba, M.M., Yurawecz, M.P., Morehouse, K., Ku, Y. (2000). Quantitative determination of conjugated linoleic acid isomers in beef fat. Eur. Lipid Science and Technology, 102: 667-672.

Marichal, A., Castro, N., Capote, J., Zamorano, M.J., Argüello, A. (2003).Effects of live weight at slaughter (6,10 and 25 kg) on kid carcass and meat quality. Livestock Production Science 83: 247-256.

Mestre Prates, J.A., Gonçalves Quaresma, M.A., Branquinho Bessa, R.J., Andrade Fontes, C.G., Mateus Alfaia, C.M.P (2006).Simultaneous HPLC quantification of total cholesterol, tocopherols and beta-carotene in Barrosã-PDO veal. Food Chemistry,94(3):469-477.

Quaresma, M.A.G., Mateus Alfaia, C.M.P., Fonseca Castro, M.L.S.D.F., Branquinho Bessa, R.J, Mestre Prates, J.A. (2004). Comparison of total lipids and total cholesterol contents in Portuguese bovine meats-PDO from four different autochthonous breeds (Barrosã, Arouquesa, Alentejana and Mertolenga). Congress Proceedings of the 50th International Congress of Meat Science and Technology. Helsinki, Filand pp 1045-1048,poster265.

Rowe, A., Macedo, F.A.F., Visentainer, J.V., Souza, N.E., Matsushita, M. (1999). Muscle composition and fatty acid profile in lambs fattned in drylot or pasture. Meat Science, 51:283-288.

Salvatori, G., Pantaleo, L., Di Cesari, C., Maiorano, G., Filetti, F., Oriani, G. (2004). Fatty acid composition and cholesterol contents of muscles as related to genotype and vitamin E treatment in crossbred lambs. Meat Science, 67:45-55.

PROXIMAL AND MINERAL COMPOSITION OF BEEF DERIVED FROM WATER BUFFALO (*BUBALUS BUBALIS*) AND ZEBU-TYPE CATTLE

L. Arenas de Moreno*[1], M. Giuffrida-Mendoza[2], S. Uzcátegui-Bracho[3], N. Huerta-Leidenz[1] and N. Jerez-Timaure[1]

[1.] Facultad de Agronomía, Universidad del Zulia, Aptdo. 15205, Maracaibo, Venezuela, [2.] Facultad de Medicina, Universidad del Zulia, Aptdo. 15131, Maracaibo, Venezuela, [3]. Facultad de Ciencias Veterinarias, Universidad del Zulia, Aptdo. 15252, Maracaibo, Venezuela. Email: lilia_arenas@cantv.net

Keywords: beef, *longissimus dorsi*, buffalo, nutritive value

Introduction

The Asiatic water buffalo (*Bubalus bubalis*) has been poorly utilized for meat production in the tropical Latin American regions. The interest in evaluating domestic buffalo as meat animals in Venezuela has grown due to the increase of buffalo herds and the breeders' organisation (Huerta-Leidenz, 2001). Little information exists about the nutrient composition of grass-fed red meats, particularly those derived from water buffalo and *Bos indicus*-influenced cattle (Zebu-types). This research focuses on the evaluation of proximal and mineral composition of *longissimus dorsi* muscle of buffalo and zebu-type cattle produced under extensive savannah conditions at two comparable ages (19 and 24 months of age), and different gender (bulls and steers).

Materials and Methods

Two groups, conformed by 32 buffaloes and 34 zebu-type cattle each, were randomly selected, and slaughtered at 19 and 24 months of age. At weaning (7 months), half of each group (16 buffalo and 17 cattle) was castrated. Details of the cow; calf operation and animal handling were previously described by Giuffrida de Mendoza *et al* (2005). Steaks (2.54 cm thick) were removed from the *L. dorsi thoracis* muscle and trimmed of surrounding fat. Proximate composition (g/100g fresh tissue) was determined according to official methods of the AOAC (1990), except for total lipids which were determined according to Folch *et al.* (1957). Mineral contents (mg/100 g fresh tissue) were determined by atomic absorption and/or atomic emission spectrophotometry (Perkin-Elmer, 1994), except for phosphorus, which was performed by using UV-VIS spectrophotometry (AOAC, 1990). Experimental data was subjected to analysis of variance (ANOVA) and least square means separation ($P < 0.05$) by using the Statistical Analysis System (SAS, 1996).

Results and Discussion

The species x age interaction affected dry matter and protein contents ($P < 0.05$) both decreased with age in samples derived from zebu-type cattle while, those from buffalo remained unchanged. Meat samples from the oldest animals (24 months) accumulated more intramuscular lipids (1.86 ± 0.05 vs. 1.43 ± 0.05 g/100g of fresh tissue). As observed by Ruiz *et al.* (2005), samples derived from the steers showed the highest intramuscular lipids content (1.73 ± 0.05 vs. 1.56 ± 0.06 g/100 fresh tissues). Except for Fe, species and gender did not affect the mineral composition of the analyzed samples. Age affected ($P < 0.05$) all minerals except P and Mg (Table 1). Phosphorus was not affected by any of the independent variables under study.

Table 1: Effect of age on mineral content in *longissimus dorsi* muscle.

Minerals (mg/100 g fresh meat)	Age (Least squares means ± SE[a])		*P* Value
	19 months (n=33)	24 months (n=33)	
Na	61.52 ± 1.05	72.07 ± 1.07	0.0001
K	353.21 ± 7.17	391.74 ± 7.17	0.0003
Ca	9.22 ± 0,25	8.98 ± 0.26	0.0001
Mg	24.38 ± 0.58	24.34 ± 0.49	NS
P	205.36 ± 2.47	198.79 ± 2.47	NS
Mn	0.02 ± 0.00	0.01 ± 0.00	0.0036
Fe	2.34 ± 0.05	2.15 ± 0.04	0.0029
Zn	3.71 ± 0.09	4.16 ± 0.10	0.0053
Cu	0.16 ± 0.01	0.09 ± 0.01	0.0001

[a]SE = Standard error
NS=Non significant ($P > 0.05$)

Variation in the individual mineral contents did not show a common tendency when comparing animals of 24 months to those of 19 months; among the minerals considered herein only Na, K and Zn concentrations increased with animals' age. These results differ from those reported by Kotula *et al.* (1982). Iron content varied with species x age interaction.

In buffalo meat, there were no differences ($P > 0.05$) in Fe content at the considered ages; in contrast, a decrease ($P < 0.05$) was observed in beef at 24 months (Figure 1). Also, Fe content varied with age x gender interaction. Although Fe content increased in bulls' meat samples at 24 months, those from steers had the highest content of Fe at the two ages considered (Figure 2).

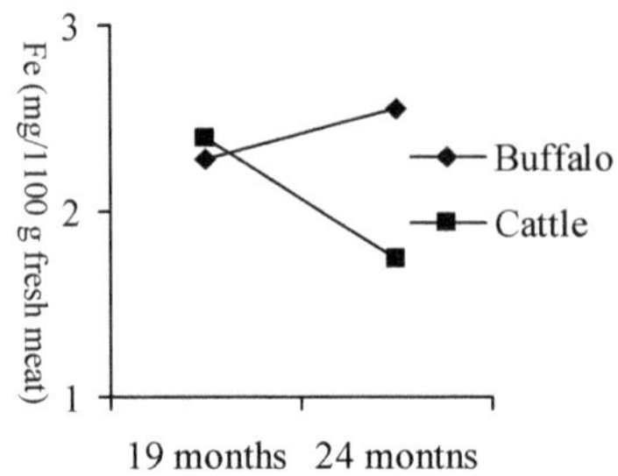

Figure 1: Effect of specie x age interaction on Fe content.

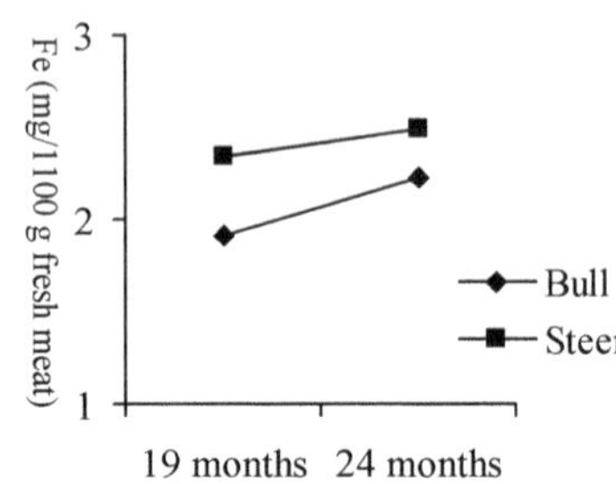

Figure 2: Effect of age x gender interaction on Fe content.

The fluctuations in the mineral content observed, could be due to the confounded effect of the diet produced by the environmental changes observed during the stocking and fattening phase of the post-weaned animal groups, described by Giuffrida de Mendoza *et al.*, (2005). This phenomenon could have produced changes in the intramuscular accumulation of some nutritional constituents like minerals. Some authors support the suggestion that pre-slaughter factors like a poor plane of nutrition during the drought period and high temperature, reduce the quality of meat (Kakkar and Pavanpreet, 2000).

Conclusions

The results showed that meat from both species have similar nutritional characteristics; the magnitude of the differences in proximate and mineral composition among age and gender groups, was not very large. The low lipid content found in the samples showed an apparent advantage of both species from a nutritional standpoint. Except for Zn, age of slaughter did not favour the accumulation of important minerals like Fe, Ca, Mg and P; however, a 100g-portion of fresh meat from both species provides nearly 25% of Fe, P, Zn and K required in the human diet.

References

AOAC. (1990). Association of Official Analytical Chemists. Official methods of analysis, Vol. II. Washington, DC.

Folch, J. M., Lees, M., Stanley, S. G. H. (1957). A simple method for the isolation and purification of total lipids from animal tissues. Journal of Biological Chemistry, 226, 497–509.

Giuffrida de Mendoza, M., Arenas de Moreno, L., Huerta-Leidenz, N., Uzcátegui-Bracho, S., Beriain, M. J., Smith, G. C. (2005). Occurrence of conjugated linoleic acid in *longissimus dorsi* muscle of water buffalo (*Bubalus bubalis*) and zebu-type cattle raised under savannah conditions. Meat Science, 69, 93–100.

Huerta-Leidenz, N.O. (2001). Venezuelan research experiences with buffalo meat production and quality. VI World Buffalo Congress Proceedings. Vol. I (pp.87-103). Maracaibo, Venezuela. N. Huerta-Leidenz, J. Vergara-López and A. Rodas-González, Editors.

Kakkar, V.K., Pavanpreet, K. (2000). Buffalo meat: Economical source of animal protein. Asian Livestock, 26(2), 11–13.

Kotula, A.W., Lusby, W.R. (1982). Mineral composition of muscles of 1 to 6 year old steers. Journal of Animal Science. 54(3),544–548.

Pelkin-Elmer. (1994). Analytical methods for atomic absorption spectrophotometry. Norwalk, Connecticut, U.S.A.

Ruiz, M.R., Matsushita, M., Visentainer, J.V., Hernández, J.A., Riveiro E.L de A., Shimokomaki, M., Reeves, J.J., de Souza, N.E. 2005. Proximate chemical composition and fatty acid profiles of *Longissimus thoracis* from pasture fed LHRH immunocastrated, castrated and intact *Bos indicus* bulls. South African Journal of Animal Science, 35(1): 13–8.

SAS. (1996). *User's guide: statistics.* Cary, N.C: Statistical Analysis System Institute, Inc

RELATIONSHIP OF FATTY ACID PROFILE AND MARBLING LEVEL IN MEAT FROM WATER BUFFALO (*BUBALUS BULALIS*) AND ZEBU-TYPE CATTLE

M. Giuffrida-Mendoza*[1], L. Arenas de Moreno[2], N. Huerta-Leidenz[2] and S. Uzcátegui-Bracho[3]

[1.] Facultad de Medicina, Universidad del Zulia, Aptdo. 15131. Maracaibo, Venezuela, [2.] Facultad de Agronomía, Universidad del Zulia, Aptdo. 15205, Maracaibo, Venezuela, [3.] Facultad de Ciencias Veterinarias, Universidad del Zulia, Aptdo. 15252. Maracaibo, Venezuela. Email: mariagvm@cantv.net

Keywords: beef, *longissimus thoracis*, buffalo meat, fatty acids, marbling

Introduction

In America, beef cattle (*Bos indicus* or *Bos taurus*) has been the traditional source of red meat. The multipurpose Asiatic water buffalo (*Bubalus bubalis*) has been mainly exploited for milk production, but poorly utilized for meat production in the Latin American region. Little information exists about the possible association between relative amounts of total fatty acid and the quantity of lipid and marbling levels in the *L. dorsi* of grass-fed buffalo and zebu-type cattle produced extensively under savannah conditions in tropical America, as well as the influence of intrinsic factors (i.e., age and gender) on the fatty acid composition of both species. The purpose of this paper was: to determine the associations between the amounts of total fatty acids (FA) with total lipids and marbling levels in the *longissimus thoracis* muscle derived from buffalo and zebu-type cattle at two comparable ages (19 and 24 months of age) and different gender (bulls and steers).

Materials and Methods

Thirty-two water buffaloes and 34 zebu-type cattle were randomly selected at weaning (7 months of age approximately) in the same cow-calf operation located at the western Venezuelan llanos. Half of each group (16 buffaloes and 17 cattle) was castrated, placed on continuous grazing of a *Brachiaria* spp pasture without supplementation other than minerals, and slaughtered at 19 and 24 months, approximately. After 24 hr chilling (0°C) carcasses were evaluated for marbling level according to Venezuelan standards (Decreto Presidencial No. 181, 1994): practically devoid (Pd), traces (Tr) and slight (L). At 48h *post-mortem* steaks (2.54 cm-thick) from the *longissimus thoracis* muscle were frozen at – 20 °C for subsequent study. Samples were partially thawed, trimmed of all surrounding fat tissue and analyzed for total intramuscular lipids (g/100g fresh tissue). Details of lipid and fatty acid profile (mg/100g fresh tissue) analyses have been described by Uzcátegui-Bracho *et al.*, (2001). Analysis of variance (ANOVA) was performed using the least squares procedure (SAS, 1996). Spearman and Pearson analyses were used to correlate the total FA with marbling level and total lipid, respectively.

Results and Discussion

Total saturated (SFA), unsaturated (UFA), monounsaturated (MUFA) and *cis*-polyunsaturated (PUFA) fatty acids increased ($P < 0.05$) with marbling levels in all animal groups (Table 1). Also, a highly significant and positive correlation ($r=0.6$; $P < 0.05$) between the already mentioned groups of FA and marbling level and total lipids was observed. PUFA content increased at the highest marbling level contrasting with the statements of Eichhorn *et al.*, (1986), for whom the increase in PUFA at lower marbling level is due to the decrease of triacilgliceride/ phospholipids ratio. This could be explained by the fact that there was very little variation in marbling levels of the analysed meat samples, probably due to the animals' genetics. Variation in the total FA and total lipids with marbling levels was similar for both species, and no interaction effect ($P > 0.05$) of age or gender was observed; the same behaviour was observed in the correlation analysis. The MUFA/SFA ratio at different marbling levels was similar for both species, and this ratio was comparable to that reported by Costa *et al.*, (2006), in pure Barrosa calves of 6 to 9 months and by Kazala *et al.*, (1999) in cattle of 16 and 18 months approximately. Uzcátegui *et al.*, (1999) reported higher MUFA/SFA ratio (1.75) in older animals when analysing meat samples from Venezuelan cattle of 2.5 to >4.0 years old. The relative low MUFA/SFA ratio observed in this work may be related to the age at slaughtered (19 and 24 months). Except for PUFA/SFA ratio (which decreased in meat samples scored "L"), there was no variation in the FA ratios considered herein. Similar results were reported by Uzcátegui *et al.*, (1999). It is assumed that the n-6/n-3 ratio in the human diet is more relevant than the PUFA/SFA ratio, since its balance plays an important role in the prevention of several chronic and auto-immune disorders (Enser *et al.*, 1996). In our study, the average value for n-6/n-3 ratio (approximately 3.5) almost triplicates that reported by Enser *et al.*,(1998), but it is within the optimal range (2.5-5.0) established by Lee *et al.*, (1989).

Conclusions

There were important changes in the content of some groups of fatty acids specially SFA, UFA, MUFA and PUFA *cis*, despite the relative low marbling levels and lipid content in the meat samples derived from the species considered. The variation observed in the total FA content according to marbling levels seemed to be independent of age and gender.

Differences in marbling levels (from Pd to L) only affected the PUFA/SFA ratio, and contributed to maintenance of optimal n-6/n-3 ratio.

Table 1: Fatty acid composition (mg/100g fresh tissue) and total lipids content (g/100g fresh tissue) in the *longissimus* muscle as affected by marbling levels (Means ± SE).

Total FA	N	Pd	Tr	L
Lipids	61	1.35 ± 0.12^{a}	1.62 ± 0.06^{ab}	1.79 ± 0.09^{b}
SFA	61	412.57 ± 29.19^{a}	548.74 ± 14.11^{b}	667.54 ± 18.40^{c}
MUFA	58	456.77 ± 27.73^{a}	550.08 ± 14.32^{b}	667.81 ± 17.54^{c}
PUFA	57	145.11 ± 5.48^{a}	173.47 ± 7.48^{ab}	187.34 ± 9.16^{b}
UFA	58	591.61 ± 36.44^{a}	721.82 ± 18.81^{b}	855.16 ± 16.04^{c}
PUFA n-6	57	109.11 ± 11.99^{a}	134.82 ± 5.79^{ab}	140.63 ± 7.09
PUFA n-3	33	31.42 ± 3.48^{a}	37.08 ± 1.78^{ab}	41.58 ± 2.12^{b}
PUFA *cis*	58	549.23 ± 34.72^{a}	669.58 ± 17.93^{b}	788.78 ± 21.96^{c}
PUFA *trans*	58	44.98 ± 4.45^{a}	52.04 ± 2.11^{a}	66.37 ± 2.63^{b}
Cis/*trans*	54	12.10 ± 0.61^{a}	12.36 ± 0.30^{a}	11.53 ± 0.38^{a}
UFA/SFA	55	1.30 ± 0.03^{a}	1.31 ± 0.01^{a}	1.28 ± 0.02^{a}
MUFA/SFA	54	1.05 ± 0.03^{a}	0.99 ± 0.01^{a}	0.99 ± 0.02^{a}
PUFA/SFA	57	0.35 ± 0.02^{a}	0.31 ± 0.01^{a}	0.28 ± 0.01^{b}
n6/n3	54	3.48 ± 0.19^{a}	3.62 ± 0.09^{a}	3.48 ± 0.12^{a}

a,b,c: Different letters in same row indicate significant differences ($P < 0.05$), NC = non correlation ($P>0.05$). SE = Standard Error.
N = total number of samples; Pd = Practically devoid, Tr = Traces, L = Slight. SFA, UFA, MUFA, PUFA = saturated, unsaturated, monounsaturated and polyunsaturated fatty acid. SFA = 14:0, 15:0, 16:0, 17:0, 18:0, 20:0. UFA = MUFA + PUFA; MUFA = 14:1, 16:1 *trans*, 16:1 *cis*, 17:1 *trans*, 18:1 *trans* (elaidic), 18:1 *cis* (oleic), 18:1 *trans* (vaccenic), 20:1 *cis*, 22:1, 24:1. PUFA = 18:2 *cis* n-6, 18:2 *trans* n-6, 18:3 γ n-6, 18:3 α n-3, 20:2 *cis*, 20:3 *cis* α n-6, 20:3 *cis* n-3, 20:4 n-6, 22:2 *cis*, 20:5 *cis* n-3, 22:4 n-6, 22:6 *cis* n-3.

References

Costa, P., Roseiro, L.C., Partidário, A., Alves, V., Bessa, R.J.B., Calkins, C.R., Santos C. (2006). Influence of slaughter, season and sex on fatty acid composition, cholesterol and α-tocopherol contents on different muscles of Barrosã-PDO veal. Meat Science, 72:130-139.

Decreto Presidencial No. 181. (1994). Ministerio de Agricultura y Cría. Faceta Oficial de la República de Venezuela. No. 30.426. Caracas, Venezuela. 5pp.

Eichhorn, J.M., Coleman, L.J., Wakayam, E.J., Blomquist, G.J., Bailey, C.M., Linkins, T.G. (1986). Effect of breed type and restricted vs ad libitum feeding on fatty acid composition and cholesterol content of muscle and adipose tissue from mature bovine females. Journal of Animal Science, 63: 781–794.

Enser, M., Hallett, K., Hweitt, B.J., Fursey, G.A., Wood, J.D. (1996). Fatty acid content and composition of English beef, lamb and pork and retail. Meat Science, 42:443-456.

Enser, M., Hallet, K.G., Wewett, B., Fursey, G.A.J., Wood, J.D., Harrington G. (1998). Fatty acid content and composition of UD beef and lamb muscle in relation to production system and implications for human nutrition. Meat Science, 49:329-341.

Kazala, E.Ch., Lozeman, F.J., Mir, P.S, Laroche, A., Bailey, D.R.C., Weselake, R.J. (1999). Relationship of fatty acid composition to intramuscular fat content in beef from crossbreed Wagyu cattle. Journal of Animal Science, 77:1717-1725

Lee J.H., Fukumoto M., Nishida H., Ikeda L., Sugano, M. (1989). The interrelated effects of n-6/n-3 and polyunsaturated/saturated ratio of dietary fats in the regulation of lipid metabolism in rat. Journal of Nutrition, 119:1893–1899.

S.A.S. (1996). Statistical Analysis System, Release 6.12. 1989-1996. SAS Institute, Inc; Cary, NC. USA.

Uzcátegui, S., Huerta-Leidenz, N., Arenas De Moreno, L., Colina G., Jérez-Timaure, N. (1999). Contenido de humedad, lípidos totales y ácidos grasos del músculo *longissimus* crudo de bovinos en Venezuela. Archivos Latinoamericanos de Nutrición, 49(2):171-180.

Uzcátegui-Bracho, S., Arenas de Moreno, L., Huerta-Leidenz, N., Buscema, I., Valero-Leal, K., Romero, S. (2001). Intramuscular content of lipids and fatty acids of grazing buffaloes and zebu-type cattle and implications for human nutrition. VI World Buffalo Congress Proceedings, Vol. II (pp.83-89). N. Huerta-Leidenz, J. Vergara-López and A. Rodas-González, Editors.

FUNCTIONAL EFFECTS OF MEAT PRODUCTS WITH ADDED WALNUT

B. Olmedilla-Alonso[1], F. Sánchez-Muniz[2], C. Ruiz-Capillas*[3] and F. Jiménez-Colmenero[3]

[1]*Unidad de Vitaminas. Hospital Universitario Puerta de Hierro de Madrid.* [2] *Depto de Nutrición. Facultad de Farmacia. UCM de Madrid.* [3] *Instituto del Frío (CSIC). Ciudad Universitaria. 28040-Madrid. Spain. Email: claudia@if.csic.es*

Keywords: restructured beef steak, frankfurters, walnut, functional effect

Introduction

Observational epidemiological studies show an inverse association between walnut consumption frequency and risk of coronary heart disease (CHD) (Fraser *et al.*, 1992). Walnuts, as part of a heart-healthy diet may reduce cholesterol concentrations in humans and animals (Sabaté, 1993; Iwamoto *et al.*, 2000). Recently, the FDA authorised a qualified health claim indicating that eating 42.5g of walnuts per day as part of a low-saturated-fat and low-cholesterol diet with no increase in the caloric intake may reduce the risk of CHD (FDA, 2003). Moreover the WHO (2003) recommends consuming at least 400g of fruit and vegetables a day, of which at least 30g must be walnuts, pulses and seeds. However, the consumption of walnuts is generally below the recommended consumption, since very few people are able to consume walnuts in sufficient amounts on a regular basis and for long periods of time. One way of encouraging their intake would be to incorporate them as an ingredient in commonly consumed food, such as meat products, to which different bioactive compounds could be added to make them better for the heart. Restructured beef steaks and frankfurters with added walnuts have been formulated resulting in products with acceptable physicochemical and sensory properties (Jiménez Colmenero *et al.*, 2003, 2005; Serrano *et al.*, 2006). Meat products with added walnut could be considered potential functional foods in that they incorporate several biologically active components that have the potential to produce functional effects (Nus *et al.*, 2004; Serrano *et al.*, 2005). However, for a food to be considered functional it must be satisfactorily shown that its consumption has a beneficial effect on certain groups of the population. Tests are therefore necessary to support this effect, preferably in humans and using appropriate intermediary biomarkers, important for early detection and diagnosis of the disease. This research focuses on evaluating the functional effect of meat products (restructured beef steak and frankfurter sausages) formulated with 20% walnut. The evaluation of the functional effect was carried out in an intervention study on subjects with cardiovascular risk to observe the effect of meat product consumption on exposure, effect and risk biomarkers in quantities the same as in a balanced diet.

Materials and Methods

Restructured beef steak and frankfurters, without walnut (control) and with 20% added walnut were prepared as reported by Serrano *et al.*, (2005, 2006) and Jiménez Colmenero *et al.*, (2005).

The functional effect was evaluated in an intervention study with men and women who had some cardiovascular risk in accordance with WHO/FAO considerations in their report on diet, nutrition and the prevention of chronic diseases (WHO, 2003). Men 45-70 years and women 50-70 years (without any substitute hormonal therapy) and who had no known illness or did not take medicines that interfered with absorption or lipid metabolism were selected, and among the modifiable factors, excess weight, dislipemia, smoking habit and blood pressure, establishing the following criteria: overweight (IMC >25 y <34.90 kg/m^2), hypercholesterolemia (>220 mg/dl y < 300 mg/dl), and at least one of the following: blood pressure (close to 140 / 90 mm Hg) or smoking habit. An assay protocol was used similar to normal diet conditions as a sustainable long-term medium. A cross-intervention study was carried out for 5 weeks, in which the consumption of meat and meat products were substituted with the consumption of 5 meat products (4 beef steaks and 1 sausage with and without walnut) a week (150 g meat/ration). In consumption terms, this would be 30 g walnut / beef steak and 16 g walnut / ration of sausages which implies a consumption of 136 g of walnuts a week, representing a mean intake of 19.4 g walnuts/day. The consumption of 150 g of any of these products provides ca. 70% of the daily amount of walnut suggested as appropriate for decreasing the risk of cardiovascular diseases (FDA, 2003).

Serum γ-tocopherol was used as a consumption marker (exposure biomarker) of this food in the subjects, since the consumption of the meat product with walnut provides an amount of γ-tocopherol significantly greater than that habitually present in the diet (Olmedilla Alonso *et al.,* 2005). Total cholesterol, low-density lipoprotein (LDL) cholesterol and high-density lipoprotein (HDL) cholesterol, triglycerides, homocysteine and blood pressure were used as function biomarkers. The CVD risk factors of recognised analytical and biological validation which we considered were: total cholesterol, LDL-cholesterol and blood pressure.

Results and Discussion

The composition, physicochemical and sensory properties of meat products with added walnut and their potential health benefits have already been reported (Jiménez Colmenero *et al.*, 2005; Serrano *et al.*, 2005 and 2006).

The evaluation results of the functional effect after the consumption of meat product with walnut were: a) the exposure biomarker response, the serum γ-tocopherol, increased after the consumption of meat with walnut, and b) the decrease in effect biomarkers, with clinical importance (risk markers), total cholesterol and LDL-cholesterol after the intervention period of the potentially functional meat product (Table 1); these biomarkers did not vary when the study was done using the same meat products but without added walnut (Olmedilla Alonso *et al.*, 2006); c) even in the study phase, the results suggested that the consumption of meat with walnuts improved the antioxidant status of the consumers by modifying the concentration of some enzymes of the glutation and paraoxonase systems.

Table 1: Variation in the cholesterol and γ-tocopherol in blood after the consumption of meat products with walnut.

	Variation	Confidence interval (95%)	P
Total cholesterol (mg/dl)	↓ 10.7	4.1 -17.1	0.002
LDL-cholesterol (mg/dl)	↓ 7.6	2.2 -13.0	0.007
γ-tocopherol (μg/dl)	↑ 8.9	1.0 – 16.8	0.03

Conclusion

Meat products with walnut can be considered as functional foods in subjects with the same characteristics as in this study (men and women between 45 and 70 years, mildly hypercholesterolemic, overweight, smokers or with slight hypertension), since the decrease in the total and LDL-cholesterol mentioned earlier was significant. However, to make declarations on these meat products regarding their cardiovascular risk, the same kind of study would have to be done on subjects with the same characteristics.

Acknowledgements

This research was supported under project AGL2001-2398-C03. Plan Nacional de Investigación Científica, Desarrollo e Innovación Tecnológica (I+D+I).

References

FDA (2003). Food and Drug Administration. Office of Nutritional Products, Labeling and Dietary Supplements. Qualified Health Claims: Walnuts and Coronary Heart Disease (Docket No 02P-0292).

Fraser, G.E., Sabaté, J., Beeson, W.L. and Strahan, T.M. (1992). A possible protective effect of nut consumption on risk of coronary heart disease. *Archives of Internal Medicine*, 152(7), 1416-1424.

Iwamoto, M., Sato, M., Kono, M,. Hirooka, Y., Sakai, K., Takeshita, A. and Imaiuzumi, K. (2000). Walnuts lower serum cholesterol in Japanease men and women. *Journal of Nutrition*, 130 (2), 171-176.

Jiménez Colmenero, F., Ayo, J. and Carballo, J. (2005). Physicochemical properties of low sodium frankfurter with added walnut: Effect of transglutaminase combined with caseinate, KCl and dietary fibre as salt replacers. *Meat Science*, 69, 781-788.

Jiménez Colmenero, F., Serrano, A., Ayo, J., Solas, M.T. Cofrades, S. and Carballo, J. (2003). Physicochemical and sensory characteristics of restructured beef steak with added walnuts. *Meat Science*, 65, 1391-1397.

Nus, M., Ruperto, M. and Sánchez-Muniz, F.J. (2004). Frutos secos y riesgo cardio y cerebrovascular. Una perspectiva española. *Arch. Latinoamer. Nutr.* 54: 137-148.

Olmedilla Alonso, B., Granado Lorencio, F., Herrero Barbudo, C., Blanco Navarro, I. and Sánchez Muniz, F.J (2005). "Productos cárnicos funcionales preparados con nuez. Evaluación del efecto funcional. (Parte 3)". *CTC Alimentación*, 24: 37-41.

Olmedilla-Alonso, B., Granado-Lorencio, F., Herrero-Barbudo, C. and Blanco-Navarro, I. (2006). Effect of meat products with added walnuts consumption on biomarkers of cardiovascular risk. *Nutr. Metab. Cardiovasc. Dis.* (submitted).

Sabaté, J., Fraser, G.E., Burke, K., Knutsen, S., Bennett, H. and Lindsted, K.D. (1993). Effects of walnuts on serum lipids levels and blood pressure in normal men. *N. Eng. J. Med.* 328: 603-607.

Serrano, A.; Cofrades, S. and Jiménez-Colmenero, F. (2006). Frozen storage characteristics of restructured beef steak containing walnut. *Meat Science*, 72, 108-115.

Serrano, A., Cofrades, S., Ruiz-Capillas, C., Olmedilla-Alonso, B., Herrero-Barbudo, C. and Jiménez-Colmenero, F. (2005). Nutritional profile of restructured beef steak with added walnuts. *Meat Science*, 70, 647-654.

WHO (2003). Diet, Nutrition and the Prevention of chronic diseases. WHO Technical report Series 916.

ASSESSMENT OF ANTIOXIDATIVE CAPACITY OF ANIMAL LIVER AND BILE

R. Sakata*[1], T. Oshida[1], K.-H. Yu[2] and L.-C. Lin[2]

[1] *School of Veterinary Medicine, Azabu University, Sagamihara 229-8501, Japan,* [2] *Research Institute of Animal Science, National Chung-Hsing University, 250 Kuo-Kuan Road, Taichung 402, Taiwan. Email: lab207@ms85.url.com.tw*

Keywords: antioxidation, bile, Fe^{2+} chelation, liver, superoxide scavenging

Introduction

Nearly 10 million pigs are slaughtered in Taiwan each year. Only a small portion of the liver is used for human consumption, with the greater part used as animal feed or discarded. The pig bile is thrown away immediately. These practices are also particularly common in poultry slaughterhouses. Discarding those by-products is increasing the environmental load. Liver is a valuable source of numerous proteins, amino acids, minerals and vitamins. Animal bile is rich in various compounds, and is potentially a source of numerous nutrients following appropriate extraction, and thus had medicinal application in traditional Chinese therapies. Bezoar is one such traditional Chinese medicine whose curative effect has been reported (Liu, 1980). The major compound present in this medicine is bilirubin which is abundant in pig bile and is the major ingredient for the synthesis of bezoar. Large quantities of liver and bile may be obtained from slaughterhouses in Taiwan and Japan. Liver extracts and bile powders were examined in this study for their antioxidative capacity.

Materials and Methods

1. Sample preparation

(1) Liver extracts: Liver was cut into 1 cm^3 pieces, placed in boiling water for 1 min, and then water-flow cooled for removal of impurities. Following the addition of an equal volume of deionized water, the samples were placed in a 100°C water bath for 6 hours and then filtered through 2-layer swabs. Decompression was then carried out to reduce original volume by 75% (Eyela, Model N-1NW, Japan). A powder of the liver extract was obtained by freeze dehydration (FD-5N, Eyela; Pump, G-100D, ULVAC, Japan) and then stored at -20°C.

(2) Bile powders: Bile gall was obtained by needle puncture and filtered as above. This was followed by heat treatment at 100°C for 15min. The samples were converted into powder form by freeze dehydration and then stored at -20°C.

2. Antioxidation assessment

(1) Reducing capacity: This capacity was assessed according to Oyaizu (1986). 2.5 mL sample solution were added to 2.5 mL 0.2 M Na_2HPO_4 (pH 6.6) and 2.5 mL 1% $K_3Fe(CN)_6$, placed in a water bath at 50°C for 20 min and cooled in flowing water, followed by addition of 2.5 mL 10% TCA and centrifugation (3,000 rpm) for 10 minutes. 5 mL supernatant were added to 5 mL deionized water and 1 mL 0.1% $FeCl_3 \cdot 6H_2O$, and the reaction was allowed to proceed for 10 min. The absorbance was measured at 370 nm by spectrophotometer. Higher absorbance is an indication of greater reducing capacity.

(2) Fe^{2+} ion chelating capacity: 1 mL sample solution is added to 3.7 mL methanol and 0.1 mL 2 mM $FeCl_2 \cdot 4H_2O$ followed by 0.2 mL 5 mM ferrozine after 20 sec, and the reaction was allowed to proceed for 10 min. The absorbance was measured spectrophotometrically at 562 nm. Lower absorbance is an indication of greater chelating capacity.

Chelating capacity (%) = [1– (absorbance of sample at 562 nm)/ (absorbance of control at 562 nm)] × 100%

(3) Capacity for scavenging superoxide anions: 1 mL sample solution was added to 120μM phenazine methosulfate (PMS), 936 μM β-nicotinamide adenine dinucleotide (NADH) and 300μM nitro bluetetrazolium (NBT), and the reaction is allowed to proceed at room temperature for 5 min. The absorbance was immediately measured spectrophotometrically at 560 nm, followed by addition of superoxide dismutase (SOD, 5,000 units). A lower absorbance indicates greater scavenging capacity, which is expressed as [1 – (absorbance of sample at 560 nm) / (absorbance of control at 560 nm)] × 100%.

Results and Discussion

1. Reducing capacity

Figure 1 shows this ability for the various extracts and appeared greatest for PLE. The reducing capacity of PLE, PBL, CLE and CBL was significant and exceeded that of BHT ($p<0.01$), while that of PBP and CBP was not significant. Takahashi *et al.* (1995) indicated that taurine (Tau) inhibits lipid peroxidation in cells. Tau is the main component of PLE and CLE which may explain the better reducing capacity in liver extracts.

2. Capacity for Fe^{2+} ion chelation

Metal ions can promote lipid peroxidation and some may induce free radical production. Fe^{2+} ions strongly promote oxidation which in turn may enhance this production. However, this ability is lost subsequent to the chelation of these ions. Chelating ability may thus serve to enhance that of antioxidation, as evident from Figure 2. All extracts were found to have more than 50% chelating ability. Pierno *et al.* (1998) consider Tau to chelate Fe^{2+} ions in cells, and bilirubin has this ability to the same extent. Chelating ability may possibly be due to Tau and bilirubin.

(3) Scavenging ability of superoxide anions

Superoxide anions may possess reducing and oxidation capacity simultaneously, and under certain conditions may act as a single oxide, hydroxyl radical or hydrogen peroxide. Superoxide dismutase (SOD) cleans superoxide anions by the dismutase reaction. Many substrates having the same ability are said to be SOD-like and remove superoxide anions by a different mechanism. Figure 3 shows the scavenging ability of superoxide anions under various conditions, with PLE and CLE showing greater ability to scavenge superoxide anions, the extent differing significantly ($P<0.01$). PBP and CBP appear not to possess this ability. Liver extracts were noted in this study to contain more Tau than bile powder. This possibly may explain why liver extracts possess the ability to scavenge superoxide anions.

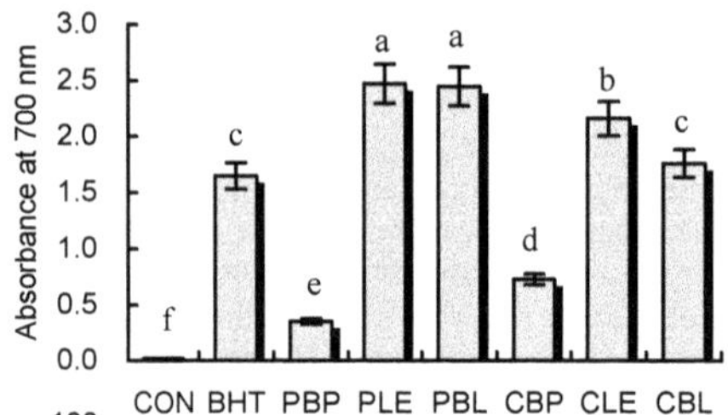

Figure 1: Reducing capacity of liver and bile extracts.
Bars indicate mean ± S.D. Different superscript letters (a-f) indicate significant differences ($P < 0.01$).
Deionized water (CON), butylated hydroxytoluene (BHT), pig bile powder (PBP) ; pig liver extract (PLE), pig bile and liver extract (PBL), chicken bile powder (CBP), chicken liver extract (CLE), chicken bile and liver extract (CBL), n=3.

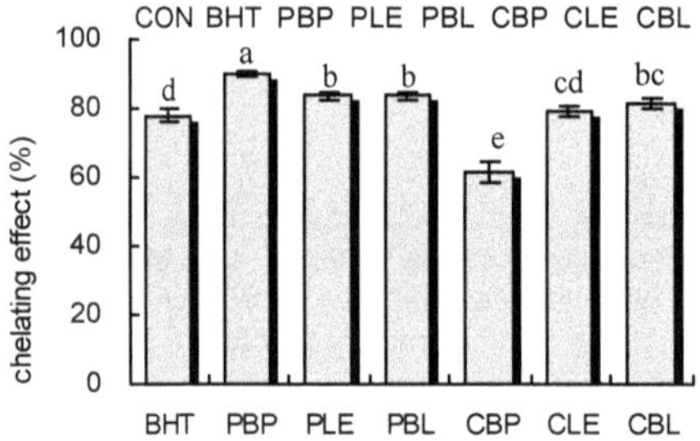

Figure 2: The Fe^{2+} ion chelating capacity of liver and bile extracts.
Bars indicate means ± S.D. Different superscript letters (a-e) indicate significant differences ($P < 0.01$).
Butylated hydroxytoluene (BHT), pig bile powder (PBP), pig liver extract (PLE), pig bile and liver extract (PBL), chicken bile powder (CBP), chicken liver extract (CLE), chicken bile and liver extract (CBL), n=3.

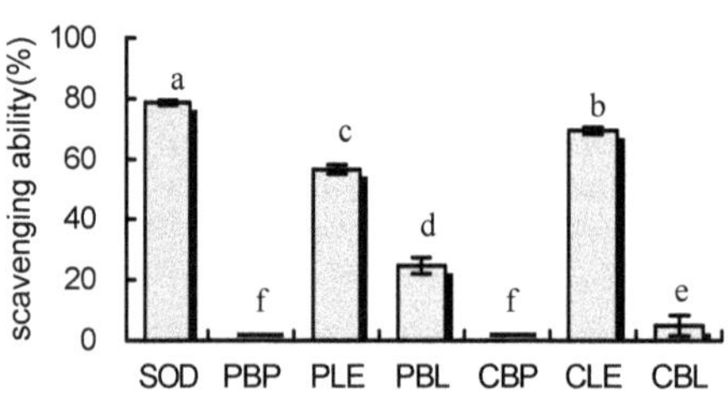

Figure 3: Superoxide anion scavenging capacity of various liver and bile extracts.
Bars indicate means ± S.D. Different superscript letters (a-f) indicate significant differences ($P<0.01$).
Superoxide dismutase (SOD), pig bile powder (PBP); pig liver extract (PLE), pig bile and liver extract (PBL), chicken bile powder (CBP), chicken liver extract (CLE), chicken bile and liver extract (CBL), n=3.

Conclusions

Pig liver and bile extracts were found to have greater capacity for reduction. Chicken and pig liver had greater capacity for scavenging superoxide anions. For pig bile extract, Fe^{2+} ion chelation capacity was greatest. These extracts are thus shown to have significantly potential as nutrients and healthful constituents in food production.

References

Liu. H.-N. (1980). Chinese Medicine. Chen Chung Book, Taiwan, R.O.C. (in Chinese)

Oyaizu, M. (1986). Antioxidative activity of browning products of glucosamine fractionated by organic solvent and thin-layer chromatography. Nippon Shokuhin Kogyo Gakkaishi, 35:771-775. (in Japanese)

Pierno, S., *et al.* (1998). Chronic administration of taurine to aged rats improves the electrical and contractile properties of skeletal muscle fibers. J. Pharmacol. Exp. Ther., 286: 1183-1190.

Takahashi, K., *et al.* (1995). Gas chromatographic separation of bile acid 3-glucosides and 3-glucuronides without prior deconjugation on a stainless- steel capillary column. J. Chromatogr. A., 689: 78-84.

DETERMINING WHETHER FAT CONSUMPTION FROM LEAN AND MARBLE STEAKS ARE WITHIN RECOMMENDED DIETARY GUIDELINES

R. Bickerstaffe*[1], A.E. Bekhit[1], K. Gately[1], J. Morton[1] and F. Carruthers[2]

[1.] *Agriculture and Life Sciences, Lincoln University, Canterbury, New Zealand.* [2.] *Beef and Lamb Marketing Bureau, Auckland, New Zealand. Email: bickerst@lincoln.ac.nz*

Keywords: steaks, fat, distribution, consumer intakes

Introduction

Meat, particularly, red meat has been associated with unfavourable publicity about its negative effects on human health due to its contribution to high levels of dietary fat and, consequently, saturated fatty acid intakes (Langesen and Swinburn, 2006). This publicity has persisted despite in New Zealand there being a reduced incidence in deaths associated with coronary heart disease. The meat industry over the last ten years has, however, undergone considerable change, both in the production of lean animals and trimming fat from retail meat cuts. The question is 'Have these changes resulted in a reduction in fat intake by consumers and have the changes been sufficient to satisfy the dietary guidelines recommended by nutritionalists?'.

The objectives of this research was to determine

(i) whether production management practices in the red meat industry have changed and had an impact on the supply of fat into the human food chain.
(ii) whether processing or retail trimming procedures have changed and contributed to a reduction in the supply of fat into the human food chain.
(iii) the amount of dietary intake of fat from steaks, varying in fat content, consumed by human volunteers.

Materials and Methods

Information in this study was obtained for lamb and beef, which represent the majority of red meat consumed by the public in New Zealand, in the following three areas:

Lamb information was obtained from a major processing company on the changes in their stock purchased and processed from 1988 to 2002.
A line of lambs with the same carcass weights, but with different external fat covers over the 12th rib, were subjected to commercial boning out procedures to yield market ready retail cuts. The yield of retail cuts and 'waste' fat were measured.
A consumer trial was conducted whereby two 2.5 cm thick *M. longissimus dorsi* (LD) steaks, A and B were presented to volunteer consumers. Steak A had a high level of external fat and steak B, a low level of external fat. The consumers were requested to return to the laboratory, in pre-packaged pre-paid containers, any fat trimmed off the steak before or after cooking the steak and any residual fat left in the pan after dry grilling the steak. All steaks provided to the consumers were pre-tested for tenderness and had kg F values between 4.5 to 5.5 kg F (tender steaks).

Results and Discussion

Lamb carcasses in New Zealand are graded according to their carcass weight and fat cover over the 12th rib. YL lambs (9.0-12.5 kg carcasses with fat cover < 6 mm fat) are considered to be lean lambs. In 1988, only 15% of the lamb kill was YL lambs. By 2002, the percentage of YL lambs had increased dramatically to 42% of the lamb kill reflecting changes in lamb schedule prices which rewarded farmers producing leaner animals and the changes in breed and management practices by producers responding to the schedule prices. The pricing policy and producer practices resulted over a 14 year period in a 173% increase in leaner or low fat animals being supplied to the market.
The boning out of large lean lambs (YM, 17.5 kg carcasses < 6 mm fat) and large fat lambs (PM, 17.8 kg carcasses > 6 mm but < 12 mm fat) showed that both carcasses yielded 50% of their weight as prime retail cuts. However, 15.7% of the carcass weight of the PM lambs was trimmed off as fat and discarded whilst only 9.6% of the carcass weight of the YM lambs was discarded as fat. These results illustrate that modern boning procedures trim up to 16% of the carcass weight of 'an overfat' lamb to meet the specifications of lean lamb cuts required by retailers and consumers. This move to large lean animals promotes a reduction in the supply of fat into the human food chain and the diet.
A total of 86 low external fat covered steaks (13-22% external fat) were paired with 86 high external fat covered steaks (22-31% external fat) and assessed in pairs by 43 separate individuals. Fat cover and intramuscular fat was determined by image analysis. In the case of the high external fat covered steaks, 52% of the individuals discarded more than 20% of the weight of the steaks as fat. Only 2% of the individuals actually retained and consumed the fat. In the case of the low external fat covered steaks, only 9% of the individuals discarded more than 20% of the weight of the steak as fat.

The results indicate that consumers do not tolerate a high fat cover on steaks and will discard the fat either before or after cooking the steak. In contrast, consumers do tolerate and will consume low levels of external fat cover.

Intramuscular fat levels varied independently of the steak fat cover. It was calculated that the fat consumed from the external fat cover and intramuscular fat by the 43 individuals consuming the high external fat covered steaks was 16.5 g fat/200 g steak and from the low external fat covered steaks was 7.9 g fat/200 g steak. The intake of 7.9 to 16.5 g fat/200 g steak is well within the recommended dietary fat intake of 75 g for a 2200 calorie diet.

Conclusion

Red meat has a legacy of providing undesirable high levels of fat which may impact on the well-being and health status of the community (Higgs, 2000). The meat industry has been accused of not responding to the needs of the community and that the industry contributes, in some way, to an increase in human metabolic disturbances which manifest themselves as clinical disorders such as coronary heart disease.

This report shows that the red meat industry has responded in a responsible way to the consumer demands for a healthy product. At the producer level, schedule prices and stock management changes have led to at least 45% of the lamb kill being lean animals in New Zealand. This trend is continuing. At the processing/retail level, boning procedures have changed and resulted in more fat being trimmed from carcasses to produce low fat trimmed retail products. At the consumer level, fat is trimmed from steaks prior to or after cooking. The net result is that the consumers consume 7.9 to 11.5 g fat/200 g steak which is well within the dietary guidelines of 75 g fat per day. The whole supply chain has contributed to a reduction in the availability of dietary fat to consumers which, when combined with additional trimming by individuals at cooking or dining, has led to a considerable reduction in dietary fat intake from red meats.

A conclusion and a recommendation to the industry is that trimming any external fat to a 2 mm fat thickness, before eating, will ensure that the consumer has a satisfactory eating experience whilst delivering a healthy product which provides fat well within the recommended daily intake of 75g/day.

References

Higgs, J.F. (2000). The changing nature of red meat: 20 years of improving quality. Trends in Food Science and Technology, 11: 85-95.

Langesen, M., Swinburn, B. (2006). The New Zealand food supply and diet-trends. New Zealand Medical Journal, 113: 311-315.

EFFECT OF LOW SALT CONTENT ON THE QUALITY OF PAINHO DE PORTALEGRE, A TRADITIONAL PORTUGUESE SAUSAGE – PHYSICAL, CHEMICAL AND MICROBIOLOGICAL ASPECTS

M. Elias*, A.C. Agulheiro-Santos and P. Nunes

Universidade de Évora, Instituto de Ciências Agrárias Mediterrânicas, Apartado 94, 7002-554 Évora, Portugal Email: elias@uevora.pt

Keywords: Portuguese sausage, low salt, microbiology, physical and chemical aspects

Introduction

The role of salt in cured sausage production is well known: aw depressor, and consequently its bacteriostatic action, as a taste enhancer, it decreases the isoelectric point of proteins and, consequently, decreases the drip loss. However, salt in excess plays an undesirable role in human health, as it contributes to an increase in heart disease. The salt content in traditional Portuguese sausages is typically 5-6%. Nowadays NaCl levels below 3% are recommended. The main objective of this work was to evaluate the influence of decreasing salt content in the production of one type of Portuguese traditional sausage, on the microbiological, physical and chemical characteristics of this product.

Materials and Methods

"Painho de Portalegre" is a traditional Portuguese sausage (length: ±30 cm; diameter: 4-5 cm; curing period: 40 days; ingredients: meat, fat, salt (NaCl), garlic and fermented *Capsicum annuum* L.) made with meat and fat from the Alentejano pig breed, a Portuguese breed. Sausages were produced with 3% and 4.5% of NaCl after their curing period. Five samples were taken from each group. Physical and chemical analyses were carried out: pH, a_w, %NaCl, chromatic coordinates L*, a* and b*. Microbiological analyses carried out were: total mesophylic aerobic bacteria, yeasts, *Micrococcaceae*, Lactic acid bacteria, *Enterobacteriaceae*, *Enterococcus*, coliforms and *E. coli*. For statistical treatment of the data, an ANOVA was performed, using Statistica 5.1 program, and means were compared using the HSD Tukey method.

Results and Discussion

Physical and chemical analyses (Table 1) showed lower pH values in the sausages with 4.5% NaCl, this maybe due to the decrease in pH often found with this salt. As expected, a_w values were lower in products with more salt. The analyses of %NaCl confirm the concentration of this salt in the final product. Higher a* values in sausages with 4.5% of NaCl probably result from the higher *Micrococcaceae* counts with nitrate reducing action (Barriere *et al.*, 1998) (Table 2).

Table 1: Physical and chemical characteristics of "Painho de Portalegre" with 3% and 4,5% NaCl – means and standard deviation.

NaCl	pH	a_w	%NaCl	L*	a*	b*
3%	6,04 a	0,88 a	3,14 a	45,8 a	11,8 a	14,1
	±0,03	±0,01	±0,06	±0,4	±0,2	±0,3
4.5%	5,90 b	0,84 b	4,54 b	43,0 b	12,9 b	14,5
	±0,03	±0,00	±0,11	±0,6	±0,5	±0,2

Different letters in the same column represent significantly different means.

Concerning microbiological analyses (Tables 2 and 3), mesophylics, yeasts and *Micrococcaceae* counts show significantly higher values in sausages with 4.5% NaCl. This reveals a good adaptation of those microrganisms to those salt concentrations in the substrate studied. The main microrganisms found at high levels in "Painho de Portalegre" were lactic acid bacteria, with more than 8 log values. Results of microrganisms related with hygiene shows that sausages with lower salt content (3%) exhibit similar values to those with higher salt content.

Table 2: Microbiological analyses of "Painho de Portalegre" (mesophylic aerobial bacteria, yeasts, *Micrococcaceae* and Lactic acid bacteria) with 3% and 4,5% of NaCl – means and standard deviation.

NaCl	Mesophylic (log ufc/g)	Yeasts (log ufc/g)	*Micrococcaceae* (log ufc/g)	Lactic acid bacteria (log ufc/g)
3%	7,24 a	2,85 a	3,15 a	8,44 a
	±0,31	±0,15	±0,12	±0,03
4.5%	7,96 b	3,95 b	4,25 b	8,33 b
	±0,03	±0,02	±0,08	±0,01

Different letters in the same column represent significantly different means.

Table 3: Microbiological analyses of "Painho de Portalegre" (*Enterobacteriaceae*, *Enterococci*, coliforms and *E. coli*) with 3% and 4,5% of NaCl – means and standard deviation.

NaCl	*Enterobacteriaceae* (log ufc/g)	*Enterococci* (log ufc/g)	Coliforms *	*E. coli* *
3%	2,97	5,65	4,50	3,50
	±0,49	±0,16	±0,50	±0,50
4.5%	2,77	5,41	4,00	3,50
	±0,07	±0,15	±0,00	±0,50

Different letters in the same column represent significantly different means.
* Results in classes: class 1 (< 1 bacteria/g); class 2 (≥ 1; < 10); class 3 (≥ 10; < 100); class 4 (≥ 100; < 1 000); class 5 (≥1 000; < 10 000).

Conclusions

Based on the results obtained for physical, chemical and microbiological analyses, it was possible to produce "Painho de Portalegre" with low salt content (3% in the end product) without quality loss.

Acknowledgements

This research work was supported by the IFADAP under Project AGRO 218.

References

Barrierre, J-C., Bacque, E., Puchalt, G., Quenet, Y., Molherat, C., Cassayre, J and Paris, J-M. (1998). Reactivity of 1-(des-3-hydroxy-picolinoyl) pristinamycin I_A. Tetrahedron 54 (42): 35-40.

Beriain, M.J., Peña, M.P. and Bello, J. (1993). A study of the chemical components which characterize Spanish saucisson. Food Chemistry, 48: 31-37.

Gimeno, O., Astiasarán, I. and Bello, J. (2001). Influence of partial replacement of NaCl with KCl and $CaCl_2$ on microbiological evolution of dry fermented sausages. Food Microbiology, 18: 329-334.

EFFECT OF LOW SALT CONTENT IN THE QUALITY OF PAINHO DE PORTALEGRE, A TRADITIONAL PORTUGUESE SAUSAGE – RHEOLOGICAL AND SENSORIAL ASPECTS

M. Elias*, A.C. Agulheiro-Santos and P. Nunes

Universidade de Évora, Instituto de Ciências Agrárias Mediterrânicas, Apartado 94, 7002-554 Évora, Portugal Email: elias@uevora.pt

Keywords: Portuguese sausage, low salt, rheology, sensory evaluation

Introduction

Despite the benefits of salt (NaCl) (mainly its role in the control of micro organisms and its influence on food taste), its abused use is more and more censured. In fact, heart disease is often correlated with the excessive consumption of NaCl (Truswell, 1994; Ibáñez *et al.,* 1995). As a result, food production with low salt content has increased over the last number of years (Gou *et al.*, 1996; Gimeno *el al.*, 2001; Muguerza *et al.*, 2004). In Portugal, in traditional sausages salt content is typically 5-6%. Nowadays NaCl levels above 3% are recommended. The objective of this work was to evaluate, the effect of reduced salt content in one type of Portuguese sausage, on the rheological and sensorial characteristics of this product.

Materials and Methods

"Painho de Portalegre" is a traditional Portuguese sausage (length: ±30 cm; diameter: 4-5 cm; curing period: 40 days; ingredients: meat, fat, salt (NaCl), garlic and fermented *Capsicum annuum* L.) made with meat and fat from the Alentejano pig breed, a Portuguese breed. Sausages with 3% and 4.5% of NaCl after the curing period were produced. From each group 5 samples were taken. The rheological analyses included: Texture Profile Analysis (parameters: hardness, cohesiveness, springiness, gumminess, chewinness) and cutting test. For sensorial analysis the attributes evaluated were: colour intensity, off colours, aroma intensity, off aromas, taste intensity, off taste, tenderness, salt intensity and global evaluation.
For statistical data treatment, an ANOVA was performed, using Statistica 5.1 program, and means were compared using the HSD Tukey method.

Results and Discussion

Rheological tests revealed, for all the characteristics studied, higher values from products with 4.5% NaCl (Table 1). Salt (NaCl) increased meat protein binding properties and consequently it interferes with textural characteristics of sausages (Fernández, 2000). For that reason the results for hardness and cohesiveness are significantly higher in sausages with 4.5% NaCl.

Table 1: Rheological characteristics of "Painho de Portalegre" with 3% and 4,5% of NaCl – means and standard deviation.

NaCl	Hardness (g)	Cohesiveness	Springiness	Gumminess (g)	Chewinness (g)	Cutting test (g)
3%	797,1 a	0,74 a	0,78	590,8	475,0	750,1
	±456,6	±0,03	±0,06	±342,0	±308,9	±434,4
4.5%	2032,5 b	0,79 b	0,81	1626,4	1349,4	1957,1
	±1295,5	±0,04	±0,10	±1117,8	±1034,4	±1269,3

Different letters in the same column represent significantly different means.

Sensorial analysis carried out (Table 2) shows that colour intensity, aroma intensity, taste intensity and global evaluation attributes of sausages with 4.5% NaCl samples had higher values. The role of NaCl as a promoter of flavour could explain those results.

Table 2: Sensorial evaluation of "Painho de Portalegre" with 3% and 4.5% of NaCl – means and standard deviation (results obtained using a scale from 0 to 100).

NaCl	Colour intensity	Off colour	Aroma intensity	Off aroma	Taste intensity	Off taste	Tenderness	Salt intensity	Global evaluation
3%	59,1	4,8	58,7 a	9,1	56,6	7,4	65,9	46,6 a	52,9
	±16,4	±14,0	±17,6	±19,7	±24,1	±21,0	±23,9	±18,1	±20,9
4,5%	68,1	3,81	67,8 b	8,8	68,4	8,2	65,9	61,1 b	63,6
	±15,4	±13,6	±9,4	±19,7	±15,1	±17,2	±22,7	±9,7	±12,7

Different letters in the same column represent significantly different means.

Conclusions

"Painho de Portalegre", a Portuguese traditional sausage, with 3% salt content showed good textural and sensorial characteristics. However, a sensorial panel revealed a preference for sausages with 4.5% salt content. Therefore, this value is acceptable and considerably lower in comparison with the values commonly used for traditional Portuguese sausages, 5%-6% salt content.

Acknowledgements

This research work was supported by the IFADAP under Project AGRO 218.

References

Fernández, J.M.B. (2000). Utilización de mohos y sus extractos enzimáticos intracelulares para potenciar la generación de substancias aromáticas y sápidas en embutidos curados. Dissertação para obtenção do grau de Doutor em Ciência e Tecnologia dos Alimentos. Universidade Complutense de Madrid.

Gimeno, O., Astiasarán, I., Bello, J. (2001). Calcium ascorbate as a potential partial substitute for NaCl in dry fermented sausages: effect on colour, texture and hygienic quality at different concentrations. Meat Science, 57: 23-29.

Gou, P., Guerrero, L., Gelabert, J., Arnau, J. (1996). Potassium chloride, potassium lactate and glycine as sodium chloride substitutes in fermented sausages and in dry-cured pork loin. Meat Science, 42: 37-48.

Ibañez, C., Quintanilla, L., Irigoyan, A., García-Jalón, I., Cid, C., Astiasáran, I., Bello, J. (1995). Partial replacement of sodium cloride with potassium cloride in dry fermented sausages: influence in carbohydrate fermentation and the nitrosation process. Meat Science, 40: 45-53.

Muguerza, E., Gimeno, O., Ansorema, D., Astiasarán, I. (2004). New formulations for healthier dry fermented sausages: a review. Trends in Food Science and Technology. Elsevier Applied Science Publishers (eds.), London, Reino Unido.

Truswell, A.S. (1994). The evolution of diets for wertern diseases. In: B. Harris-While, and R. Hoffenderg (eds.). Multidisciplinary perspectives. Oxford Basil Blackwell.

BIOAVAILABILITY OF CARNOSINE FROM BEEF MEAT IN MINI-PIGS

C. Bauchart*[1,2], I. Savary-Auzeloux[3], C. Cossoul[3], C. Buffière[1], P. Patureau Mirand[1], E. Thomas[4], M. Morzel[2] and D. Rémond[1]

[1.] Unité de Nutrition Humaine, [2.] Qualité des Produits Animaux, [3.] Unité de Recherches sur les Herbivores, Institut National de la Recherche Agronomique, Centre de Theix, 63122 Saint Genès Champanelle, France. [4.] Association pour le Développement de l'Institut de la Viande, 2 rue Chappe, 63039 Clermont-Ferrand, France. ***Email:*** *cbauchar@clermont.inra.fr.*

Keywords: carnosine, absorption, meat, mini-pig

Introduction

Carnosine (β-alanyl-*L*-histidine) is a cytoplasmic dipeptide found in skeletal muscle at concentrations ranging from 2 to 25 mM (Chan and Decker, 1994). Various physiological actions have been ascribed to carnosine including pH buffering, antioxidant activity, reactive aldehydes scavenging, prevention of protein crosslinking, and involvement in the degradation of carbonylated proteins (Guitto *et al.*, 2005). Through these activities, carnosine has several potential health benefits for humans and it could be considered as a bioactive food component. Indeed, it was reported that some dietary carnosine can cross the gut barrier and appear into human plasma after consumption of beef meat (Park *et al.*, 2005).

In this study, we have investigated the quantitative significance of carnosine absorption in mini-pigs following ingestion of beef meat coming from three different muscles differing in their carnosine concentration.

Materials and Methods

Four Pitman-Moore mini-pigs (21-22kg), 6-month old, were surgically fitted with catheters in the portal vein and the abdominal aorta, and an ultrasonic flow probe around the portal vein. Three meats were tested on each animal: grilled sirloin, grilled neck-brisket and braised shoulder containing 7.77, 2.92 and 1.93mg of carnosine/g of muscle, respectively. Sampling sessions were separated by at least 2 days. Before sampling, the mini-pigs were deprived of food from 17.00 to 11.00. Base-line arterial and portal blood samples were simultaneously withdrawn at 10.00, 10.30 and 11.00 during the post-absorptive period (PA). Then, at 11.00, the mini-pigs were offered a meal exclusively of meat (30 g of proteins per meal). Arterial and portal blood samples were then taken at different times (11.30, 12.00, 13.00, 14.00, 15.00, 16.30 and 18.00) during the post-prandial period (PP). Blood samples (5mL) were collected in cold syringes with heparin. Haemoglobin concentration and packed cell volume were immediately determined (ABL510 System, Radiometer, Copenhagen). Blood samples were centrifuged at 3000g for 10 min at 4 °C. Plasma (300 μL) spiked with homocarnosine as an internal standard (30μL, 10μM) was vortexed with sulfosalicylic acid (5% w/v, final concentration) for 1 min, then left for 15 min at room temperature and centrifuged at 10000g for 15 min at 4°C. The resulting supernatant was stored at -80°C prior analysis. Portal vein blood flow was continuously recorded during the sampling session.

Arterial plasma carnosine concentrations were determined by RP-HPLC (Perkin Elmer, France) using pre-column derivatisation with O-phthaldehyde reagent (column: C18-HDO 250 x 4.6 mm, 5 µm; Uptisphere, Interchim), adapted from the method of Maynard *et al.* (2001). The portal net flux of carnosine in plasma (μmole/h), noted PNF_{carn}, is calculated as followed: $PNF_{carn} = ([Carn_{port}] \times (Hb_{art}/Hb_{port}) - [Carn_{art}]) \times PPF$, where $[Carn_{port}]$ and $[Carn_{art}]$ are carnosine concentrations (μM) in arterial and portal plasma, Hb_{art} and Hb_{port} are haemoglobin concentrations (g/dL) in arterial and portal whole blood and PPF is portal plasma flow (L/h). Portal plasma flow (L/h) is calculated as followed: $PPF = PBF \times (100 - PCV_{port})/100$, where PBF is portal blood flow (L/h) and PCV_{port} is packed cell volume in portal blood (%).

Data were statistically analysed using repeated measures analysis of variance through SAS/PC Program.

Results and Discussion

Average portal blood flow during the PA period was 49 ± 2mL.min^{-1}.kg^{-1}. The time-variations in PBF throughout the PP period were not affected by the nature of meat.

During the PA period, average $[Carn_{art}]$ and PNF_{carn} were 3.86 ± 1.05μM and -1.9 ± 4.8μmole/h, respectively. Whereas $[Carn_{art}]$ was nearly not affected by the ingestion of shoulder, it sharply increased between 30 and 60 min after the meal of sirloin and neck-brisket (Figure 1). The maximum $[Carn_{art}]$ was reached 1h, 2h and 3h after the meal of neck-brisket, sirloin and shoulder, respectively. The maximum $[Carn_{art}]$ and the area under the curve of $[Carn_{art}]$ throughout the PP period were 2 to 3 times greater for sirloin than for neck-brisket. This increase in plasma $[Carn_{art}]$ following meat ingestion is in agreement with the observations of Park *et al.* (2005) in humans. In the present study, differences observed with the $[Carn_{art}]$ pattern according to the type of meat can be related to the carnosine concentration in the ingested meat.

Portal net flux of carnosine (Figure 2) reflects carnosine absorption by the small intestine. As for $[Carn_{art}]$, PNF_{carn} was not affected by ingestion of shoulder but it significantly increased after the meal of sirloin and neck-brisket, and the maximum PNF_{carn} was reached 1h after the meals. It returned to the basal level after 2h and 3h for neck-brisket and sirloin, respectively. Portal net release of carnosine throughout the post-prandial period, calculated from PNF_{carn} integration, was 470, 220, and 0μmole/7 h for sirloin, neck-brisket and shoulder, respectively. For sirloin and neck-brisket, it accounts for 13% of ingested carnosine.

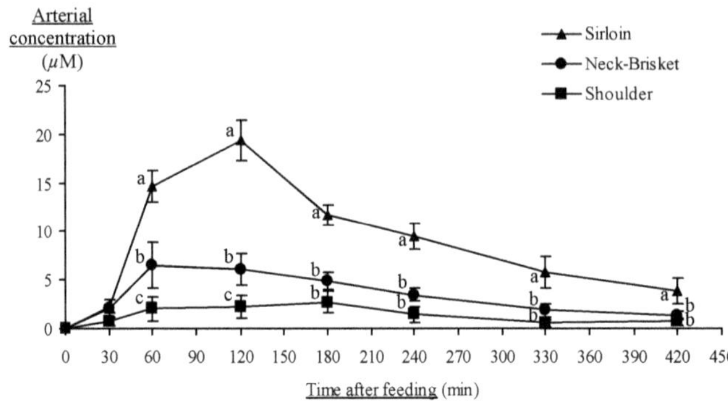

Figure 1: Increase in arterial plasma carnosine concentration (μM) in mini-pigs (n=4) after a meal of grilled sirloin, grilled neck-brisket or braised shoulder. Values are means ± standard error. For each sampling time, mean values with different letters are significantly different (P<0.05).

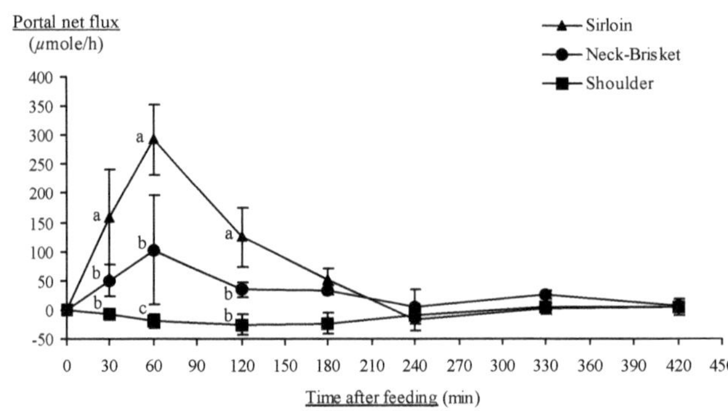

Figure 2: Increase in portal net flux of carnosine in plasma (μmole/h) in mini-pigs (n=4) after a meal of grilled sirloin, grilled neck-brisket or braised shoulder. Values are means ± standard error. For each sampling time, mean values with different letters are significantly different (P<0.05).

Conclusions

In the present study, absorption of intact carnosine was evidenced for the first time from direct measurements of portal net release. This absorption appeared highly related to carnosine concentration in food. Below 2 mg/g of meat, all dietary carnosine seems to be degraded during digestion and absorption. Above this value, carnosine bioavailability was proportional to carnosine intake.

References

Chan, K.M. and Decker, E.A. (1994). Endogenous skeletal muscle antioxidants. Critical Reviews in Food Science and Nutrition, 34: 403-426.

Guitto, A., Calderan, A., Ruzza, P. and Borin, G. (2005). Carnosine and carnosine-related antioxidants: a review. Current Medicinal Chemistry, 12: 2293-2315.

Maynard, L.M., Boissonneault, G.A., Chow, C.K. and Bruckner, G.G. (2001). High Levels of dietary carnosine are associated with increased concentrations of carnosine and histidine in rat soleus muscle. Journal of Nutrition, 131: 287-290.

Park, Y.J., Volpe, S.L. and Decker, E.A. (2005). Quantitation of carnosine in humans plasma after dietary consumption of beef. Journal of Agricultural and Food Chemistry, 53: 4736-4739.

EFFECT OF SLAUGHTER SEASON AND MUSCLE TYPE ON FATTY ACID COMPOSITION AND RELATED NUTRITIONAL INDEXES IN BARROSÃ-PDO VEAL

C.M. Alfaia[1], M.F. Castro[2], M.A.G. Quaresma*[1], S.V. Martins[1], S.P.A. Alves[3], A.P.V. Portugal[3], R.J.B. Bessa[3] and J.A.M. Prates[1]

[1.]Faculdade de Medicina Veterinária - CIISA, 1300-477 Lisboa, Portugal, [2.] Faculdade de Farmácia - CECF 1649-003 Lisboa, Portugal, [3.] Estação Zootécnica Nacional - INIAP, 2005-048 Vale de Santarém, Portugal. Email: cpmateus@fmv.utl.pt

Keywords: veal, fatty acids, slaughter season, muscle type

Introduction

Veal quality is influenced by several factors, such as breed, feeding system, gender and slaughter weight and particularly duration of milk consumption that have been reported to have an important influence on the intramuscular fat content and composition (Sañudo *et al.*, 2000; Moreno *et al.*, 2006). Barrosã-PDO (Protected Denomination of Origin) veal is a traditional meat produced in the Norwest of Portugal based on a pasture-based system according to the product specifications (Commission Regulation nº. 1263/96 of 01/07, EEC). This research focuses on the influence of two slaughter seasons (early autumn and late spring) and muscle type (*longissimus thoracis*, LT, *longissimus lumborum*, LL, and *semitendinosus*, ST) on fatty acid composition and nutritional value of Barrosã-PDO veal. This evaluation was based on the composition of main fatty acids, and related health indexes (PUFA/SFA and *n*-6/*n*-3 ratios), determined by GC-FID.

Materials and Methods

Barrosã purebred calves were raised with their dams in a semi-extensive grazing system based on natural pastures. After weaning (at 6 ± 0.5 months of age), calves were raised on a late summer grass pasture until slaughter in October 2002 (early autumn sampling; *n*=12; 8.0 ± 0.4 months of age and 192 ± 10.2 kg live body weight) or were exposed to the middle-late spring grass and slaughtered in June 2003 (late spring sampling; *n*=15; 8.0 ± 0.5 months of age and 212 ± 11.2 kg live body weight). *M. longissimus thoracis* (LT), *M. longissimus lumborum* (LL) and *M. semitendinosus* (ST) were removed and meat samples were collected 2-3 days after slaughter (+1ºC) and stored at -80°C until required for analysis. Intramuscular fat was extracted according to the method of Fritsche *et al.*, (2000). Total lipids were measured gravimetrically, in duplicate, by weighing the fatty residue obtained after solvent evaporation. Methyl esters of fatty acids (FAME) were obtained by base-catalysed transesterification (sodium methoxide in anhydrous methanol) and nonadecanoic acid (C19:0) methyl ester was used as internal standard. The FAME were separated by GC-FID, on a SP^{TM}-2560 capillary column (100 m) using a split/splitless injection system (split ratio of 1:5) and helium, as carrier gas, at flow rate of 1.5 ml/min. The column temperature was programmed to increase from 75°C to 225°C and detector and injector temperatures were set at 250°C. Fatty acids were quantified based on the internal standard technique, after adjusting the corrected response factor of each fatty acid determined using the Supelco standard mixture and according to ISO 5508 (1990). Data were analysed by analysis of variance, using the MIXED procedure of SAS (2001), considering the effect of slaughter season and muscle type (repeated measure) and its interaction.

Results and Discussion

The results concerning to the influence of slaughter season and muscle type on intramuscular fatty acid profile are summarized in Table 1. In both seasons, the predominant fatty acids in intramuscular fat were C16:0 (22-25% of total FAME) and C18:0 (13-16%) as SFA and C18:1c9 (35-37%) as monounsaturated fatty acids (MUFA). No seasonal variations (P>0.05) in the fatty acid composition of Barrosã-PDO veal were apparent, except for total lipids, which showed higher percentages (P<0.05) in meat from early autumn than in that from late spring. Total fatty acids (ranging from 996 to 1526 mg/100 g muscle), calculated as the sum of identified fatty acids, represented on average 78% of the sum of detected fatty acids (GC analysis) and 64% of total lipids (gravimetric analysis). As for total fatty acids and total lipids, the partial sums (SFA, PUFA and *n*-3 fatty acids) and the nutritional ratio PUFA/SFA were also affected by muscle type. LT and LL muscles depicted greater intramuscular fatty acid content compared with ST muscle, simultaneously with a higher proportion of SFA. In contrast, the percentages of PUFA (P<0.05) and *n*-3 fatty acids (P<0.01) were lower in LT and LL muscles than in ST muscle. These differences between muscles likely result from distinct percentages in muscle fibre type because they are reflected in differences in fatty acid composition (Wood *et al.*, 2004). Additionally, there was an interaction between slaughter season and muscle type for the fatty acids C18:0, C18:1c9, MUFA (*p*<0.05) and for total lipids (P<0.01). According to the current nutritional recommendations, the PUFA/SFA ratio in human diets should be 0.45 and, within the PUFA, the n-6/n-3 ratio should not exceed 4.0 (British Department of Health, 1994). Although the n-6/n-3 ratios obtained for Barrosã-PDO veal are inside the recommended values, which is favourable, for both slaughter seasons and muscle types (2.9-3.1), the values of PUFA/SFA ratios in veal-PDO were below, and so unfavourable, than the guideline recommended for the human diet.

Table 1: Percentages of main fatty acids (% w/w), total fatty acid content (mg/100 g muscle) and total lipids (mg/100 g muscle) of intramuscular fat in *longissimus thoracis* (LT), *longissimus lumborum* (LL) and *semitendinosus* (ST) muscles of veal from Barrosã calves reared according Barrosã-PDO specifications and slaughtered in early autumn and late spring.

	Autumn			Spring				Effects[1]		
	LT	LL	ST	LT	LL	ST	SEM	S	M	S×M
C14:0	4.62	4.68	3.93	4.35	4.25	4.09	0.298	ns	**	ns
C16:0	23.04	23.57	22.22	22.73	24.68	23.57	0.541	ns	**	ns
C16:1c9	3.71	3.71	3.77	3.78	4.19	3.92	0.165	ns	ns	ns
C18:0	14.20^{a,b}	14.52^{a,b}	13.38^{a}	15.92^{b}	13.49^{a}	14.14^{a}	0.699	ns	*	*
C18:1c9	35.49^{a}	35.15^{a}	37.00^{b}	35.64^{a,b}	34.95^{a,b}	34.91^{ab}	0.927	ns	ns	*
MUFA	41.14^{a}	40.76^{a}	42.78^{b}	42.46^{a,b}	42.31^{a,b}	42.08^{a,b}	1.036	ns	ns	*
PUFA	10.72	10.22	11.68	8.37	9.04	10.42	0.911	ns	*	ns
SFA	43.93	44.81	41.38	44.86	44.19	43.52	0.728	ns	***	ns
TFA	3.32	3.32	3.24	3.37	3.44	3.11	0.259	ns	ns	ns
PUFA/SFA ratio	0.25	0.23	0.29	0.19	0.20	0.24	0.022	ns	**	ns
n-6 PUFA	8.03	7.56	8.56	6.32	6.77	7.73	0.740	ns	ns	ns
n-3 PUFA	2.69	2.66	3.12	2.05	2.27	2.68	0.251	ns	**	ns
n-6/*n*-3 PUFA ratio	3.10	2.98	2.99	3.05	3.03	2.92	0.235	ns	ns	ns
Total fatty acids[+]	1421	1379	1226	1526	1094	996	1.409	ns	**	ns
Total lipids	2319^{a}	2133^{b}	1960^{c}	2294^{a,b}	1779^{c}	1561^{d}	1.057	*	***	**

[1]Levels of significance: ns, $p>0.05$; *, $p<0.05$; **, $p<0.01$; ***, $p<0.001$; means in the same row with different superscripts are significantly different ($p<0.05$); SEM, standard error of mean. The symbols used mean as follow: S, season; M, muscle type and MUFA, PUFA, SFA and TFA refer to monounsaturated, polyunsaturated, saturated and *trans* fatty acids, respectively.

[+] Total fatty acids is the sum of identified fatty acids.

Conclusions

As we observed in this study, Barrosã-PDO veal only showed seasonal changes in the levels of total lipids. In addition, significant differences were obtained among LT, LL and ST muscles for total lipids, some main fatty acids and total fatty acids, some partial sums of fatty acids and the PUFA/SFA index. The results indicate that veal-PDO has values of pasture-fed cattle for both slaughter seasons. The low n-6/n-3 ratio (below 2), which is mainly influenced by diet, suggests that the calves were raised on a good quality grass production system.

Acknowledgments

Sampling assistance (Eng. João Paulo Teixeira and Eng. João Paulo Costa, CAPOLIB - Cooperativa Agrícola de Boticas CRL), doctorate fellowship (SFRH/BD/2005/22566 to Susana Martins) and financial support (grants AGRO/2003/512, POCTI/CVT/2002/44750 and CIISA/2002/52) are acknowledged.

References

British Department of Health (1994). Nutritional aspects of cardiovascular disease. Report on Health and Social Subjects N° 46. HMSO: London.

Fritsche, J., Fritsche, S., Solomon, M.B., Mossoba, M.M., Yurawecz, M.P., Morehouse, K. and Ku, Y. (2000). Quantitative determination of conjugated linoleic acid isomers in beef fat. *European Journal of Lipid Science and Technology*, 102: 667-672.

Moreno, T., Varela, A., Oliete, B., Carballo, J.A., Sánchez, L. and Montserrat, L. (2006). Nutritional characteristics of veal from weaned and unweaned calves: Discriminatory ability of the fat profile. *Meat Science*, 73:209-217.

Sañudo, C., Enser, M.E., Campo, M.M., Nute, G.R., Maria, G., Sierra, I. and Wood, J.D. (2000). Fatty acid composition and sensory characteristics of lamb carcasses from Britain and Spain. *Meat Science*, 54: 339-346.

Wood, J.D., Richardson, R.I., Nute, G.R., Fisher, A.V., Campo, M.M., Kasapidou, P.R. and Enser, M. (2004). Effects of fatty acids on meat quality: a review. *Meat Science*, 66(1): 21-32.

MEDICAL-BIOLOGICAL ASPECTS OF GROUND MEAT-BASED SEMI-PREPARED PRODUCTS FOR DIETETIC THERAPY OF CARDIOVASCULAR PATIENTS

I.M. Tchernoukha*, A.V. Ustinova, S.I. Khvylya, N.E. Belyakina and I.K. Morozkina

The V.M .Gorbatov All-Russian Meat Research Institute, Talalikhina 26, 109316 Moscow, Russia Email: imche@inbox. ru

Keywords: ground semi-prepared foods, biological additives, medical-biological evaluation, blood count

Introduction

Based on formulated medical-biological requirements, with consideration of specific metabolic processes in organisms of people pre-disposed to and suffering from cardiovascular pathology, meat-plant ground semi-prepared foods for preventive nutrition were developed. The objective of the study was to develop recipe compositions based on meat raw materials added with plant components and biologically active ingredients and study of their influence on physiological state of laboratory animals, pathological morphology, general clinical blood indices and biochemical indices of blood serum of animals.

Materials and Methods

The formulations were developed using the methods of mathematical modelling (Lipatov, 1988) according to formalized medical and biological requirements. Medical and biological evaluation was carried out on the growing white male rats of Vistar line during 28 days in the vivarium of VNIIMP, using traditional experimental methods. Four groups of animals, 10 to each, were formed. The base of recipes of the studied products was ground pork and beef, oatmeal, soy protein (isolate and concentrate). The control group №1 received the ground meat as a basis, bread and rendered lard; group №2 received ground meat added with the complex of BAD№1 (wheat fibre, ascorbic and succinic acid, natural red palm oil, linseed oil, phospholipids, mineral calcium fortifier); group 3 received ground meat added with the complex of BAD№2 (iodocasein, wheat fibre, rendered lard, dihidroquercetin, ascorbic and succinic acid) and group №4 received complex of BAD №1 in addition to the diets of vivarium. The integral index of chronic intoxication (IPHI) (Snitsar *et al.*, 2000) was calculated as the ratio of the organ mass to the total mass of the body in %. General clinical blood examination and determination of glucose, cholesterol and triglycerides in the blood were carried out in veterinary clinic of MGUPB.

Results and Discussion

No clinical evidences of deviations in health were marked in animals of control and experimental groups during the whole experiment. Viability of animals in the experimental and control groups during the whole period was full (100%). The lowest gain weight (222%) and final weight were observed in group №4; use of BAD №1 in the recipe of the meat product (group №2) has led to the largest gain weight (279%) of the animals.

Pathological and anatomical examination of animals after slaughter did not reveal any external signs of pathological or inflammatory processes in internal organs. IPHI of internal organs characterizing possible negative effects of the investigated complexes of BAD are presented in Table 1.

Table1. Integral indices of state of rats having control and tested diets.

Indices	Group №1 (control)	Group №2	Group №3	Group №4
IPHI of spleen	0.36±0.03	0.34±0.03	0.30±0.08	0.29±0.05
IPHI of kidney	0.47± 0.02	0.39±0.11	0.46±0.01	0.37±0.04
IPHI of liver	3.76± 0.22	3.53±0.83	4.10 ± 0.52	4.05 ±1.17
IPHI of heart	0.51±0.19	0.54±0.28	0.30±0.18	0.46±0.07

The calculated integral index did not reveal any significant differences in the relationships between the body mass of the experimental and control animals as compared to the mass of their main important internal organs, which showed the absence of cumulative toxic effect of all the tested mixtures.

The data on hematological indices show a positive effect on hemopoietic function for rats having in their diets dihidroquercetin, which is a strong antioxidant and helps better assimilate the nutritive iron (Pilat and Ivanov, 2002). The value of hematocrit (plasma volume and blood formed elements ratio) in all the groups exceeds the control suggesting the absence of iron-deficient anemia. The colour index (hemoglobin content in one erythrocyte) reflecting the degree of saturation of erythrocytes with hemoglobin and important in diagnostics for differentiation of kinds of anemia, in test groups doesn't differ from the control.

SOE (the velocity of erythrocytes sedimentation) is the least in group №4 (BAD complex), that corresponds to the increase of blood viscosity, and is extremely undesirable for cardio-vascular patients. When BAD are used together with meat products (group 2) this index reaches the level of the control group. Dihidroquercetin allows to increase SOE 1.5 times as compared to the control. The analysis of leucotytic formula of animals' blood has shown insignificant

increase of the total number of leucocytes in groups 3 and 4 as compared to the control, that can be a concomitant symptom of immunological resistance increase of the organism.
The content of monocytes in the blood of rats consuming BAD and half-prepared products with additives is to some extent higher than in the control, but doesn't exceed the limits of the norm (0 –4%) (Batuyev, *et al.*,), which shows the absence of immune conflict in the animal's organism. Full absence of basophil lymphocytes in peripheral blood indicates a low level of food sensibilization in all the samples.

Table 2: Blood count of rats.

Indices	Group №1 (control)	Group №2	Group №3	Group№4
Erythrocytes, x 10^6/µl	4.4 ± 0.07	4.45±0.06	4.85±0.07	4.6±0.1
Mean volume of erythrocytes, µm^3	85.65±2.8	99.7±3.7	98.1±3.3	97.8±3.5
Hematocrit,%	38.0±0.0	44.5±1.7	46.0±1.4	44.0±1.8
Hemoglobin, g/dl	11.2±0.9	10.9±0.8	13.0±0.7	11.6±0.7
Colour index	0.7±0.0	0.7±0.07	0.8±0.07	0.75±0.07
Velocity of sedimentation of Erythrocytes, mm/hour	1.0±0.0	1.05±1.3	1.5±0.7	0.55±0.6
Leucocytes, x 10^5/µl	4.95±0.3	3.7±0.7	5.3±0.6	5.25±0.4
Neutrophils				
- juvenil, %	0.0±0.0	0.0±0.0	0.0±0.0	0.0±0.0
- stab, %	2.01±0.1	2.5±0.09	1.0±0.2	0.0±0.0
- segmented, %	26.01±0.1	49.0±1.4	63.0±2.3	34.0±2.6
Eosinophils, %	0.0±0.0	2.0±0.08	4.0±0,0	6.0±0.5
Monocytes, %	0.0±0.0	2.0±0.08	1.0±1.4	2.0±0.1
Basophils. %	0.0±0.0	0.0±0,0	0.0±0.0	0.0±0.0
Lymphocytes,%	72.0±2.3	34±1.4	31.0±1.2	58.0±2.1

The results of biochemical investigations of animals' blood (Table 3) have shown that use of a complex of BAD (group 2) in a meat product leads to a more rapid reduction of the level of cholesterol and triglycerides compared to other samples, and in case of use the complex of BAD only (group №4) the antiatherogenic effect was the least, which confirms the usefulness of incorporation of BAD into the recipe of a meat product with the purpose of enhancing the theraupetic effect for cardiovascular patients.
The incorporation of dihidroquercetin (group №3) to the ground meat led to less reduction of cholesterol (as compared to the control) than addition of the complex BAD №1 (group №2).

Table 3: Biochemical indices of animals' blood serum.

Indices	Group№1(control)	Group №2	Group №3	Group №4
Glucose, µmol	4.1±0.4	4.7±0.3	4.65±0.3	3.15±0.3
Cholesterol, Mmol/l	6.02 ±0.1	4.85±0.2	5.1±0.7	6.2±0.01
Triglycerides, mol/l	0.58 ±0.1	0.44±0.1	0.75±0.13	1.02 ±0.2

Conclusions

A chronic biological test of feeding meat and plant half-finished products to laboratory rats gave the following results:

- absence of cumulative toxic effect of the tested diets and, correspondingly, of the food additives in the diets;
- use of the complex BAD №1 (wheat fibre, ascorbic and succinic acids, natural red palm oil, linseed oil, phospholipids, mineral calcium enricher) in meat-plant half-finished products gives the largest antiatherogenic effect. The incorporation of dihidroquercetin (complex BAD №2) led to decrease of blood viscosity and increase of hemoglobin level, which indicates a positive effect of both versions;
- Theraupetic effect of BAD complex increases when used in combination with the meat product;
- As a result of the tests, 6 recipe compositions of meat-plant half-finished products for dietary therapy of cardio-vascular diseases have been developed.

References

Batuyev, A.S., Nikitina A.P., Zhuravlev V.L. and Sokolova N.N., (2001). A limited practical work on humans and animals physiology. A manual edited by A.S.Batuyev.- SPb.: Saint-Petersburg Publishing House, 2001.

Lipatov N.N. (1988). Principles of designing of the composition and improving the technology of multi-component meat and dairy foods. Theoretical and experimental bases. Methods and results of their implementation. M.: Thesis for a doctor' degree. 599 pages.

Pilat T.L. and Ivanov A.A. (2002). Biologically active additives to foods. M. Avvallon. 387 pp.

Snitsar A.I., Khvylya S.I. and Baryshnikova E.V. (2000). Study of toxicity of food protein-mineral additive with hydrolyzed rice shell in experiments on model animals. Vse of myase, 2000, №3

STABILITY OF PLANT POLYPHENOL SUPPLEMENT DURING COOKING AND STORAGE OF MECHANICALLY DEBONED MEAT

T. Püssa*[1], M. Rei[2], R. Pällin[2], P. Raudsepp[1] and R. Soidla[2]

[1.] *Department of Food Hygiene and Control and* [2.] *Department of Food Science, Estonian University of Life Sciences, Kreutzwaldi 58A, 51014 Tartu, Estonia. Email: tonu@chem.ut.ee.*

Keywords: mechanically deboned meat, functional food, sea buckthorn, polyphenols, LC-MS/MS

Introduction

Health-promoting functional foods on the basis of meat are becoming more and more popular. It is economically substantiated to use mechanically deboned (or recovered) meat (MDM) as a raw material for preparation of these products. Main problems connected with consumption of MDM are an elevated risk of bacterial intoxications and higher ingested doses of (per)oxidated potentially carcinogenic fatty acids comparing with the hand deboned meat. The plant polyphenolic additives should contribute an extra antibacterial as well as an antioxidant effect to the MDM.
There are papers concerning the effect of plant polyphenols on lipid oxidation (mainly estimation of TBARS) during long-time storage of MDM at subzero temperatures (Mielnik, 2003, Tang, 2002, Weisburger, 2002), but no data is available about the actual stability of added polyphenols during cooking and storage at low positive temperatures.
We have studied the dynamics of plant polyphenols content and spectrum in turkey MDM supplemented with 1,2 or 4% of dried sea buckthorn (SB) berry powder macerated in ethanol in the course of short-time cooking and storage at a low positive temperature.

Materials and Methods

SB-MDM samples preparation. Solid residue of SB berries after removal of juice by pressing was dried at 40°C, stored in polyethylene bags at room temperature and milled shortly before usage. Frozen turkey MDM was thawed overnight and respective amounts of berry residue, macerated during 20 hours in 15 ml ethanol were added to 240 g of MDM by mixing for 30 min. The products were divided into 2 parts and half of each sample was cooked during 3 minutes in microwave oven at 800 W. First samples were taken immediately, and both raw and cooked SB-MDM samples were stored during 6 days at temperature +6°C.
Chromatographic analysis. Sample preparation (Weisburger, 2002): 2 g of SB-MDM was extracted with 4 ml of methanol under shaking during 30 minutes and centrifuged. The supernatant was treated twice with 2 ml of hexane and the methanolic layer was passed through 100 mg reversed phase spe-column and kept at -40°C until analysed.
For chromatographic separation of SB polyphenols, the HPLC method on a Zorbax 300SB-C18 column (2.1×150 mm; 5μm) was used in a mobile phase gradient of 0.1% formic acid and acetonitrile at velocity 0.3 ml/min. For detection and quantitation of substances, Agilent 1100 Series LC/MSD Trap-XCT with an ESI interface (m/z interval 50-1000 in negative ion mode) and UV-Vis diode array detector were used.

Results and Discussion

SB berries contain large amounts of flavonol glucosides with UV absorption maximum at λ=370 nm as the main group of polyphenols (Guliyev, 2004). Areas under HPLC curves at 370 nm as a measure of total flavonol content are collected in Table 1. Some example chromatograms of 2% SB-MDM are presented in Fig. 1. Evidently the total flavonol glucoside content is unaffected by cooking, but is remarkably reduced during the storage of uncooked SB-MDM samples. The antioxidant content in cooked samples is only slightly reduced in the course of storage.

Table 1: Areas under curves of HPLC-DAD chromatograms (λ=370 nm) of raw and cooked SB-MDM-s on the 1st day and and after storage for 6 days.

% of SB in SB-MDM	1st day		6th day	
	raw	cooked	raw	cooked
1	241.9	292.9	109.4	236.5
2	551.3	597.2	230.5	510.6
4	1072.7	1255.0	651.4	1140.0

Table 2 and Fig.2 show that different flavonol glucosides behave differently during storage of uncooked SB-MDM, derivatives of isorhamnetin (excluded rhamnoside) and kaempferol being more stable than others.

Table 2: Extracted ion chromatogram (EIC) mass peak heights of the main polyphenols in 2% SB-MDM (t_R – retention time at LC/MS chromatogram at Figure 1, amu – atom mass units, * - not detectable).

t_R min	$[M-H]^-$ amu	EIC-peak height of raw and cooked SB-MDM-s				Flavonol glucoside
		Raw 1st day	Cooked, 1st d	Raw 6th day	Cooked, 6th d	
11.4	609	3341	4162	*	2164	Quercetin rhamnoside glucoside
12.5	623	20841	19320	14685	20941	Isorhamnetin rhamnoside glucoside
12.6	609	3020	4224	*	1550	Quercetin rutinoside
12.9	463	6478	6976	2905	3581	Quercetin glucoside
14.0	623	15351	17874	10492	16203	Isorhamnetin rutinoside
14.4	477	17531	18034	8339	18177	Isorhamnetin glucoside
15.9	447	2584	3729	*	*	Quercetin pentoside
17.8	593	3699	4533	2949	4147	Kaempferol rhamnoside glucoside
18.1	461	10967	14182	*	*	Isorhamnetin rhamnoside

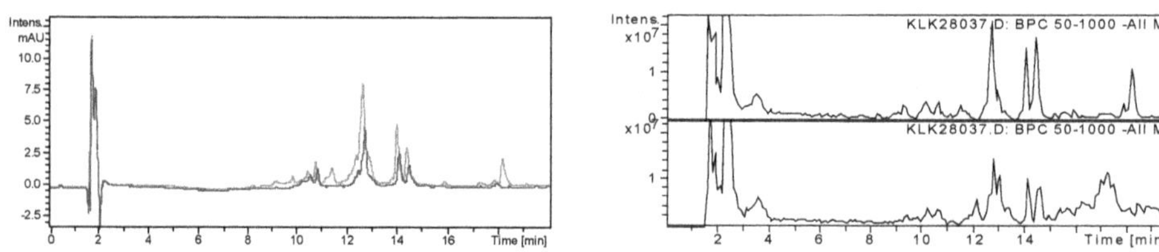

a) b)

Figure 1: HPLC chromatograms of raw 2% SB-MDM: a) UV-chromatograms, λ=370 nm at 1-st day (light line) and 6th day (dark line); b) MS base peak chromatograms (BPC) at 1st day (upper) and 6th day (lower).

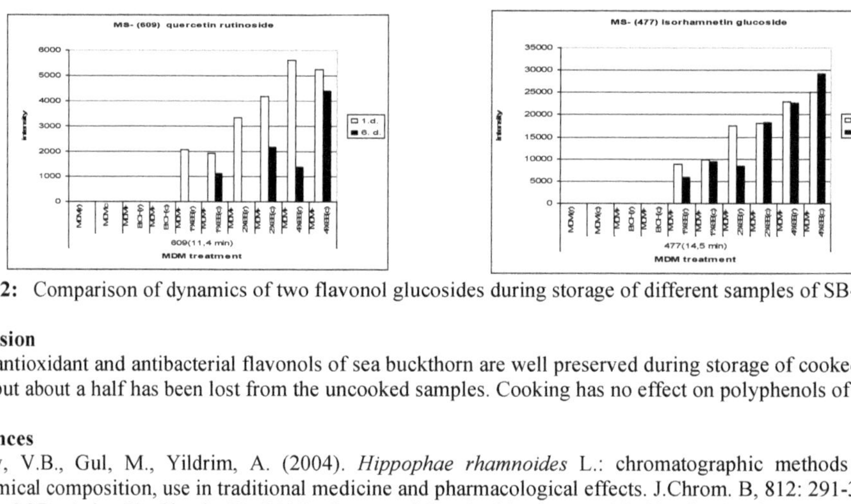

Figure 2: Comparison of dynamics of two flavonol glucosides during storage of different samples of SB-MDM

Conclusion

Potent antioxidant and antibacterial flavonols of sea buckthorn are well preserved during storage of cooked SB-MDM at +6°C, but about a half has been lost from the uncooked samples. Cooking has no effect on polyphenols of SB-MDM.

References

Guliyev, V.B., Gul, M., Yildrim, A. (2004). *Hippophae rhamnoides* L.: chromatographic methods to determine chemical composition, use in traditional medicine and pharmacological effects. J.Chrom. B, 812: 291-307.

Mielnik, M.B., Aaby, K., Skrede, K. (2003). Commercial antioxidants control lipid oxidation in mechanically deboned turkey meat. Meat Science, 65: 1147-1155.

Tang, S.Z., Kerry, J.P., Sheehan, D., Buckley, D.J. (2002). Antioxidative mechanisms of tea catechins in chicken meat systems. Food Chemistry, 76: 45-51.

Weisburger, J.H., Veliath, E., Larios, E., Pitman, B., Zang, E., Hara, Y. (2002). Tea polyphenols inhibit the formation of mutagens during the cooking of meat. Mutation Research, 516: 19-22.

FAT AND CHOLESTEROL CONTENT IN BOVINE AND PORK TISSUES

L. Turubatovic*[1], D. Vranic[1], M.B. Spasic[2] and P. Radetic[1]

[1.] *Institute of Meat Hygiene and Technology, 11 000 Belgrade, SCG, e-mail: meatinst@beotel.yu,* [2.] *Institute for biological research "Sinisa Stankovic", 11 000 Belgrade, SCG*

Keywords: meat, fat, cholesterol, pigs, bovine

Introduction

Cholesterol is susceptible to oxidation, which results in formation of oxysterols. Apart from eggs, egg products, milk and diary products, meat fried in animal fat or vegetable oil may also be important source of oxysterols in human nutrition (Valenzuela *et al.*, 2003). Interest in oxysterols is growing due to their various biological effects as well as potential cytotoxic, mutagenic and carcinogenic effect of some oxysterols. Fat and cholesterol content in human nutrition is related to the level of cholesterol in serum, increase of which poses a risk for development of cardio-vascular diseases. Therefore, the aim of the task was to determine the differences in fat and cholesterol content in the selected identical groups of beef and pork muscles, with the special emphasis on the fatty tissues of these two species. It is often the case that not only the non-professionals, but also the medical experts (nutritionists) blame pork to be a rich source of cholesterol. These experiments represent an attempt to answer numerous medical recommendations that people with increased cholesterol level in serum have to exclude from their diet even lean pork meat and especially pork fat. We also included in our investigation beef as less anathemised food or low-risk meat for cardio vascular diseases.

Materials and Methods

Meat from five Swedish Landrace pigs were used for the experiment. The examined beef meat (China Meat Research Centre, 2000) originated from the crosses of Charolais and Domestic Spotted breeds. Fat content in the beef and pork samples was determined according to JUS ISO method (1992). Cholesterol content was determined according to the method from China meat research centre (2000).

Results and Discussion

Figure 1 shows comparison of fat content (%) and Figure 2 comparison of cholesterol content (mg/100g) in the examined pork and beef samples.

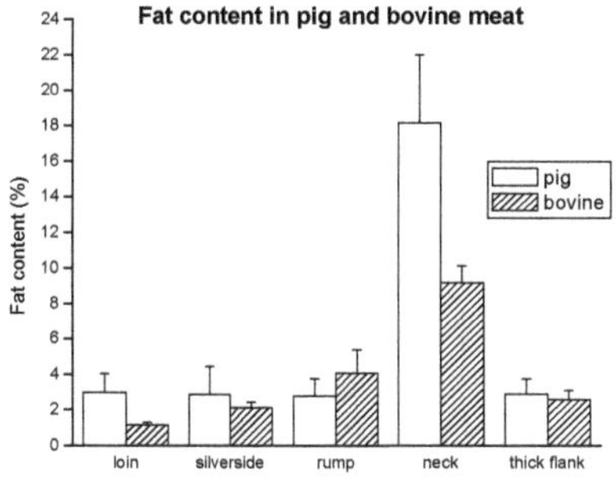

Figure 1.

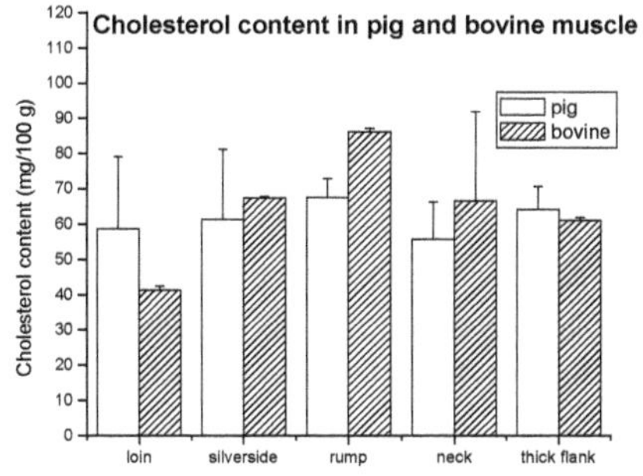

Figure 2.

Comparison of the mean values of fat content in the examined samples showed that the highest value was in pork neck (18.17±3.84%). Fat content in loin, silverside, rump and thick flank was similar and in all the examined samples was less than 3%. Both beef and pork had the highest fat content in neck muscles (9.16±0.97%). Fat content in beef rump (4.08±1.30%) was higher compared to pork rump (2.77±0.97%), while fat content in beef loin, silverside and thick flank was lower than 3% (similar to pork). Beef loin had lower fat content (1.15±0.15%) compared with pork loin (2.97±1.06%).

The highest cholesterol content was determined in pork rump (67.58±5.34 mg/100g), lower content was found in thick flank (64.18±6.40 mg/100g) and silverside (61.40±19.71 mg/100g) and the lowest content was found in loin (58.62±20.45 mg/100g) and neck (55.68±10.58 mg/100g). Comparison of the mean values of cholesterol content between beef and pork showed the highest cholesterol content in rump (in both beef and pork) and for the beef is 86.15±1.03 mg/100g. Relatively similar values of cholesterol were found in beef silverside (67.42±0.45 mg/100g), neck (66.57±25.41 mg/100g) and thick flank (61.04±0.78 mg/100g). The lowest cholesterol content was found in beef loin (41.42±0.78 mg/100g) compared to other anatomical regions of this species.

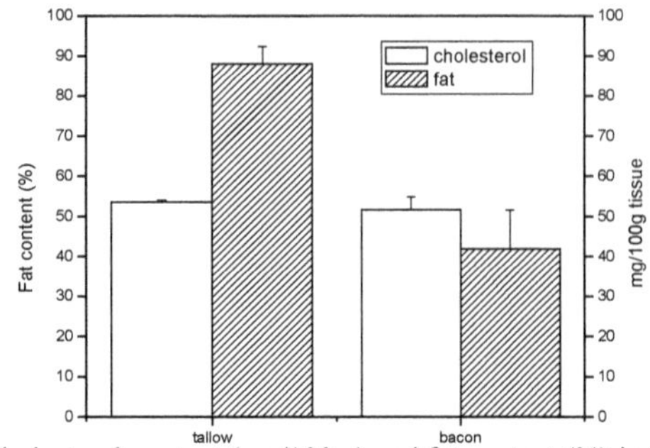

Figure 3: Cholesterol content (mg/100 g) and fat content (%) in bacon and tallow.

Figure 3 shows comparison between fat content (%) and cholesterol content (mg/100g) in the examined bacon and tallow samples. Comparison of the mean values of cholesterol content showed slightly higher cholesterol content in tallow (53.57±0.4 mg/100g) compared to Hamburg bacon (51.63 mg/100g), while fat content was more than 100% higher in tallow (88.00±4.36 mg/100g) compared to bacon (42.88±9.68 mg/100g). Our results point to the conclusion that beef and pork have similar cholesterol content. The literature data can not be used for direct comparison considering the differences in sampling procedure, applied analytical method of determination and different groups of muscles analyzed as well as lack of data considering nutrition of animals used in the experiments. The literature data (Souci *et al.*, 1981; Pearson and Dutson, 1990; Biro and Linder, 1995; Chan *et al.*, 1995; Honikel and Arneth, 1996; Zsarnoczay and Zelenak, 2003) in general showed that the cholesterol content in pork ranged between 48.60 and 94.00 mg/100g, and in beef, between 41.42 and 86.15mg/100g, what confirms our results. As for the fat content, comparison with the other authors' data (Souci *et al.*, 1994; Honikel, 1995; Elmadfa *et al.*, 2001; Piironen *et al.*, 2002; Zsarnoczay and Zelenak, 2003) shows that pork, even with the higher fat content compared to beef, like in neck, does not show increased cholesterol content.

Conclusions

- The examined samples of beef and pork (loin, silverside, rump, neck and thick flank) showed no significant difference in cholesterol content.
- There is also no significant difference in fat content, except for the analysed samples of neck (in both beef and pork). Determined fat content in neck was 18.17% and 9.16% for pork and beef, respectively.
- Results of the investigations of fat and cholesterol content in beef and pork fatty tissues showed similar values for cholesterol content in the analyzed samples of bacon and tallow. However, there is a significant difference in fat content (42.88% in Hamburg bacon and 88.00% in tallow).
- Considering the ratio of cholesterol in fat in all examined samples, thermally treated pork neck and tallow may represent a significant source of oxysterols, which is the subject of our further investigations.

References

Biro, G.Y., Linder K.(1995).Tapanyagtablazat. Medicina Konyvkiado, Budapest.

Chan, W., Brown, J., Lee, S. M., Buss, D. H. (1995). Meat, poultry and game. The Royal Society of Chemistry and Ministry of Agriculture, Fisheries and Food.

China meat research center (2000). Teaching material for international training course, Principle and method for food analysis, Analysis method of cholesterol.

Elmadfa, I., Aign,W., Muskat, E., Fritzche, D. (2001). Die grosse GU Nahrwert- und Koloren Tabele. Universitat Wien. Austria.

Honikel, K. O. (1995). Inhaltsstoffe von Fleisch und Fleischwaren, Fleisch, Gesundheit-tierschutz-umwelt, Bundesantalt fur fleischforschung: 59-87.

Honikel, K. O., Arneth, W. (1996). Cholesterin Gechalt in Fleisch und Eirn, Fleischwirtschaft, 76 (12): 1244-1253.

JUS ISO 1443/1992. Meso i proizvodi od mesa - Određivanje sadržaja ukupne masti.

Pearson, A. M., Dutson, T. R. (1990). Meat and Health. Elsevier Applied Science, London and New York.

Piironen, V., Toivo, J. Lampi A. M. (2002). New data for Cholesterol Contents in Meat, Fish, Milk, Eggs and Their Products Consumed in Finland, Journal of Food Composition and Analysis, 15: 705-713.

Souci, S. W., Fachmann, W., Kraut, H. (1981). Food composition and nutrition tables 1981/82. Wissenschaftliche Verlagsgesellschaft mbH, Stuttgart.

Souci, S. W., Fachmann, W., Kraut, H. (1994). Food Composition and Nutrition Tables, 5th ed. Stuttgart, Wissenschaftlicheverlagsgesellschaft.

Valenzuela, A., Sanhueza, J., Nieto, S. (2003). Cholesterol oxidation: Health hazard and the role of antioxidants in prevention, Biological Research, 36: 291-302.

Zsarnoczay, G., Zelenak, L. (2003). Hus-es baromfiipari alapanyagok zsir-es koleszterintartalmanak vizsgalata, a HUS, 1: 23-28.

REPLACING BACK FAT WITH MELON SEED MEAL IN LOW-FAT BEEF SAUSAGE

S.S. Abiola*, T.O. Oyelowo and B.F. Alo

Department of Animal Production and Health, University of Agriculture, P.M.B. 2240, Abeokuta, Ogun State, Nigeria
E-mail: abiolass@yahoo.com

Keywords: melon seed meal, low-fat, beef sausage

Introduction

The meat industry worldwide is positively responding to the growing demand of consumers for low-fat meat products. The consumer concern is attributed to the positive correlation that exists between fat intake and incidence of coronary heart disease. Consequently, new processed meat products with lower caloric densities have been developed. Abiola and Adegbaju (2001) replaced pork back fat with rind in pork sausage without adverse effect on processing yield. Non-meat ingredients such as mushroom (Akobundu and Eluchie,1992) and olive oil (Bloukas *et al.* , 1997) were also used in the manufacture of low-fat meat products. In a related study, Abiola *et al.*, (2004) incorporated Melon Seed Meal (MSM) in low fat chicken sausages. Melon seed is a protein-rich oil seed which serves as a popular condiment in soup and stew particularly in developing countries. Earlier studies indicated that the seed contains high amount of unsaturated fatty acid and linoleic acid thereby indicating possible hypocholesteronic effect. Very limited research has been conducted on the use of MSM in meat products. The objective of this study was to determine the effect of replacing back fat with MSM on quality characteristics of low-fat beef sausages.

Materials and Methods

The melon seeds used were purchased from the market . The seeds were shelled and oven dried in an electric oven (Model 38060 KAYSERI/TURKEY TE-5604) at 50^0C for 30 minutes. The dried seeds were air-cooled for 30 minutes after which they were ground in a fabricated milling machine (2.00mm screen) to form MSM. Thigh muscle and back-fat of freshly slaughtered cattle were obtained from a slaughter slab. The meat and fat were cut into pieces and later ground separately in a Kenwood Chef food processor Model A901. Four batches of beef sausages (3kg/batch) were prepared at room temperature. The back fat in each batch of sausage was replaced with MSM at 0, 33, 66 and 100% levels. Equal amounts of wheat flour, seasoning, and water were added to each batch of sausage.

Samples of the sausages were stuffed separately into 35mm diameter cellulose casing and tied at the ends. After labelling, the samples were refrigerated at 2°C for 24 hours while some samples were cooked in a water bath at 70°C for 20 minutes and later air cooled for 30 minutes. Treatment in both cases was replicated three times and the study was repeated twice.

Cooking losses were based on pre- and post-cooking weights of samples while refrigeration losses were also based on the pre- and post-refrigeration weights of samples. The cooked beef sausages were organoleptically evaluated by 5 trained panelists for colour, flavour, juiciness, tenderness and overall acceptability on a 1-7 hedonic scale (1 = unacceptable, 2 =very poor, 3 =poor, 4 =fair, 5 =good, 6 = very good and 7 = excellent).

Data collected were subjected to analysis of variance while means were compared using the Duncan's New Multiple Range Test (Steel and Torrie, 1980).

Results and Discussion

Results of weight losses during cooking and refrigeration are shown in Table 1. Cooking weight losses were generally low and the values obtained ranged between 0.69%-1.77% in all the products. Lower cooking weight losses were obtained on MSM sausages compared with the control. Troutt *et al.*, (1992) also recorded lower cooking losses compared with the control when texture-modifying ingredients were incorporated in low-fat ground beef. In the present study, refrigeration weight losses ranged between 4.51%-8.77%. Lower refrigeration weight losses were also recorded for MSM sausages compared with the control. These findings corroborate the report of Abiola *et al.*, (2004) which indicated that MSM decreased refrigeration weight losses in chicken sausages. In a related study, Abiola and Adegbaju (2001) observed that replacement of pork back fat with rind decreased both refrigeration and cooking weight losses in pork sausage.

All the MSM sausages were superior in all sensory traits evaluated (Table 2). Highest scores of 5.93, 5.47, 5.27 and 5.53 were recorded for colour, flavour, juiceness and overall acceptability respectively. In a similar study, Abiola *et al.*, (2004) indicated that MSM improved overall acceptability of the finished products.

Conclusion

MSM can replace 100% fat in beef sausage to produce lower cooking and refrigeration weight losses and improve overall acceptability. Such formulation will satisfy the demand of consumers for low-fat meat products.

Table 1: Cooking and Refrigeration Weight Losses in Beef Sausages.

	Cooking loss				Refrigeration loss			
	Replacement levels of melon seed meal				Replacement levels of melon seed meal			
	0%	33%	66%	100%	0%	33%	66%	100%
Initial Weight (g)	50.00	50.00	50.00	50.00	50.00	50.00	50.00	50.00
Final weight (g)	49.12	49.65	49.56	49.65	45.62	47.36	47.74	46.86
Weight Loss (%)	1.77^{a}	0.69^{a}	0.89^{b}	0.69^{b}	8.77^{b}	5.29^{b}	4.51^{a}	6.28^{a}

a, b Means in the same row with different superscripts are significantly different (P< 0.05)

Table 2: Sensory Evaluation of Cooked Beef Sausage.

	Replacement levels of melon seed meal			
PARAMETERS	0%	33%	66%	100%
Colour	4.27^{b}	5.67^{a}	5.27^{a}	5.93^{a}
Flavour	4.53	4.87	4.93	5.47
Juiciness	4.80	5.13	5.13	5.27
Tenderness	5.13	5.27	5.27	4.80
Acceptability	4.73	5.20	5.47	5.53

a, b Means in the same row with different superscripts are significantly different (P<0.05)

References

Abiola, S. S. and Adegbaju, S.W.(2001). Effect of substituting pork back fat with rind on quality characteristics of pork sausage. Meat Science. 58: 409-412.

Abiola, S. S, Kadiri, E.I. and Kareem, T. T. (2004). Effect of melon seed meal addition on some quality characteristics of chicken sausages. Journal of the Science of Food and Agriculture. 84: 423-426.

Akobundu, E.N.T. and Eluchie, G.U. (1992). Quality characteristics of pork sausage containing mushroom (Pleurotus tuber-regium) and spices. Journal of Food Science and Technology. Vol. 29, No3, 159-161.

Bloukas, J.G; Paneras, E.D. and Fournitzis, G.C. (1997). Effect of replacing pork back fat with olive oil on processing and quality characteristics of fermented sausages. Meat Science. 45 (2): 133-144.

Steel, R.G.D. and Torrie, J.H. (1980). Principles and procedures of statistics. 2nd ed. McGraw-Hill NewYork.

Troutt, E.S., Hunt, M.C., Johnson, D.E., Claus, J.R., Kastner, C.L and Kropf, D.H. (1992). Characteristics of low-fat ground beef containing texture-modifying ingredients. Journal of Food Science. 57(1): 19-24.

A STUDY ON MINERAL COMPOSITION OF BEEF

P. Barge*, A. Brugiapaglia, M.T. Barge and G. Destefanis

Dip. di Scienze Zootecniche, Università di Torino, 10095 Grugliasco, Italy. Email: paolo.barge@unito.it

Keywords: beef, minerals, category, muscle

Introduction

Knowing the mineral content of food is interesting from a nutritional and environmental point of view. Even though the need for studying the mineral content in meat has been long acknowledged (Doornebal and Murray, 1981) the data on this subject are rather lacking and sometimes contradictory. The aim of this investigation is to study the content of some minerals in commonly marketed beef belonging to 2 categories (cow and young bull). Considering the importance of muscle factor for the characteristics of beef (Barge *et al.*, 2001), we analysed the *longissimus dorsi* (LD), the *supraspinatus* (SS) and the *semitendinosus* (SS).

Materials and Methods

In some commercial abattoirs of the Piedmont, from 22 carcasses-10 young bulls (range 12-19 months old) and 12 cows (range 39-120 months old) we collected samples of 3 muscles (LD, ST and SS), which were freeze-dried, digested with nitric acid and analysed in duplicate for Fe, Zn, Ca, Na, K and Mg by an atomic absorption spectrophotometer (3030 Perkin-Elmer). Data were analysed by Anova (SPSS), considering as sources of variation the muscle, the animal category and their interaction. We also studied the correlations (not reported) of each mineral with the age of the animal, with protein content and with lipid content of the meat.

Results and Discussion

Table 1: Minerals: mean value (ppm on dry matter) and coefficient of variation (%).

		Fe		Zn		Ca		Na		K		Mg	
		mean	cv	mean	cv	mean	cv	mean	cv	mean	cv	Mean	cv
Cow	LD	88,3	22,99	166,6	13,80	459,2	10,45	1912,0	19,68	13957,3	11,14	964,7	13,50
	ST	81,8	19,86	129,5	23,25	379,7	11,18	1741,7	16,95	17540,6	6,37	1262,9	4,77
	SS	115,9	16,08	254,1	9,80	424,7	16,39	2752,7	12,92	14471,8	11,06	961,4	14,58
	all	95,5^{w}	24,51	181,8	32,36	418,9^{z}	14,92	2139,3	26,65	15019,4	15,47	1068,8	17,14
Young bulls	LD	63,1	23,82	160,4	29,19	533,3	15,68	1649,6	18,05	14729,1	7,51	996,8	16,48
	ST	54,1	22,46	126,8	14,65	394,6	6,83	1803,0	15,35	17129,3	11,86	1273,8	5,51
	SS	76,8	9,96	218,4	14,54	481,6	21,75	4425,5	20,29	14469,9	5,91	1047,6	18,09
	all	65,2^{z}	22,82	166,8	30,24	467,2^{w}	20,36	2528,9	53,76	15467,4	11,97	1106,1	17,17
Muscle	LD	77,7^{b}	28,19	163,7^{b}	21,70	494,3^{a}	15,28	1754,6^{b}	19,66	14304,6^{b}	9,75	980,7^{b}	14,80
	ST	70,8^{b}	28,32	128,3^{c}	19,48	386,5^{b}	9,37	1769,6^{b}	15,93	17353,6^{a}	9,05	1267,9^{a}	5,02
	SS	99,1^{a}	24,86	238,0^{a}	13,82	449,1^{a}	19,78	3496,2^{a}	30,41	13925,5^{b}	9,94	1000,6^{b}	16,61

$^{a,b,c\ (w,z)}$ Means of muscle (of category) within a column with different superscripts are significantly different at P<.05.

The 2 factors accounted for more than 50% of the variability (R^2 ranging from 0.53 for Mg to 0.87 for Na); only for Ca the R^2 was equal to 0.40. The data confirmed the great importance of muscle for meat characteristics: this factor significantly influenced all minerals, the size of the differences being not at all negligible. Among the 3 muscles analysed, ST has a low content of iron (as in Von Seggern *et al.*, 2005), zinc and calcium; on the other hand it was the richest in K and Mg. SS was the poorest for K and Ca, but it was definitely the richest not only in sodium, but also in 2 elements as important as iron and zinc. About iron, it is well-known that beef is an important source, either for quantity or for bioavailability; this could have a practical importance, considering that iron deficiency is the most common and widespread nutritional disorder in the world (Biesalski, 2004). Our data confirm that LD was poorer in iron in comparison with less valuable muscle: f.i. *vs diaphragm* (Doornebal and Murray, 1981); *vs triceps brachii* (Purchas and Busboom, 2005). Referring to zinc, Biesalski (2004) underlined that a low intake of this mineral is associated with a weakened immune system; according to Lombardi-Boccia *et al.*, (2005), meat is the richest source of zinc in the Italian total diet. In our study, Zn content in SS was 1.5 times higher than in LD; a figure similar to 163% that can be reckoned from the data of Künhe, (1976). Our results and particularly the low content of ST agree only in part with a statement reported by Doornebal and Murray (1981); it was suggested that Zn would have a higher concentration in those muscles involved in movement.

LD was intermediate between the other muscles for zinc and iron, similar to ST for Na (as in Kotula and Lusby, 1982) and to SS for the K; on the other hand it was the highest for the calcium content.

Recently, Ruusunen and Puolanne (2005) observed that consumers have become more aware of the relationship between sodium and hypertension and review the possibilities to reduce the sodium content in meat products. Our

results indicate that, although we considered only 3 muscles, the variability between muscles is so high (LD sodium content is about half as compared to SS) that the choice of muscles could be a valid way to reduce the sodium intake.
On the contrary, the category only influenced some minerals: cow meat was richer in iron and tended to have more zinc, but was poorer in calcium. It interacted with muscle only in one case: the meat of young bulls contains less Na than cow in LD, whereas it contains more in the other muscles.
Referring to the correlations, we found one positive (significant in 2 out of 3 muscles) between iron and age, as for other authors (Kotula and Lusby, 1982). Age was in negatively related (although not significant in LD) to calcium content, as found in steers (1-6 years) by Kotula and Lusby; whereas in Doornenbal and Murray's cow group (3-12 years) the Ca content increases with age. The protein content was negatively correlated with sodium (not significantly in SS); positively with LD content in K; with Mg content in ST; negatively in zinc content (significant in LD and ST). Sim and Wellington (1976) found a high negative correlation between fat and K content. In our study, the lipid content was negatively correlated to the K content (significant in LD and SS) and Mg content. Lipid and iron content were positively correlated (significant in SS).

Conclusions

Even though obtained from a small number of samples, the present data allow us to highlight that beef normally marketed is a good source of different minerals, contributing to fulfil the RDA (INRAN, 1996) to not a negligible degree. Considering the mean of 3 muscles, 100 g of beef (about 25 g on dry matter) satisfy 11,5% of the needs of iron in adult women, but for SS of cow the figure grows to 29% in comparison to the needs of adult males. About zinc, the percentages for SS of cow was equal to 52.9% of the needs of nursing women, up to 90.7% for the other women, to 63.5% in adult males.
The value of a cut of beef has no relation with its mineral richness: in fact we found the highest content both of iron and zinc in the SS of cow.
Deepening the knowledge of mineral in more muscles of different categories of cattle –and thereby contributing to our knowledge of meat as a varying raw material (Kunhe, 1976)- would allow the upgrading of under utilised cuts and to complete the beef muscling profiling research (Von Seggern *et al.*, 2005). Knowledge of the raw material, together with studies about losses depending on cooking procedures, is the only way to allow consumers to adjust their diet to their specific requirements.

Acknowledgements

The authors thank Prof. Giuseppe Piccone and the staff of the chemical section of the Agricultural Sciences Faculty for their valuable help.

References

Barge M.T., Destefanis G., Brugiapaglia A., Barge P. (2001). Chemical composition of beef muscles in relation to the composition of *longissimus*. Proceedings of the 47th ICoMST, I:178-179.
Biesalski H.F. (2004). Meat as a component of a healthy diet – are there any risks or benefits if meat is avoided in the diet? Meat Science, 70:509-524.
Doornenbal H. and Murray A.C. (1981). Effects of age, breed, sex and muscle on certain mineral concentrations in cattle. Journal of Food Science, 47:55-58.
INRAN (1996). LARN - Livelli di Assunzione Giornalieri Raccomandati di Nutrienti per la Popolazione Italiana. Istituto Nazionale di Ricerca per gli Alimenti e la Nutrizione. *www.inran.it*
Kotula A.W. and Lusby W.R. (1982). Mineral composition of muscles of 1 to 6 year old steers. Journal of Animal Science, 54:544-548
Künhe D. (1976). Zum Mineralstoffgehalt verschiedener Muskeln von Schwein und Rind. Fleischwirtschaft, 56:570-572.
Lombardi-Boccia G., Lanzi S., Aguzzi A. (2005). Aspect of meat quality: trace elements and B vitamins in raw and cooked meats. Journal of Food composition and analysis, 18:39-46.
Purchas R.W. and Busboom J.R. (2005). The effect of production system and age on levels of iron, taurine, carnosine, *coenzyme* Q_{10}, and creatinine in beef muscle and liver. Meat Science, 70:589-596.
Ruusunen M. and Puolanne E. (2005). Reducing sodium intake from meat products. Meat Science, 70: 531-541.
Sim D.W. and Wellington G.H. (1976). Potassium concentration in bovine muscle as influenced by carcass location, breed, sex, energy intake, age and shrunk body weight. Journal of Animal Science, 42:84-91.
Von Seggern D.D., Calkins C.R. Johnson D.D., Brickler J.E., Gwartney B.L. (2005). Muscle profiling: characterizing the muscles of the beef chuck and round. Meat Science, 71:39-51.

CANONICAL CORRELATION BETWEEN MINERAL CONTENT AND MEAT QUALITY TRAITS OF BUFFALO AND ZEBU-TYPE CATTLE

N. Jerez-Timaure*[1], L. Arenas de Moreno[1], C. Colmenares[1] and Y. Navas-Sánchez[2].

[1.] *Facultad de Agronomía-Universidad del Zulia, Maracaibo Venezuela.* [2.] *División de Postgrado, Facultad de Ingeniería-Universidad del Zulia, Aptdo. 15205, Maracaibo-Venezuela. E-mail: jerez03@yahoo.com*

Keywords: tenderness, flavour, mineral content, beef, buffalo

Introduction

There is a growing interest in Venezuela and other South American countries to utilise water buffaloes (*Bubalus bubalis*) as an alternative to produce meat in tropical flooding savannas where conditions are detrimental for productivity of most cattle breed types. Recent studies (Huerta-Leidenz *et al.*, 2001a,b) have shown that buffalo raised in tropical conditions can produce carcass with excellent attributes and good-quality meat at young ages. Red meat provides essential nutrients, and it is rich in mineral of high bioavailability like Fe and Zn. Many researchers have used several minerals, such as Ca, Zn, and P in solutions injected or infused in whole cuts or carcasses as a way to improve meat tenderness. However, the relationship between natural mineral content of meat and its quality traits have not been reported. The objective of this study was to determine the relationship between the mineral content and the organoleptic characteristics in meat of buffalo and cattle.

Materials and Methods

A group of conformed by 47 buffalo and 53 cattle was randomly selected in a cow-calf operation located at the western Venezuelan llanos. Buffalo were crossbred, with a predominance of Murrah and Mediterranean breeds. Beef calves came from Brahman dams which were artificially inseminated with bulls of several breeds (Brahman, Black Angus, Red Angus, Romo Sinuano, and Charolais). Half of the animals were castrated at weaning (7mo of age). Three groups of 33 animals each (16 buffaloes and 17 cattle) were randomly selected, and serially slaughtered at 17, 19 and 24 mo of age, respectively, following industrial standard procedures. At 48 h. *postmortem*, two 2.5 cm-thick steaks from the *longissimus dorsi thoracis* muscle were taken from each animal for Warner-Bratzler shear force (WBSF) determination and trained panel evaluation of tenderness, juiciness, amount of connective tissue, general tenderness and flavour intensity. A third steak was taken for mineral content analysis. The Na, K, Ca, Fe, Mg, Cu, Zn and Mn contents (mg/100g fresh tissue)were determined by atomic absorption and /or emission spectrophotometry (Pelkin-Elmer, 1994), while P was determined by UV-VIS spectrophotometry (A.O.A.C. 1990). Canonical correlation analysis using the CANCORR procedure was used to determine the relationship between mineral concentrations and meat quality traits (SAS, 1999-2000) for each species. The analysis is a multivariate statistical method that consists on finding two composite variables (canonical variables) correctly, so that the relationship among the groups of variables is a maximum (Johnson, 2000). The canonical variate that included the mineral content (Ca, Mg, Na, P, K, Cu, Fe, Zn, and Mn) was called "Mineral". The canonical variate that included meat quality-related traits (Warner Bratzler shear force, juiciness, general tenderness, and flavour intensity) was called "Quality".

Results and Discussion

Two canonical correlations coefficients were obtained for mineral content and quality traits for both beef and buffalo groups. The first canonical correlations for beef samples explained approximately 50.53 % of the relationship (r= 0.71) between variables (P=0.01). The first canonical correlation coefficient for buffalo samples explained a tendency of (P=0.09) of 41.36 % of the association (r= 0.64) between variables. These results point out a dependent relationship between mineral content and meat quality traits. For both species, the second canonical correlation was not significant (P>0.10). Canonical correlation coefficients by gender or age within specie were not significant.

Canonical variety coefficients for the canonical variety "Mineral" (M) were as follows:

$M_{Buffalo}$ = (0.82) Ca + (-0.09) Mg + (-0.10) P + (0.14) K + (-0.22) Na + (0.03) Fe + (-0.25) Zn + (-0.07) Cu + (-0.20) Mn.

M_{Beef} = (0.90) Ca + (0.12) Mg + (0.15) P + (-0.06) K + (-0.18) Na + (-0.04) Fe + (-0.22) Zn + (-0.32) Cu + (-0,21) Mn.

Calcium content resulted as the variable within the Mineral canonical variety with the highest weight in the equation for both species.

Canonical variety coefficients for the canonical correlation variety "Quality" (Q) were as follows:

$Q_{Buffalo}$ = (1.12) WBSF + (-0.11) Juiciness + (0.46) General Tenderness + (-0.33) Flavour intensity.

Q_{Beef} = (0.84) WBSF + (0.30) Juiciness + (0.22) General Tenderness + (-0.41) Flavour intensity.

WBSF and flavour intensity and WBSF and tenderness were the variables with the largest weight in the second equation for beef and buffalo samples, respectively.

Correlations coefficients between original variables and canonical variates are shown in Table 1. Among mineral components traits, Ca concentrations showed the highest canonical coefficient for both buffalo and beef samples (0.60 and 0.63, respectively), when it is correlated with canonical variety "Quality". WBSF showed the highest canonical correlation coefficient (0.60 and 0.62) for buffalo and beef samples, respectively, when it is correlated with canonical variety "Mineral". The correlation coefficients between the original variables Tenderness and Canonical variate "Mineral" were negative for both groups of samples. The dual function of Ca^{+2} in *post-mortem* muscle is well known. The initial *post-mortem* toughening (*rigor mortis*) and the tenderisation processes (rigor resolution) are both regulated by Ca^{+2}. Results from this study showed a strong positive relationship between Ca content and WBSF, indicating that tenderness decreases as muscle's intact Ca content increases. This is in contrast to the findings of other researchers, who have demonstrated that the injection or infusion calcium chloride solutions has improved tenderness in pre-*rigor* and *post-rigor* meat (Wheeler *et al.*, 1992). However, the relationship between intact Ca content in the muscle and tenderness has not been reported. An important association between Mg and Cu in beef samples, and P and Mg in buffalo samples, with the canonical variety "Quality", was obtained, which should be considered for further studies.

Table 1: Canonical Correlation between Original variables and Canonical variates.

	Beef		Buffalo	
Original variables	Canonical Variety "Mineral"	Canonical Variety "Quality"	Canonical Variety "Mineral"	Canonical Variety "Quality"
Ca	**0.8985**	**0.6387**	**0.9388**	**0.6038**
Mg	0.3597	0.2557	0.2671	0.1718
P	-0.2000	-0.1422	-0.5721	-0.3679
K	0.2481	0.1764	0.3117	0.2004
Na	0.0819	0.0582	-0.2417	-0.1555
Fe	-0.2229	-0.1585	-0.0854	-0.0549
Zn	-0.1351	-0.0961	-0.0607	-0.0390
Cu	-0.3345	-0.2378	-0.1107	-0.0712
Mn	-0.2735	-0.1944	-0.3696	-0.2377
WB Shear force	**0.6234**	**0.8769**	**0.6038**	**0.8674**
Juiciness	0.2033	0.2860	0.1718	-0.1075
Tenderness	-0.1988	-0.2797	-0.3679	-0.2040
Flavour intensity	-0.4000	-0.5626	0.2004	-0.3268

Bold font indicates the highest correlation coefficients.

Conclusions

The mineral content and quality traits of meat from buffalo and beef are not independent. However, results are not conclusive to justify prediction analysis based on mineral content. The Ca content is the variable most highly correlated among the minerals studied. Results indicated a strong and positive correlation between Ca content and shear force values. Other nutritional components such as fatty acids profile, total lipids, and amino acids might be included in further studies to improve understanding of the relationship between quimical content of meat and its organoleptic characteristics.

References

A.O.A.C. Association of Official Analytical Chemist. (1990) Official Methods of Analysis 15th Ed. Washington, DC vol. II.

Huerta-Leidenz, N., Rodas-González, A., Vidal, A., Colina, O., Rodríguez, R. (2001a) Comparison of Water Buffaloes vs. Zebu-Type Cattle on Carcass Traits at contemporary Ages. In: The Buffalo: An Alternative for Animal Agriculture in the Third Millenium. VI Word Buffalo Congress. Maracaibo, Zulia, Venezuela. 2, 46-52.

Huerta-Leidenz, N., Rodas-González, A., Vidal, A., Colina, O. (2001b) Evaluation by Trained Panelists of Rib Steaks Derived from Zebu Type Cattle and Domestic Buffaloes. In: The Buffalo: An Alternative for Animal Agriculture in the Third Millenium. VI Word Buffalo Congress. Maracaibo, Zulia, Venezuela. 2, 53-58.

Johnson, D. E. (2000) Métodos Multivariados Aplicados al Análisis de Datos. Editorial Internacional Thomson Editores. México, México.

SAS. (1999-2000) SAS user's guide: Statistics. Versión 8.1 Edition. SAS Institute Inc., Cary, NC. USA.

FREE AMINO ACIDS AND CONSTITUTED AMINO ACIDS OF PROTEIN OF FOUR MUSCLES AND LIVER FROM RUBIA GALLEGA

D. Franco, T. Moreno, E. Bispo*, N. Pérez Seijas and L. Monserrat

Centro de Investigaciones Agrarias de Mabegondo Apartado 10 15080 La Coruña, SPAIN.
Email:daniel.franco.ruiz@xunta.es

Keywords: amino acids, beef, taurine

Introduction

Nowadays, consumers wish to have better knowledge about the nutritive quality of food; in the case of meat, this depends on nitrogenous components, basically protein of high biological value, free amino acids, dipeptides and other non-protein nitrogenous compounds (NPN), because of their contribution to taste and flavour in food (Kato *et al.*, 1989). It has been established that oxidative muscles are tastier than their glycolytic counterparts; this is due, in part, to their content of free amino acids and dipeptides (Cournet y Bousset, 1990). On the other hand, amino acids play an important biological role, for example accumulation of taurine is necessary for functional regulation of the eyes, heart, brain and central nervous system (Stapelton *et al.*, 1997). Taurine needed in the human body is synthesised from the dietary sulphur amino acids; methionine and cysteine; the metabolism of babies is not able to form it. Taurine is highly recommended in heart diseases like congestive heart failure (Azuma *et al.*, 1982). Protein content is constant in muscle, but the ageing process produces liberation of peptides, amino acids and other NPN due to protein hydrolysis. The amount of essential/non-essential (E/NE) amino acids in different muscles can vary and from a commercial point of view, better muscles have not necessarily got a favourable E/NE ratio. Amino acid determination at 24 hour *postmorten* is useful; at this point the carcass is cold and the ageing process has not started yet. The aim of this work was to determine amino acids in different commercial muscles and the liver of Rubia Gallega calves.

Materials and Methods

24 hours *post morten* steak was excised from middle of each following muscle: *Semitendinosus* (ST), *Gluteus bíceps* (GB), *Semimembranosus* (SM), *Longisimus Thoracis* (LT) and liver of two animals. External fat and connective tissue were removed and each sample was ground and freeze-drying. Samples were stored at -20 ºC until analyzed. Free amino acids were extracted according to the method of Konosu *et al.* (1974). Protein precipitate with TCA was hydrolyzed 2with 6M HCL in sealed ampoules for 24 hours in a oven at 110 ºC. Amino acids were determined by HPLC using AccQ-Tag derivatization reagent. Amino acids were separated on a Water AccQ-Tag column (3.9 mm x 150 mm with a 4 µm of particle size) and detected by fluorescence (Waters 2475) according to methodology developed by Vandelen and Cohen, (1997). Amino acids were identified by retention time using an amino acid standard (Sigma Aldrich) to which taurine (Panreac) was added.

Results and Discussion

The content of free amino acids (FAA) and constituted amino acids of protein (HAA) in the muscles studied is shown in Table 1. With regard to free amino acids, the amount of essential and non-essential are similar apart from SM and liver where the amount of non-essential amino acids are higher (59.01 and 43.57 vs 79.94 and 113.79 mg free amino acid/100 g fresh muscle respectively). It is interesting to note that the free amino acid at highest concentration in all muscles is taurine ,except LD in which 17.82 mg taurine/100 mg fresh muscle is found as against 74.26 mg/100 mg fresh muscle in SM. In the case of liver, taurine is a minor component. There is little information about taurine levels in beef and our results are in the same range as stated by Spitze *et al.*, (2003) and Purchas *et al.* (2004). However taurine values in liver were lower than these authors reported, because taurine quantity in liver is very variable (Spitze *et al.* 2003).

In general, amino acid composition (g/100 g protein) in different muscles showed a similar pattern; this result is in agreement with Schweigert (1987) who found that amino acid composition of protein remains constant in different commercial cuts. Glutamic acid and aspartic acid were quantitatively the most important non-essential amino acids in all muscles, whereas lysine and leucine were quantitatively the most important essential fraction. Limiting amino acids in dietary protein are typically the sulphur-containing amino acids, like threonine and lysine. The amount of threonine and lysine in the LD muscle were 2.5 and 1.7 times higher respectively than in hen egg protein ,commonly used as a reference protein with regard to amino acid composition in other food items (Matthews, 1999).

Table 1: Free amino acid contents (mg free amino acid/100 g fresh muscle) and amino acid composition of Rubia Gallega meat (g /100 g protein).

	ST		GB		SM		LD		Liver	
	FAA	HAA	FAA	HAA	FAA	HAA	FAA	HAA	FAA	HAA
Essential										
Histidine	37.58	1.58	41.10	1.73	20.85	1.73	9.71	1.63	14.8	1.63
Isoleucine	10.43	6.60	5.02	5.80	5.55	7.32	2.16	6.16	5.10	5.14
Leucine	5.99	6.55	2.90	9.04	2.40	7.49	1.55	10.37	2.36	9.14
Lysine	7.46	9.12	4.35	10.53	4.52	9.90	1.64	11.61	5.67	7.70
Methionine	4.00	2.97	1.55	3.38	1.77	3.52	0.87	4.21	1.81	2.74
Phenylalanine	7.57	4.40	3.77	5.04	4.09	5.05	2.46	6.69	4.16	5.80
Threonine	11.77	4.71	14.01	5.25	15.14	4.78	5.88	5.62	6.01	4.56
Valine	8.52	5.51	4.10	6.13	4.69	6.26	1.48	7.26	3.66	6.51
Total	93.32	41.44	76.8	46.9	59.01	46.05	25.75	53.55	43.57	43.22
Non-essential										
Arginine	10.90	6.39	21.81	6.98	18.92	7.04	nd	8.25	5.52	6.38
Alanine	21.18	6.12	20.83	6.54	19.01	6.46	5.22	7.77	8.33	5.80
Aspartic acid	2.84	9.80	2.61	10.98	2.39	10.62	2.36	12.52	4.58	9.53
Cystine	8.62	0.85	6.05	1.19	7.35	0.99	2.62	1.25	10.20	0.89
Glutamic acid	20.10	15.73	8.87	17.49	11.18	16.94	5.00	20.32	49.8	12.34
Glycine	9.38	4.39	6.81	4.43	7.77	4.44	3.04	5.14	21.69	4.79
Proline	5.86	3.64	3.66	3.85	4.05	3.76	2.55	4.47	3.96	4.25
Serine	10.08	3.45	5.54	4.04	5.74	3.78	3.02	4.25	6.99	3.90
Tyrosine	6.87	3.68	3.08	4.29	3.53	4.23	1.50	5.04	2.72	3.60
Other										
Taurine*	59.81	nd	69.03	nd	74.26	nd	17.82	nd	1.93	nd
Total	95.83	54.05	79.26	59.79	79.94	58.26	25.31	69.01	113.79	51.48
E/NE ratio	0.97	0.77	0.97	0.78	0.74	0.79	1.02	0.78	0.38	0.84

nd. not detected, * taurine not included in Total.

Conclusions

Taurine is the most important free amino acid in all muscles, with a maximum content of 74.26 mg per 100 g of fresh muscle Although there is not a recommended daily intake of taurine, "Ternera Gallega" meat is an important taurine source. The best E/NE ratio was found in liver so from a nutritive point of view its intake should be recommended.

References

Azuma, J., Hasegawa, H., Sawamura, N. (1982). Taurine for treatment of congestive heart failure. International Journal of Cardiology, 303-314.

Cornet, M. and Bousset, J. (1990) Free amino acids and dipeptides in porcine muscles. Proceedings of 36th ICOMST, Havana, pp. 226-231.

Kato, H., Rhue, M.R. and Nishimura, T. (1989). Role of free amino acids and peptides in food taste. In: Flavour Chemistry. Trends and Developments. Eds. R. Teranishi, R.G. Buttery and F. Shahidi, ACS Symp Series 388, pp. 158-174 American Chemical Society, Washington DC.

Konosu, S. and Yamaguchi, K. (1982). The flavour components in fish and shellfish. In: "Chemistry and Biochemistry of Marine Food Products". pp. 367-404. Martin, R. E., Flick, G. J., Hebard, E. and Ward, D. R. eds. AVI Publishing Co. Westport, CT, U.S.A.

Matthews, D. (1999). Proteins and amino acids. In: M. Shild, J. Olson, M. Shike, & A. Ross (Eds.), Modern nutrition in health and disease, pp. 11–48. Baltimore, USA. publ. Lippincott Williams and Wilkins.

Purchas, R.W., Rutherfurd, S.M., Pearce, P.D., Vather, R., Wilkinson, B.H.P. (2004) Concentrations in beef and lamb of taurine, carnosine, coenzyme Q10, and creatine. Meat Science, 66, 629–637.

Schweigert, B.S. (1987). The nutritional content and value of meat products. In: The Science of Meat and Meat Products, 4th edn. Eds J.F. Price and B.S. Schweigert. Food and Nutrition Press Inc., Westport, CT, USA, p.275.

Spitze,. A. R., Wong, D. L., Rogers, Q. R., Ascetti, A.J. (2003). Taurine concentrations in animal feed ingredients; cooking influences taurine content. Journal of Animal Physiology and Animal Nutrition, 87, 251–262.

Stapleton, P. P., Charles, R. P., Redmond, H. P. and Bouchier-Hayes, D. J. (1997). Taurine and human nutrition. Clinical Nutrition, 16, 103-108.

Vandelen, C.,. Cohen, S. (1997). Using quaternary high-performance liquid chromatography eluent systems for separating 6-minoquinolyl-N-hydroxysuccinimidyl carbamate-derivatized amino acid.

PHYSICO-CHEMICAL CHARACTERISTICS OF RESTRUCTURED BEEF STEAKS WITH ADDED WALNUT AS AFFECTED BY COOKING METHODS

A. Serrano, S. Cofrades* and F. Jiménez Colmenero

Instituto del Frío (Consejo Superior de Investigaciones Científicas). C/ José Antonio Novais, 10, 28040 Madrid, Spain.
Email: scofrades@if.csic.es

Keywords: restructured meat, walnut, texture, cooking method

Introduction

Consumers are currently more health conscious and as a result demand healthier products. Meat processing can be used to modify the composition of meat products reducing, increasing, adding and/or replacing different components (from animal and plant sources) with potential functional effect (Jiménez Colmenero *et al.*, 2006). Modification of the ingredients used for the preparation of meat products has been tested in various ways with a view to enhancing health-beneficial components (Anandh *et al.*, 2003; Fernández-Ginés *et al.*, 2005). Different studies have shown that the frequent consumption of walnuts reduces the risk of cardiovascular disease, since it lowers serum cholesterol and favourably modifies the lipoprotein profile (Sabaté, 1993). In order to promote walnut intake, restructured beef steaks with added walnuts have been formulated resulting in products with acceptable physicochemical and sensory properties (Jiménez Colmenero *et al.*, 2003; Cofrades *et al.*, 2004), frozen storage stability (Serrano *et al.*, 2006) and potential health benefits (Serrano *et al.*, 2005).
Restructured beef steaks undergo cooking prior to consumption. It is well known that meat product composition and cooking techniques are the main factors which affect the final quality attributes of the product (Berry and Leddy, 1984; Dal Bosco *et al.*, 2002). This study was conducted to determine how three different cooking techniques (oven, grill and pan) affect some physicochemical characteristics of restructured beef steak.

Materials and Methods

Preparation of two different restructured beef steaks was performed as described by Jiménez-Colmenero *et al.*, (2003) and the formulation of the these products are shown in Table 1. Three different cooking methods were evaluated: oven, grill and pan. In each case, the final cooking temperature of 70°C, (in the core of the sample) was determined beforehand by inserting thermocouples connected to a temperature recorder.

Table 1: Formulation of different restructured beef steaks[1].

Samples denomination	Beef (g)	NaCl +STP (g)	Walnut (g)	Beef fat (g)	MTG (g)	Caseinate (g)	Water (g)	Total (g)
Control (C)	2800	60+12	0	660	28	40	400	4000
20 % added walnut (20W)	2660	60+12	800	0	28	40	400	4000

[1]STP, sodium tripolyphosphate. MTG, microbial transglutaminase

Protein, moisture and fat content of the raw products were evaluated. After the cooking process and the cooling step (for 30 min at 20-22°C), the steaks were manually wiped with a paper towel to remove visible exudate. Cooking loss was calculated as weight loss (%). Bind strength (BS) and Kramer shear force (KSF) of the raw and cooked restructured steaks were performed as described by Jiménez Colmenero *et al.*, (2003). Data was analysed using Statgraphics Plus 2.1 for one-way and two-way ANOVA.

Results and Discussion

Proximate composition of raw restructured beef steak (Table 2) are consistent with the meat product formulation (Table 1). Addition of walnut reduced ($P<0.05$) cooking loss (Table 3). Similar results have been described by Jiménez Colmenero *et al.*, (2003) in restructured beef steak with added walnut. Irrespective of the type of restructuring, the effect of the cooking method seems not to be relevant (Table 3). Cooking methods and type of restructured steak had an effect ($P<0.05$) on textural parameters. BS and KSF of the raw and cooked products containing walnut were lower than those observed in the control samples (Table 3). Softer textures have been reported in restructured beef steaks containing walnut (Jiménez Colmenero *et al.*, 2003) and could be due to the addition of walnut reducing the cohesiveness among meat pieces, so less applied force (BS and KSF) was required. Generally products with high KSF and BS also result in greater cooking loss (Table 3). No clear effect was observed in the texture due to the different thermal treatments applied, however the products cooked by means of pan cooking tend to have softer textures (Table 3).

Table 2: Proximate composition (%) of the raw products.

(%)	C	20W	SEM
Moisture	63.72[1]	61.25[2]	0.58
Protein	17.16[1]	19.38[2]	0.08
Fat	13.03[1]	12.57[1]	0.3
Ash	2.97[1]	3.44[1]	0.01

[1]Different numbers in the same column indicate significant differences (P<0.05). SEM = Standard error of the mean.

Table 3: Cooking loss and textural parameters of the different restructured as affected by cooking method.

Sample		Cooking loss (%)	BS (N)	KSF (N/g)
C				
	Raw		4.28 [1]	3.54 [1]
	Oven	31.74 $_{ab}^{1}$	24.89 $_{ab}^{1}$	16.14 $_{a}^{1}$
	Grill	33.16 $_{a}^{1}$	27.23 $_{a}^{1}$	19.40 $_{b}^{1}$
	Pan	27.27 $_{b}^{1}$	20.77 $_{b}^{1}$	16.75 $_{ab}^{1}$
20W				
	Raw		2.20 [2]	2.91 [2]
	Oven	12.52 $_{a}^{2}$	19.71 $_{a}^{2}$	9.60 $_{a}^{2}$
	Grill	13.48 $_{a}^{2}$	14.32 $_{b}^{2}$	11.31 $_{a}^{2}$
	Pan	11.46 $_{a}^{2}$	13.31 $_{b}^{2}$	11.58 $_{a}^{2}$
SEM		1.27	1.23	0.77

[1]Different letters indicate significant differences (P<0.05) between cooking treatments and different numbers indicate significant differences (P<0.05) between samples. SEM = Standard error of the mean.

Conclusions

Our results suggest that the different cooking methods did not a clear influence on the physico-chemical characteristics studied. The incorporation of walnut in restructured beef steak improves, the binding properties in comparison to the incorporation of animal fat, producing products with softer textures.

Acknowledgments

This research was supported under project AGL2005-07204-C02-02 of the Plan Nacional de Investigación Científica, Desarrollo e Innovación Tecnológica (I+D+I). Ministerio de Ciencia y Tecnología.

References

Anandh, M.A., Lakshmanam, V., Anjaneyulu, A.S.R. (2003). Designer meat foods. Indian Food Industry, 22, 40-45.

Berry, BW., Leddy, K.F. (1984). Nutritional quality of rabbit meat as affected by cooking procedures and dietary vitamin E. Journal of Food Science, 49, 870-875.

Cofrades, S., Serrano, A., Ayo, J. Solas, M.T. Carballo, J. and Jiménez Colmenero, F. (2004). Restructured beef with different proportions of walnut as affected by meat particle size. *Eur. Food Res. Technol.*, 218, 230-236.

Dal Bosco, A., Castellini, C., Bernardini, M. (2002). Effect of fat level and cooking methods on sensory and textural properties of ground beef patties. Jorunal of Food Science, 67, 1047-1051.

Fernández-Ginés, J.M., Fernández-López, J., Sayas-Barberá, E., Pérez-Alvarez, J.A. (2005). Meat products as functional foods: A review. Journal of Food Science, 70, R37-R43.

Jiménez Colmenero F, Serrano A, Ayo J, Solas MT, Cofrades S, Carballo J. 2003. Physicochemical and sensory characteristics of restructured beef steak with added walnuts. Meat Science, 65:1391-7.

Jiménez-Colmenero, F., Reig, M., Toldrá, F. (2006). New approaches for the development of functional meat products. In *Advances technologies for meat processing.* Ed. Leo M.L. Nollet and Fidel Toldrá. Taylor and Francis. Boca Raton, London, New York.

Sabaté, J. and Fraser GE. (1993). The probable role of nuts in preventing coronary heart disease. Prim Cardiol., 19, 65-72.

Serrano, A., Cofrades, S., Ruiz-Capillas, C., Olmedilla-Alonso, B., Herrero-Barbudo, C. and Jiménez-Colmenero, F. 2005. Nutricional profile of restructured beef steak with added walnuts. Meat Science, 70, 647-654.

Serrano, A., Cofrades, S. and Jiménez-Colmenero, F. 2006. Characteristics of restructured beef steak with different proportions of walnut during frozen storage. Meat Science, 72, 108-115.

ANTIOXIDANT AND ANTIMICROBIAL EFFECTS OF GREEN TEA AND GRAPE SEED EXTRACTS ON BEEF PATTIES WITH A LOW DOSE OF SULPHITE

P. Díaz, A. Price, M. Rodríguez, G. Nieto*, M.D. Garrido and S. Bañón

Department of Food Technology and Science and Human Nutrition. Veterinary Faculty. University of Murcia. Campus Espinardo 30071. Murcia. Spain. E-mail: gnieto@um.es

Keywords: beef patties, natural antioxidants, sulphite, ascorbate, tea, grape

Introduction

EC Directive 95/2/EC authorises the addition of sulphites (up to 450mg/kg) in order to delay microbiological and colour spoilage in patties. The preservative action of sulphite is usually strengthened by ascorbate or other antioxidative additives. The use of sulphites in food is questionable because it is related to certain health problems, such as allergies, thiamine absorption deficit and disorders on the carbohydrate metabolism. Vegetable substances with preservative effect are an alternative for prolonging the shelf life of meat. Green tea and grape seed are rich in polyphonic compounds and their antioxidative activity on lipids has been proven in raw and pre-cooked meat products (Gil *et al.*, 2001; Tang *et al.*, 2001; Lise *et al.*, 2004; Mielnik *et al.*, 2006). Other studies suggest that they may also have a certain antimicrobial effect (Jayaprakasha *et al.*, 2003). Previous studies (Diaz *et al.*, 2005) have shown that green tea and grape seed extract do not delay the colour deterioration of raw beef patties, unless sulphites are added, although they may contribute to a reduction in the dose of sulphites necessary. The aim of this study was to determine the effect of green tea and grape seed extracts on the shelf life of patties manufactured with a low level of sulphites (100mg/kg) and stored under retail display conditions.

Materials and Methods

Five batches of beef patties (minced to 5mm and 2% salt) were prepared: the control group, S100 (100mg/kg sulphites in the form of sodium metabisulphite), S100 + A400 (100mg/kg sulphites + 400mg/kg sodium ascorbate), S100 + T300 (100mg/kg sulphites + 300 mg/kg watery green tea extract) and S100 + G300 (100mg/kg sulphites + 300 mg/kg watery grape seed extract) (sodium metabisulphite and sodium ascorbate, Juan Martínez Pérez S.L., Spain; vegetable extracts, GMBH & Co. Kg, Germany). The patties were packed in the presence of air and stored for a maximum of 9 days at 4°C under natural lighting conditions. The spoilage of the meat was determined by colour change, expressed as Chroma and Hue (Minolta Chromameter Reflectance II), the Thiobarbituric Acid Reactive Substances (TBARs) index (mg Malondialdehyde (MDA)/kg meat) and the total viable count and total coliform count (log u.f.c/g) at days 0, 3, 6 and 9 of storage. In order to determine whether the extracts used had caused adverse effects in the meat, a trained panel of 10 judges carried out sensory analysis of the patties at day 0.

Results and Discussion

According to the TBARs values (Table 1), the 3 vegetable antioxidants tested were effective in delaying lipid oxidation in comparison with the control and S100 groups. The antioxidative effect on raw meat has also been described by Gil *et al.* (2001) (tea catechins), Lise *et al.* (2004) (green tea and grape skin), Mitsumoto *et al.* (2005) (vitamin C and tea catechins) and Mielnik *et al.* (2006) (grape seed). On the other hand, the 3 vegetable antioxidants tested had no stabilising effect on the colour above that of S100 (Table 1). Neither was there a greater antimicrobial effect than with S100 (Table 2). The antimicrobial capacity of ascorbate on minced buffalo meat has been questioned by Sahoo and Anjaneyulu (1997). Sensory analysis (Table 3) showed that none of the 3 vegetable extracts tested caused adverse odours or flavours in cooked meat, which agrees with the findings of Mitsumoto *et al.* (2005). Attributes with scores of less than 2 were rancid odour, rancid, liver, acid, astringent and metallic flavours.

Table 1: Average values of TBARs (TBA), Chroma (C*) and Hue (H*) in beef patties at 0, 3, 6 and 9 days under retail display conditions.

	Day 0			Day 3			Day 6			Day 9		
	TBA	C*	H*	TBA	C*	H*	TBA	C*	H*	TBA	C*	H*
Control	0.83	17.3	22.8	2.47^{a}	14.8^{a}	28.5^{a}	3.07^{a}	13.2^{a}	40.0^{a}	3.86^{a}	10.6^{a}	44.1^{a}
S100	0.78	20.2	22.2	2.65^{a}	19.5^{b}	23.0^{ab}	2.96^{a}	15.1^{ab}	30.5^{b}	3.31^{b}	11.7^{a}	37.8^{b}
S100+A400	0.44	21.1	21.1	0.60^{b}	20.3^{b}	21.1^{b}	0.78^{b}	16.1^{ab}	22.4^{c}	0.69^{c}	15.8^{b}	26.0^{c}
S100+T300	0.51	21.5	21.0	0.84^{b}	20.7^{b}	22.4^{b}	0.63^{b}	17.4^{b}	26.4^{bc}	0.60^{c}	15.3^{b}	28.1^{c}
S100+U300	0.45	20.1	20.2	0.47^{b}	21.2^{b}	22.0^{b}	0.45^{b}	17.6^{b}	24.3^{bc}	0.47^{c}	14.9^{b}	30.6^{c}

S: sulphite; A; ascorbate; T: green tea; G grape seed.
Means with different superscripts (rows) are significantly different ($p<0.05$).

Table 2: Average total viable count (TVC) and total coliform count (TCC) (log ufc/g) in beef patties at 0, 3, 6 and 9 days under retail display conditions.

	Day 0		Day 3		Day 6		Day 9	
	TVC	TCC	TVC	TCC	TVC	TCC	TVC	TCC
Control	4.59	3.41	4.60	3.50	5.58^{a}	4.71^{a}	6.37	5.50^{a}
S100	4.46	3.29	4.44	3.50	5.18^{ab}	3.53^{b}	5.98	4.15^{ab}
S100 + A400	4.43	3.23	4.32	3.23	4.36^{b}	3.21^{b}	5.60	3.66^{b}
S100 + T300	4.31	3.34	4.55	3.28	4.49^{b}	3.25^{b}	5.81	3.82^{b}
S100 + G300	4.45	3.45	4.27	3.24	4.35^{b}	3.25^{b}	5.55	3.45^{b}

S: sulphite; A; ascorbate; T: green tea; G grape seed.
Means with different superscripts (rows) are significantly different ($p<0.05$).

Table 3: Average values of sensory attributes in beef patties (day 0).

Attributes	Control	S 100	S 100 + A 400	S 100 + U 300	S 100 + T 300
MO	3.90	3.67	3.77	4.05	3.82
MF	4.35	4.32	4.25	4.30	4.22
FP	2.70	3.02	2.87	2.85	2.87
JU	3.87	4.22	3.95	3.67	3.92
CH	3.85	3.75	3.70	4.05	3.97

S: sulphite; A; ascorbate; T: green tea; G grape seed.
Scoring scale: (1: minimum; 5: maximum)
MO: meaty odour; MF: meaty flavour; FP: persistent flavour; JU: juiciness; CH: chewiness.

Conclusions

The use of green tea and grape seed extracts delays fat oxidation and contributes to prolonging the shelf life of beef patties manufactured with reduced levels of sulphite and stored under usual retail display conditions. 300 mg/kg of green tea and grape seed extracts have a preservative action similar to that of 400 mg/kg sodium ascorbate. The doses of green tea and grape seed tested do not modify the sensory quality of cooked meat, therefore it is acceptable to add these natural antioxidants to meat products.

References

Díaz, P., Price, A., Rodríguez, M., Garrido, M.D. and Bañón, S. (2005). Empleo de té verde, semilla de uva y clara de huevo deshidratada para mejorar la salubridad de los productos cárnicos. *III Jornada de Actualización en Alimentación y Salud*, Murcia. Spain.

Gil, M.D., Buckley, J., Cayuela, J.M., Bañón, S., Granados, M.V., Laencina J. and Garrido M.D. (2001). Estudio del efecto antioxidante de la vitamina E y de la catequina de té sobre carne de cerdo. *I Congreso Nacional de Ciencia y Tecnología de Alimentos*. Granada. Spain.

Jayaprakasha, G. K., Selvi, T. and Sakariah, K. K. (2003). Antibacterial and antioxidant activities of grape (Vitis vinifera) seed extracts. *Food Research International*, 36: 117-122.

Lise, R., N., Derek, V.B., Byrne, G.B. and Leif, H.S. (2004). The antioxidative activity of plant extracts in cooked pork patties as evaluated by descriptive sensory profiling and chemical analysis. *Meat Science*, 68: 485-495.

Mielnik, M.B., Olsen, E., Vogt, G., Adeline, D. and Skrede, S. (2006). Grape seed extract as antioxidant in cooked, cold stored turkey meat. *LWT*, 39: 191-198.

Mitsumoto, M., O'Grady, M. N., Kerry, J. P. and Buckley, D. J. (2005). Addition of tea catechins and vitamin C on sensory evaluation, colour and lipid stability during chilled storage in cooked or raw beef and chicken patties. *Meat Science*, 69: 773-779.

Sahoo, J. Y. and Anjaneyulu, A. S. R. (1997). Quality improvement of ground buffalo meat by preblending with sodium ascorbate. *Meat Science*, 46: 237-247.

Tang, S., Kerry, J. P., Sheehan, D., Buckley, D. J. and Morrissey, P. A. (2001). Antioxidative effect of added tea catechins on susceptibility of cooked red meat, poultry and fish patties to lipid oxidation. *Food Research International,* 34: 651-657.

UTILIZING REGRESSION TECHNIQUES FOR ESTIMATING THE NUTRIENT COMPOSITION OF GROUND BEEF PRODUCTS

J.C. Howe*[1], J.M. Holden[1], D.R. Buege[2] and L.A. Douglass[3]

[1]*USDA, ARS, BHNRC, Nutrient Data Laboratory, 10300 Baltimore Avenue, Bldg. 005, BARC-West, Beltsville, MD 20705, USA* [2] *University of Wisconsin, Dept. of Animal Sciences, Madison, WI 53706, USA* [3] *University of Maryland, Dept. of Animal Sciences, College Park, MD, USA. Email: howej@ba.ars.usda.gov*

Keywords: ground beef, regression equations, nutrient composition

Introduction

Ground beef is a unique meat product in that a wide range of formulations for this product are available in most US retail stores. Until 2003, nutrient composition data for ground beef reported in the USDA National Nutrient Database for Standard Reference (SR) (USDA, 2003) were for products containing 27%, 21%, and 17% fat. However, in response to consumer demand for healthier foods, a much broader selection of ground beef products were marketed at that time, with product offerings as low as 5% fat. The USDA, in collaboration with America's Beef Producers, undertook a study to update the nutrient composition data for ground beef products in SR. However, providing nutrient information for such a range of products presented a unique challenge. In order to provide consumers and industry with the nutrient composition information for this variable product, the study was designed to establish the mathematical relationship between the various nutrients and the total fat content of raw ground beef through regression techniques. The ultimate aim was to use these relationships for prediction of the nutrient composition for ground beef, as purchased, and for broiled, ground beef patties. This report will provide the predictive equations used to estimate nutrient values for proximate components, and for selected vitamins and minerals disseminated in SR.

Materials and Methods

Sampling: Ground beef samples for each of 3 fat categories (label declarations of <12% fat, 12-22% fat, or >22% fat) were purchased from 24 retail outlets nationwide. In this sampling plan developed for the National Food and Nutrient Analysis Program (Pehrsson *et al.*, 2000), the country was divided into 4 regions, with 3 consolidated metropolitan statistical areas (CMSA) within each region; 2 retail stores were selected within each CMSA.

Sample preparation: Ground beef products were analyzed in raw and cooked form (broiled patty). To achieve uniform sizing for the broiled patties, 112 g of ground beef were pressed into a patty mold. Patties were broiled in a preheated conventional oven for 8.7 min at 71°C. Patties were cut in half to evaluate degree of doneness based on color, then stored at –24°C in sealed vacuum bags until homogenization and analysis.

Sample analyses: Samples from each location and each fat level (n=72) were analyzed for moisture, nitrogen, total fat, ash, and selenium. Samples were pooled by CMSA (n=36) for analyses of minerals (calcium, magnesium, potassium, manganese, iron, phosphorus, sodium, copper, zinc), niacin, thiamin, riboflavin, vitamins B_6 and B_{12}, cholesterol, and fatty acids (C8 - C22); twelve samples (pooled by region) were analyzed for total choline and vitamin K. A commercial testing laboratory using AOAC methods analyzed proximate components, vitamins, and minerals. Selenium was analyzed by isotope dilution gas chromatography/mass spectrometry (Reimer and Veillon, 1981); Vitamin K was analyzed by high performance liquid chromatography (Dumont *et al.*, 2003); choline and betaine were analyzed by liquid chromatography/electrospray ionization-isotope dilution mass spectrometry (Koc *et al.*, 2002). Quality control measures included duplicate sampling, and the use of control composites and NIST standard reference materials (SRM 1546: Meat Homogenate).

Statistics: Data were analyzed using mixed model regression analysis (SAS, 2004) to obtain a regression equation for each nutrient and preparation method (raw and broiled).

Results and Discussion

For each nutrient in the raw product, the relationship between nutrient levels and total fat content could best be defined by a simple linear expression. Simple linear equations and second-degree polynomial models appropriately described the relationship between nutrient level of broiled patties and total fat content of the raw product. Quadratic equations were the 'best fit' models for cholesterol and total fatty acids (saturated, monounsaturated and polyunsaturated) with respect to total fat content. The best regression model to explain the relationship of moisture to total fat was also a quadratic equation. Higher levels of polynomial regressions did not improve the fit of the model for any nutrient analyzed. Table 1 presents the equations for predicting nutrient values for raw ground beef and broiled ground beef patties based on the total fat content of raw ground beef. Table 1 also illustrates the use of these equations and provides the nutrient composition of 93% lean ground beef (raw and broiled patty) as derived from these equations.

Table 1: Regression equations for determining nutrient content of ground beef products and the estimated nutrient values for ground beef containing 93% lean meat/7% fat (values per 100g ground beef).

		Raw		Broiled	
Nutrient	N	Regression Equation[1]	93% lean meat	Regression Equation[1]	93% lean meat
Protein, g	72	Y=22.8261 - 0.2826X	20.85	Y=26.4707 - 0.0363X	26.22
Total Fat, g	72	Y=0.0000 + 1.0000X	7.00	Y=(-0.0369) + 1.4601X - 0.0284X^2	11.58
Ash, g	72	Y=1.1141 - 0.0137X	1.02	Y=1.0736 - 0.0036X	1.05
Moisture, g	72	Y=77.0618 - 0.7560X	71.77	Y=72.0784 - 1.3458X + 0.0273X^2	64.00
Calcium, g	36	Y=6.2241 + 0.5864X	10	Y=1.3947 + 1.1347X	9
Iron, g	36	Y=2.5329 - 0.0296X	2.33	Y=2.9465 - 0.0232X	2.78
Magnesium, mg	36	Y=23.2122 - 0.3216X	21	Y=23.1760 - 0.1453X	22
Phosphorus, mg	36	Y=210.7459 - 2.6304X	192	Y=210.5172 - 0.8434X	205
Potassium, g	36	Y=372.0656 - 5.1254X	336	Y=362.0251 - 2.9055X	342
Sodium, mg	36	Y=65.3488 + 0.0699X	66	Y=61.7564 + 0.6534X	66
Zinc, mg	36	Y=5.3958 - 0.0609X	4.97	Y=6.4899 - 0.0118X	6.41
Copper, mg	36	Y=0.0829 - 0.0011X	0.075	Y=0.1011 - 0.0011X	0.093
Manganese, mg	36	Y=0.0107 + 0.0000X	0.011	Y=0.0153 - 0.0002X	0.014
Selenium, mg	72	Y=18.167 - 0.1565X	17.1	Y=21.7719 - 0.0139X	21.7
Thiamin, mg	36	Y=0.0406 + 0.0001X	0.041	Y=0.0404 + 0.0004X	0.043
Riboflavin, mg	36	Y=0.1647 - 0.0008X	0.159	Y=0.1714 + 0.0003X	0.174
Niacin, mg	36	Y=5.9167 - 0.0845X	5.325	Y=6.2196 - 0.0561X	5.827
Vitamin B_6, mg	36	Y=0.4150 - 0.0046X	0.383	Y=0.4275 - 0.0030X	0.407
Vitamin B_{12}, mg	36	Y=2.2786 - 0.0069X	2.23	Y=2.3861 + 0.0171X	2.51
Choline, mg	12	Y=75.6921 - 0.9630X	68.95	Y=86.9261 - 0.3047X	84.79
Vitamin K, mg	12	Y=(-0.2039) + 0.1024X	0.5	Y=1.6773 - 0.1048X + 0.0050X^2	1.2
Betaine, mg	12	Y=7.6230 + 0.0299X	7.83	Y=6.8812 + 0.1065X	7.63
Cholesterol, mg	36	Y=58.2332 + 0.6534X	63	Y=62.6671 + 3.0539X - 0.0806X^2	80
SFA[2], g	35	Y=0.4424 + 0.3615X	2.973	Y=0.9357 + 0.4471X - 0.0078X^2	3.683
MUFA[3], g	35	Y=(-0.0509) + 0.4404X	3.032	Y=(-0.0013) + 0.5942X - 0.0100X^2	3.668
PUFA[4], g	35	Y=0.1686 + 0.0176X	0.292	Y=0.2026 + 0.0278X - 0.0006X^2	0.368

[1]X=% total fat; [2]total saturated fatty acids; [3]total monounsaturated fatty acids; [4]total polyunsaturated fatty acids

Conclusion

USDA has been developing food composition tables and electronic databases for more than 100 years (Atwater and Woods, 1891; Nutrient Data Laboratory, 2003). SR is provided to the scientific community and national population free of charge, and is widely used for many purposes, including nutrition monitoring, nutrition policy development, health research and food product development. The results of this study provide a mechanism by which the nutrient content of ground beef can be estimated for any product formulated with a fat content ranging from 5% to 30%. Nutrient information derived from these equations will be used by consumers to estimate dietary intakes. These data will also be used by industry to meet anticipated mandatory labeling requirements.

References

Atwater, W. O. and Woods, C. D. (1891). Investigations upon the chemistry and economy of foods. Report of the Storrs School Experiment Station for 1891; U.S. Department of Agriculture, Storrs, Connecticut.

Dumont JF, Peterson JW, Haytowitz DB, Booth SL. (2003). Phylloquinone and dihydrophylloquinone contents of mixed dishes, processed meats, soups and cheeses. Journal of Food Composition and Analysis 16:595-603.

Koc, H., Mar, M. H., Ranasinghe, A., Swenberg, J. A. and Zeisel, S. H. (2002). Quantitation of choline and its metabolites in tissues and foods by liquid chromatography/electrospray ionization-isotope dilution mass spectrometry. Analytical Chemistry 74: 4734-4740.

NDL (Nutrient Data Laboratory). (2003). Compiling food composition data for 110 years. Nutrient Data Laboratory Web site, http://www.ars.usda.gov/nutrientdata.

Pehrsson, P. R., Hatowitz, D.B., Holden, J.M., Perry, C.R. and Beckler, D.G. (2000). USDA's National Food and Nutrient Analysis Program: food sampling. Journal of Food Composition and Analysis 13:379-389.

Reimer, D. C. and Veillon, C. (1981). Determination of selenium in biological materials by stable isotope dilution gas chromatography-mass spectrometry. Anal. Chem. 53 (14): 2166-2169

SAS Institute Inc. (2004). SAS OnlineDoc® 9.1.2. Cary, NC: SAS Institute Inc.

U.S. Department of Agriculture, Agricultural Research Service. (2003). USDA National Nutrient Database for Standard Reference, Release 15. Nutrient Data Laboratory Web site, http://www.ars.usda.gov/nutrientdata.

DIETARY SUPPLEMENTATION AND ADDITION OF TEA CATHECHINS: ASSESSMENT OF THE EFFECTS OF CATECHIN LEVEL AND PH ON ANTIOXIDANT ACTIVITY IN FRESH BEEF

M.N. O'Grady*[1], M. Maher[1], D.J. Troy[2], A.P. Moloney[3] and J.P. Kerry[1]

[1]*Department of Food and Nutritional Sciences, University College Cork, Cork, Ireland.* [2]*Ashtown Food Research Centre, Teagasc, Ashtown, Dublin 15, Ireland.* [3]*Teagasc, Grange Research Centre, Co. Meath, Ireland. Email: Michael.ogrady@ucc.ie*

Keywords: tea catechins, colour, lipid oxidation

Introduction

Colour and lipid oxidation are important factors influencing the quality of fresh beef. The muscle pigment myoglobin is responsible for the colour of fresh beef. The principal catechins found in green tea (*Camellia sinensis*) are (-)-epicatechin (-EC), (-)-epigallocatechin (-EGC), (-)-epicatechin gallate (-ECG), and (-)-epigallocatechin gallate (-EGCG). Addition of tea catechins (TC) to poultry diets (up to 300 mg/kg feed for 42 days pre-slaughter) significantly reduced lipid oxidation in poultry meat (Tang et al., 2001). In contrast, supplementation of beef cattle diets with TC (1000 mg TC/animal/day for 100 days pre-slaughter) did not significantly improve beef quality in terms of colour and lipid stability (O'Grady *et al.*, 2006). It was concluded that the lack of an effect of dietary TC may be attributed to the level of TC included in the diet, pH sensitivity of TC isomers (Zhu *et al.*, 1997) to alkaline conditions present in the bovine rumen and intestine, or simply to excretion of TC in the urine. The aim of this study was to assess the effect of increasing levels of TC (1000, 4000 and 10,000 mg TC/animal/day) on fresh beef quality. The pH sensitivity of the TC supplement, employed in the dietary trial, was also assessed in minced beef.

Materials and Methods

Twenty eight continental cross bred heifers were randomly assigned to one of four diets for 90 days pre-slaughter. The control group were offered a diet consisting of a barley-based concentrate and the three remaining groups were fed the control diet plus incremental amounts of TC, 1000 (+ TC 1000), 4000 (+ TC 4000) and 10,000 (+TC 10,000mg/animal/day). The TC supplement (81.43%) extracted from green tea (*Camellia sinensis* L. variety *assamica*) was supplied by Kinglong Natural Plant Products Industry Ltd., China. *M. longissimus dorsi* (LD) samples were vacuum packaged and stored at 0°C (~ 1 week) prior to analysis. Muscle TC levels and composition of the dietary TC supplement were determined by HPLC following the extraction procedure of Tsuchiya *et al.* (1998), with modifications. The TC supplement contained 10.29% -EGC, 8.23% -EC, 44.15% -EGCG and 17.65% -ECG. LD samples were cut into steaks (~ 25mm thickness), placed in retail display trays and flushed with 80% O_2 : 20% CO_2 for storage in modified atmosphere packs (MAP) for up to 8 days at 4°C under simulated retail display conditions (616 lux fluorescent lighting).

Portions of LD samples from animals fed the control diet were pooled and minced through a plate with 4 mm holes. Following mincing, pH adjusted TC (5.5 – 8.0) dissolved in 60 mM sodium phosphate buffers (Na_2HPO_4) (pH 5.5, 6.0, 6.5, 7.0, 7.5 and 8.0) were added to minced LD samples (10 mg TC/kg beef) which were subsequently formed into meat patties. Meat patties were stored in MAP (80% O_2 : 20% CO_2) for up to 10 days at 4°C (616 lux). Colour measurements were made at 2 day intervals using a Cr-300 Chromameter (Minolta Co. Ltd., Japan) and reported as CIE 'a' redness values. Lipid oxidation was measured at 2 day intervals by the distillation method of Tarladgis et al. (1960), as modified by Ke *et al.* (1977), and results were expressed as 2-thiobarbituric acid reactive substances (TBARS) in mg malondialdehyde (MDA)/kg muscle. Data was analysed by repeated measures ANOVA using the SPSS 11.0 for Windows software package.

Results and Discussion

Dietary supplementation with increasing concentrations of TC (1000, 4000 and 10,000 mg TC/animal/day) did not significantly improve the colour or lipid stability of LD steaks, relative to controls (Table 1). Similarly, TC isomers were not detected in LD samples (chromatograms not shown). Zhu et al. (1997) reported that stability of TC was pH dependent with -EGC and -EGCG demonstrating least stability in alkaline conditions. Both -EGC and -EGCG represent >50% of the total TC concentration of the dietary supplement used in the present study. Therefore lack of a dietary effect of TC on fresh beef quality may have been due to degradation of TC in the alkaline conditions of the bovine rumen and intestine. In meat patties containing pH adjusted TC, surface colour was unaffected by TC pH (Figure 1A). An upward trend in lipid oxidation was observed from day 4 to day 10 at pH 7.0 and higher (Figure 1B) which indicated lower antioxidant activity of TC above pH 7.0.

Conclusions

Supplementation of beef cattle diets with TC at levels up to 10,000 mg/animal/day did not affect beef quality. This may be attributed to pH sensitivity of certain TC isomers to alkaline pH conditions.

Table 1: Effect of dietary TC (0, 1000, 4000 and 10,000 mg TC/animal/day) on surface redness ('a' value) and lipid oxidation (TBARS) in *M. longissimus dorsi* (LD) steaks stored in modified atmosphere packs (MAP) (80% O_2 : 20% CO_2).

Treatment	Storage time at 4°C, d									
	Surface redness, 'a' value					TBARS, mg MDA/kg muscle				
	0	2	4	6	8	0	2	4	6	8
Control	20.86[a]	18.03	17.17	15.31	12.68	0.07[a]	0.43	0.43	1.26	1.80
+ TC 1000	21.41	18.08	16.82	13.77	12.26	0.07	0.43	0.74	1.10	2.16
+ TC 4000	21.59	18.23	16.16	15.56	11.60	0.07	0.33	0.52	1.17	2.07
+ TC 10,000	21.67	18.07	15.96	14.44	12.09	0.07	0.35	0.65	1.35	2.14

[a]Within each day, no significant effects were observed for dietary treatment, $P > 0.05$.

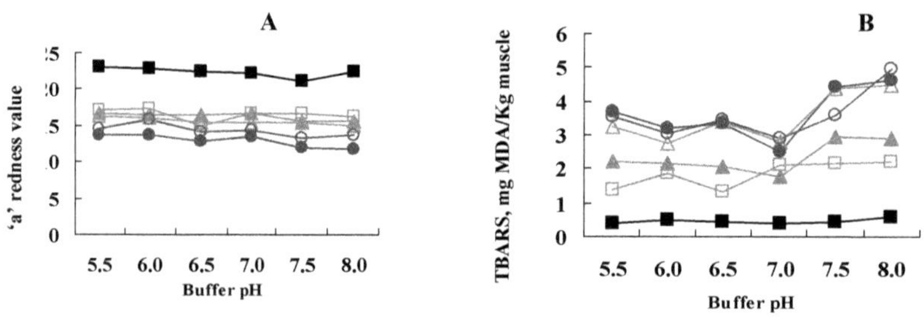

Figure 1: Effect of pH adjusted TC (5.5 – 8.0) (10 mg TC/kg beef) on surface redness ('a' value) (A) and lipid oxidation (TBARS) (B) in minced meat patties stored in MAP (80% O_2 : 20% CO_2) for 10 days at 4°C.
■, day 0; □, day 2; ▲, day 4; △, day 6; ○, day 8; ●, day 10.

References

Ke, P.J., Ackman, R.G., Linke, B.A. and Nash, D.M. (1977). Differential lipid oxidation in various parts of frozen mackerel. Journal of Food Technology, 12, 37-47.

O'Grady, M.N., Maher, M., Troy, D.J., Moloney, A.P. and Kerry, J.P. (2006). An assessment of dietary supplementation with tea catechins and rosemary extract on the quality of fresh beef. Meat Science, 73, 132-143.

Tang, S.Z., Kerry, J.P., Sheehan, D., Buckley, D.J. and Morrissey, P.A. (2001). Antioxidative effect of dietary tea catechins on lipid oxidation of long-term frozen stored chicken meat. Meat Science, 57, 331-336.

Tarladgis, B.G., Watts, B.M. and Younathan, M.T. and Dudan, L.R. Jr. (1960). A distillation method for the quantitative determination of malonaldehyde in rancid foods. Journal of the American Oil Chemists Society, 37, 44-48.

Tsuchiya, H., Sato, M., Kato, H., Kureshiro, H. and Takagi, N. (1998). Nanoscale analysis of pharmacologically active catechins in body fluids by HPLC using borate complex extraction pre-treatment. Talanta, 46, 717-726.

Zhu, Q.Y., Zhang, A., Tsang, D., Huang, Y. and Chen, Z.-Y. (1997). Stability of green tea catechins. Journal of Agricultural and Food Chemistry, 45, 4624-4628.

Acknowledgements

Funded by the Irish Government under the National Development Plan 2000 - 2006.

KEYWORD INDEX

AUTHORS INDEX

www.ingramcontent.com/pod-product-compliance
Ingram Content Group UK Ltd.
Pitfield, Milton Keynes, MK11 3LW, UK
UKHW051448290926
15471UKWH00008B/1083

9 789086 860104